MATHEMATICS DICTIONARY

FIFTH EDITION

JAMES / JAMES

CONTRIBUTORS

ARMEN A. ALCHIAN
University of California at Los Angeles

EDWIN F. BECKENBACH
University of California at Los Angeles

CLIFFORD BELL
University of California at Los Angeles

HOMER V. CRAIG
University of Texas

GLENN JAMES
University of California at Los Angeles

ROBERT C. JAMES
Claremont Graduate School

ARISTOTLE D. MICHAL
California Institute of Technology

IVAN S. SOKOLNIKOFF
University of California at Los Angeles

FIFTH EDITION

JAMES / JAMES

VNR VAN NOSTRAND REINHOLD
New York

Copyright © 1992 by Van Nostrand Reinhold

Library of Congress Catalog Card Number 92-6757
ISBN 0-442-00741-8
ISBN 0-442-01241-1 (pbk.)

Manufactured in the United States of America

Published by Van Nostrand Reinhold
115 Fifth Avenue
New York, NY 10003

Chapman and Hall
2-6 Boundary Row
London, SE 1 8HN, England

Thomas Nelson Australia
102 Dodds Street
South Melbourne 3205
Victoria, Australia

Nelson Canada
1120 Birchmount Road
Scarborough, Ontario M1K 5G4, Canada

16 15 14 13 12 11 10 9 8 7 6 5 4 3 2 1

Library of Congress Cataloging-in-Publication Data
James. Robert C., 1918–
  Mathematics dictionary / Robert C. James.—5th ed.
    p. cm.
  Includes index
    ISBN 0-442-00741-8—ISBN 0-442-01241-1 (pbk.)
    1. Mathematics—Dictionaries. I. Title.
QA5.J33 1992
510′.3—dc20   92-6757
                                      CIP

CONTENTS

PREFACE

Throughout the preparation of each edition of this dictionary, the guiding objective has been to make it useful for students, scientists, engineers, and others interested in the meaning of mathematical terms and concepts. It is intended to be essentially complete in the coverage of topics that occur in precollege or undergraduate college mathematics courses, as well as covering many topics from beginning graduate-level courses. In addition, many other interesting and important mathematical concepts are included.

When preparing this edition, emphasis was given on revising and updating topics included in previous editions and to introducing many new topics, so as to reflect recent interesting and exciting developments in mathematics. An important feature continued and extended in this edition is the multilingual index in French, German, Russian, and Spanish. The English equivalents of mathematical terms in these languages enable the reader not only to learn the English meaning of a foreign-language mathematical term, but also to find its definition in the body of this book.

Although this is by no means a mere word dictionary, neither is it an encyclopedia. It is a correlated condensation of mathematical concepts, designed for time-saving reference work. Nevertheless, the general reader can come to an understanding of a new concept by looking up unfamiliar terms in the definition at hand and following this procedure down to familiar concepts.

Main headings are printed in boldface capitals beginning at the left margin. Each main heading that is also a proper name is followed by the appropriate given names, birth and death dates, and a brief biographical statement, to the extent that these have been determined. Subheadings are printed in boldface type at the beginning of paragraphs. Citations to subheadings under other main headings give the main heading in small capitals, followed by a dash and the subheading (if giving the subheading seems useful) as: ANGLE—adjacent angle.

The first publication of this dictionary (1942) was due primarily to Glenn James, assisted by Robert C. James. The D. Van Nostrand edition (1949) benefitted from contributions by Armen A. Alchian, Edwin F. Beckenbach, Clifford Bell, Homer V. Craig, Aristotle D. Michal, and Ivan S. Sokolnikoff. Professor Beckenbach also made major contributions to the second (1959), third (1968), and fourth (1976) editions published by Van Nostrand Reinhold. The multilingual index was introduced in the second (1959) edition. Translators were J. George Adashko (Russian), Aaron Bakst (French), Samuel Gitler (Spanish), and Kuno Lorenz (German). The index has been extended in this edition with help from Gilles Pisier (French), Albert Fässler (German), Zvi Ruder (Russian), and Helga Fetter (Spanish).

Comments on definitions as well as discussions of any phase of this dictionary are invited. Such comments have been very helpful in preparing this edition, especially those from John K. Baumgart, Ralph P. Boas, Patrick W. Kearney, and Edward Wakeling.

ROBERT C. JAMES

A

A.E. An abbreviation for *almost everywhere.* See MEASURE —measure zero.

AB′A-CUS, *n.* [*pl.* **abaci** or **abacuses**]. A counting frame to aid in arithmetic computation; an instructive plaything for children, used as an aid in teaching place value; a primitive predecessor of the modern computing machine. One form consists of a rectangular frame carrying as many parallel wires as there are digits in the largest number to be dealt with. Each wire contains nine beads free to slide on it. A bead on the lowest wire counts unity, on the next higher wire 10, on the next higher 100, etc. Two beads slid to the right on the lowest wire, three on the next higher, five on the next and four on the next denote 4532.

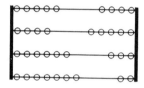

ABEL, Niels Henrik (1802–1829). Norwegian algebraist and analyst. When about nineteen, he proved the general quintic equation in one variable cannot be solved by a finite number of algebraic operations (see RUFFINI). Made fundamental contributions to the theories of infinite series, transcendental functions, groups, and elliptic functions.

Abel identity. The identity

$$\sum_{i=1}^{n} a_i u_i \equiv s_1(a_1 - a_2) + s_2(a_2 - a_3) + \cdots$$

$$+ s_{n-1}(a_{n-1} - a_n) + s_n a_n,$$

where

$$s_n = \sum_{i=1}^{n} u_i.$$

This is easily obtained from the evident identity:

$$\sum_{i=1}^{n} a_i u_i \equiv a_1 s_1 + a_2(s_2 - s_1) + \cdots$$

$$+ a_n(s_n - s_{n-1}).$$

Abel inequality. If $u_n \ge u_{n+1} > 0$ for all positive integers n, then $\left| \sum_{n=1}^{p} a_n u_n \right| \le L u_1$, where L is the largest of the quantities: $|a_1|$, $|a_1 + a_2|$, $|a_1 + a_2 + a_3|, \cdots, |a_1 + a_2 + \cdots + a_p|$. This inequality can be easily deduced from the Abel identity.

Abel method of summation. The method of *summation* for which a series $\sum_{0}^{\infty} a_n$ is *summable* and has *sum S* if $\lim_{x \to 1-} \sum_{0}^{\infty} a_n x^n = S$ exists. A convergent series is summable by this method [see below, Abel theorem on power series (2)]. Also called *Euler method of summation.* See SUMMATION —summation of a divergent series.

Abel problem. Suppose a particle is constrained (without friction) to move along a certain path in a vertical plane under the force of gravity. The **Abel problem** is to find the path for which the time of descent is a given function f of x, where the x-axis is the horizontal axis and the particle starts from rest. This reduces to the problem of finding a solution $s(x)$ of the *Volterra integral equation of the first kind*

$$f(x) = \int_0^x \frac{s(t)}{\sqrt{2g(x-t)}} \, dt, \quad \text{where } s(x) \text{ is the}$$

length of the path. If f' is continuous, a solution is

$$s(x) = \frac{\sqrt{2g}}{\pi} \frac{d}{dx} \int_0^x \frac{f(t)}{(x-t)^{1/2}} \, dt.$$

Abel tests for convergence. (1) If the series $\sum u_n$ converges and $\{a_n\}$ is a bounded monotonic sequence, then $\sum a_n u_n$ converges. (2) If $\left| \sum_{n=1}^{k} u_n \right|$ is equal to or less than a properly chosen constant for all k and $\{a_n\}$ is a positive, monotonic decreasing sequence which approaches zero as a limit, then $\sum a_n u_n$ converges. (3) If a series of complex numbers $\sum a_n$ is convergent, and the series $\sum(v_n - v_{n+1})$ is absolutely convergent, then $\sum a_n v_n$ is convergent. (4) If the series $\sum a_n(x)$ is uniformly convergent in an interval (a, b), $v_n(x)$ is positive and monotonic decreasing for any value of x in the interval, and there is a number k such that $v_0(x) < k$ for all x in the interval, then $\sum a_n(x) v_n(x)$ is uniformly convergent (this is the **Abel test for uniform convergence**).

Abel theorem on power series. (1) If a power series, $a_0 + a_1 x + a_2 x^2 + \cdots + a_n x^n + \cdots$, converges for $x = c$, it converges absolutely for $|x| < |c|$. (2) If $\sum_0^\infty a_n$ is convergent, then $\lim_{t \to 1-} \sum_0^\infty a_n t^n = \sum_0^\infty a_n$, where the limit is the limit on the left at $+1$. An equivalent statement is that if $\sum_0^\infty a_n x^n$ converges when $x = R$, then S is continuous if $S(x)$ is defined as $\sum_0^\infty a_n x^n$ when x is in the closed interval with end points 0 and R. This theorem is designated in various ways, most explicitly by "Abel's theorem on continuity up to the circle of convergence."

A-BEL′IAN, *adj.* **Abelian group.** See GROUP.

A-BRIDGED′, *adj.* **abridged multiplication.** See MULTIPLICATION.

Plücker's abridged notation. A notation used for studying curves. Consists of the use of a single symbol to designate the expression (function) which, equated to zero, has a given curve for its locus; hence reduces the study of curves to the study of polynomials of the first degree. *E.g.*, if $L_1 = 0$ denotes $2x + 3y - 5 = 0$ and $L_2 = 0$ denotes $x + y - 2 = 0$, then $k_1 L_1 + k_2 L_2 = 0$ denotes the family of lines passing through their common point $(1, 1)$. See PENCIL—pencil of lines through a point.

AB-SCIS′SA, *n.* [*pl.* **abscissas** or **abscissae**]. The horizontal coordinates in a two-dimensional system of rectangular coordinates; usually denoted by x. Also used in a similar sense in systems of oblique coordinates. See CARTESIAN —Cartesian coordinates.

AB′SO-LUTE, *adj.* **absolute constant, continuity, convergence, error, inequality, maximum (minimum), symmetry.** See CONSTANT, CONTIN-UOUS, CONVERGENCE, ERROR, INEQUALITY, MAXI-MUM, SYMMETRIC —symmetric function.

absolute moment. For a random variable X or the associated distribution function, the kth absolute moment about a is the expected value of $|X - a|^k$, whenever this exists. See MO-MENT—moment of a distribution.

absolute number. A number represented by figures such as 2, 3, or $\sqrt{2}$, rather than by letters as in algebra.

absolute property of a surface. Same as IN-TRINSIC PROPERTY OF A SURFACE.

absolute term in an expression. A term which does not contain a variable. *Syn.* constant term. In the expression $ax^2 + bx + c$, c is the only absolute term.

absolute value of a complex number. See MODULUS —modulus of a complex number.

absolute value of a real number. The absolute value of a, written $|a|$, is the nonnegative number which is equal to a if a is nonnegative and equal to $-a$ if a is negative; *e.g.*, $3 = |3|$, $0 = |0|$, and $3 = |-3|$. Useful properties of the absolute value are that $|xy| = |x| \, |y|$ and $|x + y| \leq |x| + |y|$ for all real numbers x and y. *Syn.* numerical value. See TRIANGLE —triangle inequality.

absolute value of a vector. See VECTOR —absolute value of a vector.

AB-SORB′, *v.* A subset A of a vector space X is said to **absorb** a subset B if there exists a positive number ϵ such that $aB \subset A$ if $0 < |a| \leq \epsilon$ (sometimes it is only required that there exist at least one positive number a for which $aB \subset A$). A subset is **absorbing** or **absorbent** if it absorbs each point of X.

AB-SORP′TION, *adj.* **absorption property.** For a Boolean algebra or for set theory, either of the properties $A \cup (A \cap B) = A$ or $A \cap (A \cup B) = A$. See BOOLE—Boolean algebra.

AB′STRACT, *adj.* **abstract mathematics.** See MATHEMATICS —pure mathematics.

abstract number. Any number as such, simply as a number, without reference to any particular objects whatever except in so far as these objects possess the number property. Used to emphasize the distinction between a number, as such, and concrete numbers. See CONCRETE, NUMBER, DENOMINATE.

abstract space. A formal mathematical system consisting of undefined objects and axioms of a geometric nature. Examples are Euclidean spaces, metric spaces, topological spaces, and vector spaces.

abstract word or symbol. (1) A word or symbol that is not concrete; a word or symbol denoting a concept built up from consideration of many special cases; a word or symbol denoting a property common to many individuals or individual sets, as yellow, hard, two, three, etc. (2) A word or symbol which has no specific reference in the sense that the concept it represents exists quite independently of any specific cases whatever and may or may not have specific reference.

A′BUN-DANT, *adj.* **abundant number.** See NUMBER —perfect number.

AC-CEL'ER-A'TION, *n.* The time rate of change of velocity. Since velocity is a directed quantity, the acceleration **a** is a vector equal to $\lim_{\Delta t \to 0} \dfrac{\Delta \mathbf{v}}{\Delta t} = \dfrac{d\mathbf{v}}{dt}$, where $\Delta \mathbf{v}$ is the increment in the velocity **v** which the moving object acquires in t units of time. Thus, if an airplane moving in a straight line with the speed of 2 miles per minute increases its speed until it is flying at the rate of 5 miles per minute at the end of the next minute, its **average acceleration** during that minute is 3 miles per minute per minute. If the increase in speed during this one minute interval of time is uniform, the average acceleration is equal to the actual acceleration. If the increase in speed in this example is not uniform, the instantaneous acceleration at the time t_1 is determined by evaluating the limit of the quotient $\dfrac{\Delta \mathbf{v}}{\Delta t}$ as the time interval $\Delta t = t - t_1$ is made to approach zero by making t approach t_1. For a particle moving along a curved path, the velocity **v** is directed along the tangent to the path and the acceleration **a** can be shown to be given by the formula

$$\mathbf{a} = \frac{dv}{dt}\boldsymbol{\tau} + v^2 c \boldsymbol{\nu},$$

where $\dfrac{dv}{dt}$ is the derivative of speed v along the path, c is the curvature of the path at the point, and $\boldsymbol{\tau}$ and $\boldsymbol{\nu}$ are vectors of unit lengths directed along the tangent and normal to the path. The first of these terms, $\dfrac{dv}{dt}$, is called the **tangential component**, and the second, $v^2 c$, the **normal** (or **centripetal**) **component** of acceleration. If the

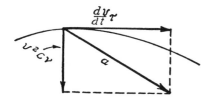

path is a straight line, the curvature c is zero, and hence the acceleration vector will be directed along the path of motion. If the path is not rectilinear, the direction of the acceleration vector is determined by its tangential and normal components as shown in the figure.

acceleration of Coriolis. If S' is a reference frame rotating with the angular velocity $\boldsymbol{\omega}$ about a fixed point in another reference frame S, the acceleration **a** of a particle, as measured by the observer fixed in the reference frame S, is given by the sum of three terms: $\mathbf{a} = \mathbf{a}' + \mathbf{a}_t + \mathbf{a}_c$, where $\mathbf{a}'$ is the acceleration of the particle relative to S', $\mathbf{a}_t$ is the acceleration of the moving space, and $\mathbf{a}_c = 2\boldsymbol{\omega} \times \mathbf{v}'$ is the *acceleration of Coriolis*. The symbol $\boldsymbol{\omega} \times \mathbf{v}'$ denotes the vector product of the angular velocity $\boldsymbol{\omega}$, and the velocity $\mathbf{v}'$ relative to S', so that the acceleration of Coriolis is normal to the plane determined by the vectors $\boldsymbol{\omega}$ and $\mathbf{v}'$ and has the magnitude $2v' \sin(\boldsymbol{\omega}, \mathbf{v}')$. The acceleration of Coriolis is also called the **complementary acceleration.**

acceleration of gravity. The acceleration with which a body falls *in vacuo* at a given point on or near a given point on the earth's surface. This acceleration, frequently denoted by g, varies by less than one percent over the entire surface of the earth. Its "average value" has been defined by the International Commission of Weights and Measures as 9.80665 meters (or 32.174− feet) per second per second. Its value at the poles is 9.8321 and at the equator 9.7799.

angular acceleration. The time rate of change of angular velocity. If the angular velocity is represented by a vector $\boldsymbol{\omega}$ directed along the axis of rotation, then the angular acceleration $\boldsymbol{\alpha}$, in the symbolism of calculus, is given by $\boldsymbol{\alpha} = \dfrac{d\boldsymbol{\omega}}{dt}$. See VELOCITY —angular velocity.

centripetal, normal, and tangential components of acceleration. See above, ACCELERATION.

uniform acceleration. Acceleration in which there are equal changes in the velocity in equal intervals of time. *Syn.* constant acceleration.

AC'CENT, *n.* A mark above and to the right of a quantity (or letter), as in a' or x'; the mark used in denoting that a letter is primed. See PRIME —prime as a symbol.

AC-CEPT'ANCE, *adj.* acceptance region. See HYPOTHESIS —test of a hypothesis.

AC-CU'MU-LAT'ED, *adj.* accumualated value. Same as AMOUNT at simple or compound interest. The accumulated value (or **amount**) of an **annuity** at a given date is the sum of the compound amounts of the annuity payments to that date.

AC-CU'MU-LA'TION, *adj., n.* Same as ACCUMULATED VALUE.

accumulation of discount on a bond. Writing up the book value of a bond on each dividend date by an amount equal to the interest on the investment (interest on book value at yield rate) minus the dividend. See VALUE —book value.

accumulation factor. The name sometimes given to the binomial $(1 + r)$, or $(1 + i)$, where r, or i, is the rate of interest. The formula for compound interest is $A = P(1 + r)^n$, where A is the amount accumulated at the end of n periods from an original principal P at a rate r per period. See COMPOUND —compound amount.

accumulation point. (1) An accumulation point of a **set of points** is a point P such that there is at least one point of the set distinct from P in any neighborhood of the given point; a point which is the limit of a sequence of points of the set (for spaces which satisfy the first axiom of countability). A point of P that is not an accumulation point of P is an **isolated point** of P. An **adherent point** of P is any point that either is in P or is an accumulation point of P. See BOLZANO—Bolzano-Weierstrass theorem, CONDENSATION—condensation point, POINT— isolated point. (2) See SEQUENCE —accumulation point of a sequence. *Syn.* cluster point, limit point.

accumulation problem. The determination of the amount when the principal, or principals, interest rate, and time for which each principal is invested are given.

accumulation schedule of bond discount. A table showing the accumulation of bond discounts on successive dates. Interest and book values are usually listed also.

AC-CU′MU-LA′TOR, *n.* In a computing machine, an adder or counter that augments its stored number by each successive number it receives.

AC′CU-RA-CY, *n.* Correctness, usually referring to numerical computations. The **accuracy of a table** may mean either: (1) The number of significant digits appearing in the numbers in the table (*e.g.*, in the mantissas of a logarithm table); (2) the number of correct places in computations made with the table. (This number of places varies with the form of computation, since errors may repeatedly combine so as to become of any size whatever.)

AC′CU-RATE, *adj.* Exact, precise, without error. One speaks of an accurate statement in the sense that it is correct or true and of an accurate computation in the sense that it contains no numerical error. **Accurate to a certain decimal place** means that all digits preceding and including the given place are correct and the next place has been made zero if less than 5 and 10 if greater than 5 (if it is equal to 5, the most usual convention is to call it zero or 10 as is necessary to leave the last digit even). *E.g.*, 1.26 is accu-rate to two places if obtained from 1.264 or 1.256 or 1.255. See ROUNDING —rounding off.

AC′NODE, *n.* See POINT—isolated point.

A-COUS′TI-CAL, *adj.* **acoustical property of conics.** See ELLIPSE—focal property of ellipse, HYPERBOLA—focal property of hyperbola, PARABOLA —focal property of parabola.

A′CRE, *n.* The unit commonly used in the United States in measuring land; contains 43,560 square feet, 4,840 square yards, or 160 square rods.

AC′TION, *n.* A concept in advanced dynamics defined by the line integral $A = \int_{P_1}^{P_2} m\mathbf{v} \cdot d\mathbf{r}$, called the **action integral**, where m is the mass of the particle, $\mathbf{v}$ is its velocity, and $d\mathbf{r}$ is the vector element of the arc of the trajectory joining the points P_1 and P_2. The dot in the integrand denotes the scalar product of the momentum vector $m\mathbf{v}$ and $d\mathbf{r}$. The *action A* plays an important part in the development of dynamics from variational principles. See below, principle of least action.

law of action and reaction. The basic law of mechanics asserting that two particles interact so that the forces exerted by one on another are equal in magnitude, act along the line joining the particles, and are opposite in direction. See NEWTON —Newton's laws of motion.

principle of least action. Of all curves passing through two fixed points in the neighborhood of the natural trajectory, and which are traversed by the particle at a rate such that for each (at every instant of time) the sum of the kinetic and potential energies is a constant, that one for which the action integral has an extremal value is the natural trajectory of the particle. See ACTION.

A-CUTE′, *adj.* **acute angle.** An angle numerically smaller than a right angle (usually a positive angle less than a right angle).

acute triangle. See TRIANGLE.

AD′DEND, *n.* One of a set of numbers to be added, as 2 or 3 in the sum 2 + 3.

AD′DER, *n.* In a computing machine, any arithmetic component that performs the operation of addition of positive numbers. An arithmetic component that performs the operations of addition and subtraction is said to be an **algebraic adder.** See ACCUMULATOR, COUNTER.

AD-DI′TION, *n.* **addition of angles, directed line segments, integers, fractions, irrational numbers, mixed numbers, matrices, and vectors.** See various headings under SUM.

addition of complex numbers. See COMPLEX —complex numbers.

addition of decimals. The usual procedure for adding decimals is to place digits with like place value under one another, *i.e.*, place decimal points under decimal points, and add as with integers, putting the decimal point of the sum directly below those of the addends. See SUM—sum of real numbers.

addition formulas of trigonometry. See TRIGONOMETRY.

addition of series. See SERIES.

addition of similar terms in algebra. The process of adding the coefficents of terms which are alike as regards their other factors: $2x + 3x = 5x$, $3x^2y - 2x^2y = x^2y$ and $ax + bx = (a + b)x$. See DISSIMILAR —dissimilar terms.

addition of tensors. See TENSOR.

algebraic addition. See SUM—algebraic sum, sum of real numbers.

arithmetic addition. See SUM—arithmetic sum.

proportion by addition (and **addition and subtraction**). See PROPORTION.

ADD′I-TIVE, *adj.* **additive function.** A function f which has the property that $f(x + y)$ is defined and equals $f(x) + f(y)$ whenever $f(x)$ and $f(y)$ are defined. A *continuous* additive function is necessarily *homogeneous*. A function f is **subadditive** or **superadditive** according as

$$f(x_1 + x_2) \leq f(x_1) + f(x_2),$$

or

$$f(x_1 + x_2) \geq f(x_1) + f(x_2),$$

for all x_1, x_2, and $x_1 + x_2$ in the domain of f (this domain is usually taken to be an interval of the form $0 \leq x \leq a$).

additive inverse. See INVERSE —inverse of an element.

additive set function. A function which assigns a number $\phi(X)$ to each set X of a family F of sets is **additive** (or **finitely additive**) if the *union* of any two members of F is a member of F and

$$\phi(X \cup Y) = \phi(X) + \phi(Y)$$

for all disjoint members X and Y of F. The function ϕ is **countably additive** (or **completely additive**) if the union of any finite or countable

set of members of F is a member of F and

$$\phi\left(\bigcup X_i \right) = \sum \phi(X_i)$$

for each finite or countable collection of sets $\{X_t\}$ which are *pairwise disjoint* and belong to F. If $\phi(\cup X_i) \leq \Sigma\phi(x_i)$, then ϕ is said to be **subadditive** (it is then not necessary to assume the sets are pairwise disjoint). See MEASURE —measure of a set.

AD-HER′ENT, *adj.* **adherent point.** See ACCU-MULATION —accumulation point.

AD′I-A-BAT′IC, *adj.* **adiabatic curves.** Curves showing the relation between pressure and volume of substances which are assumed to have adiabatic expansion and contraction.

adiabatic expansion (or contraction). (*Thermodynamics*) A change in volume without loss or gain of heat.

AD IN′FI-NI′TUM. Continuing without end (according to some law); denoted by three dots, as $\cdots$; used, principally, in writing infinite series, infinite sequences, and infinite products.

AD-JA′CENT, *adj.* **adjacent angles.** Two angles having a common side and common vertex and lying on opposite sides of their common side. In the figure, AOB and BOC are adjacent angles.

AD-JOINED′, *adj.* **adjoined number.** See FIELD —number field.

AD′JOINT, *adj., n.* **adjoint of a differential equation.** For a homogeneous differential equation

$$L(y) \equiv p_0 \frac{d^n y}{dx^n} + p_1 \frac{d^{n-1}y}{dx^{n-1}} + \cdots$$

$$+ p_{n-1} \frac{dy}{dx} + p_n y = 0,$$

the adjoint is the differential equation

$$\overline{L}(y) \equiv (-1)^n \frac{d^n(p_0 y)}{dx^n} + (-1)^{n-1} \frac{d^{n-1}(p_1 y)}{dx^{n-1}}$$

$$+ \cdots - \frac{d(p_{n-1} y)}{dx} + p_n y = 0.$$

This relation is symmetric, $L = 0$ being the adjoint of $\overline{L} = 0$. A function is a solution of one of these equations if and only if it is an *integrating factor* of the other. There is an expression $P(u, v)$ for which

$$v L(u) = u \overline{L}(v) \equiv \frac{dP(u, v)}{dx}.$$

$P(u, v)$ is linear and homogeneous in u, $u', \cdots, u^{(n-1)}$, and in $v, v', \cdots, v^{(n-1)}$. It is known as the **bilinear concomitant**. An equation is **self-adjoint** if $L(y) \equiv \overline{L}(y)$. E.g., Sturm-Liouville differential equations and Legendre differential equations are self-adjoint.

adjoint of a matrix. The *transpose* of the matrix obtained by replacing each element of the original matrix by its *cofactor*; the matrix obtained by replacing each element a_{rs} (in row r and column s) by the cofactor of the element a_{sr} (in row s and column r). The adjoint is defined only for square matrices. Sometimes (rarely) the adjoint is called the **adjugate**, although adjugate has also been used for the square matrix of order $\binom{n}{r}$ formed from a square matrix of order n by arranging all rth-order minors in some specified order. The *Hermitian conjugate* matrix is frequently called the adjoint matrix by writers on quantum mechanics.

adjoint of a transformation. For a bounded linear transformation T which maps a **Hilbert space** H into H (with domain of T equal to H), there is a unique bounded linear transformation T^*, the *adjoint* of T, such that the inner products (Tx, y) and (x, T^*y) are equal for all x and y of H. It follows that $\|T\| = \|T^*\|$. Two linear transformations T_1 and T_2 are said to be *adjoint* if $(T_1 x, y) = (x, T_2 y)$ for each x in the domain of T_1 and y in the domain of T_2. If T is a linear transformation whose domain is dense in H, there is a unique transformation T^* (called the *adjoint* of T) such that T and T^* are adjoint and, if S is any other transformation adjoint to T, then the domain of S is contained in the domain of T^* and S and T^* coincide on the domain of S. For a **finite dimensional space** and a transformation T which maps vectors $x = (x_1, x_2, \cdots, x_n)$ into $Tx = (y_1, y_2, \cdots, y_n)$ with

$y_i = \sum_j a_{ij} x_j$ (for each i), the adjoint of T is the transformation for which $T^* x = (y_1, y_2, \cdots, y_n)$ with $y_i = \sum_j \overline{a}_{ji} x_i$ and the matrices of the coefficients of T and of T^* are *Hermitian conjugates* of each other. If T is a bounded linear transformation which maps a **Banach space** X into a Banach space Y, and X^* and Y^* are the *first conjugate spaces* of X and Y, then the adjoint of T is the linear transformation T^* which maps Y^* into X^* and is such that $T^*(g) = f$ (for f and g members of X^* and Y^*, respectively) if f is the continuous linear functional defined by $f(x) = g[T(x)]$. For two bounded linear transformations T_1 and T_2, the adjoints of $T_1 + T_2$ and $T_1 \cdot T_2$ are $T_1^* + T_2^*$ and $T_2^* \cdot T_1^*$, respectively. If T has an inverse whose domain is all of H (or Y), then $(T^*)^{-1} = (T^{-1})^*$. For Banach spaces, the adjoint T^{**} of T^* is a mapping of X^{**} into Y^{**} which is a norm-preserving extension of T (T maps a subset of X^{**}, which is isometric with X, into Y^{**}). For Hilbert space, T^{**} is identical with T if T is bounded with domain H; T^{**} is a linear extension of T otherwise. See SELF—self-adjoint transformation.

adjoint space. See CONJUGATE—conjugate space.

AD′JU-GATE, *n.* See ADJOINT—adjoint of a matrix.

AD-MIS′SI-BLE, *adj.* **admissible hypothesis.** See HYPOTHESIS.

AF-FINE′, *adj.* **affine algebraic variety.** See VARIETY.

affine space. A set S for which there is an associated vector space V with scalars in some field F and an operation (denoted by addition) that satisfies the three conditions: (i) $s + v \in S$ if $s \in S$ and $v \in V$; (ii) $(s + u) + v = s + (u + v)$ whenever $s \in S$, $u \in V$, and $v \in V$; (iii) if s and σ are members of S, then there is a unique $v \in V$ such that $s = \sigma + v$. It follows that $s + \mathbf{0} = s$. Also, if σ is an arbitrary member of S, then $s = \sigma + v$ defines a one-to-one correspondence $s \leftrightarrow v$ between S and V. If $\sigma \in S$ and U is a vector subspace of V, then the set of all $s \in S$ for which $s = \sigma + u$ for some $u \in U$ is an **affine subspace** of S and it is a **line**, **plane**, or **hyperplane** according as U is one-dimensional, two-dimensional, or a hyperplane of V. *Example:* Given a vector space V and a nonsingular linear transformation T of V onto V, let $S = V$ and define $s + v$ to be $T(v) + s$, where the last "$+$" denotes vector addition. See VECTOR—vector space (2). When studying properties preserved by affine transformations on a vector space, one often refers to the vector space as an **affine space**.

affine transformation. *In the plane*, a transformation of the form

$$x' = a_1 x + b_1 y + c_1, \qquad y' = a_2 x + b_2 y + c_2.$$

It is **singular** or **nonsingular** according as the determinant $\begin{vmatrix} a_1 & b_1 \\ a_2 & b_2 \end{vmatrix} = a_1 b_2 - a_2 b_1$ is nonzero or zero. The following are important special cases of nonsingular affine transformations of the plane: (a) **translations** ($x' = x + a$, $y' = y + b$); (b) **rotations** ($x' = x \cos \theta + y \sin \theta$, $y' = -x \sin \theta + y \cos \theta$); (c) **stretchings** ($x' = kx$, $y' = ky$ with $k > 1$), and **shrinkings** ($x' = kx$, $y' = ky$ with $0 < k < 1$), called *transformations of similitude* or *homothetic transformations*; (d) **reflections in the x-axis** ($x' = x$, $y' = -y$) and **in the y-axis** ($x' = -x$, $y' = y$); (e) **reflections in the origin** ($x' = -x$, $y' = -y$); (f) **reflections in the line** $y = x$ ($x' = y$, $y' = x$); (g) **simple shear transformations** ($x' = x + ky$, $y' = y$) or ($x' = x$, $y' = kx + y$). The affine transformation carries parallel lines into parallel lines, finite points into finite points, and leaves the line at infinity fixed. An affine transformation can always be factored into the product of transformations belonging to the above special cases. An **homogeneous affine transformation** is an affine transformation in which the constant terms are zero; an affine transformation which does not contain a translation as a factor. Its form is

$$x' = a_1 x + b_1 y, \qquad y' = a_2 x + b_2 y,$$

$$\Delta = a_1 b_2 - a_2 b_1 \neq 0.$$

An **isogonal affine transformation** is an affine transformation which does not change the size of angles. It has the form

$$x' = a_1 x + b_1 y + c_1,$$

$$y' = a_2 x + b_2 y + c_2,$$

where either $a_1 = b_2$ and $a_2 = -b_1$ or $-a_1 = b_2$ and $a_2 = b_1$. *For an n-dimensional vector space V*, an affine transformation is a transformation T of V into V of type $T(\mathbf{x}) = L(\mathbf{x}) + \boldsymbol\alpha$, where L is a linear transformation of V into V and $\boldsymbol\alpha$ is a fixed member of V. It is **singular** or **nonsingular** (or **regular**) according as L is singular or nonsingular. See LINEAR —linear transformation.

AGE, *n.* (*Life Insurance*) The **age at issue** is the age of the insured at his birthday nearest the policy date. The **age year** is a year in the lives of a group of people of a certain age. The age year l_x refers to the year from x to $x + 1$, the year during which the group is x years old.

AG′GRE-GA′TION, *n.* **signs of aggregation: Parenthesis**, ( ); **bracket**, [ ]; **brace**, { } and **vinculum** or **bar**, ——. Each means that the terms enclosed are to be treated as a single term. *E.g.*, $3(2 - 1 + 4)$ means 3 times 5, or 15. See various headings under DISTRIBUTIVE.

AGNESI, Maria Gaetana (1718–1799). Distinguished Italian mathematician. See WITCH.

AHLFORS, Lars Valerian (1907–). Finnish-American mathematician and Fields Medal recipient (1936). Research in complex-variable theory and theory of quasiconformal mappings, with important contributions to Riemann surfaces and meromorphic functions.

AHMES (RHYND or RHIND) PAPYRUS. Probably the oldest mathematical book known, written 2000 to 1800 B.C. and copied by the Egyptian scribe Ahmes about 1650 B.C. See PI, RHIND.

ALBERT, Abraham Adrian (1905–1972). American algebraist. Made fundamental contributions to the theory of Riemannian matrices and to the structure theory of associative and nonassociative algebras, Jordan algebras, quasigroups, and division rings.

ALBERTI, Leone Battista (1404–1472). Italian mathematician and architect. Wrote on art, discussing perspective and raising questions that pointed toward the development of projective geometry.

A′LEPH, *n.* The first letter of the Hebrew alphabet, written ℵ.

aleph-null or **aleph-zero.** The *cardinal number* of countably infinite sets, written $\aleph_0$. See CARDINAL —cardinal number.

ALEXANDER, James Waddell (1888–1971). American algebraic topologist who did research in complex-variable theory, homology and ring theory, fixed points, and the theory of knots.

Alexander's subbase theorem. A topological space is *compact* if and only if there is a subbase S for its topology which has the property that, whenever the union of a collection of members of S contains X, then X is contained in the union of a finite number of members of this collection.

ALEXANDROFF, Pavel Sergeevich (1896–).
Alexandroff compactification. See COMPACTIFI-
CATION.

AL'GE-BRA, *n.* (1) A generalization of arith-
metic. *E.g.*, the arithmetic facts that $2 + 2 + 2 = 3 \times 2$, $4 + 4 + 4 = 3 \times 4$, etc., are all special
cases of the (general) algebraic statement that
$x + x + x = 3x$, where x is any number. Letters
denoting any number, or any one of a certain set
of numbers, such as all real numbers, are related
by laws that hold for any numbers in the set;
e.g., $x + x = 2x$ for all x (all numbers). On the
other hand, conditions may be imposed upon a
letter, representing any one of a set, so that it
can take on but one value, as in the study of
equations; *e.g.*, if $2x + 1 = 9$, then x is restricted to 4. Equations are met in arithmetic,
although not so named. For instance, in percentage one has to find one of the unknowns in the
equation, interest = principal $\times$ rate, or $I = p \times r$,
when the other two are given. (2) A system of
logic expressed in algebraic symbols, or a Boolean
algebra (see BOOLE—Boolean algebra). (3) See
below, algebra over a field.

    algebra over a field. An **algebra** (or **linear
algebra**) over a field F is a ring R that is also a
vector space with members of F as scalars and
satisfies $(ax)(by) = (ab)(xy)$ for all scalars a and
b and all members x and y of R. The dimension
of the vector space is the **order** of R. The algebra is a **commutative algebra**, or an **algebra with
unit element**, according as the ring is a commutative ring, or a ring with unit element. A **division
algebra** is an algebra that is also a division ring.
A **simple algebra** is an algebra that is a simple
ring. The set of real numbers is a commutative
division algebra over the field of rational numbers; for any positive integer n, the set of all
square matrices of order n with complex numbers (or real numbers) as elements is an algebra
(noncommutative) over the field of real numbers.
Any algebra consisting of all n by n matrices
with elements in a given field is a simple algebra.
An algebra of order n with a unit element is
isomorphic to an algebra of n by n square matrices.

    algebra of propositions. See
BOOLE—Boolean algebra.

    algebra of subsets. An algebra of subsets of a
set X is a class of subsets of X which contains
the *complement* of each of its members and the
union of any two of its members (or the *intersection* of any two of its members). It is called a
σ-**algebra** if it also contains the union of any
sequence of its members. An algebra of subsets
of a *Boolean algebra* relative to the operations of
union and intersection. A *ring* of subsets of a set

X is an algebra of subsets of X if and only if it
contains X as a member. For any class C of
subsets of a set X, the intersection of all algebras (or σ-algebras) which contain C is the
smallest algebra (σ-algebra) which contains C
and is said to be the algebra (σ-algebra) generated by C. For the real line (or n-dimensional
space) examples of σ-algebras are the system of
all measurable sets, the system of all Borel sets,
and the system of all sets having the property of
Baire. See RING—ring of sets.

    Banach algebra. An algebra over the field of
real numbers (or complex numbers) which is also
a real (or complex) Banach space for which $\|xy\|
\le \|x\| \cdot \|y\|$ for all x and y. It is called a **real** or a
complex Banach algebra according as the field is
the real or the complex number field. The set of
all functions which are continuous on the closed
interval $[0, 1]$ is a Banach algebra over the field
of real numbers if $\|f\|$ is defined to be the
largest value of $|f(x)|$ for $0 \le x \le 1$. *Syn.* normed
vector ring.

    Boolean algebra. See BOOLE.

    fundamental theorem of algebra. See FUNDA-
MENTAL—fundamental theorem of algebra.

    homological algebra. See HOMOLOGICAL.

    measure algebra. See MEASURE—measure
ring and measure algebra.

AL'GE-BRA-IC, *adj.* **algebraic adder.** See
ADDER.

    algebraic addition. See SUM—algebraic sum,
sum of real numbers.

    algebraic curve. See CURVE.

    algebraic deviation. See DEVIATION.

    **algebraic expression, equation, function, oper-
ation, etc.** An expression, etc., containing or
using only algebraic symbols and operations, such
as $2x + 3$, $x^2 + 2x + 4$, or $\sqrt{2} - x + y = 3$. **Algebraic operations** are the operations of addition,
subtraction, multiplication, division, and raising
to integral or fractional powers. A **rational algebraic expression** is an expression that can be
written as a quotient of polynomials. An **irrational algebraic expression** is one that is not
rational, as $\sqrt{x + 4}$. See FUNCTION—algebraic
function, and various headings under RATIONAL.

    algebraic extension of a field. See EXTEN-
SION.

    algebraic hypersurface. See HYPERSURFACE.

    algebraic multiplication. See MULTIPLICA-
TION.

    algebraic number. (1) Any ordinary positive
or negative number; any real directed number.
(2) Any number which is a root of a polynomial
equation with rational coefficients; the degree of
the polynomial is said to be the **degree** of the

algebraic number α, and the equation is the **minimal equation** of α, if α is not a root of such an equation of lower degree. An **algebraic integer** is an algebraic number which satisfies some *monic* equation,

$$x^n + a_1 x^{n-1} + \cdots + a_n = 0,$$

with *integers* as coefficients. The minimal equation of an algebraic integer is also monic. A rational number is an algebraic integer if and only if it is an ordinary integer. The set of all algebraic numbers is an integral domain (see DOMAIN—integral domain). (3) Let F^* be a field and F a subfield of F^*. A member c of F^* is **algebraic** with respect to F if c is a zero of a polynomial with coefficients in F; otherwise, c is **transcendental** with respect to F. See BAKER, GELFOND—Gelfond-Schneider theorem.

   algebraic proofs and solutions. Proofs and solutions which use algebraic symbols and no operations other than those which are algebraic. See above, algebraic expression.

   algebraic subtraction. See SUBTRACTION.

   algebraic symbols. Letters representing numbers, and the various operational symbols indicating *algebraic operations*. See MATHEMATICAL SYMBOLS in the appendix.

   algebraic variety. See VARIETY.

   irrational algebraic surface. See IRRATIONAL.

AL-GE-BRA′IC-AL-LY, *a.* **algebraically complete field.** A field F which has the property that every polynomial equation with coefficients in F has a root in F. The field of algebraic numbers and the field of complex numbers are algebraically complete. Every field has an extension that is algebraically complete. *Syn.* algebraically closed field.

AL′GO-RITHM, *n.* Some special process of solving a certain type of problem, particularly a method that continually repeats some basic process.

   division algorithm. See DIVISION.

   Euclidean algorithm. A method of finding the greatest common divisor (G.C.D.) of two numbers—one number is divided by the other, then the second by the remainder, the first remainder by the second remainder, the second by the third, etc. When exact division is finally reached, the last divisor is the greatest common divisor of the given numbers (integers). In algebra, the same process can be applied to polynomials. *E.g.*, to find the greatest common divisor of 12 and 20, we have $20 \div 12$ is 1 with remain-

der 8; $12 \div 8$ is 1 with remainder 4; and $8 \div 4 = 2$; hence 4 is the G.C.D.

A-LIGN′MENT, *adj.* **alignment chart.** Same as NOMOGRAM.

AL′I-QUOT PART. Any exact divisor of a quantity; any factor of a quantity; used almost entirely when dealing with integers. *E.g.*, 2 and 3 are *aliquot parts* of 6.

AL′MOST, *adj.* **almost all** and **almost everywhere.** See MEASURE—measure zero.

   almost periodic. See PERIODIC.

AL′PHA, *n.* The first letter in the Greek alphabet: lower case, α; capital, A.

AL-TER′NANT, *n.* A determinant for which there are n functions $f_1, f_2, \cdots, f_n$ (if the determinant is of order n) and n quantities $r_1, r_2, \cdots, r_n$ for which the element in the ith column and jth row is $f_i(r_j)$ for each i and j (this determinant with rows and columns interchanged is also called an alternant). The Vandermonde determinant is an alternant (see DETERMINANT).

AL′TER-NATE, *adj.* The **alternate angles** are angles on opposite sides of a transversal cutting two lines, each having one of the lines for one of its sides. They are **alternate exterior angles** if neither lies between the two lines cut by the transversal. They are **alternate interior angles** if both lie between the two lines. See ANGLE—angles made by a transversal.

AL-TER-NAT′ING, *adj.* **alternating function.** A function f of more than one variable for which $f(x_1, x_2, \cdots, x_n)$ changes sign if two of the variables are interchanged. If f is an alternating multilinear function of n vectors in an n-dimensional vector space and $\{e_1, e_2, \cdots, e_n\}$ is a basis for V, then

$$f(v_1, \cdots, v_n) = \det(v_{ij}) f(e_1, \cdots, e_n),$$

where $v_k = \sum_{i=1}^{n} v_{ik} e_i$ and $\det(v_{ij})$ is the determinant with v_{ij} in the ith row and jth column.

   alternating group. See PERMUTATION—permutation group.

   alternating series. A series whose terms are alternately positive and negative, as

$$1 - \tfrac{1}{2} + \tfrac{1}{3} - \tfrac{1}{4} + \cdots + (-1)^{n-1}/n + \cdots.$$

An alternating series converges if the absolute values of its terms decreases monotonically with limit zero (the *Leibniz test for convergence*). This is a sufficient but not a necessary condition for convergence of an alternating series. If one convergent series has only positive terms and another only negative terms, then the series obtained by alternating terms from these series is convergent, but the absolute values of its terms may not be monotonically decreasing. The series

$$1 - \tfrac{1}{2} + \tfrac{1}{3} - \tfrac{1}{4} + \tfrac{1}{9} - \tfrac{1}{8} + \tfrac{1}{27} - \tfrac{1}{16} + \cdots$$

is such a series. See NECESSARY —necessary condition for convergence.

AL′TER-NA′TION, *n.* (1) In logic, the same as DISJUNCTION. (2) See PROPORTION.

AL-TER′NA-TIVE, *adj.* **alternative hypothesis.** See HYPOTHESIS —test of a hypothesis.

AL′TI-TUDE, *n.* A line segment indicating the height of a figure in some sense (or the length of such a line segment). See CONE, CYLINDER, PARABOLIC—parabolic segment, PARALLELO-GRAM, PARALLELEPIPED, PRISM, PYRAMID, RECT-ANGLE, SEGMENT —spherical segment, TRAPE-ZOID, TRIANGLE, ZONE.

   altitude of a celestial point. Its angular distance above, or below, the observer's horizon, measured along a great celestial circle (vertical circle) passing through the point, the zenith, and the nadir. The altitude is taken positive when the object is above the horizon and negative when below. See figure under HOUR—hour-angle and hour-circle.

AM-BIG′U-OUS, *adj.* Not uniquely determinable.

   ambiguous case in the solution of triangles. For a **plane triangle**, the case in which two sides and the angle opposite one of them is given. One of the other angles is then found by use of the law of sines; but there are always two angles less than 180° corresponding to any given value of the sine (unless the sine be unity, in which case the angle is 90° and the triangle is a right triangle). When the sine gives two distinct values of the angle, two triangles result if the side opposite is less than the side adjacent to the given angle (assuming the data are not such that there is no triangle possible, a situation that may arise in any case, ambiguous or nonambiguous). In the figure, angle A and sides a and b are given ($a < b$); triangles AB_1C and AB_2C are both solutions. If $a = b \sin A$, the right triangle ABC

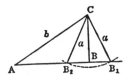

is the unique solution. For a **spherical triangle**, the ambiguous case is the case in which a side and the opposite angle are given (the given parts may then be either two sides and an angle opposite one side, or two angles and the side opposite one angle).

A-MER′I-CAN, *adj.* **American experience table of mortality.** See MORTALITY.

AM′I-CA-BLE, *adj.* **amicable numbers.** Two numbers, each of which is equal to the sum of all the exact divisors of the other except the number itself. *E.g*, 220 and 284 are amicable numbers, for 220 has the exact divisors 1, 2, 4, 5, 10, 11, 20, 22, 44, 55 and 110, whose sum is 284; and 284 has the exact divisors 1, 2, 4, 71, and 142, whose sum is 220. There are 236 known *amicable pairs* for which the smaller of the two amicable numbers is less than 10^8.

A-MOR-TI-ZA′TION, *n.* **amortization of a debt.** The discharge of the debt, including interest, by periodic payments, usually equal, which continue until the debt is paid without any renewal of the contract. The mathematical principles are the same as those used for annuities.

   amortization equation. An equation relating the amount of an obligation to be amortized, the interest rate, and the amount of the period payments.

   amortization of a premium on a bond. Writing down (decreasing) the book value of the bond on each dividend date by an amount equal to the difference between the dividend and the interest on the investment (interest on the book value at the yield rate). See VALUE—book value.

   amortization schedule. A table giving the annual payment, the amount applied to principal, the amount applied to interest, and the balance of principal due.

A-MOUNT′, *n.* **amount of a sum of money at a given date.** The sum of the principal and interest (simple or compound) to the date; designated as *amount at simple interest* or *amount at compound interest* (or *compound amount*), according as interest is simple or compound. In practice, the word *amount* without any qualification usually refers to amount at compound interest. The amount at some date in the future is the **future value**.

    amount of annuity. See ACCUMULATED — accumulated value of an annuity at a given date.

    compound amount. See INTEREST.

AM'PERE, *n.* The unit of electric current in the International System of Units (SI). The current in each of two long parallel wires which carry equal currents and for which there is a force of $2 \cdot 10^{-7}$ newton per meter acting on each wire. The legal standard of current before 1950 was the **international ampere**, which is the current which when passed through a standard solution of silver nitrate deposits silver at the rate of .001118 gram per sec. Our international ampere equals 0.999835 absolute ampere. See COULOMB, OHM.

AM'PLI-TUDE, *n.* **amplitude of a complex number.** The angle that the vector representing the complex number makes with the positive horizontal axis. *E.g.*, the *amplitude* of $2 + 2i$ is $45°$. See POLAR—polar form of a complex number.

    amplitude of a curve. Half the difference between the greatest and the least values of the ordinates of a periodic curve. The *amplitude* of $y = \sin x$ is 1; of $y = 2\sin x$ is 2.

    amplitude of a point. See POLAR—polar coordinates in the plane.

    amplitude of simple harmonic motion. See HARMONIC —simple harmonic motion.

AN'A-LOG, *adj.* **analog computer.** See COMPUTER.

A-NAL'O-GY, *n.* A form of inference sometimes used in mathematics to set up new theorems. It is reasoned that, if two or more things agree in some respects, they will probably agree in others. Exact proofs must, of course, be made to determine the validity of any theorems set up by this method.

A-NAL'Y-SIS, *n.* [*pl.* **analyses**]. That part of mathematics which uses, for the most part, algebraic and calculus methods—as distinguished from such subjects as synthetic geometry, number theory, and group theory.

    analysis of a problem. The exposition of the principles involved; a listing, in mathematical language, of the data given in the statement of the problem, other related data, the end sought, and the steps to be taken.

    analysis of variance. See VARIANCE.

    analysis situs. The field of mathematics now called *topology*.

    diophantine analysis. See DIOPHANTUS.

    proof by analysis. Proceeding from the thing to be proved to some known truth, as opposed to synthesis which proceeds from the true to that which is to be proved. The most common method of *proof by analysis* is, in fact, by *analysis* and *synthesis*, in that the steps in the analysis are required to be reversible.

    unitary analysis. A system of analysis that proceeds from a given number of units to a unit, then to the required number of units. Consider the problem of finding the cost of 7 tons of hay if $2\frac{1}{2}$ tons cost \$25.00. Analysis: If $2\frac{1}{2}$ tons cost \$25.00, 1 *ton costs* \$10.00. Hence 7 tons cost \$70.00.

AN-A-LYT'IC, *adj.* **analytic continuation.** If f is given to be a single-valued analytic function of a complex variable in a domain D, then possibly there is a function F analytic in a domain of which D is a proper subdomain, and such that

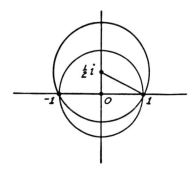

$F(z) = f(z)$ in D. If so, the function F is necessarily unique. The process of obtaining F from f is called **analytic continuation**. *E.g.*, the function f defined by $f(z) = 1 + z + z^2 + z^3 + \cdots$ is thereby defined only for $|z| < 1$, the radius of convergence of the series being 1 and the circle of convergence having center at 0. The series represents the function $1/(1 - z)$, but if this function is given a new representation, say by

$$\sum_{n=0}^{\infty} \frac{f^{(n)}(\tfrac{1}{2}i)}{n!} \left(z - \frac{1}{2}i\right)^n,$$

where the coefficients are determined from the original series, the new circle of convergence extends outside the old one (see the figure). The given function f, usually given as a power series (but not necessarily so), is called a **function-element** of F. The analytic continuation might well lead to a many-sheeted Riemann surface of definition of F. See MONOGENIC —monogenic analytic function.

analytic curve. A curve in n-dimensional Euclidean space which, in the neighborhood of each of its points, admits a representation of the form $x_j = x_j(t)$, $j = 1, 2, \cdots, n$, where the $x_j(t)$ are real analytic functions of the real variable t. If in addition we have $\sum_{j=1}^{n} (x_i')^2 \neq 0$, the curve is said to be a **regular analytic curve** and the parameter t is a **regular parameter** for the curve. For three-dimensional space, an analytic curve is a curve which has a parametric representation $x = x(t)$, $y = y(t)$, $z = z(t)$, for which each of these functions is an analytic function of the real variable t; it is a regular analytic curve if dx/dt, dy/dt and dz/dt do not vanish simultaneously. See POINT—ordinary point of a curve.

analytic function. (1) A **function of a single variable** x is analytic at h if it can be represented by a Taylor series in powers of $x - h$ whose sum is equal to the value of the function at each x in some neighborhood of h. The function is said to be analytic in the interval (a, b) if this is true for each h in the interval. See TAYLOR—Taylor's theorem. (2) A **function of r variables** is analytic at $p = (h_1, \cdots, h_r)$ if there is a neighbor of p in which the function is equal to the sum of an infinite series whose $(n + 1)$st term is a sum of terms of type

$$c_n (x_1 - h_1)^{a_1} (x_2 - h_2)^{a_2} \cdots (x_r - h_r)^{a_r},$$

where $a_1 + a_2 + \cdots + a_r = n$. (3) A **function of a complex variable** z is analytic at z_0 if there is a neighborhood U of z_0 such that the function is differentiable at each point of U. *Syn.* holomorphic, regular, or monogenic at z_0. See CAUCHY —Cauchy-Riemann partial differential equations. A single-valued function of z, or a multiple-valued function considered as a single-valued function on its Riemann surface, is analytic if it is differentiable at each point of its domain D (a non-null connected open set). Such a function is sometimes said to be analytic in D if it is continuous in D and has a derivative at all except at most a finite number of points of D. If f is differentiable at all points of D, it is a **regular function**, a **regular analytic function**, or a **holomorphic function** in D. If a function f of a complex variable is analytic at a point z_0, then f has continuous derivatives of all orders at z_0 and can be expanded as a Taylor series in a neighborhood of z_0.

analytic proof or solution. A proof or solution which depends upon that sort of procedure in mathematics called analysis; a proof which consists, essentially, of algebraic (rather than geometric) methods and/or of methods based on

limiting processes (such as the methods of differential and integral calculus).

analytic set. Let X be the real line, or any space that is homeomorphic to a separable complete metric space. A subset S of X is **analytic** if S is the continuous image of a Borel set in X, or (equivalently) if S is a continuous image of the space of irrational numbers. If both S and $X - S$ are analytic, then S is a Borel set (**Souslin's theorem**).

analytic structure for a space. A covering of a *locally Euclidean space* of dimension n by a set of open sets each of which is homeomorphic to an open set in n-dimensional Euclidean space E_n and which are such that whenever any two of these open sets overlap, the coordinate transformation (in both directions) is given by analytic functions (*i.e.*, functions which can be expanded in power series in some neighborhood of any point). If neighborhoods U and V overlap and P is in their intersection, then the homeomorphisms of U and V with open sets of n-dimensional Euclidean space define coordinates $(x_1, x_2, \cdots, x_n)$ and $(y_1, y_2, \cdots, y_n)$ for P, and the functions $x_i = x_i(y_1, \cdots, y_n)$ and $y_i = y_i(x_1, \cdots, x_n)$ are the functions required to be analytic. The analytic structure is **real** or **complex** according as the coordinates of points in E_n are taken as real or complex numbers. See EUCLIDEAN—locally Euclidean space, MANIFOLD.

a-point of an analytic function. An a-point of the analytic function $f(z)$ of the complex variable z is a *zero point* of the analytic function $f(z) - a$. The order of an a-point is the order of the zero of $f(z) - a$ at the point. See ZERO—zero point of an analytic function of a complex variable.

normal family of analytic functions. See NORMAL.

quasi-analytic function. For a sequence of positive numbers $\{M_1, M_2, \cdots\}$ and a closed interval $[a, b] = I$, the *class of quasi-analytic functions* is the set of all functions which possess derivatives of all orders on I and which are such that for each function f there is a constant k such that

$$|f^{(n)}(x)| < k^n M_n \quad \text{for} \quad n \geq 1 \quad \text{and} \quad x \in I,$$

provided this set of functions has the property that if f is a member of the set and $f^{(n)}(x_0) = 0$ for $n \geq 0$ and $x_0 \in I$, then $f(x) \equiv 0$ on I. If $M_n = n!$, or $M_n = n^n$, then the corresponding class of functions is precisely the class of all analytic functions on I. Every function which possesses derivatives of all orders on I (*e.g.*, e^{-1/x^2} on $[0, 1]$) is the sum of two functions each of which belongs to a quasi-analytic class. Even if the class defined by $M_1, \cdots$ and I is not quasi-analytic, certain subclasses are sometimes said to be quasi-analytic if they do not contain a nonzero function f for which $f^{(n)}(x_0) = 0$ for $n \geq 0$ and $x_0 \in I$. Quasi-analyticity is one of the most important properties of analytic functions, but there exist classes of quasi-analytic functions which contain nonanalytic functions.

singular point of an analytic function. See SINGULAR.

AN'A-LYT'I-CAL-LY, *a.* Performed by analysis, by analytic methods, as opposed to synthetic methods.

AN'A-LY-TIC'I-TY, *n.* **point of analyticity.** A point at which a function of the complex variable z is analytic.

ANCHOR RING or TORUS. A surface in the shape of a doughnut with a hole in it; a surface generated by the rotation, in space, of a circle about an axis in its plane but not cutting the circle. If r is the radius of the circle, k the distance from the center to the axis of revolution, in this case the z-axis, and the equation of the generating circle is $(y - k)^2 + z^2 = r^2$, then the equation of the anchor ring is

$$\left(\sqrt{x^2 + y^2} - k\right)^2 + z^2 = r^2.$$

Its volume is $2\pi^2 k r^2$ and the area of its surface is $4\pi^2 k r$.

AN'GLE, *n* A **geometric angle** (or simply **angle**) is a set of points consisting of a point P and two rays extending from P (sometimes it is required that the rays do not lie along the same straight line). The point P is the **vertex** and the rays are the **sides** (or **rays**) of the angle. Two geometric angles are **equal** if and only if they are congruent. When the two rays of an angle do not extend along the same line in opposite directions from the vertex, the set of points between the rays is the **interior** of the angle. The **exterior** of an angle is the set of all points in the plane that are not in the union of the angle and its interior. A **directed angle** is an angle for which one ray is designated as the initial side and the other as the terminal side. There are two commonly used signed measures of directed angles. If a circle is drawn with unit radius and center at the vertex of a directed angle, then a **radian measure** of the angle is the length of an arc that extends counterclockwise along the circle from the initial side to the terminal side of the angle, or the negative of the length of an arc that extends clockwise

along the circle from the initial side to the terminal side. The arc may wrap around the circle any number of times. For example, if an angle has radian measure $\frac{1}{2}\pi$, it also has radian measure $\frac{1}{2}\pi + 2\pi$, $\frac{1}{2}\pi + 4\pi$, etc., or $\frac{1}{2}\pi - 2\pi$, $\frac{1}{2}\pi - 4\pi$, etc. **Degree measure** of an angle is defined so that $360°$ corresponds to radian measure of 2π

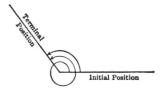

[see SEXAGESIMAL —sexagesimal measure of an angle]. A **rotation angle** consists of a directed angle *and* a signed measure of the angle. The angle is a **positive angle** or a **negative angle** according as the measure is positive or negative. **Equal** rotation angles are rotation angles that have the same measure. Usually, **angle** means rotation angle (*e.g.*, see below, angle of depression, angle of inclination, obtuse angle). A rotation angle can be thought of as being a directed angle together with a description of how the angle is formed by rotating a ray from an initial position (on the initial side) to a terminal position (on the terminal side).

acute angle. See ACUTE.

addition of angles. See SUM—sum of angles.

adjacent angles. See ADJACENT.

angle of depression. The angle between the horizontal plane and the oblique line joining the observer's eye to some object lower than (beneath) the line of his eye.

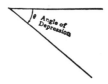

angle of elevation. The angle between the horizontal plane and the oblique line from the observer's eye to a given point above his eye.

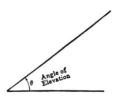

angle of friction. See FRICTION.

angle of inclination of a line. The positive angle, less than 180°, measured from the positive x-axis to the given line.

angle of intersection. The angle of intersection of **two lines in a plane** is defined thus: The angle from line L_1, say, to line L_2 is the smallest positive angle that has L_1 as initial side and L_2

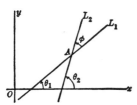

as terminal side, angle ϕ in the figure. The tangent of the angle from L_1 to L_2 is given by

$$\tan \phi = \tan(\theta_2 - \theta_1) = \frac{m_2 - m_1}{1 + m_1 m_2},$$

where $m_1 = \tan\theta_1$ and $m_2 = \tan\theta_2$. The **angle between lines** L_1 and L_2 is the least positive angle between the two lines (the angle between two parallel lines is defined to be of measure 0). The **angle between two lines in space** (whether or not they intersect) is the angle between two intersecting lines which are parallel respectively to the two given lines. The cosine of this angle is equal to the sum of the products in pairs of the corresponding direction cosines of the lines (see DIRECTION —direction cosines). The angle between two **intersecting curves** is the angle between the tangents to the curves at their point of intersection. The angle between **a line and a plane** is the smaller (acute) angle which the line makes with its projection in the plane. The angle between **two planes** is the dihedral angle which they form (see DIHEDRAL); this is equal to the angle between the normals to the planes (see above, angle between two lines in space). When the equations of the planes are in normal form, the cosine of the angle between the planes is equal to the sum of the products of the corresponding coefficients (coefficients of the same variables) in their equations.

angle of a lune. See LUNE.

angles made by a transversal. The angles made by a line (the transversal) which cuts two or more other lines. In the figure, the transversal t cuts the lines m and n. The angles a, b, c', d' are **interior angles**; a', b', c, d are **exterior angles**; a and c', and b and d' are the pairs of **alternate-interior angles**; b' and d, a' and c are

the pairs of **alternate-exterior angles**; a' and a, b' and b, c' and c, d' and d are the **exterior-interior** or **corresponding angles**.

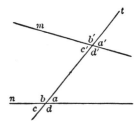

angle of a polygon. An **interior angle** of a polygon is an angle whose vertex is a vertex of the polygon, whose sides contain the sides of the polygon that meet at this vertex, and whose measure is equal to the smallest positive measure that describes a rotation from one side through the interior of the polygon to the other side. An **exterior angle** is an angle whose vertex is a vertex of the polygon, whose sides contain one side (of the polygon) with an endpoint at this vertex and the other such side of the polygon extended through the vertex, and whose measure is equal to the least positive measure that describes a rotation from one side of the angle to the other through the exterior of the polygon. At each vertex of a polygon, there is one interior angle and there are two exterior angles. These definitions suffice for any polygon for which no side contains points of more than two other sides (in other cases, the sides must be ordered in some way so that the angles between them can be defined uniquely).

angle of reflection. See REFLECTION.

angle of refraction. See REFRACTION.

base angles of a triangle. The angles in the triangle having the base of the triangle for their common side.

central angle. See CENTRAL.

complementary angles. See COMPLEMENTARY.

conjugate angles. Two angles whose sum is 360°. Such angles are sometimes said to be *explements* of each other.

coterminal angles. See COTERMINAL.

dihedral angle. See DIHEDRAL.

direction angles. See DIRECTION —direction angles.

eccentric angle. See ELLIPSE.

Euler angles. See EULER—Euler angles.

explementary angles. See above, conjugate angles.

face angles. See below, polyhedral angle.

flat angle. Same as STRAIGHT ANGLE.

hour angle of a celestial point. The angle between the plane of the meridian of the observer and the plane of the hour circle of the star, measured westward from the plane of the meridian. See HOUR—hour angle and hour circle.

interior angle. See above, angle of a polygon, angles made by a transversal.

measure of an angle. See ANGLE, MIL, RADIAN, SEXAGESIMAL —sexagesimal measure of an angle.

obtuse angle. An angle numerically **greater** than a right angle and less than a straight angle; sometimes used for all angles numerically greater than a right angle.

opposite angle. See OPPOSITE.

plane angle. See above, ANGLE.

polar angle. See POLAR—polar coordinates in the plane.

polyhedral angle. The configuration formed by the lateral faces of a polyhedron which have a common vertex (*A-BCDEF* in the figure); the positional relation of a set of planes determined by a point and the sides of some polygon whose plane does not contain the point. The planes

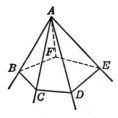

(*ABC*, etc.) are **faces** of the angle; the lines of intersection of the planes are **edges** of the *polyhedral angle*. Their point of intersection (*A*) is the **vertex**. The angles (*BAC, CAD*, etc.) between two successive edges are **face angles**. A **section** of a polyhedral angle is the polygon formed by cutting all the edges of the angle by a plane not passing through the vertex.

quadrant angles. See QUADRANT.

quadrantal angles. See QUADRANTAL.

reflex angle. An angle greater than a straight angle and less than two straight angles; an angle between 180° and 360°.

related angle. See RELATED.

right angle. Half of a straight angle; an angle of 90° or $\frac{1}{2}\pi$ radians.

solid angle. See SOLID.

spherical angle. The figure formed at the intersection of two great circles on a sphere; the difference in direction of the arcs of two great circles at their point of intersection. In the figure, the spherical angle is *APB*. It is equal to

the plane angles *A'PB'* and *AOB*. See SPHERICAL—spherical degree.

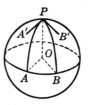

straight angle. An angle whose sides lie on the same straight line, but extend in opposite directions from the vertex; an angle of 180° or π radians. *Syn.* flat angle.

supplementary angles. See SUPPLEMENTARY.

tetrahedral angle. A polyhedral angle having four faces.

trihedral angle. A polyhedral angle having three faces.

trisection of an angle. See TRISECTION.

vertex angle. The angle opposite the base of a triangle.

vertical angles. Two angles such that each side of one is a prolongation, through the vertex, of a side of the other.

zero angle. The figure formed by two rays drawn from the same point in the same direction (so as to coincide); an angle whose measure in degrees is 0.

AN'GU-LAR, *adj.* Pertaining to an angle; circular; around a circle.

angular acceleration. See ACCELERATION — angular acceleration.

angular distance. See DISTANCE—angular distance between two points.

angular momentum. See MOMENTUM— moment of momentum.

angular speed. See SPEED.

angular velocity. See VELOCITY.

AN'HAR-MON'IC RATIO. See RATIO— anharmonic ratio.

AN-NI'HI-LA'TOR, *n.* The annihilator of a set *S* is the class of all functions of a certain type which *annihilate S* in the sense of being zero at each point of *S*. *E.g.*, if the functions are continuous linear functionals and *S* is a subset of a normed linear space *N*, then the annihilator of *S* is the linear subset *S'* of the first conjugate space *N** which consists of all continuous linear functionals which are zero at each point of *S*. Analogously, the annihilator of a linear subset *S* of Hilbert space is the orthogonal complement of *S*.

AN′NU-AL, *adj.* Yearly.

annual premium (net annual premium). See PREMIUM.

annual rent. Rent, when the payment period is a year. See RENT.

AN-NU′I-TANT, *n.* The life (person) upon whose existence each payment of a life annuity is contingent, *i.e.*, the beneficiary of an annuity.

AN-NU′I-TY, *n.* A series of payments at regular intervals. An **annuity contract** is a written agreement setting forth the amount of the annuity, its cost, and the conditions under which it is to be paid (sometimes called an **annuity policy**, when the annuity is a temporary annuity). The **payment interval** of an annuity is the time between successive payment dates; the **term** is the time from the beginning of the first payment interval to the end of the last one. An annuity is a **simple annuity**, or a **general annuity**, according as the payment intervals do, or do not, coincide with the interest conversion periods. A **deferred annuity** (or **intercepted annuity**) is an annuity in which the first payment period begins after a certain length of time has lapsed; it is an **immediate annuity** if the term begins immediately. An **annuity due** is an annuity in which the payments are made at the beginning of each period. If the payments are made at the end of the periods, the annuity is called an **ordinary annuity**. An **annuity certain** is an annuity that provides for a definite number of payments, as contrasted to a **life annuity**, which is a series of payments at regular intervals during the life of an individual (a **single-life annuity**) or of a group of individuals (a **joint-life annuity**). A **last-survivor annuity** is an annuity payable until the last of two (or more) lives end. An annuity whose payments continue forever is called a **perpetuity**. A **temporary annuity** is an annuity extending over a given period of years, provided the recipient continues to live throughout that period, otherwise terminating at his death. A **reversionary annuity** is an annuity to be paid during the life of one person, beginning with the death of another. An annuity whose payments depend upon certain conditions, such as some person (not necessarily the beneficiary) being alive, is called a **contingent annuity**. A **forborne annuity** is a life annuity whose term began sometime in the past; *i.e.*, the payments have been allowed to accumulate with the insurance company for a stated period. In case a group contributes to a fund over a stated period and at the end of the period the accumulated fund is converted into annuities for each of the survivors, the annuity is also called a *forborne annuity*. A life annuity is **curtate**, or **complete**,

according as a proportionate amount of a payment is not made, or is made, for the partial period from the last payment before death of the beneficiary to the time of death. A complete annuity is also called an **apportionate annuity** and a **whole-life annuity**. An annuity is **increasing** if each payment after the first is larger than the preceding payment; it is **decreasing** if each payment except the last is larger than the next payment. Also see TONTINE—tontine annuity.

accumulated value of an annuity. The accumulated value (or **amount**) of an annuity at a given date is the sum of the compound amounts of the annuity payments to that date. The **amount** of an annuity is the accumulated value at the end of the term of the annuity.

annuity bond. See BOND.

cash equivalent (or **present value**) **of an annuity.** See VALUE—present value.

ANN′U-LUS, *n.* [*pl.* **annuli** or **annuluses**]. The portion of a plane bounded by two concentric circles in the plane. The area of an annulus is the difference between the areas of the two circles, namely $\pi(R^2 - r^2)$, where R is the radius of the larger circle and r is the radius of the smaller.

A-NOM′A-LY, *n.* anomaly of a point. See PO-LAR—polar coordinates in the plane.

AN-TE-CED′ENT, *n.* (1) The first term (or numerator) of a ratio; that term of a ratio which is compared with the other term. In the ratio 2/3, 2 is the *antecedent* and 3 is the *consequent*. (2) See IMPLICATION.

AN′TI-AU-TO-MOR′PHISM, *n.* See ISOMOR-PHISM.

AN′TI-COM-MU′TA-TIVE, *adj.* A method of combining two objects, $a \cdot b$, is *anticommutative* if $a \cdot b = -b \cdot a$. E.g., see MULTIPLICATION — multiplication of vectors (3).

AN-TI-DE-RIV′A-TIVE, *n.* antiderivative of a function. Same as the PRIMITIVE or INDEFINITE INTEGRAL of the function. See INTEGRAL—indefinite integral.

AN-TI-HY-PER-BOL′IC functions. Same as IN-VERSE HYPERBOLIC functions. See HYPERBOL-IC—inverse hyperbolic functions.

AN-TI-I-SO-MOR′PHISM, *n.* See ISOMOR-PHISM.

AN-TI-LOG′A-RITHM, *n.* antilogarithm of a given number. The number whose logarithm is the given number; *e.g.*, antilog$_{10}$ 2 = 100. *Syn.* inverse logarithm. To find an antilogarithm corresponding to a given logarithm that is not in the tables, one can use interpolation. See INTERPOLATION.

AN′TI-PAR′AL-LEL, *adj.* **antiparallel lines.** Two lines which make, with two given lines,

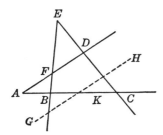

angles that are equal in opposite order. In the figure, the lines *AC* and *AD* are antiparallel with respect to the lines *EB* and *EC*, since $\angle EFD = \angle BCD$ and $\angle ADE = \angle EBC$. Two parallel lines have a similar property. The parallel lines *AD* and *GH* also make equal angles with the lines *EB* and *EC*, but in the same order; *i.e.*, $\angle EFD = \angle BGH$ and $\angle ADE = \angle GHD$.
    antiparallel vectors. See PARALLEL—parallel vectors.

AN-TIP′O-DAL, *adj.* **antipodal points.** Points on a sphere at opposite ends of a diameter.

AN′TI-SYM-MET′RIC, *adj.* **antisymmetric dyadic.** See DYAD.
    antisymmetric relation. See SYMMETRIC—symmetric relation.

AN′TI-TRIG′O-NO-MET′RIC, *adj.* **antitrigonometric function.** Same as INVERSE TRIGONOMETRIC FUNCTION. See TRIGONOMETRIC —inverse trigonometric function.

A′PEX, *n.* [*pl.* **apexes** or **apices**]. A highest point relative to some line or plane. The apex of a triangle is the vertex opposite the side which is considered as the base; the apex of a cone is its vertex.

A-PHE′LI-ON, *n.* See PERIHELION.

APOLLONIUS of Perga (c. 255–170 B.C.). A great Greek geometer.
    Problem of Apollonius. To construct a circle tangent to three given circles.

A POS-TE′RI-O′RI, *adj.* **a posteriori knowledge.** Knowledge from experience. *Syn.* Empirical knowledge.
    a posteriori probability. See PROBABILITY—empirical or a posteriori probability.

A-POTH′E-CAR-Y, *n.* **apothecaries′ weight.** The system of weights used by druggists. The pound and the ounce are the same as in troy weight, but the subdivisions are different. See DENOMINATE NUMBERS in the appendix.

AP′O-THEM, *n.* The perpendicular distance from the center of a regular polygon to a side. *Syn.* short radius.

AP-PAR′ENT, *adj.* **apparent distance.** See DISTANCE—angular distance between two points.
    apparent time. Same as APPARENT SOLAR TIME. See TIME.

APPLIED MATHEMATICS. See MATHEMATICS.

AP-POR′TION-A-BLE, *adj.* **apportionable annuity.** See ANNUITY.

AP-PROACH′, *v.* **approach a limit.** See LIMIT.

AP-PROX′I-MATE, *adj., v.* To calculate nearer and nearer to a correct value; used mostly for numerical calculations. *E.g.*, one *approximates* the square root of 2 when he finds, in succession, the numbers 1.4, 1.41, 1.414, whose successive squares are nearer and nearer to 2. Finding any one of these decimals is also called approximating the root; that is, to *approximate* may mean either to secure one result near a desired result, or to secure a succession of results approaching a desired result.
    approximate result, value, answer, root, etc. One that is nearly but not exactly correct. Sometimes used of results either nearly or exactly correct. See ROOT—root of an equation.

AP-PROX′I-MA′TION, *n.* (1) A result that is not exact, but is accurate enough for some specific purpose. (2) The process of obtaining such a result.
    approximation by differentials. See DIFFERENTIAL.
    approximation property. A Banach space X having the **approximation property** means that, for any $\epsilon > 0$ and any compact set K in X, there is a continuous linear transformation L from X onto a finite-dimensional subspace of X for

which $\|L(x) - x\| \le \epsilon$ if $x \in K$. The space X has the approximation property if, and only if, for each $\epsilon > 0$ and each continuous linear transformation T from any Banach space Y into X for which the range of T is compact, there is a continuous linear transformation L from Y onto a finite-dimensional subspace of X such that $\|T - L\| < \epsilon$. If a Banach space X has a basis, then X has the approximation property. It was shown by Per Enflo in 1973 that there exists a separable Banach space that does not have the approximation property and therefore has no basis. See LINEAR—linear transformation.

successive approximation. The successive steps taken in working toward a desired result or calculation. See APPROXIMATE.

A PRI-O'RI, *adj.* **a priori fact.** Used in about the same sense as axiomatic or self-evident fact.

a priori knowledge. Knowledge obtained from pure reasoning from cause to effect, as contrasted to empirical knowledge (knowledge obtained from experience); knowledge which has its origin in the mind and is (supposed to be) quite independent of experience.

a priori reasoning. Reasoning which arrives at conclusions from definitions and assumed axioms or principles; deductive reasoning.

AR'A-BIC, *adj.* **Arabic numerals:** 1, 2, 3, 4, 5, 6, 7, 8, 9, 0; introduced into Europe from Arabia, probably originating in India. Also called HINDU-ARABIC NUMERALS.

AR'BI-LOS, *n.* A plane figure A bounded by a semicircle C of diameter d and two smaller semicircles with diameters d_1 and d_2 for which $d_1 + d_2 = d$ and the smaller semicircles lie inside C with their diameters along the diameter of C. The area of A is $\frac{1}{4}\pi d_1 d_2$. Archimedes and other Greeks studied the arbilos in great detail. *Syn.* shoemaker's knife. See SALINON.

AR'BI-TRAR'Y, *adj.* **arbitrary assumption.** An assumption constructed at the pleasure of the individual without regard to its being consistent either with the laws of nature or (sometimes) with accepted mathematical principles.

arbitrary constant. See CONSTANT.

arbitrary ϵ. A statement is true for *arbitrary* ϵ if it is true for any numerical value (usually restricted to be positive) which may be assigned to ϵ. This idiom usually occurs in situations where small values of ϵ are of the most interest.

arbitrary function in the solution of partial differential equations. A symbol that stands for an unspecified function in an expression that satisfies the differential equation whatever func-

tion (of some specified type) may be substituted for the symbol. *E.g.*, $z = xf(y)$ is a solution of $x(\partial z/\partial x) - z = 0$, if f is any differentiable function.

arbitrary parameter. Same as *parameter* in its most commonly used sense. The addition of the attribute *arbitrary* places emphasis upon the fact that this particular parameter is not determined, but may be any member of some set (*e.g.*, any real number).

ARC, *n.* A segment, or piece, of a curve. *Tech.* (1) The image of a closed interval $[a, b]$ under a one-to-one continuous transformation; *i.e.*, a simple curve that is not closed. (2) A curve that is not a closed curve. If a curve is the continuous image of the interval $[a, b]$, then an **arc of the curve** is any arc that is the image of an interval $[c, d]$ contained in $[a, b]$.

arc length. The length of an arc. See LENGTH—length of a curve.

degree of arc. An arc of a circle is an arc of one degree if it subtends an angle of one degree at the center of the circle. The measure of an arc in degrees is the measure of the angle subtended at the center of the circle.

differential (or element) of arc. See ELEMENT —element of integration.

limit of the ratio of an arc to its chord. See LIMIT—limit of the ratio of an arc to its chord.

minor arc of a circle. See SECTOR —sector of a circle. *Syn.* short arc.

ARC CO-SE'CANT, *n.* The arc cosecant of a number x is an angle (or number) whose cose-

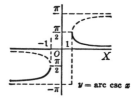

$y = \text{arc csc } x$

cant is x, written $\csc^{-1} x$ or **arc csc** x. *E.g.*, $\csc^{-1} 2$ is $30°$, $150°$, or in general, $n180° + (-1)^n 30°$. See TRIGONOMETRIC—inverse trigonometric function. *Syn.* inverse cosecant, anticosecant. (In the figure, y is in radians.)

ARC CO'SINE, *n.* The arc cosine of a number x is an angle (or number) whose cosine is x, written $\cos^{-1} x$ or **arc cos** x. *E.g.*, $\arccos \frac{1}{2}$ is $60°$, $300°$, or in general $n360° \pm 60°$. See TRIGONOMETRIC—inverse trigonometric function. *Syn.* inverse cosine, anticosine. The figure shows the graph of $y = \cos^{-1} x$ (y in radians).

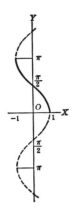

ARC COTANGENT, *n.* The arc cotangent of a number x is an angle (or number) whose cotangent is x, written $\cot^{-1} x$, $\operatorname{ctn}^{-1} x$ or $\operatorname{arc cot} x$. *E.g.*, arc cot 1 is 45°, 225°, or in general $n180° + 45°$. See TRIGONOMETRIC —inverse trigonometric function. *Syn.* inverse cotangent, anticotangent. (In the figure, y is in radians.)

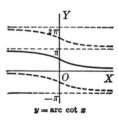

$y = \operatorname{arc cot} x$

ARCHIMEDES (c. 287–212 B.C.). Greek geometer, analyst and physicist. He used the limiting processes of both differential and integral calculus. One of the greatest mathematicians of all time.

    Archimedean property. The property of real numbers that for any positive numbers a and b there is a positive integer n such that $a < nb$.

    Archimedean solid. A polyhedron whose faces are regular polygons (not necessarily of the same type) and whose polyhedral angles are all congruent. There are 13 such solids, of which only 5 are regular. One of the simplest that is not regular is the **cubooctahedron**, formed from a cube by joining the mid-points of edges in each face and removing the 8 corner pyramids; it has 14 sides and 12 vertices. *Syn.* semiregular solid.

    method of exhaustion. See EXHAUSTION.

    spiral of Archimedes. See SPIRAL.

ARC-HY-PER-BOL'IC, *adj.* **arc-hyperbolic sine, cosine, etc.** See HYPERBOLIC —inverse hyperbolic functions.

ARC SE-CANT', *n.* The arc secant of a number x is an angle (or number) whose secant is x, written $\sec^{-1} x$ or $\operatorname{arc sec} x$. *E.g.*, arc sec 2 is $-60°$, 300°, or in general, $n360° \pm 60°$. See TRIGONOMETRIC—inverse trigonometric function. *Syn.* inverse secant, anti-secant. (In the figure, y is in radians.)

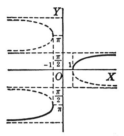

ARC SINE, *n.* The arc sine of a number x is an angle (or number) whose sine is x, written

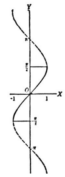

$\sin^{-1} x$ or $\operatorname{arc sin} x$. *E.g.*, arc sin $\frac{1}{2}$ is 30°, 150°, or in general $n180° + (-1)^n 30°$. See TRIGONOMETRIC—inverse trigonometric function. *Syn.* inverse sine, antisine. The figure shows the graph of $y = \sin^{-1} x$ (y in radians).

ARC TAN-GENT', *n.* The arc tangent of a number x is an angle (or number) whose tangent is x, written $\tan^{-1} x$ or $\operatorname{arc tan} x$.

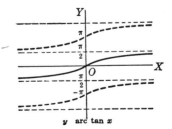

y arc tan x

E.g., arc tan 1 is 45°, 225°, or in general, $n180° + 45°$. See TRIGONOMETRIC —inverse trigonometric function. *Syn.* inverse tangent, antitangent. The figure shows the graph of $y = \tan^{-1} x$ (*y* in radians.)

ARC′ WISE, *adv.* **arcwise-connected set.** See CONNECTED.

ARE, *n.* A unit of measure of area in the metric system, equal to 100 square meters or 119.6 square yards. See HECTARE.

A′ RE-A, *n.* **area of a lune.** See LUNE.
  area of a plane set. A rectangle with adjacent edges of length *a* and *b* has area *ab*. The area of any bounded plane set is the least upper bound α of the sum of the areas of a finite collection of nonoverlapping rectangles contained in the set, or the greatest lower bound β of the sum of the areas of a finite collection of rectangles which together completely cover the set, provided $\alpha = \beta$ (if $\alpha = \beta = 0$, the set has area and the area is zero; if $\alpha \neq \beta$, the set does not have area). An unbounded set with area is an unbounded set S for which there is a number m such that $R \cap S$ has area not greater than m whenever R is a rectangle; the area of S is then the least upper bound of the areas of $R \cap S$ for rectangles R. This definition can be used to prove the usual formulas for area (see the specific configuration: circle, triangle, etc.) Calculus is very useful for computing area (see INTEGRAL —definite integral). The "method of exhaustion" is related to the methods of calculus (see EXHAUSTION). *Syn.* two-dimensional content. See CONTENT, DIDO'S PROBLEM, MEASURABLE —measurable set, PAPPUS.
  area of a surface. See SURFACE —surface area.
  differential (or element) of area. See ELEMENT—element of integration, SURFACE— surface area, surface of revolution.
  lateral area of a cone, cylinder, parallelepiped, etc. See the specific configuration.
  relations between areas of similar surfaces. Areas of similar surfaces have the same ratio (vary as) the squares of corresponding lines. *E.g.*, (1) the areas of two circles are in the same ratio as the squares of their radii, (2) the areas of two similar triangles are in the same ratio as the squares of corresponding sides or altitudes.

ARGAND, Jean Robert (1768–1822). Swiss mathematician. One of the first to publish (1806) an account of the graphical representation of complex numbers. See GAUSS—Gauss plane, WALLIS, WESSEL.

Argand diagram. Two perpendicular axes on one of which real numbers are represented and on the other pure imaginaries, thus providing a

frame of reference for graphing complex numbers. These axes are called the REAL AXIS and the IMAGINARY AXIS or the AXIS OF REALS and the AXIS OF IMAGINARIES.

AR′ GU-MENT, *n.* **argument of a complex number.** Same as AMPLITUDE. See AMPLITUDE — amplitude of a complex number.
  argument of a function. Same as the INDEPENDENT VARIABLE. See FUNCTION.
  arguments in a table of values of a function. The values in the domain of the function for which the values in the range are tabulated. The arguments in a trigonometric table are the angles for which the functions are tabulated; in a table of logarithms, the numbers for which the logarithms are tabulated.

A-RITH′ ME-TIC, *n.* The study of the positive integers $1, 2, 3, 4, 5, \cdots$ under the operations of addition, subtraction, multiplication, and division, and the use of the results of these studies in everyday life.
  arithmetic modulo *n.* See CONGRUENCE.
  fundamental operations of arithmetic. Addition, subtraction, multiplication, and division.

AR-ITH-MET′ IC or AR-ITH-MET′ I-CAL, *adj.* Employing the principles and symbols of arithmetic.
  arithmetic component. In a computing machine, any component that is used in performing arithmetic, logical, or other similar operations.
  arithmetic mean (or average). See MEAN.
  arithmetic means between two numbers. The other terms of an *arithmetic sequence* of which the given numbers are the first and last terms; a single mean between two numbers *x* and *y* is their **mean,** $\frac{1}{2}(x + y)$. See MEAN, and below, arithmetic sequence.
  arithmetic number. See NUMBER —arithmetic numbers.
  arithmetic progression. Same as ARITHMETIC SEQUENCE.
  arithmetic sequence. A sequence, each term of which is equal to the sum of the preceding term and a constant; written: $a, a + d, a +$

$2d, \cdots, a + (n-1)d$, where a is the **first term**, d is the **common difference**, or simply the **difference**, and $a + (n-1)d = l$ is the **last**, or n**th**, **term**. The positive integers, $1, 2, 3, \cdots$, form an arithmetic sequence. *Syn.* arithmetic progression. See below, arithmetic series.

arithmetic series. The indicated sum of the terms of an arithmetic sequence. The sum of the arithmetic sequence described above is equal to

$$\tfrac{1}{2}n(a+l) \quad \text{or} \quad \tfrac{1}{2}n[2a + (n-1)d].$$

AR-ITH-MOM′E-TER, *n.* A computing machine.

ARM, *n.* **arm of an angle.** A side of the angle.

AR-RANGE′MENT, *n.* Same as PERMUTATION (1).

AR-RAY′, *n.* A display of objects in some regular arrangement, as a *rectangular array* or *matrix* in which numbers are displayed in rows and columns, or an arrangement of *statistical data* in order of increasing (or decreasing) magnitude.

ARTIN, Emil (1898–1962). German algebraist and group theorist who spent many years in the U.S.
    Artinian ring. See CHAIN—chain conditions on rings.

AS-CEND′ING, *adj.* **ascending powers of a variable in a polynomial.** Powers of the variable that increase as the terms are counted from left to right, as in the polynomial

$$a + bx + cx^2 + dx^3 + \cdots.$$

    ascending chain condition on rings. See CHAIN—chain conditions on rings.

ASCOLI, Giulio (1843–1896). Italian analyst.
    Ascoli's theorem. Let A be an infinite set of functions whose domains are the same closed bounded set D in a finite-dimensional Euclidean space (*e.g.*, a closed bounded interval) and whose ranges are sets of real numbers. If these functions are *pointwise equicontinuous* and there is a number M such that $|f(x)| \leq M$ for all f in A and all x in D, then there is a sequence $\{f_n\}$ of distinct members of A that converges uniformly to a continuous function. The following stronger theorem also is true: Let A be a set of functions whose domains are the same separable metric space X and whose ranges are in a metric space Y. If these functions are *pointwise equicontinuous* and for each x in a dense subset of X the set

$\{f(x): f \in A\}$ is compact, then there is a sequence $\{f_n\}$ of distinct members of A which converges pointwise to a continuous function f, and the convergence is uniform on each compact subset of X.

AS-SESSED′, *adj.* **assessed value.** A value set upon property for the purpose of taxation.

AS-SES′SOR, *n.* One who estimates the value of (evaluates) property as a basis for taxation.

AS′SETS, *n.* **assets of an individual or firm.** All of his (or its) goods, money, collectable accounts, etc., which have value; the opposite of **liabilities**.
    depreciating assets. Assets whose book value decreases. See DEPRECIATION—depreciation charge. *Syn.* wasting assets.
    fixed assets. Assets represented by equipment for use but not for sale—such as factories, buildings, machinery, and tools.

AS-SO′CI-ATE, *adj.*, *n.* In a commutative semigroup or ring, **associates** are members a and b for which there exist members x and y such that $a = bx$ and $b = ay$.
    associate matrix. See HERMITIAN—Hermitian conjugate of a matrix.

AS-SO′CI-AT′ED, *adj.* **associated radius of convergence.** If the power series

$$\sum a_{k_1, k_2, \ldots, k_n} z_1^{k_1} z_2^{k_2} \cdots z_n^{k_n}$$

converges for $|z_j| < r_j$, $j = 1, 2, \cdots, n$, and diverges for $|z_j| > r_j$, $j = 1, 2, \cdots, n$, where r_j is positive, then the set $r_1, r_2, \cdots, r_n$ is called a set of associated radii of convergence for the series. *E.g.*, for the series

$$1 + z_1 z_2 + z_1^2 z_2^2 + \cdots \equiv \frac{1}{1 - z_1 z_2}$$

associated radii are any positive numbers r_1, r_2 with $r_1 r_2 = 1$.

AS-SO′CI-A′TIVE, *adj.* A method of combining objects two at a time is *associative* if the result of the combination of three objects (order being preserved) does not depend on the way in which the objects are grouped. If the operation is denoted by $\circ$ and the result of combining x and y by $x \circ y$, then

$$(x \circ y) \circ z = x \circ (y \circ z)$$

for any x, y and z for which the "products" are

defined. For ordinary **addition** of numbers, the associative law states that $(a + b) + c = a + (b + c)$ for any numbers a, b, c. This law can be extended to state that in any sum of several terms any method of grouping may be used (*i.e.*, at any stage of the addition one may add two adjacent terms). The associative law for **multiplication** states that

$$(ab)c = a(bc)$$

for any numbers a, b, c. This law can be extended to state that in any product of several factors any method of grouping may be used (*i.e.*, at any stage of the multiplication one may multiply any two adjacent factors). See GROUP. For a type of non-associative multiplication, see CAYLEY —Cayley algebra.

AS-SUMP'TION, *n.* See AXIOM, and below, fundamental assumptions of a subject.

empirical assumption. See EMPIRICAL—empirical formula, assumption, or rule.

fundamental assumptions of a subject. A set of assumptions upon which the subject is built. For instance, in algebra the commutative and associative laws are *fundamental assumptions*. Sets of fundamental assumptions for the same subject vary more or less with different writers.

AS-SUR'ANCE, *n.* Same as INSURANCE.

AS'TROID, *n.* The hypocycloid of four cusps.

AS-TRO-NOM'I-CAL, *adj.* **astronomical unit.** The mean distance between the sun and the earth. Approx. 92,955,807 miles or $1.4959787 \cdot 10^{13}$ cm.

A'SYM-MET'RIC, *adj.* **asymmetric relation.** See SYMMETRIC —symmetric relation.

AS'YMP'TOTE, *n.* For a plane curve, an **asymptote** is a line which has the property that the distance from a point P on the curve to the line approaches zero as the distance from P to the origin increases without bound and P is on a suitable piece of the curve. Often it is required that the curve not oscillate about the line. See below, asymptote to the hyperbola.

asymptote to the hyperbola. When the equation of the hyperbola is in the standard form $x^2/a^2 - y^2/b^2 = 1$, the lines $y = bx/a$ and $y = -bx/a$ are its asymptotes. This can be sensed by writing the above equation in the form $y = \pm(bx/a)\sqrt{1 - a^2/x^2}$ and noting that a^2/x^2 approaches zero as x increases without limit. *Tech.*

The numerical difference between the corresponding ordinates of the lines and the hyperbola is

$$|bx/a|\left(1 - \sqrt{1 - a^2/x^2}\right)$$

$$= |ab/x|\Big/\left(1 + \sqrt{1 - a^2/x^2}\right),$$

which approaches zero as x increases, and the distances from the hyperbola to the lines are the product of this infinitesimal by the cosines of the angles the lines make with the x-axis; hence the distances between the lines and the hyperbola each approach zero as x increases. See above, ASYMPTOTE, and the figure under HYPERBOLA.

AS'YMP-TOT'IC, *adj.* **asymptotic directions on a surface at a point.** Directions at a point P on a surface S for which $D\,du^2 + 2D'\,du\,dv + D''\,dv^2 = 0$. See SURFACE —fundamental coefficients of a surface. Asymptotic directions at P on S are the directions at P on S in which the tangent plane at P has contact of at least the third order. See DISTANCE —distance from a surface to a tangent plane. Asymptotic directions are also the directions in which the *normal curvature* vanishes. At a planar point, all directions are asymptotic directions; otherwise there are exactly two asymptotic directions, which are real and distinct, real and coincident, or conjugate imaginary, according as the point on the real surface S is hyperbolic, parabolic, or elliptic.

asymptotic cone of an hyperboloid. If either of the hyperboloids

$$\frac{x^2}{a^2} + \frac{y^2}{b^2} - \frac{z^2}{c^2} = 1,$$

and

$$-\frac{x^2}{a^2} - \frac{y^2}{b^2} + \frac{z^2}{c^2} = 1,$$

is cut by the plane $y = mx$, hyperbolas are formed whose asymptotes pass through the origin. The cone described by these lines as m varies is the **asymptotic cone of the hyperboloid under consideration.**

asymptotic distribution. If a distribution $F(x)$ of a random variable x is a function of a parameter n (*e.g.*, n may be the size of a sample and x the mean), the limit of $F(x)$ as $n \to \infty$ is the asymptotic distribution function of x. In particular, if two quantities u and σ can be obtained so that the distribution function of $\left(\dfrac{x-u}{\sigma}\right) = y_n$ will be, in the limit as $n \to \infty$, equal to

$$\lim_{n \to \infty} p(y_n < t) = \frac{1}{\sqrt{2\pi}} \int_{-\infty}^{t} e^{-x^2/2} \, dx,$$

then $F(x)$ is *asymptotically* normally distributed. This means that x is asymptotically normally distributed in the sense that the limit, as $n \to \infty$, of the probability of

$$\left(\frac{x-u}{\sigma} = y_n\right) < t,$$

is given by the normal distribution regardless of whether or not x has a mean and variance of u and σ. Whatever the distribution of x, the probability of the variable y is given in the limit by the normal distribution, if x can be so transformed as to be asymptotically normal.

asymptotic expansion. A divergent series of the form $a_0 + (a_1/z) + (a_2/z^2) + (a_3/z^3) + \cdots + (a_n/z^n) + \cdots$, where the numbers a_i are constants, is an *asymptotic expansion* of a function f if $\lim_{z \to \infty} z^n[f(z) - S_n(z)] = 0$ for any fixed value of n, where $S_n(z)$ is the sum $a_0 + (a_1/z) + \cdots + (a_n/z^n)$. E.g.,

$$\int_x^\infty t^{-1} e^{x-t} \, dt = (1/x) - (1/x^2) + (2!/x^3)$$

$$- \cdots + (-1)^{n-1}(n-1)!/x^n$$

$$+ (-1)^n n! \int_x^\infty \frac{e^{x-t}}{t^{n+1}} \, dt.$$

For n fixed,

$$\lim_{x \to \infty} x^n \int_x^\infty \frac{e^{x-t}}{t^{n+1}} \, dt = 0.$$

Hence the series with general (nth) term $(-1)^{n-1}(n-1)!/x^n$ is the asymptotic expansion of the function of x given by the integral. This fact is written:

$$\int_x^\infty t^{-1} e^{x-t} \, dt \sim (1/x) - (1/x^2)$$

$$+ (2!/x^3) - (3!/x^4) + \cdots.$$

asymptotic line on a surface. A curve C on a surface S such that the direction of C at each of its points in an asymptotic direction on S; curves defined by the differential equation $D \, du^2 + 2D' \, du \, dv + D'' \, dv^2 = 0$. In general there are two such curves through each point of S. See above, asymptotic directions on a surface at a point.

AS'YMP-TOT'I-CAL-LY, *adv.* **asymptotically equal.** See MAGNITUDE —order of magnitude.

asymptotically unbiased. See UNBIASED— unbiased estimator.

ATIYAH, Michael Francis (1929–   ). English mathematician and Fields Medal recipient (1966). Research in K-theory, theory of indices, fixed-point theory, algebraic geometry and theory of cobordism.

AT'MOS-PHERE, *n.* (Meteorology) The pressure of 76 cm of mercury at $0°$ C. Approximately one bar, approximately 14.7 lb/sq. in., and approximately the air pressure at mean sea level. *Syn.* standard atmosphere.

AT'OM, *n.* For a lattice (or a ring of sets) R, an **atom** is a nonzero (nonempty) member U for which there is no nonzero (nonempty) X in R which precedes (is properly contained in) U. Any finite Boolean algebra is isomorphic to the Boolean algebra of all sets of its atoms. See JOIN —join-irreducible.

AT'TO. A prefix meaning 10^{-18}, as in **attometer.**

AT-TRAC'TION, *n.* **center of attraction.** See CENTER —center of mass.

gravitational attraction. See GRAVITATION.

AT'TRI-BUTE, *n.* (*Statistics*) An inherent property, usually one that is measured by a finite number or a discrete set of values. Sometimes used as the outcome of one of a set of mutually exclusive events (*e.g.*, the sex of a person, or the weight to the nearest 10 lb); when used in this sense, **attribute space** is the same as **sample space.** Often **attribute sampling** refers to an experiment with only two outcomes; *e.g.*, classifying an object as defective or nondefective.

AUG-MEN'TED, *adj.* **augmented matrix.** See MATRIX.

AU-TO-MOR'PHIC, *adj.* **automorphic function.** A single-valued function f, analytic except for poles in a domain D of the complex plane, is **automorphic with respect to a group of linear transformations** provided for each transformation T of the group it is true that, if z lies in D, then also $T(z)$ lies in D, and $f[T(z)] \equiv f(z)$.

AU-TO-MOR′PHISM, *n.* See ISOMORPHISM, CATEGORY.

AU′TO-RE-GRES′SIVE, *n.* **autoregressive series.** If the variable $y = f(t)$, written y_t, is of the form

$$y_t = a_1 y_{t-1} + a_2 y_{t-2} + \cdots + a_m y_{t-m} + k,$$

the variable y forms an *autoregressive series.* Specifically, a difference equation of the variable y forms an *autoregressive series.*

AUX-IL′IA-RY, *adj.* **auxiliary circle of an ellipse.** The larger of the two eccentric circles of the ellipse. See ELLIPSE.

    auxiliary circle of a hyperbola. See HYPERBOLA—parametric equations of a hyperbola.

    auxiliary equation. See DIFFERENTIAL —linear differential equations.

AV′ER-AGE, *adj., n.* Same as MEAN. See various headings under MEAN.

    average speed and velocity. See SPEED, VELOCITY.

    moving average. Given a sequence of random variables $\{X_n\}$, the **moving average of order** k is the sequence of random variables $\{Y_n\}$ for which Y_n is $(X_n + X_{n+1} + \cdots + X_{n+k-1})/k$. E.g., if daily maximum temperatures are $\{T_1, T_2, \cdots\}$, the three-day moving average would have values $\{\frac{1}{3}(T_1 + T_2 + T_3), \frac{1}{3}(T_2 + T_3 + T_4), \frac{1}{3}(T_3 + T_4 + T_5), \cdots\}$ Weighted averages may be used.

AV′ER-AG-ING, *n.* **averaging an account.** Finding the *average date.* See EQUATED— equated date.

AV′OIR-DU-POIS′, *adj.* **avoirdupois weight.** A system of weights using the pound as its basic unit, the pound being equal to 16 ounces. See DENOMINATE NUMBERS in the appendix.

AX′I-AL, *adj.* **axial symmetry.** Symmetry with respect to a line. The line is the **axis of symmetry.**

AX′I-OM, *n.* A statement that is accepted without proof. The axioms of a mathematical system are the basic propositions from which all other propositions can be derived. A set of axioms is **inconsistent** if it is possible to deduce from these axioms that some statement is both true and false. An axiom is **independent** of the others of a set of axioms if it is not a consequence of the others, *e.g.*, if there exists a model that satisfies all the axioms except the one in question. See DEDUCTIVE.

    axiom of choice. See CHOICE.

    axiom of continuity. To every point on the real axis there corresponds a real number (rational or irrational); the assumption that there exist numbers such as those indicated by the *Cauchy necessary and sufficient condition for convergence*, and the *Dedekind cut postulate. Syn.* principle of continuity.

    axiom of countability. See SEPARABLE —separable space.

    axiom of superposition. Any figure may be moved about in space without changing either its shape or size.

    Euclid's axioms or "common notions." (1) Things equal to the same thing are equal to each other. (2) If equals are added to equals, the results are equal. (3) If equals are subtracted from equals the differences are equal. (4) Things which coincide with one another are equal. (5) The whole is greater than any of its parts. Axioms (4) and (5) are not universally attributed to Euclid.

AX′IS, *n.* [*pl.* **axes**]. See CONE, CYLINDER, ELLIPSE, ELLIPSOID, HYPERBOLA, PARABOLA, PENCIL —pencil of planes, RADICAL —radical axis, REFERENCE —axis of reference, SYMMETRIC —symmetric geometric configurations.

    axis of a curve or surface. Same as an AXIS OF SYMMETRY. See SYMMETRIC —symmetric geometric configurations.

    axis of perspective. See PERSPECTIVE —perspective position.

    axis of revolution. See REVOLUTION —solid of revolution.

    axis of symmetry. A geometric configuration is symmetric with respect to a line, the **axis of symmetry**, if for any point P there is a point Q such that the axis of symmetry is the perpendicular bisector of the line segment PQ. See SYMMETRIC—symmetric geometric configurations.

    coordinate axis. A line along which (or parallel to which) a coordinate is measured. See CARTESIAN —Cartesian coordinates.

    major and minor axis of an ellipse. See ELLIPSE.

    polar axis. See POLAR—polar coordinates in the plane.

    principal axes of inertia. See MOMENT— moment of inertia.

    real and imaginary axes. See ARGAND —Argand diagram.

    transverse and conjugate axes of the hyperbola. See HYPERBOLA.

AZ′I-MUTH, *n.* **azimuth of a point in a plane.** See POLAR—POLAR COORDINATES IN THE PLANE.

azimuth of a celestial point. See HOUR—hour angle and hour circle.

AZ′I-MUTH′AL, *adj.* **azimuthal map.** A map of a spherical surface S in which the points of S are projected onto a tangent plane from a point on that diameter of S that is perpendicular to the plane. An azimuthal map is said to be a **gnomic** map if the point of projection is the center of the sphere; it is an **orthographic map** if the point of projection is at infinite distance. See PROJECTION—stereographic projection of a sphere on a plane.

B

B.T.U. See BRITISH—British thermal unit.

BABBAGE, Charles (1792–1871). English analyst, statistician, and inventor. Prophet of the modern digital computing machine, envisaging a mechanical device using the basic principles of arithmetic operations and information storage and recall for astronomical and navigational computations.

BAB-Y-LO′NI-AN, *adj.* **Babylonian numerals.** A sexagesimal system in which numbers less than 60 were expressed using cuneiform marks to denote 1, 10, and subtraction. A number was expressed by repetition of the symbol for 1 and the symbol for 10, with subtraction used to simplify (*e.g.*, 38 could be represented by four "tens" with two "ones" subtracted). No place markers such as zero were used and place values had to be determined from the context.

BAC-TE′RI-AL, *adj.* **law of bacterial growth.** The rate of increase of bacteria growing freely in the presence of unlimited food is proportional to the number present. It is defined by the equation $dN/dt = kN$, where k is a constant, t the time, N the number of bacteria present, and kN the rate of increase. The solution of this equation is $N = ce^{kt}$, where c is the value of N when $t = 0$. This is also called the **law of organic growth.**

BAIRE, Louis René (1874–1932). French analyst.

Baire category theorem. See CATEGORY.

Baire function. A real-valued function f which has the property that for any real number a the set of all x for which $f(x) > a$ is a *Borel set*. Equivalent definitions result if the set of all x satisfying $f(x) \geq a$, or the set of all x satisfying $a \leq f(x) \leq b$ for arbitrary a and b, are required to be Borel sets (and either or both of the signs $\leq$ could be replaced by $<$). Any Baire function is *measurable*. The Baire functions can be classified as follows. The set of continuous functions are of the **first Baire class.** In general, a function is of Baire class α if it is not of Baire class β for any $\beta < \alpha$ and is a point-wise limit of functions which belong to Baire classes corresponding to numbers preceding α. By transfinite induction, these classes are defined for all ordinal numbers corresponding to denumerable well-ordered sets. No additional functions are obtained by further extensions. *Syn.* Borel measurable functions. To every measurable function there corresponds a Borel measurable function which differs from f only on a set of *measure zero*.

Baire space. A topological space X for which (i) every nonempty open subset is of the second category in X, or (equivalently) (ii) every countable intersection of dense open subsets is dense in X. Every complete metric space and every locally compact topological space is a Baire space. See CATEGORY —category of sets.

property of Baire. A set S contained in a set T has the property of Baire if each non-empty open set U contains a point where either S or the complement of S is of *first category*. A set has the property of Baire if and only if it can be made into an open (or a closed) set by adjoining and taking away suitable sets of the first category, or if and only if it can be represented as a G_δ set plus a set of first category, or if and only if it can be represented as an F_σ set minus a set of first category. The class of all sets having the property of Baire is the σ-algebra generated by the open sets together with the sets of first category. See BOREL—Borel set, MEASURABLE—measurable set.

BAKER, Allan (1939–   ). English mathematician and Fields Medal recipient (1970). Extended the Gelfond-Schneider theorem, showing that $\alpha_1^{\beta_1}\alpha_2^{\beta_2} \cdots \alpha_k^{\beta_k}$ is transcendental if $\alpha_1, \alpha_2, \cdots, \alpha_k$ are algebraic numbers (not 0 or 1) and $\beta_1, \beta_2, \cdots, \beta_k$ are linearly independent, algebraic and irrational. See GELFOND —Gelfond-Schneider theorem.

BALL, *n.* See SPHERE.

BANACH, Stefan (1892–1945). Polish algebraist, analyst, and topologist.

Banach algebra. See ALGEBRA –Banach algebra.

Banach category theorem. See under CATEGORY.

Banach fixed-point theorem. If C is a nonempty complete metric space and T is a contraction of C into C, then for any x_0 in C the sequence $\{x_0, T(x_0), T(T(x_0)), \cdots\}$ converges to the unique x^* in C for which $T(x^*) = x^*$. Also,

$$\mathrm{dist}(x^*, x_n) \leq \frac{\theta^n}{1 - \theta} \mathrm{dist}(x_1, x_0),$$

where $\theta < 1$ is the number for which $\mathrm{dist}[T(x), T(y)] \leq \sigma \, \mathrm{dist}(x, y)$ if x and y are in C. Also known as the Caccioppoli-Banach principle. See CONTRACTION.

Banach space. A vector space whose scalar multipliers are the real numbers (or the complex numbers) and which has associated with each element x a real number $\|x\|$, the **norm** of x, satisfying the postulates: (1) $\|x\| > 0$ if $x \neq 0$; (2) $\|ax\| = |a| \cdot \|x\|$ for all real numbers a; (3) $\|x + y\| \leq \|x\| + \|y\|$ for all x and y; (4) the space is complete, a *neighborhood* of an element x being the set of all y satisfying $\|x - y\| < \epsilon$ for some fixed ϵ. Without postulate (4), the space is a **normed linear space** or **normed vector space**. The Banach space is a **real Banach space** or a **complex Banach space** according as the scalar multipliers are real numbers or complex numbers. Examples of Banach spaces are *Hardy spaces*, *Hilbert space*, the spaces $l^{(r)}$ ($r \geq 1$) of all sequences $x = (x_1, x_2, \cdots)$ for which $\sum_{i=1}^{\infty} |x_i|^r$ is finite and $\|x\| = [\sum_{i=1}^{n} |x_i|^r]^{1/r}$, the spaces $L^{(r)}$ ($r \geq 1$) of measurable functions f on some set T for which $\int_T |f|^r \, dm$ is finite and $\|f\| = [\int_T |f|^r \, dm]^{1/r}$ (see FUNCTION —function of class L_p, INTEGRABLE —integrable function), and the space **C** of all continuous functions f defined on the interval $[0, 1]$ with $\|f\| = \max |f(x)|$ for $0 \leq x \leq 1$.

Banach-Steinhaus theorem. Let X and Y be Banach spaces and let Φ be a collection of continuous linear transformations from X into Y. If the set $\{\|T(x)\| : T \in \Phi\}$ is bounded for each $x \in X$, then there is a number M such that $\|T(x)\| \leq M \|x\|$ for each x in X and each T in Φ (*i.e.*, $\|T\| \leq M$ for each T in Φ). More generally, suppose X and Y are linear topological spaces, Φ is a collection of continuous linear transformations from X into Y, and S is the set of all x in X for which $\{T(x): T \in \Phi\}$ is a bounded set in Y. If S is of second category in X, then $S = X$ and Φ is uniformly equicontinuous. *Syn.* uniform-boundedness principle.

Banach-Tarski paradox. The theorem of Banach and Tarski which states that if A and B

are bounded sets in a Euclidean space of dimension at least 3 and if both A and B have interior points, then A can be decomposed into a finite number of pieces and reassembled by moving the pieces by rigid motions (translations and rotations) to form a set congruent to B. In particular, it is possible to decompose a solid sphere into a finite number of pieces and to reassemble these pieces to form two solid spheres the same size as the original sphere. No estimate of the number of pieces needed in this case was given by Banach and Tarski, but R. M. Robinson has proved that the smallest possible number of pieces is 5 and that one of these pieces can be a single point; he also proved that the surface S of a sphere can be separated into two pieces each of which can be separated into two pieces congruent to itself (thus only four pieces are needed to make from S two identical copies of S). The situation is very different in the plane. A circle and its interior can be decomposed into fewer than 10^{50} pieces that can be translated to form a square, but the square must have the same area as the circle. More generally, a Jordan curve and its interior can be decomposed into finitely many pieces that can be translated to form a square. Less can be done if the "pieces" are required to have area, but any polygon P can be decomposed into triangles that can be used to form another polygon Q if and only if P and Q have the same area. See HAUSDORFF—Hausdorff paradox.

Hahn-Banach theorem. See HAHN.

Mazur-Banach game. See MAZUR.

BANK, n. **bank discount.** See DISCOUNT.

bank note. A note given by a bank and used for currency. It usually has the shape and general appearance of government paper money.

mutual savings bank. See MUTUAL.

BAR, n. (1) In the sense of being a "stripe" that is much longer than it is wide, the word **bar** is used in many ways in mathematics. *E.g.*, to indicate the *conjugate* of a complex number, to indicate the *absolute value* of a number, and to separate the numerator and denominator of a fraction. See AGGREGATION, GRAPH—bar graph. (2) In *Meteorology*, a **bar** (or **barye**) is a unit of pressure equal to 10^5 newtons per square meter, 10^6 dynes per square cm, or 10^5 pascals. In *physics*, **bar** often means a pressure of 1 dyne per square cm.

BARROW, Isaac (1630–1677). English theologian, geometer, and analyst. Though highly talented and original, he is remembered primarily as Newton's teacher.

BAR'Y-CEN'TER, *n.* Same as CENTER OF MASS. See CENTER —center of mass.

BAR'Y-CEN'TRIC, *adj.* **barycentric coordinates.** Let $p_0, p_1, \cdots, p_n$ be $n + 1$ points of n-dimensional Euclidean (or vector) space E_n that are not in the same hyperplane of E_n. Then for each point x of E_n, there is one and only one set $\{\lambda_0, \cdots, \lambda_n\}$ of real numbers for which

$$x = \lambda_0 p_0 + \lambda_1 p_1 + \cdots + \lambda_n p_n$$

and $\lambda_0 + \lambda_1 + \cdots + \lambda_n = 1$. The point x is (by definition) the center of mass of point masses $\lambda_0, \lambda_1, \cdots, \lambda_n$ at the points $p_0, \cdots, p_n$, respectively, and the numbers $\lambda_0, \lambda_1, \cdots, \lambda_n$ are said to be *barycentric coordinates* of the point x. The motivation for this definition is that if three objects have weights $\lambda_0, \lambda_1, \lambda_2$ with $\lambda_0 + \lambda_1 + \lambda_2 = 1$, and their centers of mass are at the points $p_0 = (x_0, y_0, z_0)$, $p_1 = (x_1, y_1, z_1)$, $p_2 = (x_2, y_2, z_2)$, then the center of mass of the three objects together is the point

$$\bar{p} = \lambda_0 p_0 + \lambda_1 p_1 + \lambda_2 p_2$$
$$= (\lambda_0 x_0 + \lambda_1 x_1 + \lambda_2 x_2, \lambda_0 y_0 + \lambda_1 y_1 + \lambda_2 y_2,$$
$$\lambda_0 z_0 + \lambda_1 z_1 + \lambda_2 z_2).$$

BASE, *n.* A base of a geometric configuration is usually a side (or face) upon which (perpendicular to which) an altitude is constructed, or is thought of as being constructed. See the particular geometric configuration. For an expression such as a^n, the quantity a is called the *base* and n the exponent. Also see the various headings below.

base angles of a triangle. The two angles which have the base of the triangle for a common side.

base for a topology. A collection B of open sets is a **base for the topology** of a topological space T if each open set is the union of some of the members of B. A **subbase for a topology** is a collection S of open sets such that the collection of all finite intersections of members of S is a base for the topology. A collection N of open sets is a **base for the neighborhood system** of a point x (or a **local base** at X) if x belongs to each member of N and any open set which contains x also contains a member of N. A **subbase** for the neighborhood system of a point x (or a **local subbase** at x) is a collection S of sets such that the collection of all finite intersections of members of S is a base for the neighborhood system of x. A topological space is said to satisfy the **first axiom of countability** if each point has a countable base for its neighborhood system; it satisfies the **second axiom of countability** or is **perfectly separable** if its topology has a

countable base. A metric space satisfies the second axiom of countability if and only if it is *separable*. *Syn.* basis for a topology. See SEPARABLE.

base in mathematics of finance. A number, usually a sum of money, of which some per cent is to be taken; a sum of money upon which interest is to be calculated.

base of a logarithmic system. See LOGARITHM.

base of a number system. The number of units, in a given digit's place or decimal place, which must be taken to denote 1 in the next higher place. *E.g.*, if the base is ten, ten units in units place are denoted by 1 in the next higher place, which is ten's place; if the base is twelve, twelve units in units place are denoted by 1 in the next higher place, which is twelve's place. For example, when the base is twelve, 23 means $2 \times$ twelve $+ 3$. *Tech.*, an integer to any base is of form $d_0 + d_1(\text{base}) + d_2(\text{base})^2 + d_3(\text{base})^3 + \cdots$, where $d_0, d_1, d_2, d_3, \cdots$, are each nonnegative integers less than the base. A number between 0 and 1 can be written as

$$0.d_1 d_2 d_3 \cdots = \frac{d_1}{\text{base}} + \frac{d_2}{(\text{base})^2}$$
$$+ \frac{d_3}{(\text{base})^3} + \cdots.$$

See BINARY, DECIMAL —decimal number system, DUODECIMAL, HEXADECIMAL, OCTAL, SEXADECIMAL, SEXAGESIMAL, VIGESIMAL.

BA'SIS, *n.* **basis of a vector space.** (1) A set of linearly independent vectors such that every vector of the space is equal to some finite linear combination of vectors of the basis. *Syn.* Hamel basis (see HAMEL). (2) For an infinite dimensional and (separable) vector space with a vector length (or *norm*) defined, a basis usually means a sequence of elements $\{x_1, x_2, \cdots\}$ such that every x is uniquely expressible in the form $x = \sum_{i=1}^{\infty} a_i x_i$ (meaning that the limit as n becomes infinite of the length of $x - \sum_{i=1}^{n} a_i x_i$ is zero). If the vectors of the basis are mutually orthogonal, the basis is an **orthogonal basis**; if they are also all of unit length, the basis is a **normal** (or **normalized**) **orthogonal basis**, or an **orthonormal basis**. If there is a finite number of vectors in the basis, the space is said to be **finite-dimensional** and its dimension is equal to the number of vectors in its basis. Otherwise, it is **infinite-dimensional**. The examples given of *Banach spaces* (see BANACH –Banach space) possess such a basis, but not all separable Banach spaces have bases. See APPROXIMATION—approximation property, INNER—inner-product space.

dual basis. (1) For a finite-dimensional linear space V with a basis $\{x_1, \cdots, x_n\}$, the **dual basis** is the set of linear functionals $\{f_1, \cdots, f_n\}$ defined by $f_k(\Sigma a_i x_i) = a_k$. The dual basis is a basis for the first conjugate space V^*. If V is taken as the dual of V^* with x the linear functional on V^* defined by $x(f) = f(x)$ for all x in V and all f in V^*, then the dual of $\{f_1, \cdots, f_n\}$ is $\{x_1, \cdots, x_n\}$. (2) If a Banach space B has a basis $\{x_1, x_2, \cdots\}$, then the sequence $\{f_1, f_2, \cdots\}$ defined by

$$f_k\left(\sum_1^\infty a_i x_i\right) = a_k$$

is a sequence of continuous linear functionals and it is a basis (a **dual basis**) for the first conjugate space if and only if it is **shrinking** in the sense that $\lim_{n\to\infty}\|f\|_n = 0$ for each continuous linear functional f, where $\|f\|_n$ is the norm of f as a continuous linear functional with domain the linear span of $\{x_{n+1}, x_{n+2}, \cdots\}$. This condition is satisfied by all bases in reflexive spaces. If $\{x_\alpha\}$ is a complete orthonormal set for an inner product space T, then $\{f_\alpha\}$ is a complete orthonormal set for the first conjugate space of T, where $f_\beta(\Sigma a_\alpha x_\alpha) = a_\beta$. Analogously to (1), each of the orthonormal bases $\{x_\alpha\}$ and $\{f_\alpha\}$ is dual to the other. See INNER—inner-product space.

BAYES, Thomas (1702–1761). English theologian and probabilist.

Bayes' theorem. Suppose A and B_1, $B_2, \cdots, B_n$ are events for which the probability $P(A)$ of A is not 0, $\sum_{i=1}^n P(B_i) = 1$, and $P(B_i$ and $B_j) = 0$ if $i \neq j$. Then the conditional probability $P(B_j|A)$ of B_j given that A has occurred is given by

$$P(B_j|A) = \frac{P(B_j)P(A|B_j)}{\sum_{i=1}^n P(B_i)P(A|B_i)}.$$

Sometimes $P(B_j|A)$ is called the *inverse probability* of the event B_j. E.g., suppose 4 urns are equally likely to be sampled. Number 1 contains 1 white and 2 red balls, number 2 has 1 white and 3 red, number 3 has 1 white and 4 red, and number 4 has 1 white and 5 red. The probability of an urn being sampled is $\frac{1}{4} = P(B_i)$; $P(A|B_i)$ equals $\frac{1}{3}, \frac{1}{4}, \frac{1}{5}$, and $\frac{1}{6}$, respectively, for $i = 1, \cdots, 4$, where A is the draw of a white ball. Application

of Bayes' formula yields

$$P(B_2|A) = \frac{\frac{1}{4}\cdot\frac{1}{4}}{\frac{1}{4}\cdot\frac{1}{3} + \frac{1}{4}\cdot\frac{1}{4} + \frac{1}{4}\cdot\frac{1}{5} + \frac{1}{4}\cdot\frac{1}{6}} = \frac{15}{57}.$$

See PROBABILITY—conditional probability.

BEAR'ING, *n.* **bearing of a line.** The angle which the line makes with the north and south line; its direction relative to the north-south line.

bearing of a point, with reference to another point. The angle the line through the points makes with the north-south line.

BECKENBACH, Edwin Ford (1906–1982). American mathematician who wrote extensively on inequalities and complex analysis. Received the Mathematical Association of America Award for Distinguished Service to Mathematics for his work on behalf of professional organizations and his expository writing, including valuable contributions to several editions of this dictionary.

BEHRENS, Walter Ulrich. German agricultural statistician.

Behrens-Fisher problem. The problem of determining confidence intervals for the difference of the means of two normal populations when the variances of the populations are unknown, and the means of two random samples are known.

BEI, *adj.* **bei function.** See BER—ber function.

BEND, *adj.* **bend point.** A point on a plane curve where the ordinate is a maximum or minimum.

BENDING MOMENT. See MOMENT.

BEN'E-FI'CI-AR-Y, *n.* *(Insurance)* The one to whom the amount guaranteed by the policy is to be paid.

BEN'E-FIT, *n.* **benefits of an insurance policy.** The sum or sums which the company promises to pay provided a specified event occurs, such as the death of the insured or his attainment of a certain age.

BER, *adj.* **ber function.** The **ber, bei, her, hei, ker,** and **kei** functions are defined by the relations:

$$\mathrm{ber}_n(z) \pm i\,\mathrm{bei}_n(z) = J_n(ze^{\pm 3\pi i/4}),$$

$$\mathrm{her}_n(z) + i\,\mathrm{hei}_n(z) = H_n^{(1)}(ze^{3\pi i/4}),$$

$$\mathrm{her}_n(z) - i\,\mathrm{hei}_n(z) = H_n^{(2)}(ze^{-3\pi i/4}),$$

$$\mathrm{ker}_n(z) \pm i\,\mathrm{kei}_n(z) = i^{\mp n}K_n(ze^{\pm \pi i/4}),$$

where J_n is a *Bessel function*, $H_n^{(1)}$ and $H_n^{(2)}$ are *Hankel functions*, and K_n is a *modified Bessel function* of the second kind. The following conventions are also used: $\mathrm{ber}_0(z) = \mathrm{ber}(z)$, $\mathrm{bei}_0(z) = \mathrm{bei}(z)$, etc. It follows that:

$$2\ker_n(z) = -\pi\,\mathrm{hei}_n(z);$$

$$2\mathrm{kei}_n(z) = \pi\,\mathrm{her}_n(z).$$

These six functions are real when n is real and z is real and positive. In particular,

$$\mathrm{ber}\,x = 1 - \frac{x^4}{2^2\cdot 4^2} + \frac{x^8}{2^2\cdot 4^2\cdot 6^2\cdot 8^2} - \cdots,$$

$$\mathrm{bei}\,x = \frac{x^2}{2^2} - \frac{x^6}{2^2\cdot 4^2\cdot 6^2}$$

$$+ \frac{x^{10}}{2^2\cdot 4^2\cdot 6^2\cdot 8^2\cdot 10^2} - \cdots.$$

Also, $\int_0^x t\,\mathrm{ber}(t)\,dt = x\,\mathrm{bei}'(x)$, $\int_0^x t\,\mathrm{bei}(t)\,dt = -x\,\mathrm{ber}'(x)$, these still being valid if ber is replaced by ker and bei by kei.

BERNOULLI, Daniel (1700–1782). Swiss anatomist, botanist, hydrodynamicist, analyst, and probabilist. The most celebrated of his generation of Bernoullis. See PETERSBERG PARADOX, RICCATI —Riccati equation.

Bernoulli polynomials. (1) The polynomials B_n defined by

$$\frac{te^{zt}}{e^t - 1} = \sum_1^\infty B_n(z)t^n.$$

The first four Bernoulli polynomials are $B_1(z) = z - \frac{1}{2}$, $B_2(z) = (z^2/2) - (z/2) + \frac{1}{12}$, $B_3(z) = (z^3/3!) - (z^2/4) + (z/12)$, $B_4(z) = (z^4/4!) - (z^3/12) + (z^2/24) - \frac{1}{720}$. It follows that $B'_{n+1}(z) = B_n(z)$, $B_n(z+1) - B_n(z) = nz^{n-1}(n > 1)$,

$$B_{2n}(z) = (-1)^{n-1}\sum_{r=1}^\infty \frac{2\cos 2r\pi z}{(2r\pi)^{2n}},$$

and

$$B_{2n+1}(z) = (-1)^{n-1}\sum_{r=1}^\infty \frac{2\sin 2r\pi z}{(2r\pi)^{2n+1}}$$

$$(n \ge 1).$$

(2) The polynomials ϕ_n defined by

$$t\frac{e^{zt} - 1}{e^t - 1} = \sum_{n=1}^\infty \frac{\phi_n(z)t^n}{n!}.$$

It follows that $\phi_n = n!(B_n - B'_n)$, and that $\phi(0) = 0$. Trivial variations of these definitions are sometimes given.

BERNOULLI, James (or **Jacques** or **Jakob**) (1654–1705). Swiss physicist, analyst, combinatorist, probabilist, and statistician. First and perhaps most famous of the Bernoulli family of mathematicians. See BRACHISTOCHRONE.

Bernoulli distribution. A random variable X has a **Bernoulli distribution**, or is a **Bernoulli random variable**, if there is a number p such that X is the number of successes in a single Bernoulli experiment with probability of success p. The range of X is the set $\{0, 1\}$ and the probability of k successes is

$$P(X = k) = p^k q^{1-k} \quad \text{if } k \text{ is } 0 \text{ or } 1,$$

where $q = 1 - p$. The *mean* is p and the *variance* is pq. See BINOMIAL —binomial distribution.

Bernoulli equation. A differential equation of the form

$$\frac{dy}{dx} + yf(x) = y^n g(x).$$

Bernoulli experiment. (*Statistics*) An experiment or trial for which there are two possible outcomes, such as "heads" or "tails" when tossing a coin, or "A" or "B" when asking if candidate A or candidate B is favored. See above, Bernoulli distribution. *Syn.* Bernoulli trial.

Bernoulli inequality. The inequality

$$(1 + x)^n > 1 + nx$$

for $x > -1$, $x \ne 0$, and n an integer greater than 1.

Bernoulli numbers. (1) The numerical values of the coefficients of $x^2/2!, x^4/4!, \ldots, x^{2n}/(2n)!, \cdots$ in the expansion of $x/(1 - e^{-x})$, or $xe^x/(e^x - 1)$. Substituting the exponential series for e^x and starting the division by the expansion of $(e^x - 1)$ one obtains, for the first four terms of this quotient,

$$1 + \left(\tfrac{1}{2}\right)x + \left(\tfrac{1}{6}\right)x^2/2! - \left(\tfrac{1}{30}\right)x^4/4!.$$

The odd terms all drop out after the term $(\tfrac{1}{2})x$. Some authors denote the Bernoulli numbers by B_1, B_2, etc. Others use B_2, B_4, etc. With the first notation: $B_1 = \frac{1}{6}$, $B_2 = \frac{1}{30}$, $B_3 = \frac{1}{42}$, $B_4 = \frac{1}{30}$, $B_5 = \frac{5}{66}$, $B_6 = \frac{691}{2730}$, $B_7 = \frac{7}{6}$, $B_8 = \frac{3617}{510}$. In general,

$$B_n = \frac{(2n)!}{2^{2n-1}\pi^{2n}} \sum_{i=1}^{\infty} (1/i)^{2n}.$$

(2) The numbers defined by the relation $\dfrac{t}{e^t - 1} = \sum_{t=1}^{\infty} B'_n \dfrac{t^n}{n!}$. It follows that $B'_{2n} = B_n$ except possibly for sign, that $B'_{2n+1} = 0$, for all $n > 1$ ($B'_1 = -\frac{1}{2}$), and that $n! B'_n = B_n(0)$, where $B_n(z)$ is the nth Bernoulli polynomial. See BERNOULLI, (Daniel)—Bernoulli polynomials. STIRLING—Stirling's series. Various trivial variations of these definitions are sometimes given.

Bernoulli theorem. (*Statistics*) See LARGE—law of large numbers.

lemniscate of Bernoulli. See LEMNISCATE.

BERNOULLI, John (or **Jean** or **Johann**) (1667–1748). Swiss mathematician. Student, rival, and perhaps equal of his older brother James. Gave impetus to the study of the calculus of variations through posing his famous *brachistochrone problem*. See BRACHISTOCHRONE, EULER—Euler's equation.

BERNOULLI, Nikolaus (or **Nicolaus**) (1623–1708). Progenitor of celebrated Swiss family of mathematicians, the best known of whom are listed in the following chart. The family came originally from Antwerp, but because of religious persecution fled to Basel in 1583.

BERNOULLI, Nikolaus (or **Nicolaus**), **II** (1687–1759). Swiss mathematician. Educated by his uncles, James and John, he solved many of their problems.

BERNSTEIN, Sergei Natanovich (1880–1968). Russian analyst and approximation theorist.

Bernstein polynomials. If f is a real-valued function with domain the closed interval $[0, 1]$, then

$$B_n(f) = \sum_{i=0}^{n} f\left(\frac{i}{n}\right)\binom{n}{i} x^i (1-x)^{n-i},$$

$$n = 1, 2, \cdots,$$

are **Bernstein polynomials**. If f is continuous, then $B_n(f)$ converges uniformly to f on $[0, 1]$.

Schröder–Bernstein theorem. See SCHRÖDER.

BERTRAND, Joseph Louis François (1822–1903). French analyst, differential geometer, and probabilist.

Bertrand curve. A curve whose principal normals are the principal normals of a second curve. *Syn.* conjugate curve.

Bertrand's postulate. There is always at least one prime number between n and $(2n - 2)$, provided n is greater than 3. *E.g.*, if n is 4, $2n - 2 = 6$ and the prime 5 is between 4 and 6. Bertrand's "postulate" is a true theorem [P. L. Chebyshev (1852)].

BESICOVITCH, Abram Samollevitch (1891–1970). Soviet mathematician who worked with

Nikolaus
(1623–1708)

James I
(1654–1705)

Nikolaus I
(1662–1716)

John I
(1667–1748)

Nikolaus II
(1687–1759)

Nikolaus III
(1695–1726)

Daniel I
(1700–1782)

John II
(1710–1790)

John III
(1746–1807)

Daniel II
(1751–1834)

James II
(1759–1789)

Christoph
(1782–1863)

John Gustave
(1811–1863)

probability, almost periodic functions, complex analysis, topology, and number theory.

Hausdorff-Besocovitch dimension. See HAUS-DORFF—Hausdorff dimension.

BESSEL, Friedrich Wilhelm (1784–1846). German astronomer and mathematician.

Bessel functions. For n a positive or negative integer, the nth Bessel function, $J_n(z)$, is the coefficient of t^n in the expansion of $e^{z[t-1/t]/2}$ in powers of t and $1/t$. In general,

$$J_n(z) = \frac{1}{\pi} \int_0^\pi \cos(nt - z \sin t) \, dt$$

$$= \sum_{r=0}^\infty \frac{(-1)^r}{r!\Gamma(n+r+1)} \left(\frac{z}{2}\right)^{n+2r},$$

the second form being valid if $n \neq -1, -2, \cdots$.
$$J_{1/2}(z) = \sqrt{\frac{2}{\pi z}} \sin z, \text{ and for all } n,$$

$$2[dJ_n(z)/dz] = J_{n-1}(z) - J_{n+1}(z),$$

$$(2n/z)J_n(z) = J_{n-1}(z) + J_{n+1}(z),$$

and $J_n(z)$ is a solution of Bessel's differential equation. Sometimes called *Bessel functions of the first kind.* See HANKEL—Hankel function, NEUMANN (Karl)—Neumann function, STEEPEST—method of steepest descent (2), and below, modified Bessel functions.

Bessel's differential equation. The differential equation

$$z^2 \frac{d^2y}{dz^2} + z \frac{dy}{dz} + (z^2 - n^2)y = 0.$$

Bessel's inequality. (1) For any real function F and an *orthogonal normalized* system of real functions $f_1, f_2, \cdots$ on an interval (a, b), Bessel's inequality is

$$\int_a^b [F(x)]^2 \, dx \geq \sum_{n=1}^p \left[\int_a^b F(x)f_n(x) \, dx\right]^2,$$

or for complex valued functions,

$$\int_a^b |F(x)|^2 \, dx \geq \sum_{n=1}^p \left|\int_a^b F(x)\overline{f_n(x)} \, dx\right|^2.$$

These are valid for all p if the functions F, $f_1, f_2, \cdots$ are assumed to be Riemann integrable

(or, more generally, if they are Lebesgue measurable and their squares are Lebesgue integrable). For the Fourier coefficients of any (measurable) real function whose square is Riemann (or Lebesgue) integrable, Bessel's inequality becomes:

$$\frac{1}{\pi} \int_0^{2n} [F(x)]^2 \, dx \geq (a_0/2)^2 + \sum_{k=1}^n \left(a_k^2 + b_k^2\right),$$

for all n, where $a_k = \frac{1}{\pi} \int_0^{2n} F(x) \cos kx \, dx$,

$$b_k = \frac{1}{\pi} \int_0^{2\pi} F(x) \sin kx \, dx \quad (k = 0, 1, 2, \cdots).$$

(2) For a vector space with an inner product $(\mathbf{x}, \mathbf{y})$ and an orthogonal normalized set of vectors $\mathbf{x}_1, \mathbf{x}_2, \cdots, \mathbf{x}_n$, Bessel's inequality is

$$(\mathbf{u}, \mathbf{u}) = |\mathbf{u}|^2 \geq \sum_{k=1}^n |(\mathbf{u}, \mathbf{x}_k)|^2.$$

See RIESZ—Riesz–Fischer theorem, VECTOR—vector space, PARSEVAL—Parseval's theorem.

modified Bessel functions. The *modified Bessel functions of the first kind* and *of the second kind* are the functions $I_n(z) = i^{-n}J_n(iz)$ and

$$K_n(z) = \tfrac{1}{2}\pi(\sin n\pi)^{-1}[I_{-n}(z) - I_n(z)];$$

$K_n(z)$ is the limit of this expression if n is an integer. These functions are real when n is real and z is positive. Also, I_n is a solution of the **modified Bessel differential equation,**

$$z^2 \frac{d^2y}{dz^2} + z \frac{dy}{dz} - (z^2 + n^2)y = 0,$$

and $I_n(z) = \sum_{r=0}^\infty \frac{1}{n!\Gamma(n+r+1)} \left(\frac{z}{2}\right)^{n+2r}$. The functions I_n and I_{-n} are independent solutions of this differential equation when n is not an integer, while the limit of K_n is a second solution when n is an integer. These functions satisfy various recurrence relations, such as $I_{n-1}(z) - I_{n+1}(z) = (2n/z)I_n(z)$ and $K_{n-1}(z) - K_{n+1}(z) = -(2n/z)K_n(z)$. The definition of K_n is sometimes taken as the product of $\cos n\pi$ and the above value (I_n and K_n then satisfy the same recurrence formulae). See BER—ber function.

BE′TA, n. The second letter in the Greek alphabet: lower case, β; capital, B.

beta coefficient. See CORRELATION —multiple correlation.

beta distribution. A random variable X has a **beta distribution** or is a **beta random variable** if X has the interval $(0, 1)$ as range and there are positive numbers α and β for which the *probability density function* f satisfies

$$f(x) = \frac{\Gamma(\alpha + \beta)}{\Gamma(\alpha)\Gamma(\beta)} x^{\alpha-1}(1-x)^{\beta-1}$$

$$= \frac{x^{\alpha-1}(1-x)^{\beta-1}}{B(\alpha, \beta)},$$

where Γ is the *gamma function* and B is the *beta function*. The *mean* is $\alpha/(\alpha + \beta)$ and the *variance* is $\alpha\beta/[(\alpha + \beta)^2(\alpha + \beta + 1)]$. The kth *moment about zero* is $B(\alpha + k, \beta)/B(\alpha, \beta)$. If F is an *F random variable* with (m, n) degrees of freedom, then $X = nF/(m + nF)$ is a beta random variable with $\alpha = \frac{1}{2}n$, $\beta = \frac{1}{2}m$.

beta function. The function β defined by

$$\beta(m, n) = \int_0^1 x^{m-1}(1-x)^{n-1} \, dx,$$

for m and n positive. In terms of the Γ function,

$$\beta(m, n) = \frac{\Gamma(m)\Gamma(n)}{\Gamma(m + n)}.$$

See GAMMA—gamma function. The **incomplete beta function** is defined by

$$\beta_x(m, n) = \int_0^x t^{m-1}(1-t)^{n-1} \, dt,$$

which is equal to $m^{-1}x^m F(m, 1 - n; m + 1; x)$, where F is the hypergeometric function.

BETTI, Enrico (1823–1892). Italian algebraist, analyst, topologist, and politician. See HOMOLOGY—homology group.

Betti number. Let H_r be the r-dimensional *homology group* (of a *simplician complex* K) formed by using the group G. If G is the group of integers modulo π, where π is a prime, then G is a field, H_r is a linear (vector) space, and the dimension of H_r is the **r-dimensional Betti number (modulo π)** of K. If G is the group of integers, then H_r is a commutative group with a finite number of generators and is the *Cartesian product* of infinite cyclic groups $E_1, \cdots, E_m$ and cyclic groups $F_1, \cdots, F_n$ of finite orders $r_1, \cdots, r_n$ (see TORSION—torsion coefficients of a group). The number m is the **r-dimensional Betti num-**

ber and $r_1, \cdots, r_n$ are the r-dimensional **torsion coefficients** of K. The Betti numbers (especially the 1-dimensional Betti number modulo 2, or 1 plus this number) are sometimes called *connectivity numbers* (see CONNECTIVITY). For an ordinary closed surface, $\chi = 2 - B_2^1$, where χ is the Euler characteristic and B_2^1 is the 1-dimensional Betti number modulo 2. If the surface is not closed (has boundary curves), then $\chi = 1 - B_2^1$. If the surface is orientable, then the *genus* of the surface is equal to $\frac{1}{2}B_2^1$.

BE-TWEEN, *prep.* Used in mathematics in many senses that are roughly consistent with English usage. *E.g.*, if a and c are real numbers and $a < c$, then b being between a and c means that $a < b < c$; for a point B on a line to be between distinct points A and C on the line means that B is on the same side of A as C and on the same side of C as A; for a set S to be between sets R and T means that $R \subset S \subset T$ (usually it is not required that S not be equal to either R or T).

BÉZOUT, Étienne (1730–1783). French analyst and geometer.

Bézout's theorem. If two algebraic plane curves of degrees m and n do not have a common component, then they have exactly mn points of intersection [points of intersection are counted to the degree of their multiplicity and include points of intersection at infinity (see COORDINATE —homogeneous coordinates, and PROJECTIVE —projective plane)]. For n-dimensional Euclidean space, if p algebraic hypersurfaces have degrees $d_1, d_2, \cdots, d_p$ and have only a finite number of common points, then there are at most $d_1 d_2 \cdots d_p$ common points (exactly $d_1 d_2 \cdots d_p$, if points at infinity are counted and if multiplicities are defined suitably and points of intersection are counted to the degrees of their multiplicities).

BI-AN'NU-AL, *adj.* Twice a year. *Syn.* semiannual.

BI'ASED or BI'ASSED, *adj.* **biased estimator.** See UNBIASED —unbiased estimator.

biased test. See HYPOTHESIS —test of a hypothesis.

BI-COM-PACT', *adj.* See COMPACT.

BI-COM-PAC'TUM, *n.* Same as COMPACTUM.

BI'CON-DI'TION-AL, *adj.* See EQUIVALENCE —equivalence of propositions.

BIEBERBACH, Ludwig (1886–1982). German complex analyst.

Bieberbach conjecture. Let f be a function of a complex variable for which f has domain the set D of all z for which $|z| < 1$, f is analytic at all points of D, f is one-to-one on D, $f(0) = 0$, and $f'(0) = 1$. The Taylor series for f,

$$f(z) = z + a_2 z^2 + a_3 z^3 + \cdots = z + \sum_{n=2}^{\infty} a_n z^n,$$

has the property that $|a_n| \leq n$ for each $n \geq 2$, and, if there is an $n \geq 2$ for which $|a_n| = n$, then there is a complex number c for which $|c| = 1$ and

$$f(z) = z + 2cz^2 + 3c^2 z^3 + \cdots = z / (1 - cz)^2.$$

This was conjectured in 1916 by Bieberbach, who proved only the case $n = 2$. By 1972, proofs were known for $n \leq 6$. A complete proof was given in 1984 by Louis de Branges. See KOEBE—Koebe function, RIEMANN —Riemann mapping theorem.

BIENAYMÉ, Irénée Jules (1796–1878). French probabilist.

Bienaymé-Chebyshev inequality (*Statistics*). See CHEBYSHEV —Chebyshev's inequality.

BI-EN′NI-AL, *adj.* Once in two years; every two years.

BI-FUR-CA′TION, *n.* A branching or division into two branches or parts; a splitting apart.

bifurcation theory. Study of the behavior of a solution of a nonlinear problem in the neighborhood of a known solution, particularly as a parameter varies. *E.g.*, the equation

$$y'' + \left(\lambda - \frac{1}{\pi} \int_0^\pi y^2 \, dx \right) y = 0$$

has the trivial solution $y \equiv 0$. It has the solution $y = A \sin nx$ if $A^2 = 2(\lambda - n^2)$. A new solution $y = \sqrt{2(\lambda - 1)} \sin x$ appears when $\lambda > 1$, another solution $y = \sqrt{2(\lambda - 4)} \sin 2x$ when $\lambda \geq 4$, etc. The values $1, 4, 9, \cdots$ of λ are **bifurcation points** or **branching points**. Also see CHAOS.

BIG′IT, *n.* A digit in a binary representation of a number.

BI′HAR-MON′IC, *adj.* **biharmonic boundary value problem.** See BOUNDARY.

biharmonic function. A solution of the fourth order partial differential equation $\Delta \Delta u = 0$,

where Δ is the Laplace operator $\partial^2/\partial x^2 + \partial^2/\partial y^2 + \partial^2/\partial z^2$; thus, a solution $u(x, y, z)$ of the equation

$$\frac{\partial^4 u}{\partial x^4} + \frac{\partial^4 u}{dy^4} + \frac{\partial^4 u}{dz^4} + 2\frac{\partial^4 u}{\partial x^2 \partial y^2} + 2\frac{\partial^4 u}{\partial y^2 \partial z^2}$$

$$+ 2\frac{\partial^4 u}{dz^2 \partial x^2} = 0.$$

The definition applies equally well to functions of two, four, or any other number of independent variables. Biharmonic functions occur in the study of electrostatic boundary value problems and elsewhere in mathematical physics.

BI-JEC′TION, *n.* A **bijection** from a set A to a set B is a one-to-one correspondence between A and B, *i.e.*, a function from A into B that is both an injection and a surjection. *Syn.* bijective function. See INJECTION, SURJECTION.

BI-JEC′TIVE, *adj.* See BIJECTION.

BI-LIN′E-AR, *adj.* A mathematical expression is *bilinear* if it is linear with respect to each of two variables or positions. *E.g.*: The function $f(x, y) = 3xy$ is linear in x and y, since

$$f(x_1 + x_2, y) = 3(x_1 + x_2)y = 3x_1 y + 3x_2 y$$

$$= f(x_1, y) + f(x_2, y)$$

and $f(x, y_1 + y_2) = f(x, y_1) + f(x, y_2)$. The *scalar product* of vectors $\mathbf{x} = x_1\mathbf{i} + x_2\mathbf{j} + x_3\mathbf{k}$ and $\mathbf{y} = y_1\mathbf{i} + y_2\mathbf{j} + y_3\mathbf{k}$ is $\mathbf{x} \cdot \mathbf{y} = x_1 y_1 + x_2 y_2 + x_3 y_3$, which is bilinear since $(\mathbf{u} + \mathbf{v}) \cdot \mathbf{y} = \mathbf{u} \cdot \mathbf{y} + \mathbf{v} \cdot \mathbf{y}$ and $\mathbf{x} \cdot (\mathbf{u} + \mathbf{v}) = \mathbf{x} \cdot \mathbf{u} + \mathbf{x} \cdot \mathbf{v}$. The scalar product and the function $3xy$ are **bilinear forms** (see FORM). The function F for which $F(u, v)$ is a function whose value at x is

$$\int_0^1 t^2 u(t, x) v(t, x) \, dt$$

is a bilinear function of u and v, where u and v are functions of two variables.

bilinear concomitant. See ADJOINT—adjoint of a differential equation.

BILL, *n.* A statement of money due, usually containing an itemized statement of the goods or services for which payment is asked.

BIL′LION, *n.* (1) In the U.S. and France, a thousand millions (1,000,000,000). (2) In England and Germany, a million millions (1,000,000,000,000).

BI-MO′DAL, *adj.* **bimodal distribution.** A distribution with two modes; *i.e.*, there are two different values which are conspicuously more frequent than neighboring values.

BI'NA-RY, *adj.* **binary number system.** A system of numerals for representing real numbers that uses the base 2 instead of the base 10. Only the digits 0 and 1 are needed. For example, 101110 in binary notation is equal to $1 \cdot 2^5 + 0 \cdot 2^4 + 1 \cdot 2^3 + 1 \cdot 2^2 + 1 \cdot 2^1 + 0 \cdot 2^0$, or 46 in decimal notation. This example illustrates the process of changing from binary notation to decimal notation. The reverse process is accomplished by successive division by 2. *E.g.*, 29 in decimal notation is equal to

$$2 \cdot 14 + 1 = 2^2 \cdot 7 + 1 = 2^2(2 \cdot 3 + 1) + 1$$

$$= 2^3 \cdot 3 + 2^2 + 1$$

$$= 2^3(2 + 1) + 2^2 + 1$$

$$= 2^4 + 2^3 + 2^2 + 1,$$

which is 11101 in binary notation. As with decimal notation, a real number has a repeating infinite sequence representation in binary notation if and only if it is rational. *E.g.*, $0.1100000 \cdots$ and $0.10, 10, 10, \cdots$ in binary notation are equal to $\frac{1}{2} + \frac{1}{4} = \frac{3}{4}$ and $\frac{2}{3}$, respectively, in decimal notation. Binary arithmetic is useful in connection with electronic computers, since the digits 0 and 1 can be described electrically as "off" and "on." *Syn.* dyadic number system. See BASE—base of a number system, BIGIT.

binary operation. An operation which is applied to two objects. For example, any two numbers can be added, two numbers can be multiplied, the intersection of two sets is a set, the product of a matrix with n columns and a matrix with n rows is a matrix, and the composition of two functions produces a function. A set is **closed** with respect to a particular binary operation if, whenever the operation is applied to a pair of members of the set, it gives a member of the set. The set of positive integers is closed under addition and multiplication. It is not closed under division or subtraction, since $2/3$ is not an integer and $2 - 3$ is not a positive integer. The set $\{1, 2, 3\}$ is not closed under addition, since $2 + 3$ is not a member of the set. *Tech.* A **binary operation** is a function f whose domain is a set of ordered pairs of members of a set S; the set S is **closed** with respect to the binary operation f if and only if $f(x, y)$ belongs to S whenever x and y belong to S. See TERNARY—ternary operation.

binary quantic. See QUANTIC.

BI-NO'MI-AL, *n.* A polynomial of two terms, such as $2x + 5y$ or $2 - (a + b)$. See TRINOMIAL.

binomial coefficients. The coefficients in the expansion of $(x + y)^n$. For example,

$$(x + y)^2 = x^2 + 2xy + y^2,$$

so that the binomial coefficients of order 2 are 1, 2, and 1. The $(r + 1)$st binomial coefficient of order n (n a positive integer) is $n!/[r!(n - r)!]$, the number of combinations of n things r at a time, and is denoted by $\binom{n}{r}$, $_nC_r$, $C(n, r)$ or C_r^n. The sum of the binomial coefficients is equal to 2^n, shown by putting 1 for each of x and y in $(x + y)^n$. See PASCAL—Pascal's triangle, and below, binomial theorem.

binomial differential. A differential of the form $x^m(a + bx^n)^p \, dx$, where a and b are any constants and the exponents m, n, and p are rational numbers.

binomial distribution. A random variable X is **binomially distributed** or is a **binomial random variable** if there is an integer n and a number p such that X is the number of successes in n independent Bernoulli experiments, where the probability of success in a single experiment is p. The range of X is the set $\{0, 1, \cdots, n\}$ and the probability of k successes is

$$P(X = k) = \binom{n}{k} p^k q^{n-k},$$

where $q = 1 - p$. *E.g.*, if three coins are thrown, then $p = \frac{1}{2}$ and the probabilities of 0, 1, 2 or 3 heads are $\frac{1}{8}, \frac{3}{8}, \frac{3}{8}, \frac{1}{8}$; these are the terms in the expansion of $(\frac{1}{2} + \frac{1}{2})^3$ by the binomial theorem. In general,

$$(p + q)^n = \sum_{k=0}^{n} \binom{n}{k} p^k q^{n-k} = \sum_{k=0}^{n} P(X = k).$$

The *mean* of the binomial distribution is np, the *variance* is npq, and the *moment generating function* is $M(t) = (q + pe^t)^n$. When n is large, the binomial distribution can be approximated by a normal distribution with mean np and variance npq. The binomial distribution can be approximated by a Poisson distribution mean np if n is large. See BERNOULLI (James)—Bernoulli distribution, Bernoulli experiment, CENTRAL—central limit theorem, MOMENT—moment generating function, MULTINOMIAL—multinomial distribution, NORMAL—normal distribution, POISSON—Poisson distribution.

binomial equation. An equation of the form $x^n - a = 0$.

binomial expansion. The expansion given by the binomial theorem.

binomial formula. The formula given by the binomial theorem.

binomial series. A binomial expansion which contains infinitely many terms. That is, the expansion of $(x + y)^n$, where n is not a positive integer or zero. Such an expansion converges and its sum is $(x + y)^n$ if $|y| < |x|$, or if $x = y \neq 0$ and $n > -1$, or if $x = -y \neq 0$ and $n > 0$. E.g.,

$$\sqrt{3} = (2 + 1)^{1/2} = 2^{1/2} + \tfrac{1}{2}(2)^{-1/2}$$

$$-\left(\tfrac{1}{2}\right)^3 (2)^{-3/2} + \cdots.$$

binomial surd. See SURD.

binomial theorem. A theorem (or rule) for the expansion of a power of a binomial. The theorem can be stated thus: The first term in the expansion of $(x + y)^n$ is x^n; the second term has n for its coefficient, and the other factors are x^{n-1} and y; in subsequent terms the powers of x decrease by 1 for each term and those of y increase by 1, while the next coefficient can be obtained from a given coefficient by multiplying by the exponent of x and dividing by one more than the exponent of y. E.g., $(x + y)^3 = x^3 + 3x^2 y + 3xy^2 + y^3$. In general, if n is a positive integer,

$$(x + y)^n = x^n + nx^{n-1}y$$

$$+ \frac{n(n - 1)}{2!} x^{n-2} y^2 + \cdots + y^n.$$

The coefficient of $x^{n-r} y^r$ is $n!/[r!(n - r)!]$. See above, binomial coefficients, binomial series.

negative binomial distribution. A random variable S has a **negative binomial distribution** or is **a negative binomial random variable** if there are numbers r and p such that X is the number of repeated independent Bernoulli trials with probability of success p that are performed to obtain r successes. The range of X is the infinite set $\{r, r + 1, r + 2, \cdots\}$ and the probability of n trials is

$$P(X = n) = \binom{n - 1}{r - 1} p^r q^{n-r} \text{ if } n \geq r,$$

where $q = 1 - p$. The *mean is* r/p, the *variance* is rq/p^2, and the *moment generating function* is

$$M(t) = e^{tr} p^r (1 - qe^t)^{-r}.$$

Syn. Pascal distribution. If $r = 1$, X has a **geometric distribution** or is a **geometric random variable**. Then $P(X = n)$ is pq^{n-1} if $n \geq 1$, the *mean* is $1/p$, and the *variance* is q/p^2. Sometimes $Y = X - 1$, the number of trials before the first success, is called a geometric random variable; then $P(X = n)$ is pq^n if $n \geq 0$ and the mean is q/p.

BI-NOR′MAL, *n.* See NORMAL—normal to a curve or surface.

BI-PAR′TITE, *n.* **bipartite cubic.** The locus of the equation

$$y^2 = x(x - a)(x - b), \qquad 0 < a < b.$$

The curve is symmetric about the x-axis and intersects the x-axis at the origin and at the points $(a, 0)$ and $(b, 0)$. It is said to be bipartite because it has two entirely separate branches.

bipartite graph. See COLORING—graph coloring.

BI-QUAD-RAT′IC, *adj.* **biquadratic equation.** An algebraic equation of the fourth degree. *Syn.* quartic.

BI-REC-TANG′U-LAR, *adj.* Having two right angles. *E.g.*, a **birectangular spherical triangle** is a spherical triangle with two right angles.

BIRKHOFF, George David (1884–1944). Leading American topologist, analyst, and applied mathematician of his time. Worked on coloring of maps, calculus of variations, and dynamical systems. Proved Poincare's last theorem, concerning fixed points in a ring. See ERGODIC—ergodic theory, POINCARÉ—Poincaré-Birkhoff fixed-point theorem.

BI-SECT′, *v.* To divide in half.

bisect an angle. To draw a line through the vertex dividing the angle into two equal angles.

bisect a line segment. To find the point on the line segment and equally distant from the ends. *Analytically*, the Cartesian coordinates of the midpoint can be found as the arithmetic means or averages of the corresponding coordinates of the two end points. See POINT—point of division. If $P_1(x_1, y_1)$ and $P_2(x_2, y_2)$ are the end points of a line segment, the coordinates of the midpoint are

$$x = (x_1 + x_2)/2, \qquad y = (y_1 + y_2)/2.$$

BI-SECT′ING, *adj.* **bisecting point of a line segment.** Same as MIDPOINT of a line segment.

BI-SEC′TOR, *n.* **bisector of an angle.** The straight line which divides the angle into two equal angles. The equation of an angle bisector can be obtained by equating the distances of a variable point from the two lines. See DISTANCE —distance from a line to a point.

bisector of the angle between two intersecting planes. A plane containing all the points equidistant from the two planes. There are two such bisectors for any two such planes. Their equations are obtained by equating the distances of a variable point from the two planes—first giving these distances like signs, and then unlike signs. See DISTANCE —distance from a plane to a point.

BI-VAR′I-ATE, *adj.* Involving two variables.

bivariate distribution. See DISTRIBUTION — distribution function.

bivariate normal distribution. The vector random variable (X, Y) has a **bivariate normal distribution** if its probability density function is given by

$$f(x, y) = \frac{1}{2\pi\sigma_X\sigma_Y(1 - r^2)^{1/2}} e^{-\frac{1}{2}w/(1-r^2)},$$

$$w = \left(\frac{x - \mu_X}{\sigma_X}\right)^2 - 2r\frac{(x - \mu_X)(y - \mu_Y)}{\sigma_X\sigma_Y}$$

$$+ \left(\frac{y - \mu_Y}{\sigma_Y}\right)^2,$$

where $-1 \le r \le 1$, μ_X and μ_Y are the *means* of X and Y, and σ_X^2 and σ_Y^2 are the *variances* of X and Y. The conditional distribution of X given Y (or of Y given X) is normal. The conditional mean of X given that $Y = y$ is $\mu_X + r(\sigma_X/\sigma_Y) \cdot (y - \mu_Y)$. The parameter r is the **correlation parameter** and is equal to the *correlation coefficient* between the random variables X and Y.

BLASCHKE, Wilhelm (1885–1962). Austrian-German analyst and geometer.

Blaschke product. A product of type

$$B(z) = z^k \prod_{n=1}^{\infty} \frac{(a_n - z)|a_n|}{(1 - \bar{a}_n z)a_n},$$

where $0 < |a_n| < 1$ for each n, $\sum_1^{\infty}(1 - |a_n|)$ converges, and k is a nonnegative integer. The function B is bounded and analytic on the set of all complex numbers z such that $|z| < 1$. The zeros of B are the numbers $\{a_n\}$ and 0 (if $k > 0$).

Blaschke's theorem. Each bounded closed convex plane set of *width* 1 contains a circle of radius $\frac{1}{3}$. See JUNG—Jung's theorem.

BLISS, Gilbert Ames (1876–1951). American analyst. Best known for his work in the calculus of variations, he also contributed to the study of algebraic functions of a complex variable and to the study of exterior ballistics.

BLOCK, *n.* **randomized blocks.** An experimental design for which an experiment is repeated for each of several situations, called **blocks.** *E.g.,* the yields of three types of corn might be tested in several fields, the *blocks*, by planting each type of corn in a plot in each field, assuming all plots in a given field have equal fertility. When studying the quality of a product, the machines might be grouped into several types, the *blocks*, and the operators chosen randomly. See VARIANCE —analysis of variance.

BOARD MEASURE. The system of measuring used for measuring lumber. See MEASURE — board measure.

BOCHNER, Salomon (1899–1982). Polish-born analyst who obtained a doctoral degree at the University of Berlin and came to the U.S. in 1933.

Bochner integral. See INTEGRABLE —integrable function.

BOD′Y, *n.* **convex body.** See CONVEX—convex set.

BOHR, Harold (1887–1951). Danish analyst and number theorist. Brother of physicist Niels Bohr. Worked on theory of summability, Dirichlet series, and zeta function. Founded theory of almost periodic functions. See PERIODIC —almost periodic function.

BOLYAI, Janos (1802–1860). Hungarian geometer. Invented non-Euclidean geometry independently of Lobachevski. See GEOMETRY— non-Euclidean geometry, LOBACHEVSKI.

BOLYAI, Farkas (1775–1856). Hungarian mathematician; friend of Gauss and father of Janos Bolyai.

BOLZA, Oskar (1857–1942). German analyst who spent many years in the U.S. Best known for his work in the calculus of variations, he also contributed to the study of elliptic and hyperelliptic functions.

problem of Bolza. In the calculus of variations, the general problem of determining, in a class of curves subject to constraints of the form $Q_j(x, y, y') = 0$ and

$$g_k[x_1, y(x_1), x_2, y(x_2)]$$

$$+ \int_{x_1}^{x_2} f_k(x, y, y') \, dx = 0,$$

an arc that minimizes a function of the form

$$l = g[x_1, y(x_1), x_2, y(x_2)] + \int_{x_1}^{x_2} f(x, y, y') \, dx.$$

BOLZANO, Bernhard (1781–1848). Czechoslovakian analyst whose work helped establish the importance of rigorous proof. He became a Catholic priest and in 1805 was given a chair in Philosophy at the University of Prague. Because he spoke out on educational reform, the right of individual conscience, the absurdity of war and militarism, and refused to recant his statements, he was forced to retire in 1824 on a small pension.

Bolzano's theorem. A real-valued function f of a real variable x is zero for at least one value of x between a and b if it is continuous on the closed interval $[a, b]$ and $f(a)$ and $f(b)$ have opposite signs.

Bolzano-Weierstrass theorem. If E is a bounded set containing infinitely many points, there is a point x which is an accumulation point of E. The set E may be a set of real numbers, a set in a plane, or a set in n-dimensional Euclidean space. An equivalent statement of the theorem is that for any (finite dimensional) Euclidean space the concepts of bounded closed sets and sets with the Bolzano-Weierstrass property are equivalent (see COMPACT). This theorem is frequently credited to Weierstrass, but was proved by Bolzano in 1817 and seems to have been known to Cauchy.

BOMBIERI, Enrico (1940–   ). Italian mathematician awarded a Fields Medal (1974) for his contributions to number theory and the theory of minimal surfaces.

BOND, *n.* A written agreement to pay interest (dividends) on a certain sum of money and to pay the sum in some specified manner, unless it be a **perpetual bond** (which draws interest, but whose principal need never be paid). **Callable** (or **optional**) bonds are redeemable prior to maturity at the option of the issuing corporation, usually under certain specified conditions and at certain specified times. An **annuity bond** is redeemed in equal installments which include the interest on the unpaid balance and sufficient payment of the face of the bond to redeem it by the end of a specified time. **Coupon bonds** are bonds for which the interest is paid by means of coupons (in effect, the coupons are post-dated checks, attached to the bond, which may be detached and used at the specified date); **registered bonds** are bonds whose ownership is registered with the debtor, the interest being paid by check directly to the registered owner. If an issue of bonds is such that part of the bonds mature on a certain date and part of the bonds mature at each of certain dates thereafter (usually each year), the bonds are said to be **serial bonds**. **Collateral trust bonds** are bonds issued by corporations whose assets consist primarily of securities of subsidiaries and of other corporations (the securities are deposited with a trust company as trustee); **guaranteed bonds** are bonds for which some corporation (in addition to the one which issues the bonds) guarantees payment of principal or interest or both; **debenture bonds** are unsecured and usually protected only by the credit and earning power of the issuer; **mortgage bonds** have the highest priority in case of liquidation of the corporation (they are called *first mortgage bonds, second mortgage bonds,* etc.).

"and interest price," purchase price, and **redemption price** of a bond. See PRICE.

bond rate. See DIVIDEND —dividend on a bond.

bond table. A table showing the values of a bond at a given bond rate for various investment rates, and for various periods. Most tables are based on interest computed semiannually (the usual practice) and on the assumption that the bonds will be redeemed at par.

par value of a bond. The principal named in the bond. *Syn.* face value. See PAR.

premium bonds. See PREMIUM.

valuation of bonds. Computing the *present value*, at the investor's rate of interest, of the face value of the bond and of the interest payments (an annuity whose rental is equal to the dividend payments on the bond).

$$P = C(1 + i)^{-n} + R[1 - (1 + i)^{-n}]/i,$$

where P denotes the value of the bond, C its

redemption value, R the interest payments (coupon value of a coupon bond), n the number of periods before redemption, and i the investor's (purchaser's) rate per period.

   yield of a bond.   See YIELD.

BONNET, Pierre Ossian (1819–1892). French analyst and differential geometer.

   Bonnet's mean-value theorem. See MEAN–mean-value theorems (or laws of the mean) for integrals.

BO′NUS, *n.*   A sum paid in addition to a sum that is paid periodically, as bonuses added to dividends, wages, etc. See INSURANCE—participating insurance policy.

BOOK, *n.* **book value.**   See VALUE.

BOOLE, George (1815–1864). Pioneering British logician. He also worked in algebra, analysis, calculus of variations, and probability theory.

   Boolean algebra.   A *ring* which has the properties that $x \cdot x = x$ for each x, and there is an element I such that $x \cdot I = x$ for each x. If the members of the ring are sets, then *addition* and *multiplication* for the ring correspond to *symmetric difference* and *intersection* of sets and I is a set which contains all sets belonging to the ring. If a collection of subsets of a set S contains the complement of each of its members and the union of any two of its members, then it is a Boolean algebra if the ring operations of *addition* and *multiplication* are taken to be *symmetric difference* and *intersection*. Conversely, any Boolean algebra is an algebra of subsets of a collection of subsets of some set (see ALGEBRA—algebra of subsets). If, for any Boolean algebra, the operators $\cup$ and $\cap$, and the concept of inclusion, are defined by

$$A \cup B = (A + B) + (A \cdot B),$$

$$B \cap B = A \cdot B,$$

$$A \subset B \quad \text{if and only if} \quad A \cap B = A,$$

then these correspond to the union, intersection, and inclusion concepts for sets and the following statements can easily be proved. ($A + A$ can be proved to have the same value for all elements A of the Boolean algebra and this common value

is denoted by θ):

$$A \cup (B \cup C) = (A \cup B) \cup C,$$

$$A \cap (B \cap C) = (A \cap B) \cap C,$$

$$A \cup B = B \cup A,$$

$$A \cap B = B \cap A,$$

$$A \cap (B \cup C) = (A \cap B) \cup (A \cap C),$$

$$A \cup (B \cap C) = (A \cup B) \cap (A \cup C),$$

$$A \cup A = A \cap A = A, \, \theta \cup A = I \cap A = A,$$

$$\theta \subset A \subset I,$$

$$A = B \text{ if } A \subset B \text{ and } B \subset A,$$

$$A \subset C \text{ if } A \subset B \text{ and } B \subset C.$$

If the complement A' of A is defined to be $A + I$, then

$$(A \cap B)' = A' \cup B', (A \cup B)' = A' \cap B',$$

$$A \cup A' = I, \, A \cap A' = \theta,$$

$$(A')' = A, \, I' = \theta, \, \theta' = I.$$

The simplest Boolean algebra is the one whose elements are the empty set and the set of one point, θ and I. Then $A \cup B = I$ if and only if one (or both) of A and B is I, and $A \cap B = \theta$ if and only if one (or both) of A and B is θ. As well as being interpreted as an algebra of sets, a Boolean algebra can also be interpreted as an algebra of elementary logical properties of statements (propositions). The statement $p = q$ means that the statements denoted by "p" and "q" are logically equivalent; $p \cup q$ denotes the statement "p or q"; $p \cap Q$ denotes the statement "p and q"; and p' denotes the statement "not p." If p is the statement "triangle x is isosceles", and "q" is the statement "triangle x is equilateral", then $p \cup q$ is the statement "triangle x is isosceles or triangle x is equilateral"; $p \cap q$ is the statement "triangle x is isosceles and triangle x is equilateral", and $q \subset p$ is the statement "$q \cap p = q$" (*i.e.,* "for any triangle x, x is isosceles if x is equilateral"). See ATOM, LATTICE.

BOR′DER-ING, *v.* **bordering a determinant.** Annexing a column and a row. Usually refers to annexing a column and a row which have 1 as a common element—all the other elements of either the column or the row being zero. This increases the order of the determinant by 1 but does not change its value.

BOREL, Félix Édouard Justin Émile (1871–1956). French mathematician and politician. Founded modern theory of measure, divergent series; contributed to probability, game theory, etc.

Borel covering theorem. Same as the HEINE–BOREL THEOREM.

Borel definition of the sum of a divergent series. If $\sum a_n$ is the series to be summed, the sum is (1)

$$S = \lim_{\alpha \to \infty} \lim_{n \to \infty} \frac{s_0 + s_1\alpha + s_2\alpha^2/2! + \cdots + s_n\alpha^n/n!}{1 + \alpha + \alpha^2/2! + \cdots + \alpha^n/n!}$$

$$= \lim_{\alpha \to \infty} \left(e^{-\alpha} \sum_{n=0}^{\infty} \frac{s_n}{n!} \alpha^n \right),$$

where $s_i = \sum_{j=0}^{i} a_j$. (2) The sum of $\sum a_n$ is defined to be $\int_0^{\infty} e^{-x} \sum_0^{\infty} a_n \frac{x^n}{n!} \, dx$, where x is real, if this limit exists. Both definitions are *regular*. See SUMMATION —summation of divergent series.

Borel measurable function. See MEASURABLE —measurable function (2).

Borel set. Let X be a topological space (*e.g.*, X could be the real line or any Euclidean space). The Borel sets are the members of the smallest σ-algebra that contains all open sets in X (or that contains all closed sets in X). Examples of Borel sets are F_σ sets, which are countable unions of closed sets, and G_δ sets, which are countable intersections of open sets. A Borel set is sometimes called a **Borel measurable set**. If X is Euclidean space, then all Borel sets are Lebesgue measurable.

BOUND, n. A **lower bound** of a set of numbers is a number which is less than or equal to every number in the set; an **upper bound** is a number which is greater than or equal to every number in the set. The **greatest lower bound** or **infimum** (abbreviated **g.l.b.** or **inf**) of a set of numbers is the largest of its lower bounds. This is either the smallest number in the set or the largest number that is less than all the numbers in the set. The **least upper bound** or **supremum** (abbreviated **l.u.b.** or **sup**) of a set of numbers is the smallest of its upper bounds. This is either the largest number in the set or the smallest number that is greater than all the numbers in the set. If a g.l.b. or a l.u.b. of a set does not belong to the set (and sometimes if it does), then it is an accumulation point of the set. *E.g.*, the set $\{0.3, 0.33, 0.333, \cdots\}$ has the l.u.b. $\frac{1}{3}$, which also is an accumulation point. These concepts can be extended to anypartially ordered set. *E.g.*, an upper bound of a collection of sets is a set U that contains each of the given sets. See LATTICE.

bound of a function. A bound of a function f on a set S is a bound for the set of numbers $f(x)$ for which x is in S.

least-upper-bound-axiom. The statement: "A set of real numbers that has an upper bound has a least upper bound." This often is taken as one of the axioms for the real number system, but otherwise is proved. It is equivalent to the **greatest-lower-bound axiom**: "A set of real numbers that has a lower bound has a greatest lower bound."

BOUND'A-RY, n. biharmonic boundary-value problem. For a region R with boundary surface S, the biharmonic boundary-value problem is the problem of determining a function $U(x, y, z)$ that is biharmonic in R and is such that its first-order partial derivatives coincide with prescribed boundary value functions on S. This problem, along with the Dirichlet problem, arises in particular problems concerning elastic bodies.

boundary of a set. See INTERIOR —interior of a set.

boundary of a simplex and a chain and boundary operator. See CHAIN —chain of simplexes.

boundary of half-line, half-plane, and **half-space.** See HALF —half-plane, half-space, and RAY.

boundary-value problem. The problem of finding a solution to a given differential equation or set of equations which will meet certain specified requirements for a given set of values of the independent variables—the boundary points. Many of the problems of mathematical physics are of this type.

first boundary-value problem of potential theory (the Dirichlet problem). Given a region R, its boundary surface S, and a function f defined and continuous over S, to determine a solution U of Laplace's equation $\nabla^2 U = 0$ which is regular in R, is continuous in $R + S$, and satisfies the equation $U = f$ on the boundary. This problem occurs in electrostatics and heat flow. It has at most one solution. See GREEN —Green's function.

second boundary-value problem of potential theory (the Neumann problem). Given a region R, its boundary surface S, and a function f defined and continuous over S and such that $\iint f \, dS$ over S vanishes, to find a solution of Laplace's equation $\nabla^2 U = 0$ which is regular in R, which together with its normal derivative is continuous in $R + S$, and which has a normal derivative equal to f on the boundary S. This problem occurs in fluid dynamics. Any two of its solutions differ at most by a constant. See NEUMANN (Karl)—Neumann's function.

third boundary-value problem of potential theory. As in the two above problems, except the function U is required to satisfy the equation $k \, \partial U / \partial n + hU = f$ on the boundary, where k, h, and f are prescribed functions that are continuous on S. This problem includes the other two and is of importance in heat flow and fluid mechanics. If $h/k > 0$, it has at most one solution. See ROBIN—Robin's function.

BOUND'ED, *adj.* **bounded convergence theorem.** Suppose m is a countably additive measure on a σ-algebra of subsets of a set T for which $m(T) < +\infty$, and that $\{S_n\}$ is a sequence of measurable functions for which there is a number M such that $|S_n(x)| \leq M$ for all n and all x in T. Then each S_n is integrable and, if there is a function S for which $\lim_{n \to +\infty} S_n(x) = S(x)$ a.e. on T, then S is integrable and

$$\int_T S \, dm = \lim_{n \to \infty} \int_T S_n \, dm.$$

For Riemann integration, the theorem can be stated as follows: Suppose that for a sequence of functions $\{S_n\}$ and an interval I there is a number M such that $|S_n(x)| \leq M$ for all n and all x in I. Also suppose that each S_n is Riemann integrable on I and that there is a function S which is Riemann integrable on I and for which $\lim_{n \to \infty} S_n(x) = S(x)$ a.e. on I. Then the integral of S over I is equal to the limit of the integral of S_n over I as $n \to \infty$. See LEBESGUE—Lebesgue convergence theorem, MONOTONE—monotone convergence theorem, and SERIES—integration of an infinite series.

bounded linear transformation. See LINEAR—linear transformation.

bounded quantity, or function. A quantity whose numerical value is always less than or equal to some properly chosen constant. The ratio of a leg of a right triangle to the hypotenuse is a bounded quantity since it is always less than or equal to 1; that is, the functions $\sin x$ and $\cos x$ are bounded functions since their numerical values are always less than or equal to 1. The function $\tan x$ in the interval $(0, \frac{1}{2}\pi)$ is not bounded.

bounded sequence. See SEQUENCE—bound to a sequence.

bounded set of numbers. A set of numbers which has both a lower bound and an upper bound; a set of numbers for which there are numbers A and B such that $A \leq x \leq B$ for each number x of the set.

bounded set of points. (1) In a *metric space*, a **bounded set** is a set of points for which the set of distances between pairs of points is a bounded set. The least upper bound of such distances is called the **diameter** of the set. A set T is **totally bounded** if, for any $\epsilon > 0$, there is a finite set of points in T such that each point of T is at distance less than ϵ from at least one of these points. A metric space is *compact* if and only if it is *complete* and *totally bounded*. (2) In a *linear topological space*, a **bounded set** is a subset S which has the property that, for each neighborhood U of O, there exists a positive number a such that $S \subset aU$.

bounded variation. See VARIATION—variation of a function.

essentially bounded function. A function f for which there is a number K such that the set of all x for which $|f(x)| > K$ is of measure zero. The greatest lower bound of such numbers K is the **essential supremum** of $|f(x)|$.

BOURBAKI, Nicholas. The pseudonym of a group of expert mathematicians, almost all French, who since the late 1930s have been writing the multivolume *Elements of Mathematics*, a survey of all "important" mathematics. The members of the group change now and then and has deliberately been kept mysterious. The original group was H. Cartan, C. Chevalley, J. Dieudonné, and A. Weil. Usually there have been ten to twenty members.

BOX, *n.* **three-boxes game.** A game in which there are three boxes marked 1, 2, and 3. For a given play of the game, player A removes the bottom of one of the boxes, but player B does not know which one it is. Player B then puts an amount of money equal to the number marked on the box in each of two of the three boxes. He loses the money put in the box with no bottom and wins the money put in the others. This is a *zero-sum* game with *imperfect information*. The *payoff matrix* does not have *a saddle point* and the solutions are *mixed strategies*. The solutions are $(0, \frac{1}{2}, \frac{1}{2})$ for A and $(\frac{3}{5}, \frac{2}{5}, 0)$ for B, meaning that A removes the bottoms of boxes 2 and 3, each with probability $\frac{1}{2}$, player B puts money into boxes 1 and 2, or 1 and 3, with respective probabilities $\frac{3}{5}$ and $\frac{2}{5}$ (never in 2 and 3). The *value* of this game is 1 (with B the *maximizing player*).

BOYLE, Robert (1627–1691). British chemist and natural philosopher.

Boyle's law. At a given temperature, the product of the volume of a gas and the pressure

(pv) is constant. Also called **Boyle and Mariott's law.** Approximately true for moderate pressures.

BRACE, *n.* See AGGREGATION.

BRA-CHIS'TO-CHRONE, *adj.* **brachistochrone problem.** The calculus of variations problem of finding the equation of the path down which a particle will fall from one point to another in the shortest time. Proposed by John Bernoulli in 1696 as a challenge to the mathematicians of Europe. The time required for a particle with the initial velocity v_0 to fall along a path $y = f(x)$ from a point $(x_1, 0)$ to a point (x_2, y_2) is

$$t = \frac{1}{\sqrt{2g}} \int_{x_1}^{x_2} \sqrt{\frac{1 + (y')^2}{y + a}} \, dx,$$

where $a = v_0^2 / 2g$. The solution of the problem then requires the determination of y that minimizes the integral. See CALCULUS—calculus of variations. Newton, Leibniz, l'Hôpital, and James and John Bernoulli all found the correct solution, which is a *cycloid* through the two points. See CYCLOID.

BRACK'ET, *n.* See AGGREGATION.

BRAID, *n.* Let s_1 and s_2 be two parallel lines in a plane P and let $\{A_1, \cdots, A_n\}$ and $\{B_1, \cdots, B_n\}$ be two sets of points placed consecutively on s_1 and s_2, respectively. Let $\{\lambda 1, \lambda 2, \cdots, \lambda n\}$ be a permutation of $\{1, 2, \cdots, n\}$ and, for each i, let L_i be a broken line above P connecting A_i and $B_{\lambda i}$ so that L_i and L_j are disjoint if $i \neq j$, the projection L_i^P of L_i onto P lies between s_1 and s_2, any line that is parallel to s_1 intersects L_i^P in at most one point, and $L_i^P \cap L_j^P$ is a finite set of points—all on different lines parallel to s_1. This is a **braid of order** *n* and L_i is the *i*th **string.** A **closed braid** is obtained by deleting the original lines and, for each i, joining A_i and B_i by a line in the plane P. A closed braid consists of several knots and any knot is equivalent to a closed braid. See KNOT—knot in topology.

BRANCH, *n.* **branch of a curve.** Any section of a curve separated from the other sections of the curve by discontinuities or special points such as vertices, maximum or minimum points, cusps, nodes, etc. One would speak of the two *branches* of an hyperbola, or even of four *branches* of an hyperbola; or of two *branches* of a semi-cubical parabola, or of the *branch* of a curve above (or below) the *x*-axis.

branch cut of a Riemann surface. A line or curve C on a Riemann surface such that on crossing C a variable point is considered as passing from one sheet to another.

branch of a multiple-valued analytic function. The single-valued analytic function $w = f(z)$ corresponding to values of z on a single sheet of the Riemann surface of definition.

branch point of a Riemann surface. A point of the Riemann surface at which two or more sheets of the surface hang together.

infinite branch. See INFINITE.

BRANGES, Louis de (1932–   ). American complex analyst who proved the Bieberbach conjecture.

BREADTH, *n.* Same as WIDTH.

BRIANCHON, Charles Julien (1783–1864). French geometer.

Brianchon's theorem. If a hexagon is circumscribed about a conic section, the three diagonals (lines through opposite vertices) are concurrent. This is the dual of *Pascal's theorem.* See DUALITY—principle of duality of projective geometry, PASCAL—Pascal's theorem.

BRIDG'ING, *v.* **bridging in addition.** In adding a one-place number to a second number, *bridging* is said to occur if the sum is in a decade different from that in which the second number lies. Thus bridging occurs in $14 + 9 = 23$ but not in $14 + 3 = 17$. See DECADE.

bridging in subtraction. If the difference obtained by subtracting a number from a second number (the minuend) is in a decade different from that in which the minuend lies, bridging is said to have occurred. Thus bridging occurs in the examples $64 - 9 = 55$, $34 - 27 = 7$, but not in $64 - 3 = 61$.

BRIGGS, Henry (1561–1630). English astronomer, geometer, and numerical-table maker.

Briggsian logarithms. Logarithms using 10 as a base. *Syn.* common logarithms. See LOGARITHM.

BRITISH, *adj.* **British thermal unit** or **B.T.U.** The amount of heat required to raise the temperature of 1 lb. of water 1° F., when the water is at its maximum density, which is at 4° C. or 39.2° F.

BRO'KEN, *adj.* **broken line.** A curve consisting of segments of lines joined end to end and not forming a single straight line segment. When defining the length of a curve, it is customary to approximate the curve by a broken line **inscribed in the curve** (i.e., having its vertices on the curve).

BRO'KER, *n.* One who buys and sells stocks and bonds on commission, that is, for pay equal to a given percentage of the value of the paper. *Broker* is sometimes applied to those who sell any kind of goods on commission, but commission merchant, or commission man, is more commonly applied to those who deal in staple goods.

BRO'KER-AGE, *n.* A commission charged for selling or buying stocks, bonds, notes, mortgages, and other financial contracts. See BROKER.

BROUWER, Luitzen Egbertus Jan (1881–1966). Dutch topologist and logician. Founder of modern intuitionism, wherein the positive integers furnish the prototype for the intellectual construction of all mathematical objects. In accordance with this philosophy, he objected to the unrestricted use of Aristotelian logic (including the law of the excluded middle), particularly in dealing with infinite sets. Thus the question of whether there exists an integer with a certain property P can be settled only if one constructs such an integer or shows "constructively" that no such integer exists. It cannot be settled by proving that it is false that no such integer exists.

Brouwer's fixed-point theorem. If C is a closed disc and T is a continuous mapping of C into C, then T has a fixed point—*i.e.*, there is a point x in C for which $T(x) = x$. This theorem also is true for closed intervals and for spheres with their interiors. It was extended by Schauder to the case for which C is a compact convex subset of a normed vector space. Tychonoff extended this result from normed vector spaces to locally convex linear topological spaces. See VECTOR —vector space.

BROWN, Robert (1773–1858). Scottish botanist.
Brownian motion process. Same as WIENER PROCESS.

BUDAN DE BOIS LAURENT, Ferdinand François Désiré (c. 1800–1853 or later). French physician and amateur mathematician.
Budan's theorem. The number of real roots of $f(x) = 0$ between a and $b(a < b)$, where $f(x)$ is a polynomial of degree n, is $V(a) - V(b)$, or less by an even number, $V(a)$ and $V(b)$ being the numbers of variations in sign of the sequence

$$f(x), f'(x), f''(x), \cdots, f^{(n)}(x),$$

when $x = a$ and $x = b$, respectively. (Vanishing terms on the sequence are not counted and

m-tuple roots are counted as m roots.) *E.g.*, to find the number of roots of $x^3 - 5x + 1 = 0$ between 0 and 1, we form the sequence $x^3 - 5x + 1$, $3x^2 - 5$, $6x$, 6, then substitute 0 and 1 for x, successively. This gives the sequences 1, -5, 0, 6 and -3, -2, 6, 6, whence $V(0) - V(1) = 2 - 1 = 1$. Thus there is one root between 0 and 1. Similarly the other roots can be located between 2 and 3 and between -3 and -2.

BUFF'ER, *n.* In a computing machine, a switch that transmits a signal if any one of several signals is received by the switch; thus a buffer is the machine equivalent of the logical "or". See DISJUNCTION, GATE. *Syn.* inverse gate.

BUFFON, George Louis Leclerc, Comte de (1707–1788). French naturalist and probabilist.
Buffon needle problem. Suppose a board is ruled with equidistant parallel lines and that a needle fine enough to be considered a segment of length δ less than the distance d between consecutive lines is thrown on the board. What is the probability it will hit one of the lines? The answer is $2\delta/(\pi d)$. It is possible to approximate the value of π by throwing such a needle a large number of times.

BUILD'ING, *n.* **building and loan association.** A financial organization whose objective is to loan money for building homes. One plan, called the *individual-account plan*, is essentially as follows: Members may buy shares purely as an investment, usually paying for them in monthly installments at an annual nominal rate; or they may borrow money (shares) from the company with which to build, securing (guaranteeing) this money with mortgages on their homes. In both cases the monthly payments are called dues. Failure to meet monthly payments on time is sometimes subjected to a fine which goes into the profits of the company. The profits of the company are distributed to the share purchasers, thus helping to mature (complete the payments on) their shares. In practice, the interest rate is usually figured so that it returns all profits automatically. A *serial plan* is a plan under which shares are issued at different times to accommodate new members. Monthly dues are paid and profits distributed to all share holders. This plan naturally resolves into the INDIVIDUAL-ACCOUNT PLAN. A *guaranteed-stock plan* is a plan in which certain investors provide certain funds and guarantee the payment of certain dividends on all shares, any surplus over this guarantee being divided among these basic stockholders. A *terminating plan* is a plan under which the members pay dues for a certain number of years to facilitate their building homes, the highest bidder getting the use of the money, since there is not enough to go around. New members coming in have to pay back-dues and back-earnings. This is the earliest plan of building and loan association and is not usually practiced now.

BULK, *n.* **bulk modulus.** See MODULUS.

BUN'DLE, *n.* **bundle of planes.** See SHEAF.

BUNIAKOVSKI (or BOUNIAKOWSKY), Victor Jakowlewitsch (1804–1899). Russian probabilist.
  Buniakovski inequality. See SCHWARZ–Schwarz inequality.

BURALI-FORTI, Cesare (1861–1931). Italian mathematician.
  Burali-Forti paradox. The "set of all ordinal numbers" (each of which is an *order type* of a well-ordered set) is a well-ordered set. However, the order type Y of this set is then a largest ordinal number. This is impossible, since $Y + 1$ is a larger ordinal number (Y is the order type of a certain well-ordered set and $Y + 1$ is the order type of the well-ordered set obtained by introducing a single new element to follow every member of this set).

BURNSIDE, William (1852–1927). **Burnside conjecture.** See THOMPSON.

BUSH, Vannevar (1890–1974). American electrical engineer. Starting about 1925, he built the first large-scale mechanical but electrically powered analog computer.

C

C.G.S. UNITS. Units of the centimeter—gram—second system. Centimeter is the unit of distance (length); gram, mass; and second, time. See ERG, FORCE-unit of force.

CABLE, *n.* **parabolic cable.** See PARABOLIC.

CAL'CU-LATE, *v.* To carry out some mathematical process; to supply theory or formula and secure the results (numerical or otherwise) that are required; a looser and less technical term than compute. One may say, "Calculate the volume of a cylinder with radius 4' and altitude 5'"; he may also say, "Calculate the derivative of $\sin(2x + 6)$." *Syn.* compute.

CALCULATING MACHINE. Same as COMPUTING MACHINE, but see COMPUTER.

CAL'CU-LUS, *n.* The field of mathematics which deals with differentiation and integration of functions and related concepts and applications. Sometimes called the **infinitesimal calculus** because of the prominence of the use of *infinitesimal* in the early development of the subject. See below, differential calculus, integral calculus.
  calculus of variations. The study of the theory of maxima and minima of definite integrals whose integrand is a known function of one or more independent variables and of one or more dependent variables and their derivatives, the problem being to determine the dependent variables so that the integral will be a maximum or a minimum. The simplest such integral is of the form

$$I = \int_a^b f(x, y, dy/dx)\, dx,$$

where y is to be determined to make I a maximum or a minimum (whichever is desired). The name *calculus of variations* originated as a result of notations introduced by Lagrange in about 1760 (see VARIATION). Other integrals studied are of the form

$$I = \int_a^b f(x, y_1, \cdots, y_n, y_1', \cdots, y_n')\, dx,$$

where $y_1, \cdots, y_n$ are unknown functions of x, or multiple integrals such as

$$I = \int_a^b \int_a^b f\left(x, y, z, \frac{\partial z}{\partial x}, \frac{\partial z}{\partial y}\right) dx\, dy,$$

where z is an unknown function of x and y, or multiple integrals of higher order or of various numbers of dependent variables (the integrand may also be a function of derivatives of higher order than the first). See BRACHISTOCHRONE, EULER—Euler equation (2), FUNDAMENTAL — fundamental lemma of the calculus of variations, ISOPERIMETRIC, VARIATION.
  differential calculus. The study of the variation of a function with respect to changes in the independent variable, or variables, by means of the concepts of derivative and differential; in particular, the study of slopes of curves, nonuniform velocities, accelerations, forces, approximations to the values of a function, maximum and minimum values of quantities, etc. See DERIVATIVE.

fundamental lemma of the calculus of variations. See FUNDAMENTAL.

fundamental theorem of calculus. See FUNDAMENTAL —fundamental theorem of calculus.

integral calculus. The study of integration as such and its application to finding areas, volumes, centroids, equations of curves, solutions of differential equations, etc.

CALL'A-BLE, *adj.* **callable bonds.** See BOND.

CAL'O-RIE (or CAL'O-RY), *n.* The amount of heat required to raise the temperature of one gram of water one degree Centigrade. The calorie thus defined varies slightly for different temperatures. A standard calorie is usually defined as the amount of heat required to raise the temperature of one gram of water from 14.5° to 15.5° C. This unit is about the average amount required to raise the temperature of one gram of water one degree at any point between 0° and 100° C. A more exact definition (generally accepted in the U.S.) is that one calorie equals 4.1840 absolute joules.

CAN'CEL, *v.* (1) To divide numbers (or factors) out of the numerator and denominator of a fraction;

$$\frac{6}{8} = \frac{2 \times 3}{2 \times 4} = \frac{3}{4},$$

the number 2 having been *canceled out*. (2) Two quantities of opposite sign but numerically equal are said to *cancel* when added; $2x + 3y - 2x$ reduces to $3y$, the terms $2x$ and $-2x$ having *canceled* out.

CAN'CEL-LA'TION, *n.* The act of dividing like factors out of numerator and denominator of a fraction; sometimes used of two quantities of different signs which cancel each other in addition. Also used for the process of eliminating z when replacing $x + z = y + z$ by $x = y$, or $xz = yz$ by $x = y$ (if $z \neq 0$). See DOMAIN—integral domain, SEMI—semigroup.

CAN-DEL'A, *n.* The unit of luminous intensity which is 1/60 of the intensity of a black-body radiator (or ideal radiator) at the temperature at which platinum solidifies (2047 K).

CAN'DLE-POW'ER, *n.* Luminous intensity given in candelas.

CA-NON'I-CAL, *adj.* **canonical random variables.** Given a set S of p random variables and a set T of q random variables, there are sets of random variables $\{X_1, \cdots, X_p\}$ and $\{Y_1, \cdots, Y_q\}$, called **canonical random variables**, such that any two members of the same set have zero correlation, each X_i has zero correlation with each Y_j for which $i \neq j$, each X_i is a linear combination of members of S, each Y_i is a linear combination of members of T, and each X_i and each Y_i has mean 0 and variance 1. The correlations of X_i with Y_i for $1 \leq i \leq$ (minimum of p and q) are **canonical correlations**.

canonical form of a matrix. That which has been considered the simplest and most convenient form to which square matrices of a certain class can be reduced by a certain type of transformation. *Syn.* normal form. *E.g.*: (1) Any square matrix can be reduced by *elementary operations* or an *equivalent transformation* to the canonical form having nonzero elements only in the principal diagonal; or when the elements are polynomials (or integers, etc.) to **Smith's canonical form** having zeros except in the principal diagonal and each diagonal element being a factor of the next lower (if not zero). (2) Any matrix can be reduced by a *collineatory transformation* to the **Jacobi canonical form** having zeros below the principal diagonal and characteristic roots as elements of the principal diagonal, or to the **classical canonical form** having zeros except for a sequence of *Jordan matrices* situated along the principal diagonal. The exact type of the classical canonical matrix is specified by its **Segre characteristic**—a set of integers which are the orders of the Jordan submatrices, those integers which correspond to submatrices containing the same characteristic root being bracketed together. When the characteristic roots are distinct, the classical canonical form is a diagonal matrix. (3) A *symmetric matrix* can be reduced to a diagonal matrix by a *congruent transformation*. (4) A *normal matrix* (and hence a *Hermitian* or a *unitary matrix*) can be reduced by a *unitary transformation* to a diagonal matrix having characteristic roots along the principal diagonal.

canonical representation of a space curve in the neighborhood of a point. Representation of the curve in the neighborhood of the point P_0, with the arc length from the point as parameter and the axes of the moving trihedral as coordinates axes. The representation has the form

$$x = s - \frac{1}{6}\frac{1}{\rho_0^2}s^3 + \cdots,$$

$$y = \frac{1}{2\rho_0}s^2 + \frac{1}{6}\frac{d}{ds}\left(\frac{1}{\rho}\right)_0 s^3 + \cdots,$$

$$z = -\frac{1}{6}\frac{1}{\rho_0\tau_0}s^3 + \cdots,$$

where ρ_0 and τ_0 are the radii of curvature and torsion, respectively, at P_0.

CAN'TI-LE'VER, *adj.* **cantilever beam.** A projecting beam supported at one end only.

CANTOR, Georg Ferdinand Ludwig Philipp (1845–1918). German set theorist (born in Russia, whither his father had migrated from Denmark before moving to Germany). In his day, his theories concerning infinite sets seemed revolutionary and created much controversy.

Cantor function. The function f defined on the closed interval $[0, 1]$ by the following. If $x = .d_1d_2 \cdots$ belongs to the Cantor set, then $f(x) = .c_1c_2 \cdots$, where c_i is 0 or 1 according as d_i is 0 or 2. If x does not belong to the Cantor set, then $f(x)$ is the common value of f at the end points of the deleted open interval that contains x. This function is *continuous* (but not absolutely continuous), constant on each deleted interval, *monotone increasing*, maps the Cantor set onto $[0, 1]$, and its derivative is zero almost everywhere (*i.e.*, except on a set of measure zero). *Syn.* Cantor ternary function. See below, Cantor set.

Cantor set. The set of numbers in the closed interval $[0, 1]$ that are not deleted if the open middle third of $[0, 1]$ is deleted, the open middle third of each of the two remaining intervals is deleted, the open middle third of each of the four remaining intervals is deleted, etc. A number belongs to the Cantor set if and only if it has a representation in the ternary system of type $.d_1d_2 \cdots$, where each d_n is either 0 or 2. The Cantor set is *non-denumerable, perfect, non-dense*, and all its points are *frontier points*. Each neighborhood of a point in the Cantor set contains a set that is similar to the entire set. Also called the **Cantor discontinuum** and the **Cantor ternary set.** Other **Cantor-type** sets can be obtained in many ways. *E.g.*, for any odd integer p, we could divide $[0, 1]$ into p subintervals and delete the open middle interval, then repeat this for each of the remaining $p - 1$ subintervals, and continue this *ad infinitum*; or one could divide a square into 9 congruent squares and delete the middle square, then do this for each of the remaining squares, and continue this *ad infinitum*. See FRACTAL.

CAP, *n.* The symbol $\cap$ used to denote **intersection** or **greatest lower bound.** See INTERSECTION, LATTICE.

CAP'I-TAL, *adj., n.* **capital stock.** The money invested by a corporation to carry on its business; wealth used in production, manufacturing, or business of any sort, which, having been so used, is available for use again. Capital stock may be

disseminated by losses but is not consumed in the routine process of a business.

circulating capital. Capital consumed, or changed in form, in the process of production or of operating a business—such as that used to purchase raw materials. Capital invested permanently—such as that invested in buildings, machinery, etc., is **fixed capital**.

CAP'I-TAL-IZED, *adj.* **capitalized cost.** The sum of the first cost of an asset and the present value of replacements to be made perpetually at the ends of given periods.

CARATHÉODORY, Constantin (1873–1950). German analyst who worked largely in complex-variable theory and the calculus of variations.

Carathéodory measure. See MEASURE.

Carathéodory theorem. If S is a subset of n-dimensional space, then each point of the convex span of S is a convex combination of $n + 1$ (or fewer) points of S. See related theorems under HELLY, RADON, STEINITZ.

CARDAN, Jerome (Girolamo Cardano) (1501–1576). Italian physician and mathematician.

Cardan solution of the cubic. A solution of the *reduced cubic* (see REDUCED —reduced cubic),

$$x^3 + ax + b = 0,$$

by the substitution $x = u + v$ [$x = u + v$ will be a root of the equation if $u^3 + v^3 = -b$ and $uv = -\frac{1}{3}a$, or if u^3 is a root of the quadratic equation in u^3,

$$\left(u^3\right)^2 + b\left(u^3\right) - a^3/27 = 0, \quad \text{and} \quad uv = -\tfrac{1}{3}a\,].$$

If u_1 is a cube root of $\frac{1}{2}(-b + \sqrt{b^2 + 4a^3/27}\,)$, and $v_1 = -\frac{1}{3}a/u_1$, then the three roots of the reduced cubic are

$$z_1 = u_1 + v_1, \qquad z_2 = \omega u_1 + \omega^2 v_1,$$

$$z_3 = \omega^2 u_1 + \omega v_1,$$

where $\omega = -\frac{1}{2} + \frac{1}{2}i\sqrt{3}$ is a cube root of unity. This is equivalent to the formula

$$x = \left[-\tfrac{1}{2}b + \sqrt{R}\,\right]^{1/3} + \left[-\tfrac{1}{2}b - \sqrt{R}\,\right]^{1/3},$$

where $R = (\frac{1}{2}b)^2 + a^3/27$ and the cube roots are to be chosen so that their product is $-\frac{1}{3}a$. The number R is negative if and only if the three roots of the cubic are real and distinct; this is called the *irreducible case*, since the formulas (although still correct) involve the cube roots of complex numbers. This general solution of the reduced cubic was completed by Tartaglia, who showed it to Cardan. Cardan gave an oath of secrecy, but published the solution (giving credit to Tartaglia).

CAR′DI-NAL, *adj.* **cardinal number.** A number which designates the manyness of a set of things; the number of units, but not the order in which they are arranged; used in distinction to signed numbers. *E.g.*, when one says 3 dolls, the 3 is a cardinal number. *Tech.* Two sets are said to have the same cardinal number if their elements can be put into one-to-one correspondence with each other. Thus a symbol or *cardinal number* can be associated with any set. The cardinal number of a set is also called the **potency** of the set and the **power** of the set (*e.g.*, a set whose elements can be put into 1–1 correspondence with the real numbers is said to have the **power of the continuum**). The cardinal number of the set $\{a_1, a_2, \cdots, a_n\}$ is denoted by n. The cardinal number of all countably infinite sets is called **Aleph-null** or **Aleph-zero** and is designated by $\aleph_0$, and the cardinal number of all real numbers is designated by c. The cardinal number 2^c is the cardinal number of the set of all subsets of the real numbers (*i.e.*, the set of all real-valued functions defined for all real numbers) and is greater than c in the sense that the real numbers can be put into one-to-one correspondence with a subset of the real functions but not conversely. See ORDINAL—ordinal number, EQUIVALENT-equivalent sets.

CAR′DI-OID, *n.* The locus (in a plane) of a fixed point on a given circle which rolls on an equal but fixed circle. If a is the radius of the fixed circle, ϕ the vectorial angle, and r the radius vector—where the pole is on the fixed circle and the polar axis is on a diameter of the fixed circle—the polar equation of the cardioid is $r = 2a \sin^2 \frac{1}{2}\phi = a(1 - \cos \phi)$. A *cardioid* is an epicycloid of one loop and a special case of the limaçon.

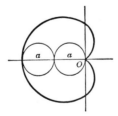

CARTAN, Élie Joseph (1869–1951). French algebrist, group theorist, differential geometer,

and relativist. Worked on classification of Lie algebras and Lie groups, differential geometry in the large, and stability theory. Introduced exterior differential forms and spinors.

CARTAN, Henri Paul (1904–). French analyst, algebraist, and topologist. Son of Élie Cartan. Worked particularly in algebraic topology, theory of analytic functions of one and several variables, theory of sheaves, and potential theory.

CAR-TE′SIAN, *adj.* **Cartesian coordinates. In the plane,** a point can be located by its distances from two intersecting straight lines, the distance from one line being measured along a parallel to the other line. The two intersecting lines are called **axes** (*x*-axis and *y*-axis), **oblique axes** when they are not perpendicular, and **rectangular axes** when they are perpendicular. The coordinates

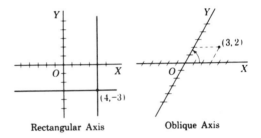

Rectangular Axis   Oblique Axis

are then called **oblique coordinates** and **rectangular coordinates,** respectively. The coordinate measured from the *y*-axis parallel to the *x*-axis is called the **abscissa** and the other coordinate is called the **ordinate. In space,** three planes (*XOY*, *XOZ*, and *YOZ* in the figure) can be used to locate points by giving their distance from each of the planes along a line parallel to the intersection of the other two. If the planes are mutually perpendicular, these distances are called the **rectangular Cartesian coordinates** of the point in space, or the **rectangular** or **Cartesian coordinates.** The three intersections of these three planes are called the **axes of coordinates** and are usually labeled the *x*-axis, *y*-axis, and *z*-axis. Their common point is called the origin. The three axes are called a **coordinate trihedral** (see TRIHEDRAL). The coordinate planes separate space into eight compartments, called octants. The octant containing the three positive axes as edges is called the **1st octant** (or **coordinate trihedral**). The other octants are usually numbered 2, 3, 4, 5, 6, 7, 8; 2, 3, and 4 are reckoned counterclockwise around the positive z-axis (or clockwise if the coordinate system is left-handed), then the quadrant vertically beneath the first quadrant is labeled 5, and the remaining quad-

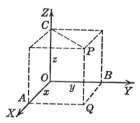

rants 6, 7, and 8, taken in counterclockwise (or clockwise) order as before. A rectangular space coordinate is quite commonly thought of as the projection of the line from the origin to the point upon the axis perpendicular to the plane from which the coordinate is measured; i.e., $x = OA$, $y = OB$, and $z = OC$ in the figure.

Cartesian product. See PRODUCT—Cartesian product.

Cartesian space. Same as EUCLIDEAN SPACE.

CASH, *n.* Money of any kind; usually coin or paper money, but frequently includes checks, drafts, notes, and other sorts of commercial paper, which are immediately convertible into currency.

cash equivalent of an annuity. Same as PRESENT VALUE. See SURRENDER —surrender value of an insurance policy, VALUE —present value.

CASSINI, Jean Dominique (1625–1712). French astronomer, geographer, and geometer.

Ovals of Cassini. The locus of the vertex of a triangle when the product of the sides adjacent to the vertex is a constant and the length of the opposite side is fixed. When the constant is equal to one-fourth of the square of the fixed side, the curve is called a **lemniscate.** If k^2 denotes the constant and a one-half the length of the fixed side, the Cartesian equation takes the form

$$\left[(x+a)^2 + y^2\right]\left[(x-a)^2 + y^2\right] = k^4.$$

If k^2 is less than a^2, the curve consists of two distinct ovals; if k^2 is greater than a^2, it consists of one, and if k^2 is equal to a^2, it reduces to the lemniscate. The figure illustrates the case in which $k^2 > a^2$.

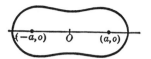

CAST′ING, *n.* **casting out nines.** A method used to check multiplication (and sometimes division); based on the fact that the *excess of nines* in the product equals the excess in the product of the excesses in the multiplier and multiplicand. See EXCESS. *E.g.*, to check the multiplication $832 \times 736 = 612,352$, add the digits in 612,352, subtracting 9 as the sum reaches or exceeds 9. This gives 1. Do the same for 832, and for 736; the results are 4 and 7. Now multiply 4 by 7, getting 28. Then add 2 and 8 and subtract 9. This leaves 1—which is the same excess that was obtained for the product. This method can also be used to check addition (or subtraction), since the *excess of nines* in a sum is equal to the excess in the sum of the excesses of the addends.

CATALAN, Eugène Charles (1814–1894). Belgian mathematician.

Catalan conjecture. It was conjectured by Catalan (1844) that the only pair of consecutive positive integers which are powers of smaller integers is $(8, 9)$. It is known now that there are at most only finitely many such pairs and each member of a pair is less than the 500th power of 10^{10}.

Catalan numbers. The numbers $1, 1, 2, 5, 14, 42, 132, \cdots$, where $C_n = \dfrac{(2n)!}{n!(n+1)!}$ if $n \geq 0$. These numbers occur in many contexts. *E.g.*, C_n is the middle *binomial coefficient* divided by $n + 1$; C_n is the number of ways the $n + 1$ terms of the sequence $\{x_1, \cdots, x_{n+1}\}$ can be combined (without changing the order of the subscripts) by a binary nonassociative product [*e.g.*, for $C_2 = 2$, we have $(x_1 x_2)x_3$ and $x_1(x_2 x_3)$]; C_n is the number of plane connected *graphs* that have $n + 2$ vertices, $n + 1$ edges, no loops, and a designated vertex with valence one.

CA-TAS′TRO-PHE, *n.* **catastrophe theory.** The study of singularities in certain types of mathematical models, together with applications of this to the study of discontinuous events in nature; the study of classifying singularities of systems whose behavior usually is smooth, but which sometimes (or in some places) exhibits discontinuities, which is then used to propose mathematical models for discontinuous events in other situations. *E.g.*, if a sphere is projected perpendicularly onto a plane that contains the equator of the sphere, then each point of the sphere not on the equator has a neighborhood that maps in a one-to-one way onto a neighborhood of the image of the point, but at points of the equator there is a singularity called a **fold**—a neighborhood of the point is "folded" by the projection. The singularities of a smooth map of a surface onto a plane are only folds and cusps. The figure

shows how a **cusp** can result from projecting a surface onto a plane: the cusp is at the intersection P of two curves consisting of fold points. There are seven "elementary" catastrophes: fold, cusp, butterfly, swallowtail, elliptic umbilic, hyperbolic umbilic, and parabolic umbilic. There are applications to elastic stability, geometric optics, and wave propagation, as well as controversial applications, *e.g.*, to economics, psychology, political elections, mental disorders, and the outbreak of war.

CAT′E-GO-RY, *adj., n.* (1) See below, category of sets. (2) A category K consists of two classes, O_K and M_K, for which the members of O_K are called **objects**, the members of M_K are called **morphisms**, and the following conditions are satisfied: (i) With each ordered pair (a, b) of objects there is associated a set $M_K(a, b)$ of morphisms such that each member of M_K belongs to exactly one of these sets; (ii) if f is in $M_K(a, b)$ and g is in $M_K(b, c)$, then the product or composite $g \circ f$ of f and g is defined uniquely and is a member of $M_K(a, c)$; (iii) if f, g, and h are members of $M_K(a, b)$, $M_K(b, c)$, and $M_K(c, d)$, respectively, so that $(h \circ g) \circ f$ and $h \circ (g \circ f)$ are defined, then $(h \circ g) \circ f = h \circ (g \circ f)$; (iv) for each object a, there is a morphism e_a in $M_K(a, a)$, called the **identity morphism**, such that $f \circ e_a = f$ and $e_a \circ g = g$ if there are objects b and c for which f is in $M_K(b, a)$ and g is in $M_K(a, c)$. The concept of category provides an abstract general model for many situations in which sets with certain structures are studied along with a class

of mappings that preserve these structures. Examples of such categories are: (i) O_K is the collection of all subsets of a set T and $M_K(a, b)$ is the set of all functions which have domain a and whose ranges are contained in the set b; (ii) O_K is a collection of groups and $M_K(a, b)$ is the set of all homomorphisms from the group a into the group b; (iii) O_K is a collection of topological spaces and $M_K(a, b)$ is the set of all continuous functions which have domain a and whose ranges are contained in b. A **zero** in a category is an object 0 with the property that, for any object a, both $M_K(0, a)$ and $M_K(a, 0)$ have exactly one member. If a category has a zero, then the **zero morphism** in $M_K(a, b)$ is $g_{0b} \circ f_{a0}$, where f_{a0} and g_{0b} are the unique members of $M_K(a, 0)$ and $M_K(0, b)$, respectively. An **isomorphism** or **equivalence** in $M_K(a, b)$ is a morphism f in $M_K(a, b)$ which has the property that there is a morphism g in $M_K(b, a)$ such that $f \circ g$ and $g \circ f$ are the identity morphisms in $M_K(b, b)$ and $M_K(a, a)$, respectively. An isomorphism that belongs to $M_K(a, a)$ is an **automorphism** and a morphism that belongs to $M_K(a, a)$ is an **endomorphism**. See FUNCTOR.

Baire category theorem. The theorem which states that a *complete metric space* is of second category in itself. An equivalent statement is that the intersection of any sequence of dense open sets in a complete metric space is dense. *E.g.*, the space C of all functions which are continuous on the closed interval $[0, 1]$ is a complete metric space if the distance $d(f, g)$ is defined to be the least upper bound of $|f(x) - g(x)|$. The set of all members of C which are differentiable at one or more points of $[0, 1]$ can be shown to be of *first category* in C, so that the set of continuous functions not differentiable at any point of $[0, 1]$ is of *second category*.

Banach category theorem. The theorem which states that if a set S contained in a topological space T (of type T_1) is of second category in T, then there is a nonempty open set U in T such that S is of second category at every point of U. It follows from this theorem that a subset of T is of first category in T if it is of first category at each point of T.

category of sets. A set S is of **first category** in a set T if it can be represented as a countable union of sets each of which is *nowhere dense* in T. Any set which is not of the first category is said to be of **second category**. A set S is of **first category at the point** x if there is a neighborhood U of x such that the intersection of U and S is of first category. The complement of a set of first category in T is called a **residual set** of T (sometimes residual set is used only for complements of sets of first category in sets T which have the property that every nonempty open set in T is of second category). A subset S of the real line is of the first category if and only if there is a one-to-one transformation of the line onto itself for which S corresponds to a set of *measure zero* which is also an F_σ set (see BOREL —Borel set).

CAT'E-NA-RY, *n.* The plane curve in which a uniform flexible cable hangs when suspended from two points. Its equation in rectangular co-ordinates is

$$y = (a/2)(e^{x/a} + e^{-x/a}),$$

where a is the y-intercept.

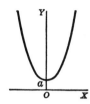

CAT'E-NOID, *n.* The surface of revolution generated by the rotation of a catenary about its axis. The only minimal surface of revolution is the catenoid. See CATENARY.

CAUCHY, Augustin Louis (1789–1857). Great French analyst, applied mathematician, and group theorist. After Euler, the most prolific mathematician in history. Worked in wave, elasticity, and group theory; introduced modern rigor into calculus; founded theory of functions of a complex variable; inaugurated modern era in differential equations with his existence theorems. He left his family and went into exile in 1830 when he refused the required loyalty oath. The exiled king Charles X made him a baron. In 1848 Napoleon III exempted him from the oath, and he became professor of mathematical astronomy at the Sorbonne.

Cauchy condensation test for convergence. If Σa_n is a series of positive monotonic decreasing terms and p is any positive integer, then the series $a_1 + a_2 + a_3 + \cdots$ and $pa_p + p^2 a_{p^2} + p^3 a_{p^3} + \cdots$ are either both convergent or both divergent.

Cauchy condition for convergence of a sequence. An infinite sequence converges if, and only if, the numerical difference between two of its terms is as small as desired, provided both terms are sufficiently far out in the sequence. *Tech.* The infinite sequence $s_1, s_2, s_3, \cdots, s_n, \cdots$ converges if, and only if, for every $\epsilon > 0$ there

exists an N such that

$$|s_{n+h} - s_n| < \epsilon$$

for all $n > N$ and all $h > 0$. See SEQUENCE — Cauchy sequence. Same as CAUCHY CONDITION FOR CONVERGENCE of a series when s_n is looked upon as the sum to n terms of the series

$$s_1 + (s_2 - s_1) + (s_3 - s_2) + \cdots$$
$$+ (s_n - s_{n-1}) + \cdots.$$

Cauchy condition for convergence of a series. The sum of any number of terms can be made as small as desired by starting sufficiently far out in the series. *Tech.* A necessary and sufficient condition for convergence of a series $\sum_1^\infty a_n$ is that, for any $\epsilon > 0$, there exists an N such that

$$|a_n + a_{n+1} + \cdots + a_{n+h}| < \epsilon$$

for all $n > N$ and all $h > 0$. See COMPLETE — complete space.

Cauchy distribution. A random variable has a **Cauchy distribution** or is a **Cauchy random variable** if there are numbers u and L for which the *probability density function f* satisfies

$$f(x) = \frac{L}{\pi [L^2 + (x-u)^2]}.$$

The distribution is symmetric about u, but it has no mean or variance since it has no finite moments of any order. The *distribution function* is $F(x) = \frac{1}{2} + \pi^{-1} \arctan[(x-u)/L]$. The means of random samples of order n of a Cauchy random variable X have the same distribution as X for all n. The *t distribution* with one degree of freedom is a Cauchy distribution with $L = 1$ and $u = 0$.

Cauchy form of the remainder for Taylor's theorem. See TAYLOR —Taylor's theorem.

Cauchy-Hadamard theorem. The theorem states that the radius of convergence of the Taylor series $a_0 + a_1 z + a_2 z^2 + \cdots$ in the complex variable z is given by

$$r = \frac{1}{\overline{\lim_{n \to \infty}} \sqrt[n]{|a_n|}}.$$

Cauchy inequality. The inequality

$$\left| \sum_1^n a_i b_i \right|^2 \leq \sum_1^n |a_i|^2 \cdot \sum_1^n |b_i|^2.$$

Also see SCHWARZ—Schwarz inequality.

Cauchy integral formula. The formula

$$f(z) = \frac{1}{2\pi i} \int_C \frac{f(\zeta)}{\zeta - z} \, d\zeta,$$

where f is an analytic function of the complex variable z in a finite simply connected domain D, C is a *simple closed rectifiable curve* in D, and z is a point in the finite domain bounded by C. This formula can be extended to the following, for n any positive integer:

$$f^{(n)}(z) = \frac{n!}{2\pi i} \int_C \frac{f(\zeta)}{(\zeta - z)^{n+1}} \, d\zeta.$$

Cauchy integral test for convergence of an infinite series. Suppose that for the series Σa_n there is a function f which has the properties: (i) there is a number N such that f is a monotonically decreasing positive function on the interval consisting of all numbers greater than N, and (ii) $f(n) = a_n$ for all n sufficiently large. Then a necessary and sufficient condition for convergence of the series, Σa_n, is that there exists a number a such that

$$\int_a^\infty f(x) \, dx \quad \text{converges}.$$

In the case of the p series,

$$\sum 1/n^p, \quad f(x) = 1/x^p,$$

$$\int_1^\infty x^{-p} \, dx = x^{1-p}/(1-p)\big]_1^\infty \quad \text{if} \quad p \neq 1,$$

$$= \log x\big]_1^\infty \quad \text{if } p = 1,$$

$$\lim_{x \to \infty} \frac{x^{1-p}}{1-p} = 0 \quad \text{if} \quad p > 1,$$

$$= \infty \quad \text{if} \quad p < 1$$

and

$$\lim_{x \to \infty} \log x = \infty.$$

Hence the p series converges for $p > 1$ and diverges for $p \leq 1$.

Cauchy integral theorem. If f is analytic in a finite simply connected domain D of the complex plane, and C is a closed rectifiable curve in D, then

$$\int_C f(z) \, dz = 0.$$

Cauchy-Kovalevski theorem. In its simplest form, the theorem that the equation

$$\frac{\partial z}{\partial x} = f\left(x, y, z, \frac{\partial z}{\partial y}\right),$$

for which f is analytic at $[x_0, y_0, z_0, (\partial z/\partial y)_0]$, has exactly one solution $z(x, y)$ which is analytic at (x_0, y_0) and for which $z(x_0, y) = g(y)$ defines a function g with $g(y_0) = z_0$ and $g'(y_0) = (\partial z/\partial y)_0$. This can be generalized to functions of more than two independent variables, to derivatives of higher order, and to systems of equations. See ANALYTIC—analytic function.

Cauchy mean-value formula. See MEAN—mean-value theorems (or laws of the mean) for derivatives.

Cauchy ratio and root tests. See RATIO—ratio test, ROOT—root test.

Cauchy-Riemann partial differential equations. For functions u and v of x and y, the **Cauchy-Riemann equations** are $\partial u/\partial x = \partial v/\partial y$ and $\partial u/\partial y = -\partial v/\partial x$. The equations characterize analytic functions $u + iv$ of the complex variable $z = x + iy$, and are satisfied if and only if the map T defined by $T(z) = u + iv$ is *conformal* except at points where all four partial derivatives vanish.

Cauchy sequence. See SEQUENCE—Cauchy sequence.

CAVALIERI, Francesco Bonaventura (1598–1647). Italian physicist and mathematician. Further developed Archimedes' method of exhaustion, thus anticipating the invention of the integral calculus.

Cavalieri's theorem. If two solids have equal altitudes and all plane sections parallel to their bases and at equal distances from their bases have equal areas, then the solids have the same volume.

CAYLEY, Arthur (1821–1895). English algebraist, geometer and analyst. Contributed especially to the theory of algebraic invariants and higher-dimensional geometry. See SYLVESTER.

Cayley algebra. The set of symbols of type $A + Be$, where A and B are quaternions and addition and multiplication are defined by

$$(A + Be) + (C + De) = (A + C) + (B + D)e,$$

$$(A + Be)(C + De) = (AC - B\overline{D})$$

$$+ (AD + B\overline{C})e,$$

with $\overline{C}$ and $\overline{D}$ the conjugates of the quaternions C and D. Except for multiplication not being associative, the Cayley algebra satisfies all axioms for a division algebra with unit element. As a vector space over the field of real numbers, it is of dimension 8 with the basis $\{1, i, j, k, e, ie, je, ke\}$. The definition of multiplication implies that $e^2 = -1$, $ie = -ei$, $je = -ej$, $ke = -ek$, but that $(ij)e = ke$ and $i(je) = -ke$. The members of the Cayley algebra are **Cayley numbers.** See FROBENIUS —Frobenius' theorem.

Cayley's theorem. Any group is isomorphic to a group of transformations. In particular, a group G is isomorphic to a permutation group on the set G.

ČECH, Eduard (1893–1960). Czechoslovakian topologist and projective differential geometer.
Stone-Čech compactification. See COMPACTIFICATION.

CE·LES'TIAL, *adj.* Of or pertaining to the skies or heavens.
altitude of a celestial point. See ALTITUDE.
celestial equator. See HOUR—hour angle and hour circle, EQUATOR.
celestial horizon, meridian, and pole. See HOUR—hour angle and hour circle.
celestial sphere. The conceptual sphere on which all the celestial objects are seen in projection and appear to move.

CELL, *n.* An n-dimensional cell (n-cell) is a set which is homeomorphic either with the set of points $(x_1, \cdots, x_n)$ of n-dimensional Euclidean space for which $\sum x_i^2 < 1$, or with the set for which $\sum x_i^2 \leq 1$ (it is an **open n-cell** in the first case, a **closed n-cell** in the other). A 0-cell is a point; a 1-cell is an open or a closed interval or a continuous deformation of an open or a closed interval. Circles (or simply polygons) and their interiors are examples of closed 2-cells; spheres (or simple polyhedrons) and their interiors are closed 3-cells. A closed n-cell is sometimes called a *solid n-sphere*, or an *n-disk*.

CELSIUS, Anders (1701–1744). Swedish astronomer.

Celsius temperature scale. The temperature scale established in 1742 by Anders Celsius for which the difference in temperature between the freezing and boiling points of water is divided into 100 degrees. The freezing point is 0° Celsius and the boiling point is 100° Celsius. Widely known as the centigrade temperature scale because of this difference of 100°. Formulas for converting from a temperature T_c measured on the Celsius scale to the corresponding temperature T_f as measured on the Fahrenheit scale, and conversely, are $T_f = \frac{9}{5}T_c + 32$, $T_c = \frac{5}{9}(T_f - 32)$.

CEN'TER, *n.* Usually the *center of symmetry*, such as the center of a circle, or the center of a regular polygon as the center of the inscribed circle. See CIRCLE, ELLIPSE, ELLIPSOID, HYPERBOLOID, SYMMETRIC —symmetric geometric configurations.
center of attraction. Same as CENTER OF MASS. See below.
center of curvature. See CURVATURE —curvature of a curve, curvature of a surface, GEODESIC —geodesic curvature.
center of a curve. The point (if it exists) about which the curve is symmetrical. Curves such as the hyperbola, which are not closed, but are symmetrical about a given point, are said to have this point as a center, but the term center commonly refers to closed curves such as circles and ellipses. *Syn.* center of symmetry. See SYMMETRIC—symmetric geometric configurations.
center of gravity. Same as CENTER OF MASS.
center of mass. The point at which a mass (body) can be considered as being concentrated without altering the effect of the attraction that the earth has upon it; the point in a body through which the resultant of the gravitational forces, acting on all its particles, passes regardless of the orientation of the body; the point about which the body is in equilibrium; the point such that the moment about any line is the same as it would be if the body were concentrated at that point. See MOMENT—moment of mass. The center of mass is the point of the body which has the same motion that a particle having the mass of the whole body would have if the resultant of all the forces acting on the body were applied to it. If the body consists of a **set of particles**, the center of mass is the point determined by the vector $\mathbf{r} = \dfrac{\Sigma_i \mathbf{r}_i m_i}{\Sigma_i m_i}$, where $\mathbf{r}_i$ is the position vector of the mass m_i in the system of particles $m_1, m_2, \cdots, m_n$. In case of a continuous distribution of mass, the vector $\bar{\mathbf{r}}$ locating the center of

mass of a body is given by $\bar{\mathbf{r}} = \dfrac{\int_s \mathbf{r}\, dm}{\int_s dm}$, where the integration is carried out throughout the space s occupied by the body. The coordinates $\bar{x}$, $\bar{y}$, and $\bar{z}$ of the center of mass are given by

$$\bar{x} = (1/m)\int_s x\, dm, \qquad \bar{y} = (1/m)\int_s y\, dm,$$

$$\bar{z} = (1/m)\int_s z\, dm,$$

where m is the total mass of the body, x, y, z are the coordinates of some point in the element of mass, dm, and $\int_s$ indicates that the integration is to be taken over the entire body, the integration being single, double, or triple depending upon the form of dm; dm may, for instance, be one of the forms: $\rho\, ds$, $\rho x\, dy$, $\rho\, dy\, dx$, $\rho\, dz\, dy\, dx$, where ρ is density. If elements such as $y\, dx$ or $x\, dy$ are used, one must take for the point in the element of mass the approximate center of mass of these strips (elements). *Syn.* center of attraction, center of gravity. See CENTROID.

center of pressure of a surface submerged in a liquid. That point at which all the force could be applied and produce the same effect as when the force is distributed.

center of a sheaf. See SHEAF—sheaf of planes.

center of similarity (or similitude) of two configurations. See RADIALLY—radially related figures.

radical center. See RADICAL.

CEN-TES'I-MAL, *adj.* **centesimal system of measuring angles.** The system in which the right angle is divided into 100 equal parts, called *degrees,* a *degree* into 100 *minutes* and a *minute* into 100 *seconds.* Not in common use.

CEN'TI. A prefix meaning $1/100$, as in **centimeter.**

CEN'TI-GRADE, *adj.* **centigrade temperature scale.** See CELSIUS—Celsius temperature scale.

CEN'TI-GRAM, *n.* One hundredth of a gram. See DENOMINATE NUMBERS in the appendix.

CEN'TI-ME'TER, *n.* One hundredth part of a meter. See DENOMINATE NUMBERS in the appendix.

centimeter-gram-second system. See CGS SYSTEM.

CEN'TRAL, *adj., n.* **central angle in a circle.** An angle whose sides are radii. An angle with its vertex at the center. See figure under CIRCLE.

central conics. Ellipses and hyperbolas.

central death rate. See DEATH.

central of a group. The set of all elements of the group which commute with every element of the group. The central is an *invariant subgroup,* but may be contained properly in an invariant subgroup. See GROUP.

central limit theorem. Given a sequence $\{X_1, X_2, \cdots\}$ of independent random variables, a *central limit theorem* is any theorem which gives conditions for $\left(\sum\limits_{i=1}^{n} X_i - m_n\right)/s_n$ to have approximately a normal distribution when n is large, where m_n and s_n^2 are the mean and variance of $\sum\limits_{i=1}^{n} X_i$. For example, if all the random variables have the same distribution function with finite mean μ and finite variance σ^2, then the distribution of $\left(\sum\limits_{i=1}^{n} X_i - n\mu\right)/(\sigma\sqrt{n})$ approaches uniformly the normal distribution with mean 0 and variance 1 as $n \to \infty$. If in particular each X_n is a Bernoulli experiment with probability p of success, then $[s(n) - np]/[np(1-p)]^{1/2}$ approaches uniformly the normal distribution with mean and variance 1, where $s(n)$ is the number of successes in n experiments (see BINOMIAL—binomial distribution).

central moment. See MOMENT—moment of a distribution.

central plane and point of a ruling on a ruled surface. See RULING.

central projection. See PROJECTION.

central quadrics. Quadrics having centers—ellipsoids and hyperboloids.

measures of central tendency. See MEASURE—measure of central tendency.

CEN-TRIF'U-GAL, *adj.* **centrifugal force.** (1) The force which a mass, constrained to move in a path, exerts on the constraint in a direction along the radius of curvature. (2) A particle of mass m, rotating with the angular velocity ω about a point O at a distance r from the particle, is subjected to a force, called **centrifugal force,** of magnitude $m\omega^2 r$ (or mv^2/r, where v is the speed of the particle relative to O). The direction of this force on the particle is away from the center of rotation. The equal and oppositely directed force is called **centripetal force.**

CEN-TRIP'E-TAL, *adj.* **centripetal acceleration.** See ACCELERATION.

centripetal force. The force which restrains a body, in motion, from going in a straight line. It is directed toward the center of curvature. A force equal, but opposite in sign, to the **centrifugal force.**

CEN′TROID, *n.* **centroid of a set.** The point whose coordinates are the *mean values* of the coordinates of the points in the set. The center is the centroid of a circle; the centroid of a triangle is the point of intersection of its medians. For sets in space over which integration can be performed, the coordinates of the centroid, $\bar{x}$, $\bar{y}$, and $\bar{z}$, are given by

$$\bar{x} = \left[\int_s x\,ds\right]/s,$$

$$\bar{y} = \left[\int_s y\,ds\right]/s,$$

and

$$\bar{z} = \left[\int_s z\,ds\right]/s;$$

where $\int_s$ denotes the integral over the set, ds denotes an element of area, arc length, or volume, and s denotes the area, arc length, or volume of the set. The centroid is the same as the center of mass if the set is regarded as having constant density (constant mass per unit area, length, or volume). See CENTER —center of mass, INTEGRAL —definite integral, MEAN —mean value of a function.

centroid of a triangle. See MEDIAN —median of a triangle.

CESÀRO, Ernesto (1859–1906). Italian geometer and analyst.

Cesàro summation formula. A specific method of attributing a sum to certain divergent series. A sequence of partial sums $\{S_n\}$, where $S_n = \sum_{i=0}^{n} a_i$, is replaced by the sequence $\{S_n^{(k)}/A_n^{(k)}\}$, where

$$S_n^{(k)} = \binom{n+k-1}{n} S_0$$

$$+ \binom{n+k-2}{n-1} S_1 + \cdots + S_n$$

and

$$A_n^{(k)} = \binom{k+n}{n} = \sum_{i=0}^{n} \binom{n+k-1-i}{n-i},$$

$\binom{n}{r}$ being the rth binomial coefficient of order n. If the sequence $\{S_n^{(k)}/A_n^{(k)}\}$ has a limit, the series Σa_n is **summable** C_k, or (C, k), to this limit. In terms of the a_i of the original series,

$$S_n^{(k)}/A_n^{(k)} = a_0 + \frac{n}{k+n} a_1$$

$$+ \frac{n(n-1)}{(k+n-1)(k+n)} a_2 + \cdots$$

$$+ \frac{n!}{(k+1)(k+2)\cdots(k+n)} a_n.$$

For $k = 1$, this becomes $(\Sigma_0^n S_k)/(n+1)$, or

$$a_0 + \frac{n}{n+1} a_1 + \frac{n-1}{n+1} a_2 + \cdots + \frac{1}{n+1} a_n.$$

In general, $S_n^{(k)}/A_n^{(k)} = (\Sigma_0^n S_n^{(k-1)}/A_n^{k-1})/(n+1)$. The Cesàro summation formula is *regular*. See SUMMATION —summation of divergent series.

CEULEN, Ludolph van (1540–1610). Dutch mathematician who worked most of his life calculating π to 35 digits; π was the entire epitaph on his tombstone. See PI.

CEVA, Giovanni (1647–1734). **Ceva's theorem.** If, in the triangle ABC, lines L_1, L_2, L_3 pass through the vertices A, B, C, respectively, and intersect lines containing the opposite sides at points Q_1, Q_2, Q_3, then these lines are concurrent (or parallel) if and only if

$$\frac{AQ_3}{Q_3 B} \frac{BQ_1}{Q_1 C} \frac{CQ_2}{Q_2 A} = 1.$$

It is assumed that none of Q_1, Q_2, Q_3 is one of A, B, C, and that the lines containing the sides of the triangle are directed. See DIRECTED— directed line, MENELAUS —Menelaus' theorem.

CGS SYSTEM. A system of units for which the units of length, time, and mass are the centimeter, second, and gram. See METRIC—metric system, MKS SYSTEM, SI.

CHAIN, *adj., n.* (1) A linearly ordered set. See ORDERED —ordered set, NESTED —nested sets. (2) See below, chain of simplexes, etc.

chain conditions on rings. A ring R satisfies the **descending chain condition** on right ideals (or is **Artinian** on right ideals) if every nonempty set of right ideals has a minimal member, or, equivalently, if no sequence of right ideals $\{I_n\}$ for which $I_k \supset I_{k+1}$ for each k has more than a finite number of different members; R satisfies the **ascending chain condition** on right ideals (or is **Noetherian** on right ideals) if every nonempty set of right ideals has a maximal member, or, equivalently, if no sequence of right ideals $\{I_n\}$ for which $I_k \subset I_{k+1}$ for each k has more than a finite number of different members. Similar definitions can be given for left ideals. See WEDDERBURN.

chain discounts. See DISCOUNT—discount series.

chain of simplexes. Let G be a commutative group with the group operation indicated as addition. Let $S_1^r, S_2^r, \cdots, S_n^r$ be oriented r-dimensional simplexes of a simplicial complex K. Then

$$x = g_1 S_1^r + g_2 S_2^r + \cdots + g_n S_n^r$$

is an **r-dimensional chain**, or an **r-chain**. It is understood that if $*S^r$ is the simplex S^r with its orientation changed, then $g(*S^r) = (-g)S^r$ for any g of G. The set of all r-chains is a group if chains are added in the natural way, i.e., by adding coefficients of each oriented simplex. The group G is usually taken as either the group I of integers or one of the finite groups I_n of integers modulo an integer n. Of the latter, the group I_2 of integers modulo 2 is especially useful. If G is one of these groups of integers, then the **boundary of an r-simplex** S^r is defined to be the $(r-1)$-chain

$$\Delta(S^r) = \epsilon_0 B_0^{r-1} + \epsilon_1 B_1^{r-1} + \cdots + \epsilon_n B_n^{r-1},$$

where $B_0^{r-1}, \cdots, B_n^{r-1}$ is the set of all $(r-1)$-dimensional faces of S^r and ϵ_k is $+1$ or -1 according as S^r and B_k^{r-1} are *coherently oriented* or *noncoherently oriented*. If $r = 0$, the boundary ΔS^0 is defined to be 0. The **boundary of the chain** x is defined to be

$$\Delta(x) = g_1 \Delta S_1^r + g_2 \Delta S_2^r + \cdots + g_n \Delta S_n^r.$$

It follows that the boundary of a boundary is 0, i.e., $\Delta(\Delta x) = 0$ for x any chain. A chain whose boundary is 0 is called a **cycle** (any boundary is a cycle). E.g., a chain of "edges" $S_1^1, S_2^1, \cdots, S_n^1$ is a cycle if the "edges" are joined so as to form a closed oriented path. See HOMOLOGY—homology group.

chain rule. For **ordinary differentiation**, the rule of differentiation which states that, if F is the composite function of f and u defined by $F(x) = f[u(x)]$ for all x in the domain of u for which $u(x)$ is in the domain of f, then

$$\frac{dF}{dx} = \frac{df}{du}\frac{du}{dx}.$$

For example, the derivative of u^3 with $u = x^2 + 1$ is $3u^2 \cdot du/dx$, or $3(x^2 + 1)^2(2x)$. Sufficient conditions for the chain rule to be valid at x are that u be differentiable at x, f be differentiable at $u(x)$, and each neighborhood of x contain points in the domain of F other than x. This rule can be used repeatedly, e.g., as $D_x u[v(w)] = D_v u \cdot D_w v \cdot D_x w$, or along with other differentiation formulas in an explicit differentiation, such as

$$D_x\left[\left(x^2 + 1\right)^3 + 3\right]^2$$

$$= 2\left[\left(x^2 + 1\right)^3 + 3\right] \cdot 3\left(x^2 + 1\right)^2 \cdot 2x.$$

The chain rule can also be used to change variables; e.g., if y is replaced by $z = 1/y$ in the differential equation $D_x y + y^2 = 0$, one uses the formula

$$D_x z = D_y z \cdot D_x y = \left(-1/y^2\right) D_x y$$

to obtain $-y^2 D_x z + y^2 = 0$, or $D_x z = 1$. Now let F be a function of one or more variables $u_1, u_2, \cdots, u_n$ and each of these variables be a function of one or more variables $x_1, x_2, \cdots$. The **chain rule for partial differentiation** is

$$\frac{\partial F}{\partial x_p} = \sum_{i=1}^{n} \frac{\partial F}{\partial u_i}\frac{\partial u_i}{\partial x_p}.$$

This formula can be applied at a point $P_0 = (x_1^0, \cdots, x_n^0)$ if P_0 is an interior point of the domains of each of $u_1, \cdots, u_n$, each of these functions is differentiable with respect to x_p at P_0, and F is *differentiable* at the point $(u_1^0, \cdots, u_n^0)$ obtained by evaluating each u_i at P_0. If each of the variables $u_1, u_2, \cdots, u_n$ is a function of one variable x, then the formula becomes

$$\frac{dF}{dx} = \sum_{i=1}^{n} \frac{\partial F}{\partial u_i}\frac{du_i}{dx}.$$

This is the **derivative** of F with respect to x. E.g., if $z = f(x, y)$, $x = \phi(t)$ and $y = \theta(t)$, the total derivative of z with respect to t is given by

$$\frac{dz}{dt} = f_x(x, y)\phi'(t) + f_y(x, y)\theta'(t).$$

ε-chain. See EPSILON —epsilon chain.

surveyor's chain. A chain 66 feet long containing 100 links, each link 7.92 inches long. Ten square chains equal one acre. See DENOMINATE NUMBERS in the appendix.

CHANCE, *adj.*, *n.* Same as PROBABILITY. Has considerable popular, but little technical, usage.

chance variable. Same as RANDOM VARIABLE. See RANDOM.

CHANGE, *n.* **change of base in logarithms.** See LOGARITHM.

change of coordinates. See TRANSFORMATION —transformation of coordinates.

change of variable in integration. See INTEGRATION —change of variables in integration.

cyclic change of variables. Same as CYCLIC PERMUTATION. See PERMUTATION.

CHA'OS, *n.* In a very general sense, a **dynamic system** is a map of a set into itself. The theory of **chaos** is basically the study of the complicated behavior (mysterious behavior?) of some dynamic systems under repeated iteration and of changes in this behavior that result from small changes in initial conditions of parameters in the definition of the map. More specifically, a dynamic system might be said to be **chaotic** if: (i) there is sensitive dependence on initial conditions, meaning that there is a $\delta > 0$ such that, for "many" x and any $\varepsilon > 0$, there is a y and an n for which x and y differ by less than ε, but the nth term in the orbit of x and the nth term in the orbit of y differ by more than δ (see ORBIT); (ii) many points have periodic orbits; (iii) many orbits are dense. The following three examples would be said to exhibit chaotic behavior. (1) Suppose T maps each point p on the unit circle with center at the origin onto the point determined by doubling the polar-coordinate angle $\theta(p)$ between the positive x-axis and the segment from the origin to p. Let the orbit of p be defined with respect to successive iterations of T. If $\theta(p) = 2\pi\rho$ in radians, then the orbit of p eventually becomes periodic if ρ is rational. For a dense set of irrational values of ρ, each orbit is dense in the circle, but for another dense set of irrational values of ρ, each orbit eventually is attracted by (becomes close to) the **stable points** $\pi, \frac{1}{2}\pi, \frac{1}{4}\pi, \frac{1}{2}\pi, \cdots$: first π, then $\frac{1}{2}\pi$ and π, then $\frac{1}{4}\pi, \frac{1}{2}\pi$, and π, etc. Thus an arbitrarily small change in p can cause the orbit to have a dense set of accumulation points, even if for p itself there are only finitely many. (2) Let T be the one-to-one map T of the plane into itself for which $T(x, y) = (x^*, y^*)$ with $x^* = y + 1 - ax^2$

and $y^* = bx$. If $a = 1.3$ and $b = 0.3$, then there are points $p_1, p_2, \cdots, p_7$ in the plane (called **periodic attractors**) for which some points (x, y) have the property that their orbits tend to get close to p_1, then close to p_2, then close to $p_3, \cdots$, then close to p_7, and repeat this indefinitely, getting closer to p_1 again, then p_2, etc. But there are two fixed points for T and other points whose orbits are unbounded. However, if $a = 1.4$ and $b = 0.3$, then there are still 2 fixed points, but there is a complicated set of curves (called **strange attractors**) and certain points for which the orbits tend to cluster near these curves. (3) Maps of type $T(x) = r(x - x^2)$ have been used, *e.g.*, to study turbulence in fluids, changing biological populations, and fluctuation of economic prices. If $0 \le r \le 4$, then T maps the closed interval $[0, 1]$ into itself. If $0 \le r \le 1$, then the orbit of any x converges to 0. If $1 \le r \le 4$, then $\xi = 1 - 1/r$ is a fixed point; if $1 < r \le 3$, then the orbits of all x other than 0 or 1 converge to ξ and ξ is said to be **stable**. If $3 < r \le 4$, then in addition to the fixed point $\xi = 1 - 1/r$, there is a set of 2 periodic points $\frac{1}{2}[r + 1 \pm \sqrt{(r+1)(r-3)}]/r$ that are stable for $r > 3$ and sufficiently near 3. As r becomes larger and approaches approximately 3.57, the number of stable points in a set may be 4, 8, 16, 32, etc. For r approximately 3.83, there is a set of 3 stable points, then sets of 6, 12, 24, etc. Similarly, for each positive integer p there is an r for which there is a set of p stable points, then $2p, 4p$, etc. The number ρ is a **bifurcation point** if there is an integer n such that for $r < \rho$ and near ρ there is a set of $m \cdot 2^n$ stable points and no set of $m \cdot 2^{n+1}$ stable points, but for $r > \rho$ and near ρ there is a set of $m \cdot 2^{n+1}$ stable points. When $r = 4$, the nth term in the orbit of x is $\sin^2(2^n \sin^{-1}\sqrt{x})$. The set of points S is dense in $[0, 1]$ if S is the set of points whose orbits eventually repeat periodically, or if S is the set of points whose orbits contain a fixed point, or if S is the set of points whose orbits are dense in $[0, 1]$. If T of this example is considered to be a map of the complex plane into itself, then with simple changes of variables it becomes the map $z^2 + c$ used to defined the *Mandelbrot set*. These examples illustrate that when infinitely many operations are involved, the results may be very complicated, very interesting, and extremely surprising. For other examples of "chaotic behavior", see JULIA, MANDELBROT —Mandelbrot set.

CHAR'AC-TER, *n.* **finite character.** A collection *A* **of sets** is of finite character if *A* contains any set all of whose finite subsets belong to *A* and each finite subset of a member of *A* belongs to A. A **property of subsets** of a set is of finite character if a subset S has the property if and only if each nonempty finite subset of S has the property. E.g., the property of being *simply ordered* is of finite character, while the property of being *well-ordered* is not. If a property is of finite character, then the collection of all sets with this property is of finite character. If a collection A of sets is of finite character, then the property of belonging to A is of finite character. See ZORN —Zorn's lemma.

group character. A *character* of a group G is a homomorphism of G into the group of complex numbers of absolute value 1; *i.e.*, it is a continuous function f defined on G for which $f(x)$ is a complex number with $|f(x)| = 1$ and $f(x)f(y) = f(x \cdot y)$ for all x and y of G (the group operation of G is indicated here by multiplication). The set of all characters of G is called a **character group**, the "product" of characters f and g being defined to be the character h defined by $h(x) = f(x)g(x)$ for each x of G. If G is commutative and locally compact, then G is algebraically isomorphic with the character group of its character group. The character group can be given a topology by defining neighborhoods of a point so that U is a neighborhood of a character f if there are elements $x_1, \cdots, x_n$ of G and a positive number ϵ such that U is the set of all characters G for which

$$|f(x_k) - g(x_k)| < \epsilon \quad \text{for} \quad k = 1, \cdots, n.$$

It then follows that the character group is a topological group and is locally compact if G is locally compact; it is discrete if G is compact. If G is the group of translations of the real line, then the character group of G is isomorphic with G.

CHAR'AC-TER-IS'TIC, *adj.*, *n.* **characteristic curves of a surface.** That conjugate system of curves on a surface S such that the directions of the tangents of the two curves of the system through any point P of S are the characteristic directions at P on S. See CONJUGATE —conjugate system of curves on a surface, and below, characteristic directions on a surface. The characteristic curves are parametric if and only if $D : D'' = E : G$ and $D' = 0$. See SURFACE —fundamental coefficients of a surface.

characteristic directions on a surface. The pair of conjugate directions on a surface S at a point P of S which are symmetric with respect to the directions of the lines of curvature on S through P. The characteristic directions on S at P are unique except at umbilical points, and are the directions which minimize the angle between pairs of conjugate directions on S at P.

characteristic equation of a matrix. Let I be the unit matrix of the same order as the square matrix A. If $d(xI - A)$ is the determinant of the matrix $xI - A$, then $d(xI - A) = 0$ is the characteristic equation of A and $d(xI - A)$ is the **characteristic function** of A. Thus the characteristic equation of the matrix $M = \begin{pmatrix} 2 & 1 \\ 2 & 3 \end{pmatrix}$ is the equation $\begin{vmatrix} x - 2 & -1 \\ -2 & x - 3 \end{vmatrix} = 0$, or $x^2 - 5x + 4 = 0$. It is known that every matrix satisfies its characteristic equation (the Hamiltonian-Cayley Theorem). E.g., the matrix $M \cdot M - 5M + 4I$ is zero. The **reduced characteristic equation** of a matrix is the equation of lowest degree which is satisfied by the matrix. If A is a matrix of order n and I is the unit matrix of order n, then the reduced characteristic equation is $f(x)/g(x) = 0$, where $f(x)$ is the determinant of the matrix $xI - A$ and $g(x)$ is the greatest common divisor of the $(n - 1)$ - rowed minor determinants of $xI - A$. The **reduced characteristic function** is $f(x)/g(x)$. The reduced characteristic equation is also called the **minimal** (or **minimum**) **equation**, and the matrix **derogatory** of its order is greater than that of its reduced characteristic equation. See EIGENVALUE.

characteristic function. For a random variable X or the associated distribution function, the characteristic function ϕ is the function for which $\phi(t)$ is the expected value of e^{itX} for real numbers t. For a discrete random variable with values $\{x_n\}$ and probability function p, $\phi(t) = \Sigma e^{itx_n} p(x_n)$; for a continuous random variable with probability density function f,

$$\phi(t) = \int_{-\infty}^{\infty} e^{itx} f(x)\, dx.$$

If $\phi^{(n)}$ is the nth derivative of ϕ, then $(-i)^n \phi^{(n)}(0)$ is the nth moment if the nth moment exists. The characteristic function ϕ of a vector random variable $(X_1, \cdots, X_n)$ is defined by letting $\phi(a_1, \cdots, a_n)$ be the expected value of $e^{i(a_1 X_1 + \cdots + a_n X_n)}$. Then the random variables $X_1, \cdots, X_n$ are independent if and only if $\phi(a_1, \cdots, a_n) = \prod_{k=1}^{n} \phi_k(a_k)$, where ϕ_k is the characteristic function of X_k. If $X_1, \cdots, X_n$ are independent, then the characteristic function for the random variable $\sum_{k=1}^{n} X_k$ is $\prod_{k=1}^{n} \phi_k$. See CUMULANTS, FOURIER —Fourier transform, INVARIANT—semi-invariant, MOMENT—moment generating function.

characteristic function of a matrix. See above, characteristic equation of a matrix.

characteristic function of a set. The function f which is defined by $f(x) = 1$ for each point x in the set, and $f(x) = 0$ if x is not in the set.

characteristic of the logarithm of a number. See LOGARITHM —characteristic and mantissa of a logarithm.

characteristic number of a matrix. Same as CHARACTERISTIC ROOT. See below.

characteristic number, functions, and vector in the study of symmetric operators. See EIGENVALUE.

characteristic of a one-parameter family of surfaces. The limiting curve of intersection of two neighboring members of the family as they approach coincidence—*i.e.*, as the two values of the parameter determining the two members of the family of surfaces approach a common value. The equations of a given characteristic curve are the equation of the family taken with the partial derivative of this equation with respect to the parameter, each equation being evaluated for a particular value of the parameter. The locus of the characteristic curves, as the parameter varies, is the envelope of the family of surfaces. *E.g.*, if the family of surfaces consists of all spheres of a given, fixed radius with their centers on a given line, the characteristic curves are circles having their centers on the line, and the envelope is the cylinder generated by these circles.

characteristic of a ring or field. If there is a smallest positive integer n such that $nx = 0$ for all x in the ring, then n is the **characteristic** of the ring; otherwise, the characteristic is zero. If the ring is an integral domain (*e.g.*, a field), then the characteristic is a prime if it is not zero. Sometimes "characteristic ∞" is used instead of "characteristic zero." See PERFECT—perfect field.

characteristic root of a matrix. A root of the characteristic equation of the matrix. *Syn.* characteristic number, latent root. See EIGENVALUE.

Euler characteristic. See EULER—Euler characteristic.

Segre characteristic of a matrix. See CANONICAL—canonical form of a matrix.

CHARGE, *n.* **Coulomb's law for point-charges.** See COULOMB.

density of charge. See various headings under DENSITY.

electrostatic unit of charge. See ELECTROSTATIC.

point-charge. A point endowed with electrical charge. The electrical counterpart of point-mass or particle, *i.e.*, electrical charge considered as concentrated at a point.

set (or complex) of point-charges. A collection of charges located at definite points of space. Sometimes the term *complex* carries the connotation that the maximum distance between the various pairs of charges is small in comparison to the distance to the field-points at which the electrical effects are to be determined.

surrender charge. See SURRENDER.

CHARLIER, Carl Vilhelm Ludvig (1862–1934). Swedish astronomer.

Gram-Charlier series. See GRAM, J. P.

CHART, *n.* **flow chart.** In machine computation, a diagram with labeled boxes, arrows, etc., showing the logical pattern of a problem, but not ordinarily including machine-language instructions and commands. See CODING, PROGRAMMING—programming for a computing machine.

CHEBYSHEV, Pafnuti Lvovich (1821–1894). Russian mathematician who worked in algebra, analysis, geometry, number theory, and probability theory. Numerous other transliterations are used, *e.g.*, Tchebycheff, Tchebychev. See BERTRAND—Bertrand's postulate, MOMENT—moment problem.

Chebyshev differential equation. The differential equation

$$(1 - x^2)\frac{d^2y}{dx^2} - x\frac{dy}{dx} + n^2y = 0.$$

Chebyshev inequality. If X is a random variable, f is a nonnegative real-valued function, and $k > 0$, then

$$P[f(X) > k] \leq E[f(X)]/k,$$

where $P[f(X) > k]$ is the probability of $f(X) > k$ and $E[f(X)]$ is the *mean* or *expected value* of $f(X)$. The special case for which $k = t^2\sigma^2$ and $f(X) = (X - \mu)^2$ is the **Bienaymé-Chebyshev inequality** (often called simply the **Chebyshev inequality**) which was discovered by Bienaymé in 1853 and rediscovered by Chebyshev in 1867:

$$P[|X - \mu| > \sigma t] \leq 1/t^2,$$

where σ^2 and μ are the *variance* and *mean* of X, and $t > 0$. If σ^2 is finite and $\epsilon > 0$, this can be written as

$$P[|X - \mu| > \epsilon] \leq \sigma^2/\epsilon^2.$$

Chebyshev net of parametric curves on a surface. See PARAMETRIC —equidistant system of parametric curves on a surface.

Chebyshev polynomials. The polynomials defined by $T_0(x) = 1$ and

$$T_n(x) = 2^{1-n}\cos(n\arccos x) \quad \text{for } n \geq 1,$$

or by

$$\frac{1 - t^2}{1 - 2tx + t^2} = \sum_{n=0}^{\infty} T_n(x)(2t)^n.$$

For $n \geq 2$, $T_{n+1}(x) - xT_n(x) + \frac{1}{4}T_{n-1}(x) = 0$, while $T_2 - xT_1 + \frac{1}{4}T_0 = -\frac{1}{4}$ and $T_1 - xT_0 = 0$. $T_n(x)$ is a solution of Chebyshev's differential equation. Also,

$$T_n(x) = 2^{1-n}\frac{(x^2 - 1)^{1/2}}{1 \cdot 3 \cdots (2n-1)}$$

$$\times \frac{d^n(x^2 - 1)^{n-1/2}}{dx^n}.$$

$T_n(x)$ is sometimes defined as 2^{n-1} times the above value. See JACOBI —Jacobi polynomials.

CHECK, *n.*, *v.* (1) A draft upon a bank, usually drawn by an individual. (2) To verify by repetition or some other device. Any process used to increase the probability of correctness of a solution.

CHI, *adj.*, *n.* The twenty-second letter of the Greek alphabet: lower case, χ; capital, X.

chi-square distribution. A random variable has a chi-square (or χ^2) distribution or is a chi-square (or χ^2) random variable with n degrees of freedom if it has a *probability density function* f for which $f(x) = 0$ if $x \leq 0$ and

$$f(x) = \frac{1}{2^{\frac{1}{2}n}\Gamma(\frac{1}{2}n)}x^{\frac{1}{2}n-1}e^{-\frac{1}{2}x} \quad \text{if } x > 0,$$

where Γ is the gamma function. The *mean* is n, the *variance* is $2n$, and the *moment generating function* is $M(t) = (1 - 2t)^{-\frac{1}{2}n}$. If $X_1, \cdots, X_n$ are independent normal random variables with means 0 and variances 1, then $\chi^2 = \sum_{i=1}^{n} X_i^2$ is a chi-square random variable with n degrees of freedom. The χ^2 distribution is the same as the *gamma distribution* with its parameters given the values $\lambda = \frac{1}{2}$ and $r = \frac{1}{2}n$. For large n, $(2\chi^2)^{1/2}$

has approximately a *normal distribution* with mean $\sqrt{2n - 1}$ and unit variance. Chi-square random variables have the *additive property* that $\sum_{i=1}^{k} X_i$ is a χ^2 random variable with $\sum_{i=1}^{k} n_i$ degrees of freedom if $X_1, \cdots, X_k$ are independent χ^2 random variables with $n_1, \cdots, n_k$ degrees of freedom. The preceding are sometimes called **central chi-square distributions**; if $X_1, \cdots, X_n$ are independent normal random variables with means $m_1, \cdots, m_n$ and variances 1, then $\chi^2 = \sum_{i=1}^{n} X_i^2$ has a **noncentral chi-square distribution** which has mean $n + \lambda$ and moment generating function $M(t) = (1 - 2t)^{-\frac{1}{2}n} \cdot \epsilon^{\lambda t/(1-2t)}$, where $\lambda = \sum_{i=1}^{n} m_i^2$ is the **noncentrality parameter**. Chi-square distributions are used widely for testing hypotheses; *e.g.*, concerning contingency tables, goodness of fit, estimates of variances, and parameters of random samples. See COCHRAN —Cochran's theorem, F—F distribution, GAMMA —gamma distribution, and below, chi-square test.

chi-square test. Suppose one wishes to test the hypothesis that a certain random variable X with a finite number of values $\{x_1, \cdots, x_k\}$ has a specified probability function p. For a random sample of size n, let

$$Y = \sum_{i=1}^{k} \frac{[n_i - n \cdot p(x_i)]^2}{n \cdot p(x_i)},$$

where n_i is the number of times x_i occurs in the sample. If n is sufficiently large (one empirical rule is that $np(x_i) \geq 5$ for all i), then Y has approximately the *chi-square distribution* with $k - 1$ degrees of freedom; using a significance level α, one accepts or rejects the hypothesis "p is the probability function of X" according as $Y \leq t_\alpha$ or $Y > t_\alpha$, where $1 - F(t_\alpha) = \alpha$ for the distribution function F of this chi-square distribution. This test can be used for random variables with infinitely many values by grouping the values, *e.g.*, by using intervals. The chi-square test can be adapted to many other situations, *e.g.*, when the distribution is only partially specified or when there are unknown parameters.

CHI′NESE′, *adj.* **Chinese-Japanese numerals.** A number system which adopted symbols for $1, 2, \cdots, 9$ and for $10, 10^2, 10^3, \cdots$. Then multiplication is used to indicate how many greater numerals are wanted. *E.g.*, if we use Arabic numerals for $1, 2, \cdots, 9$ and $a, b, c, \cdots$

for $10, 10^2, 10^3, \cdots$, then $7492 = 7c4b9a2$, but this would have been written vertically.

Chinese remainder theorem. If the integers $\{m_i\}$ are relatively prime in pairs and $\{b_i\}$ are any n integers, then there is an integer x which satisfies all the congruences $x \equiv b_i \pmod{m_i}$, $i = 1, 2, 3, \cdots, n$, and any two solutions of these congruences are congruent modulo $\prod_1^n m_i$.

CHOICE, n. An alternative elected by one of the players, or determined by a random device, for a move in the play of a game. See GAME, MOVE, PLAY.

axiom of choice. Given any collection of sets, there exists a "method" of designating a particular element of each set as a "special" element of that set; for any collection A of sets there exists a function f such that $f(S)$ is an element of S for each set S of the collection A. See ORDERED —well-ordered set, ZORN—Zorn's lemma. *Syn.* Zermelo's axiom.

finite axiom of choice. The *axiom of choice* for the special case that the collection of sets is a finite collection.

CHORD, n. A chord of any curve (or surface) is a segment of a straight line between two designated points of intersection of the line and the curve (surface). See CIRCLE, SPHERE, etc.

chord of contact with reference to a point outside a circle. The chord joining the points of contact of the tangents to the circle from the given point.

focal chords of conics. See FOCAL.

supplemental or supplementary chords in a circle. See SUPPLEMENTAL.

CHRISTOFFEL, Elwin Bruno (1829–1900). German differential geometer. Invented process of covariant differentiation.

Christoffel symbols. Certain symbols representing particular functions of the coefficients, and of the first-order derivatives of the coefficients, of a quadratic differential form. The differential form is usually the first fundamental quadratic differential form of a surface. For the quadratic differential form $g_{11}\, dx_1^2 + 2g_{12}\, dx_1\, dx_2 + g_{22}\, dx_2^2$, the **Christoffel symbols of the first kind** are

$$\begin{bmatrix} ij \\ k \end{bmatrix} = \frac{1}{2}\left(\frac{\partial g_{ik}}{\partial x_j} + \frac{\partial g_{jk}}{\partial x_i} - \frac{\partial g_{ij}}{\partial x_k} \right),$$

where $i, j, k = 1, 2$, separately. For a quadratic form in n variables, we have $i, j, k = 1, 2, \cdots, n$, separately. The symbol $\begin{bmatrix} ij \\ k \end{bmatrix}$ is sometimes replaced by $[ij, k]$, C_{ij}^k, or Γ_{ijk}. The symbols are symmetric in i and j. For the quadratic differential form

$$g_{11}\, dx_1^2 + 2g_{12}\, dx_1\, dx_2 + g_{22}\, dx_2^2,$$

the **Christoffel symbols of the second kind** are $\begin{Bmatrix} ij \\ k \end{Bmatrix} = g^{k1}\begin{bmatrix} ij \\ 1 \end{bmatrix} + g^{k2}\begin{bmatrix} ij \\ 2 \end{bmatrix}$, where $i, j, k = 1, 2$, separately, g^{ki} is the factor of g_{ki} in the determinant $\Delta = \begin{vmatrix} g_{11} & g_{12} \\ g_{12} & g_{22} \end{vmatrix}$ divided by Δ, and $\begin{bmatrix} ij \\ k \end{bmatrix}$ are the Christoffel symbols of the first kind. The symbol $\begin{Bmatrix} ij \\ k \end{Bmatrix}$ is sometimes replaced by $\begin{Bmatrix} k \\ ij \end{Bmatrix}$ in keeping with the summation convention, or by Γ_{ij}^k. The symbols are symmetric in i and j. For the quadratic differential form $g_{ij}\, dx^i\, dx^j$ in n variables $x^1, x^2, \cdots, x^n$ (where the *summation convention* applies), the **Christoffel symbols of the first kind** are

$$\begin{bmatrix} ij \\ k \end{bmatrix} = \frac{1}{2}\left(\frac{\partial g_{ik}}{\partial x^j} + \frac{\partial g_{jk}}{\partial x^i} - \frac{\partial g_{ij}}{\partial x^k} \right),$$

it being assumed that $g_{ij} = g_{ji}$ for all i and j. The **Christoffel symbols of the second kind** are $\begin{Bmatrix} k \\ ij \end{Bmatrix} = g^{k\sigma}\begin{bmatrix} ij \\ \sigma \end{bmatrix}$, where g is the determinant having g_{ji} in the ith row and jth column and $g^{ij} = [\text{cofactor of } g_{ji}]/g$. Neither kind of Christoffel symbol is a *tensor*. The law connecting Christoffel symbols of the second kind in two systems of coordinates x^i and $\bar{x}^i$ is

$$\overline{\begin{Bmatrix} i \\ jk \end{Bmatrix}} = \begin{Bmatrix} \lambda \\ \mu\nu \end{Bmatrix} \frac{\partial x^\mu}{\partial \bar{x}^j} \frac{\partial x^\nu}{\partial \bar{x}^k} \frac{\partial \bar{x}^i}{\partial x^\lambda} + \frac{\partial^2 x^\lambda}{\partial \bar{x}^j \partial \bar{x}^k} \frac{\partial \bar{x}^i}{\partial x^\lambda},$$

where $\begin{Bmatrix} i \\ jk \end{Bmatrix}$ and $\overline{\begin{Bmatrix} i \\ jk \end{Bmatrix}}$ are the Christoffel symbols of the second kind in the x^i and $\bar{x}^i$ coordinate systems, respectively. See below, Euclidean Christoffel symbols.

Euclidean Christoffel symbols. Christoffel symbols in an Euclidean space (*i.e.*, where rectangular Cartesian coordinates $y_1, y_2, \cdots, y_n$ exist such that the element of arc length ds is given by $ds^2 = \Sigma\, dy_i^2$). The Euclidean symbols of the second kind are all identically zero in rectangular Cartesian coordinates. However, the Euclidean Christoffel symbols are not all zero in general coordinates and are given by the alternative expression

$$\begin{Bmatrix} i \\ jk \end{Bmatrix} = \frac{\partial^2 y^\lambda}{\partial x^j \partial x^k} \frac{\partial x^i}{\partial y^\lambda},$$

in terms of the transformation functions and their inverses taking the rectangular Cartesian coordinates y^i into the general coordinates x^i. Since the Euclidean Christoffel symbols of the second kind are all identically zero in rectangular Cartesian coordinates, it follows that *covariant differentiation* is ordinary partial differentiation in rectangular Cartesian coordinates. Hence successive covariant differentiation in Euclidean space is a commutative operation even in general coordinates as long as partial differentiation is commutative.

Riemann-Christoffel curvature tensor. See RIEMANN.

CHRO-MAT'IC, *adj.* **chromatic number.** See FOUR—four-color problem.

CI'PHER(or CYPHER), *n.* The symbol 0, denoting *zero.* *Syn.* zero, naught.

CI'PHER, *v.* To compute with numbers; to carry out one or more of the fundamental operations of arithmetic.

CIR'CLE, *n.* (1) See PATH (2). (2) A plane curve consisting of all points at a given distance (called the **radius**) from a fixed point in the plane, called the **center**. The **diameter** is twice the radius ("diameter" can mean any chord passing through the center; see DIAMETER —diameter of a conic). An **arc** is one of the two pieces bounded by two points on the circle. The **circumference** is the length of the circle, which is $2\pi r$ if r is the radius (see PI); sometimes "circumference" is used to mean a circle itself rather than its interior. The **area** of a circle (i.e., the area of the interior) is πr^2, or in terms of the diameter d,

Arc
Chord
Diameter
Central Angle
O Radius

$\frac{1}{4}\pi d^2$. A **unit circle** is a circle with radius 1; it is has circumference 2π and area π. A **null circle** is a circle with radius 0. A **secant** of a circle is any line that is not tangent to the circle and intersects the circle; a **chord** is the segment of a secant joining the two points of intersection with the circle.

circle of convergence. See CONVERGENCE — circle of convergence of a power series.

circle of curvature. See CURVATURE— curvature of a curve, curvature of a surface.

circumscribed circle. See CIRCUMSCRIBED.

eccentric circles of an ellipse. See ELLIPSE —parametric equations of an ellipse.

equation of a circle in the plane. *In rectangular Cartesian coordinates*, $(x - h)^2 + (y - k)^2 = r^2$, where r is the radius of the circle and the center is at the point (h, k). When the center is at the origin, this becomes $x^2 + y^2 = r^2$. (See DISTANCE —distance between two points.) *In polar coordinates*, the equation is

$$\rho^2 + \rho_1^2 - 2\rho\rho_1 \cos(\phi - \phi_1) = r^2,$$

where ρ is the radius vector, ϕ the vectorial angle, (ρ_1, ϕ_1) the polar coordinates of the center, and r the radius. When the center of the circle is at the origin, this equation becomes $\rho = r$. **The parametric equations of a circle** are $x = a \cos \theta$, $y = a \sin \theta$, where θ is the angle between the positive x-axis and the radius from the origin to the given point, and a is the radius of the circle.

equations of a circle in space. The equations of any two surfaces whose intersection is the circle; a sphere and a plane, each containing the circle, would suffice.

escribed circle. See ESCRIBED.

family of circles. All the circles whose equations can be obtained by assigning particular values to an essential constant in the equation of a circle. *E.g.*, $x^2 + y^2 = r^2$ describes the family of circles with their centers at the origin, r being the essential constant in this case. See PENCIL —pencil of circles.

great circle. A section of a sphere by a plane which passes through its center; a circle (on a sphere) which has its diameter equal to that of the sphere.

hour circle of a celestial point. The great circle on the celestial sphere that passes through the point and the north and south celestial poles. See HOUR —hour angle and hour circle.

imaginary circle. The name given to the set of points which satisfy the equation $x^2 + y^2 = -r^2$, or $(x - h)^2 + (y - k)^2 = -r^2$, $r \neq 0$. Both coordinates of such a point can not be real. Although no points in the real plane have such coordinates, this terminology is desirable because of the algebraic properties common to these imaginary coordinates and the real coordinates of points on real circles.

inscribed circle. See INSCRIBED.

nine-point circle. The circle through the midpoints of the sides of a triangle, the feet of the perpendiculars from the vertices upon the sides, and the midpoints of the line segments between the vertices and the point of intersection of the altitudes.

osculating circle. See CURVATURE—curvature of a curve.

parallel circle. See SURFACE —surface of revolution.

small circle. A section of a sphere by a plane that does not pass through the center of the sphere.

squaring the circle. See SQUARING.

CIR′CUIT, *n.* (*Graph theory*) A closed path for which each node and each edge in the path occurs exactly once. A circuit is topologically a circle. See GRAPH THEORY, HAMILTON —Hamiltonian graph, PATH (2).

flip-flop circuit. In a computing machine, any bistable circuit that remains in either of its two stable states until receiving a signal changing it to the other state.

CIR′CU-LANT, *n.* A determinant in which the elements of each row are the elements of the previous row slid one place to the right (with the last element put first). The elements of the main diagonal are all identical.

CIR′CU-LAR, *adj.* **circular argument or reasoning.** An argument which is incorrect because it makes use of the theorem to be proved, or makes use of a theorem that is a consequence of the theorem to be proved but is not known to be true. One flagrant type of circular reasoning is to use the theorem to be proved as a reason for a step in its "proof."

circular cone and cylinder. See CONE, CYLINDER.

circular functions. The trigonometric functions.

circular (or cyclic) permutation. See PERMUTATION.

circular point of a surface. An elliptic point of the surface at which $D = kE, D' = kF, D'' = kG, k \neq 0$. See SURFACE —fundamental coefficients of a surface, and ELLIPTIC—elliptic point of a surface. For a circular point, the principal radii of normal curvature are equal, and the *Dupin indicatrix* is a circle. A surface is a sphere if and only if all its points are circular points. The points where an ellipsoid of revolution cuts its axis of revolution are circular points. See PLANAR—planar point of a surface, UMBILICAL —umbilical point of a surface.

circular region. See REGION.

uniform circular motion. See UNIFORM.

CIR′CUM-CEN′TER, *n.* **circumcenter of a triangle.** The center of the circumscribed circle (the circle passing through the three vertices of the triangle); the point of intersection of the perpendicular bisectors of the sides; point O in the figure.

CIR′CUM-CIR′CLE, *n.* Same as CIRCUMSCRIBED CIRCLE.

CIR-CUM′FER-ENCE, *n.* (1) See CIRCLE. (2) The boundary of any region whose boundary is a simple closed curve (*e.g.*, a polygon). *Syn.*, periphery, perimeter.

circumference of a sphere. the circumference of any great circle on the sphere.

CIR-CUM-PO′LAR, *adj.* **circumpolar** star. A star whose path lies entirely above the horizon. The angle between the star and the *north celestial pole* is less than the latitude of the observer. See HOUR—hour angle and hour circle.

CIR′CUM-SCRIBED′, *adj.* A configuration composed of lines, curves, or surfaces, is said to be **circumscribed** about a polygon (or polyhedron) if every vertex of the latter is incident upon the former and the polygon (or polyhedron) is contained in the configuration. A polygon (or polyhedron) is circumscribed about a configuration if every side of the polygon (or face of the polyhedron) is tangent to the configuration and the configuration is contained in the polygon (or polyhedron). If one figure is circumscribed about another, the latter is said to be **inscribed** in the former. In particular, the **circumscribed circle of a polygon** is a circle which passes through the vertices of the polygon. The polygon is then an **inscribed polygon of the circle.** If the polygon is regular and s is the length of a side and n the number of sides, the radius of the circle is

$$r = \frac{s}{2} \csc \frac{180°}{n}.$$

If the polygon is a triangle with sides a, b, c, and $s = \frac{1}{2}(a + b + c)$, then

$$r = \frac{abc}{4\sqrt{s(s - a)(s - b)(s - c)}}.$$

If the polygon is *regular* and has n sides, its area is $\frac{1}{2}r^2 \sin 360°/n$ and its perimeter is $2rn \sin 180°/n$, where r is the radius of the circumscribed circle. A **circumscribed polygon of a circle** is a polygon which has its sides tangent to the circle. The circle is then an **inscribed circle of the polygon** (see INSCRIBED —inscribed circle of a triangle). If the polygon is regular, its area is

$$nr^2 \tan \frac{180°}{n},$$

and its perimeter is

$$2nr \tan \frac{180°}{n},$$

where r is the radius of the inscribed circle and n the number of sides of the polygon. If s is the length of a side of the polygon and n the number of its sides, the radius is $\frac{1}{2}s \cot 180°/n$. A **circumscribed sphere** of a polyhedron is a sphere which passes through all the vertices of the polyhedron (the polyhedron is then said to be *inscribed in the sphere*). An **inscribed sphere** of a polyhedron is a sphere which is tangent to all the faces of the polyhedron (the polyhedron is then said to be *circumscribed about the sphere*). A **circumscribed pyramid of a cone** is a pyramid having its base circumscribed about the base of the cone and its vertex coincident with that of the cone. The cone is then an **inscribed cone of the pyramid.**

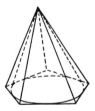

A **circumscribed cone of a pyramid** is a cone whose base is circumscribed about the base of the pyramid and whose vertex coincides with the vertex of the cone. The pyramid is then an inscribed pyramid of the cone.

A **circumscribed prism of a cylinder** is a prism whose bases are coplanar with, and circumscribed about the bases of the cylinder. The lateral faces of the prism are then tangent to the cylindrical surface and the cylinder is an **inscribed cylinder of the prism**. A **circumscribed cylinder of a prism** is a cylinder whose bases are coplanar with, and circumscribed about, the bases of the cylinder. The lateral edges of the prism are then elements of the cylinder and the prism is an **inscribed prism of the cylinder**.

CIS'SOID (cissoid of Diocles), *n.* The plane locus of a variable point on a variable line passing through a fixed point on a circle, where the distance of the variable point from the fixed point is equal to the distance from the line's intersection with the circle to its intersection with a fixed tangent to the circle at the extremity of the diameter through the fixed point; the locus of the foot of the perpendicular from the vertex of a parabola to a variable tangent. If a is

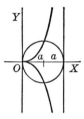

taken as the radius of the circle in the first definition, the polar equation of the cissoid is $r = 2a \tan \phi \sin \phi$; its Cartesian equation is $y^2(2a - x) = x^3$. The curve has a cusp of the first kind at the origin, the x-axis being the double tangent. The cissoid was first studied by Diocles about 200 B.C., who gave it the name "Cissoid" (meaning *like ivy*).

CIV'IL, *adj.* **civil year.** See YEAR.

CLAIRAUT (or CLAIRAULT), Alexis Claude (1713–1765). French analyst, differential geometer, and astronomer.

Clairaut's differential equation. A differential equation which is of the form $y = xy' + f(y')$ for some function f. The *general solution* is $y = cx + f(c)$, and a *singular solution* is given by the parametric equations $y = -pf'(p) + f(p)$ and $x = -f'(p)$.

CLASS, *n.* (*Statistics*) Often the set of all observations of a random variable are grouped into **classes** by divisions of the range of the variable. *E.g.*, a variable whose range is the interval $[0, 100]$ may be grouped into **class intervals** ten units wide with $0 \le x \le 10$ for x in the first interval, $10 < x \le 20$ in the second, etc. The **class limits** or **class bounds** are the upper and lower bounds of the values in a class interval. The **class frequency** is the frequency with which the random variable assumes a value in a given class interval.

class of a plane algebraic curve. The greatest number of tangents that can be drawn to it from any point in the plane not on the curve.

equivalence class. See EQUIVALENCE.

subclass. Same as SUBSET. See SET, NUMBER.

CLAS'SI-FI-CA'TION, *n*. **one-way classification.** A classification of values of a random variable according to several classes of a single factor, *e.g.*, the factor might be *sex* and the classes *female* and *male*.

CLIFFORD, William Kingdom (1845–1879).
Clifford algebra. An *algebra with unit element* which is generated by members $\{e_1, \ldots, e_n\}$ with $e_i e_j = -e_j e_i$ if $i \ne j$ and $e_i^2 = 1$. See ALGEBRA—algebra over a field, DIRAC—Dirac matrix.

CLOCK'WISE, *adj*. In the same direction of rotation as that in which the hands of a clock move around the dial.

counterclockwise. In the direction of rotation opposite to that in which the hands of a clock move around the dial.

CLOSED, *adj*. **closed curve.** A curve which has no end points. *Tech*. A set of points which is the image of a circle under a continuous transformation. See CONTINUOUS—continuous function, CURVE, PATH (2), SIMPLE—simple closed curve.

closed function, mapping, or transformation. (1) See OPEN—open mapping. (2) A **linear transformation** T is said to **be closed** if it has the property that, if $\lim x_n = x_0$ and $\lim T(x_n) = y_0$ exist, where x_n is in the domain D of T for each n, then x_0 is in D and $T(x_0) = y_0$. This is equivalent to stating that the set of points of type $[x, T(x)]$ is closed in the *Cartesian product* $\overline{D} \times \overline{R}$ of the closure of the domain D and the closure of R, the range of T.

closed interval. See INTERVAL.

closed region. See REGION.

closed set. A set of points U such that every accumulation point of U is a point of U; the complement of an open set. The set of points on and within a circle is a closed set.

closed surface. A surface with no boundary curves; a space such that each point has a neighborhood topologically equivalent to the interior of a circle. See SURFACE.

closed with respect to a binary operation. See BINARY—binary operation.

CLO'SURE, *n*. **closure of a set of points.** The set which contains the given set and all *accumulation points* of the given set. The closure of a closed set is the set itself, while the closure of any set is closed. The set of all accumulation points of a given set is the **derived set.** The closure of a set U is usually denoted by $\overline{U}$ and the derived set by U'.

Kuratowski closure-complementation problem. See KURATOWSKI.

CLUS'TER, *adj*. **cluster point.** Same as ACCUMULATION POINT.

CO'A-LI'TION, *n*. In an n-person game, a set of players, more than one in number, who coordinate their strategies, presumably for mutual benefit. See GAME—cooperative game.

CO-AL'TI-TUDE, *n*. **coaltitude of a celestial point.** Same as ZENITH DISTANCE.

COARS'ER, *adj*. See PARTITION—partition of a set.

CO-AX'I-AL, *adj*. **coaxial circles.** Circles such that all pairs of the circles have the same radical axis.

coaxial planes. Planes which pass through the same straight line. The line is called the **axis.**

CO-BOUND'A-RY, *n*. See COHOMOLOGY—cohomology group.

COCHRAN, William Gemmell (1909–1980). Scottish statistician, who has spent many years in the U.S.

Cochran's theorem. If X_i $(i = 1, \cdots, n)$ are independently and normally distributed random variables with zero mean and unit variance, and if $q_1, q_2, \cdots, q_k$ are k quadratic forms in the variables X_i with ranks $r_1, r_2, \cdots, r_k$ and $\sum\limits_{j=1}^{k} q_j = \sum\limits_{i=1}^{n} X_i^2$, then a necessary and sufficient condition that q_j be each independently distributed with the χ^2 distribution with r_j degrees of freedom is that $\sum\limits_{j=1}^{k} r_j = n$. On the basis of this theorem it follows that if $\{X_1, \cdots, X_n\}$ is a random sample from a normal distribution with mean u and variance σ^2, then $\sum\limits_{i=1}^{n} \dfrac{(X_i - \overline{X})^2}{\sigma^2}$ is distributed as χ^2 with $n - 1$ degrees of freedom, where $\overline{X}$ is the mean of the sample. This theorem is useful, for example, in establishing the independence of the mean and sum of squares of deviations around the mean in random samples from a normal population.

CO-CY'CLE, *n.* See COHOMOLOGY —cohomology group.

CODAZZI, Delfino (1824–1873). Italian differential geometer.

Codazzi equations. The equations

$$\frac{\partial D}{\partial v} - \frac{\partial D'}{\partial u} - \left\{ \begin{matrix} 1 \\ & 1 \end{matrix} \begin{matrix} 2 \end{matrix} \right\} D$$

$$+ \left(\left\{ \begin{matrix} 1 \\ & 1 \end{matrix} \begin{matrix} 1 \end{matrix} \right\} - \left\{ \begin{matrix} 1 \\ & 2 \end{matrix} \begin{matrix} 2 \end{matrix} \right\} \right) D'$$

$$+ \left\{ \begin{matrix} 1 \\ & 2 \end{matrix} \begin{matrix} 1 \end{matrix} \right\} D'' = 0$$

and

$$\frac{\partial D''}{\partial u} - \frac{\partial D'}{\partial v} + \left\{ \begin{matrix} 2 \\ & 1 \end{matrix} \begin{matrix} 2 \end{matrix} \right\} D$$

$$+ \left(\left\{ \begin{matrix} 2 \\ & 2 \end{matrix} \begin{matrix} 2 \end{matrix} \right\} - \left\{ \begin{matrix} 1 \\ & 1 \end{matrix} \begin{matrix} 2 \end{matrix} \right\} \right) D'$$

$$- \left\{ \begin{matrix} 1 \\ & 2 \end{matrix} \begin{matrix} 2 \end{matrix} \right\} D'' = 0,$$

involving the fundamental coefficients of the first and second orders of a surface. In tensor notation: $d_{\alpha\alpha,\beta} - d_{\alpha\beta,\alpha} = 0$, $\alpha \neq \beta$. There are no relations between these fundamental coefficients and their derivatives which cannot be derived from the Gauss equation and the Codazzi equations, for these three equations uniquely determine a surface to within its position in space. See CHRISTOFFEL —Christoffel symbols.

CO-DEC'LI-NA'TION, *n.* **codeclination of a celestial point.** Ninety degrees minus the declination; the complementary angle of the declination. See HOUR—hour angle and hour circle. *Syn.* polar distance.

COD'ING, *n.* In machine computation, the detailed preparation from the programmer's instructions or flow charts, of machine commands that will lead to the solution of the problem at hand. See PROBLEM —problem formulation, PROGRAMMING —programming for a computing machine.

CO'EF-FI'CIENT, *n.* In elementary algebra, the numerical part of a term, usually written before the literal part, as 2 in $2x$ or $2(x + y)$. (See PARENTHESIS.) In general, the product of all the factors of a term except a certain one (or a certain set), of which the product is said to be the *coefficient*. *E.g.*, in $2axyz$, $2axy$ is the coefficient of z, $2ayz$ the coefficient of x, $2ax$ the coefficient of yz, etc. Most commonly used in algebra for the constant factors, as distinguished from the variables.

binomial coefficients. See BINOMIAL.

coefficient of correlation. See CORRELATION —correlation coefficient.

coefficient of friction. See FRICTION.

coefficient of linear expansion. (1) The quotient of the change in length of a rod, due to one degree change of temperature, and the original length (not the same at all temperatures). (2) The change in length of a unit rod when the temperature changes one degree centigrade beginning at $0°$ C.

coefficient of strain. See STRAIN—one-dimensional strains.

coefficient of thermal expansion. A term used to designate both the *coefficient of linear expansion* and the *coefficient of volume expansion*.

coefficient of variation. See VARIATION.

coefficient of volume (or cubical) expansion. (1) The change in volume of a unit cube when the temperature changes one degree. (The coefficient thus defined is different at different temperatures.) (2) The change in unit volume due to a change of $1°$ C beginning at $0°$ C.

coefficients in an equation. (1) The coefficients of the variables. (2) The constant term and the coefficients of all the terms containing variables. If the constant term is not included, the phrase *coefficients of the variables* in the equation is often used.

confidence coefficient. See CONFIDENCE.

correlation coefficient. See CORRELATION.

detached coefficients, multiplication and division by means of. Abbreviations of the ordinary multiplication and division processes used in algebra. The coefficients alone (with their signs) are used, the powers of the variable occurring in the various terms being understood from the order in which the coefficients are written, missing powers being assumed to be present with zero coefficients. *E.g.*, $(x^3 + 2x + 1)$ is multiplied by $(3x - 1)$ by using the expressions $(1 + 0 + 2 + 1)$ and $(3 - 1)$. See SYNTHETIC —synthetic division.

determinant of the coefficients. See DETERMINANT—determinant of the coefficients.

differential coefficient. Same as DERIVATIVE.

leading coefficient. See LEADING.

Legendre's coefficients. See LEGENDRE—Legendre's polynomials.

matrix of the coefficients. See MATRIX.

phi coefficient. (*Statistics*) A coefficient obtained from a four-fold table, in which the two variables are essentially dichotomous. The phi coefficient ϕ is defined by

$$\phi = \sqrt{\chi^2/n}\,,$$

where χ^2 is computed from the cell entries and the marginal totals, which are the basis of the expected values. See CHI-SQUARE.

regression coefficient. See REGRESSION.

relation between the roots and coefficients of a polynomial equation. See ROOT—root of an equation.

undetermined coefficients. See UNDETERMINED.

CO-FAC'TOR, *n.* **cofactor of an element of a determinant.** See MINOR—minor of an element of a determinant.

cofactor of an element of a matrix. This is defined only for square matrices, and is the same as the cofactor of the same element in the determinant of the matrix.

CO-FIN'AL, *adj.* **cofinal subset.** See MOORE (E. H.)—Moore-Smith convergence.

CO-FUNC'TION, *n.* See TRIGONOMETRIC—trigonometric cofunctions.

COHEN, Paul Joseph (1934–   ). American analyst, topological group theorist, and logician. Fields Medal recipient (1966). Proved the independence in set theory of the axiom of choice, and the independence of the continuum hypothesis from the axiom of choice, thus completing the negative solution of the continuum problem (Hilbert's first problem), the first part of the solution having been given by Kurt Gödel in 1938–40.

CO-HER'ENT-LY, *adv.* **coherently oriented.** See MANIFOLD, SIMPLEX.

CO'HO-MOL'O-GY, *adj.* **cohomology group.** Let K be an n-dimensional simplicial complex and let Δ be the boundary operator, so that the boundary of a p-chain $x = \Sigma g_i S_i^p$ is $\Delta x = \Sigma g_i \Delta S_i^p$. Then Δx is a $(p-1)$-chain and the boundary operator maps the group of p-chains

into the group of $(p-1)$-chains. In particular,

$$\Delta\sigma_i^p = \sum_j g_i^j \sigma_j^{p-1}$$

for each p-simplex σ_i^p, where $\{\sigma_j^{p-1}\}$ are $(p-1)$-simplexes and g_i^j are elements of the group G whose elements are used as coefficients in forming chains. If the matrix (g_i^j) is $r \times s$, its transpose is $s \times r$ and can be used to define a mapping

$$\nabla\sigma_j^{p-1} = \sum_i g_i^j \sigma_i^p,$$

the chain $\nabla\sigma_j^{p-1}$ being called the **coboundary** of σ_j^{p-1}. This operator can be extended to all $(p-1)$-chains by the definition

$$\nabla x = \sum_i g_i \nabla S_i^{p-1} \quad \text{if} \quad x = \sum_i g_i S_i^{p-1}.$$

A chain is called a **cocycle** if its coboundary is zero. The **r-dimensional cohomology group** is the quotient group, T^r/H^r, where T^r is the group of all r-dimensional cocycles of K and H^r is the group of all cycles which are 0 or are coboundaries of an $(r-1)$-chain of K. The concepts of homology and cohomology can be defined for certain generalizations of simplicial complexes (called *complexes*) for which each *complex* has a *dual complex* such that the homology group of one is the co-homology group of the other.

COIN, *n.* **coin-matching game.** See GAME.

CO'IN-CIDE', *v.* To be coincident.

CO-IN'CI-DENT, *adj.* **coincident configurations.** Two configurations which are such that any point of either one lies on the other. Two lines (or curves or surfaces) which have the same equation are coincident.

CO-LAT'I-TUDE, *n.* **colatitude of a point on the earth.** Ninety degrees minus the latitude; the complementary angle of the latitude.

COL-LAT'E-RAL, *adj.* **collateral security.** Assets deposited to guarantee the fulfillment of some contract to pay, and to be returned upon the fulfillment of the contract.

collateral trust bonds. See BOND.

COL-LECT'ING, *p.* **collecting terms.** Grouping terms in a parenthesis or adding like terms. *E.g.*, to collect terms in $2 + ax + bx$, we write it in the form $2 + x(a + b)$; to collect terms in $2x + 3y - x + y$, we write it in the form $x + 4y$.

COL-LIN′E-AR, *adj.* **collinear planes.** Planes having a common line. *Syn.* coaxial planes. Three planes are either *collinear* or *parallel* if the equation of any one of them is a linear combination of the equations of the others. See CONSIS-TENCY—consistency of linear equations.

    collinear points. Points lying on the same line. Two points in the plane are **collinear with the origin** if and only if their corresponding rectangular Cartesian coordinates are proportional, or the determinant, whose first row is composed of the Cartesian coordinates of one of the points and the second of those of the other, is zero; *i.e.*, $x_1 y_2 - x_2 y_1 = 0$, where the points are (x_1, y_1) and (x_2, y_2). Two points in space are *collinear with the origin* if, and only if, their corresponding Cartesian coordinates are proportional, *i.e.*, the *matrix*

$$\begin{Vmatrix} x_1 & y_1 & z_1 \\ x_2 & y_2 & z_2 \end{Vmatrix}$$

whose columns are composed of the coordinates of the points, is of rank one. Three points in a plane are collinear if, and only if, the third-order determinant whose rows are $x_1, y_1, 1$; $x_2, y_2, 1$; and $x_3, y_3, 1$, where the x's and y's are the coordinates of the three points, is zero. Three points in space are collinear if, and only if, lines through different pairs of the points have their direction numbers proportional, or if, and only if, the coordinates of any one of them can be expressed as a linear combination of the other two, in which the constants of the linear combination have their sum equal to unity.

COL-LIN-E-A′TION, *n.* A transformation of the plane or space which carries points into points, lines into lines, and planes into planes. See TRANSFORMATION —collineatory transformation.

COL-LIN′E-A-TO-RY, *adj.* **collineatory transformation.** See TRANSFORMATION —collineatory transformation.

CO-LOG′A-RITHM, *n.* The **cologarithm** of a number is the logarithm of the reciprocal of the number (*i.e.*, the negative of the logarithm), expressed with the decimal part positive. Used in computations to avoid subtracting mantissas and the confusion of dealing with the negatives of mantissas. *E.g.*, to evaluate $\frac{641}{1246}$ by use of logarithms, write $\log \frac{641}{1246} = \log 641 + \text{colog } 1246$, where colog $1246 = 10 - \log 1246 - 10 = 10 - (3.0955) - 10 = 6.9045 - 10$.

"COLONEL BLOTTO" GAME. See GAME.

COL′OR-ING, *adj.* **graph coloring.** A graph is **n-colorable** if each node can be colored using one of n colors in such a way that different colors are assigned to any pair of nodes belonging to the same edge. The problem of whether every planar graph is 4-colorable is equivalent to the *four-color problem* (see under FOUR). A 2-colorable graph is said to be **bipartite.**

    map-coloring problems. See FOUR—four-color problem.

COL′UMN, *n.* A vertical array of terms, used in addition and subtraction and in determinants and matrices.

    column in a determinant. See DETERMINANT.

COMBESCURE, Jean Joseph Antoine Éduard (1824–1889). **Combescure transformation of a curve.** A one-to-one continuous mapping of the points of one space curve on another in such a way that the tangents at corresponding points are parallel. It follows that the principal normals and the binormals, respectively, at corresponding points must also be parallel.

    Combescure transformation of a triply orthogonal system of surfaces. A one-to-one continuous mapping of the points of three-dimensional Euclidean space on itself, such that the normals to the members of one triply orthogonal system of surfaces are parallel to the normals to the members of another system at points corresponding under the transformation.

COM′BI-NA′TION, *n.* A *combination* of a set of objects is any subset *without regard to order*. The **number of combinations of n things, r at a time,** is the number of sets that can be made up from the n things, each set containing r different things and no two sets containing exactly the same r things. This is equal to the number of permutations of the n things, taken r at a time, divided by the number of permutations of r things taken r at a time; that is,

$$_nP_r/r! = n!/[(n-r)!r!],$$

which is denoted by $_nC_r$, $C(n, r)$, or $\binom{n}{r}$. *E.g.*, the combinations of a, b, c, two at a time, are ab, ac, bc. $_nC_r$ is also the coefficient in the $(r+1)$th term of the binomial expansion, $(x-a)^n = (x-a_1)(x-a_2)\cdots(x-a_n)$, with the a's all equal. That is, the coefficient of x^{n-r} is a^r times the number of combinations of the n different a's r at a time. See BINOMIAL —binomial coefficients. The **number of combinations of n** things, *r* at a time, when repetitions are allowed is the number of sets which can be made up of *r* things chosen from the given *n*, each being used as many times as desired. The number of such combinations is the same as the number of combinations of $n + r - 1$ different things taken *r* at a time, repetitions not allowed; *i.e.*,

$$\frac{(n + r - 1)!}{(n - 1)! r!}.$$

The combinations of a, b, c two at a time, repetitions allowed, are: aa, bb, cc, ab, ac, bc. The **total number of combinations of** *n* **different things** (repetitions not allowed) is the sum of the number of combinations taken $0, 1, 2, \cdots, n$ at a time, *i.e.*, $\sum_{r=0}^{n} {}_nC_r$, which is the sum of the binomial coefficients in $(x + y)^n$, or 2^n.

linear combination. See LINEAR.

COM'BI-NA-TO'RI-AL, *adj.* **combinatorial topology.** See TOPOLOGY.

COM-MAND', *n.* An instruction, in machine language, for a computing machine to perform a certain operation.

COM-MEN'SU-RA-BLE, *adj.* **commensurable quantities.** Two quantities which have a common measure; *i.e.*, there is a measure which is contained an integral number of times in each of them. A rule, a yard long, is commensurable with a rod, for they each contain, for instance, 6 inches an integral number of times. Two real numbers are commensurable if and only if their ratio is a rational number.

COM-MER'CIAL, *adj.* **commercial bank.** A bank that carries checking accounts.

commercial draft. See DRAFT.

commercial paper. Negotiable paper used in transacting business, such as drafts, negotiable notes, endorsed checks, etc.

commercial year. See YEAR.

COM-MIS'SION, *n.* A fee charged for transacting business for another person.

commission man or merchant. See BROKER.

COM'MON, *adj.* **common denominator.** For two or more fractions, a common denominator is a common multiple of the denominators. For example, a common denominator for the fractions $\frac{2}{3}$, $\frac{3}{4}$, and $\frac{1}{6}$ is 12, or any multiple of 12. The **least common denominator** (L.C.D.) is the smallest of all common denominators, *i.e.*, the least

common multiple of the denominators. In the example, 12 is the least common denominator.

common difference in an arithmetic progression. The difference between any term and the preceding term, usually denoted by *d*.

common divisor or **common factor.** See DIVISOR—common divisor.

common fraction. See FRACTION.

common logarithms. Logarithms having 10 for their base. See LOGARITHM.

common multiple. See MULTIPLE—common multiple.

common stock. Stock upon which the dividends paid are determined by the net profits of the corporation after all other costs, including dividends on preferred stock, have been paid.

common tangent of two circles. A line which is tangent to each of the circles. If each circle is in the exterior of the other, there are four common tangents. The two tangents that separate the circles are **internal tangents**. The other two tangents are **external tangents**; both circles are on the same side of each external tangent.

COM-MU-TA'TION, *adj.* **commutation symbols in life insurance.** Symbols denoting the nature of the numbers in the columns of a commutation table. For instance, D_x and N_x. See below, commutation tables.

commutation tables (columns). Tables from which the values of certain types of insurance can be quickly computed. *E.g.*, suppose that one has a commutation table with the values of D_x and N_x for all ages appearing in the mortality tables, where D_x is the product of the number of persons who attain the age *x* in any year and the present value of a sum of money *x* years hence, at some given rate, and N_x is the sum of the series $(D_x + D_{x+1} + \cdots$ to end of table). The value of an immediate annuity of \$1.00 at age *x* is the quotient N_{x+1}/D_x, and that of an annuity due is N_x/D_x. Sometimes (following Davies) N_x is defined as the sum of the series $(D_{x+1} + D_{x+2} + \cdots)$. With this definition, the annuity values must be N_x/D_x and N_{x-1}/D_x, respectively. Commutation tables based on the latter definition of N_x are called the **terminal form**, while those based on the former definition of N_x are called the **initial form**.

COM-MU'TA-TIVE, *adj.* A method of combining objects two at a time is *commutative* if the result of the combination of two objects does not depend on the order in which the objects are given. For example, the **commutative law of addition** states that the order of addition does not affect the sum: $a + b = b + a$ for any numbers *a* and *b* (*e.g.*, $2 + 3 = 3 + 2$). The **commutative law**

of multiplication states that the order of multiplication does not affect the product: $a \cdot b = b \cdot a$ for any numbers a and b (e.g., $3 \cdot 5 = 5 \cdot 3$). There are many mathematical systems that satisfy commutative laws, but many others that do not. For example, (vector) multiplication of vectors and multiplication of matrices are not commutative. See GROUP, RING.

    commutative group. See GROUP.

COM′MU-TA-TOR, *n.* **commutator of elements of a group.** The commutator of two elements a and b of a group is the element $a^{-1}b^{-1}ab$, or the element c such that $bac = ab$. The group of all elements of the form $c_1 c_2 \cdots c_n$, where each c_i is the commutator of some pair of elements, is called the **commutator subgroup**. The commutator subgroup of an Abelian group contains only the identity element. A group is said to be **perfect** if it is identical with its commutator subgroup. A commutator subgroup is an *invariant subgroup* and the *factor group* formed with it is Abelian.

COM-MUT′ING, *p.* **commuting obligations.** Exchanging one set of obligations to pay a certain sum (or sums) at various times for another to pay according to some other plan. The common date of comparison at which the two sets are equivalent (equal in value at that time) is the **focal date**.

COM′PACT, *adj.* A topological space E being **compact** (originally, **bicompact**) means that it has the property that, for any union of open sets that contains E, there is a finite number of these open sets whose union contains E. A space E is compact if and only if each collection of closed sets contained in E has a nonempty intersection if it has the *finite intersection property*. All closed subsets of a compact space are compact; all compact subsets of a Hausdorff topological space are closed. A **locally compact space** is a space E with the property that each point of E has a neighborhood that is contained in a compact subset of E. The set $\{0, 1, \frac{1}{2}, \frac{1}{3}, \frac{1}{4}, \cdots \}$ is compact. The set R of real numbers is not compact, since R is contained in the union of all open intervals of unit length, but R is not contained in any finite union of such intervals (see HEINE—Heine-Borel theorem). A **countably compact space** is a space E with the property that, for any union of a countable number of open sets that contains E, there is a finite number of these open sets whose union contains E. A space E is countably compact if and only if each sequence in E has an accumulation point in E. A **sequentially compact** (originally, **compact**) space is a

space E with the property that each sequence in E contains a subsequence that converges to a point in E. For a space to have the **Bolzano-Weierstrass property** means that each infinite subset has at least one accumulation point (see BOLZANO —Bolzano-Weierstrass theorem). For a Lindelöf space (and therefore for a metric space), compactness, countable compactness, sequential compactness, and the Bolzano-Weierstrass property are equivalent. All compact spaces are countably compact, and all countably compact spaces have the Bolzano-Weierstrass property. All T_1-spaces with the Bolzano-Weierstrass property are countably compact. All sequential compact spaces are countably compact, and all countably compact spaces satisfying the first countability axiom are sequentially compact. See METACOMPACT, PARACOMPACT, WEAK—weak compactness.

    compact support. See SUPPORT —support of a function.

COM-PACT′I-FI-CA′TION, *n.* A compactification of a topological space T is a *compact* topological space W which contains T (or is such that T is homeomorphic with a subset of W). The *complex plane* (or *sphere*) is the compactification of the *Euclidean plane* obtained by adjoining a single point (usually designated by the symbol ∞) and defining the neighborhoods of ∞ to be sets containing ∞ and the complement of a bounded, closed (i.e., compact) subset of the plane. Likewise, a *locally compact Hausdorff space H* has a **one-point** (or **Alexandroff**) **compactification** (also a Hausdorff space) obtained by adjoining a single point, which can be designated by the symbol ∞, whose neighborhoods are sets containing ∞ and the complement of a compact subset of H. The **Stone-Čech compactification** of a *Tychonoff space T* is the closure of the image of T in the space I^{ϕ}, where I^{ϕ} is the Cartesian product of the closed unit interval I (taken ϕ times) and ϕ is the cardinal number of the family F of all continuous functions from T to I (the image of a point x of T in I^{ϕ} is the member of I^{ϕ} whose f "component" is $f(x)$ for each member f of F). The Stone-Čech compactification is (in a certain real sense) a maximal compactification. The entire space I^{ϕ} is compact, a consequence of the Tychonoff theorem (see PRODUCT—Cartesian product).

COM-PAC′TUM, *n.* A topological space which is *compact* and *metrizable*. Examples of compacta are closed intervals, closed spheres (with or without their interiors), and closed polyhedra. See COMPACT —compact set.

COM'PA-RA-BLE, *adj.* **comparable functions.** Functions f and g which have real-number values, which have a common domain of definition D, and which are such that either $f(x) \le g(x)$ for all x in D or $f(x) \ge g(x)$ for all x in D.

COM-PAR'I-SON, *adj. n.* **comparison date.** See EQUATION —equation of payments.

comparison property of the real numbers. The property that exactly one of the statements $x < y$, $x = y$, and $x > y$ is true for any numbers x and y. See TRICHOTOMY.

comparison test for convergence of an infinite series. If, after some chosen term of a series, the *absolute value* of each term is equal to or less than the value of the corresponding term of some convergent series of positive terms, the series converges (and converges absolutely); if each term is equal to or greater than the corresponding term of some divergent series of positive terms, the series diverges.

COM'PASS, *n.* An instrument for describing circles or for measuring distances between two points. Usually used in the plural, as **compasses.**

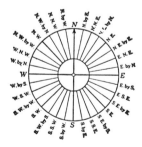

mariner's compass. A magnetic needle that rotates about an axis perpendicular to a card (see figure) on which the directions are indicated. The needle always indicates the direction of the magnetic meridian.

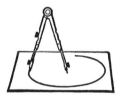

COM-PAT'I-BIL'I-TY, *adj., n.* Same as CONSISTENCY.

compatibility equations. (*Elasticity*) The differential equations connecting the components of the strain tensor which guarantee that the state of strain be possible in a continuous body.

COM'PLE-MENT, *n.* **complement of an angle.** The complement of an angle A is the angle $90° - A$. See COMPLEMENTARY —complementary angles.

complement of a set. The set of all objects that do not belong to the given set U, but belong to a given whole space (or set) that contains U. The complement of the set of positive numbers with respect to the space of all real numbers is the set containing all negative numbers and zero. See LATTICE.

COM'PLE-MEN'TA-RY, *adj.* **complementary acceleration.** See ACCELERATION —acceleration of Coriolis.

complementary angles. Two angles whose sum is a right angle. The two acute angles in a right triangle are always complementary. See TRIGONOMETRIC —trigonometric cofunctions.

complementary function. See DIFFERENTIAL —linear differential equations.

complementary minor. See MINOR —minor of an element in a determinant.

complementary trigonometric functions. Same as TRIGONOMETRIC COFUNCTIONS. See TRIGONOMETRIC.

surface complementary to a given surface. Given a surface S, there is an infinitude of parallel surfaces such that S is a *surface of center* relative to each of them. See SURFACE —surfaces of center relative to a given surface, PARALLEL —parallel surfaces. The other common surface of center of the family of parallel surfaces is said to be **complementary** to S.

COM-PLETE', *adj.* **complete annuity.** See ANNUITY.

complete field. See FIELD —ordered field.

complete graph. See GRAPH THEORY.

complete induction. See INDUCTION —mathematical induction.

complete lattice. See LATTICE.

complete scale. See SCALE —number scale.

complete space. A **complete metric space** is a metric space such that every *Cauchy sequence* converges to a point of the space (see SEQUENCE —Cauchy sequence). The space of all real numbers (or all complex numbers) is complete. The space of all continuous functions defined on the interval $[0, 1]$ is not complete if the distance between f and g is defined as $\int_0^1 |f - g|\, dx$, since the sequence $f_1, f_2, \cdots$ does not then converge to a continuous function if $f_n(x) = 0$ for $0 \le x \le \frac{1}{2}$ and $f_n(x) = (x - \frac{1}{2})^{1/n}$ for $\frac{1}{2} \le x \le 1$. A **topologically complete topological space** is a topological space that is homeomorphic to some complete metric space. A subset of a complete metric space is topologically complete if and only if it is a G_δ subset (see BOREL—Borel set). A **complete linear topological space** is a linear topological space for which each Cauchy net converges to some point in the space. A **Cauchy net** is a net $\{x_\alpha\}$ for which $\{x_\alpha - x_\beta\}$ converges to zero, where x_α is a member of the space for each α in the given directed set [see MOORE (E. H.)—Moore-Smith convergence].

complete system of functions. See ORTHOGONAL—orthogonal functions.

complete system of representations for a group. See REPRESENTATION—representation of a group.

weakly complete space. See WEAK—weak completeness.

COM-PLET′ING, v. **completing the square.** A process used in solving quadratic equations. It consists of transposing all terms to the left side of the equation, dividing through by the coefficient of the square term, then adding to the constant (and to the right side) a number that will make the left member a perfect trinomial square. This method is sometimes modified by first multiplying through by a number chosen to make the coefficient of the square of the variable a perfect square, then adding a constant to both sides of the equation, as before, to make the left side a perfect trinomial square. *E.g.*, to complete the square in $2x^2 + 8x + 2 = 0$, divide both members of the equation by 2, obtaining $x^2 + 4x + 1 = 0$. Now add 3 to both sides, obtaining

$$x^2 + 4x + 4 = (x + 2)^2 = 3.$$

Frequently, *completing* the square refers to writing any polynomial of the form $a_1 x^2 + b_1 x + c_1$ in the form $a_1(x + b_2)^2 + c_2$, a procedure used, for example, in reducing the equations of conics to their standard form.

COM′PLEX, *adj., n.* As a *noun*, a complex may mean simply a SET (also see below, simplicial complex).

absolute value of a complex number. See MODULUS—modulus of a complex number.

amplitude, or argument, of a complex number. See POLAR—polar form of a complex number.

complex conjugate of a matrix. See MATRIX—complex conjugate of a matrix.

complex domain (field). The set of all complex numbers. See FIELD.

complex fraction. See FRACTION.

complex integration. See CONTOUR—contour integral.

complex measure. See MEASURE—measure of a set.

complex number. Any number, real or imaginary, of the form $a + bi$, where a and b are real numbers and $i^2 = -1$. Called **imaginary numbers** when $b \neq 0$, and **pure imaginary** when $a = 0$ and $b \neq 0$ (although complex numbers are not imaginary in the usual sense). Two complex numbers are defined to be **equal** if and only if they are identical. *I.e.*, $a + bi = c + di$ means $a = c$ and $b = d$. A complex number $x + yi$ can be represented in the plane by the vector with components x and y, or by the point (x, y) (see the figure below, and ARGAND—Argand diagram).

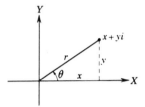

Thus two complex numbers are equal if and only if they are represented by the same vectors, or by the same points. In the above figure, $x = r \cos \theta$ and $y = r \sin \theta$. Therefore

$$x + yi = r(\cos \theta + i \sin \theta),$$

which is the **polar form** of $x + yi$ (see POLAR). The **sum of complex numbers** is obtained by adding the real parts and the coefficients of i separately; *e.g.*, $(2 - 3i) + (1 + 5i) = 3 + 2i$. Geometrically, this is the same as the addition of the corresponding vectors in the plane. In the figure below,

$$OP_1 + OP_2 = OP_3 \ (OP_2 = P_1 P_3).$$

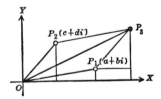

The **product of complex numbers** is computed by treating the numbers as polynomials in i with the special property $i^2 = -1$. Thus:

$$(a + bi)(c + di) = ac + (ad + bc)i + bdi^2$$

$$= ac - bd + (ad + bc)i.$$

If the complex numbers are in the form $r_1(\cos A + i \sin A)$ and $r_2(\cos B + i \sin B)$, their product is $r_1 r_2[\cos(A + B) + i \sin(A + B)]$; *i.e.*, to multiply two complex numbers, multiply their moduli and add their amplitudes (see DE MOIVRE —De Moivre's theorem). Similarly, the **quotient of two complex numbers** is the complex number whose modulus is the quotient of the modulus of the dividend by that of the divisor and whose amplitude is the amplitude of the dividend minus that of the divisor; that is,

$$r_1(\cos \theta_1 + i \sin \theta_1) \div r_2(\cos \theta_2 + i \sin \theta_2)$$
$$= \frac{r_1}{r_2}[\cos(\theta_1 - \theta_2) + i \sin(\theta_1 - \theta_2)].$$

When the numbers are not in polar form, the quotient can be computed by multiplying dividend and divisor by the conjugate of the divisor, as illustrated in the following example:

$$\frac{2+i}{1+i} = \frac{(2+i)(1-i)}{(1+i)(1-i)} = \frac{3-i}{2}.$$

Tech. The system of complex numbers is the set of ordered pairs (a, b) of real numbers in which two pairs are considered equal if, and only if, they are identical [$(a, b) = (c, d)$ if, and only if, $a = c$ and $b = d$], and in which addition and multiplication are defined by

$$(a, b) + (c, d) = (a + c, b + d),$$
$$(a, b)(c, d) = (ac - bd, ad + bc).$$

The system satisfies most of the fundamental algebraic laws, such as the associative and commutative laws for addition and multiplication. It is a *field*, but not an *ordered field*. A remarkable consequence of these definitions is:

$$(0, 1)(0, 1) = (-1, 0);$$
$$(0, -1)(0, -1) = (-1, 0).$$

That is, the number $(-1, 0)$, or -1, has the two square roots $(0, 1)$ and $(0, -1)$. See FUNDAMENTAL—fundamental theorem of algebra.

complex plane. The plane (of complex numbers) with a single point at infinity whose neighborhoods are exteriors of circles with center at 0. The complex plane is topologically (and conformally) equivalent to a sphere. See PROJECTION—stereographic projection. *Syn.* Gauss plane.

complex roots of a quadratic equation ($ax^2 + bx + c = 0$). Used in contrasting roots of the form $a + bi$ with real roots, although the latter are special cases of the former for which $b = 0$. See DISCRIMINANT —discriminant of a quadratic equation in one variable.

complex sphere. A unit sphere on which the complex plane is represented by a stereographic projection. The complex plane is usually the equatorial plane relative to the pole of projection, or the tangent plane at the point diametrically opposite the pole of projection.

conjugate complex numbers. Two numbers of type $a + bi$ and $a - bi$, where a and b are real numbers. Frequently called **conjugate imaginaries**. If $\bar{z}$ denotes the conjugate of z, then $\overline{z_1 z_2} = \bar{z}_1 \bar{z}_2$, $\overline{z_1 + z_2} = \bar{z}_1 + \bar{z}_2$, $z\bar{z} = x^2 + y^2$ if $z = x + iy$ and x and y are real, and $\bar{z}$ is a root of any polynomial equation with real coefficients which has z as a root.

modulus of a complex number. See MODULUS.

real and imaginary parts of a complex number. See REAL, IMAGINARY.

root of a complex number. See ROOT.

simplicial complex. A set which consists of a finite number of simplexes (not necessarily all of the same dimension) with the property that the intersection of any two of the simplexes either is empty or is a face of each of them. This definition is sometimes modified in various ways, *e.g.*, by requiring that each simplex be oriented. A simplicial complex is sometimes called a complex, but a **complex** is sometimes defined with fewer restrictions (*e.g.*, it may be a countable set of simplexes such that the intersection of any two of the simplexes either is empty or is a face of each of them, and no vertex of a simplex belongs to more than a finite number of the simplexes). The **dimension of a simplicial complex** is the largest of the dimensions of the simplexes making up the simplicial complex. The class of all simplexes which belong to a simplicial complex K and have dimension less than that of K is called the **skeleton** of K. A finite set K of elements $c_0, c_1, \cdots, c_n$ is called an **abstract simplicial complex** (or an **abstract complex** or a **skeleton complex**), and the elements $c_0, \cdots, c_n$ are called **vertices**, if certain nonempty subsets of K, which are called **abstract simplexes** (or **skeletons**) are such that each subset (called a face) of an abstract simplex is an abstract simplex and each of the vertices is an abstract simplex. The **dimension of an abstract simplex** of $r + 1$ points is r, and the **dimension of an abstract complex** is the largest of the dimensions of its abstract simplexes. An abstract complex of dimension n can always be represented by a simplicial complex imbedded in the Euclidean space of dimension $2n + 1$. A simplicial complex is sometimes called a **geometric complex** (or simply a **complex**), or a **triangulation**. The set of all those points which belong to simplexes of a simplicial complex is called a **polyhedron**. A topological space is said to be **triangulable**, and is sometimes called a **polyhedron** or a **topological simplicial complex**, if it is homeomorphic to the set of points belonging to simplexes of a simplicial complex K; the homeomorphism together with the complex K is a **triangulation** of the polyhedron. A simplicial complex is **oriented** if each of its simplexes is oriented. See CHAIN—chain of simplexes, MANIFOLD, SIMPLEX, SURFACE, TRIANGULATION.

unit complex number. A complex number whose modulus is 1; a complex number of the form $\cos \theta + i \sin \theta$. Unit complex numbers are represented by points on the unit circle in the plane. The products and quotients of unit complex numbers are unit complex numbers.

COM-PO'NENT, *n.* **component of acceleration, force, or velocity.** See VECTOR.

component of a computing machine. Any physical mechanism or abstract concept having a distinct role in automatic computation. See headings below and under ARITHMETIC, CONTROL, INPUT, OUTPUT, STORAGE.

component of a graph. See GRAPH THEORY.

component of an algebraic plane curve. See CURVE—algebraic plane curve.

component of a set of points. A subset which is *connected* and is not contained in any other connected subset of the given set of points. A component is necessarily a *closed* subset relative to the set.

component of a vector. See VECTOR.

component of the stress tensor. In linear theory of elasticity, a set of six functions determining the state of stress at any point of the substance.

direction components. See DIRECTION—direction numbers.

elementary potential digital computing component. In a computing machine, any component that can assume any one of a fixed discrete set of stable states, and that can influence and/or be influenced by other components of the machine. See CIRCUIT—flip-flop circuit.

COM-POS'ITE, *adj.* **composite function.** (1) See COMPOSITION —composition of functions. (2) A function which is factorable (can be written as the product of two or more functions), as $x^2 - y^2$ or $x^2 - 1$ (usually refers only to polynomial functions which are factorable relative to some specified field).

composite hypothesis. See HYPOTHESIS.

composite life of a plant. The time required for the total annual depreciation charge to accumulate, at a given rate of interest, to the original wearing value.

composite number. A number that has two or more prime factors, as 4, 6, or 10, in distinction to ± 1 and prime numbers such as 3, 5, or 7. Refers only to integers, not to rational or irrational numbers.

composite quantity. A quantity that is factorable.

derivative and differential of a composite function. See CHAIN—chain rule, DIFFERENTIAL.

COM'PO-SI'TION, *n.* **composition in a proportion.** Passing from the statement of the proportion to the statement that the sum of the first antecedent and its consequent is to its consequent as the sum of the second antecedent and its consequent is to its consequent; *i.e.,* passing from

$$a/b = c/d \quad \text{to} \quad (a + b)/b = (c + d)/d.$$

composition and division in a proportion. Passing from the statement of the proportion to the statement that the sum of the first antecedent and its consequent is to the difference between the first antecedent and its consequent as the sum of the second antecedent and its consequent is to the difference between the second antecedent and its consequent; *i.e.,* passing from

$$a/b = c/d$$

to

$$(a + b)/(a - b) = (c + d)/(c - d).$$

See above, composition in a proportion, and DIVISION —division in a proportion.

composition of functions. Forming a new function h (the **composite function**) from given functions g and f by the rule that $h(x) = g[f(x)]$ for all x in the domain of f for which $f(x)$ is in the domain of g. For example, if $g(x) = \sqrt{x}$ when $x \geq 0$ and $f(x) = x + 3$ for all real numbers, then the domain of the composite h of g and f is the set of all x with $x \geq -3$, and $h(x) = \sqrt{x + 3}$. In particular, $g[f(-2)] = g(1) = 1$. This composite h of g and f often is written as $g \circ f$ or gf, but sometimes as $f \circ g$ or fg (as for the composition of relations, of which functions are special cases). The order in which functions are combined is important. For example, if $g(x) = x + 1$ and $f(x) = x^2$, then $g[f(x)] =$

$g(x^2) = x^2 + 1$, but $f[g(x)] = f(x + 1) = (x + 1)^2$. The derivative of the composite of two functions can be computed by use of the *chain rule*.

composition of relations. Given a relation R and a relation S, the **composite relation** $R \circ S$ is the relation for which x is related to z if and only if there is an object y for which xRy and ySz. For example, if r, s and t denote positive integers, rRs means $r < s$, and rSs means "r divides s," then $r(R \circ S)t$ means "there is an integer s greater than r that divides t," and $r(S \circ R)t$ means "there is a positive integer s less than t that is divisible by r." See RELATION, and above, composition of functions.

composition of tensors. See INNER—inner product of tensors.

composition of vectors. The same process as *addition of vectors*. but the term composition of vectors is used more when speaking of adding vectors which denote forces, velocities, or accelerations; finding the vector which represents the resultant of forces, velocities, accelerations, etc., represented by the given vectors. See SUM—sum of vectors.

graphing by composition of ordinates. See GRAPHING —graphing by composition.

COM′POUND, *adj.* **compound amount** and **compound interest.** See INTEREST.

compound event. See EVENT—compound event.

compound number. The sum of two or more denominations of a certain kind of denominate number; *e.g.*, 5 feet, 7 inches, or 6 pounds, 3 ounces.

compound survivorship life insurance. See INSURANCE —life insurance.

COM-PRES′SION, *n.* See TENSION.

modulus of compression. See MODULUS— bulk modulus.

simple or one-dimensional compressions. Same as ONE-DIMENSIONAL STRAINS. See STRAIN.

COM′PU-TA′TION, *n.* The act of carrying out mathematical processes; used mostly with reference to arithmetic rather than algebraic work. One might say, "Find the formula for the number of gallons in a sphere of radius r and *compute* the result for $r = 5$"; or "*Compute* the square root of 3." Frequently used to designate long arithmetic or analytic processes that give numerical results, as *computing* the orbit of a planet.

computation by logarithms. See LOGARITHM.

numerical computation. A computation involving numbers only, not letters representing numbers.

COM-PUTE′, *v.* To make a computation. *Syn.* calculate.

COM-PUT′ER, *n.* Any instrument which performs numerical mathematical operations. A mechanical machine which primarily performs combinations of addition, subtraction, multiplication and division is sometimes called a *calculating machine* in distinction from such more versatile instruments as electronic computers. *Syn.* computing machine.

analog computer. A computing machine in which numbers are converted into measurable quantities, such as lengths or voltages, that can be combined in accordance with the desired arithmetic operations; *e.g.*, a slide rule. Generally, if two physical systems have corresponding behavior, and one is chosen for study in place of the other (because of greater familiarity, economy, feasibility, or other factors), then the first is called an analog device, analog machine, or analog computer. See BUSH.

digital computer. A computing machine that performs mathematical operations on numbers expressed by means of digits. *Syn.* digital device. See BABBAGE, ENIAC, LEIBNIZ.

CON′CAVE, *adj.* **concave toward a point (or line or plane).** A curve is **concave toward a point (or line)** if every segment of the arc cut off by a chord lies on the chord or on the opposite side of the chord from the point (or line). If there exists a horizontal line such that the curve lies above it and is concave toward it (lies below it and is concave toward it) the curve is said to be *concave downward* (*upward*). A curve is concave upward if and only if it is the graph of a convex function (see CONVEX —convex function). A circle with center on the x-axis is concave toward that axis, the upper half being concave down and the lower half concave up.

concave function. The negative of a convex function. See CONVEX —convex function.

concave polygon and polyhedron. See POLYGON, POLYHEDRON.

CON-CAV′I-TY, *n.* The state or property of being concave.

CON′CEN-TRA′TION, *n.* **concentration method for the potential of a complex.** See POTENTIAL.

CON-CEN′TRIC, *adj.* **concentric circles.** Circles lying in the same plane and having a common center. *Concentric* is applied to any two figures which have centers (that is, are symmetric about some point) when their centers are coincident. *Concentric* is opposed to *eccentric*, meaning not concentric.

CON′CHOID, *n.* The locus of one end point of a segment of constant length, on a line which rotates about a fixed point (*O* in figure), the other end point of the segment being at the

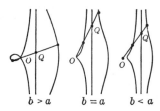

$$b > a \qquad b = a \qquad b < a$$

intersection *Q* of this line with a fixed line not containing the fixed point. If the polar axis is taken through the fixed point and perpendicular to the fixed line, the length of the segment is taken as *b*, and the distance from the fixed point to the fixed line as *a*, the polar equation of the conchoid is $r = b + a \sec \theta$. Its Cartesian equation is

$$(x - a)^2 (x^2 + y^2) = b^2 x^2.$$

The curve is asymptotic to the fixed line in both directions, and on both sides of it. If the length of the line segment is greater than the perpendicular distance from the pole to the fixed line, the curve forms a loop with a node at the pole. If these two distances are equal, it forms a cusp at the pole. *Syn.* conchoid of Nicomedes.

CON-CLU′SION, *n.* **conclusion of a theorem.** The statement which follows (or is to be proved to follow) as a consequence of the hypothesis of the theorem. See IMPLICATION.

CON-CORD′ANT-LY, *adv.* **concordantly oriented.** See MANIFOLD, SIMPLEX.

CON′CRETE, *adj.* **concrete number.** A number referring to specific objects or units, as 3 people, or 3 houses. The number and its references are denoted by *concrete number*.

CON-CUR′RENT, *adj.* Having a point in common. A collection of sets is **concurrent** if there is a point that belongs to each of the sets. For example, the medians of a triangle are concur-

rent; these three lines all contain the point $\frac{2}{3}$ of the distance from a vertex to the opposite side along a median. In space, the planes that contain the origin are concurrent (they are a bundle of planes). See CONSISTENCY —consistency of linear equations.

CON-CYC′LIC, *adj.* **concyclic points.** Points which lie on a common circle.

CON′DEN-SA′TION, *adj.* **condensation point.** A point *P* is a *condensation point* of a set *S* if each neighborhood of *P* contains uncountably many points of *S*. See ACCUMULATION —accumulation point, COUNTABLE —countable set.

CON-DI′TION, *n.* A mathematical assumption or truth that suffices to assure the truth of a certain statement, or which must be true if this statement is true. A condition from which a given statement logically follows is a **sufficient condition**; a condition which is a logical consequence of a given statement is a **necessary condition**. A **necessary and sufficient condition** is a condition that is both necessary and sufficient. A condition may be necessary but not sufficient, or sufficient but not necessary. It is necessary that a substance be sweet in order that it be called sugar, but it may be sweet and be arsenic; it is sufficient that it be granulated and have the chemical properties of sugar, but it can be sugar without being granulated. In order for a quadrilateral to be a parallelogram, it is necessary, but not sufficient, that two opposite sides be equal, and sufficient, although not necessary, that all of its sides be equal; but it is *necessary and sufficient* that two opposite sides be equal and parallel. See IMPLICATION.

CON-DI′TION-AL, *adj.* **conditional convergence of series.** See CONVERGENCE —conditional convergence.
    conditional equation and inequality. See EQUATION, INEQUALITY.
    conditional probability. See PROBABILITY.
    conditional statement. Same as IMPLICATION.

CON-DUC′TOR, *adj., n.* **conductor potential.** For a region *R* with boundary *S*, the conductor potential is the function harmonic in the interior of *R*, continuous on $R \cup S$, and taking on the constant value 1 on *S*. It describes the potential of an electric charge in equilibrium on the surface of a conductor.

CONE, *n.* (1) A conical surface (see CONICAL —conical surface). (2) A solid bounded by a region (the **base**) in a plane and the surface formed by the straight line segments (the **elements**) which join points of the boundary of the base to a fixed point (the **vertex**) not in the plane of the base (the surface bounding this solid is also called a *cone*). The perpendicular distance from the vertex to the plane of the base is the **altitude** of the cone. If the base has a center, the line passing through the center of the base and the vertex is the **axis** of the cone. The cone is **circular** or **elliptic** in the cases its base is a circle or ellipse (sometimes a circular cone is defined to be a cone whose intersections with planes perpendicular to the axis, but not intersecting the base, are circles). An **oblique circular cone** is a circular cone whose axis is not perpendicular to its base. A **right circular cone** (or **cone of revolution**) is a circular cone whose base is perpendicular to its axis (sometimes called simply a *circular cone*). A right circular cone can be generated by revolving a right triangle about one of its legs, or an isosceles triangle about its altitude. The **slant height** of a right circular cone is the length of an element of the cone. The **lateral area** of a cone is the area of the surface formed by the elements (for a right circular cone this is equal to $\pi r h$, where r is the radius of the base and h is the slant height). The **volume** of a cone is equal to one-third of the product of the area of the base and the altitude. If the cone has a circular base, the volume is $\frac{1}{3}\pi r^2 s$, where r is the radius of the base and s is the altitude.

    frustum of a cone. The part of the cone bounded by the base and a plane parallel to the base (see figure). The volume of a frustum of a cone equals one-third the **altitude** (the distance between the planes) times the sum of the areas

of the bases and the square root of the product of the areas of the bases; *i.e.*,

$$\tfrac{1}{3}h\left(B_1 + B_2 + \sqrt{B_1 B_2}\right).$$

The **lateral area** of a frustum of a right circular cone (the area of the curved surface) is equal to $\pi l (r + r')$, where l is the slant height and r and r' are the radii of the bases.

    ruling of a cone. See RULING.

    spherical cone. A surface composed of the spherical surface of a spherical segment and the conical surface defined by the bounding circle of

the segment and the center of the sphere (see CONICAL—conical surface); a spherical sector whose curved base is a zone of one base. The volume of a spherical cone is $\frac{2}{3}\pi r^2 h$, where r is the radius of the sphere and h is the altitude of the zone base.

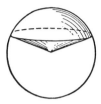

    tangent cone of a quadric surface. A cone whose elements are each tangent to the quadric. In particular, a **tangent cone of a sphere** is any circular cone all of whose elements are tangent to the sphere. If a ball is dropped into a circular cone, the cone is tangent to the ball.

    truncated cone. The portion of a cone included between two nonparallel planes whose line of intersection does not pierce the cone. The two plane sections of the cone are the **bases** of the *truncated cone*.

CON′FI-DENCE, *adj.* **confidence interval.** An interval which is believed, with a preassigned degree of confidence, to include the particular value of some parameter being estimated. *Tech.* For a random variable X whose distribution depends on an unknown parameter θ, a $(100\alpha)\%$ **confidence interval** is an interval (T_1, T_2) for which T_1 and T_2 are *statistics, i.e.,* functions of random samples $(X_1, \cdots, X_n)$, and

$$\mathrm{Prob}[T_1 \le \theta \le T_2] \ge \alpha.$$

If repeated samples are taken and T_1 and T_2 are computed for each sample, we would find, on the average, that at least the proportion α of the intervals did include θ and not more than the proportion $1 - \alpha$ did not. For example, for a normal distribution with unknown mean μ, known variance σ^2, and a random sample $(X_1, \cdots, X_n)$ with $\sum\limits_{i=1}^{n} X_i/n = \overline{X}$, we have

$$\mathrm{Prob}\left[\overline{X} - z_{0.95}\frac{\sigma}{\sqrt{n}} < \mu < \overline{X} + z_{0.95}\frac{\sigma}{\sqrt{n}}\right] = 0.95$$

and $(\overline{X} - z_{0.95}\sigma/\sqrt{n}, \overline{X} + z_{0.95}\sigma/\sqrt{n})$ is a 95% confidence interval if $z_{0.95}$ is chosen as 1.96 + so that, for a normal distribution with mean 0 and variance 1, $\mathrm{Prob}[-z_{0.95} < X < z_{0.95}] = 0.95$. This can also be written as

$$\text{Prob}\left[\mu - z_{0.95}\frac{\sigma}{\sqrt{n}} < \overline{X} < \mu + z_{0.95}\frac{\sigma}{\sqrt{n}}\right] = 0.95.$$

If the variance also is unknown, the 95% confidence interval is

$$\left(\overline{X} - t_{n-1}(0.975)\frac{s}{\sqrt{n}},\ \overline{X} + t_{n-1}(0.975)\frac{s}{\sqrt{n}}\right),$$

where s is the *sample standard deviation*,

$$\left[\sum_{i=1}^{n}\left(X_i - \overline{X}\right)^2/(n-1)\right]^{1/2},\ \text{and } t_{n-1}(0.975) \text{ is}$$

the 97.5 percentile of the t-distribution with $n-1$ degrees of freedom.

confidence region. For a random variable X whose distribution depends on unknown parameters $\theta_1,\cdots,\theta_n$, a $(100\alpha)\%$ **confidence region** is a set S in the n-dimensional space of possible values for $(\theta_1,\cdots,\theta_n)$ that is determined in some way by sample values $(X_1,\ldots,X_n)$ and satisfies

$$\text{Prob}\left[(\theta_1,\cdots,\theta_n)\in S\right]\geq\alpha.$$

most selective confidence interval. A $(100\alpha)\%$ confidence interval (T_1,T_2) (see above, confidence interval) which has the property that, if (S_1,S_2) is any other $(100\alpha)\%$ confidence interval, then, for any incorrect value θ_1 of the unknown parameter, we have

$$\text{Prob}\left[\theta_1\in(T_1,T_2)\right]$$

$$\leq\text{Prob}\left[\theta_1\in(S_1,S_2)\right].$$

That is, the most selective confidence intervals cover false values of the parameters with minimal probability. Sometimes called SHORTEST CONFIDENCE INTERVAL, but the most selective confidence interval is not necessarily shortest in the sense of minimizing $T_2 - T_1$.

unbiased confidence interval. A confidence interval (T_1,T_2) (see above, confidence interval) which has the property that the conditional probability of "$\theta\in(T_1,T_2)$," given that θ is the true value of the unknown parameter, is never less than the conditional probability of "$\theta\in(T_1,T_2)$," given that the parameter has some other value.

CON-FIG'U-RA'TION, *n.* A general term for any geometrical figure, or any combination of geometrical elements, such as points, lines, curves, and surfaces.

CON-FO'CAL, *adj.* **confocal conics.** Conics having their foci coincident. *E.g.*, the ellipses and hyperbolas represented by the equation

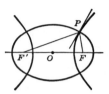

$x^2/(a^2 - k^2) + y^2/(b^2 - k^2) = 1$, where $b^2 < a^2$, $k^2 \neq b^2$, and k takes on all other real values for which $k^2 < a^2$, are confocal. These conics intersect at right angles, forming an orthogonal system. (See point P in figure.)

confocal quadrics. Quadrics whose *principal planes* are the same and whose sections by any one of these planes are *confocal conics*. *E.g.*, if k is a parameter and a, b, and c are fixed, the equation

$$\frac{x^2}{a^2 - k} + \frac{y^2}{b^2 - k} + \frac{z^2}{c^2 - k} = 1,$$

$a^2 > b^2 > c^2$, represents a triply orthogonal system of confocal quadrics: For $c^2 > k > -\infty$, the equation represents a family of **confocal ellipsoids**; for $b^2 > k > c^2$, it represents a family of **confocal hyperboloids of one sheet**; and for $a^2 > k > b^2$, it represents a family of **confocal hyperboloids of two sheets**. Each member of one family is confocal and orthogonal to each member of other families. See ORTHOGONAL —triply orthogonal system of surfaces. For $k = c^2$ we get, by a limiting process, the elliptic portion of the (x, y)-plane (counted twice) bounded by (1) $\frac{x^2}{a^2 - c^2} + \frac{y^2}{b^2 - c^2} = 1$; similarly for $k = b^2$ we get the hyperbolic portion of the (x, z)-plane (counted twice) bounded by (2) $\frac{x^2}{a^2 - b^2} - \frac{z^2}{b^2 - c^2} = 1$. Equations (1) and (2) define the **focal ellipse** and the **focal hyperbola**, respectively, of the system. Through each point (x, y, z) of space there pass three quadrics of the system. The corresponding values k_1, k_2, k_3 of k are called the **ellipsoidal coordinates** of (x, y, z). See COORDINATE —ellipsoidal coordinates.

CON-FORM'A-BLE, *adj.* **conformable matrices.** Two matrices A and B such that the number of columns in A is equal to the number of rows in

B. It is possible to form the product *AB* if, and only if, *A* and *B* are conformable. Being conformable is not a symmetric relation. See PRODUCT—product of matrices.

CON-FORM'AL, *adj.* **conformal-conjugate representation of one surface on another.** A representation which both is conformal and is such that each conjugate system on one surface corresponds to a conjugate system on the other. *Syn.* isothermal-conjugate representation of one surface on another.

conformal map or conformal transformation. A map which preserves angles; *i.e.*, a map such that if two curves intersect at an angle θ, then the images of the two curves in the map also intersect at the same angle θ. The functions $x = x(u, v)$, $y = y(u, v)$, $z = z(u, v)$ map the (u, v)-domain of definition conformally on a surface S if and only if the fundamental quantities of the first order satisfy $E = G = \lambda(u, v) \neq 0$, $F = 0$. See ISOTHERMIC—isothermic map. The coordinates u, v are called **conformal parameters**. The correspondence between surfaces S and $\bar{S}$ determined by $x = x(u, v)$, $y = y(u, v)$, $z = z(u, v)$ and $x = \bar{x}(u, v)$, $y = \bar{y}(u, v)$, $z = \bar{z}(u, v)$ is conformal at regular points if, and only if, the fundamental quantities of the first order satisfy $E : F : G = \bar{E} : \bar{F} : \bar{G}$. The only conformal correspondences between open sets in three-dimensional Euclidean space are obtained by inversions in spheres, reflections in planes, translations, and magnifications. See CAUCHY—Cauchy-Riemann partial differential equations.

conformal parameters. See above, conformal map.

CON'GRU-ENCE, *n.* A statement of type $x \equiv y \pmod{w}$, which is read: "x is congruent to y modulus (or modulo) w"; w is the **modulus** of the congruence. When x, y, and w are integers, the congruence is equivalent to the statement "$x - y$ is divisible by w," or "there is an integer k for which $x - y = kw$." *E.g.*, $23 \equiv 9 \pmod{7}$, since $23 - 9 = 14$ and 14 is divisible by 7. For a given positive integer n, a **modular arithmetic** or **arithmetic modulo n** is obtained by using only the integers $0, 1, 2, \cdots, n - 1$ and defining addition and multiplication by letting the sum $a + b$ and the product ab be the remainder after division by n of the ordinary sum and product of a and b. *E.g.*, if $n = 7$, then $2 + 5 \equiv 0$, $3 \cdot 6 \equiv 4$, and the multiplicative inverse of 2 is 4, since $2 \cdot 4 \equiv 1$. If $n = 15$, then 3 has no multiplicative inverse, since a multiplicative inverse a would have the property that $3 \cdot a - 1 = k \cdot 15$ for some integer k and that $3(a - 5k) = 1$. Arithmetic modulo n is a *commutative ring with unit ele-*

ment; if n is a prime, then arithmetic modulo n is a *field* [see FIELD, and RING]. Congruences may be used in many situations. If polynomials are used, then with the modulus chosen as the polynomial $x^2 - 1$ the congruence $f \equiv g \pmod{x^2 - 1}$ means that $f - g$ is divisible by $x^2 - 1$. *E.g.*, $x^3 + 5x^2 - 1 \equiv 3x^2 + x + 1$, since $(x^3 + 5x^2 - 1) - (3x^2 + x + 1) = (x^2 - 1) \cdot (x + 2)$. If x and y are members of a group and W is a subgroup, then $x \equiv y \pmod{W}$ means that $x \cdot y^{-1}$ belongs to W. *E.g.*, if x and y are complex numbers and W is the set of real numbers, then $x \equiv y \pmod{W}$ could be defined as meaning that x/y is a real number, or as meaning that $x - y$ is a real number. See FERMAT—Fermat's theorem.

linear congruence. A congruence in which all terms are of the first degree in the variables involved. *E.g.*, $12x + 10y - 6 \equiv 0 \pmod{42}$ is a linear congruence.

quadratic congruence. A congruence of the second degree. Thus its general form is $ax^2 + bx + c \equiv 0 \pmod{n}$, where $a \neq 0$.

CON'GRU-ENT, *adj.* **congruent figures.** In **plane geometry**, it is customary to say that two figures are congruent if one of them can be made to coincide with the other by a rigid motion in space (*i.e.*, by translations and rotations in space). Thus it might be said that two figures are congruent if they "differ only in location." Two line segments of equal length are congruent and two circles of equal radii are congruent. Each of the following is a necessary and sufficient condition for two triangles to be congruent: (i) There is a one-to-one correspondence between the sides of one triangle and the sides of the other for which corresponding sides are equal (abbrev.: SSS); (ii) there is a one-to-one correspondence between the sides of one triangle and the sides of the other for which two sides and the angle determined by these sides are equal, respectively, to the corresponding sides of the other triangle and the angle determined by these sides (SAS); (iii) there is a one-to-one correspondence between the angles of one triangle and the angles of the other for which two angles and the side between the vertices of these angles are equal, respectively, to the corresponding angles of the other triangle and the side between the vertices of these angles (ASA). If we change the definition of congruence to allow only rigid motions in the plane, a different concept of congruence results. In **solid geometry**, two figures are congruent if one of them can be made to coincide with the other by a rigid motion in space. Sometimes such figures are said to be **directly congruent** and two figures for which one is directly congruent to the reflection of the other through a plane are **oppo-**

sitely congruent (then two figures are either directly or oppositely congruent if and only if one can be made to coincide with the other by a rigid motion in four-dimensional space). Often when giving axioms for a geometric system, congruence is taken as an undefined concept restricted by suitable axioms.

congruent matrices. See TRANSFORMATION — congruent transformation.

congruent numbers, or quantities. See CONGRUENCE.

congruent transformation. See TRANSFORMATION—congruent transformation.

CON′IC, *n.* Any curve which is the locus of a point which moves so that the ratio of its distance from a fixed point to its distance from a fixed line is constant. The ratio is the **eccentricity of the curve**, the fixed point the **focus**, and the fixed line the **directrix**. The *eccentricity* is always denoted by *e*. When $e = 1$, the conic is a

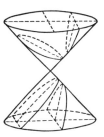

parabola; when $e < 1$, an ellipse; and when $e > 1$, a hyperbola. These are called conics, or **conic sections**, since they can always be gotten by taking plane sections of a conical surface (see DANDELIN). The **equation of a conic** can be given in various forms. If the eccentricity is *e*, the focus is at the pole, and the directrix perpendicular to the polar axis and at a distance *q* from the pole, the equation in polar coordinates is

$$\rho = (eq)/(1 + e\cos\theta).$$

This is equivalent to the following equation in Cartesian coordinates (the focus being at the origin and the directrix perpendicular to the *x*-axis at a distance *q* from the focus):

$$(1 - e^2)x^2 + 2e^2qx + y^2 = e^2q.$$

The general algebraic equation of the second degree in two variables always represents a conic (including here degenerate conics); *i.e.*, an ellipse, hyperbola, parabola, a straight line, a pair of straight lines, or a point, provided it is satisfied by any real points. See DISCRIMINANT —dis-

criminant of a quadratic equation in two variables, ELLIPSE, HYPERBOLA, PARABOLA.

acoustical, optical, or focal property of conics. See ELLIPSE—focal property of ellipse, HYPERBOLA—focal property of hyperbola, and PARABOLA —focal property of parabola.

central conics. Conics which have centers—ellipses and hyperbolas. See CENTER.

confocal conics. See CONFOCAL.

degenerate conic. A point, a straight line, or a pair of straight lines, which is a limiting form of a conic. *E.g.*, the parabola approaches a straight line, counted twice, as the plane, whose intersection with a conical surface defines the parabola, moves into a position in which it contains a single element of the conical surface, and the parabola approaches a pair of parallel lines as the vertex of the cone recedes infinitely far; the ellipse becomes a point when the cutting plane passes through the vertex of the cone but does not contain an element; the hyperbola becomes a pair of intersecting lines when the cutting plane contains the vertex of the conical surface. All these limiting cases can be obtained algebraically by variation of the parameters in their several equations. See DISCRIMINANT —discriminant of the general quadratic.

diameter of a conic. See DIAMETER.

focal chords of conics. See FOCAL.

similarly placed conics. Conics of the same type (both ellipses, both hyperbolas, or both parabolas) which have their corresponding axes parallel.

tangent to a general conic. (1) If the equation of the conic in *Cartesian coordinates* is $ax^2 + 2bxy + cy^2 + 2dx + 2ey + f = 0$, then the equation of the tangent at the point (x_1, y_1) is

$$ax_1x + b(xy_1 + x_1y) + cy_1y$$

$$+ d(x + x_1) + e(y + y_1) + f = 0.$$

(2) If the equation of the conic in *homogeneous* Cartesian coordinates is written

$$\sum_{i,j=1}^{3} a_{ij}x_ix_j = 0, \quad \text{where } a_{ij} = a_{ji},$$

then the equation of the tangent at the point (b_1, b_2, b_3) is

$$\sum_{i,j=1}^{3} a_{ij}b_ix_j = 0.$$

See COORDINATE —homogeneous coordinates.

CON′I-CAL, *adj.* **conical surface.** A surface which is the union of all lines that pass through a fixed point and intersect a fixed curve. The fixed point is the **vertex,** or **apex,** of the conical surface, the curve the **directrix,** and each of the lines is a **generator** or **generatrix.** Any homogeneous equation in rectangular Cartesian coordinates is the equation of a conical surface with vertex at the origin.

circular conical surface. A conical surface whose directrix is a circle and whose vertex is on the line perpendicular to the plane of the circle and passing through the center of the circle. If the vertex is at the origin and the directrix in a plane perpendicular to the z-axis, its equation in rectangular Cartesian coordinates is $x^2 + y^2 = k^2 z^2$.

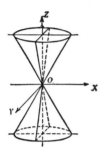

quadric conical surfaces. Conical surfaces whose directrices are conics.

CON′I-COID, *n.* An ellipsoid, hyperboloid, or paraboloid; usually does not refer to limiting (degenerate) cases.

CON′JU-GATE, *adj.* **complex conjugate of a matrix.** See MATRIX—complex conjugate of a matrix.

conjugate algebraic numbers. Any set of numbers that are roots of the same irreducible algebraic equation with rational coefficients, an equation of the form:

$$a_0 x^n + a_1 x^{n-1} + \cdots + a_n = 0.$$

E.g., the roots of $x^2 + x + 1 = 0$, which are $\frac{1}{2}(-1 + i\sqrt{3})$ and $\frac{1}{2}(-1 - i\sqrt{3})$, are conjugate algebraic numbers (in this case conjugate imaginary numbers).

conjugate angles. See ANGLE—conjugate angles.

conjugate arcs. Two nonoverlapping arcs whose union is a complete circle.

conjugate axis of a hyperbola. See HYPERBOLA.

conjugate complex numbers. See COMPLEX—conjugate complex numbers.

conjugate convex functions. See CONVEX—conjugate convex functions.

conjugate curves. Two curves each of which is a *Bertrand curve* with respect to the other. The only curves having more than one conjugate are plane curves and circular helices.

conjugate diameters. A diameter and the diameter which occurs among the parallel chords that define the given diameter. The conjugate diameters in a circle are perpendicular. The axes of an ellipse are conjugate diameters. But, in general, conjugate diameters are not perpendicular. See DIAMETER —diameter of a conic.

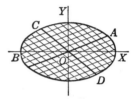

conjugate diameter of a diametral plane of a central quadric. The diameter which contains the centers of all sections of the central quadric by planes parallel to a given diameter. The diametral plane is likewise said to be conjugate to the diameter.

conjugate directions on a surface at a point. The directions of a pair of conjugate diameters of the *Dupin indicatrix* at an elliptic or hyperbolic point P of a surface S. There is a unique direction conjugate to any given direction on S through P, so that there are infinitely many pairs of conjugate directions on S at P. Two conjugate directions which are mutually perpendicular are necessarily principal directions. Conjugate directions are not defined at parabolic or planar points. The characteristic of the tangent plane to S, as the point of contact moves along a curve C on S, is the tangent to the surface in the direction conjugate to the direction of C. See below, conjugate system of curves on a surface.

conjugate dyads and dyadics. See DYAD.

conjugate elements and conjugate subgroups of a group. See TRANSFORM —transform of an element of a group.

conjugate elements of a determinant. Elements which are interchanged if the rows and columns of the determinant are interchanged; *e.g.,* the element in the second row and third column is the conjugate of the element in the third row and second column. In general, the elements a_{ij} and a_{ji} are conjugate elements, a_{ij} being the element in the ith row and jth column and a_{ji} the element in the jth row and ith column. See DETERMINANT.

conjugate harmonic functions. See HARMONIC —harmonic functions.

conjugate hyperboloids. Hyperboloids which with suitable choice of coordinate axes have equations

$$\frac{x^2}{a^2} + \frac{y^2}{b^2} - \frac{z^2}{c^2} = 1$$

and

$$-\frac{x^2}{a^2} - \frac{y^2}{b^2} + \frac{z^2}{c^2} = 1.$$

Any plane containing the common axis $z = 0$ cuts the two hyperboloids in conjugate hyperbolas. See HYPERBOLOID—hyperboloid of one sheet, hyperboloid of two sheets.

conjugate imaginaries. See COMPLEX—conjugate complex numbers.

conjugate points relative to a conic. Two points such that one of them lies on the line joining the points of contact of the two tangents drawn to the conic from the other; two points that are harmonic conjugates of the two points of intersection of the conic and the line drawn through the points; a point and any point on the polar of the point. *Tech.* If the conic is written in the form

$$\sum_{i,j=1}^{3} a_{ij} x_i x_j, \quad \text{where} \quad x_1, x_2, \text{and } x_3$$

are *homogeneous rectangular Cartesian coordinates* and $a_{ij} = a_{ji}$, then two points, (x_1, x_2, x_3) and (y_1, y_2, y_3), are conjugate points if, and only if,

$$\sum_{i,j=1}^{3} a_{ij} x_i y_{ij} = 0.$$

See below, harmonic conjugates with respect to two points.

conjugate quaternions. See QUATERNION.

conjugate radicals. (1) Conjugate binomial surds (see SURD). (2) Radicals that are conjugate algebraic numbers.

conjugate roots. See above, conjugate algebraic numbers.

conjugate ruled surfaces. See RULED.

conjugate space. If V is a *vector space* with scalars in a field F, then the **conjugate space** of V is the vector space V^* whose members are linear functions (functionals) with domain V and range contained in F. If V is finite-dimensional, then V and V^* have the same dimension and V is isomorphic to its second conjugate space $(V^*)^*$, the member. F_x of $(V^*)^*$ corresponding to the member x of V being defined by $F_x(f) = f(x)$ for all f in V^*. If N is a *normed vector space* (real or complex) and f is a continuous linear function (functional) with domain N and range in the set of scalars (real or complex), then there is a least number, called the **norm** of f and written $\|f\|$, such that $|f(x)| \le \|f\| \cdot \|x\|$ for each x of N. The set of all such functionals is a complete normed linear space, or *Banach space*, which is the **first conjugate space** of N. The first conjugate space of this space is the **second conjugate space** of N, etc. If N is finite-dimensional, then N and its second conjugate space are identical (*i.e.*, isometric). For any normed linear space N, N is isometric with a subspace of its second conjugate space (see REFLEXIVE —reflexive Banach space). If N is a Hilbert space with a complete orthonormal sequence $u_1, u_2, \cdots$, then the sequence of functions $f_n(x) = (x, u_n)$, $n = 1, 2, \cdots$, is a complete orthonormal sequence in the first conjugate space and the correspondence $\sum_1^\infty a_i u_i \leftrightarrow \sum_1^\infty \bar{a}_i f_i$ is an isometric correspondence between the two spaces. *Syn.* adjoint space; dual space.

conjugate subgroups. See ISOMORPHISM.

conjugate system of curves on a surface. Two one-parameter families of curves on a surface S such that through each point P of S there passes a unique curve of each family, and such that the directions of the tangents to these two curves at P are conjugate directions on S at P. See above, conjugate directions on a surface at a point. The parametric curves form a conjugate system if, and only if, $D' \equiv 0$ on S. See SURFACE —fundamental coefficients of a surface. The lines of curvature form a conjugate system, and the only orthogonal conjugate system.

harmonic conjugates with respect to two points. Any two points that separate the line through the two points internally and externally in the same numerical ratio; two points (the 3rd and 4th) which with the given two (the 1st and 2nd) have a *cross ratio* equal to -1 (see RATIO). If two points are harmonic conjugates with respect to two others, the latter two are harmonic conjugates with respect to the first two.

isogonal conjugate lines. See ISOGONAL—isogonal lines.

mean-conjugate curve on a surface. A curve C on a surface S such that C is tangent to a *mean-conjugate direction* on S at each point of C. See below, mean-conjugate directions on a surface.

mean-conjugate directions on a surface. Conjugate directions on the surface S at the point P of S such that the directions make equal angles with the lines of curvature of S at P. The mean-conjugate directions are real if the Gaussian curvature of S is positive at P, and the radius of normal curvature R of S in each of these two directions is the mean of the principal radii there: $R = (\rho_1 + \rho_2)/2$. See above, mean-conjugate curves on a surface.

method of conjugate directions. A generalization of the *method of conjugate gradients* for solving a system of n linear equations in n unknowns. In the method of conjugate directions, special restrictions on the conjugate directions to be used do not need to be specified.

method of conjugate gradients. An iterative method, terminating in n steps if there is no round-off error, for solving a system of n equations in n unknowns, $x = (x_1, x_2, \cdots, x_n)$. Starting with an initial estimate x_0 of the solution vector x, the correction steps are in directions that are conjugate to each other relative to the matrix of coefficients, and (to within this constraint) they are successively chosen to be in gradient directions relative to an associated quadratic function that assumes its minimum value 0 at the solution x of the original problem. The sets of residuals are mutually orthogonal. See above, conjugate points relative to a conic.

method of successive conjugates. In complex variable theory, an iterative method for the approximate evaluation of an analytic function that maps a given nearly circular domain conformally onto the interior of a circle. This mapping might be considered as the second step in a two-step process of mapping a given simply connected domain conformally on the interior of a circle, the mapping of the given domain onto a nearly circular domain having previously been attained through known functions or through a catalogue of conformal maps.

CON-JUNC'TION, *n.* **conjunction of propositions.** The proposition formed from two given propositions by connecting them with the word *and*. E.g., the conjunction of "Today is Wednesday" and "My name is Harry" is the proposition "Today is Wednesday *and* my name is Harry." The conjunction of propositions p and q is usually written as $p \wedge q$, or $p \cdot q$, and read "p *and* q." The conjunction of p and q is true if and only if both p and q are true. See DISJUNCTION.

CON-JUNC'TIVE, *adj.* **conjunctive transformation.** See TRANSFORMATION —conjunctive transformation.

CON-NECT'ED, *adj.* **arcwise-connected set.** A set such that each pair of its points can be joined by a curve all of whose points are in the set. *Syn.* path-connected set, pathwise-connected set. The set is **locally arcwise-connected** if each neighborhood of any point contains an arcwise-connected neighborhood of x.

connected graph. See GRAPH THEORY.

connected relation. A relation such that if $a \neq b$, then either a is related to b or b is related to a. The relation $a < b$ for real numbers is *connected*.

connected set. A set that cannot be separated into two sets U and V which have no points in common and which are such that no accumulation point of U belongs to V and no accumulation point of V belongs to U (see DISCONNECTED —disconnected set). The set of all rational numbers is not connected, since the set of rational numbers less than $\sqrt{5}$ and the set greater than $\sqrt{5}$ are both closed in the set of all rational numbers. An *arcwise-connected* set is connected, but a connected set need not be either *arcwise-connected* or *simply connected*. A **locally connected set** is a set S such that, for any point x of S and neighborhood U of x, there is a neighborhood V of x such that the intersection of S and V is connected and contained in U.

simply connected set. An arcwise-connected set such that any closed curve within it can be deformed continuously to a point of the set without leaving the set. For a plane set, this means the set is arcwise connected and no closed curve lying entirely within the set encloses a boundary point of the set. An arcwise-connected set that is not simply connected is **multiply connected**. See CONNECTIVITY —connectivity number.

CON'NEC-TIV'I-TY, *adj.* **connectivity number.** The connectivity number of a curve is 1 *plus* the maximum number of points that can be deleted without separating the curve into more than one piece (this is $2 - \chi$, where χ is the Euler characteristic). The connectivity number of a (connected) surface is 1 *plus* the largest number of closed cuts (or cuts joining points of previous cuts, or joining points of the boundary or a point of the boundary to a point of a previous cut, if the surface is not closed) which can be made without separating the surface. This is equal to $3 - \chi$ for a closed surface and to $2 - \chi$ for a surface with boundary curves. A *simply connected* curve or surface then has connectivity number 1; a curve or surface is **doubly connected**, **triply connected**, etc., according as its connectivity number is 2, 3, etc. The region between two concentric circles in a plane is *doubly* *connected*; the surface of a doughnut (a torus) is *triply connected*. In the above sense, the connectivity number of a connected *simplicial complex* (which may be a curve or a surface) is 1 *plus* the 1-*dimensional Betti number* (modulo 2). However, the connectivity number is sometimes defined to be equal to this Betti number.

CONNES, Alain (1947–   ). French mathematician awarded a Fields Medal (1983) for his contributions to the classification of von Neumann algebras and his results on the connections between operator algebras, foliations, and index theorems.

CO′NOID, *n.*   (1) A surface that is the union of all straight lines parallel to a given plane, intersecting a given line, and intersecting a given curve. (2) A paraboloid of revolution, a hyperboloid of revolution, or an ellipsoid of revolution. (3) The general paraboloid and hyperboloid, but not the general ellipsoid.
   right conoid.   A conoid for which the given plane and the given line are mutually orthogonal.

CON-SEC′U-TIVE, *adj.*   Following in order without jumping. *E.g.*, in the sum $1 + x + x^2 + x^3 + x^4$, the terms x and x^2 are **consecutive terms**; the sets $\{3, 4\}$ and $\{5, 6, 7\}$ are sets of **consecutive integers**; and $\{3, 5, 7, 9\}$ is a set of **consecutive odd integers**. Such a concept can not be applied to the rational numbers, since for no rational number x is there a first rational number larger than x. *Syn.* successive.
   consecutive angles and sides.   Two **consecutive angles** of a polygon are two angles with a common side; two **consecutive sides** are two sides with a common vertex.

CON′SE-QUENT, *n.*   (1) The second term of a ratio; the quantity to which the first term is compared, *i.e.*, the divisor. *E.g.*, in the ratio $\frac{2}{3}$, 3 is the *consequent*, and 2 the *antecedent*. (2) See IMPLICATION.

CON′SER-VA′TION, *n.* **conservation of energy.** See ENERGY.

CON-SER′VA-TIVE, *adj.* **conservative field of force.**   A force field such that the work done in displacing a particle from one position to another is independent of the path along which the particle is displaced. In a *conservative field* the work done in moving a particle around any closed path is zero. If the work done on the particle is

represented by a line integral,

$$\int_C F_x \, dx + F_y \, dy + F_z \, dz,$$

where F_x, F_y, and F_z are the Cartesian components of force in a *conservative field*, then the integrand is an exact differential. The gravitational and electrostatic fields of force are examples of conservative fields, whereas the magnetic field due to current flowing in a wire and fields of force involving frictional effects are nonconservative.

CON-SIGN′, *v.* **to consign goods, or any property.**   To send it to someone to sell, usually at a fixed fee, in contrast to selling on commission.

CON′SIGN-EE′, *n.*   A person to whom goods are consigned.

CON-SIGN′OR, *n.*   A person who sends goods to another for him to sell; a person who consigns goods.

CON-SIST′EN-CY, *n.* **consistency of systems of equations.**   The property possessed by a system of equations when there is at least one set of values of the variables that satisfies each equation, *i.e.*, the solution sets have one or more common points. If they are not satisfied by any one set of values of the variables, they are **inconsistent**. *E.g.*, the equations $x + y = 4$ and $x + y = 5$ are inconsistent; the equations $x + y = 4$ and $2x + 2y = 8$ are consistent, but are not independent (see INDEPENDENT); and the equations $x + y = 4$ and $x - y = 2$ are *consistent and independent*. The first pair of equations represents two parallel lines, the second represents two coincident lines, and the third represents two distinct lines intersecting in a point, the point whose coordinates are $(3, 1)$.
   consistency of linear equations.   A linear equation in two variables is the equation of a line in the plane. Therefore a single equation has an unlimited number of solutions. Two equations have a unique simultaneous solution if the lines they represent intersect and are not coincident; there is no solution if the lines are parallel and not coincident; there is an unlimited number of solutions if the lines are coincident. These correspond to the three cases of the following discussion. Consider the equations: $a_1 x + b_1 y = c_1$, $a_2 x + b_2 y = c_2$, where at least one of a_1, b_1 and at least one of a_2, b_2 is not zero. Multiply the first equation by b_2 and the second by b_1, then

subtract. This gives $(a_1b_2 - a_2b_1)x = b_2c_1 - b_1c_2$. Similarly, $(a_1b_2 - a_2b_1)y = a_1c_2 - a_2c_1$,

or

$$x\begin{vmatrix} a_1 & b_1 \\ a_2 & b_2 \end{vmatrix} = \begin{vmatrix} c_1 & b_1 \\ c_2 & b_2 \end{vmatrix},$$

and

$$y\begin{vmatrix} a_1 & b_1 \\ a_2 & b_2 \end{vmatrix} = \begin{vmatrix} a_1 & c_1 \\ a_2 & c_2 \end{vmatrix}.$$

Three cases follow: I. If the determinant of the coefficients

$$\begin{vmatrix} a_1 & b_1 \\ a_2 & b_2 \end{vmatrix}$$

is not zero, one can divide by it and secure unique values for x and y. The equations are then consistent and independent. The equations $2x - y = 1$ and $x + y = 3$ reduce in the above way to

$$3x = 4 \quad \text{and} \quad 3y = 5$$

and have the unique simultaneous solution $x = \frac{4}{3}$, $y = \frac{5}{3}$. II. If the determinant of the coefficients is zero and one of the determinants formed by replacing the coefficients of x (or of y) by the constant terms is not zero, there is no solution; *i.e.*, the equations are *inconsistent*. The equation $2x - y = 1$ and $4x - 2y = 3$ reduce to

$$0 \cdot x = 1 \quad \text{and} \quad 0 \cdot y = 2,$$

which have no solution. III. If all three determinants entering are zero, there results $0 \cdot x = 0$ and $0 \cdot y = 0$. The equations are then consistent but not independent. This is the situation for the equations $x - y = 1$ and $2x - 2y = 2$. An infinite number of pairs of values of x and y can be found that satisfy both of these equations. **A linear equation in three variables** is the equation of a plane in space. Therefore a single equation has an unlimited number of solutions. Two equations either represent parallel planes and have no common solution or else represent planes which intersect in a line or coincide and the equations have an unlimited number of solutions. Eliminating the variables, two at a time, from the equations

$$a_1 x + b_1 y + c_1 z = d_1,$$

$$a_2 x + b_2 y + c_2 z = d_2,$$

$$a_3 x + b_3 y + c_3 z = d_3,$$

gives $Dx = K_1$, $Dy = K_2$, and $Dz = K_3$, where K_1, K_2, and K_3 are the determinants resulting from substituting the d's in the determinant of the coefficients, D, in place of the a's, b's, and c's, respectively. Three cases arise: I. If $D \neq 0$, it can be divided out and a unique set of values for x, y, and z obtained; *i.e.*, the three planes representing the three equations then intersect at a point and the equations are *consistent* (and also independent). II. If $D = 0$ and at least one of K_1, K_2, and K_2 is not zero, there is no solution; the three planes do not have any point in common and the three equations are *inconsistent*. III. If $D = 0$ and $K_1 = K_2 = K_3 = 0$, three cases arise: *a*). Some second-order determinant in D is not zero, in which case the equations have infinitely many points in common; the planes (the loci of the equations) intersect in a line and the equations are *consistent*. *b*). Every second-order minor in D is zero and a second-order minor in K_1, K_2, or K_3 is not zero. The planes are then parallel but at least one pair is not coincide; the equations are *inconsistent*. *c*). All the second-order minors in D, K_1, K_2, and K_3 are zero. The three planes then coincide and the equations are *consistent* (but not independent). The general situation of **m linear equations in n unknowns** is best handled by consideration of matrix rank (see MATRIX—rank of a matrix): the equations are consistent if and only if the rank of the matrix of the coefficients is equal to the rank of the augmented matrix. If the constant terms in a system of linear equations are all zero (the equations are homogeneous), then the equations have a trivial solution (each variable equal to zero). For **n homogeneous linear equations in m unknowns**: (1) If $n < m$, the equations have a nontrivial solution (not all variables zero). (2) If $n = m$, the equations have a nontrivial solution if, and only if, the determinant of the coefficients is equal to zero. (3) If $n > m$, the equations have a nontrivial solution if, and only if, the *rank* of the *matrix* of the coefficients is less than m. These are simply the special case of the results for n linear equations in m variables when the constant terms are all zero.

CON·SIST′ENT, *adj*. **consistent assumptions, hypotheses, postulates.** Assumptions, hypotheses, postulates that do not contradict each other.

  consistent estimator. An estimator Φ for a parameter ϕ such that, for each positive number ϵ, the probability that $|\Phi(X_1, \cdots, X_n) - \phi| > \epsilon$ approaches 0 as $n \to \infty$. If an estimator is *asymptotically unbiased*, *i.e.*, the expected value of $\Phi(X_1, \cdots, X_n)$ approaches ϕ as $n \to \infty$, and if the *variance* of $\Phi(X_1, \cdots, X_n)$ approaches 0 as $n \to \infty$, then Φ is a consistent estimator for ϕ. *E.g.*, if $\{X_1, \cdots, X_n\}$ is a random sample from a normal population and

$$\Phi(X_1, X_2, \cdots, X_n) = \frac{1}{n} \sum_{i=1}^{n} (X_i - \bar{X})^2,$$

where $\bar{X} = \sum_{i=1}^{n} X_i/n$, then $E(\Phi) = (n-1)\sigma^2/n$, which approaches σ^2 as $n \to \infty$, and the variance of Φ is $2(n-1)\sigma^4/n^2$, which approaches 0 as $n \to \infty$. Thus Φ is a consistent estimator for σ^2. See UNBIASED —unbiased estimator, VARIANCE.

consistent system of equations. See CONSISTENCY.

CON'STANT, *adj., n.* A particular object or number. A symbol that represents the same object throughout a certain discussion or sequence of mathematical operations. A variable that can assume only one value. See VARIABLE.

absolute constant. A constant that never changes in value, such as numbers in arithmetic.

arbitrary constant. A symbol that stands for an unspecified constant. *E.g.,* in the quadratic equation $ax^2 + bx + c = 0$, the symbols a, b, and c are arbitrary constants. See below, constant of integration.

constant function. A function whose range has only one member; a function f for which there is an object a such that $f(x) = a$ for all x in the domain of f.

constant of integration. An arbitrary constant that must be added to any function arising from integration to obtain all the antiderivatives. The integral, $\int 3x^2\,dx$, can have any of the values $x^3 + c$, where c is a constant, because the derivative of a constant is zero; further, it follows from the mean-value theorem that there are no other values for the integral. See MEAN—mean value theorems for derivatives.

constant of proportionality or variation. See VARIATION —direct variation.

constant speed and velocity. An object is said to have **constant speed** if it passes over equal distances in equal intervals of time (although the object need not move in a straight line). It has **constant velocity** if it passes over equal distances *in the same direction* in equal intervals of time (this means that the *instantaneous velocity* is the same *vector* at each point of the path; see VELOCITY). Constant velocity is also sometimes called **uniform (rectilinear) velocity** and **uniform motion** (although uniform motion is sometimes used in such a sense as uniform circular motion, meaning motion around a circle with constant speed).

constant term in an equation or function. A term which does not contain a variable. *Syn.* absolute term.

essential constant. A set of essential constants in an equation is a set of *arbitrary constants* which: (1) cannot be replaced by a smaller number (changing the form of the equation if desired) so as to have a new equation which represents essentially the same family of curves, or (2) are equal in number to the number of points needed to determine a unique member of the family of curves represented by the equation, or (3) are arbitrary constants in an equation $y = f(x)$ for which the number of arbitrary constants is equal to the minimum order of a differential equation which has $y = f(x)$ as a solution. The linear equation $y = Ax + B$ defines a family of straight lines; it has 2 essential constants, since 2 points (not in a vertical line) determine a unique line of the family, and $y = Ax + B$ is a solution of the differential equation $y'' = 0$, which is of order 2. The equation $ax + by + c = 0$ does not have 3 essential constants, since 2 points determine a line of the family of lines it represents; also, it represents the same family of curves as $y = Ax + B$, except for the lines $x = $ constant. The equation $y^2 + bxy - abx - (a - c)y - ac = 0$ has 4 arbitrary constants; these are not essential, since the equation can be factored as $(y - a)(y + bx + c) = 0$ and has the same family of curves as the equation $y = mx + b$. The **number of essential constants** in an equation is the number of essential constants to which the arbitrary constants can be reduced. *E.g.,* the number of essential constants in $y^2 + bxy - abx - (a - c)y - ac = 0$ is 2. The constants $A_1, \cdots, A_n$ in the equation

$$y = A_1 u_1(x) + A_2 u_2(x) + \cdots + A_n u_n(x)$$

are essential if and only if the functions $u_1, \cdots, u_n$ are *linearly independent*.

gravitational constant. See GRAVITATION—law of universal gravitation.

Lamé's constants. See LAMÉ—Lamé's constants.

CON-STRAIN'ING, *p., adj.* **constraining forces (constraints).** (1) Those forces that tend to prevent a particle's remaining at rest or moving at a uniform velocity in a straight line (according to Newton's first law of motion). (2) Those forces that are exerted perpendicularly to the direction of motion of a particle.

CON-STRUCT', *v.* To draw a figure so that it meets certain requirements; usually consists of drawing the figure and proving that it meets the requirements. *E.g.*, to construct a line perpendicular to another line, or to construct a triangle having three given sides.

CON-STRUC'TION, *n.* (1) The process of drawing a figure that will satisfy certain given conditions. See CONSTRUCT. (2) Construction in proving a theorem; drawing the figure indicated by the theorem and adding to the figure any additional parts that are needed in the proof. Such "additional" lines, points, etc., are usually called **construction lines, points,** etc.

CON-STRUCT'IVE, *adj.* **constructive mathematics.** The constructive method has its roots in the intuitionism of Brouwer, but there are fundamental differences. There also are differences between various concepts of constructivism, but each of the following is used by some or all constructivists in some sense. (i) Only finitely representable objects are to be used and operations on such objects must be executed in a finite number of steps. (ii) It is not permissible to define x by using a collection X of which x is a member (*e.g.*, it is not permissible to use the set of all upper bounds of a set to define the least upper bound). (iii) A proof of the existence of x must explain how the construction of x is to be carried out; *e.g.*, the existence of x cannot be proved by showing that the assumption "x does not exist" leads to a contradiction. (iv) Infinite sequences may be given constructively by giving some process to determine the nth term; a real number can be specified by giving a sequence of rationals and a specific description of a Cauchy modulus.

CON'TACT, *n.* **chord of contact.** See CHORD.
   order of contact. See ORDER—order of contact of two curves.
   point of contact. See TANGENT—tangent to a curve.

CON'TENT, *n.* **content of a set of points.** The **exterior content** (or **outer content**) of a set of points E is the greatest lower bound of the sums of the lengths of a finite number of intervals (open or closed) such that each point of E is in one of the intervals, for all such sets of intervals. The **interior content** (or **inner content**) is the least upper bound of the sums of the lengths of a finite number of nonoverlapping intervals such that each interval is completely contained in E, for all such sets of intervals; or (equivalently) the difference between the length of an interval I containing E and the exterior content of the complement of E in I. Also called the **exterior**

Jordan content and **interior Jordan content.** If the exterior content is equal to the interior content, their common value is the **(Jordan) content.** If the exterior content is zero, then the interior content is also, and the set is said to have **(Jordan) content zero.** The set of rational numbers in $(0, 1)$ has exterior content 1 and interior content zero; the set $(1, \frac{1}{2}, \frac{1}{3}, \frac{1}{4}, \cdots)$ has content zero. This definition is for sets of points on a line. A similar definition holds for sets in the plane, or in n-dimensional Euclidean space. See VOLUME.

CON-TIN'GENCE, *n.* **angle of contingence.** The angle between the positive directions of the tangents to a given plane curve at two given points of the curve.
   angle of geodesic contingence. For two points P_1 and P_2 of a curve C on a surface, the angle of geodesic contingence is the angle of intersection of the geodesics tangent to C at P_1 and P_2. See above, angle of contingence.

CON-TIN'GEN-CY, *n.* **contingency table.** (*Statistics*). If each member of a population can be classified according to each of two criteria and the numbers of categories are p and q, respectively, then there are pq distinct classifications and a table with pq cells showing in each cell the number of members of the population in that classification is a p by q or $p \times q$ contingency table. This concept can be extended to cases for which there are more than two criteria. Following is a **two-by-three contingency table** with six cells resulting from

	Male	Female
Favor	234	195
Opposed	108	124
Undecided	58	81

classifying a set of 800 persons according to sex and according to opinion on a certain political question.

CON-TIN'GENT, *adj.* **contingent annuity and life insurance.** See ANNUITY, INSURANCE —life insurance.

CON-TIN'U-A'TION, *adj., n.* **analytic continuation of an analytic function of a complex variable.** See ANALYTIC —analytic continuation.
   continuation notation. Three center dots or dashes following a few indicated terms. In case

there is an infinite number of terms, the most common usage is to indicate a few terms at the beginning of the set, follow these with three center dots, write the general term, and add three center dots as follows:

$$1 + x + x^2 + \cdots + x^n + \cdots.$$

continuation of sign in a polynomial. Repetition of the same algebraic sign before successive terms.

CON-TIN′UED, *adj.* **continued equality.** See EQUALITY.

continued fraction. See FRACTION —continued fraction.

continued product. A product of an infinite number of factors, or a product such as $(2 \times 3) \times 4$ of more than two factors; denoted by Π, that is, capital pi, with appropriate indices.

$$E.g., \left(\tfrac{1}{2}\right)\left(\tfrac{2}{3}\right)\left(\tfrac{3}{4}\right) \cdots [n/(n+1)] \cdots$$

$$= \prod_{n=1}^{\infty} [n/(n+1)]$$

is a continued product.

CON′TI-NU′I-TY, *n.* The property of being continuous.

axiom of continuity. See AXIOM.

equation of continuity. A fundamental equation of fluid mechanics, namely, $d\rho/dt + \rho\nabla \cdot \eta = 0$, where ρ is the density of the fluid and η is the velocity vector. A more general equation takes account of sources and sinks at which fluid is created and destroyed.

principle of continuity. See AXIOM—axiom of continuity.

CON-TIN′U-OUS, *adj.* **absolutely continuous function.** With respect to an interval I, an **absolutely continuous function** is a function f whose domain contains I and which has the property that, for any positive number ϵ, there is a positive number η such that if (a_1, b_1), $(a_2, b_2), \cdots, (a_n, b_n)$ is any finite set of nonoverlapping intervals such that the sum of the lengths of the intervals is less than η, then

$$\sum_{i=1}^{n} |f(a_i) - f(b_i)| < \epsilon.$$

The definition remains equivalent to this if it is changed to allow a countable number of intervals. An absolutely continuous function is *continuous* and of *bounded variation*. A function f is

absolutely continuous on a bounded closed interval $[a, b]$ if and only if there is a function g such that

$$f(x) = \int_a^x g(t)\, dt$$

if x is in $[a, b]$, where the integral may be a Lebesgue integral.

absolutely continuous measure. See RADON —Radon-Nikodým theorem.

continuous annuity. An annuity payable continuously. Such an annuity cannot occur, but has theoretical value. Formulas for this sort of annuity are limiting forms of the formulas for noncontinuous annuities, when the number of payments per year increases without limit while the nominal rate and annual rental remain fixed. The results differ very little from annuities having a very large number of payments per year. Approximate present values for a single life continuous annuity of one dollar is that of a single life annuity payable annually at the end of the year plus $\frac{1}{2}$ of a dollar; or that of a single life annuity payable annually at the beginning of the year minus $\frac{1}{2}$ of a dollar.

continuous conversion of compound interest. See CONVERSION.

continuous function. A function f whose domain and range are topological spaces is **continuous at a point** x if for any neighborhood W of $f(x)$ there is a neighborhood U of x such that W contains all points $f(u)$ for which u is in U. Such a function f is **continuous** if it is continuous at each point of D. It then can be proved that f is continuous if and only if the inverse of each open set in R is open in D (or if and only if the inverse of each closed set in R is closed in D) [see OPEN—open mapping]. For a function f whose domain and range are sets of real or complex numbers, this means that f is continuous at x_0 if $f(x)$ can be made as nearly equal to $f(x_0)$ as one might wish by restricting x to be sufficiently close to x_0; *i.e.*, for any positive number ϵ there is a positive number δ such that $|f(x) - f(x_0)| < \epsilon$ if $|x - x_0| < \delta$ and x is in the domain of f. This implies that f is continuous at x_0 if x_0 is an isolated point of the domain of f (*i.e.*, there is a $\delta > 0$ such that there is no x in the domain of f for which $x \neq x_0$ and $|x - x_0| < \delta$), and that f is continuous at a non-isolated point x_0 if and only if $\lim_{x \to x_0} f(x) = x_0$. A function f is **continuous on a set** S if it is continuous at each point of S. All polynomial, trigonometric, exponential, and logarithmic functions are continuous at all points of their domains. All functions are continuous at all points at which they are differentiable. A function

$f(x, y)$ of two real variables x and y, or of a point (x, y), is continuous at (a, b) if for each positive number ϵ there is a positive number δ such that $|f(x, y) - f(a, b)| < \epsilon$ if the distance between (x, y) and (a, b) is less than δ and (x, y) is in the domain of f. If (a, b) is not an isolated point of the domain, this is equivalent to requiring that

$$\lim_{(x, y) \to (a, b)} f(x, y) = f(a, b).$$

See DISCONTINUITY, UNIFORM —uniform continuity.

continuous game. See GAME.

continuous on the left or right. A real-valued function f is **continuous on the right** at a point x_0 if for each $\epsilon > 0$ there is a $\delta > 0$ such that $|f(x) - f(x_0)| < \epsilon$ if $x_0 < x < x_0 + \delta$, and to **be continuous on the left** if for any $\epsilon > 0$ there is a $\delta > 0$ such that $|f(x) - f(x_0)| < \epsilon$ if $x_0 - \delta < x < x_0$. A function is continuous on the right (or left) on an interval (a, b) if it is continuous on the right (or left) at each point of (a, b). See LIMIT—limit on the right.

continuous in the neighborhood of a point. A function is *continuous in the neighborhood of a point* if there exists a neighborhood of the point such that the function is continuous at each point of the neighborhood. Thus $f(x_1, x_2, \cdots, x_n)$ is continuous in the neighborhood of $(a_1, \cdots, a_n)$ if there exists a positive number ϵ such that f is continuous at $(x_1, \cdots, x_n)$ if $|x_i - a_i| < \epsilon$ for each i, or if

$$\left[\sum_{i=1}^{n} |x_i - a_i|^2 \right]^{1/2} < \epsilon.$$

continuous random variable. See RANDOM—random variable.

continuous surface in a given region. The graph of a continuous function of two variables; the locus of the points whose rectangular coordinates satisfy an equation of the form $z = f(x, y)$, where f is a continuous function of x and y in the region of the (x, y)-plane which is the projection of the surface on that plane. E.g., a sphere about the origin is a continuous surface, for $z = \sqrt{r^2 - (x^2 + y^2)}$ is a continuous function on, and within, the circle $x^2 + y^2 = r^2$. To determine the entire sphere, both signs of the radical must be considered. Thought of in this way, the sphere is a multiple (two) valued surface.

continuous transformation. See above, continuous function.

piecewise-continuous function. See PIECEWISE.

semicontinuous function. If for any arbitrary positive number ϵ a real-valued function f satisfies the relation $f(x) < f(x_0) + \epsilon$ for all x in some neighborhood of x_0, then f is **upper semicontinuous** at x_0; if $f(x) > f(x_0) - \epsilon$ for all x in some neighborhood of x_0, then f is **lower semicontinuous** at x_0. Equivalent conditions are, respectively, that the *limit superior* of $f(x)$ as $x \to x_0$ be $\leq f(x_0)$ and that the limit inferior be $\geq f(x_0)$. A function is upper semicontinuous (or lower semicontinuous) on an interval or region R if, and only if, it is so at each point of R. The function defined by $f(x) = \sin x$ if $x \neq 0$ and $f(0) = 1$ is upper semicontinuous, but not lower semicontinuous, at $x = 0$.

CON-TIN'U-UM, *n.* [*pl.* **continua** or **continuums**]. A *compact connected* set. It is usually required that the set contain at least two points, which implies that it contains an infinite number of points. The set of all real numbers (rational and irrational) is called the **continuum of real numbers**. Any closed interval of real numbers is a continuum. A continuum is topologically equivalent to a closed interval of real numbers if and only if it does not contain more than two noncut points (see CUT).

continuum hypothesis. See HYPOTHESIS—continuum hypothesis.

continuum of real numbers. The totality of rational and irrational real numbers.

CON'TOUR, *adj.* **contour integral.** For a complex-valued function f of complex numbers z and a curve C joining points p and q in the complex plane (or on a Riemann surface), let $z_0 = p, z_1, \cdots, z_n = q$ be $n + 1$ arbitrary points on C which separate C into n consecutive segments, ζ_i be a point on the closed segment of C which joins z_{i-1} to z_i, and δ be the largest of the numbers $|z_i - z_{i-1}|$. Then the contour integral

$$\int_p^q f(z)\, dz$$

is the limit of $\sum_{i=1}^{n} f(\zeta_i)(z_i - z_{i-1})$ as δ approaches zero, if this limit exists. If f is continuous on C and C is *rectifiable*, this contour integral exists; if it is also true that F is a function such that $dF(z)/dz = f(z)$ at each point of C, then $\int_p^q f(z)\, dz = F(q) - F(p)$. With suitable restrictions on the nature of C, this contour integral can be evaluated as either of the line integrals

$$\int f(z) z'(t)\, dt,$$

$$\int (u\, dx - v\, dy) + i \int (v\, dx + u\, dy),$$

where $z = z(t)$ is an equation for C and

$$f(z) = u(x, y) + iv(x, y)$$

with $z = x + iy$ and $u(x, y)$ and $v(x, y)$ real. See CAUCHY—Cauchy's integral formula, Cauchy's integral theorem.

    contour lines. (1) Projections on a plane of all the sections of a surface by planes parallel to this given plane and equidistant apart; (2) lines on a map which pass through points of equal elevation. Useful in showing the rate of ascent of the surface, since the contour lines are thicker where the surface rises more steeply. *Syn.* level lines.

CON-TRAC′TION, *n.* Let T be a mapping of a metric space M into itself for which there is a number θ such that

$$\text{dist}[T(x), T(y)] \le \theta \, \text{dist}(x, y)$$

if x and y are in M. Then T is a **contraction**, a **nonexpansive mapping**, or a **Lipschitz mapping** (and to satisfy a **Lipschitz condition**) according as $0 \le \theta < 1$, $\theta = 1$, or $\theta > 0$. See BANACH—Banach fixed-point theorem.

    contraction of a tensor. The operation of putting one contravariant index equal to a covariant index and then summing with respect to that index. The resultant tensor is called the **contracted tensor**.

CON′TRA-DIC′TION, *n.* **law of contradiction.** The principle of logic which states that a proposition and its negation cannot both be true; *i.e.*, a proposition can not be both true and false. *E.g.*, for no number x are both the statements "$x^2 = 4$" and "$x^2 \ne 4$" true. This also is called the **law of the excluded middle**, which states that a proposition is true, or its negation is true, but not both are true. See BROUWER, CONSTRUCTIVE —constructive mathematics (iii), DICHOTOMY.

    proof by contradiction. Same as REDUCTIO AD ABSURDUM PROOF and INDIRECT PROOF. See PROOF—direct and indirect proofs.

CON′TRA-POS′I-TIVE, *n.* **contrapositive of an implication.** The implication which results from replacing the antecedent by the negation of the consequent and the consequent by the negation of the antecedent. *E.g.*, the contrapositive of "If x is divisible by 4, then x is divisible by 2" is "If x is not divisible by 2, then x is not divisible by 4." An implication and its contrapositive are *equivalent*—they are either both true or both false. The contrapositive of an implication is the *converse* of the *inverse* (or the *inverse* of the *converse*) of the implication.

CON′-TRA-VA′RI-ANT, *adj.* **contravariant derivative of a tensor.** The *contravariant derivative* of a tensor $t_{b_1 \cdots b_q}^{a_1 \cdots a_p}$ is the tensor

$$t_{b_1 \cdots b_q}^{a_1 \cdots a_p, j} = g^{j\sigma} t_{b_1 \cdots b_q, \sigma}^{a_1 \cdots a_p},$$

where the *summation covention* applies, g^{ij} is $1/g$ times the cofactor of g_{ij} in the determinant $g = \{g_{ij}\}$, and $t_{b_1 \cdots b_q, \sigma}^{a_1 \cdots a_p}$ is the *covariant derivative*. See COVARIANT—covariant derivative of a tensor, CHRISTOFFEL —Christoffel symbols.

    contravariant functor, indices, tensor, vector field. See FUNCTOR, TENSOR, TENSOR—contravariant tensor.

CON-TROL′, *adj., n.* **control chart.** A graph of the results of sampling the product of a process; usually consists of a horizontal line indicating expected mean value of some characteristic of quality and two lines on either side indicating the allowable extent of sampling and/or random production deviations. Usually used for control of quality of production.

    control component. In a computing machine, any component that is used in manual operation, for starting, testing, etc.

    control group. In estimating the effect of a given factor, it may be necessary to compare the result with another situation in which the tested factor is absent (or held constant). The *control group* is that sample in which the factor is absent (or held constant).

    quality control. A statistical method of testing the output of a production process in order to detect major, nonchance variations in quality of output.

    statistical control. A state of *statistical control* exists if a process of obtaining data under essentially the same conditions is such that the variations in the values of the data are random and cannot be attributed to any assignable causes, and the mean values of the subgroups show no trend. Values that conform to what would be expected under a random sampling scheme from a hypothetical normal population also are frequently regarded as characterizing a state of statistical control.

CON-VERGE′, *v.* To draw near to. (1) A series is said to converge when the sum of the first n of its terms approaches a limit as n increases without bound (see LIMIT). (2) A curve is said to converge to its asymptote, or to a point, when the distance from the curve to the asymptote, or point, approaches zero; *e.g.*, the polar spiral, $r = 1/\theta$, converges to the origin; the curve $xy = 1$ converges to the x-axis as x increases and to the y-axis as y increases. (3) A variable is said to converge to its limit. See various headings under CONVERGENCE.

CON-VER′GENCE, *n.* **absolute convergence of an infinite product.** See PRODUCT—infinite product.

absolute convergence of an infinite series. The property that the sum of the absolute values of the terms of the series form a convergent series. Such a series is said to **converge absolutely** and to be **absolutely convergent**; $1 - \frac{1}{2} + \frac{1}{2}^2 - \frac{1}{2}^3 + \cdots + (-1)^{n-1}\frac{1}{2}^{n-1} + \cdots$ is absolutely convergent. See SUM—sum of an infinite series; and below, conditional convergence.

circle of convergence. For a power series, $c_0 + c_1(z - a) + c_2(z - a)^2 + \cdots + c_n(z - a)^n + \cdots$, there is an R such that the series converges (absolutely) if $|z - a| < R$ and diverges if $|z - a| > R$. The circle of radius R with center at a in the complex plane is the **circle of convergence** (its equation is $|z - a| = R$); R is the **radius of convergence** (R may be zero or infinite). The series converges uniformly in any circle with center at a and radius less than R. The series may either converge or diverge on the circumference of the circle. *E.g.*,

$$\sum_{n=1}^{\infty} (3z)^n / n$$

converges absolutely within the circle whose radius is $\frac{1}{3}$ and whose center is the origin, and diverges outside this circle. It converges for $z = -\frac{1}{3}$, but diverges for $z = +\frac{1}{3}$. See below, interval of convergence.

conditional convergence. An infinite series is *conditionally convergent* if it is convergent and there is another series which is divergent and which is such that each term of each series is also a term of the other series (the second series is said to be derived from the first by a *rearrangement* of terms); *i.e.*, an infinite series is *conditionally convergent* if its convergence depends on the order in which the terms are written. A convergent series is conditionally convergent if and only if it is not absolutely convergent. *E.g.*, the series $1 - \frac{1}{2} + \frac{1}{3} - \frac{1}{4} + \cdots$ is conditionally convergent because it converges and the series $1 + \frac{1}{2} + \frac{1}{3} + \cdots$ diverges.

convergence of an infinite product. See PRODUCT—infinite product.

convergence of an infinite sequence. See SEQUENCE—limit of a sequence.

convergence of an infinite series. See SUM—sum of an infinite series.

convergence of an integral. The existence of an *improper integral*. *E.g.*, the integral $\int_2^{\infty} (1/x^2)\, dx$ is convergent and equals $\frac{1}{2}$, since

$$\int_2^{y} (1/x^2)\, dx = -1/y + \frac{1}{2}$$

approaches $\frac{1}{2}$ as y increases without bound.

convergence in the mean. A sequence of functions f_n is said to **converge in the mean of order p** to F on the interval or region Ω if

$$\lim_{n \to \infty} \int_{\Omega} |F(x) - f_n(x)|^p\, dx = 0.$$

When the term **convergence in the mean** is used without qualification, it is sometimes understood to mean "convergence in the mean of order two" and sometimes "convergence in the mean of order one."

convergence in measure. A sequence $\{f_n\}$ of measurable functions is said to **converge in measure** to F on the set S if, for any pair (ϵ, η) of positive numbers, there is a number N such that the measure of E_n is less than η when $n > N$, where E_n is the set of all x for which

$$|F(x) - f_n(x)| > \epsilon.$$

If S is of finite measure, then a sequence $\{f_n\}$ of measurable functions converges in measure to F if

$$\lim_{n \to \infty} f_n(x) = F(x)$$

for all x except a set of measure zero.

convergence in probability. See PROBABILITY.

interval of convergence. A power series,

$$c_0 + c_1(x - a) + c_2(x - a)^2 + \cdots$$

$$+ c_n(x - a)^n + \cdots,$$

either converges for all values of x, or there is a

number R such that the series converges if $|x - a| < R$ and diverges if $|x - a| > R$. The interval $(a - R, a + R)$ is the *interval of convergence* (R may be zero). The series converges *absolutely* if $|x - a| < R$ and converges *uniformly* in any interval (A, B) with $a - R < A \leq B < a + R$. See ABEL—Abel's theorem on power series; and above, circle of convergence.

tests for convergence of an infinite series. See ABEL, ALTERNATING, CAUCHY—Cauchy's integral test for convergence of an infinite series, COMPARISON, DIRICHLET, KUMMER, NECESSARY—necessary condition for convergence, RATIO, ROOT—root test.

uniform convergence of a series. An infinite series whose terms are functions with domain D is **uniformly convergent** on D if the numerical value of the remainder after the first n terms is as small as desired *throughout* D for n greater than a sufficiently large chosen number. *Tech.* If the sum of the first n terms of a series is $s_n(x)$, the series converges uniformly to $f(x)$ on D if for arbitrary positive ϵ there exists an N (dependent on ϵ) such that

$$|f(x) - s_n(x)| < \epsilon$$

for all n greater than N and all x in the set D. Equivalently, the series converges uniformly on D if, and only if, for arbitrary positive ϵ, there exists an N (dependent on ϵ) such that $|s_{n+p}(x) - s_n(x)| < \epsilon$ for all $n > N$, for all positive p, and for all x in the set D. E.g., the series

$$1 + x/2 + (x/2)^2 + \cdots + (x/2)^{n-1} + \cdots$$

converges uniformly for x in any closed interval contained in the interval $(-2, 2)$; but does not converge uniformly for $-2 < x < 2$, since the absolute value of the difference of

$$f(x) = 1/(1 - x/2) \quad \text{and}$$

$$s_n(x) = \left[1 - (x/2)^n\right] / (1 - x/2)$$

is $|(x/2)^n / (1 - x/2)|$, which (for any fixed n) becomes infinite as x approaches 2. See ABEL, DINI, DIRICHLET, WEIERSTRASS for tests of uniform convergence.

uniform convergence of a set of functions. See UNIFORM.

CON-VER'GENT, *adj., n.* Possessing the property of convergence. See SEQUENCE—limit of a sequence, SUM—sum of an infinite series, and various headings under CONVERGENCE.

convergent of a continued fraction. The fraction terminated at one of the quotients. See FRACTION—continued fraction.

permanently convergent series. Series which are convergent for all values of the variable, or variables, involved in its terms; *e.g.*, the exponential series, $1 + x + x^2/2! + x^3/3! + \cdots$, is equal to e^x for all values of x; hence the series is *permanently* convergent.

CON'VERSE, *n.* **converse of a theorem (or implication).** The theorem (or implication) resulting from interchanging the hypothesis and conclusion. If only a part of the conclusion makes up the new hypothesis, or only a part of the old hypothesis makes up the new conclusion, the new statement is sometimes spoken of as *a* converse of the old. E.g., the converse of "If x is divisible by 4, then x is divisible by 2" is the false statement "If x is divisible by 2, then x is divisible by 4." If an implication is true, its converse may be either true or false. If an implication $p \to q$ and its converse $q \to p$ are both true, then the *equivalence* $p \leftrightarrow q$ is true. See INVERSE—inverse of an implication.

CON-VER'SION, *adj., n.* **continuous conversion of compound interest.** Finding the limit of the amount, at the given rate of interest, as the length of the period approaches zero. For one year, this is the product of the initial principal and $\lim_{m \to \infty}(1 + j/m)^m$, where j is the fixed nominal rate and m the number of interest periods per year. This limit is e^j. See e.

conversion interval or period. See INTEREST.

conversion tables. Tables such as those giving the insurance premiums (annual or single), at various rates of interest, which are equivalent to a given annuity.

frequency of conversion of compound interest. The number of times a year that interest is compounded.

CON'VEX, *adj.* **convex curve in a plane.** A curve such that any straight line cutting the curve cuts it in just two points.

conjugate convex functions. If the function f with $f(0) = 0$ is *strictly increasing* for $x \geq 0$, and g is its *inverse*, then the convex functions $F(x) = \int_0^x f(t)\, dt$ and $G(y) = \int_0^y g(t)\, dt$ are **conjugate**. More generally, for a convex function

$$F(x_1, x_2, \cdots, x_n) = F(\mathbf{x})$$

defined in a domain D, the conjugate convex function is defined by

$$G(y_1, y_2, \cdots, y_n) = \text{l.u.b.}\left[\sum_{t=1}^{n} x_i y_i - F(\mathbf{x})\right]$$

for $\mathbf{x}$ in D. See YOUNG, W. H.

convex function. A real-valued function f whose domain contains an interval I is **convex on** I if, whenever $a < b < c$ and the numbers a, b and c are in I, we have $f(b) \leq l(x)$, where l is the linear function coinciding with f at a and b. *I.e.*, f is convex on I if and only if each chord joining two points of the graph of f lies on or above the graph of f. A necessary and sufficient condition that f be convex on I is that for any points x_1 and x_2 in I and any number λ with $0 < \lambda < 1$, we have

$$f[\lambda x_1 + (1 - \lambda)x_2] \leq \lambda f(x_1) + (1 - \lambda)f(x_2).$$

A convex function is necessarily continuous; but see below, convex in the sense of Jensen. If the second derivative of f is continuous on I, then f is convex on I if and only if $f''(x) \geq 0$ at each point of I.

convex in the sense of Jensen. A real-valued function f whose domain contains the interval I is **convex in the sense of Jensen** provided that, for any points x_1 and x_2 in I, we have

$$f\left(\frac{x_1 + x_2}{2}\right) \leq \frac{1}{2}[f(x_1) + f(x_2)].$$

A function convex in the sense of Jensen is not necessarily continuous, but, if such a function is bounded in any subinterval of I, it is necessarily continuous in I. See above, convex function.

convex linear combination. See LINEAR —linear combination.

convex polygon and polyhedron. See POLYGON, POLYHEDRON.

convex sequence. A sequence of numbers $\{a_1, a_2, a_3, \cdots\}$ is **convex** if $a_{i+1} \leq \frac{1}{2}(a_i + a_{i+2})$ for all i [or for $1 \leq i \leq n - 2$ if the sequence is the finite sequence $\{a_1, a_2, \cdots, a_n\}$]. If the inequality is reversed, the sequence is **concave**.

convex set. A set that contains the line segment joining any two of its points; in a *vector space*, a set such that $rx + (1 - r)y$ is in the set for $0 < r < 1$ if x and y are in the set. A set is **locally convex** if for any point x of the set and any neighborhood U of x there is a neighborhood V of x which is convex and contained in U. A convex set is a **convex body** if it has an interior point (it is sometimes also required that a convex body be *closed* or *compact*).

convex span. See SPAN.

convex surface. A surface such that any plane section of it is a convex curve.

convex toward a point (or line or plane). A curve is **convex toward a point (or line)** when every segment of it, cut off by a chord, lies on the chord or on the same side of the chord as does the point (or line). If there exists a horizontal line such that a curve lies above (below) it and is convex toward it, the curve is said to be *convex downward* (*upward*). A curve is convex downward if and only if it is the graph of a convex function (see above, convex function). A surface is said to be **convex toward** (or **away from**) a **plane** when every plane perpendicular to this plane cuts it in a curve which is convex toward (or away from) the line of intersection of the two planes.

generalized convex function. Let $\{F\}$ be a family of functions which are continuous on an interval (a, b) and such that, for any two points (x_1, y_1) and (x_2, y_2) with x_1 and x_2 two different numbers of the interval (a, b), there is a unique member F of $\{F\}$ satisfying

$$F(x_1) = y_1; \qquad F(x_2) = y_2.$$

A function f is a **generalized convex function** relative to $\{F\}$, or a **sub-F** function in an interval I provided that, for any numbers x_1, ξ, x_2 in I with ξ between x_1 and x_2, we have $f(\xi) \leq F(\xi)$, where F is the member of $\{F\}$ for which $F(x_1) = f(x_1)$ and $F(x_2) = f(x_2)$.

logarithmically convex function. A function whose logarithm is convex. The *gamma function* is the only logarithmically convex function which is defined and positive for $x > 0$, which satisfies the functional equation $\Gamma(x + 1) = x\Gamma(x)$, and for which $\Gamma(1) = 1$.

strictly convex space. A normed linear space which has the property that if $\|x + y\| = \|x\| + \|y\|$ and $y \neq 0$, then there is a number t such that $x = ty$. A finite-dimensional space is *strictly convex* if and only if it is *uniformly convex*, but an infinite-dimensional space can be strictly convex without being uniformly convex. *Syn.* rotund space.

uniformly convex space. A normed linear space is *uniformly convex* if for any $\epsilon > 0$ there is a $\delta > 0$ such that $\|x - y\| < \epsilon$ if $\|x\| \leq 1$, $\|y\| \leq 1$, and $\|x + y\| > 2 - \delta$. A finite-dimensional space is uniformly convex if and only if elements x and y are proportional whenever $\|x + y\| = \|x\| + \|y\|$. Hilbert space is uniformly convex. Any uniformly convex Banach space is reflexive, but there are reflexive Banach spaces which are not isomorphic with any uniformly convex space. A Banach space is isomorphic with a uniformly convex space if and only if it is super-reflexive. See SUPER-REFLEXIVE. *Syn.* uniformly rotund space.

CON-VO-LU'TION, *n.* **convolution of two functions.** The function h defined by

$$h(x) = \int_0^x f(t)g(x-t)\,dt = \int_0^x g(t)f(x-t)\,dt$$

is the convolution of f and g. The function $H(x) = \int_{-\infty}^{\infty} f(t)g(x-t)\,dt$ is sometimes also called a convolution of f and g, but is also called a **bilateral convolution.** *Syn.* faltung (German), resultant.

convolution of two power series. The convolution of two series of the form

$$\sum_{n=-\infty}^{\infty} a_n z^n \quad \text{and} \quad \sum_{n=-\infty}^{\infty} b_n z^n$$

is the series

$$\sum_{n=-\infty}^{\infty} c_n z^n, \quad \text{where} \quad c_n = \sum_{p=-\infty}^{\infty} a_p b_{n-p}.$$

This is the formal term-by-term product of the series.

CO-OP'ER-A-TIVE, *adj.* **cooperative game.** See GAME.

CO-OR'DI-NATE, *n.* One of a set of numbers which locate a point in space. If the point is known to be on a given line, only one coordinate is needed; if in a plane, two are required; if in space, three. See CARTESIAN, POLAR.

barycentric coordinates. See BARYCENTRIC.

Cartesian coordinates. See CARTESIAN.

complex coordinates. (1) Coordinates which are complex numbers. (2) Coordinates used in representing complex numbers in the plane (see COMPLEX —complex numbers).

coordinate geometry. See GEOMETRY—analytic geometry.

coordinate paper. Paper ruled with graduated rulings to aid in plotting points and drawing the loci of equations. See CROSS—cross-section paper, LOGARITHMIC —logarithmic coordinate paper.

coordinate planes. See CARTESIAN —Cartesian coordinates.

coordinate system. Any method by which a set of numbers is used to represent a point, line, or any geometric object. See COORDINATE, CARTESIAN, POLAR.

coordinate trihedral. See TRIHEDRAL.

curvilinear coordinates. See CURVILINEAR.

cylindrical coordinates. See CYLINDRICAL—cylindrical coordinates.

ellipsoidal coordinates. Through each point in space there passes just one of each of the families of confocal quadrics whose equations are the following, if $a^2 > b^2 > c^2$:

$$\frac{x^2}{a^2-k} + \frac{y^2}{b^2-k} + \frac{z^2}{c^2-k} = 1, \quad k < c^2,$$

$$\frac{x^2}{a^2-l} + \frac{y^2}{b^2-l} - \frac{z^2}{l-c^2} = 1, \quad c^2 < l < b^2,$$

$$\frac{x^2}{a^2-m} - \frac{y^2}{m-b^2} - \frac{z^2}{m-c^2} = 1, \quad b^2 < m < a^2.$$

The values of k, l, m, which determine these three quadrics, are the **ellipsoidal coordinates** of the given point. However, three such quadrics intersect in eight points, so further restrictions are necessary to determine a point from a given set of quadrics, such perhaps as the octant in which the point shall lie. See CONFOCAL —confocal quadrics.

geodesic coordinates. See GEODESIC.

geographical coordinates. See SPHERICAL—spherical coordinates.

homogeneous coordinates. In a plane, the homogeneous coordinates of a point whose Cartesian coordinates are x and y are any three numbers (x_1, x_2, x_3) for which $x_1/x_3 = x$ and $x_2/x_3 = y$. The coordinates are called homogeneous since any polynomial equation in Cartesian coordinates becomes homogeneous when the transformation to homogeneous coordinates is made; *e.g.*, $x^3 + xy^2 + 9 = 0$ becomes $x_1^3 + x_1 x_2^2 + 9x_3^3 = 0$. Homogeneous coordinates are defined analogously for spaces of three or more dimensions. See LINE— ideal line, PROJECTIVE —projective plane.

inertial coordinate system. In mechanics, any system of coordinate axes moving with constant velocity with respect to a system of axes fixed in space relative to the positions of "fixed" stars. The latter system is the **primary inertial system.**

left-handed coordinate system. A coordinate system in which the positive directions of the axes form a left-handed trihedral. See TRIHEDRAL —directed trihedral.

logarithmic coordinates. Coordinates using the logarithmic scale; used in plotting points on logarithmic paper. See LOGARITHMIC —logarithmic coordinate paper.

normal coordinates. Coordinates y^i such that the parametric equations of any geodesic going through the origin $y^i = 0$ have the linear form $y^i = \xi^i s$ in terms of the arc-length parameters. Normal coordinates are special kinds of *geodesic coordinates*. Also see ORTHOGONAL —orthogonal transformation.

oblique coordinates. See CARTESIAN —Cartesian coordinates.

polar-coordinate paper. Paper ruled with concentric circles about the point which is to serve as the pole and with radial lines through this point at graduated angular distances from the initial line. Used to graph functions in polar coordinates. See POLAR—polar coordinates in the plane.

polar coordinates. See POLAR, SPHERICAL — spherical coordinates.

rectangular Cartesian coordinates. See CARTESIAN, and above, coordinate planes.

right-handed coordinate system. A coordinate system in which the positive directions of the axes form a right-handed trihedral. See TRIHEDRAL —directed trihedral.

spherical coordinates. See SPHERICAL— spherical coordinates.

symmetric coordinates. Coordinates u, v of a surface S: $x = x(u, v)$, $y = y(u, v)$, $z = z(u, v)$, such that the element of length is given by $ds^2 = F\,du\,dv$; that is, such that $E = G = 0$. See LINEAR—linear element of a surface. We have symmetric coordinates if, and only if, the parametric curves are minimal. See PARAMETRIC —parametric curves of a surface, MINIMAL —minimal curve.

tangential coordinates of a surface. Let X, Y, Z denote the direction cosines of the normal to a surface S: $x = x(u, v)$ $y = y(u, v)$, $z = z(u, v)$, and let W denote the algebraic distance from the origin to the plane tangent to S at the point P: (x, y, z) of S, $W = xX + yY + zZ$. The surface S is uniquely determined by the functions X, Y, Z, W, which are called the **tangential coordinates** of S.

transformation of coordinates. See TRANSLATION, TRANSFORMATION —transformation of coordinates.

CO-PLA'NAR, *adj.* Lying in the same plane. *E.g.*, **coplanar lines** are lines which lie in the same plane; **coplanar points** are points which lie in the same plane. Three points are necessarily coplanar; four points described by their rectangular Cartesian coordinates are coplanar if and only if the following determinant is zero (otherwise, the absolute value of the determinant is the volume of a parallelpiped with the four points as four of the eight vertices and three of the points

adjacent to the fourth):

$$\begin{vmatrix} x_1 & y_1 & z_1 & 1 \\ x_2 & y_2 & z_2 & 1 \\ x_3 & y_3 & z_3 & 1 \\ x_4 & y_4 & z_4 & 1 \end{vmatrix}.$$

CO-PUNC'TAL, *adj.* **copunctal planes.** Three or more planes having a point in common.

CORIOLIS, Gaspard Gustave de (1792–1843). French mathematician and physicist.

Coriolis acceleration. See ACCELERATION— acceleration of Coriolis, and below, Coriolis force.

Coriolis force. A force on terrestrial particles arising from the rotation of the earth about its axis. Its magnitude is $2m\omega v$, where ω is the angular velocity of rotation of the earth and v is the speed of the particle of mass m relative to the earth. Because of the small angular velocity of the earth (2π radians per day, or 7.27×10^{-5} radians per second) the effects of the *Coriolis force* are negligible in most technical applications, but are important in meteorological and geographical considerations since they account for the trade winds.

COR'OL-LA'RY, *n.* A theorem that follows so obviously from the proof of some other theorem that no, or almost no, proof is necessary; a by-product of another theorem. See THEOREM.

COR-REC'TION, *n.* **Sheppard's correction, Yates' correction.** See the respective names.

COR'RE-LA'TION, *n.* (1) In pure mathematics: A linear transformation which, in the plane, carries points into lines and lines into points and, in space, carries points into planes and planes into points. (2) (*Statistics*) Most generally, an interdependence between random variables or between sets of numbers. Thus, random variables might be said to be correlated if they are not *independent* (see INDEPENDENT —independent random variables). Most commonly, correlation is used in a more restrictive sense as measured by correlation coefficients (see various headings below).

canonical correlation. See CANONICAL— canonical random variables.

correlation coefficient. For two random variables X and Y with finite nonzero *variances* σ_X^2 and σ_Y^2, the **correlation coefficient** is

$$r = \frac{\sigma_{X,Y}}{\sigma_X \sigma_Y},$$

where $\sigma_{X,Y}$ is the *covariance* of X and Y. The number r indicates the degree of linear relationship between X and Y; it satisfies $-1 \le r \le 1$. For two sets of numbers $(x_1, x_2, \cdots, x_n)$ and $(y_1, y_2, \cdots, y_n)$, the **correlation coefficient** is

$$r = \frac{\sum_{i=1}^{n} (x_i - \bar{x})(y_i - \bar{y})}{\sqrt{\sum_{i=1}^{n} (x_i - \bar{x})^2 \sum_{i=1}^{n} (y_i - \bar{y})^2}},$$

where $\bar{x}$ and $\bar{y}$ are the corresponding *means*. It measures how near the points (x_1, y_1), $(x_2, y_2), \cdots, (x_n, y_n)$ are to lying on a straight line. If $r = 1$, the points lie on a line and the two sets of data are said to be in **perfect correlation**: $r = 1$ if and only if there is a positive number A such that $y_i - \bar{y} = A(x_i - \bar{x})$ for all i. There is a negative number B such that $y_i - \bar{y} = B(x_i - \bar{x})$ for all i if and only if $r = -1$. *Syn.* coefficient of linear correlation, Pearson's coefficient, product moment correlation coefficient.

correlation ratio. The **correlation ratio** $\eta_{X,Y}$ of X on Y for random variables X and Y in a bivariate distribution is $\sigma_{X,Y}/\sigma_X$, where σ_X^2 is the *variance* of X and $\sigma_{X,Y}^2$ is the *expected value* of $[\mu(Y) - \bar{X}]^2$, where $\mu(Y)$ is the conditional expectation (mean) of X given Y and $\bar{X}$ is the mean of X. If X and Y take on values $(x_1, x_2, \cdots, x_n)$ with probability function p and $(y_1, y_2, \cdots, y_n)$ with probability function q, respectively, the correlation ratio of X on Y is the square root of

$$\frac{\sum [\mu(y_i) - \bar{x}]^2 q(y_i)}{\sum (x_i - \bar{x})^2 p(x_i)},$$

where $\mu(y_i)$ is the conditional expectation (mean) of X given Y has the value y_i. The correlation ratio is equal to the correlation coefficient if and only if there is a number β such that $\mu(y_i) = \beta y_i$ for each i.

illusory correlation. See ILLUSORY.

multiple correlation. Given a set of n random variables, $X_1, \cdots, X_n$, let

$$X' = a + b_{12}X_2 + b_{13}X_3 + b_{14}X_4$$
$$+ \cdots + b_{1n}X_n$$

be the linear function of $X_2, \cdots, X_n$ which minimizes the *expected value* of $(X_1 - X')^2$. The **multiple correlation coefficient** of X_1 with respect to $X_2, \cdots, X_n$ is the ordinary *correlation coefficient* between X_1 and X'. It is the maximum correlation between X_1 and linear functions of $X_2, \cdots, X_n$.

nonsense correlation. See ILLU-SORY—illusory correlation.

partial correlation. (1) Suppose two random variables X_1 and X_2 are considered along with $n - 2$ other variables $X_3, \cdots, X_n$. Let Y_1 and Y_2 be the random variables obtained by subtracting from X_1 and X_2 the respective linear functions of $X_3, \cdots, X_n$ which maximize the *multiple correlations* of X_1 and X_2 with respect to $X_3, \cdots, X_n$. The correlation coefficient between Y_1 and Y_2 is the **partial correlation coefficient** of X_1 and X_2 with respect to $X_3, \cdots, X_n$. The partial correlation coefficient can be expressed in terms of lower order coefficients by

$$r_{12:34\cdots k}$$

$$= \frac{r_{12:34\cdots k-1} - r_{1k:34\cdots k-1}r_{2k:34\cdots k-1}}{\sqrt{\left(1 - r_{1k:34\cdots k-1}^2\right)\left(1 - r_{2k:34\cdots k-1}^2\right)}}.$$

(2) Given random variables $X_1, X_2, \cdots, X_n$, another type of **partial correlation coefficient** is the (conditional) correlation coefficient between X_1 and X_2, given that $X_3, \cdots, X_n$ have specified values, *i.e.*, the correlation coefficient between X_1 and X_2 computed using their conditional distributions given that $X_3, \cdots, X_n$ have specified values.

product-moment correlation coefficient. Same as CORRELATION COEFFICIENT. See above.

rank correlation. The degree of association between two rankings of the same objects, as the rankings of students according to weight and according to height. Given two sets $(x_1, x_2, \cdots, x_n)$ and $(y_1, y_2, \cdots, y_n)$, each of which consists of the integers $1, 2, \cdots, n$ in some order, one measure of rank correlation is the correlation coefficient of these two sets:

$$r = \frac{\sum (x_i - \bar{x})(y_i - \bar{y})}{\left[\sum (x_i - \bar{x})^2 \sum (y_i - \bar{y})^2\right]^{1/2}}$$

$$= 1 - \frac{6 \sum (x_i - y_i)^2}{n^3 - n}.$$

This is often called *Spearman's rank correlation*. There are numerous other measures of rank correlation.

COR′RE-SPOND′ENCE, *n.* Same as RELATION. See RELATION.

one-to-one correspondence. A correspondence between two sets for which each member of either set is paired with exactly one member of the other set. For example, a one-to-one correspondence between the sets $\{a,b,c,d\}$ and $\{1,2,3,4\}$ is given by the pairing $\{(a,1),(b,2),(c,3),(d,4)\}$. *Tech.* A one-to-one correspondence between sets A and B is a collection S of ordered pairs (x,y) whose first members are the members of A, whose second members are the members of B, and for which (x_1,y_1) and (x_2,y_2) are identical if $x_1 = x_2$ or $y_1 = y_2$. *Syn.* bijection, one-to-one function, one-to-one mapping, one-to-one transformation.

COR′RE-SPOND′ING, *adj.* **corresponding angles, lines, points, etc.** Points, angles, lines, etc., in different figures, similarly related to the rest of the figures. *E.g.*, in two right triangles the hypotenuses are *corresponding* sides.

corresponding angles of two lines cut by a transversal. See ANGLE—angles made by a transversal.

corresponding rates. Rates producing the same amount on the same principal in the same time with different conversion periods. The nominal rate of 6%, money being converted semiannually, and the effective rate of 6.09%, are corresponding rates. *Syn.* equivalent rates.

CO-SE′CANT, *adj., n.* See TRIGONOMETRIC — trigonometric functions.

cosecant curve. The graph of $y = \text{cosec } x$; the same as the curve obtained by moving the secant curve $\pi/2$ radians to the right, since $\text{cosec } x = \sec(x - \pi/2)$. See SECANT —secant curve.

CO-SET, *n.* **coset of a subgroup of a group.** A set consisting of all products hx, or of all products xh, of elements h of the subgroup by a fixed element x of the group. If the multiplication by x is on the right, the set is a **right coset.** If the multiplication by x is on the left, the set is a **left coset.** Two cosets either are identical or have no elements in common. Each element of the group belongs to one of the left cosets and to one of the right cosets. See GROUP.

CO′SINE, *n.* See TRIGONOMETRIC —trigonometric functions.

cosine curve. The graph of $y = \cos x$ (see figure). The curve has a y-intercept 1, is concave toward the x-axis, and cuts this axis at odd multiples of $\frac{1}{2}\pi$ (radians).

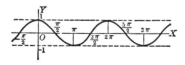

direction cosine (in space). See DIRECTION — direction cosines.

law of cosines. For a **plane triangle**, if a,b,c are the sides and C the angle opposite c, the law of cosines is

$$c^2 = a^2 + b^2 - 2ab\cos C.$$

This formula is useful for solving a triangle when two sides and an angle, or three sides, are given. For a **spherical triangle**, the cosine laws are:

$$\cos a = \cos b \cos c + \sin b \sin c \cos A,$$

$$\cos A = -\cos B \cos C + \sin B \sin C \cos a,$$

where a,b,c are the sides of the spherical triangle, and A,B,C are the corresponding opposite angles.

CO-TAN′GENT, *n.* See TRIGONOMETRIC —trigonometric functions.

cotangent curve. The graph of $y = \cot x$. It is asymptotic to the lines $x = 0$ and $x = n\pi$ and cuts the x-axis at odd multiples of $\pi/2$ (radians). It can be obtained by rotating the tangent curve about the Y-axis and translating so the Y-axis is an asymptote. See TANGENT —tangent curve.

CO-TER′MI-NAL, *adj.* **coterminal angles.** Angles (rotation angles) having the same terminal line and the same initial line; *e.g.*, 30°, 390°, and $-330°$ are coterminal angles. See ANGLE.

COTES, Roger (1682–1716). English mathematician. Assisted Newton in preparation of the second edition of Principia. Worked on computational tables. See NEWTON—Newton-Cotes integration formulas.

COULOMB, Charles Augustin de (1736–1806) French physicist.

coulomb. A unit of electrical charge, abbreviation coul or Cb. The **absolute coulomb** is defined as the amount of electrical charge which crosses a surface in one second if a steady current of one **absolute ampere** is flowing across the surface. The absolute coulomb has been the legal standard of quantity of electricty since 1950. The **international coulomb,** the legal standard before 1950, is the quantity of electricity which, when passed through a solution of silver nitrate in water, in accordance with certain definite specifications, depositis 0.00111800 gm of sliver.

$$1 \text{ Int. coul} = 0.999835 \text{ abs. coul.}$$

Coulomb's law for point-charges. A point-charge of magnitude e_1 located at a point P_1 exerts a force on a point-charge of magnitude e_2 located at a point P_2, which is given by the expression $ke_1e_2r^{-2}\boldsymbol{\rho}_1$, where k is a positive constant depending on the units used, r is the distance between the charges, and $\boldsymbol{\rho}_1$ is a unit vector having the direction of the displacement P_1 to P_2. Thus the force is a repulsion or attraction according as the charges are of the same or opposite sign. If we replace the positive constant k with the negative constant $-G$ and replace e_1 and e_2 with m_1 and m_2, the magnitudes of two point-masses, the foregoing formula becomes *Newton's law of gravitation for particles.*

COUNT, $v.$ To name a set of consecutive integers in their natural order, usually beginning with 1.

  count by twos (threes, fours, fives, etc.). To name, in order, a set of integers that have the difference 2 (3, 4, 5, etc.); *e.g.*, when counting by two's, one says "2, 4, 6, 8, $\cdots$"; when counting by three's, "3, 6, 9, 12, $\cdots$."

COUNT-A-BIL'I-TY, $n.$ **first and second axioms of countability.** See BASE—base for a topology.

COUNT'A-BLE, *adj.* **countable set.** (1) A set of objects whose members can be put into one-to-one correspondence with the positive integers; a set whose members can be arranged in an infinite sequence $p_1, p_2, p_3, \cdots$ in such a way that every member occurs in one and only one position. *Syn.* countably infinite. (2) A set of objects which either has a finite number of members or can be put into one-to-one correspondence with the set of positive integers. The sets of all integers and of all rational numbers are countable, but the set of all real numbers is not. See DENUMERABLE, ENUMERABLE.

COUNT'A-BLY, *adv.* **countably compact.** See COMPACT.

COUN'TER, *adj., n.* In a computing machine, an adder that receives only unit addends, or addends of amount one. Counters are usually built up by means of simple modulo 2 counters, a **modulo 2 counter** being a simple arithmetic component that is in one of its two steady states according as the number of impulses it has received is even or odd. See ADDER.

  counter life. See INSURANCE —life insurance.

COUN'TER-CLOCK'WISE, *adj.* In the direction of rotation opposite to that in which the hands move around the dial of a clock.

COUNT'ING, *adj.* **counting number.** The numbers used in counting objects. This may mean the set of positive integers: $1, 2, 3, \cdots$. It may also mean the set of positive integers and the number 0, since 0 is the number of members in the empty set.

COU'PON, $n.$ **bond coupons.** See BOND.

COURANT, Richard (1888–1972). German-American analyst and applied mathematician. Contributed especially to potential theory, complex-variable theory, and the calculus of variations.

  Courant maximum-minimum and minimum-maximum principles. Theorems which describe the nth eigenvalue for certain eigenvalue problems without using the previous eigenvalues or associated eigenvectors. Such theorems apply to a symmetric operator T on an inner-product space H for which there is a complete orthonormal sequence $\{\phi_n\}$ such that each ϕ_n is an eigenvector of T with the eigenvalue λ_n [*i.e.*, $T(\phi_n) = \lambda_n\phi_n$]. If the sequence $\{\lambda_1, \lambda_2, \cdots\}$ is monotone increasing, the **maximum-minimum theorem** states that λ_n is the maximum value of $m(f_1, f_2, \cdots, f_{n-1})$, where $m(f_1, f_2, \cdots, f_{n-1})$ is the minimum value of the inner product (Ty, y) for y in the domain of T with $\|y\| = 1$ and y orthogonal to f_k if $k < n$. If $\{\lambda_1, \lambda_2, \cdots\}$ is monotone decreasing, the **minimum-maximum** (or **minimax**) **theorem** states that λ_n is the minimum value of $M(f_1, f_2, \cdots, f_{n-1})$, where $M(f_1, f_2, \cdots, f_{n-1})$ is the maximum value of (Ty, y) for y in the domain of T with $\|y\| = 1$ and y orthogonal to f_k if $k < n$.

COURSE, $n.$ **course of a ship.** See SAILING—plane sailing.

CO-VAR'I-ANCE, $n.$ The first *product moment* about the means of two random variables X and Y. See CORRELATION—correlation coefficient, MOMENT —product moment, VARIANCE.

  analysis of covariance. Statistical analysis of variance of a variable (as in the analysis of variance) which is affected by other variables. *E.g.*, the effect of variation in carbon content on the tensile strength of steel may be removed at the same time that other nonlinearity related factors, such as different manufacturers or blast furnaces, are also controlled. In this example, the intent is (1) to determine if there is any relationship between tensile strength and carbon content, (2) to determine if that relationship varies from manufacturer to manufacturer, and (3) to remove the effect of carbon content in determining the relation between tensile strength and manufacturer.

CO-VAR′I-ANT, *adj.* **covariant derivative of a tensor.** The *covariant derivative* of a tensor $t_{b_1 \cdots b_q}^{a_1 \cdots a_p}$ is the tensor

$$t_{b_1 \cdots b_1, j}^{a_1 \cdots a_p} = \frac{\partial t_{b_1 \cdots b_q}^{a_1 \cdots a_p}}{\partial x_j}$$

$$- \sum_{r=1}^{q} t_{b_1 \cdots b_r - 1 i b_{r+1} \cdots b_q}^{a_1 \cdots a_p} \left\{ \begin{matrix} i \\ b_r \ j \end{matrix} \right\}$$

$$+ \sum_{r=1}^{p} t_{b_1 \cdots b_q - 1 i a_{r+1} \cdots a_p}^{a_1 \cdots a_r} \left\{ \begin{matrix} a_r \\ i \ j \end{matrix} \right\},$$

where the summation convention applies and $\left\{ \begin{matrix} i \\ j \ k \end{matrix} \right\}$ is the *Christoffel symbol* of the second kind. This tensor is contravariant of rank p and covariant of rank $q + 1$. Covariant differentiation is not commutative. E.g., $t_{,j,k}^i \neq t_{,k,j}^i$ in general, since $t_{,j,k}^i - t_{,k,j}^i = R_{rjk}^i t^r$, where R_{jkl}^i is the *Riemann-Christoffel tensor.* If $t_i(x^1, x^2, \cdots, x^n)$ is a covariant tensor of rank one (*i.e.*, a *covariant field*), then the covariant derivative of t_i is

$$t_{i,j} = \frac{\partial t_i}{\partial x^j} - \left\{ \begin{matrix} \sigma \\ i \ j \end{matrix} \right\} t_\sigma,$$

a covariant tensor of rank two. If $t^i(x^1, \cdots, x^n)$ is a contravariant tensor of rank one (*i.e.*, a *contravariant vector field*), then the covariant derivative of t^i is

$$t_{,j}^i = \frac{\partial t^i}{\partial x^j} + \left\{ \begin{matrix} i \\ \sigma \ j \end{matrix} \right\} t^\sigma,$$

which is contravariant of rank one and covariant of rank one. In Cartesian coordinates, or for *scalar fields*, covariant differentiation is ordinary differentiation. See CONTRAVARIANT—contravariant derivative of a tensor.

covariant functor, indices, tensor, vector field. See FUNCTOR, TENSOR—covariant tensor, VECTOR—covariant vector field.

Stokian covariant derivative. If

$$t_{a_1 a_2 \cdots a_p}(x^1, \cdots, x^n)$$

is an alternating covariant tensor field, then the covariant tensor field $t_{a_1 a_2 \cdots a_p | \beta}$ of rank $p + 1$ defined by

$$t_{a_1 a_2 \cdots a_p | \beta} = \frac{\partial t_{a_1 \cdots a_p}}{\partial x^\beta} - \sum_{r=1}^{n} \frac{\partial t_{a_1 \cdots a_{r-1} \beta a_{r+1} \cdots a_p}}{\partial x^{a_r}}$$

is an alternating tensor, the **Stokian covariant derivative** of $t_{a_1 \cdots a_p}$. The terminology is appro-priate, since in the generalized Stokes' theorem on multiple integrals we have

$$\int_{B_p} \cdots \int t_{a_1 \cdots a_p} \, dx^{a_1} \cdots dx^{a_p}$$

$$= \int_{V_{p+1}} \cdots \int t_{a_1 \cdots a_p | \beta} \, dx^{a_1} \cdots dx^{a_p} \, dx^\beta.$$

It is to be observed that Stokian covariant dif-ferentiation does not depend on the machinery put at one's disposal by a metric tensor field g_{ij}.

COV′ER, *n.* A **cover** of a set T is a collection of sets such that each point of T belongs to at least one of the sets. A cover is **closed** (or **open**) according as each of the covering sets is closed (or each is open). An **ϵ-cover** of a metric space M is a cover of M by a finite number of sets each of which is such that the distance between any two of its points is less than ϵ. An ϵ-cover is of **order n** if there is a point which is contained in n of the sets of the covering, but no point is contained in $n + 1$ of these sets. *Syn.* covering. Also see VITALI—Vitali covering.

CO′VERSED, *adj.* **coversed sine.** *One minus* the *sine* of an angle; geometrically the difference between the radius and the sine of an angle constructed in a unit circle. See TRIGONOMETRIC —trigonometric functions.

CO′VER-SINE′. Same as COVERSED SINE.

CRAMER, Gabriel (1704–1752). Swiss mathe-matician and physicist.

Cramer's rule. A simple rule using determi-nants to express the solution of a system of linear algebraic equations for which the number of equations is equal to the number of variables. The rule for n equations is: The solution value for each variable is equal to the fraction in which the denominator is the determinant of the coef-ficients of the n variables and the numerator is the same determinant, except that the coeffi-cients of the variable whose solution value is being found are replaced by the constant terms if these appear as the righthand members of the system of equations and by their negatives if they appear in the left members. *E.g.*, the values of x

and y which satisfy

$$x + 2y = 5$$

$$2x + 3y = 0 \quad \text{are}$$

$$x = \begin{vmatrix} 5 & 2 \\ 0 & 3 \end{vmatrix} \div \begin{vmatrix} 1 & 2 \\ 2 & 3 \end{vmatrix} = -15,$$

$$y = \begin{vmatrix} 1 & 5 \\ 2 & 0 \end{vmatrix} \div \begin{vmatrix} 1 & 2 \\ 2 & 3 \end{vmatrix} = 10.$$

This rule gives the solution of equations for which there is a unique solution, that is, for which the determinant of the coefficients is not zero. See DETERMINANT —determinant of the second order, CONSISTENCY.

CRAMÉR, Harald (1893–1985). Swedish statistician.

Cramér-Rao inequality. Suppose a random variable X has a probability density function $f(X;\theta)$ depending on a parameter θ and that $T(X_1, \cdots, X_n)$ is an estimator for θ whose *bias* is $b_T(\theta) = E(T) - \theta$, where $E(T)$ is the expected value of T. If suitable regularity conditions are satisfied, then the expected value of $(T - \theta)^2$, the *mean-square error* of T, satisfies

$$E\left[(T - \theta)^2\right] \geq \frac{\left[1 + b_T'(\theta)\right]^2}{n \cdot E\left[(\partial \ln f / \partial \theta)^2\right]},$$

the numerator reducing to 1 if the estimator is unbiased. A similar result holds for the discrete case, with f the frequency function.

CRED'IT, *adj.* **credit business.** A business in which goods are sold without immediate payment, but with a promise to pay later, generally at some specified time.

CRED'I-TOR, *n.* One who accepts a promise to pay in the future in place of immediate payment; a term most commonly applied to retail merchants who do a *credit* business.

CRISP, *n.* **crisp set.** See FUZZY.

CRI-TE'RI-ON, *n.* [*pl.* cri-te'ri-a.] A law or principle by which a proposition can be tested.

CRIT'I-CAL, *adj.* **critical point.** A STATIONARY POINT. Sometimes a point at which the graph of a function has a vertical tangent also is called a critical point.

critical region. See HYPOTHESIS —test of a hypothesis.

CROSS, *adj.* **cross-cap.** A Möbius strip has a boundary which is a simple closed curve. It can

be deformed into a circle, although in the process the strip must be allowed to intersect itself (the curve of intersection is regarded as two different curves, each belonging to exactly one of the two parts of the surface which "cross" along the curve). The surface which results is a nonorientable surface called a **cross-cap.** It can be described as a hemisphere which has been pinched together along a short vertical line starting at the pole; the surface then appears to intersect along this line and the line is to be regarded as two lines, each belonging to one of the two parts of the surface which cross along the line. See GENUS—genus of a surface.

cross product. See MULTIPLICATION —multiplication of vectors.

cross ratio. See RATIO—cross ratio.

cross-section of an area or solid. A plane section perpendicular to the axis of symmetry or to the longest axis, if there be more than one; rarely used except in cases where all cross sections are equal, as in the case of a circular cylinder or rectangular parallelepiped.

cross-section paper. Paper ruled with vertical and horizontal lines equally spaced; used in graphing equations in rectangular coordinates. *Syn.* ruled paper, squared paper.

CRU'CI-FORM, *adj.* **cruciform curve.** The locus of the equation

$$x^2 y^2 - a^2 x^2 - a^2 y^2 = 0.$$

The curve is symmetric about the origin and the coordinate axes, has four branches, one in each quadrant, and is asymptotic to each of the four lines $x = \pm a$, $y = \pm a$. It is called the *cruciform curve* because of its resemblance to a cross.

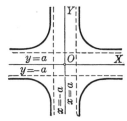

CRU'NODE, *n.* A point on a curve through which there are two branches of the curve with distinct tangents.

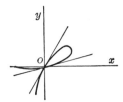

CUBE, *n.* (1) A polyhedron bounded by six plane faces with its twelve edges all equal and its face angles all right angles. (2) An **n-dimensional cube** with edges of length k is a figure in n-dimensional Euclidean space that can, after an appropriate rigid motion, be represented as the set of all those points $(x_1, x_2, \cdots, x_n)$ for which $0 \leq x_i \leq k$ for each i. The *volume* (or *measure*) of the cube is equal to k^n. Such a cube is the *Cartesian product* of n closed intervals, each of length k. The four-dimensional cube sometimes is called a **tesseract**; it has 16 vertices, 32 edges, 24 faces, and 8 "three-dimensional faces".

cube of a number. The third power of the number. *E.g.,* the cube of 2 is $2 \times 2 \times 2$, written 2^3.

cube of a quantity. The third power of the quantity; *e.g.,* the cube of $(x + y)$ is

$$(x + y)(x + y)(x + y),$$

written $(x + y)^3$ or $x^3 + 3x^2y + 3xy^2 + y^3$.

cube root of a given quantity. A quantity whose cube is the given quantity. See ROOT—root of a number.

duplication of the cube. See DUPLICATION.

CU′BIC, *adj., n.* Of the third degree. For example, a **cubic equation** is an equation of the third degree, such as

$$2x^3 + 3x^2 + x + 5 = 0.$$

A **cubic polynomial** is a polynomial of the third degree.

Cardan's solution of the cubic. See CARDAN.

bipartite cubic. See BIPARTITE.

cubic curve. See CURVE—algebraic plane curve.

reduced cubic. See REDUCED.

resolvent cubic. See FERRARI—solution of the quartic.

twisted cubic. A curve which cuts each plane in three points, real or imaginary, distinct or not. *E.g.,* the equations $x = at$, $y = bt^2$, $z = ct^3$, $abc \neq 0$, represent such a curve.

CU′BI-CAL, *adj.* **cubical and semicubical parabola.** See PARABOLA —cubical parabola.

CU′BOID, *n.* A polyhedron with six faces all of which are rectangles. It is a **cube** if each rectangle is a square. *Syn.* rectangular parallelepiped.

CU′BO-OC-TA-HE′DRON or **CU′BOC-TA-HE′DRON,** *n.* See ARCHIMEDES —Archimedean solid.

CU′MU-LANTS, *n.* For a random variable or the associated distribution function, the nth cu-mulant is the evaluation at 0 of the nth derivative of $\ln M(t)$, where M is the *moment-generating function*; $\ln M(t)$ is the *cumulant-generating function*. The first three cumulants are equal to the corresponding central moments. See INVARIANT—semi-invariant.

CU′MU-LA′TIVE, *adj.* **cumulative frequency.** See FREQUENCY.

CU-NE′I-FORM, *adj.* **cuneiform symbols.** Symbols shaped like a wedge. See BABYLONIAN.

CUP, *n.* The symbol $\cup$ used to denote **union** or **least upper bound**. See LATTICE, UNION.

CURL, *n.* For a vector-valued function $\boldsymbol{F}$ of x, y, and z, the **curl** is denoted by $\nabla \times \boldsymbol{F}$ and defined as

$$\boldsymbol{i} \times \frac{\partial \boldsymbol{F}}{\partial x} + \boldsymbol{j} \times \frac{\partial \boldsymbol{F}}{\partial y} + \boldsymbol{k} \times \frac{\partial \boldsymbol{F}}{\partial z},$$

where ∇ is the operator,

$$\boldsymbol{i} \frac{\partial}{\partial x} + \boldsymbol{j} \frac{\partial}{\partial y} + \boldsymbol{k} \frac{\partial}{\partial z}.$$

E.g., if $\boldsymbol{F}$ is the velocity at a point $P(x, y, z)$ in a moving fluid, $\frac{1}{2}\nabla \times \boldsymbol{F}$ is the vector angular velocity of an infinitesimal portion of the fluid about P. Also known as **rotation**. See VECTOR—vector components.

CUR′RENT, *adj.* **current rate.** Same as PREVAILING INTEREST RATE. See INTEREST.

current yield rate. The ratio of the dividend (bond interest) to the purchase price.

CUR′TATE, *adj.* **curtate annuity.** See ANNUITY.

curtate expectation of life. See EXPECTATION —expectation of life.

CUR′VA-TURE, *n.* **curvature of a curve.** For a **circle**, the curvature is the reciprocal of the radius. For other plane curves, the curvature at a point can be thought of as the curvature of the circle which approximates the curve most closely near the point. Explicitly, for a **plane curve**, the curvature is the absolute value of the rate of change of the angle of inclination of the tangent line with respect to distance along the curve, *i.e.,* the absolute value of the derivative of $\tan^{-1}(dy/dx)$ with respect to distance along the curve. In rectangular Cartesian coordinates, the curvature K is given by

$$K = |d^2y/dx^2| \Big/ \left[1 + (dy/dx)^2\right]^{3/2};$$

in parametric coordinates it is

$$\left| \frac{(dx/dt)(d^2y/dt^2) - (dy/dt)(d^2x/dt^2)}{\left[(dx/dt)^2 + (dy/dt)^2\right]^{3/2}} \right|,$$

where x and y are functions of a parameter t; and in polar coordinates it is

$$\left| \frac{r^2 + 2(dr/d\theta)^2 - r(d^2r/d\theta^2)}{\left[r^2 + (dr/d\theta)^2\right]^{3/2}} \right|.$$

The absolute value of the ratio of the change in angle of inclination of the tangent line along a given arc to the length of the arc is the **average curvature** along the arc. The limit of the average curvature as the length of the arc approaches zero is the curvature. The circle tangent to a curve on the concave side and having the same

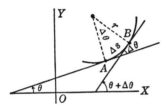

curvature at the point of tangency is called the **circle of curvature** of the curve (at that point). Its radius is the numerical value of the **radius of curvature**, and its center is the **center of curvature**. The curvature at A is the limit of $|\Delta\theta/\Delta s|$ as Δs approaches zero (where Δs is the length of the arc AB and θ is measured in radians). Some writers define the curvature with the absolute value signs removed in the preceding formulas, in which case the sign of K depends upon that of $\Delta\theta$, which is positive or negative according as the curve is concave up or concave down (according as d^2y/dx^2 is positive or negative). For a **space curve** C, let P be a fixed point, P' a variable point, s the length of arc on C from P to P', and $\Delta\theta$ the angle between the positive directions of the tangents to C at P and P'. Then the curvature $K = 1/\rho$ of C at P is defined by $K = \dfrac{1}{\rho} = \lim\limits_{\Delta s \to 0}\left|\dfrac{\Delta\theta}{\Delta s}\right|$. Again, the curvature is a measure of the rate of turning of the tangent to C relative to the distance along the curve. The number ρ is the **radius of curvature** (the **circle of curvature** is the osculating circle; see OSCULATING). Also called the **first curvature** (see TORSION). For a **plane curve** or a **space curve**, the curvature is equal to $|d\mathbf{T}/ds|$, where $\mathbf{T}$ is the unit tangent vector and s denotes distance along the curve. For the position vector $P = x\mathbf{i} +$ $y\mathbf{j} + z\mathbf{k}$, where x, y, and z are functions of a parameter t and (x, y, z) is a point on the curve, the curvature is equal to $|\mathbf{V} \times \mathbf{A}|/v^3$, where $\mathbf{V} = d\mathbf{P}/dt$, $\mathbf{A} = d\mathbf{V}/dt$, $v = |\mathbf{V}|$, and the multiplication is vector multiplication of the vectors $\mathbf{V}$ and $\mathbf{A}$.

curvature of a surface. The **normal curvature** of a surface at a point in a given direction is the curvature, with proper choice of sign, of the normal section C of the surface S at the point and in the given direction. The sign is positive if the positive direction of the principal normal of C coincides with the positive direction of the normal to S, otherwise negative. The normal curvature $1/R$ is given by

$$\frac{1}{R} = \frac{D\,du^2 + 2D'\,du\,dv + D''\,dv^2}{E\,du^2 + 2F\,du\,dv + G\,dv^2}.$$

The reciprocal R of the normal curvature is called the **radius of normal curvature** of the surface at the point in the given direction and the **center of normal curvature** is the center of curvature of the corresponding normal section C. [See MEUSNIER, SURFACE —fundamental coefficients of a surface.] At an ordinary point of a surface there are directions in which the radius of normal curvature attains its absolute maximum and its absolute minimum. These directions are at right angles to each other (unless the radius of normal curvature is the same for all directions at the point), and are called the **principal directions** on the surface at the point [see UMBILICAL —umbilical point on a surface]. The **principal curvatures** of a surface at a point are the *normal curvatures* $1/\rho_1$ and $1/\rho_2$ in the *principal directions* at the point. The numbers ρ_1 and ρ_2 are the **principal radii of normal curvature** of the surface at the point and the **centers of principal curvature** are the centers of normal curvature in the two principal directions. The **mean curvature** (or **mean normal curvature**) of a surface is the sum of the principal curvatures:

$$K_m = \frac{1}{\rho_1} + \frac{1}{\rho_2} = \frac{ED'' + GD - 2FD'}{EG - F^2}.$$

The **total curvature** (or **total normal curvature** or **Gaussian curvature**) of a surface at a point is the product of the *principal curvatures* of the surface at the point:

$$K = \frac{1}{\rho_1\rho_2} = \frac{DD'' - D'^2}{EG - F^2},$$

In tensor notation, this is the scalar field

$$K = \frac{R_{1221}}{g_{11}g_{22} - g_{12}^2},$$

if $ds^2 = g_{\mu\gamma} dx^\mu dx^\gamma$ is the line element of the surface as a two-dimensional Riemannian space and, within sign, R_{1221} is the only nonzero component of the covariant Riemann-Christoffel tensor $R_{\alpha\beta\gamma\dot{o}}$. Since R_{1221} is the determinant of the coefficients of the second fundamental differential form of the surface, it follows that the total curvature of a surface is the ratio of the determinant of its second fundamental differential form to the determinant of its first fundamental differential form $ds^2 = g_{\alpha\beta} dx^\alpha dx^\beta$. See RADIUS — radius of total curvature of a surface.

Gaussian curvature of a surface. Same as TOTAL CURVATURE. See above, curvature of a surface.

integral curvature. For a region on a surface, the **integral curvature** is the integral of the Gaussian curvature over the region: $\int\int K dA$. *Syn.* curvatura integra. For a **geodesic triangle** on a surface, the integral curvature is π less than the sum of the angles of the triangle. With proper account of sign, the integral curvature is equal to the area of the part of the unit sphere covered by the spherical image of the region.

lines of curvature of a surface. The curves on a surface S: $x = x(u, v)$, $y = y(u, v)$, $z = z(u, v)$ defined by $(ED' - FD) du^2 + (ED'' - GD) du\,dv + (FD'' - GD') dv^2 = 0$. See SURFACE —fundamental coefficients of a surface. The curves form an orthogonal system on S, and the two curves of the system through a point P of S determine the principal directions on S at P. See above, curvature of a surface.

radius of curvature. See RADIUS —radius of total curvature, GEODESIC —geodesic curvature, and above, curvature of a curve, curvature of a surface.

Riemannian curvature. See RIEMANN.

second curvature of a space curve. Same as TORSION.

surface of negative curvature. A surface on which the *total curvature* is negative at every point. Such a surface lies part on one side and part on the other side of the tangent plane in the neighborhood of a point; *e.g.*, the inner surface of a torus, and the hyperboloid of one sheet.

surface of positive total curvature. A surface on which the *total curvature* is positive at every point, such as a sphere or an ellipsoid.

surface of zero total curvature. A surface on which the total curvature is zero at every point,

such as a cylinder or, in fact, any developable surface.

CURVE, *n.* The locus of a point which has one degree of freedom. *E.g.*, in a plane, a straight line is the locus of points whose coordinates satisfy a linear equation, and a circle with radius 1 is the locus of points which satisfy $x^2 + y^2 = 1$. *Tech.* A set of points C and a continuous transformation T for which C is the image of a closed interval $[a, b]$ under the continuous transformation T. The image of a is the **initial point** and the image of b is the **terminal point** of the curve. These are **end points** if they are not coincident. *E.g.*, a **plane curve** is the graph of parametric equations, $x = f(t)$, $y = g(t)$, for which the functions f and g are continuous and the domain of each is a closed interval $[a, b]$; a special case is the graph of $y = f(x)$, where f is a continuous function on $[a, b]$. If the images of a and b coincide, the curve is a **closed curve**. A **simple curve** is a curve with the property that, with the possible exception of a and b, no two unequal numbers in $[a, b]$ determine the same point on the curve. A simple closed curve is a **Jordan curve**.

algebraic plane curve. A plane curve which has an equation in Cartesian coordinates of the form $f(x, y) = 0$, where f is a polynomial in x and y. If f has degree n, the curve is an **algebraic curve of degree** n. If n is one, the curve is a **straight line**, if n is two, the curve is a **quadratic** or **conic**; and if it is of the third, fourth, fifth, sixth degree, etc., it is a **cubic**, **quartic**, **quintic**, **sextic**, etc. When n is greater than 2, the curve is a **higher plane curve**. For an algebraic plane curve with equation $f(x, y) = 0$, a **component** is an algebraic plane curve with equation $g(x, y) = 0$ for which there exists a polynomial $h(x, y)$, possibly a constant, for which $f(x, y) \equiv g(x, y)h(x, y)$. An algebraic plane curve with only one component is **irreducible**. *E.g.*, the circle with equation $x^2 + y^2 - 9 = 0$ is irreducible, but the graph of $(y - x)(2x + y - 1) = 0$ is reducible and has components whose equations are $y - x = 0$ and $2x + y - 1 = 0$. See BÉZOUT—Bézout's theorem, PROJECTIVE —projective plane curve.

analytic curve. See ANALYTIC.

curve fitting. Determining empirical curves. See EMPIRICAL.

curve in a plane, or plane curve. A curve all points of which lie in a plane. See above, CURVE.

curve of constant width. See REULEAUX — Reuleaux "triangle."

curve of zero length. Same as MINIMAL CURVE. See MINIMAL.

curve tracing. Plotting or graphing a curve by finding points on the curve and, in a more advanced way, by investigating such matters as symmetry, extent, and asymptotes and using the derivatives to determine critical points, slope, change of slope, and concavity and convexity.

derived curve. See DERIVED.

empirical curves. See EMPIRICAL.

family of curves. See FAMILY.

growth curve. See GROWTH.

integral curves. See INTEGRAL—integral curves.

length of a curve. See LENGTH—length of a curve.

normal frequency curve. See FREQUENCY.

parabolic curve. An algebraic curve which has an equation in Cartesian coordinates of the following type:

$$y = a_0 + a_1 x + \cdots + a_n x^n.$$

parallel curves (in a plane). Two curves which have their points paired on the same normals, always cutting off segments of the same length on these normals. Their tangents at points where they cut a common normal are parallel. See INVOLUTE.

path curves. See PATH.

pedal curve. See PEDAL.

periodic curves. See PERIODIC.

primitive curve. See PRIMITIVE.

quadric (or quadratic) curve. A curve whose equation is of the 2nd degree. *Syn.* conic, conic section.

skew curve. Same as TWISTED CURVE. See below.

smooth curve. See SMOOTH.

space curve. A curve in space, but not necessarily a twisted curve. See below, twisted curve, and above, CURVE.

spherical curve. A curve that lies wholly on the surface of a sphere.

turning points on a curve. See TURNING.

twisted curve. A space curve that does not lie in a plane. *Syn.* skew curve. A twisted curve is of the **nth order** if it cuts each plane in n points, real or imaginary, distinct or not. See CUBIC—twisted cubic.

u-curves on a surface. See PARAMETRIC—parametric curves on a surface.

CUR′VI-LIN′E-AR, *adj.* **curvilinear coordinates of a point in space.** The surfaces of a triply orthogonal system of surfaces may be determined by three parameters. The values of these three parameters which determine the three surfaces of the system through a given point P in space are the **curvilinear coordinates** of P. See ORTHOGONAL —triply orthogonal system of surfaces, CONFOCAL —confocal conics.

curvilinear coordinates of a point on a surface. Parametric coordinates u, v of a point on a surface S: $x = x(u, v)$, $y = y(u, v)$, $z = z(u, v)$. See PARAMETRIC —parametric equations of a surface.

curvilinear motion. See MOTION.

CUSP, *n.* A *double point* at which the two tangents to the curve are coincident. *Syn.* spinode. A **cusp of the first kind** (or **simple cusp**) is a cusp in which there is a branch of the curve on each side of the double tangent in the neighborhood of the point of tangency; *e.g.*, the semicubical parabola, $y^2 = x^3$, has a cusp of the *first*

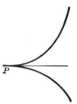

kind at the origin. A **cusp of the second kind** is a cusp for which the two branches of the curve lie on the same side of the tangent in the neighborhood of the point of tangency; the curve $y = x^2 \pm \sqrt{x^5}$ has a cusp of the second kind at the origin. See RAMPHOID.

A **double cusp** is the same as a *point of osculation* (see OSCULATION). For a given family of curves, a **cusp locus** is a set of points each of which is a cusp for one of the members of the family. See CATASTROPHE, DISCRIMINANT —discriminant of a differential equation.

hypocycloid of four cusps. See HYPOCYCLOID.

CUT, *n.* A **cut** (or **cutting**) of a set T is a subset C of T such that $T - C$ is not connected. If a cut C is a point or a line, then C is a **cut point** or a **cut line**. A point which is not a cut point is a **noncut point**.

Dedekind cut. See DEDEKIND.

CY′BER-NET′ICS, *n.* A field of science developed by N. Wiener which generalizes common properties of varied systems such as automatic factories, computers, and living organisms. It builds a common theory for certain aspects of biology and engineering, and is concerned with problems of information and noise, communications, control, and electronic computers.

CY′CLE, *n.* See CHAIN —chain of simplexes, PERMUTATION (2).

CY'CLIC, *adj.* **cyclic change, interchange, or permutation.** See PERMUTATION (2).

cyclic group and module. See GROUP, MODULE.

cyclic polygon. A polygon whose vertices lie on a circle. A convex quadrilateral is cyclic if and only if the opposite angles are supplementary. See PTOLEMY—Ptolemy's theorem.

CYC'LI-DES, *n.* **cyclides of Dupin.** The envelope of a family of spheres tangent to three fixed spheres.

CY'CLOID, *n.* The plane locus of a point which is fixed on the circumference of a circle, as the circle rolls upon a straight line. *E.g.*, the path described by a point on the rim of a wheel. The cycloid is a special case of the **trochoid**, although the two words are sometimes used synonymously. If *a* is the radius of the rolling circle and θ is the central angle of this circle, which is subtended by the arc *OP* that has contacted the line upon

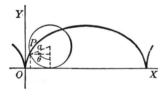

which the circle rolls, the parametric equations of the cycloid are:

$$x = a(\theta - \sin\theta), \qquad y = a(1 - \cos\theta).$$

The cycloid has a cusp at every point where it touches the *base line*. The distance along the base line between successive cusps is $2\pi a$; the length of the cycloid between cusps (not necessarily adjacent) is $4L/\pi$, where *L* is the distance along the base line between the cusps. The **pendulum property** of the cycloid is that if the cycloid is inverted and a simple pendulum of length $4a$ is hung at a cusp and compelled to stay between the branches and wrap around the cycloid as it swings, then the end of the pendulum describes another cycloid and the period of oscillation is independent of the amplitude. It was shown by Huygens that the cycloid (inverted) has the **isochronous property**, *i.e.*, the time of descent of a particle sliding without friction to a lowest point is the same wherever the particle starts on the cycloid. See BRACHISTOCHRONE, TAUTOCHRONE.

curtate and prolate cycloids. See TROCHOID.

CY'CLO-SYM'ME-TRY, *n.* See SYMMETRIC — symmetric function.

CY'CLO-TOM'IC, *adj.* **cyclotomic integer.** If *z* is a primitive *n*th root of unity and each a_i is an ordinary integer, then the number

$$a_0 + a_1 z + a_2 z^2 + \cdots + a_{n-1} z^{n-1}$$

is a **cyclotomic integer.** For each *n*, the corresponding set of cyclotomic integers is an integral domain. See DOMAIN—integral domain.

cyclotomic polynomial. The *n*th cyclotomic polynomial Φ_n is the monic polynomial of degree $\phi(n)$ whose zeros are the primitive *n*th roots of unity. This polynomial is irreducible (in the field of real numbers). If *n* is a prime, then

$$\Phi_n(x) = x^{n-1} + x^{n-2} + x^{n-3} + \cdots + x + 1,$$

but $\Phi_4(x) = x^2 + 1$ and $\Phi_{12}(x) = x^4 - x^2 + 1$. See EULER—Euler ϕ-function, ROOT—root of unity.

CYL'IN-DER, *n.* (1) A cylindrical surface (see CYLINDRICAL). (2) Suppose we are given two parallel planes and two simple closed curves C_1 and C_2 in these planes for which lines joining corresponding points of C_1 and C_2 are parallel to a given line *L*. A **cylinder** is a closed surface consisting of two bases which are plane regions bounded by such curves C_1 and C_2 and a **lateral surface** which is the union of all line segments joining corresponding points of C_1 and C_2. Each of the curves C_1 and C_2 is a **directrix** of the cylinder and the line segments joining corresponding points of C_1 and C_2 are **elements** (or **generators** or **rulings**). The cylinder is **circular** or **elliptic** if a directrix is a circle or an ellipse, respectively. Sometimes a **circular cylinder** is defined to be a cylinder whose intersections with planes perpendicular to the elements are circles. The cylinder is a **right cylinder** or an **oblique cylinder** according as *L* is perpendicular to the planes containing the bases or not perpendicular to these planes. The **altitude** of a cylinder is the perpendicular distance between the planes containing the bases and a **right section** is the intersection of the cylinder and a plane perpendicular to the elements that crosses the cylinder between the bases. The **volume** of a cylinder is equal to the product of the area of a base and the altitude; the lateral area (the area of the lateral surface) is equal to the product of the length of an element and the perimeter of a right section. Sometimes a cylinder is permitted to have a directrix that is not a closed curve, *e.g.*, a parabola or an hyperbola. The figure shows a parabolic cylinder whose equation is $x^2 = 2py$, where $\frac{1}{2}p$ is the distance from the *z*-axis of the

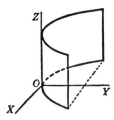

focus of a cross section. If the bases of a cylinder have centers, then the **axis** is the line segment joining the centers. A **right circular cylinder** (or **cylinder of revolution**) is a circular cylinder whose bases are perpendicular to the axis. This is the surface generated by revolving a rectangle about

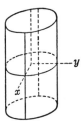

one of its sides. Its volume is $\pi a^2 h$ and its lateral area is $2\pi a h$, where h is its altitude and a the radius of the base. A **solid cylinder** has bases B_1 and B_2 which are congruent sets in parallel planes for which line segments joining corresponding points of B_1 and B_2 are parallel to some given line L. The cylinder is the union of all line segments joining corresponding points in B_1 and B_2.

   similar right circular cylinders. Right circular cylinders for which the ratio between the radius of the base and the length of an element in any one cylinder is the same as the corresponding ratio in any of the other cylinders.

CY-LIN′DRI-CAL, *adj.* **cylindrical coordinates.** Space coordinates making use of polar coordinates in one coordinate plane, usually the (x, y)-plane, the third coordinate being simply the rectangular coordinate measured from this plane. These are called cylindrical coordinates because when r is fixed and z and θ vary, they develop a cylinder; *i.e.*, $r = c$ is the equation of a cylinder. The locus of points for which θ has a fixed value is a plane, *PNO*, containing the z-axis; the points for which z is constant define a plane parallel to the x, y plane. The three surfaces for r, θ, and z constant, respectively, locate the point $P(r, \theta, z)$ as their intersection. The **transformation of cylindrical into rectangular coordi-**

nates is given by the formulas $x = r \cos \theta$, $y = r \sin \theta$, $z = z$.

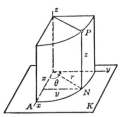

   cylindrical function. Any solution of Bessel's differential equation. Sometimes taken to be synonymous with Bessel function.

   cylindrical map. For a spherical surface S with longitude and latitude denoted by θ and ϕ, respectively, a **cylindrical map** is a continuous one-to-one map of the points of S onto a set of points of a (u, v)-plane given by formulas of the type $u = \theta$ and $v = v(\phi)$, with $v(0) = 0$ and $v(\phi) > 0$ for $\phi > 0$. A cylindrical map that is given by the formulas $u = \theta$, $v = \tan \phi$ is said to be a **central cylindrical projection**. This is a projection of a sphere from its center onto a tangent right circular cylinder that is then slit along one of its elements and spread out on a plane. A cylindrical map that is given by the formulas $u = \theta$, $v = \phi$ is said to be an **even-spaced map**. Lines of longitude and latitude with equal angular increments appear as squares, such as the squares on a checkerboard. See MERCATOR — Mercator's projection.

   cylindrical surface. A surface consisting of all straight lines parallel to a given line and intersecting a given curve (if the curve is a plane curve with its plane parallel to the given line, the cylinder is in the plane). Each of the lines is an **element** (or **generator** or **generatrix**). The curve is the **directrix**. A cylindrical surface is not necessarily closed, since the directrix is not restricted to being a closed curve. A cylindrical surface is named after its right sections; *e.g.*, if the right section is an ellipse, it is an **elliptical cylindrical surface**, or simply an **elliptic cylinder**. The **equation of a cylindrical surface**, when one of the coordinate planes is perpendicular to the elements, is the equation of the trace of the cylinder in this plane. *E.g.*, the equation $x^2 + y^2 = 1$ is the equation of a right circular cylindrical surface since, for every pair, (x, y), of numbers that satisfies this equation, z may take all values; similarly, $y^2 = 2x$ is the equation of a parabolic cylindrical surface with its elements parallel to the z-axis; and

$$\frac{x^2}{a^2} + \frac{y^2}{b^2} = 1$$

is the equation of an elliptical cylindrical surface with its elements parallel to the z-axis. See CYLINDER.

CYL′IN-DROID, *n.* (1) A cylindrical surface whose sections perpendicular to the elements are ellipses. (2) A surface that is the union of all straight lines that intersect each of two given curves and are parallel to a given plane.

D

D'ALEMBERT, Jean Le Rond (1717–1783). French mathematician, philosopher, and physicist.

D'Alembert's test for convergence (or divergence) of an infinite series. Same as the *generalized ratio test*. See RATIO—ratio test.

DAMPED, *p.* **damped harmonic motion.** Harmonic motion having its amplitude continually reduced. See HARMONIC —harmonic motion.

damped oscillations. See OSCILLATION.

DANDELIN, Germinal Pierre (1794–1847). French-Belgian geometer. Noted for the discovery (shared by Quetelot) of the beautiful relationship between a conic section and tangent spheres in its circular cone (see below, **Dandelin spheres**). It seems remarkable that this was not noticed by the ancient Greeks, who knew so much about conic sections.

Dandelin sphere. When a conic is represented as the intersection of a plane and a circular cone, the **Dandelin spheres** are spheres that are tangent to the plane and also tangent to the cone along a circle. If the intersection of the plane and cone is a parabola, there is one such sphere; if it is an ellipse or hyperbola, there are two. The Dandelin spheres are tangent to the plane at a focus of the conic.

DARBOUX, Jean Gaston (1842–1917). French differential geometer and analyst.

Darboux's monodromy theorem. The theorem that if the function f of the complex variable z is analytic in the finite domain D bounded by the simple closed curve C, is continuous in the closed region $D + C$, and takes on no value more than once for z on C, then f takes on no value more than once for z in D. See MONODROMY —monodromy theorem.

Darboux's theorem. If f is bounded on the closed interval $[a, b]$, the numbers

$M_1, M_2, \cdots, M_n$ and $m_1, m_2, \cdots, m_n$ are the least upper bounds and greatest lower bounds of $f(x)$ on the intervals $[a, x_1], [x_1, x_2], \cdots, [x_{n-1}, b]$, and δ is the length of the largest of these subintervals, then

$$\lim_{\delta \to 0} \left[M_1(x_1 - a) + M_2(x_2 - x_1) + \cdots \right.$$
$$\left. + M_n(b - x_{n-1}) \right]$$

and

$$\lim_{\delta \to 0} \left[m_1(x_1 - a) + m_2(x_2 - x_1) + \cdots \right.$$
$$\left. + m_n(b - x_{n-1}) \right]$$

both exist. The former is the **upper Darboux integral** of f and is written

$$\int_a^{\overline{b}} f(x) \, dx;$$

the latter is the **lower Darboux integral** of f and is written

$$\int_{\underline{a}}^b f(x) \, dx.$$

A necessary and sufficient condition that f be Riemann integrable is that these two integrals be equal. See INTEGRAL —definite integral.

DATE, *n.* **after-date draft.** See DRAFT.

average date. Same as EQUATED DATE.

dividend date. See DIVIDEND —dividend date.

equated date. See EQUATED.

focal date. See COMMUTING—commuting obligations.

DEATH, *n.* **death rate.** Same as RATE OF MORTALITY. See MORTALITY.

central death rate, during one year. The ratio of the number of persons dying during that year to the number living at some particular time during the year; denoted by M_x, where x is the year. Usually M_x is defined as $d_x / [\frac{1}{2}(l_x + l_{x+1})]$, where d_x is the number of the group dying during the year x, l_x is the number living at the beginning of the year, and l_{x+1} the number living at the end. Compare MORTALITY —rate of mortality.

DE-BEN′TURE, *adj., n.* A written recognition of a debt or loan; usually carries the seal of a corporation or other firm and represents funds raised in addition to ordinary stocks and bonds. Such a **debenture bond** is usually unsecured and protected only by the credit and earning power of the issuer.

DEBT, *n.* An obligation to pay a certain sum of money.

DEBT′OR, *n.* One who owes a debt.

DEC′A (or **DEK′A**). A prefix meaning 10, as in **decameter**.

DEC′ADE, *n.* (1) A division or group of ten. Thus the numbers from 1 to 10 inclusive form one decade, those from 11 to 20 inclusive another, etc. (2) Ten years.

DEC′A-GON, *n.* A polygon having ten sides. It is a **regular decagon** if the polygon is a regular polygon.

DEC′A-ME′TER, *n.* A term used in the metric system; 10 meters or approximately 32.808 feet. See DENOMINATE NUMBERS in the appendix.

DE-CEL′ER-A′TION, *n.* Negative acceleration. See ACCELERATION.

DEC′I. A prefix meaning 1/10, as in **decimeter**.

DEC′I-MAL, *adj., n.* Any numeral in the *decimal number system* (see below), but sometimes restricted to a **decimal fraction** (*i.e.*, a number that in decimal notation has no digits other than zeros to the left of the decimal point. A **mixed decimal** is a decimal plus an integer, as 23.35. Decimals having the same number of decimal places, as 2.361 and 0.253, are **similar**. Any decimals can be made similar by annexing the proper number of zeros; *e.g.*, 0.36 can be made similar to 0.321 by writing it as 0.360 (see DIGIT—significant digits).

    accurate to a certain decimal place. See ACCURATE.

    addition and multiplication of decimals. See ADDITION—addition of decimals, PRODUCT—product of real numbers.

    decimal equivalent of a common fraction. A decimal fraction equal to the common fraction. *E.g.*, $\frac{1}{8} = 0.125$; $\frac{1}{3} = 0.333\cdots$.

    decimal expansion. For a real number, the representation of the number using the decimal number system.

    decimal measure. Any system of measuring in which each unit is derived from some standard unit by multiplying or dividing the latter by some power of 10. See METRIC—metric system.

    decimal number. The representation of a number using the decimal number system.

    decimal number system. A system of numerals for representing real numbers that uses the base 10 (see BASE—base of a number system); the common system of numerals for which each real number is represented by a sequence of the digits 0, 1, 2, 3, 4, 5, 6, 7, 8, 9, and a single period (**decimal point**) in some order (see DECIMAL). A **finite decimal** (or **terminating decimal**) is a decimal that contains a finite number of digits; to be able to represent all real numbers, it is necessary to use **infinite decimals** (or **nonterminating decimals**) as well, *i.e.*, decimals that have an unending string of digits to the right of the decimal point. A **repeating decimal** (or **periodic decimal**) is a decimal that either is finite or is infinite and has a finite block of digits that eventually repeats indefinitely. For example, $\frac{15}{28} =$ 0.53 571428 571428 $\cdots$ is a repeating decimal with the repeating block 571428. A number has a repeating decimal representation if and only if it is a rational number. A decimal that is not repeating is **nonrepeating** (or **nonperiodic**) and represents an irrational number. See LIOUVILLE —Liouville number, NORMAL—normal number.

    decimal place. The position of a digit in a decimal. *E.g.*, in 123.456, 1 is in the *hundred's place*, 2 is in the *ten's place*, 3 is in the *unit's place*, 4 is in the *first decimal place*, 5 in the *second decimal place*, and 6 in the *third decimal place*.

    decimal point. See above, decimal number system.

    decimal system. (1) The decimal number system (see above). (2) Any system of decimal measure, as the *metric system*.

    floating decimal point. See FLOATING.

DEC′I-ME′TER, *n.* A term used in the metric system; one-tenth of a meter; approximately 3.937 inches. See DENOMINATE NUMBERS in the appendix.

DEC′LI-NA′TION, *n.* **declination of a celestial point.** Its angular distance north or south of the celestial equator, measured along the hour circle passing through the point. See HOUR—hour angle and hour circle.

    north declination. The celestial declination of a point which is north of the celestial equator. It is always regarded as *positive*.

    south declination. The celestial declination of a point which is south of the celestial equator. It is always regarded as *negative*.

DE′COM-PO-SI′TION, *n.* **decomposition of a fraction.** Breaking a fraction up into **partial fractions**.

decomposition of a set. Representation of the set as the union of (usually finitely many) pairwise disjoint subsets.

DE′CREASE, *n.* **percent decrease.** See PERCENT.

DE-CREAS′ING, *adj.* **decreasing function of one variable.** A function whose value decreases as the independent variable increases; a function whose graph falls as the abscissa increases. If a function is differentiable on an interval I, then the function is decreasing on I if the derivative is nonpositive throughout I and is not identically zero in any interval. A decreasing function is often said to be **strictly decreasing**, to distinguish it from a *monotone decreasing* function. *Tech.* A function f is **strictly decreasing** on an interval (a, b) if

$$f(y) < f(x)$$

for all numbers x and y of I with $x < y$; f is **monotone decreasing** on I if $f(y) \le f(x)$ for all numbers x and y of I with $x < y$. See MONOTONE.

decreasing the roots of an equation. See ROOT—root of an equation.

decreasing sequence. A sequence $x_1, x_2, \cdots$, for which $x_i > x_j$ if $i < j$. The sequence is **monotone decreasing** if $x_i \ge x_j$ if $i < j$. See MONOTONE.

DEC′RE-MENT, *n.* The decrease, at a given age, of the number of lives in a given group—such as the number in the service of a given company.

DEDEKIND, Julius Wilhelm Richard (1831–1916). German mathematician who worked in number theory and analysis, particularly with algebraic integers, algebraic functions, and ideals. He defined real numbers by *Dedekind cuts* of the sets of rational numbers and thus clarified notions concerning the real number system. See EUDOXUS.

Dedekind cut. A subdivision of the rational numbers into two nonempty disjoint sets A and B such that: (a) if x belongs to A and y to B, then $x < y$; (b) set A has no largest member (this condition can be replaced by the requirement that B have no least member). *E.g.*, A might be the set of all rational numbers less than 3, and B the set of all rational numbers greater than or equal to 3; or A might be the set whose members are the negative rational numbers, 0, and all positive rational numbers x for which $x^2 < 2$, and B the set of all positive rational numbers x for which $x^2 > 2$. In the first exam-

ple, B has a least member; in the second example, it does not. The **real numbers** can be defined as the set of all Dedekind cuts. It is then convenient to use the notation (A, B) for the real number or cut consisting of the sets A and B. Inequality, addition, and multiplication of real numbers can then be defined as follows: **Inequality**: $(A_1, B_1) > (A_2, B_2)$ if there is an x which belongs to A_1 but does not belong to A_2. **Addition**: If (A_1, B_1) and (A_2, B_2) are real numbers, their sum is the real number (A, B) for which A is the set of all $x + y$ where x belongs to A_1 and y belongs to A_2. **Multiplication**: If (A_1, B_1) and (A_2, B_2) are real numbers, their product is the real number (A, B) for which A is the set of all rational numbers x with the property that, for any positive number ϵ, there are numbers a_1, b_1, a_2, b_2, belonging to A_1, B_1, A_2, B_2, respectively, such that $b_1 - a_1 < \epsilon$, $b_2 - a_2 < \epsilon$, and x is less than each of $a_1 a_2, a_1 b_2, b_1 a_2, b_1 b_2$ (note that if A_1 and A_2 each contain positive numbers, then A is the set of all nonpositive rational numbers and all products xy of positive numbers x and y which belong to A_1 and A_2, respectively. A real number (A, B) for which there is a rational number a which is the least upper bound of A is usually identified with a and called a rational number, since the correspondence $a \leftrightarrow (A, B)$ preserves order, sums, and products. See IRRATIONAL—irrational number. Dedekind cuts of the real numbers might now be similarly defined as subdivisions of the real numbers into two subsets, but this would yield a set of objects isomorphic with the real numbers themselves and would not lead to a further extension of the number system.

DE-DUC′TIVE, *adj.* **deductive method or theory.** A formal structure based on a set of *unproved* axioms and a set of *undefined* objects. New terms are defined in terms of the given undefined objects and new statements, or *theorems*, are derived from the axioms by proof. A *model* for a deductive theory is a set of objects that have the properties stated in the axioms. The deductive theory can be used to prove theorems that are true for all of its models. *Syn.* axiomatic method. See AXIOM.

DE-FAULT′ED, *adj.* **defaulted payments.** (1) Payments made on principal after the due date; occurs most frequently in installment plan paying. (2) Payments never made.

DE-FEC′TIVE, *adj.* **defective equation.** See EQUATION—defective equation.

defective (or deficient) number. See NUMBER—perfect number.

DE-FERRED′, *adj.* **deferred annuity and life insurance.** See ANNUITY, INSURANCE —life insurance.

DE-FI′CIEN-CY, *n.* **premium deficiency reserve.** See RESERVE.

DE-FI′CIENT, *adj.* **deficient number.** See NUMBER—perfect number.

DEF′I-NITE, *adj.* **definite integral.** See INTEGRAL—definite integral.
    definite integration. The process of finding definite integrals.
    partial definite integral. One of the definite integrals constituting an ITERATED INTEGRAL.
    positive definite and semidefinite quadratic form. See FORM.

DEF-I-NI′TION, *n.* An agreement to use something (*e.g.*, a symbol or set of words) as a substitute for something else, usually for some expression that is too lengthy to write easily or conveniently. For example, the definition "a *square* is a rectangle whose angles are all right angles and whose sides are equal in length" is an agreement to substitute "square" for "rectangle whose angles are all right angles and whose sides are equal in length."

DEF′OR-MA′TION, *adj.*, *n.* **continuous deformation.** A transformation which shrinks, twists, etc., in any way without tearing. *Tech.* **A continuous deformation** of an object *A* into an object *B* is a continuous mapping $T(p)$ of *A* onto *B* for which there is a function $F(p, t)$ which is defined and continuous (simultaneously in *p* and *t*) for real numbers *t* with $0 \leq t \leq 1$ and points *p* of *A* and for which $F(p, 0)$ is the identity mapping of *A* onto *A*, $F(p, 0) \equiv p$, and $F(p, 1)$ is identical with $T(p)$. With this definition, a circle in the plane can be continuously deformed to a point (although a circle around the outer circumference of a torus cannot be deformed continuously into a point, or into one of the small circles around the body of the torus, without leaving the torus—*i.e.*, with all values of $F(p, t)$ being points on the torus). It is frequently required that a continuous deformation not bring points together; *i.e.*, that the above function $F(p, t)$ be a one-to-one correspondence for each value of *t*. Then a circle in the plane can be continuously deformed into a square, but not into a point or a figure "8"; a sphere with one hole can be continuously deformed into a disc (a circle and its interior), but not into a cylinder or a sphere. Two mappings T_1 and T_2 of a topological space *A* into a topological space *B* can be continuously

deformed into each other if there is a function $F(x, t)$ which has values in *B*, which is continuous simultaneously in *x* and *t* for *x* in *A* and $0 \leq t \leq 1$, and for which $F(x, 0) \equiv T_1(x)$ and $F(x, 1) \equiv T_2(x)$ for each *x* of *A*. Two mappings are **homotopic** if they can be continuously deformed into each other. If *A* is contained in *B* and T_1 is the identity mapping of *A* onto *A*, then T_2 is a continuous deformation (in the above sense) of *A* into the range of T_2 if T_1 can be continuously deformed into T_2. See INESSENTIAL.

deformation (*Elasticity*) A change in the position of the points of a body accompanied by a change in the distance between them. See STRAIN.

deformation ratio. In a conformal map, the magnification is the same in all directions at a point, *i.e.*, $ds^2 = [M(x, y)]^2(dx^2 + dy^2)$. The function $M(x, y)$ is the **linear deformation ratio**. The **area deformation ratio** is $[M(x, y)]^2$. If the map is given by the analytic function $w = f(z)$ of the complex variable *z*, then

$$M = |f'(z)|.$$

Syn. ratio of magnification.

DE-GEN′ER-ATE, *adj.* **degenerate conics.** See CONIC—degenerate conic.

DE-GREE′, *n.* (1) A unit of angular measure. See SEXAGESIMAL —sexagesimal measure of an angle. (2) A unit of measure of temperature. (3) Sometimes used in the same sense as *period* in arithmetic. (4) See the various headings below.
    degree of an alternating group or a permutation group. See PERMUTATION—permutation group.
    degree of arc. See ARC.
    degree of curve. See CURVE—algebraic plane curve.
    degree of a differential equation. The degree of the equation with respect to its highest-order derivative; the greatest power to which the highest-order derivative occurs. The degree of

$$\left(\frac{d^4 y}{dx^4}\right)^2 + 2\left(\frac{dy}{dx}\right)^3 = 0$$

is two. See DIFFERENTIAL —differential equation (ordinary).
    degree of an extension of a field. See EXTENSION—extension of a field.
    degree of freedom. See FREEDOM.
    degree of a polynomial, or equation. The degree of its highest-degree term. The degree of a

term in one variable is the exponent of that variable; a term in several variables has degree equal to the sum of the exponents of its variables (or one may speak of the degree with respect to a certain variable, meaning the exponent of that variable). *E.g.*, $3x^4$ is of degree four; $7x^2yz^3$ is of degree six, but of degree two in x. The equation $3x^4 + 7x^2yz^3 = 0$ is of degree six, but is of degree four in x, of degree one in y, and of degree three in z.

general equation of the *n*th degree. See EQUATION —polynomial equation.

spherical degree. See SPHERICAL.

DEL, *n.* The operator $\mathbf{i}\dfrac{\partial}{\partial x} + \mathbf{j}\dfrac{\partial}{\partial y} + \mathbf{k}\dfrac{\partial}{\partial z}$, denoted by ∇. Also known as **nabla.** See GRADIENT —gradient of a function. DIVERGENCE —divergence of a vector function.

DELAMBRE, Jean Baptiste Joseph (1749–1822). French astronomer and geometer.

Delambre's analogies. Same as GAUSS'S FORMULAS.

DE LA VALLÉE POUSSIN, Charles Jean Gustave Nicolas (1866–1962). Belgian analyst and number theorist. See PRIME —prime number theorem.

DELIGNE, Pierre Jacques (1944—   ). French mathematician awarded a Fields Medal (1978) for his work on the relations between the cohomological structure of algebraic varieties over the complex numbers and the diophantine structure of algebraic varieties over finite fields.

DEL'TA, *adj., n.* The fourth letter of the Greek alphabet: lower case, δ; capital, Δ.

delta distribution. See DISTRIBUTION (2).

DEL'TA-HE'DRON, *n.* A polyhedron whose faces are congruent equilateral triangles, but whose polyhedral angles are not necessarily congruent. There are exactly eight such convex polyhedrons.

DEL'TOID, *n.* A non-convex quadrilateral which has two pairs of adjacent equal sides. It is symmetric about a diagonal.

DE-MAND', *adj.* **demand note.** See NOTE.

DE MOIVRE, Abraham (1667–1754). Statistician, probabilist, and analyst; born in France,

studied in Belgium, settled in England. Remembered principally for *De Moivre's theorem* and his work in probability theory.

De Moivre's theorem. A rule for raising a complex number to a power when the number is expressed in polar form. The rule is: Raise the modulus to the given power and multiply the amplitude by the given power; *i.e.*

$$\left[r(\cos\theta + i\sin\theta) \right]^n = r^n(\cos n\theta + i\sin n\theta).$$

E.g.,

$$\left(\sqrt{2} + i\sqrt{2} \right)^2 = \left[2(\cos 45° + i\sin 45°) \right]^2$$

$$= 4(\cos 90° + i\sin 90°)$$

$$= 4i.$$

DE MORGAN, Augustus (1806–1871). British analyst, logician, and probabilist.

De Morgan formulas. For sets A and B, the formulas

$$(A \cup B)' = A' \cap B'$$

and

$$(A \cap B)' = A' \cup B',$$

where the complement of a set S is indicated by S'. These formulas are also valid for any collection of subset $\{A_\alpha\}$ of S, and may be stated symbolically as

$$\left(\cup A_\alpha \right)' = \cap A'_\alpha \quad \text{and} \quad \left(\cap A_\alpha \right)' = \cup A'_\alpha.$$

DE-NI'AL, *n.* Same as NEGATION.

DE-NOM'I-NATE, *adj.* **denominate number.** A number whose unit represents a unit of measure —such as 3 inches, 2 pounds, or 5 gallons.

addition, subtraction, or multiplication of denominate numbers. The process of reducing them to the same denomination and then proceeding as with ordinary (abstract) numbers. *E.g.*, to find the number of square yards in a room $17'\,6''$ by $12'\,4''$, the length in yards is $5\frac{5}{6}$ and the width is $4\frac{1}{9}$. The required number of square yards is $5\frac{5}{6} \times 4\frac{1}{9}$. See PRODUCT —product of real numbers.

DE-NOM'I-NA'TION, *n.* **denomination of a bond.** Its par value.

denomination of a number. See NUMBER — denominate number.

DE-NOM′I-NA′TOR, *n.* The term below the line in a fraction; the term that divides the numerator. The denominator of $\frac{2}{3}$ is 3. See CON- SEQUENT (1).

common denominator. See COMMON.

DENSE, *adj.* **dense set.** A set E in a space M is **dense** (or **dense in** M, or **everywhere dense**) if every point of M is a point of E or a limit point of E, or (equivalently) if the closure of E is M, or if every neighborhood in M contains a point of E. A set E is **dense in itself** if every point of E is an accumulation point of E; *i.e.*, if each neighborhood of any point of E contains another point of E. A set E is **nondense** (or **nowhere dense**) relative to M if no neighborhood in M is contained in the closure of E, or (equivalently) if the complement of the closure of E is dense in M. The set of rational numbers is dense in itself and dense in the set R of all real numbers, as is also the set of irrational numbers. This is equivalent to the fact that between any two real numbers (either rational or irrational) there are both rational numbers and irrational numbers. However, the set $S = \{0, 1, \frac{1}{2}, \frac{1}{3}, \frac{1}{4}, \cdots\}$ is nowhere dense in R, since 0 is the only accumulation point of S and therefore no neighborhood in R contains only accumulation points of S.

DEN′SI-TY, *n.* The mass or amount of matter per unit volume. Since the mass of 1 cc. of water at 4° C. is one gram, density in the *metric system* is the same as *specific gravity*. See SPECIFIC.

density of a sequence of integers. Let $\{a_1, a_2, \cdots\}$ be an increasing sequence A of integers. Let $F(n)$ be the number of integers which are in this sequence and not larger than n. Then $0 \le F(n)/n \le 1$. The **density** $d(A)$ of A is the *greatest lower bound* of $F(n)/n$; the **asymptotic density** is the *limit inferior* of $F(n)/n$; the **upper density** is the *limit superior* of $F(n)/n$; and the **natural density** is $\lim_{n \to \infty} F(n)/n$, if this limit exists (see SEQUENCE — accumulation point of a sequence). The density $d(A)$ is 0 if $a_1 \ne 1$, or if A contains "very few" of the integers; *e.g.*, if A is a geometric sequence, a sequence of primes, or a sequence of perfect squares. Let the sum of two sequences A and B of the above type be defined as the sequence of all numbers (arranged in order of size) which can be represented as a sum of a term of one sequence and a term of the other sequence. It can be proved that $d(A + B) \ge d(A) + d(B)$ if $d(A) + d(B) \le 1$. A sequence has density 1 if and only if it contains all nonzero integers.

mean density. The mass divided by the volume. *Tech.*

$$\int_v \rho \, dv \div \int_v dv,$$

where ρ is the density and $\int_v$ denotes the integral taken over the total volume.

metric density. See METRIC.

surface density of charge. Charge per unit area. It is sometimes advantageous to think of a body as having a skin of definite thickness at its boundary. If we think of all of the charge in the skin as being shifted and concentrated on the outer surface of the skin, then so far as total charge is concerned we may replace the original charge per unit volume of the skin with the charge per unit area obtained by the shift. The volume integral of the former density would equal the surface integral of the latter.

surface density (or moment per unit area) of a double layer (or dipole distribution). Polarization per unit area. Instead of charge being concentrated on the surface, we may consider that we have a continuous distribution of dipoles spread over the surface.

volume density of charge. Charge per unit volume. The most fundamental property of density of charge is that its volume integral taken throughout any given volume V gives the total charge in V. If we start with point charge as the fundamental concept, instead of density, it is found that we may approximate the electric field at points external to the complex as closely as we please by introducing a sufficiently complicated density function. See POTENTIAL — concentration method for the potential of a complex. In terms of total charge, density may be defined as the limit of e_i/V_i, where V_i is a sequence of regions having the property that each V_i is within a sphere of radius r_i with center at P and $\lim r_i = 0$, e_i is the total charge in V_i, and it is required that the limit of e_i/V_i be independent of the sequence of regions V_i.

DE-NU′MER-A-BLE, *adj.* **denumerable set.** Same as COUNTABLE SET. *Denumerably infinite* is used in the same sense as *countably infinite*, to denote an infinite set whose elements can be put into one-to-one correspondence with the positive integers.

DE-PAR′TURE, *n.* **departure between two meridians on the earth's surface.** The length of the arc of the parallel of latitude subtended by the two meridians. This grows shorter the nearer the parallel of latitude is to a pole. Used in parallel sailing.

DE-PEND′ENCE, *n.* **domain of dependence.** For an initial-value problem for a partial differential equation, the value of the solution at a point p and time t might be determined by the initial values on only a portion of their entire range, called the *domain of dependence.* *E.g.*, for the wave equation $(1/c^2)u_{tt} = u_{xx}$, with initial conditions $u(x, 0) = f(x)$, $u_t(x, 0) = g(x)$, the value of the solution at the point x and time t depends only on the initial values in the interval $[x - ct, x + ct]$. See HUYGENS —Huygens principle.

DE-PEND′ENT, *adj.* **dependent equations.** An equation is **dependent** on a set of equations if it is satisfied by every set of values of the unknowns that satisfy all the other equations. One of three linear equations in two unknowns is dependent on the other two if the graphs of these two are not coincident and the three graphs are concurrent lines. A set of equations (perhaps not all different—but distinguished by indices or in some other way), is **dependent** if some member of the set is dependent on the others. When solving a system of equations, any dependent equation can be discarded.

    dependent event. See EVENT.

    dependent functions. A set of functions (perhaps not all different—but distinguished by indices or in some other way), one of which can be expressed as a function of the others. *E.g.*, $v = \sin u$ and thus $\{u, v\}$ is a set of dependent functions if

$$u(x, y) = \frac{x + 1}{y + 1}, \qquad v(x, y) = \sin \frac{x + 1}{y + 1}.$$

Also, f_1 and f_2 are dependent if $f_1(x) = x^2$ and $f_2(x) = x^2$ for all x, since $f_1 = f_2$. *Syn.* interdependent functions. Functions that are not *independent* are *dependent* [see INDEPENDENT].

    dependent variable. See FUNCTION —function of one variable.

    linearly dependent. A set of objects z_1, $z_2, \cdots, z_n$ (vectors, matrices, polynomials, etc.) is **linearly dependent** on a given set if there is a linear combination of them,

$$a_1 z_1 + a_2 z_2 + \cdots + a_n z_n,$$

which is zero (with the coefficients in the given set), but at least one of the coefficients is nonzero. The dependence is relative to the nature of the coefficients allowed (see the examples to follow).

A set of objects is **linearly independent** if it is not linearly dependent. The binomials $x + 2y$ and $3x + 6y$ are linearly dependent, since

$$-3(x + 2y) + (3x + 6y) \equiv 0.$$

The numbers 3 and π are linearly independent with respect to rational numbers, since $a_1 \cdot 3 + a_2 \cdot \pi$ cannot be zero if a_1 and a_2 are rational numbers, not both zero. Since

$$-1 \cdot 3 + (3/\pi)\pi = 0,$$

3 and π are linearly dependent with respect to real numbers. Similarly, $1 + i$ and $3 - 5i$ are *linearly independent* with respect to the field of real numbers and *linearly dependent* with respect to the field of complex numbers. If $v^k = (x_1^k, x_2^k, \cdots, x_n^k)$, $k = 1, 2, \cdots, r$, are vectors (or points) of n-dimensional space, then these vectors are linearly dependent if there exist numbers $\lambda_1, \lambda_2, \cdots, \lambda_r$, not all zero, such that $\lambda_1 v^1 + \lambda_2 v^2 + \cdots + \lambda_r v^r = 0$. This means that a similar equation holds for each component: $\lambda_1 x_p^1 + \lambda_2 x_p^2 + \cdots + \lambda_r x_p^r = 0$ for each p. See GRAMIAN, WRONSKI —Wronskian.

DE-PRE′CI-A′TION, *adj., n.* The loss in value of equipment; the difference between the *cost value* and the *book value.*

    depreciation charge. A decrease in the book value, usually annual, such that the total of these decreases, without interest, will equal the original book value (or cost) minus the scrap value at the end of a certain number of years (the **total depreciation**). There are various methods for computing depreciation charges. For the **straight line method**, equal depreciation charges are made each year. For the **declining balance method** (or **constant percentage method**) each depreciation charge is computed as a constant percent of the book value at the time of computation; this percent is equal to

$$\left(1 - \sqrt[n]{R/C}\right) \times 100,$$

where C is the cost, R the scrap value, and n the number of years.

DE-PRESSED′, *adj.* **depressed equation.** The equation resulting from reducing the number of roots in an equation, *i.e.*, by dividing out the difference of the unknown and a root; *e.g.*, $x^2 - 2x + 2 = 0$ is the *depressed* equation obtained from $x^3 - 3x^2 + 4x - 2 = 0$ by dividing the left member of the latter by $x - 1$.

DE-PRES′SION, *n.* **angle of depression.** See ANGLE—angle of depression.

DE-RIV′A-TIVE, *n.* The instantaneous rate of change of a function with respect to the variable. Let f be a given **function of one variable** and let Δx denote a number (positive or negative) to be added to the number x. Let Δf denote the corresponding increment of f:

$$\Delta f = f(x + \Delta x) - f(x).$$

Form the increment ratio

$$\frac{\Delta f}{\Delta x} = \frac{f(x + \Delta x) - f(x)}{\Delta x}.$$

Then let Δx approach zero. If $\Delta f/\Delta x$ approaches a *limit* as Δx approaches zero, this limit is the **derivative** of f at the point x. If $f(x) = x^2$, the process gives:

$$f(x + \Delta x) = (x + \Delta x)^2;$$

$$f(x + \Delta x) - f(x) = (x + \Delta x)^2 - x^2;$$

$$\frac{f(x + \Delta x) - f(x)}{\Delta x} = \frac{(x + \Delta x)^2 - x^2}{\Delta x}$$

$$= 2x + \Delta x;$$

$$\lim_{\Delta x \to 0} (2x + \Delta x) = 2x = \frac{dx^2}{dx}.$$

The derivative of a function f is a function; this function is denoted by symbols such as

$$f'(x), D_x f, \frac{df}{dx}, f'(x), D_x f(x),$$

$$\frac{d}{dx} f(x) \quad \text{or} \quad \frac{df(x)}{dx}.$$

The derivative, evaluated at a point a, is written

$$f'(a), D_x f(x)_{x=a}, \left[\frac{df(x)}{dx}\right]_{x=a},$$

$$f'(x)|_{x=a}, \quad \text{etc.}$$

The definition of the derivative of f at the point a can also be written in the form

$$f'(a) = \lim_{x \to a} \frac{f(x) - f(a)}{x - a}.$$

Two special interpretations of the derivative are: (1) As the **slope** of a curve: In the figure, the ratio $\Delta y/\Delta x$ is the slope of the line PP'. Therefore the limit of this ratio as Δx approaches zero is the slope of the tangent PT. It follows that a

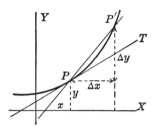

function is **increasing** in intervals where the derivative is positive and **decreasing** in intervals where the derivative is negative. If the derivative is zero, the function may have a **maximum** or **minimum** at the point (see MAXIMUM). (2) As the **speed** and **acceleration** of a moving particle: If $f(t)$ is the distance traversed by the particle in time t, then the derivative of f at t_1 is the *speed* of the particle at the time t_1; the increment ratio $\Delta f/\Delta t$ is the *average speed* during the time interval Δt. The derivative of the speed (the second derivative of the distance) at t_1 is the *acceleration* of the particle at the time t_1. There are many powerful and useful formulas for evaluating derivatives (see DIFFERENTIATION FORMULAS in the appendix). *E.g.*, the **derivative of a sum** is the sum of the derivatives, the derivative of x^n is nx^{n-1}, and the derivative of a function $F(u)$, where u is a function of x, is given by the following formula (**chain rule**):

$$\frac{dF(u)}{dx} = \frac{dF(u)}{du} \frac{du}{dx}.$$

From these rules, it follows that

$$D_x(x^3 + x^2) = 3x^2 + 2x,$$

$$D_x\left[x^{1/2} + (x^2 + 7)^\pi\right]$$

$$= \frac{1}{2}x^{-1/2} + \pi(x^2 + 7)^{\pi-1}(2x),$$

etc. *Syn.* differential coefficient. See various headings under ACCELERATION, DIFFERENTIAL, DIFFERENTIATION, LEIBNIZ, TANGENT, VELOCITY. Similar definitions of the derivative are used when the function is of a different type (see below, derivative of a function of a complex variable, derivative of a vector).

chain rule for derivatives. See CHAIN.

covariant derivative of a tensor. See COVARI-ANT.

derivative of a distribution. See GENERAL-IZED—generalized function.

derivative of a function of a complex variable. The complex-valued function f whose domain contains a neighborhood of the complex number z_0 is differentiable at z_0 if and only if

$$\lim_{z \to z_0} \frac{f(z) - f(z_0)}{z - z_0}$$

exist. The limit is the **derivative** of f at z_0 and is denoted by $f'(z)|_{z=z_0}, f'(z_0), \frac{df}{dz}\Big|_{z=z_0}$, etc. See ANALYTIC —analytic function of a complex variable.

derivative of higher order. Derivatives of other derivatives, the latter being considered as functions of the independent variable just as was the function of which the first derivative was taken. E.g., $y = x^3$ has the first derivative $y' = 3x^2$ and the second derivative $y'' = 6x$, obtained by taking the derivative of $3x^2$; similarly $y''' = 6$, and $y^{[4]} = 0$.

derivative of an integral. (1) The derivative of $\int_a^x f(t)\,dt$ at the point x_0 exists and is equal to $f(x_0)$, provided f is integrable on the interval (a,b) and is continuous at x_0, where x_0 is in the open interval (a, b). See FUNDAMENTAL —fundamental theorem of calculus. (2) If $f(t, x)$ has a partial derivative $\partial f/\partial t = f_t(t, x)$ which is continuous in both x and t for t in the closed interval $[a, b]$ and x in an interval that contains x_0 as an interior point, and $\int_a^b f(t, x)\,dx = F(t)$ exists, then dF/dt exists and equals $\int_a^b f_t(t, x)\,dx$. This is sometimes called *Leibniz's rule*, although he did not specify the conditions on $f(t, x)$. (3) Combining (1) and (2) by using the *chain rule* for partial differentiation gives the formula:

$$D_t \int_u^v f(t, x)\,dx = D_t v \cdot f(t, v) - D_t u \cdot f(t, u)$$

$$+ \int_u^v f_t(t, x)\,dx.$$

E.g., the derivative of $\int_1^2 (x^2 + y)\,dx$, with respect to y, is $\int_1^2 dx$, and the derivative, with respect to y, of $\int_y^{y^2} (x^2 + y)\,dx$ is

$$\int_y^{y^2} dx + (y^4 + y)2y - (y^2 + y).$$

derivative from parametric equations. See PARAMETRIC—differentiation of parametric equations.

derivative of a vector. Let t be the parameter of a curve and suppose that corresponding to each point of the curve there is a vector. Let $v(t)$ denote the vector at the point of the curve for which the parameter has the value t. Then

$$\lim_{\Delta t \to 0} \frac{v(t + \Delta t) - v(t)}{\Delta t}$$

is the **derivative of the vector**, relative to parameter of the curve, at the point t, provided this limit exists. See ACCELERATION, VELOCITY.

directional derivative. See DIRECTIONAL.

formal derivative. See FORMAL.

normal derivative. See NORMAL—normal derivative.

partial derivative. See PARTIAL.

total derivative. See CHAIN—chain rule (for partial differentiation).

DE-RIVED′, *adj.*, *n.* **derived curve.** The **first derived curve** of a given curve is the curve each of whose ordinates is equal to the slope of the given curve for the same value of the abscissas. E.g., the curve whose equation is $y = 3x^2$ is the derived curve of the curve whose equation is $y = x^3$. The derived curve of the first derived curve is the **second derived curve**, etc.

derived equation. (1) *In algebra*, an equation obtained from another one by adding terms to both sides, powering both, or multiplying or dividing by some quantity. A derived equation is not always equivalent to the original, *i.e.*, does not always have the same set of roots. (2) *In calculus*, the equation resulting from differentiating the given equation. See above, derived curve.

derived set. See CLOSURE.

DESARGUES, Girard (1591–1661). French geometer who initiated formal study of projective geometry.

Desargues' theorem on perspective triangles. The lines joining corresponding vertices of two triangles are concurrent if and only if the intersections of the three pairs of corresponding sides are collinear.

DESCARTES, René (1596–1650). Philosopher and mathematician; born in France, lived in several countries of Western Europe, settled in Holland. Founded analytic geometry (an honor shared with Fermat), which is commonly called *Cartesian geometry* after the latinized version of his name, Cartesius. For this reason, he and Fermat are often considered to be the first modern mathematicians.

Descartes' rule of signs. A rule determining an upper bound to the number of positive zeros and to the number of negative zeros of a polynomial. The rule states that the number of positive zeros of a polynomial $p(x)$ either is equal to the number of its variations of sign or is less than that number by an even integer, a zero of multiplicity m being counted as m zeros. The polynomial can be tested for negative zeros by applying this rule to the polynomial $p(-x)$. For example, the polynomial

$$x^4 - x^3 - x^2 + x - 1$$

has three variations in sign and hence has either three or one positive zeros. Since replacing x by $-x$ gives $x^4 + x^3 - x^2 - x - 1$, which has only one variation of sign, the original polynomial has exactly one negative zero. See VARIATION —variation of sign in a polynomial.

folium of Descartes. See FOLIUM.

DE-SCEND′ING, *adj.* **descending chain condition on rings.** See CHAIN—chain conditions on rings.

DE-SCENT′, *n.* **method of steepest descent.** See STEEPEST.

proof by descent or **method of infinite descent.** Same as MATHEMATICAL INDUCTION. See INDUCTION.

DE-TACHED′, *adj.* **detached coefficient.** See COEFFICIENT —detached coefficients.

DE-TACH′MENT, *n.* **rule of detachment.** If an implication is true and the antecedent is true, then the consequent is true. *E.g.,* if the statements "If my team lost, I will eat my hat" and "My team lost" are both true, then the statement "I will eat my hat" is true.

DE-TER′MI-NANT, *n.* A square array of quantities, called **elements**, symbolizing the sum of certain products of these **elements**. The number of rows (or **columns**) is the **order** of the determinant. The diagonal, from the upper left corner to the lower right corner, is the **principal (or leading) diagonal**. The diagonal from the lower left corner to the upper right corner is the **secondary diagonal**. A determinant of the second order is a square array of type

$$\begin{vmatrix} a_1, & b_1 \\ a_2, & b_2 \end{vmatrix},$$

whose value is $a_1 b_2 - a_2 b_1$. A determinant of the third order is a square array of type

$$\begin{vmatrix} a_1 & b_1 & c_1 \\ a_2 & b_2 & c_2 \\ a_3 & b_3 & c_3 \end{vmatrix},$$

whose value is $(a_1 b_2 c_3 + a_2 b_3 c_1 + a_3 b_1 c_2 - a_3 b_2 c_1 - a_2 b_1 c_3 - a_1 b_3 c_2)$. This expression is equal to the sum of the products of the elements in a given column (or row) by their cofactors (see below, expansion of a determinant by minors). The element in row i and column j of a determinant is usually indicated by some such symbol as a_{ij}, where i is called the row index and j the column index. The value of the determinant is then the algebraic sum of all products obtained by taking one and only one factor from each row and each column and attaching the positive or negative sign to each product according as the column (or row) indices form an even or an odd permutation when the row (or column) indices are in natural order $(1, 2, 3,$ etc.$)$. *E.g.,* the term $a_{13} a_{21} a_{34} a_{42}$ of the expansion of a determinant of order four has the column indices in order $(3, 1, 4, 2)$. This term should have a negative sign attached, since three successive interchanges will change the column indices to $(1, 3, 4, 2), (1, 3, 2, 4)$, and $(1, 2, 3, 4)$, the last being in natural order. In practice, a determinant is usually evaluated by using minors (see below, expansion of a determinant by minors, Laplace's expansion of a determinant) after simplifying the determinant by use of certain properties of determinants. Some of the simple properties of determinants are: (1) If all the elements of a column (or row) are zero, the value of the determinant is zero. (2) Multiplying all the elements of a column (or row) by the same quantity is equivalent to multiplying the value of the determinant by this quantity. (3) If two columns (or rows) have their corresponding elements equal, the determinant is zero. (4) The value of a determinant is unaltered if the same multiple of the elements of any column (row) are added to the corresponding elements of any other column (row). (5) If two columns (or rows) of a determinant are interchanged, the sign of the determinant is changed. (6) The value of a determinant is unaltered when all the corresponding rows and columns are interchanged. See ALTERNATING —alternating function.

cofactor of an element in a determinant. See MINOR—minor of an element in a determinant.

conjugate elements of a determinant. See CONJUGATE—conjugate elements of a determinant.

determinant of the coefficients of a set of linear equations. For n equations in n unknowns, the determinant whose element in the ith row and jth column is the coefficient of the jth variable in the ith equation (the variables being written in the same order in each equation). This determinant is not defined if the number of equations is not equal to the number of variables (see MATRIX—matrix of the coefficients). The determinant of the coefficients of the variables of

$$\begin{cases} 2x + 3y - 1 = 0 \\ 4x - 7y + 5 = 0 \end{cases} \text{is} \begin{vmatrix} 2 & 3 \\ 4 & -7 \end{vmatrix}.$$

determinant of a matrix. See MATRIX.

elementary operations on determinants. See ELEMENTARY.

expansion of a determinant by minors. The expansion of the determinant by writing it in terms of determinants of one lower order, using the elements of a selected row (or column) as coefficients. The determinant is equal to the sum of the products of the elements of that row (or column) by their *signed minors* (or *cofactors*) (see MINOR—minor of an element in a determinant). *E.g.*,

$$\begin{vmatrix} a_1 & b_1 & c_1 \\ a_2 & b_2 & c_2 \\ a_3 & b_3 & c_3 \end{vmatrix} = a_1 \begin{vmatrix} b_2 & c_2 \\ b_3 & c_3 \end{vmatrix} - a_2 \begin{vmatrix} b_1 & c_1 \\ b_3 & c_3 \end{vmatrix}$$
$$+ a_3 \begin{vmatrix} b_1 & c_1 \\ b_2 & c_2 \end{vmatrix}.$$

Fredholm's determinant. See FREDHOLM.

functional determinant. Same as JACOBIAN.

Gram determinant. Same as GRAMIAN.

Laplace's expansion of a determinant. Let A be a determinant of order n and $A_{s_1 s_2 \cdots s_k}^{r_1 r_2 \cdots r_k}$ be the determinant formed from A by using the elements in rows $r_1, r_2, \cdots, r_k$ and columns $s_1, s_2, \cdots, s_k$. Laplace's expansion is

$$A = \sum (-1)^h \left(A_{i_1 \cdots i_k}^{r_1 \cdots r_k} \right) \left(A_{i_{k+1} \cdots i_n}^{r_{k+1} \cdots r_n} \right),$$

where $(r_1, r_2, \cdots, r_n)$ and $(i_1, i_2, \cdots, i_n)$ are permutations of the integers $(1, 2, \cdots, n)$, h is the number of inversions necessary to bring the order $(i_1, i_2, \cdots, i_n)$ into the order $(r_1, r_2, \cdots, r_n)$, and the summation is taken over

the $n!/[k!(n-k)!]$ ways of choosing the combinations $(i_1, i_2, \cdots, i_k)$ from the integers $(1, 2, \cdots, n)$. See above, expansion of a determinant by minors, which is the special case of Laplace's expansion for $k = 1$.

minor of an element in a determinant. See MINOR.

multiplication of determinants. See MULTIPLICATION—multiplication of determinants.

numerical determinant. A determinant whose elements are numbers.

skew-symmetric determinant. See SKEW.

symmetric determinant. See SYMMETRIC—symmetric determinant.

Vandermonde determinant. A determinant (or its transpose) that has 1 in each place of the first row, the second row has n arbitrary elements $x_1, x_2, \cdots, x_n$, and the ith row ($2 \leq i \leq n$) has the elements $x_1^{i-1}, x_2^{i-1}, \cdots, x_n^{i-1}$. The value of the determinant is the product of all $x_i - x_k$ for which $i > k$.

DE-VEL′OP-A-BLE, *adj.*, *n.* **developable surface.** The envelope of a one-parameter family of planes; a surface that can be developed, or rolled out, on a plane without stretching or shrinking; a surface for which the total curvature vanishes identically. See below, rectifying developable of a curve, polar developable of a curve, and TANGENT—tangent surface of a curve.

polar developable of a space curve. The envelope of the normal planes of the space curve; the totality of points on the polar lines of the curve. See above, developable surface, and NORMAL—normal to a curve or surface.

rectifying developable of a space curve. The envelope S of the rectifying planes of the space curve C. This developable surface S is called the **rectifying developable** of C because the process of developing S on a plane results in rolling C out along a straight line. See above, developable surface, and RECTIFYING—rectifying plane of a space curve at a point.

DE′VI-A′TION, *n.* The difference between the value of a random variable and some standard value, usually the mean. However, see various headings below.

mean deviation. The *mean* or *expected value* of $|X - \mu|$, where μ is the mean. For a continuous random variable with probability density function f, the **mean deviation** is

$$\int_{-\infty}^{\infty} |x - \mu| f(x) \, dx,$$

if this integral converges. For a discrete random

variable with values $\{x_n\}$ and probability function p, the **mean deviation** is $\sum |x_i - \mu| p(x_i)$, if this sum converges; sometimes the median is used for μ. If a random variable has a finite number of values with equal probabilites, then the sum of deviations, with regard to sign, is zero about the mean and, without regard to sign, it is minimal about the median. See ABSOLUTE —absolute moment.

mean-square deviation. The second moment about some number a. If a is the mean, the mean-square deviation is the *variance*. See MOMENT—moment of a distribution.

probable deviation. The product of approximately 0.6745 and the *standard error*; for a normal random variable, the probability of a deviation with absolute value greater than this is $\frac{1}{2}$. No longer generally used, being replaced by the standard error. *Syn.* probable error.

quartile deviation. One-half of the difference between the upper and lower quartiles: $\frac{1}{2}(Q_3 - Q_1)$.

standard deviation. For a random variable (or the associated distribution function), the positive square root of the *variance* (see VARIANCE). *Syn.* root mean square deviation.

DEXTROROSUM [*Latin*] or **DEX'TRORSE,** *adj.* **dextrorse curve.** Same as RIGHT-HANDED CURVE. See RIGHT.

DI-AG'O-NAL, *adj.*, *n.* **diagonal of a determinant.** See DETERMINANT.

diagonal of a matrix. See MATRIX.

diagonal of a polygon. A line connecting two nonadjacent vertices. In *elementary geometry* it is thought of as the **line segment** between nonadjacent vertices; in *projective geometry* it is the straight line (of infinite length) passing through two nonadjacent vertices.

diagonal of a polyhedron. A line segment between any two vertices that do not lie in the same face. See PARALLELEPIPED.

diagonal scale, for a rule. A scale in which the rule is divided crosswise and diagonally by systems of parallel lines. *E.g.*, suppose that there are 11 longitudinal (lengthwise of the ruler) lines per inch (counting the lines at the beginning and end of the inch interval) and one diagonal line per inch. Then the intersections of the diagonal lines with the longitudinal lines are $\frac{1}{10}$ inch apart longitudinally, for the 10 segments cut off on any one diagonal by the horizontal lines are equal, and hence the 10 corresponding distances measured along the longitudinal lines must be equal. Thus the inch is divided into 10 equal parts.

Similarly one diagonal per $\frac{1}{10}$ inch scales the ruler in $\frac{1}{100}$ inch, etc.

DI-AG'ON-AL'IZE, *v.* To transform a matrix or transformation to diagonal form. For any Hermitian matrix M there is a unitary matrix P such that PMP^{-1} is a diagonal matrix. If M is a real symmetric matrix, then P can be an orthogonal matrix (see ORTHOGONAL —orthogonal matrix). There are similar statements for linear transformations of finite-dimensional spaces (see MATRIX —matrix of a linear transformation) and for linear transformations of Hilbert space (see SPECTRAL—spectral theorem).

DI'A-GRAM, *n.* A drawing representing certain data and, perhaps, conclusions drawn from the data; a drawing representing pictorially (graphically) a statement or a proof; used to aid readers in understanding algebraic explanations.

DI'A-LYT'IC, *adj.* **Sylvester's dialytic method.** See SYLVESTER.

DI-AM'E-TER, *n.* **conjugate diameters.** See CONJUGATE —conjugate diameters.

diameter of a central quadric surface. The locus of the centers of parallel sections of the central quadric. This locus is a straight line.

diameter of a circle. See CIRCLE.

diameter of a conic. Any straight line which is the locus of the midpoints of a family of parallel chords. Any conic has infinitely many diameters. In the central conics, ellipses and hyperbolas, they form a pencil of lines through the center of the conic. See CONJUGATE —conjugate diameters.

diameter of a set of points. See BOUNDED —bounded set of points.

DI-AM'E-TRAL, *adj.* **conjugate diametral planes.** Two diametral planes, each of which is parallel to the set of chords defining the other.

diametral line in a conic (ellipse, hyperbola or parabola). Same as DIAMETER.

diametral plane of a quadric surface. A plane containing the middle points of a set of parallel chords.

DI-CHOT'O-MY, *n.* A classification into two classes. The principle of dichotomy in logic states that a proposition is either true or false, but not both [see CONTRADICTION —law of contradiction]. *E.g.*, for two numbers x and y, exactly one of the statements $x = y$ and $x \neq y$ is true. See TRICHOTOMY.

DICKSON, Leonard Eugene (1874–1954). American mathematician who made important contributions to the theory of finite linear groups, finite fields, and linear associative algebras. Wrote a monumental three-volume *History of the Theory of Numbers*, a nearly complete guide to the subject from antiquity to about 1920.

DIDO'S PROBLEM. The problem of finding the curve, with a given perimeter, which encloses the maximum area. The required curve is a circle. If part of the boundary is freely given as a straight-line segment of arbitrary length, as along a river, then the solution is a semicircle. The solution is reported to have been known by Queen Dido of Carthage, who was given as much land as she could enclose with a cowhide. She cut it into thin strips and formed a semi-circle along the coast of North Africa, within which was established the State of Carthage.

DIF′FE-O-MOR′PHISM, *n.* A function f defined on a subset S of an Euclidean space with range in another (or the same) Euclidean space is **differentiable** (or **smooth**) if, for each x in S, f can be extended to an open set U containing x such that f has continuous mixed partial derivatives of all orders at all points of U. Let T be a map of Euclidean m-space to Euclidean n-space, with $T(x) = (f_1(x), f_2(x), \cdots, f_n(x))$ if $x = (x_1, x_2, \cdots, x_m)$. Then T is a **differentiable** (or **smooth) map** if each f_i is differentiable. A **diffeomorphism** is a one-to-one map T for which both T and T^{-1} are differentiable.

DIF′FER-ENCE, *adj., n.* The result of subtracting one quantity from another. *Syn.* remainder.

difference equation. See below, ordinary difference equation, and partial difference equation.

difference of like powers. An algebraic expression of type $x^n - y^n$. If n is odd, the difference $x^n - y^n$ is divisible by $x - y$; whereas if n is even, the difference is divisible by both $x + y$ and $x - y$. E.g.,

$$x^3 - y^3 = (x - y)(x^2 + xy + y^2),$$

while

$$x^4 - y^4 = (x - y)(x + y)(x^2 + y^2).$$

A difference of squares is the particular case $x^2 - y^2$, which is equal to $(x + y)(x - y)$. See SUM—sum of like powers.

difference quotient. For a function f, the increment ratio $[f(x + \Delta x) - f(x)]/\Delta x$; *e.g.*, if the function f is defined by $f(x) = x^2$, the *difference quotient* is

$$\frac{f(x + \Delta x) - f(x)}{\Delta x} = \frac{(x + \Delta x)^2 - x^2}{\Delta x}$$

$$= 2x + \Delta x.$$

See DERIVATIVE.

difference of sets. The difference $A - B$ of two sets A and B is the set of all objects that belong to A and do not belong to B. The **symmetric difference** of two sets A and B is the set which contains all the objects that belong to either of the sets but not to both; *i.e.*, the symmetric difference of A and B is the union of the sets $A - B$ and $B - A$. Some of the notations used for the symmetric difference of A and B are $A \ominus B$, $A \nabla B$, $A + B$. See RING—ring of sets.

differences of the first order or **first-order differences.** The sequence formed by subtracting each term of a sequence from the next succeeding term. The first-order differences of the sequence $(1, 3, 5, 7, \cdots)$ would be $(2, 2, 2, \cdots)$.

differences of the second order or **second-order differences.** The *first-order differences* of the first-order differences; *e.g.*, the first-order differences of the sequence $(1, 2, 4, 7, 11, \cdots)$ are $(1, 2, 3, 4, \cdots)$, while the second-order differences are $(1, 1, 1, \cdots)$. Similarly, the *third-order differences* are the first-order differences of the second-order differences; and, in general, the rth-*order differences* are the first-order differences of the $(r - 1)$th order differences. If the sequence is $(a_1, a_2, a_3, \cdots, a_n, \cdots)$, the first-order differences are $a_2 - a_1, a_3 - a_2, a_4 - a_3, \cdots$, the second-order are $a_3 - 2a_2 + a_1, a_4 - 2a_3 + a_2, \cdots$, and the rth-order are:

$$\left[a_{r+1} - ra_r + \{r(r-1)/2!\}a_{r-1} - \cdots \pm a_1 \right],$$

$$\left[a_{r+2} - ra_{r+1} \right.$$

$$\left. + \{r(r-1)/2!\}a_r - \cdots \pm a_2 \right], \cdots.$$

finite differences. The differences derived from the sequence of values obtained from a given function by letting the variable change by arithmetic progression. If f is the given function, the arithmetic progression

$$(a, a + h, a + 2h, \cdots)$$

gives the sequence of values: $f(a), f(a + h), f(a + 2h), \cdots$. The differences may be of any given order. The first-order differences are $f(a + h) - f(a), f(a + 2h) - f(a + h), \cdots$. The successive differences of order one, two, three, etc., are written: $\Delta f(x), \Delta^2 f(x), \Delta^3 f(x)$, etc. In the study of difference equations, it is sometimes understood that $\Delta f(x) = f(x + 1) - f(x)$, $\Delta^2 f(x) = \Delta \, \Delta f(x) = f(x + 2) - 2f(x + 1) + f(x)$, etc.

ordinary difference equation. An expressed relation between an independent variable x and one or more dependent variables or functions $f, g, \cdots$, and any successive differences of f, g, etc., as $\Delta f(x) = f(x + h) - f(x)$, $\Delta^2 f(x) = f(x + 2h) - 2f(x + h) + f(x)$, etc., or equivalently, the results of any successive applications of the operator E, where $Ef(x) = f(x + h)$. The **order** of a difference equation is the order of the highest difference (or exponent of the highest power of E), and the **degree** is the highest power to which the highest difference is involved. A difference equation is **linear** if it is of the first degree with respect to all of the quantities $f(x), \Delta f(x), \Delta^2 f(x)$, etc.; or $f(x), Ef(x)$, etc. The equation $f(x + 1) = xf(x)$ is a linear difference equation. See below, partial difference equation.

partial difference equation. An expressed relation between two or more independent variables, $x, y, z, \cdots$, one or more dependent variables $f(x, y, z, \cdots)$, $g(x, y, z, \cdots), \cdots$, and partial differences of these dependent variables.

partial differences. Partial differences of a function $f(x, y, z, \cdots)$ of two or more variables are any of the expressions arising from successive derivation of ordinary differences, holding all the variables but one fixed at each step.

tabular difference. See TABULAR.

DIF'FER-ENC-ING, *p.* **differencing a function.** Taking the successive differences. See DIFFERENCE—finite differences.

DIF'FER-EN'TI-A-BLE, *adj.* A function f of one variable is **differentiable** at a point x if x is in the domain of the derivative of f; f is **differentiable on a set** D if it is differentiable at each point of D. For a function of several variables, see DIFFEOMORPHISM, DIFFERENTIAL.

DIF'FER-EN'TIAL, *adj., n.* Let f be a **function of one variable** which is differentiable at x. Then the **differential** of f is

$$df = f'(x)\,dx,$$

where dx is an independent variable. Thus df is a function of the two variables x and dx. Since the derivative of x is 1, it follows that the differential of x is dx. The differential df has the property that, if x is changed by Δx, and the resulting change in f is denoted by Δf, then df is a good approximation for Δf when Δx is small, in the sense that $(\Delta f - df)/\Delta x$ approaches zero as $\Delta x \to 0$. For since

$$\lim_{\Delta x \to 0} \frac{\Delta f}{\Delta x} = f'(x), \qquad \frac{\Delta f}{\Delta x} = f'(x) + \epsilon,$$

where $\lim_{x \to 0} \epsilon = 0$. Hence

$$\Delta f = f'(x)\,\Delta x + \epsilon \, \Delta x,$$

or $\Delta f = dx + \epsilon \, \Delta x$. E.g., to find the approximate change in the area of a circle of radius 2 feet

when the radius increases 0.01 foot, we have $A = \pi r^2$, from which $dA = 2\pi r \, dr = 2\pi \times 2 \times \frac{1}{100} = \frac{1}{25}\pi$ square feet, which is the approximate increase in area. The **differential** (the **total differential) of a function of several variables,** $f(x_1, x_2, \cdots, x_n)$, is the function

$$df = \frac{\partial f}{\partial x_1}\,dx_1 + \frac{\partial f}{\partial x_2}\,dx_2 + \cdots + \frac{\partial f}{\partial x_n}\,dx_n,$$

which is a function of the independent variables $x_1, \cdots, x_n, dx_1, \cdots, dx_n$. Each of the terms $\frac{\partial f}{\partial x_i}\,dx_i$ is a **partial differential.** If $u = f(x, y, z)$ and z is a function of x and y, then

$$du = \left(\frac{\partial f}{\partial x} + \frac{\partial f}{\partial z}\frac{\partial z}{\partial x}\right) dx + \left(\frac{\partial f}{\partial y} + \frac{\partial f}{\partial z}\frac{\partial z}{\partial y}\right) dy.$$

Each term on the right is a partial differential, but is sometimes called an **intermediate differential** in cases such as the above where at least one of the variables of f is dependent on the others. The formulas for the differentials of functions of one or more variables hold when the functions are **composite.** One may, in that case, replace the differentials of the variables by their total differentials in terms of the variables of which they are functions. E.g., if $z = f(x, y)$, $x = u(s, t)$, and $y = v(s, t)$, then

$$dz = \frac{\partial f}{\partial x}\,dx + \frac{\partial f}{\partial y}\,dy$$

$$= \frac{\partial f}{\partial x}\left[\frac{\partial u}{\partial s}\,ds + \frac{\partial u}{\partial t}\,dt\right]$$

$$+ \frac{\partial f}{\partial y}\left[\frac{\partial v}{\partial s}\,ds + \frac{\partial v}{\partial t}\,dt\right].$$

For a function $f(x, y)$ of two variables,

$$df = \frac{\partial f}{\partial x}\, dx + \frac{\partial f}{\partial y}\, dy.$$

The function f is **differentiable** at (x, y) if, for any $\epsilon > 0$, there is a $\delta > 0$ such that if Δx and Δy are numbers with $|\Delta x|$ and $|\Delta y|$ each less than δ, then

$$\left| \Delta f - \left(\frac{\partial f}{\partial x} \cdot \Delta x + \frac{\partial f}{\partial y}\, \Delta y \right) \right| < \epsilon (|\Delta x| + |\Delta y|),$$

where $\Delta f = f(x + \Delta x, y + \Delta y) - f(x, y)$. Thus if the independent variables are changed by small amounts, the change in a *differentiable function* can be approximated by its differential with an error which is small relative to changes in the variables. A function with continuous partial derivatives is differentiable. This concept of approximation can be used to define differentials for more general situations (see FUNCTIONAL — differential of a functional). *Syn.* total differential. See INCREMENT —increment of a function, ELEMENT —element of integration.

adjoint of a differential equation. See ADJOINT.

binomial differential. See BINOMIAL.

differential analyzer. An instrument for solving differential equations (or system of differential equations) by mechanical means. The **Bush differential analyzer**, designed in the 1920's by Vannevar Bush, was the first differential analyzer ever built. It was based on the two fundamental operations of addition and integration, performed respectively by differential gear boxes and wheel and disc mechanisms.

differential of arc, area, attraction, mass, moment, moment of inertia, pressure, volume, and work. Same as ELEMENT OF ARC, AREA, ATTRACTION, etc. See ELEMENT —element of integration.

differential calculus. See CALCULUS.

differential coefficient. Same as DERIVATIVE.

differential equation (ordinary). An equation containing at most two variables, and derivatives of the first or higher order of one of the variables with respect to the other, such as $y(dy/dx) + 2x = 0$. The **order** of a differential equation is the order of the highest derivative which appears. When an equation contains only derivatives of the first order it is frequently written in terms of differentials. This is permissible because the first derivative may be treated as the quotient of the differentials. Thus the equation above may

be written $y\, dy + 2x\, dx = 0$. See the headings below, and PARTIAL —partial differential equation.

differential equation with variables separable. An ordinary differential equation which can be written in the form $P(x)\, dx + Q(y)\, dy = 0$, by means of algebraic operations performed on the given equation. Its general solution is obtainable directly by integration.

differential equations of Bernoulli (James), Bessel, Chebyshev, Clairaut, Gauss, Hermite, Laguerre, Laplace, Legendre, Mathieu, Sturm-Liouville. See the respective names.

differential form. A homogeneous polynomial in differentials. *E.g.*, if $g_{i_1 i_2 \cdots i_n}$ is a symmetric covariant tensor field and $t_{\beta_1 \beta_2 \cdots \beta_q}$ is an alternating covariant tensor field, then

$$g_{i_1 i_2 \cdots i_r}\, dx^{i_1}\, dx^{i_2} \cdots dx^{i_r}$$

and

$$t_{\beta_1 \beta_2 \cdots \beta_q}\, dx^{\beta_1}\, dx^{\beta_2} \cdots dx^{\beta_q}$$

transform like scalar fields and are a **symmetric differential form** and **alternating differential form**, respectively.

differential geometry. The theory of the properties of configurations in the neighborhood of one of its general elements. See GEOMETRY—metric differential geometry; and below, projective differential geometry.

differential operator. A polynomial in the operator D, where D stands for d/dx and Dy for dy/dx. *E.g.*, $(D^2 + xD + 5)y = d^2y/dx^2 + x(dy/dx) + 5y$. Symbols of the form $1/f(D)$, where $f(D)$ is a polynomial in D, are **inverse differential operators**. *E.g.*, the symbol $1/(D - a)$ arises from the equation $dy/dx - ay = f(x)$. The equation is written in the form $(D - a)y = f(x)$. Then

$$y = \frac{1}{(D - a)} f(x)$$

is a solution, where

$$\frac{1}{(D - a)} f(x) = Ce^{ax} + e^{ax} \int e^{-ax} f(x)\, dx.$$

differential parameter of a surface. For a given function $f(u, v)$ and a given surface S:

$x = x(u, v)$, $y = y(u, v)$, $z = z(u, v)$, the function

$$\Delta_1 f \equiv \left(\frac{df}{ds} \right)^2$$

$$= \frac{E \left(\dfrac{\partial f}{\partial v} \right)^2 - 2F \dfrac{\partial f}{\partial u} \dfrac{\partial f}{\partial v} + G \left(\dfrac{\partial f}{\partial v} \right)^2}{EG - F^2},$$

where the derivative df/ds is evaluated in the direction perpendicular to the curve $f = $ const. on S, is invariant under change of parameters: $u = u(u_1, v_1)$, $v = v(u_1, v_1)$. See VARIATION — variation of a function on a surface. The invariant $\Delta_1 f$ is the **differential parameter of the first order** for the function f relative to the surface S. See below, mixed differential parameter of the first order. The **differential parameter of the second order** is the invariant

$\Delta_2 f$

$$\equiv \frac{\dfrac{\partial}{\partial u} \left(\dfrac{G \dfrac{\partial f}{\partial u} - F \dfrac{\partial f}{\partial v}}{(EG - F^2)^{1/2}} \right) + \dfrac{\partial}{\partial v} \left(\dfrac{E \dfrac{\partial f}{\partial v} - F \dfrac{\partial f}{\partial u}}{(EG - F^2)^{1/2}} \right)}{(EG - F^2)^{1/2}}.$$

For a conformal map of the (u, v)-domain of defintion on the surface S, which is a map with $E = G = \sigma(u, v) \neq 0$, $F = 0$, the numerator of $\Delta_2 f$ reduces to the Laplacian of f:

$$\Delta_2 f = \left(\frac{\partial^2 f}{\partial u^2} + \frac{\partial^2 f}{\partial v^2} \right) \bigg/ \sigma.$$

There are other differential invariants of the second and higher orders, such as $\Delta_1 \Delta_1 (f, g), \Delta_1 \Delta_2 f$, etc. See below, mixed differential parameters of the first order.

    exact differential equation. A differential equation which is obtained by setting the total differential of some function equal to zero. An exact differential equation in two variables can be put in the form

$$[\partial f(x, y)/\partial x] \, dx + [\partial f(x, y)/\partial y] \, dy = 0.$$

A necessary and sufficient condition that an equation of the form $M \, dx + N \, dy = 0$, where M and N have continuous first-order partial derivatives, be exact is that the partial derivative of M with respect to y be equal to the partial derivative of N with respect to x; i.e., $D_y M = D_x N$.

The equation

$$(2x + 3y) \, dx + (3x + 5y) \, dy = 0$$

is exact. If a differential equation in three variables is of the form

$$P \, dx + Q \, dy + R \, dz = 0,$$

where P, Q, and R have continuous first-order partial derivatives, then a necessary and sufficient condition that it be exact is that $D_y P = D_x Q$, $D_z Q = D_y R$ and $D_x R = D_z P$, where $D_y P$, etc., denote partial derivatives. This can be generalized to any number of variables. See FACTOR —integrating factor.

    homogeneous differential equation. A name usually given to a differential equation of the first degree and first order which is homogeneous in the variables (the derivatives not being considered) such as

$$y^2 + \left(xy + x^2 \right) \frac{dy}{dx} = 0$$

and

$$\frac{x}{y} + \left(\sin \frac{x}{y} \right) \frac{dy}{dx} = 0.$$

Such equations are solvable by use of the substitution $y = ux$. An equation of the type

$$\frac{dy}{dx} = \frac{ax + by + c}{dx + ey + f}$$

can be reduced to a homogeneous equation by the substitutions: $x = x' + h$, $y = y' + k$, where h and k are to be chosen so as to remove the constant terms in the numerator and denominator of the fraction.

    homogeneous linear differential equation. A *linear differential equation* which does not contain a term involving only the independent variable. E.g., $y' + yf(x) = 0$.

    integrable differential equation. A differential equation that is exact, or that can be made so by multiplying through by an integrating factor.

    linear differential equation. A linear differential equation of first order is an equation of the form

$$\frac{dy}{dx} + P(x)y = Q(x).$$

Such an equation has an integrating factor of the form $e^{\int P \, dx}$. See example above, under differential operator. The general linear differential equation is an equation of the first degree in y and its derivatives, the coefficients of y and its derivatives being functions of x alone. *I.e.*, an equation of the form

$$L(y) \equiv p_0 \frac{d^n y}{dx^n} + p_1 \frac{d^{n-1} y}{dx^{n-1}} + \cdots + p_n y = Q(x).$$

The *general solution* can be found by finding n linearly independent particular solutions of the homogeneous equation $L(y) = 0$, multiplying each of these functions by an arbitrary parameter, and adding to the sum of these products (called the **complementary function**) some particular solution of the original differential equation. The equation $L(y) = 0$ is the **auxiliary equation** or **reduced equation**. The original equation $L(y) = Q$ is the **complete equation**. For methods of finding a general solution, after having the complementary function, see UNDETERMINED —undetermined coefficients, VARIATION —variation of parameters.

metric differential geometry. See GEOMETRY.

mixed differential parameter of the first order. The invariant

$$\Delta_1(f, g)$$

$$\equiv \frac{E \dfrac{\partial f}{\partial v} \dfrac{\partial g}{\partial v} - F\left(\dfrac{\partial f}{\partial u} \dfrac{\partial g}{\partial v} + \dfrac{\partial f}{\partial v} \dfrac{\partial g}{\partial u}\right) + G \dfrac{\partial f}{\partial u} \dfrac{\partial g}{\partial u}}{EG - F^2},$$

for given functions $f(u, v)$ and $g(u, v)$ and a given surface S: $x = x(u, v)$, $y = y(u, v)$, $z = z(u, v)$. See above, differential parameter of the first order. The invariance of $\Delta_1(f, g)$ under change of parameters u, v follows from its geometrical significance:

$$\cos \theta = \frac{\Delta_1(f, g)}{[\Delta_1 f]^{1/2} [\Delta_1 g]^{1/2}},$$

where θ is the angle between the curves $f = $ const. and $g = $ const. through a point of S. Another mixed differential parameter of the first order is

$$\Theta(f, g) \equiv \frac{\partial(f, g)}{\partial(u, v)} \bigg/ (EG - F^2)^{1/2}.$$

We have

$$\Delta_1^2(f, g) + \Theta^2(f, g) = [\Delta_1 f][\Delta_1 g].$$

partial differential equation. See PARTIAL —partial differential equations

primitive of a differential equation. See PRIMITIVE —primitive of a differential equation.

projective differential geometry. The theory of the differential properties of configurations, which are invariant under projective transformations.

simultaneous (or systems of) differential equations. Two or more differential equations involving the same number of dependent variables, taken as a system in the sense that solutions are sought which will satisfy them simultaneously.

solution of a differential equation. Any function which reduces the differential equation to an identity when substituted for the dependent variable; $y = x^2 + cx$ is a solution of

$$x \frac{dy}{dx} - x^2 - y = 0, \quad \text{for} \quad \frac{dy}{dx} = 2x + c$$

and substituting $2x + c$ for $\dfrac{dy}{dx}$, and $x^2 + cx$ for y, in the differential equation reduces it to the identity $0 = 0$. *Syn.* primitive integral. The constant c in the solution $y = x^2 + cx$ is an **arbitrary constant** in the sense that $y = x^2 + cx$ is a solution whatever value is given to c. The **general solution** of a differential equation is a solution in which the number of *essential* arbitrary constants is equal to the order of the differential equation. A **particular solution** is a solution obtained from the general solution by giving particular values to the arbitrary constants. A **singular solution** is a solution not obtainable by assigning particular values to the parameters in the general solution; it is the equation of an envelope of the family of curves represented by the general solution. This envelope satisfies the differential equation because at every one of its points its slope and the coordinates of the point are the same as those of some member of the family of curves representing the general solution. See DISCRIMINANT —discriminant of a differential equation, PICARD, RUNGE.

total differential. See DIFFERENTIAL.

DIF'FER-EN'TI-A'TION, *adj.*, *n.* The process of finding the derivative, or differential coefficient. See DERIVATIVE.

differentiation formulas. Formulas that give the derivatives of functions or enable one to reduce the finding of their derivatives to the problem of finding the derivatives of simpler functions. See DIFFERENTIATION FORMULAS in the appendix, CHAIN —chain rule, DERIVATIVE.

differentiation of an infinite series. See SERIES —differentiation of an infinite series.

differentiation of parametric equations. See PARAMETRIC—differentiation of parametric equations.

differentiation of an integral. See DERIVATIVE—derivative of an integral.

implicit differentiation. The process of finding the derivative of one of two variables with respect to the other by differentiating all the terms of a given equation in the two variables, leaving the derivative of the dependent variable (with respect to the independent variable) in indicated form, and solving the resulting identity to find this derivative. *E.g.*, if

$$x^3 + x + y + y^3 = 4,$$

then

$$3x^2 + 1 + y' + 3y^2 y' = 0,$$

whence

$$y' = -(3x^2 + 1)/(3y^2 + 1).$$

In cases where an equation cannot be solved for one of the variables, this method is indispensable. It generally facilitates the work even when the equation can be so solved. For the equation $f(x, y) = 0$ with $D_y f(x, y) \neq 0$, one may also use the formula

$$dy/dx = -D_x f(x, y)/D_y f(x, y).$$

This is easily seen to be equivalent to the above method. See DIFFERENTIAL [if z is constant, $dz = df(x, y)$ is zero and the above formula results].

logarithmic differentiation. Finding derivatives by the use of logarithms. Consists of taking the logarithm of both sides of an equation and then differentiating. It is used for finding the derivatives of variable powers of variable bases, such as x^x, and to simplify certain differentiation processes. *E.g.*, if $y = x^x$, one can write log $y = x$ log x and find the derivative of y with respect to x from the latter equation by means of the usual method of *implicit differentiation*.

successive differentiation. The process of finding higher-order derivatives by differentiating lower-order derivatives.

DI-GAM'MA, *adj.* **diagamma function.** The derivative ψ of $\log_e \Gamma(x)$. For $x > 0$, $\psi(x + 1) - \psi(x) = 1/x$ and $\psi'(x) = \sum_{n=0}^{\infty} (z + n)^{-2}$; ψ' is the **trigamma function.** See GAMMA—gamma function.

DIG'IT, *n.* For the decimal number system, any of the symbols $0, 1, 2, 3, 4, 5, 6, 7, 8, 9$. The number 23 has the digits 2 and 3.

significant digits. (1) The digits which determine the mantissa of the logarithm of the number; the digits of a number beginning with the first digit on the left that is not zero and ending with the last digit on the right that is not zero. (2) The digits of a number which have significance; the digits of a number beginning with the first nonzero digit on the left of the decimal point, or with the first digit after the decimal point if there is no nonzero digit to the left of the decimal point, and ending with the last digit to the right. *E.g.*, the significant digits in 230 are 2, 3, and 0. The significant digits in 0.230 are 2, 3, and 0, the 0 meaning that, to three-place accuracy, the number is 0.230. In 0.23, the 0 is not significant, but in 0.023 the second zero is significant.

DIG'IT-AL, *adj.* **digital device.** See COMPUTER—digital computer.

DI-HE'DRAL, *adj.* **dihedral angle.** The union of a line and two half-planes which have this line as a common edge. The line is the **edge** of the dihedral angle and the union of the line and one of the planes is a **face.** A **plane angle** of a dihedral angle is an angle formed by the two rays which are the intersections of the faces of the dihedral angle and a plane perpendicular to the edge. Any two plane angles (as at A and A' in the figure) are congruent. A MEASURE of a dihedral angle is a measure of one of its plane angles; a dihedral angle is **acute**, **right**, or **obtuse** according as its plane angles are acute, right, or obtuse.

dihedral group. The group of motions (symmetries) of 3-dimensional space that preserve a regular polygon. If the polygon has n sides, the group has $2n$ members and is generated by two members x and y for which $x^n = e$, $y^2 = e$, and $xy = yx$. See GROUP, SYMMETRY—group of symmetries.

DIL'A-TA'TION, *n.* The change in volume per unit volume of the element of a deformed substance. If the principal strains are denoted by e_1, e_2, and e_3, the dilatation is

$$\vartheta = (1 + e_1)(1 + e_2)(1 + e_3) - 1,$$

and for small strains $\vartheta = e_1 + e_2 + e_3$, approximately.

DI-MEN′SION, *n.* Refers to those properties called length, area, and volume. A configuration having length only is said to be of one dimension; area and not volume, two dimensions; volume, three dimensions. A geometric configuration is of dimension n if n is the least number of real-valued parameters which can be used to (continuously) determine the points of the configuration; *i.e.*, if there are n degrees of freedom, or the configuration is (locally) topologically equivalent to a subspace of n-dimensional Euclidean space. See BASIS—basis of a vector space, SIMPLEX. Definitions of dimension can be given for more general sets of points. See headings below.

fractal (or **Mandelbrot) dimension.** Let X be a metric space. For each positive number ϵ, let $N(X, \epsilon)$ be the least positive integer M for which there are M balls of radius less than ϵ whose union contains X. Then the **fractal dimension** of X is

$$D = \lim_{\epsilon \to 0} \frac{\log N(X, \epsilon)}{\log 1/\epsilon},$$

which also is the greatest lower bound of positive numbers d such that

$$\limsup_{\epsilon \to 0} \epsilon^d N(X, \epsilon) = 0.$$

If X is a *fractal* that can be divided into N congruent subsets each of which can be made congruent to X by magnifying it by a factor of r, then the fractal dimension of X is equal to $(\log N)/(\log r)$. The Cantor set has fractal dimension $(\log 2)/(\log 3)$. For any metric space, the *Hausdorff dimension* is less than or equal to the *fractal dimension*. See FRACTAL.

Hausdorff dimension. Let X be a metric space. For positive numbers ϵ and p, define $m_p^\epsilon(X)$ to be the greatest lower bound of

$$\sum_{k=1}^{\infty} [\text{diameter of } A_k]^p,$$

where $\bigcup_{k=1}^{\infty} A_k = X$ and the diameter of each A_k is less than ϵ (see BOUNDED—bounded set of points). If X is compact, one need use only

finitely many such A_k. Now define $m_p(X)$ to be $\lim_{\epsilon \to 0} m_p^\epsilon(X)$. The **Hausdorff dimension** of X is the greatest lower bound of all p for which $m_p(X) = 0$, or equivalently, the least upper bound of all p for which $m_p(X) = \infty$. Also called the **Hausdorff-Besicovitch dimension.** The Cantor set has Hausdorff dimension $\log 2/\log 3$. A subset of ordinary n-dimensional Euclidean space which contains the interior of a sphere has dimension n. For any metric space, the *topological dimension* is less than or equal to the *Hausdorff dimension*. See FRACTAL, JULIA—Julia set.

topological dimension. An inductive definition of the dimension of a **topological space** X can be given as follows. Write D for *dimension* and define D of the empty set to be -1. Suppose D has been defined for all positive integers less than n. Define the statement "$D_p(X) \leq n$" to be true if and only if each neighborhood of p contains an open neighborhood of p whose boundary has $D \leq n - 1$, and then define "$D_p(X) = n$" to be true if and only if $D_p(X) \leq n$ and it is false that $D_p(X) \leq n - 1$. Finally, define $D(X) = n$ to mean that $D(X) \leq n$ and it is false that $D(X) \leq n - 1$, where "$D(X) \leq n - 1$" is true if and only if $D_p(X) \leq n$ for all p in X. There are other concepts of dimension for topological spaces. *E.g.*, if M is a **metric space**, then the dimension of M is n if (i) for each positive number ϵ there is a closed ϵ-cover of *order* less than or equal to $n + 1$, and (ii) there is a positive number ϵ for which each closed ϵ-cover of M is of *order* greater than n (see COVER). This definition of dimension of a metric space is topologically invariant. The Cantor set has dimension 0. A subset of ordinary n-dimensional Euclidean space which contains the interior of a sphere has dimension n. Two topological spaces are said to be of the same **dimension type** if each is homeomorphic to a subset of the other.

dimensions of a rectangular figure. The length and width of a rectangle, or the length, width, and height of a box (a rectangular parallelepiped).

DI-MEN′SION-AL′I-TY, *n* The number of dimensions. See DIMENSION.

DINI, Ulisse (1845–1918). Italian analyst.

Dini condition for convergence of Fourier series. See FOURIER—Fourier's theorem.

Dini theorem on uniform convergence. If $\{S_n\}$ is a monotone sequence of real-valued continuous functions that converges to a continuous function S on a compact set D, then $\{S_n\}$ converges uniformly to S on D.

DIOPHANTUS (c. 250 A.D.) Ancient Greek arithmetician and algebraist; lived in Egypt.

Diophantine analysis. Finding *integral* solutions of certain algebraic equations. Depends mostly on ingenious use of arbitrary parameters.

Diophantine equations See EQUATION—indeterminate equation.

DI'POLE, *n.* See POTENTIAL—concentration method for the potential of a complex.

DIRAC, Paul Adrien Maurice (1902–1984).

Dirac distribution or **Dirac δ-function.** See GENERALIZED —generalized function.

Dirac matrix. Any one of four 4×4 matrices, γ_i $(i = 1, 2, 3, 4)$, for which $\gamma_i \gamma_j = -\gamma_j \gamma_i$ if $i \neq j$ and γ_i^2 is the identity matrix. See CLIFFORD —Clifford algebra.

DI-RECT', *adj.* **direct product (or sum).** See headings under PRODUCT.

direct proof. See PROOF—direct and indirect proof.

direct proportion or variation. See VARIATION—direct variation.

direct trigonometric functions. The trigonometric functions *sine, cosine, tangent,* etc., as distinguished from the *inverse trigonometric functions.* See TRIGONOMETRIC —trigonometric functions.

DI-RECT'ED, *adj.* **directed angle.** See ANGLE.

directed line, or line segment. A line (or line segment) on which the direction from one end to the other has been indicated as positive, and the reverse direction as negative. The direction can be described by specifying two distinct points on the line (for a line segment, usually the end points) and stating which precedes the other. Two such points determine a vector with initial point at the first point and terminal point at the other.

directed numbers. Numbers having signs, positive or negative, indicating that the negative numbers are to be measured, geometrically, in the direction opposite to that in which the positive are measured when the numbers are considered to be points on the number line. *Syn.* signed numbers, algebraic numbers. See POSITIVE —positive number.

directed set. See MOORE, E. H.—Moore-Smith convergence.

DI-REC'TION, *adj., n.* The **direction of a line** can be taken as any vector parallel to the line, or as a set of direction angles or cosines, or for a line in a plane as an angle of inclination of the line. The **direction of a curve** is the direction of the tangent line.

characteristic directions on a surface. See CHARACTERISTIC.

direction angle. For a line in a plane, the smallest positive (or zero) angle that the line makes with the positive x-axis. For a line in space, any one of the three positive angles which the line makes with the positive directions of the coordinate axes. There are two such sets for an undirected line, one for each direction which can be assigned to the line. Direction angles are not independent (see PYTHAGORAS —Pythagorean relation between direction cosines). In the figure, direction angles of the line L are the angles, α, β, and γ, which the parallel line L' makes with the coordinate axes.

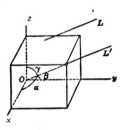

direction components of the normal to a surface For a surface S given in parametric representation, $x = x(u, v)$, $y = y(u, v)$, $z = z(u, v)$, **direction components** of the normal at a regular point are any three numbers having the ratio $A : B : C$, where

$$A = \begin{vmatrix} \dfrac{\partial y}{\partial u} & \dfrac{\partial z}{\partial u} \\ \dfrac{\partial y}{\partial v} & \dfrac{\partial z}{\partial v} \end{vmatrix}, \quad B = \begin{vmatrix} \dfrac{\partial z}{\partial u} & \dfrac{\partial x}{\partial u} \\ \dfrac{\partial z}{\partial v} & \dfrac{\partial x}{\partial v} \end{vmatrix},$$

$$C = \begin{vmatrix} \dfrac{\partial x}{\partial u} & \dfrac{\partial y}{\partial u} \\ \dfrac{\partial x}{\partial v} & \dfrac{\partial y}{\partial v} \end{vmatrix}.$$

The positive direction of the normal is taken to be the direction for which the direction cosines are $X = A/H$, $Y = B/H$, $Z = C/H$, where $H = \sqrt{A^2 + B^2 + C^2}$. Thus the orientation of the normal depends on the choice of parameters.

direction cosines. The cosines of the *direction angles*. They are usually denoted by l, m, and n, where, if α, β, γ are the direction angles with respect to the x-axis, y-axis, and z-axis, respectively, $l = \cos \alpha$, $m = \cos \beta$, and $n = \cos \gamma$.

Direction cosines are not independent. When two of them are given, the third can be found, except for sign, by use of the *Pythagorean relation*, $\cos^2 \alpha + \cos^2 \beta + \cos^2 \gamma = 1$. See below, direction numbers.

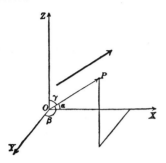

direction numbers (or ratios) of a line in space. Any three numbers, not all zero, proportional to the *direction cosines* of the line. *Syn.* direction components. If a line passes through the points (x_1, y_1, z_1) and (x_2, y_2, z_2), its direction numbers are proportional to $x_2 - x_1, y_2 - y_1, z_2 - z_1$, and its direction cosines are

$$\frac{x_2 - x_1}{D}, \frac{y_2 - y_1}{D}, \frac{z_2 - z_1}{D},$$

where

$$D = \sqrt{(x_2 - x_1)^2 + (y_2 - y_1)^2 + (z_2 - z_1)^2},$$

the distance between the points.
    principal direction of strain. See STRAIN
    principal direction on a surface. See CURVATURE—curvature of a surface.

DI-REC′TION-AL, *adj.* **directional derivative.** The rate of change of a function with respect to arc length in a given direction (*i.e.*, along a line or curve in that direction). This is equal to the sum of the directed projections, upon the tangent line to the path, of the rates of change of the function in directions parallel to the three axes. Explicitly, for a function F of x, y, z, the directional derivative in the direction of a curve whose parametric equations are $x = x(s)$, $y = y(s)$, $z = z(s)$, where s is arc length, is given by

$$\frac{du}{ds} = F_x(x, y, z)\frac{dx}{ds} + F_y(x, y, z)\frac{dy}{ds}$$

$$+ F_z(x, y, z)\frac{dz}{ds}$$

$$= lF_x(x, y, z) + mF_y(x, y, z) + nF_z(x, y, z),$$

where l, m, and n are the direction cosines of the tangent to the curve. For a function f of two variables, this can be written as

$$f_x(x, y)\cos \theta + f_y(x, y)\sin \theta,$$

where θ is the angle which the tangent to the curve (directed in the direction of motion) makes with the directed x-axis. See CHAIN—chain rule.

DI-REC′TOR, *adj.* **director circle of an ellipse (or hyperbola).** The locus of the intersection of pairs of perpendicular tangents to the ellipse (or hyperbola). In the figure, the circle is the director circle of the ellipse, being the locus of points P which are the intersections of perpendicular tangents like (1) and (2).

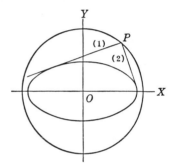

    director cone of a ruled surface. A cone formed by lines through a fixed point in space and parallel to the rectangular generators of the given ruled surface. See INDICATRIX —spherical indicatrix of a ruled surface.

DIR-REC″TRIX, *n.* [*pl.* **directrices** or **directrixes**]. See CONIC, CONICAL—conical surface, CYLINDRICAL—cylindrical surface, PYRAMIDAL—pyramidal surface, RULED—ruled surface.
    directrix planes of a hyperbolic paraboloid. The two lines of intersection of $z = 0$ with the hyperbolic paraboloid

$$\frac{x^2}{a^2} - \frac{y^2}{b^2} = 2z,$$

each taken with the z-axis, determine two planes which are called the directrix planes of the hyperbolic paraboloid.

DIRICHLET, Peter Gustav Lejeune (1805–1859). German number theorist, analyst, and applied mathematician.

Dirichlet characteristic properties of the potential function

$$\iiint \rho/r \, dV.$$

Assume that ρ and its first partial derivatives are piecewise continuous and that the set of points at which ρ is not zero may be enclosed in a sphere of finite radius. The **Dirichlet properties** of the potential function

$$U = \iiint \rho/r \, dV$$

are: (1) U is of class C^1 throughout space; (2) U is of class C^2 except on surfaces of discontinuity of ρ, $\partial\rho/\partial x$, $\partial\rho/\partial y$, and $\partial\rho/\partial z$; (3) at points external to the body ($\rho = 0$), U satisfies Laplace's equation

$$\partial^2 U/\partial x^2 + \partial^2 U/\partial y^2 + \partial^2 U/\partial z^2 = 0,$$

while at points internal to the body but not on the boundary, U satisfies the more general Poisson equation

$$\partial^2 U/\partial x^2 + \partial^2 U/\partial y^2 + \partial^2 U/\partial z^2 = \pm 4\pi\rho$$

(the sign is plus in the electrostatic case and plus or minus in the gravitational case depending on the conventions adopted); (4) if $M = \iiint \rho \, dV$ and $R^2 = x^2 + y^2 + z^2$, then as $R \to \infty$, $R(U - M/R) \to 0$, while each of $R^3 \, \partial(U - M/R)/\partial x$, $R^3 \, \partial(U - M/R)/\partial y$, $R^3 \, \partial(U - M/R)/\partial z$ remains bounded. See POTENTIAL — potential function for a volume distribution of charge or mass.

Dirichlet conditions for convergence of Fourier series. See FOURIER — Fourier's theorem.

Dirichlet drawer principle. If $1 \le p < n$ and a set with n members is expressed as the union of p disjoint subsets, then at last one subset has more than one member. Often called the **Dirichlet pigeon-hole principle**, or the **pigeon-hole principle**, and stated: "If n objects are put into p boxes ($1 \le p < n$), then some box contains at least two of the objects." Sometimes p is taken to be $n - 1$.

Dirichlet integral. The integral

$$\iint_A \left[\left(\frac{\partial w}{\partial x}\right)^2 + \left(\frac{\partial w}{\partial y}\right)^2 \right] dx \, dy,$$

or its analogue for a function of any number of independent variables. The **Dirichlet principle** states that if the *Dirichlet integral* is minimized

in the class of functions continuously assuming a given boundary value function on the boundary of A, then the minimizing function is *harmonic* on the interior of A.

Dirichlet kernel. See KERNEL.

Dirichlet problem. Same as the first boundary value problem of potential theory. See BOUNDARY.

Dirichlet product. For a given domain R and a given nonnegative function $p(x, y, z)$, the Dirichlet product $D[u, v]$ of functions $u(x, y, z)$ and $v(x, y, z)$ is defined by

$$D[u, v] = \iiint_R (\nabla u \cdot \nabla v + puv) \, dx \, dy \, dz,$$

where $\nabla u \cdot \nabla v = \dfrac{\partial u}{\partial x}\dfrac{\partial v}{\partial x} + \dfrac{\partial u}{\partial y}\dfrac{\partial v}{\partial y} + \dfrac{\partial u}{\partial z}\dfrac{\partial v}{\partial z}$. See above, Dirichlet integral.

Dirichlet series. An infinite series of type $\sum_1^\infty a_n/n^z$, where z and a_n can be complex numbers. See ZETA—Riemann zeta function.

Dirichlet test for convergence of a series. Let $a_1, a_2, \cdots$ be a sequence for which there is a number K with $\left| \sum_{n=1}^p a_n \right| < K$ for all p. Then $\sum_{n=1}^\infty a_n u_n$ converges if $u_n \ge u_{n+1}$ for all n and $\lim_{n \to \infty} u_n = 0$. This test is easily deduced from the *Abel inequality*.

Dirichlet test for uniform convergence of a series. If $a_1, a_2, \cdots$ are functions for which there is a K for which $\left| \sum_{n=1}^p a_n(x) \right| < K$ (where K is independent of p and x), and if $u_n(x) \ge u_{n+1}(x)$ and $u_n(x) \to 0$ *uniformly* as $n \to \infty$, then $\sum_{n=1}^\infty a_n(x)u_n(x)$ converges uniformly. Sometimes called *Hardy's test*.

Dirichlet theorem. If a and r are relatively prime integers, then the infinite sequence $\{a, a + r, a + 2r, a + 3r, \cdots\}$ contains infinitely many primes.

DISC (or DISK), *n.* A circle and its interior. *Tech.* An **open disc** is the interior of a circle; a **closed disc** is a circle and its interior.

DIS′CON-NECT′ED, *adj.* **disconnected set.** A set which can be separated into two sets U and V which have no points in common and which are such that no accumulation point of U belongs to V and no accumulation point of V belongs to U. A set is **totally disconnected** if no subset of more than one point is connected; *e.g.*, the set of rational numbers is totally disconnected. A set is **extremally disconnected** if the closure of each open set is open, or (equivalently) if the closures of two disjoint open sets are disjoint. An extremally disconnected Hausdorff space is *totally disconnected*.

DIS-CON′TI-NU′I-TY, *n.* (1) The property of being noncontinuous. (2) A point in the domain of a function at which the function is not continuous (also called a **point of discontinuity**). Sometimes a point x not in the domain of a function f is called a discontinuity if f is discontinuous at x whatever value is assigned to $f(x)$ (*e.g.*, 0 for the function $1/x$). Points of discontinuity of a real-valued function are classified as follows. If the function can be made continuous at the point by being given a new value at the point, then the discontinuity is **removable** (this is the case if the limits from the right and left exist and are equal); *e.g.*, $x \sin 1/x$ has a removable discontinuity at the origin. It approaches zero as x approaches zero either from the left or right, although it is not defined for $x = 0$. A **nonremovable discontinuity** is any discontinuity which is not removable. An **ordinary discontinuity** (or **jump discontinuity**) is a discontinuity at which the limits of the function from the right and left exist, but are not equal; *e.g.*, the limits (at $x = 0$) from the right and left of $1/(1 + 2^{1/x})$ are 0 and 1, respectively. The difference between the right and left limits is called the **jump** of the function (sometimes a jump discontinuity is called simply a jump). A **finite discontinuity** is a discontinuity (removable, ordinary, or neither) such that there is an interval about the point in which the function is bounded; *e.g.*, $\sin 1/x$ has a finite discontinuity at $x = 0$, which is neither removable nor ordinary. An **infinite discontinuity** of a function f is a discontinuity such that $|f(x)|$ has arbitrarily large values arbitrarily near the point. See SINGULAR—singular point of an analytic function.

DIS′CON-TIN′U-OUS, *adj.* **discontinuous function.** A function that is not continuous. Ordinarily one refers to a function as being discontinuous at a point, or at (all) points of some set, although it is sometimes said that a function is discontinuous on an interval when it is discontinuous at some point or points on the interval. See DISCONTINUITY.

DIS′COUNT, *adj., n.* **bank discount.** A discount equal to the simple interest on the obligation; interest paid in advance on a note or other obligation (strictly speaking the interest is made part of the face of the note and paid when the note is paid). *E.g.*, a note for \$100, discounted by the *bank rule* at 6%, would leave \$94 (paying the face value of \$100 at the end of the year is equivalent to paying 6.38% interest, or if the interest is to be 6%, the *true discount* would be \$5.66 and the discount rate 5.66%, not 6%).

bond discount. The difference between the redemption value and the purchase price when the bond is bought below par.

cash discount, or discount for cash. A reduction in price made by the seller because the buyer is paying cash for the purchase.

chain discount. Same as DISCOUNT SERIES.

commercial discount. A reduction in the price of goods, or in the amount of a bill or debt, often given to secure payment before the due date. This discount may be computed either by means of discount rate or by means of interest rate; the former is used in discounting prices and the latter, usually, in discounting interest bearing contracts. The discount in the latter case is the face of the contract minus its present value at the given rate. When simple interest is used, *present value* is $S/(1 + ni)$; when compound interest is used, it is $S/(1 + i)^n$; where n in both cases is the number of interest periods and S the face of the contract. See below, discount rate, simple discount, compound discount.

compound discount. Discount under compound interest; the difference between face value and the present value at the given rate after a given number of years:

$$D = S - S/(1 + i)^n,$$

or, in terms of *discount rate*,

$$D = S - A, \qquad \text{where } A = S(1 - d)^n.$$

discount factor. The factor which, when multiplied into a sum, gives the present value over a period of n years; *i.e.*, gives the principal which would amount to the sum at the end of n years at the given interest rate. For compound interest, this factor is $(1 + i)^{-n}$, where i is the interest rate. In terms of the *discount rate*, d, this factor is $(1 - d)^n$. See below, discount rate.

discount on a note. The difference between the selling value and the present value of a note.

discount on stocks. The difference between the selling value and the face value of stocks, when the former is lower than the latter.

discount problem under compound interest. Finding the present value of a given sum, at a given rate of compound interest; *i.e.*, solving the equation $S = P(1 + r)^n$ for P.

discount rate. The percentage used to compute the discount. This is never the same as the interest rate on the contract. See above, bank discount, and below, true discount. If d is the discount rate and i the interest rate, the discount on a sum, S, for one interest period is $S - S/(1 + i)$ or Sd, where

$$d = 1 - 1/(1 + i) = i/(1 + i).$$

discount series. A sequence of discounts consisting of a discount, a discount upon the discounted face value, a discount upon the discounted, discounted face value, etc. The successive discount rates may or may not be the same. *E.g.*, if \$100 is discounted at a discount rate of 10%, the new principal is \$90; if this principal is discounted 5%, the new principal is \$85.50, and the discounts \$10 and \$4.50 are called a discount series. *Syn.* chain discount.

simple discount. Discount proportional to the time (on the basis of simple interest). If S is the amount due in the future (after n years), P the present value, and i the interest rate, the discount D is equal to $S - P$, where $P = S/(1 + ni)$.

time discount. Discount allowed if payment is made within a prescribed time; usually called *cash discount*. A vendor who does credit business often prices his goods high enough to make his credit sales cover losses due to bad accounts, then *discounts* for cash or for cash within a certain period.

trade discount. A reduction from the list price to adjust prices to prevailing prices or to secure the patronage of certain purchasers, especially purchasers of large amounts.

true discount. The reduction of the face value of an agreement to pay, by the simple interest on the reduced amount at a given rate; *e.g.*, the *true discount* on \$100 for one year at 6% is \$5.66, because 6% of \$94.34 is \$5.66. The formula for true discount is:

$$D = S - S/(1 + ni),$$

where S is face value, n time in years, and i interest rate.

DIS-CRETE′, *adj.* **discrete Fourier transform.** See FOURIER —discrete Fourier transform.

discrete mathematics. Mathematics that does not involve calculus or limits. Essentially the same as finite mathematics, but courses in "discrete mathematics" often give more emphasis to providing a background for computer science. See FINITE —finite mathematics.

discrete set. A set that has no accumulation points; *i.e.*, each point has a neighborhood that contains no other points of the set. *E.g.*, the set of integers is a discrete set. The set of rational numbers is not discrete, since any interval of nonzero length that contains a rational number contains other rational numbers. See ISOLATED —isolated set.

discrete topology. For a set S, the discrete topology is the set of all subsets of S. Each subset is both open and closed and each subset is a neighborhood of each of its points. See TOPOLOGY —topology of a space.

discrete variable. A variable whose possible values form a discrete set. See RANDOM —random variable.

DIS-CRIM′I-NANT, *adj., n.* **discriminant function.** (*Statistics*) A linear combination of a set of n variables that will classify (into two different classes) the events or items for which the measurements of the n variables are available, with the smallest possible proportion of misclassifications. Useful, for example, in the taxonomic problems of classifying individuals of a plant into the various species.

discriminant of a differential equation. For a differential equation of type $F(x, y, p) = 0$, where $p = dy/dx$, the **p-discriminant** is the result of eliminating p between the equations $F(x, y, p) = 0$ and $\dfrac{\partial F(x, y, p)}{\partial p} = 0$. If the solution of the differential equation is

$$u(x, y, c) = 0,$$

the **c-discriminant** is the result of eliminating c between the equations $u(x, y, c) = 0$ and $\dfrac{\partial u(x, y, c)}{\partial c} = 0$. The curve whose equation is obtained by setting the p-discriminant equal to zero contains all *envelopes* of solutions, but also may contain a *cusp locus*, a *tac-lacus*, or a *particular solution* (in general, the equation of the tac-locus will be squared and the equation of the particular solution will be cubed). The curve whose equation is obtained by setting the c-discriminant equal to zero contains all *envelopes* of solutions, but also may contain a *cusp locus*, a *node locus*, or a *particular solution* (in general, the equation of the node locus will be squared and the equation of the cusp locus will be cubed). In general, the cusp locus, node locus, and tac-locus are not solutions of the differential equation. *E.g.*, the differential equation $(dy/dx)^2 (2 - 3y)^2 = 4(1 - y)$ has the general solution $(x - c)^2 = y^2(1 - y)$ and the p-discriminant and

c-discriminant equations are, respectively,

$$(2 - 3y)^2(1 - y) = 0 \quad \text{and} \quad y^2(1 - y) = 0.$$

The line $1 - y = 0$ is an envelope; $2 - 3y$ is a tac-locus; $y = 0$ is a node locus.

discriminant of a polynomial equation, $x^n + a_1x^{n-1} + \cdots + a_n = 0$. The product of the squares for all the differences of the roots taken in pairs. The discriminant is equal to the *resultant* of the equation and its derived equation, except possibly in sign. The discriminant is $(-1)^{n(n-1)/2}$ times this resultant. If the leading coefficient is a_0 instead of 1, the factor a_0^{2n-2} is introduced in the discriminant and the discriminant is $(-1)^{n(n-1)/2}/a_0$ times the resultant. The discriminant is zero if and only if the polynomial equation has a double root. For a **quadratic equation**, $ax^2 + bx + c = 0$, the discriminant is $b^2 - 4ac$. If a, b, and c are real, the discriminant is zero when and only when the roots are equal, and negative or positive according as the roots are imaginary or real. *E.g.*, the discriminant of $x^2 + 2x + 1 = 0$ is 0 and the two roots are equal; the discriminant of $x^2 + x + 1 = 0$ is -3 and the roots are imaginary; the discriminant of $x^2 - 3x + 2 = 0$ is 1 and the roots, 1 and 2, are real and unequal. See QUADRATIC —quadratic formula. For a real **cubic equation**, $x^3 + ax^2 + bx + c = 0$, the discriminant is equal to

$$a^2b^2 + 18abc - 4b^3 - 4a^3c - 27c^2.$$

This discriminant is positive if the equation has three real, distinct roots; it is negative if there is a single real root and two conjugate imaginary roots; and it is zero if the roots are all real and at least two of them are equal. See RESULTANT — resultant of a set of polynomial equations.

discriminant of a quadratic equation in two variables. If the equation is

$$ax^2 + bxy + cy^2 + dx + ey + f = 0,$$

then the discriminant is the quantity $\Delta = (4acf - b^2f - ae^2 - cd^2 + bde)$, which can be written

$$\Delta = \frac{1}{2} \begin{vmatrix} 2a & b & d \\ b & 2c & e \\ d & e & 2f \end{vmatrix}.$$

The discriminant is also equal to the product of $-(b^2 - 4ac)$ and the constant term in the equation obtained by translating the axes so as to remove the first degree terms—namely,

$$a'x^2 + b'xy + c'y^2 - \Delta/(b^2 - 4ac) = 0.$$

The discriminant and the invariant, $b^2 - 4ac$, provide the following criteria concerning the locus of the general quadratic in two variables. If $\Delta \neq 0$ and $b^2 - 4ac < 0$, the locus of the general quadratic is a real or imaginary ellipse; if $\Delta \neq 0$ and $b^2 - 4ac > 0$, a hyperbola; if $\Delta \neq 0$ and $b^2 - 4ac = 0$, a parabola. If $\Delta = 0$ and $b^2 - 4ac < 0$, the locus is a point ellipse; if $\Delta = 0$ and $b^2 - 4ac > 0$, two intersecting lines; and if $\Delta = 0$ and $b^2 - 4ac = 0$, two parallel or coincident lines or no (real) locus. The discriminant Δ is defined differently by different writers, but all the forms are the same except for multiplication by some constant.

discriminant of a quadratic form. The determinant with a_{ij} in row i and column j, where the quadratic form $Q = \sum\limits_{i,j=1}^{n} a_{ij}x_ix_j$ is written so that $a_{ij} = a_{ji}$ for all i and j. If Δ_m is the discriminant of the quadratic form obtained from Q by discarding all terms but those involving only $x_1, x_2, \cdots, x_m$, then there is a linear transformation of the form $x_i = y_i + \sum\limits_{j=1}^{n} b_{ij}y_j$ such that $\sum\limits_{i,j=1}^{n} a_{ij}x_ix_j = \sum\limits_{i=1}^{n} \alpha_iy_i^2$, where $\alpha_1 = \Delta_1$, $\alpha_2 = \Delta_2/\Delta_1$, $\alpha_3 = \Delta_3/\Delta_2, \cdots, \alpha_n = \Delta_n/\Delta_{n-1}$. See TRANSFORMATION—congruent transformation, INDEX—index of a quadratic form.

DIS-JOINT, *adj.* Two sets are disjoint if there is no object which belongs to each of the sets (*i.e.*, if the intersection of the sets is the null set). A system of more than two sets is **pairwise disjoint** (sometimes simply **disjoint**) if each pair of sets belonging to the system is disjoint.

DIS-JUNC'TION, *n.* **disjunction of propositions.** The proposition formed from two given propositions by connecting them with the word *or*, thereby asserting the truth of one or both of the given propositions. The disjunction of two propositions is false if and only if both the propositions are false. *E.g.*, the disjunction of "$2 \cdot 3 = 7$" and "Chicago is in Illinois" is the true statement "$2 \cdot 3 = 7$ *or* Chicago is in Illinois." The disjunction of "Today is Tuesday" and "Today is Christmas" is the statement "Today is Tuesday *or* today is Christmas," which is true unless today is neither Tuesday nor Christmas. The disjunction of propositions p and q is usually written $p \vee q$ and read "p *or* q." This is the *inclusive disjunction*, which ordinarily is used in mathematics. The *exclusive disjunction* of p and q is true if and only if exactly one of p and q is true. Both types of disjunctions are common in ordinary English usage. See CONJUNCTION. *Syn.* alternation.

DIS-PER'SION, *n.* Scattering or dispersion of data. Dispersion is measured in various ways, *e.g.*, by mean deviation, standard deviation, quartile deviation.

DIS-SIM'I-LAR, *adj.* **dissimilar terms.** Terms that do not contain the same powers or the same unknown factors. *E.g.*, $2x$ and $5y$, or $2x$ and $2x^2$, are *dissimilar terms*. See ADDITION—addition of similar terms in algebra.

DIS'TANCE, *n.* **angular distance between two points.** The angle between the two rays drawn from the point of observation (point of reference) through the two points. *Syn.* apparent distance.
  distance between lines. For two **parallel lines**, the length of a common perpendicular joining the lines; the distance from a point of one line to the other line. For two **skew lines**, the length of the line segment that joins the lines and is perpendicular to both.
  distance between planes. For two parallel planes, the length of the segment which they cut off on a common perpendicular; the distance from one of them to a point on the other.
  distance between points. The length of the line segment joining two points. In analytic geometry, it is found by taking the square root of the sum of the squares of the differences of the corresponding rectangular Cartesian coordinates of the two points. In the plane, this is $\sqrt{(x_2 - x_1)^2 + (y_2 - y_1)^2}$, where the points are (x_1, y_1) and (x_2, y_2); for points (x_1, y_1, z_1) and (x_2, y_2, z_2) in space, it is

$$\sqrt{(x_2 - x_1)^2 + (y_2 - y_1)^2 + (z_2 - z_1)^2},$$

  distance from a point to a line or a plane. The perpendicular distance from the point to the line or plane. The distance from a point to a line in the (x, y)-plane is equal to the left member of the *normal form* of the equation of the line when the coordinates of the point are substituted (see LINE—equation of a line), or it can be found by finding the foot of the perpendicular from the point to the line and then finding the distance between these two points. The distance from a point to a plane is equal to the left member of the *normal form* of the equation of the plane when the coordinates of the point are substituted (see PLANE—equation of a plane).

distance-rate-time formula. The formula which states that the distance passed over by a body, moving at a fixed rate for a given time, is equal to the product of the rate and time, written $d = rt$.
  distance from a surface to a tangent plane. For a surface S whose parametric equations are $x = x(u, v)$, $y = y(u, v)$, $z = z(u, v)$, the distance between the point corresponding to $(u + du, v + dv)$ and the plane tangent to S at (u, v) is given by $\frac{1}{2}(dx\,dX + dy\,dY + dz\,dZ) + e = \frac{1}{2}(D\,du^2 + 2D'\,du\,dv + D''\,dv^2) + e = \frac{1}{2}\Phi + e$, where X, Y, Z are the direction cosines of the normal to S, e denotes terms of the third and higher orders in du and dv, and

$$D = X\frac{\partial^2 x}{\partial u^2} + Y\frac{\partial^2 y}{\partial u^2} + Z\frac{\partial^2 z}{\partial u^2}$$

$$= \sum X\frac{\partial^2 x}{\partial u^2} = -\sum \frac{\partial X}{\partial u}\frac{\partial x}{\partial u},$$

$$D' = \sum \frac{\partial^2 x}{\partial u\,\partial v} = -\sum \frac{\partial X}{\partial u}\frac{\partial x}{\partial v} = -\sum \frac{\partial X}{\partial v}\frac{\partial x}{\partial v},$$

$$D'' = \sum X\frac{\partial^2 x}{\partial v^2} = -\sum \frac{\partial X}{\partial v}\frac{\partial x}{\partial v}.$$

See SURFACE —fundamental quadratic form of a surface.
  polar distance. Same as CODECLINATION.

DIS'TRI-BU'TION, *n.* (1) See GENERALIZED —generalized function. (2) The relative arrangement of a set of numbers; a set of values of a variable and the frequencies of each value. *Tech.* A *random variable* together with its *probability density function, probability function,* or *distribution function.* For examples, see under BERNOULLI, BINOMIAL, BIVARIATE, CAUCHY, CHI, F, GAMMA, GEOMETRIC, HYPERGEOMETRIC, LOG-NORMAL, MULTINOMIAL, NORMAL, PEARSON, POISSON, T, UNIFORM.
  distribution function. The distribution function of a random variable X is the function F_X defined for each real number t by letting $F_X(t)$ be the probability of the event "$X \leq t$." The distribution function is **discrete** or **continuous** if X is discrete or continuous, respectively. If a discrete random variable X has possible values $x_1, x_2, \cdots$, and if $f(x_i)$ is the probability of x_i, then $F_X(t)$ is the sum of all $f(x_i)$ for which $x_i \leq t$. If a continuous random variable X has the probability density function f, then $F_X(t) = \int_{-\infty}^{t} f(x)\,dx$. For a *vector random variable* (X, Y), the **joint distribution function**

$F_{(X,Y)}$ is defined by letting $F_{(X,Y)}(a,b)$ be the probability of the event "$X \leq a$ and $Y \leq b$," for arbitrary real numbers a and b. Random variables X and Y are independent if and only if $F_{(X,Y)}(a,b) = F_X(a)F_Y(b)$ for all a and b (see INDEPENDENT —independent random variables). The function $F_{(X,Y)}$ also is called a **bivariate distribution; joint-distribution functions** or **multivariate distribution functions** are defined similarly for arbitrary vector random variables. *Syn.* cumulative distribution function, probability distribution function. Sometimes "distribution function" is used in the sense of any function that describes the distribution of a population, such as the *frequency function* or the *probability function*. For examples, see CAUCHY—Cauchy distribution, GAMMA—gamma distribution, UNIFORM —uniform distribution.

relative distribution function. See PROBABILITY—probability density function.

skew distribution. A nonsymmetric distribution. If the third moment about the mean is nonzero, the distribution is skew (but not conversely). A distribution is said to be **skewed to the left (right)**—or to have **negative (positive) skewness**—if the graph of the probability density function (or the probability function) has a bigger tail (greater values) far to the left (right) of the mean than far to the right (left). Sometimes the third moment about the mean μ_3 is used as a measure of skewness and the distribution is said to be skewed to the left or right as μ_3 is negative or positive.

symmetric distribution. A distribution for which there is a number m about which the probability density function (or the probability function) is symmetric. This number m is then both the *mean* and the *median*, although the mean and median can be equal for nonsymmetric distributions.

DIS-TRIB'U-TIVE, *adj.* An operation is distributive relative to a rule of combination if performing the operation upon the combination of a set of quantities is equivalent to performing the operation upon each member of the set and then combining the results by the same rule of combination. *E.g.*,

$$\frac{d(u+v)}{dx} = \frac{du}{dx} + \frac{dv}{dx},$$

the rule of combination being addition. The function $\sin x$ is not distributive, since $\sin(x+y) \neq \sin x + \sin y$. See FIELD, and below, distributive property of arithmetic and algebra.

distributive lattice. See LATTICE.

distributive properties of logic. Conjunction ($\wedge$) distributes over disjunction ($\vee$),

$$p \wedge (q \vee r) \Leftrightarrow (p \wedge q) \vee (p \wedge r),$$

and disjunction distributes over conjunction,

$$p \vee (q \wedge r) \Leftrightarrow (p \vee q) \wedge (p \vee r).$$

See CONJUNCTION, DISJUNCTION.

distributive properties of set theory. Intersection ($\cap$) distributes over union ($\cup$),

$$A \cap (B \cup C) = (A \cap B) \cup (A \cap C),$$

and union distributes over intersection,

$$A \cup (B \cap C) = (A \cup B) \cap (A \cup C).$$

See INTERSECTION, UNION.

distributive property of arithmetic and algebra. The property that

$$a(b+c) = ab + ac$$

for any numbers a, b, and c; *e.g.*, $2(3+5) = 2 \cdot 3 + 2 \cdot 5$, each expression being equal to 16. This can be extended to state that the product of a monomial by a polynomial is equal to the sum of the products of the monomial by each term of the polynomial; *e.g.*,

$$2(3 + x + 2y) = 6 + 2x + 4y.$$

When two polynomials are multiplied together, one is first treated as a monomial and multiplied by the individual terms of the other, then the results multiplied out according to the above law (or one may multiply each term of one polynomial by each term of the other and add the results). *E.g.*,

$$(x+y)(2x+3) = x(2x+3) + y(2x+3)$$

$$= 2x^2 + 3x + 2xy + 3y.$$

For noncommutative operations, it is necessary to distinguish between **left distributivity**,

$$a(b+c) = ab + ac,$$

and **right distributivity**,

$$(b+c)a = ba + ca.$$

DI-VER'GENCE, *adj., n.* **divergence of a sequence or series.** The property of being divergent; the property of not being convergent. See DIVERGENT.

divergence of a vector function. For a vector-valued function F, the **divergence** is denoted by $\nabla \cdot F$, where ∇ is the operator,

$$i\frac{\partial}{\partial x} + j\frac{\partial}{\partial y} + k\frac{\partial}{\partial z},$$

and is defined as

$$\nabla \cdot F = i \cdot \frac{\partial F}{\partial x} + j \cdot \frac{\partial F}{\partial y} + k \cdot \frac{\partial F}{\partial z}$$

$$= \frac{\partial F_x}{\partial x} + \frac{\partial F_y}{\partial y} + \frac{\partial F_z}{\partial z},$$

where $F = iF_x + jF_y + kF_z$. If, for instance, $F(x, y, z)$ is the velocity of a fluid at the point P: (x, y, z), then $\nabla \cdot F$ is the rate of change of volume per unit volume of an infinitesimal portion of the fluid containing P. See below, divergence of a tensor.

divergence of a tensor. The divergence of a contravariant tensor T^i of order one (*i.e.*, **a contravariant vector field**) is $T^i_{,i}$, or

$$\frac{1}{\sqrt{g}}\frac{\partial(T^i\sqrt{g})}{\partial x^i},$$

where the *summation convention* applies, g is the determinant having g_{ij} in the ith row and jth column (g_{ij} being the *fundamental metric tensor*), and $T^i_{,j}$ is the covariant derivative of T^i. The divergence of a covariant tensor T_i of order one (*i.e.*, a *covariant vector field*) is $g^{ij}T_{i,j}$, or $T^i_{,i}$, where $T^i = g^{ij}T_j$ and g^{ij} is $1/g$ times the cofactor of g_{ji} in g.

divergence theorem. Let V be a bounded open three-dimensional set whose boundary S is a surface composed of a finite number of smooth surface elements. Then the *divergence theorem* is a theorem which states that, with certain specific conditions on a vector-valued function $\mathbf{F}$,

$$\int_S (\mathbf{F} \cdot \mathbf{n}) \, d\sigma = \int_V (\nabla \cdot \mathbf{F}) \, dV,$$

where $\mathbf{n}$ is the unit vector normal to S and pointing out of V and $\nabla \cdot \mathbf{F}$ is the divergence of $\mathbf{F}$. A sufficient condition on $\mathbf{F}$ is that $\mathbf{F}$ be continuous on the union of V and S and the first-order partial derivatives of the components of $\mathbf{F}$ be bounded and continuous on V. See GREEN—Green's formulas, SURFACE—surface integral. *Syn.* Gauss' theorem, Green's theorem in space, Ostrogradski's theorem.

DI-VER′GENT, *adj.* **divergent sequence.** A sequence which does not converge. It might either be **properly divergent**, or **oscillate**, in the sense described below for the sequence of partial sums of a divergent series.

divergent series. A series which does not converge. The sequence of partial sums $\{S_1, S_2, \cdots\}$ of a *divergent series* (S_n is the sum of the first n terms of the series) is a *divergent sequence*. The series is **properly divergent** if the partial sums become arbitrarily large for large values of n in the sense that, for any number M, $S_n > M$ for all but a finite number of values of n, or if they become arbitrarily small (algebraically) in the sense that, for any number M, $S_n < M$ for all but a finite number of values of n. In these two cases, one writes $\lim_{n \to \infty} S_n = +\infty$ and $\lim_{n \to \infty} S_n = -\infty$. All other types of divergent series are called **oscillating divergent series**. The series $1 + 2 + 3 + \cdots$, $1 + \frac{1}{2} + \frac{1}{3} + \cdots$, and $-1 - 1 - 1 \cdots$ are *properly divergent*, while $1 - 1 + 1 - 1 + \cdots$ and $1 - 2 + 3 - 4 + \cdots$ are *oscillating divergent series*. For the last example, the partial sums are $1, -1, 2, -2, 3, -3, 4, -4, \cdots$. This sequence is divergent in the sense that, for any number M, $|S_n| > M$ for all but a finite number of values of n.

summation of divergent series. See SUMMATION—summation of divergent series.

DI-VIDE′, *v.* To perform a division.

DIV′I-DEND, *n.* (1) A quantity which is to be divided by another quantity. See DIVISION—division algorithm. (2) *In finance*, profits of a company which are to be distributed among the shareholders. For a bond, the dividend is the periodic, usually semiannual, interest paid on the bond. The **dividend date** is the date upon which the dividend is due; the interest rate named in the bond is the **dividend rate** or **bond rate**. An **accrued dividend** is a partial dividend; the interest on the face value of a bond from the nearest preceding dividend date to the purchase date. In bond market parlance, **accrued interest** is used synonymously with *accrued dividends*.

DI-VIS′I-BLE, *adj.* In general, an object x is divisible by an object y if there exists an object q of a certain specified type for which $x = yq$. For example, an integer m is divisible by an integer n if there is an integer q for which $m = nq$; a polynomial F is divisible by a polynomial G if there is a polynomial Q for which $F = GQ$. There are many special **tests for divisibility** of integers when written in decimal notation: *Divisibility by 2:* the last digit divisible by 2. *Divisibility by 3 (or 9):* the sum of the digits divisible by 3 (or 9) [*e.g.*,

35,712 is divisible by both 3 and 9, since the sum of the digits is 18]. *Divisibility by* 4: the number represented by the last two digits divisible by 4. *Divisibility by* 5: the last digit is either 0 or 5. *Divisibility by* 11: the sum of the digits in even places minus the sum in odd places is divisible by 11.

DI-VI′SION, *n.* (1) Finding the quotient and remainder in the division algorithm (see below, division algorithm). (2) The *inverse operation* to multiplication. The result of dividing one number (the **dividend**) by another (the **divisor**) is their **quotient.** The quotient a/b of two numbers a and b is that number c such that $b \cdot c = a$, provided c exists and has only one possible value (if $b = 0$, then c does not exist if $a \neq 0$, and c is not unique if $a = 0$; *i.e.*, $a/0$ is meaningless for all a, and **division by zero is meaningless**); the quotient a/b can also be defined as the product of a and the inverse of b (see GROUP). *E.g.*, $6/3 = 2$, because $3 \cdot 2 = 6$; $(3 + i)/(2 - i) = 1 + i$, because

$$3 + i = (2 - i)(1 + i).$$

The **division of a fraction by an integer** can be accomplished by dividing the numerator (or multiplying the denominator) by the integer ($4/5 \div 2 = 2/5$ or $4/10$); **division by a fraction** can be accomplished by inverting the fraction and multiplying by it, or by writing the quotient as a complex fraction and simplifying

$$\frac{\frac{7}{5}}{\frac{2}{3}} = \frac{7}{5} \cdot \frac{3}{2} = \frac{21}{10},$$

or

$$\frac{\frac{7}{5}}{\frac{2}{3}} = \frac{\frac{7}{5} \cdot 15}{\frac{2}{3} \cdot 15} = \frac{7 \cdot 3}{2 \cdot 5} = \frac{21}{10}$$

(see FRACTION —complex fraction); **division of mixed numbers** is accomplished by reducing the mixed numbers to fractions and dividing these results ($1\frac{2}{3} \div 3\frac{1}{2} = 5/3 \div 7/2 = 10/21$).

division algebra. See ALGEBRA —algebra over a field.

division algorithm. For integers, the theorem that, for any integer m and any positive integer n, there exist unique integers q and r such that

$$m = nq + r, \qquad 0 \leq r < n.$$

The integer m is the *dividend*, n is the *divisor*, q is the *quotient*, and r is the *remainder*. For polynomials, the division algorithm states that,

for any polynomial f and any nonconstant polynomial g, there exist unique polynomials q and r such that

$$f(x) \equiv g(x)q(x) + r(x),$$

where either $r = 0$ or the degree of r is less than the degree of g. The polynomials f, g, q, and r are the *dividend*, *divisor*, *quotient*, and *remainder*. See REMAINDER —remainder theorem.

division by a decimal. See above, DIVISION. Accomplished by multiplying dividend and divisor by a power of 10 that makes the divisor a whole number (*i.e.*, moving the decimal point to the right in the dividend as many places as there are decimal places in the divisor), then dividing as with whole numbers, placing a decimal point in the quotient in the place arrived at before using the first digit after the decimal place in the dividend. *E.g.*,

$$28.7405 \div 23.5 = 287.405 \div 235 = 1.223.$$

division modulo *p.* If, in the process of performing division, $f(x) = q(x) \cdot d(x) + r(x)$, any of the coefficients are increased or diminished by multiples of p, the process is **division modulo** *p* and is written $f(x) = q(x) \cdot d(x) + r(x) \pmod{p}$. This definition applies only when each coefficient is an integer. Each coefficient is usually written as one of the integers $0, 1, 2, \cdots, p - 1$, two integers being regarded as equal (or equivalent) if they differ by a multiple of p.

division in a proportion. Passing from a proportion to the statement that the first antecedent minus its consequent is to its consequent as the second antecedent minus its consequent is to its consequent, *i.e.*, from $a/b = c/d$ to $(a - b)/b = (c - d)/d$. See COMPOSITION —composition in a proportion.

division ring. See RING.

division transformation. The relation dividend = quotient × divisor + remainder. Rarely used.

division by use of logarithms. See LOGARITHM.

harmonic division of a line. See HARMONIC.

point of division. See POINT —point of division.

ratio of division or **division ratio.** See POINT —point of division.

short division and long division. (1) Division is called short (or long) according as the process can (or cannot) be carried out mentally. It is customary to discriminate between long and short division solely on the basis of the complexity of the problem. When the steps in the division must be written down, it is called **long division**; otherwise it is **short division**. (2) Division is **short** (or **long**) if the divisor contains one digit (or more than one digit); in *algebra*, if the divisor contains one term (or more than one term).

synthetic division. See SYNTHETIC.

DI-VI′SOR, *n.* The quantity by which the dividend is to be divided. See DIVISION, and DIVISION—division algorithm.

common divisor of two or more quantities. A quantity which is a factor of each of the quantities. A common divisor of 10, 15, and 75 is 5; a common divisor of $x^2 - y^2$ and $x^2 - 2xy + y^2$ is $x - y$, since $x^2 - y^2 = (x - y) \times (x + y)$ and $x^2 - 2xy + y^2 = (x - y)^2$. *Syn.* common factor, common measure.

greatest common divisor of two or more quantities. A common divisor that is divisible by all other common divisors, often abbreviated as G.C.D. For positive integers, the greatest common divisor is the largest of all common divisors; the common divisors of 30 and 42 are 2, 3, and 6, the largest being the greatest common divisor 6. *Syn.* greatest common factor, greatest common measure.

normal divisor of a group. See NORMAL—normal subgroup.

DO-DEC′A-GON, *n.* A polygon having twelve sides, See POLYGON.

DO′DEC-A-HE′DRON, *n.* [*pl.* **dodecahedrons** or **dodecahedra**]. A polyhedron having twelve faces. A **regular dodecahedron** has regular pentagons as faces. See POLYHEDRON.

DO-MAIN′, *n.* (1) A *field*, as the *number domain* of all rational numbers, or of all real numbers (see FIELD). (2) Any *open connected set* that contains at least one point. Also used for any open set containing at least one point. Sometimes called a **region**. (3) The **domain of a function** is the set of values which the independent variable may take on, or the *range of the independent variable*. See FUNCTION. (4) See below, integral domain.

domain of dependence for a partial differential equation. See DEPENDENCE.

integral domain. A *commutative ring with unit element* which has no proper divisors of zero (proper divisors of zero are nonzero members x and y for which $x \cdot y = 0$, where 0 is the additive identity). The assumption that there are no proper divisors of zero is equivalent to the cancellation law: $x = y$ whenever $xz = yz$ and $z \neq 0$. The set of ordinary integers (positive, negative, and 0) and the set of all algebraic integers are integral domains. An **ordered integral domain** is

an integral domain D that contains a set of "positive" members satisfying the conditions: (1) the sum and product of two positive members is positive; (2) for a given member x of D, one and only one of the following is true: x is positive, $x = 0$, $-x$ is positive. A **unique-factorization domain** is an integral domain for which each member is a unit, or is a prime, or can be expressed as the product of a finite number of primes and this expression is unique except for unit factors and the order of the factors. If D is a unique-factorization domain, then so is the set of all polynomials with coefficients in D. In particular, the set of all polynomials with coefficients in a field F is a unique-factorization domain. See ALGEBRAIC—algebraic number, RING.

DOM′I-NANT, *adj.* **dominant strategy**. See STRATEGY.

dominant vector. A vector $\mathbf{a} = (a_1, a_2, \cdots, a_m)$ such that, relative to a second vector $\mathbf{b} = (b_1, b_2, \cdots, b_m)$, the inequality $a_i \geq b_i$ holds for each i ($i = 1, 2, \cdots, m$). If the strict inequality $a_i > b_i$ holds for each i, the dominance is said to be **strict**.

DOM′I-NO, *n.* See POLYOMINO.

DONALDSON, Simon Kirwan (1957–). American mathematician awarded a Fields Medal (1986) for showing the existence of "exotic" four-spaces; *i.e.*, four-dimensional manifolds that are topologically but not differentially equivalent to the usual R^4.

DOT, *n.* **dot product.** See MULTIPLICATION — multiplication of vectors.

DOU′BLE, *adj.* **double-angle formulas.** See TRIGONOMETRY—double angle formulas of trigonometry.

double integral. See INTEGRAL —iterated integral, multiple integral.

double law of the mean. See MEAN—mean value theorem for derivatives.

double ordinate. See ORDINATE.

double point. See POINT—multiple point.

double root of an algebraic equation. A root that is repeated, or occurs exactly twice in the equation; a root such that $(x - r)^2$, where r is the root, is a factor of the left member of the equation when the right is zero, but $(x - r)^3$ is not such a factor. *Syn.* repeated root, root of multiplicity two, coincident roots, equal roots. See MULTIPLE —multiple root of an equation.

double rule of three. See THREE—rule of three.

double tangent. (1) A tangent which has two *noncoincident* points of tangency with the curve. (2) Two coincident tangents, as the tangents at a cusp. See POINT—double point.

DOU′BLET, *n.* See POTENTIAL —concentration method for the potential of a complex.

DOUGLAS, Jesse (1897–1965). American analyst and Fields Medal recipient (1936). Solved the *Plateau problem* in 1931, about the same time as Tibor Radó.

DRAFT, *n.* An order written by one person and directing another to pay a certain amount of money. A **bank draft** is a draft drawn by one bank upon another; a **commercial draft** is a draft made by one firm on another to secure the settlement of a debt. A draft is **after-date** or **after-sight** according as the time during which discount is reckoned (if there is any) begins on the date of the draft or begins with the date of acceptance of the draft.

DRAG, *n.* If the total force F that is applied to a body B gives B a motion with velocity vector v, then the component of F in the direction opposite to v is called *drag*. In exterior ballistics, the drag F_v is given approximately by the formula

$$F_v = \rho d^2 v^2 K,$$

where ρ is the density of air, d is the diameter of the shell, v is the speed at which it is traveling, and K is the constant called the **drag coefficient** of the shell. See LIFT.

axial drag. In exterior ballistics, *axial drag* is the component, in the direction opposite to that of the advancing axis of a shell, of the total force acting on the shell. It is found that the axial drag F_a is given approximately by the formula

$$F_a = \rho d^2 v_a^2 K_a,$$

where ρ is the density of air, d is the diameter of the shell, v_a is the component of the velocity in the direction of the axis of the shell, and K_a is a constant. The constant K_a is the **axial-drag coefficient** of the shell; it depends mostly on the shape of the shell, but also somewhat on its size.

DRINFEL′D, Vladimir (1954–). Soviet mathematician awarded a Fields Medal (1990) for his contributions to algebraic geometry, number theory, and the theory of quantum groups.

DU′AL, *adj.* **dual formulas.** Formulas related in the same way as dual theorems.

dual space. See CONJUGATE—conjugate space.

dual theorems. See DUALITY—principle of duality of projective geometry, and principle of duality in a spherical triangle. Sometimes called RECIPROCAL THEOREMS.

DU-AL′I-TY, *n.* **Poincaré duality theorem.** The p-dimensional Betti numbers B_G^p of an orientable manifold which is homeomorphic to the set of points of an n-dimensional simplicial complex satisfy

$$B_G^p = B_G^{n-p},$$

where G is the group for which chains and homology groups are defined. Poincaré proved the theorem for the case G is the group of rational numbers; the proof for G the group of integers $mod\, 2$ was given by Veblen and the proof for G the group of integers $mod\, p$ (p a prime) was given by Alexander. See BETTI—Betti number, HOMOLOGY —homology group.

principle of duality of projective geometry. The principle that if one of two dual theorems is true the other is also. In a *plane*: point and line are **dual elements**; the drawing of a line through a point and the marking of a point on a line are **dual operations**, as are also the drawing of two lines through a point and the marking of two points on a line, or the bringing of two lines to intersect in a point and the joining of two points by a line; figures which can be obtained from one another by replacing each element by the dual element and each operation by the dual operation are **dual figures**, as three lines passing through a point and three points lying on a line (three *concurrent* lines and three *collinear* points). Theorems which can be obtained from one another by replacing each element in one by the dual element and each operation by the dual operation are **dual theorems**. In *space*: the point and plane are dual elements (called **space duals**), the definitions of dual operations, figures, and theorems being analogous to those in the plane. Some writers state dual theorems in such terms that they are interchanged merely by interchanging the words point and line (or point and plane), as, for instance, two points determine a line—two lines determine a point, or two points on a line —two lines on a point. *E.g.*, the two following statements are plane duals: (a) one and only one line is determined by a point and the point common to two lines; (b) one and only one point is determined by a line and the line common to two points.

principle of duality in a spherical triangle. In any formula involving the sides and the supplements of the angles opposite the sides, another true formula may be obtained by interchanging each of the sides with the supplement of the angle opposite it. The new formula is the **dual formula**.

DU′EL, *n*. A two-person zero-sum game involving the timing of decisions. Delay of action increases accuracy but also increases the likelihood that the opponent will have acted first. A duel is a **noisy duel** if each player knows at all times whether or not the opponent has taken action; it is a **silent duel** if the players never know whether or not their opponent has taken action.

DUHAMEL, Jean Marie Constant (1797–1872). French analyst and applied mathematician.

Duhamel's theorem. If the sum of *n* infinitesimals (each a function of *n*) approaches a limit as *n* increases (becomes infinite), then the same limit is approached by the sum of the infinitesimals formed by adding to each of these infinitesimals other infinitesimals which are *uniformly* of higher order than the ones to which they are added. *E.g.*, the sum of *n* terms, each equal to $1/n$, is equal to 1 for all *n*. Hence this sum approaches (is) 1 as *n* increases. The sum of *n* terms, each equal to $1/n + 1/n^2$, must, by *Duhamel's Theorem*, also approach 1 as *n* increases. This is seen to be true, from the fact that this sum is $1 + 1/n$, which certainly approaches 1 as *n* increases. See INTEGRAL —definite integral. *Tech.* If

$$\lim_{n \to \infty} \sum_{i=1}^{n} \alpha_i(n) = L,$$

then

$$\lim_{n \to \infty} \sum_{i=1}^{n} \left[\alpha_i(n) + \beta_i(n) \right] = L,$$

provided that for any $\epsilon > 0$ there exists an N such that $|\beta_i(n)/\alpha_i(n)| < \epsilon$ for all *i*'s and for all $n > N$. [When β_i/α_i satisfy this restriction, the ratios β_i/α_i $(i = 1, 2, 3, \cdots)$ are said to *converge uniformly* to zero.] This is a sufficient, but not a necessary, condition for the two limits to be the same. All that is necessary is that the sum of the *n* betas approach zero as *n* becomes infinite, which can happen, for example, when any finite number of the betas are larger than the alphas, provided each of them approaches zero and the other betas are such that all β_i/α_i formed from them converge uniformly to zero. See UNIFORM —uniform convergence of a set of functions.

DUM′MY, *adj.* **dummy variable.** A symbol which could be replaced by any other arbitrary symbol without changing the meaning of some given expression. *E.g.*, $\sum_{i=1}^{n} a_i$ has exactly the same meaning as $\sum_{j=1}^{n} a_j$, being simply the sum of $a_1, a_2, \cdots, a_n$ [in such cases, *i* (or *j*) also is called a **dummy index**]. In $\int_a^b f(x)\, dx$, *x* is a dummy variable, since any other letter could be substituted for *x*; this integral also is equal to $\int_a^b f(s)\, ds$, and its value is determined when *a*, *b*, and the function *f* are specified. See SUMMATION —summation convention.

DU′O-DEC′I-MAL, *adj.* **duodecimal number system.** A system of numerals for representing real numbers for which twelve is the base, instead of ten. *E.g.*, in the *duodecimal system*, 24 would mean $2 \cdot 12 + 4$, which is 28 in the decimal system. Since 12 has many factors, arithmetic operations can be performed more simply in the duodecimal system than in the decimal system. For example, $\frac{1}{2}, \frac{1}{3}, \frac{1}{4}, \frac{1}{6}, \frac{1}{8}$ and $\frac{1}{9}$ can be written in the duodecimal system as 0.6, 0.4, 0.3, 0.2, 0.16 and 0.14, respectively. However, $\frac{1}{5}$ is the infinite repeating sequence $0.2497, 2497, \cdots$. See BASE—base of a number system.

DUPIN, François Pierre Charles (1784–1873). French differential geometer and physicist.

Dupin indicatrix of a surface at a point. If the tangents to the lines of curvature at the point *P* of the surface *S* are taken as ξ, η coordinate axes, and ρ_1 and ρ_2 are the corresponding radii of principal curvature of *S* at *P*, then the **Dupin indicatrix** of *S* at *P* is $\dfrac{\xi^2}{|\rho_1|} + \dfrac{\eta^2}{|\rho_2|} = 1$, or $\dfrac{\xi^2}{\rho_1} + \dfrac{\eta^2}{\rho_2} = \pm 1$, or $\xi^2 = |\rho_1|$, according as the total curvature of *S* at *P* is positive, negative, or zero $(1/\rho_2 = 0)$. The curve of intersection of *S* and a nearby plane parallel to the tangent plane at *P* is approximately similar to the Dupin indicatrix of *S* at *P*, or, if the curvature of *S* is negative at *P*, to one of the hyperbolas constituting the Dupin indicatrix. Accordingly, a point of the surface is said to be an **elliptic, hyperbolic,** or **parabolic point** according as the total curvature is positive, negative, or zero there.

DU′PLI-CA′TION, *n.* **duplication formula.** For the gamma function, the formula

$$\Gamma(2z) = \pi^{-\frac{1}{2}} \Gamma(z) \Gamma\left(z + \tfrac{1}{2}\right) 2^{2z-1}.$$

Also called the *Legendre duplication formula*.

duplication of the cube. Finding the edge of a cube whose volume is twice that of a given cube, using only straight-edge and compass; the problem of solving the equation $y^3 = 2a^3$ for y, using only straightedge and compass. This is impossible, since the cube root of 2 cannot be expressed in terms of radicals of index 2, and only such numbers can be evaluated by means of straightedge and compasses alone.

DÜRER, Albrecht (1471–1528). German artist and mathematician. Gave first description of epicycloid. Displayed mathematical objects in his wood engravings.

DU′TY, *n.* A tax levied by a government on imported (sometimes on exported) merchandise at the time of entering (or leaving) the country.
   ad valorem duty. A duty which is a certain per cent of the value of the goods.

DY′AD, *n.* The juxtaposition of two vectors, without either scalar or vector multiplication being indicated, as $\mathbf{AB} = \mathbf{\Phi}$. A dyad is thought of as an operator which may operate on a vector by either *scalar* or *vector multiplication*, and in either order: $\mathbf{\Phi} \cdot \mathbf{F} = \mathbf{A}(\mathbf{B} \cdot \mathbf{F})$, $\mathbf{F} \cdot \mathbf{\Phi} = (\mathbf{F} \cdot \mathbf{A})\mathbf{B}$, $\mathbf{\Phi} \times \mathbf{F} = (\mathbf{A} \times \mathbf{B}) \times \mathbf{F}$, $\mathbf{F} \times \mathbf{\Phi} = \mathbf{F} \times (\mathbf{A} \times \mathbf{B})$. The first vector is the **antecedent**, the second the **consequent**. The sum of two or more dyads is a **dyadic**. If the order of the factors in each term of a dyadic is changed, *i.e.*, $\mathbf{A}_1\mathbf{B}_1 + \mathbf{A}_2\mathbf{B}_2 + \mathbf{A}_3\mathbf{B}_3$ is written $\mathbf{B}_1\mathbf{A}_1 + \mathbf{B}_2\mathbf{A}_2 + \mathbf{B}_3\mathbf{A}_3$, the two dyadics are **conjugate dyadics**. Two dyadics $\mathbf{\Phi}_1$ and $\mathbf{\Phi}_2$ are defined to be equal if $\mathbf{r} \cdot \mathbf{\Phi}_1 = \mathbf{r} \cdot \mathbf{\Phi}_2$ and $\mathbf{\Phi}_1 \cdot \mathbf{r} = \mathbf{\Phi}_2 \cdot \mathbf{r}$ for all $\mathbf{r}$. If a dyadic is equal to its conjugate, it is **symmetric**. If it is equal to the negative of its conjugate, it is **antisymmetric**. The (direct) product of dyads $\mathbf{AB}$ and $\mathbf{CD}$ is defined as the dyad $(\mathbf{B} \cdot \mathbf{C})\mathbf{AD}$.

DY-AD′IC, *n.* See DYAD.
   dyadic number system Same as BINARY NUMBER SYSTEM. See BINARY.
   dyadic rational A real number of the form $p/2^n$, where p is an integer and n is a positive integer.

DY-NAM′ICS, *n.* A branch of mechanics studying the effects of forces on rigid and deformable bodies. It is usually treated under two heads—statics and kinematics. See STATICS, KINEMATICS, KINETICS.

DYNE, *n.* The unit of force in the c.g.s. system of units (centimeter-gram-second system); the force required to give a mass of one gram an acceleration of 1 cm per sec per sec. It is equal to a little more than a milligram. See BAR, FORCE —unit force.

E

e. The base of the natural system of logarithms; the limit of $(1 + 1/n)^n$ as n increases without limit. Its numerical value is $2.718281828459045 + $. Also, e is the sum of the infinite series $1 + 1/1! + 1/2! + 1/3! + 1/4! + \cdots$. It is related to the important numbers 1, π, and i by $e^{\pi i} = -1$. In 1737 Euler showed, in effect, that e is irrational; in 1873 Hermite proved that e is a transcendental number.

EC-CEN′TRIC, *adj.* **eccentric angle and circles.** See ELLIPSE, HYPERBOLA —parametric equations of a hyperbola.
   eccentric, or excentric, configuration. Configurations with centers which are not coincident. The term is used mostly with reference to two circles.

EC′CEN-TRIC′I-TY, *n.* **eccentricity of a parabola, ellipse, or hyperbola.** See CONIC.

ECH′E-LON, *adj.* A matrix is an **echelon matrix** or is in **echelon form** if it has the property that the rows with some nonzero terms precede all of those whose terms are zero, the first nonzero term in a row is 1, and the 1 appears in a column to the right of the first nonzero term in any preceding row. An echelon matrix is a **reduced echelon matrix** or is in **reduced echelon form** if the first nonzero term in a row is the only nonzero term in its column. Any matrix can be changed to reduced echelon form by using only elementary row transformations. See ELEMENTARY—elementary operations on determinants or matrices.

E-CLIP′TIC, *n.* The great circle in which the plane of the earth's orbit cuts the celestial sphere; the path in which the sun appears to move. The intersection of the plane of the equator and the plane of the ecliptic is the **line of nodes**. The points where this line intersects the celestial sphere are the **equinoxes**. Also, if any object orbits the sun, then each of the two points where its orbit crosses the ecliptic is called a **node**.

EDGE, *n.* A line or a line segment which is the intersection of two plane faces of a geometric figure, or which is in the boundary of a plane figure. Examples are the edges of a polyhedron and of a polyhedral angle, and the lateral edge of a prism. See ANGLE—polyhedral angle, DIHEDRAL—dihedral angle, HALF—half-plane.

EDVAC. A computing machine built at the University of Pennsylvania for the Ballistic Research Laboratories, Aberdeen Proving Ground. EDVAC is an acronym for *Electronic Discrete Variable Automatic Computer.*

EF-FEC′TIVE, *adj.* **effective interest rate.** See INTEREST.

EF-FI′CIENT, *adj.* **efficient estimator.** (1) An *unbiased estimator* $T(X_1, X_2, \cdots, X_n)$ for a parameter θ which has the property that the variance of T, *i.e.*, the expected value of $(T - \theta)^2$, is the least for all such estimators. Sometimes defined as an unbiased estimator for which

$$E\left[(T - \theta)^2\right] = \frac{1}{n \cdot E\left[(\partial \ln f / \partial \theta)^2\right]},$$

the last expression being a lower bound for $E[(T - \theta)^2]$ (see CRAMÉR—Cramér-Rao inequality). If T is an efficient estimator and t another estimator, the efficiency of t is $E[(T - \theta)^2] / E[(t - \theta)^2] \leq 1$. (2) Given a sequence of estimators $\{T_n\}$ with T_n a function of a random sample $(X_1, X_2, \cdots, X_n)$ for $n = 1, 2, \cdots$, $\{T_n\}$ is asymptotically efficient, or simply efficient, if the distribution of $\sqrt{n}\,(T_n - \theta)$ approaches the normal distribution with mean 0 and variance σ^2 as n increases (*i.e.*, $\{T_n\}$ is *asymptotically normal*) and σ^2 is least for all such estimators.

E-GYP′TIAN, *adj.* **Egyptian numerals.** Numerals used in Egyptian hieroglyphics as early as 3400 B.C. Symbols (pictures) were adopted for $1, 10, 10^2, 10^3 \cdots$. Numbers were expressed by repeated use of these symbols.

EIGENFUNCTION, *n.* See CHARACTERISTIC—Characteristic equation of a matrix, EIGENVALUE.

EIGENVALUE, *n.* For a linear transformation T on a vector space V [see LINEAR—linear transformation (2)], an **eigenvalue** is a scalar λ for which there is a nonzero member $\mathbf{v}$ of V for which $T(\mathbf{v}) = \lambda\mathbf{v}$. The vector $\mathbf{v}$ is an **eigenvector** (or **characteristic vector**). For a **matrix** A, the eigenvalues are the roots of the characteristic equation of the matrix (they are also called **characteristic roots** and **latent roots**); the number λ being an eigenvalue means there is a nonzero vector $\mathbf{x} = (x_1, x_2, \cdots, x_n)$ for which $A\mathbf{x} = \lambda\mathbf{x}$, where multiplication is matrix multiplication and $\mathbf{x}$ is considered to be a one-column matrix. For the **homogeneous integral equation,**

$$\lambda y(x) = \int_a^b K(x, t) y(t)\, dt,$$

the number λ being an eigenvalue means that there is a nonzero solution y, the eigenfunction corresponding to λ. *E.g.*, if $K(x, t) = xt$ and $y(x) \equiv x$, then $\int_a^b K(x, t) y(t)\, dt = \int_a^b xt^2\, dt = \frac{1}{3}x(b^3 - a^3)$, so that $\frac{1}{3}(b^3 - a^3)$ is an eigenvalue and x is an eigenfunction. *Syn.* characteristic number (or constant or value), fundamental number. See HILBERT—Hilbert-Schmidt theory of integral equations with symmetric kernels, SPECTRUM, STURM—Sturm-Liouville differential equation.

EIGENVECTOR, *n.* See EIGENVALUE.

EILENBERG, Samuel (1913–). Polish-American topologist, group theorist, and algebraist. Co-founder of *category theory* with Saunders MacLane.

EINSTEIN, Albert (1879–1955). Superb German-American theoretical physicist and philosopher. Increased interest and activity of mathematicians in Riemannian geometry and tensor analysis with his origination of special and general relativity theory.

    Einstein tensor. See RICCI—Ricci tensor.

EISENSTEIN, Ferdinand Gotthold Max (1823–1852). German algebraist, analyst, and number theorist.

    Eisenstein irreducibility criterion. Let f be the polynomial $a_n x^n + a_{n-1} x^{n-1} + \cdots + a_1 x + a_0$ with integer coefficients. If there is a prime p such that p divides each of $a_0, a_1, \cdots, a_{n-1}$, but p does not divide a_n and p^2 does not divide a_0, then f is irreducible in the field of rational numbers.

E-LAS′TIC, *adj.* **elastic bodies.** Bodies possessing the property of recovering their size and shape when the forces producing deformations are removed.

    elastic constants. See HOOKE—generalized Hooke's law, MODULUS—Young's modulus, POISSON—Poisson's ratio, LAMÉ—Lamé's constants.

E′LAS-TIC′I-TY, *n.* (1) (*Economics*) The ratio of the percent change in one variable to the percent change in another. Suppose X is a function of Y, where $X = f(Y)$. The **point elasticity** of X with respect to Y is

$$\frac{f'(Y) \cdot Y}{f(Y)} = \frac{\partial X}{\partial Y} \cdot \frac{Y}{X}.$$

The **arc elasticity** of X with respect to Y, between Y_1 and Y_2, is

$$\frac{f(Y_1) - f(Y_2)}{\frac{1}{2}[f(Y_1) + f(Y_2)]} \bigg/ \frac{Y_1 - Y_2}{\frac{1}{2}(Y_1 + Y_2)}$$

(2) The property possessed by substances of recovering their size and shape when the forces producing deformations are removed. (3) The mathematical theory concerned with the study of the behavior of elastic bodies. It deals with the calculation of stresses and strains in elastic substances subjected to the action of prescribed forces or deformations. The theory of elasticity of small displacements is the **linear theory**. The **first fundamental problem of elasticity** is the problem of the determination of the state of stress and deformation in the interior of a body when its surface is deformed in a known way. The **second fundamental problem of elasticity** is the problem of the determination of the state of stress and deformation in the interior of a body when its surface is subjected to a specified distribution of external forces.

volume elasticity, or bulk modulus. The quotient of the increase in pressure and the change in unit volume; the negative of the product of the volume and the rate of change of the pressure with respect to the volume, *i.e.*, $E = -V \, dp/dV$.

Young's modulus of elasticity. A measure of the elasticity of stretching or compression; the ratio of the stress to the resulting strain.

E-LEC′TRO-MO′TIVE, *adj.* **electromotive force.** Denoted by E. M. F. (1) That which causes current to flow. (2) The energy added per unit charge due to the mechanical (or chemical) action producing the current. (3) The open circuit difference in potential between the terminals of a cell or generator.

E-LEC′TRO-STAT′IC, *adj.* **electrostatic intensity.** The force a unit positive charge would experience if placed at the point in question, assuming this is done without altering the positions of the other charges in the universe. This assumption should be regarded as a convenient mathematical fiction rather than as a physical possibility. If $e\mathbf{E}$ is the electrostatic force experienced by a charge e when placed at the point P, then the vector $\mathbf{E}$ is the electrostatic intensity at P. Dimensionally, $\mathbf{E}$ is force per unit charge. The electric intensity due to a single charge e is given by the expression $er^{-2}\rho_1$. See CHARGE—point charge, and COULOMB—Coulomb's law for point charges. Here r is the distance from the charge

point to the field point and ρ_1 is the unit vector pointing from charge point to field point.

electrostatic potential of a complex of charges. The scalar point function $\Sigma e_i/r_i$ ($= e_1/r_1 + e_2/r_2 + \cdots + e_n/r_n$). Here $r_1, r_2, \cdots, r_n$ are the distances from the charges $e_1, e_2, \cdots, e_n$ to the field point—the point, supposedly free of charge, at which we are computing the potential. Thus the potential in this case is a point function which is defined at all points excepting the charge points. In rectangular Cartesian coordinates,

$$e_i/r_i = e_i \left[(x - x_i)^2 + (y - y_i)^2 + (z - z_i)^2 \right]^{-1/2},$$

where x, y, z and x_i, y_i, z_i are the coordinates of the field point and of the ith charge point. The potential at a field point P is equal to the work done by the field in repelling a unit positive charge from P to infinity, or to the work that must be done against the field to being the unit charge from infinity to rest at the point P. This work is given by the line integral of the tangential component of the electric intensity E —namely, $\int_p^\infty \mathbf{E} \cdot \boldsymbol{\tau} \, ds$, where $\boldsymbol{\tau}$ is the unit tangent vector to the curve employed and s is the arc length. See COULOMB—Coulomb's law for point charges; FORCE—field of force; and above, electrostatic intensity. This line integral is independent of the path. The negative *gradient* of this potential function considering the charge points as fixed and the field point as variable is equal to the *electric intensity*. This potential function satisfies Laplace's partial differential equation at field points and is of the order of $1/r$ for r large.

electrostatic unit of charge. A charge of such magnitude that when placed one centimeter away from a duplicate charge will repel it with a force of one dyne. Evidently, if force, distance, and charge are measured in dynes, centimeters, and electrostatic units, respectively, the constant k in Coulomb's law for point charges will assume the value unity. See COULOMB. In this definition, the charges are to be in free space—otherwise the dielectric constant of the medium must also be involved.

Gauss's fundamental theorem of electrostatics. See GAUSS.

superposition principle for electrostatic intensity. The principle that the electrostatic intensity due to a complex (or set) of charges e_1 at P_1, e_2 at P_2, etc., is equal to the vector sum of the electric intensities due to the separate charges: $\Sigma e_i r_i^{-2} \rho_i$.

EL′E-MENT, *n.* See CONE, CYLINDER, CYLINDRICAL—cylindrical surface, DETERMINANT.

element of an analytic function of a complex variable. See ANALYTIC—analytic continuation.

element of integration. The expression following the integral sign (or signs) in a definite integral (or multiple integral). If the integral is being used to determine area (or volume, mass, etc.), the element is the **element** (or **differential**) of area (or volume, mass, etc.). It can then be interpreted as an approximation to the area (or volume, mass, etc.) of small pieces, the limit of whose sum as the pieces decrease in size in a suitable way is the value of the area (or volume, mass, etc.). See INTEGRAL—definite integral, multiple integral. Following are some particular examples of elements of integration: The **element of arc length** (or **linear element**) of a curve is an approximation to the length of the curve (see LENGTH), between two points, which (for a curve in the plane) is equal to

$$ds = \sqrt{(dx)^2 + (dy)^2} = \sqrt{1 + (dy/dx)^2}\ dx$$

$$= \sqrt{(dx/dy)^2 + 1}\ dy,$$

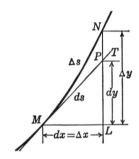

where dy/dx is to be determined in terms of x before integrating, and dx/dy in terms of y, from the equation of the curve. It can be seen from the figure that $ds = MP$ is an approximation of the arc-length $MN = \Delta s$, which results from an increase of Δx in the independent variable. **In polar form:**

$$ds = \sqrt{\rho^2 + (d\rho/d\theta)^2}\ d\theta.$$

If the equation of a space curve is in the parametric form, $x = f(t)$, $y = g(t)$, $z = h(t)$, the element of length is

$$\sqrt{(dx/dt)^2 + (dy/dt)^2 + (dz/dt)^2}\ dt.$$

The **element of plane area** (denoted by dA) for an area bounded by the curve $y = f(x)$, the x-

axis, and the lines $x = a$ and $x = b$, is usually taken as $f(x)\,dx$. The area is then equal to

$$\int_a^b f(x)\ dx.$$

In polar coordinates, dA is taken as $\frac{1}{2}r^2\,d\theta$ or $\frac{1}{2}\rho^2\,d\theta$. Then

$$A = \frac{1}{2}\int_{\theta_1}^{\theta_2}\rho^2\ d\theta$$

is the area bounded by the two rays $\theta = \theta_1$, $\theta = \theta_2$ and by the given curve for which ρ is expressed as a function of θ. In double integration, the element of area in rectangular Cartesian coordinates is $dx\,dy$, and in polar coordinates it is $\rho\,d\rho\,d\theta$ (also see SURFACE—surface area, surface of revolution). The **element of volume** can be taken as $A(h)\,dh$, where $A(h)$ is the area of a plane section perpendicular to the h axis (for particular examples of this, see REVOLUTION—solid of revolution). For triple integration in Cartesian coordinates, the element of volume is $dx\,dy\,dz$. The volume then equals

$$\int_{z_1}^{z_2}\int_{y_1}^{y_2}\int_{x_1}^{x_2} dx\,dy\,dz,$$

where z_1 and z_2 are constants, y_1 and y_2 may be functions of z, and x_1 and x_2 may be functions of y, or z, or of both y and z, these functions depending on the particular shape of the surface that bounds the volume. The order of integration may, of course, be changed (the proper change in limits being made) to best suit the volume under consideration. The figure shows an element of volume in rectangular coordinates and illustrates the process of finding the volume by an integral of the form

$$\int_{x_1}^{x_2}\int_{y_1}^{y_2}\int_{z_1}^{z_2} dz\,dy\,dx.$$

In cylindrical coordinates, the element of volume is $dv = r\,dr\,d\theta\,dz$, and in polar (spherical) coordinates it is $dv = r^2\sin\theta\,dr\,d\theta\,d\phi$. The **element of mass** is $dm = \rho\,dV$, where dV is an element of arc, area, or volume and ρ is the density (mass per unit length, area, or volume). Also see AREA, VOLUME, MOMENT—moment of a mass, moment of inertia, PRESSURE —fluid pressure, WORK.

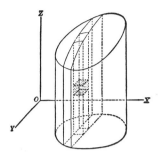

element of a set. Any member of the set. See MEMBER.

elements of geometry, calculus, etc. The fundamental assumptions and propositions of the subject.

geometric element. (1) A point, line, or plane. (2) Any of the parts of a configuration, as the sides and angles of a triangle.

EL-E-MEN′TA-RY, *adj.* **elementary divisor of a matrix.** See INVARIANT —invariant factor of a matrix.

elementary operations on determinants or matrices. The operations: (1) Interchange of two rows, or of two columns; (2) addition to a row of a multiple of another row, or addition to a column of a multiple of another column; (3) multiplication of a row or of a column by a nonzero constant. Operation (2) leaves the value of a determinant unchanged, (1) leaves the numerical value unchanged but changes the sign, and (3) is equivalent to multiplying the determinant by the constant. See EQUIVALENT —equivalent matrices.

elementary symmetric functions. See SYMMETRIC.

EL′E-VA′TION, *n.* **elevation of a given point.** The height of the point above a given plane, above sea level unless otherwise indicated.

angle of elevation. See ANGLE —angle of elevation.

E-LEV′EN, *n.* **divisibility by eleven.** See DIVISIBLE.

E-LIM′I-NANT, *n.* See RESULTANT.

E-LIM′I-NA′TION, *n.* **elimination of an unknown** from a set of simultaneous equations. The process of deriving from these equations another set of equations which does not contain the unknown that was to be eliminated and is satisfied by any values of the remaining unknowns which satisfy the original equations. This can be

done in various ways. **Elimination by addition or subtraction** is the process of putting a set of equations in such a form that when they are added or subtracted in pairs one or more of the variables disappears, then adding or subtracting them as the case may require to secure a system (or perhaps one equation) containing at least one less variable. *E.g.*, (a) given $2x + 3y + 4 = 0$ and $x + y - 1 = 0$, x can be eliminated by multiplying the latter equation by 2 and subtracting the result from the first equation, giving $y + 6 = 0$; (*b*) given

$$(1) \qquad 4x + 6y - z - 9 = 0,$$

$$(2) \qquad x - 3y + z + 1 = 0,$$

$$(3) \qquad x + 2y + z - 4 = 0,$$

y can be eliminated by multiplying (2) by 2 and adding the result to (1), and (3) by -3 and adding to (1). The results are $6x + z - 7 = 0$ and $x - 4z + 3 = 0$. **Elimination by comparison** is the process of putting two equations in such forms that their left (or right) members are identical and the other members do not contain one of the variables, then equating the right (or left) members. *E.g.*, $x + y = 1$ and $2x + y = 5$ can be written $x + y = 1$ and $x + y = 5 - x$, respectively. Hence $5 - x = 1$. **Elimination by substitution** is the process of solving one of a set of equations for one of the variables (in terms of the other variables), then substituting this expression in place of this variable in the other equations. *E.g.*, in solving $x - y = 2$ and $x + 3y = 4$, one might solve the first equation for x, getting $x = y + 2$, and substitute in the second, getting

$$y + 2 + 3y = 4 \quad \text{or} \quad y = \tfrac{1}{2}.$$

See RESULTANT —resultant of a set of polynomial equations.

EL-LIPSE′, *n.* A sort of elongated circle that is the intersection of a circular conical surface and a plane that cuts the surface in a single closed curve; a plane curve which is the set of all points such that the sum of the distances of one of the points from two fixed points (called the **foci**) is constant; a conic whose eccentricity is less than 1. The ellipse is symmetric with respect to two lines, its **axes**. *Axes* usually refer to the segments cut off on these lines by the ellipse, and are the *major* (longer) and *minor* (shorter) axes. If the major and minor axes lie on the *x*- and *y*-axes, respectively, the center is then at the origin and the equation of the ellipse, in Cartesian coordinates, is

$$\frac{x^2}{a^2} + \frac{y^2}{b^2} = 1$$

where a and b are the lengths of the *semimajor* and *semiminor* axes. This is the *standard form* of the equation of the ellipse, and is the equation of the ellipse in the position illustrated. The dis-

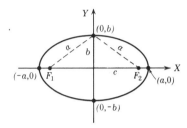

tance from an end of the minor axis to a focus is a. If c is the distance from the center to a focus, then the ratio c/a is the **eccentricity** of the ellipse (see CONIC). Two ellipses are **similar** if they have the same eccentricity. The intersection of the axes is the **center** of the ellipse, the points where the ellipse cuts its major axis are its **vertices**, and the chords through its foci and perpendicular to its major axis are the **latera recta** (plural of **latus rectum**). If the center of the ellipse is at the point (h, k) and its axes are parallel to the coordinate axes, its Cartesian equation is

$$\frac{(x-h)^2}{a^2} + \frac{(y-k)^2}{b^2} = 1.$$

If the 1 in the right member of this equation is replaced by 0, the equation is the equation of **a point ellipse**, since the equation has the form of the equation of an ellipse but is satisfied by the coordinates of only one real point. If 1 is replaced by -1, the equation is said to be the equation of an **imaginary ellipse**, since no point with real coordinates satisfies the equation. When the ellipse has its center at the origin and its axes on the coordinate axes, the **parametric equations** are

$$x = a\cos\alpha, \qquad y = b\sin\alpha,$$

where a and b are the lengths of the semimajor and semiminor axes, and α is the angle (at the origin) in the right triangle whose legs are the abscissa, OA, of the point $P(x, y)$ on the ellipse, and the ordinate, AB, to the circle with radius a and center at the origin. The angle α is the **eccentric angle** of the ellipse. The two circles in

the figure are the **eccentric circles** of the ellipse. A circle is an extreme case of an ellipse with eccentricity zero, with its major and minor axes equal, and with coincident foci. See CONIC, DISCRIMINANT —discriminant of a quadratic equation in two variables.

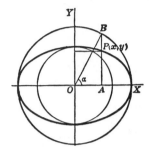

area of an ellipse. The product of π and the lengths of the semimajor and semiminor axes (*i.e.,* πab). This reduces to the formula for the area of a circle (πr^2) when the major and minor axes of the ellipse are equal, *i.e.,* when the ellipse is a circle.

diameter of an ellipse. The locus of the midpoints of a set of parallel chords. Any diameter must pass through the center of the ellipse and always belongs to a set of parallel chords defining some other diameter. Two diameters in this relation to each other are **conjugate diameters**.

director circle of an ellipse. See DIRECTOR.

focal property of an ellipse. Lines drawn from the foci of an ellipse to any point on the ellipse make equal angles with the tangent (and normal) to the ellipse at the point (see figure). Hence, if

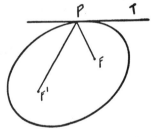

the ellipse is constructed from a strip of polished metal, rays of light emanating from one focus will come together at the other focus. This is the **optical** or **reflection property** of the ellipse. When reflection of sound instead of light is being considered it is the **acoustical property** of the ellipse.

EL·LIP'SOID, *n.* A surface all of whose plane sections are either ellipses or circles. An ellipsoid is symmetrical with respect to three mutually perpendicular lines (the **axes**), and with respect to the three planes determined by these lines. The intersection of these lines is the **center**. Any chord through the center is **a diameter**. The standard equation of the ellipsoid, with center at the origin and intercepts on the axes,

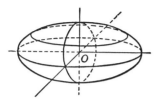

$a, -a, b, -b$, and $c, -c$, is

$$\frac{x^2}{a^2} + \frac{y^2}{b^2} + \frac{z^2}{c^2} = 1.$$

If $a > b > c$, a is the **semimajor axis**, b the **semimean axis**, and c the **semiminor axis**. If $a = b = c$, the equation becomes the equation of a sphere. The volume is equal to $\frac{4}{3}\pi abc$, which becomes the usual formula $\frac{4}{3}\pi r^3$ for the volume of a sphere when $a = b = c = r$. If the 1 in the right member of the above equation is replaced by 0, the equation is the equation of a **point ellipsoid** (since it is satisfied by only one point with real coordinates); if the 1 is replaced by -1, then no real values of the coordinates satisfy the equation and it is the equation of an **imaginary ellipsoid**. The **center of an ellipsoid** is the point of symmetry of the ellipsoid. This point is the intersection of the three principal planes of the ellipsoid. An **ellipsoid of revolution** (or **spheroid**) is an ellipsoid generated by revolving an ellipse about one of its axes (see SURFACE—surface of revolution). This is an ellipsoid whose sections by planes perpendicular to one of its axes are all circles. The axis passing through the centers of these circular sections is the **axis of revolution**. The largest circular section is the **equator** of the ellipsoid of revolution. The extremities of the axis of revolution are the **poles** of the ellipsoid of revolution. The ellipsoid of revolution is **prolate** if the diameter of its equatorial circle is less than the length of the axis of revolution, and **oblate** if this diameter is greater than the length of the axis of revolution. See GEOID.

confocal ellipsoids. See CONFOCAL—confocal quadrics.

similar ellipsoids. See SIMILAR.

EL′LIP-SOI′DAL, *adj.* **ellipsoidal coordinates.** See COORDINATE.

EL-LIP′TIC, or EL-LIP′TI-CAL, *adj.* **elliptic conical surface.** A conical surface whose directrix is an ellipse. When the vertex is at the origin and the axis coincident with the z-axis in a system of rectangular Cartesian coordinates, its equation is

$$x^2/a^2 + y^2/b^2 - z^2/c^2 = 0.$$

When $a = b$, this is a right circular cone.

elliptic coordinates of a point. Coordinates in the plane determined by confocal conics (ellipses and hyperbolas) or coordinates in space determined by confocal quadrics (in the latter case, usually called *ellipsoidal coordinates*). See CONFOCAL—confocal quadrics, CURVILINEAR—curvilinear coordinates of a point in space.

elliptic cylinder. See CYLINDER.

elliptic function. The inverse $x = \phi(y)$ of an elliptic integral y with limits of integration x_0 and x. See below, Jacobian elliptic functions, and Weierstrassian elliptic functions. An **elliptic function of a complex variable** is defined as a doubly periodic single-valued function f of the complex variable z such that f has no singularities other than poles in the finite plane. A doubly periodic function cannot be an *entire* function, unless it is a constant.

elliptic integral. Any integral of the type $\int R(x, \sqrt{S})\, dx$, where $S = a_0 x^4 + a_1 x^3 + a_2 x^2 + a_3 x + a_4$ has no multiple roots and a_0 and a_1 are not both zero, and $R(x, \sqrt{S})$ is a *rational function* of x and $\sqrt{S}$. Integrals of the form

$$I_1 = \int_0^x \frac{dt}{(1-t^2)^{1/2}(1-k^2t^2)^{1/2}}$$

$$= \int_0^\phi \frac{d\psi}{(1 - k^2 \sin^2 \psi)^{1/2}},$$

$$I_2 = \int_0^x \frac{(1 - k^2 t^2)^{1/2}}{(1 - t^2)^{1/2}}\, dt$$

$$= \int_0^\phi (1 - k^2 \sin^2 \psi)^{1/2}\, d\psi,$$

$$I_3 = \int_0^x \frac{dt}{(t^2 - a)(1 - t^2)^{1/2}(1 - k^2 t^2)^{1/2}}$$

$$= \int_0^\phi \frac{d\psi}{(\sin^2 \psi - a)(1 - k^2 \sin^2 \psi)^{1/2}},$$

where $\sin \phi = x$, were called (by Legendre) **incomplete elliptic integrals of the first, second, and third kinds,** respectively. The **modulus** of one of these elliptic integrals is k and the **complementary modulus** is $k' = (1 - k^2)^{1/2}$, it being usual to take $0 < k^2 < 1$. The integrals are **complete** if $x = 1$ ($\phi = \frac{1}{2}\pi$). Also, $I_1 = \beta$,

$$I_2 = \int_0^\beta \mathrm{dn}^2 \, t \, dt,$$

and

$$I_3 = \int_0^\beta (\mathrm{sn}^2 t - \mathrm{sn}^2 \alpha)^{-1} \, dt,$$

where $x = \mathrm{sn}\,\beta$, $a = \mathrm{sn}^2\,\alpha$, and $\mathrm{sn}\,t$ and $\mathrm{dn}\,t$ are *Jacobian elliptic functions.* The incomplete elliptic integral of the second kind is sometimes taken to be of the form

$$\int_0^x t^2 (1 - t^2)^{-1/2} (1 - k^2 t^2)^{-1/2} \, dt.$$

Elliptic integrals are so named because they were first encountered in the problem of finding the circumference of an ellipse.

elliptic modular function. See MODULAR.

elliptic paraboloid. See PARABOLOID.

elliptic partial differential equation. A real second-order partial differential equation of the form

$$\sum_{i,j=1}^n a_{ij} \frac{\partial^2 u}{\partial x_i \, \partial x_j}$$

$$+ F\left(x_1, \cdots, x_n, u, \frac{\partial u}{\partial x_1}, \cdots, \frac{\partial u}{\partial x_n}\right) = 0,$$

such that the quadratic form $\displaystyle\sum_{i,j=1}^n a_{ij} x_i x_j$ is non-singular and definite; *i.e.,* by means of a real linear transformation this quadratic form can be reduced to a sum of n squares all of the same sign. Typical examples are the Laplace and Poisson equations. See INDEX—index of a quadratic form.

elliptic point on a surface. A point whose *Dupin indicatrix* is an ellipse. See DUPIN.

elliptic Riemann surface. See RIEMANN—Riemann surface.

Jacobian elliptic functions. The functions $\mathrm{sn}\,z$, $\mathrm{cn}\,z$, $\mathrm{dn}\,z$ defined by $y = \mathrm{sn}(z, k) = \mathrm{sn}\,z$ if

$$z = \int_0^y (1 - t^2)^{-1/2} (1 - k^2 t^2)^{-1/2} \, dt,$$

and $\mathrm{sn}^2 z + \mathrm{cn}^2 z = 1$, $k^2 \mathrm{sn}^2 z + \mathrm{dn}^2 z = 1$, where the sign of $\mathrm{cn}\,z$ and $\mathrm{dn}\,z$ are chosen so that $\mathrm{cn}(0) = \mathrm{dn}(0) = 1$. The number k is the **modulus** of the functions and $k' = \sqrt{1 - k^2}$ is the **complementary modulus**. If

$$K = \int_0^1 (1 - t^2)^{-1/2} (1 - k^2 t^2)^{-1/2} \, dt$$

and

$$K' = \int_0^1 (1 - t^2)^{-1/2} (1 - k'^2 t^2)^{-1/2} \, dt,$$

then $\mathrm{sn}\,z$, $\mathrm{cn}\,z$, and $\mathrm{dn}\,z$ are *doubly periodic* functions with periods $(4K, 2iK')$, $(4K, 2K + 2iK')$, and $(2K, 4iK')$, respectively. Also,

$$\frac{d\,\mathrm{sn}\,z}{dz} = \mathrm{cn}\,z\,\mathrm{dn}\,z, \qquad \frac{d\,\mathrm{cn}\,z}{dz} = -\mathrm{sn}\,z\,\mathrm{dn}\,z,$$

$$\frac{d\,\mathrm{dn}\,z}{dz} = -k^2\,\mathrm{sn}\,z\,\mathrm{cn}\,z.$$

Jacobi's notation for these functions was $\mathrm{sinam}\,z$, $\mathrm{cosam}\,z$, $\Delta\mathrm{am}\,z$. He also wrote $\mathrm{tanam}\,z$ for $\mathrm{sn}\,z/\mathrm{cn}\,z$. See above, elliptic integrals.

Weierstrass elliptic functions. The function $p(z)$ defined by $y = p(z)$ if $z = \displaystyle\int_y^\infty S^{-1/2} \, dt$, where

$$S = 4t^3 - g_2 t - g_3 = 4(t - e_1)(t - e_2)(t - e_3),$$

and the function $p'(z) = \sqrt{4p^3 - g_2 p - g_3}$. These are *doubly periodic* functions with periods $2\omega_1, 2\omega_2$, where $\omega_1 = K(e_1 - e_3)^{-1/2}$ and $\omega_2 = iK'(e_1 - e_3)^{-1/2}$, and K and K' are as defined above under *Jacobian elliptic functions.* Any *elliptic function* $f(z)$ can be expressed as a product of $p'(z)$ and a rational function of $p(z)$, where $p(z)$ and $p'(z)$ have the same periods as $f(z)$. Also, $p(z) = e_3 + (e_1 - e_3)[\mathrm{sn}\{z(e_1 - e_3)^{1/2}\}]^{-2}$, where $\mathrm{sn}\,z$ is a *Jacobian elliptic function,* and

$$p(z) = \frac{1}{z^2} + \sum_{m,n} \left\{ \frac{1}{(z - \Omega_{m,n})^2} - \frac{1}{\Omega_{m,n}^2} \right\},$$

where $\Omega_{m,n} = 2m\omega_1 + 2n\omega_2$ and the summation is over all integral values of m and n except $m = n = 0$.

E′LON·GA′TION, *n.* (1) The limit of the ratio of the increment Δl in length l of a vector, joining two points of a body (this increment resulting from the body being subjected to a deformation), to its undeformed length l as l is allowed to approach zero. In symbols, $e = \lim_{l \to 0} \Delta l / l$. This limit has in general different values depending on the direction of the vector in the deformed medium. (2) The change in length per unit length of a vector in a deformed medium.

elongations and compressions. Same as ONE-DIMENSIONAL STRAINS. See STRAIN.

EM-PIR′I-CAL, *adj.* **empirical formula, assumption, or rule.** A statement whose reliability is based upon observation and experimental evidence (such as laboratory experiments) and is not necessarily supported by any established theory or laws; formulas based upon experience rather than logical (or mathematical) conclusions.

empirical curve. A curve that is drawn to fit approximately a set of statistical data. It is usually assumed to represent, approximately, additional data of the same kind. See LEAST—method of least squares, GRAPH—broken-line graph.

EMP′TY, *adj.* **empty set.** The set with no members. *Syn.* null set.

END, *adj.* **end point.** See CURVE, and INTERVAL.

EN′DO-MOR′PHISM, *n.* See HOMOMORPHISM, CATEGORY.

EN′DORSE′, *v.* Same as INDORSE.

EN-DOW′MENT, *adj.* **endowment insurance.** See INSURANCE —life insurance.

EN′ER-GY, *n.* The capacity for doing work.

conservation of energy. A principle asserting that energy can neither be created nor destroyed. In mechanics this principle asserts that in a conservative field of force the sum of the kinetic and potential energies is a constant.

energy integral. (1) An integral that arises in the solution of the particular differential equation of motion, $d^2 s / dt^2 = \pm k^2 s$, describing simple harmonic motion. The integral is $v^2/2 = \pm k^2$ $\int s \, ds$ and is called the energy integral, because when it is multiplied by m it is equal to the *kinetic energy*, $\frac{1}{2} m v^2$. (2) An integral stating that the sum of the potential and kinetic energies is constant, in a dynamic system in which this is true.

kinetic energy. The energy a body possesses by virtue of its motion. A particle of mass m moving with velocity v has kinetic energy of amount $\frac{1}{2} m v^2$. In a conservative field of force, the work done by the forces in displacing the particle from one position to another is equal to the change in the kinetic energy. A body rotating about an axis and having angular velocity ω and moment of inertia I about the axis has kinetic energy of amount $\frac{1}{2} I \omega^2$.

potential energy. The energy a body possesses by virtue of its position. A term applicable to conservative fields of force only; potential energy is defined as the negative of the work done in displacing a particle from its standard position to any other position. See ENERGY—conservation of energy.

principle of energy. A principle in mechanics asserting that the increase in kinetic energy is equal to the work done by the force.

ENIAC, *n.* First fully electronic general-purpose computing machine (Electronic Numerical Integrator and Computer). Built at the University of Penn., demonstrated publicly in 1946, and moved to the Ballistic Research Laboratories, Aberdeen Proving Ground, in 1947.

ENNEPER, Alfred (1830–1885). German differential geometer.

equations of Enneper. Integral equations for the coordinate functions of a minimal surface referred to its minimal curves as parametric curves:

$$x = \frac{1}{2} \int (1 - u^2) \phi(u) \, du + \frac{1}{2} \int (1 - v^2) \psi(v) \, dv,$$

$$y = \frac{i}{2} \int (1 + u^2) \phi(u) \, du$$

$$- \frac{i}{2} \int (1 + v^2) \psi(v) \, dv,$$

$$z = \int u \phi(u) \, du + \int v \psi(v) \, dv,$$

where ϕ and ψ are arbitrary analytic functions. See WEIERSTRASS —equations of Weierstrass.

surface of Enneper. See SURFACE.

EN-TIRE, *adj.* **entire function.** A function which can be expanded in a Maclaurin's series, valid for all finite values of the variable; a function of a complex variable which is analytic for all finite values of the variable. *Syn.* integral function. An entire function f is of **order** ρ for

$\theta_1 < \theta < \theta_2$ provided

$$\limsup_{r \to \infty} \frac{\log|f(re^{i\theta})|}{r^{\rho + \epsilon}} = 0,$$

uniformly in $\theta, \theta_1 < \theta < \theta_2$, for each $\epsilon > 0$, but not for any $\epsilon < 0$. See LIOUVILLE—Liouville's theorem, PHRAGMÉN—Phragmén-Lindelöf function, PICARD—Picard's theorems.

entire series. A power series which converges for all values of the variable. The exponential series, $1 + x + x^2/2! + x^3/3! + \cdots + x^n/n! + \cdots$, is an entire series.

EN'TRO-PY, *n.* The amount of disorder or randomness of errors relative to energy or information. Suppose a countably additive measure (or probability measure) is defined on a σ-algebra of subsets of a set X. For a finite or countably infinite partition ρ of X into measurable sets $S_1, S_2, \cdots$, the **entropy** of the partition is

$$E(\rho) = -\Sigma_n m(S_n) \log_e [m(S_n)].$$

If φ is a measure-preserving transformation on X, let φ^{-1} denote the partition consisting of all $\varphi^{-1}(S)$ for S a member of ρ. The **entropy** of φ is the least upper bound with respect to all ρ with $E(\rho) < \infty$ of

$$\lim_{n \to \infty} \frac{1}{n} [E(n, \rho, \varphi)],$$

where $E(n, \rho, \varphi)$ is the entropy of the coarsest partition that is finer than each of $\rho, \varphi^{-1}\rho, \cdots, \varphi^{-(n-1)}\rho$. This is an isomorphism invariant of φ. In communication and information theory, logarithms to the base 2 often are used. See INFORMATION—information theory.

E-NU'MER-A'BLE, *adj.* **enumerable set.** Same as COUNTABLE SET. *Enumerably infinite* is used in the same sense as *countably infinite*, to denote an infinite set whose members can be put into one-to-one correspondence with the positive integers.

EN'VE-LOPE, *n.* The **envelope of a one-parameter family of curves** is a curve that is tangent to (has a common tangent with) every curve of the family. Its equation is obtained by eliminating the parameter between the equation of the curve and the partial derivative of this equation with respect to this parameter. The envelope of the circles $(x-a)^2 + y^2 - 1 = 0$ is $y = \pm 1$. See DIFFERENTIAL—solution of a differential equation. In particular, the **envelope of a one-parameter family of straight lines** is a

curve that is tangent to every member of the family of lines; *e.g.*, the curve $4(x-2)^3 = 27y^2$ is the envelope of the family of lines $y = -\frac{1}{2}cx + c + \frac{1}{8}c^3$ and is the result of eliminating c between this equation and the equation $0 = -\frac{1}{2}x + 1 + \frac{3}{8}c^2$. The **envelope of a one-parameter family of surfaces** is the surface that is tangent to (has a common tangent plane with) each of the surfaces of the family along their characteristics; the locus of the characteristic curves of the family. See CHARACTERISTIC—characteristic of a one-parameter family of surfaces.

EP-I-CY'CLOID, *n.* The plane locus of a point P on a circle as the circle rolls on the outside of a given fixed circle. If a (OB in the figure) is the

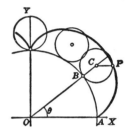

radius of the fixed circle with center at O, b (BC) is the radius of the rolling circle, and θ is the angle from OA to the ray from O through the center C of the rolling circle, then the parametric equations of the curve are

$$x = (a+b)\cos\theta - b\cos[(a+b)\theta/b],$$

$$y = (a+b)\sin\theta - b\sin[(a+b)\theta/b].$$

The curve has one arch when $a = b$, two arches, when $a = 2b$, and n arches when $a = nb$. It has a cusp of the first kind at every point at which it touches the fixed circle. See HYPOCYCLOID.

EP-I-TRO'CHOID, *n.* A generalization of an epicycloid, in which the describing point may be at any fixed point on the radius of the rolling circle, or this radius extended. If h denotes the distance from the center of the rolling circle to the describing point, and a, b, and θ are as in the discussion of the epicycloid, then the parametric equations of the *epitrochoid* are

$$x = (a+b)\cos\theta - h\cos[(a+b)\theta/b],$$

$$y = (a+b)\sin\theta - h\sin[(a+b)\theta/b].$$

The cases for $h < b$ and $h > b$ are analogous to the corresponding cases ($b < a$ and $b > a$) in the discussion of the *trochoid*. See figure under TRO-CHOID.

EP′I-TRO-CHOI′DAL, *adj.* **epitrochoidal curve.** The locus of a point in the plane of a circle which rolls without slipping on another circle in such a way that the planes of the two circles meet under constant angle. All epitrochoidal curves are spherical curves. See CURVE—spherical curve.

EP′SI-LON, *adj., n.* The fifth letter of the Greek alphabet: lower case, ϵ; capital, **E**.

epsilon-chain. A finite succession of points $p_1, p_2, \cdots, p_n$ such that the distance between any two successive points is less than epsilon (ϵ), ϵ being some positive real number. Any two points of a *connected* set can be joined by an ϵ-chain for any $\epsilon > 0$, while a compact set is connected if every pair of its elements can be joined by an ϵ-chain for any $\epsilon > 0$.

epsilon symbols. The symbols $\epsilon^{i_1 i_2 \cdots i_k}$ and $\epsilon_{i_1 i_2 \cdots i_k}$, which are defined as being zero unless the integers $i_1, i_2, \cdots, i_k$ consist of the integers $1, 2, \cdots, k$ in some order, and to be $+1$ or -1 according as $i_1, i_2, \cdots, i_k$ is obtained from $1, 2 \cdots, k$ by an even or an odd permutation. If $\delta_{i_1 i_2 \cdots i_k}^{i_1 i_2 \cdots i_k}$ is the *generalized Kronecker delta*, then $\epsilon^{i_1 i_2 \cdots i_k} = \delta_{1,2,\cdots k}^{i_1,i_2 \cdots i_k} = \delta_{i_1 i_2 \cdots i_k}^{1,2,\cdots k} = \epsilon_{i_1 i_2 \cdots i_k}$. The two epsilon symbols are *relative numerical tensor fields* of weight $+1$ and -1, respectively.

E′QUAL, *adj.* See EQUIVALENCE —equivalence relation.

equal roots of an equation. See MULTIPLE —multiple root of an equation, DISCRIMINANT —discriminant of a polynomial equation.

E-QUAL′I-TY, *n.* The *relation* of being *equal*; the statement, usually in the form of an equation, that two things are equal.

continued equality. Three or more quantities set equal by means of two or more equality signs in a continuous expression, as $a = b = c = d$, or $f(x, y) = g(x, y) = h(x, y)$. The last expression is equivalent to the equations $f(x, y) = g(x, y)$ and $g(x, y) = h(x, y)$.

equality of two complex numbers. The property of having their real parts equal and their pure imaginary parts equal ($a + bi = c + di$ means $a = c$ and $b = d$); the property of having equal moduli and having amplitudes which differ by integral multiples of 2π.

E-QUATE′, *v.* **to equate one expression to another.** To form the algebraic statement of equality which states that the two expressions are equal. The statement may be either an *identity*

or a *conditional equation* (commonly called simply an *equation*). *E.g.*, one may *equate* $(x + 1)^2$ to $x^2 + 2x + 1$, getting the identity $(x + 1)^2 = x^2 + 2x + 1$; or one may *equate* $\sin x$ and $2x + 1$; or one may *equate* coefficients in $ax + b$ and $2x + 3$, getting $a = 2$, $b = 3$.

E-QUAT′ED, *adj.* **equated date (for a set of payments).** The date upon which they could all be discharged by a single payment equal to the sum of their values when due, taking into account accumulations of payments due prior to that date and present values, at that date, of future payments. *Syn.* average date.

equated time (*Finance*). The time from the present to the equated date.

E-QUA′TION, *n.* A statement of equality between two expressions. Equations are of two types, **identities** and **conditional equations** (or usually simple *equations*). A *conditional equation* is true only for certain values of the variables involved (see IDENTITY); *e.g.*, $x + 2 = 5$ is a true statement only when $x = 3$; and $xy + y - 3 = 0$ is true when $x = 2$ and $y = 1$, and for many other pairs of values of x and y; but for still other pairs it is false. A **solution** (or **root**) of a conditional equation is a value of the variable (or a set of values of the variables if there is more than one variable) for which the equation is a true statement (see MULTIPLE —multiple root). Equations often are named according to the types of functions used. For example, an **irrational** (or **radical**) equation is an equation for which the variable or variables occur under radical signs or with fractional exponents, as for

$$\sqrt{x^2 + 1} = x + 2 \quad \text{and} \quad x^{1/2} + 1 = 3x;$$

a trigonometric equation contains the variables in trigonometric functions, as for $\cos x - \sin x = \frac{1}{2}$; in an **exponential equation**, variables occur in exponents, as for $2^x - 5 = 0$. See below, **polynomial equation.** An equation of a curve, cylinder, plane, etc., is any equation, or set of simultaneous equations, that is satisfied by those, and only those, values of the variables that are coordinates of points on the curve, cylinder, etc. See CURVE, LINE—equations of a straight line, PARAMETRIC—parametric equations, SURFACE.

auxiliary equation. See DIFFERENTIAL —linear differential equation.

compatibility equations. See COMPATIBILITY.

defective equation. An equation which has fewer roots than some equation from which it has been obtained. Roots may be lost, for instance, by dividing both members of an equation

Equation 148 **Equation**

by a function of the variable. If $x^2 + x = 0$ is divided by x, the result, $x + 1 = 0$, is defective; it lacks the root 0.

difference and differential equations. See DIFFERENCE, DIFFERENTIAL —differential equation.

differential equations of Bessel, Hermite, Laguerre, Legendre, etc. See the respective names.

equation of continuity (*Hydrodynamics*). The equation div $q = \nabla \cdot q = 0$, where q represents the flux of some fluid. If there are no sources or sinks in a fluid, this equation states that the fluid does not concentrate toward or expand from any point. If this equation holds at every point in a body of liquid, then the lines of vector flux must be closed or infinite. Such a distribution of vectors is called *solenoidal*.

equation of motion. An equation, usually a differential equation, stating the law by which a particle moves.

equation in p-form. A polynomial equation in one variable, in which the coefficient of the highest degree term is unity and the other coefficients are all integers.

equation of payments. An equation stating the equivalence of two sets of payments on a certain date, each payment in each set having been accumulated or discounted to that certain date, called the **comparison,** or **focal, date.**

equation of value. An equation of payments stating that a set of payments is equivalent to a certain single payment, at a given time. Some authors use only the term **equation of value** and others use only **equation of payments**, making no distinction between the cases in which there are two sets of several payments and those in which one set is replaced by a single payment. See above, equation of payments.

homogeneous equation. See HOMOGENEOUS.

inconsistent equations. See CONSISTENCY.

indeterminate equation. A single equation is **indeterminate** if it has more than one variable, such as $x + 2y = 4$, and has an unlimited number of solutions. Historically, this kind of equation has been of particular interest when the coefficients are integers and it is required to find expressions for the sets of integral values of the variables that satisfy the given equation. Under these restrictions, the equations are **Diophantine equations.** An **indeterminate system of linear equations** is a system of linear equations having an infinite number of solutions. See CONSISTENCY.

integral equations. See various headings under INTEGRAL.

locus of an equation. See LOCUS.

logarithmic equation. An equation containing the logarithm of the variable. It is usually called logarithmic only when the variable occurs only in the arguments of logarithms; $\log x + 2 \log 2x + 4 = 0$ is a logarithmic equation.

minimal (or **minimum**) **equation.** See ALGEBRAIC—algebraic number, CHARACTERISTIC—characteristic equation of a matrix.

numerical equation. An equation in which the coefficients of the variables and the constant term are numbers, not literal constants. The equation $2x^2 + 5x + 3 = 0$ is a numerical equation.

parametric equations. See PARAMETRIC.

polynomial equation. A polynomial in one or more variables, set equal to zero. The **degree** of the equation is the degree of the polynomial (see DEGREE—degree of a polynomial, or equation). The **general equation of the second degree in two variables** is

$$ax^2 + by^2 + cxy + dx + ey + f = 0,$$

in which a, b and c are not all zero. (See DISCRIMINANT—discriminant of the general quadratic.) The **general equation of the nth degree in one variable** is a polynomial equation of the nth degree whose coefficients are literal constants, such as

$$a_0 x^n + a_1 x^{n-1} + \cdots + a_n = 0.$$

A polynomial equation of the nth degree is **complete** if none of its coefficients are zero; it is **incomplete** if one or more coefficients (other than the coefficient of x^n) are zero. A polynomial equation is **linear, quadratic, cubic, quartic** (or **biquadratic**), or **quintic,** according as its degree is 1, 2, 3, 4, or 5. A **root** of a polynomial equation in one variable is a value of the variable which reduces the equation to a true equality. Sometimes a solution can be found by factoring the equation; *e.g.*, the equation $x^2 + x - 6 = 0$ has roots 2 and -3, since $x^2 + x - 6 = (x - 2)(x + 3)$. If the equation can not be factored, the method of solution usually consists of some method of successive approximations. Horner's and Newton's methods offer systematic approximations of this type. For the determination of imaginary roots, one can substitute $u + iv$ for the variable, equate the real and imaginary parts to zero, and solve these equations for u and v (usually by some method of successive approximations). See RATIONAL—rational root theorem, QUADRATIC—quadratic formula, CARDAN—Cardan's solution of the cubic, FERRARI, FUNDAMENTAL—fundamental theorem of algebra.

reciprocal equation. See RECIPROCAL.

Equation 149 **Equivalence**

redundant equation. An equation containing roots that have been introduced by operating upon a given equation; *e.g.*, such roots may be introduced by multiplying both members by the same function of the unknown, or by raising both members to a power. The introduced roots are **extraneous roots.** The equation $x - 1 = \sqrt{x + 1}$, when squared and simplified, becomes $x^2 - 3x = 0$, which has the roots 0 and 3, hence is *redundant*, because 0 does not satisfy the original equation.

simultaneous equations. See SIMULTANEOUS.

theory of equations. See THEORY.

transformation of an equation. See TRANSFORMATION.

E-QUA'TOR, *n.* **celestial equator.** The great circle in which the plane of the earth's equator cuts the celestial sphere. See HOUR—hour angle and hour circle.

equator of an ellipsoid of revolution. See ELLIPSOID—ellipsoid of revolution.

geographic equator (earth's equator). The great circle that is the section of the earth's surface by the plane through the center of the earth and perpendicular to the earth's axis. See ELLIPSOID—ellipsoid of revolution.

E'QUI-AN'GU-LAR, *adj.* **equiangular hyperbola.** Same as RECTANGULAR HYPERBOLA. See HYPERBOLA.

equiangular polygon. A polygon having all of its interior angles equal. An equiangular triangle is necessarily equilateral, but an equiangular polygon of more than three sides need not be equilateral. Two polygons are **mutually equiangular** if their corresponding angles are equal.

equiangular spiral. Same as *logarithmic* spiral. Called equiangular because the angle between the tangent and radius vector is a constant. See LOGARITHMIC—logarithmic spiral.

equiangular transformation. See ISOGONAL—isogonal transformation.

E-QUI-A'RE-AL MAP. Same as AREA-PRESERVING MAP. See MAP.

E-QUI-CON-TIN'U-OUS, *adj.* A collection F of functions with domain S is **equicontinuous at a point** x of S if, for arbitrary $\epsilon > 0$, there exists a $\delta > 0$ such that $|f(y) - f(x)| < \epsilon$ for all f in F whenever the distance between x and y is less than δ and y is in S. The collection F is **pointwise equicontinuous** if it is equicontinuous at each point of S; it is **uniformly equicontinuous** if, for arbitrary $\epsilon > 0$, there exists a $\delta > 0$ such that $|f(r) - f(s)| < \epsilon$ for all f in F whenever the distance between r and s is less than δ and r and s are members of S. See ASCOLI—Ascoli's theorem.

E'QUI-DIS'TANT, *adj.* At the same distance.

equidistant system of parametric curves. See PARAMETRIC.

E'QUI-LAT'ER-AL, *adj.* **equilateral hyperbola.** See HYPERBOLA—rectangular hyperbola.

equilateral polygon. A polygon having all of its sides equal. An **equilateral triangle** is necessarily equiangular, but an equilateral polygon of more than three sides need not be equiangular. Two polygons are **mutually equilateral** if their corresponding sides are equal.

equilateral spherical polygon. A spherical polygon which has all of its sides equal.

E'QUI-LIB'RI-UM, *n.* **equilibrium of forces.** The property of having their resultant force and the sum of their torques about any axis equal to zero. See RESULTANT.

equilibrium of a particle or a body. A particle is in equilibrium when the resultant of all forces acting on it is zero. A body is in equilibrium when it has no acceleration, either of translation or rotation; a rigid body is in equilibrium when its center of mass has no acceleration and the body has no angular acceleration. The conditions for a body to be in equilibrium are: (1) that the resultant of the forces acting on it be equal to zero; (2) that the sum of the moments of these forces about every axis be equal to zero (about each of three mutually perpendicular axes suffices).

E'QUI-NOX, *n.* See ECLIPTIC.

E-QUI-NU'MER-A-BLE, *adj.* See EQUIVALENT—equivalent sets.

E-QUI-PO'TENT, *adj.* See EQUIVALENT—equivalent sets.

E'QUI-PO-TEN'TIAL, *adj.* **equipotential surface.** A surface on which a potential function U maintains a constant value. More generally, if U is any point function, then an equipotential surface relative to U is a surface on which U is constant.

E-QUIV'A-LENCE, *adj., n.* **equivalence class.** If an *equivalence relation* is defined on a set, then the set can be separated into classes by the convention that two elements belong to the same class if and only if they are equivalent. These classes are **equivalence classes.** Two equivalence classes are identical if they have an element in common. Each element belongs to one of the equivalence classes. *E.g.*, if one says that *a* is equivalent to *b* if $a - b$ is a rational number, this is an equivalence relation for the real numbers; the equivalence class which contains a number *a* is then the class which contains all numbers which can be obtained by adding rational numbers to *a*. If a set *T* has been divided into nonoverlapping subsets, then an equivalence relation can be defined by letting "*x* is related to *y*" mean that *x* and *y* are in the same subset.

equivalence of propositions. An **equivalence** is a proposition formed from two given propositions by connecting them by "*if, and only if*." An equivalence is true if both propositions are true, or if both are false. The proposition "For all triangles *x*, *x* is equilateral if, and only if, *x* is equilateral" is true, since any particular triangle is either both equilateral and equiangular, or is neither equilateral nor equiangular. The equivalence formed from propositions *p* and *q* is usually denoted by $p \leftrightarrow q$, or $p \equiv q$. The equivalence $p \leftrightarrow q$ is the same as the statements "*p* is a *necessary and sufficient* condition for *q*," or "*p* if, and only if, *q*"; it is equivalent to the *conjunction* of the implications $p \rightarrow q$ and $q \rightarrow p$. Propositions *p* and *q* are **equivalent** if $p \leftrightarrow q$ is true. An equivalence is also called a **biconditional** statement (or proposition). Two statements are **logically equivalent** if they are equivalent because of their logical form rather than because of mathematical content. *E.g.*, $\sim (p \wedge q)$ is equivalent to $(\sim p) \vee (\sim q)$ whatever the statements *p* and *q* [see CONJUNCTION, DISJUNCTION, NEGATION.] See EQUIVALENT—equivalent propositional functions.

equivalence relation. A relation between elements of a given set which is a *reflexive, symmetric*, and *transitive* relation and which is such that any two elements of the set are either equivalent or not equivalent. An equivalence relation also is called an **equals relation**, and one says that two objects so related are **equivalent** or **equal** with respect to that equivalence relation. Ordinary equality of numbers is an equivalence relation. In geometry, "equal" often is used to denote agreement with respect to some particular property. *E.g.*, triangles may be said to be equal if they are similar, or if they have the same area, or if they are congruent. See above, equivalence class.

E-QUIV'A-LENT, *adj., n.* Given an equivalence relation, one speaks of objects as being **equivalent** if they have this relation to each other. See EQUIVALENCE —equivalence relation, and various headings below.

cash equivalent of an annuity. Same as PRESENT VALUE. See VALUE.

equivalent angles. Two angles (*i.e.*, rotation angles) that have the same measure. The corresponding geometric angles are then congruent. See ANGLE.

equivalent equations, inequalities, etc. Equations, inequalities, etc., that have the same solution sets. *E.g.*, the equations $x^2 = 1$ and $x^4 = 2x^2 - 1$ are equivalent, since each of their solution sets is the set whose members are -1 and $+1$. The system of equations $\{2x + 3y = 3, x + 3y = 9\}$ and the system $\{x = -6, x + 3y = 9\}$ are equivalent, since the solution set of each is the ordered pair $(-6, 5)$. The inequalities $|x - 3| < 2$ and $1 < x < 5$ are equivalent, since the solution set of each is the open interval $(1, 5)$.

equivalent geometric figures. See EQUIVALENCE —equivalence relation.

equivalent matrices. Two square matrices *A* and *B* for which there exist nonsingular square matrices *P* and *Q* such that $A = PBQ$. Two square matrices are equivalent if, and only if, one can be derived from the other by a finite number of operations of the types: (a) interchange of two rows, or of two columns; (b) addition to a row of a multiple of another row, or addition to a column of a multiple of another column; (c) multiplication of a row or of a column by a nonzero constant. Every matrix is equivalent to some diagonal matrix. This transformation PBQ of the matrix *B* is an **equivalent transformation**. If $P = Q^{-1}$, it is a **collineatory** (or **similarity**) **transformation**; if *P* is the transpose of *Q*, it is a **congruent transformation**; if *P* is the Hermitian conjugate of *Q*, it is a **conjunctive transformation**; if $P = Q^{-1}$ and *Q* is orthogonal, it is an **orthogonal transformation**; if $P = Q^{-1}$ and *Q* is unitary, it is a **unitary transformation**. Also, see under TRANSFORMATION.

equivalent propositional functions (or **open sentences** or **statement functions**). See PROPOSITIONAL —propositional function.

equivalent propositions. See EQUIVALENCE —equivalence of propositions.

equivalent sets. Sets that can be put into one-to-one correspondence. *Syn.* equinumerable, equipotent. See CARDINAL —cardinal number, CORRESPONDENCE.

topologically equivalent spaces. See TOPOLOGICAL —topological transformation.

ERATOSTHENES of Alexandria (c. 276–c. 194 B.C.). Greek astronomer, geographer, philosopher and mathematician.

sieve of Eratosthenes. The process of determining all the primes not greater than a number *N* by writing down all the numbers from 2 to *N*, removing those after 2 which are multiples of 2, those after 3 which are multiples of 3, and continuing until all multiples of primes not greater than $\sqrt{N}$, except the primes themselves, have been removed. Only prime numbers will remain.

ERDÖS, Paul (1913–   ). Prolific Hungarian mathematician who has made contributions to algebra, analysis, combinatorics, geometry, topology, number theory, and graph theory. Noted no less for his cosmopolitan life style and stimulating conversations, conjectures, and collaborations than for his own brilliant contributions. See PRIME—prime-number theorem.

ERG, *n.* A unit of work; the work done by a force of one dyne operating over a distance of one centimeter.

ER-GOD'IC, *adj.* **ergodic theory.** The study of measure-preserving transformations. In particular, the study of theorems concerning the limits of probability and weighted means. *E.g.*, the following is a theorem of this type: If T is a measure-preserving one-to-one transformation of a bounded open region of n-dimensional space onto itself, then there is a set M of *measure zero* such that if x is a point not in M and U is a neighborhood of x, then the points $T(x)$, $T^2(x), T^3(x), \cdots$ are in U with a definite positive limiting frequency; *i.e.*, $\lim_{n \to \infty} [\sum_1^n \phi_k(x)]/n$ exists and is positive, where $\phi_k(x)$ is $+1$ or 0 according as $T^k(x)$ belongs to U or does not belong to U [$T^k(x)$ is the result of applying the transformation T to x successively k times]. The **ergodic theorem of Birkhoff** states that if T is a measure-preserving point transformation of the interval $(0, 1)$ onto itself, and if the function f is Lebesgue integrable over $(0, 1)$, then there is a function f^* which is Lebesgue integrable over $(0, 1)$ and is such that we have

$$f^*(x) = \lim \frac{f(x) + f(Tx) + \cdots + f(T^n x)}{n + 1}$$

almost everywhere on $(0, 1)$. The **mean ergodic theorem** (a weaker result than that of Birkhoff's ergodic theorem) states that, under the same hypotheses as in Birkhoff's theorem, the same conclusion is true with *point-wise convergence almost everywhere* replaced by *convergence in the mean of order two.*

ERLANG, A. K. (1878–1929). Danish statistician and engineer.

Erlang distribution. Same as GAMMA DISTRIBUTION. See under GAMMA.

ERLANGEN PROGRAM. See KLEIN.

ER'ROR, *n.* The difference between a number and the number it approximates; the difference $E = A - X$, where A is an approximation of X. The **absolute error** is $|A - X|$; *i.e.*, the absolute value of the error. The relative error is E/X, or sometimes $|E/X|$, and the percent error is the relative error expressed in percent, *i.e.*, multiplied by 100. *E.g.*, if a distance of 10 ft. is measured as 10.3 ft., then the error is 0.3, the relative error is 0.03, and the percent error is 3%. In *statistics*, an error of observation is the difference between an observation and the "true value" or the expected value due to uncontrollable factors (either human or instrumental) in the method of observation. If the uncontrollable factors are independent and additive in their effect on the variation around some "true" or expected value, the deviations will be normally distributed around this value. Measurements are presumably affected by such a set of factors—hence the name error curve for the normal distribution. **A sampling error** is the part of the difference between a "true value" or an expected value and an estimate obtained from a random sample that is due to the fact that only a sample of values is being used. When testing a hypothesis, an **error of type I** is the error of rejecting a true hypothesis; an **error of type II** is the acceptance of an alternative to the hypothesis when the hypothesis is true. See HYPOTHESIS —test of a hypothesis.

error function. Any of the functions

$$Erf(x) = \int_0^x e^{-t^2} \, dt = \tfrac{1}{2}\gamma\left(\tfrac{1}{2}, x^2\right),$$

$$Erfc(x) = \int_x^\infty e^{-t^2} \, dt = \tfrac{1}{2}\Gamma\left(\tfrac{1}{2}, x^2\right),$$

$$Erfi(x) = \int_0^x e^{t^2} \, dt = -i \cdot Erf(ix).$$

mean-square error. The *expected value* of $(t - \theta)^2$, where t is an *estimator* for the parameter θ. This is equal to $\sigma^2 + b^2$, where σ^2 and b are the *variance* and *bias* of t.

standard error. For an *unbiased estimator*, the standard deviation of the estimator, obtained by replacing any unknown moments involved by the moments computed from the sample. *E.g.*, $\sum_1^n X_i/n$ is an estimator for the *mean*, and its standard deviation is $\sigma/\sqrt{n}$, so the standard error is $\left[\sum_{1}^{n}\left(X_i - \bar{X}\right)^2/n\right]^{1/2}$, where $\bar{X} = \sum_{1}^{n} X_i/n$.

ES-CRIBED′, *adj.* **escribed circle of a triangle.** A circle tangent to one side of the triangle and to the extensions of the other sides. *Syn.* excircle. In the figure, the circle is an excircle of the triangle ABC, being tangent to BC at L and to AB and AC, extended, at N and M, respectively. The bisector of angle BAC passes through the center of the circle.

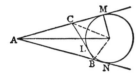

ES-SEN′TIAL, *adj.* **essential constant.** See CONSTANT —essential constant.
   essential map. See INESSENTIAL.

ES-SEN′TIAL-LY, *adv.* **essentially bounded function.** See BOUNDED.

ES′TI-MATE, *n., v.* (1) A numerical value assigned to a parameter of a distribution function of a random variable on the basis of evidence from random samples of the variable. *Tech.* A particular value of an *estimator* given a particular sample (see ESTIMATOR). (2) A value of a certain variable or a description of any mathematical concept, such as a property of a function, made on the basis of evidence.
   estimate a desired quantity. To pass judgment based on very general considerations, as contrasted to finding the quantity by exact mathematical procedure. One might *estimate* the square root of any number to the nearest integer, but one would *compute* it systematically, by some rule for extracting roots, if accuracy to three or four decimal places was required.

ES′TI-MA′TOR, *n.* A function of the terms $\{X_1, X_2, \cdots, X_n\}$ of a *random sample* that is used to estimate some parameter of a distribution. An estimator is itself a random variable; the value of the estimator computed from a given set of sample values is a number that is an estimate of the parameter. For example, $\left(\sum_{1}^{n} X_i\right)/n$ is an estimator for the *mean*; if $n = 2$ and X_1 and X_2 have the values 3 and 5, then $(3 + 5)/2 = 4$ is an

estimate of the mean. See CONSISTENT —consistent estimator, EFFICIENT, LIKELIHOOD, STATISTIC—sufficient statistic, UNBIASED, VARIANCE.

E′TA, *n.* The seventh letter of the Greek alphabet: lower case, η; capital, H.

EUCLID (fl. about 300 B.C.). Greek geometer, number theorist, astronomer and physicist. His axiomatic *Elements* has been the most enduring and widely used mathematical work of all time, standing alone and unquestioned until the nineteenth century.
   Euclid's algorithm, axioms, postulates. See ALGORITHM, AXIOM, POSTULATE.

EU-CLID′E-AN, *adj.* **Euclidean algorithm.** See ALGORITHM.
   Euclidean geometry. See GEOMETRY— Euclidean geometry.
   Euclidean ring. A commutative ring R for which there is a function n whose domain is R with 0 deleted, whose range is a set of nonnegative integers, and which satisfies the conditions: (i) $n(xy) \geq n(x)$ if $xy \neq 0$; (ii) for any members x and y of R with $x \neq 0$, there are members q and r such that $y = qx + r$ and either $r = 0$ or $n(r) < q(x)$. A Euclidean ring has a unit element and is a principal ideal ring. A ring of polynomials over a field is a Euclidean ring if $n(p)$ is the degree of p.
   Euclidean space. (1) Ordinary two- or three-dimensional space. (2) A space consisting of all sets (points) of n numbers $(x_1, x_2, \cdots, x_n)$, where the distance $\rho(x, y)$ between $x = (x_1, \cdots, x_n)$ and $y = (y_1, \cdots, y_n)$ is defined as $\rho(x, y) = \left[\sum_{i=1}^{n} |x_i - y_i|^2\right]^{1/2}$. This is an n-dimensional Euclidean space; it is **real** or **complex** according as the coordinates $x_1, \cdots, x_n$ of $x = (x_1, \cdots, x_n)$ are real or complex numbers, although sometimes components are required to be real and the space for which components are complex is called a *unitary space*. See INNER —inner product space.
   locally Euclidean space. A topological space T for which there is an integer n such that each point of T has a neighborhood which is homeomorphic to an open set in n-dimensional Euclidean space. The space T is then said to be of dimension n. It has been proved that any locally Euclidean topological group is isomorphic to a Lie group (the fifth problem of Hilbert). See MANIFOLD.

EUDOXUS (c. 408–355 B.C.). Greek astronomer and mathematician. His definition of equal ratios anticipated the modern theory of real numbers; the *method of exhaustion* for finding areas and volumes probably is due to him. See DEDEKIND, EXHAUSTION.

EULER, Leonhard (1707–1783). Born in Basel, Switzerland. In 1727, he accepted a chair at the new St. Petersburg Academy. Later, he was head of the Prussian Academy in Berlin for 25 years, but then returned to St. Petersburg. Although he was blind the last 17 years of his life, this was one of his most productive periods. Euler was the most prolific mathematician in history, and the first modern mathematical universalist. It has been said that he wrote mathematics as effortlessly as most men breathe. He had a fantastically flawless memory.

equation of Euler. When the lines of curvature of a surface S are parametric, the equation for the normal curvature $1/R$ for a given direction at a point of S becomes

$$\frac{1}{R} = \frac{\cos^2 \theta}{\rho_1} + \frac{\sin^2 \theta}{\rho_2},$$

where θ is the angle between the directions whose normal curvatures are $1/\rho_1$ and $1/\rho_2$. The above equation is called the **equation of Euler.** See CURVATURE —normal curvature of a surface, CURVATURE —principal curvatures of a surface at a point.

Euler angles. The three angles usually chosen to fix the directions of a new set of rectangular space coordinate axes with reference to an old set. They are the angle between the old and the new z-axis, the angle between the new x-axis and the intersection of the new xy-plane with the old xy-plane, and the angle between this intersection and the old x-axis. This intersection is called the **nodal line** of the transformation. Euler angles are often defined in other ways. Another common usage is to take the angles between the old and new z-axes, between the old y-axis and the normal to the plane of the two z-axes, and between this normal and the new y-axis.

Euler characteristic. For a **curve,** the Euler characteristic is the difference between the number of vertices and the number of segments when the curve is divided into segments by points (vertices) such that each segment (together with its end points) is topologically equivalent to a closed straight line segment (can be continuously deformed into a closed interval). The Euler characteristic of a surface is equal to the number of vertices *minus* the number of edges *plus* the number of faces if the surface is divided into faces by means of vertices and edges in such a way that each face is topologically equivalent to a plane polygon. For both curves and surfaces, the Euler characteristic is independent of the method of subdivision. A surface has Euler characteristic 2 if and only if it is topologically equivalent to a sphere; Euler characteristic 1 if and only if it is topologically equivalent to the projective plane or to a disc (a circle and its interior); Euler characteristic zero if and only if it is topologically equivalent to a cylinder, torus, Möbius strip, or Klein bottle. See GENUS —genus of a surface, SURFACE. For an n-dimensional simplicial complex K, the Euler characteristic is the number $\chi = \sum_{r=0}^{n} (-1)^r s(r)$, where $s(r)$ is the number of r-simplexes of K; χ is also equal to

$$\sum_{r=0}^{n} (-1)^r B_m^r,$$

where B_m^r is the *r-dimensional Betti number modulo m* (m a prime), and to $\sum_0^n (-1)^r B^r$, where B^r is the *r-dimensional Betti number.* Sometimes called the *Euler-Poincaré characteristic.*

Euler constant (or **Mascheroni constant**). The constant defined as

$$\lim_{n \to \infty} \left(1 + \tfrac{1}{2} + \tfrac{1}{3} + \cdots + 1/n - \log n \right)$$

$$= 0.5772156649015328606 + .$$

It is not known whether this number is irrational.

Euler criterion for residues. See RESIDUE.

Euler equation. (1) An ordinary differential equation of type

$$a_0 x^n \frac{d^n y}{dx^n} + a_1 x^{n-1} \frac{d^{n-1} y}{dx^{n-1}} + \cdots$$

$$+ a_{n-1} x \frac{dy}{dx} + a_n y = f(x),$$

where $a_0, \cdots, a_n$ are constants. Such equations were studied by Euler about 1740, but the general solution was known to John Bernoulli by 1700. (2) (*Calculus of Variations*) The differential equation

$$\frac{\partial f(x, y, y')}{\partial y} - \frac{d}{dx} \left(\frac{\partial f(x, y, y')}{\partial y'} \right) = 0,$$

where

$$y' = \frac{dy}{dx}.$$

A necessary condition that y minimize the integral $\int_a^b f(x, y, y')\, dx$ is that y satisfy Euler's equation. This condition, and the more general necessary condition

$$\frac{\partial f}{\partial y} + \sum_{r=1}^{n} (-1)^r \frac{d^r}{dx^r} \left\{ \frac{\partial f}{\partial y^{(r)}} \right\} = 0,$$

where $y^{(r)} = \dfrac{d^r y}{dx^r}$, for y to minimize the integral $\int_a^b f(x, y, y', \cdots, y^{(n)})\, dx$, were first discovered by Euler in 1744. For the double integral $\iint_S f(x, y, z, z_x, z_y)\, dx\, dy$, the Euler equation becomes

$$\frac{\partial f}{\partial z} - \frac{\partial}{\partial x}\left(\frac{\partial f}{\partial z_x}\right) - \frac{\partial}{\partial y}\left(\frac{\partial f}{\partial z_y}\right) = 0,$$

where

$$z_x = \frac{\partial z(x, y)}{\partial x}, \quad z_y = \frac{\partial z(x, y)}{\partial y}.$$

Also called the **Euler-Lagrange equation**. See CALCULUS —calculus of variations.

Euler formula. The formula $e^{ix} = \cos x + i \sin x$. This formula can be taken as the definition of e^{ix} for x a real number; if e^z is defined as $\sum_0^{\infty} z^n / n!$ for z any complex number, the Euler formula can be proved by using the Maclaurin series for $\cos x$ and $\sin x$. Interesting special cases are those in which $x = \pi$ and 2π, for which $e^{\pi i} = -1$ and $e^{2\pi i} = 1$, respectively.

Eulerian graph. See GRAPH THEORY.

Euler-Maclaurin sum formula. A formula for approximating a definite integral, say $\int_a^b f(x)\, dx$, where f has continuous derivatives of all orders up to the highest order used for all points of $[a, b]$, and $b - a = m$ is an integer. The formula

is

$$\int_a^b f(x)\, dx$$

$$= \frac{1}{2}[f(a) + f(b)] + \sum_{r=1}^{m} f(a + r)$$

$$- \sum_{r=1}^{n-1} \frac{B_r}{(2r)!} \left[f^{[2r-1]}(b) - f^{[2r-1]}(a) \right]$$

$$- f^{[2n]}(\theta m) \frac{mB_n}{(2n)!},$$

where θ is some number satisfying $0 \le \theta \le 1$ and B_n is a Bernoulli number. See BERNOULLI (James)—Bernoulli's numbers (1).

Euler method. A method for obtaining an approximate solution of a differential equation of type $dy/dx = f(x, y)$. To determine an approximate solution that passes through the point (x_0, y_0), we let $x_1 = x_0 + h$, $y_1 = y_0 + hf(x_0, y_0)$; $x_2 = x_1 + h$, $y_2 = y_1 + hf(x_1, y_1)$; etc. See RUNGE —Runge-Kutta method.

Euler pentagonal-number theorem. The equality

$$\prod_{n=1}^{\infty} (1 - x^n)$$

$$= 1 + \sum_{n=1}^{\infty} (-1)^n \left(x^{\frac{1}{2}n(3n-1)} + x^{\frac{1}{2}n(3n+1)} \right),$$

stated by Euler as "quite certain, although I cannot prove it", but he proved it ten years later. This theorem has been extremely important for number theory, especially for relations between number theory and elliptic functions. The numbers $\frac{1}{2}n(3n - 1)$, $n = 1, 2, 3, \cdots$, are called **pentagonal numbers** because of their relation to certain "pentagonal arrays" of points.

Euler ϕ-function of an integer. The number of integers not greater than the given integer and relatively prime to it. If the number is $n = a^p b^q c^r \cdots$, where $a, b, c, \cdots$ are distinct primes, then the *Euler ϕ-function* of n, written $\phi(n)$, is equal to

$$n(1 - 1/a)(1 - 1/b)(1 - 1/c) \cdots.$$

The values of $\phi(n)$, for $n = 1, 2, 3$ and 4, are 1, 1, 2, and 2, respectively, while $\phi(12) = 12(1 - \frac{1}{2})(1 - \frac{1}{3}) = 4$. *Syn.* indicator, totient, ϕ-function.

Euler theorem on homogeneous functions. A homogeneous function of degree n in the variables $x_1, x_2, x_3, \cdots, x_n$, multiplied by n, is equal to x_1 times the partial derivative of the function with respect to x_1, plus x_2 times the partial derivative of the function with respect to x_2, etc. E.g., if $f(x, y, z) = x^2 + xy + z^2$, then

$$2(x^2 + xy + z^2) = x(2x + y) + y(x) + z(2z).$$

Euler theorey for polyhedrons. For any simple polyhedron, $V - E + F = 2$, where V is the number of vertices, E the number of edges, and F the number of faces. See above, Euler characteristic.

Euler transformation of series. A transformation for oscillating series which increases the rate of convergence of convergent series and sometimes defines sums for divergent series. Consider the series $a_0 - a_1 + a_2 - a_3 + \cdots$. The Euler transformation carries this into

$$\frac{a_0}{2} + \frac{a_0 - a_1}{2^2} + \frac{a_0 - 2a_1 + a_2}{2^3}$$

$$+ \cdots = \sum \frac{\Delta^n a_0}{2^n},$$

where $\Delta^n a_0$ is the nth difference of the sequence $a_0, a_1, a_2, \cdots$ (i.e., $\Delta^n a_0 = a_0 - \binom{n}{1} a_1 + \binom{n}{2} a_2 - \cdots + (-1)^n a_n$, where $\binom{n}{r}$ is the rth binomial coefficient of order n). E.g., this transformation carries the series $1 - \frac{1}{2} + \frac{1}{3} - \cdots$ into $\frac{1}{1 \cdot 2} + \frac{1}{2 \cdot 2^2} + \frac{1}{3 \cdot 2^3} + \cdots$ and the series $1 - 1 + 1 - 1 + \cdots$ into $\frac{1}{2} + 0 + 0 + 0 + \cdots$.

E-VAL′U-ATE, *v.* To find the value of. E.g., to evaluate $8 + 3 - 4$ means to reduce it to 7; to evalute $x^2 + 2x + 2$ for $x = 3$ means to replace x by 3 and collect the results (giving 17); to evaluate an integral means to carry out the integration and, if it is a definite integral, substitute the limits of integration. Also, see DETERMINANT — evaluation of a determinant.

E-VAL′U-A′TION, *n.* The act of evaluating.

E′VEN, *adj.* **even number.** An integer that is divisible by 2. All even numbers can be written in the form $2n$, where n is an integer.

    even function. See FUNCTION —even function.

    even permutation. See PERMUTATION (2).

    even-spaced map. See CYLINDRICAL —cylindrical map.

E-VENT′, *n.* For an experiment with a finite or countably infinite number of outcomes, an event is any subset of the possible outcomes of the experiment, *i.e.*, any subset of the sample space of the experiment (see EXPERIMENT). The event is said to occur if the observed outcome is a member of the subset. *E.g.*, if two dice are thrown, then the set $\{(3, 6), (4, 5), (5, 4), (6, 3)\}$ is an event (this event also can be described as "sum of 9"); the events are the subset of the set of all ordered pairs (m, n) for which each of m and n is one of the integers $1, 2, 3, 4, 5, 6$. An **elementary event** (or **simple event**) is a single outcome of an experiment, *i.e.*, an event that is a set with only one member. If a black die and a red die are thrown, then $\{(3, 6)\}$ is an elementary event consisting of the single outcome for which the black die shows 3 and the red die shows 6. Now suppose an experiment has a noncountable number of outcomes (the sample space S is not countable) and let $\mathscr{E}$ be a σ-algebra of subsets of S, *i.e.*, a collection of subsets of S which has the properties: (i) S is a member of $\mathscr{E}$, (ii) if A is in $\mathscr{E}$, then the complement of A is in $\mathscr{E}$, (iii) if $\{A_1, A_2, \cdots\}$ is a sequence of members of $\mathscr{E}$, then the union of these members of $\mathscr{E}$ is in $\mathscr{E}$ (see ALGEBRA —algebra of subsets). Then an event is any subset of S that is a member of the collection $\mathscr{E}$. *E.g.*, suppose an experiment consists of spinning a pointer and the sample space is the set S of all angles x such that $0 \leq x < 360°$, where x is the angle from a fixed ray to the final position of the pointer. Then one might designate as *events* all measurable subsets of S, or all Borel subsets of S. In either case, any interval would be an event; thus "x falling between 90° and 180°" would be an event. See PROBABILITY —probability function.

    compound event. Suppose S_1 and S_2 are sample spaces for the outcomes of two experiments and that E_1 and E_2 are events contained in S_1 and S_2, respectively. Then the Cartesian product $E_1 \times E_2$ is a compound event. *E.g.*, the event "head on each of two tosses of a coin" is a compound event relative to sample spaces S_1 and S_2 that are each $\{H, T\}$ – the set of outcomes H and T of a toss of a single coin.

    dependent and independent events. Two events are independent if the occurrence or nonoccurrence of one of them does not change the probability of the occurrence of the other event. Two events are dependent if they are not independent. If A and B are events whose probabilities, $P(A)$ and $P(B)$, are not zero, then A and B are independent if and only if $P(A$ and $B) = P(A) \cdot P(B)$. If neither $P(A)$ nor $P(B)$ is zero, then A and B are independent if and only if $P(A)$ is equal to the *conditional probability* of

A given *B* and *P*(*B*) is equal to the *conditional probability* of *B* given *A*. *E.g.*, if a bag contains 3 white and 5 black balls, and two balls are drawn, then the probability $P(W2)$ of the second ball being white is $\frac{3}{8}$. But if the first ball is known to be white, then the probability of the second being white is $\frac{2}{7}$; also $P(W1 \text{ and } W2) = \frac{3}{28}$ and $P(W_1) \cdot P(W2) = \frac{9}{64} \neq \frac{3}{28}$. Thus the events "first ball white" and "second ball white" are not independent. See INDEPENDENT—independent random variables.

mutually exclusive events. Two or more events such that the occurrence of any one precludes the occurrence of any of the others; events $\{E_1, \cdots, E_n\}$ such that $E_i \cap E_j$ is empty for all *i* and *j* (see EVENT, p. 155). If a coin is tossed, the coming up of heads and the coming up of tails on a given throw are mutually exclusive events.

simple event. See EVENT, above.

EV′O-LUTE, *n.* **evolute of a curve.** See INVOLUTE—involute of a curve.

evolute of a surface. The two surfaces of center relative to the given surface *S*. See SURFACE—surfaces of center relative to a given surface. If we choose the normals to *S* as normals to the lines of curvature of *S*, we obtain the surfaces of center as loci of the evolutes of the lines of curvature of *S*. See above, evolute of a curve. The evolute of *S* is also the evolute of any surface parallel to *S*. See PARALLEL—parallel surfaces, and INVOLUTE—involute of a surface.

mean evolute of a surface. The envelope of the planes orthogonal to the normals of a surface *S* and cutting the normals midway between the centers of principal curvature of *S*.

EV′O-LU′TION, *n.* The extraction of a root of a quantity; *e.g.*, finding a square root of 25. *Evolution* is the inverse of finding a power of a number, or **involution.**

EX-ACT′, *adj.* **exact differential equation.** See DIFFERENTIAL.

exact division. Division in which the remainder is zero. It is then said that the divisor is an **exact divisor**. The quotient is required to be of some specified type. *E.g.*, 7 is not exactly divisible by 2 if the quotient must be an integer, but 7 is exactly divisible by 2 if the quotient can be a rational number $(3\frac{1}{2})$.

exact interest. See INTEREST.

EX-CEN′TER, *n.* **excenter of a triangle.** The center of an *escribed* circle; the intersection of the bisectors of two exterior angles of the triangle.

EX-CESS′, *n.* **excess of nines.** The remainder left when any positive integer is divided by nine; the remainder when the greatest possible number of nines have been subtracted from it. It is equal to the remainder determined by dividing the sum of the digits by 9. It is customary, but not necessary, to restrict the process to positive integers. The excess of nines in 237 is 3, since $237 = 26 \times 9 + 3$ (or $2 + 3 + 7 = 9 + 3$). See CASTING—casting out nines.

spherical excess. See SPHERICAL—spherical excess.

EX-CHANGE′, *n.* Payment of obligations other than by direct use of money; by use of checks, drafts, money orders, exchange of accounts, etc.

foreign exchange. Exchange carried on with other countries (between countries). The **rate of foreign exchange** is the value of the foreign money in terms of the money of one's own country (or vice versa).

EX-CIR′CLE, *n.* **excircle of a triangle.** See ESCRIBED—escribed circle of a triangle.

EX-CLUD′ED, *adj.* **law of the excluded middle.** See CONTRADICTION—law of contradiction.

EX-CLU′SIVE, *adj.* **mutually exclusive events.** See EVENT.

EX-HAUS′TION, *n.* **method of exhaustion.** A method probably invented by Eudoxus and used (notably by Archimedes and Eudoxus) to find areas (*e.g.*, of circles, ellipses, and segments of parabolas) and volumes (pyramids and cones). For areas, it consists of finding an increasing (or decreasing) sequence of sets whose areas are known and less than (greater than) the desired area and increasing (decreasing); then showing that the area approaches the area of the given set because the region between the boundary of the given set and the approximating set is "exhausted."

EX-IST′ENCE, *adj.* **existence theorem.** A theorem that asserts the existence of at least one object of some specified type. Examples are: (1) The fundamental theorem of algebra, which asserts that, for any polynomial *p* of degree at least 1 with complex coefficients, there *exists* at least one complex number *z* for which $p(z) = 0$; (2) "There *exists* a solution for a system of *n* linear equations in *n* unknowns if the determinant of the coefficients is nonzero"; (3) "If *f*, *g* and *h* are continuous on the closed interval $[a, b]$ and y_0 and y_1 are two real numbers, then there *exists* a solution *y* of the differential equation $y'' + f(x)y' + g(x)y = h(x)$ for which y'' is continuous on $[a, b]$, $y(a) = y_0$, and $y'(a) = y_1$." An argument that establishes an existence theorem is an *existence proof*. See UNIQUENESS — uniqueness theorem.

EX'IS-TEN'TIAL, *adj.* **existential quantifier.** See QUANTIFIER.

EX-OT'IC, *adj.* **exotic four-space.** See DONALD-SON.

exotic sphere. Any manifold that has a differential structure of class C^∞ and is homeomorphic, but not diffeomorphic, to a natural sphere. See MANIFOLD.

EX-PAND'ED, *adj.* **expanded notation.** The number represented by 537.2 in decimal notation can also be represented as $(5 \cdot 10^2) + (3 \cdot 10) + (7 \cdot 1) + (2 \cdot \frac{1}{10})$, which is an **expanded numeral** and an example of the use of **expanded notation.** Writing numerals with positional value understood, as for 537.2, is using **positional notation.**

EX-PAN'SION, *n.* (1) The form a quantity takes when written as a sum of terms, or as a continued product, or in general in any type of expanded (extended) form. See FOURIER —Fourier series, TAYLOR —Taylor's theorem. (2) The act, or process, of obtaining the expanded form of a quantity. (3) Increase in size.
   binomial expansion. The expansion given by the binomial theorem. See BINOMIAL —binomial theorem.
   coefficient of linear expansion, thermal expansion, and volume (or cubical) expansion. See COEFFICIENT.
   expansion of a determinant. See DETERMINANT—expansion of a determinant.
   expansion (of a function) in a series. Writing a series which converges to the function for certain values of the variables (or which "represents" the function in some other sense). The series itself is also spoken of as the *expansion* of the function.

EX-PEC-TA'TION, *n.* **expectation of life.** The average number of years that members of a given group may be expected to live after attaining a certain age, according to a mortality table. Also called **complete expectation of life** as distinguished from **curtate expectation of life,** which is the average number of *entire years* that members of a given group may be expected to live. **Joint expectation of life** is the average number of years that two (or more) persons at a given age may both (all) be expected to live, according to a mortality table.

mathematical expectation. Same as EX-PECTED VALUE. See below.

EX-PECT'ED, *adj.* **expected value.** For a discrete random variable X with values $\{x_n\}$ and probability function p, the expected value of X is $\Sigma x_i p(x_i)$, if this sum is finite or converges absolutely. For example, if two coins are tossed and Paul receives \$5 if they both show heads, \$1 if one shows heads and one shows tails, and pays \$6 if both show tails, then Paul's expectation is $5 \cdot \frac{1}{4} + 1 \cdot \frac{1}{2} + (-6) \cdot \frac{1}{4} = \frac{1}{4}$ dollars. Most generally, if p is a probability measure defined on a set S, then the **expected value** of a function f is

$$\int_S f \, dp \text{ [see INTEGRABLE —integrable function]. In}$$

particular, if f is the probability density function (relative frequency function) of the random variable X, then

$$E(X) = \bar{X} = \int_{-\infty}^{+\infty} t f(t) \, dt$$

is the expected value of X. The quantity

$$\sigma^2 = \int_{-\infty}^{+\infty} (t - \bar{X})^2 f(t) \, dt$$

is the **variance** of X, or the expected value of the square of the deviation of X from its expected value (or mean). The expected value of the product XY of two independent variables X and Y is equal to the product of their expected values. If ϕ is a function of the random variable X whose probability density function is f, then

$$E[\phi(X)] = \int_{-\infty}^{+\infty} \phi(t) f(t) \, dt$$

is the **expected value** of ϕ. *Syn.* (arithmetic) average, expectation, mathematical expectation, (arithmetic) mean. See MEAN, MOMENT —moment of a distribution, SAMPLE —sample moment.

EX-PENS'ES, *n.* **overhead expenses.** Administrative expenses, such as salaries of officers and employees, cost of supplies, losses by credit, and rent and plant depreciation. (It sometimes includes some of the selling expenses.)
   selling expenses. Expenses such as insurance, taxes, advertising and salesmen's wages.

EX-PER'I-MENT, *n.* (*Statistics*) Any operation or process of doing or observing something happen under certain conditions, resulting in some final "outcome." The collection of all possible outcomes is the **sample space** of the experiment. E.g., the toss of a coin has two "outcomes," H (heads) and T (tails); the sample space is the set $\{H, T\}$. *Syn.* chance experiment, experiment of chance, random experiment, trial.

EX-PLE-MEN′TA-RY, *adj.* **explementary angles.** See ANGLE—conjugate angles.

EX-PLIC′IT, *adj.* **explicit function.** See IMPLICIT—implicit function.

EX-PO′NENT, *n.* A number placed at the right of and above a symbol. The value assigned to the symbol with this exponent is called a **power** of the symbol, although *power* is sometimes used in the same sense as *exponent*. If the exponent is a positive integer and x denotes the symbol, then x^n means x if $n = 1$, and it means the product of n factors each equal to x if $n > 1$. *E.g.*, $3^1 = 3$; $3^2 = 3 \times 3 = 9$ (the second power of 3 is 9); $x^3 = x \times x \times x$. If x is a nonzero number, the value of x^0 is defined to be 1 (if $x \neq 0$, x^0 can be thought of as the result of subtracting exponents when dividing a quantity by itself: $x^2/x^2 = x^0 = 1$). A negative exponent indicates that in addition to the operations indicated by the numerical value of the exponent, the quantity is to be reciprocated. Whether the reciprocating is done before or after the other exponential operations have been carried out is immaterial. *E.g.*, $3^{-2} = (3^2)^{-1} = (9)^{-1} = \frac{1}{9}$, or $3^{-2} = (3^{-1})^2 = (\frac{1}{3})^2 = \frac{1}{9}$. The following **laws of exponents** are valid when m and n are any integers (positive, negative, or zero) and a and b are real or complex numbers (not zero if in a denominator or if the exponent is 0 or negative):

(1) $a^n a^m = a^{n+m}$; (2) $a^m/a^n = a^{m-n}$;
(3) $(a^m)^n = a^{mn}$; (4) $(ab)^n = a^n b^n$;
(5) $(a/b)^n = a^n/b^n$, if $b \neq 0$.

If the exponent on a symbol x is a **fraction** p/q, then $x^{p/q}$ is defined as $(x^{1/q})^p$, where $x^{1/q}$ is the positive qth root of x if x is positive, and the (negative) qth root if x is negative and q is odd. If follows that $x^{p/q} = (x^p)^{1/q}$ and that the above five laws are valid if m and n are either fractions or integers (*i.e.*, **rational numbers**), provided a and b are positive numbers. If the exponent is **irrational**, the power is defined to be the quantity approximated by using rational exponents which approximate the irrational exponent; *e.g.*, 3 with exponent $\sqrt{2}$ denotes the limit of the sequence $3^{1.4}, 3^{1.41}, 3^{1.414}, \cdots$. *Tech.* if the sequence $a_1, a_2, \cdots, a_n, \cdots$ of rational numbers approaches an irrational limit a, then c^a denotes

the limit of the sequence $c^{a_1}, c^{a_2}, c^{a_3}, \cdots,$ $c^{a_n}, \cdots$. The above laws of exponents are valid for m and n any real numbers (rational or irrational) if a and b are positive. If x is a complex number, then x^m is defined to be $e^{m(\log x)}$, where this is computed by substituting $m(\log x)$ for t in the Taylor's series

$$e^t = 1 + t + t^2/2! + t^3/3! + t^4/4! + \cdots$$

(see EXPONENTIAL —exponential series, LOGARITHM—logarithm of a complex number). With this definition, x^m is multiple valued and the laws of exponents are valid only in the sense that, if one member of one of the equations is computed, then this is one of the values of the other member. *E.g.*, $(2/-3)^{1/2} = (-\frac{2}{3})^{1/2} = i\sqrt{\frac{2}{3}}$ is not equal to $2^{1/2}/(-3)^{1/2} = \sqrt{2}/(i\sqrt{3}) = -i\sqrt{\frac{2}{3}}$ if $(-3)^{1/2}$ is taken as $i\sqrt{3}$, but these are equal if one takes $(-3)^{1/2}$ to be $-i\sqrt{3}$. See DE MOIVRE —De Moivre's theorem.

EX′PO-NEN′TIAL, *adj.* **derivative of an exponential.** See DIFFERENTIATION FORMULAS in the appendix.

exponential curve. The plane locus of $y = a^x$ (or, what is the same, $x = \log_a y$). It can be obtained geometrically by revolving the logarithmic curve, $y = \log_a x$, about the line $y = x$, *i.e.*,

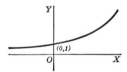

reflecting it in this line. The curve is asymptotic to the negative x-axis, when $a > 1$ as in the figure, and has its y intercept unity.

exponential distribution. See GAMMA—gamma distribution.

exponential equation. See EQUATION—exponential equation.

exponential function. See FUNCTION —exponential function.

exponential series. The series

$$1 + x + x^2/2! + x^3/3! + \cdots + x^n/n! + \cdots.$$

This is the Maclaurin expansion of e^x and converges to e^x for all real numbers x. The series can be used to define e^z for z a complex number, or the series can be proved to converge to e^z if e^{x+iy} is defined to be $e^x e^{iy}$ with $e^{iy} = \cos y + i \sin y$. See EULER —Euler's formula.

exponential values of sin x and cos x.

$$\sin x = \frac{e^{ix} - e^{-ix}}{2i}$$

and

$$\cos x = \frac{e^{ix} + e^{-ix}}{2},$$

where $i^2 = -1$. These can be proved by use of EULER'S FORMULA.

EX-PRES'SION, *n.* A very general term used to designate any symbolic mathematical form, such, for instance, as a polynomial.

EX-SE'CANT, *n.* See TRIGONOMETRIC—trigonometric functions.

EX-TEND'ED, *adj.* **extended mean-value theorem.** (1) Same as TAYLOR'S THEOREM. (2) Same as the SECOND MEAN-VALUE THEOREM. See MEAN—mean-value theorems for derivatives.

extended real-number system. The real-number system together with the symbols $+\infty$ (or ∞) and $-\infty$. The following definitions are used: $-\infty < a < +\infty$ if a is a real number, $a + \infty = \infty + a = \infty$ if $a \neq -\infty$, $a + (-\infty) = (-\infty) + a = -\infty$ if $a \neq +\infty$, $a \cdot \infty = \infty \cdot a = \infty$ and $a(-\infty) = (-\infty)a = -\infty$ if $0 < a \leq +\infty$, $a \cdot \infty = \infty \cdot a = -\infty$ and $a(-\infty) = (-\infty)a = +\infty$ if $-\infty \leq a < 0$, $a/\infty = a/(-\infty) = 0$ if a is a real number.

EX-TEN'SION, *n.* **extension of a field.** Any field F^* that contains a field F is an **extension** of F. The **degree** of the extension is the dimension of F^* as a vector space with scalars in F [see VECTOR—vector space (2)]. A **finite extension** is an extension whose degree is finite. An **algebraic extension** of F is an extension whose members all satisfy polynomial equations with coefficients in F. A **simple extension** F^* of F is an extension which has a member c such that F^* is the set of all quotients $p(c)/q(c)$, where p and q are polynomials with coefficients in F and $q(c) \neq 0$. A simple extension is finite if and only if c is algebraic with respect to F. A **normal extension** F^* of F is an extension which has one of the following equivalent properties: (i) F is the set of all members of F^* such that $a(x) = x$ for all automorphisms a of F^* for which $a(x) = x$ whenever x is in F; (ii) F^* is the Galois field of some polynomial with coefficients in F; (iii) if p is an irreducible polynomial with coefficients in

F and a zero in F^*, then all zeros of p are in F^*. See SEPARABLE —separable extension of a field.

EX-TE'RI-OR, *adj., n.* The exterior of a circle, polygon, sphere, triangle, etc., is the set of all points neither on nor inside the circle, polygon, etc. For this definition, we think of the circle as being a disk including the circumference and the interior of the circle, and similarly for the polygon, etc. In general, the **exterior** of a set E is the set of all points which have a neighborhood having no points in common with E. Same as the *interior* of the complement of E. Each such point is an **exterior point** of E.

alternate exterior, exterior, and **exterior-interior angles.** See ANGLE—angles made by a transversal.

exterior angle of a polygon. See ANGLE—angle of a polygon.

exterior angle of a triangle. The angle between one side produced and the adjacent side (not produced). A triangle has six exterior angles.

exterior content. See CONTENT—content of a set of points.

exterior measure. See MEASURE—exterior measure.

exterior of a simple closed curve. See JORDAN—Jordan curve theorem.

EX-TER'NAL, *adj.* **external operation.** See OPERATION.

external ratio. See POINT—point of division.

external tangent of two circles. See COMMON—common tangent of two circles.

EX-TER'NAL-LY, *a.* **externally tangent circles.** See TANGENT—tangent circles.

EX-TRACT', *v.* **extract a root of a number.** To find a root of the number; usually refers to finding the positive real root, or the real negative root if it be an odd root of a negative number. *E.g.*, one extracts the square root of 2 when it is found to be $1.4142\cdots$, or the cube root of -8 when it is found to be -2. *Syn.* find, compute, or estimate a root.

EX-TRA'NE-OUS, *adj.* **extraneous root.** A number obtained in the process of solving an equation, which is not a root of the equation given to be solved. It is generally introduced either by squaring the original equation, or clearing it of fractions. *E.g.*, (1) the equation $(x^3 - 3x + 2)/(x - 2) = 0$ has only one root, 1; but if one multiplies through by $x - 2$, the resulting equation has the root 2 also; (2) the equation $1 - \sqrt{x - 1} = x$ has only one root, 1; but if one adds -1 to both members and squares, thus getting rid of the radical, the resulting equation is $x^2 - 3x + 2 = 0$, which has the two roots 1 and 2. The root, 2, is an extraneous root of the original equation, since its substitution in that equation gives $1 - 1 = 2$.

EX′TRA-PO-LA′TION, *n.* Estimating (approximating) the value of a function (quantity) for a value of the argument which is either greater than, or less than, all the values of the argument which are being used in the estimating (approximating). Using log 2 and log 3 one might find an approximate value of log 3.1 by extrapolation, using the formula

$$\log 3.1 = \log 3 + \tfrac{1}{10}(\log 3 - \log 2).$$

See INTERPOLATION.

EX′TREM-AL-LY, *adv.* **extremally disconnected.** See DISCONNECTED —disconnected set.

EX′TREME, *n.* **extremes in a proportion.** See PROPORTION.

extreme or extremum of a function. A *maximum* or *minimum* value of the function. See MAXIMUM.

extreme point. For a convex set K, a point P such that K with P deleted is convex; equivalently, a point that does not belong to any segment joining two other points of K. A compact convex subset of a finite-dimensional space is the convex span of its extreme points. Also see KREIN—Krein-Milman theorem.

F

F, *adj., n.* **F distribution.** A random variable has an **F distribution** or is an **F random variable** with (m, n) degrees of freedom if it has a *probability density function* f for which $f(x) = 0$ if $x \leq 0$ and, if $x > 0$,

$$f(x) = \frac{\Gamma\left[\tfrac{1}{2}(m + n)\right] m^{\frac{1}{2}m} n^{\frac{1}{2}n} x^{\frac{1}{2}m - 1}}{\Gamma\left(\tfrac{1}{2}m\right)\Gamma\left(\tfrac{1}{2}n\right)(mx + n)^{\frac{1}{2}(m + n)}},$$

where Γ is the *gamma function*. The *mean* is

$n/(n - 2)$ if $n > 2$ and the *variance is*

$$\frac{2n(m + n - 2)}{m(n - 2)^2(n - 4)} \quad \text{if } n > 4.$$

If X and Y are independent *chi-square random variables* with m and n degrees of freedom, respectively, then $\mathbf{F} = nX/(mY)$ is an **F** random variable with (m, n) degrees of freedom; equivalently, $\mathbf{F}$ is the ratio of two independent estimates of the variance of a normal distribution with mean 0 and variance 1: $\left(\displaystyle\sum_{i=1}^{m} X_i^2/m\right) \div \left(\displaystyle\sum_{i=1}^{n} X_i^2/n\right)$. If X is an **F** random variable with (m, n) degrees of freedom, then $z = \tfrac{1}{2}\ln X$ is a *Fisher z distribution* with (m, n) degrees of freedom. *Syn.* Snedecor F. See BETA—beta distribution, CHI—chi-square distribution, T—t distribution.

F_σ set. See BOREL—Borel set.

FACE, *n.* See ANGLE—polyhedral angle, DIHEDRAL—dihedral angle, HALF—half space, POLYHEDRON, POLYTOPE, PRISM, PYRAMID.

face amount of an insurance policy. The amount that the company is contracted to pay when the exigency stated in the policy occurs.

face value. See PAR.

FAC′ET, *n.* See POLYTOPE.

FAC′TOR, *adj., n., v.* As a *verb*, to resolve into factors. One factors 6 when he writes it in the form 2×3. A **factor** of an object is any object (perhaps of some specified type) that divides the given object. See PRIME—prime factor, and various headings below.

accumulation factor. See ACCUMULATION.

common factor. Same as COMMON DIVISOR. See DIVISOR.

factor analysis. A branch of multivariate analysis for which observed random variables X_i $(i = 1, \cdots, n)$ are assumed to be representable in terms of other random variables; *e.g.*, as

$$X_i = \sum_{j=1}^{m} a_{ij} U_j + b_i e_i,$$

where $n > m$. The random variables $\{U_j\}$ are **factors** of the $\{X_i\}$ and the $\{e_i\}$ are error terms. Problems incident to this method involve estimation of the a_{ij} on the basis of observations of the variables X_i, and the existence of meaningful interpretations that may be assigned to the U_j factors. *E.g.*, *nr* scores on *n* different psychological tests may be obtained from a set of *r* persons. These scores may be related to a set of *factors*, *e.g.*, verbal facility, arithmetic ability, and form recognition, which are interpretations given to the factors U_i.

factor of an integer. An integer whose product with some integer is the given integer. *E.g.*, 3 and 4 are factors of 12, since $3 \cdot 4 = 12$; the positive factors of 12 are $1, 2, 3, 4, 6, 12$, and the negative factors are $-1, -2, -3, -4, -6, -12$. For tests as to whether certain integers are factors, see DIVISIBLE.

factor modulo *p*. If $r(x) \equiv 0 \pmod{p}$ in the congruence $f(x) \equiv g(x) \cdot d(x) + r(x) \pmod{p}$, then $d(x)$ is said to be a *factor modulo p* of $f(x)$.

factor of a polynomial. One of two or more polynomials whose product is the given polynomial. Sometimes one of the polynomials is allowed to be the constant 1, but usually in elementary algebra a polynomial with rational coefficients is considered factorable if and only if it has two or more nonconstant polynomial factors whose coefficients are rational (sometimes it is required that the coefficients be integers). *Tech.* One of a set of polynomials whose product gives the polynomial to be factored and whose coefficients lie in a given *field* (*domain*). Unless a field is specified, the field of the coefficients of the given polynomial is understood. *Factor* is sometimes used of any quantity whatever that divides a given quantity. *E.g.*, $(x^2 - y^2)$ has the factors $(x - y)$ and $(x + y)$ in the ordinary (elementary) sense; $(x^2 - 2y^2)$ has the factors $(x - \sqrt{2}y)$ and $(x + \sqrt{2}y)$ in the field of real numbers; $(x^2 + y^2)$ has the factors $(x - iy)$ and $(x + iy)$ in the complex field. See FACTORING —type forms for factoring, IRREDUCIBLE —irreducible polynomial.

factor of proportionality. See VARIATION —direct variation.

factor of a term. Any divisor of the term. *E.g.*, $x + 1$ is a factor of $3x^{1/2}(x + 1)$.

factor space (group, ring, vector space, etc.). See QUOTIENT —quotient space.

factor theorem. A polynomial in x is divisible by $(x - a)$ if it reduces to zero when a is substituted for x. See REMAINDER —remainder theorem. The **converse of the factor theorem** also is true: "If $(x - a)$ is a factor of the polynomial $p(x)$, then $p(a) = 0$." See REMAINDER —remainder theorem.

integrating factor. A factor which, when multiplied into a differential equation, with right-hand member zero, makes the left-hand member an exact differential, or makes it an exact derivative. *E.g.*, if the differential equation

$$\frac{dy}{x} + \frac{y}{x^2} dx = 0$$

is multiplied by x^2, there results

$$x \, dy + y \, dx = 0,$$

which has the solution $xy = c$. The differential equation

$$xy'' + (3 - x^3)y' - 5x^2 y + 4x = 0$$

has the integrating factor x^2; when multiplied by x^2 the equation becomes

$$\frac{d}{dx}\left(x^3 y' - x^5 y + x^4 \right) = 0.$$

See ADJOINT —adjoint of a differential equation.

monomial factor. See MONOMIAL.

FAC'TOR-A-BLE, *adj.* *In arithmetic*, containing factors other than unity and itself (referring to integers). *In algebra*, containing factors other than constants and itself (referring to polynomials); $x^2 - y^2$ is factorable in the domain of real numbers, while $x^2 + y^2$ is not.

FAC-TO'RI-AL, *adj.*, *n.* For a positive integer *n*, the product of all the positive integers less than or equal to *n*. *Factorial n* is denoted by *n*! or (rarely) by $\underline{|n}$. *E.g.*, $1! = 1$, $2! = 1 \cdot 2$, $3! = 1 \cdot 2 \cdot 3$, and in general, $n! = 1 \cdot 2 \cdot 3 \cdots n$. This definition of factorial leaves the case when *n* is zero meaningless. In order to make certain formulas valid in all cases, **factorial zero** is arbitrarily defined to be 1. Despite the fact that this is the value of factorial 1, there is considerable to be gained by using this definition; *e.g.*, this makes the formula for the general binomial coefficient, $n!/[r!(n - r)!]$, valid for the first and last terms, which are the terms for which $r = 0$ and $r = n$, respectively.

factorial moment. See MOMENT —moment of a distribution, moment generating function.

factorial series. See SERIES —factorial series.

FAC'TOR-ING, *p.* type forms for factoring:

(1) $x^2 + xy = x(x + y)$;
(2) $x^2 - y^2 = (x + y)(x - y)$;
(3) $x^2 + 2xy + y^2 = (x + y)^2$;
(4) $x^2 - 2xy + y^2 = (x - y)^2$;
(5) $x^2 + (a + b)x + ab = (x + a)(x + b)$;

(6) $acx^2 + (bc + ad)x + bd = (ax + b) \times (cx + d)$;

(7) $x^3 \pm 3x^2y + 3xy^2 \pm y^3 = (x \pm y)^3$;

(8) $x^3 \pm y^3 = (x \pm y)(x^2 \mp xy + y^2)$.

[In (7) and (8) either the upper, or the lower, signs are to be used throughout.]

FAC-TO-RI-ZA'TION, *n.* The process of factoring.

factorization of a transformation. Finding two or more transformations which, when made successively, have the same effect as the given transformation. For an example, see AFFINE — affine transformation.

unique-factorization theorem. The *fundamental theorem of arithmetic* (see FUNDAMENTAL) or similar theorems for integral domains, *e.g.*, for polynomials. See DOMAIN — integral domain, IRREDUCIBLE — irreducible polynomial.

FAH'REN-HEIT, Gabriel Daniel (1686–1736). Physicist; born in Poland, lived in England and Holland.

Fahrenheit temperature scale. A temperature scale for which water freezes at 32° and boils at 212°. See CELSIUS — Celsius temperature scale.

FALSE, *adj.* **method of false position.** (1) Same as REGULA FALSI. (2) A method for approximating the roots of an algebraic equation. Consists of making a fairly close estimate, say r, then substituting $(r + h)$ in the equation, dropping the terms in h of higher degree than the first (since they are relatively small), and solving the resulting linear equation for h. This process is then repeated, using the new approximation $(r + h)$ in place of r. *E.g.*, the equation $x^3 - 2x^2 - x + 1 = 0$ has a root near 2 (between 2 and 3). Hence we substitute $(2 + h)$ for x. This gives (when the terms in h^2 and h^3 have been dropped) the equation $3h - 1 = 0$; whence $h = \frac{1}{3}$. The next estimate will then be $2 + \frac{1}{3}$ or $\frac{7}{3}$.

FALTINGS, Gerd (1954–). German mathematician awarded a Fields Medal (1986) for his proof of Mordell's conjecture.

FALTUNG, *n.* German for CONVOLUTION.

FAM'I-LY, *n.* **family of curves.** A set of curves whose equations can be obtained from a given equation by varying n essential constants which occur in the given equation is an *n*-**parameter family of curves**; *e.g.*, a set of curves whose equations are nonsingular solutions (special cases

of the general solution) of a differential equation of order n. A set of concentric circles constitutes a **one-parameter family of curves**, the radius being the arbitrary parameter. The set of circles in the plane having a given radius is a **two-parameter family of curves**, the two coordinates of the center being the parameters. All the circles in a plane constitute a three-parameter family, and all the conics in a plane a five-parameter family. The set of all lines tangent to a given circle is a one-parameter family of lines, while the set of all lines in the plane is a two-parameter family.

family of surfaces. A set of surfaces whose equations can be obtained from a given equation by varying n essential constants which occur in the given equation is an *n*-**parameter family of surfaces**. The set of all spheres with a given center is a one-parameter family, while the set of all spheres is a four-parameter family.

FAREY, John (1766–1826). English civil engineer and mathematician.

Farey sequence. The **Farey sequence of order** n is the increasing sequence of all fractions p/q for which $0 \leq p/q \leq 1$, $q \leq n$, and p and q are nonnegative integers with no common divisors other than 1. *E.g.*, the Farey sequence of order 5 is

$$\frac{0}{1}, \frac{1}{5}, \frac{1}{4}, \frac{1}{3}, \frac{2}{5}, \frac{1}{2}, \frac{3}{5}, \frac{2}{3}, \frac{3}{4}, \frac{4}{5}, \frac{1}{1}.$$

If a/b, c/d, e/f are three consecutive terms of a Farey sequence, then $bc - ad = 1$ and $c/d = (a + e)/(b + f)$. These facts were stated by Farey without proof in 1816 and proved later by Cauchy, although they had been stated and proved by C. Haros in 1802.

FAST, *n.* **fast Fourier transform.** See FOURIER — fast Fourier transform.

FATOU, Pierre (1878–1929). French analyst.

Fatou's theorem (or lemma). Let μ be a countably additive measure on subsets of a measurable set E and let $\{f_n\}$ be a sequence of measurable functions on E with ranges in the extended real number system. Then $\liminf f_n$ and $\limsup f_n$ are measurable and (1) if there is a measurable function g for which $\int_E g \, d\mu \neq +\infty$ and $f_n(x) \leq g(x)$ for all n and all x in E, then

$$\limsup \int_E f_n \, d\mu \leq \int_E (\limsup f_n) \, d\mu;$$

(2) if there is a measurable function h for which $\int_E h \, d\mu \neq -\infty$ and $f_n(x) \geq h(x)$ for all n and all x in E, then

$$\int_E (\liminf f_n) \, d\mu \leq \liminf \int_E f_n \, d\mu.$$

FEFFERMAN, Charles (1949–   ). American mathematician awarded a Fields Medal (1978) for his discovery of the dual of the Hardy space H^1 and his contributions to the theory of partial differential equations, harmonic analysis, and functions of several complex variables.

FEIT, Walter (1930–   ). **Feit-Thompson theorem**. See THOMPSON.

FEJÉR, Leopold (1880–1959). Hungarian mathematician who worked in the theory of complex variables and summability of series.

   Fejér kernel. See KERNEL—Fejér kernel.

   Fejér's theorem. Let f be a function of the real variable x, defined arbitrarily when $-\pi < x \leq \pi$, and defined by $f(x + 2\pi) = f(x)$ for all other values of x. By Fejér's theorem is meant one of the following: (1) If $\int_{-\pi}^{\pi} f(x) \, dx$ exists, and, if this is an improper integral, $\int_{-\pi}^{\pi} |f(x)| \, dx$ exists, then the *Fourier series* associated with f is *summable* (C1) at all points x for which the limits from the right and left, $f(x + 0)$ and $f(x - 0)$, exist, and its sum (C1) is $\frac{1}{2}(f(x + 0) + f(x - 0))$. (2) If, in addition, f is continuous everywhere in an interval (a, b), then the *first Cesàro sums* converge uniformly to $f(x)$ in any interval (α, β) with $a < \alpha < \beta < b$. Both of these theorems were published by Fejér in 1904. See CESÀRO—Cesàro's summation formula.

FEM'TO. A prefix meaning 10^{-5}, as in **femtometer**.

FERMAT, Pierre de (1601–1665). A French lawyer, who was a brilliant and versatile amateur mathematician. An unexcelled number theorist, Fermat also conceived and applied the leading ideas of the differential calculus before either Newton or Leibniz was born. He is considered, along with Pascal, as the founder of probability theory. As the coinventor of analytic geometry, he is considered, along with Descartes, as one of the first modern mathematicians.

   Fermat numbers. Numbers of the type $F_n = 2^{2^n} + 1$, $n = 1, 2, 3, \cdots$. Fermat thought that these numbers might all be primes. Actually, F_5 is not a prime,

$$F_5 = (641)(6{,}700{,}417) = 4{,}294{,}967{,}297,$$

and F_n is not a prime if $5 \leq n \leq 16$. At least 19 values of $n > 16$ are known for which F_n is not a prime (*e.g.*, 18, 23, 36, 38, 39, 55) and there is no value of $n > 4$ for which F_n is known to be prime. The largest value of n for which F_n actually has been factored is 9: F_9 is the product of 3 primes, with 7 digits, 49 digits, and 99 digits. A regular polygon with p sides, where p is a prime, can be constructed with straight edge and compass alone if and only if p is a Fermat number. See MERSENNE—Mersenne prime.

   Fermat's last theorem. Fermat had studied the book *Arithmetica* by Diophantus carefully and made many marginal notes, one of which was: "*... it is impossible ... for any number that is a power greater than the second to be the sum of two like powers. I have discovered a truly marvelous demonstration of this proposition that this margin is too narrow to contain.*" The conjecture that $x^n + y^n = z^n$, where $n > 2$, has no solution with x, y, and z positive integers is called **Fermat's last theorem**, although this "theorem" has never been proved. It is known that $x^n + y^n = z^n$ has no solution with $n > 2$ unless $n > 1{,}000{,}000$. If a prime $p > 125{,}100$ divides x, then $x^p > (125{,}100)^{125{,}100}$. If none of x, y, or z has a factor in common with n, there is no solution unless $n > 7.568 \cdot 10^{17}$ and the least of x, y, and z is greater than $\frac{111}{77} n(2n^2 + 1)^n$. Although $x^2 + y^2 = 1$ has infinitely many rational solutions, for no value of n greater than 2 are there infinitely many rational solutions of $x^n + y^n = 1$. See MORDELL—Mordell conjecture, PYTHAGORAS—Pythagorean triple.

   Fermat's principle. The principle that a ray of light requires less time along its actual path than it would along any other path having the same end points. This principle was used by John Bernoulli in solving the brachistochrone problem. See BRACHISTOCHRONE.

   Fermat's spiral. See PARABOLIC—parabolic spiral.

   Fermat's theorem. If p and a are positive integers, p is a prime, and a is prime to p, then a^{p-1} divided by p leaves a remainder of 1. *Tech.* $a^{p-1} \equiv 1 \pmod{p}$; *e.g.*, $2^4 \equiv 1 \pmod{5}$, where $p = 5$ and $a = 2$. See CONGRUENCE.

FERRARI (or FERRARO), Ludovico (1522–1565). Italian mathematician. First to solve the general quartic equation in one variable.

Ferrari's solution of the quartic. The solution of the quartic equation

$$x^4 + px^3 + qx^2 + rx + s = 0$$

by showing that the roots of this equation are also the roots of the two equations

$$x^2 + \tfrac{1}{2}px + k = \pm(ax + b),$$

where $a = (2k + \tfrac{1}{4}p^2 - q)^{1/2}$, $b = (kp - r)/(2a)$, and k is obtained from the following cubic equation (the RESOLVENT CUBIC):

$$k^3 - \tfrac{1}{2}qk^2 + \tfrac{1}{4}(pr - 4s)k$$

$$+ \tfrac{1}{8}(4qs - p^2s - r^2) = 0.$$

FERRO, Scipione del (c. 1465–1526). Italian mathematician. He had a secret solution for $x^3 + mx = n$ with m and n positive. See CARDAN —Cardan's solution of the cubic, TARTAGLIA.

FIBONACCI, Leonardo. Nickname of the Italian number theorist and algebraist *Leonardo da Pisa or Leonardo Pisano* (c. 1175–c. 1250). *Fibonacci* is a short form of *Filius Bonacci*, meaning *son of Bonacci*. Fibonacci's publication of *Liber abaci* in 1202 greatly influenced the replacement in Europe of the Roman numerals by the Hindu-Arabic system of numbers.

Fibonacci sequence. The sequence of numbers $1, 1, 2, 3, 5, 8, 13, 21, \cdots$, each of which, after the second, is the sum of the two previous ones (these numbers are also called **Fibonacci numbers**). The ratio of one Fibonacci number to the preceding one is a *convergent* of the continued fraction.

$$1 + \cfrac{1}{1 + \cfrac{1}{1 + \cfrac{1}{1 + \cfrac{1}{1 +}}}} \cdots,$$

which satisfies the equation $x = 1 + 1/x$ and is equal to $\tfrac{1}{2}(\sqrt{5} + 1)$. Thus the sequence $\tfrac{1}{1}, \tfrac{2}{1}, \tfrac{3}{2}, \tfrac{5}{3}, \tfrac{8}{5}, \tfrac{13}{8}, \cdots$ has the limit $\tfrac{1}{2}(\sqrt{5} + 1)$. See FAREY— Farey sequence, GOLDEN—Golden section.

FIELD, *n.* A set for which two operations, called *addition* and *multiplication*, are defined and have the properties: (i) the set is a *commutative group* with addition as the group operation; (ii) multiplication is commutative and the set, with the identity (0) of the additive group omitted, is a group with multiplication as the group operation; (iii) $a(b + c) = ab + ac$ for all a, b, and c in the set. An **ordered field** is a field that contains a set of positive members satisfying the

conditions: (1) the sum and product of two positive members is positive; (2) for a given member x of the field, one and only one of the following possibilities is valid: (a) x is positive, (b) $x = 0$, $(c) - x$ is positive. An ordered field is **complete** if every nonempty subset has a least upper bound if it has an upper bound. The real numbers form a complete ordered field. See below, number field, for examples of fields. Also see DOMAIN— integral domain, RING.

algebraically complete field. See ALGEBRAICALLY.

characteristic of a field. See CHARACTERISTIC —characteristic of a ring or field.

extension of a field. See EXTENSION.

field of force. See FORCE.

field plan. (*Statistics*). The spatial arrangement of experimental trials when the repetitions of different factors must be located at different points in space, *e.g.*, a *Latin square*, or a *randomized-block experiment* in agricultural experiments.

field of study. A group of subjects that deal with closely related material, such as the field of analysis, the field of pure mathematics, or the field of applied mathematics.

Galois field. See GALOIS.

number field. Any set of real or complex numbers such that the sum, difference, product, and quotient (except by 0) of any two members of the set is in the set. *Syn.* number domain. The set of all rational numbers, and of all numbers of the form $a + b\sqrt{2}$ with a and b rational, are number fields. A number field is necessarily a *field*. A number field F can be enlarged by **adjoining** a number z to it, the new field consisting of all numbers which can be derived from z and the numbers of F by the operations of addition, subtraction, multiplication, and division.

p-adic field. See *p*-ADIC.

perfect field. A field for which no irreducible polynomial with coefficients in the field has a multiple zero. See SEPARABLE —separable extension of a field.

tensor field. See TENSOR.

FIELDS, John Charles (1863–1932). Canadian analyst. As President of the International Congress of Mathematicians held in Toronto in 1924, he suggested funds remaining at the end of the Congress be used for awarding medals at each International Congress for outstanding discoveries in mathematics. The **Fields medals** are also intended as encouragement for further achievements on the part of the recipients and have been confined to persons under 40 years of age. These awards were established at the Congress in Zürich (1932). Medals have been presented at each Congress since then: AHLFORS and DOUGLAS (Oslo, 1936); SCHWARTZ and SELBERG (Cambridge, Mass., 1950); KODAIRA and SERRE (Amsterdam, 1954); ROTH and THOM (Edinburgh, 1958); HÖRMANDER and MILNOR (Stockholm, 1962); ATIYAH, COHEN, GROTHENDIECK, and SMALE (Moscow, 1966); BAKER, HIRONAKA, NOVIKOV, and THOMPSON (Nice, 1970); BOMBIERI and MUMFORD (Vancouver, 1974); DELIGNE, FEFFERMAN, MARGULIS, QUILLEN (Helsinki, 1978); CONNES, THURSTON, YAU (Warsaw, 1983); DONALDSON, FALTINGS, and FREEDMAN (Berkeley, 1986); DRINFEL'D, JONES, MORI, and WITTEN (Berkeley, 1990).

FIG'URE, *n.* (1) A character or symbol denoting a number, as 1, 5, 12; sometimes used in the same sense as digit. (2) A drawing, diagram or cut used to aid in presenting subject matter in text books and scientific papers.

    geometric figure. See GEOMETRIC.

    plane figure. See PLANE.

FIL'TER, *n.* A family F of nonempty subsets of a set X, for which the intersection of any two members of F is a member of F and for which any subset B of X which contains a member of F is itself a member of F. An **ultrafilter** (or **maximal filter**) is a filter which is not a proper subset of any filter (if F is an ultrafilter and A is a subset of X, then either A or the complement of A is a member of F). It follows from Zorn's lemma that if F is a filter of subsets of X, then there is an ultrafilter that contains F. A **free ultrafilter** is an ultrafilter that contains A whenever the complement of A is finite. If X is a topological space, a filter F is said to **converge** to a point x if each neighborhood of x is a member of F. See NON-nonstandard numbers.

    filter base. A family F of nonempty subsets of a set X which has the property that the intersection of any two members of F contains another member. If F is a filter base, then F^* is a filter if F^* contains all subsets of X that contain a member of F.

FINE'NESS, *n.* **fineness of a partition.** See PARTITION —partition of an interval, partition of a set.

FI'NITE, *adj.* **finite character.** See CHARACTER.

    finite decimal. See DECIMAL —decimal number system.

    finite differences. See DIFFERENCE—finite differences.

finite discontinuity. See DISCONTINUITY.

finite extension of a field. See EXTENSION — extension of a field.

finite-intersection property. For a collection of sets to have the **finite-intersection property** means that each nonempty finite subcollection of these sets has a nonempty intersection.

finite mathematics. Mathematics that does not involve calculus or limits. Courses in "finite mathematics" usually treat introductory topics from such fields as matrices and linear algebra, number theory, mathematics of finance, set theory and logic, probability, linear programming, theory of games, operations research, combinatorial analysis, graph theory, and computing.

finite projective plane. See PROJECTIVE — projective plane (2).

    finite quantity. (1) Any quantity which is *bounded*; *e.g.*, a function is *finite* on an interval if it is *bounded* on the interval. However, one also says a function is *finite* on a set if it has only finite values (not $+\infty$, $-\infty$, or ∞); *e.g.*, $1/x$ is finite but not bounded for $x > 0$. (2) A real or complex number may be said to be finite to distinguish it from an ideal number $+\infty$, $-\infty$, ∞. See COMPLEX —complex plane, EXTENDED —extended real-number system.

    finite set. A set which contains a finite (limited) number of members; a set which has, for some integer n, just n members. *Tech.* A set which cannot be put into one-to-one correspondence with a proper part of itself. All the integers between 0 and 100 constitute a finite set. See INFINITE —infinite set.

    locally finite family of sets. A family of subsets of a topological space T is **locally finite** if each point of T has a neighborhood which intersects only a finite number of these subsets. A family of subsets is **point finite** if each point of T is in at most finitely many of these subsets.

FI'NITE-LY, *adv.* **finitely representable.** A Banach space X being **finitely representable** in a Banach space Y means that any finite-dimensional subspace X_n of X is "nearly isometric" to a subspace of Y; *i.e.*, for any positive numbers $c < 1$ and $d > 1$, there is an isomorphism between X_n and a subspace of Y for which $c\|x\| \le \|x^*\| \le d\|x\|$ if x in X_n corresponds to x^* in Y. See ISOMORPHISM, SUPER-REFLEXIVE.

FISCHER, Ernst Sigismund (1875–1954). Algebraist and analyst; born in Austria and lived in Czechoslovakia and Germany.

    Riesz-Fischer theorem. See RIESZ.

FISHER, Ronald Aylmer (1890–1962). British statistician.

Fisher's *z.* The correlation-coefficient transformation

$$z(r) = \tfrac{1}{2}\log_e \frac{1+r}{1-r} = \tanh^{-1} r,$$

where *r* is the correlation coefficient. If the random samples are from a bivariate normal population, the distribution of *z* approaches normality more rapidly than that of the correlation coefficient itself. The *mean* of *z* is approximately $z(\rho)$ and the *variance* approximately $1/(n-3)$ if the size *n* of the sample is large, where ρ is the population correlation coefficient.

Fisher's *z* **distribution.** See F–F distribution.

FIT'TING, *n.* **curve fitting.** See EMPIRICAL — empirical curve, LEAST — method of least squares.

FIVE, *n.* **divisibility by five.** See DIVISIBLE.

FIXED, *adj.* **fixed assets.** See ASSETS.

    fixed point. A point which is not moved by a given transformation or mapping. *E.g.,* $x = 3$ is a fixed point of the transformation $T(x) = 4x - 9$.

    fixed-point theorem. A theorem which states that under certain conditions a mapping *T* of a subset of *E* into *E* has at least one fixed point. For examples, see under BANACH, BROUWER, POINCARÉ.

    fixed value of a letter or quantity. A value that does not change during a given discussion or series of discussions; not arbitrary. In an expression containing several letters, some may be *fixed* and others subjected to being assigned certain values or to taking on certain values by virtue of their place in the expression; *e.g.,* if in $y = mx + b$, *b* is *fixed*, *m* arbitrary and *x* and *y* variables, the equation represents the pencil of lines through the point whose coordinates are $(0, b)$. When *m* has also been assigned a particular value, *x* and *y* are then thought of as taking on all pairs of values which are coordinates of points on a particular line.

FLAT, *adj.* **flat angle.** Same as *straight angle*, that is, an angle of $180°$.

    flat price of a bond. See PRICE.

FLEC'NODE, *n.* A *node* which is also a point of inflection on one of the branches of the curve that touch each other at the node.

FLEX, *n.* Same as INFLECTION.

FLEX'ION, *n.* A name sometimes used for the rate of change of the slope of a curve; the second derivative of a function.

FLIP-FLOP, *adj.* See CIRCUIT.

FLOAT'ING, *p.* **floating decimal point.** A term applied in machine computation when the decimal point is not fixed at a certain machine position throughout the computation but is placed by the machine as each operation is performed.

FLOW, *adj.* **flow chart.** See CHART.

FLUC'TU-A'TION, *n.* Same as VARIATION.

FLU'ENT, *n.* Newton's name for a varying, or "flowing," function.

FLU'ID, *adj., n.* **fluid pressure.** See PRESSURE —fluid pressure.

    mechanics of fluids. See MECHANICS.

FLUX'ION, *n.* Newton's name for the rate of change or derivative of a "fluent." It is denoted in Newton's writing by a letter with a dot over it.

FO'CAL, *adj.* **focal chord of a conic.** A chord passing through a focus of the conic. A **focal radius** is a line segment joining a focus to a point on the conic.

    focal point. For an integral $I = \int_{x_1}^{x_2} \cdot f(x, y, y')\,dx$ and a curve *C*, the *focal point* of *C* on the *transversal T* is the point of contact of *T* with the envelope of transversals of *C*. In order for an arc $[(x_1, y_1), (x_2, y_2)]$ of *T*, with (x_2, y_2) on *C*, to minimize *I*, the focal point of *C* on *T* must not lie between (x_1, y_1) and (x_2, y_2) on *T*. See TRANSVERSALITY —transversality condition.

    focal property of conics. See under ELLIPSE, HYPERBOLA, PARABOLA.

FO'CUS, *n.* [*pl.* **foci** or **focuses**]. See ELLIPSE, HYPERBOLA, and PARABOLA.

FOLD, *n.* See CATASTROPHE.

FO'LI-UM, *n.* **folium of Descartes.** A plane cubic curve consisting of a single loop, a node, and two branches asymptotic to the same line. Its rectangular Cartesian equation is $x^3 + y^3 = 3axy$, where the curve passes through the origin and is asymptotic to the line $x + y + a = 0$.

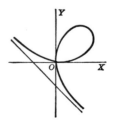

FONTANA, Niccolò. Same as TARTAGLIA. See TARTAGLIA.

FOOT, *n.* (1) A unit of linear measure, equal to 12 inches or 0.3048 meters. See DENOMINATE NUMBERS in the appendix. (2) The point of intersection of a line with another line or a plane. In particular, the *foot of a perpendicular to a line* is the point of intersection of the line and the perpendicular; the *foot of a perpendicular to a plane* is the point in which the perpendicular cuts the plane.

    foot-pound. A unit of work; the work done when a body weighing one pound is lifted one foot. More precisely, the work done when a force of one pound produces a displacement of one foot in the direction of the force. See HORSE-POWER.

FORCE, *n.* That which pushes, pulls, compresses, distends, or distorts in any way; that which changes the state of rest or state of motion of a body. *Tech.* The time rate of change of momentum of a body. If the mass of the body does not change with time, the force $\mathbf{F}$ is proportional to the product of the mass m by its vector acceleration $\mathbf{a}$. See NEWTON—Newton's laws of motion (2).

    centrifugal force. See CENTRIFUGAL.
    centripetal force. See CENTRIPETAL.
    conservative force. See CONSERVATIVE.
    electromotive force. See ELECTROMOTIVE.
    field of force. A region of space endowed with the property that a physical object of the proper sort will experience a force acting on it if placed at any point of the region. *E.g.*, if a stationary electric charge would experience a force if placed at any point of a certain region, we should speak of the region as bearing an *electrostatic field*. Similarly, if it is a point-mass or isolated magnetic pole that is acted on, then we have a *gravitational field* or a *magnetic field*, as the case may be.
    force of interest. See INTEREST.
    force of mortality. See MORTALITY.
    force vector. A vector equal in magnitude to the magnitude of a given force and having its direction parallel to the line of action of the force. See PARALLELOGRAM —parallelogram of forces.
    moment of a force. See MOMENT.
    parallelogram of forces. See PARALLELO-GRAM—parallelogram of forces.
    projection of a force. See PROJECTION —orthogonal projection.
    tube of force. A tube whose boundary surface is made up of lines of force. In general, if C is a closed curve, no part of which is a line of force,

and if through each point of C there passes a line of force, then the collection of these lines will make up the boundary of a *tube of force.*

    unit force. The force which will give unit acceleration to a unit mass. The force which, acting on a mass of one gram for one second, increases its velocity by one centimeter per second is called a force of one *dyne*. The force which, acting on a mass of one pound for one second, will increase the velocity of the mass one foot per second is called a force of one *poundal.*

FORCED, *p.* **forced oscillations and vibrations.** See OSCILLATION.

FOR′EIGN, *adj.* **foreign exchange.** See EXCHANGE.

FORM, *n.* (1) A mathematical expression of a certain type; see STANDARD —standard form of an equation. (2) A homogeneous polynomial expression in two or more variables. In particular, a **bilinear form** is a polynomial of the second degree which is homogeneous of the first degree in variables $x_1, x_2, \cdots, x_n$ and in variables $y_1, y_2, \cdots, y_n$; a polynomial of the form

$$P(x, y) = \sum_{i,j=1}^{n} a_{ij} x_i y_j.$$

If $x_1, \cdots, x_n$; $y_1, \cdots, y_n$; $\cdots$; $z_1, \cdots, z_n$ are m sets of variables, then an expression of the form $\sum a_{ij\cdots k} x_i y_j \cdots z_k$ is a **multilinear form** of order m (the form is **linear, bilinear, trilinear**, etc., according as $m = 1, 2, 3$, etc.). A **quadratic form** is a homogeneous polynomial of the second degree, a polynomial of the form $\sum_{i,j=1}^{n} a_{ij} x_i x_j$. If it is positive for all real values (not all zero) of the variables $\{x_i\}$ it is a **positive definite** quadratic form; if it is positive or zero it is **positive semidefinite**. See DISCRIMINANT—discriminant of a quadratic form, TRANSFORMATION —congruent transformation.

    standard form. See STANDARD—standard form of an equation.

FORM′AL, *adj.* **formal derivative.** For a polynomial $a_0 x^n + a_1 x^{n-1} + \cdots + a_{n-1} x + a_n$ with coefficients in some ring (or in any structure for which multiplication by integers is defined), the formal derivative is the polynomial $n a_0 x^{n-1} + (n-1) a_1 x^{n-2} + \cdots + a_{n-1}$. See SEPARABLE —separable polynomial.

    formal power series. See SERIES —power series.

FOR'MU-LA, *n.* [*pl.* **formulas** or **formulae**]. A general answer, rule, or principle stated in mathematical language. See EMPIRICAL—empirical formula, INTEGRATION, and various headings under SPHERICAL and TRIGONOMETRY.

FOUR, *adj., n.* **four-color problem.** The problem of answering the question raised about 1852 of whether any map on a sphere or plane can be colored with four colors so that no two countries having a common boundary line have the same color. It can be assumed that the boundary of each country is a simple closed curve (or even a polygon), which implies that each country is connected; *i.e.*, that it is possible to go between any two points of a given country without leaving that country. If two countries share only a finite number of points, they may use the same color. With the help of many hours of computing on a very large computer, Kenneth Appel and Wolfgang Haken proved in 1976 that four colors suffice for all such maps. It had been known since 1890 that 5 colors suffice and that there are maps for which 3 colors are insufficient. The **chromatic number** of a surface is the least number of colors that suffice to color any map on that surface. Except for the Klein bottle, the chromatic number of a surface is equal to the greatest integer not greater than $\frac{1}{2}(7 + \sqrt{49 - 24\chi})$, where χ is the *Euler characteristic* of the surface. Since $\chi = 0$ for the cylinder, Möbius strip, and torus, their chromatic numbers are 7. But $\chi = 0$ for the Klein bottle and its chromatic number is 6. For the projective plane, $\chi = 1$ and the chromatic number is 6. For the plane or sphere, $\chi = 2$ and the chromatic number is 4. For the plane or sphere, if each country has at most one "colony" (which must be colored the same color as the country), then at most 12 colors are needed and there is a map which requires 12 colors (the **twelve-color problem**).

four-group. The non-cyclic group of order 4. It is a commutative group and is the group of motions (symmetries) of 3-dimensional space that preserve a rectangle. It has members x and y for which $x^2 = y^2 = e$ and $xy = yx$. See GROUP, SYMMETRY —group of symmetries.

FOURIER, Jean Baptiste Joseph, Baron de (1768–1830). French analyst and mathematical physicist. Famous for the widely applicable series that bears his name. Initiated an important phase of mathematical physics with investigations of heat conduction.

discrete Fourier transform. A function that assigns to a sequence $z = (z_0, z_1, \cdots, z_{n-1})$ of complex numbers the sequence

$$w = (w_0, w_1, \cdots, w_{n-1})$$

for which

$$w_s = \alpha n^{-1} \sum_{n=0}^{n-1} z_r e^{2\pi i r s / n} \qquad \text{for each } s.$$

The inverse is the transformation

$$z_r = \alpha^{-1} \sum_{n=0}^{m-1} w_s e^{-2\pi i r s / n}.$$

Sometimes α is taken to be 1, sometimes n, sometimes $n^{1/2}$. These transformations can be written as $w = Fz$ and $z = F^{-1}w$, where F is the product of αn^{-1} and the $n \times n$ *Vandermonde matrix* whose element in row r and column s is ω^{rs}, where rows and columns are labeled from 0 to $n - 1$ and $\omega = e^{2\pi i / n}$ is an nth root of unity; F^{-1} is the product of α^{-1} and the Vandermonde matrix that has ω^{-rs} in row r and column s.

fast Fourier transform. The calculation of a *discrete Fourier transform* by making clever use of certain algebraic properties of the roots of unity that are elements of the Vandermonde matrix that describes the transform (see above, discrete Fourier transform). If the matrix is $n \times n$, direct calculation of the transform uses "of the order of" n^2 multiplications. If $n = n_1 n_2 \cdots n_p$, then factoring the matrix can reduce the number of multiplications to "of the order of" $n(n_1 + \cdots + n_p)$, which gives the estimate $2n \log_2 n$ if n is a power of 2. As with other transforms, there are problems that can be transformed into problems that can be solved more easily; then, using the inverse transform, the solution can be brought back to the desired solution. *E.g.*, the multiplication of large integers, with applications to primality testing and cryptography.

Fourier half-range series. *A Fourier series* of the form

$$\frac{1}{2}a_0 + a_1 \cos x + a_2 \cos 2x + \cdots$$

$$= \frac{1}{2}a_0 + \sum_{n=1}^{\infty} a_n \cos nx,$$

or

$$b_1 \sin x + b_2 \sin 2x + \cdots = \sum_{n=1}^{\infty} b_n \sin nx.$$

These are also called the **cosine** and **sine** series, respectively. Since the cosine is an *even* function, the cosine series can represent a function f on the whole interval $-\pi < x < \pi$ only if f is an even function, i.e., $f(-x) = f(x)$. Likewise, since the sine is an *odd* function, the sine series can represent a function f on the whole interval $-\pi < x < \pi$ only if f is an odd function, i.e., $f(-x) = -f(x)$.

Fourier integral theorem. If f has at most a finite number of infinite discontinuities, if f is integrable on any finite interval not containing one of these discontinuities, and if $\int_{-\infty}^{\infty} |f(x)| \, dx$ exists, then f can be represented as

$$f(x) \approx \frac{1}{2\pi} \int_{-\infty}^{\infty} dt \int_{-\infty}^{\infty} f(s) \cos t(x-s) \, ds,$$

where the right member of this expression has the value

$$\lim_{h \to 0} \tfrac{1}{2}[f(x+h) + f(x-h)]$$

if f is of bounded variation in a neighborhood of x.

Fourier series. A series of the form

$$\tfrac{1}{2}a_0 + (a_1 \cos x + b_1 \sin x)$$

$$+ (a_2 \cos 2x + b_2 \sin 2x) + \cdots$$

$$= \tfrac{1}{2}a_0 + \sum_{n=1}^{\infty} (a_n \cos nx + b_n \sin nx),$$

for which there is a function f such that for $n \geq 0$

$$a_n = \frac{1}{\pi} \int_{-\pi}^{\pi} f(x) \cos nx \, dx,$$

and for $n \geq 1$

$$b_n = \frac{1}{\pi} \int_{-n}^{n} f(x) \sin nx \, dx.$$

The marked characteristic of a Fourier series is that it can be used to represent functions that ordinarily are represented by different expressions in different parts of the interval, the functions being subject only to certain very general restrictions (see below, Fourier's theorem). Since the sine and cosine each have a period of 2π, the Fourier series has a period of 2π. E.g., if f is defined by the relations $f(x) = 1$ when $-\pi \leq x \leq 0$, $f(x) = 2$ when $0 < x \leq \pi$, then

$$\pi a_0 = \int_{-\pi}^{\pi} f(x) \, dx = \int_{-\pi}^{0} dx + \int_{0}^{\pi} 2 \, dx = 3\pi,$$

from which $a_0 = 3$. Similarly, $a_n = 0$ for all n, $b_n = 0$ for n even and $b_n = 2/(n\pi)$ for n odd. Whence

$$f(x) = \tfrac{3}{2} + \frac{2}{\pi}\left(\sin x + \tfrac{1}{3}\sin 3x\right.$$

$$\left. + \tfrac{1}{5}\sin 5x + \cdots\right).$$

The **complex Fourier series** for a function f that is defined on the interval $(-\pi, \pi)$ is

$$\sum_{-\infty}^{\infty} c_n e^{inx}, \qquad c_n = \frac{1}{2\pi} \int_{-\pi}^{\pi} f(x) e^{-inx} \, dx.$$

This is related to the above Fourier series by

$$c_n = \tfrac{1}{2}(a_n - ib_n), \qquad c_{-n} = \tfrac{1}{2}(a_n + ib_n),$$

$$\text{for } n \geq 0.$$

The sequence $\{e^{inx}: n = 0, \pm 1, \pm 2, \cdots\}$ is an *orthogonal system* on the interval $(-\pi, \pi)$. Series on intervals other than $(-\pi, \pi)$, derived from the above types of Fourier series, are also called Fourier series. See ORTHOGONAL —orthogonal functions, and below, Fourier's theorem.

Fourier's theorem. Suppose f and $|f|$ are integrable on $[-\pi, \pi]$ and f is extended outside the interval containing all x with $-\pi < x \leq \pi$ so as to be periodic with period 2π. If at least one of the following conditions (i) through (v) is satisfied, and if a_n and b_n are defined by

$$\pi a_n = \int_{-\pi}^{\pi} f(x) \cos nx \, dx,$$

$$\pi b_n = \int_{-\pi}^{\pi} f(x) \sin nx \, dx,$$

then the series

$$\tfrac{1}{2}a_0 + \sum_{n=1}^{\infty} (a_n \cos nx + b_n \sin nx)$$

converges to $f(x)$ if f is continuous at x, and converges to $\tfrac{1}{2}[f(x+) + f(x-)]$ whether or not f is continuous at x, where $f(x+)$ and $f(x-)$ denote the limits of f at x on the right and left, respectively. (i) **(Dirichlet condition)** f is bounded and has only a finite number of maxima and minima and a finite number of discontinuities on the interval $[-\pi, \pi]$; (ii) there is an interval I with x as midpoint such that f is bounded on I and monotone on each of the open halves of I; (iii) **(Jordan condition)** there is a neighborhood of x on which f is of bounded variation; (iv) (**Dini condition**) both $f(x+)$ and $f(x-)$ exist and there is a positive number δ such that the following function is integrable on $[-\delta, \delta]$:

$$\left| \frac{f(x+t) - f(x+)}{t} + \frac{f(x-t) - f(x-)}{t} \right|;$$

(v) f is differentiable on the right and left at x. There are functions that are integrable (belong to L^1), but whose Fourier series diverge everywhere. However, if f is square-integrable (belongs to L^2), then the Fourier series for f converges to $f(x)$ almost everywhere (*i.e.*, except for a set of measure zero). See BANACH—Banach space, FEJÉR—Fejér's theorem, KERNEL—Dirichlet kernel, Fejér kernel, LOCALIZATION —localization principle.

Fourier transform. A function f is the **Fourier transform** of g if

$$f(x) = \frac{1}{\sqrt{2\pi}} \int_{-\infty}^{\infty} g(t) e^{itx} \, dt$$

(the factor $1/\sqrt{2\pi}$ is sometimes omitted). Under suitable conditions on g (*e.g.*, those given above for the Fourier integral theorem), it then follows that

$$g(x) \approx \frac{1}{\sqrt{2\pi}} \int_{-\infty}^{\infty} f(t) e^{-itx} \, dt,$$

where the value of the right member is

$$\lim_{h \to 0} \tfrac{1}{2} [g(x+h) + g(x-h)]$$

if g is of *bounded variation* in the neighborhood of x. Such functions f and g are sometimes said to be a *pair of Fourier transforms*. A function f is the *Fourier cosine transform* of g, or the *Fourier sine transform* of g, according as

$$f(x) = \sqrt{\frac{2}{\pi}} \int_0^{\infty} g(t) \cos xt \, dt,$$

or

$$f(x) = \sqrt{\frac{2}{\pi}} \int_0^{\infty} g(t) \sin xt \, dt.$$

These transformations of g are inverses of themselves.

FRAC'TAL, *n.* Originally, "fractal" was used to describe any irregular set (or pattern) whose *Hausdorff dimension* is not an integer and therefore is greater than its *topological dimension*. It has come to mean a set that is self-similar under magnification. *I.e.*, a **fractal** is a set F which has the property that F is the union of sets $\{F_i\}$ for which each F_i is similar to F and $F_i \cap F_j$ ($i \neq j$) is empty (or negligible in some sense), each F_i also can be decomposed in this way, and this can be continued indefinitely (see SIMILAR). The *Cantor set* C is a fractal; its topological dimension is 0 and its Hausdorff dimension is $\log 2/\log 3$. Also, the part of C in the interval $[0, \frac{1}{3}]$ is congruent to the part in $[\frac{2}{3}, 1]$ and each is similar to the whole. Each neighborhood of a point of the Cantor set contains a set that is similar to the entire Cantor set (see CANTOR—Cantor set). Variations of the Cantor-set process can produce other interesting fractals. *E.g.*, replace each deleted third when forming the Cantor set by a circle with that diameter; or divide a square into nine congruent squares, replace the middle square by an inscribed circle (or some other object), repeat this for each of the other eight squares, and continue this ad infinitum—perhaps making it more "artistic" by coloring the circles in some way. For other examples of fractals, see JULIA—Julia set, MANDELBROT —Mandelbrot set.

fractal dimension. See DIMENSION.

FRAC'TION, *n.* An indicated quotient of two quantities. The dividend is the **numerator** and the divisor the **denominator** (the numerator of $\frac{3}{4}$ is 3 and the denominator is 4). A **simple fraction** (or **common fraction** or **vulgar fraction**) is a fraction whose numerator and denominator are both integers, as contrasted to a **complex fraction** which has a fraction for the numerator or denominator or both. If the numerator of a simple fraction is 1, it is a **unit fraction**. A complex fraction can be changed to a simple fraction by inverting the denominator and multiplying (*i.e.*, multiplying both numerator and denominator by the reciprocal of the denominator), or by multiplying numerator and denominator by the least common multiple of all denominators in the complex fraction; *e.g.*,

$$\tfrac{1}{3} \Big/ \Big(\tfrac{1}{2} + \tfrac{1}{4} \Big) = \tfrac{1}{3} \Big/ \tfrac{3}{4} = \tfrac{1}{3} \times \tfrac{4}{3} = \tfrac{4}{9},$$

$$\tfrac{1}{3} \Big/ \Big(\tfrac{1}{2} + \tfrac{1}{4} \Big) = \Big(12 \times \tfrac{1}{3} \Big) \Big/ \Big[12 \Big(\tfrac{1}{2} + \tfrac{1}{4} \Big) \Big] = \tfrac{4}{9}.$$

Two simple fractions are **similar** if they have the same denominators. A fraction whose numerator and denominator are real numbers is a **rational**

fraction if the numerator and denominator are rational numbers; it is a **proper fraction** if its numerator is less in absolute value than its denominator, and an **improper fraction** if the numerator is greater than (or equal to) the denominator in absolute value ($\frac{2}{3}$ is a proper fraction and $\frac{4}{3}$ is an improper fraction). A fraction whose numerator and denominator are both polynomials is a **rational fraction** (or **rational function**); it is a **proper fraction** if the numerator is of lower degree than the denominator, and an **improper fraction** otherwise; $x/(x^2 + 1)$ is a proper fraction and $x^2/(x + 1)$ is an improper fraction, while $(x - y)/xy$ is a proper fraction in x and y together, but not in either x or y alone. See SOLIDUS.

addition, subtraction, multiplication, and division of fractions. See SUM—sum of real numbers, SUBTRACTION, PRODUCT—product of real numbers, DIVISION.

clearing of fractions. Multiplying both members of an equation by a common denominator of the fractions. See EXTRANEOUS —extraneous root.

continued fraction. A number plus a fraction whose denominator is a number plus a fraction, etc., such as:

$$a_1 + b_2 \over a_2 + b_3 \over a_3 + b_4 \over a_4 + b_5 \over a_5, \text{ etc.}$$

A continued fraction may have either a finite or an infinite number of terms. In the former case it is said to be a **terminating** continued fraction; in the latter, **nonterminating**. If a certain sequence of the a's and b's occurs periodically, the continued fraction is **recurring** or **periodic**. The terminating continued fractions

$$a_1, \quad a_1 + b_2 \over a_2, \quad a_1 + b_2 \over a_2 + b_3 \over a_3, \quad \text{etc.},$$

are **convergents** of the continued fraction. The quotients b_2/a_2, b_3/a_3, etc., are **partial quotients**.

decimal fraction. See DECIMAL.

decomposition of a fraction. Writing the fraction as a sum of partial fractions.

partial fraction. See PARTIAL—partial fractions.

FRAC′TION-AL, *adj.* **fractional equation.** (1) An equation containing fractions of any sort, such as $\frac{1}{2}x + 2x = 1$. (2) An equation containing fractions with the variable in the denominator, such as $x^2 + 2x + 1/x^2 = 0$.

fractional exponent. See EXPONENT.

FRAME, *n.* **frame of reference.** Any set of lines or curves *in a plane*, by means of which the position of any point *in the plane* may be uniquely described; any set of planes or surfaces by means of which the position of a point *in space* may be uniquely described.

FRANCESCA, Piero della (1416–1492). Italian painter and mathematician. Wrote extensively on perspective and was regarded by his contemporaries as the scientific painter *par excellence*.

FRÉCHET, René Maurice (1878–1973). French analyst, probabilist, and topologist.

Fréchet space. (1) A T_1-space (see TOPOLOGICAL—topological space). (2) A linear topological space that is locally convex, metrizable, and complete (sometimes it is not required that the space be locally convex). See VECTOR —vector space.

FREDHOLM, Erik Ivar (1866–1927). Swedish analyst and physicist.

Fredholm determinant. The Fredholm determinant for the kernel K is the power series in λ defined as

$$D(\lambda) = 1 - \lambda \int_a^b K(t,t)\,dt$$

$$+ \frac{\lambda^2}{2!} \int_a^b \int_a^b \begin{vmatrix} K(t_1,t_1) & K(t_1,t_2) \\ K(t_2,t_1) & K(t_2,t_2) \end{vmatrix} dt_1\,dt_2$$

$$- \frac{\lambda^3}{3!} \int_a^b \int_a^b \int_a^b \begin{vmatrix} K(t_1,t_1) \cdots K(t_1,t_3) \\ \cdots\cdots\cdots\cdots\cdots\cdots \\ K(t_3,t_1) \cdots K(t_3,t_3) \end{vmatrix}$$

$$\times dt_1\,dt_2\,dt_3 + \cdots .$$

Fredholm integral equations. The *Fredholm integral equation of the first kind* is the equation $f(x) = \int_a^b K(x,t)y(t)\,dt$, and the *Fredholm integral equation of the second kind* is $y(x) = f(x) + \lambda \int_a^b K(x,t)y(t)\,dt$, in which f and K are two given functions and y is the unknown function. The function K is the **kernel** or **nucleus** of the equation. The Fredholm equation of the second kind is **homogeneous** if $f(x) \equiv 0$. Also called the **integral equations of the first and second kind**, respectively. The above is sometimes modified by letting $\lambda = 1$.

Fredholm minors. The *first Fredholm minor* $D(x, y; \lambda)$ for the *kernel* $K(x, y)$ is

$$D(x, y, \lambda)$$

$$= \lambda K(x, y) - \lambda^2 \int_a^b \begin{vmatrix} K(x, y) K(x, t) \\ K(t, y) K(t, t) \end{vmatrix} dt$$

$$+ \frac{\lambda^3}{2!} \int_a^b \int_a^b \begin{vmatrix} K(x, y) K(x, t_1) K(x, t_2) \\ K(t_1, y) K(t_1, t_1) K(t_1, t_2) \\ K(t_2, y) K(t_2, t_1) K(t_2, t_2) \end{vmatrix}$$

$$\times dt_1 dt_2 + \cdots .$$

The higher Fredholm minors are defined in a similar way. This definition is sometimes modified by letting $\lambda = 1$.

Fredholm solution of the Fredholm integral equation of the second kind. If f is a continuous function of x for $a \le x \le b$ and K is a continuous function of x and t for $a \le x \le b$, $a \le t \le b$, and if the *Fredholm determinant* $D(\lambda)$ of the *kernel* $K(x, t)$ is not zero, then the *Fredholm integral equation*

$$y(x) = f(x) + \lambda \int_a^b K(x, t) y(t) \, dt$$

has a unique continuous solution y given by

$$y(x) = f(x) + \frac{1}{D(\lambda)} \int_a^b D(x, t; \lambda) f(t) \, dt,$$

where $D(x, t; \lambda)$ is the *first Fredholm minor* for the kernel $K(x, t)$ and $D(\lambda)$ is the *Fredholm determinant* for $K(x, t)$. See LIOUVILLE—Liouville-Neumann series, HILBERT—Hilbert-Schmidt theory of integral equations with symmetric kernels, VOLTERRA —Volterra's reciprocal functions.

FREE, *adj.* **free group.** A group is **free** if the group has a set of generators such that no product of generators and inverses of generators is equal to the identity unless it can be written as a product of expressions of type $a \cdot a^{-1}$. *E.g.*, if a free group has generators a and b, then expressions of type ab, aba, $a^{-1}babbab^{-1}$, etc., are all distinct members of the group. An **Abelian group** is **free** if no product of generators and inverses of generators is equal to the identity unless it can be reduced to a product of expressions of type $a \cdot a^{-1}$ by use of the commutative law. If an Abelian group has a finite number of generators,

then it is free if and only if no element is of finite period (the group is then a direct product of infinite cyclic groups). A **free element** of a group is an element that is not of finite period.

free ultrafilter. See FILTER.

FREEDMAN, Michael (1951–). American mathematician awarded a Fields Medal (1986) for his proof of the Poincaré conjecture for dimension 4.

FREE'DOM, *n.* **degrees of freedom.** The number of values of coordinates or parameters that are necessary to determine an object or system. *E.g.*, a point on a line has one degree of freedom, a point in a plane or on the surface of a sphere has two degrees of freedom, and a moving point in space has six degrees of freedom (three coordinates are needed to specify its position and three components of velocity are needed to specify its velocity). In *statistics*, a statistic based on n independent random variables is said to have n degrees of freedom; *e.g.*, the chi-square random variable that is the sum of squares of n independent normal random variables with means 0 and variances 1. A random sample of size n has n degree of freedom and a statistic calculated from it has n degrees of freedom. However, the statistic $\sum_{i=1}^{n} (x_i - \bar{x})^2$, where $\bar{x}$ is the sample mean, is said to have $n - 1$ degrees of freedom since the n random variables $x_i - \bar{x}$ are related by $\sum_{i=1}^{n} (x_i - \bar{x}) = 0$ and thus $\sum_{i=1}^{n} (x_i - \bar{x})^2$ can be expressed as a function of $n - 1$ random variables [it also is a function of the single random variable $\sum_{i=1}^{n} (x_i - \bar{x})^2$]— this might be defended on the grounds the sample mean is nearly constant. See CHI—chi-square distribution, F—F distribution, T—t distribution.

FRENCH, *adj.* **French horsepower.** See HORSEPOWER.

FRENET, Jean Frédéric (1816–1900). French differential geometer.

Frenet-Serret formulas. The central formulas in the theory of space curves. If α, β, γ denote the unit vectors along the tangent, principal normal, and binormal, respectively, of a space curve C, and ρ and τ its radii of curvature and torsion, then the formulas are $\dfrac{d\alpha}{ds} = \dfrac{\beta}{\rho}$, $\dfrac{d\beta}{ds} = -\dfrac{\alpha}{\rho} - \dfrac{\gamma}{\tau}$, $\dfrac{d\gamma}{ds} = \dfrac{\beta}{\tau}$, where s denotes the arclength.

FRE′QUEN-CY, *adj., n.* For a collection of data, the number of items in a given category. *E.g.,* if the number of grades on an examination in the ranges 0–24, 25–49, 50–74, and 75–100 are 2, 10, 20, and 8, respectively, then the absolute frequencies for these groups are 2, 10, 20, and 8. The relative frequencies of $\frac{1}{20}, \frac{1}{4}, \frac{1}{2},$ and $\frac{1}{5}$ are obtained by dividing the frequencies by 40, the total number of examinations. When a collection of data is separated into several categories, the number of items in a given category is the **absolute frequency** and the absolute frequency divided by the total number of items is the **relative frequency.** The **cumulative frequency** is the sum of all preceding frequencies, a certain order having been established. *E.g.,* for the absolute frequencies of 2, 10, 20, and 8, the cumulative frequencies are 2, 12, 32, and 40. For a variable x that can have values x_i, the sum of the absolute (or relative) frequencies of values of x equal to or less than x_i is the upward cumulative absolute (or relative) frequency of x. It may also be cumulated in a downward direction. The upward cumulative relative frequency is called a **distribution function.** See DISTRIBUTION —distribution function, PROBABILITY —probability function.

frequency curve or diagram. A graphic picture of a frequency distribution, or of a set of frequencies for the various values of a variable. The ordinates of a **frequency curve** are proportional to the frequencies for the various values of the variable which are noted on the horizontal axis. Customarily, the area under the curve depicts the total frequency, while the ratio of the area over an interval to the total area is the relative frequency for that interval. *E.g.,* suppose a variable has values between 0 and 100, indicated on a horizontal axis. For each of the intervals $0 \le x < 10$, $10 \le x < 20, \cdots, 90 \le x < 100$, called **class intervals,** construct a vertical bar or rectangle with that interval as base and area proportional to the frequency of that class. This diagram is a **histogram.** More generally, to construct a **histogram** one divides the range of a variable into successive intervals (not necessarily of equal lengths) and draws rectangles with the intervals as bases and areas proportional to the frequencies for the classes described by the intervals. If the intervals are equal, then the altitudes of the rectangles are proportional to the frequencies and a **frequency polygon** is obtained

if one joins by line segments the midpoints of the top sides of the rectangles.

frequency function. For a variable x that has only a finite or denumerably infinite number of values, the **absolute frequency function** is the function f for which $f(x_i)$ is the absolute frequency of x_i; the **relative frequency function** is the function g for which $g(x_i)$ is the relative frequency of x_i. For a *random variable* with possible values $X_1, X_2, \cdots, X_n$, the **frequency function** is the function p for which $p(X_i)$ is the probability of X_i; sometimes this is called a probability function, but strictly it is the restriction of the probability function to the events $X_1, X_2, \cdots, X_n$. See PROBABILITY—probability function.

frequency of a periodic function. See PERIODIC—periodic function of a real variable.

frequency polygon. See above, frequency curve or diagram.

FRESNEL, Augustin Jean (1788–1827). French physicist and engineer.

Fresnel integrals. (1) The integrals

$$\int_0^x \sin t^2 \, dt \quad \text{and} \quad \int_0^x \cos t^2 \, dt,$$

called the **Fresnel sine** and the **Fresnel cosine integrals.** These are equal, respectively, to

$$\frac{x^3}{3} - \frac{x^7}{7.3!} + \frac{x^{11}}{11 \cdot 5!} - \cdots$$

and

$$x - \frac{x^5}{5 \cdot 2!} + \frac{x^9}{9 \cdot 4!} - \cdots,$$

which converge for all values of x. (2) The integrals

$$\int_x^\infty \frac{\cos t}{t^{1/2}} \, dt \quad \text{and} \quad \int_x^\infty \frac{\sin t}{t^{1/2}} \, dt,$$

which are equal, respectively, to

$$(U \cos x - V \sin x)$$

and

$$(U \sin x + V \cos x),$$

where

$$U = \frac{1}{x} \left(\frac{1}{x} - \frac{3!}{x^3} + \frac{5!}{x^5} - \cdots \right)$$

and

$$V = \frac{1}{x}\left(1 - \frac{2!}{x^2} + \frac{4!}{x^4} - \cdots\right).$$

FRIC′TION, *n.* **angle of friction.** See below, force of friction.

coefficient of friction. See below, force of friction.

force of friction. If two bodies are in contact and one of these is at rest, or in motion without acceleration, relative to the other, then the external forces acting on A are balanced by a normal reaction force **N** perpendicular to the plane of contact and a force of friction **F** in the plane of contact. When A is on the verge of moving, the acute angle α is the **angle of friction,** and $\tan \alpha = \dfrac{F}{N} = \mu$ is the **coefficient of static friction.** When A is moving without acceleration relative to the other body, μ is the **coefficient of kinetic (sliding) friction.**

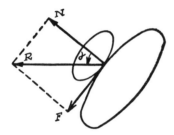

FROBENIUS, Ferdinand Georg (1849–1917). German algebraist, analyst, and group theorist. See SYLOW—Sylow's theorem.

Frobenius theorem. If D is a division algebra over the field of real numbers for which each member of D satisfies a polynomial equation with real coefficients, then D is isomorphic to the field of *real numbers*, the field of *complex numbers*, or the division algebra of *quaternions*. The theorem can be generalized in the sense that if the restrictions on D are reduced by removing the assumption that multiplication is associative, then the only additional possibility for D is the *Cayley algebra*. See CAYLEY.

FRON-TIER′, *n.* **frontier of a set.** See INTERIOR —interior of a set.

FRUS′TUM, *n.* **frustum of any solid.** The part of the solid between two parallel planes cutting the solid. See PYRAMID, CONE.

FUBINI, Guido (1879–1943). Italian analyst, algebraist, and projective differential geometer.

Fubini theorem. Let m_1 and m_2 be *measures* defined on spaces X and Y, and $m_1 \times m_2$ be the product measure defined on the *Cartesian product* $X \times Y$. Fubini's theorem states that if h is integrable on $X \times Y$, then the subset of Y for which g is not integrable on X, if $g(x) = h(x, y)$, is of *measure zero*; the subset of X for which f is not integrable on Y, if $f(y) = h(x, y)$, is of *measure zero*; and

$$\int h\, d(m_1 \times m_2) = \int F\, dm_1 = \int G\, dm_2,$$

where

$$F(x) = \int h(x, y)\, dm_2$$

and

$$G(y) = \int h(x, y)\, dm_1.$$

Such results as the following are sometimes referred to as parts of the Fubini theorem: "If S is a measurable subset of $X \times Y$, then the set of points y in Y for which S_y is not measurable is of measure zero (S_y is the set of all points x of X for which (x, y) is in S)." See SIERPINSKI —Sierpinski set.

FUL′CRUM, *n.* The point at which a lever is supported. See LEVER.

FUNC′TION, *n.* An association of exactly one object from one set (the **range**) with each object from another set (the **domain**). E.g., it can be said that a person's age is a function of the person, and that the domain of this function is the set of all human beings and the range is the set of all integers which are ages of persons presently living. The area of a circle is a function of the radius; the sine of an angle is a function of the angle; the logarithm of a number is a function of the number. The expression $y = 3x^2 + 7$ defines y as a function of x when it is specified that the domain is (for example) the set of real numbers; y is then a function of x, a value of y is associated with each real-number value of x by multiplying the square of x by 3 and adding 7 (the range of this function is the set of all real numbers not less than 7), and x is said to be the **independent variable** or **argument** of the function (y is called the **dependent variable** or **value of the function**). If the equation $y = 3x^2 + 7$ is denoted by $y = f(x)$, then the value of y when

$x = 2$ is $f(2) = 3 \cdot 2^2 + 7 = 19$. Symbols such as f, F, ϕ, etc., are used to denote a function, the function values corresponding to x being denoted by $f(x)$, $F(x)$, $\phi(x)$, etc. and read as "f of x" or "the f function of x," etc. If a function f associates with each pair (x, y) of certain objects a single object $z = f(x, y)$, then f is said to be a **function of two variables** and both x and y are called **independent variables**. Similarly, a **function of several variables** is a function that associates with a set of objects $\{x_1, x_2, \cdots\}$ a single object $f(x_1, x_2, \cdots)$. E.g., the equation $z = 2x + xy$ defines z as a function of x and y. This function can be regarded either as a function of two variables x and y, or as a function of points (x, y). In this way, any function of several variables can be regarded as a function of one variable; e.g., a function $z = f(x_1, \cdots, x_n)$, where $x_1, \cdots, x_n$ can have any real number values, can be regarded as a function whose *domain* consists of the objects which are sequences $(x_1, \cdots, x_n)$ of n real numbers. A function can be described as a *mapping* from its domain D to its range R and one can say that the function *maps* a member x of D onto its *image y* in R. A function can also be regarded as being a set of ordered pairs (x, y). Each ordered pair is said to be an **element** of the function. The domain of the function is the collection of all objects which occur as the first member of some element, and the range is the collection of all objects which occur as the second member of some element. For a function f, a symbol such as x is used to denote a member of the domain, the symbol $f(x)$ denotes the corresponding member of the range, and f denotes the function itself—i.e., the rule associating $f(x)$ with x or simply the collection of all ordered pairs $(x, f(x))$. A function is **complex-valued, real-valued**, or **vector-valued** according as the range is a subset of the complex numbers, the real numbers, or a vector space. *Syn.* functional relation, map, mapping, operator, transformation. See CONTINUOUS —continuous function, GRAPH, HOMOMORPHISM, IMAGE, ISOMETRY, ISOMORPHISM, LINEAR—linear transformation and various headings below, particularly "multiple-valued function."

algebraic function. A function which can be generated by algebraic operations alone. *Tech.* (1) A function f such that, for some polynomial $P(x, y)$, it is true that $P[x, f(x)] \equiv 0$. Any polynomial is algebraic, but (for example) $\log x$ is not. (2) A many-valued analytic function w such that, for some irreducible polynomial $P(z, w)$ with complex coefficients, the values of $w(z)$ are the values of w for which $P(z, w) = 0$.

analytic function. See ANALYTIC—analytic function.

automorphic function. See AUTOMORPHIC.
Bessel function. See BESSEL.
beta function. See BETA.
characteristic function. See CHARACTERISTIC.
complementary function. See DIFFERENTIAL —linear differential equation.
composite function. See COMPOSITE.
continuous function. See CONTINUOUS.
decreasing function. See DECREASING.
dependent functions. See DEPENDENT.
entire function. See ENTIRE.
Euler's ϕ-function. See EULER.
even function. A function whose value does not change when the sign of the independent variable is changed; *i.e.*, a function f such that $f(-x) = f(x)$ for all x in the domain of f; x^2 and $\cos x$ are even functions, for $(-x)^2 = x^2$ and $\cos(-x) = \cos x$.

exponential function. (1) The function e^x [see e]. (2) The function a^x, where a is a positive constant. If $a \neq 1$, the function a^x is the inverse of the logarithmic function $\log_a x$. (3) A function in which the variable or variables appear in exponents and possibly also as a base, such as 2^{x+1} or x^x. For a complex number z, the function e^z may be defined either by $e^z = e^x(\cos y + i \sin y)$, where $z = x + iy$, or by $e^z = 1 + z + z^2/2! + z^3/3! + \cdots$. Two important properties of the exponential function are that $de^z/dz = e^z$ and $e^u e^v = e^{u+v}$. In fact, the exponential functions a^x restricted to real numbers are the only continuous functions which satisfy the functional equation $f(u + v) = f(u)f(v)$ for all real numbers u and v.

function of class C_n. A function f is of class C_n (or $C^{(n)}$) on a set S if, at each point of S, f is continuous and has continuous derivatives of all orders up to and including the nth. The functions of class C_0 are the continuous functions.

function of class L_p. A function f is of class L_p (or $L^{(p)}$) on a measurable set Ω if it is *measurable* and the integral of $|f(x)|^p$ over Ω is finite. The space L_p ($p \geq 1$) of all functions of class L_p is a complete normed vector space (*i.e.*, a *Banach space*) if addition and multiplication by scalars are taken as ordinary addition and multiplication,

$$\|f\| = \left[\int_\Omega |f|^p \, d\Omega \right]^{1/p}$$

is defined as the "length" or norm of f, and $f = g$ means $\|f - g\| = 0$ (or equivalently, $f(x) = g(x)$ except on a set of measure zero). Minkowksi's inequality is then equivalent to $\|f + g\| \leq \|f\| + \|g\|$, and Hölder's inequality is equivalent to

$$\int_{\Omega} |fg| \, d\Omega \le \|f\| \cdot \|g\|$$

if f is of class L_p and g of class L_q and $p + q = pq$ ($p > 1$, $q > 1$). See INTEGRABLE — integrable function, LEBESGUE —Lebesgue integral, MEASURE —measure of a set.

function element of an analytic function of a complex variable. See ANALYTIC —analytic continuation.

    function theory. See THEORY.

    gamma function. See GAMMA.

    Hamiltonian function. The sum of *kinetic energy* and *potential energy*.

    harmonic function. See HARMONIC.

    holomorphic function. See ANALYTIC —analytic function of a complex variable.

    hyperbolic functions. See HYPERBOLIC.

    implicit function. See IMPLICIT.

    increasing function. See INCREASING.

    integrable function. See INTEGRABLE.

    integral function. Same as ENTIRE FUNCTION.

    inverse of a function. See INVERSE.

    logarithmic function. A function defined by an expression of the form $\log f(x)$.

    measurable function. See MEASURABLE.

    meromorphic function. See MEROMORPHIC.

    monogenic function. See MONOGENIC.

    monotonic functions. Functions which are either monotonic increasing or monotonic decreasing. See MONOTONIC.

    multiple-valued function. An association of one or more objects from one set (the range) with each object from another set (the domain). A multiple-valued function can be regarded as being a set of ordered pairs (x, y). Then the domain is the set of all objects that are used as first members and the range is the set of all objects used as second members. Actually, a multiple-valued function is a *relation*, but is not a function unless it is single-valued, *i.e.*, to each point in the domain there corresponds exactly one point in the range. The relation defined by $x^2 + y^2 = 1$ is a **double-valued function** if y is regarded as a function of x, since

$$y = \pm (1 - x^2)^{1/2} \quad \text{if } |x| \le 1.$$

The relation defined by $x = \sin y$ is multiple-valued if y is regarded as the dependent variable, since if for particular numbers x and y we have $x = \sin y$, then

$$x = \sin\left[(-1)^n y + n\pi \right]$$

for any positive integer n. See FUNCTION, RELATION.

    odd function. A function whose sign changes, but whose absolute value does not change, when the sign of the independent variable is changed; *i.e.*, a function such that $f(-x) = -f(x)$ for all x in the domain of f; x^3 and $\sin x$ are odd functions, for

$$(-x)^3 = -x^3 \quad \text{and} \quad \sin(-x) = -\sin x.$$

    orthogonal functions. See ORTHOGONAL.

    periodic function. See PERIODIC.

    rational integral function of one variable. Same as POLYNOMIAL IN ONE VARIABLE. See POLYNOMIAL.

    regular function. See ANALYTIC—analytic function of a complex variable.

    step function. See STEP—step function.

    stream function. Let $f(x, y)$ denote the flux, in an incompressible fluid, across some curve AP, where A is a fixed point and $P = (x, y)$ is variable. The value of f is dependent only on the position of P, since the flux across two curves joining A and P must be the same; otherwise fluid would be added or taken away from the space between the two curves. When P moves in such a way as to keep f constant, P traces curves across which there is no flux. These curves are **stream lines**, and if the equation of such a stream line is $F(x, y) = 0$ then $F(x, y)$ is a **stream function**.

    subadditive and subharmonic functions. See ADDITIVE —additive set function, SUBHARMONIC.

    transcendental function. See TRANSCENDENTAL.

    trigonometric function. See TRIGONOMETRIC —trigonometric functions.

    unbounded function. See UNBOUNDED.

    vector function. A function whose values are vectors. The expression

$$F = f_1 i + f_2 j + f_3 k,$$

where f_1, f_2, and f_3 are scalar functions, defines a vector (vector-valued) function.

FUNC′TION-AL, *adj., n.* As an *adjective*: Of, relating to, or affecting a function. As a *noun*: A function whose domain is a set C_1 of functions and whose range is contained in another set C_2 of functions, not necessarily distinct from C_1. The "degenerate" case in which C_2 is a set of numbers is also included as a possible case. Many authors in the modern theories of abstract spaces use the term only in the case in which C_2 is a set of numbers, but in these generalizations the elements of the set C_1 need not be functions. E.g., $\int_a^b \alpha(x)y(x)\,dx$, $\max |y(x)|$, $dy(x)/dx$, and $\alpha(x)y(x) + \int_a^b \beta(x,s)y(s)\,ds$ are *functionals* of y. In each of these examples, the set C_1 is a suitably restricted set of real functions y of a real variable x, while C_2 is the set of real numbers in the first two examples and a suitably restricted set of real functions of a real variable in the last two examples. Two simple examples of a functional in which both C_1 and C_2 are sets of real functions of two real variables are $\dfrac{\partial^2 y(s,r)}{\partial s^2} + \dfrac{\partial^2 y(s,r)}{\partial r^2}$ and $\int_r^s y(s,t)y(t,r)\,dt$. A functional f defined on a vector space is a **linear functional** if $f(x+y) = f(x) + f(y)$ and $f(ax) = af(x)$ for any vectors x and y and scalar a. If f has real or complex number values, then f is **continuous** for each x if and only if there is a number M such that $|f(x)| \leq M \cdot \|x\|$ for each x. The least such number M is called the **norm** of f. See CONJUGATE —conjugate space.

    differential of a functional. If f is a *functional* from a set C_1 of functions to a set C_2 of functions, then a differential of f at y_0 with increment δy is a continuous, additive functional $\delta f(y_0, \delta y)$ from C_1 to C_2 such that $f(y_0 + \delta y) - f(y_0) = \delta f(y_0, \delta y) +$ "higher order terms in δy" for all δy in some neighborhood of the "zero function" in C_1. In order that this definition be applicable, the sets C_1 and C_2 must be such that the notions of addition, subtraction, zero function, neighborhood, and continuity are meaningful, and the meaning of "higher order terms in $\delta y(x)$" must be specified. E.g., if C_1 and C_2 are both the *Banach space* of all real continuous functions y of a real variable x in $a \leq x \leq b$ with

$$\|f\| = \max_{a \leq x \leq b} |f(x)| \quad \text{and} \quad \rho(f,g) = \|f - g\|$$

the *distance* between f and g, then $\delta f(y_0, \delta y)$ is a *Fréchet differential* of f at y_0 if it is a continuous additive functional from C_1 to C_2 and

$$f(y_0 + \delta y) - f(y_0)$$

$$= \delta f(y_0, \delta y) + \|\delta y\| \epsilon(y_0, \delta y),$$

where $\|\epsilon(y_0, \delta y)\|$ tends to zero with $\|\delta y\|$ uniformly for all functions δy continuous in $a \leq x \leq b$. If $\alpha(x)$ and $\beta(x,s)$ are fixed continuous functions, then the *Fréchet differential* of $f(y) = \alpha(x)y(x) + \int_a^b \beta(x,s)y^2(s)\,ds$ exists for each y_0

in C_1 and is given by $\delta f(y_0, \delta y) = \alpha(x)\delta y(x) + 2\int_a^b \beta(x,s)y_0(s)\delta y(s)\,ds$.

    functional determinant. See JACOBIAN.
    functional notation. See FUNCTION.

FUNC′TOR, *n.* Let K and L be two categories with O_K, M_K and O_L, M_L denoting, respectively, the sets of objects and morphisms in K, and the sets of objects and morphisms in L [see CATEGORY (2)]. A **covariant functor** of K into L is a function F whose domain is the set of all objects and morphisms in K; which maps O_K into O_L and, for each a and b in O_K, maps $M_K(a,b)$ into $M_L[F(a), F(b)]$; and which has the properties: (i) If e_a is the identity morphism in $M_K(a,a)$, then $F(e_a)$ is the identity morphism in $M_L[F(a), F(a)]$; (ii) if f and g are morphisms in $M_K(a,b)$ and $M_K(b,c)$, respectively, then $F(g \circ f) = F(g) \circ F(f)$. A **contravariant functor** is defined similarly, except that F maps $M_K(a,b)$ into $M_L[F(b), F(a)]$ and the equality in (ii) is replaced by $F(g \circ f) = F(f) \circ F(g)$. An **isomorphism** between categories K and L is a one-to-one correspondence between M_K and M_L with the property that if f and g corrrespond to f' and g', respectively, then $f \circ g$ is defined if and only if $f' \circ g'$ is defined, and then $(f \circ g)' = f' \circ g'$. For any isomprphism, it follows that identity morphisms correspond, so that a one-to-one correspondence between O_K and O_L is induced. Also, an isomorphism is a covariant functor. Similarly, for an **antiisomorphism** it is required that $f \circ g$ be defined if and only if $g' \circ f'$ is defined, and then that $(f \circ g)' = g' \circ f'$. An anti-isomorphism is a contravariant functor.

FUND, *n.* Money (sometimes other assets immediately convertible into money) which is held ready for immediate demands.
    endowment fund. A fund permanently appropriated for some objective, such as carrying on a school or church.
    reserve fund. (*Insurance*) See RESERVE. Used to take care of additional costs of policies in later years. (*Business*) A sum held ready to meet emergencies or take advantage of opportunities.
    sinking fund. A fund accumulated by periodic investments for some specific purpose such as retiring bonds, replacing equipment, providing pensions, etc. (The amount of the sinking fund is the amount of the annuity formed by the payments.)

FUN′DA-MEN′TAL, *adj.* **fundamental assumption.** See ASSUMPTION.

fundamental coefficients and quadratic forms of a surface. See under SURFACE.

fundamental group. Let S be a set which has the property that any two of its points can be connected by a *path*, a *path* being the image (for a continuous mapping) of a (directed) interval. The fundamental group of S is the *quotient group* of the group of all paths, with initial and terminal point at a designated base point P, and the subgroup of all paths which are homotopic to the path consisting of the single point P (using the quotient group is equivalent to defining two paths to be *equal* whenever they are homotopic to each other). The *product* of paths f and g is the path obtained by attaching g to the end of f; the *inverse* of f is the path obtained by reversing the direction assigned to f. Two fundamental groups with different base points are isomorphic. If the fundamental group contains only the identity, then S is *simply connected*. The fundamental group of a circle is an infinite cyclic group. For a torus, it is the commutative group generated by two elements a and b. A closed orientable surface of genus p has a fundamental group generated by $2p$ elements a_i, b_i which satisfy the relation

$$a_1 b_1 a_1^{-1} b_1^{-1} a_2 b_2 a_2^{-1} b_2^{-1} \cdots a_p b_p a_p^{-1} b_p^{-1} = 1$$

and is not commutative unless $p = 1$ (the surface is then a torus). A closed nonorientable surface has a fundamental group generated by q elements a_i with the relation

$$a_1 a_1 a_2 a_2 \cdots a_q a_q = 1,$$

which for $q = 1$ is a group of order two generated by a single element (the surface is then the projective plane). The commutators of the fundamental group generate a group which is isomorphic with the 1-dimensional *homology group* (based on the integers).

fundamental identities of trigonometry. See TRIGONOMETRIC —trigonometric functions.

fundamental lemma of the calculus of variations. If α is continuous for $a \leq x \leq b$ and $\int_a^b \alpha(x)\phi(x)\,dx = 0$ for all ϕ which have continuous first derivatives in $a \leq x \leq b$ and have $\phi(a) = \phi(b) = 0$, then $\alpha(x) \equiv 0$ throughout the interval $a \leq x \leq b$.

fundamental numbers and functions. Same as EIGENVALUES and EIGENFUNCTIONS.

fundamental operations of arithmetic. Addition, subtraction, multiplication, and division.

fundamental period of a periodic function of a complex variable. Same as PRIMITIVE PERIOD.

See PERIODIC —periodic function of a complex variable, and various headings under PERIOD.

fundamental sequence. Same as CAUCHY SEQUENCE. See SEQUENCE.

fundamental theorem of algebra. Every polynomial equation of degree $n \geq 1$ with complex coefficients has at least one root, which is a complex number (real or imaginary). A simple proof of the fundamental theorem of algebra can be given by using the minimum value theorem to show that, for any polynomial p, the absolute minimum of $|p(z)|$ is attained at some point, and then showing that this absolute minimum is zero unless p is a constant. Also see GAUSS —Gauss' proof of the fundamental theorem of algebra, WINDING —winding number.

fundamental theorem of arithmetic. Any positive integer greater than 1 is a prime or can be expressed as a product of primes. Except for the order of the factors, this expression is unique. *E.g.*, $60 = 2 \cdot 2 \cdot 3 \cdot 5$ (or $5 \cdot 2 \cdot 3 \cdot 2$, etc.). *Syn.* unique factorization theorem.

fundamental theorem of calculus. The relation between differentiation and integration, which can be expressed as follows: (1) If $\int_a^b f(x)\,dx$ exists and there is a function F such that $F'(x) = f(x)$ for all x in the closed interval $[a, b]$, then

$$\int_a^b f(x)\,dx = F(b) - F(a).$$

(2) If $\int_a^b f(x)\,dx$ exists and F is defined by $F(x) = \int_a^x f(x)\,dx$ for x in the closed interval $[a, b]$, then F is differentiable at x_0 and $F'(x_0) = f(x_0)$ if x_0 is in $[a, b]$ and f is continuous at x_0. See DARBOUX —Darboux's theorem, INTEGRAL —definite (Riemann) integral, ELEMENT —element of integration.

FU′TURE, *adj.* **future value of a sum of money.** See AMOUNT.

FUZZ′Y, *adj.* A body of concepts and methods intended to provide a systematic framework for dealing with vagueness and imprecision is called **fuzzy mathematics.** A **crisp set** divides the universe under consideration into those objects that belong to the set and those that do not belong. But for most sets of objects, the transition from membership to non-membership is gradual; *e.g.*, the set of tall women, democratic countries, intelligent students, sunny days. A **fuzzy set** can be defined by assigning to each member of the universe under consideration a number that represents its degree of membership in the fuzzy set. This number often is chosen in the closed interval $[0, 1]$. *E.g.*, a 60-year-old man might be said to be 70% old and 30% young, or that 0.7 is his degree of membership in the fuzzy set of old men. Classical logic may be used with fuzzy sets treated as mathematical objects, but **fuzzy logic** may treat truth-values themselves as fuzzy sets and the rules of inference may be approximate rather than exact. Not only are fuzzy assertions made about fuzzy sets, but the truth-values and the rules of inference may be fuzzy. Conventional computers use Boolean logic, which requires a statement to be true or false. Fuzzy logic allows a statement to be partly true or partly false at the same time. *E.g.*, a normal furnace thermostat turns heat on at a prescribed temperature and off at another temperature. A "fuzzy" thermostat allows the furnace to run at different power levels—if it senses the temperature is closer to the lower limit than the upper, it turns the furnace up an appropriate amount until the temperature is in the "about right" range.

G

G$_\delta$ set. See BOREL —Borel set.

GALILEI, Galileo (1564–1642). Italian astronomer, mathematician, and physicist who inaugurated the age of modern physics; discarded theory that heavier bodies fall faster than lighter ones, established the formula $s = \frac{1}{2}gt^2$ for freely falling bodies, and showed that projectiles move in parabolic curves. Realized that the squares of integers can be put in one-to-one correspondence with the integers, but concluded incorrectly that one cannot say one infinite number is greater than another. Persecuted for supporting Copernican theory of the solar system.

GAL′LON, *n*. A measure of volume equal to 4 quarts, 231 cubic inches, or 3.7853 liters (this is the old English **wine gallon**). The English **imperial gallon** is equal to 277.418 cubic inches, or 4.5460 liters.

GALOIS, Évariste (1811–1832). Brilliantly innovative algebraist responsible for *Galois theory* (see below). Killed in a political dual at the age of twenty.

   Galois field. For a polynomial p with coefficients in a field F, the **Galois field** F^* of p relative to F is the minimal field that contains F and has the property that p can be factored into linear factors with coefficients in F^*. If p has degree n, then p has n zeros in F^* if each zero is counted to the degree of its multiplicity, and F^* has degree at most $n!$ as an extension of F. See EXTENSION —extension of a field. *Syn.* root field, splitting field.

   Galois group. If F^* is the Galois field of the polynomial p relative to the field F, then the **Galois group** of p relative to F is the group of all automorphisms a of F^* for which $a(x) = x$ whenever x is in F. The Galois group is isomorphic to a group of permutations of the zeros of p.

   Galois theory. A theory of the Galois field F^* and the Galois group of a polynomial p with coefficients in a field F. The theory involves a one-to-one correspondence between the subfields of F^* that contain F and the subgroups of the Galois group (the field K corresponds to the group G if and only if K is the set of members x of F^* for which $a(x) = x$ if a is in G, or if and only if G is the set of all automorphisms a of F^* for which $a(x) = x$ if x is in K). This leads to the theorem that the Galois group of a polynomial p with respect to a field F is *solvable* if the equation $p(x) = 0$ is solvable in F by radicals, from which it follows that there is a real quintic equation that is not solvable by radicals.

GAME, *n*. A set of rules, for individuals or groups of individuals involved in a competitive situation, giving their permissible actions, the amount of information each receives as the competition progresses, the probabilities associated with the chance events that might occur during the competition, the circumstances under which the competition ends, and the amount each individual pays or receives as a consequence. An *n***-person game** is a game in which there are exactly n players or interests (a **two-person** game is a game in which there are exactly two players or conflicting interests). The **theory of games** is the mathematical theory, founded largely by the mathematician John von Neumann, of optimal behavior in situations involving conflict of interest. See BOX, DUEL, HER, MAZUR, MINIMAX — minimax theorem, MORRA, NIM, PAYOFF, PLAYER, STRATEGY, and the headings below.

   "Colonel Blotto" game. In the theory of games, the problem of dividing attacking and defending forces between fortresses under the assumption that at each fortress each side loses a number of men equal to the number of men in the smaller force involved at the fortress, and that the fortress is then occupied by the side having survivors; occupation of a fortress is considered as being equivalent to having a certain number of survivors, and the payoff is then measured in terms of the total number of survivors at the fortresses.

circular symmetric game. A finite two-person zero-sum game whose matrix is a *circulant* (*i.e.*, the elements of each row are the elements of the previous row slid one place to the right, with the last element put first).

coin-matching game. A two-person zero-sum game in which each of the two players tosses a coin of like value. If the two coins show like faces—either both "heads" or both "tails"—the first player wins, while if they show unlike faces the second player wins.

completely mixed game. A game having a unique solution that is also a simple solution; equivalently, a game such that every possible strategy has positive probability in the solution. See below, solution of a game.

concave and convex games. A concave game is a two-person zero-sum game for which the payoff function $M(x, y)$ is a *concave function* of the strategy x of the maximizing player (the dual of the convex game with payoff function $-M(y, x)$). A **convex game** is a two-person zero-sum game for which the payoff function $M(x, y)$ is a *convex function* of the strategy y of the minimizing player (the dual of the concave game with payoff function $-M(y, x)$). A **concave-convex game** is a two-person zero-sum game for which the payoff function $M(x, y)$ is a *concave function* of the strategy x of the maximizing player and is a *convex function* of the strategy y of the minimizing player.

continuous game. An infinite game in which each player has a closed and bounded continuum of pure strategies, usually taken (without loss of generality) to be identified with the numbers of the closed interval $[0, 1]$. See below, finite and infinite games.

cooperative game. A game in which the formation of coalitions is possible and permissible. A game is **noncooperative** if the formation of coalitions is either not possible or not permissible. See COALITION.

extensive form of a game. The general description of a game in terms of its moves, information patterns, etc. See below, normal form of a game.

finite and infinite games. A game is **finite** if each player has only a finite number of possible pure strategies; it is **infinite** if at least one player has an infinite number of possible pure strategies (*e.g.*, a pure strategy might ideally consist of choosing an instant from a given interval of time at which to fire a gun). See STRATEGY.

game of survival. A two-person zero-sum

game that continues until one player loses all.

game with perfect information. A game such that at each move each player knows the outcome of all previous moves of the game. Such a game of necessity has a saddle point, and accordingly each player has an optimal pure strategy. A game with imperfect information is a game in which at least one move is made by a player not knowing the outcome of all previous moves.

normal form of a game. A description of a game in terms of its strategies and associated payoff matrix or function. See above, extensive form of a game.

polynomial game. A continuous game having payoff function of the form $M(x, y) = \sum_{i, j=0}^{m, n} a_{ij} x^i y^j$, where the strategies x, y range over the closed interval $[0, 1]$. See below, separable game.

positional game. A game with simultaneous moves by the players, in which each player knows at all times the outcome of all previous moves. See above, game with perfect information.

saddle point of a game. See SADDLE.

separable game. A continuous game having payoff function of the form

$$M(x, y) = \sum_{i, j=0}^{m, n} a_{ij} f_i(x) g_j(y),$$

where the strategies x, y range over the closed interval $[0, 1]$, the a_{ij} are constants, and the functions f_i and g_j are continuous. A *polynomial game* is a particular instance of a separable game.

solution of a two-person zero-sum game. A pair of optimal mixed (or pure) strategies, one for each of the players of the game. A **simple solution** is a solution X, Y such that each pure strategy for the minimizing player, when used against the maximizing player's optimal mixed strategy X, and each pure strategy for the maximizing player, when used against the minimizing player's optimal mixed strategy Y, yields an expected value of the payoff exactly equal to the value of the game. A game might have a solution without having a simple solution. The *coin-matching game* is an example of a game having a simple solution. A set of **basic solutions** of a game is a finite set S of solutions of the game, such that each solution can be written as a convex linear combination of the members of S, but such that there is no proper subset of S in terms of which each solution can be so written. See PLAYER, STRATEGY.

symmetric game. A finite two-person zero-sum game with a (square) skew-symmetric payoff matrix, *i.e.*, a payoff matrix for which $a_{ij} = -a_{ji}$ for all i and j. More generally, a two-person zero-sum game with payoff function $M(x, y)$ satisfying $M(x, y) = -M(y, x)$ for all x and y. The *value* of such a game is zero, and both players have the same *optimal strategies*.

value of a game. The number v associated with any two-person zero-sum game for which the minimax theorem holds. See MINIMAX —minimax theorem, SADDLE —saddle point of a game.

zero-sum game. A game in which the sum of the winnings of the various players is always zero. Thus games like poker, in which we consider only the financial payoff, are zero-sum games unless there is a house charge for playing. A **non-zero-sum game** is a game in which, for at least one play, the sum of the winnings of the various players is not zero.

GAM'MA, *n.* The third letter of the Greek alphabet: lower case, γ; capital, Γ.

gamma distribution. A random variable X has a **gamma** distribution or is a **gamma random variable** if the range of X is the set of positive real numbers and there are positive numbers r and λ for which the *probability-density function f* satisfies

$$f(x) = \frac{\lambda}{\Gamma(r)} (\lambda x)^{r-1} e^{-\lambda x} \quad \text{if} \quad x > 0,$$

where Γ is the *gamma function*. The *mean* is r/λ, the *variance* is r/λ^2, and the *moment-generating function* is $M(t) = (1 - t/\lambda)^{-r}$. If r is an integer, then X is the length of the interval $[t_0, t]$, where t_0 is an arbitrary point and t is the first point such that, for a Poisson process with parameter λ, r events occur in the interval $[t_0, t]$. If $r = 1$, the gamma random variable is called the **exponential random variable** with parameter λ. Then $f(x) = \lambda e^{-\lambda x}$, where f is the probability-density function, and the *distribution function F* satisfies $F(x) = 0$ if $x \leq 0$ and $F(x) = 1 - e^{-\lambda x}$ if $x \geq 0$. *Syn.* Erlang distribution. See CHI—χ^2 distribution, POISSON —Poisson process.

gamma function. The function Γ defined by

$$\Gamma(x) = \int_0^\infty t^{x-1} e^{-t} \, dt$$

for $x > 0$, or the real part of x greater than zero if x is complex. It follows that $\Gamma(x + 1) = x \Gamma(x)$, that $\Gamma(1) = 1$, and that $\Gamma(n) = (n - 1)!$, when n is any positive integer. Also, $\Gamma(\frac{1}{2}) = \pi^{1/2}$, $\Gamma(\frac{3}{2}) = \frac{1}{2}\pi^{1/2}$, etc., and Γ has an analytic extension with domain the set of all complex numbers except the negative integers and 0. This exten-

sion can be defined by using $\Gamma(z + 1) = z \Gamma(z)$ or

$$\Gamma(z) = \left[z e^{\gamma z} \prod_{n=1}^\infty \left\{ \left(1 + \frac{z}{n} \right) e^{-z/n} \right\} \right]^{-1}$$

where γ is *Euler's constant*. The **incomplete gamma functions**, γ and Γ, are defined by

$$\gamma(a, x) = \int_0^x t^{a-1} e^{-1} \, dt,$$

$$\Gamma(a, x) = \int_x^\infty t^{a-1} e^{-1} \, dt.$$

It follows that $\Gamma(a) = \gamma(a, x) + \Gamma(a, x)$ and that

$$\gamma(a + 1, x) = a\gamma(a, x) - x^a e^{-x},$$

$$\Gamma(a + 1, x) = a\Gamma(a, x) + x^a e^{-x},$$

$$\gamma(a, x) = \sum_0^\infty \frac{(-1)^n x^{a+n}}{n!(a+n)}.$$

See DIGAMMA, DUPLICATION.

GATE, *n.* In a computing machine, a switch that allows the passage of a signal if and only if one or more other signals are present; thus a gate is the machine equivalent of the logical "and". See CONJUNCTION, BUFFER.

inverse gate. Same as BUFFER.

GAUSS, Carl Friedrich (1777–1855). German mathematician, usually considered along with Archimedes and Newton to be one of the three greatest mathematicians of all time. Made important contributions to algebra, analysis, geometry, number theory, numerical analysis, probability, and statistics, as well as astronomy and physics. See PRIME —prime-number theorem.

Gauss-Bonnet theorem. (*Differential geometry*) For a simply connected portion S of a surface for which the curvature K is continuous and the geodesic curvature k of the contour C is continuous,

$$\int_C k \, ds + \int\int_S K \, dA = 2\pi - \sum_{i=1}^n \theta_i,$$

where $\theta_1, \cdots, \theta_n$ are the exterior angles (if any) at the vertices of C.

Gauss' differential equation. See HYPERGEOMETRIC —hypergeometric differential equation.

Gauss' equation. (*Differential Geometry*) An equation expressing the total curvature $K = \dfrac{DD'' - D'^2}{EG - F^2}$ in terms of the fundamental coefficients of the first order, E, F, G, and their partial derivatives of the first and second orders:

$$K = \frac{1}{2H}\left\{ \frac{\partial}{\partial u}\left[\frac{F}{EH}\frac{\partial E}{\partial v} - \frac{1}{H}\frac{\partial G}{\partial u}\right] \right.$$

$$\left. + \frac{\partial}{\partial v}\left[\frac{2}{H}\frac{\partial F}{\partial u} - \frac{1}{H}\frac{\partial E}{\partial v} - \frac{F}{EH}\frac{\partial E}{\partial u}\right] \right\},$$

where $H = \sqrt{EG - F^2}$, or in terms of Christoffel symbols,

$$K = \frac{1}{H}\left\{ \frac{\partial}{\partial u}\left(\frac{H}{G}\begin{bmatrix} 2 & & 2 \\ & 1 & \end{bmatrix}\right) \right.$$

$$\left. - \frac{\partial}{\partial v}\left(\frac{H}{G}\begin{bmatrix} 1 & & 2 \\ & 1 & \end{bmatrix}\right)\right\}$$

$$= \frac{1}{H}\left\{ \frac{\partial}{\partial v}\left(\frac{H}{E}\begin{bmatrix} 1 & & 1 \\ & 2 & \end{bmatrix}\right) \right.$$

$$\left. - \frac{\partial}{\partial u}\left(\frac{H}{E}\begin{bmatrix} 1 & & 2 \\ & 2 & \end{bmatrix}\right)\right\}.$$

In tensor notation:

$$x^i_{,\alpha\beta} = d_{\alpha\beta}X^i.$$

For isothermic parameters,

$$E = G = \lambda(u, v), \qquad F = 0,$$

the formula reduces to

$$K = -\frac{1}{2\lambda}\left[\frac{\partial^2 \log \lambda}{\partial u^2} + \frac{\partial^2 \log \lambda}{\partial v^2}\right];$$

and for geodesic parameters, $E = 1$, $F = 0$, $G = [\mu(u, v)]^2$, $\mu \geq 0$, the formula reduces to

$$K = -\frac{1}{\mu}\frac{\partial^2 \mu}{\partial u^2}.$$

See CODAZZI —Codazzi equations, and below, theorem of Gauss.

Gauss's formulas (or Delambre's analogies). Formulas stating the relations between the sine (or cosine) of half of the sum (or difference) of two angles of a spherical triangle and the other angle and the three sides. If the angles of the triangle are A, B, and C, and the sides opposite these angles are a, b, and c, respectively, then Gauss' formulas are:

$$\cos\tfrac{1}{2}c\,\sin\tfrac{1}{2}(A + B) = \cos\tfrac{1}{2}C\,\cos\tfrac{1}{2}(a - b),$$

$$\cos\tfrac{1}{2}c\,\cos\tfrac{1}{2}(A + B) = \sin\tfrac{1}{2}C\,\cos\tfrac{1}{2}(a + b),$$

$$\sin\tfrac{1}{2}c\,\sin\tfrac{1}{2}(A - B) = \cos\tfrac{1}{2}C\,\sin\tfrac{1}{2}(a - b),$$

$$\sin\tfrac{1}{2}c\,\cos\tfrac{1}{2}(A - B) = \sin\tfrac{1}{2}C\,\sin\tfrac{1}{2}(a + b).$$

Gauss's fundamental theorem of electrostatics. The surface integral of the exterior normal component of the electric intensity over any closed surface all of whose points are free of charge is equal to 4π times the total charge enclosed by the surface. In the corresponding theorem for gravitational matter, the constant is -4π.

Gauss's mean-value theorem. Let u be a regular harmonic function in a region R. Let P be a point in R, S a sphere with center at P and lying entirely (boundary and interior points) within R, and A the area of S; then $u(P) = (1/A)\int\int_S u\,dS$. For R a plane region and C a circle with perimeter c, $u(P) = \dfrac{1}{c}\int_C u\,ds$.

Gauss plane. Same as COMPLEX PLANE.

Gauss' proof of the fundamental theorem of algebra. The first known proof of this theorem. A geometrical proof consisting essentially of substituting a complex number, $a + bi$, for the unknown of the equation, separating the real and imaginary parts of the result, and then showing that the two resulting functions of a and b are zero for some pair of values of a and b.

Gaussian distribution. Same as normal distribution. See under NORMAL.

Gaussian integer. See INTEGER.

Gaussian representation of a surface. See SPHERICAL —spherical image (or representation) of curves and surfaces.

theorem of Gauss. The famous theorem that the total curvature of a surface is a function of the fundamental coefficients of the first order of the surface and their partial derivatives of the first and second orders. See DIVERGENCE —divergence theorem, and above, Gauss' equation.

GEL′FOND, Alexander Osipovič (1906–1968). Russian analyst.

Gelfond-Schneider theorem. If α and β are algebraic numbers, α is not 0 or 1, and if β is not a rational number, then any value of α^β is transcendental (*i.e.*, is a real or complex number which is not a root of a polynomial equation whose coefficients are integers). This theorem was proved independently by Gelfond (1934) and Schneider (1935). It is an affirmative solution of Hilbert's seventh problem. See BAKER.

GEN′ER-AL, *adj.* Not specific or specialized; covering all special cases. Some examples are the **general polynomial equation** (see EQUATION —polynomial equation), **general term** (see TERM—general term), and **general solution of a differential equation** (see DIFFERENTIAL—solution of a differential equation).

GEN′ER-AL-IZED, *adj.* **generalized function.** A concept introduced by L. Schwartz in 1945 that provides a mathematical foundation for some formal methods of mathematical physics and powerful tools for many fields of mathematics. (1) For the one-dimensional case, let the **test functions** be the linear space Φ of all functions whose supports are bounded and which are differentiable everywhere to all orders. A **distribution** or **generalized function** is a linear functional T defined on Φ which is continuous in the sense that $\lim_{n \to \infty} T(\varphi_n) = 0$ whenever the sequence $\{\varphi_n\}$ has the property that there is a bounded interval that contains the support of each φ_n and each sequence $\{\varphi_n^{(k)}\}$ of kth order derivatives converges uniformly to zero. *E.g.*, if f is measurable and $|f|$ has a finite integral on each bounded interval, then

$$T_f(\varphi) = \int_{-\infty}^{\infty} \varphi(t) f(t) \, dt, \qquad \varphi \in \Phi,$$

defines a distribution T_f. Two functions determine the same distribution if and only if they are equal almost everywhere. Each distribution T has a derivative T' defined by $T'(\varphi) = -T(\varphi')$. (2) More generally, let the **test functions** be the linear space of all complex-valued functions with compact support in n-dimensional Euclidean space R^n which have mixed partial derivatives of all orders. A sequence $\{\varphi_n\}$ of members of Φ is said to converge to 0 if there is a compact set that contains the support of each φ_n and if the sequences $\{\varphi_n\}$ and $\{D\varphi_n\}$ converge uniformly to 0 for each mixed partial derivative D. A **distribution** or **generalized function (in the sense of L. Schwartz)** is a linear functional T defined on Φ which is continuous in the sense that $\lim_{n \to \infty} T(\varphi_n) = 0$ whenever the sequence $\{\varphi_n\}$ converges to 0. *E.g.*, if f is a locally integrable function with domain in R^n, then

$$T_f(\varphi) = \int \varphi(t) f(t) \, dt, \qquad \varphi \in \Phi,$$

defines a distribution T_f. Similarly, if μ is a complex-valued measure for R^n and μ is finite on compact sets, then

$$T_\mu(\varphi) = \int \varphi(t) \, d\mu, \qquad \varphi \in \Phi,$$

defines a distribution T_μ. If $\mu(E)$ is 1 or 0 according as the origin belongs to E or does not belong to E and if δ is the corresponding distribution, then $\delta(\varphi) = \int \varphi(t) \, d\mu = \varphi(0)$. This δ is

the **Dirac distribution**, sometimes called the **Dirac δ-function** (although it is not a function). Let D^p with $p = (p_1, p_2, \cdots, p_n)$ denote mixed partial differentiation, where there are p_i differentiations with respect to the ith variable. Also, let $|p| = p_1 + \cdots + p_n$. Then for any distribution T the mixed partial derivative $D^p T$ exists and is defined by

$$D^p T(\varphi) = (-1)^{|p|} T(D^p \varphi), \qquad \varphi \in \Phi.$$

If T is a distribution and g has mixed partial derivatives of all orders, then the product gT is defined by $(g(T)(\varphi) = T(g\varphi)$. Sometimes other spaces of test functions are used when they are more useful for the problems considered.

generalized mean-value theorem. (1) Same as TAYLOR'S THEOREM. (2) Same as SECOND MEAN-VALUE THEOREM. See MEAN—mean-value theorems for derivatives.

generalized ratio test. See RATIO—ratio test.

generalized Riemann integral. See INTEGRAL—generalized Riemann integral.

GEN′ER-AT′ING, *p.* **generating function.** A function F that, through its representation by means of an infinite series of some sort, gives rise to a certain sequence of constants or functions as coefficients in the series. *E.g.*, the function $(1 - 2ux + u^2)^{-1/2}$ is a generating function of the *Legendre polynomials P_n* by means of the expansion

$$(1 - 2ux + u^2)^{-1/2} = \sum_0^\infty P_n(x) u^n.$$

GEN′ER-A′TOR, *n.* Same as GENERATRIX (the feminine form of *generator*).

generator of a surface of translation. See SURFACE—surface of translation.

generators of a group. A set of generators of a group G is a subset S of G such that each member of G can be represented (using the group operations) in terms of members of S, repetitions of members of S being allowed. The set S of generators is **independent** if no member of S is in the group generated by the other members of S.

rectilinear generators. See RULED—ruled surface.

GEN′ER-A′TRIX, *n.* **generatrix of a ruled surface.** A straight line which forms the surface by moving according to some law. The elements of a cone are different positions of its generatrix. See RULED—ruled surface.

GE′NUS, *n.* **genus of a surface.** For a closed surface, the **genus** is the maximum number of disjoint circles along which the surface can be cut without disconnecting it. A *closed orientable surface* is topologically equivalent to a sphere with an even number $2p$ of holes (made by removing discs) which have been connected in pairs by p *handles* (shaped like the surface of half of a doughnut). A *closed nonorientable surface* is topologically equivalent to a sphere which has had a certain number q of discs replaced by *cross-caps*. The numbers p and q are said to be the *genus* of the surface. In either of the above cases, the surface not being closed means that some discs have been removed and the hole left open. A torus is a sphere with one handle; a Möbius strip is a sphere with one cross-cap and one "hole"; a Klein bottle is a sphere with two cross-caps; a cylinder is a sphere with two "holes". In general, the *Euler characteristic* of a surface is equal to $2 - 2p - q - r$, where p is the number of handles (which can be zero for a nonorientable surface), q is the number of cross-caps (zero for an orientable surface), and r is the number of holes (or boundary curves).

GE′O-DES′IC, *adj., n.* A curve C on a surface S such that for any two points of C an arc of C between these points is the shortest curve on S that joins these points. A geodesic C has the properties that at each point of C the principal normal coincides with the normal to S and that the geodesic curvature vanishes identically. See below, geodesic curvature of a curve on a surface. If a straight line lies on a surface, then the line is a geodesic for the surface. A geodesic yields a stationary value of the length integral

$$s = \int_{t_0}^{t_1} \sqrt{g_{ij} \frac{dx^i}{dt} \frac{dx^j}{dt}} \, dt.$$

In terms of the arc-length parameter s, the Euler-Lagrange equations for this calculus of variations problem are the system of second-order differential equations

$$\frac{d^2 x^i(s)}{ds^2} + \Gamma_{\alpha\beta}^i \left(x^1(s), \cdots, x^n(s) \right)$$

$$\times \frac{dx^\alpha(s)}{ds} \frac{dx^\beta(s)}{ds} = 0,$$

where the $\Gamma_{\alpha\beta}^i$ are the Christoffel symbols of the second kind based on the metric tensor

$g_{ij}(x^1, \cdots, x^n)$. See RIEMANN—Riemannian spaces.

geodesic circle on a surface. If equal lengths are laid off from a point P of a surface S along the geodesics through P on S, the locus of the end points is an orthogonal trajectory of the geodesics. The locus of end points is called a **geodesic circle** with center at P and radius r. The "radius" r is a **geodesic radius**; it is the geodesic distance on the surface from the "center" P to the circle. See below, geodesic polar coordinates.

geodesic coordinates in Riemannian space. Coordinates y^i such that the Christoffel symbols

$$\Gamma_{\alpha\theta}^i(y^1, \cdots, y^n)$$

are all zero when evaluated at the point in question, which we take to be the origin $y^1 = y^2 = \cdots = y^n = 0$. Thus the coordinate system is locally Cartesian. If the x^i are general coordinates, then the coordinate transformation

$$x^i = q^i + y^i - \frac{1}{2!} \left[\Gamma_{\alpha\beta}^i(x^1, \cdots, x^n) \right]_{x^j = q^j} y^\alpha y^\beta$$

defines implicitly geodesic coordinates y^i. See below, geodesic parameters (coordinates), geodesic polar coordinates.

geodesic curvature of a curve on a surface. For a curve C with parametric equations $u = u(s)$, $v = v(s)$, on a surface S with parametric equations $x = x(u,v)$, $y = y(u,v)$, $z = z(u,v)$, let π be the plane tangent to S at a point P of C, and let C' be the orthogonal projection of C on π. Let the positive direction of the normal to the cylinder K projecting C on C' be determined so that the positive directions of the tangent to C, the normal to K, and the normal to S, at P, have the same mutual orientation as the positive x, y, and z axes; and let ψ be the angle between the positive directions of the principal normal to C and the normal to K at P. Then the geodesic curvature $1/\rho_g$ of the curve C on the surface S at the point P is defined by

$$\frac{1}{\rho_g} = \frac{\cos \psi}{\rho},$$

where ρ is the curvature of C at P. Thus the geodesic curvature of C is numerically equal to the curvature of C' and is positive or negative according as the positive directions of the principal normal to C and the normal to K at P lie on the same or opposite sides of the normal to S. If the positive direction on C is reversed, the geodesic curvature changes sign. The **radius of geodesic curvature** is the reciprocal of the geodesic curvature. The **center of geodesic curvature** is the center of curvature, relative to P, of the curve C'.

geodesic ellipses and hyperbolas on a surface. Let P_1 and P_2 be distinct points on a surface S (or let C_1 and C_2 be curves on S such that C_1 and C_2 are not geodesic parallels of each other on S). Let u and v measure geodesic distances on S from P_1 and P_2 (or from C_1 and C_2), respectively. Then the curves $u' = $ const. and $v' = $ const., where $u' = \frac{1}{2}(u + v)$, $v' = \frac{1}{2}(u - v)$, are **geodesic ellipses** and **hyperbolas**, respectively, on S relative to P_1 and P_2 (or to C_1 and C_2). These names are given because, for instance, the sum of the geodesic distances from P_1 and P_2 (or from C_1 and C_2) to a variable point of a fixed geodesic ellipse has a constant value.

geodesic parallels on a surface. Given a smooth curve C_0 on a surface S, there exists a unique family of geodesics on S intersecting C_0 orthogonally; if segments of equal length s be measured along the geodesics from C_0, then the locus of their end points is an orthogonal trajectory C_s of the geodesics. The curves C_s are **geodesic parallels** on S. See below, geodesic parameters for a surface.

geodesic parameters (coordinates). Parameters u, v for a surface S such that the curves $u = $ const. are the members of a family of geodesic parallels, while the curves $v = $ const. $= v_0$ are members of the corresponding orthogonal family of geodesics, of length $u_2 - u_1$ between the points (u_1, v_0) and (u_2, v_0). See above, geodesic parallels on a surface. A necessary and sufficient condition that u, v be geodesic parameters is that the first fundamental form of S reduce to $ds^2 = du^2 + G\, dv^2$. See below, geodesic polar coordinates.

geodesic polar coordinates. These are geodesic parameters u, v for a surface S, except that the curves $u = $ const. $= u_0$, instead of being geodesic parallels, are concentric geodesic circles, of **radius** u_0, and center, or **pole**, P corresponding to $u = 0$; the curves $v = v_0$ are the geodesic radii; and, for each v_0, v_0 is the angle at P between the tangents to $v = 0$ and $v = v_0$. Necessary and sufficient conditions that u, v be geodesic polar coordinates are that the first fundamental quadratic form of S reduce to $ds^2 = du^2 + \mu^2\, dv^2$, $\mu \geq 0$, and that at $u = 0$ we have $\mu = 0$ and $\partial \mu / \partial u = 1$. All points on $u = 0$ are singular points corresponding to P. See above, geodesic parameters for a surface.

geodesic representation of one surface on another. A representation such that each geodesic on one surface corresponds to a geodesic on the other.

geodesic torsion. The **geodesic torsion of a surface** at a point P in a given direction is the torsion of the geodesic through P in the given direction. The **geodesic torsion of a curve on a surface** is the geodesic torsion of the surface at the given point in the direction of the curve. See TORSION —torsion of a curve.

geodesic triangle on a surface. A triangle formed by three geodesics, intersecting by pairs on the surface. See CURVATURE —integral curvature.

umbilical geodesic. See UMBILICAL.

GE'O-GRAPH'IC, *adj.* Pertaining to the surface of the earth.

geographic coordinates. Same as SPHERICAL COORDINATES in the sense of coordinates on a sphere. Spherical coordinates use the longitude and colatitude of a point on a sphere of radius r.

geographic equator. See EQUATOR.

GE'OID, *n.* The solid object that would result if the earth were modified by removing all parts above mean sea level and filling all places below mean sea level. It is almost exactly an ellipsoid and nearly an oblate spheroid. See ELLIPSOID.

GE'O-MET'RIC, or GE'O-MET'RI-CAL, *adj.* Pertaining to geometry; according to rules or principles of geometry; done by geometry.

geometric average. The geometric average of n positive numbers is the positive nth root of their product. The geometric average of two numbers is the middle term in a geometric progression of three terms including the two given numbers. There are always two such means, but it is common to use the positive root of the product, unless otherwise indicated. The *geometric means* of 2 and 8 are $\pm \sqrt{16}$ or ± 4. See AVERAGE. *Syn.* geometric mean.

geometric construction. In elementary geometry, a construction made using only a straight edge and compass. Simple examples are bisecting an angle and circumscribing a circle about a triangle. For constructions that can not be made in this way, see DUPLICATION —duplication of the cube, FERMAT —Fermat numbers, SQUARING —squaring the circle, TRISECTION —trisection of an angle.

geometric distribution. See BINOMIAL —negative binomial distribution.

geometric figure. Any combination of points, lines, planes, circles, etc.

geometric locus. Any system of points, curves, or surfaces defined by certain general conditions or equations, such as the *locus of points* equidistant from a given point, or the *locus* of the *equation* $y = x$.

geometric means between two numbers. The other terms of a *geometric sequence* of which the given numbers are the first and last terms; a single geometric mean between two positive numbers x and y is their **geometric mean**: $\sqrt{xy}$. The **geometric mean** of n positive numbers is the positive nth root of their product. See MEAN, and below, geometric sequence.

geometric sequence. A sequence for which the ratio of a term to its predecessor is the same for all terms. The general form of a finite geometric sequence is $\{a, ar, ar^2, ar^3, \cdots, ar^{n-1}\}$, where a is the **first term**, r is the **common ratio** (or simply **ratio**), and ar^{n-1} is the **last term**. The sum of the terms is $a(1 - r^n)/(1 - r)$. *Syn.* geometric progression.

geometric series. The indicated sum of the terms of a geometric sequence. The sum of the first n terms of the infinite geometric sequence

$$a + ar + ar^2 + ar^3 + \cdots + ar^{n-1} + \cdots$$

is $a(1 - r^n)/(1 - r)$, which approaches $a/(1 - r)$ as n increases if $|r| < 1$ and $\lim_{n \to \infty} r^n = 0$. Thus, if $|r| < 1$, the series converges and its sum is $a/(1 - r)$. *E.g.*, the sum of $1 + \frac{1}{2} + (\frac{1}{2})^2 + \cdots + (\frac{1}{2})^{n-1}$ is $[1 - (\frac{1}{2})^n]/(1 - \frac{1}{2}) = 2 - (\frac{1}{2})^{n-1}$. Since

$$\lim_{n \to \infty} \left[2 - \left(\tfrac{1}{2} \right)^{n-1} \right] = 2,$$

the sum of the geometric series $1 + \frac{1}{2} + (\frac{1}{2})^2 + \cdots$ is 2. Any repeating decimal is a geometric series. *E.g.*, $3.575757 \cdots$ is the sum of the geometric series

$$3 + 57\left(\frac{1}{100} \right) + 57\left(\frac{1}{100} \right)^2 + 57\left(\frac{1}{100} \right)^3 + \cdots,$$

which is $3 + 57(\frac{1}{100})/(1 - \frac{1}{100}) = 118/33$. See SUM—sum of an infinite series.

geometric solid. Any portion of space which is occupied conceptually by a physical solid; *e.g.*, a cube or a sphere.

geometric solution. The solution of a problem by strictly geometric methods, as contrasted to algebraic or analytic solutions.

geometric surface. Same as SURFACE.

GE-OM′E-TRY, *n.* The science that treats of the shape and size of things. *Tech.* The study of invariant properties of given elements under specified groups of transformations.

analytic geometry. The geometry in which position is represented analytically (by coordinates) and algebraic methods of reasoning are used for the most part.

Euclidean geometry. The study of geometry based on the assumptions of Euclid. The *Elements* of Euclid (*c.* 300 B.C.) contains a systematic development of the basic propositions of elementary geometry (as well as propositions about numbers), but has defects (*e.g.*, the need of a postulate of type: "If a line cuts one side of a triangle and does not contain a vertex, then it cuts another side"). In modern usage, a **Euclidean space** is any finite-dimensional vector space in which the distance between points (vectors) is given by an extension of the usual formula for three-dimensional space. See EUCLIDEAN—Euclidean space.

metric differential geometry. The study, by means of differential calculus, of properties of general elements of curves and surfaces which are invariant under rigid motions.

non-Euclidean geometry. A geometry for which the parallel postulate of Euclid is not satisfied, or more generally any geometry not based on the postulates of Euclid. See BOLYAI, LOBACHEVSKI.

plane analytic geometry. Analytic geometry in the plane (in two dimensions), devoted primarily to the graphing of equations in two variables and finding the equations of loci in the plane.

plane (elementary) geometry. The branch of geometry that treats of the properties and relations of plane figures (such as angles, triangles, polygons, circles) which can be drawn with ruler and compasses.

projective geometry. See PROJECTIVE.

solid analytic geometry. Analytic geometry in three dimensions; devoted primarily to the graphing of equations (in three variables) and finding the equations of loci in space.

solid (elementary) geometry. The branch of geometry which studies figures in space (three dimensions) whose plane sections are the figures studied in plane elementary geometry, such as cubes, spheres, polyhedrons, and angles between planes.

synthetic geometry. See SYNTHETIC—synthetic geometry.

GERGONNE, Joseph Diaz (1771–1859). French analytic and projective geometer. Shares priority with Poncelet in formulation of principle of duality of projective geometry.

GIBBS, Josiah Willard (1839–1903). American mathematical physicist. Initiated the development of vector analysis and contributed to statistical mechanics.

Gibbs phenomenon. Quite generally, for a sequence of transformations $\{T_n\}$, $n = 1, 2, \cdots$, of a function f, if the interval

$$\left[\liminf_{x \to x_0, \, n \to \infty} T_n(x), \; \limsup_{x \to x_0, \, n \to \infty} T_n(x) \right]$$

contains points outside the interval

$$\left[\liminf_{x \to x_0} f(x), \; \limsup_{x \to x_0} f(x) \right],$$

then the sequence is said to exhibit a Gibbs phenomenon at $x = x_0$. This is particularly important for Fourier series. If f has a jump discontinuity at a and the limits on the left and right are $f(a-)$ and $f(a+)$, and if f' is continuous in an interval I around a except at a, then as n increases the approximating curve $y = S_n(x)$ closely approximates in I a curve consisting of the graph of f and a vertical line segment of length $|f(a+) - f(a-)|(1 + 2\lambda/\pi)$ with center at $(a, \frac{1}{2}[f(a-) + f(a+)])$, where S_n is the nth partial sum of the Fourier series and

$$\lambda = -\int_n^\infty (\sin x)/x \, dx = 0.28 + .$$

GIG′A. A prefix meaning 10^9, as in **gigameter**.

GIRTH, *n.* The length of the perimeter of a cross section of a surface when that length is the same for all right cross sections in planes parallel to the plane of that cross section.

GLO′BAL, *adj.* **global property.** See LOCAL—local property.

GÖDEL, Kurt (1906–1978). Czechoslovakian-American logician and philosopher. Proved the axiom of choice and the continuum hypothesis are consistent with the usual axioms of set theory; also proved that the consistency of a logical system cannot be proved within the system.

GOLDBACH, Christian (1690–1764). Number theorist and analyst; born in Prussia, lived in various West European countries, settled in Russia.

Goldbach conjecture. The conjecture (unproved) that every even number (except 2) is equal to the sum of two prime numbers. This conjecture has been verified up to $2 \cdot 10^{10}$.

GOLD′EN, *adj.* **golden section.** The division of a line segment AB by an interior point P in "extreme and mean ratio," *i.e.*, so that $AB/AP = AP/PB$. It follows that $AP/PB = \frac{1}{2}(1 + \sqrt{5})$, which is a root of $x^2 - x - 1 = 0$. Such a division appears in sculpture, painting, architecture, anatomy, and patterns of nature. It is considered to be "esthetically pleasing". *Syn.* divine proportion, golden mean, golden ratio. See FIBONACCI —Fibonacci sequence.

golden rectangle. A rectangle R with the property that it can be divided into a square and a rectangle similar to R; a rectangle whose sides have the ratio $\frac{1}{2}(1 + \sqrt{5})$.

GOMPERTZ, Benjamin (1779–1865). English actuary, analyst, and astronomer. He was largely self-educated, since Jews were barred from universities when he was young.

Gompertz curve. A curve whose equation is of the form $\log y = \log k + (\log a)b^x$, or $y = ka^{bx}$, where $0 < a < 1$, $0 < b < 1$. The value of y at $x = 0$ is ka, and, as $x \to \infty$, $y \to k$. The rate of change of $\log y$ is proportional to $\log(y/k)$. This is a type of **growth curve**.

Gompertz's law. The force of mortality (risk of dying) increases geometrically; is equal to a constant multiple of a power of a constant, the exponent being the age for which the force of mortality is being determined. See MAKEHAM —Makeham's law.

GOO′GOL, *n.* (1) The digit 1 followed by 100 zeros; *i.e.*, 10^{100}. (2) Any extremely large number.

GOSSET, William Sealy (1876–1937). English industrial statistician. Wrote under pseu. Student. See $T - t$ distribution.

GRAD, *n.* One-hundredth part of a right-angle in the centesimal system of measuring angles, also called a **grade** or **degree**.

GRADE, *n.* (1) The slope of a path or curve. (2) The inclination of a path or curve, the angle it makes with the horizontal. (3) The sine of the inclination of the path, vertical rise divided by the length of the path. (4) An inclined path. (5) A class of things relatively equal. (6) A division or class in an elementary school. (7) A rating, given students on their work in a given course. (8) One-hundredth part of a right angle. See GRAD.

GRA'DI-ENT, *n.* (*Physics*) The rate at which a variable quantity, such as temperature or pressure, changes in value; in these instances, called *thermometric gradient*, and *barometric gradient*, respectively.

gradient of a function. The vector whose components along the coordinate axes are the partial derivatives of the function with respect to the variables. Written: $\operatorname{grad} f = \nabla f = i f_x + j f_y + k f_z$, where f_x, f_y and f_z are the partial derivatives of f, a function of x, y, and z. The gradient of f is a vector whose component in any direction is the derivative of f in that direction. Its direction is that in which the derivative of f has its maximum, and its absolute value is equal to that maximum. Grad f, evaluated at a point P: (x_1, y_1, z_1), is normal to the surface $f(x, y, z) = c$ at P, where c is the constant $f(x_1, y_1, z_1)$. See VARIATION—variation of a function on a surface.

method of conjugate gradients. See CONJUGATE.

GRAD'U-AT'ED, *adj.* Divided into intervals, by rulings or other marks, such as the graduations on a ruler, a thermometer or a protractor.

GRÄFFE (or GRAEFFE), Karl Heinrich (1799–1873). German-Swiss analyst.

Gräffe's method for approximating the roots of an algebraic equation with numerical coefficients. The method consists of replacing the equation by an equation whose roots are the 2^kth power of the roots of the original equation. If the roots $r_1, r_2, r_3, \cdots$ are real and such that $|r_1| > |r_2| > |r_3| > \cdots$, then k can be made large enough that the ratio of $r_1^{2^k}$ to the coefficient of the next to the highest degree term is numerically as near 1 as one desires and also the ratio of $r_1^{2^k} r_2^{2^k}$ to the coefficient of the third highest degree term is numerically as near 1 as desired, etc. From these relations, $|r_1|, |r_2|, \cdots$ can be determined. If the roots are complex or equal, variations of this method can be used to obtain them.

GRAM, *n.* A unit of weight in the metric system; one thousandth of a standard kilogram of platinum preserved in Paris. It was intended to be the weight of one cubic centimeter of water at 4° C (the temperature at which its density is a maximum), and this is very nearly true. See DE-NOMINATE NUMBERS in the appendix.

GRAM, Jörgen Pedersen (1850–1916). Danish number theorist and analyst.

Gram-Charlier series. (1) **Type A:** a series used in a certain system of deriving frequency functions based on a Fourier integral theorem. In particular, the frequency function

$$f(x) = \frac{1}{\sqrt{2\pi}} e^{-(x^2/2)} \left[1 + \frac{1}{3!} u_3 H_3 \right.$$
$$\left. + \frac{u_4 - 3}{4!} H_4 + \cdots \right],$$

where x is in standard deviation units, u_i is the ith moment, and H_i are *Hermite polynomials*. Successive terms in the series do not necessarily diminish monotonically. Thus a satisfactory approximation may not be obtained by the first few terms. This is essentially a system of representing a given function by means of a series of derivatives of the normal distribution curve. (2) **Type B:** A *Poisson distribution*, instead of the normal, is used as a base for the series.

Gram-Schmidt process. The process of forming an orthogonal sequence $\{y_n\}$ from a linearly independent sequence $\{x_n\}$ of members of an inner-product space by defining y_n inductively as follows:

$$y_1 = x_1, \quad y_n = x_n - \sum_{k=1}^{n-1} \frac{(x_n, y_k)}{\|y_k\|^2} y_k, \quad n \geq 2.$$

To obtain an orthonormal sequence, one can replace each y_n by $y_n / \|y_n\|$ or use the auxiliary sequence $\{u_n\}$ and the formulas $u_1 = x_1$, $y_1 = u_1 / \|u_1\|$,

$$u_n = x_n - \sum_{k=1}^{n-1} (x_n, y_k) y_k, \qquad y_n = \frac{u_n}{\|u_n\|}.$$

If the inner product is complex-valued and $(ax, y) = \bar{a}(x, y)$, then (x_n, y_k) should be replaced by $\overline{(x_n, y_k)}$ in these formulas. See INNER—inner-product space.

GRAM'I-AN, *n.* (1) For n vectors $u_1, u_2, \cdots, u_n$ in n dimensions, the determinant with $u_i \cdot u_j$ as the element in the ith row and jth column, where $u_i \cdot u_j$ is the *scalar product* of u_i and u_j (the *Hermitian scalar product* if u_i and u_j have complex components). This determinant being zero is a necessary and sufficient condition for the linear dependence of $u_1, \cdots, u_n$. (2) For n functions $\phi_1, \phi_2, \cdots, \phi_n$, the determinant with $\int_\Omega \phi_i \phi_j \, d\Omega$ as the element in the ith row and jth column. This determinant is zero if and only if the functions ϕ_i are linearly dependent in the interval or region of integration Ω if suitable restrictions are satisfied by the functions. *E.g.,* (a) that each ϕ_i be continuous; or (b) that each ϕ_i be (Lebesgue) measurable and $|\phi_i|$ be (Lebesgue) integrable (linearly dependent here meaning that there exist $a_1, \cdots, a_n$, not all zero, such that

$$\sum_{i=1}^{n} a_i \phi_i = 0$$

almost everywhere in Ω). Under condition (b), the Gramians (1) and (2) become equivalent when the vectors and functions are regarded as elements of *Hilbert space. Syn.* Gram determinant.

GRAPH, *n., v.* (1) As a verb, "to draw the graph of". See below, bar graph, broken-line graph, etc. (2) See GRAPH THEORY. (3) A drawing which shows the relation between certain sets of numbers (see below, bar graph, broken line graph, circular graph). Used to convey a better idea of the meaning of the data than is evident directly from the numbers. (4) A representation of some quantity by a geometric object, such as the representation of a complex number by a point in the plane (see COMPLEX —complex number). (5) A drawing which depicts a functional relation. *E.g.,* the **graph of an equation in two variables** is (*in the plane*) the curve which contains those points, and only those points, whose coordinates satisfy the given equation. *In space,* it is the cylinder which contains those points, and only those points, whose coordinates satisfy the given equation (*i.e.,* whose right section is the graph in a plane of the given equation). The **graph of an equation in three variables** is a surface which contains those points and only those points whose coordinates satisfy the equation. The graph of a first-degree linear equation in Cartesian coordinates is a straight line in the plane or a plane in three dimensions. The graph of a set of simultaneous equations is either: (1) The graphs of all the equations, showing their intersections, or: (2) The intersection of the graphs of the equations. The **graph of a function** f is the set of all ordered pairs $[x, f(x)]$ and sometimes is not distinguished from the function [see FUNCTION]. Thus the graph of f is the same as the graph of the equation $y = f(x)$. Also see INEQUALITY —graph of an inequality.

bar graph. A graph consisting of parallel bars (see figure) whose lengths are proportional to certain quantities given in a set of data.

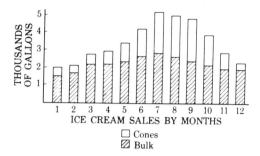

ICE CREAM SALES BY MONTHS
☐ Cones
▨ Bulk

broken-line graph. A graph formed by segments of straight lines which join the points representing certain data. The days during a certain period of time might be indicated by successive, equally spaced points on the *x*-axis and ordinates drawn at each point proportional in length to the highest temperature on those days. If the upper ends of these ordinates are connected by line segments, a broken line graph results. *Syn.* line graph.

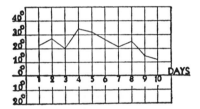

circular graph. A compact scheme for geometrically comparing parts of a whole to the whole. The whole is represented by the area of a circle, while the parts are represented by the areas of sectors of the circle.

graph theory. See GRAPH THEORY.

GRAPH'IC-AL or GRAPH'IC, *adj.* Pertaining to graphs, or scale drawings; working with scale drawings rather than with algebraic tools.

graphical solution. A solution obtained (approximately) by graphical or geometric methods. For example, the real roots of an equation $f(x) = 0$ can be found by sketching the graph and estimating the points where the graph crosses the *x*-axis; solutions of an equation such as $e^x = 5 + \ln x$ can be estimated by drawing graphs of $y = e^x$ and $y = 5 + \ln x$ and estimating the points of intersection.

GRAPH'ING, *n.* Drawing the graph of an equation or the graph representing a set of data. See CURVE —curve tracing.

graphing by composition. A method of graphing which consists of writing the given function as the sum of several functions whose graphs are easier to draw, plotting each of these functions, then adding the corresponding ordinates. The graph of $y = e^x - \sin x$ can be readily drawn by drawing the graphs of each of the equations $y = e^x$ and $y = -\sin x$, then adding the ordinates of these two curves, which correspond to the same values of x. *Syn.* graphing by composition of ordinates.

graph theory. The study of graphs, where a **graph** is an abstract mathematical system that consists of a set of objects called **nodes** (or **vertices** or **points**), a set of objects called **edges** (or **arcs, lines,** or **segments**), and a function f (the **incidence function**) defined on the set of edges that assigns to each directed edge exactly one ordered pair of nodes and to each undirected edge exactly one unordered pair. If a node is assigned to an edge by the function f, then the node and edge are said to be **incident** with each other. A graph is **directed** or **undirected** according as all its edges are directed or all are undirected. Edges assigned the same pair of nodes are **parallel edges** or **multiple edges.** A **loop** is an edge that joins a node to itself [*i.e.*, its pair of nodes is of type (x, x)]. Other edges are called **links.** The **valence** of a node is the number of edges incident with the node, with loops counted twice. A graph is **complete** if any two distinct nodes are joined by exactly one edge; it is **connected** if any two nodes can be joined by moving along edges. A **component** of a graph is a maximal connected subgraph. Graph-theoretical methods were used first by Euler in 1736 (see KÖNIGSBERG-BRIDGE PROBLEM), who showed that a connected graph can be traversed in such a way as to end at the starting point and cover each edge exactly once if and only if each node has even valence. Such a path is said to be an **Eulerian path** and the graph is said to be an **Eulerian graph.** See CIRCUIT, COLORING—graph coloring, HAMILTON—Hamiltonian graph, PATH (2), PLANAR—planar graph, TREE.

GRAV′I-TA′TION, *n.* **law of universal gravitation.** The law of attraction, formulated by Newton, in accordance with which two particles of masses m_1 and m_2 interact so that the force of attraction is proportional to the product of the masses and varies inversely as the square of the distance between the particles. In symbols, $F = k\dfrac{m_1 m_2}{r^2}$, where r is the distance between the particles and k is the universal **constant of gravitation** whose value, determined by experiments, in the c.g.s. system of units, is 6.675×10^{-8} cm.^{3} per gram sec^2. See GRAVITY—acceleration of gravity.

GRAV′I-TY, *n.* **acceleration of gravity.** See AC-CELERATION—acceleration of gravity.

center of gravity. See CENTER—center of mass.

GREAT, *adj.* **great circle.** See CIRCLE—great circle.

GREAT′ER, *adj.* One **cardinal number** is greater than a second when the set of units represented by the second is a part of that represented by the first, but not conversely; one cardinal number is greater than a second if the units of the first can be paired one-to-one with a subset of the units of the second, but not conversely. *E.g.*, 5 is greater than 3, since any set of 5 objects contains a set of 3 objects, but no set of 3 objects contains a set of 5 objects. One **real number** is greater than a second when a positive number can be added to the second to yield a number equal to the first; one real number is greater than a second when it is to the right of the second in the number scale: $\cdots -4, -3, -2, -1, 0, 1, 2, 3, 4, \cdots$. Thus 3 is greater than 2 (written $3 > 2$); and $-2 > -3$, because 1 must be added to -3 to yield -2. For **ordinal numbers** α and β which have ordinal types corresponding to well-ordered sets, α is greater than β if $\alpha \neq \beta$ and any set of ordinal type β can be put into a one-to-one, order-preserving correspondence with an initial segment of any set of ordinal type α. For any numbers A and B, the statements "A is **less than** B" and "B is **greater than** A" are equivalent.

GREAT′EST, *adj.* **greatest common divisor, factor, or measure.** See DIVISOR—greatest common divisor.

GREEK, *adj.* **Greek alphabet.** See the appendix.

Greek numerals. (1) An early system of numerals used symbols for $1, 10, 10^2, 10^3, 10^4$, with a special symbol to indicate five of a kind. *E.g.*, 754 would have been written using the 5-symbol with a 100-symbol attached, then two 100-symbols, then a 10-symbol with a 5-symbol attached, then four 1-symbols. (2) The *Ionic*, or *alphabetic*, system also was based on 10. It assigned the 27 letters of the alphabet (including 3 that now are obsolete) to indicate the 27 numbers $1, 2, \cdots, 9$; $10, 20, \cdots, 90$; $100, 200, \cdots, 900$. *E.g.*, $732 = \psi\lambda\beta$, $884 = \omega\pi\delta$. Larger numbers were indicated by primes or other signs, as $2000 = \beta'$. The number 10,000 was indicated by M (for **myriad**); multiplication was used for multiples of 10,000, as $20,000 = \beta M$ and $4,000,000 = \nu M$ ($\nu = 400$). Archimedes developed a further system that could represent fantastically large numbers.

GREEN, George (1793–1841). English analyst and applied mathematician.

Green's formulas:

(1)
$$\int_V u\,\nabla^2 u\,dV + \int_V \nabla u \cdot \nabla u\,dV$$

$$= \int_S u\,\partial u/\partial n\,d\sigma;$$

(2)
$$\int_V u\,\nabla^2 v\,dV + \int_V \nabla u \cdot \nabla v\,dV$$

$$= \int_S u\,\partial v/\partial n\,d\sigma;$$

(3)
$$\int_V u\,\nabla^2 v\,dV - \int_V v\,\nabla^2 u\,dV$$

$$= \int_S (u\,\partial v/\partial n - v\,\partial u/\partial n)\,d\sigma.$$

The second of these may be obtained from Green's theorem:

$$\int_V \nabla \cdot \phi\,dV = \int_S \phi \cdot \mathbf{n}\,d\sigma$$

by taking ϕ to be $u\nabla v$ so that $\nabla \cdot \phi = u\nabla^2 v + \nabla u \cdot \nabla v$. The first is the special case of the second with $v = u$, and the third may be obtained from the second by permuting u and v and subtracting. The volume integrals are taken over a set V which meets the requirements of the divergence theorem, while the surface integrals are taken over the boundary S of V. The symbol $\partial u/\partial n$ denotes the directional derivative of u in the direction of the exterior normal to the surface, i.e., $\partial u/\partial n = \nabla u \cdot \mathbf{n}$ if $\mathbf{n}$ is the unit exterior normal.

Green's function. For a region R with boundary surface S, and for a point Q interior to R, the Green's function $G(P, Q)$ is a function of the form $G(P, Q) = 1/(4\pi r) + V(P)$, where r is the distance PQ, $V(P)$ is harmonic, and G vanishes on S. The solution $U(Q)$ of the Dirichlet problem can be represented in the form

$$U(Q) = -\int_S f(P)\,\frac{\partial G(P, Q)}{\partial n}\,d\sigma_P.$$

Green's functions, Neumann's functions, and Robin's functions are sometimes called *Green's functions of the first, second, and third kinds*, respectively. See BOUNDARY—first boundary value problem of potential theory (the Dirichlet problem).

Green's theorem. (1) *In the plane*: Let R be a bounded open set in the plane whose boundary C is a rectifiable simple closed curve. Then *Green's theorem* is any theorem which states that, with certain specific conditions on L and M, the line integral of $L\mathbf{i} + M\mathbf{j}$ around C in the positive direction is equal to the integral over R of $\dfrac{\partial M}{\partial x} - \dfrac{\partial L}{\partial y}$ (the positive direction for C is the direction for which the winding number of C is $+1$ with respect to each point of R; intuitively, R is to be on the left as one moves around C). Sufficient conditions on L and M are that L and M be continuous on the union of R and C and $\partial L/\partial y$ and $\partial M/\partial x$ be bounded and continuous on R. Green's theorem is the special case of Stokes' theorem when the surface lies in the (x, y)-plane. If C is piece-wise smooth, Green's theorem can be written as

$$\int_C L\,dx + M\,dy = \int_R \left(\frac{\partial M}{\partial x} - \frac{\partial L}{\partial y} \right) dA.$$

See INTEGRAL —line integral. (2) *In space*: Same as DIVERGENCE THEOREM.

GREGORY, James (1638–1675). Scottish astronomer, algebraist, and analyst. First to distinguish between convergent and divergent series, employed power series to determine function values, and used the limiting processes of differential and integral calculus and understood the relation between them. His methods and results were little known or appreciated by his contemporaries.

Gregory-Newton formula. A formula for interpolation. If $x_0, x_1, x_2, x_3, \cdots$ are successive values of the argument, and $y_0, y_1, y_2, y_3, \cdots$ the corresponding values of the function, the formula is:

$$y = y_0 + k\Delta_0 + \frac{k(k-1)}{2!}\Delta_0^2$$

$$+ \frac{k(k-1)(k-2)}{3!}\Delta_0^3 + \cdots,$$

where $k = (x - x_0)/(x_1 - x_0)$, x (the value of the argument for which y is being computed) lies between x_0 and x_1,

$$\Delta_0 = y_1 - y_0,$$

$$\Delta_0^2 = y_2 - 2y_1 + y_0,$$

$$\Delta_0^3 = y_3 - 3y_2 + 3y_1 - y_0,$$

etc., the coefficients in Δ_0^n being the binomial coefficients of order n. If all the terms in the formula except the first two are dropped, the result is the ordinary interpolation formula used with logarithmic and trigonometric tables and in approximating roots of an equation, namely

$$y = y_0 + [(x-x_0)/(x_1-x_0)](y_1-y_0).$$

(This is, incidentally, the two-point form of the equation of a straight line.)

GROSS, *adj.*, *n.* Twelve dozen; 12×12.

gross capacity, price, profit, etc. The totality before certain parts have been deducted to leave the balance designated by the term *net*. *E.g.*, the *gross profit* is the sale price minus the initial or first cost. When the overhead charges have been deducted the remainder is the *net profit*.

gross premium. See PREMIUM.

GROTHENDIECK, Alexandre (1928–). French mathematician and Fields Medal recipient (1966), noted for his work in algebraic geometry, functional analysis, group theory, and *K*-theory.

GROUP, *n.* A set G with a binary operation (usually called multiplication) whose domain is the set of all ordered pairs of members of G, whose range is contained in G, and which satisfies the conditions: (1) there is a member of G (called the **identity** or **unit element**) such that its product with any member, in either order, is that same member; (2) for each member of G there is a member (called the **inverse**) such that the product of the two, in either order, is the identity; (3) the *associative law* holds. The cube roots of unity form a group under ordinary multiplication. The positive and negative integers and zero form a group under ordinary addition, the identity being zero and the inverse of an integer its negative. A group is **Abelian** (or **commutative**) if (in addition to the four assumptions listed above) it satisfies the commutative law. *I.e.*, $ab = ba$, where a and b are any two members of the group. A group for which all members are powers of one member is **cyclic**. A cyclic group is necessarily Abelian. The cube roots of unity form

a cyclic, Abelian group. A group having only a finite number of members is a **finite group**; otherwise it is an **infinite group** (the set of all integers, with ordinary addition, is an infinite group). The number of members of a finite group is the **order** of the group (see PERIOD—period of a member of a group). A group whose members are members of another group (and subject to the same rule of combination in both groups) is a **subgroup** of the latter [see LAGRANGE—Lagrange's theorem, and SYLOW—Sylow's theorem]. The group consisting of the cube roots of unity is a subgroup of the group consisting of the sixth roots of unity, the binary operation being ordinary multiplication. The product of any two members of a subgroup is in the subgroup, but the product of a member of the subgroup and a member of the group not in the subgroup is not in the subgroup.

alternating group. See PERMUTATION —permutation group.

composite group. See below, simple group.

direct product of groups. See PRODUCT—direct product of groups.

free group. See FREE—free group.

full linear group. The full linear group (of dimension n) is the group of all non-singular matrices of order n with complex numbers as elements and matrix multiplication as the group operation.

fundamental group. See FUNDAMENTAL.

group character. See CHARACTER.

group of symmetries. See SYMMETRY.

group without small subgroups. A topological group for which there is a neighborhood U of the identity such that the only subgroup completely contained in U is the subgroup consisting of the identity alone.

Lie group. See LIE.

modular group. See MODULAR.

order of a finite group. See above, GROUP.

perfect group. See COMMUTATOR.

permutation group. See PERMUTATION —permutation group.

quotient (or factor) group. See QUOTIENT—quotient space.

real linear group. The real linear group (of dimension n) is the group of all nonsingular matrices of order n with real numbers as elements and matrix multiplication as the group operation. See above, full linear group.

representation of a group. See REPRESENTATION.

semigroup. See SEMI.

simple group. A group that has no *normal subgroups* other than the identity alone and the whole group. A group which is not simple is said to be **composite**. The only cyclic simple groups are the cyclic groups of prime order. All alternating groups of degree 5 or more are simple. All finite non-cyclic simple groups have even order (see THOMPSON). All finite simple groups can now be described explicitly (but not easily!).

solvable group. A group G that contains a finite sequence of subgroups, $N_0, N_1, \cdots, N_k$, for which $N_0 = g$, N_k contains only the identity, each N_i is a normal subgroup of N_{i-1}, and each quotient group N_{i-1}/N_i is abelian. The meaning of the definition is not changed if "abelian" is replaced by "cyclic" or by "of prime order." All finite groups whose orders are odd or less than 60 are solvable.

symmetric group. A group of all permutations on n letters. See PERMUTATION —permutation group.

topological group. See TOPOLOGICAL —topological group.

GROUPING TERMS. A method of factoring consisting of rearranging terms, when necessary, inserting parentheses, and taking out a factor; e.g.,

$$x^3 + 4x^2 - 8 - 2x = x^3 + 4x^2 - 2x - 8$$
$$= x^2(x+4) - 2(x+4)$$
$$= (x^2 - 2)(x+4).$$

GROUP'OID, *n.* A set S with a binary operation whose domain is the set of all ordered pairs of members of S and whose range is contained in S. For example, the set V of ordinary vectors is a groupoid if the operation is *vector multiplication* (since such multiplication is not associative, V is not a *semigroup*). A **unit element** is a member e for which $x \circ e = e \circ x = x$ for all x, where the binary operation is denoted by $\circ$. See GROUP, SEMIGROUP.

GROWTH, *adj.* **growth curve** (*Statistics*). A curve designed to indicate the general pattern of growth of some variable. These are of several types. See GOMPERTZ —Gompertz curve, LOGISTIC—logistic curve.

GUDERMANN, Christof (1798–1852). German analyst and geometer.

Gudermannian. The function u of the variable x defined by the relation $\tan u = \sinh x$; u and x also satisfy the relations $\cos u = \operatorname{sech} x$ and $\sin u = \tanh x$. The Gudermannian of x is written gd x.

GUNTER, Edmund (1581–1626). English

mathematician and astronomer. Invented logarithmic scale on the line, leading to the development of the slide rule.

GY-RA'TION, *n.* **radius of gyration.** See RADIUS.

H

HAAR, Alfréd (1885–1933). Hungarian analyst.

Haar functions. The functions $\{y_n\}$ for which $y_1 \equiv 1$ on the interval $[0, 1]$ and, for each pair of positive integers (r, k) with $1 \le k \le 2^{r-1}$, $y_{2^{r-1}+k}$ is identically $2^{\frac{1}{2}(r-1)}$ on I^r_{2k-1}, identically $-2^{\frac{1}{2}(r-1)}$ on I^r_{2k}, and identically 0 on the rest of $[0, 1]$, where $[0, 1]$ is partitioned into 2^r intervals $\{I^r_j : 1 \le j \le 2^r\}$ of equal lengths. If $1 \le p < \infty$, the sequence of Haar functions is a basis for L^p of $[0, 1]$. If $1 < p < \infty$, this basis is unconditional; it is orthonormal if $p = 2$. See RADEMACHER — Rademacher functions.

Haar measure. See MEASURE —Haar measure.

HADAMARD, Jacques Salomon (1865–1963). Great French analyst, functional analyst, algebraist, number theorist, and mathematical physicist. Defended Alfred Dreyfus (a relative) and was active in peace movements after World War II. See PRIME —prime-number theorem.

Hadamard conjecture. The wave equation for $3, 5, \cdots$ space dimensions satisfies the Huygens principle, while that for 1 or an even number of space dimensions does not. The Hadamard conjecture is that no equation essentially different from the *wave equation* satisfies Huygens principle. See HUYGENS —Huygens principle.

Hadamard inequality. For a determinant of order n and value D, with real or complex elements a_{ij}, the inequality

$$|D|^2 \le \prod_{i=1}^{n} \left(\sum_{j=1}^{n} |a_{ij}|^2 \right).$$

Hadamard three-circles theorem. The theorem that if the complex function f is analytic in the ring $a < |z| < b$, and $m(r)$ denotes the maximum of $|f(z)|$ on a concentric circle of radius r in the ring, then $\log m(r)$ is a convex function of $\log r$. The name of the theorem, coined by Landau, reflects the fact that three radii are needed in order to express the convexity inequality. The result has been extended by Hardy to mean-value functions $m_t(r)$ of arbitrary nonnegative order t, of which the maximum-value function $m(r)$ is the limiting case as $t \to +\infty$.

HAHN, Hans (1879–1934). Austrian analyst and topologist.

Hahn-Banach theorem. Let L be a linear subset contained in a Banach space B. Let f be a real-valued continuous linear functional defined on L. Then there is a real-valued continuous linear functional F defined on all of B such that $f(x) = F(x)$ if x is in L and the norm of f on L is equal to the norm of F on B. If B is a complex Banach space, then f and F can be complex-valued. See CONJUGATE—conjugate space.

HALF, *adj.* **half-angle and half-side formulas.** See TRIGONOMETRY.

half-line. See RAY.

half-plane. The part of a plane which lies on one side of a line in the plane. It is an **open half-plane** or a **closed half-plane** according as the line is not included or is included. The line is the **boundary** or **edge** of the half-plane in either case. *Syn.* side of a line.

half-space. The part of space that lies on one side of a plane. It is an **open half-space** or a **closed half-space** according as the plane is not included or is included. The plane is the **boundary** or **face** of the half-space in either case.

HAM, *adj.* **ham-sandwich theorem.** (1) If, with respect to some concept of limit, the functions f and h have the same limit L, and if $f(x) \le g(x) \le h(x)$ for all x, then g also has the limit L. (2) If X, Y, and Z are three bounded, connected, open sets in space, then there is a plane which cuts each set into two sets with equal volume (or Lebesgue measure); thus, a ham sandwich can be cut with one stroke of a knife so that the ham and each slice of bread are cut in half.

HAMEL, Georg Karl Wilhelm (1877–1954). German analyst and applied mathematician.

Hamel basis. If L is a vector space whose scalar multipliers are the elements of a field F, then it can be shown (using Zorn's Lemma) that there exists a set B of elements of L (a **Hamel basis for** L) which has the properties that the elements of a finite subset of B are *linearly independent* and each element of L can be written as a *finite linear combination* (with coefficients in F) of elements of B. E.g., there is a Hamel basis B (necessarily noncountable) for the real numbers regarded as a vector space with

rational numbers as scalar multipliers; each nonzero real number x can be written as $\sum_{1}^{n} x_i b_{a_i}$ in exactly one way, with the x_i's nonzero rational numbers and the b_{a_i}'s in B.

HAMILTON, William Rowan (1805–1865). Great Irish algebraist, astronomer, and physicist. See QUATERNION.

Hamilton-Cayley theorem. The theorem that every matrix satisfies its characteristic equation. See CHARACTERISTIC —characteristic equation of a matrix.

Hamiltonian. (1) In *classical particle mechanics*, a *function* of n generalized coordinates q_i and momenta p_i, commonly symbolized by H and defined by

$$H = \sum_{i=1}^{n} p_i \dot{q}_i - L,$$

where p_i is the generalized momentum associated with q_i ($p_i = \partial L / \partial \dot{q}_i$), $\dot{q}_i$ is the first time derivative of the ith **generalized coordinate**, L is a **Lagrangian function**. If the Lagrangian function does not contain time explicitly, then H is equal to the total energy of the system. H satisfies the canonical equations of motion

$$\frac{\partial H}{\partial p_i} = \dot{q}_i, \quad \frac{\partial H}{\partial q_i} = -\dot{p}_i \quad (i = 1, \cdots, n).$$

(2) In *quantum theory*, an *operator* H which gives the equation of motion for the wave function ψ in the form

$$i\hbar \frac{\partial \psi}{\partial t} = H\psi.$$

Hamiltonian graph. A graph which has a closed path that passes each node exactly once. Such a path is a **Hamiltonian path** or **Hamiltonian circuit**. See CIRCUIT, GRAPH THEORY, PATH (2).

Hamilton's principle. The principle that over short intervals of time, and in a conservative field of force, a particle moves in such a way as to minimize the action integral $\int_{t_1}^{t_2} (T - U)\, dt$, where $T = \frac{1}{2} m \Sigma \dot{q}_i \cdot \dot{q}_i$ denotes kinetic energy and $U = U(q_1, q_2, q_3)$ is the potential function satisfying $m\ddot{q}_i = -U_{q_i}$. Thus (in a conservative field of force) trajectories are extremals of the action integral.

HAND, *n.* A unit of linear measure equal to four inches.

HAN′DLE, *n.* **handle of a surface.** See GENUS —genus of a surface.

HANKEL, Hermann (1839–1873). German analyst and geometer. Proved that the system of complex numbers cannot be enlarged if all field properties are to be preserved.

    Hankel function. A function of one of the types

$$H_n^{(1)}(z) = \frac{i}{\sin n\pi}\left[e^{-n\pi i}J_n(z) - J_{-n}(z)\right]$$

$$= J_n(z) + iN_n(z),$$

$$H_n^{(2)}(z) = \frac{-i}{\sin n\pi}\left[e^{n\pi i}J_n(z) - J_{-n}(z)\right]$$

$$= J_n(z) - iN_n(z),$$

where J_n and N_n are Bessel and Neumann functions (limits of these expressions are used when n is a nonzero integer). The Hankel functions are solutions of Bessel's differential equation (if n is not an integer). Both $H_n^{(1)}$ and $H_n^{(2)}$ are unbounded near zero; they behave exponentially at ∞ (like e^{iz} and e^{-iz}, respectively). Also called *Bessel functions of the third kind.*

HARDY, Godfrey Harold (1877–1947). Outstanding English analyst and number theorist. Famous for his collaborations with Littlewood. Did research on geometry of numbers, Diophantine approximations, Waring's problem, Fourier series, real and complex variable theory, and inequalities. See RIEMANN—Riemann hypothesis about the zeros of the zeta function.

    Hardy space. If $0 < p \le \infty$, the **Hardy space** H^p is the set of all functions f of a complex variable z that are analytic for $|z| < 1$ and for which

$$\|f\| = \lim_{r\to 1}\left\{\frac{1}{2\pi}\int_{-\pi}^{\pi}\left|f(re^{i\theta})\right|^p d\theta\right\}^{1/p} < \infty$$

if $0 < p < \infty$, and $\|f\| = \sup_{|z|<1}|f(z)| < \infty$ if $p = \infty$. If $1 \le p \le \infty$, H^p is a Banach space. If $1 < p < \infty$ and $1/p + 1/q = 1$, then H^p is the conjugate space of H^q. The conjugate space of H^1 is the space BMO of functions of bounded mean oscillation; *i.e.*, the space of functions which on each interval differs in the mean from its mean value by a bounded amount. See ANALYTIC—analytic function (3), CONJUGATE —conjugate space.

HAR-MON′IC, *adj.* **damped harmonic motion.** The motion of a body which would have simple harmonic motion except that it is subjected to a resistance proportional to its velocity. The equation of motion is

$$x = ae^{-ct}\cos(kt + \phi).$$

The exponential factor continuously reduces the amplitude. The differential equation of the motion is

$$\frac{d^2x}{dt^2} = -(c^2 + k^2)x - 2c\frac{dx}{dt}.$$

    harmonic analysis. The study of the representation of functions by means of linear operations (*summation* or *integration*) on characteristic sets of functions; in particular, the representation by means of *Fourier series.*

    harmonic average. See MEAN.

    harmonic conjugates of two points. See CONJUGATE—harmonic conjugates with respect to two points.

    harmonic division of a line. A line segment is divided *harmonically* when it is divided externally and internally in the same ratio. See RATIO —harmonic ratio.

    harmonic function. (1) A function u which satisfies Laplace's equation in two variables:

$$\frac{\partial^2 u}{\partial x^2} + \frac{\partial^2 u}{\partial y^2} = 0.$$

Some kind of regularity condition is usually assumed, such as that u has continuous partial derivatives of the first and second order in some given region. Two harmonic functions u and v are said to be **conjugate harmonic functions** if they satisfy the *Cauchy-Riemann partial-differential equations*; *i.e.*, if and only if $u + iv$ is an *analytic function* (it is assumed here that u and v have continuous first-order partial derivatives). The conjugate of a harmonic function can be found by integration, using the Cauchy-Riemann equations. (2) A solution u of Laplace's equation in three variables:

$$\frac{\partial^2 u}{\partial x^2} + \frac{\partial^2 u}{\partial y^2} + \frac{\partial^2 u}{\partial z^2} = 0.$$

Some kind of regularity condition is usually assumed, such as that u has continuous partial derivatives of the first and second order in some given region. (3) Sometimes a function of type $A \cdot \cos(kt + \phi)$ or $A \cdot \sin(kt + \phi)$ is called a harmonic function, or a **simple harmonic function** (see below, simple harmonic motion). Then a sum such as $3\cos x + \cos 2x + 7\sin 2x$ is called a **compound harmonic function**.

harmonic means between two numbers. The other terms of a *harmonic sequence* of which the given numbers are the first and last terms; a single harmonic mean between two numbers x and y is their **harmonic mean**: $1/[\frac{1}{2}(1/x + 1/y)]$. The **harmonic mean** of n numbers is the reciprocal of the arithmetic mean of their reciprocals. See MEAN, and below, harmonic sequence.

harmonic ratio. See RATIO—harmonic ratio.

harmonic sequence. A sequence whose reciprocals form an arithmetic sequence. In *music*, strings of the same material, same diameter, and same torsion, whose lengths are proportional to terms in a harmonic sequence, produce harmonic tones. The sequence $\{1, \frac{1}{2}, \frac{1}{3}, \cdots, 1/n\}$ is a harmonic sequence. *Syn.* harmonic progression.

harmonic series. A series whose terms form a harmonic sequence.

simple harmonic motion. Motion like that of the projection upon a diameter of a circle of a point moving with uniform speed around the circumference; the motion of a particle moving on a straight line under a force proportional to the particle's distance from a fixed point and directed toward that point. If the fixed point is the origin and the x-axis the line, the acceleration of the particle is $-k^2x$, where k is a constant. *I.e.*, the equation of motion of the particle is

$$\frac{d^2x}{dt^2} = -k^2x.$$

The general solution of this equation is $x = a\cos(kt + \phi)$. The particle moves back and forth (oscillates) between points at a distance a on either side of the origin. The time for a complete oscillation is $2\pi/k$. The number a is the **amplitude** and $2\pi/k$ the **period**. The angle $\phi + kt$ is the **phase** and ϕ the **initial phase**.

spherical harmonic. A spherical harmonic of degree n is an expression of type

$$r^n \left\{ a_n P_n(\cos\theta) + \sum_{m=1}^{n} \left[a_n^m \cos m\phi + b_n^m \sin m\phi \right] \right.$$

$$\left. \times P_n^m(\cos\theta) \right\},$$

where P_n is a *Legendre polynomial* and P_n^m is an *associated Legendre function*. Any spherical harmonic is a homogeneous polynomial of degree n in x, y, and z and is a particular solution of

Laplace's equation (in spherical coordinates); any solution of Laplace's equation which is analytic near the origin is the sum of an infinite series $\sum_0^\infty H_n$, where H_n is a spherical harmonic of degree n.

surface harmonic. A surface harmonic of degree n is an expression of type

$$a_n P_n(\cos\theta) + \sum_{m=1}^{n} \left[a_n^m \cos m\phi \right.$$

$$\left. + b_n^m \sin m\phi \right] P_n^m(\cos\theta),$$

where P_n is a *Legendre polynomial* and P_n^m is an *associated Legendre function*. A surface harmonic of type $(\cos m\phi)P_n^m(\cos\theta)$ or $(\sin m\phi)P_n^m(\cos\theta)$ is a **tesseral harmonic** if $m < n$, and a **sectoral harmonic** if $m = n$; it is a solution of the differential equation

$$\frac{1}{\sin\theta}\frac{\partial}{\partial\theta}\left(\sin\theta\frac{\partial y}{\partial\theta}\right) + \frac{1}{\sin^2\theta}\frac{\partial^2 y}{\partial\theta^2}$$

$$+ n(n+1)y = 0.$$

A tesseral harmonic is zero on $n - m$ parallels of latitude and $2m$ meridians (on a sphere with center at the origin of spherical coordinates); a sectoral harmonic is zero along $2n$ meridians (which divide the surface of the sphere into sectors).

zonal harmonic. A function $P_n(\cos\theta)$, where P_n is the *Legendre polynomial* of degree n. The function $P_n(\cos\theta)$ is zero along n great circles on a sphere with center at the origin of a system of spherical coordinates (these circles pass through the poles and divide the sphere into n zones). See above, spherical harmonic.

HARVARD MARK I, II, III, IV. Certain automatic digital computing machines built at Harvard University.

HAUSDORFF, Felix (1868–1942). German analyst and general topologist.

Hausdorff dimension. See DIMENSION.

Hausdorff maximal principle. See ZORN—Zorn's lemma.

Hausdorff paradox. The theorem which states that it is possible to represent the surface S of a sphere as the union of four disjoint sets A, B, C, and D such that D is a countable set and A is congruent to each of the three sets B, C, and $B \cup C$. Thus, except for the countable set

D, A is both a half and a third of S. See BANACH —Banach-Tarski paradox.

Hausdorff space. See TOPOLOGICAL —topological space.

HA′VER-SINE, *n.* See TRIGONOMETRIC —trigonometric functions.

HEAT, *adj., n.* **heat equation.** The parabolic second-order partial differential equation

$$\frac{\partial u}{\partial t} = \frac{k}{c\rho}\left(\frac{\partial^2 u}{\partial x^2} + \frac{\partial^2 u}{\partial y^2} + \frac{\partial^2 u}{\partial z^2}\right),$$

where $u = u(x, y, z; t)$ denotes temperature, (x, y, z) are space coordinates, and t is the time variable; the constant k is the thermal conductivity of the body, c its specific heat, and ρ its density.

HEAVISIDE, Oliver (1850–1925). English electrical engineer. His methods of studying operators were developed experimentally rather than logically, but many of them have been rigorously justified by other investigators.

HEC′TARE, *n.* A unit of measure of area in the metric system, equal to 10,000 square meters or 2.471 acres.

HEC′TO. A prefix meaning 100, as in **hectometer.**

HEI, *adj.* **hei function.** See BER—ber function.

HEINE, Heinrich Eduard (1821–1881). German analyst.

Heine-Borel theorem. If S is a subset of a finite-dimensional Euclidean space, then S is compact if S is bounded and closed. The converse also is true: S is bounded and closed if S is compact. See COMPACT—compact set.

HEL′I-COID, *n.* A surface generated by a plane curve or a twisted curve which is rotated about a fixed line as axis and also is translated in the direction of the axis in such a way that the ratio of the two rates is constant. A helicoid can be represented parametrically by equations $x = u \cos v$, $y = u \sin v$, $z = f(u) + mv$. For $m = 0$, the helicoid is a surface of revolution; and for $f(u) = $ const., the surface is a special right conoid called a right helicoid. See below, right helicoid.

right helicoid. A surface that can be represented parametrically by the equations $x = u \cos v$, $y = u \sin v$, $z = mv$. It is shaped rather like a propeller screw. If u is held fixed, $u \neq 0$,

the equations define a helix (the intersection of the helicoid and the cylinder $x^2 + y^2 = u^2$). The right helicoid is the one and only real ruled minimal surface.

HE′LIX, *n.* A curve which lies on a cylinder or cone and cuts the elements under constant angle. It is a **cylindrical helix** or a **conical helix** according as it lies on a cylinder or a cone. If the cone is a right circular cylinder, the helix is a **circular helix** and its equations, in parametric form, are $x = a \sin \theta$, $y = a \cos \theta$, $z = b\theta$, where a and b are constants and θ is the parameter. The thread of a bolt may be a circular helix.

HELLY, Eduard (1884–1943). Austrian analyst, geometer, topologist, and physicist.

Helly's condition. Let $\{c_1, \cdots, c_n\}$ be n numbers, M a positive number, and $\{f_1, \cdots, f_n\}$ continuous linear functionals defined on a normed linear space X. Then

$$\left|\sum_1^n k_i c_i\right| \leq M\left\|\sum_1^n k_i f_i\right\| \quad \text{for all numbers}$$

$\{k_1, \cdots, k_n\}$ is a necessary and sufficient condition that, for each $\epsilon > 0$, there is an x in X such that $\|x\| \leq M + \epsilon$ and $f_i(x) = c_i$ for each i. If X is finite-dimensional (or if X is a reflexive Banach space), then ϵ can be 0. Helly's condition implies the usual condition for consistency of m linear equations in n unknowns. See CONSISTENCY—consistency of systems of equations.

Helly's theorem. There is a point that belongs to each member of a collection of bounded closed sets in n-dimensional Euclidean space if the collection has at least $n + 1$ members and any $n + 1$ members of the collection have a common point. See related theorems under CARATHÉODORY, RADON, STEINITZ.

HELMHOLTZ, Hermann Ludwig Ferdinand von (1812–1894). German physiologist, physician, and physicist.

Helmholtz' differential equation. The equation

$$L\frac{dI}{dt} + RI = E.$$

The equation is satisfied by the current I in a circuit which has resistance R and inductance L, where E is the impressed or external electromotive force.

HEM′I-SPHERE, *n.* A half of a sphere bounded by a great circle.

HENNEBERG, Ernst Lebrecht (1850–1933). German engineer and differential geometer. See SURFACE —surface of Henneberg.

HEP'TA-GON, *n.* A polygon having seven sides. A heptagon whose sides are all equal and whose interior angles are all equal is a **regular heptagon**.

HEP-TA-HE'DRON, *n.* [*pl.* **heptahedrons, heptahedra**]. A polyhedron with seven faces. There are 34 types of heptahedrons.

HER, *adj.* **her function.** See BER—ber function.

HER, *n.* A game in which one player deals one card to his opponent and one to himself, at random from an ordinary deck of cards. Each looks only at his own card. The dealer's opponent may elect to keep his own card or to exchange cards with the dealer, except that the dealer is not compelled to relinquish a king. Thereafter, the dealer may elect to keep the card he then has or to exchange it for a new card dealt from the deck, except that if the new card is a king he must keep the card he already has. High card wins. This is a game with both personal moves and chance moves. See MOVE.

HERMITE, Charles (1822–1901). French algebraist, analyst, and number theorist. Solved the general quintic equation in one variable by means of elliptic functions. Widely influential; trained many distinguished mathematicians. See *e*.

Hermite polynomials. The polynomials

$$H_n(x) = (-1)^n e^{x^2} \frac{d^n e^{-x^2}}{dx^n}.$$

The functions $e^{-1/2 x^2} H_n(x)$ are *orthogonal functions* on the interval $(-\infty, \infty)$, with

$$\int_{-\infty}^{\infty} \left[e^{-1/2 x^2} H_n(x) \right]^2 dx = 2^n n! \sqrt{\pi}.$$

The Hermite polynomial H_n is a solution of *Hermite's differential equation* with the constant $\alpha = n$. For all n, $H_n'(x) = 2nH_{n-1}(x)$ and

$$e^{x^2 - (t-x)^2} = \sum_{n=1}^{\infty} H_n(x) t^n / n!.$$

Hermite differential equation. The differential equation $y'' - 2xy' + 2\alpha y = 0$, where α is a constant. Any solution of the Hermite equation, multiplied by $e^{-1/2 x^2}$, is a solution of $y'' + (1 - x^2 + 2\alpha) y = 0$.

Hermite's formula of interpolation. An interpolation formula for functions of period 2π. The

formula, which is a trigonometric analogue of Lagrange's formula, is

$$f(x) = \frac{f(x_1) \sin(x - x_2) \cdots \sin(x - x_n)}{\sin(x_1 - x_2) \cdots \sin(x_1 - x_n)} + \cdots$$

to n terms. See LAGRANGE —Lagrange's formula of interpolation.

HERMITIAN, *adj.* **Hermitian conjugate of a matrix.** The transpose of the *complex conjugate* of the matrix. Called the **adjoint** of the matrix by some writers on quantum mechanics. *Syn.* associate matrix. See ADJOINT —adjoint of a transformation.

Hermitian form. A bilinear form in conjugate complex variables whose matrix is Hermitian; an expression of the form

$$\sum_{i,j=1}^{n} a_{ij} x_i \bar{x}_j,$$

where $a_{ij} = \bar{a}_{ji}$. See TRANSFORMATION—conjunctive transformation.

Hermitian matrix. A matrix which is its own Hermitian conjugate; a square matrix such that a_{ij} is the complex conjugate of a_{ji} for all i and j, where a_{ij} is the element in the ith row and jth column.

Hermitian transformation. For bounded linear transformations (which include any linear transformation of a finite-dimensional space), same as SELF-ADJOINT TRANSFORMATION, or SYMMETRIC TRANSFORMATION (see SELF, SYMMETRIC). For unbounded linear transformations, *Hermitian* usually means *self-adjoint*.

skew Hermitian matrix. A matrix which is the negative of its *Hermitian conjugate*; a square matrix such that a_{ij} is the complex conjugate of $-a_{ji}$ for all i and j, where a_{ij} is the element in the ith row and jth column.

HERON (or HERO) of Alexandria (1st century A.D.). Greek mathematician and physicist.

Hero's (or Heron's) formula. A formula expressing the area of a triangle in terms of the sides, a, b, c. It is

$$A = \sqrt{s(s-a)(s-b)(s-c)},$$

where $s = \frac{1}{2}(a + b + c)$.

HESSE, Ludwig Otto (1811–1874). German differential geometer.

Hessian. For a function f of n variables $x_1, x_2, \cdots, x_n$, the Hessian of f is the nth-order determinant whose element in the ith row and jth column is $\dfrac{\partial^2 f}{\partial x_i \partial x_j}$. The Hessian is analogous to the second derivative of a function of one variable, as a Jacobian is analogous to the first derivative. E.g., the Hessian of a function f of two variables x and y is

$$\frac{\partial^2 f}{\partial x^2} \frac{\partial^2 f}{\partial y^2} - \left(\frac{\partial^2 f}{\partial x \, \partial y} \right)^2,$$

which is useful in determining maxima, minima, and saddle points (see MAXIMUM, SADDLE — saddle point).

HEU-RIS′TIC, *adj.* **heuristic method.** A method of gaining better understanding of a problem by trying several approaches or techniques and evaluating the progress toward a solution after each attempt. A method of education or discovery that uses empirical methods.

HEX, *n.* A game first introduced as *polygon* in Denmark in 1942. The game is played with buttons of two colors on a board of n^2 hexagons with n hexagons along each edge (usually $n = 11$). The first player places a button in any hexagon except the center one (sometimes he is not allowed to play on the small diagonal). Thereafter, the players take turns placing a button on any unoccupied hexagon. The winner is the first player to complete a chain of hexagons joining his opposite sides. A game of hex can not end without a winner.

HEX-A-DEC′I-MAL, *adj.* **hexadecimal number system.** A system of numerals for representing real numbers for which 16 is the base. *Syn.* sexadecimal number system. See BASE — base of a number system.

HEX′A-GON, *n.* A polygon of six sides. It is **regular** if its sides are all equal and its interior angles are all equal. See PASCAL — Pascal's theorem.

simple hexagon. Six points, no three of which are collinear, and the six lines determined by joining consecutive vertices.

HEX-AG′O-NAL, *adj.* **hexagonal prism.** A prism having hexagons for bases. See PRISM.

HEX′A-HE′DRON, *n.* [*pl.* **hexahedrons** or **hexahedra**]. A polyhedron having six faces. A **regular hexahedron** is a cube. There are seven types of convex hexahedrons. See POLYHEDRON.

HIGH′ER, *adj.* **higher plane curve.** An algebraic plane curve of degree higher than the second. Sometimes includes transcendental curves.

HIGH′EST, *adj.* **highest common factor.** Same as GREATEST COMMON DIVISOR. See DIVISOR.

HILBERT, David (1862–1943). Great German mathematical universalist and philosopher. Leading mathematician of the twentieth century. Contributed to the theory of algebraic invariants, algebraic manifolds, number fields, class fields, integral equations, functional analysis, and applied mathematics. In 1899, proposed establishment of postulational basis for all mathematics and began with the foundations of geometry. In 1900, proposed 23 problems at Paris International Congress that have influenced mathematical research throughout this century. Professor at the Univ. of Göttingen from 1895 until his death. See WARING — Waring's problem.

Hilbert parallelotope. See PARALLELOTOPE.

Hilbert-Schmidt theory of integral equations with symmetric kernels. A theory built on the orthogonality of *eigenfunctions* corresponding to distinct *eigenvalues.* Some characteristic results are: (1) $K(x,t)$ has at least one eigenvalue, all eigenvalues are real, and $\phi_1(x)$ and $\phi_2(x)$ are orthogonal if they are eigenfunctions corresponding to unequal eigenvalues; (2) there is an *orthonormal sequence* of eigenfunctions ϕ_i corresponding to the eigenvalues λ_i (not necessarily all distinct) such that (a) if $\sum\limits_{n=1}^{\infty} \lambda_n \phi_n(x) \phi_n(t)$ is uniformly convergent for $a \leq x \leq b$, $a \leq t \leq b$, then it is equal to $K(x,t)$, (b) if for the function f there is a continuous function g such that

$$f(x) = \int_a^b K(x,t) g(t) \, dt,$$

then $f(x) = \sum\limits_{t=1}^{\infty} \lambda_i a_i \phi_i(x)$, where

$$a_i = \int_a^b g(t) \phi_i(t) \, dt,$$

the series converging absolutely and uniformly; (4) if f is continuous and λ is not an eigenvalue, then

$$\theta(x) = f(x) + \frac{1}{\lambda} \int_a^b K(x,t) \theta(t) \, dt$$

has a unique continuous solution θ given by

$$\theta(x) = f(x)$$

$$+ \sum_{n=1}^{\infty} \left[\frac{\lambda_n}{\lambda - \lambda_n} \int_a^b f(t)\phi_n(t)\,dt \right] \phi_n(x),$$

the series being absolutely and uniformly convergent; (5) if λ is an eigenvalue, there is a solution if, and only if, f is orthogonal to each eigenfunction corresponding to λ and the solution is given by the above (with the terms for which $\lambda_n = \lambda$ omitted) plus any linear combination of these eigenfunctions.

Hilbert space. A complete inner-product space [see INNER—inner-product space]. Examples of Hilbert space are: (1) The set of all sequences $x = (x_1, x_2, \cdots)$ of complex numbers, where $\sum_{i=1}^{\infty} |x_i|^2$ is finite. The sum $x + y$ is defined as $(x_1 + y_1, x_2 + y_2, \cdots)$, the product ax as $(ax_1, ax_2, \cdots)$, and the **inner product** as $(x, y) = \sum_{i=1}^{\infty} \bar{x}_i y_i$, where $x = (x_1, x_2, \cdots)$ and $y = (y_1, y_2, \cdots)$. (2) The set of all (Lebesgue) measurable functions f on an interval $[a, b]$ for which the integral of $|f|^2$ is finite. Two functions are considered identical if they are equal *almost everywhere* on (a, b); the operations of addition and multiplication by complex numbers are defined as ordinary addition and multiplication; and (f, g) is defined as $\int_a^b f(x)g(x)\,dx$. If ordinary (Riemann) integration is used, all postulates are satisfied except that of completeness. Each Hilbert space H contains a complete orthonormal set $B = \{x_\alpha\}$ which has the property that, if x is in H, then $x = \sum(x, x_\alpha)x_\alpha$ where all but a countable number of the coefficients (x, x_α) are zero and the order of summation is immaterial. Two Hilbert spaces are isomorphic if and only if they use the same scalars and their complete orthonormal sets have the same cardinal numbers. The isomorphism is required to preserve inner products as well as the operations of addition and multiplication by scalars, so one can say that there is essentially only one Hilbert space of a specified dimension that uses real scalars and one that uses complex scalars. See OPERATOR—linear operator, SPECTRAL—spectral theorem.

HIN'DU, *n.* **Hindu-Arabic numerals.** Same as ARABIC NUMERALS. See ARABIC.

HIPPOCRATES OF CHIOS (c. 440 B.C.). **lunes of Hippocrates.** Hippocrates "squared" certain lunes, perhaps adding fuel to the hopeless search for a technique of squaring the circle (see SQUARING). Following are examples. (1) Let ABC be an isosceles triangle with the angle at C a right angle. Form a lune by drawing two arcs of circles that connect A and B, one with center at C and the other with diameter AB. The area of the lune is equal to the area of the triangle. (2) Let ABC be a triangle inscribed in a semicircle. For each of the sides not on a diameter, form the lune bounded by an arc of the semicircle and by the semicircle (outside the triangle) which has that side as diameter. The sum of the areas of these two lunes is the area of the triangle. (3) Let $T = ABCD$ be half of a regular hexagon, with $AB = BC = CD$. For each of the three equal sides, form the lune bounded by an arc of the circumscribed circle of T and the semicircle outside T with that side as diameter. The area of T is equal to the sum of the areas of these three lunes plus the area of one of the semicircles that has one of the three equal edges as diameter.

HIRONAKA, Heisouke (1931–   ). Japanese-American algebraic geometer and Fields Medal recipient (1970). Research on algebraic varieties and resolution of singularities.

HIS'TO-GRAM, *n.* See FREQUENCY —frequency curve or diagram.

HITCHCOCK, Frank Lauren (1875–1957). American vector analyst and physicist. Formulated transportation problem in 1941.

Hitchcock transportation problem. See TRANSPORTATION.

HOD'O-GRAPH, *n.* If the velocity vectors of a moving particle be laid off from a fixed point, the extremities of these vectors trace out a curve called the *hodograph* of the moving particle. The *hodograph* of uniform motion in a straight line is a point. The *hodograph* of uniform motion in a circle is another circle whose radius is equal to the speed of the particle. If $\bar{v} = \bar{f}(t)$ is the vector equation of the path of the particle, $d\bar{v}/dt = \bar{f}'(t)$ is the equation of the *hodograph*.

HOLD'ER, *n.* (*Finance*) The one who owns a note, not necessarily the payee named in the note. See NEGOTIABLE.

HÖLDER, Ludwig Otto (1859–1937). German group theorist who also worked on summability of series.

Hölder condition. See LIPSCHITZ —Lipschitz condition.

Hölder definition of the sum of a divergent series. If the series is $\sum a_n$, Hölder's definition gives the sum as

$$\lim_{n \to \infty} s'_n = \lim_{n \to \infty} \frac{s_1 + \cdots + s_n}{n},$$

where $s_n = \sum_{i=1}^{n} a_i$, or

$$\lim_{n \to \infty} \frac{s'_1 + \cdots + s'_n}{n},$$

where $s'_n = \frac{1}{n} \sum_{i=1}^{n} s_i$, etc. This is the repeated application of the process of taking the average of the first n partial sums until a stage is reached where the limit of this average exits. This sum is *regular*. See REGULAR —regular definition of the sum of a divergent series.

Hölder inequality. Either of the inequalities

$$(1) \quad \sum_{1}^{n} |a_i b_i| \le \left[\sum_{1}^{n} |a_i|^p \right]^{1/p} \left[\sum_{1}^{n} |b_i|^q \right]^{1/q},$$

where n may be $+\infty$, or

$$(2) \quad \int_{\Omega} |fg| \, d\mu \le \left[\int_{\Omega} |f|^p \, d\mu \right]^{1/p} \left[\int_{\Omega} |g|^q \, d\mu \right]^{1/q},$$

which are valid if $p > 1$, $p + q = pq$, and both $|f|^p$ and $|g|^q$ are integrable on Ω. The integrals may be Riemann integrals, or μ may be a measure defined on a σ-algebra of subsets of Ω (see INTEGRABLE —integrable function). The numbers in (1) or the functions in (2) may be real or complex. Either of these inequalities is easily deduced from the other. If $p = q = 2$, they become Schwarz's inequalities. See MINKOWSKI —Minkowski's inequality, SCHWARZ.

HOL-O-MOR′PHIC, *adj.* **holomorphic function.** See ANALYTIC —analytic function of a complex variable.

HO′ME-O-MOR′PHISM, *n.* Same as TOPO-LOGICAL TRANSFORMATION.

HO′MO-GE′NE-OUS, *adj.* **homogeneous affine transformation.** See AFFINE —affine transformation.

homogeneous coordinates. See COORDINATE —homogeneous coordinates.

homogeneous differential equation. See DIFFERENTIAL.

homogeneous equation. An equation such that, if it is written with zero as the right-hand member, the left-hand member is a *homogeneous function* of the variables involved. For the solution of homogeneous linear equations, see CONSISTENCY —consistency of linear equations.

homogeneous function. A function such that if each of the variables is replaced by t times the variable, t can be completely factored out of the function whenever $t \neq 0$. The power of t which can be factored out of the function is the **degree of homogeneity** of the function. The functions $\sin x/y + x/y$ and $x^2 \log x/y + y^2$ are homogeneous. See below, homogeneous polynomial.

homogeneous integral equation. An integral equation which is homogeneous of the first degree in the unknown function. See FREDHOLM —Fredholm integral equations, and VOLTERRA —Volterra integral equation.

homogeneous polynomial. A polynomial whose terms are all of the same degree with respect to all the variables taken together; $x^2 + 3xy + 4y^2$ is homogeneous.

homogeneous solid. (1) A solid whose density is the same at all points. (2) A solid such that if congruent pieces be taken from different parts of it they will be alike in all respects.

homogeneous strains. See STRAIN.

homogeneous transformation. See TRANSFORMATION —homogeneous transformation.

HO-MO-GE-NE′I-TY, *n.* (*Statistics*) (1) k populations are **homogeneous** if the distribution functions are identical. (2) In a two-by-two table, the **test for homogeneity** is a test for the equality of the proportions in the two classifications. This test is also called a **test of independence**. No interaction is present if independence exists.

HO-MO-LOG′IC-AL, *adj.* **homological algebra.** A branch of algebra that uses tools from algebraic topology, especially homology theory. Particularly emphasized are the structure of modules and the use of categories and functors.

HO′MOL′O-GOUS, *adj.* **homologous elements** (such as terms, points, lines, angles). Elements that play similar roles in distinct figures or functions. The numerators or the denominators of two equal fractions are *homologous terms*. The vertices of a polygon and those of a projection of the polygon on a plane are *homologous points* and the sides and their projections are *homologous lines*. *Syn.* corresponding. Also see HOMOLOGY —homology group.

HO-MOL′O-GY, *adj.* **homology group.** Let K be an n-dimensional simplicial complex [or a topological simplicial complex or a complex in a suitable more general sense (see COHOMOLOGY —cohomology group)] and T^r be the set of all r-dimensional *cycles* of K, defined by use of a group G (see CHAIN—chain of simplexes). An r-dimensional cycle is **homologous to zero** if it is 0 or is the *boundary* of an $(r + 1)$-dimensional chain of K, while two r-dimensional cycles are **homologous** if their difference is homologous to zero. The commutative quotient group T^r/H^r, where H^r is the group of all cycles which are homologous to zero, is called an **r-dimensional homology group** or **Betti group**. The elements of a homology group are therefore classes of mutually homologous cycles. This definition depends on the particular group G whose elements are used as coefficients in forming chains. However, if the homology groups over the groups of integers are known, the homology groups over any group G can be determined. The zero-dimensional homology group is isomorphic with the group G. If G is the group of integers, the 1-dimensional homology group of the torus has two generators of infinite order (a small circle around the torus and a large circle around the "hole"); the 1-dimensional homology group of the surface of an ordinary sphere contains only the identity (any two 1-cycles are homologous, any 1-cycle being a boundary of a 2-chain). See FUNDAMENTAL —fundamental group.

HO′MO-MOR′PHISM, *n.* A function with domain D and range R which is subject to certain restrictions depending on the nature of D and R. It is an **endomorphism** if R is D or a subset of D. If D and R are **topological spaces**, it is required that the correspondence be continuous (see CONTINUOUS —continuous function). If operations such as multiplication, addition, or multiplication by scalars are defined for D and R, it is required that these correspond as described in the following. If D and R are **groups** (or semigroups) with the operation denoted by $\cdot$, and x corresponds to x^* and y to y^*, then $x \cdot y$ must correspond to $x^* \cdot y^*$. If D and R are **rings** (or integral domains or fields) and x corresponds to x^* and y to y^*, then xy must correspond to $x^* y^*$ and $x + y$ to $x^* + y^*$. If D and R are **vector spaces**, multiplication and addition must correspond as for rings and scalar multiplication must correspond in the sense that if a is a scalar and x corresponds to x^*, then ax corresponds to ax^*. If the vector space is normed (*e.g.*, if it is a Banach space or Hilbert space), then the correspondence must be continuous. This is equivalent to requiring that there be a number M such

that $\|x^*\| \le M\|x\|$ if x corresponds to x^* (a homomorphism of normed vector spaces is also a *bounded linear transformation*). See ISOMORPHISM, ISOMETRY, IDEAL.

HO′MO-SCE-DAS′TIC, *adj.* Having equal variance. *E.g.*, several *distributions* are homoscedastic if their variances are equal. In a *bivariate distribution*, one of the variables is homoscedastic if, for given values of the second variable, the variance of the first variable is the same regardless of the values of the second variable. In a *multi-variance distribution*, one of the variables is homoscedastic if its conditional distribution function has a constant variance regardless of the particular set of values of the other variable.

HO-MO-THET′IC, *adj.* **homothetic figures.** Figures so related that lines joining corresponding points pass through a point and are divided in a constant ratio by this point. See SIMILAR — similar sets of points.

   homothetic ratio. See RATIO—ratio of similitude.

   homothetic transformation. See SIMILITUDE —transformation of similitude.

HO-MO-TOP′IC, *adj.* See DEFORMATION— continuous deformation.

HOOKE, Robert (1635–1703). English physicist, chemist, and mathematician.

   Hooke's law. The basic law of proportionality of stress and strain published by Robert Hooke in 1678. In its simple form it states that within elastic limits of materials the elongation produced by the tensile force is proportional to the tensile force. If the elongation is denoted by e and the tensile stress by T, then $T = Ee$, where E is a constant depending on the properties of the material. The constant E is called the modulus of tension. This law is found experimentally to be valid for many substances when the forces and the deformations produced by them are not too great. See MODULUS—Young's modulus, and below, generalized Hooke's law.

   generalized Hooke's law. The law in the theory of elasticity asserting that, for sufficiently small strains, each component of the stress tensor is a linear function of the other components of this tensor. The coefficients of the linear forms connecting the components of these tensors are elastic constants. It is known that the general elastic medium requires 21 such constants for its complete characterization. A homogeneous, isotropic elastic medium is characterized by two constants, Young's modulus and Poisson's ratio. See MODULUS—Young's modulus, POISSON —Poisson's ratio.

HO-RI′ZON, *n.* **horizon of an observer on the earth.** The circle in which the earth, looked upon as a plane, appears to meet the sky; the great circle on the celestial sphere which has its pole at the observer's zenith. See HOUR—hour angle and hour circle.

HOR′I-ZON′TAL, *adj.* Parallel to the earth's surface looked upon as a plane; parallel to the plane of the horizon. *Tech.* In a plane perpendicular to the plumb line.

HÖRMANDER, Lars (1931–   ). Swedish analyst and Fields Medal recipient (1962). Research in the theory of partial differential equations and linear differential operators.

HORNER, William George (1786–1837). English algebraist.
    Horner's method. A method for approximating the real roots of an algebraic equation. Its essential steps are as follows: (1) Isolate a positive root between two successive integers (if the equation has only negative real roots, *transform* it to one whose roots are the negatives of those of the given equation). (2) *Transform* the equation into an equation whose roots are decreased by the lesser of the integers between which the root lies, by the substitution $x' = x - a$. The root of the new equation will lie between 0 and 1. (3) Isolate the root of the new equation between successive tenths. (4) *Transform* the last equation into an equation whose roots are decreased by the smaller of these tenths and isolate the root of this equation between hundredths. Continue this process to one decimal place more than the place to which the answer is to be correct. The root sought is then the total amount by which the roots of the original equation were reduced, the last decimal being rounded off to make the result accurate to the desired decimal place. Fractions may be avoided in locating the roots by transforming the equation in step (4) to one whose roots are ten times as large, and repeating the same transformation each time another digit in the root is sought. Synthetic division is generally used to expedite the work of substituting values for the variable. Roots often can be approximated quickly after the first step by solving the equation obtained by dropping the terms of higher degree than the first. See REMAINDER — remainder theorem, INTERPOLATION, REGULA FALSI.

HORSEPOWER. A unit of power; a measure of how fast work is being done. Several values have been assigned to this unit. The one used in England and America is the *Watts horsepower*. It is defined as 550 foot-pounds per second, at sea level and 50° latitude. The Watts horsepower is 1.0139 times the French horsepower. See FOOT —foot-pound.

HOUR, *n.* Approximately one twenty-fourth of a *mean solar* day; 3600 seconds. See SECOND, TIME.
    hour angle and hour circle. In the figure, let *O* be the place of the observer, *NESW* the circle in which the plane of the observer's horizon cuts the celestial sphere, *EKW* the circle in which the plane of the earth's equator cuts the celestial sphere, *NS* the north-and-south line, and *EW* the east-and-west line. The circles *NESW* and *EKW* are, respectively, the astronomical **horizon** and the **celestial equator.** *Z* is the zenith and *P* the **north celestial pole.** *SZPN* is the **celestial meridian** or **meridian** of *O*. Let *M* be any celestial object, and draw great circles *ZR* and *PL* which pass through *M* and are perpendicular to the horizon and the equator, respectively. *RM* is the **altitude** of *M*, and *NR* its **azimuth.** *LM* is the **declination** of *M*, and *KOL* is its **hour angle.** *LP* is the **hour circle** of *M*. If *M* is north of the equator the declination is taken as positive; if

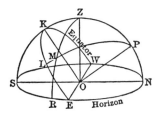

south, as negative. The hour angle is positive if *M* is west of the meridian, negative if it is east of the meridian. The hour angle of a celestial object changes at the rate of 15° an hour or 360° a day (sidereal time), the hour circle passing through it appearing to rotate in the opposite direction to that of the earth, that is, to the west.
    kilowatt-hour. See KILOWATT.

HULL, *n.* Same as SPAN. See SPAN.

HUN′DRED, *n.* **hundred's place.** See PLACE— place value.

HUN′DREDTH, *adj.* **hundredth part of a number.** The quotient of the number and 100, or $\frac{1}{100}$ times the number.

HUYGENS (or HUYGHENS), Christian (1629–1695). Dutch physical scientist, astronomer, and mathematician. Noted for his pioneering work on continued fractions, the tautochrone, probability, and analysis leading toward the invention of calculus. See CYCLOID.

Huygens formula. The length of an arc of a circle is *approximately* equal to twice the chord subtending half the arc plus one-third of the difference between twice this chord and the chord subtending the entire arc, or eight-thirds of the chord subtending half the arc minus one-third of the chord subtending the whole arc.

Huygens principle. If, for an initial-value problem in a space of n dimensions, the domain of dependence of each point is a manifold of at most $n - 1$ dimensions, then the problem is said to satisfy Huygens' principle. See DEPENDENCE —domain of dependence, HADAMARD —Hadamard conjecture.

HY-PER′BO-LA, *n.* A curve with two branches which is the intersection of a plane and a circular conical surface; the set of points whose distances from two fixed points, the **foci,** have a constant difference. The standard form of the equation in rectangular Cartesian coordinates is

$$\frac{x^2}{a^2} - \frac{y^2}{b^2} = 1,$$

where the hyperbola is symmetric about the x- and y-axes and cuts the x-axis in the points whose coordinates are $(a, 0)$ and $(-a, 0)$, as in the figure below. The intercepts on the y-axis are imaginary. The axes of symmetry of the hyperbola are the **axes of the hyperbola** (regardless of whether they coincide with the coordinate axes). The segment (of length $2a$) of the axis which cuts the hyperbola is the **transverse** (real) axis and the **conjugate axis** is the line segment of length $2b$ (as illustrated). The line segments a and b are the **semitransverse** and **semiconjugate** axes, respectively. (Transverse and conjugate axes are also used in speaking of the entire axes of symmetry.) If c is the distance from the center to a focus, then $c^2 = a^2 + b^2$ and the **eccentricity** of the hyperbola is c/a. Two hyperbolas are **similar** if they have the same eccentricity. The extremities of the transverse axes are the **vertices** of the hyperbola; the double ordinate at a focus, *i.e.*, the chord through a focus and perpendicular

to the transverse axis, is a **latus rectum.** See CONIC.

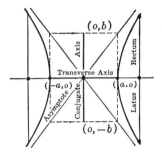

asymptote to a hyperbola. See ASYMPTOTE.

auxiliary circle of a hyperbola. See below, parametric equations of a hyperbola.

conjugate hyperbolas. Hyperbolas for which the real (transverse) and conjugate axes of one are, respectively, the conjugate and real axes of the other. Their standard equations are $x^2/a^2 - y^2/b^2 = 1$ and $x^2/a^2 - y^2/b^2 = -1$. They have the same asymptotes.

diameter of a hyperbola. See DIAMETER — diameter of a conic.

director circle of a hyperbola. See DIRECTOR.

equiangular (or equilateral) hyperbola. Same as RECTANGULAR HYPERBOLA.

focal (or reflection) property of a hyperbola. The angle formed by the focal radii drawn from

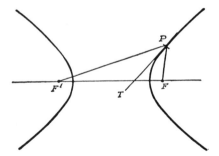

any point P (see figure) on the hyperbola is bisected by the tangent line drawn to the hyperbola at P. If the hyperbola be constructed from a polished strip of metal, a ray of light emanating from one focus (F) is reflected along a line whose extension passes through the other focus (F').

parametric equations of a hyperbola. With the origin as center draw two circles with radii equal to the semiconjugate and semitransverse axes of the required hyperbola (see figure). These are the **eccentric circles** of the hyperbola. Draw OR intersecting the eccentric circle of radius OA (the **auxiliary circle**) in R. Draw a tangent line to the auxiliary circle at R. It crosses the x-axis at

S. Draw a tangent line to the other eccentric circle at *L*. This line intersects *OR* at *Q*. Through *Q* and *S* draw lines parallel, respectively, to the *x*- and *y*-axes. The intersection of these lines gives a point *P* on the hyperbola. The angle *LOQ*, designated by ϕ, is the **eccentric angle** of the hyperbola. Letting *a* and *b* represent *OA* and *OB*, respectively, we find that the rectangular Cartesian coordinates of *P* are $x = a \sec \phi$, $y = b \tan \phi$. These equations are **parametric equations** of the hyperbola.

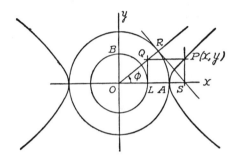

rectangular hyperbola. A hyperbola whose major and minor axes are equal. Its equation, in standard form, is $x^2 - y^2 = a^2$. The equations of the asymptotes are $y = x$ and $y = -x$. *Syn.* equiangular hyperbola, equilateral hyperbola.

HY′PER-BOL′IC, *adj.* **hyperbolic cylinder.** See CYLINDER.

hyperbolic functions. The functions *hyperbolic sine* of *z*, *hyperbolic cosine* of *z*, etc., written sinh *z*, cosh *z*, etc. They are defined by the relations

$$\sinh z = \tfrac{1}{2}(e^z - e^{-z}), \qquad \cosh z = \tfrac{1}{2}(e^z + e^{-z}),$$

$$\tanh z = \frac{\sinh z}{\cosh z}, \qquad \coth z = \frac{\cosh z}{\sinh z},$$

$$\operatorname{sech} z = \frac{1}{\cosh z}, \qquad \operatorname{csch} z = \frac{1}{\sinh z}.$$

The Taylor's series for sinh *z* and cosh *z* are

$$\sinh z = z + z^3/3! + z^5/5! + \cdots,$$

$$\cosh z = 1 + z^2/2! + z^4/4! + \cdots.$$

The hyperbolic functions for *z* real are related to the hyperbola in a manner somewhat similar to the way the trigonometric functions are related to the circle. The hyperbolic and trigonometric functions are connected by the relations: $\sinh iz = i \sin z$, $\cosh iz = \cos z$, $\tanh iz = i \tan z$, where

$i^2 = -1$. Some of the properties of the hyperbolic functions are:

$$\sinh(-z) = -\sinh z,$$

$$\cosh(-z) = \cosh z,$$

$$\cosh^2 z - \sinh^2 z = 1,$$

$$\operatorname{sech}^2 z + \tanh^2 z = 1,$$

$$\coth^2 z - \operatorname{csch}^2 z = 1.$$

See EXPONENTIAL —exponential values of sin *x* and cos *x*.

hyperbolic logarithms. Another name for *natural logarithms*. See LOGARITHM.

hyperbolic paraboloid. See PARABOLOID.

hyperbolic partial differential equation. A real second-order partial differential equation of the form

$$\sum_{i,j=1}^{n} a_{ij} \frac{\partial^2 u}{\partial x_i \, \partial x_j}$$

$$+ F\left(x_1, \cdots, x_n, u, \frac{\partial u}{\partial x_1}, \cdots, \frac{\partial u}{\partial x_n}\right) = 0,$$

for which the quadratic form $\displaystyle\sum_{i,j=1}^{n} a_{ij} y_i y_j$ is *non-singular* and *indefinite*; *i.e.*, by means of a real linear transformation the quadratic form can be reduced to a sum of *n* squares, but not all of the same sign. The term is often reserved for the case where all but one of the squares are of the same sign, although this case sometimes is called **normal hyperbolic.** A typical example is the *wave equation*. See INDEX—index of a quadratic form.

hyperbolic point of a surface. A point of the surface at which the total curvature is negative; a point at which the Dupin indicatrix is a hyperbola.

hyperbolic (or reciprocal) spiral. A plane curve whose radius vector varies inversely with

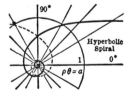

the vectorial angle. Its polar equation is $\rho\theta = a$, where *a* is the constant of proportionality. It is asymptotic to a straight line parallel to the polar axis and at a distance a above it.

hyperbolic Riemann surface. See RIE-MANN—Riemann surface.

inverse hyperbolic functions. The inverses of the hyperbolic functions; written $\sinh^{-1} z$, $\cosh^{-1} z$, etc., and read inverse hyperbolic sine of z, etc. Also called **arc-hyperbolic functions**. The explicit forms of the functions can be derived from the definitions of the hyperbolic functions; they are

$$\sinh^{-1} z = \log\left(z + \sqrt{z^2 + 1}\right), \qquad -\infty < z < \infty,$$

$$\cosh^{-1} z = \log\left(z \pm \sqrt{z^2 - 1}\right), \qquad z \geq 1,$$

$$\tanh^{-1} z = \tfrac{1}{2} \log \frac{1 + z}{1 - z}, \qquad |z| < 1,$$

$$\operatorname{ctnh}^{-1} z = \tfrac{1}{2} \log \frac{z + 1}{z - 1}, \qquad |z| > 1,$$

$$\operatorname{sech}^{-1} z = \log \frac{1 \pm \sqrt{1 - z^2}}{z}, \qquad 0 < z \leq 1,$$

$$\operatorname{csch}^{-1} z = \log \frac{1 + \sqrt{1 + z^2}}{|z|}, \qquad z \neq 0.$$

In each case, log means $\log_e$. The inverses of cosh and sech are relations and not functions, since they are not one-to-one. It is customary to reduce the domain of cosh to be the nonnegative real numbers, which are the **principal values** of $\cosh^{-1}$. The same choice of domain is made for sech. This corresponds to using the upper signs in the preceding formulas. See INVERSE—inverse of a function.

HY-PER′BO-LOID, n. A term referring to the so-called hyperboloids of one sheet and of two sheets. See below.

asymptotic cone of a hyperboloid. See ASYMPTOTIC.

center of a hyperboloid. The point of symmetry of the hyperboloid. This is the intersection of the three principal planes of the hyperboloid.

confocal hyperboloids. See CONFOCAL—confocal quadrics.

conjugate hyperboloids. See CONJUGATE.

hyperboloid of one sheet. A hyperboloid that is cut in an ellipse or hyperbola by every plane parallel to a coordinate plane. If its equation is written in the form

$$\frac{x^2}{a^2} + \frac{y^2}{b^2} - \frac{z^2}{c^2} = 1,$$

planes parallel to the xy-plane cut the surface in ellipses, while planes parallel to the xz- or yz-plane cut it in hyperbolas (see figure). The surface is a *ruled surface*. It contains two sets of rulings (two families of straight lines), and through each point of the surface there passes one member of each family. The equations of the two families of lines are:

$$\frac{x}{a} - \frac{z}{c} = p\left(1 - \frac{y}{b}\right),$$

$$p\left(\frac{x}{a} + \frac{z}{c}\right) = 1 + \frac{y}{b},$$

and

$$\frac{x}{a} - \frac{z}{c} = p\left(1 + \frac{y}{b}\right),$$

$$p\left(\frac{x}{a} + \frac{z}{c}\right) = 1 - \frac{y}{b},$$

where p is an arbitrary parameter. The product of the two equations in either set gives the original equation of the hyperboloid. Therefore the lines represented by these sets must lie on the

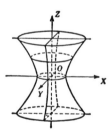

hyperboloid. Either set of rulings is a set of **rectilinear generators**, since it may be used to generate the surface (see RULED—ruled surface). A **hyperboloid of revolution of one sheet** is a hyperboloid of one sheet whose elliptical sections parallel to the xy-plane, when in the position illustrated above, are circles. The parameters a and b are equal in this case and the surface can be generated by revolving the hyperbola, $x^2/a^2 - z^2/c^2 = 1$, about the z-axis.

hyperboloid of two sheets. A surface whose sections by planes parallel to two of the three coordinate planes (see figure) are hyperbolas and whose sections by planes parallel to the third plane are ellipses, except for a finite interval where there is no intersection (the intersection is imaginary). When the surface is in the position

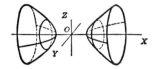

illustrated, its equation is of the form $x^2/a^2 - y^2/b^2 - z^2/c^2 = 1$. A **hyperboloid of revolution of two sheets** is a hyperboloid of two sheets whose elliptical sections are circles. The parameters b and c (in the equation of the hyperboloid of two sheets) are equal in this case, and the *hyperboloid of revolution of two sheets* can be generated by revolving the hyperbola, $x^2/a^2 - y^2/c^2 = 1$, about the x-axis.

    similar hyperboloids. See SIMILAR.

HY'PER-COM'PLEX, *adj.* **hypercomplex numbers.** The quaternions, but sometimes used more generally for members of any division algebra. See ALGEBRA —algebra over a field.

HY'PER-GE'O-MET'RIC, *adj.* **hypergeometric differential equation.** The differential equation

$$x(1-x)\frac{d^2y}{dx^2} + [c - (a+b+1)x]\frac{dy}{dx} - aby = 0.$$

When $c \neq 1,2,3,\cdots$, the general solution (for $|x| < 1$) is

$$y = c_1 F(a,b;c;x)$$

$$+ c_2 x^{1-c} F(a-c+1, b-c+1; 2-c; x),$$

where $F(a,b;c;x)$ is the *hypergeometric function*. *Syn.* Gauss' differential equation.

    hypergeometric distribution. Suppose a box contains M balls with colors $c_1, c_2, \cdots, c_k$ and that n balls are drawn randomly in succession *without replacement*. Let X be the vector random variable $(X_1, X_2, \cdots, X_k)$, where X_i is the number of balls drawn of color c_i. Then X is a **hypergeometric (vector) random variable** and X has a **hypergeometric distribution.** The range of X is the set of all k-tuples of nonnegative integers $(n_1, n_2, \cdots, n_k)$ for which $\sum_{i=1}^{k} n_i = n$. The probability function P satisfies

$$P(n_1, n_2, \cdots, n_k)$$

$$= \left[\binom{M_1}{n_1}\binom{M_2}{n_2}\cdots\binom{M_k}{n_k}\right]\binom{M}{n}^{-1},$$

where M_i is the number of balls of color c_i. For each i, the *mean* for the random variable X_i is np_i and the *variance* for X_i is $np_i q_i (M-n)/(M-1)$, where $p_i = M_i/M$ and $q_i = 1 - p_i$. The probability function P is approximated closely by that of the multinomial distribution when M is large relative to n (see MULTINOMIAL —multinomial distribution).

    hypergeometric function. For $|z| < 1$, the hypergeometric function $F(a,b;c;z)$ is the sum of the hypergeometric series (see below). This function has an analytic continuation which is analytic in the complex plane with the line from $+1$ to $+\infty$ omitted. For $|x| > 1$ and $a - b$ not an integer or zero, the hypergeometric function can be expressed in the form:

$$F(a,b;c;z)$$

$$= \frac{\Gamma(c)\Gamma(a-b)}{\Gamma(b)\Gamma(a-c)}(-z)^{-a}$$

$$\times F(a, 1-c+a; 1-b+a; z^{-1})$$

$$+ \frac{\Gamma(c)\Gamma(b-a)}{\Gamma(a)\Gamma(b-c)}(-z)^{-b}$$

$$\times F(b, 1-c+b; 1-a+b; z^{-1}),$$

where z is not real and $\Gamma(z)$ is the *gamma function*. See JACOBI —Jacobi polynomials, and above, hypergeometric differential equation.

    hypergeometric series. A series of the form

$$1 + \sum_{n=1}^{\infty} \frac{\left[\begin{array}{c}a(a+1)\cdots(a+n-1)b(b+1)\cdots \\ (b+n-1)z^n\end{array}\right]}{n!c(c+1)(c+2)\cdots(c+n-1)}$$

where c is not a negative integer. The series converges absolutely for $|z| < 1$. A necessary and sufficient condition for it to converge when $z = 1$ is that $a + b - c$ be negative (or that its real part be negative if it is complex). Also called *Gaussian series*. See above, hypergeometric differential equation.

HY'PER-PLANE, *n.* A subset H of a linear space L such that H contains all x for which there are numbers $\lambda_1, \lambda_2, \cdots, \lambda_n$, and elements $h_1, h_2, \cdots, h_n$ of H, satisfying $x = \sum \lambda_i h_i$ and $\sum \lambda_i = 1$; it is also usually required that H be a *maximal* proper subset of this type (see BARYCENTRIC —barycentric coordinates). Equivalently, H is a hyperplane if there is a maximal linear subset M of L such that, for any element h of H, H consists precisely of all sums of type

$x + h$ with x belonging to M. If L is a normed linear space, it is usually required that H be closed; this is equivalent to requiring the existence of a continuous linear functional f and a number c for which H is the set of all x with $f(x) = c$.

hyperplane of support. See SUPPORT —plane (and hyperplane) of support.

HY′PER-RE′AL, *adj.* **hyperreal numbers.** See NON —nonstandard numbers.

HY-PER-SUR′FACE, *n.* A generalization of the concept of surface in three-dimensional Euclidean space to surface in n-dimensional Euclidean space. *E.g.*, an **algebraic hypersurface** is the graph in n-dimensional Euclidean space of an equation of type $f(x_1, x_2, \cdots, x_n) = 0$, where f is a polynomial in $x_1, x_2, \cdots, x_n$.

HY-PER-VOL′UME, *n.* The n-dimensional content of a set in n-dimensional Euclidean space. See CONTENT.

HY-PO-CY′CLOID, *n.* The plane locus of a point P on a circle as the circle rolls on the inside of a given fixed circle. If a is the radius of

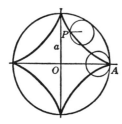

the fixed circle, b is the radius of the rolling circle, and θ is the angle from OA to the ray from O through the center of the rolling circle, then the parametric equations of the hypocycloid are:

$$x = (a - b)\cos\theta + b\cos[(a - b)\theta/b],$$

$$y = (a - b)\sin\theta - b\sin[(a - b)\theta/b].$$

The criteria for the number of arches are the same as for the *epicycloid*. The rectangular Cartesian equation of the hypocycloid of four cusps (which is the case shown in the figure) is $x^{2/3} + y^{2/3} = a^{2/3}$. The hypocycloid has a cusp of the first kind at every point at which it touches the fixed circle. See EPICYCLOID.

HY-POT′E-NUSE, *n.* The side opposite the right angle in a plane right triangle. See TRIANGLE.

HY-POTH′E-SIS, *n.* [*pl.* **hypotheses**]. (1) An assumed proposition used as a premise in proving something else; a condition; that from which something follows. See IMPLICATION. (2) A proposition held to be probably true because its consequences, according to known general principles, are found to be true. (3) (*Statistics*) A statement about the distribution of a *random variable*; *e.g.*, about the parameters or the type of distribution.

admissible hypothesis. (*Statistics*) Any hypothesis that is regarded as possibly true.

alternative hypothesis. (*Statistics*) See below, test of a hypothesis.

composite hypothesis. (*Statistics*) Any hypothesis that is not *simple*. *E.g.*, the hypothesis "X is normally distributed with zero mean" is *composite* since the variance must also be known to determine the distribution.

continuum hypothesis. The conjecture, advanced by Cantor in 1878, that every infinite subset of the set of real numbers (the *real continuum*) has the cardinal number either of the positive integers or of the entire set of real numbers. The **general continuum hypothesis** is that $2^{\aleph}$ is the least cardinal number greater than $\aleph$, where $\aleph$ is any infinite cardinal number. See COHEN.

linear hypothesis. (*Statistics*) Logically, any hypothesis concerning the parameters of a distribution that can be expressed linearly might be called a linear hypothesis; *e.g.*, given a distribution with parameters μ and σ, the hypothesis $\mu = 2\sigma - 1$ is linear. However, *linear hypothesis* usually refers to the following general type of hypothesis. Suppose n independent random variables are normally distributed with equal variances and that their means $\mu_1, \cdots, \mu_n$ are related to parameters $\theta_1, \cdots, \theta_k$ by linear equations

$$\mu_i = \sum_{j=1}^{k} c_{ij}\theta_j, \qquad k \leq n.$$

A **linear hypothesis** is a hypothesis that asserts the parameters $\theta_1, \cdots, \theta_k$ satisfy certain linear conditions. *E.g.*, the hypothesis "$\mu_1 - \mu_2 = 3$" concerning the means of two particular normal random variables with equal variances is a linear hypothesis. The weight w of a child might be assumed to depend linearly on height and age, for heights $h_1, \cdots, h_r$ and ages $y_1, \cdots, y_s$. That is,

$$E\big[w|(h_i, y_j)\big] = \theta_1 + \theta_2 h_i + \theta_3 y_j, \ i \leq r, \ j \leq s,$$

where $E[w|(h_i, y_j)]$ is the mean weight of a child of height h_i and weight y_j. The *linear hypothesis* could be a statement specifying the values of θ_1, θ_2 and θ_3.

null hypothesis. See below, test of a hypothesis.

parametric hypothesis. (*Statistics*) A hypothesis that assumes a specific type of distribution for a random variable and states conditions on the parameters of the distribution.

simple hypothesis. (*Statistics*) A hypothesis that specifies the distribution exactly. *E.g.*, the hypothesis "X is normally distributed with mean 0 and variance 1" is a simple hypothesis.

test of a hypothesis. (*Statistics*) A rule for reaching the decision to accept a stated hypothesis, or to reject the hypothesis (or sometimes not to make a decision yet, but to take another sample). Usually the rule is based on random samples. The stated hypothesis is the **null hypothesis**, and one has in mind an **alternative hypothesis**; either the null hypothesis is accepted and the alternative hypothesis is rejected, or the null hypothesis is rejected and the alternative hypothesis is accepted. The sample space is divided into an **acceptance region** A and a **rejection region**, or **critical region**, C. Given a random sample, the null hypothesis is accepted if the sample is in A and it is rejected if the sample is in C. A test of a hypothesis has the possibility of two types of errors: an **error of type I** is the rejection of the null hypothesis when it is true; an **error of type II** is the acceptance of the null hypothesis when it is false. The **size of a test** or the **size of the critical region** for a simple null hypothesis is the probability of an error of type **I**. The **significance level of a test** (simple or composite) is the least upper bound of the probability of an error of type **I** for all distributions consistent with the null hypothesis. The smaller this size or level, the more probable it is that the null hypothesis will be accepted when it should be accepted. The **power of a test** is a function p for which $p(\theta)$ is the probability of accepting the alternative hypothesis when the distribution is θ (θ is a distribution consistent with either the null or the alternative hypothesis—sometimes, particularly when the hypotheses are simple, θ is restricted to being consistent with the alternative hypothesis). An **unbiased test** is a test for which $p(\theta) \geq \alpha$ if θ is a distribution consistent with the alternative hypothesis and α is the significance level of the test; a **biased test** is a test that is not unbiased. If T_1 and T_2 are tests of significance level α with power functions p_1 and p_2, then T_1 is **more powerful** than T_2 with respect to α if $p_1(\theta) \geq p_2(\theta)$ whenever θ is consistent with the alternative hypothesis. A test T is a **uniformly**

most powerful test of significance level α if it has significance level α and is more powerful than any other test of that significance level. See CHI —chi-square test, CONFIDENCE —confidence interval, NEYMAN —Neyman-Pearson test, SEQUENTIAL—sequential analysis, TEST—test statistic.

HY-PO-TRO′CHOID, *n.* Same as *hypocycloid*, except that the describing point may lie within the rolling circle or on its radius extended. If h is the distance from the center of the rolling circle to the describing point, and the other parameters are the same as for the *hypocycloid*, the parametric equations are:

$$x = (a - b)\cos\theta + h\cos[(a - b)\theta/b],$$

$$y = (a - b)\sin\theta - h\sin[(a - b)\theta/b].$$

The cases when h is less than b, or greater than b, are similar to the corresponding cases for the *trochoid* (see TROCHOID).

I

I-CO-SA-HE′DRAL, *adj.* **icosahedral group.** The group of motions (symmetries) of 3-dimensional space that preserve the regular icosahedron. It also is the alternating group of order 60 that is the set of even permutations on five objects. See GROUP, SYMMETRY —group of symmetries.

I′CO-SA-HE′DRON, *n.* [*pl.* **icosahedrons** or **icosahedra**]. A polyhedron having twenty faces. A **regular icosahedron** is an icosahedron whose faces are congruent equilateral triangles and whose polyhedral angles are congruent. See POLYHEDRON.

I-DE′AL, *adj., n.* Let R be a set which is a ring with respect to operations called addition and multiplication (it may also be an integral domain, algebra, etc.). A subset I which is an additive *group* (or, equivalently, a subset I which is such that $x - y$ belongs to I whenever x and y belong to I) is a **left ideal** if cx belongs to I whenever c belongs to R and x belongs to I (it is a **right ideal** if xc belongs to I whenever c belongs to R and x belongs to I). It is a **two-sided ideal** (or simply an **ideal**) if cx and xc belongs to I whenever c belongs to R and x belongs to I. For any ideal I in a ring R, there is a homomorphism which maps R onto the *quotient ring* R/I. This homomorphism maps each member of I

onto zero. Also, R/I is isomorphic to the image of R under any homomorphism for which I is the set of those elements which map onto zero. A subset I of a ring is an ideal if there is a homomorphism of the ring for which I is the set of those elements which map onto zero. A **principal ideal** is an ideal that contains an element such that all elements of the ideal are multiples of this element. *E.g.*, the set of even integers is a principal ideal in the integral domain of all integers. The product AB of two ideals A and B is the ideal obtained by multiplying every member of A by every member of B and then forming all possible sums of these products. If D is the integral domain of all algebraic integers and a **prime ideal** in D is one which is not the ideal I of all multiples of 1 and has no factors other than itself and 1, then any ideal of D can be represented uniquely (except for order of the factors) as a product of prime ideals. See ALGEBRAIC—algebraic number, MODULE, QUOTIENT—quotient space.

ideal point. A point at infinity; a term used to complete the terminology of certain subjects (*e.g.*, projective geometry) so that it is not necessary to state exceptions to certain theorems. Instead of saying that two straight lines in the same plane intersect except when they are parallel, it is said that two straight lines in a plane always intersect, intersecting in the ideal point being synonymous with being parallel. Thus an *ideal point* is thought of as a direction, the direction of a certain set of parallel lines. When expressed in *homogeneous coordinates*, x_1, x_2, x_3, the ideal points are the points $(x_1, x_2, 0)$, where x_1 and x_2 are not both zero. The point $(x_1, x_2, 0)$ lies on any line whose slope is equal to x_2/x_1. See COORDINATE —homogeneous coordinates in the plane, INFINITY—point at infinity.

nilpotent ideal. See NILPOTENT.

radical of an ideal. See RADICAL—radical of an ideal.

I-DEM-FAC′TOR, *n.* The dyadic $\mathbf{ii} + \mathbf{jj} + \mathbf{kk}$, called *idemfactor* because its *scalar product* with any vector in either order does not change the vector. See DYAD.

I-DEM′PO-TENT, *adj.* An idempotent quantity is one unchanged under multiplication by itself. Unity and the matrix

$$\begin{vmatrix} 1 & 0 & 0 \\ 0 & 1 & 0 \\ 1 & 0 & 0 \end{vmatrix}$$

are idempotent.

idempotent property. The property for some sets of objects and operations $\circ$ that $x \circ x = x$. *E.g.*, for a Boolean algebra or for set theory, either of the properties $A \cup A = A$ or $A \cap A = A$. See BOOLE—Boolean algebra.

I-DEN′TI-CAL, *adj.* **identical figures.** Figures that are exactly alike in form and size; two triangles with three sides of one equal to three sides of the other are identical. *Syn.* congruent.

identical quantities. Quantities which are alike in form as well as value. Quantities which form the left and right members of an *identity* are not necessarily identical, usually being different in form although always having the same values for all values of the variables.

I-DEN′TI-TY, *adj., n.* A statement of equality, usually denoted by $\equiv$, which is true for all values of the variables, with the exception of values of the variables for which each member of the statement of equality does not have meaning. *E.g.*, $2/(x^2 - 1) \equiv 1/(x - 1) - 1/(x + 1)$ and $(x + y)^2 \equiv x^2 + 2xy + y^2$ are identities. The equality sign, $=$, is quite commonly used in place of the identity sign, $\equiv$. Two functions f and g are identical if they have the same domains and $f(x) = g(x)$ for all x in the domains; this identity is written either as $f(x) \equiv g(x)$ or $f = g$. See EQUATION, TRIGONOMETRY —identities of plane trigonometry.

identity element. For a set S and a binary operation $x \circ y$, an **identity** is a member e of S for which $x \circ e = e \circ x = x$ for all x in S (a **right identity** is a member e of S such that $x \circ e = x$ for all x in S; a **left identity** is a member e of S such that $e \circ x = x$ for all x in S). For example, the **identity for addition** of numbers is 0, since $x + 0 = 0 + x = x$ for all x; the **identity for multiplication** of numbers is 1, since $x \cdot 1 = 1 \cdot x = x$ for all x. If S is the set of all subsets of a set T, then the identity for *union* is the empty set $\varnothing$ and the identity for *intersection* is T, since $A \cup \varnothing = \varnothing \cup A = A$ and $A \cap T = T \cap A = A$ for all subsets A of T. See GROUP, INVERSE—inverse of an element.

identity function. The function from a set into itself that maps each member of the set to itself. For a set S, the **identity function** is the function f defined by $f(x) = x$ for each x in S. For example, the identity function for the set of real numbers is the function f whose graph is the line $y + x$, *i.e.*, the function that associates with each number x the number itself. *Syn.* identity operator, identity transformation. See FUNCTION —inverse of a function, TRANSFORMATION—inverse transformation.

identity matrix. See MATRIX.

Pythagorean identities and other **trigonometric identities**. See TRIGONOMETRY —identities of plane trigonometry.

ILLIAC. An automatic digital computing machine built at the University of Illinois.

IL-LU'SO-RY, *adj.* **illusory correlation.** Correlation between two variables that is significant without implying a causal relation. *E.g.*, the population of South Africa and the number of kilowatts of electrical energy consumed in California are correlated, since both are positively correlated with time. The correlation observed in a random sample, if assignable to random sampling fluctuations, is illusory correlation. *Syn.* nonsense correlation.

IM'AGE, *n.* For a function f, the **image of a point** x is the functional value $f(x)$ corresponding to x. If A is a subset of the domain of f, then the **image** of A is denoted by $f(A)$ and is the set of all images of members of A. The **pre-image** or **inverse image** of a set B contained in the range of f is denoted by $f^{-1}(B)$ and is the subset of the domain whose members have images in B. In particular, the inverse image of a point y in the range is the set of all x for which $f(x) = y$.

spherical image. See SPHERICAL.

IM-AG'I-NA'RY, *adj.* **imaginary axis.** See COMPLEX —complex numbers, ARGAND —Argand diagram.

imaginary curve (surface). A term used to provide continuity in speaking of loci of equations, the imaginary part of the curve (surface) corresponding to imaginary values of the variable which satisfy the equation. The equation $x^2 + y^2 + z^2 = 1$ has for its real locus the sphere with radius one and center at the origin, but is also satisfied by $(1, 1, i)$ and many other points whose coordinates are not all real. See CIRCLE —imaginary circle, ELLIPSE, ELLIPSOID, INTERSECTION.

imaginary number. See COMPLEX —complex number.

imaginary part of a complex number. If $z = x + iy$ (where x and y are real), the *imaginary part* of z is y, denoted by I(z), Im(z), or $\Im(z)$.

imaginary roots. Roots of an equation or number which are complex numbers whose imaginary part is not zero. *E.g.*, the roots of $x^2 + x + 1 = 0$ are $-\frac{1}{2} \pm \frac{1}{2}\sqrt{3} \cdot i$. See COMPLEX —complex number, FUNDAMENTAL —fundamental theorem of algebra, ROOT—root of a number.

IM-BED', *v.* See SPACE —enveloping space.

IM GROSSEN. *German* for IN THE LARGE. See SMALL —in the small.

IM KLEINEN. *German* for IN THE SMALL. See SMALL —in the small.

IM-ME'DI-ATE, *adj.* **immediate annuity.** See ANNUITY.

IM'PLI-CA'TION, *n.* (1) A statement that follows from other given statements. *E.g.*, an **implication** of $x = -1$ is that $x^2 = 1$. (2) A proposition formed from two given propositions by connecting them in the form "If $\cdots$, then $\cdots$." The first statement is the **antecedent** (or **hypothesis**) and the second the **consequent** (or **conclusion**). An implication is true in all cases except when the antecedent is true and the consequent is false. *E.g.*, the following implications are true: "If $2 \cdot 3 = 7$, then $2 \cdot 3 = 8$"; "If $2 \cdot 3 = 7$, then $2 \cdot 3 = 6$"; "If $2 \cdot 3 = 6$, then $3 \cdot 4 = 12$." Such a proposition as "If a quadrilateral is a square, then it is a parallelogram" can be written as "For any quadrilateral x, if x is a square, then x is a parallelogram," which is true, since when x is a specific quadrilateral the expression "If x is a square, then x is a parallelogram" is a true proposition. For propositions p and q, the implication "*if p, then q*" is usually written as $p \to q$, or $p \subset q$, and read "*p implies q.*" The implication $p \to q$ has the same meaning as the propositions "*p is a *sufficient condition* for q,*" or "*q is a *necessary condition* for p.*" *Syn.* conditional statement (or proposition). See CONVERSE— converse of an implication, EQUIVALENCE — equivalence of propositions.

IM-PLIC'IT, *adj.* **implicit differentiation.** See DIFFERENTIATION —implicit differentiation.

implicit function. A function defined by an equation of the form $f(x, y) = 0$ [in general, $f(x_1, x_2, \cdots, x_n) = 0$]. If y is thought of as the dependent variable, $f(x, y) = 0$ is said to define y as an *implicit* function of x. Sometimes an equality of type $y = F(x)$ can be derived; when this has been done, y is called an **explicit** function of x. In $x + y^3 + 2x^2y + xy = 0$, y is an *implicit* function of x, while in $y = x^2 + 1$, y is an *explicit* function of x. The equality $x^2 + y^2 = 4$ defines a *relation* between x and y; solving gives $y = +(4 - x^2)^{1/2}$ or $y = -(4 - x^2)^{1/2}$, each of which gives y as an explicit function of x.

implicit-function theorem. A theorem stating conditions under which an equation, or a system of equations, can be solved for certain dependent variables. For a **function of two variables**, the implicit-function theorem states conditions under which an equation in two variables possesses a unique solution for one of the variables in a neighborhood of a point whose coordinates satisfy the given equation. *Tech.* If F and $D_y F$, the partial derivative of F with respect to y, are continuous in the neighborhood of the point (x_0, y_0) and if $F(x_0, y_0) = 0$ and $D_y F(x_0, y_0) \neq 0$, then there is a number $\epsilon > 0$ such that there exists one and only one function f which is such that $y_0 = f(x_0)$ and which is continuous and satisfies $F[x, f(x)] = 0$ for $|x - x_0| < \epsilon$. *E.g.*, $x^2 + xy^2 + y - 1$ and its partial derivative with respect to y, namely, $2xy + 1$, are both continuous in the neighborhood of $(1, 0)$, and $x^2 + xy^2 + y - 1 = 0$ while $2xy + 1 \neq 0$ when $x = 1$, $y = 0$. Hence there exists a unique solution for y, in the neighborhood of $(1, 0)$, which gives $y = 0$ for $x = 1$. That solution is

$$y = \frac{-1 + \sqrt{1 - 4x(x^2 - 1)}}{2x}.$$

The **general implicit-function theorem** states conditions under which a system of $n + p$ equations in n dependent variables and p independent variables possesses solutions for the dependent variables in a neighborhood of a point whose coordinates satisfy the given equations. Consider a system of n equations between the $n + p$ variables

$$u_1, u_2, \cdots, u_n, \quad \text{and} \quad x_1, x_2, \cdots, x_p,$$

namely

$$f_1(x_1, x_2, \cdots, x_p; u_1, u_2, \cdots, u_n) = 0,$$

$$f_2(x_1, x_2, \cdots, x_p; u_1, u_2, \cdots, u_n) = 0,$$

$$\cdots\cdots\cdots\cdots\cdots\cdots\cdots\cdots\cdots$$

$$f_n(x_1, x_2, \cdots, x_p; u_1, u_2, \cdots, u_n) = 0.$$

Suppose that these equations are satisfied for the

values $x_1 = x_1^0, \cdots, x_p = x_p^0, u_1 = u_1^0, \cdots, u_n = u_n^0$, that the functions f_i are continuous in the neighborhood of this set of values and possess first partial derivatives which are continuous for this set of values of the variables and, finally, that the Jacobian of these functions does not vanish for $x_i = x_i^0$, $u_k = u_k^0$ ($i = 1, 2, \cdots, p$; $k = 1, 2, \cdots, n$). Under these conditions there exists one and only one system of continuous functions,

$$u_1 = \phi_1(x_1, x_2, \cdots, x_p),$$

$$\cdots\cdots\cdots\cdots\cdots\cdots\cdots\cdots$$

$$u_n = \phi_n(x_1, x_2, \cdots, x_p),$$

defined in some neighborhood of

$$\left(x_1^0, x_2^0, \cdots, x_p^0\right),$$

which satisfy the above equations and which reduce to $u_1^0, u_2^0, \cdots, u_n^0$ for $x_1 = x_1^0, x_2 = x_2^0, \cdots, x_p = x_p^0$.

IM-PROP'ER, *adj.* **improper fraction.** See FRACTION.

   improper integral. See INTEGRAL.

IN'CEN'TER, *n.* **incenter of a triangle.** The center of the inscribed circle; the intersection of the bisectors of the interior angles of the triangle.

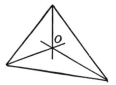

INCH, *n.* A unit of measure of length, or distance; one twelfth of one foot, approximately 2.54 centimeters. See DENOMINATE NUMBERS in the appendix.

   inch of mercury. (*Meteorology*) The pressure due to a one-inch-high column of mercury when the *acceleration of gravity* has the standard value. Equal to 3386.3886 + newtons per square meter.

IN′CI-DENCE, *adj.* **incidence function.** See GRAPH THEORY.

IN-CIR′CLE, *n.* Same as INSCRIBED CIRCLE OF A TRIANGLE.

IN′CLI-NA′TION, *n.* The **inclination of a line in a plane** is an angle from the positive direction of the *x*-axis to the line, the angle usually being taken to be greater than or equal to 0° and less than 180° (see ANGLE—angle of intersection, and SLOPE—slope of a line). The **inclination of a line in space** with respect to a plane is the smaller angle the line makes with its orthogonal projection in the plane. The **inclination of a plane** with respect to a given plane is the smaller of the dihedral angles which it makes with the given plane.

IN-CLU′SION, *adj.* **inclusion relation.** The relation usually denoted by $\subset$ such that, if A and B are sets, then $A \subset B$ means that each member of A is a member of B. To indicate that a particular x is a member of B, one usually uses the notation $x \in B$.

IN′COME, *n.* **income rate.** Same as YIELD.

income tax. A tax on incomes (salaries or profits), levied by the federal or other governments. The tax is determined by taking a certain percent of the remainder of the income after certain deductions (exemptions) have been made.

IN′COM-MEN′SU-RA-BLE, *adj.* Not having a common measure, *i.e.*, not having a common unit of which both are integral multiples. Two **incommensurable numbers** are numbers that are not both integral multiples of the same number. *E.g.*, the numbers $\sqrt{2}$ and 3 are incommensurable, since if there were integers m and n and a number x for which $\sqrt{2} = mx$ and $3 = nx$, then it would follow that $\sqrt{2} = 3m/n$. This is impossible, since $\sqrt{2}$ is not a rational number. Two **incommensurable line segments** are line segments that are not both integral multiples of another segment, *i.e.*, their lengths are incommensurable numbers.

IN′COM-PAT′I-BLE, *adj.* **incompatible equations.** Same as INCONSISTENT EQUATIONS.

IN′COM-PLETE′, *adj.* **incomplete beta and gamma functions.** See BETA—beta function, GAMMA—gamma function.

incomplete induction. See INDUCTION—mathematical induction.

IN-CON-SIS′TENT, *adj.* **inconsistent axioms.** See AXIOM.

inconsistent equations. Two (or more) equations that are not satisfied by any one set of values of the variables; equations that are not consistent. *E.g.*, $x + y = 2$ and $x + y = 3$ are inconsistent. See CONSISTENCY. *Syn.* incompatible.

IN′CREASE, *n.* **percent increase.** See PERCENT.

IN-CREAS′ING, *adj.* **increasing function.** A function of real numbers whose value increases as the independent variable increases; a function whose graph, in Cartesian coordinates, rises as the abscissa increases. If a function is differentiable on an interval I, then the function is increasing on I if the derivative is non-negative throughout I and is not identically zero in any interval. An increasing function often is said to be **strictly increasing**, to distinguish it from a *monotone increasing* function. *Tech.* A function f is **strictly increasing** on an interval I if

$$f(x) < f(y)$$

for all numbers x and y of I with $x < y$; f is **monotone increasing** on I if $f(x) \leq f(y)$ for all numbers x and y of I with $x < y$. See MONOTONE.

increasing premium policy. An insurance policy on which the early premiums are smaller than the later ones.

increasing sequence. A sequence $\{x_1, x_2, \cdots\}$ for which $x_i < x_j$ if $i < j$. The sequence is **monotone increasing** if $x_i \leq x_j$ for $i < j$. See MONOTONE.

IN′CRE-MENT, *n.* A change in a variable; an amount added to a given value of a variable, usually thought of as a small amount (positive or negative).

increment of a function. The change in the function due to changes in the values of the independent variables (these changes in the independent variables are **increments of the independent variables** and may be either positive or negative). If the function is f and the change in the independent variable x is Δx, then the increment in $f(x)$ is $\Delta f = f(x + \Delta x) - f(x)$. If f is differentiable at x, then

$$f(x + \Delta x) - f(x) = f'(x)\Delta x + \epsilon \cdot \Delta x,$$

where ϵ approaches zero as Δx approaches zero; $f'(x)\Delta x$ is the **principal part** of Δf, or the **differential** of f. If u is a function of x and y, the increment of u, written Δu, is equal to $u(x + \Delta x, y + \Delta y) - u(x, y)$. By use of the *mean* *value theorem* for a function of two variables this expression can be written as the sum $(D_x u\, \Delta x + D_y u\, \Delta y) + (\epsilon_1 \Delta x + \epsilon_2 \Delta y)$, where $D_x u$ and $D_y u$ denote the partial derivatives of u with respect to x and y, respectively, and ϵ_1 and ϵ_2 approach zero as Δx and Δy approach zero, provided there is a neighborhood of the point (x, y) in which $D_x u$ and $D_y u$ both exist and one (at least) is continuous. The sum of the two terms in the first parentheses is the **principal part** of Δu, or **total differential** of u, written $du = D_x u\, dx + D_y u\, dy$. The increments of x and y, Δx and Δy, have been written as dx and dy, in order to avoid confusion of notation in the cases $u = x$ and $u = y$. The increments dx and dy are also the *differentials* of the *independent* variables x and y. See DIFFERENTIAL.

IN-DEF′I-NITE, *adj.* **indefinite integral.** See INTEGRAL —indefinite integral.

IN′DE-PEND′ENCE, *n.* **statistical (or stochastic) independence.** See EVENT—dependent and independent events, INDEPENDENT—independent random variables.

IN′DE-PEND′ENT, *adj.* **independent axioms.** See AXIOM.

independent equations. A system of equations such that no one of them is necessarily satisfied by a set of values of the independent variables which satisfy all the others. See CONSISTENCY, DEPENDENT —dependent equations.

independent events. See EVENT.

independent functions. A set of functions $u_1, u_2, \cdots, u_n$, where $x_1, x_2, \cdots, x_n$ are the independent variables, such that there does not exist a relation $F(u_1, u_2, \cdots, u_n) \equiv 0$, where not all $\partial F/\partial u_i$ are identically zero. The functions are independent if and only if the *Jacobian*

$$\frac{D(u_1, u_2, \cdots, u_n)}{D(x_1, x_2, \cdots, x_n)}$$

does not vanish identically provided not all of the functions $\partial F/\partial u_i$ vanish at any point under consideration and each u_i has continuous first partial derivatives. The functions $2x + 3y$ and $4x + 6y + 8$ are dependent, since $4x + 6y + 8 = 2(2x + 3y) + 8$. The functions

$$u_1 = 2x + 3y + z, \quad u_2 = x + y - z, \quad u_3 = x + y$$

are independent. Their Jacobian does not vanish; it is

$$\begin{vmatrix} 2 & 3 & 1 \\ 1 & 1 & -1 \\ 1 & 1 & 0 \end{vmatrix} = -1.$$

independent random variables. Random variables X and Y such that, whenever A and B are events associated with X and Y, respectively, the probability $P(A \text{ and } B)$ of both A and B is equal to $P(A)P(B)$. For discrete random variables with values $\{x_i\}$ and $\{y_i\}$, this is equivalent to $P(x_i \text{ and } y_j) = P(x_i)P(y_j)$ for all i and j. Random variables X and Y with distribution functions F_X and F_Y and joint distribution function $F_{(X,Y)}$ are independent if and only if $F_{(X,Y)}(a, b) = F_X(a)F_Y(b)$ for all numbers a and b; if X and Y have probability density functions g and h and joint probability density function f, then X and Y are independent if and only if $f(a, b) = g(a)g(b)$. The preceding generalizes naturally to a set $\{X_1, \cdots, X_n\}$ of random variables: they are independent if and only if, whenever A_i is an event associated with X_i for $i = 1, 2, \cdots, n$, the probability of *all* these events is $\prod_{i=1}^{n} P(A_i)$. If X and Y are independent random variables with nonzero standard deviation, then their correlation coefficient is 0; if the correlation coefficient of X and Y is 0 and X and Y are normally distributed with nonzero standard deviations, then X and Y are independent. See RANDOM—random variable, vector random variable.

independent variable. See FUNCTION.

linearly independent quantities. Quantities which are not linearly dependent. See DEPENDENT—linearly dependent.

IN′DE-TER′MI-NATE, *adj.* **indeterminate equation.** See EQUATION —indeterminate equation.

indeterminate form. An expression of type $\infty - \infty, 0/0, \infty/\infty, 0 \cdot \infty, \infty^0, 0^0, 1^\infty$, which are undefined. These may arise from replacing different members of composite functions by their limits before combining the members properly. The correct procedure is to find the limits of the difference, quotient, etc., not the difference, quotient, etc., of the limits. See L'HÔPITAL.

IN′DEX, *n.* [*pl.* **indexes** or **indices**]. A number used to point out a specific characteristic or operation.

contravariant and covariant indices. See TENSOR.

dummy (or umbral) index. See SUMMATION —summation convention.

free index. See SUMMATION—summation convention.

index of a matrix. See below, index of a quadratic form.

index of a point relative to a curve. See WINDING —winding number.

index of precision. See PRECISION —modulus of precision.

index of a quadratic form. The number of positive terms when the quadratic form is reduced to a sum of squares by means of a linear transformation. Likewise, the **index of a Hermitian form** is the number of terms with positive coefficients when the form is reduced to the type $\sum_{i=1}^{n} a_i z_i \bar{z}_i$ by means of a linear transformation. See TRANSFORMATION —congruent transformation, conjunctive transformation. Analogously, the **index of a symmetric or of a Hermitian matrix** is the number of positive elements when it is transformed to diagonal form. For forms or matrices, the number of positive terms diminished by the number of negative terms is the **signature** and the total number of nonzero terms is the **rank**. Two quadratic forms (or two symmetric or Hermitian matrices) have the same rank and the same index if and only if they are congruent, *i.e.*, can be transformed into each other by an invertible linear transformation (this is **Sylvester's law of inertia**).

index of a radical. An integer placed above and to the left of a radical to indicate what root is sought; *e.g.*, $\sqrt[3]{64} = 4$. The index is omitted when it would be 2; $\sqrt{x}$ rather than $\sqrt[2]{x}$ indicates the nonnegative square root of x.

index of refraction. See REFRACTION.

index of a subgroup. The quotient of the order of the group by the order of the subgroup. See GROUP, LAGRANGE —Lagrange's theorem.

IN′DI-CA′TOR, *n.* **indicator diagram.** A diagram in which the ordinates of a curve represent a varying force, the abscissas the distance passed over, and the area beneath the curve the work done.

indicator of an integer. See EULER —Euler's ϕ-function.

IN′DI-CA′TRIX, *n.* **binormal indicatrix of a space curve.** The locus of the extremities of the radii of the unit sphere parallel to the positive directions of the binormals of the given space curve. See below, spherical indicatrix of a space curve. *Syn.* spherical indicatrix of the binormal to a space curve.

principal normal indicatrix of a space curve. The locus of the extremities of the radii of the unit sphere parallel to the positive directions of the principal normals of the given space curve. *Syn.* spherical indicatrix of the principal normal to a space curve.

spherical indicatrix of the principal normal to a space curve. See above, principal normal indicatrix of a space curve.

spherical indicatrix of a ruled surface. The intersection of the *director cone* of the ruled surface with the unit sphere, when the vertex of the cone is at the origin. See DIRECTOR —director cone of a ruled surface.

spherical indicatrix of a space curve. The curve traced out on a unit sphere by the end of a radius which is always parallel to a tangent that moves along the curve. If the curve is a plane curve, its spherical indicatrix lies on a great circle of the sphere. Hence, the amount of deviation of the spherical indicatrix from a great circle gives some idea of the amount of deviation of the curve from being a plane curve, *i.e.*, of the amount of torsion of the curve. *Syn.* spherical representation of a space curve. See above, principal normal indicatrix of a space curve, and binomial indicatrix of a space curve.

tangent indicatrix of a space curve. Same as the SPHERICAL INDICATRIX. See above, principal normal indicatrix of a space curve, and binormal indicatrix of a space curve.

IN′DI-CES, *n.* Plural of **index.** See INDEX.

IN′DI-RECT′, *adj.* **indirect differentiation.** The differentiation of a function of a function by use of the formula

$$df(u)/dx = (df(u)/du)(du/dx).$$

See CHAIN —chain rule.

indirect proof. (1) See PROOF —direct and indirect proof. (2) Proving a proposition by first proving another theorem from which the given proposition follows.

IN-DIS-CRETE, *adj.* **indiscrete topology.** See TRIVIAL —trivial topology.

IN-DORSE′, *v.* **to indorse a note or other financial arrangement.** To accept the responsibility for carrying out the obligation of the maker of the paper provided the maker does not meet his own obligation; to assign a paper to a new payee by signing it on the back (this makes the indorser liable for the maker's obligation unless the signature is accompanied by the statement *without recourse*).

IN-DORSE′MENT, *n.* The act of indorsing; that which is written when indorsing. See INDORSE.

IN-DUC'TION, *n.* **mathematical induction.** A method of proving a law or theorem by showing that it holds in the first case and showing that, if it holds for all the cases preceding a given one, then it also holds for this case. Before the method can be applied it is necessary that the different cases of the law depend upon a parameter which takes on the values $1, 2, 3, \cdots$. The essential steps of the proof are as follows: (1) Prove the theorem for the first case. (2) Prove that if the theorem is true for the nth case (or for the first through the nth cases), then it is true for the $(n + 1)$th case. (3) Conclude that it must then be true for all cases. For, if there were a case for which it is not true, there must be a *first* case for which it is not true. Because of (1), this is not the first case. But because of (2), it cannot be any other case [since the previous case could not be true without the next case (known to be false) being true; it could not be false because the next case is the false case]. *E.g.,* (1) since the sun rose today, if it can be shown that if it rises any one day, it will rise the next day, then it follows by mathematical induction that it will always rise. (2) To prove $1 + 2 + 3 + \cdots + n = \frac{1}{2}n(n + 1)$. If $n = 1$, the right member becomes 1, which completes step (1). Adding $(n + 1)$ to both members gives

$$1 + 2 + 3 + \cdots + n + (n + 1)$$

$$= \tfrac{1}{2}n(n + 1) + n + 1 = \tfrac{1}{2}(n + 1)(n + 2),$$

which completes step (2). Therefore, the statement is true for all values of n. Mathematical induction is called **complete induction**, in contrast to **incomplete induction** which draws a "conclusion" from the examination of a finite number of cases. *Syn.* method of infinite descent, proof by descent.

transfinite induction. See TRANSFINITE.

IN-DUC'TIVE, *adj.* **inductive methods.** Drawing conclusions from several known cases; reasoning from the particular to the general. See INDUCTION —mathematical induction.

IN'E-QUAL'I-TY, *n.* A statement that one quantity is less than (or greater than) another. If a is less than b, their relation is denoted symbolically by $a < b$; the relation a greater than b is written $a > b$. Inequalities have many important properties (see ORDER —order properties of real numbers). An inequality which is not true for all values of the variables involved is a **conditional inequality**; an inequality which is true for all values of the variables (or contains no variables) is an **unconditional inequality** (or **absolute inequality**). *E.g.,* $(x + 2) > 3$ is a conditional inequality, because it is true only for x greater than 1, while $(x + 1) > x$, $3 > 2$, and $(x - 1)^2 + 3 > 2$ are unconditional inequalities. A **polynomial inequality** is an inequality for which the expression on each side of the inequality sign is a polynomial [see QUADRATIC —quadratic inequality]. The direction (*greater than* or *less than*) in which the inequality sign points is the **sense** of the inequality. This is used in such phrases as *same sense* and *opposite sense*. The inequalities $a < b$ and $c < d$, or $b > a$, and $d > c$, are said to have the *same sense*; the inequalities $a < b$ and $d > c$ are said to have *opposite senses*.

graph of an inequality. The set of all points that satisfy the inequality. *E.g.,* the graph of the inequality $y < x$ is the set of all points below the line $y = x$, the graph of $2x - y < 3$ is the set of all points above the line $y = 2x - 3$, and the set of solutions of the simultaneous inequalities, $y < x$, $2x - y < 3$, is the set of all points that are below the line $y = x$ and above the line $y = 2x - 3$. The graph of $x^2 + y^2 + z^2 < 4$ is the interior of the sphere $x^2 + y^2 + z^2 = 4$. *Syn.* solution set. See QUADRATIC —quadratic inequality.

inequalities of Abel, Bernoulli, Bessel, Cauchy, Chebyshev, Hadamard, Hölder, Jensen, Minkowski, Newton, Schwarz, and Young. See the respective names.

simultaneous inequalities. See SIMULTANEOUS.

triangle inequality. See TRIANGLE.

IN-ER'TI-A, *n.* **moment of inertia, principal axes of inertia, products of inertia.** See MOMENT —moment of inertia.

inertia of a body. Its resistance to change of its state of motion, or rest; the property of a body which necessitates exertion of force upon the body to give it acceleration. *Syn.* mass.

law of inertia. A law of mechanics stating that material bodies not subjected to action by forces either remain at rest or move in a straight line with constant speed. This law was deduced by Galileo in 1638 and incorporated by Isaac Newton in the *Principia* (1867) as one of the postulates of mechanics. This law is also known as *Newton's First Law of Motion*.

IN-ER'TIAL, *adj.* **inertial coordinate system.** See COORDINATE.

IN'ES-SEN'TIAL, *adj.* A map of a topological space X into a topological space Y is **inessential** if it is *homotopic* to a map whose range is a single point (see DEFORMATION —continuous deformation). A map is **essential** if it is not inessential. A map into a circle (or an n-sphere) whose range is not the entire circle (or sphere) is *inessential*. A map of an interval (or an *n*-cell) into a circle (or an *n*-sphere) is *inessential*. A map of a circle into a circle is *essential* if and only if the winding number of the image of the circle (relative to its center) is not zero.

INF, *n.* See BOUND.

IN′FER-ENCE, *n.* **statistical inference.** The making of statements, or the process of drawing judgments, about a population on the basis of random samples in such a manner that the probability of making correct inferences is determinable under various alternative hypotheses about the population being sampled. See HYPOTHESIS —test of a hypothesis.

IN-FE′RI-OR, *adj.* **limit inferior.** (1) See SEQUENCE —accumulation point of a sequence. (2) The **limit inferior** of a function f at a point x_0 is the least number L such that for any $\epsilon > 0$ and neighborhood U of x_0 there is a point $x \neq x_0$ of U for which $f(x) < L + \epsilon$. This limit is denoted by $\liminf_{x \to x_0} f(x)$ or $\underline{\lim}_{x \to x_0} f(x)$. The limit inferior of $f(x)$ as $x \to x_0$ is equal to the limit as $\epsilon \to 0$ of the g.l.b. of $f(x)$ for $|x - x_0| < \epsilon$ and $x \neq x_0$, and may be positively or negatively infinite. (3) The **limit inferior** of a sequence of sets $\{U_1, U_2, \cdots\}$ is the set consisting of all points belonging to all but a finite number of the sets. It is equal to the union over p of the intersection of $U_p, U_{p+1}, \cdots$, or $\bigcup_{p=1}^{\infty} \bigcap_{n=p}^{\infty} U_n$, and written $\liminf_{n \to \infty} U_n$ or $\underline{\lim}_{n \to \infty} U_n$. In this sense (3), it also is called the $\overline{\text{restricted limit}}$. See SUPERIOR —limit superior. *Syn.* lower limit.

IN′FI-MUM, *n.* See BOUND.

IN′FI-NITE, *adj.* (1) Becoming large beyond any fixed bound. *Tech.* Let f be a function and a be a point such that each neighborhood of a contains a point of the domain of f other than a. Then f **becomes infinite** as x approaches a if, for any number C, there is a neighborhood U of a such that $|f(x)| > C$ if x is in the domain of f, $x \in U$, and $x \neq a$; f becomes **positively infinite** (or **approaches plus infinity**) as x approaches a if, for any number C, there is a neighborhood U of a such that $f(x) > C$ if x is in the domain of f, $x \in U$, and $x \neq a$; f becomes **negatively infinite** (or **approaches minus infinity**) as x approaches a if, for any number C, there is a neighborhood U of a such that $f(x) < C$ if x is in the domain of f, $x \in U$, and $x \neq a$. The above

are written, respectively, as

$$\lim_{x \to a} |f(x)| = \infty, \quad \lim_{x \to a} f(x) = \infty,$$

and

$$\lim_{x \to a} f(x) = -\infty.$$

The above definitions also apply when a is one of the symbols $+\infty$ or $-\infty$, if a "neighborhood of $+\infty$" is the set of x satisfying $x > M$ for a specified number M, and "neighborhood of $-\infty$" is similarly defined. See EXTENDED —extended real-number system, UNBOUNDED —unbounded function. (2) See NON—nonstandard numbers.

   infinite branch of a curve. A part of the curve which cannot be enclosed in any (finite) circle.

   infinite decimal. See DECIMAL—decimal number system.

   infinite integral. An integral at least one of whose limits of integration is infinite. The value of the integral is the limit it approaches as its limit (or limits) of integration becomes infinite. The integral exists only if this limit exists. An infinite integral is a type of **improper integral**. *E.g.,*

$$\int_1^{\infty} \frac{dp}{p^2} = \lim_{h \to \infty} \int_1^{h} \frac{dp}{p^2} = \lim_{h \to \infty} \left[-\frac{1}{h} + 1 \right] = 1.$$

   infinite limit. See INFINITE.
   infinite point. Same as IDEAL POINT.
   infinite product. See PRODUCT.
   infinite root. See ROOT—infinite root of an equation.
   infinite sequence. See SEQUENCE.
   infinite series. See SERIES.
   infinite set. A set which is not finite; a set which has an unlimited number of members; a set which can be put into one-to-one correspondence with a proper part of itself. *E.g.,* all positive integers constitute an infinite set. It can be put into one-to-one correspondence, for instance, with the set of positive even integers. The rational numbers between 0 and 1 constitute an infinite set. It can be put into one-to-one correspondence with the set of all positive integers.

   method of infinite descent. Same as MATHEMATICAL INDUCTION. See INDUCTION.

IN′FIN-I-TES′I-MAL, *adj., n.* (1) A variable which approaches zero as a limit, usually meaning a function f such that $\lim_{x \to 0} f(x) = 0$. (2) See NON—nonstandard numbers.

infinitesimal analysis. The study of differentials, and of integration as a process of summing a set of n infinitesimals with $n \to \infty$ (see INTEGRAL—definite integral). Sometimes used for the calculus and all subjects making use of the calculus.

infinitesimal calculus. Ordinary calculus; so called because it is based on the study of infinitesimal quantities.

order of an infinitesimal. If two infinitesimals u and v are functions of x that approach 0 as x approaches 0, then u and v are of the **same order** if there are positive numbers A, B, and ϵ such that $A < |u/v| < B$ if $0 < |x| < \epsilon$; if $\lim_{x \to 0} u(x)/v(x) = 0$, or if $\lim_{x \to 0} |v(x)/u(x)| = +\infty$, u is of **higher order** than v and v is of **lower order** than u. If u^n is of the same order as v, then v is **of the nth order** relative to u; $(1 - \cos x)$ is of the second order relative to x, since $x^2/(1 - \cos x)$ approaches 2 as x approaches zero.

IN-FIN'I-TY, n. **approach infinity.** See INFINITE.

line at infinity. See LINE—line at infinity.

order of infinities. See MAGNITUDE —order of magnitude.

point at infinity. (1) See IDEAL—ideal point. (2) See EXTENDED —extended real number system. (3) The **point at infinity in the complex plane** is a single point adjoined to the complex plane to make the complex plane compact (the Euclidean plane is not). The complex plane may be thought of as a sphere—for instance, a sphere which is mapped conformally on the complex plane by *stereographic projection*. The *pole* of the projection corresponds to the point at infinity.

IN-FLEC'TION, n. **point of inflection.** A point at which a plane curve changes from concavity toward any fixed line to convexity toward it; a point at which a curve has a *stationary tangent* and at which the tangent is changing from rotating in one direction to rotating in the opposite direction. The vanishing of the 2nd derivative, if it is continuous, is a *necessary* but not a *sufficient* condition for a point of inflection, because the second derivative may be zero without changing signs at the point. *E.g.*, the curve $y = x^3$ has its second derivative zero at the origin, and has a *point of inflection* there; the curve $y = x^4$ also has its second derivative zero at the origin, but has a minimum there; the curve $y = x^4 + x$ has its 2nd derivative zero at the origin, but has neither a point of inflection nor a maximum or minimum there. A necessary and sufficient condition that a point be a point of inflection is that the second derivative change sign at this point, *i.e.*, have different signs for values of the independent variable slightly less and slightly greater than at the point.

IN-FLEC'TION-AL, *adj.* **inflectional tangent to a curve.** A tangent at a point of inflection. Such a tangent has contact of order 3, since dy/dx and d^2y/dx^2 each have the same value (at the point of inflection) for the curve as for the tangent. See CONTACT—order of contact.

IN'FLU-ENCE, n. **range of influence.** See RADIATION.

IN'FOR-MA'TION, *adj.*, n. **information theory.** The branch of probability theory, founded in 1948 by C. E. Shannon, that is concerned with the likelihood of the transmission of messages, accurate to within specified limits, when the bits of information comprising the messages are subject to certain probabilities of transmission failure, distortion, and accidental additions called **noise**. If there are k pieces of information, one of which is to be chosen for transmission, then these pieces of information are called **messages**. The individual receiving the message is the **receiver** and the individual transmitting the message is the **sender**. A **channel** is a means of communication. Mathematically, it is described by the **input set** (the totality of possible elements from which the sender selects one for the purpose of transmitting a message to the receiver in accordance with a prearranged code), the **output set** (the totality of elements of which the receiver can observe just one), and the probability law that gives, for each element a of the input set and each element b of the output set, the probability that if a is sent then b is received. If there is a total of k messages that might be sent, a **code** is a sequence of k elements $a_1, \cdots, a_k$ of the input set and a division of the entire output set into k disjoint sets $E_1, \cdots, E_k$. Observing message b, the receiver determines the set E_j of which b is a member and concludes that the jth message was sent. A **probability law** is a rule specifying, for a given channel, for each element a of the input set, and for each element b of the output set, the probability that if element a is sent then element b is received. For a given code, if the sender transmits element a_i representing the ith message, and the probability of the receiver's observing message b when a_i is transmitted is $p(b|a_i)$, then the probability of error when the ith message is sent is

$$P_e(i) = \sum_{b \text{ not in } E_i} p(b|a_i).$$

The **maximum-error probability** for the given code is

$$\max_i P_e(i).$$

The **entropy** of a set of messages, each with a known probability of being sent, is roughly the number of binary digits needed to encode all long sequences of messages, except for the messages comprising a set of sufficiently small total probability of being sent. *Tech.* If the random variable X has k different possible values, with respective probabilities $p_1, \cdots, p_k$, then the entropy of X is

$$H(X) = -\sum_{i=1}^{k} p_i \log_2 p_i.$$

For random variables X and Y with entropies $H(X)$ and $H(Y)$, respectively, the **conditional entropy** of X when Y is given is $H(X|Y) = H(X,Y) - H(Y)$. It represents the additional number of bits of information (binary digits) required to identify an element of X when Y is given. For two sets of random numbers X and Y with entropies $H(X)$ and $H(Y)$, respectively, the **mutual information** is the number $R(X,Y) = H(X) + H(Y) - H(X,Y)$. It represents the number of bits of information, that is, the number of binary digits, of information that can be obtained concerning X by observing Y. For a channel with input set A and output set B, if X is the input variable and Y the output variable, with associated entropies $H(X)$, $H(Y)$, and $H(X,Y)$, and mutual information $R(X,Y) = H(X) + H(Y) - H(X,Y)$, the **capacity C of the channel** is defined to be the maximum of $R(X,Y)$ over all input distributions on A. The **fundamental theorem of information theory** states, roughly, that if a channel has capacity C, then in using the channel a sufficiently large number N of times, any one of about 2^{CN} messages can be transmitted with small error probability.

IN-I′TIAL, *adj.* **initial point.** See CURVE, DIRECTED —directed line.

initial side of an angle. See ANGLE.

IN-JEC′TION, *n.* An **injection** from a set A to a set B is a function that is one-to-one and whose domain is A and whose range is contained in B. *Syn.* injective function. See BIJECTION, SURJECTION.

IN-JEC′TIVE, *adj.* See INJECTION.

IN′NER, *adj.* **inner automorphism.** See ISOMORPHISM.

inner measure. Same as INTERIOR MEASURE. See MEASURE —exterior and interior measure.

inner-product space. A vector space V for which there is a function (called the **inner product** or **scalar product**) whose domain is the set of all ordered pairs of members of V, whose range is in the set of scalars (real or complex numbers), and which satisfies the following axioms, with (x, y) denoting the inner product of x and y: $(ax, y) = \bar{a}(x, y)$, $(x + y, z) = (x, z) + (y, z)$, $(x, y) = \overline{(y, x)}$, (x, x) is real and positive if $x \neq 0$ [sometimes the first of these axioms is replaced by $(ax, y) = a(x, y)$, in which case $(x, ay) = \bar{a}(x, y)$]. If a norm is defined by letting $\|x\| = (x, x)^{1/2}$, then the space becomes a *normed vector space*. An inner-product space also is called a **generalized Euclidean space** or a **pre-Hilbert space**. A complete inner-product space is a **Hilbert space**, although sometimes a Hilbert space is required to be separable and not finite-dimensional. A finite-dimensional inner-product space with real scalars is an **Euclidean space** and a finite-dimensional inner-product space with complex scalars is an **unitary space**. See HILBERT —Hilbert space, MULTIPLICATION —multiplication of vectors, VECTOR —vector space.

inner product of tensors. The *inner product* of two tensors $A^{a_1 \cdots a_n}_{i_1 \cdots i_m}$ and $B^{b_1 \cdots b_p}_{j_1 \cdots j_q}$ is the *contracted tensor* obtained from the product

$$C^{a_1 \cdots a_n b_1 \cdots b_p}_{i_1 \cdots i_m j_1 \cdots j_q} = A^{a_1 \cdots a_n}_{i_1 \cdots i_m} B^{b_1 \cdots b_p}_{j_1 \cdots j_q}$$

by putting a contravariant index of one equal to a covariant index of the other and summing with respect to that index. This inner multiplication of two tensors is also called **composition**, and the inner product is spoken of as the tensor **compounded** from the two given tensors.

inner product of two functions. For real-valued functions f and g on a set E, the inner product is $(f, g) = \int_E fg\, dx$. If f and g are complex-valued, then (f, g) is $\int_E \bar{f}g\, dx$ or $\int_E f\bar{g}\, dx$. See HILBERT —Hilbert space, ORTHOGONAL —orthogonal functions, and above, inner-product space.

inner product of two vectors. See MULTIPLICATION —multiplication of vectors, VECTOR —vector space, HILBERT —Hilbert space, and above, inner-product space.

IN′PUT, *adj., n.* **input component.** In a computing machine, any component that is used in introducing problems into the machine; for example, a numerical keyboard, typewriter, punched-card machine, or tape might be used for this purpose.

IN-SCRIBED′, *adj.* A polygon (or polyhedron) is said to be inscribed in a closed configuration composed of lines, curves, or surfaces, when it is contained in the configuration and every vertex of the polygon (or polyhedron) is incident upon the configuration; a closed configuration is said to be inscribed in a polygon (or polyhedron) when every side (in the polygon) or every face (in the polyhedron) is tangent to it and the configuration is contained in the polygon (or polyhedron). If one configuration is inscribed in another, the latter is said to be **circumscribed** about the former (see CIRCUMSCRIBED). The figure shows a polygon inscribed in a circle, the circle inscribed in another polygon. See BROKEN — broken line.

inscribed angle. An angle formed by two chords which intersect on the curve, each with an end point at the vertex of the angle. An angle inscribed in a circle of unit radius has radian measure equal to half the length of the intercepted arc; in particular, an angle inscribed in a semicircle is a right angle.

inscribed circle of a triangle. The circle tangent to the sides of the triangle. Its center is the intersection of the bisectors of the angles of the triangle, and is the **incenter** of the triangle. Its radius is

$$\sqrt{\frac{(s-a)(s-b)(s-c)}{s}}$$

where a, b, c are the lengths of the sides of the triangle and $s = \frac{1}{2}(a + b + c)$. *Syn.* incircle.

inscribed cone, cylinder, polygon, prism, pyramid. See CIRCUMSCRIBED.

IN-STALL′MENT (or IN-STAL′MENT), *n.* **installment payments.** Payments on notes, accounts, or mortgages at regular periods. Each payment may include a fixed payment on principal and interest on the balance over the preceding period, or the interest on the entire amount

may have been added to the principal at the outset and the payments made equal. The latter practice makes the actual interest practically double the quoted rate. The former method is sometimes called the **long-end interest plan**. Sometimes fixed payments plus the interest on the amount of the payment from the beginning of the contract are made periodically (monthly). This is sometimes called the **short-end interest plan.** See AMORTIZATION.

installment plan of buying. Any plan under which a debt is paid in installments.

installment policies. Insurance policies whose benefits are payable in installments beginning at the death (or end of the death year) of the insured, instead of in a single payment.

installment premium. See PREMIUM.

IN′STAN-TA′NE-OUS, *adj.* **instantaneous acceleration, speed, and velocity.** See ACCELERATION, SPEED, VELOCITY.

IN′SUR′ANCE, *adj., n.* **extended insurance.** Insurance extended (because of default in payment or some other exigency) over a period of such length that the *net single premium* is equal to the *surrender value* of the policy.

insurance policy. A contract under which an insurance company agrees to pay the beneficiary a certain sum of money in case of death, fire, sickness, accident, or whatever contingency the policy provides against.

life insurance. A contract to pay a specified benefit under specified conditions when certain lives end (or periodically thereafter), or when the lives end if that is within a certain period. The former is **whole-life insurance**; the latter, **temporary (or term) insurance.** The money paid by the insured for his protection is the **premium** and the formal contract between him and the company is the **policy. Deferred life insurance** can refer to any type of life insurance which has the provision that benefits are not payable if the insured lives end before a specified future date. **Ordinary life insurance** is whole life insurance with premiums payable annually throughout life (a **limited-payment policy** is one for which the annual payment is to be paid only for a stated number of years—it is **insurance with return of premiums** if the company returns all premiums after payments are completed). Term insurance which permits the insured to reinsure (usually at the end of the period) without another medical examination is **option-term insurance.** Insurance payable upon the end of the life of the insured (the policy life) provided some other life (the counter life) continues is **contingent-life insurance.** This is a simple case of **compound-survi-**

vorship insurance which is payable at the end of a certain life provided a group of lives, including this one, end in a certain order. Insurance payable when the last of two or more lives ends is **last-survivor insurance**; insurance payable when the first of two (or more) lives ends is **joint-life insurance.** An **endowment insurance** policy provides that the benefits be payable (usually to the insured) after a definite period or (to the beneficiary) at the death of the insured if that occurs before the end of the period; **double-endowment insurance** provides twice the benefits at the close of the period as would be paid if the insured life ends during the period; **pure-endowment insurance** provides payment of a benefit at the end of a given period if, and only if, the insured survives that period. Any type of life insurance policy is an **installment policy** if the benefits are payable in installments instead of in a single payment. An endowment policy is a **continuous-installment policy** if it provides that if the insured survives the period of the contract an annuity will be paid for a certain period and so long thereafter as the insured lives. Also see TONTINE—tontine insurance, DEBENTURE, and various headings above and below.

    loan value of an insurance policy. See LOAN —loan value.

    mutual insurance company. See MUTUAL.

    participating insurance policy. A policy which entitles the holder to participate in the profits of the insurance company. The profits are usually paid in the form of a **dividend** or **bonus** at the time of payment of the policy, or at periods when the company invoices its insurance contracts. A bonus paid when the policy is paid is a **reversionary bonus.** A reversionary bonus is a **uniform reversionary bonus** if it is determined as a percentage of the original amount of the insurance; it is a **compound reversionary bonus** if it is a sum of a percentage of the original amount of the insurance and any bonus already credited to the policy. A **discounted bonus policy** is a policy in which the premiums are reduced by the estimated (anticipated) future bonuses. A **guaranteed-bonus policy** is a policy which carries a guarantee of a certain rate of profit-sharing (bonus) (really nonparticipating since the bonuses assured are determined at the policy date). See MUTUAL—mutual insurance company.

    reserve of an insurance policy. See RESERVE.

IN′TE-GER, *n.* Any of the numbers $\cdots, -4, -3, -2, -1, 0, 1, 2, \cdots$. The **positive integers** (or **natural numbers**) are $1, 2, 3, \cdots$ and the **negative integers** are $-1, -2, -3, \cdots$. The entire class of integers consists of $0, \pm 1, \pm 2, \cdots$.

Peano defined the positive integers as a set of elements which satisfies the following postulates: (1) There is a positive integer 1; (2) every positive integer a has a *consequent* a^+ (a is called the *antecedent* of a^+); (3) the integer 1 has no antecedent; (4) if $a^+ = b^+$, then $a = b$; (5) every set of positive integers which contains 1 and the consequent of every number of the set contains all the positive integers. A positive integer (or zero) can also be thought of as describing the "many-ness" of a set of objects in the sense of being a symbol denoting that property of a set of individuals which is independent of the natures of the individuals. That is, it is a symbol associated with a set and with all other sets which can be put into one-to-one correspondence with this set. This collection of sets is a **number class.** *Syn.* whole number. See CARDINAL—cardinal number, PRODUCT—product of real numbers, SUM—sum of real numbers.

    algebraic integer. See ALGEBRAIC—algebraic number.

    cyclotomic integer. See CYCLOTOMIC.

    Gaussian integer. Any complex number of the form $a + bi$, where a and b are ordinary (real) integers. *Syn.* complex integer.

    indicator, totient, or ϕ-function of an integer. See EULER—Euler's ϕ-function.

    p-adic integer. See p-ADIC.

IN′TE-GRA-BLE, *adj.* **integrable differential equation.** See DIFFERENTIAL—integrable differential equation.

    integrable (or **summable**) **function.** A function whose integral exists in some specified sense and is finite (sometimes a function is said to be integrable if the integral exists, even though the integral may be allowed to be $+\infty$ or $-\infty$). See DARBOUX—Darboux's theorem, and various headings under INTEGRAL. If m is a measure on a σ-algebra of subsets of a set T, then a **simple function** is a measurable function s whose range is a finite set. If $\{q_1, \cdots, q_n\}$ are the nonzero members of the range of a real-valued simple function s, and Q_i is the inverse image of q_i, then the integral of s over T is

$$\int_T s \, dm = \sum_{i=1}^{n} q_i \cdot m(Q_i),$$

provided it is not true that one term on the right is $+\infty$ and another is $-\infty$ (sometimes it is required that all terms be finite). The integral of a nonnegative measurable function f is defined as the least upper bound of $\int_T s \, dm$, where s is a simple function with $s(x) \le f(x)$ on T, or (equivalently) as the least upper bound of sums of type

$$\sum_{i=1}^{n-1} y_i \cdot m(A_i) + y_n \cdot m(A_n), \quad \text{where} \quad 0 < y_1 <$$

$y_2 \cdots < y_n$, A_i is the inverse image of the interval $[y_i, y_{i+1}]$ if $i < n$, and A_n is the inverse image of the set of all y for which $y \geq y_n$. For any measurable function f, let f^+ and f^- be defined by letting $f^+(x)$ be the larger of $f(x)$ and 0 and $f^-(x)$ be the larger of $-f(x)$ and 0, so that $f = f^+ - f^-$. Then f is defined to be integrable and $\int_T f \, dm$ to be

$$\int_T f^+ \, dm - \int_T f^- \, dm,$$

if these two integrals exist and not both are $+\infty$ (sometimes it is required that neither be $+\infty$). This definition can be extended to functions with domain T and range in a Banach space X. Such a function f is **measurable** if the inverse image of open sets in X is measurable. If also, for any positive number ϵ, there is a simple function s for which $\int_T \|f - s\| dm < \epsilon$, then f is said to be **integrable** and $\int_T f \, dm$ is $\lim_{\epsilon \to 0} F_\epsilon$, where F_ϵ is the set of all $\int_T s \, dm$ with $\int_T \|f - s\| \, dm < \epsilon$. This is the **Bochner integral** of f. A measurable function is Bochner integrable if and only if $\int_T \|f\| \, dm < \infty$. See LEBESGUE —Lebesgue integral, MEASURE —measure of a set.

   locally integrable function. See LOCALLY — locally integrable function.

IN′TE-GRAL, *adj., n.* **definite integral** (or **Riemann integral**). The *definite integral* is the fundamental concept of the integral calculus. It is written as

$$\int_a^b f(x) \, dx,$$

where $f(x)$ is the integrand, a and b are the lower and upper limits of integration, and x is the variable of integration. Geometrically, if $a < b$, the integral exists if and only if the region W between the closed interval $[a, b]$ and the graph of f has area; the integral then is equal to the area of the part of W above the x-axis *minus* the area of the part of W below the x-axis. There are many other interpretations of the definite integral. *E.g.*, if $v(t)$ is the velocity of a particle

at time t moving along a straight line, then $\int_a^b v(t) \, dt$ is the distance traveled by the particle between times a and b. Suppose the interval $[a, b]$ is divided into n equal subintervals of length $\Delta x = (b - a)/n$, as indicated in the figure. Then the definite integral is the limit, as Δx

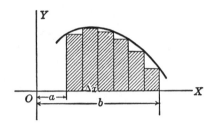

approaches zero, of the sum of the areas of the rectangles, Δx in width, formed with successive ordinates taken Δx apart through the interval between a and b. *Tech.* Choose numbers $x_1, x_2, \cdots, x_{n+1}$ in order from a to b, with $x_1 = a$ and $x_{n+1} = b$. Also, let $\Delta_i x$ denote $x_{i+1} - x_i$ and let ξ_i be any number in the closed interval with end points x_i and x_{i+1}. Then the sum

$$R(x_1, \cdots, x_{n+1}; \xi_1, \cdots, \xi_n) = \sum_{i=1}^{n} f(\xi_i) \Delta_i x$$

is called a **Riemann sum**. Let δ be the largest of the numbers $|x_{i+1} - x_i|$. Then the **Riemann integral** of f on the interval with end points a and b is

$$\int_a^b f(x) \, dx = \lim_{\delta \to 0} R(x_1, \cdots, x_{n+1}; \xi_1, \cdots, \xi_n),$$

provided this limit exists. The definite integral always exists for a function which is continuous in the closed interval defined by the limits of integration; continuity is here a sufficient but not a necessary condition. A necessary and sufficient condition that a bounded function have a (Riemann) integral on a given interval is that the function be continuous *almost everywhere*. See DARBOUX —Dauboux's theorem. Some elementary properties of a definite integral are: (1) Interchanging the limits of integration simply changes the sign of the integral. (2) For any number c,

$$\int_a^b cf(x) \, dx = c \int_a^b f(x) \, dx.$$

(3) If the first two integrals exist, then the other

exists and

$$\int_a^b f(x)\, dx + \int_b^c f(x)\, dx = \int_a^c f(x)\, dx.$$

(4) If the first two integrals exist, then the other exists and

$$\int_a^b f(x)\, dx + \int_a^b g(x)\, dx = \int_a^b [f(x) + g(x)]\, dx.$$

(5) If $a < b$ and $m \le f(x) \le M$ when $a \le x \le b$, then

$$m(b - a) \le \int_a^b f(x)\, dx \le M(b - a).$$

(6) If f is continuous, then there is a number ξ between a and b for which

$$\int_a^b f(x)\, dx = (b - a)f(\xi).$$

See FUNDAMENTAL —fundamental theorem of calculus, INTEGRABLE —integrable function, INTEGRATION —change of variables in integration, MEAN—mean value of a function, mean-value theorems for integrals, and various headings below.

    derivative of an integral. See DERIVATIVE.

    elliptic integral. See ELLIPTIC.

    energy integral. See ENERGY.

    Fredholm integral equations and their solutions. See various headings under FREDHOLM.

    Fresnel integrals. See FRESNEL.

    generalized Riemann integral. Let f be a real-valued function with domain the interval $[a, b]$. Then GI is the **generalized Riemann integral** of f on $[a, b]$ if, for each $\epsilon > 0$, there exists a function δ with domain $[a, b]$ and range in the positive reals for which

$$\left| GI - \sum_{i=1}^{n} f(t_i)(x_{i+1} - x_i) \right| < \epsilon$$

whenever $\{x_1 = a,\ x_2, \cdots, x_{n+1} = b\}$ defines a partition of $[a, b]$, $t_i \in [x_i, x_{i+1}]$ and $|x_{i+1} - x_i| < \delta(t_i)$ for each i. If there is a real-valued continuous function F on $[a, b]$ for which $F'(x) = f(x)$ except for at most a countably infinite subset of $[a, b]$, then the generalized Riemann integral of f on $[a, b]$ exists and is equal to $F(b) - F(a)$. Also, any function whose Lebesgue integral exists on $[a, b]$ has a generalized Riemann integral and these integrals are equal. See above, definite integral.

Hilbert-Schmidt theory of integral equations. See HILBERT.

    homogeneous integral equation. See HOMOGENEOUS.

    improper integral. An integral for which the interval (or region) of integration is not bounded, or the integrand is not bounded, or neither is bounded.

    indefinite integral of a function of a single variable. Any function whose derivative is the given function. If g is an indefinite integral of f, then $g + c$, where c is an *arbitrary constant*, is also an integral of f; c is the **constant of integration**. The indefinite integral of f with respect to x is written $\int f(x)\, dx$; $f(x)$ is the **integrand**. Many basic formulas for finding integrals are listed in the appendix. More extensive tables have been published, but the list of integrals is inexhaustible. *Syn.* antiderivative.

    infinite integral. See INFINITE —infinite integral.

    integral calculus. See CALCULUS.

    integral curves. The family of curves whose equations are solutions of a particular differential equation; the *integral curves* of the differential equation $y' = -x/y$ are the family of circles $x^2 + y^2 = c$, where c is an arbitrary parameter. See DIFFERENTIAL —solution of a differential equation.

    integral domain. See DOMAIN.

    integral equation. An equation in which the unknown function occurs under an integral sign. The (Fourier) equation

$$f(x) = \int_{-\infty}^{\infty} \cos(xt)\phi(t)\, dt,$$

where f is an even function, is the first integral equation that was solved. Under certain conditions, a solution is

$$\phi(x) = \frac{2}{\pi} \int_0^{\infty} \cos(ux)f(u)\, du.$$

An integral equation of the **third kind** is an integral equation of type

$$g(x)y(x) = f(x) + \lambda \int_a^b K(x, t)y(t)\, dt,$$

where f, g, and K are given functions and y is the unknown function. The function K is the **kernel** or **nucleus** of the equation. The (Fredholm) integral equations of the **first and second kind** are special cases of this equation. See FREDHOLM —Fredholm integral equations.

integral expression. An algebraic expression in which no variables appear in any denominator when the expression is written in a form having only positive exponents.

integral function. Same as ENTIRE FUNCTION.

integral of a function of a complex variable. See CONTOUR —contour integral.

integral number. An INTEGER (positive, negative, or zero).

integral tables. Tables giving the primitives (indefinite integrals) of the more common functions and sometimes some of the more important definite integrals. See INTEGRATION TABLES in the appendix.

integral test for convergence. See CAUCHY— Cauchy integral test.

iterated integral. An indicated succession of integrals in which integration is to be performed first with respect to one variable, the others being held constant, then with respect to a second, the remaining ones being held constant, etc.; the inverse of successive partial differentiation, if the integration is *indefinite integration*. When the integration is *definite integration*, the limits may be either constants or variables, the latter usually being functions of variables with respect to which integration is yet to be performed. (Some writers use the term *multiple integral* for *iterated integral*.) E.g., (1) The **iterated integral**,

$$\int \int xy \, dy \, dx,$$

may be written

$$\int \left\{ \int xy \, dy \right\} dx.$$

Integrating the inner integral gives

$$\left(\tfrac{1}{2} xy^2 + C_1 \right)$$

where C_1 is any function of x, only. Integrating again gives

$$\tfrac{1}{4} x^2 y^2 + \int C_1 \, dx + C_2$$

where C_2 is any function of y. The result may be written in the form $\tfrac{1}{4} x^2 y^2 + \phi_1(x) + \phi_2(y)$, where $\phi_1(x)$ and $\phi_2(y)$ are any differentiable functions of x and y, respectively. The order of integration is usually from the inner differential out, as taken here. The orders are not always inter-

changeable. (2) The **definite iterated integral**,

$$\int_a^b \int_x^{x+1} x \, dy \, dx,$$

is equivalent to

$$\int_a^b \left\{ \int_x^{x+1} x \, dy \right\} dx$$

which is equal to

$$\int_a^b \left\{ x(x+1) - x^2 \right\} dx = \tfrac{1}{2}(b^2 - a^2).$$

See below, multiple integral.

Lebesgue integral. See LEBESGUE.

Lebesgue-Stieltjes integral. See STIELTJES.

line integral. Let C be a rectifiable curve defined by parametric equations on the closed interval $[a, b]$, so that the point $(x(t), y(t), z(t))$ has the position vector $\mathbf{P}(t) = x(t)\mathbf{i} + y(t)\mathbf{j} + z(t)\mathbf{k}$. Also, let $\mathbf{F}$ be a vector-valued function whose domain contains $[a, b]$ and let $\{a = t_1, t_2, \cdots, t_{n+1} = b\}$ be a partition of $[a, b]$. Let τ_i be a point of the interval $[t_1, t_{i+1}]$ and form the sum

$$\sum_1^n \mathbf{F}(\tau_i) \cdot \Delta_i \mathbf{P},$$

where $\Delta_i \mathbf{P} = \mathbf{P}(t_{i+1}) - \mathbf{P}(t_i)$ and the dot indicates the scalar product. If this sum has a limit as the fineness of the partition approaches zero, the limit is the **line integral** of $\mathbf{F}$ over C, denoted by $\int_C \mathbf{F}(t) \cdot d\mathbf{P}$. A sufficient condition for existence of the integral is that $\mathbf{F}$ be continuous on C. If $\mathbf{P}$ is differentiable and $\mathbf{F} = L\mathbf{i} + M\mathbf{j} + N\mathbf{k}$, then the line integral can be written as $\int_a^b (Lx' + My' + Nz') \, dt$ or as $\int_C L \, dx + \int_C M \, dy + \int_C N \, dz$. If the parameter is the length of an arc with a fixed end point, then the integral can be written as $\int_C F_t \, ds$, where F_t is the tangential component of $\mathbf{F}$. Similarly, if $A = P_1, P_2, \cdots, P_n = B$ are successive points on a curve, then

$$\int_C f(x, y, z) \, ds$$

is the limit as $\delta \to 0$ of a sum of type

$$\sum_1^n f(x_i, y_i, z_i) \Delta_i s,$$

where (x_i, y_i, z_i) is a point on the arc from P_i to P_{i+1}, $\Delta_i s$ is the length of this arc, and δ is the largest of the numbers $\Delta_i s$. If **F** is continuous on an open set S, then **F** is the gradient $\nabla\Phi$ of a function Φ if and only if **F** is conservative, *i.e.*,

$\int_{C_1} \mathbf{F}\cdot d\mathbf{P} = \int_{C_2} \mathbf{F}\cdot d\mathbf{P}$ whenever C_1 and C_2 have

the same initial and terminal points [in this case, the integral from **P** to **Q** is equal to $\Phi(\mathbf{Q}) - \Phi(\mathbf{P})$]. If, on an open set S in the plane, $\mathbf{F} = L\mathbf{i} + M\mathbf{j}$ with $\partial L/\partial y$ and $\partial M/\partial x$ continuous, then **F** is conservative only if $\partial L/\partial y = \partial M/\partial x$; if S is simply connected, then **F** is conservative if $\partial L/\partial y = \partial M/\partial x$. See CONSERVATIVE —conservative field of force. If $\partial M/\partial x - \partial L/\partial y = 1$, then the line integral of $L\mathbf{i} + M\mathbf{j}$ around a simple closed curve C is the area of the region bounded by C; in

particular, the area is given by $\frac{1}{2}\int_C (x\,dy - y\,dx)$.

multiple integral. A generalization of the integral of a function of a single variable as the limit of a sum. The **double integral** of a function f over a set R that has area and is contained in the domain of f is defined as follows: Divide the set k into n nonoverlapping subsets whose areas are $\Delta_i A$, $i = 1, 2, 3, \cdots, n$. Let Δ be the area of the smallest square in which each of these subsets can be embedded. Let (x_i, y_i) be a point in the ith subset. Form the sum

$$\sum_1^n f(x_i, y_i)\Delta_i A.$$

The double integral of f over R is defined as the limit of this sum as Δ approaches zero, if this limit exists, and is written:

$$\int_R f(x, y)\,dA.$$

The double integral exists if and only if the cylindrical region W, that is perpendicular to the (x, y)-plane and between R and the graph of f, has volume; if the double integral exists, it is equal to the volume of the part of W above the (x, y)-plane *minus* the volume of the part of W below the (x, y)-plane. If f is continuous and bounded on R, the double integral exists and is equal to the iterated integral of f over R. See above, iterated integral. The **triple integral** of a function f over a region R of space is defined in essentially the same way (Δ is then the volume of the smallest cube in which each subset can be embedded) and is also equal to a *triple iterated integral* if f is continuous. Multiple integrals of

higher order can be similarly defined. See FUBINI —Fubini theorem.

rationalization of integrals. See RATIONALIZE.

Riemann integral. See above, definite integral, generalized Riemann integral.

Riemann-Stieltjes integral. See STIELTJES.

surface integral. See SURFACE —surface integral.

triple integral. See above, iterated integral, multiple integral.

Volterra integral equations and their solutions. See various headings under VOLTERRA.

IN′TE-GRAND, *n.* See INTEGRAL —definite integral.

exact integrand. An integrand which is an exact differential. See DIFFERENTIAL —exact differential.

IN′TE-GRAPH, *n.* A mechanical device for finding areas under curves; a mechanical device for performing *definite integration*.

IN′TE-GRAT′ING, *adj.* **integrating factor.** See FACTOR.

integrating machines. Mechanical instruments for use in evaluating definite integrals; such instruments as the *integraph* and *polar planimeter*.

IN′TE-GRA′TION, *adj., n.* The process of finding an indefinite or definite integral. See INTEGRAL, and the INTEGRAL TABLES in the appendix.

change of variables in integration. For an integral with a single variable of integration, this is also called *integration by substitution*, since it is ordinarily used to change the integral into a form that is more easily evaluated. Typical substitutions are those which replace the variable by the square of a variable in order to rationalize indicated square roots of the variable, and the *trigonometric substitutions* by which the square roots of quadratic binomials are reduced to monomials. *E.g.*, if one substitutes $\sin u$ for x in $\sqrt{1 - x^2}$ the result is $\cos u$; similarly $\sqrt{1 + x^2}$ and $\sqrt{x^2 - 1}$ lend themselves to the substitutions $\tan u = x$, and $\sec u = x$, respectively. In all cases when a substitution is made in the integrand, it must also be made in the differential, and proper changes must be made in the limits if the integrals are definite integrals (an equivalent and quite common practice is to integrate the new integrand as an indefinite integral, then make the inverse substitution and use the original limits). *E.g.,*

$$\int_0^1 \sqrt{1-x^2}\, dx,$$

transformed by the substitution $x = \sin u$, becomes

$$\int_0^{\pi/2} \cos^2 u\, du = \left[\tfrac{1}{2}u + \tfrac{1}{4}\sin 2u\right]_0^{\pi/2} = \tfrac{1}{4}\pi,$$

or one may write

$$\int \cos^2 u\, du = \tfrac{1}{2}u + \tfrac{1}{4}\sin 2u$$

$$= \tfrac{1}{2}\sin^{-1} x + \tfrac{1}{2}x\sqrt{1-x^2},$$

whence

$$\int_0^1 \sqrt{1-x^2}\, dx$$

$$= \left[\tfrac{1}{2}\sin^{-1} x + \tfrac{1}{2}x\sqrt{1-x^2}\right]_0^1 = \tfrac{1}{4}\pi.$$

The rule for change of variables is of the same form for multiple integrals of all orders when expressed in terms of a Jacobian and will therefore be given only for triple integrals (the conditions given are sufficient, but not necessary). Let T be a transformation which maps an open set W in xyz-space onto an open set W^* in uvw-space and let D be a subset of W which is the image in xyz-space of a closed bounded set D^* in uvw-space. If $f(x, y, z)$ is continuous on D; x, y, and z have continuous first-order partial derivatives with respect to u, v, and w; and the Jacobian

$$J = \frac{\partial(x, y, z)}{\partial(u, v, w)}$$

is not zero on D^*, then

$$\int\int\int_D f(x, y, z)\, dx\, dy\, dz$$

$$= \int\int\int_{D^*} f(x, y, z)|J|\, du\, dv\, dw,$$

where x, y, and z in the second integral are functions of u, v, and w as determined by the transformation T. For spherical coordinates, this becomes:

$$\int\int\int f(x, y, z)\, dx\, dy\, dz$$

$$= \int\int\int f(x, y, z)\rho^2 \sin\phi\, d\rho\, d\phi\, d\theta.$$

For a single integral and the substitution $x = x(u)$, the Jacobian is dx/du and the analogous formula is

$$\int f(x)\, dx = \int f[x(u)]\frac{dx}{du}\, du;$$

this was used in the above illustrations of integration by substitution.

definite integration. The process of finding definite integrals. See INTEGRAL —definite integral.

element of integration. See ELEMENT —element of integration.

formulas of integration. Formulas giving the indefinite integrals and certain definite integrals of a few of the most commonly met functions.

integration by partial fractions. A specific method of integration used when the integrand is a rational function with denominator of higher degree than the first. Consists of breaking the integrand up into *partial fractions* with simpler denominators. *E.g.,*

$$\int \frac{dx}{1-x^2} = \frac{1}{2}\int \frac{dx}{1-x} + \frac{1}{2}\int \frac{dx}{1+x}.$$

See PARTIAL —partial fractions.

integration by parts. A process of integrating by use of the formula for the differential of a product. The formula $d(uv) = u\, dv + v\, du$ is written $u\, dv = d(uv) - v\, du$; integrating both sides of this equation gives $\int u\, dv = uv - \int v\, du$. This last formula enables one to modify the form of an integrand and simplify the process of integration, or actually integrate functions whose exact integral could not otherwise be found directly. It is especially useful in integrating such functions as xe^x, $\log x$, $x \sin x$, etc.; *e.g.,*

$$\int xe^x\, dx = xe^x - \int e^x\, dx,$$

where $x = u$, $e^x\, dx = dv$, and $v = e^x$.

integration of sequences and series. See BOUNDED—bounded convergence theorem, LEBESGUE—Lebesgue convergence theorem, MONOTONE—monotone convergence theorem, SERIES —integration of an infinite series.

integration by substitution. See above, change of variables in integration.

integration by use of series. The expanding of the integrand in a series and integrating term by term. An upper bound to the numerical value of the remainder of the series, after any given number of terms, can be integrated to find limits to the error. See INTEGRAL —definite integral (4).

mechanical integration. Determining the area bounded by a curve without the use of its equation, by the use of some specific mechanical device such as the *integraph* or *polar planimeter*.

reduction formulas in integration. Formulas expressing an integral as the sum of certain functions and a simpler integral. Such formulas are most commonly derived by *integration by parts*. See INTEGRAL TABLES in the appendix, formulas 54, 66, 74, etc.

IN'TE-GRA'TOR, *n.* An instrument which approximates definite integrals mechanically, such as a *planimeter* for measuring areas. In a computing machine, an integrator is any *arithmetic component* that performs the operation of integration.

IN-TEN'SI-TY, *n.* **electrostatic intensity.** See ELECTROSTATIC.

IN'TER-AC'TION, *n.* (*Statistics*) When the outcomes of experiments are grouped according to several factors, there is **interaction** if these factors are not independent. *E.g.*, suppose three fields are each divided into two plots and one of these plots is planted with corn of type C_1, the other with type C_2. The yields are observed for the six plots. There is no interaction between fertility of the fields and type of corn if the difference in yields between the two types of corn is the same for each field.

IN'TER-CEPT', *v., adj., n.* To cut off or bound some part of a line, plane, surface, or solid. Two radii **intercept** arcs of the circumference of a circle. An angle **intercepts** an arc if the arc (except for its end points) lies in the interior of the angle with one end point on each side of the angle. If a line L is cut at points A and B by other lines, planes, etc., then the lines, planes, etc., are said to **intercept** the segment AB on L. The **intercept of a straight line, curve, or surface on an axis of coordinates** is the distance from the origin to the point where the line, curve, or surface cuts the axis. The intercept on the axis of abscissas, or x-axis, is the **x-intercept**, and that on the axis of ordinates, or y-axis, the **y-intercept**. (In space, the intercept on the z-axis is the z-intercept.) The intercepts of the line $2x + 3y = 6$ on the x-axis and y-axis, respectively, are 3 and 2.

intercept form of the equation of a plane. See PLANE—equation of a plane.

intercept form of the equation of a straight line. See LINE—equation of a straight line.

IN'TER-DE'PEND'ENT, *adj.* **interdependent functions.** Same as DEPENDENT FUNCTIONS.

IN'TER-EST, *n.* Money paid for the use of money. The interest due at the end of a certain period is **simple interest** if the interest is computed on the original principal during the entire period. In this case, the interest is equal to the product of the *time, rate of interest*, and *original principal*; *e.g.*, the interest on $100 at 6% for 5 years is $5 \cdot (6/100) \cdot \$100 = \$30$. If the interest when due is added to the principal and thereafter earns interest, the interest (calculated in this way) is **compound interest**; the interest is computed on the principal for the first period, on the principal and the first period's interest for the second period, on the new principal and the second period's interest for the third period, etc. Thus at 6%, the interest plus principal at the end of the first, second, and nth years is respectively $P(1.06)$, $P(1.06)^2$, and $P(1.06)^n$, where P denotes the principal (see CONVERSION —continuous conversion of compound interest). The interval of time between successive conversions of interest into principal is the **interest** or **conversion period**; the total amount due at any time is the **compound amount**. If 6% is the rate, the compound amount of $1.00 at the end of 1 year is $1.06; at the end of 2 years, $(1.06)^2$; at the end of n years, $(1.06)^n$. The **nominal rate of interest** is the stated yearly rate when interest is compounded over periods of less than a year. When interest is computed at the rate of 3% semiannually, the nominal rate would be 6%. The annual rate which gives the same yield as the nominal computed over fractions of the year is the **effective rate**. The effective rate for 6% nominal compounded semiannually is 6.09%.

exact interest. Interest computed on the basis of the exact number of days in a year (365 days except for leap year, which has 366 days). Interest for 90 days at 6% would be 90/365 of 6% of the principal. In counting days between dates, the last, but not the first, date is usually included. The number of days from Dec. 25th to Feb. 2nd would be counted as 39 under the customary practice. See below, ordinary interest.

force of interest. The nominal rate which converted *continuously* is equivalent to a certain effective rate.

interest rate. The ratio of interest to principal, times 100%. *Syn.* rate of interest, rate, rate per cent.

long-end and short-end interest plan. See INSTALLMENT —installment payments.

ordinary interest. Interest computed on the basis of the commercial year of 360 days, with 30 days to the month. Interest for 2 months at 6% would be 60/360 of 6% times the principal; when the time of a note is expressed in days, the exact number of days is counted. *E.g.*, a note dated July 26th and due in 30 days would be due Aug. 25th, whereas if due in one month it would be due Aug. 26th. See above, exact interest.

prevailing interest rate (for any given investment). The rate which is common, or generally accepted, for that particular type of investment at the time under consideration. *Syn.* current rate.

six-per-cent method. A method of computing simple interest by computing it first for 6%, then for the given rate, if it is different. At 6%, the interest on $1 for one year is $0.06, for one month $0.005, and for one day $\frac{1}{30}(0.005)$. When the rate is other than 6%, one computes the interest for 6% as above and then takes the proper part of the result; for instance, if the rate is 5%, the interest is $\frac{5}{6}$ of the result obtained for 6%.

sixty-day method for computing interest. A method for computing simple interest at 6%. The rate for 60 days is $(\frac{60}{360})\frac{6}{100}$ or 1%, so the interest for 60 days is $\frac{1}{100}$ of the principal and for 6 days $\frac{1}{1000}$ of the principal. The time over which interest is to be computed is expressed in terms of 6 days and 60 days and fractional parts thereof. Also used when the rate is other than 6%. See above six-per-cent method.

IN-TE′RI-OR, *adj., n.* The interior of a circle, polygon, sphere, triangle, etc., is the set of all points inside the circle, polygon, etc. To make this consistent with the following general definition, we must think of the circle as being a disk including the circumference and the interior of the circle, and similarly for the polygon, etc. In general, the **interior** of a set E is the set of all points of E that have a neighborhood contained in E. Each such point is an **interior point**. The set of all points which belong to the closure of E and to the closure of the complement $C(E)$ of E is the **boundary** (or frontier) of E and of $C(E)$. It contains all points which are not interior points of E or of $C(E)$. See EXTERIOR —exterior of a set.

alternate interior, interior, and exterior-interior angles. See ANGLE—angles made by a transversal.

interior of an angle. See ANGLE.

interior angle of a polygon. See ANGLE—angle of a polygon.

interior content. See CONTENT —content of a set of points.

interior mapping (or transformation). Same as OPEN MAPPING. See OPEN.

interior measure. See MEASURE—exterior measure.

interior of a simple closed curve. See JORDAN —Jordan curve theorem.

IN-TER-ME′DI-ATE, *adj.* **intermediate differential.** See DIFFERENTIAL.

intermediate-value theorem. The theorem which states that if a function f is continuous for $a \le x \le b$, $f(a) \ne f(b)$, and k is between $f(a)$ and $f(b)$, then there is a number ξ between a and b for which $f(\xi) = k$.

IN-TER′NAL, *adj.* **internal operation.** See OPERATION.

internal ratio. See POINT—point of division.

internal tangent of two circles. See COMMON —common tangent of two circles.

IN-TER′NAL-LY, *a.* **internally tangent circles.** See TANGENT —tangent circles.

IN-TER-NA′TION-AL, *adj.* **international system of units.** See SI.

IN-TER′PO-LA′TION, *adj., n.* The process of finding a value of a function between two known values by a procedure other than the law which is given by the function itself. *E.g.*, in **linear interpolation** the procedure is based on the assumption that the three points having these values of the function for ordinates lie on a straight line. This is approximately true when the values of the arguments are close together and the graph of the function is smooth (has a tangent that varies continuously). If the function is f and its value is known at x_1 and x_2, then the formula for linear interpolation is

$$f(x) = f(x_1) + [f(x_2) - f(x_1)]\frac{x - x_1}{x_2 - x_1}.$$

Before the coming of calculators and computers, interpolation was used a great deal to estimate the value of a logarithm (or trigonometric function) of a number (or angle) which was not in the table. See EXTRAPOLATION.

interpolation formulas of Gregory-Newton, Hermite, and Lagrange. See the respective names.

IN′TER-QUAR′TILE, *adj.* **interquartile range.** The difference between the first and third *quartiles* of a distribution. This covers the middle half of the values in the frequency distribution.

IN′TER-SEC′TION, *n.* The point, or set of points, common to two (or more) geometric configurations. The **intersection of two curves** usually consists of a finite number of points, but may even contain an arc of one of the curves if this also is part of the other curve; two distinct straight lines either have an empty intersection or an intersection that is a single point. If two **surfaces** do not have an empty intersection, their intersection usually consists of curves, but may contain isolated points or pieces of the surfaces; the intersection of two distinct planes is either empty or a straight line. The term **imaginary intersection** is used to complete the analogy between discussions of equations and their graphs; it consists of the sets of imaginary values of the variables which are common solutions of the equations. The **intersection of two sets** consists of all the points that belong to each of the sets. The intersection of sets U and V usually is denoted by $U \cap V$, but sometimes by UV or $U \cdot V$, and is also called the **product,** or **meet,** of U and V.

angle of intersection. See ANGLE—angle of intersection.

IN′TER-VAL, *n.* (1) an **interval of real numbers** is the set containing all numbers between two given numbers (the **end points** of the interval) and one, both, or neither end point. An interval that contains none of its end points is an **open interval** and is denoted by (a, b), where a and b are the end points. A **closed interval** is an interval that contains all of its end points. A closed interval with end points a and b often is denoted by $[a, b]$, but sometimes by $]a, b[$ or (a, b). Sometimes an interval of real numbers is defined as any set of real numbers with the property that it contains all numbers between any two of its members. Then, in addition to intervals with two end points, the set of all real numbers is an interval with no end points that is both closed and open, and we have open intervals defined by inequalities of types $x < a$ or $x > a$, and closed intervals defined by inequalities of types $x \leq a$ or $x \geq a$. (2) In n-dimensional space, a **closed interval** is a set of points containing those and only those points x whose coordinates satisfy inequalities $a_i \leq x_i \leq b_i$ (for each i), for some fixed numbers $a_1, a_2, \cdots, a_n$ and $b_1, b_2, \cdots, b_n$ with $a_i < b_i$ for all i. The set of points x satisfying the inequalities $a_i < x_i < b_i$ (for each i) is an **open interval**. An interval may be an open interval, a closed interval, or one that

is partly open and partly closed (*i.e.*, some of the signs $\leq$ may be replaced by $<$).

confidence interval. See CONFIDENCE.

interval of convergence. See CONVERGENCE.

open interval. See above, closed interval.

IN′TO, *prep.* See ONTO.

IN-TRANS′SI-TIVE, *adj.* **intransitive relation.** See TRANSITIVE.

IN-TRIN′SIC, *adj.* **intrinsic equations of a space curve.** Since a space curve is determined to within its position in space by its radii of curvature and torsion as functions of the arc length, $\rho = f(s)$, $\tau = g(s)$, these equations are called the intrinsic equations of the curve. *Syn.* natural equations of a space curve.

intrinsic properties of a curve. Properties which are not altered by any change of coordinate systems. Some of the intrinsic properties of the conics are their eccentricity, distances from foci to directrices, length of latus rectum, length of the axes (of an ellipse or hyperbola), and their reflection properties.

intrinsic property of a surface. A property which pertains merely to the surface, not to the surrounding space; a property which is preserved under isometric transformations; a property expressible in terms of the coefficients of the first fundamental quadratic form alone. *Syn.* absolute property of a surface.

IN-TU-I′TION-ISM, *n.* See BROUWER.

IN-VA′RI-ANT, *adj., n.* **invariant factor of a matrix.** One of the diagonal elements when the matrix is reduced to *Smith's canonical form*, the elements of the matrix being polynomials. See CANONICAL—canonical form of a matrix. The invariant factors are unchanged by multiplication on either side by a matrix whose determinant does not involve the variable (λ). Each invariant factor is a product of the type $E_j(\lambda) = (\lambda - \lambda_1)^{p_{1j}} (\lambda - \lambda_2)^{p_{2j}} \cdots$, where $\lambda_1, \lambda_2, \cdots$ are distinct. Each factor $(\lambda - \lambda_i)^{p_{ij}}$ is called an **elementary divisor** of the matrix.

invariant of an algebraic equation. An algebraic expression involving the coefficients which remains unaltered in value when any translation or rotation of the axes is made. For the general quadratic, $ax^2 + bxy + cy^2 + dx + ey + f = 0$, $a + c$, $b^2 - 4ac$, and the **discriminant** are invariants. See DISCRIMINANT—discriminant of a quadratic equation in two variables.

invariant property. A property of a function, configuration, or equation that is not altered by a particular transformation. The invariant property is used with reference to a particular transformation or type of transformation. *E.g.*, the value of a cross ratio is not changed by a projection, and hence is said to be invariant under projective transformations. See TENSOR.

invariant subgroup. Same as NORMAL SUBGROUP. See NORMAL.

invariant-subspace problem. The problem of determining which bounded linear operators T on a Banach space X have a nontrivial ($\neq \{0\}$) proper **invariant subspace** (*i.e.*, a closed linear subspace L such that $Tx \in L$ if $x \in L$). It is known that T has such an invariant subspace if there exists a compact operator that commutes with T and does not map all of X onto 0, but there is a nonreflexive Banach space X and a bounded linear operator T on X that has no nontrivial proper invariant subspaces.

semi-invariant. If, for each transformation of type $Y = a + bX$ of a random variable X, a sequence of parameters $\{v_n\}$ associated with X are related to the same parameters $\{w_n\}$ associated with Y by having $w_n = b^n v_n$ if $n \geq 2$, the parameters are **semi-invariants**. *E.g.*, central moments and cumulants about the mean are semi-invariants.

IN-VERSE′, *adj.*, *n.* The **additive inverse** of a number a is the number $-a$ for which $a + (-a) = 0$; the **multiplicative inverse** (or **reciprocal**) of a nonzero number a is the number $1/a$ for which $a \cdot (1/a) = 1$. More generally, suppose that for a set S there is a binary operation $x \circ y$ and an identity e such that $x \circ e = e \circ x$ for all x in S. Then an **inverse** of x is a member x^* of S for which $x \circ x^* = x^* \circ x = e$. A **right inverse** is an x^* for which $x \circ x^* = e$; a **left inverse** is an x^* for which $x^* \circ x = e$. See GROUP, IDENTITY—identity element, TRANSFORMATION—inverse transformation, and below, inverse of a function.

inverse of a function. If $y = f(x)$ is equivalent to $x = g(y)$, then f is the **inverse** of g (and vice versa). It is customary to interchange the variables in the latter, writing $y = g(x)$ as the inverse. A function f has a **right inverse** g for which $f[g(x)] = x$ for all x in the range of f if and only if f has a **left inverse** h for which $h[f(x)] = x$ for all x in the domain of f. If either g or h exists, then $g = h$ and g is the inverse of f. A function has an inverse if and only if it is one-to-one. *E.g.*, the inverse of $y = \sin x$ is $y = \sin^{-1} x$ if the domains of these functions are $[-\frac{1}{2}\pi, +\frac{1}{2}\pi]$ and $[-1, +1]$, respectively, and their ranges are $[-1, +1]$ and $[-\frac{1}{2}\pi, +\frac{1}{2}\pi]$, respectively. The inverse of a continuous function f is continuous if the domain of f is compact. If the domain of f is an interval and f is either increasing throughout the domain or decreasing

throughout the domain, then f has an inverse and the inverse is continuous, and moreover, if $f'(x)$ exists and is nonzero and g is the inverse of f then $g'(y)$ exists and $f'(x)g'(y) = 1$ if $y = f(x)$. See IMAGE.

inverse hyperbolic functions. See HYPERBOLIC—inverse hyperbolic functions.

inverse image. See IMAGE

inverse logarithm of a given number. The number whose logarithm is the given number. Log 100 is 2; hence 100 is inverse log 2. *Syn.* antilogarithm.

inverse-mapping theorem. See OPEN—open-mapping theorem.

inverse of an implication. The implication which results from replacing both the antecedent and the consequent by their negations. *E.g.*, the inverse of "If x is divisible by 4, then x is divisible by 2" is the (false) statement "If x is not divisible by 4, then x is not divisible by 2." The *converse* and the *inverse* of an implication are equivalent—they are either both true or both false.

inverse of a number. One divided by the number. *Syn.* reciprocal of a number.

inverse of an operation. That operation which, when performed after a given operation, annuls the given operation. Subtraction of a quantity is the inverse of addition of that quantity. Addition is likewise the inverse of subtraction. Since an operation is a function and a transformation, see above, inverse of a function, TRANSFORMATION —inverse transformation.

inverse of a point or curve. See INVERSION.

inverse of a relation. See RELATION.

inverse or reciprocal proportion. A proportion containing one reciprocal ratio. See INVERSELY —inversely proportional quantities.

inverse or reciprocal ratio of two numbers. The ratio of the reciprocals of the numbers.

inverse trigonometric function. See TRIGONOMETRIC.

inverse variation. See VARIATION —inverse variation.

IN-VERSE′LY, *adv.* **inversely proportional quantities.** (1) Two variables having their product constant; *i.e.*, either of them is equal to a constant times the inverse (reciprocal) of the other. (2) The numbers $(a_1, a_2, \cdots)$ are **inversely proportional** to the numbers $(b_1, b_2, \cdots)$ if and only if $a_1 b_1 = a_2 b_2 = \cdots$. *E.g.*, $(1, 2)$ and $(6, 3)$ are inversely proportional, since $1 \cdot 6 = 2 \cdot 3$.

IN-VER′SION, *adj.*, *n.* **inversion formulas.** Such formulas as *Fourier transforms*, *Laplace transforms*, and the *Mellin inversion formulas* which give a pair of linear transformations T_1, T_2

such that T_2 applied to $T_1(f)$ produces f for any function of a certain class. See FOURIER, LAPLACE, MELLIN.

inversion of a point with respect to a circle. Finding the point on the radial line through the given point such that the product of the distances of the two points from the center of the circle is equal to the square of the radius. Either of the points is called the **inverse** of the other and the center of the circle is called the **center of inversion.** Any curve whose points are the inverses of the points of a given curve is called the inverse of the given curve. *E.g.*, the inverse of a circle which passes through the center of inversion is a straight line; the inverse of any other circle is a circle. If the equation of a curve is $f(x, y) = 0$, the equation of its inverse relative to a circle with center at the origin is

$$f\left(\frac{k^2 x}{x^2 + y^2}, \frac{k^2 y}{x^2 + y^2} \right) = 0,$$

where k is the radius of the circle.

inversion of a point with respect to a sphere. Finding the point on the radial line through the given point such that the product of the distances of the two points from the center of the sphere is equal to the square of the radius of the sphere. *E.g.*, the inverse of every sphere with respect to a fixed sphere is another sphere, except that the inverse of a sphere passing through the center of the fixed sphere is a plane.

inversion of a sequence of objects. The interchange of two adjacent objects. The *number of inversions* in a sequence is the minimum number of inversions which can be performed in order to put the objects in a certain *normal* order. The permutation $1, 3, 2, 4, 5$ has one inversion, if the normal order is $1, 2, 3, 4, 5$, whereas the permutation $1, 4, 3, 2, 5$ has three. A permutation is **odd** or **even** according as it contains an *odd* or an *even* number of inversions.

proportion by inversion. See PROPORTION.

IN-VER'SOR, *n.* A mechanical device which simultaneously traces out a curve and its *inverse*. A rhombus, with its sides pivoted at the vertices and a pair of opposite vertices each linked to a fixed point (the *center of inversion*) by equal links, is such a mechanism, called a *Peaucellier cell.* When one of the unlinked vertices traces out a curve, the other traces out the inverse of the curve. See INVERSION.

IN-VERT'I-BLE, *adj.* A **right-invertible** member of a groupoid (or a ring) with a unit element e is a member x for which there is an x^* with

$x \cdot x^* = e$; a **left-invertible** member is a member x for which there is an x' with $x' \cdot x = e$; and an **invertible** member is a member x for which there is an x^* such that $x \cdot x^* = x^* \cdot x = e$. A matrix is invertible if and only if it is nonsingular; a transformation is invertible if and only if it is one-to-one (or for linear transformations of finite-dimensional spaces, if and only if its matrix is nonsingular). See INVERSE —inverse of a function, MATRIX —inverse of a matrix.

IN-VEST'MENT, *n.* Money used to buy notes, bonds, etc., or put into any enterprise for the purpose of making profit.

fixed investment. An investment which yields a fixed income; the amount which must be put into a sinking fund at the end of each year to accumulate to a given sum at the end of a given term.

investment (or investor's) rate. See YIELD.

mathematics of investment. Same as MATHEMATICS OF FINANCE.

IN'VO-LUTE, *n.* **involute of a curve.** For a **plane curve,** the locus of a fixed point on a nonflexible string as it is unwound, under tension, from the curve; the locus of any fixed point on a tangent line as this line rolls, but does not slide, around the curve; a curve orthogonal to the family of tangents to a given curve. Any two involutes of the same curve are parallel, *i.e.*, the segments cut off by the two involutes on a common normal are always of the same length. Also,

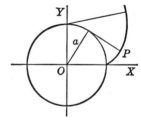

an involute of a given curve is any curve whose *evolute* is the given curve. The **evolute** of a curve is the locus of the centers of curvature of the given curve. The family of straight lines normal to a given curve are tangent to the evolute of this curve, and the change in length of the radius of curvature is equal to the change in length of arc of the evolute as the point on the curve moves continuously in one direction along the curve. The equation of the *evolute* is obtained by eliminating the coordinates of the point on the curve between the equation of the curve and equations expressing the coordinates of the center of curvature in terms of the coordinates of the point on the curve. The **involute of a circle** is the curve described by the end of a thread as it is unwound from a stationary spool. The parametric equations of the involute of a circle in the position illustrated, where θ is the angle from the x-axis to the radius marked a, are:

$$x = a(\cos\theta + \theta\sin\theta),$$

$$y = a(\sin\theta - \theta\cos\theta).$$

An **involute of a space curve** is a curve orthogonal to the tangents of the given curve. The involutes of a space curve lie on its tangent surface. A given space curve has an infinitude of involutes; they constitute a family of geodesic parallels on the tangent surface (see GEODESIC —geodesic parallels on a surface). An **evolute of a space curve** is a curve of which the given curve is an involute. A given space curve C admits an infinitude of evolutes; when all the normals to C which are tangent to one of the evolutes are turned through the same angle in their planes normal to C, the resulting normals are tangent to another evolute of C.

involute of a surface. A surface of which the given surface is one of the two branches of the evolute. See EVOLUTE —evolute of a surface.

IN′VO-LU′TION, *n.* (1) Raising to a power; multiplying a quantity by itself a given number of times; the inverse of *evolution*. The process of squaring 2 is *involution*, of finding a square root of 4 is *evolution*. (2) A function which is its own inverse, *e.g.*, $x = 1/x'$ is an involution. See INVERSE —inverse of a function. (3) An *automorphism* that satisfies $(x^*)^* = x$ for all x, where for each x the image of x under the automorphism is x^*. See ISOMORPHISM.

involution on a line. A projective correspondence between the points of a line, which is its own inverse. Algebraically, the transformation

$$x' = \frac{ax + b}{cx - a},$$

where $a^2 + bc \neq 0$. If $c \neq 0$, this can be written $x' = k/x$, by a proper choice of the origin.

involution of lines of a pencil. A correspondence between the lines, which is such that corresponding lines pass through corresponding points of an involution of points on a line which does not pass through the vertex of the pencil.

I-O′TA, *n.* The ninth letter of the Greek alphabet: lower case, ι; capital, I.

IR-RA′TION-AL, *adj.* **irrational algebraic surface.** The graph of an algebraic function in which the variable (or variables) appear irreducibly under a radical sign. The loci of $z = \sqrt{y + x^2}$ and $z = x^{1/2} + xy$ are irrational algebraic surfaces.

irrational equation. See EQUATION—irrational equation.

irrational exponent. See EXPONENT.

irrational number. A real number not expressible as an integer or quotient of integers; a nonrational number. The irrational numbers are those numbers defined by sets (A, B) of a *Dedekind cut* such that A has no greatest member and B has no least member. Also, the irrational numbers are precisely those infinite decimals which are not repeating. The irrational numbers are of two types, **algebraic irrational numbers** (irrational numbers which are roots of polynomial equations with rational coefficients; see ALGEBRAIC —algebraic number) and **transcendental numbers**. The numbers e and π are transcendental, as well as trigonometric and hyperbolic functions of any nonzero algebraic number and any power α^β, where α and β are algebraic numbers, α is not 0 or 1, and β is not a real rational number (see GELFOND —Gelfond-Schneider theorem). The algebraic numbers (including the rational numbers) are less numerous than the transcendental numbers in two senses, of which the second is a consequence of the first: (1) the algebraic numbers are *countable*, while the transcendental numbers are not; (2) the set of algebraic numbers is of *measure zero*, while the *measure* of the transcendental numbers on an interval is the length of the interval. Also see LIOUVILLE —Liouville number, NORMAL —normal number.

IR′RE-DUC′I-BLE, *adj.* **irreducible case** in Cardan's formula for the roots of a cubic. See CARDAN—Cardan solution of the cubic.

irreducible equation. A rational integral equation of the form $f(x) = 0$, where $f(x)$ is a polynomial irreducible in a certain field, usually the field of all rational numbers. See below, polynomials irreducible in a given field (domain).

irreducible matrices and transformations. See under REDUCIBLE.

irreducible module. See MODULE.

irreducible polynomial. A polynomial that cannot be written as the product of two polynomials with degrees at least 1 and having coefficients in some given domain or field. Unless otherwise stated, *irreducible* means irreducible in the field of the coefficients of the polynomial under consideration. The binomial $x^2 + 1$ is irreducible in the field of real numbers, although in the field of complex numbers it can be factored as $(x + i)(x - i)$, where $i^2 = -1$. In elementary algebra, it is understood that an irreducible polynomial is a polynomial that cannot be factored into factors having rational coefficients. If complex numbers are used as coefficients, it follows from the *fundamental theorem of algebra* that the only irreducible polynomials are of the first degree. If p is a polynomial with coefficients in a field F, then either p is irreducible or p can be written as a product of irreducible polynomials; except for constant factors and the order of the factors, this product is unique. See DOMAIN—integral domain, EISENSTEIN.

irreducible radical. A radical which cannot be written in an equivalent rational form. The radicals $\sqrt{6}$ and $\sqrt{x}$ are irreducible, whereas $\sqrt{4}$ and $\sqrt[3]{x^3}$ are reducible since they are equal, respectively, to 2 and x.

join- and **meet-irreducible.** See JOIN.

IR-RO-TA′TION-AL, *n.* **irrotational vector in a region.** A vector whose integral around every *reducible* closed curve in the region is zero. The *curl* of a vector is zero at each point of a region if, and only if, the vector is irrotational, or if, and only if, it is the *gradient* of a scalar function (called a **scalar potential**); *i.e.*, $\nabla \times \mathbf{F} \equiv 0$ if and only if $\mathbf{F} = -\nabla\Phi$ for some scalar potential Φ. See CURL, GRADIENT.

I-SOCH′RO-NOUS (or **I-SOCH′RO-NAL**), *adj.* **isochronous curve.** A curve with the property that a particle sliding along the curve without friction will reach a lowest point on the curve in a time that is the same for all starting points. See CYCLOID.

I-SOG′O-NAL, *adj.* Having equal angles.
isogonal affine transformation. See AFFINE—affine transformation.
isogonal conjugate lines. See below, isogonal lines.
isogonal lines. Lines through the vertex of an angle and symmetric with respect to (making equal angles with) the bisector of the angle. The lines are **isogonal conjugates.**
isogonal transformation. A transformation which leaves all the angles in any configuration unchanged. *E.g.*, the general similarity transformation is an isogonal transformation. *Syn.* equiangular (or conformal) transformation.

I′SO-LATE, *v.* **isolate a root.** To find two numbers between which the root (and usually no other root) lies. See ROOT—root of an equation.

I′SO-LA-TED, *adj.* **isolated point.** See ACCUMULATION—accumulation point, POINT—isolated point.

isolated set. A set which contains none of its accumulation points; a set consisting entirely of isolated points, a point of a set E being an isolated point of E if it has a neighborhood which contains no other point of E. A **discrete set** is a set that has no accumulation points. A discrete set is an isolated set, but the set $(1, \frac{1}{2}, \frac{1}{3}, \frac{1}{4}, \cdots)$ is isolated and not discrete.

isolated singular point of an analytic function. See SINGULAR—singular point of an analytic function.

I′SO-MET′RIC, *adj.* **isometric family of curves on a surface.** A one-parameter family of curves on the surface such that the family together with its orthogonal trajectories forms an *isometric system* of curves on the surface.

isometric map and **parameters.** See ISOMETRY.

isometric surfaces. Surfaces on which corresponding distances are equal and angles between corresponding lines are equal.

isometric system of curves on a surface. A system of two one-parameter families of curves (on the surface) which might be taken as parametric curves in an isometric map.

I-SOM′E-TRY, *n.* (1) An isothermic map. (2) A length preserving map. In the map given by $x = x(u, v)$, $y = y(u, v)$, $z = z(u, v)$, lengths are preserved if, and only if, the fundamental coefficients of the first order satisfy $E = G = 1$, $F = 0$. The coordinates u, v are **isometric parameters.** The above functions and the functions $x = \bar{x}(u, v)$, $y = \bar{y}(u, v)$, $z = \bar{z}(u, v)$ give length preserving maps between corresponding surfaces S and $\bar{S}$ if, and only if, the corresponding fundamental coefficients of the first order satisfy $E = \bar{E}$, $F = \bar{F}$, $G = \bar{G}$. Then the surfaces S and $\bar{S}$ are said to be **applicable.** *Syn.* isometric map. (3) A one-to-one correspondence of a metric space A with a metric space B such that if x corresponds to x^* and y to y^*, then the "distances" $d(x, y)$ and $d(x^*, y^*)$ are equal. It is then said that A and B are **isometric.** If A and B are vector spaces with a norm, it is also required that the correspondence be an *isomorphism*. The preservation of distance is then equivalent to $\|x\| = \|x^*\|$ whenever x and x^* correspond. If A and B are Hilbert spaces, this is equivalent to the equality of the inner products (x, y) and (x^*, y^*) whenever x and x^* correspond and y and y^* correspond (see UNITARY—unitary transformation).

I-SO-MOR′PHISM, *n.* (1) See CATEGORY (2). (2) A one-to-one correspondence of a set A with a set B (the sets A and B are then said to be **equinumerable, equipotent,** or **equivalent**). If operations such as multiplication, addition, or multiplication by scalars are defined for A and B, it is required that these correspond between A and B in the ways described in the following. If A and B are **groups** (or semigroups) with the operation denoted by $\cdot$, and x corresponds to x^* and y to y^*, then $x \cdot y$ must correspond to $x^* \cdot y^*$. An isomorphism of a set with itself is an **automorphism.** An automorphism of a group is an **inner automorphism** if there is an element t such that x corresponds to x^* if and only if $x^* = t^{-1}xt$; it is an **outer automorphism** if it is not an inner automorphism. The correspondence $\omega_1 \to \omega_2$, $\omega_2 \to \omega_1$, $1 \to 1$, is an outer automorphism of the group consisting of the cube roots of unity $(1, \omega_1, \omega_2)$. If a set S^* corresponds to a subgroup S by an automorphism (meaning that the elements of S^* and S correspond in pairs), then S^* is also a subgroup (if the automorphism is an inner automorphism, then S and S^* are said to be **conjugate subgroups**). If A and B are **rings** (or integral domains or fields) and x corresponds to x^* and y to y^*, then xy must correspond to x^*y^* and $x + y$ to $x^* + y^*$. A correspondence that would be an isomorphism or an automorphism, except that xy corresponds to y^*x^*, is an **anti-isomorphism** or an **anti-automorphism** (*e.g.*, the *transpose* operation for matrices). If A and B are **vector spaces**, addition must correspond as for groups and scalar multiplication must also correspond in the sense that if a is a scalar and x corresponds to x^*, then ax corresponds to ax^*. If the vector space is normed (*e.g.*, if it is a Banach or Hilbert space), then the correspondence must be continuous in both directions. This is equivalent to requiring that there be positive numbers c and d such that $c\|x\| \le \|x^*\| \le d\|x\|$ if x and x^* correspond. See HOMOMORPHISM, ISOMETRY.

I′SO-PER′I-MET′RIC or **I′SO-PER′I-MET′RI-CAL,** *adj.* Having equal perimeters.

isoperimetric inequality. The inequality $A \le \dfrac{1}{4\pi}L^2$ between the area A of a plane region and the length L of its boundary curve. The equality sign holds if and only if the region is a circle. The inequality holds also for regions on surfaces of nonpositive total curvature, and actually characterizes these surfaces. See below, isoperimetric problem in the calculus of variations.

isoperimetric problem in the calculus of variations. A problem in which it is required to make one integral a maximum or minimum while keeping constant the integral of a second given function (both integrals being of the general type indicated under CALCULUS —calculus of variations). An example is the problem of finding the closed plane curve of given perimeter and maximum area (the solution is a circle). In polar coordinates, the problem is that of finding the curve $r = f(\phi)$ for which $A = \int_0^{2\pi} \frac{1}{2}r^2\, d\phi$ is maximum and $P = \int_0^{2\pi}(r^2 + r'^2)^{1/2}\, d\phi$ is constant. The solution can be found by maximizing the integral

$$A + \lambda P = \int_0^{2\pi}\left[\tfrac{1}{2}r^2 + \lambda\left(r^2 + r'^2\right)^{1/2}\right] d\phi$$

and determining the constant λ by the condition that P is a given constant. Most isoperimetric problems can be similarly reduced to the usual type of calculus of variations problem, a solution of the given problem necessarily being a solution of the reduced problem.

I-SOS′CE-LES, *adj.* **isosceles triangle.** See SPHERICAL —spherical triangle, TRIANGLE.

isosceles trapezoid. See TRAPEZOID.

I′SO-THERM, *n.* A line drawn on a map through places having equal temperatures.

I′SO-THER′MAL, *adj.* Relating to equal temperatures.

isothermal change. A change in the volume and pressure of a substance which takes place at constant temperature.

isothermal-conjugate parameters. Parameters such that the parametric curves form an isothermal-conjugate system on the surface. See below, isothermal-conjugate system of curves on a surface.

isothermal-conjugate representation of one surface on another. See CONFORMAL —conformal-conjugate representation of one surface on another.

isothermal-conjugate system of curves on a surface. A system of two one-parameter families of curves on the surface S such that, when the curves are taken as parametric curves, the second fundamental quadratic form of S reduces to $\mu(u, v)(du^2 \pm dv^2)$. In particular, then, an isothermal-conjugate system is a conjugate system. See CONJUGATE —conjugate system of curves on a surface. An isothermal-conjugate system has a relation to the second fundamental quadratic form of S similar to that which an isothermic system has to the first fundamental quadratic form. See ISOTHERMIC—isothermic family of curves on a surface.

isothermal lines. Lines on a map connecting points which have the same mean (annual) temperatures. In *physics*, curves obtained by plotting pressure against volume for a gas kept at constant temperature.

I′SO-THER′MIC, *adj.* **isothermic family of curves on a surface.** A one-parameter family of curves on the surface such that the family together with its orthogonal trajectories forms an isothermic system of curves on the surface.

isothermic map. A map of a (u, v)-domain on a surface S in which the fundamental quantities of the first order satisfy $E = G = \lambda(u, v)$, $F = 0$. The map is conformal except at the singular points where $\lambda = 0$. The coordinates u, v are **isothermic parameters.** See CONFORMAL—conformal map, and above, isothermic family of curves on a surface.

isothermic parameters. See above, isothermic map.

isothermic surface. A surface whose lines of curvature form an *isothermal system*. All surfaces of revolution are isothermic surfaces.

isothermic system of curves on a surface. A system of two one-parameter families of curves on a surface S such that there exist parameters u, v for which the curves of the system are the parametric curves of the surface, and for which the first fundamental quadratic form reduces to $\lambda(u, v)(du^2 + dv^2)$. See ISOTHERMAL—isothermal-conjugate system of curves on a surface, and above, isothermic map.

I′SO-TROP′IC, *adj.* **isotropic curve.** Same as MINIMAL CURVE.

isotropic developable. An imaginary surface for which $EG - F^2$ vanishes identically. Such a surface is the tangent surface of a minimal curve. See SURFACE—fundamental coefficients of a surface.

isotropic elastic substances. Substances whose elastic properties are independent of the direction in the substances are said to be *isotropic*. This means that the elastic properties are the same in all directions.

isotropic matter. Matter which, at any point, has the same properties in any direction (such properties, for instance, as elasticity, density, and conductivity of heat or electricity).

isotropic plane. An imaginary plane with equation $ax + by + cz + d = 0$, where $a^2 + b^2 + c^2 = 0$. E.g., the osculating planes of minimal curves are isotropic.

IS′SUE, *n.* **issue of bonds, bank notes, money, or stock.** A set of bonds, bank notes, etc., which is (or has been) issued at a certain time.

IT′ER-ATE, *v.* **iterated integral.** See INTEGRAL—iterated integral.

J

JACOBI, Karl Gustav Jacob (1804–1851). German algebraist and analyst.

Jacobi polynomials. The polynomials $J_n(p, q; x) = F(-n, p + n; q; x)$, where $F(a, b; c; x)$ is the *hypergeometric function* and n is a positive integer. It follows that $J_n[1, 1, \frac{1}{2}(1 - x)] = P_n(x)$ and

$$2^{1-n} J_n[0, \tfrac{1}{2}, \tfrac{1}{2}(1 - x)] = T_n(x),$$

where P_n and T_n are Legendre and Chebyshev polynomials, respectively.

Jacobian elliptic functions. See ELLIPTIC.

Jacobian of two or more functions in as many variables. For the n functions $f_i(x_1, x_2, \cdots, x_n)$, $i = 1, 2, 3, \cdots, n$, the Jacobian is the determinant

$$\begin{vmatrix} \dfrac{\partial f_1}{\partial x_1} & \dfrac{\partial f_1}{\partial x_2} & \dfrac{\partial f_1}{\partial x_3} & \cdots & \dfrac{\partial f_1}{\partial x_n} \\[2mm] \dfrac{\partial f_2}{\partial x_1} & \dfrac{\partial f_2}{\partial x_2} & \dfrac{\partial f_2}{\partial x_3} & \cdots & \dfrac{\partial f_2}{\partial x_n} \\[2mm] \cdots & & & & \\[2mm] \dfrac{\partial f_n}{\partial x_1} & \dfrac{\partial f_n}{\partial x_2} & \dfrac{\partial f_n}{\partial x_3} & \cdots & \dfrac{\partial f_n}{\partial x_n} \end{vmatrix}$$

which is often designated by one of

$$\frac{D(f_1, f_2, f_3, \cdots, f_n)}{D(x_1, x_2, x_3, \cdots, x_n)}, \quad \frac{\partial(f_1, f_2, f_3, \cdots, f_n)}{\partial(x_1, x_2, x_3, \cdots, x_n)}.$$

For two functions, $f(x, y)$ and $g(x, y)$, the Jacobian is the determinant

$$\begin{vmatrix} \dfrac{\partial f}{\partial x} & \dfrac{\partial f}{\partial y} \\[2mm] \dfrac{\partial g}{\partial x} & \dfrac{\partial g}{\partial y} \end{vmatrix}, \text{ designated by } \frac{D(f, g)}{D(x, y)}.$$

See IMPLICIT—implicit function theorem, INDEPENDENT—independent functions. *Syn.* functional determinant.

Jacobi's theorem. See PERIODIC—periodic function of a complex variable.

JENSEN, Johan Ludvig William Valdemar (1859–1925). Danish analyst, algebraist, and engineer. Introduced formal study of convex functions.

Jensen's formula. The formula in the conclusion of Jensen's theorem. This formula is basic in the modern theory of entire functions. See below, Jensen's theorem.

Jensen's inequality. For convex functions f, the inequality

$$f\left(\sum_1^n \lambda_i x_i\right) \le \sum_1^n \lambda_i f(x_i),$$

where the x_i are arbitrary values in the region on which f is convex and the λ_i are nonnegative numbers satisfying $\sum_1^n \lambda_i = 1$. The term "Jensen's inequality" is also applied to the inequality expressing the fact that for $t > 0$ the *sum of order t* is a nonincreasing function of t; *i.e.*, for positive numbers a_i and positive numbers s and t with $s > t$,

$$\left(\sum_1^n a_i^s\right)^{1/s} \le \sum_1^n \left(a_i^t\right)^{1/t}.$$

Jensen's theorem. If f is analytic in the disk $|z| \le R < \infty$, if the zeros of f in this disk are $a_1, a_2, \cdots, a_n$, where each zero is repeated as often as is indicated by its multiplicity, and if $f(0)$ is not 0, then

$$\frac{1}{2\pi}\int_0^{2\pi} \ln|f(Re^{i\theta})|\,d\theta$$

$$= \ln|f(0)| + \sum_{j=1}^n \ln\frac{R}{|a_j|}.$$

JOACHIMSTHAL, Ferdinand (1818–1861). German analyst and geometer.

surface of Joachimsthal. See under SURFACE.

JOHNIAC. An automatic digital computing machine at the RAND Corporation. Named for the mathematician John von Neumann, the machine was similar to one at the Institute for Advanced Study.

JOIN, *n.* See LATTICE, UNION—union of sets.

join-irreducible. For a lattice or a ring of sets, a **join-irreducible** member is a member W having the property that $X = W$ or $Y = W$ if X and Y are members such that $X \cup Y = W$. A member of a Boolean algebra is join-irreducible if and only if it is an atom or 0. Every member of a finite lattice is the join of join-irreducible members. **Meet-irreducible** is defined similarly, with $\cup$ replaced by $\cap$.

JOINT, *adj.* **joint-distribution function.** See DISTRIBUTION —distribution function.

joint expectation of life. See EXPECTATION —expectation of life.

joint-life annuity. See ANNUITY.

joint-life insurance. See INSURANCE —life insurance.

joint variation. See VARIATION —joint variation.

JONES, Vaughan F. R. (1952–   ). American mathematician awarded a Fields Medal (1990) for his contributions to von Neumann algebras.

JORDAN, Camille (1838–1922). French algebraist, group theorist, analyst, geometer, and topologist.

Jordan condition for convergence of Fourier series. See FOURIER —Fourier's theorem.

Jordan content. See CONTENT —content of a set of points.

Jordan curve. Same as SIMPLE CLOSED CURVE. See SIMPLE.

Jordan-curve theorem. A simple closed curve C in a plane determines two regions, of which it is the common boundary. One of these regions is bounded and is the **interior** of C; the other region is the **exterior** of C. Each point of the plane is on C, belongs to the interior of C, or belongs to the exterior of C. Any two points in the interior (or the exterior) can be joined by a curve that contains no points of C. Any curve that joins a point of the interior and a point of the exterior of C contains a point of C. The Jordan-curve theorem was proved incorrectly by Jordan; the first correct proof was given by Veblen in 1905.

Jordan matrix. A matrix having the elements of its principal diagonal equal and not zero, the elements immediately above those in the diagonal unity, and all other elements zero. *Syn.* simple classical matrix. Such a matrix is also said to be in **Jordan form.** See CANONICAL —canonical form of a matrix.

JOUKOWSKI, Nikolai Jegórowitch (1847–1921). Russian applied mathematician and aerodynamicist.

Joukowski transformation. In complex variable theory, the transformation

$$w = z + 1/z.$$

It maps the points z and $1/z$ into the same point, so that the image of the exterior of the unit circle $|z| = 1$ is the same as the image of the interior of this circle. There are simple zeros of dw/dz at $z = \pm 1$, and otherwise $dw/dz \neq 0$; accordingly, the map is conformal except at these two points. The upper half of the z-plane, with its half of the unit circle deleted, is mapped on the upper half of the w-plane. Under the Joukowski transformation, the exterior of a circle through the point $z = -1$ and having $z = +1$ in its interior is mapped on the exterior of a contour that, for some positions of the circle, bears a striking resemblance to the profile of an airplane wing. Such a contour is a **Joukowski airfoil profile**.

JOULE, James Prescott (1835–1889). English physicist.

joule. A unit of energy or work; the work done when a force of one *newton* produces a displacement of one meter in the direction of the force.

$$1 \text{ J} = 10^7 \text{ erg} = .2390 \text{ calorie}.$$

JULIA, Gaston Maurice (1892–1978). **Julia set.** Given a polynomial f with degree greater than 1, the **Julia set** of f is the boundary of the set of all complex numbers z whose orbits with respect to the sequence of functions $\{f, f^2, \cdots, f^n, \cdots\}$ are bounded, where $f^2(z) = f[f(z)]$, etc. (see ORBIT). Many interesting *fractals* can be computer-generated by using suitable polynomials. The most famous are generated by polynomials of type $z^2 + c$, where c is a constant complex number. The Julia set for $z^2 + c$ with $|c| < 1$ has *topological dimension* 0 and *Hausdorff dimension* $1 + |c|^2 (4 \log 2)^{-1}$, plus higher order terms. By simple changes of variable, one can reduce any quadratic polynomial to a function of type $z^2 + c$. Thus all Julia sets generated by quadratic polynomials can be described by using a polynomial of type $z^2 + c$. See CHAOS, FRACTAL, MANDELBROT—Mandelbrot set.

JUMP, *n.* See DISCONTINUITY.

JUNG, Wilhelm Ewald (1876–1953). German analyst and geometer.

Jung's theorem. A set of diameter 1 in n-dimensional Euclidean space can be enclosed in a closed ball of radius $[\frac{1}{2}n/(n+1)]^{1/2}$. In particular, a plane set of diameter 1 can be enclosed in a circle of radius $1/\sqrt{3}$. See BLASCHKE—Blaschke's theorem.

K

KAKEYA, Sôichi (1886–1947). Japanese analyst and geometer.

Kakeya problem. The problem of finding a plane set S of minimum area such that a unit line segment can be moved continuously in S so as to return to its original position with its ends reversed. There is no solution, since for each positive number ϵ there exists such a set S with area less than ϵ. Moreover, S can be simply connected and contained in a circle of radius 1.

KAP′PA, *adj.*, *n.* The tenth letter of the Greek alphabet: lower case, κ; capital K.

kapa curve. The graph of the rectangular equation $x^4 + x^2 y^2 = a^2 y^2$. The curve has the lines $x = \pm a$ as asymptotes, is symmetrical about the coordinate axes and the origin, and has a double cusp at the origin. It is called the kappa curve because of its resemblance to the Greek letter K.

KEI, *adj.* **kei function.** See BER—ber function.

KEL′VIN, *adj.*, *n.* The name in the International System of Units (SI) for the degree of temperature used in the Kelvin temperature scale, for which $0°$ K is absolute zero (-273.16 degrees celsius) and the kelvin degree is the same as the celsius degree.

KEPLER, Johann (1571–1630). Astronomer, mathematician, and philosopher; born in Würtemberg, lived in various parts of Eastern Europe and finally in Silesia. His laws of planetary motion were determined empirically, based on over twenty years of ingenious and laborious calculations.

Kepler's laws of planetary motion The three laws: (1) The orbits of the planets are ellipses having the sun at one focus. (2) The areas described by the radius vectors of a planet in equal times are equal. (3) The square of the period of revolution of a planet is proportional to the cube of its mean distance from the sun. These laws can be derived directly from the law of gravitation and Newton's laws of motion as applied to the sun and one planet, but were in fact determined earlier and helped direct Newton in his work.

KER, *adj.* **ker function.** See BER—ber function.

KER′NEL, *n*. **Dirichlet kernel.** The function $D_n(t) = \sum_{-n}^{n} e^{ikt}$, which is $2n + 1$ if $e^{it} = 1$ and

$$D_n(t) = \frac{\sin\left(n + \frac{1}{2}t\right)}{\sin\frac{1}{2}t},$$

otherwise, but sometimes with a coefficient of $\frac{1}{2}$ or $1/2\pi$. For the *complex form* of the Fourier series of a function f and $s_n(x) = \sum_{-n}^{n} c_n e^{inx}$,

$$s_n(x) = \frac{1}{2\pi} \int_{-\pi}^{\pi} f(x - t) D_n(t)\, dt.$$

See FOURIER —Fourier series.
   Fejér kernel. The function

$$K_n(t) = (n + 1)^{-1} \sum_{0}^{n} D_k(t),$$

which is $2n + 1$ if $e^{it} = 1$ and

$$K_n(t) = \frac{1}{n+1} \frac{1 - \cos(n+1)t}{1 - \cos t},$$

otherwise. If s_n is as defined above under *Dirichlet kernel* and if $\sigma_n = \sum_{0}^{n} s_k/(n + 1)$, then

$$\sigma_n(x) = \frac{1}{2\pi} \int_{-\pi}^{\pi} f(x - t) K_n(t)\, dt.$$

See CESÀRO —Cesàro summation formula, FEJÉR —Fejér's theorem, and above, Dirichlet kernel.
   iterated kernels. The functions K_n defined by $K_1(x, y) = K(x, y)$ and

$$K_{n+1}(x, y) = \int_{a}^{b} K(x, t) K_n(t, y)\, dt$$

$(n = 1, 2, \cdots)$, where K is a given *kernel*. It follows that the *resolvent kernel* $k(x, t; \lambda)$ is equal to $(-1) \cdot \sum_{n=0}^{\infty} \lambda^n K_{n+1}(x, t)$.
   kernel of a homomorphism. If a homomorphism maps a **group** G onto a group G^*, then the kernel of the homomorphism is the set N of all elements which map onto the identity element of G^*. Then N is a *normal subgroup* of G and G^* is isomorphic with the *quotient group* G/N. If a homomorphism maps a **ring** R onto a ring R^*, then the kernel of the homomorphism is the set I of elements which map onto the zero element of R^*. The kernel I is an *ideal* and R^* is isomorphic with the *quotient ring* R/I (see IDEAL).

kernel of an integral equation. See VOLTERRA —Volterra integral equations, INTEGRAL—integral equation of the third kind. *Syn.* nucleus.
   resolvent kernel. See VOLTERRA —Volterra reciprocal functions, and above, iterated kernels.

KHINCHINE (or KHINTCHINE), Aleksandr Iakovlevich (1894–1959). Soviet analyst and probabilist.
   Khinchine's theorem. Let $x_1, x_2, \cdots$ be independent random variables having equivalent distribution functions $F(x)$, with mean u. Then the variable

$$\bar{x} = \sum_{i=1}^{n} x_i/n$$

converges in probability to u as $n \to \infty$.

KIL′O-GRAM, *n*. One thousand grams; the weight of a platinum rod preserved in Paris as the standard unit of the metric system of weights; approximately 2.2 lbs. avoirdupois. See DENOMINATE NUMBERS in the appendix.

KIL′O-ME′TER, *n*. One thousand meters; approximately 3280 feet. See DENOMINATE NUMBERS in the appendix.

KIL′O-WATT, *n*. A unit of measure of electrical power; 1000 watts. See WATT.
   kilowatt-hour. A unit of energy; 1000 watt-hours; a kilowatt of power used for one hour; approximately $\frac{4}{3}$ horsepower acting for one hour.

KIN′E-MAT′ICS, *n*. A branch of mechanics dealing with the motion of rigid bodies without reference to their masses or forces producing the motion. The ingredients of kinematics are the concepts of space and time.

KI-NET′IC, *adj*. **kinetic energy.** See ENERGY.

KI-NET′ICS, *n*. That part of *mechanics* which treats the effect of forces in changing the motion of bodies.

KLEIN, Christian Felix (1849–1925). German algebraist, analyst, geometer, topologist, mathematical historian, and physicist. In Erlangen, development program (the *Erlangen program*) for classifying geometries according to properties left invariant under groups of transformations.
   Klein bottle. A one-sided surface with no edges and no "inside" or "outside," which is formed by pulling the small end of a tapering

tube through one side of the tube and spreading it so as to join with the other end.

KNOT, *n.* Nautical miles per hour. "A ship sails 20 knots" means it sails at a speed of 20 nautical miles per hour.

knot in a spline. See SPLINE.

knot in topology. A curve in space formed by looping and interlacing a piece of string in any way and then joining the ends together. Any two knots are topologically equivalent, but it may not be possible to continuously deform one into the other (*i.e.*, deform without breaking the string). *Tech.* A knot is a set of points in space which is topologically equivalent to a circle. The *theory of knots* consists of the mathematical analysis of possible types of knots and of methods for determining whether two knots can be continuously deformed into each other. See BRAID.

KODAIRA, Kunihiko (1915–   ). Japanese-American mathematician and Fields Medal recipient (1954). Research on harmonic integrals and harmonic forms, with applications to Kählerian and algebraic varieties.

KOEBE (or Köbe), Paul (1882–1945). German complex analyst who gave the first correct proof of the Riemann mapping theorem.

Koebe function. (1) The function f for which

$$f(z) = z/(1-z)^2 = z + 2z^2 + 3z^3 + \cdots,$$

if z is a complex number with $|z| < 1$. (2) Any function f for which there is a complex number c with $|c| = 1$ and

$$f(z) = z/(1-cz)^2 = z + 2cz^2 + 3c^2z^3 + \cdots,$$

if z is a complex number with $|z| < 1$.

KOLMOGOROV, Andrei Nikolaevich (1903–1987). Soviet analyst, probabilist, and topologist. Laid the set-theoretic foundation of probability theory in 1933.

Kolmogorov space. Same as T_0-SPACE. See TOPOLOGICAL —topological space.

KÖNIGSBERG–BRIDGE PROBLEM. The Prussian city of Königsberg had 7 bridges, as indicated in the figure. The problem is to show that it is impossible to cross all 7 bridges without crossing at least one bridge more than once. This was proved by Euler in 1736. To analyze the problem, one can replace it by the network of vertices and segments indicated in the figure by the dashed lines. It can be shown that there is a path that traverses such a network without crossing any segment more than once if and only if there are fewer than 3 vertices that belong to an odd number of segments (there are 4 such vertices for the Königsberg-bridge problem), but there is such a path that returns to its starting point if and only if no vertex belongs to an odd number of segments. See GRAPH THEORY.

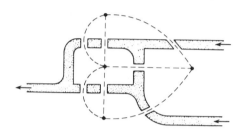

KOVALEVSKI, Sonya Vasilyevna (or **KOVALEVSKAYA, Sofya Vasilyevna**) (1850–1891). Born in Moscow. Studied under Weierstrass in Berlin (1871–74), but because of sex discrimination was barred from courses. Given a Ph.D. in 1874 from the University of Göttingen. In 1889, Mittag-Leffler secured for her a "tenured" professorship at the University of Stockholm. Published important mathematics about Abelian integrals, partial differential equations, and rotation of a solid body about a point.

Cauchy-Kovalevski theorem. See under CAUCHY.

KREIN, Mark Grigorievich (1907–1989). Soviet functional analyst and applied mathematician.

Krein-Milman property. The property of some linear topological spaces that each bounded closed convex subset is the closure of the convex span of its extreme points. All finite-dimensional spaces and many Banach spaces (including all reflexive spaces) have this property. A Banach space has the property if and only if each bounded closed convex subset has an extreme point. It has been conjectured that the Banach space has the Krein-Milman property if and only if it has the Radon-Nikodým property. See below, Krein-Milman theorem.

Krein-Milman theorem. Any compact convex subset of a locally convex linear topological space is the closure of the convex span of the set of its extreme points. See EXTREME —extreme point.

KREMER, Gerhard. See MERCATOR.

KRONECKER, Leopold (1823–1891). German algebraist, algebraic number theorist, and intuitionist. Rejected irrational numbers, insisting that mathematical reasoning be based only on the integers and finite processes.

Kronecker delta. The function δ_j^i of two variables i and j defined by $\delta_j^i = 1$ if $i = j$, and $\delta_j^i = 0$ if $i \neq j$. The *generalized Kronecker delta* $(\delta_{j_1 j_2 \cdots j_k}^{i_1 i_2 \cdots i_k})$ has k superscripts and k subscripts. If no two superscripts are equal and the subscripts are the same set of numbers as the superscripts, the value is said to be $+1$ or -1 according as an even or odd permutation is needed to arrange the subscripts in the same order as the superscripts. In all other cases its value is zero. See EPSILON. All of the Kronecker deltas are *numerical tensors*.

KUMMER, Ernst Eduard (1810–1893). German analyst, geometer, number theorist and physicist. Originated notion of ideal number and often is considered the father of modern arithmetic.

Kummer's test for convergence. Let $\sum a_n$ be a series of positive numbers and $\{p_n\}$ be a sequence of positive numbers. If c_n denotes the number $(a_n/a_{n+1})p_n - p_{n+1}$, then the series $\sum a_n$ converges if there is a positive number δ and a number N such that $c_n > \delta$ if $n > N$, and the series diverges if $\sum 1/p_n$ diverges and there is a number N such that $c_n \leq 0$ if $n > N$.

KURATOWSKI, Kazimierz (1896–1980). Polish analyst and topologist. See PLANAR —planar graph.

Kuratowski closure-complementation problem. The problem solved by Kuratowski of showing that if S is a subset of a topological space, then at most 14 sets can be obtained from S by closure and complementation. There is a subset of the real line from which 14 different sets can be so obtained.

Kurtowski lemma. See ZORN —Zorn's lemma.

KUR-TO'SIS, *n.* A descriptive property of distributions designed to indicate the general form of concentration around the mean. It is sometimes defined by the ratio $B_2 = u_4/u_2^2$, where u_2 and u_4 are the 2nd and 4th moments about the

mean. In a normal distribution, $B_2 = 3$. It is mesokurtic, platykurtic or leptokurtic according as $B_2 = 3$, $B_2 < 3$ or $B_2 > 3$. A platykurtic distribution often appears to be less heavily concentrated about the mean, a leptokurtic distribution to be more heavily concentrated, than the normal distribution.

KUTTA, Wilhelm Martin (1867–1944). German applied mathematician. See RUNGE —Runge-Kutta method.

L

LAC′U-NAR′Y, *adj.* **lacunary space relative to a monogenic analytic function.** A domain in the z-plane no point of which is covered by the domain of existence of the given function. See MONOGENIC —monogenic analytic function.

LAGRANGE, Joseph Louis (1736–1813). Born **Giuseppe Lodovico Lagrangia** in Turin, Italy. In 1766, he accepted a post he held for 20 years at the court of Frederick the Great in Berlin, after which he moved to France. He was a superb analyst, algebraist, number theorist, probabilist, physicist, and astronomer. Contributed especially to the calculus of variations, analytic mechanics, and astronomy.

Lagrange form of the remainder for Taylor's theorem. See TAYLOR —Taylor's theorem.

Lagrange interpolation formula. A formula for finding an approximation of an additional value of a function within a given interval of the independent variable, when certain values of the function within that interval are known. It is based upon the assumption that a polynomial of degree one less than the number of given values of the independent variable can be determined which will approximate the given function to the accuracy desired for the value sought. If $x_1, x_2, \cdots, x_n$ are the values of x for which the values of the function f are known, the formula is

$$f(x) = \frac{f(x_1)(x-x_2)(x-x_3) \cdots (x-x_n)}{(x_1-x_2)(x_1-x_3) \cdots (x_1-x_n)}$$

$$+ \frac{f(x_2)(x-x_1)(x-x_3) \cdots (x-x_n)}{(x_2-x_1)(x_2-x_3) \cdots (x_2-x_n)}$$

$$+ \cdots, \text{ to } n \text{ terms.}$$

Lagrange method of multipliers. A method for finding the maximum and minimum values of a function of several variables when relations between the variables are given. If it is desired to find the maximum area of a rectangle whose perimeter is a constant, k, it is necessary to find the maximum value of xy for $2x + 2y - k = 0$. Lagrange's method of multipliers is to solve the three equations $2x + 2y - k = 0$, $\partial u/\partial x = 0$, and $\partial u/\partial y = 0$ for x and y, where $u = xy + t(2x + 2y - k)$ and t is to be treated as an unknown to be eliminated. In general, given a function $f(x_1, x_2, \cdots, x_n)$ of n variables connected by h distinct relations, $\phi_1 = 0$, $\phi_2 = 0, \cdots, \phi_h = 0$, in order to find the values of $x_1, x_2, \cdots, x_n$, for which this function may have a maximum or minimum, equate to zero the partial derivatives of the auxiliary function $f + t_1\phi_1 + \cdots + t_h\phi_h$, with respect to $x_1, x_2, \cdots, x_n$, regarding $t_1, t_2, \cdots, t_h$ as constants, and solve these n equations simultaneously with the given h relations, treating the t's as unknowns to be eliminated.

Lagrange's theorem. If G is a subgroup of a group H of finite order, then the order of G divides the order of H.

Lagrangian function. See POTENTIAL— kinetic potential.

LAGUERRE, Edmond Nicolas (1834–1886). French geometer and analyst.

associated Laguerre functions. The functions $y = e^{-\frac{1}{2}x} x^{\frac{1}{2}(k-1)} L_n^k(x)$, where L_n^k is an *associated Laguerre polynomial*. This function is a solution of the differential equation

$$xy'' + 2y' + \left[n - \tfrac{1}{2}(k-1) - \tfrac{1}{4}x \right.$$

$$\left. - \tfrac{1}{4}(k^2 - 1)/x \right] y = 0.$$

associated Laguerre polynomials. The polynomials L_n^k defined by

$$L_n^k(x) = \frac{d^k}{dx^k} L_n(x),$$

where L_n is a *Laguerre polynomial*. The differential equation $xy'' + (k+1-x)y' + (n-k)y = 0$ is satisfied by L_n^k.

Laguerre polynomials. The polynomials L_n defined by

$$L_n(x) = e^x \frac{d^n}{dx^n} (x^n e^{-x}).$$

For all n, $(1 + 2n - x)L_n - n^2 L_{n-1} - L_{n+1} = 0$, and

$$(1 - t)^{-1} e^{-xt/(1-t)} = \sum_{n=1}^{\infty} L_n(x) t^n / n!.$$

The Laguerre polynomial L_n is a solution of *Laguerre's differential equation* with the constant $\alpha = n$. The functions $e^{-x} L_n(x)$ are *orthogonal functions* on the interval $(0, \infty)$.

Laguerre's differential equation. The differential equation $xy'' + (1-x)y' + \alpha y = 0$, where α is a constant.

LA HIRE, Phillipe de (1640–1718). French geometer. Applied methods of projective geometry to prove theorems of Apollonius on conics.

LAMB′DA, n. The eleventh letter of the Greek alphabet: lower case, λ; capital, Λ.

LAMBERT, Johann Heinrich (1728–1777). German analyst, number theorist, astronomer, physicist, and philosopher. Proved π irrational; introduced hyperbolic functions.

LAMÉ, Gabriel (1795–1870). French engineer and applied mathematician.

Lamé's constants. Two positive constants λ and μ, introduced by Lamé, which completely characterize the elastic properties of an isotropic body. They are related to Young's modulus E and Poisson's ratio σ by the formulas

$$\lambda = \frac{E\sigma}{(1+\sigma)(1-2\sigma)}, \qquad \mu = \frac{E}{2(1+\sigma)}.$$

The constant μ is the **modulus of rigidity** (or **shearing modulus**), and it is equal to the ratio of the shearing stress to the change in angle produced by the shearing stress.

LAM′I-NA, n. [*pl.* **laminas** or **laminae**]. A thin plate or sheet of uniform thickness and of constant density.

LAPLACE, Pierre Simon, Marquis de (1749–1827). French analyst, probabilist, astronomer, and physicist. Best known for his monumental work on celestial mechanics, for his great contributions to probability theory, and for the differential equation that bears his name.

Laplace's differential equation. The partial differential equation

$$\frac{\partial^2 V}{\partial x^2} + \frac{\partial^2 V}{\partial y^2} + \frac{\partial^2 V}{\partial z^2} = 0.$$

Under certain conditions, gravitational, electrostatic, magnetic, electric, and velocity potentials satisfy Laplace's equation. In general coordinates with the *fundamental metric tensor* g_{ij}, Laplace's equation takes the form

$$g^{ij}V_{,i,j} = 0 \quad \text{or} \quad \frac{1}{\sqrt{g}} \frac{\partial\left(\sqrt{g}\, g^{ij} \frac{\partial V}{\partial x^j}\right)}{\partial x^i} = 0,$$

where g is the determinant $|g_{ij}|$, g^{ij} is $1/g$ times the cofactor of g_{ji} in g, $V_{,i,j}$ is the second *covariant derivative* of the scalar V, and the *summation convention* applies. In cylindrical and spherical coordinates, Laplace's equation takes the respective forms:

$$\frac{\partial^2 V}{\partial r^2} + \frac{1}{r}\frac{\partial V}{\partial r} + \frac{\partial^2 V}{\partial z^2} + \frac{1}{r^2}\frac{\partial^2 V}{\partial \theta^2} = 0,$$

$$\frac{1}{r^2}\frac{\partial}{\partial r}\left(r^2 \frac{\partial V}{\partial r}\right) + \frac{1}{r^2 \sin\theta}\frac{\partial}{\partial\theta}\left(\sin\theta \frac{\partial V}{\partial\theta}\right)$$

$$+ \frac{1}{r^2 \sin^2\theta}\frac{\partial^2 V}{\partial\phi^2} = 0.$$

See DIRICHLET —Dirichlet characteristic properties of the potential function.

Laplace's expansion of a determinant. See DETERMINANT —Laplace's expansion of a determinant.

Laplace transform. The function f is the *Laplace transform* of g if

$$f(x) = \int e^{-xt} g(t)\, dt,$$

where the path of integration is some curve in the complex plane. It has become customary to restrict the path of integration to the real axis from 0 to $+\infty$. Suppose that $g(x)$ is defined for $x > 0$, has only a finite number of infinite discontinuities, $\int |g(t)|\, dt$ exists for any finite interval, and $f(x) = \int_0^\infty e^{-xt} g(t)\, dt$, where this integral converges absolutely for $x > a$. Then this Laplace transformation has an inverse given by

$$g(x) = \frac{1}{2\pi i} \int_{\alpha-i\infty}^{\alpha+i\infty} e^{xt} f(t)\, dt,$$

where the value of the integral is

$$\lim_{h\to 0} \tfrac{1}{2}\left[g(x+h) + g(x-h) \right]$$

if g is of *bounded variation* in the neighborhood of x and if $\alpha > a$. See FOURIER —Fourier transform.

LARGE, *adj., n.* **in the large.** See SMALL —in the small.

law of large numbers. Given a sequence $\{X_1, X_2, \cdots\}$ of independent random variables with means $\{\mu_1, \mu_2, \cdots\}$, **a strong law of large numbers** is a theorem which gives conditions that $\sum_{i=1}^{n} (X_i - \mu_i)/n$ approach zero with probability 1 as $n \to \infty$. For example, if the random variables all have the same distribution with mean μ and finite variance, then for each positive ϵ the probability $\left| \mu - \sum_{i=1}^{k} X_i/k \right| > \epsilon$ for some $k > n$ approaches 0 as $n \to \infty$. A **weak law of large numbers** is a theorem which gives conditions that the probability $\left| \sum_{i=1}^{n} (X_i - \mu_i)/n \right| > \epsilon$ approach 0 as $n \to \infty$, whatever ϵ may be. A sufficient condition is that there be a number A such that $\sigma_n^2 < A$ for all n, where σ_n is the variance of X_n. The *Bernoulli theorem* is a specialized weak law of large numbers which states that, if $s(n)$ is the number of successes in n independent Bernoulli experiments all of which have probability p of success, then for each positive number ϵ the probability $|p - s(n)/n| < \epsilon$ approaches 1 as $n \to \infty$.

LA'TENT, *adj.* **latent root of a matrix.** See EIGENVALUE —eigenvalue of a matrix.

LAT'ER-AL, *adj.* **lateral surface and area.** The **lateral surface** of a surface with bases, such as a cone or cylinder, is the surface with the bases removed. The **lateral area** is the area of the lateral surface. See CONE, CYLINDER, PRISM, PYRAMID.

lateral edge and face. For a prism or a pyramid, an edge or face that is not part of a base.

LA'TIN, *adj.* **Latin square.** A Latin square of order n is an n by n square array which has the property that each row and each column consists of some permutation of the same n symbols; it is a **diagonal Latin square** if the diagonals also have this property. In statistics, such arrays are used to give a method for controlling sources of variability. *E.g.*, suppose one wishes to test five makes of typewriters using five typists and five work-samples. A 5 by 5 Latin square with work-samples as entries describes how each sample can be done by each typist and also be done on each machine, if rows in the square are assigned to typists and columns to machines. Two Latin squares A and B of the same order are **orthogonal** if no two members of P are identical, where P is the set of all ordered pairs (a_{ij}, b_{ij}) for which a_{ij} and b_{ij} are the symbols in row i and column j of A and B, respectively. If there is a set of k mutually orthogonal Latin squares of order n, then $k \leq n - 1$. If $k = n - 1$, the set is **complete**. See PROJECTIVE —projective plane (2).

LAT′I-TUDE, *n.* **latitude of a point on the earth's surface.** The number of degrees in the arc of a meridian from the equator to the point; the angle which the plane of the horizon makes with the earth's axis; the elevation of the pole of the heavens; the angle which a plumb line at the point makes with a plumb line on the same meridian at the equator.

    middle latitude of two places. The arithmetic mean between the latitudes of the two places; one-half the sum of their latitudes if they are on the same side of the equator, one-half the difference (taken north or south according to which latitude was the larger) if the places are on different sides of the equator.

    middle latitude sailing. See SAILING.

LAT′TICE, *n.* A partially ordered set in which any two elements have a greatest lower bound (g.l.b.) and a least upper bound (l.u.b.), the g.l.b. of a and b being an element c such that $c \leq a$, $c \leq b$, and there is no d for which $c < d \leq a$ and $d \leq b$, and the l.u.b. being defined analogously. The g.l.b. and the l.u.b. of a and b are denoted by $a \cap b$ and $a \cup b$, respectively, and called the **meet** and **join**, respectively, of a and b. The set of all subsets $U, V, \cdots$ of a given set is a lattice if $U \leq V$ means that each element of U is contained in V. Then $U \cap V$ is the intersection of the sets U and V, and $U \cup V$ is the union of U and V. A lattice is **complete** if every subset has a g.l.b. and a l.u.b., **modular** if it has the property that $x \geq z$ implies $x \cap (y \cup z) = (x \cap y) \cup z$ for all y, and **distributive** if the distributive laws $x \cup (y \cap z) = (x \cup y) \cap (x \cup z)$ and $x \cap (y \cup z) = (x \cap y) \cup (x \cap z)$ hold (actually, either of these laws implies the other). A lattice is a *Boolean algebra* if and only if it is distributive and **complemented** (*i.e.*, there are members O and I such that, for all x, $O \leq x \leq I$ and there is a "complement" x' for which $x \cap x' = O$ and $x \cup x' = I$). See JOIN—join-irreducible.

LA′TUS, *adj.* **latus rectum.** [*pl.* **latera recta.**] See PARABOLA, ELLIPSE, HYPERBOLA.

LAURENT, Paul Matthieu Hermann (1841–1908). French analyst. Best known for *Laurent series*, which generalizes Taylor series.

    Laurent expansion. Let f be a function of a complex variable that is analytic in the circular annulus $a < |z - z_0| < b$. Then f can be represented by a two-way power series in the annulus; *i.e.*,

$$f(z) = \sum_{-\infty}^{\infty} a_n (z - z_0)^n.$$

The series is the **Laurent expansion** or the **Laurent series** of f about z_0. The coefficients a_n are given by

$$a_n = \frac{1}{2\pi i} \int_C (\zeta - z_0)^{-n-1} f(\zeta) \, d\zeta,$$

where C is a simple closed rectifiable curve lying in the annulus and enclosing the inner circle $|z - z_0| = a$. If there are only finitely many negative values of n for which $a_n \neq 0$, then f has a **pole** at z_0 and $\sum_{-\infty}^{-1} a_n (z - z_0)^n$ is the **principal part** of f at z_0. See SINGULAR —singular point of an analytic function.

    Laurent series. See above, Laurent expansion.

LA VALLÉE POUSSIN. See DE LA VALLÉE POUSSIN.

LAW, *n.* A general principle or rule. For examples, see under ASSOCIATIVE, BOYLE, COMMUTATIVE, CONTRADICTION, EXPONENT, GOMPERTZ, BACTERIAL, COSINE, INERTIA, KEPLER, LARGE, LOGARITHM, MAKEHAM, NEWTON, QUADRANT, QUADRATIC, SINE, SPECIES, TANGENT.

LEAD′ING, *p.* **leading coefficient.** For a polynomial in one variable, the coefficient of the term of highest degree.

LEAST, *adj.* **least common denominator.** See COMMON —common denominator.

    least common multiple. See MULTIPLE—least common multiple.

    least upper bound. See BOUND.

    method of least squares. A method by which an estimation procedure is determined by minimizing the sum of squares of differences between observed and estimated values. If one is estimating a single number on the basis of a set of observations, the method of least squares gives the *mean*. If random variables X and Y are considered to be related by an equation of type

$Y = mX + b$ and if, for $X = 1, 2, 3$, the observed values of Y are $2, 4, 7$, the method of least squares would determine $m = \frac{5}{2}$ and $b = -\frac{2}{3}$, the numbers which minimize

$$(m + b - 2)^2 + (2m + b - 4)^2 + (3m + b - 7)^2.$$

If X has values $Y_1, \cdots, Y_n$ for values $X_1, \cdots, X_n$ of X, the least-squares method for determining the equation $Y = mX + b$ gives

$$Y = \bar{Y} + r(\sigma_Y / \sigma_X)(X - \bar{X}),$$

where $\bar{X}$ and $\bar{Y}$ are sample means, σ_X and σ_Y are sample standard deviations, and r is the sample correlation coefficient. More generally, Y might be considered to be a function of p random variables $X_1, \cdots, X_p$; if the jth of n observations gives the value Y_j for Y when X_i has the value $X_{i,j}$ for $i = 1, \cdots, p$, then the method of least squares consists of determining the parameters $\{a_i\}$ so as to minimize

$$\sum_{j=1}^{n} \left(Y_j - \sum_{i=1}^{p} a_i X_{i,j} \right)^2.$$

This does not exclude the use of other functions to estimate Y, since each X_i could be a function of other random variables. The least-squares estimators of the parameters $\{a_i\}$ are *minimum variance unbiased estimators* in the class of unbiased linear estimators if the random variables $\{Y_j\}$ are independent, their means are linear functions of known mathematical variables $\{X_i\}$, and their variances are equal. If also the $\{Y_j\}$ are normally distributed, the least-squares estimators become *maximum likelihood estimators*. See NORMAL—normal equations, REGRESSION —regression line.

LEBESGUE, Henri Léon (1875–1941). French analyst who greatly influenced mathematics, particularly through his theory of measure and integration, but also through his work on trigonometric series.

Lebesgue convergence theorem. Suppose m is a countably additive measure on a σ-algebra of subsets of a set T, that g is nonnegative and measurable with $\int_T g\,dm < +\infty$, and that $\{S_n\}$ is a sequence of measurable functions such that $|S_n(x)| \le g(x)$ on T. Then each S_n in integrable

and, if there is a function S for which $\lim_{n \to \infty} S_n(x) = S(x)$ a.e. on T, then

$$\int_T S\,dm = \lim_{n \to \infty} \int_T S_n\,dm.$$

If we replace the hypothesis $\int_T g\,dm < +\infty$ by $\int_T g^p\,dm < +\infty$, then the conclusion becomes: "$|S_n|^p$ is integrable for each n, $|S|^p$ is integrable, and $\lim_{n \to \infty} \int_T |S - S_n|^p\,dm = 0$." Syn. Lebesgue dominated convergence theorem. See BOUNDED —bounded convergence theorem, MONOTONE — monotone convergence theorem, and SERIES — integration of an infinite series.

Lebesgue integral. First suppose that f is a *bounded* measurable function defined over a (Lebesgue) measurable set E of finite measure. If L and U are lower and upper bounds of f, then the Lebesgue integral $\int_\Omega f(x)\,dx$ of f over Ω is defined as the limit of

$$\sum_{i=1}^{n} y_{i-1} m(e_i), \quad \text{or of} \quad \sum_{i=1}^{n} y_i m(e_i),$$

as the greatest of the numbers $y_i - y_{i-1}$ approaches zero, where the interval $[L, U]$ is divided into n parts by the increasing sequence of numbers, $y_0 = L$, $y_1, y_2, \cdots, y_n = U$, and where $m(e_1)$ is the measure of the set e_i, e_i consisting of all points x for which $y_{i-1} \le f(x) < y_i$ ($i = 1, 2, \cdots, n - 1$) and e_n of all x satisfying $y_{n-1} \le f(x) \le y_n$. If f is unbounded and f_n^m is defined by $f_n^m(x) = f(x)$ if $n \le f(x) \le m$, $f_n^m(x) = m$ if $f(x) > m$, and $f_n^m(x) = n$ if $f(x) < n$, then f has the Lebesgue integral

$$\int_\Omega f(x)\,dx = \lim_{\substack{m \to \infty \\ n \to -\infty}} \int_\Omega f_n^m(x)\,dx,$$

provided this limit exists. If the set Ω does not have finite measure and

$$\int_{\Omega \cap I} f(x)\,dx$$

approaches a limit as the boundaries of an interval I all increase indefinitely, in any manner, then that limit is defined as $\int_\Omega f(x)\,dx$. A function ϕ defined on a set E contained in an interval I has a Lebesgue integral over E if, and only if, there exists a sequence of step functions (or of continuous functions) $\{f_n\}$ such that

$$\lim_{n \to \infty} f_n(x) = \phi(x)$$

for almost all x and I [where $\phi(x)$ is taken as zero for points x not in E] and

$$\lim_{m,n \to \infty} \int_I |f_n(x) - f_m(x)|\, dx = 0.$$

In this case, $\lim_{n \to \infty} \int_I f_n(x)\, dx$ exists and is the Lebesgue integral of ϕ over E. See INTEGRABLE —integrable function, MEASURABLE —measurable function. A function which has a Riemann integral necessarily has a Lebesgue integral, but not conversely.

Lebesgue measure. See MEASURE —Lebesgue measure.

LEFSCHETZ, Solomon (1884–1972). Russian–American theoretical engineer, algebraic geometer, and topologist. Worked also in differential equations, control theory and nonlinear mechanics.

LEFT, *adj.* **left coset, ideal, identity, inverse.** See COSET, IDEAL, IDENTITY —identity element, INVERSE —inverse of an function.

left-handed coordinate system. See COORDINATE —left-handed coordinate system, TRIHEDRAL.

left-handed curve. If the torsion of a directed curve C at a point P is positive, then a variable point moving through the position P in the positive direction along C goes from the positive to the negative side of the osculating plane at P. See CANONICAL —canonical representation of a space curve in the neighborhood of a point. Accordingly, C is said to be **left-handed** at P. See RIGHT-HANDED-CURVE. *Syn.* sinistrorsum [*Latin*], or sinistrorse curve.

left-handed trihedral. See TRIHEDRAL.

LEG, *n.* **leg of a right triangle.** Either one of the sides adjacent to the right angle.

LEGENDRE, Adrien Marie (1752–1833). French analyst and number theorist.

associated Legendre functions. The functions

$$P_n^m(x) = \left(1 - x^2\right)^{m/2} \frac{d^m}{dx^m} P_n(x),$$

where P_n is a *Legendre polynomial*. The function

P_n^m is a solution of the differential equation

$$(1 - x^2)y'' - 2xy'$$
$$+ \left[n(n+1) - m^2/(1 - x^2) \right] y = 0.$$

See HARMONIC —spherical harmonic, zonal harmonic.

Legendre differential equation. A differential equation of type $(1 - x^2)y'' - 2xy' + n(n+1)y = 0$. See below, Legendre polynomials.

Legendre duplication formula. See DUPLICATION —duplication formula.

Legendre functions of the second kind. See NEUMANN, F. E.

Legendre necessary condition. (*Calculus of Variations*) A condition, namely $f_{y'y'} \geq 0$, that must be satisfied if the function y is to minimize $\int_{x_1}^{x_2} f(x, y, y')\, dx$. See CALCULUS —calculus of variations, EULER —Euler equation, WEIERSTRASS —Weierstrass necessary condition.

Legendre polynomials. The coefficients $P_n(x)$ in the expansion

$$\left(1 - 2xh + h^2\right)^{-1/2} = \sum_{n=0}^{\infty} P_n(x)h^n.$$

Thus $P_0(x) = 1$, $P_1(x) = x$, $P_2(x) = \frac{1}{2}(3x^2 - 1)$, $P_3(x) = \frac{1}{2}(5x^3 - 3x)$, $P_4(x) = \frac{1}{8}(35x^4 - 30x^2 + 3)$. The function P_n is a solution of the *Legendre differential equation*. For all n,

$$P'_{n+1}(x) - xP'_n(x) = (n+1)P_n(x),$$

$$(n+1)P_{n+1}(x) - (2n+1)xP_n(x)$$
$$+ nP_{n-1}(x) = 0,$$

and

$$P_n(\cos\theta) = \frac{(-1)^n}{n!} r^{n+1} \frac{\partial^n}{\partial z^n}\left(\frac{1}{r}\right),$$

where $\cos\theta = z/r$ and $r^2 = x^2 + y^2 + z^2$. The set of Legendre polynomials is a *complete* set of *orthogonal functions* on the interval $(-1, 1)$. Also called LEGENDRE COEFFICIENTS. See RODRIQUES —Rodriques formula, SCHLÄFLI —Schläfli integral.

Legendre symbol $(c|p)$. The symbol $(c|p)$ is equal to 1 if the integer c is a quadratic residue of the odd prime p, and is equal to -1 if c is a quadratic nonresidue of p. E.g., $(6|19) = 1$ since the congruence $x^2 \equiv 6 \pmod{19}$ has a solution, and $(39|47) = -1$ since the congruence $x^2 \equiv 39$ (mod 47) has no solutions.

LEIBNIZ, Gottfried Wilhelm von (1646–1716). Born in Leipzig, Germany, he was one of the last to master most major fields of knowledge. He was awarded a doctor of law degree in 1667 and had a career in law and international politics. He invented calculus (independently of Newton) and introduced many of the present-day symbols. See BRACHISTOCHRONE.

Leibniz test for convergence. An alternating series converges if the absolute values of its terms decrease monotonically with limit zero. See ALTERNATING —alternating series.

Leibniz theorem (or formula). A formula for finding the nth derivative of the product of two functions; the formula is as follows:

$$D^n(uv) = vD^n u + nD^{n-1}uDv$$

$$+ \tfrac{1}{2}n(n-1)D^{n-2}uD^2 v$$

$$+ \cdots + uD^n v,$$

the numerical coefficients being the coefficients in the expansion of $(u + v)^n$ and the indicated derivatives being of the same order as the corresponding powers in this expansion. Analogously, the nth derivative of the product of k functions can be written out from the multinomial expansion of the nth power of the sum of k quantities.

LEM′MA, n. A theorem proved for use in the proof of another theorem. See THEOREM.

LEM-NIS′CATE, n. The plane locus of the foot of the perpendicular from the origin to a variable tangent to the equilateral hyperbola; the locus of the vertex of a triangle when the product of the two adjacent sides is kept equal to one-fourth the square of the third side (which is fixed in

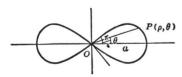

length). In terms of polar coordinates, if the node (see figure) is taken as the pole, the axis of symmetry as the initial line, and the greatest distance from the pole to the curve as a, the equation of the lemniscate is $\rho^2 = a^2 \cos 2\theta$. Its

corresponding rectangular Cartesian equation is

$$\left(x^2 + y^2\right) = a^2\left(x^2 - y^2\right).$$

This curve was first studied by Jacques Bernoulli, hence is frequently called the *lemniscate of Bernoulli*. See CASSINI.

LENGTH, n. **length of a curve.** Let A and B be points on a curve and choose points $A = P_1, P_2, P_3, \cdots, P_n = B$, starting at A and moving along the curve to B. If the sum of the lengths of the chords joining successive points,

$$\overline{P_1P_2} + \overline{P_2P_3} + \overline{P_3P_4} + \cdots + \overline{P_{n-1}P_n},$$

has an upper bound, then the least upper bound of such sums is the length of the curve between A and B (otherwise the length is not defined, *i.e.*, does not exist). If a simple curve C has parametric equations $x = f(t)$, $y = g(t)$, $z = h(t)$, where $a \leq t \leq b$, then C has length if f, g, and h are differentiable on the interval $[a, b]$ and f', g', and h' are bounded on $[a, b]$. Moreover, if f', g', and h' are continuous on $[a, b]$, then the length of C is equal to

$$\int_a^b \left[f'(t)^2 + g'(t)^2 + h'(t)^2\right]^{1/2} dt.$$

If a plane curve has an equation $y = f(x)$ for $x_1 \leq x \leq x_2$ and if dy/dx is continuous, then the length is

$$\int_{x_1}^{x_2} \left[1 + \left(\frac{dy}{dx}\right)^2\right]^{1/2} dx.$$

For polar coordinates, the length is

$$\int_{\theta_1}^{\theta_2} \left[r^2 + \left(\frac{dr}{d\theta}\right)^2\right]^{1/2} d\theta.$$

length of a line segment or of a **set of points on a line.** Same as CONTENT (see CONTENT), so that the length of a line segment is the absolute value of the difference of the coordinates of its endpoints (whether or not they belong to the segment). An equivalent definition is that the length is the number of times a unit interval will fit in the line segment, this being defined as the sum of the number of complete unit intervals that can be embedded in the line, $\tfrac{1}{2}$ the number of intervals of length $\tfrac{1}{2}$ that can be embedded in the remainder, $\tfrac{1}{4}$ the number of intervals of length $\tfrac{1}{4}$ that can be embedded in the remainder, etc. (an interval of length $\frac{1}{2}$ is one of two intervals which results from bisecting the unit interval, etc.).

length of a rectangle or of a **rectangular parallelepiped.** The length of one of its longest edges. See WIDTH.

LEONARDO DA VINCI (1452–1519). The outstanding intellect of the Italian Renaissance. Best known mathematically for his applications of mathematics to perspective (on which he wrote a treatise) and to science.

LEONARDO of Pisa. Same as FIBONACCI. See FIBONACCI.

LEPTOKURTIC, *adj.* **leptokurtic distribution.** See KURTOSIS.

LESS, *adj.* See GREATER.

LEV'EL, *adj.* **level lines.** See CONTOUR —contour lines.
  net level premiums. See PREMIUM.

LE'VER, *adj., n.* A rigid bar used to lift weights by placing the bar against a support called the *fulcrum*, and applying a force or weight. A lever is said to be of the *first*, *second* or *third* type, respectively, when the fulcrum is under the bar and between the weights, under the bar at one end, or above the bar at one end.
  law of the lever. If there is equilibrium for two weights (forces), the weights (forces) are to each other inversely as their level arms, or, what is equivalent, the products of the weights by their lever arms are equal, or the algebraic sum of the moments of all the forces about the fulcrum is equal to zero.
  lever arm. The distance of a weight (or line of action of a force) from the fulcrum of the lever.

LEVI-CIVITA, Tullio (1873–1941). Italian geometer, analyst, and physicist. Collaborator of Ricci. Einstein used his absolute differential calculus in developing the theory of relativity.

LEX'I-CO-GRAPH'I-CAL-LY, *a.* If a set S is ordered in some way, *e.g.*, numbers by size or letters alphabetically, then a collection of sequences of members of S is **ordered lexicographically** if two sequences with different first members are given the order of the first terms, then two sequences with the same first terms and different second terms are given the order of the second term, etc.

L'HÔPITAL (*also* **L'Hospital, Lhospital**), **Guillaume François Antoine de [Marquis de St. Mesme]** (1661–1704). French analyst and geometer. Writer of first textbook on differential calculus. See BRACHISTOCHRONE.
  L'Hôpital's rule. A rule (actually discovered by John Bernoulli and given to L'Hôpital in return for salary) to evaluate *indeterminate forms*: If

$$\lim_{x \to a} f(x) = \lim_{x \to a} F(x) = 0,$$

or

$$\lim_{x \to a} |f(x)| = \lim_{x \to a} |F(x)| = +\infty,$$

and $f'(x)/F'(x)$, where f' and F' are the derivatives of f and F, approaches a limit as x approaches a, then $f(x)/F(x)$ approaches the same limit. *E.g.*, if $f(x) = x^2 - 1$, $F(x) = x - 1$, and $a = 1$, $f(a)/F(a)$ takes the form $0/0$ and

$$\lim_{x \to 1} f'(x)/F'(x) = \lim_{x \to 1} 2x = 2,$$

which is $\lim_{x \to 1}(x^2 - 1)/(x - 1)$. L'Hôpital's rule can be proved under the assumption that there is a neighborhood U of a such that f and F are differentiable in U except possibly at a and there is no point of U at which f' and F' are both zero. See MEAN—mean-value theorems for derivatives.

L'HUILIER, Simon Antoine Jean (1750–1840). Swiss geometer.
  L'Huilier's theorem. A theorem relating the spherical excess E of a spherical triangle to the sides: $\tan \frac{1}{2}E$ is equal to

$$\left[\tan \tfrac{1}{2}s \tan \tfrac{1}{2}(s - a)\tan \tfrac{1}{2}(s - b)\tan \tfrac{1}{2}(s - c)\right]^{1/2},$$

where a, b, and c are the sides of the triangle, and $s = \frac{1}{2}(a + b + c)$.

LI'A-BIL'I-TY, *n.* See ASSETS.

LIE, Marius Sophus (1842–1899). Norwegian analyst, geometer, and group theorist. Did research on transformation invariants and developed theory of transformation groups.
  Lie group. A topological group which can be given an *analytic structure* for which the coordinates of a product xy are analytic functions of the coordinates of the elements x and y, and the coordinates of the inverse x^{-1} of an element x are analytic functions of x. See EUCLIDEAN —locally Euclidean space.

LIFE, *adj., n.* (1) (*Life Insurance*). The difference between a policy date and the death of the insured. (2) The period during which something under consideration is effective, useful, or efficient, such as the life of a lease or contract, the life of an enterprise, or the life of a machine.

   expectation of life. See EXPECTATION.

   life annuity. See ANNUITY.

   life insurance. See INSURANCE —life insurance.

LIFT, *n.* In aerodynamics, if the total force **F** that is applied to a body *B* gives *B* a motion with velocity vector **v**, then the component of **F** perpendicular to **v** is called *lift*. See DRAG.

LIGHT, *adj.* **light year.** The distance light travels in one (mean solar) year; approximately $5.8785 \cdot 10^{12}$ miles or $9.46053 \cdot 10^{17}$ cm.

LIKE'LI-HOOD, *adj.* **likelihood function.** A *frequency function* or *probability-density function* $f(\mathbf{X}; \theta_1, \theta_2, \cdots, \theta_n)$ considered as a function of the parameters $\{\theta_1, \theta_2, \cdots, \theta_n\}$ for fixed **X**. In particular, suppose $\{\mathbf{X}_1, \mathbf{X}_2, \cdots, \mathbf{X}_k\}$ is a *random sample* from a population whose distribution is known to be of a certain general form or type, but certain parameters are not known. Let $f(\mathbf{X}; \theta_1, \theta_2, \cdots, \theta_n)$ be the frequency function or the probability density function when the parameters have values $\theta_1, \theta_2, \cdots, \theta_n$. The **likelihood function** L of the random sample satisfies

$$L(\mathbf{X}_1, \cdots, \mathbf{X}_k; \theta_1, \cdots, \theta_n) = \prod_{i=1}^{k} f(\mathbf{X}_i, \theta_1, \cdots, \theta_n)$$

and is considered to be a function of $\{\theta_1, \theta_2, \cdots, \theta_n\}$ with $\{\mathbf{X}_1, \cdots, \mathbf{X}_k\}$ fixed. The **maximum-likelihood estimates** for the parameters are the values that maximize the likelihood function. The resulting functions, $\theta_1(\mathbf{X}_1, \cdots, \mathbf{X}_k), \cdots, \theta_n(\mathbf{X}_1, \cdots, \mathbf{X}_k)$, are **maximum-likelihood estimators** for the parameters. For example, suppose a bag contains 4 balls, θ being black and the rest red. On successive draws, a black ball, a red ball, and a black ball are drawn (each being returned before the next draw). The frequency function (or probability function) has values $\frac{1}{4}\theta$ and $1 - \frac{1}{4}\theta$ for a black draw and a red draw, so the likelihood function is $(\frac{1}{4}\theta)(1 - \frac{1}{4}\theta)(\frac{1}{4}\theta) = (4\theta^2 - \theta^3)/64$, which for θ having values 0, 1, 2, 3, 4 has values $0, 3/64, 1/8, 9/64, 0$. The largest is $9/64$ for $\theta = 3$, and the *maximum-likelihood estimate* of θ is 3. See ESTIMATOR, STATISTIC —sufficient statistic, VARIANCE, and below, likelihood ratio.

likelihood ratio. The ratio L_0/L_1, where L_0 is the maximum value of the *likelihood function*, for the parameters satisfying certain conditions H_0, and L_1 is the maximum of the *likelihood function* for all possible values of the parameters; L_0/L_1 is a function of sample values $\{\mathbf{X}_1, \mathbf{X}_2, \cdots, \mathbf{X}_k\}$. The **likelihood-ratio test** with significance $(100\alpha)\%$ rejects the hypothesis H_0 if $L_0/L_1 < a$, where a is chosen so that the probability of $L_0/L_1 < a$ is equal to α. See NEYMAN —Neyman-Pearson test, and above, likelihood function.

LIM'A-ÇON, *n.* The locus of a point on a line, at a fixed distance from the intersection of this line with a fixed circle, as the line revolves about a point on the circle. If the diameter of the circle is taken as *a* (see figure), the fixed distance as *b*, the fixed point as the pole, the moving line as the radius vector, and the diameter through the fixed circle as the polar axis, the equation of the limaçon is $r = a \cos \theta + b$. This curve was first studied and named by **Étienne Pascal** (1588–1640), the father of Blaise Pascal, hence is usually called the **limaçon of Pascal.** When *b* is less than the diameter of the fixed circle, the curve consists of two loops, one within the other; the outside loop is heart-shaped and the inside loop

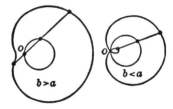

is pear-shaped, the curve having a node at the origin. When *b* is equal to *a*, there is one heart-shaped loop, the **cardioid.** When *b* is greater than *a*, there is one loop, whose shape tends toward that of a circle as *b* increases.

LIM'IT, *n.* See various headings below for specific types of limits. A general description of the concept of limit can be given by using the concept of **system of stages**, a system of stages being a collection S of sets such that no member of S is empty and $A \cap B$ belongs to S whenever A and B belong to S. For a function f, the existence of the limit with respect to a system of stages means that each stage contains points in the domain of f and there is a number ℓ (the limit) such that, for each neighborhood W of ℓ, there is a stage A such that $f(x)$ is in W if x is in the intersection of A and the domain of f.

The limit l can be proved to be unique if the range of f has the property that, for any two distinct members l_1 and l_2, there are neighborhoods U and V of l_1 and l_2 with $U \cap V$ empty [see TOPOLOGICAL —topological space]. If the range of f is a set of real or complex numbers, then the usual theorems on limits can be proved [see below, fundamental theorems on limits]. All other types of limits are special cases of the preceding. E.g., $\lim_{x \to a} f(x)$ can be defined using stages for which a positive number δ corresponds to the stage that is the set of all x for which $0 < |x - a| < \delta$ [see below, limit of a function]; $\lim_{n \to \infty} x_n$ uses stages for which a number N corresponds to a stage that is the set of all positive integers n for which $n > N$ [see SEQUENCE —limit of a sequence]; *Moore-Smith convergence* can be described in terms of stages by letting a stage correspond to a member a of the directed set D and consist of all x in D with $x \geq a$ [see MOORE, E. H.]; a *Riemann integral* on $[a, b]$ can be defined as a limit of Riemann sums by letting a stage correspond to a positive number δ and be the set of all pairs $\{(x_1, \cdots, x_{n+1}), (\xi_1, \cdots, \xi_n)\}$ for which n is arbitrary, $a = x_1$, $b = x_{n+1}$, the numbers $x_1, \cdots, x_{n+1}$ are either increasing or decreasing, and for each k the number ξ_k is in the closed interval with end points x_k and x_{k+1} [see INTEGRAL —definite integral]. See FILTER.

central limit theorem. See CENTRAL.

fundamental theorems on limits. (1) If a function u has a limit l and c is a number, then cu has the limit $c l$. (2) If u and v have the limits l and m, respectively, then $u + v$ has the limit $l + m$. (3) If u and v have the limits l and m, respectively, then uv has the limit lm. (4) If u and v have the limits l and m, respectively, and if m is not zero, then u/v has the limit l/m. (5) If u never decreases and there is a number A such that u never is greater than A, then u has a limit which is not greater than A. (6) If u never increases and there is a number B such that u never is less than B, then u has a limit which is not less than B.

inferior and superior limits. See INFERIOR, SEQUENCE —accumulation point of a sequence, SUPERIOR.

limit inferior or **lim inf.** See INFERIOR —limit inferior.

limit of a function. Strictly, all limits involve functions [see above, LIMIT]. Here, we shall be concerned with the limit of a function whose domain and range are sets of numbers. Roughly, a limit then is a number that the function approximates closely when a suitable restriction is placed on the independent variable. E.g., the limit of $1/x$ is 0, as x increases beyond all

bounds; the same is true if x takes on numerically large negative values, and also if x takes on, alternately, large positive and numerically large negative values, such as $10, -10, 100, -100, 1000, -1000, \cdots,$. A function is said to *approach its limit* or to *approach* a certain number as *a limit*. The fact that a function f approaches the limit k as x approaches a given number a is written

$$\lim_{x \to a} f(x) = k$$

and stated "limit of $f(x)$, as x approaches a, is k." *Tech.* A function f has the limit k as x approaches a if, for every positive number ϵ, there is a number δ such that $|f(x) - k| < \epsilon$ if $0 < |x - a| < \delta$; f has the limit k as x *becomes infinite* if, for every positive number ϵ, there is a number δ such that $|f(x) - k| < \epsilon$ if $x > \delta$; f has the limit k as $|x|$ becomes infinite if, for every positive number ϵ, there is a number δ such that $|f(x) - k| < \epsilon$ if $|x| > \delta$. Sometimes $+\infty$ is allowed as a limit; then

$$\lim_{x \to +\infty} f(x) = +\infty$$

means that for any positive number ϵ there is a number δ such that $f(x) > \epsilon$ if $x > \delta$ [see EXTENDED —extended real-number system]. There are many other limits of these types, e.g., see below, limit on the left or right.

limit on the left or right. The **limit on the right** of a function f at a point a is a number M such that for any $\epsilon > 0$ there exists a $\delta > 0$ such that $|M - f(x)| < \epsilon$ if

$$a < x < a + \delta,$$

and the **limit on the left** is a number N such that for any $\epsilon > 0$ there exists a $\delta > 0$ such that $|N - f(x)| < \epsilon$ if $a - \delta < x < a$. A function is continuous on the right (or left) at a if, and only if, the limit on the right (or left) exists and is equal to $f(a)$. Such limits are denoted in various ways, e.g., by $\lim_{x \to a+} f(x)$, $f(a +)$, $f(a + 0)$. *Syn.* right-hand limit, left-hand limit.

limit point. Same as ACCUMULATION POINT.

limit of a product, quotient, sum. See above, fundamental theorems on limits.

limit of the ratio of an arc to its chord. Refers to the limit of this ratio when the chord (or arc) approaches zero. If the curve is a *circle*, this limit is 1, and it is also 1 for smooth curves.

limit of a sequence. See various headings under SEQUENCE, especially *limit of a sequence*.

limits of a class interval. (*Statistics*) The upper and lower limits of the values of a class interval. *Syn.* class bounds.

limits of integration. See INTEGRAL—definite integral.

limit superior or **lim sup.** See SUPERIOR—limit superior.

problems of limit analysis and design. The problem of determining the carrying capacity, for a given type of loading, of a structure of which the geometry and the fully plastic moments of the members are known, is said to be a *problem of limit analysis*. A *problem of limit design* is the problem of determining the fully plastic moments of the members of a structure, of which the geometry and the loads it has to carry are known, in such a way as to minimize its weight.

LIM′IT-ING, *adj.* **limiting age in a mortality table.** The age which the last survivor of the group upon which the table is based would have attained had he lived to the end of the year during which he died.

LINDELÖF, Ernst Leonard (1870–1946). Finnish analyst and topologist.

Lindelöf space. A topological space T such that, for any class C of open sets whose union contains T, there is a countable class C^* of sets whose union contains T and such that each member of C^* is a member of C. A topological space which satisfies the *second axiom of countability* is a Lindelöf space (Lindelöf's theorem).

LINDEMANN, Carl Louis Ferdinand von (1852–1939). German analyst and geometer. See PI.

LINE, *n.* (1) Same as CURVE. (2) Same as STRAIGHT LINE. See below.

addition of line segments. See SUM—sum of directed line segments.

angle between two lines, or **between a line and a plane.** See ANGLE—angle of intersection.

bisection point of a line segment. Same as MIDPOINT of the line segment.

broken line. A figure composed entirely of segments of straight lines, laid end to end.

concurrent lines. See CONCURRENT.

contour lines. See CONTOUR.

curved line. A curve which is neither a broken nor a straight line; a curve that continually turns (changes direction).

directed line. See DIRECTED.

direction of a line. See DIRECTION—direction of a line.

equation of a line. A relation between the coordinates of a point which holds when, and

only when, the point is on the line. The following are forms of the **equation of a line in the plane**: (1) **Slope-intercept form.** The equation (in rectangular Cartesian coordinates) $y = mx + b$, where m is the slope of the line and b its intercept on the y-axis. (2) **Intercept form.** The equation $x/a + y/b = 1$, where a and b are the x and y intercepts, respectively. (3) **Point-slope form.**. The equation $y - y_1 = m(x - x_1)$, where m is the slope and (x_1, y_1) is a point on the line. (4) **Two-point form.** The equation

$$(y - y_1)/(y_2 - y_1) = (x - x_1)/(x_2 - x_1)$$

where (x_1, y_1) and (x_2, y_2) are two points through which the line passes. This form can be written more elegantly by equating to zero the third-order determinant whose first, second, and third rows contain, in order, the sets of elements $x, y, 1; x_1, y_1, 1; x_2, y_2, 1$. (5) **Normal form.** The equation $x \cos \omega + y \sin \omega - p = 0$, where ω is the angle from the x-axis to the perpendicular from the origin to the line and p is the length of the perpendicular from the origin to the line; any equation $ax + by + c = 0$ with $a^2 + b^2 = 1$ (some authors require that the sign of a be chosen so that $a = \cos \omega$). If $a^2 + b^2 = 1$, then $ax + by + c$ is equal to the distance from the point (x, y) to the line $ax + by + c = 0$ (positive on one side of

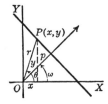

the line and negative on the other). An equation $ax + by + c = 0$ can be changed to normal form by dividing all coefficients by $\pm (a^2 + b^2)^{1/2}$, the sign being the opposite of the constant term c (it is sometimes required that the angle in the normal form be less than 180°, in which case the sign of the coefficient of y is taken as positive, or the coefficient of x positive if y doesn't appear.) To reduce $3x - 4y + 5 = 0$ to the normal form, multiply the equation by $-\frac{1}{5}$, getting $-\frac{3}{5}x + \frac{4}{5}y - 1 = 0$. The distance from $(-1, 5)$ to the line is then $(-\frac{3}{5})(-1) + (\frac{4}{5})(5) - 1 = 3\frac{3}{5}$; the distance from $(0, 0)$ to the line is -1. (6) **General form** in rectangular Cartesian coordinates. The form which includes all other forms in this system of coordinates as special cases. It is written $Ax + By + C = 0$, where A and B are not both zero. (7) **Polar form.** The equation $r = p \sec(\theta - \omega)$, where p is the perpendicular distance from the pole to the line, ω is the inclination of this perpendicular to the polar axis, and r and θ are the polar coordinates of a variable point on the line; see figure above. The **equation of a line in space** may be of the following types: (1) The equations of any two planes which intersect in the given line. (2) Equations of planes parallel to the coordinate axes are used as the **symmetric (standard) form** of the equation of a line, the equation being written

$$(x - x_1)/l = (y - y_1)/m = (z - z_1)/n,$$

where l, m, and n are *direction numbers* of the line and x_1, y_1, and z_1 are the coordinates of a point on it. (3) The **two-point form** of the equations of a line is

$$\frac{x - x_1}{x_2 - x_1} = \frac{y - y_1}{y_2 - y_1} = \frac{z - z_1}{z_2 - z_1},$$

where (x_1, y_1, z_1) and (x_2, y_2, z_2) are two points on the line. (4) The **parametric form** is derived by equating each fraction in the *symmetric form* to a parameter, say t, and solving for x, y, and z. This gives $x = x_1 + lt$, $y = y_1 + mt$, $z = z_1 + nt$. The points determined by giving t any values desired lie on the line. If l, m, n are the direction cosines of the line, t is the distance between the points (x, y, z) and (x_1, y_1, z_1).

half-line See RAY.

ideal line, or **line at infinity.** *Algebraically,* the locus of the equation $x_3 = 0$ in the system of *homogeneous coordinates* related to the Cartesian coordinates $x_1/x_3 = x$, $x_2/x_3 = y$. See CO-ORDINATE —homogeneous coordinates. *Geometrically,* the aggregate of all *ideal* points in the place. *Syn.* ideal line.

level lines. See CONTOUR —contour lines.

line graph. See GRAPH —broken line graph.

line of best fit. Usually the line determined by the method of least squares. See LEAST— method of least squares.

line integral. See INTEGRAL —line integral.

line segment. The part of a straight line between two points on the line (the points may, or may not, belong to the line segment).

material line. See MATERIAL.

midpoint of a line. See MIDPOINT.

nodal line. See NODAL.

oblique and parallel line (relative to another line or to a plane). See OBLIQUE, PARALLEL.

perpendicular line (relative to another line or to a plane). See PERPENDICULAR.

plumb line. (1) The line in which a string hangs, when supporting a weight. (2) The string itself.

polar line (and **pole of a line**). See POLE —pole and polar of a conic.

projection of a line. See PROJECTION.

straight line. A curve such that if any part of it is placed so as to have two points in common with any other part, it will lie along the other part; a straight line is usually called simply a line. *Tech.* (1) The set of all "points" (x, y) which satisfy a given linear equation, $ax + by + c = 0$, where a and b are not both zero. (2) An object called "line" in an axiomatic structure called "geometry." This may be an undefined element, which, taken with some other element or elements, such as points, satisfies certain assumptions; *e.g.*, two lines determine a point (including the ideal point), and two points determine a line.

trace of a line. See TRACE.

LIN′E-AL, *adj.* **lineal element.** A directed line segment through a point, whose slope, together with the coordinates of the point, satisfies a given first-order differential equation.

LIN′E-AR, *adj.* (1) In a straight line. (2) Along or pertaining to a curve. (3) Having only one dimension.

coefficient of linear expansion. See COEFFICIENT —coefficient of linear expansion.

consistency of a system of linear equations. See CONSISTENCY.

linear algebra. See ALGEBRA —algebra over a field.

linear combination. A linear combination of two or more **quantities** is a sum of the quantities, each multiplied by a constant (not all the constants being zero). See DEPENDENT —linearly dependent. For **equations** $f(x, y) = 0$ and $F(x, y) = 0$, a linear combination is $kf(x, y) + hF(x, y) = 0$, where k and h are not both zero. The graph of the linear combination of any two equations passes through the points of intersection on their graphs and cuts neither in any other point. A **convex linear combination** of quantities x_i $(i = 1, 2, \cdots, n)$ is an expression of the form $\sum_1^n \lambda_i x_i$, where $\sum_1^n \lambda_i = 1$ and each λ_i is a nonnegative real number. See BARYCENTRIC— barycentric coordinates.

linear congruence. See CONGRUENCE —linear congruence.

linear differential equation. See DIFFERENTIAL—linear differential equation.

linear element. See ELEMENT —element of integration. For a surface S: $x = x(u, v)$, $y =$

$y(u, v)$, $z = z(u, v)$, and a curve $f(u, v) = 0$ on S, the linear element ds is given by $ds^2 = dx^2 + dy^2 + dz^2 = E\, du^2 + 2F\, du\, dv + G\, dv^2$, where

$$E = \left(\frac{\partial x}{\partial u}\right)^2 + \left(\frac{\partial y}{\partial u}\right)^2 + \left(\frac{\partial z}{\partial u}\right)^2,$$

$$F = \frac{\partial x}{\partial u}\frac{\partial x}{\partial v} + \frac{\partial y}{\partial u}\frac{\partial y}{\partial v} + \frac{\partial z}{\partial u}\frac{\partial z}{\partial v},$$

$$G = \left(\frac{\partial x}{\partial v}\right)^2 + \left(\frac{\partial y}{\partial v}\right)^2 + \left(\frac{\partial z}{\partial v}\right)^2,$$

and du and dv satisfy $\dfrac{\partial f}{\partial u}\, du + \dfrac{\partial f}{\partial v}\, dv = 0$. Also written $ds^2 = g_{\alpha\beta}\, du^\alpha du^\beta$, in tensor notation. The **linear element of a surface** is the element of length ds, given by $ds^2 = E\, du^2 + 2F\, du\, dv + G\, dv^2$, without necessary reference to any particular curve on the surface. In an n-dimensional Euclidean space there exist rectangular Cartesian coordinates y^i (not so in other Riemannian spaces) so that the linear element has the form

$$ds^2 = \left(dy^1\right)^2 + \left(dy^2\right)^2 + \cdots + \left(dy^n\right)^2.$$

In other words, the fundamental Euclidean metric tensor $g_{ij}(x^1, \cdots, x^n)$ has the components $^*g_{ij}(y^1, \cdots, y^n) = \delta_{ij}$ in rectangular coordinates y^i, where δ_{ij} is Kronecker's delta. If $y^i = f(x^1, \cdots, x^n)$ is the transformation of coordinates from any general coordinates x^i to rectangular coordinates y^i, then the components $g_{ij}(x^1, \cdots, x^n)$ in general coordinates can be computed by

$$g_{ij}(x^1, \cdots, x^n) = \frac{\partial y^\alpha}{\partial x^i}\frac{\partial y^\alpha}{\partial x^j}.$$

Also called the **line element** and **element of length**.

linear equation or expression. An algebraic equation or expression which is of the first degree in its variable (or variables); *i.e.*, its highest degree term in the variable (or variables) is of the first degree. The equations $x + 2 = 0$ and $x + y + 3 = 0$ are linear. An equation or expression is **linear in a certain variable** if it is of the first degree in that variable. The equation $x + y^2 = 0$ is linear in x, but not in y.

linear expansion. Expansion in a straight line; expansion in one direction. The longitudinal expansion of a rod that is being heated is a linear expansion.

linear function. Same as LINEAR TRANSFORMATION (2). See below.

linear group. See GROUP—full linear group, real linear group.

linear hypothesis. See HYPOTHESIS—linear hypothesis.

linear interpolation. See INTERPOLATION.

linear programming. See PROGRAMMING.

linear regression. See REGRESSION—regression line.

linear space. Same as VECTOR SPACE.

linear span. See SPAN.

linear theory of elasticity. See ELASTICITY.

linear topological space. See VECTOR—vector space.

linear transformation. (1) A transformation effected by an equality which is a linear algebraic equation in the old variables and in the new variables. The **general linear transformation** in *one dimension* is of the form

$$x' = (ax + b)/(cx + d),$$

or $\rho x_1' = ax_1 + bx_2$, $\rho x_2' = cx_1 + dx_2$, where ρ is an arbitrary nonzero constant and x_1, x_2 are *homogeneous coordinates* defined by $x_1/x_2 = x$. In *two dimensions* the general linear transformation is

$$x' = (a_1 x + b_1 y + c_1)/(d_1 x + e_1 y + f_1),$$

$$y' = (a_2 x + b_2 y + c_2)/(d_1 x + e_1 y + f_1),$$

or in *homogeneous coordinates*

$$\rho x_1' = a_1 x_1 + b_1 x_2 + c_1 x_3,$$

$$\rho x_2' = a_2 x_1 + b_2 x_2 + c_2 x_3,$$

$$\rho x_3' = a_3 x_1 + b_3 x_2 + c_3 x_3.$$

General linear transformations in more than two dimensions are defined similarly, and are **singular** or **nonsingular** according as the determinant of the coefficients on the right side is or is not zero. (2) A transformation which takes $ax + by$ into $ax' + by'$ for all a and b if it takes vectors x and y into x' and y'. It is sometimes also required that the transformation be continuous. Here x and y may be vectors in n-dimensional Euclidean space or in a vector space, or in particular they may be ordinary real or complex numbers. The numbers a and b may be real, complex, or of any field for which multiplication with elements of the vector space is defined. In an n-dimensional vector space (or Euclidean space of dimension n), such a transformation is of the form

$$y_i = \sum_{j=1}^{n} a_{ij} x_j \qquad (i = 1, 2, \cdots, n),$$

or $y = Ax$, where x and y are one-column matrices (vectors) with elements $(x_1, x_2, \cdots, x_n)$ and $(y_1, y_2, \cdots, y_n)$, A is the matrix (a_{ij}), and multiplication is matrix multiplication (see MATRIX—matrix of a linear transformation); such a linear transformation has an inverse (is **invertible** or **nonsingular**) if and only if A is nonsingular, or if and only if its range is V. In general, a linear transformation T is invertible if and only if there is no x for which $T(x) = 0$. A linear transformation T between normed vector spaces is **bounded** if there exists a constant M such that

$$\|T(x)\| \le M \|x\|$$

for each x. The least such number M is the **norm** of the linear transformation and is denoted by $\|T\|$. If such a number M does not exist, the linear transformation is **unbounded**. A linear transformation is continuous if and only if it is bounded. See OPERATOR —linear operator.

linear velocity. See VELOCITY.

solution of a system of linear equations. See CONSISTENCY —consistency of linear equations, CRAMER —Cramer's rule, ELIMINATION.

LIN′E·AR·LY, *adv.* **linearly dependent and linearly independent.** See DEPENDENT —linearly dependent.

linearly ordered set. See ORDERED —ordered set.

LINK, *n.* See GRAPH THEORY.

LIOUVILLE, Joseph (1809–1882). French analyst and geometer. First to prove existence of transcendental numbers.

Liouville function. The function λ of the positive integers defined by $\lambda(1) = 1$ and $\lambda(n) = (-1)^{a_1 + \cdots + a_r}$ if $n = p_1^{a_1} \cdots p_r^{a_r}$, where $p_1, \cdots, p_r$ are prime numbers.

Liouville-Neumann series. The series

$$y(x) = f(x) + \sum_{n=1}^{\infty} \lambda^n \phi_n(x),$$

where

$$\phi_1(x) = \int_a^b K(x, t) f(t) \, dt$$

and

$$\phi_n(x) = \int_a^b K(x, t) \phi_{n-1}(t) \, dt$$

$$(n = 2, 3, \cdots).$$

This function y is a solution of the equation

$$y(x) = f(x) + \lambda \int_a^b K(x, t) y(t) \, dt$$

if (1) $K(x, y)$ is real, continuous, and not identically zero in the square $a \le x \le b$, $a \le y \le b$; (2) $|\lambda| < 1/[M(b-a)]$, where M is the l.u.b. of $|K(x, y)|$ in this square; (3) $f(x) \not\equiv 0$ and is real and continuous for $a \le x \le b$. See KERNEL —iterated kernels.

Liouville number. An irrational number X such that, for any integer n, there is a rational number p/q such that $q > 1$ and

$$|X - p/q| < 1/q^n.$$

All Liouville numbers are *transcendental* (see IRRATIONAL —irrational number). For any *irrational number I*, there exist infinitely many rational numbers p/q such that

$$|I - p/q| < 1/(\sqrt{5}\, q^2),$$

but $\sqrt{5}$ is the largest number that can be used for every I; for an *algebraic number* of degree n, there is a positive number c for which there are infinitely many rational numbers p/q with

$$|A - p/q| < c/q^n,$$

but the degree n is the largest exponent that can be used on q. There is a Liouville number between any two real numbers. In fact, the set of Liouville numbers is a set of *second category* (although it is of *measure zero*).

Liouville's theorem. If f is an *entire analytic function* of the complex variable z and is bounded, then f is a constant.

LIPSCHITZ, Rudolph Otto Sigismund (1832–1903). German analyst, algebraist, number theorist, and physicist.

Lipschitz condition. A function f satisfies a **Lipschitz condition** (with constant K) at a point x_0 if $|f(x) - f(x_0)| \le K|x - x_0|$ for all x in some neighborhood of x_0. It satisfies a **Hölder condition** of order p at x_0 if

$$|f(x) - f(x_0)| \le K|x - x_0|^p$$

for all x in some neighborhood of x_0 (this is sometimes called a Lipschitz condition of order p). A function f satisfies a Lipschitz condition on the interval $[a, b]$ if

$$|f(x_2) - f(x_1)| \le K|x_2 - x_1|$$

for all x_1 and x_2 on $[a, b]$. A function having a continuous derivative at each point of a closed interval satisfies a Lipschitz condition. Also see CONTRACTION.

LI′TER (LI′TRE), *n.* One cubic decimeter. Approximately equal to 61.026 cubic inches or 1.056 quarts. See DENOMINATE NUMBERS in the appendix.

LIT′ER-AL, *adj.* **literal constant.** A letter which denotes any one of certain constants (say any real number, or any rational number), as contrasted to a specific constant like 1, 2, or 3. Letters from the first part of the alphabet are ordinarily used (however, see SUBSCRIPT).

 literal expression, or equation. An expression or equation in which the constants are represented by letters; $ax^2 + bx + c = 0$ and $ax + by + cz = 0$ are *literal* equations, whereas $3x + 5 = 7$ is a *numerical* equation.

 literal notation. The use of letters to denote numbers, either unknown numbers or any of a set of numbers under discussion. *E.g.*, algebra uses letters in discussing the fundamental operations of arithmetic in order to make statements regarding all numbers, such as $a + a = 2a$.

LITTLEWOOD, John Edensor (1885–1977). Great English analyst and number theorist. Famous for his collaborations with Hardy.

LIT′U-US, *n.* [*pl.* **litui**]. A plane curve shaped like a trumpet, from which it gets its name; the locus of a point such that the square of the radius vector varies inversely as the vectorial angle. Its equation in polar coordinates is $r^2 = a/\theta$. The curve is asymptotic to the polar axis and winds around infinitely close to the pole but

never touches it. Only positive values of r are used in the figure. Negative values would give an identical branch of the curve in such a position that the two branches would be symmetrical about the pole.

LOAD′ING, *v.* (*Insurance*) The amount added to the net insurance premiums to cover agents' fees, company expenses, etc.

LOAN, *adj., n.* **building and loan association.** See BUILDING.

 loan value. A term used in connection with an insurance policy. It is an amount, usually somewhat less than the cash surrender value, which the insurance company agrees to loan the policy holder at a stipulated rate as long as the policy is in force.

LOBACHEVSKI, Nikolai Ivanovich (1793–1856). Russian geometer. Independently of Bolyai, published first system of non-Euclidean geometry. See BOLYAI, GEOMETRY —non-Euclidean geometry.

LO′CAL, *adj.* **local maximum or minimum.** See MAXIMUM.

 local property. A property that is possessed by a space, function, etc., if it is satisfied in suitable neighborhoods of certain points. *E.g.*, a line being tangent to a curve at a point P of the curve, or a system of equations being solvable near a point (see IMPLICIT —implicit-function theorem). Also see the properties listed under LOCALLY. A **global property** is defined with respect to a whole figure or a whole space. *E.g.*, the properties of being connected or compact, the properties defining a geodesic or a minimal surface, and the Gauss-Bonnet theorem or Green's theorem.

 local value. Same as PLACE VALUE.

LO′CAL-LY, *adv.* **locally arcwise connected, locally compact, locally connected, locally convex, locally Euclidean,** and **locally finite.** See COMPACT, CONNECTED, CONVEX —convex set, EUCLIDEAN —Euclidean space, and FINITE —locally finite family of sets.

 locally integrable function. A function f is locally integrable on a set S if f is measurable on S and f has a finite integral on each compact subset of S.

LO′CAL-I-ZA′TION, *adj.* **localization principle.** Let f and $|f|$ be integrable on $[-\pi, \pi]$ (1) if f vanishes on the interval $(a, b) \subset (-\pi, \pi)$, then the Fourier series for f converges uniformly to 0 in any interval (c, d) for which $a < c < d < b$. (2) The convergence of the Fourier series of f at a point x depends only on the behavior of f in some (arbitrarily small) neighborhood of x. See FOURIER —Fourier series, Fourier's theorem.

LO-CA'TION, *adj., n.* **location theorem** (or **principle**) for the roots of an equation. See ROOT—root of an equation.

LO'CUS, *n.* [*pl.* **loci**]. Any system of points, lines, or curves which satisfies one or more given conditions. If a set of points consists of those points (and only those points) whose coordinates satisfy a given equation, then the set of points is the **locus of the equation** and the equation is the **equation of the locus.** *E.g.*, the locus of the equation $2x + 3y = 6$ is a straight line, the line which contains the points $(0, 2)$ and $(3, 0)$. The locus of points which satisfy a given condition is the set which contains all the points which satisfy the condition and none which do not; *e.g.*, the locus of points equidistant from two parallel lines is a line parallel to the two lines and midway between them; the locus of points at a given distance r from a given point P is the circle of radius r with center at P. The **locus of an inequality** consists of those points whose coordinates satisfy the given inequality. Thus in a one-dimensional space the locus of the inequality $x > 2$ is the part of the x-axis to the right of the point 2. In a two-dimensional space, the locus of the inequality $2x + 3y - 6 < 0$ is that portion of the (x, y)-plane which is below the line $2x + 3y - 6 = 0$.

LOG'A-RITHM, *n* The logarithm of a positive number (see below, logarithm of a complex number) is the exponent indicating the power to which it is necessary to raise a given number, the **base**, to produce the number; the logarithm, with base a, of M is equal to x if $a^x = M$. Since $10^2 = 100$, 2 is the logarithm of 100 to the base 10, written $\log_{10} 100 = 2$; likewise, $\log_{10} 0.01 = -2$ and $\log_9 27 = \frac{3}{2}$. Logarithms which use 10 as a base are **common** (or **Briggsian**) **logarithms.** Logarithms which use the base $e = 2.71828 +$ are **natural logarithms** (sometimes called **hyperbolic logarithms** or **Napierian logarithms,** but see NAPIER); $\log_e x$ often is written as $\ln x$ (see *e*). Common logarithms are particularly useful for performing multiplication, division, evolution, and involution, because of the following **fundamental laws of logarithms** (valid for logarithms to any base) together with the fact that shifting the decimal point n places to the right (or left) changes the common logarithm by the addition (or subtraction) of the integer n (see below, characteristic and mantissa of a logarithm). (1) The logarithm of the product of two numbers is the sum of the logarithms of the numbers $[\log(4 \times 7) = \log 4 + \log 7 = 0.60206 + 0.84510 = 1.44716$. (2) The logarithm of the quotient of two numbers is equal to the logarithm of the dividend minus the logarithm of the divisor $[\log \frac{4}{7} = \log 4 - \log 7 = 10.60206 - 10 - 0.84510 = 9.75696 - 10]$. (3) The logarithm of a power of a number is equal to the product of the exponent and the logarithm of the number $[\log 7^2 = 2 \cdot \log 7 = 1.69020]$. (4) The logarithm of a root of a number is equal to the quotient of the logarithm of the number and the index of the root $[\log \sqrt{49} = \log(49)^{1/2} = \frac{1}{2} \log 49 = \frac{1}{2}(1.69020) = 0.84510]$. *Natural logarithms* are particularly adapted to analytical work. This originates from the fact that the *derivative* of $\log_e x$ is equal to $1/x$, while the derivative of $\log_a x$ is $(1/x)\log_a e$. **Change of base** of logarithms can be accomplished by use of the formula:

$$\log_b N = \log_a N \cdot \log_b a.$$

In particular, $\log_{10} N = \log_e N \cdot \log_{10} e$ and $\log_e N = \log_{10} N \cdot \log_e 10$. The number by which logarithms in one system are multiplied to give logarithms in a second system is the **modulus** of the second system with respect to the first. Thus the *modulus of common logarithms* (with respect to natural logarithms) is $\log_{10} e = 0.434294 \cdots$ and the *modulus of natural logarithms* (with respect to common logarithms) is $\log_e 10 = 2.302585 \cdots$. The calculation of logarithmic tables is usually based on infinite series, such as:

$$\log_e(N + 1) = \log_e N + 2\left[\frac{1}{2N + 1}\right.$$

$$\left. + \frac{1}{3}\frac{1}{(2N + 1)^3} + \frac{1}{5}\frac{1}{(2N + 1)^5} + \cdots \right],$$

which is convergent for all values of N. See MERCATOR (Nicolaus).

characteristic and mantissa of logarithms. Due to the fundamental laws of logarithms (see above, LOGARITHM) and the fact that $\log_{10} 10 = 1$, common logarithms have the property that

$$\log_{10}(10^n \cdot K) = \log_{10} 10^n + \log_{10} K$$

$$= n + \log_{10} K.$$

I.e., the common logarithm of a number is changed by adding (or subtracting) n if the decimal point is moved n places to the right (or left). Thus when the logarithm is written as the sum of an integer (the *characteristic*) and a positive decimal (the *mantissa*), the characteristic serves to locate the decimal point and the mantissa determines the digits in the number. The characteristic of the logarithm of a number can be determined by either of the following rules: (1) The characteristic is the number of places the decimal point is to the right of standard position, or the negative of the number of places the decimal point is to the left of standard position (**standard position** of the decimal point is the position to the right of the first nonzero digit of the number. (2) When the number is greater than or equal to 1, the characteristic is always one less than the number of digits to the left of the decimal point. When the number is less than 1, the characteristic is **negative** and numerically one greater than the number of zeros immediately following the decimal point; *e.g.*, 0.1 has the characteristic -1, and 0.01 has the characteristic -2. The mantissa of a common logarithm is the same regardless of where the decimal point is located in the number. Only mantissas are put in tables of logarithms, since characteristics can be found by the above rules. See below, proportional parts in a table of logarithms.

logarithm of a complex number. The number w is the **logarithm** of z to base e if $z = e^w$. If we write z in the form

$$z = x + iy = r(\cos \theta + i \sin \theta) = re^{i\theta},$$

it is seen that $\log(x + iy) = i\theta + \log r$, where θ is an argument of z and r is the absolute value of z; *i.e.*, $\log z = \log|z| + i \arg z$. See COMPLEX — polar form of a complex number, and EULER — Euler formula. The logarithm of a complex number is a many-valued function since the argument of a complex number is many-valued. Since $e^{i\pi} = \cos \pi + i \sin \pi = -1$, $\log(-1) = i\pi$. For any number, $-n$, $\log(-n) = i\pi + \log n$, thus providing a definition for the **logarithm of a negative number**. More generally, $\log(-n) = (2k + 1)\pi i + \log n$, where k is any integer. When $\log_e z$ is known, the logarithm of z to any base can be found. See LOGARITHMIC —logarithmic function of a complex variable, and above, LOGARITHM.

logarithm of a negative number. See above, logarithm of a complex number.

proportional parts in a table of logarithms. The numbers to be added to the next smaller mantissa to produce a desired mantissa. The proportional parts are the products of the differences between successive mantissas (written above them) and the numbers $0.1, 0.2, \cdots, 0.9$ (written in the table without decimal points). These *proportional parts tables* are multiplication (and division) tables to aid in interpolating for logarithms of numbers not in the tables (and for numbers whose logarithms are not in the tables).

LOG′A-RITH′MIC, *adj.* **logarithmic convexity.** See CONVEX —logarithmically convex function.

logarithmic coordinates. See COORDINATE — logarithmic coordinates.

logarithmic coordinate paper. Coordinate paper on which the rulings corresponding (for instance) to the numbers 1, 2, 3, etc., are at distances from the coordinate axes proportional to the logarithms of these numbers; *i.e.*, the markings on the graph are not the distances from the axes, but the antilogarithms of the actual distances. This scale is a **logarithmic scale**, while the ordinary scale, which marks the actual distance, is a **uniform scale**. See below, semilogarithmic coordinate paper.

logarithmic curve. The plane locus of the rectangular Cartesian equation $y = \log_a x$, $a > 1$. This curve passes through the point whose coordinates are $(1, 0)$ and is asymptotic to the negative y-axis. The ordinates of the curve increase arithmetically while the abscissas increase geometrically; *i.e.*, if the ordinates of three points are $1, 2, 3$, respectively, the corresponding abscissas are a, a^2, a^3. When the base a of the logarithmic system is given different values, the general characteristics of the curve are not altered. The figure shows the graph of $y = \log_2 x$.

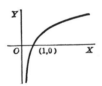

logarithmic derivative of a function. The ration $f'(z)/f(z)$. *I.e.*, $d \log f(z)/dz$.

logarithmic differentiation. see DIFFERENTIATION —logarithmic differentiation.

logarithmic equation. See EQUATION —logarithmic equation.

logarithmic function of a complex variable. The function $\log z$ can be defined as the inverse of the exponential function; *i.e.*, if $z = e^w$, then by definition $w = \log z$. It can also be defined by $\log z = \int_1^z \frac{d\zeta}{\zeta}$ with the path of integration restricted away from the branch-point $z = 0$, or by the function-element

$$f(z) = (z - 1) - \tfrac{1}{2}(z - 1)^2$$

$$+ \cdots + \frac{(-1)^{n-1}}{n}(z - 1)^n + \cdots$$

and its analytic continuations. The logarithmic function is infinitely multiple-valued; if its principal branch is denoted by $\log z$, then all of its values are given by $\log z = \log z + 2k\pi i$, $k = 0, \pm 1, \pm 2, \cdots$. The **principal branch** of the function $\log z$ is the single-valued analytic function of the complex variable $z = x + iy$ defined in the z-plane cut along the negative real axis, and coinciding with the real function $\log x$ along the positive real axis.

logarithmic plotting (or graphing). A system of graphing such that curves whose equations are of the form $y = kx^n$ are graphed as straight lines. The logarithms of both sides of the equation are taken, giving an equation of the form $\log y = \log k + n \log x$; $\log y$ and $\log x$ are then treated as the variables, and a straight line plotted whose abscissas are $\log x$ and ordinates $\log y$. Points can be found on this straight line whose coordinates are $(\log x, \log y)$, just as the coordinates of points on any line are found. It is then a matter of taking antilogarithms of the coordinates to find x and y, and even this is not necessary if *logarithmic coordinate paper* is used.

logarithmic potential. Potential based on a force which varies inversely as the first power of the distance instead of inversely as the square, as is the case in the Newtonian law of gravitation, Coulomb's law for point charges, and the law of force for isolated magnetic poles. An example of such a force field is furnished by a uniformly charged straight wire of infinite length. If we take the z-axis along this wire, then the force experienced by a unit charge at a point r units from the wire is given by $(k/r)\boldsymbol{\rho}_1$, where k is a constant and $\boldsymbol{\rho}_1$ is a unit vector having the direction of the perpendicular from the wire to the point. In this case the field depends on two variables only (say r and θ). Hence we are dealing with a two-dimensional situation. Consequently, we may replace the uniformly charged wire with a particle that supposedly exerts an attractive or repulsive force which is inversely proportional to the first power of the distance r. The potential corresponding to such a particle (the *logarithmic potential*) is given by $a \log r + b$, where a and b are constants.

logarithmic sine, cosine, tangent, cotangent, secant, or cosecant. The logarithms of the corresponding sine, cosine, etc.

logarithmic solution of triangles. Solutions using logarithms and formulas adapted to the use of logarithms, formulas which essentially involve only multiplication and division.

logarithmic spiral. The plane curve whose vectorial angle is proportional to the logarithm of the radius vector. Its polar equation is $\log r =$

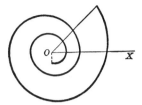

$a\phi$. The angle between the radius vector to a point on the spiral and the tangent at this point is always equal to the modulus of the system of logarithms being used. *Syn.* logistic spiral, equiangular spiral.

logarithmic transformation. (*Statistics*) Sometimes the *logarithm* of a random variable X is normally distributed (where X is not). Hence the transformation which replaces X by the logarithm of X may be used to permit application of normal distribution theory. See LOGNORMAL —lognormal distribution.

semilogarithmic coordinate paper. Coordinate paper on which the logarithmic scale is used on one axis and the uniform scale on the other. It is adapted to graphing equations of the type $y = ck^x$. When the logarithms of both sides are taken, the equation takes the form

$$\log y = \log c + x \log k.$$

Log y is now treated as one variable, say u, and the linear equation $u = \log c + x \log k$ graphed. Useful in *statistics* for showing a series in which fluctuations in the rate of change are of interest and for comparing two or more greatly divergent series, or one series which fluctuates widely. *Syn.* ratio paper. See above, logarithmic coordinate paper.

LOG'IC, *n.* **fuzzy logic.** See FUZZY.

LO-GIS'TIC, *adj. n.* (1) Logical. (2) Skilled in or pertaining to computation and calculation. (3) Proportional; pertaining to proportions. (4) The art of calculation. (5) Sexagesimal arithmetic.

logistic curve. A curve whose equation is of the form $y = k/(1 + e^{a+bx})$, where $b < 0$. The value of y at $x = 0$ is

$$k/(1 + e^a),$$

and as $x \to \infty$, $y \to k$. The increments in y as x increases are such that the difference of increments of $1/y$ is proportional to the corresponding difference in $1/y$. Also known as the **Pearl-Reed curve.** This is one of the types of curves known as **growth curves.**

logistic spiral. Same as LOGARITHMIC SPIRAL.

LO-GIS'TICS, *n.* The military art (and science) of transport, supply, and quartering. Studied in the mathematical disciplines of linear programming and the theory of games.

LOG-NOR'MAL, *adj.* **lognormal distribution.** A random variable X has a **lognormal distribution** or is a **lognormal random variable** if ln X is a *normal random variable*; *i.e.*, if there is a normal random variable Y for which $X = e^Y$. If Y has mean μ and variance σ^2, then the *probability density function* f for X satisfies $f(x) = 0$ if $x \le 0$ and

$$f(x) = \frac{1}{\sigma x \sqrt{2\pi}} e^{-1/2(\ln x - \mu)^2/\sigma^2} \quad \text{if } x > 0.$$

The *mean* is $e^{\mu + \frac{1}{2}\sigma^2}$ and the *variance* is $(e^{\sigma^2} - 1)(e^{2\mu + \sigma^2})$. *Syn.* log normal distribution.

LONG, *adj.* **long division.** See DIVISION —short and long division.

LON'GI-TUDE, *n.* The number of degrees in the arc of the equator cut off by the meridian through the place under consideration and the meridian through some established point (Greenwich, England, unless otherwise stated). See MERIDIAN —principal meridian.

LOOP, *n.* (1) **Loop of a curve.** A section of a plane curve that is the boundary of a bounded set. (2) See GRAPH THEORY.

LOW'ER, *adj.* **lower bound.** See BOUND.
  lower limit of an integral. See INTEGRAL — definite integral.

LOW'EST, *adj.* **fraction in lowest terms.** A fraction in which all common factors have been divided out of numerator and denominator; $\frac{1}{2}$, $\frac{2}{3}$ and $1/(x + 1)$ are in their lowest terms, but $\frac{2}{4}$, $\frac{6}{9}$ and

$$(x - 1)/(x^2 - 1)$$

are not.
  lowest common multiple. *Syn.* least common multiple. See MULTIPLE —least common multiple.

LOX'O-DROME, *n.* Same as LOXODROMIC SPIRAL.

LOX-O-DROM'IC, *adj.* **loxodromic spiral.** The path of a ship which cuts the meridians at a constant angle not equal to a right angle; more generally, any curve on a surface of revolution which cuts the meridians at a constant angle. See SURFACE—surface of revolution. *Syn.* rhumb-line, loxodromic line, loxodromic curve.

LUNE, *n.* (1) Any crescent-shaped figure. (2) A portion of a sphere bounded by two great semicircles. The angle at which the great circles intersect is the **angle of the lune.** The area of a lune is $4\pi r^2 A/360$, where r is the radius of the sphere and A is the degree measure of the angle of the lune.
  lunes of Hippocrates. See HIPPOCRATES OF CHIOS.

LUZIN (OR LUSIN), Nikolai Nikolaevich (1883–1950). Russian analyst, topologist, and logician.
  Luzin's theorem. Let f be defined on the real line (or an n-dimensional space), finite *almost everywhere*, and *measurable*. Then, for any positive number ϵ, there is a function g which is continuous on the line (or space) and is such that $f(x) = g(x)$ except for points of a set of measure less than ϵ.

M

MACH, Ernst (1838–1916). Austrian philosopher and physicist.
  mach number. The ratio v/a of v, the speed at which a body is traveling, to a, the local velocity of sound in air.

MACHIN, John (1685–1751). **Machin's formula.** The equality

$$\frac{\pi}{4} = 4 \arctan \tfrac{1}{5} - \arctan \frac{1}{239},$$

which was used by Machin, together with the Taylor series

$$\arctan x = x - \tfrac{1}{3}x^3 + \tfrac{1}{5}x^5 - \tfrac{1}{7}x^7 + \cdots,$$

to compute 100 digits of π in 1706. The same method was used by William Shanks in 1873 to compute 707 digits, 527 correctly.

MACLANE, Saunders (1909–   ). American algebraist, algebraic topologist, geometer, and set theorist. Cofounder of category theory with S. Eilenberg.

MACLAURIN, Colin (1698–1746). Scottish mathematician and physicist.

Maclaurin series (theorem). See TAYLOR—Taylor's theorem.

trisectrix of Maclaurin. See TRISECTRIX.

MAG'IC, *adj.* **magic square.** A square array of integers such that the sum of the numbers in each row, each column, and each diagonal are all the same, such as

$$
\begin{array}{|ccc|}
\hline
17 & 3 & 13 \\
7 & 11 & 15 \\
9 & 19 & 5 \\
\hline
\end{array}
\ \text{and}\
\begin{array}{|cccc|}
\hline
1 & 15 & 14 & 4 \\
12 & 6 & 7 & 9 \\
8 & 10 & 11 & 5 \\
13 & 3 & 2 & 16 \\
\hline
\end{array}
$$

MAG'NI-FI-CA'TION, *adj.* **magnification ratio.** Same as DEFORMATION RATIO.

MAG'NI-TUDE, *n.* (1) Greatness, vastness. (2) Size, or the property of having size; length, area, or volume. (3) The *absolute value* of a real number or a complex number, or the *length* of a vector.

magnitude of a star. Two stars differ by one magnitude if one is $(100)^{1/5}$, or $2.512 +$, times as bright as the other. The faintest stars seen with the naked eye on a clear moonless night are said to be of the 6th magnitude. The pole star (Polaris) is nearly of the 2nd magnitude.

order of magnitude. For two functions u and v to be of the **same order** of magnitude near t_0 means there are positive numbers ϵ, A, and B such that

$$
A < \left| \frac{u(t)}{v(t)} \right| < B \qquad \text{if } 0 < |t - t_0| < \epsilon.
$$

If this is true for $v(t)$ replaced by $[v(t)]^r$, then u is of the rth order with respect to v. If $\lim_{t \to t_0} u(t)/v(t) = 0$, u is of **lower order** than v and one writes $u = o(v)$; if $\lim_{t \to t_0} |u(t)/v(t)| = \infty$, u is of **higher order** than v. Also, one says that u is **of the order of** v on a set S (usually a deleted neighborhood of some t_0) and writes $u = O(v)$ if there is a positive number B such that

$|u(t)/v(t)| < B$ if $t \in S$. The statements $u = o(1)$ and $u \to 0$ are equivalent; $u = O(1)$ on S is equivalent to "u is bounded on S." If $\lim_{t \to t_0} u(t)/v(t) = 1$, then u and v are **asymptotically equal** at t_0. These concepts also are used when $t_0 = \pm\infty$, with appropriate changes. For example, if $\{u_n\}$ and $\{v_n\}$ are sequences and there are numbers B and N such that

$$
\left| \frac{u_n}{v_n} \right| < B \qquad \text{if } n > N,
$$

then u_n is **of the order of** v_n and $u_n = O(v_n)$.

MAGNUS, Heinrich Gustav (1802–1870). German chemist and physicist.

magnus effects. In aerodynamics, those forces and moments on a rotating shell that account for such phenomena as right-hand drift, etc.

MA'JOR, *adj.* **major arc.** The longer of two arcs in a circle, subtended by a secant. See SECTOR—sector of a circle.

major axis See ELLIPSE, ELLIPSOID.

major and minor segments of a circle. See SEGMENT—segment of a circle.

MAKEHAM, William Matthew (c. 1860–1892). British statistician.

Makeham's formula for bonds. The price to be paid for a bond n periods before redemption equals $Cv^n + (j/i)F(1 - v^n)$, where C is the redemption price, F the par value, j the dividend rate, i the investment rate, and $v = (1 + i)^{-1}$.

Makeham's law. The force of mortality (risk of dying) is equal to the sum of a constant and a multiple of a constant raised to a power equal to the age, x, of the life: $M = A + Be^x$. Makeham's law is a closer approximation to statistical findings than Gompertz's law. From the age of 20 to the end of life it very nearly represents the data of most tables.

MANDELBROT, Benoit B. (1924–   ). **Mandelbrot dimension.** See DIMENSION—fractal (or Mandelbrot) dimension.

Mandelbrot set. Let $f_c(z) = z^2 + c$, where c and z are complex numbers. For each c, let B_c be the set of all z whose orbits with respect to the sequence $\{f_c, f_c^2, \cdots\}$ are bounded (see ORBIT). The **Mandelbrot set** $\mathcal{M}$ is the set of all complex numbers c for which B_c is connected. For a particular c, B_c is totally disconnected (actually, a Cantor-type set) if the orbit of 0 is not bounded, and B_c is connected if the orbit of 0 is bounded. In any case, the boundary J_c of B_c is a *fractal* and is a *Julia set*. The boundary of $\mathcal{M}$ also is a fractal. If U is a neighborhood of a point in J_c, then every point in the complex plane belongs to the orbit of some point in U. See CHAOS, FRACTAL, JULIA.

MAN'I-FOLD, *n*. In general, manifold may mean any collection or sets of objects. *E.g.*, a Riemannian space is also called a Riemannian manifold; a subset of a vector space is said to be a *linear set* or a *linear manifold* if it contains all linear combinations of its members. However, manifold frequently has technical meaning beyond being a mere set, as illustrated by the following definitions. A **topological manifold** of dimension n (frequently called simply an *n*-**manifold**) is a topological space such that each point has a neighborhood which is homeomorphic to the interior of a sphere in Euclidean space of dimension n. Such a manifold M is **differentiable of order r** (or has a **differentiable structure of class C^r**) if there is a family of neighborhoods which cover M and which are such that each neighborhood is homeomorphic to the interior of a sphere in Euclidean space of dimension n; no point of M belongs to more than a finite number of the neighborhoods; and when x belongs to two neighborhoods U and V the $2n$ functions $u_k = u_k(v_1, v_2, \cdots, v_n)$ and $v_k = v_k(u_1, u_2, \cdots, u_n)$, $k = 1, \cdots, n$, have continuous partial derivatives of order r, where $(u_1, \cdots, u_n)$ and $(v_1, \cdots, v_n)$ are coordinates given to the same point in the intersection of U and V. A manifold which is compact and differentiable of order 1 is a *polyhedron* (*i.e.*, homeomorphic to the point-set union of the simplexes of a simplicial complex). A *manifold* is sometimes defined to be a *topological manifold* which is also a *polyhedron*. A connected manifold of this type is also a **manifold** (sometimes called a **pseudomanifold**) in the sense that it is an *n*-dimensional simplicial complex ($n \geq 1$) such that (i) each k-simplex ($k < n$) is a face of at least one *n*-simplex; (ii) each $(n - 1)$-simplex is a face of exactly two *n*-simplexes; and (iii) any two *n*-simplexes can be connected by a sequence whose members are alternatively *n*-simplexes and $(n - 1)$-simplexes, each $(n - 1)$-simplex being a face of the two adjacent *n*-simplexes. Such a manifold is **orientable** if its *n*-simplexes can be *coherently oriented*; *i.e.*, oriented so that no $(n - 1)$-simplex can be oriented so as to be coherently oriented with each of the *n*-simplexes of which it is a face (see SIMPLEX). Otherwise it is **nonorientable**. Any topological space which is homeomorphic to a manifold is also called a manifold and is orientable or nonorientable according as the manifold is orientable or nonorientable. A one-dimensional manifold is a simple closed curve. A two-dimensional manifold is also called a closed surface. The closed surfaces can be classified by use of certain topological invariants (see SURFACE). No such classification is known for three-dimensional manifolds.

MAN-TIS'SA, *n*. See LOGARITHM —characteristic and mantissa of logarithms.

MAN'Y, *adj.* **many-valued function.** See MULTIPLE—multiple-valued function.

MAP, *n*. Same as FUNCTION.
 angle-preserving map. See CONFORMAL—conformal map or conformal transformation.
 area-preserving map. A *map* which preserves areas. The map $x = x(u, v)$, $y = y(u, v)$, $z = z(u, v)$ of the (u, v)-domain of definition D on a surface S is area-preserving if, and only if, the *fundamental quantities of the first order* satisfy $E \cdot G - F^2 \equiv 1$. The induced map between the above surface S and the surface $\bar{S}$: $x = \bar{x}(u, v)$, $y = \bar{y}(u, v)$, $z = \bar{z}(u, v)$ is area-preserving if, and only if, $EG - F^2 \equiv \bar{E}\bar{G} - \bar{F}^2$. *Syn.* equivalent map, equiareal map.
 cylindrical map. See CYLINDRICAL.
 map-coloring problem. See FOUR—four-color problem.

MAP'PING, *n*. Same as MAP.

MAR'GIN, *n*. (1) The difference between the selling price and the cost of goods. (2) A sum of money deposited with a broker by a client to cover any losses that may occur in the broker's dealings for him.

MARGULIS, Gregori Aleksandrovitch. Soviet mathematician awarded a Fields Medal (1978) for his contributions to the theory of discrete subgroups of real and p-adic Lie groups.

MARIOTTE, Edme (1620–1684). French physicist.
 Mariotte's law. The name used in France for BOYLE'S LAW.

MARK, *n*. (*Statistics*) The value or name given to a particular class interval. Often the midvalue, or the integral value nearest the midpoint.

MAR'KET, *n*. **market value.** The amount a commodity sells for on the open market. *Syn.* market price.

MARKOV (or MARKOFF), Andreĭ Andreevich (1856–1922). Russian probabilist, algorist, algebraist, and topologist.

Markov process. A *stochastic process* in which the "future" is determined by the "present" and is independent of the "past." *Tech.* A *stochastic process* $\{X(t): t \in T\}$ with the property that if $t_1 < t_2 < \cdots < t_n$ are in the index set T, then the *conditional probability* of "$X(t_n) \leq x_n$", given that $X(t_i) = x_i$ for $i < n$, is equal to the conditional probability of "$X(t_n) \leq x_n$", given that $X(t_{n-1}) = x_{n-1}$. The probability $P_t(t_0, a; E)$ that $X(t)$ has a value in the interval (or Borel set) E, given that $X(t_0) = a$, is the **transition probability function**; the process is **stationary** if there is a function $\bar{P}$ such that $P_t(t_0, a; E) = \bar{P}(t - t_0, a; E)$ if $t > t_0$. A Markov process for which there is a discrete set that contains the ranges of all the random variables involved is a **Markov chain**; if T also is discrete, the process is a **discrete Markov chain**. Suppose T is the set of positive integers and the range of each $X(t)$ is contained in the set $\{s_1, s_2, \cdots\}$. If the process is stationary, then there is a set of transition probabilities $\{p_{ij}\}$ such that p_{ij} is the probability that $X(n) = s_j$, given that $X(n - 1) = s_i$.

MAR′TIN-GALE, *n.* A *stochastic process* $\{X(t): t \in T\}$ such that the expected value of $|X(t)|$ is finite for each t and, if $t_1 < t_2 < \cdots < t_n$ are in the index set T, then (with probability 1) the conditional expected value of $X(t_n)$, given $X(t_i) = a_i$ if $i < n$, is equal to a_{n-1}. If T is the set of positive integers, it is sufficient to require that the conditional expected value of $X(n)$, given $X(i) = a_i$ if $i < n$, is a_{n-1}. If, for each n, $G(n)$ is a fair game and $X(n)$ is the amount a player wins if he invests in $G(n)$ the winnings from the preceding game, then $\{X(n)\}$ is a martingale; if a_n is the amount a player has at time n, then the expected amount at time $n + 1$ is a_n —whatever his past fortune has been. If $\{X_n\}$ is a sequence of independent random variables with zero means, then $S(n)$ is a martingale if $S(n) = \sum_{i=1}^{n} X_i$ for each n. See WIENER —Wiener process.

MASCHERONI, Lorenzo (1750–1800). Italian geometer and analyst. Proved that all ruler-and-compass constructions can be accomplished by using compass alone.

    Mascheroni constant. See EULER—Euler constant.

MASS, *n.* The measure of the tendency of a body to oppose changes in its velocity. Mass can be defined, with the aid of Newton's Second Law of Motion, as the ratio of the magnitudes of the force and acceleration which the force produces.

This amounts to defining the mass in terms of force. At speeds small compared with the speed of light, the masses m_1 and m_2 of two bodies may be compared by allowing the two bodies to interact. Then

$$m_1/m_2 = |a_2|/|a_1|,$$

where $|a_1|$ and $|a_2|$ are the magnitudes of the respective accelerations of the two bodies as a result of the interaction. This permits the measurement of the mass of any particle with respect to a standard particle (for example, the standard kilogram). At higher speeds, the mass of a body depends on its speed relative to the observer according to the relation

$$m = m_0/\sqrt{1 - v^2/c^2},$$

where m_0 is the mass of the body as found by an observer at rest with respect to the body, v is the speed of the body relative to the observer who finds its mass to be m, and c is the speed of light in empty space (*theory of relativity*). Equal masses at the same location in a gravitational field have equal weights. Because of this, masses may be compared by weighing. Mass is particularly important because it is a conserved quantity, which can neither be created nor be destroyed. Thus, the mass of any isolated system is a constant. When relativistic mechanics is appropriate, *e.g.*, when speeds comparable to the speed of light are involved, mass may be converted into energy and *vice versa*; hence the energy of the system must be converted into mass through the Einstein equation

$$E = mc^2,$$

where c is the speed of light in empty space, before the conservation law may be applied.

    center of mass. See CENTER —center of mass, CENTROID.

    differential (or element) of mass. See ELEMENT—element of integration.

    moment of mass. See MOMENT—moment of mass.

    point-mass. Same as PARTICLE.

    unit mass. The standard unit of mass, or some multiple of this unit chosen for convenience. There are several such standard units. In the c.g.s. system, one gram-mass is defined as $\frac{1}{1000}$ part of the mass of a certain block of platinum-iridium alloy preserved in the Bureau of Weights and Measures at Sèvres, France. The corresponding unit in the British system is the standard pound of mass which is a block of platinum alloy preserved in the Standards Office, London.

MATCHED, *adj.* **matched samples.** For two random samples, **matched pairs** are samples paired by consideration of some variable other than the one under immediate study; *e.g.*, in a study of heights of two groups of ten persons each, the individuals may be paired one from each group so that two persons in a pair have the same age. Similarly, given a set of random samples, a **matched group** or a set of **matched samples** is a set of samples consisting of one sample chosen from each random sample by consideration of some variable other than those under immediate study. The objective is to control variation due to some extraneous factor.

MA-TE′RI-AL, *adj.* **material point, line, or surface.** A point, line, or surface thought of as having mass. (If one thinks of a lamina with a fixed mass whose thickness approaches zero and density increases proportionally, the limiting situation can be thought of as a surface with the fixed mass.)

MATH′E-MAT′I-CAL, *adj.* **mathematical expectation.** See EXPECTATION.
　mathematical induction. See INDUCTION.
　mathematical system. One or more sets of undefined objects, some undefined and defined concepts, and a set of axioms concerning these objects and concepts. One of the simplest and most important mathematical systems is the *group.* More complicated systems are the real numbers and their axioms (properties) and the plane Euclidean geometry system of points, lines, suitable undefined and defined concepts such as "on a line," "between," "triangle," and "parallel," and axioms relating these objects and concepts. A mathematical system is an abstract deductive theory that can be applied in other mathematical situations when the axioms can be verified. The success of application in other fields of knowledge depends on how well the mathematical system describes the situation at hand.

MATH′E-MAT′ICS, *n.* The logical study of shape, arrangement, quantity, and many related concepts. Mathematics often is divided into three fields: *algebra, analysis,* and *geometry.* However, no clear divisions can be made, since these branches have become thoroughly intermingled. Roughly, algebra involves numbers and their abstractions, analysis involves continuity and limits, and geometry is concerned with space and related concepts. *Tech.* The postulational science

in which necessary conclusions are drawn from specified premises.
　applied mathematics. A branch of mathematics concerned with the study of the physical, biological, and sociological worlds. It includes mechanics of rigid and deformable bodies (elasticity, plasticity, mechanics of fluids), theory of electricity and magnetism, relativity, theory of potential, thermodynamics, biomathematics, and statistics. Broadly speaking, a mathematical structure utilizing, in addition to the purely mathematical concepts of space and number, the notions of time and matter belongs to the domain of applied mathematics. In a restricted sense, the term refers to the use of mathematical principles as tools in the fields of physics, chemistry, engineering, biology, and social studies.
　mathematics of finance. The study of the mathematical practices in brokerage, banking, and insurance. *Syn.* mathematics of investment.
　pure mathematics. The study and development of the principles of mathematics for their own sake and possible future usefulness, rather than for their immediate usefulness in other fields of science or knowledge. The study of mathematics independently of experience in other scholarly disciplines. Often the study of problems in applied mathematics leads to new developments in pure mathematics and theories developed as pure mathematics often find applications later. Thus no sharp line can be drawn between applied and pure mathematics. *Syn.* abstract mathematics.

MATHIEU, Émile Léonard (1835–1890). French mathematical physicist.
　Mathieu differential equation. A differential equation of the form

$$y'' + (a + b \cos 2x) y = 0.$$

The general solution can be written in the form $y = A e^{rx} \phi(x) + B e^{-rx} \phi(-x)$, for some constant r and function ϕ which is periodic with period 2π. There are periodic solutions for some *characteristic values* of a, but no Mathieu equation (with $b \neq 0$) can have two independent periodic solutions.
　Mathieu function. Any solution of Mathieu's differential equation which is periodic and is either an even or an odd function, the solution being multiplied by an appropriate constant. The solution which reduces to $\cos nx$ when $b \to 0$ and $a = n^2$, and for which the coefficient of $\cos nx$ in its Fourier expansion is unity, is denoted by $ce_n(x)$; the solution which reduces to $\sin nx$ when $b \to 0$, and in which the coefficient of $\sin nx$ in its Fourier expansion is unity, is denoted by $se_n(x)$.

MA'TRIX, *adj., n.* [pl. **matrices** or **matrixes**]. A rectangular array of terms called **elements** (written between parentheses or double lines on either side of the array), as

$$\begin{pmatrix} a_1 & b_1 & c_1 \\ a_2 & b_2 & c_2 \end{pmatrix} \quad \text{or} \quad \begin{Vmatrix} a_1 & b_1 & c_1 \\ a_2 & b_2 & c_2 \end{Vmatrix}.$$

Used to facilitate the study of problems in which the relation between these elements is fundamental, as in the study of the existence of solutions of simultaneous linear equations. Unlike determinants, a matrix does not have quantitative value. It is not the symbolic representations of some polynomial, as is a determinant (see below, rank of a matrix). The **order** or **dimension** of a matrix is given by stating the number of rows and then the number of columns in the matrix. Thus the two matrices shown above are 2-by-3, or 2×3, matrices. A **real matrix** and a **complex matrix** are matrices whose elements are real numbers or complex numbers, respectively. A **square matrix** is a matrix for which the number of rows is equal to the number of columns. The **order** of a square matrix is simply the number of rows (or columns) in the matrix; the diagonal from the upper left corner to the lower right corner is the **principal** (or **main**) **diagonal**; the diagonal from the lower left corner to the upper right corner is the **secondary diagonal**. The **determinant** of a square matrix is the determinant obtained by considering the array of elements in the matrix as a determinant. A square matrix is **singular** or **nonsingular** according as the determinant of the matrix is zero or nonzero. It is nonsingular if and only if it is invertible (see INVERTIBLE). A **diagonal matrix** is a square matrix all of whose nonzero elements are in the principal diagonal. If, in addition, all the diagonal elements are equal, the matrix is a **scalar matrix**. An **identity** (or **unit**) **matrix** is a diagonal matrix whose elements in the principal diagonal are all unity. For any square matrix A of the same order as I, $IA = AI = A$.

adjoint of a matrix. See ADJOINT.

associate matrix. See HERMITIAN —Hermitian conjugate of a matrix.

augmented matrix of a set of simultaneous linear equations. The matrix of the coefficients, with an added column consisting of the constant terms of the equations. The augmented matrix of

$$a_1 x + b_1 y + c_1 z + d_1 = 0$$
$$a_2 x + b_2 y + c_2 z + d_2 = 0$$

is

$$\begin{Vmatrix} a_1 & b_1 & c_1 & d_1 \\ a_2 & b_2 & c_2 & d_2 \end{Vmatrix}.$$

canonical (or normal) form of a matrix. See CANONICAL —canonical form of a matrix.

characteristic equation and function of a matrix. See CHARACTERISTIC.

complex conjugate of a matrix. The matrix whose elements are the complex conjugates of the corresponding elements of the given matrix.

derogatory matrix. See CHARACTERISTIC — characteristic equation of a matrix.

determinant of a matrix. See above, MATRIX.

echelon matrix. See ECHELON.

eigenvalue and eigenvector of a matrix. See EIGENVALUE.

elementary divisor of a matrix. See INVARIANT—invariant factor of a matrix.

equivalent matrices. See EQUIVALENT— equivalent matrices.

Hermitian matrix. See HERMITIAN —Hermitian matrix.

inverse of a matrix. For a nonsingular square matrix, the inverse is the quotient of the *adjoint* of the matrix and the *determinant* of the matrix; the inverse of a matrix is the transpose of the matrix obtained by replacing each element of the original matrix by its cofactor divided by the determinant of the matrix. If A^{-1} is the inverse of A, then $AA^{-1} = A^{-1}A = I$, where I is the identity matrix. The inverse is defined only for nonsingular square matrices. See INVERTIBLE.

Jordan matrix. See JORDAN.

matrix of the coefficients of a set of simultaneous linear equations. The rectangular array left by dropping the variables from the equations when they are written so that the variables are in the same order in all equations and are in such a position that the coefficients of like variables are in the same columns, zero being used as the coefficient if a term is missing. When the number of variables is the same as the number of equations, the matrix of the coefficients is a square array. The matrix of the coefficients of

$$a_1 x + b_1 y + c_1 z + d_1 = 0 \quad \text{is} \quad \begin{Vmatrix} a_1 & b_1 & c_1 \\ a_2 & b_2 & c_2 \end{Vmatrix}.$$
$$a_2 x + b_2 y + c_2 z + d_2 = 0$$

See below, rank of a matrix.

matrix of a linear transformation. The matrix of a linear transformation defined by $y_i = \sum_{j=1}^{n} a_{ij} x_j$ $(i = 1, 2, \cdots, n)$ is the matrix $A = (a_{ij})$, where a_{ij} is the element of the ith row and jth column. Two linear transformations, T_1 and T_2, applied in this order, are equivalent to the linear transformation with matrix BA, where A is the matrix of T_1 and B is the matrix of T_2.

norm of a matrix. See NORM.

normal matrix. A square matrix A such that $A^*A = AA^*$, where A^* is the *Hermitian conjugate* of A (the *transpose* if A is real). A matrix is normal if, and only if, it is the *transform* of a diagonal matrix by a *unitary matrix* (*i.e.*, it can be changed to diagonal form by a unitary transformation), whereas any nonsingular matrix can be written as the product of two normal matrices.

orthogonal matrix. A real matrix that is equal to the *inverse* of its *transpose*; a matrix such that

$$\sum_{s=1}^{n} a_{is}a_{js} = \sum_{s=1}^{n} a_{si}a_{sj} = \delta_{ij}$$ for all i and j, where

δ_{ij} is *Kronecker's delta* and a_{ij} is the element in the ith row and jth column. Thus any two distinct rows or any two distinct columns are *orthogonal vectors*. The orthogonal matrices are the unitary matrices whose terms are real. See ORTHOGONAL —orthogonal transformation.

payoff matrix. See PAYOFF.

permutation matrix. See PERMUTATION— permutation matrix.

product of matrices. See PRODUCT —product of matrices, direct product of matrices.

product of a scalar and a matrix. See PRODUCT—product of a scalar and a matrix.

rank of a matrix. The order of the nonzero determinant of greatest order that can be selected from the matrix by taking out rows and columns. The concept *rank* facilitates, for instance, the statement of the condition for consistency of simultaneous linear equations: m linear equations in n unknowns are consistent when, and only when, the *rank* of the matrix of the coefficients is equal to the *rank* of the augmented matrix. In the system of linear equations

$$x + y + z + 3 = 0$$

$$2x + y + z + 4 = 0,$$

the matrix of the coefficients is

$$\left\| \begin{matrix} 1 & 1 & 1 \\ 2 & 1 & 1 \end{matrix} \right\|$$

and the augmented matrix is

$$\left\| \begin{matrix} 1 & 1 & 1 & 3 \\ 2 & 1 & 1 & 4 \end{matrix} \right\|.$$

The rank of both is two, because the determinant

$$\left| \begin{matrix} 1 & 1 \\ 2 & 1 \end{matrix} \right|$$

is not zero. Hence these equations are satisfied

by some set of values of x and y and z. See CONSISTENCY —consistency of linear equations.

reducible matrix representation of a group. See REPRESENTATION —reducible matrix representation of a group.

skew-symmetric matrix. See SKEW—skew-symmetric matrix.

square matrix. See above, MATRIX.

sum of matrices. See SUM—sum of matrices.

symmetric matrix. See SYMMETRIC —symmetric matrix.

trace of a matrix. See TRACE—trace of a matrix.

transpose of a matrix. See TRANSPOSE — transpose of a matrix.

unitary matrix. A matrix which is equal to the inverse of its *Hermitian conjugate*; a matrix such that

$$\sum_{s=1}^{n} a_{is}\bar{a}_{js} = \sum_{s=1}^{n} a_{si}\bar{a}_{sj} = \delta_{ij}$$

for all i and j, where δ_{ij} is *Kronecker's delta* and a_{ij} is the element in the ith row and jth column. Thus any two distinct rows or any two distinct columns are *orthogonal vectors* in Hermitian vector space. For real matrices this is equivalent to being an orthogonal matrix. See UNITARY — unitary transformation.

Vandermonde matrix. An $m \times n$ matrix (or its transpose) that has 1 in each place of the first row, the second row has n arbitrary elements $x_1, x_2, \cdots, x_n$, and the ith row ($2 \le i \le m$) has the elements $x_1^{i-1}, x_2^{i-1}, \cdots, x_n^{i-1}$. See DETERMINANT—Vandermonde determinant.

MAX′I-MAL, *adj.* **maximal member of a set.** In a set which is *partially ordered*, a maximal element is any element x for which there is no y which follows x in the ordering. For a family of sets, partial ordering can be defined by means of set inclusion, and a maximal member is a set which is not properly contained in any other set. *E.g.*, a maximal *connected* subset of a set S is a subset which is connected and is not contained in any other connected subset of S.

MAX′I-MUM, *n.* [*pl.* **maxima** or **maximums**]. The **maximum (minimum)** of a function is the greatest (least) value of the function if it has such a value. A function has a **local (or relative) maximum** at a point c if there is a neighborhood of c such that $f(x) \le f(c)$ if x is in U; it has a **local (or relative) minimum** at c if there is a neighborhood U of c such that $f(x) \geq f(c)$ if x is in U. For example, suppose the graph of a function of one variable is illustrated as a path over a mountain (through a valley), a path over several mountains of equal heights (through valleys of equal depth), or as a path over mountains of different heights (through valleys of different depths). In the first case, the function is said to have an **absolute maximum (minimum)**. In the second and third cases, each hilltop (valley) is a local or relative maximum (minimum), and in the third case the highest hilltop is an **absolute maximum (minimum)**. A test for local maxima (minima) can be made by examining values of the function very near the point (value) under investigation. In cases such as we have illustrated by paths, the slope of the graph of a function changes from positive to negative (negative to positive) at a maximum (minimum) point as one passes from left to right, being zero at the point if the function is differentiable. A condition for testing for maximum (minimum) is that the first derivative be zero at the point and the second derivative be negative (positive) at the point. This rule fails when the second derivative is zero or when the curve has a cusp at the point. For instance, the functions $y = x^3$ and $y = x^4$ both

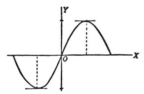

have their first and second derivatives zero at the origin, but the first has a point of inflection there and the second a minimum. For the exceptional cases in which this rule fails, a test can be made by examining the sign of the derivative on either side of the point, or finding the values of the function on either side. A general test for a local maximum (minimum) of a function that can be represented by a Taylor series is that the first derivative be zero at the point and the lowest-order derivative not zero at the point be of even order and negative (positive). For a function f of two variables, f has a local maximum (minimum) at the point (a, b) if $f(a + h, b + k) - f(a, b)$ is not greater than (not less than) zero for all sufficiently small values of h and k different from zero. A necessary condition for a maximum (or minimum) at a point (a, b) in the domains of the first-order partial derivatives of f is that these partial derivatives be zero at (a, b). If the first-order partial derivatives are zero, the second-order partial derivatives are continuous in a

neighborhood of (a, b) and the expression

$$\left(\frac{\partial^2 f}{\partial x \, \partial y} \right)^2 - \frac{\partial^2 f}{\partial x^2} \frac{\partial^2 f}{\partial y^2}$$

is greater than zero at (a, b), there is no maximum (or minimum); if this expression is less than zero at (a, b), there is a maximum if the derivatives $\partial^2 f / \partial x^2, \partial^2 f / \partial y^2$ are both negative (a minimum if both are positive). If the above expression is zero, the test fails. The function $F(x_1, x_2, \cdots, x_n)$ of the n **independent variables** $x_1, x_2, \cdots, x_n$ is said to have a maximum (minimum) at the point $P(x_1, x_2, \cdots, x_n)$ if the difference $F(x_1', x_2' \cdots, x_n') - F(x_1, x_2, \cdots, x_n)$ is not positive (not negative) for all points in a sufficiently small neighborhood of P. If F and its first partial derivatives exist in the neighborhood of a point, a necessary condition for F to have a maximum (minimum) at the point is that all of its first partial derivatives be zero at the point. For the case when the arguments of F are not independent, see LAGRANGE —Lagrange method of multipliers. See SADDLE —saddle point.

maximum-likelihood estimator. See LIKELIHOOD.

maximum-minimum principle of Courant. See COURANT.

maximum-value theorem. If a real-valued function f has a domain D that is a compact set, then there is a point x in D at which f has its greatest value. In particular, D might be a closed interval or a disk in a plane (a circle together with its interior). By using $-f$, the analogous **minimum-value theorem** follows from the maximum-value theorem.

principle of the maximum. See PRINCIPLE.

MA′YAN, *adj.* **Mayan numerals.** See VIGESIMAL.

MAZUR, Stanisław (1905–). Polish mathematician noted for important contributions to functional analysis, real analysis, and topology.

Mazur-Banach game. Let I be a given closed interval and let A and B be any two disjoint subsets whose union is I. Two players (A) and (B) alternately choose closed intervals $I_1, I_2, \cdots$, with each interval contained in the previous one. Player (A) chooses the intervals with odd subscripts, (B) those with even subscripts. Player (A) wins if there is a point which belongs to A and all of the chosen intervals; otherwise (B) wins (and there is a point which belongs to B and all of the chosen intervals). There is a strategy by which (B) can win for any strategy chosen by (A) if and only if A is of *first category* in I; there is a strategy by which (A) can win for any strategy chosen by (B) if and only if B is of *first category at some point of I*. The first of these statements can be extended to an arbitrary topological space and the second to a complete metric space, provided the players choose sets from a specified collection G of subsets which have nonempty interiors and which have the property that each nonempty open set contains a member of G. See NESTED —nested intervals.

MEA′GER, *adj.* **meager set.** A set of *first category*. See CATEGORY —*category of sets.*

MEAN, *adj., n.* A single number typifying or representing a set of numbers, usually not less than the least nor greater than the greatest. The **weighted mean** or **weighted average** of numbers $x_1, x_2, \cdots, x_n$ is

$$\bar{x} = \frac{q_1 x_1 + q_2 x_2 + \cdots + q_n x_n}{q_1 + q_2 + \cdots + q_n}.$$

The numbers q_i are the **weights**; if they are all equal, the weighted mean reduces to the **arithmetic mean** or **arithmetic average** (usually called simply the **mean** or **average**). *E.g.*, the average of the numbers 60, 70, 80, 90 is 75: their sum divided by 4. If one desired to give more preference to the grades a student makes as the semester advances, he could do so by using a weighted average. If the grades were 60, 70, 80, 90, the *mean* would be 75, but if 1, 2, 3, 4, were used as weights, the *weighted mean* would be $(60 + 140 + 240 + 360)/10$, or 80. The **harmonic mean** or **harmonic average** of nonzero numbers $x_1, x_2, \cdots, x_n$ is the reciprocal of the arithmetic mean of the reciprocals of the numbers:

$$H = \frac{n}{\dfrac{1}{x_1} + \dfrac{1}{x_2} + \cdots + \dfrac{1}{x_n}}.$$

The positive nth root of the product of a set of n positive numbers is their **geometric mean** or **geometric average**: $\sqrt[n]{x_1 x_2 \cdots x_n}$. The arithmetic mean of the logarithms of a set of numbers is the logarithm of their geometric mean. For a *random variable*, the mean is the same as the **expected value** (see EXPECTED —expected value); special cases of this are the preceding *weighted mean* (with q_i the probability of x_i) and the *mean value of a function* (see below). *Syn.* average.

arithmetic, geometric, harmonic means between two numbers. See under ARITHMETIC, GEOMETRIC, HARMONIC, and below, arithmetic-geometric mean.

arithmetic-geometric mean. For two positive numbers p, q, the arithmetic-geometric mean is the common limit of the two sequences $\{p_n\}, \{q_n\}$, where $p_1 = p$, $q_1 = q$,

$$p_n = \tfrac{1}{2}(p_{n-1} + q_{n-1}), \qquad q_n = (p_{n-1}q_{n-1})^{1/2}.$$

This mean is used in particular in Gauss' determination of the potential due to a homogeneous circular wire.

mean axis of an ellipsoid. See ELLIPSOID.

mean curvature of a surface. See CURVATURE.

mean deviation. See DEVIATION.

mean proportional. See PROPORTIONAL.

mean-square deviation. See DEVIATION.

mean-square error. See ERROR.

mean value of a function. For a function f of one variable x, the **mean value** (or **mean ordinate**) on the interval (a, b) is equal to the area of the region that is between the x-axis and the graph of f and also between lines perpendicular to the x-axis at a and b (with the area of the part of this region below the x-axis taken to be negative) *divided* by the length $b - a$ of the interval (a, b); this is the length of one side of the rectangle whose other side is of length $b - a$ and whose area is equal to the area of the region described. For any integrable function f, the **mean value** of f on the interval (a, b) is

$$\frac{1}{b - a} \int_a^b f(x)\, dx.$$

More generally, the **mean value** of a function f over a set S with respect to a measure m is $\left[\int_S f\, dm \right]/m(S)$. *E.g.*, the mean of xy over the rectangle whose vertices are $(0, 0), (2, 0), (2, 3), (0, 3)$ is

$$\frac{1}{s} \int_s xy\, ds = \frac{1}{6} \int_0^3 \int_0^2 xy\, dx\, dy = \frac{3}{2}.$$

The **mean-square ordinate** of a curve $y = f(x)$ in the interval (a, b) is the mean value of y^2 in the interval, *i.e.*, $\left[\int_a^b y^2\, dx \right]/(b - a)$. See EXPECTED —expected value.

mean-value theorems (or **laws of the mean**) for **derivatives.** For a **function of one variable**, the **mean-value theorem** states that an arc of a differentiable curve has at least one tangent parallel to its secant. When the secant is on the x-axis, this is the same as ROLLE'S THEOREM. *Tech.* If f is continuous for $a \leq x \leq b$ and if f is differentiable on the open interval (a, b), then there is a number c between a and b for which

$$f(b) - f(a) = (b - a)f'(c).$$

The **second mean-value** theorem states that if the functions f and g are continuous on the closed interval $[a, b]$ and are differentiable on the open interval (a, b), and if $g(b) - g(a) \neq 0$ and f' and g' are not both zero at any point of the open interval (a, b), then there is a number x_1 such that $a < x_1 < b$ and

$$\frac{f(b) - f(a)}{g(b) - g(a)} = \frac{f'(x_1)}{g'(x_1)}.$$

[This is also called the **double law of the mean**, the **Cauchy mean-value formula**, and **generalized** (or **extended**) **mean-value theorem**, although the generalized (or extended) mean-value theorem sometimes denotes Taylor's theorem]. The **mean-value theorem for a function of two variables** is that if f is continuous and has continuous first partial derivatives for $x_1 \leq x \leq x_2$, $y_1 \leq y \leq y_2$, there exists numbers ξ and η such that

$$f(x_2, y_2) - f(x_1, y_1)$$
$$= (x_2 - x_1)f_x(\xi, \eta) + (y_2 - y_1)f_y(\xi, \eta),$$

where f_x and f_y denote the partial derivatives of f with respect to x and y, respectively, and $x_1 < \xi < x_2$, $y_1 < \eta < y_2$ [actually, there is such a (ξ, η) that is an interior point of the line segment joining (x_1, y_1) and (x_2, y_2)]. This theorem can be extended to any number of variables. See DIFFERENTIAL.

  mean-value theorems (or **laws of the mean**) **for integrals.** The **first mean-value theorem** for integrals states that the definite integral of a continuous function over a given interval is equal to the product of the width of the interval by some value of the function within the interval. The **second mean-value theorem** for integrals means one of the following: (1) if f and g are both integrable on the interval (a, b) and $f(x)$ is always of the same sign, then

$$\int_a^b f(x)g(x)\,dx = K\int_a^b f(x)\,dx,$$

where K is between the greatest and the least values of $g(x)$, or possibly equal to one of them.

If g is continuous in the interval (a, b), then K may be replaced by $g(k)$, where k is a value of x in the interval (a, b). (2) If in addition to the above conditions, g is a *positive monotonically decreasing function*, the theorem can be written in *Bonnet's form*

$$\int_a^b f(x)g(x)\,dx = g(a)\int_a^p f(x)\,dx,$$

where $a \leq p \leq b$, or, if g is only *monotonic*, in the form

$$\int_a^b f(x)g(x)\,dx = g(a)\int_a^p f(x)\,dx$$

$$+ g(b)\int_p^b f(x)\,dx.$$

  means of a proportion. See PROPORTION.
  weighted mean. See MEAN.

MEAS′UR-A-BLE, *adj.* **measurable function.** (1) A real-valued function f is **(Lebesgue) measurable** if for any real number a the set of all x for which $f(x) > a$ is measurable. Equivalent definitions result if the set of all x satisfying $f(x) \geq a$, or the set of all x satisfying $a \leq f(x) \leq b$ for arbitrary a and b, are required to be measurable (and either of the signs $\leq$ could be replaced by $<$). A bounded function defined on a set of finite measure is integrable if it is measurable. If g is integrable and $|f(x)| \leq g(x)$ for all x, then f is integrable if it is measurable. More generally, if a measure is defined on an algebra of subsets of a set X, then a function with domain X and range contained in a topological space (*e.g.*, the real numbers or the complex numbers) is measurable if the set of all x for which $f(x) \in U$ is measurable for each open set U. (2) If X and Y are topological spaces and $\mathcal{B}$ is a σ-algebra of subsets of X, then a function f of X into Y is measurable if $f^{-1}(U)$ is a member of $\mathcal{B}$ whenever U is an open set in Y. See INTEGRABLE —integrable function, MEASURE— measure of a set.
  measurable set. (1) A set that has measure. See MEASURE —measure of a set. (2) If $\mathcal{B}$ is a σ-algebra of subsets of a topological space X, then each member of $\mathcal{B}$ is said to be measurable. See BOREL —Borel set.

MEAS′URE, *n.* Comparison to some unit recognized as a standard.
  angular measure. A system of measuring angles. See DEGREE, MIL, RADIAN, SEXAGESIMAL.

board measure. A system of measure in which boards one inch or less in thickness are measured in terms of square feet on the side, the thickness being neglected; those thicker than one inch are measured in terms of the number of square feet one inch thick to which they are equivalent.

Carathéodory measure. A function which assigns a nonnegative number $\mu^*(M)$ to each subset of a set M is a **Carathéodory outer measure** if (i) $\mu^*(R) \leq \mu^*(S)$ if R is a subset of S; (ii) $\mu^*(\cup R_i) \leq \Sigma \mu^*(R_i)$ for any sequence of sets $\{R_i\}$; (iii) $\mu^*(R \cup S) = \mu^*(R) + \mu^*(S)$ if the distance between R and S is positive. A set R is **measurable** if $\mu^*(E) = \mu^*(R \cap E) + \mu^*(R' \cap E)$ for any set E, where R' is the complement of R. A bounded set R of real numbers (or a set in n-dimensional Euclidean space) is Lebesgue measurable if and only if

$$m^*(E) = m^*(R \cap E) + m^*(R' \cap E)$$

for any bounded set E, where m^* is the exterior Lebesgue measure. This is called the *Caratheodory test* for measurability. See below, exterior measure.

circular measure. (1) Same as angular measure. (2) The measure of angles by means of radians. See RADIAN.

common measure. Same as COMMON DIVISOR.

convergence in measure. See CONVERGENCE.

cubic measure. The measurement of volumes in terms of a cube whose edge is a standard linear unit, a *unit cube*. *Syn.* volume measure.

decimal measure. See DECIMAL—decimal measure.

dry measure. The system of units used in measuring dry commodities, such as grain, fruit, etc. In the United States, the system is based on the bushel. See DENOMINATE NUMBERS in the appendix.

exterior and interior measure. Let E be a set of points and S be a finite or countably infinite set of intervals (open or closed) such that each point of E belongs to at least one of the intervals (*interval* is used in a generalized sense described below). The **exterior measure** of E is the greatest lower bound of the sum of the measures of the intervals of S, for all such sets S. Let E be contained in an interval I and let E' be the complement of E in I. Then the **interior measure** of E is the difference between the measure of I and the exterior measure of E'. The **interior measure** of a set is the least upper bound of the interior measures of its bounded subsets. If a set E is either open or closed, its

exterior and interior measures are equal and the common value is its measure (see below, Lebesgue measure). Also, the exterior measure of a set E is the greatest lower bound of the measures of open sets which contain E, and the interior measure of E is the least upper bound of the measures of closed sets which are contained in E. The measure of an interval of a straight line is its length. An **interval** I in n-dimensional space is a "generalized rectangular parallelepiped" consisting of all points $x = (x_1, x_2, \cdots, x_n)$ for which $a_i \leq x_i \leq b_i$ for each i, where a_i and b_i are given numbers. The measure of I is the product

$$(b_1 - a_1)(b_2 - a_2) \cdots (b_n - a_n),$$

the same definition being used if the interval is open, or partly open and partly closed. *Syn.* Lebesgue exterior (interior) measure, outer (inner) measure. See CONTENT, INTERVAL, and above and below. Carathéodory measure, Lebesgue measure, measure of a set.

Haar measure. Let G be a locally compact topological group. A *Haar measure* is a measure which assigns a nonnegative real number $m(E)$ to each set E of the σ-ring S generated by the *compact* subsets of G and which has the properties: (i) there is a member of S for which m is not zero; (ii) either m is left-invariant or m is right-invariant—*i.e.*, $m(aE) = m(E)$ for each element a and member E of S, or $m(Ea) = m(E)$ for each a and E, where aE and Ea are, respectively, the set of all ax for x in E and the set of all xa for x in E. Any locally compact topological group has a left-invariant Haar measure and a right-invariant Haar measure and each is unique to within a multiplicative constant.

land measure. See ACRE, and DENOMINATE NUMBERS in the appendix.

Lebesgue measure. A *bounded* set in Euclidean space is (Lebesgue) measurable if its *exterior* and *interior measures* are equal, the common value being called the **(Lebesgue) measure** of the set. For an *unbounded* set S, let W_I be the set of all points belonging to both S and a bounded interval I. The set S is (Lebesgue) measurable if, and only if, W_I is measurable for each I; then the **(Lebesgue) measure** of S is the least upper bound of the measures of the sets W_I if these are bounded, S being of infinite measure otherwise. Each of the following is a necessary and sufficient condition for a set B to be Lebesgue measurable: (1) For each positive number ϵ, there is a closed set F and an open set G such that $F \subset B \subset G$ and the measure of $G - F$

is less than ϵ; (2) for each positive number ϵ, there is an open set G such that $B \subset G$ and the exterior measure of $G - B$ is less than ϵ; (3) for each positive number ϵ, there is a closed set F such that $F \subset B$ and the exterior measure of $B - F$ is less than ϵ; (4) the measure of any interval I is the sum of the exterior measures of $I \cap B$ and $I \cap B'$, where B' is the complement of B; (5) the exterior measure of any set S is the sum of the exterior measures of $S \cap B$ and $S \cap B'$. The collection of all Lebesgue measurable sets in a given Euclidean space is a σ-ring [see RING—ring of sets]. See exterior and interior measure, measure zero, INTERVAL, MEASURABLE —measurable function.

linear measure. Measurement along a line, the line being either straight or curved.

liquid measure. The system of units ordinarily used in measuring liquids. See DENOMINATE NUMBERS in the appendix.

measure of central tendency. The *mean*, the *mode*, the *median*, and the *geometric mean* are commonly used.

measure of dispersion. Same as DEVIATION.

measure of a set. Let R be a collection of sets which constitute a *ring* (or a *semiring*) *of sets*. A **finitely additive measure** is a set function m which associates a number with each set of R, for which $m(\varnothing) = 0$, where $\varnothing$ is the empty set, and for which $m(A \cup B) = m(A) + m(B)$ if A and B are members of R for which $A \cap B = \varnothing$. The value of the measure of a set may (for example) be required to be a nonnegative real number (or $+\infty$), a real number (or $+\infty$ or $-\infty$), or a complex number (see EXTENDED —extended real-number system). A **countably additive measure** is a finitely additive measure m for which $m(\bigcup_1^\infty S_n) = \sum_1^\infty m(S_n)$ if S_n is a member of R for each positive integer n, the sets S_n and S_m have no common points if $m \neq n$, and the union $\bigcup_1^\infty S_n$ of all sets $\{S_n\}$ is a member of R. A set S in R is said to have σ-**finite measure** if there is a sequence $\{S_n\}$ of sets in R such that $S \subset \bigcup_1^\infty S_n$ and $m(S_n) \neq +\infty$ (for each n). If each set of R has σ-finite measure, the measure is said to be σ-**finite.** If R is a σ-algebra of subsets of a set X, it is customary to call a countably additive measure on R a **measure, signed measure,** or **complex measure,** according as the values of the measure are nonnegative real numbers (or $+\infty$), real numbers (or $+\infty$ or $-\infty$), or complex numbers. Then a σ-**finite measure** is a measure for which X has σ-finite measure. See above, Carathéodory measure, Lebesgue measure, and below, product measure.

measure of a spherical angle. The plane angle formed by the tangents to the sides of the spherical angle at their points of intersection.

measure ring and **measure algebra.** If a measure is defined on a σ-ring of subsets of a space X, then these measurable subsets of X are a **measure ring** if equality is defined by the definition that $A = B$ if the measure of the symmetric difference of A and B is zero (see RING—ring of sets). That is, the measure ring is the *quotient ring* of the σ-ring modulo the ideal of sets of measure zero. A measure ring is a **measure algebra** if there is a measurable set which contains all the measurable sets (it is then a Boolean algebra). The set of elements of finite measure in a measure ring is a metric space if the distance between sets A and B is defined to be the measure of the symmetric difference of A and B.

measure zero. A set is said to be of measure zero if it has measure and the measure is zero. See above, measure of a set. For Lebesgue measure and a set of points on a line or in n-dimensional Euclidean space, the set is of measure zero if and only if, for any positive number ϵ, there exists a finite or countably infinite set of intervals (open or closed) such that each point of the set is contained in at least one of the intervals, and the sum of the measures of the intervals is less than ϵ [see INTERVAL (2)]. A property is said to hold **almost everywhere, a.e.,** or for **almost all** points, if it holds for all points except those of a set of measure zero. E.g., a function is continuous *almost everywhere* if the set of points at which it is discontinuous is of measure zero. See MEASURABLE —measurable set.

probability measure. See PROBABILITY—probability function.

product measure. Let m_1 and m_2 be measures defined on σ-rings of subsets of spaces X and Y, respectively, and let $X \times Y$ be the Cartesian product whose points consist of all pairs (x, y) with x in X and y in Y. The *product measure* of m_1 and m_2 is the measure defined on the σ-ring generated by the "rectangles" $A \times B$ of $X \times Y$ for which A and B are measurable and the measure of $A \times B$ is the product of the measures of A and B.

signed measure. See above, measure of a set.

square measure. The measure of areas of surfaces in terms of a square whose side is a standard linear unit, a *unit square.* Syn. surface measure.

surveyor's measure. See DENOMINATE NUMBERS in the appendix.

wood measure. See DENOMINATE NUMBERS in the appendix.

MEAS′URE-MENT, *n.* The act of measuring.

median of a group of measurements. See MEDIAN.

ME-CHAN'IC, *n.* **mechanic's rule.** A rule for extracting square roots. The rule is as follows: Make an estimate of the root, divide the number by this estimate, and take for the approximate square root the arithmetic mean (average) of the estimate and the quotient thus obtained. If a more accurate result is desired, repeat the process. *Algebraically*, if a is the estimate, e the error, and $(a + e)$ the required number, divide $(a + e)^2$, that is $a^2 + 2ae + e^2$, by a and take the average between a and the quotient. This gives $a + e + e^2/(2a)$. The error is $e^2/(2a)$, which is very small if e is small. *E.g.*, if one estimates $\sqrt{2}$ to be 1.5, the mechanic's rule gives the root as 1.4167. The error in this is less than 0.003. The error in a second application of the rule is less than $(0.003)^2/2$ or 0.0000045. *Newton's method* applied to the equation $x^2 - B = 0$, given the approximation a for a root, gives the same next approximation as the mechanic's rule; *i.e.*, $\frac{1}{2}(a + B/a)$. See NEWTON—Newton's method of approximation.

ME-CHAN'I-CAL, *adj.* **mechanical integration.** See INTEGRATION —mechanical integration.

ME-CHAN'ICS, *n.* The mathematical theory of the motions and tendencies to motion of particles and systems under the influence of forces and constraints; the study of motions of masses and of the effect of forces in causing or modifying these motions. Usually divided into *kinematics* and *dynamics*.

  analytical mechanics. A mathematical structure of mechanics whose present-day formulation is largely due to J. L. Lagrange (1736–1813) and W. R. Hamilton (1805–1865). This structure is also known as *theoretical mechanics*. It uses differential and integral calculus as tools.

  mechanics of fluids. A body of knowledge including the theory of gases, hydrodynamics, and aerodynamics.

  theoretical mechanics. See above, analytical mechanics.

ME'DI-AN, *adj., n.* The middle measurement when items are arranged in order of size; or, if there is no middle one, then the average of the two middle ones. If five students make the grades 15, 75, 80, 95, and 100, the median is 80. For a *continuous random variable* with probability density function f, the median is the number M for which

$$\int_{-\infty}^{M} f(x)\, dx = \int_{M}^{\infty} f(x)\, dx = \frac{1}{2}.$$

median of a trapezoid. The line joining the midpoints of the nonparallel sides. *Syn.* midline.

median of a triangle. A line joining a vertex to the midpoint of the opposite side. The medians of a triangle intersect at a point which is the **centroid** of the triangle (sometimes called the **median point**). This point is two-thirds of the distance from a vertex to the opposite side along a median.

MEET, *n.* See LATTICE, INTERSECTION.

MEG- or **MEGA-.** Prefixes that signify multiplication by 1,000,000, as for *megohm* or *megavolt*. A 50 megaton bomb is one equivalent to 50,000,000 tons of TNT.

MELLIN, Robert Hjalmar (1854–1933). Finnish analyst and mathematical physicist.

  Mellin inversion formulas. The associated formulas

$$g(x) = \frac{1}{2\pi i} \int_{\sigma - i\infty}^{\sigma + i\infty} x^{-s} f(s)\, ds,$$

$$f(s) = \int_{0}^{\infty} x^{s-1} g(x)\, dx,$$

each of which gives the inverse of the other under suitable conditions of regularity. See FOURIER —Fourier transform, LAPLACE —Laplace transform.

MEM'BER, *n.* **member of an equation.** The expression on one (or the other) side of the equality sign. The two members of an equation are distinguished as the *left* or *first* and the *right* or *second* member.

  member of a set. One of the individual objects that belong to a set. One writes "$x \in S$" and "$x \notin S$" for "x is a member of S" and "x is *not* a member of S." *Syn.* element of a set.

MEM'O-RY, *adj.* **memory component.** See STORAGE —storage component.

MENELAUS of Alexandria (first cent. A.D.). **Menelaus' theorem.** If, in the triangle ABC, points P_1, P_2, P_3 are on the lines that contain the sides AB, BC, CA, respectively, then P_1, P_2, P_3 are collinear if, and only if,

$$\frac{AP_1}{P_1B} \frac{BP_2}{P_2C} \frac{CP_3}{P_3A} = -1.$$

It is assumed that none of P_1, P_2, P_3 is one of A, B, C, and that the lines containing the sides of the triangle are directed. See CEVA—Ceva's theorem, DIRECTED—directed line.

MEN'SU-RA'TION, *n.* The measuring of geometric magnitudes, such as the lengths of lines, areas of surfaces, and volumes of solids.

MER'CAN-TILE, *adj.* **mercantile rule.** Same as MERCHANT'S RULE. See MERCHANT.

MERCATOR, Gerardus (Latinized form of **Gerhard Kremer**) (1512–1594). Flemish geographer and map maker. He avoided being burnt alive for heresy by escaping, but later worked for emperor Charles V.

Mercator chart. A map made by use of *Mercator's projection.* A straight line on the plane corresponds to a curve on the sphere cutting meridians at constant angle. Magnification of area on the sphere increases with increasing distance from the equator. See below, Mercator's projection.

Mercator's projection. A correspondence between points of the (x, y)-plane and points on the surface of the sphere, given by

$$x = k\theta, \qquad y = k \operatorname{sech}^{-1}(\sin \phi) = k \log \tan \frac{\phi}{2},$$

where θ is the angle of longitude and ϕ is the angle of colatitude. The correspondence is a conformal one, except for singular points at the poles.

MERCATOR, Nicolaus (c. 1620–1687). Mathematician, astronomer, and engineer who was born Nicolaus Kaufmann in Holstein (then part of Denmark), but spent much of his life in England. He was one of several mathematicians who independently discovered the formula

$$\log_e(1 + x) = x - x^2/2 + x^3/3$$

$$-x^4/4 + x^5/5 - \cdots,$$

valid for $-1 < x \leq 1$. This formula follows from the fact that the area under the hyperbola $y = 1/(1 + x)$ between 0 and x is equal to $\log_e(1 + x)$.

MER'CHANT, *n.* **merchant's rule.** A rule for computing the balance due on a note after partial payments have been made. The method is to find the amount of each partial payment at the settlement date and subtract the sum of these from the value of the face of the note at the same date. *Syn.* mercantile rule.

ME-RID'I-AN, *n.* On the *celestial sphere*, a meridian is a great circle passing through the zenith and the north and south line in the plane of the horizon (see HOUR—hour angle and hour circle). On the *earth*, a meridian is a great circle on the surface of the earth passing through the geographic poles. The **local meridian** of a point on the earth is the meridian which passes through that point. The **principal meridian** is the meridian from which longitude is reckoned (usually the meridian through the transit-circle of the Royal Observatory of Greenwich, England, although observers frequently use the meridian through the capital of their country). The principal meridian is also called the *first, prime, zone,* and *zero meridian.*

meridian curve on a surface. A curve C on a surface such that the spherical representation of C lies on a great circle on the unit sphere.

meridian sections, or **meridians.** See SURFACE—surface of revolution.

MER-O-MOR'PHIC, *adj.* **meromorphic function.** A function of the complex variable z is said to be *meromorphic* in a domain D if it is *analytic* in D except for poles; *i.e.,* all singularities in D are poles.

MERSENNE, Marin (1588–1648). French theologian, philosopher, and number theorist.

Mersenne number. A number of the type $M_p = 2^p - 1$, where p is a prime. Mersenne asserted that the only primes for which M_p is a prime are $2, 3, 5, 7, 13, 17, 19, 31, 67, 127, 257$. Actually, M_{67} and M_{257} are not primes. There are 32 values of p for which M_p is known to be prime, which include

$$61, 89, 107, 521, 607, 1279, 2203, 2281, 3217,$$

$$\cdots, 132049, 216091, 756839.$$

See FERMAT—Fermat numbers.

MESH, *n.* See PARTITION—partition of an interval.

MES'O-KUR'TIC, *adj.* **mesokurtic distribution.** See KURTOSIS.

MET'A-COM-PACT', *adj.* **metacompact space.** A topological space T which has the property that, for any family F of open sets whose union contains T, there is a *point-finite* family F^* of open sets whose union contains T and which are such that each member of F^* is contained in a member of F; T is **countably metacompact** if F has this property whenever F is countable. A metacompact Hausdorff space need not be even countably paracompact. A T_1 topological space is compact if and only if it is countably compact and metacompact. See COMPACT, FINITE—locally finite family of sets, PARACOMPACT.

ME′TER (ME′TRE), *n.* The basic unit of linear measure of the metric system and the international system of units (see SI). Defined in 1790 as one 10-millionth of the distance from the North Pole to the equator at sea-level through Paris (supposedly the distance at 0° C between two marks on any of three platinum bars constructed in 1793, one of which is in Paris). Now defined as 1,650,763.73 wavelengths in vacuum of the orange line in the spectrum of the Krypton-86 atom. It is equal to 39.3701-inches.

METH′OD, *n.* **method of exhaustion.** See EX-HAUSTION.
    method of least squares. See LEAST.

MET′RIC, *adj., n.* **metric density.** Let E be a measurable subset of the line (or of n-dimensional Euclidean space). The metric *density* of E at a point x is the limit (if it exists) of $m(E \cap I)/m(I)$ as the length (or measure) $m(I)$ of I approaches zero, where I is an interval which contains x. The metric density of E is 1 at all points of E except a set of measure zero; it is 0 at all points of the complement of E except a set of measure zero.
    metric space. A set T such that to each pair x, y of its points there is associated a nonnegative real number $\rho(x, y)$, called their *distance*, which satisfies the conditions: (1) $\rho(x, y) = 0$ if, and only if, $x = y$; (2) $\rho(x, y) = \rho(y, x)$; (3) $\rho(x, y) + \rho(y, z) \ge \rho(x, z)$. The function $\rho(x, y)$ is said to be a **metric** for T. The plane and three-dimensional space are metric spaces with the usual distance. Hilbert space is also a metric space. A topological space is **metrizable** if a distance between points can be defined so the space is a metric space such that open sets in the original space are open sets in the metric space, and conversely; *i.e.*, there exists a topological transformation between the given space and a metric space. A compact Hausdorff space is metrizable if and only if it satisfies the *second axiom of countability*; or if and only if there is a countable set S of continuous real-valued functions such that, for any two points x and y, $f(x) \ne f(y)$ for at least one f in S. A regular T_1 topological space is metrizable if it satisfies the second axiom of countability (Urysohn's theorem). A topological space is metrizable if and only if it is a regular T_1 topological space whose topology has a base B which is the union of a

countable number of classes $\{B_n\}$ of open sets which have the property that, for each point x and class B_n, there is a neighborhood of x which intersects only a finite number of the members of B_n.
    metric system. A system of units for which the units of length, time, and mass are the meter, second, and kilogram, or multiples of these by powers of 10. It was adopted in France as part of the upheaval from the French Revolution, recommended in 1791 by a committee of the French Academy (including Lagrange and Laplace) and after much delay put into effect by the National Academy. It is now in general use in most other civilized countries, except the English-speaking countries, and is now almost universally used for scientific measurements. The unit of surface is the *are* (100 square meters) and the theoretical unit of volume is the *stere* (one cubic meter), although the *liter* (one cubic decimeter) is most commonly used. See SI and DENOMINATE NUMBERS in the appendix. The prefixes *deca-* or *deka-* (*da*), *hecto-* (*h*), *kilo-* (*k*), *myria-* (*my*), *mega-* (*M*), *giga-* (*G*), *and tera-* (*T*) are used on the above units to designate multiplication by 10, 10^2, 10^3, 10^4, 10^6, 10^9, and 10^{12}. The prefixes *deci-* (*d*), *centi-* (*c*), *milli-* (*m*), *micro-* (μ), *nano-* (*n*), *pico-* (*p*), *femto-* (*f*), *and atto-* (*a*) are used to designate multiplication by 10^{-1}, 10^{-2}, 10^{-3}, 10^{-6}, 10^{-9}, 10^{-12}, 10^{-15}, and 10^{-18}.

MET-RIZ′A-BLE, *adj.* **metrizable space.** See METRIC—metric space.

MEUSNIER, Jean Baptiste Marie (1754–1793). French engineer, chemist, physicist, and differential geometer.
    Meusnier's theorem. The theorem that the center of curvature of a curve on a surface S is the projection, on its osculating plane, of the center of curvature of the normal section tangent to the curve at the point. More graphically, if a segment equal to twice the radius of normal curvature for a given direction at a point of S is laid off from the point on the normal to S, and a sphere is drawn with the segment as diameter, then the circle of intersection of the sphere by the osculating plane of a curve on S in the given direction at the point is the circle of curvature of the curve.

MICRO or **MICR.** Prefixes indicating a millionth part, as **microhm, microsecond**.

MIDAC. An automatic digital computing machine built at the Willow Run Research Center. MIDAC is an acronym for *Michigan Digital Automatic Computer*.

MIDLINE, *n.* **midline of a trapezoid.** See ME-DIAN—median of a trapezoid.

MIDPOINT, *n.* **midpoint of a line segment.** The point that divides the line segment into two equal parts. See BISECT.

MIL, *n.* A unit of angle measure, equal to $\frac{1}{6400}$ of a complete revolution, $0.05625°$, and nearly $\frac{1}{1000}$ of a radian. Used by U.S. artillery.

MILE, *n.* A unit of linear measure equal to 5280 feet or 320 rods. Derived from an ancient Roman measure of 1000 paces.

geographical mile. See NAUTICAL—nautical mile.

nautical mile. See KNOT, NAUTICAL.

MILLI. A prefix indicating a thousandth part, as **millibar, millimeter,** or **milligram.**

MIL′LION, *n.* One thousand thousands (1,000,000).

MILNOR, John Willard (1931–). American algebraic and differential topologist and Fields Medal recipient (1962). Showed there are differential manifolds that are homeomorphic to the seven sphere but are not diffeomorphic to it.

MINAC. An automatic digital computing machine built at the California Institute of Technology. MINAC is an acronym for *Minimal Automatic Computer.*

MIN′I-MAL, *adj.* **adjoint minimal surfaces.** Two associate minimal surfaces with parameters α_1 and α_2 differing by $\pi/2$. See below, associate minimal surfaces.

associate minimal surfaces. When the minimal curves of a minimal surface are parametric, the coordinate functions are of the form $x = x_1(u) + x_2(v)$, $y = y_1(u) + y_2(v)$, $z = z_1(u) + z_2(v)$. The related equations $x = e^{i\alpha}x_1(u) + e^{-i\alpha}x_2(v)$, $y = e^{i\alpha}y_1(u) + e^{-i\alpha}y_2(v)$, $z = e^{i\alpha}z_1(u) + e^{-i\alpha}z_2(v)$ define a family of minimal surfaces, called **associate minimal surfaces,** with parameter α.

double minimal surface. A one-sided minimal surface; a minimal surface S such that, through each of its points P, there exists a closed path C on S having the property that, when a variable point traverses C returning to P, the positive direction of the normal is reversed. See SURFACE—surface of Henneberg. *Syn.* one-sided minimal surface.

minimal curve. A curve for which the linear element ds vanishes identically. With a Euclidean metric, $ds^2 = dx_1^2 + dx_2^2 + \cdots + dx_n^2$, this can occur only if the curve reduces to a point or if at least one of the coordinate functions is imaginary. See below, minimal straight line. *Syn.* isotropic curve, curve of zero length.

minimal (or minimum) equation. See ALGEBRAIC—algebraic number, CHARACTERISTIC—characteristic equation of a matrix.

minimal straight line. A minimal curve which is an imaginary straight line. There is an infinitude of them through each point of space; their direction components are $(1 - a^2)/2$, $i(1 + a^2)/2$, a, where a is arbitrary. See above, minimal curve.

minimal surface. A surface whose mean curvature vanishes identically; a surface for which the first variation of the area integral vanishes. A minimal surface does not necessarily minimize the area spanned by a given contour; but if a smooth surface S minimizes the area, then S is a minimal surface.

one-sided minimal surface. Same as DOUBLE MINIMAL SURFACE. See above.

MIN′I-MAX, *n.* Same as SADDLE POINT. See SADDLE.

minimax theorem. (1) See COURANT. (2) For a finite two-person zero-sum game, with *payoff matrix* (a_{ij}), where $i = 1, 2, \cdots, m$ and $j = 1, 2, \cdots, n$, if the maximizing player uses *mixed strategy* $X = (x_1, x_2, \cdots, x_m)$ and the minimizing player uses *mixed strategy* $Y = (y_1, y_2, \cdots, y_n)$, then the expected value of the payoff is given by the expression

$$\sum_{j=1}^{n} \sum_{i=1}^{m} a_{ij} x_i y_j = v_{X,Y} = E(X, Y).$$

It is easy to see that $\max_X(\min_Y v_{X,Y}) \leq \min_Y(\max_X v_{X,Y})$. The *minimax theorem*, fundamental in the theory of games, states that the sign of equality must hold for every finite two-person zero-sum game:

$$\max_X(\min_Y v_{X,Y}) = \min_Y(\max_X v_{X,Y}) = v.$$

This result can be extended to the class of continuous two-person zero-sum games with continuous payoff functions and to certain other classes of infinite games. In particular, if a two-person zero-sum game has a *saddle point* at (x_0, y_0) then v is the value of the payoff function at (x_0, y_0). See GAME—value of a game, SADDLE—saddle point of a game.

MIN′I-MUM, *adj., n.* [*pl.* **minima** or **mini-mums**]. See various headings under MAXIMUM.

minimum-variance unbiased estimator. See UNBIASED —unbiased estimator.

MINKOWSKI, Hermann (1864–1909). Number theorist, algebraist, analyst, and geometer. Born in Russia, lived in Switzerland and Germany. Created the geometry of numbers. Provided the four-dimensional geometric space-time frame-work for relativity theory.

Minkowski distance function. Relative to a convex body B of which the origin O is an interior point, the Minkowski *distance function F* is defined for each P (other than O) in the space by letting $F(P)$ be the greatest lower bound of the ratio $\rho(O, P)/\rho(O, Q)$, where Q is a point of B on the ray OP and $\rho(O, P)$ denotes the distance between O and P. We then define $F(O)$ to be 0 and have $F(P) < 1$ if P is interior to B, $F(P) = 1$ if P is on the boundary of B, and $F(P) > 1$ if P is exterior to B. The function F is a convex function of P. Two **polar reciprocal convex bodies** are any two convex bodies, each containing the origin in its interior, such that the *support function* of each is the *distance function* of the other. See SUPPORT —support function.

Minkowski's inequality. Either of the in-equalities

$$(1) \quad \left[\sum_1^n |a_i + b_i|^p\right]^{1/p}$$

$$\leq \left[\sum_1^n |a_i|^p\right]^{1/p} + \left[\sum_1^n |b_i|^p\right]^{1/p},$$

where n may be $+\infty$, or

$$(2)$$

$$\left[\int_\Omega |f + g|^p \, d\mu\right]^{1/p}$$

$$\leq \left[\int_\Omega |f|^p \, d\mu\right]^{1/p} + \left[\int_\Omega |g|^p \, d\mu\right]^{1/p},$$

which are valid if $p \geq 1$ and both $|f|^p$ and $|g|^q$ are integrable on Ω. The integrals may be Riemann integrals, or μ may be a measure defined on a σ-algebra of subsets of Ω (see INTEGRABLE —integrable function). The numbers in (1) or the functions in (2) may be real or complex. Either of these inequalities is easily deduced from the other or from *Hölder's inequalities*. The inequalities are reversed for $0 < p \leq 1$, if all val-

ues of a_i, b_i, f, g are nonnegative. These are also **triangle inequalities** if $p \geq 1$. *E.g.*, if l^p is the space of all sequences $\mathbf{a} = (a_1, \cdots, a_n)$ and $\|\mathbf{a}\| = \left[\sum_1^n |a_i^p|\right]^{1/p}$, then (1) becomes $\|\mathbf{a} + \mathbf{b}\| \leq \|\mathbf{a}\| + \|\mathbf{b}\|$.

MI′NOR, *adj., n.* **minor arc of a circle.** See SECTOR —sector of a circle.

minor axis of an ellipse. The smaller axis of the ellipse.

minor of an element in a determinant. The determinant, of next lower order, obtained by striking out the row and column in which the element lies. This is sometimes called the **complementary minor.** The minor, taken with a positive or negative sign according as the *sum* of the position numbers of the row and column stricken out of the original determinant is even or odd, is the **signed minor,** or **cofactor,** of the element. *E.g.*, the minor of b_1 in the determinant

$$\begin{vmatrix} a_1 & a_2 & a_3 \\ b_1 & b_2 & b_3 \\ c_1 & c_2 & c_3 \end{vmatrix}$$

is the determinant

$$\begin{vmatrix} a_2 & a_3 \\ c_2 & c_3 \end{vmatrix}$$

and the cofactor of b_1 is

$$-\begin{vmatrix} a_2 & a_3 \\ c_2 & c_3 \end{vmatrix}.$$

MIN′U-END, *n.* The quantity from which another quantity is to be subtracted. See SUBTRACTION.

MIN′US, *adj.* A word used between two quantities to state that the second is to be subtracted from the first. The statement 3 minus 2, written $3 - 2$, means 2 is to be subtracted from 3. *Minus* is also used as a synonym for *negative*.

MIN′UTE, *n.* (1) Sixty seconds. See SECOND. (2) One-sixtieth of a degree (in the sexagesimal system of measuring angles). See SEXAGESIMAL.

MITTAG-LEFFLER, Magnus Gösta (1846–1927). Swedish complex analyst who founded the journal *Acta Mathematica* and established a mathematics library and research institute in Djursholm, near Stockholm.

Mittag-Leffler theorem. A theorem about the existence of meromorphic functions with prescribed poles and principal parts. Let $\{z_1, z_2, \cdots\}$ be a sequence of complex numbers for which $\lim_{n \to \infty} |z_n| = \infty$. For each n, let P_n be a polynomial with no constant term. Then there is a function f that is meromorphic in the whole plane, f has poles only at the points $\{z_1, z_2, \cdots\}$, and the *principal part* of f at z_n is $P_n[1/(z - z_n)]$. The most general meromorphic function with these properties is of the form

$$ f(z) = \sum_{n=1}^{\infty} \left[P_n \left(\frac{1}{z - z_n} \right) + p_n(z) \right] + g(z), $$

where each p_n is a polynomial, g is an entire function, and the series converges uniformly in every bounded region where f is analytic.

MIXED, *adj.* **mixed decimal.** See DECIMAL.

    mixed number, expression. In *arithmetic*, the sum of an integer and a fraction, as $2\frac{3}{4}$. In *algebra*, the sum of a polynomial and a rational algebraic fraction, as

$$ 2x + 3 + \frac{1}{x + 1}. $$

    mixed partial derivative. See PARTIAL.

MKS SYSTEM. A system of units for which the units of length, time, and mass are the meter, second, and kilogram. See CGS SYSTEM, METRIC —metric system, SI.

MNE-MON'IC, *adj.* Assisting the memory; relating to the memory.

    mnemonics or mnemonic devices. Any scheme or device to aid in remembering certain facts. *E.g.,* (1) $\sin(x + y)$ is equal to the sum of two mixed terms in sine and cosine, whereas $\cos(x + y)$ is equal to the difference of unmixed terms; (2) the derivatives of all cofunctions (*trigonometric*) take a negative sign.

MÖBIUS, August Ferdinand (1790–1868). German geometer, topologist, number theorist, statistician, and astronomer.

    Möbius function. The function of the positive integers defined by $\mu(1) = 1$; $\mu(n) = (-1)^r$ if $n = p_1 p_2 \cdots p_r$ where $p_1, \cdots, p_r$ are distinct positive prime numbers, and $\mu(n) = 0$ for all other positive integers. It follows that $\mu(n)$ is also the sum of the primitive nth roots of unity. See RIEMANN —Riemann hypothesis.

    Möbius strip. The one-sided surface formed by taking a long rectangular strip of paper and pasting its two ends together after giving it half a twist. In addition to being one-sided, a Möbius strip has the unusual property that it remains one piece if it is "cut in two" along the center line. See SURFACE —one-sided surface.

    Möbius transformation. A transformation (in the complex plane) of the form $\omega = (az + b)/(cz + d)$, with $ad - bc \neq 0$.

MODE, *n.* The member of a series of measurements or observations that occurs most often, if there is such a member; there is no useful definition if more than one member occurs most often. If more students in a given class make 75 than any other one grade, then 75 is the mode. For a *continuous random variable* with probability density function f, the mode is the point at which f has its greatest value, if there is exactly one such point. Sometimes any point at which f has a local maximum is called a mode, but multimodality is unusual in practice. A mode need not be a mean or median.

MOD'I-FIED, *p.* **modified Bessel functions.** See BESSEL.

MOD'U-LAR, *adj.* **(elliptic) modular function.** A function which is *automorphic* with respect to the modular group (or a subgroup of the modular group) and which is single-valued and analytic in the upper half of the complex plane, except for poles. Usually suitable additional restrictions are made, so that such a function is a rational function of the modular function J defined by

$$ J(\tau) = \frac{4}{27} \frac{\left(\vartheta_3^8 - \vartheta_2^4 \vartheta_4^4 \right)^3}{\left(\vartheta_2 \vartheta_3 \vartheta_4 \right)^8}, $$

where ϑ_i denotes the function of the parameter τ obtained by setting $z = 0$ in the theta function $\vartheta_i(z)$. Also:

$$ J(\tau) = \frac{g_2^3}{g_2^3 - 27 g_3^2}, $$

where $g_2 = 60 \Sigma'(m\omega + n\omega')^{-4}$, $g_3 = 140 \Sigma'(m\omega + n\omega')^{-6}$ and $\tau = \omega'/\omega$ (the primes indicate that the summations are for all integral values of m and n except $m = n = 0$). The modular functions $\lambda(\tau) = \vartheta_2^4/\vartheta_3^4$ [also denoted by $f(\tau)$], $g(\tau)$, and $h(\tau) = -f(\tau)/g(\tau)$ are also of considerable interest; J and λ are related by

$$ 27J\lambda^2(1 - \lambda)^2 = 4(1 - \lambda + \lambda^2)^3. $$

modular arithmetic. See CONGRUENCE.

modular group. The transformation group which consists of all transformations

$$w = \frac{az + b}{cz + d}$$

with $ad - bc = 1$, where a, b, c, and d are real integers. Such transformations map the upper half (and the lower half) of the complex plane onto itself, and map real points into real points.

modular lattice. See LATTICE.

MOD'ULE, *n.* (1) Let S be a set such as a ring, integral domain, or algebra, which is a group with respect to an operation called addition (it may have other operations defined, such as multiplication and scalar multiplication). A module M of S is a subset of S which is a group with respect to addition (this is equivalent to stating that $x - y$ belongs to M whenever x and y belong to M). (2) A generalization of the concept of vector space, but with coefficients in a ring. A **left module** over a ring R (or a **left R-module**) is a set M that is a commutative group with the group operation denoted by $+$, and for which the "product" rx is a member of M whenever r is in R and x is in M and satisfies the conditions: $r(x + y) = rx + ry$, $(r_1 + r_2)x = r_1 x + r_2 x$, $r_1(r_2 x) = (r_1 r_2)x$. Similarly, a **right module** is defined by using a "product" xr and reversing the order of multiplication in each of the preceding equalities. If R is a ring with unit element 1 and if $1 \cdot x = x$ for all x in M, then M is a **unital left module.** For example, every commutative group is a unital left module over the ring of integers, and every vector space is a unital left module over the field of its coefficients. A module is **irreducible** if it contains no proper submodules except the module containing only 0. A left module is **cyclic** if it has a member x such that every member is of the form rx for r in R; it is **finitely generated** if there is a finite subset $x_1, \cdots, x_n$ such that every member is of the form $r_1 x_1 + \cdots + r_n x_n$. If R is a principal ideal ring and M is a finitely generated R-module, then M is the direct sum (Cartesian product) of a finite number of cyclic submodules.

MOD'U-LUS, *n.* [*pl.* **moduli**] **bulk modulus** (*Elasticity*) The ratio of the compressive stress to the cubical compression. It is connected with Young's modulus E and Poisson's ratio σ by the formula $k = \dfrac{E}{3(1 - 2\sigma)}$. The bulk modulus k is positive for all physical substances. *Syn.* compression modulus.

modulus of logarithms. See LOGARITHM.

modulus of a complex number. The numerical length of the vector representing the complex number (see COMPLEX —complex number). The modulus of a complex number $a + bi$ is $\sqrt{a^2 + b^2}$, written $|a + bi|$. If the number is in the form $r(\cos \beta + i \sin \beta)$ with $r \geq 0$, the modulus is r. The modulus of $4 + 3i$ is 5; the modulus of $1 + i = \sqrt{2}(\cos \frac{1}{4}\pi + i \sin \frac{1}{4}\pi)$ is $\sqrt{2}$. Important properties are that, for any complex numbers w and z, $|wz| = |w||z|$ and $|w + z| \leq |w| + |z|$. *Syn.* absolute value.

modulus of a congruence. See CONGRUENCE.

modulus of an elliptic integral and of an elliptic function. See ELLIPTIC —elliptic integral, and Jacobian elliptic functions.

modulus of rigidity. See RIGIDITY.

Young's modulus. The constant introduced in the theory of elasticity by Thomas Young in 1807. It characterizes the behavior of elastic substances. If the stress acting in the cross section of a thin rod is T and the small elongation produced by it is e, then $T = Ee$, where E is **Young's modulus** of tension. For many substances the modulus of tension differs from the corresponding compression modulus. Young's modulus and Poisson's ratio (*q.v.*) are found to be sufficient to completely characterize the elastic state of an isotropic substance.

MOEBIUS. See MÖBIUS.

MOLE, *n.* A mole of a substance is the amount whose mass in some unit is equal to the molecular weight of the substance. *E.g.*, a gram-mole (often called a mole) is the amount whose mass in grams is numerically equal to the molecular weight.

MO'MENT *n.* **method of moments.** A method of estimating the parameters of a distribution by relating the parameters to moments. If k parameters are to be estimated, equations are found to express the first k moments in terms of the k parameters. Given sample values of the moments, estimates of the parameters are then obtained from these equations. This method usually yields a simple procedure, but does not in general yield *minimum-variance unbiased estimates*. *E.g.*, for a distribution with mean μ and variance σ^2, $E(X) = \mu$ and $E(X^2) = \mu^2 + \sigma^2$, so $\sigma^2 = E(X^2) - [E(X)]^2$ and $\mu = E(X)$; σ^2 and μ are estimated from sample values $\{x_1, \cdots, x_n\}$ by replacing $E(X)$ by $(\Sigma x_i)/n$ and $E(X^2)$ by $(\Sigma x_i^2)/n$. See below, moment of a distribution.

moment of a distribution. For a random variable X or the associated distribution function, the kth moment μ_k about a is the *expected value* of $(X - a)^k$ whenever this exists. For a discrete random variable with values $\{x_n\}$ and probability function p,

$$\mu_k = \sum (x_i - a)^k p(x_i),$$

if this sum is finite or converges absolutely; for a continuous random variable with probability-density function f,

$$\mu_k = \int_{-\infty}^{\infty} (x - a)^k f(x)\, dx,$$

if this integral converges absolutely. If a is the mean, then μ_k is the **central moment** of order k. The moments about zero are usually called simply the **moments**. The first moment is the *mean* and the second moment about the mean is the *variance*. In terms of moments $\{u_n\}$, the central moments are $u_1, u_2 - u_1^2, u_3 - 3u_2 u_1 + 2u_1^3, u_4 - 4u_3 u_1 + 6u_2 u_1^2 - 3u_1^4, \cdots$. The kth **factorial moment** is the expected value of $X(X - 1)(X - 2)\cdots(X - r + 1)$. See ABSOLUTE —absolute moment, PARALLEL —parallel-axis theorem, SAMPLE —sample moment, and below, moment-generating function.

moment of a force. The moment of a force about a line is the product of the projection of the force on a plane perpendicular to the line and the perpendicular distance from the line to the line of action of the force. The **moment of a force F about a point O** is the vector product of the position vector $\mathbf{r}$ (from O to the point of application of the force) by the force. In symbols of vector analysis, the moment $\mathbf{L}$ is given by $\mathbf{L} = \mathbf{r} \times \mathbf{F}$. The magnitude of the moment of force is $L = rF \sin \theta$, where θ is the angle between the vectors $\mathbf{r}$ and $\mathbf{F}$. Numerically it is equal to the area of the parallelogram constructed on vectors $\mathbf{r}$ and $\mathbf{F}$ as sides. *Syn.* torque.

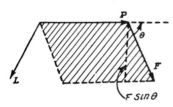

moment-generating function. For a random variable X or the associated distribution function, the moment-generating function M is defined by letting $M(t)$ be the expected value of e^{tX}, whenever this exists. For a discrete random variable with values $\{x_n\}$ and probability function p, $M(t) = \Sigma e^{tx_n} p(x_n)$ if this series converges; for

a continuous random variable with probability-density function f,

$$M(t) = \int_{-\infty}^{\infty} e^{tx} f(x)\, dx,$$

if this integral converges. Two distributions are the same if they have the same moment-generating function. For most commonly used distributions, the jth moment (about zero) is equal to the jth derivative of M evaluated at $t = 0$. E.g., for the *binomial distribution*,

$$M(t) = \sum_{k=0}^{n} e^{tk}\binom{n}{k}p^k q^{n-k} = (q + pe^t)^n.$$

Then $M'(t) = npe^t(q + pe^t)^{n-1}$ and $M'(0) = np$, the first moment. Also, $M''(0) = np + n(n - 1)p^2$, which is the second moment about zero. The kth **factorial moment-generating function** of a random variable X is the expected value of t^X. It is $M(\ln t)$, where M is the moment-generating function, and its kth derivative evaluated at 1 is the kth *factorial moment*. For various examples of specific moment-generating functions, see various types of distribution functions. See CHARACTERISTIC—characteristic function, CUMULANTS, and above, moment of a distribution.

moment of inertia. The moment of inertia of a particle about a point, line, or plane is the product of the mass and the square of the distance from the particle to the point, line or plane. The moment of inertia of a system of discrete particles about an axis is the sum of the products of the masses of the particles by the squares of their distances from the axis. Thus $I = \sum_i m_i r_i^2$, where r_i is the distance of the particle of mass m_i from the axis. For a continuous body, referred to a system of rectangular Cartesian axes x, y, z, the moments of inertia about these axes are, respectively,

$$I_x = \int_s (y^2 + z^2)\, dm, \qquad I_y = \int_s (z^2 + x^2)\, dm,$$

$$I_z = \int_s (x^2 + y^2)\, dm,$$

where the integrations are extended over the entire body. The quantities

$$I_{xy} = \int_s xy\, dm, \qquad I_{yz} = \int_s yz\, dm,$$

and $I_{xz} = \int_s xz\, dm$ are the **products of inertia**. If the coordinate axes x, y, z are so chosen that the products of inertia vanish, the axes are the **principal axes of inertia**. See PARALLEL —parallel-axis theorem.

moment of mass about a point, line, or plane. The sum of the products of the masses of each of the particles and its distance from the point, line, or plane. *Tech*. The integral, over the given mass, of the element of mass times its perpendicular distance from the point, line, or plane (this product is the **element of moment of mass**; see ELEMENT—element of integration). Algebraic (signed) distances are to be used in computing moments. A moment is essentially the sum of the moments of individual particles (elements of integration). For a set on a line (the x-axis), the moment about a point a on the line is

$$\int (x - a)\rho(x)\, dx,$$

where $\rho(x)$ is the density (mass per unit length) at the point x [this is the same as the *first moment of a frequency distribution* for which $\rho(x)$ is the frequency function (see above)]. For a set in the plane, the moment about the y-axis is

$$\int x\rho(x, y)\, dA,$$

where $\rho(x, y)$ is the density (mass per unit area) at the point (x, y) (if ρ is a function of x alone, then dA may be a strip parallel to the y-axis) [similar formulas can be given for other lines in the plane (*e.g.*, by using signed distances from the line, as given by the *normal form* of the equation of the line)]. For a set in space, the moment with respect to the (x, y)-plane is given by

$$\int z\rho(x, y, z)\, dV,$$

where $\rho(x, y, z)$ is the density (mass per unit volume) at the point (x, y, z) in the element of volume dV. **Moment of a curve** means the moment obtained by regarding the curve as having unit mass per unit length. **Moment of area** means the moment obtained by regarding the area as having unit mass per unit area.

moment of momentum. See MOMENTUM.

moment problem. The name "moment problem" was given by T. J. Stieltjes about 1894 to the problem, given a sequence of numbers $\{\mu_0, \mu_1, \mu_2, \cdots\}$, to find a monotone increasing function α such that μ_n is $\int_0^\infty t^n\, d\alpha(t)$ for all nonnegative integers n (see STIELTJES—

Riemann-Stieltjes integral). As early as 1873, a moment-type problem was solved by Chebyshev. In general, a moment problem is a problem of determining (for some set E) conditions on a sequence of numbers $\{\mu_0, \mu_1, \cdots\}$ for the existence of a function α of a specified type such that μ_n is $\int_E t^n\, d\alpha(t)$ for all nonnegative integers n, or determining the number of solutions for α, or determining all solutions. *E.g.*, if E is the interval $[0, 1]$ and α is to be monotone increasing, a necessary condition for the existence of α is that the sequence $\{|\mu_n|\}$ be bounded; a necessary and sufficient condition is that, for all nonnegative integers m and n,

$$\sum_{k=0}^m (-1)^k \binom{m}{k}\mu_{k+n} \geq 0;$$

if E is the real line and α is to be monotone increasing, a necessary and sufficient condition is that $\sum_{i=0}^n \alpha_i\mu_i \geq 0$ whenever $\{\alpha_1, \cdots, \alpha_n\}$ are such that $\sum_{i=0}^n \alpha_i t^i \geq 0$ for all t.

product moment. For a vector random variable $(X_1, \cdots, X_n)$, the **product moment** $\mu_{k_1 \cdots k_n}$ of order $(k_1, \cdots, k_n)$ about $(a_1, \cdots, a_n)$ is the *expected value* of $\prod_{i=1}^n (X_i - a_i)^{k_i}$. If $n = 2$, $k_1 = k_2 = 1$, and a_1 and a_2 are the means $\bar{X}$ and $\bar{Y}$ of X and Y, the product moment is the expected value of $(X - \bar{X})(Y - \bar{Y})$ and is called the **covariance** of X and Y; if X and Y have values $\{x_i\}$ and $\{y_i\}$ and joint probability function p, the covariance is

$$\mu_{11} = \sum (x_i - \bar{X})(y_i - \bar{Y})p(x_i, y_i);$$

if X and Y are continuous random variables with joint probability density function f,

$$\mu_{11} = \int_{-\infty}^\infty \int_{-\infty}^\infty (x - \bar{X})(y - \bar{Y})$$
$$\cdot f(x, y)\, dx\, dy.$$

second moment. See above, moment of inertia.

MO-MEN'TUM, *n*. The "quantity of motion" measured by the product of the mass and the velocity of the body. Sometimes called the *linear momentum* to distinguish it from the angular momentum or moment of momentum (*q.v.*). The *linear momentum* of a particle of mass m moving with the velocity $\mathbf{v}$ is the vector $m\mathbf{v}$. The linear momentum of a system of particles of masses $m_1, m_2, \cdots, m_n$, moving with the velocities $\mathbf{v}_1, \mathbf{v}_2, \cdots, \mathbf{v}_n$, is the vector

$$\mathbf{M} = \sum_i m_i \mathbf{v}_i,$$

representing the vector sum of the linear momenta of individual particles. The momentum of a continuous distribution of mass is defined by the integral

$$\mathbf{M} = \int_s \mathbf{v} \, dm,$$

evaluated over the body occupied by the mass.

angular momentum. See below, moment of momentum.

moment of momentum. The moment of momentum relative to a point O of a particle of mass m moving with velocity $\mathbf{v}$ is the vector product of the position vector $\mathbf{r}$ of the particle relative to O and the momentum $m\mathbf{v}$. In symbols of vector analysis, where a cross denotes the vector product, the moment of momentum is $\mathbf{H} = \mathbf{r} \times m\mathbf{v}$. The moment of momentum relative to O of a system of particles is defined as the sum of the moments of momentum of the individual particles. For a continuous distribution of mass,

$$\mathbf{H} = \int_s (\mathbf{r} \times \mathbf{v}) \, dm,$$

where the integration is extended over the entire body. *Syn.* angular momentum.

principle of linear momentum. A theorem in mechanics stating that the time rate of change of linear momentum of a system of particles is equal to the vector sum of external forces.

MONGE, Gaspard (1746–1818). French analyst and geometer. Invented descriptive geometry.

surface of Monge. See SURFACE.

MON′IC, *adj.* **monic polynomial.** A polynomial whose coefficients are integers, the coefficient of the term of highest degree being $+1$.

MON-O-DROM′Y, *n.* **monodromy theorem.** If the function f of the complex variable z is analytic at z_0 and can be continued analytically along every curve issuing from z_0 in a finite simply connected domain D, then f is a function element of an analytic function which is single valued in D; in other words, analytic continua-

tion around any closed curve in D leads to the original function element. See DARBOUX—Darboux's monodromy theorem.

MON′O-GEN′IC, *adj.* **monogenic analytic function.** The totality of all pairs $z_0, f(z)$, where

$$f(z) = \sum a_n (z - z_0)^n,$$

which can theoretically be obtained directly or indirectly by analytic continuation from a given function element f_0. The function f_0 is the **primitive element** of the monogenic function. The Riemann surface of the values z_0 is the **domain of existence** of the monogenic function, and the boundary of the domain of existence is the **natural boundary** of the monogenic analytic function. For example, the unit circle $|z| = 1$ is a natural boundary for the function

$$f(z) = \sum_{n=1}^{\infty} z^{n!}.$$

See ANALYTIC—analytic continuation of an analytic function of a complex variable.

MON′OID, *n.* A semigroup with an identity element.

MO-NO′MI-AL, *n.* An algebraic expression consisting of a single term which is a product of numbers and variables.

monomial factor. A single term that may be divided out of every member of an expression; $3x$ is a monomial factor of $6x + 9xy + 3x^2$.

MON′O-TONE MON′O-TON′IC, *adj.* A **monotonic** (or **monotone**) **increasing** quantity is a quantity which never decreases (the quantity may be a function, sequence, etc., which either increases or remains the same, but never decreases). A sequence of sets $\{E_1, E_2, \cdots\}$ is monotonic increasing if E_n is contained in E_{n+1} for each n. A **monotonic** (or **monotone**) **decreasing** quantity is a quantity which never increases (the quantity may be a function, sequence, etc., which either decreases or remains the same, but never increases). A sequence of sets $\{E_1, E_2, \cdots\}$ is monotonic decreasing if E_n contains E_{n+1} for each n. A **monotonic** (or **monotone**) **system of sets** is a system of sets such that, for any two sets of the system, one of the sets is contained in the other. A mapping of a topological space A onto a topological space B is **monotone** if the inverse image of each point of B is connected. A mapping of an ordered set A onto an ordered set B is **monotone** provided x^* precedes (or equals)

y^* whenever x^* and y^* are the images in B of points x and y of A for which x precedes y.

monotone convergence theorem. Suppose m is a countably additive measure on a σ-algebra of subsets of a set T and that $\{S_n\}$ is a monotone increasing sequence of nonnegative measurable functions. If there is a function S such that $\lim_{n \to \infty} S_n(x) = S(x)$ a.e. on T, then S is measurable and

$$\int_T S\, dm = \lim_{n \to \infty} \int_T S_n\, dm,$$

if it is understood that each member of the equality can be $+\infty$ if and only if the other is $+\infty$. See BOUNDED—bounded convergence theorem, LEBESGUE—Lebesgue convergence theorem, SERIES—integration of an infinite series.

MONTE-CARLO METHOD. Any procedure that involves statistical sampling techniques in obtaining a probabilistic approximation to the solution of a mathematical or physical problem. Such methods have been used for evaluation of definite integrals, solution of systems of linear algebraic equations and ordinary and partial differential equations, and the study of neutron diffusion. See BUFFON—Buffon needle problem.

MOORE, Eliakim Hastings (1862–1932). American general analyst, algebraist, and group theorist. Many of his doctoral students (including R. L. Moore, no kin) became eminent.

Moore-Smith convergence. A **directed set** (also called a **directed system** or a **Moore-Smith set**) is a set D which is ordered in the sense that there is a relation which holds for some of the pairs (a, b) of D (one then writes $a \geq b$) and for which (i) if $a \geq b$ and $b \geq c$, then $a \geq c$; (ii) $a \geq a$ for all a of D; (iii) if a and b are members of D, there is a c of D such that $c \geq a$ and $c \geq b$. A **net** (also called a **Moore-Smith sequence**) of a set S is a mapping of a directed set into S (onto a subset of S). E.g., the set of positive integers is a directed set and a sequence $\{x_1, x_2, x_3, \cdots\}$ is a net; the set of all open subsets of a space T is a directed set if $U \geq V$ means $U \subset V$. Let D be a directed set and ϕ a net, which is a mapping of D into a topological space T. Then ϕ is **eventually in** a subset U of T if there is an a of D such that if b belongs to D and $b \geq a$, then $\phi(b)$ is in U; ϕ is **frequently in** U if for each a in D there is a b in D such that $b \geq a$ and $\phi(b)$ is in U. The set E of all elements a of D for which $\phi(a)$ is in U is then a **cofinal subset** of D, meaning that, if b is in D, there is an a in E for which $a \geq b$.

The net ϕ **converges to a point** $\mathbf{x}$ of D if and only if it is eventually in each neighborhood of x. It follows that a point x is an *accumulation point* of a set V if and only if there is a net of V which converges to x. Also, a topological space is a *Hausdorff space* if and only if no net in the space converges to more than one point. See COMPLETE—complete space, FILTER.

MOORE, Robert Lee (1882–1974). American general topologist. No less famed for his inspiring "prove-it-yourself" teaching methods and for the success of his many doctoral students than for his fundamental topological results.

Moore space. A topological space S for which there is a sequence $\{G_n\}$ with the following properties: (1) Each G_n is a collection of open sets whose union is S. (2) For each n, G_{n+1} is a subcollection of G_n. (3) If x and y are points of an open set R and $x \neq y$, then there is an n such that if U is any member of G_n that contains x then the closure of U is contained in R and does not contain y.

MORDELL, Louis Joel (1888–1972). British number theorist.

Mordell conjecture. The conjecture (1922) that if a plane curve defined by a polynomial equation in two variables with rational coefficients has *genus* greater than or equal to 2, then there are at most finitely many points on the curve with rational coefficients. In 1983, Faltings proved that any projective plane curve of genus greater than or equal to 2 defined over a number field K has at most a finite number of K-points. The Mordell conjecture is a consequence of this. If a *smooth projective plane curve* is defined by a homogeneous polynomial of degree n, then its genus is $\frac{1}{2}(n-1)(n-2)$. Therefore the genus of the curve defined by $x^n + y^n = z^n$ is $\frac{1}{2}(n-1)(n-2)$. It follows that the equation $x^n + y^n = z^n$ has at most finitely many integral solutions if $n \geq 4$. See FERMAT—Fermat's last theorem. PROJECTIVE—projective plane curve.

MORERA, Giacinto (1856–1909). Italian analyst and mathematical physicist.

Morera's theorem. The theorem states that if the function f of the complex variable z is continuous in a finite simply connected domain D, and satisfies $\int_C f(z)\, dz = 0$ for all closed rectifiable curves in D, then f is an analytic function of z in D. This theorem is the converse of the Cauchy integral theorem.

MORI, Shigefumi (1951–   ). Japanese mathematician awarded a Fields Medal (1990) for his contributions to the theory of three-dimensional algebraic varieties.

MOR′PHISM, *n.* See CATEGORY (2).

MOR′RA, *n.* A game in which each of two players shows one, two, or three fingers and at the same time states his guess as to the number of fingers simultaneously being shown by his opponent. A player guessing correctly wins an amount proportional to the sum of the numbers of fingers shown by the two players, and the opponent loses this amount. Morra is an example of a *two-person zero-sum game* with chance moves. See MOVE.

MOR-TAL′I-TY, *adj., n.* **American experience table of mortality.** (1) A table of mortality (mortality table) based on the lives of Americans and constructed from insurance records about 1860. (2) A mortality table constructed from data obtained from American insurance companies and census records.

force of mortality. The annual rate of mortality under the assumption that the intensity of mortality is constant throughout the age-year which is under consideration and has the value it had at the moment after the beginning of the age-year.

mortality table. A table showing the number of deaths likely to occur during a given year among a group of persons of the same age (the table being based on past statistics). The number of lives at the age with which the table starts is the **radix** of the mortality table. If a mortality table is based on lives of people who have had medical examinations, or are chosen from special groups, or otherwise selected so as to constitute a better risk than the general run of persons, it is a **select mortality table**; the period during which the selection has effect upon the table is the **select period** of the table (the effect of selection often wears off with the passing of years). An **ultimate-life table** is a mortality table either based on the years after the select period (the latter not entering into consideration in the table), or using as a basis all the lives of a given age that are available.

rate of mortality. The probability a person will die within one year after attaining a certain age; d_x/l_x, where d_x is the number dying during the year x and l_x is the number attaining the age x in the group on which the mortality table is based.

MORT′GAGE, *n.* A conditional conveyance of, or lien upon, property as a security for money lent.

MO′TION, *n.* **constant** (or **uniform**) **motion.** See CONSTANT —constant speed and velocity.

curvilinear motion. Motion along a curve; motion which is not in a straight line.

curvilinear motion about a center of force. Motion such as that of celestial bodies about the sun; the motion of a particle whose initial velocity was not directed toward the center of force and which is attracted by a force at the given center. If this force is gravitational the path is a conic whose focus (or one of whose foci) is at the center of the force.

Newtonian laws of motion. See NEWTON — Newton's laws of motion.

rigid motion. See RIGID.

simple-harmonic motion. See HARMONIC.

MOVE, *n.* A component element of a game; a particular performance made at the choice of one of the players, or determined by a random device. A **personal move** is a move elected by one of the players, as contrasted with a **chance move** determined by a random device. See GAME, HER, MORRA.

MU, *n.* The twelfth letter of the Greek alphabet: lower case, μ; capital, M.

MUL′TI-AD-DRESS′, *adj.* **multiaddress system.** A method of coding problems for machine solution, whereby a single instruction might involve more than one address, memory position, or command. See SINGLE —single-address system.

MUL′TI-FOIL. A plane figure made of congruent arcs of a circle arranged on a regular polygon so that the figure is symmetrical about the center of the polygon and the ends of the arcs are on the polygon. The name is sometimes restricted to the cases in which the polygon has six or more sides. When the polygon is a square, the figure is called a *quatrefoil* (as illustrated); when it is a regular hexagon, a *hexafoil*; when a triangle, a *trefoil.*

MUL′TI-LIN′E-AR, *adj.* **multilinear form.** See FORM.

multilinear function. A function F of vectors $v_1, v_2, \cdots, v_n$ that is linear with respect to each vector when the others are given fixed values. See ALTERNATING —alternating function, and LINEAR —linear transformation.

MUL-TI-NO′MI-AL, *adj., n.* An algebraic expression which is the sum of more than one term. Compare POLYNOMIAL.

multinomial distribution. Suppose an experiment has k possible outcomes with respective probabilities $p_1, p_2, \cdots, p_k$ and that n such independent experiments are performed. Let X be the vector random variable $(X_1, X_2, \cdots, X_k)$, where X_i is the number of times the ith outcome occurs. Then X is a **multinomial (vector) random variable** and X has a **multinomial distribution**. The range of X is the set of all k-tuples of nonnegative integers $(n_1, n_2, \cdots, n_k)$ for which $\sum_{i=1}^{k} n_i = n$. The *mean* is the vector $(np_1, np_2, \cdots, np_k)$ and the probability function P satisfies

$$P(n_1, n_2, \cdots, n_k)$$
$$= \frac{n!}{n_1! n_2! \cdots n_k!} p_1^{n_1} p_2^{n_2} \cdots p_k^{n_k}.$$

See BINOMIAL —binomial distribution, HYPERGEOMETRIC —hypergeometric distribution, and below, multinomial theorem.

multinomial theorem. A theorem for the expansion of powers of multinomials. It includes the *binomial theorem* as a special case. The formula for the expansion is

$$(x_1 + x_2 + \cdots + x_m)^n$$
$$= \sum \frac{n!}{a_1! a_2! \cdots a_m!} x_1^{a_1} x_2^{a_2} \cdots x_m^{a_m}$$

where $\{a_1, a_2, \cdots, a_n\}$ is any selection of m numbers from $0, 1, 2, \cdots, n$ such that $a_1 + a_2 + \cdots + a_m = n$ and $0! = 1$.

MUL′TI-PLE, *adj., n.* In *arithmetic*, a number which is the product of a given integer and another integer; 12 is a *multiple* of 2, 3, 4, 6, and trivially of 1 and 12. In general, a product, no matter whether it be arithmetic or algebraic, is said to be a *multiple* of any of its factors.

common multiple. A quantity which is a multiple of each of two or more given quantities. The number 6 is a common multiple of 2 and 3; $x^2 - 1$ is a common multiple of $x - 1$ and $x + 1$. The **least common multiple** (denoted by l.c.m. or L.C.M.) is the least quantity that is exactly divisible by each of the given quantities; 12 is the l.c.m. of 2, 3, 4, and 6. The l.c.m. of a set of algebraic quantities is the product of all their distinct prime factors, each taken the greatest number of times it occurs in any one of the quantities; the l.c.m. of

$$x^2 - 1 \quad \text{and} \quad x^2 - 2x + 1$$

is

$$(x - 1)^2 (x + 1).$$

Tech. The l.c.m. of a set of quantities is a common multiple of the quantities which divides every common multiple of them. Also called *lowest common multiple*.

multiple correlation. See CORRELATION.

multiple edge in a graph. See GRAPH THEORY.

multiple integral. See INTEGRAL.

multiple point or tangent. See POINT —multiple (or *n*-tuple) point.

multiple regression. See REGRESSION —regression function.

multiple root of an equation. For a polynomial equation, $f(x) = 0$, a root a such that $(x - a)^n$ is a factor of $f(x)$ for some integer $n > 1$. A root which is not a multiple root is **a simple root**. If $(x - a)^n$ is the highest power of $(x - a)$ which is a factor of $f(x)$, a is a **double** or **triple** root when n is 2 or 3, respectively, and an ***n*-tuple root** in general. A multiple root is a root of $f(x) = 0$ and $f'(x) = 0$, where f' is the derivative of f. In general, a root is of order n, or an n-tuple root, if and only if it is a common root of $f(x) = 0$, $f'(x) = 0$, $f''(x) = 0, \cdots$, to $f^{[n-1]}(x) = 0$, but not of $f^{[n]}(x) = 0$. Analogously, when f is not a polynomial, a root is said to be of order n if the nth derivative is the derivative of lowest order of which it is not a root; the order of multiplicity of the root is then said to be n. *Syn.* repeated root.

multiple-valued function. (1) See FUNCTION —multiple-valued function. (2) If the Riemann surface of a monogenic analytic function f of the complex variable z covers any part of the z-plane more than once, the function is **multiple-valued**; *i.e.*, the function is multiple-valued if to any value z there corresponds more than one value of $f(z)$. A multiple-valued function may be considered as a single-valued function for z in a subdomain lying on a single sheet of its Riemann surface of existence.

MUL′TI-PLI-CAND′, *n.* The number to be multiplied by another number, called the **multiplier**. Because of the commutative property of multiplication, the product of numbers a and b is the same whether one considers a to be multiplied by b, or b to be multiplied by a.

MUL′TI-PLI-CA′TION, *n.* See various headings below and under PRODUCT.

abridged multiplication. The process of multiplying and dropping, after each multiplication by a digit of the multiplier, those digits which do not affect the degree of accuracy desired. If, in the product 235×7.1624, two-decimal place accuracy is desired (which requires the retention of only the third place throughout the multiplication), the abridged multiplication would be performed as follows:

235×7.1624

$$= 5 \times 7.1624 + 30 \times 7.1624 + 200 \times 7.1624$$

$$= 35.812 + 214.872 + 1432.480$$

$$\doteq 1683.164 \doteq 1683.16.$$

multiplication of determinants. The product of determinants (or of a determinant and a scalar) is equal to the product of the values of the determinants (or of the value of the determinant and the scalar). The **multiplication of a determinant by a scalar** can be accomplished by multiplying each element of any one row (or any one column) by the scalar. The product of **two determinants of the same order** is another determinant of the same order, in which an element in the ith row and jth column is the sum of the products of the elements in the ith row of the first determinant by the corresponding elements of the jth column of the second determinant (or vice versa). For second-order determinants,

$$\begin{vmatrix} a & b \\ c & d \end{vmatrix} \cdot \begin{vmatrix} A & B \\ C & D \end{vmatrix} = \begin{vmatrix} aA + bC & aB + bD \\ cA + dC & cB + dD \end{vmatrix}.$$

See PRODUCT—product of matrices.

multiplication of polynomials. See DISTRIBUTIVE—distributive property of arithmetic and algebra.

multiplication of the roots of an equation. The process of deriving an equation whose roots are each the same multiple of a corresponding root of the given equation. This is effected by the substitution (transformation) $x = x'/k$. The roots of the equation in x' are each k times one of the roots of the equation in x.

multiplication of series. See SERIES—multiplication of infinite series.

multiplication of vectors. (1) **Multiplication by a scalar.** The product of a scalar a and a vector $\mathbf{v}$ is the vector having the same direction as $\mathbf{v}$ and of length equal to the product of a and the length of $\mathbf{v}$; *i.e.*, the vector obtained by multiplying each component of $\mathbf{v}$ by a. (2) **Scalar multiplication of two vectors.** The scalar product of two vectors is the *scalar* which is the product of the lengths of the vectors and the cosine of the angle between them. This is frequently called the **dot product**, denoted by $\mathbf{A} \cdot \mathbf{B}$, or the **inner product**. It is equal to the sum of the products of corresponding components of the vectors. See VECTOR—vector space. (3) **Vector multiplication of two vectors.** The vector product of two vectors $\mathbf{A}$ and $\mathbf{B}$ is the vector $\mathbf{C}$ whose length is the product of the lengths of $\mathbf{A}$ and $\mathbf{B}$ and the sine of the angle between them (the angle from the first to the second), and which is perpendicular to the plane of the given vectors and directed so that the three vectors in order $\mathbf{A}, \mathbf{B}, \mathbf{C}$ form a *positively oriented trihedral* (see TRIHEDRAL). The product is also called the **cross product**. If the product of $\mathbf{A}$ by $\mathbf{B}$ is $\mathbf{C}$, one writes $\mathbf{A} \times \mathbf{B} = \mathbf{C}$. Scalar multiplication is commutative, but vector multiplication is anticommutative, for $\mathbf{B} \times \mathbf{A} = -\mathbf{A} \times \mathbf{B}$. The scalar product of $2i + 3j + 5k$ and $3i - 4j + 6k$ is $2 \cdot 3 - 3 \cdot 4 + 5 \cdot 6$ or 24, whereas the vector product is $38i + 3j - 17k$ if the vectors are multiplied in the given order, and the negative of this if they are multiplied in reverse order.

multiplication property of one and zero. For the number one, the property that $a \cdot 1 = 1 \cdot a = a$ for every number a. For zero, the property that $a \cdot 0 = 0 \cdot a = 0$ for every number a. The converse of the multiplication property of zero also is true, *i.e.*, if $ab = 0$, then at least one of a or b is zero. This is true for any field (*e.g.*, the field of complex numbers, or arithmetic modulo a prime n) and for any integral domain (*e.g.*, the set of integers), but not for all rings (in arithmetic modulo 6, $2 \cdot 3 \equiv 0$; and in the ring of 2×2 real matrices,

$$\begin{bmatrix} 1 & 2 \\ 0 & 0 \end{bmatrix} \begin{bmatrix} 0 & 2 \\ 0 & -1 \end{bmatrix} = \begin{bmatrix} 0 & 0 \\ 0 & 0 \end{bmatrix},$$

but neither factor in the left member is the zero matrix). See CONGRUENCE, DOMAIN—integral domain, RING.

MUL′TI-PLI-CA′TIVE, *adj.* **multiplicative inverse.** See INVERSE—inverse of an element.

MUL'TI-PLIC'I-TY, *n.* **multiplicity of a root of an equation.** See MULTIPLE—multiple root of an equation.

MUL'TI-PLI'ER, *n.* (1) The number which is to multiply another number, called the **multiplicand.** (2) In a computing machine, any arithmetic component that performs the operation of multiplication.

Lagrange method of multipliers. See LAGRANGE.

MUL'TI-PLY, *adv.* **multiply connected set.** See CONNECTED —simply connected set.

MUL'TI-PLY, *v.* To perform the process of *multiplication*.

MUL'TI-VAR'I-ATE, *adj.* Involving more than one variable.

multivariate distribution. See DISTRIBUTION —distribution function.

MUMFORD, David Bryant (1937–   ). British-American algebraic geometer and Fields Medal recipient (1974). Research in the theory of moduli and geometric invariants.

MUTATIS MUTANDIS. "With the necessary changes having been made."

MU'TU-AL, *adj.* **mutual fund method.** (*Insurance*) A method of computing the present value (single premium) for a *whole life annuity immediate*. See COMMUTATION—commutation tables.

mutual insurance company. A cooperative association of policyholders who divide the profits of the company among themselves. The policies, which usually provide for such division, are called *participating policies*.

mutual savings bank. A bank whose capital is that of the depositors who own the bank.

MU'TU-AL-LY, *adv.* **mutually equiangular polygons.** Polygons whose corresponding angles are equal.

mutually equilateral polygons. Polygons whose corresponding sides are equal.

mutually exclusive events. See EVENT.

MYR'I-A. A prefix meaning 10,000, as myriameter = 10,000 meters.

MYR'I-AD, *n.* An immense number (but see GREEK—Greek numerals).

N

NAB'LA, *n.* See DEL.

NA'DIR, *n.* The point in the celestial sphere diametrically opposite to the zenith; the point where a plumb line at the observer's position on the earth, extended downward, would pierce the celestial sphere.

NAN'O. A prefix meaning 10^{-9}, as in **nanometer.**

NAPIER, John (1550–1617). Innovative Scottish amateur mathematician. With trigonometric motivation, Napier defined "logarithm" (which we will call Naplog) as follows. Let S be the line segment $[0, 10^7]$ and let R be the ray with initial point 0. Let p on S and x on R start simultaneously at 0, with p moving at velocity $10^7 - p = y$ and x moving uniformly with speed 10^7. Then $x = $ Naplog y. Napier published (1614) along with this definition a table of "logarithms" of sines of angles for successive minutes of arc and proved that, if $a/b = x/y$, then

$$\text{Naplog } a - \text{Naplog } b = \text{Naplog } x - \text{Naplog } y.$$

With calculus, one can prove that Naplog $y = 10^7(\log e \, 10^7 - \log_e y)$. In 1615, Briggs and Napier worked together to make changes leading to logarithms to the base 10, *i.e.*, **common** or **Briggsian logarithms** (see LOGARITHM).

Napier's analogies. Formulas for use in solving a spherical triangle. They are as follows (a, b, c representing the sides of a spherical triangle and A, B, C the angles opposite a, b, c, respectively):

$$\frac{\sin \frac{1}{2}(A - B)}{\sin \frac{1}{2}(A + B)} = \frac{\tan \frac{1}{2}(a - b)}{\tan \frac{1}{2}c}$$

$$\frac{\cos \frac{1}{2}(A - B)}{\cos \frac{1}{2}(A + B)} = \frac{\tan \frac{1}{2}(a + b)}{\tan \frac{1}{2}c}$$

$$\frac{\sin \frac{1}{2}(a - b)}{\sin \frac{1}{2}(a + b)} = \frac{\tan \frac{1}{2}(A - B)}{\cot \frac{1}{2}C}$$

$$\frac{\cos \frac{1}{2}(a - b)}{\cos \frac{1}{2}(a + b)} = \frac{\tan \frac{1}{2}(A + B)}{\cot \frac{1}{2}C}.$$

Napier's rules of circular parts. Two ingenious rules by which one can write out the ten formulas needed in the solution of right spherical triangles. Omitting the right angle, one can think of the complements of the other two angles, the complement of the hypotenuse, and the other two sides as arranged on a circle in the same order as on the triangle. Any one of these is a middle point in the sense that there are two on either side of it. The two points nearest to a given point are called *adjacent parts*, and the two farthest are called *opposite parts*. Napier's rules are then stated as follows: I. The sine of any part is equal to the product of the tangents of the adjacent parts. II. The sine of any part is equal to the product of the cosines of the opposite parts.

Napierian logarithms. A name commonly used for *natural logarithms*. Napier did not originate natural logarithms, but see NAPIER, above.

NAPPE, *n.* One of the two parts of a conical surface into which the surface is separated by the vertex.

NAT′U-RAL, *adj.* **natural equations of a space curve.** See INTRINSIC —intrinsic equations of a space curve.

natural logarithms. Logarithms using the base e (2.71828183 +). *Syn.* Napierian logarithms. See e, and LOGARITHM.

natural numbers. The numbers $1, 2, 3, 4$, etc. Same as POSITIVE INTEGERS. See INTEGER.

NAUGHT, *n.* Same as ZERO.

NAU′TI-CAL, *adj.* **nautical mile.** The length of one minute of arc on a great circle on a sphere with the same area as the earth; approx. 6080.27 feet. The geographical mile is the length of one minute of arc on the earth's equator, or approx. 6087.15 feet. The British **admiralty mile** is 6080 feet. See KNOT.

NEC′ES-SARY, *adj.* **necessary condition.** See CONDITION.

necessary condition for convergence of an infinite series: That the terms approach zero as one goes farther out in the series; that the nth term approaches zero as n becomes infinite. This is not a sufficient condition for convergence; *e.g.*, the series

$$1 + \frac{1}{2} + \frac{1}{3} + \cdots + \frac{1}{n} + \cdots$$

is divergent, although $1/n$ approaches zero as n becomes infinite. See CAUCHY —Cauchy's condition for convergence of a series.

NE-GA′TION, *n.* **negation of a proposition.** The proposition formed from the given proposition by prefixing "It is false that," or simply "not." *E.g.*, "Today is Wednesday" has the negation "It is false that today is Wednesday." "All cows are brown" has the negation "It is false that all cows are brown," which might be written as "There is at least one cow which is not brown." The negation of a proposition p is frequently written as $\sim p$ and read "*not p.*" The negation of a proposition is true if and only if the proposition is false. *Syn.* denial. See QUANTIFIER.

NEG′A-TIVE, *adj.* **negative angle, exponent, number.** See ANGLE, EXPONENT, POSITIVE—positive number.

negative direction. The direction opposite the direction that has been chosen as positive.

negative part of a function. See POSITIVE —positive part of a function.

negative sign. The mark, $-$, denoting the negative of a number. *Syn.* minus sign. See POSITIVE—positive number.

NE-GO′TI-A-BLE, *adj.* **negotiable paper.** An evidence of debt which may be transferred by endorsement or delivery, so that the transferee or holder may sue on it in his own name with like effect as if it had been made to him originally; *e.g.*, bills of exchange, promissory notes, drafts, and checks payable to the order of a payee or to bearer.

NEIGH′BOR-HOOD, *n.* **neighborhood of a point.** The ϵ-**neighborhood** of a point P is the set of all points whose distance from P is less than ϵ (*e.g.*, an *open interval* on the line or the interior of a circle in the plane, with the point as center). A **neighborhood** of P is any set that contains an ϵ-neighborhood of P, or, for topological spaces, any set that contains an open set that contains P. Sometimes *neighborhood* and *open set* are taken as synonymous. One speaks of a property as holding **in the neighborhood** of a point if there exists a neighborhood of the point in which the property holds, or of a quantity (*e.g.*, curvature) depending on the nature of a curve or surface **in the neighborhood of a point** P if for each neighborhood of P the value of the quantity at P can be determined from knowledge of the portion of the curve or surface in this neighborhood of the point. See SMALL—in the small, TOPOLOGICAL —topological space.

NERVE, *n.* **nerve of a family of sets.** Let $S_0, S_1, \cdots, S_n$ be a finite family of sets. To each set S_k assign a symbol p_k. A *nerve* of this system of sets is the *abstract simplicial complex* whose vertices are the symbols $p_0, p_1, \cdots, p_n$ and whose *abstract simplexes* are all the subsets $p_{i_0}, p_{i_1}, \cdots, p_{i_r}$ whose corresponding sets have a nonempty intersection. E.g., if S_0, S_1, S_2, S_3 are the four faces of a tetrahedron, the nerve is the abstract simplicial complex with vertices p_0, p_1, p_2, p_3 whose abstract simplexes are all sets of three or less vertices—it has a geometric realization as a tetrahedron.

NEST, *n.* See NESTED—nested sets.

NEST'ED, *adj.* **nested intervals.** A sequence of intervals such that each is contained in the preceding. It is sometimes required that the lengths of the intervals approach zero as one goes out in the sequence. The **nested-interval theorem** states that for any sequence of nested intervals, each of which is *bounded* and *closed*, there is at least one point which belongs to each of the intervals (if the lengths of the intervals approach zero, there is exactly one such point). This theorem is true for intervals in n-dimensional Euclidean space, as well as for intervals on the line. See INTERVAL, MONOTONE.

nested sets. A collection of sets is nested if, for any two members A and B of the collection, either A is contained in B or B is contained in A. A nested collection of sets is also called a **nest, tower,** or **chain.**

NET, *adj.* Clear of all deductions (such as charges, cost, loss). E.g., **net proceeds** is the amount remaining of the sum received from the sale of goods after all expenses except the original cost have been deducted. Also see PREMIUM, PRICE, PROFIT.

NET, *n.* See MORE (E.H.)—Moore-Smith convergence.

NEUMANN, Franz Ernst (1798–1895). German mathematical physicist and crystallographer.

Neumann formula for Legendre functions of the second kind. The formula

$$Q_n(z) = \tfrac{1}{2} \int_{-1}^{1} \frac{P_n(t)}{z - 1} \, dt,$$

where P_n is a Legendre polynomial. The function Q_n is a solution of Legendre's differential equation and can be expressed in terms of the

hypergeometric function F as

$$\frac{\pi^{1/2} \Gamma(n+1)}{(2z)^{n+1} \Gamma(n + \frac{3}{2})}$$

$$\times F\left(\tfrac{1}{2}n + \tfrac{1}{2}, \tfrac{1}{2}n + 1; n + \tfrac{3}{2}; z^{-2}\right).$$

NEUMANN, Karl Gottfried (1832–1925). German analyst and potential theorist.

Neumann function. (1) A function N_n defined by

$$N_n(z) = \frac{1}{\sin n\pi} \left[\cos n\pi J_n(z) - J_{-n}(z) \right],$$

where J_n is a Bessel function. This function is a solution of Bessel's differential equation (if n is not an integer) and is also called a *Bessel function of the second kind.* See HANKEL—Hankel function. (2) (*Potential Theory*) For a region R with boundary surface S, and for a point Q interior to R, the Neumann function N is a function of the form

$$N(P, Q) = 1/(4\pi r) + V(P),$$

where r is the distance $PQ, V(P)$ is harmonic, $\partial N / \partial n$ is constant on S, and $\iint_S N \, d\sigma_P = 0$. The solution $U(Q)$ of the Neumann problem can be represented in the form

$$U(Q) = \iint_S f(P) N(P, Q) \, d\sigma_P.$$

See GREEN—Green's function, BOUNDARY—second boundary-value problem of potential theory (the *Neumann problem*).

NEVANLINNA, Rolf (1895–1980). Finnish complex analyst who made major contributions to the theory of entire functions, meromorphic functions, and computer science. The **Nevanlinna prize** is funded by the University of Helsinki and is awarded at the International Congress of Mathematicians in recognition of work in the mathematical aspects of information sciences. Awarded first in 1983.

NEW-TON, *n.* The force which will give an acceleration of one meter per second per second to a mass of one kilogram; equal to 10^5 dynes.

NEWTON, Sir Isaac (1642–1727). Superb English mathematician, physicist, and astronomer.

With Archimedes and Gauss, one of the three greatest mathematicians of all time. His masterpiece, *Philosophae Naturalis Principia Mathematica*, is considered to be the most important scientific book ever written. He and Leibniz invented calculus independently. See BRACHISTOCHRONE.

Gregory-Newton interpolation formula. See GREGORY.

Newton-Cotes integration formulas. The approximation formulas

$$\int_{x_0}^{x_0+h} y\,dx = \frac{h}{2}(y_0+y_1) - \frac{h^3}{12}y''(\xi),$$

$$\int_{x_0}^{x_0+2h} y\,dx = \frac{h}{3}(y_0+4y_1+y_2) - \frac{h^5}{90}y^{iv}(\xi),$$

$$\int_{x_0}^{x_0+3h} y\,dx = \frac{3h}{8}(y_0+3y_1+3y_2+y_3)$$

$$- \frac{3h^5}{90}y^{iv}(\xi),$$

etc., where y_k is the value of y at x_0+kh and, in each formula, ξ is an intermediate value of x. The correction term involves the sixth derivative in the next two formulas after those shown, etc. Since the foregoing formulas involve the values of y at the limits of integration, they are said to be of **closed type**. The Newton-Cotes formulas of **open type** are

$$\int_{x_0}^{x_0+3h} y\,dx = \frac{3h}{2}(y_1+y_2) + \frac{h^3}{4}y''(\xi),$$

etc. The open-type formulas are used particularly in the numerical solution of differential equations.

Newtonian potential. See POTENTIAL —gravitational potential.

Newton identities. Certain relations between sums of powers of all the roots of a polynomial equation and its coefficients. For the equation

$$x^n + a_1 x^{n-1} + a_2 x^{n-2} + \cdots + a_n = 0,$$

with the roots $r_1, r_2, \cdots, r_n$, Newton's identities are:

$$s_k + a_1 s_{k-1} + a_2 s_{k-2} + \cdots + a_{k-1}s_1 + ka_k = 0,$$

for $k \le n-1$, and

$$s_k + a_1 s_{k-1} + a_2 s_{k-2} + \cdots + a_n s_{k-n} = 0,$$

for $k \ge n$, where

$$s_k = r_1^k + r_2^k + \cdots + r_n^k.$$

Newton's inequality. The logarithmic-convexity inequality

$$p_{r-1}p_{r+1} \le p_r^2, \qquad 1 \le r < n,$$

where $p_r = b_r / \binom{n}{r}$ is the average value of the $\binom{n}{r}$ terms comprising the rth *elementary symmetric function* b_r of a set of numbers $a_1, a_2, \cdots, a_n$.

Newton's laws of motion. *First Law:* Every particle continues in its state of rest or of uniform motion in a straight line except in so far as it is compelled by forces to change that state. *Second Law:* The time rate of change of momentum is proportional to the motive force and takes place in the direction of the straight line in which the force acts. *Third Law:* The interaction between two particles is represented by two forces equal in magnitude but oppositely directed along the line joining the particles.

Newton's method of approximation. A method of approximating roots of equations. For a function of one real variable, it is based on the fact that the tangent of a curve very nearly coincides with a small arc of the curve. Suppose the equation is $f(x) = 0$, and a_1 is an approximation to one of the roots. The next approximation, a_2, is the abscissa of the point of intersection of the x-axis and the tangent to the curve $y = f(x)$ at the point whose abscissa is a_1, i.e., $a_2 = a_1 - f(a_1)/f'(a_1)$, where f' is the derivative of f. This is equivalent to using the first two terms (dropping all higher-degree terms) in the Taylor's expansion of f about a_1 and assuming that $f(a_2) = 0$. If a_1 and a_2 are in an interval I that contains the desired root c and if L is a lower bound for $|f'(x)|$ on I and U is an upper bound for $|f''(x)|$ on I, then

$$|a_1 - c| \le \frac{|f(a_1)|}{L}, \qquad |a_2 - c| \le \frac{U}{2L}|a_1 - c|^2.$$

A **generalized Newton's method** can be used to find successive approximations to a vector **u** for which $T(\mathbf{u}) = \mathbf{v}$, where **v** is a given vector in n-dimensional Euclidean space and T is a transformation of the space. Given an approximation $\mathbf{a}_1$, the next approximation $\mathbf{a}_2$ is

$$\mathbf{a}_1 - J^{-1}(\mathbf{a}_1)[T(\mathbf{a}_1) - \mathbf{v}],$$

where J^{-1} is the inverse of the Jacobian matrix of T.

Newton's three-eighths rule. An alternative to Simpson's rule for approximating the area bounded by a curve $y = f(x)$, the x-axis, and the lines $x = a$ and $x = b$. The interval (a, b) is divided into $3n$ equal parts, and the formula is

$$A = \frac{(b-a)}{8n} [y_a + 3y_1 + 3y_2 + 2y_3 + 3y_4 + 3y_5$$

$$+ 2y_6 + \cdots + 3y_{3n-1} + y_b].$$

The name of the rule comes from the fact that the coefficient $(b-a)/(8n)$ is equal to $(3/8)h$, where $h = (b-a)/(3n)$ is the length of each of the equal subintervals. For an analogous reason, Simpson's rule is sometimes called Simpson's one-third rule. The error in Simpson's rule is $-nk^5/90$ times the fourth derivative at an intermediate point, where $k = (b-a)/(2n)$, while that in Newton's rule is $-3nh^5/80$ times the fourth derivative at an intermediate point. See SIMPSON —Simpson's rule, TRAPEZOID —trapezoid rule, WEDDLE —Weddle's rule.

trident of Newton. See TRIDENT.

NEYMAN, Jerzy (1894–1981). Statistician who was born in Bessarabia, lived in Poland until age 40, and moved to the U.S. after four years in London. Noted both for important contributions to statistics and for applications of statistics in many other fields.

Neyman-Pearson test. Let X be a random variable whose distribution depends on an unknown parameter θ, where either or both of X and θ may be vectors. Let H_0 and H_1 be the hypotheses $\theta = \theta_0$ and $\theta = \theta_1$, respectively. To test H_0 versus H_1, given a random sample $(X_1, X_2, \cdots, X_n)$, let the *acceptance region R* for H_0 be defined by

$$[(X_1, \cdots, X_n) \in R] \Leftrightarrow \frac{L(\theta_1)}{L(\theta_0)} < k,$$

where L is the *likelihood function*. Let α be the conditional probability that $(X_1, \cdots, X_n)$ not belong to R, given $\theta = \theta_0$; i.e., α is the probability of *type I error*. Then this test has the least value for the conditional probability that $(X_1, \cdots, X_n)$ belong to R, given $\theta = \theta_1$, among all tests with probability α of type I error. See HYPOTHESIS — test of a hypothesis, LIKELIHOOD —likelihood function.

NICOMEDES (2nd century B.C.). Greek mathematician.

conchoid of Nicomedes. See CONCHOID.

NIKODÝM, Otton Martin (1887–1974). Analyst and topologist who came to the U.S. as a Polish refugee shortly after World War II. See RADON.

NIL′PO-TENT, *adj.* Vanishing upon being raised to some power. The matrix

$$A = \begin{pmatrix} 2 & 0 & -4 \\ 3 & 0 & 0 \\ 1 & 0 & -2 \end{pmatrix}$$

is nilpotent, since $A^3 = 0$.

nilpotent ideal. An ideal I of a ring R for which there is a positive integer n such that $x_1 x_2 \cdots x_n = 0$ for all choices of $x_1, x_2, \cdots, x_n$ in I. See IDEAL, RADICAL —radical of a ring.

NIL-SEGMENT, *n.* See SEGMENT —segment of a line.

NIM, *n.* **game of nim.** A game in which two players draw articles from several piles, each player in turn taking as many as he pleases from one pile, the player who draws the last article winning. To win, a player must, with each draw, make the sum of the corresponding digits of the numbers of articles, written in the binary scale, either 2 or 0. If the numbers of articles are $17, 6, 5$, they are written $10001, 110, 101$ in the binary scale, which we write with digits having the same place value in the same column as follows:

$$10001$$
$$110$$
$$101$$

To win, the first player, A, must take fourteen articles from the large pile, leaving the columns

$$11$$
$$110$$
$$101$$

The other player, B, has no choice but to make the sum, in at least one column, odd. A then makes it even and they finally come to the situation where it is B's draw and the numbers are 1 and 1, so that A wins. The above strategy assures a win for the first player who can make the sum of the digits even in each column. There are other variations of *nim*. E.g., with one pile, each player can be required to pick up at least one and less than k articles, the *loser* being the person picking up the last article. The player first able to leave $nk + 1$ articles (for some n) can win by leaving $n(k - 1) + 1$ at the next play, etc.

NINE, *adj.*, *n.* **casting out nines.** See CASTING.
divisibility by nine. See DIVISIBLE.
nine-point circle. See CIRCLE—nine-point circle.

NOD'AL, *adj.* **nodal line.** A line in a configuration which remains fixed while the configuration is being rotated or deformed in a certain manner; a line in an elastic plate which remains fixed while the plate vibrates. See EULER—Euler angles.

NODE, *n.* **node in astronomy.** See ECLIPTIC
node of a curve. A point at which two parts of a curve cross and have different tangents (sometimes a double point which is an *acnode* is also called a node). A set of nodes of curves which belong to a given family of curves is said to be a **node-locus.** See CRUNODE, and DISCRIMINANT—discriminant of a differential equation.
node of a graph. See GRAPH THEORY.

NOETHER, Amalie Emmy (1882–1935). German-American algebraist. Contributed to the theory of invariants, abstract axiomatic algebras, axiomatic theory of ideals, noncommutative and cyclic algebras.
Noetherian ring. See CHAIN—chain conditions on rings.

NOM'I-NAL, *adj.* **nominal rate of interest.** See INTEREST.

NOM'O-GRAM, *n.* A graph consisting of three lines or possibly curves (usually parallel) graduated for different variables in such a way that a straight edge cutting the three lines gives the related values of the three variables. *E.g.*, when considering automobile tires, one line might be graduated with the price, another with the cost per mile, and the other the mileage life of the tire, in such a way that a straight edge through a certain price point and mileage life point would cross the other line at the cost per mile. *Syn.* alignment chart.

NON, *prefix.* Negating prefix.
nondense set. See DENSE—dense set.
non-Euclidean geometry. See GEOMETRY — Euclidean geometry.
nonexpansive mapping. See CONTRACTION.
nonlinear. Not linear. See LINEAR—linear equation or expression, linear transformation.
nonperiodic, or nonrepeating, decimal. See DECIMAL—decimal number system.
nonremovable discontinuity. See DISCONTINUITY.
nonresidue. See RESIDUE.

nonsingular linear transformation. A linear transformation T of a vector space V into V for which there is no nonzero x such that $T(x) = 0$. If V is a finite-dimensional space, then T is nonsingular if and only if the determinant of the coefficients is not zero, or if and only if the range of T is V. See LINEAR—linear transformation.

nonsquare. A **nonsquare Banach space** is a Banach space for which there are no members x and y for which $\|x\| = \|y\| = \|\frac{1}{2}(x+y)\| = \|\frac{1}{2}(x-y)\| = 1$; it is **uniformly nonsquare** if there is a positive number ϵ such that there are no members x and y for which $\|x\| = \|y\| = 1$, $\|\frac{1}{2}(x+y)\| > 1 - \epsilon$, and $\frac{1}{2}(x-y)\| > 1 - \epsilon$. See SUPERREFLEXIVE.

nonstandard numbers. Let R be the set of real numbers and let U be a *free ultrafilter* of subsets of the positive integers (see FILTER). Let R^* be the set of equivalence classes of the collection of all sequences $\{r_i\} = \{r_1, r_2, r_3, \cdots\}$ of real numbers for which: (i) $\{r_i\} \equiv \{s_i\}$ means that the set of all i for which $r_i = s_i$ is a member of U; (ii) $\{r_i\} < \{s_i\}$ means that the set of all i for which $r_i < s_i$ is a member of U; (iii) $\{r_i\} + \{s_i\} \equiv \{r_i + s_i\}$; (iv) $\{r_i\} \cdot \{s_i\} \equiv \{r_i \cdot s_i\}$. The *absolute value* of $\{r_i\}$ is $\{|r_i|\}$. The function ρ is an isomorphism between R and a subset of R^*, where $\rho(r)$ is the equivalence class that contains $\{r_i\}$ with each $r_i = r$. If $\{r_i\}$ **positive** is defined to mean that $\rho(0) < \{r_i\}$, then R^* is an *ordered field*, but is not *complete* (see FIELD). Members of R^* are called **nonstandard** or **hyperreal numbers.** An **infinitesimal** is an $\{r_i\}$ for which $\rho(0) < \{|r_i|\} < \rho(r)$ for each positive real number r (*e.g.*, $\{r_i\}$ with each $r_i = 1/i$). An **infinite number** is an $\{r_i\}$ for which $\rho(r) < \{|r_i|\}$ for each real number r (*e.g.*, r_i with each $r_i = i$). The reciprocal of an infinitesimal is an infinite number, and conversely. If $x \in R^*$ and x is not an infinite number, then there is a member r of R for which $x - \rho(r)$ is an infinitesimal.

nonterminating. See DECIMAL—decimal number system, TERMINATING.

NON'A-GON, *n.* A polygon having nine sides. It is a **regular nonagon** if its sides are all equal and its interior angles are all equal.

NORM, *n.* (1) Mean; average. (2) Customary degree or condition. (3) Established pattern or form. (4) See VECTOR—vector space.
norm of a matrix. The square root of the sum of the squares of the absolute values of the elements; the square root of the *trace* of A^*A, where A is the given matrix and A^* is the *Hermitian conjugate* of A (or the *transpose* if A is real). The norm of a matrix is unchanged by

multiplication on either side by a *unitary* matrix, from which it follows that the norm of a *normal matrix* (or a Hermitian symmetric matrix) is equal to the sum of the squares of the *eigenvalues* of the matrix.

norm of a functional, quaternion, transformation, or vector. See CONJUGATE—conjugate space, LINEAR—linear transformation, QUATERNION—conjugate quaternions, VECTOR—vector space.

norm of a partition. See PARTITION—partition of an interval.

NOR'MAL, *adj., n.* (1) Perpendicular. (2) According to rule or pattern; usual or natural. (3) Possessing a property which is commonly designated by the word *normal* (*e.g.*, see below, normal number, normal transformation).

normal acceleration. See ACCELERATION.

normal curvature of a surface. See CURVATURE—curvature of a surface.

normal derivative. The *directional derivative* of a function in the direction of the normal at the point where the derivative is taken; the rate of change of a function in the direction of the normal to a curve or surface. See DIRECTIONAL.

normal distribution. A random variable X is **normally distributed** or is a **normal random variable** if there are numbers m and σ for which X has the probability density function

$$f(t) = \frac{1}{\sigma\sqrt{2\pi}} e^{-1/2(t-m)^2/\sigma^2}, \quad -\infty < t < \infty.$$

The *mean* is m, the *variance* is σ^2, and the *moment-generating function* is $M(t) = e^{im + \frac{1}{2}t^2\sigma^2}$. Such a function f is a **normal probability-density function** or **normal frequency function**; its graph is a **normal frequency curve** (or **error curve** or **normal-frequency distribution curve** or **normal-probability curve**). As shown in the figure, it is a bell-shaped curve, symmetric about the mean, with points of inflection at $\pm\sigma$ from the mean. Approximately 68% of the distribution is included in the interval $A \pm \sigma$, while 99.7% is included in the interval $A \pm 3\sigma$. The normal distribution occurs frequently in practical problems. It also is important because of the *central-limit theorem* and because it is the limiting form of the *binomial distribution* as n increases, in the sense that if X is a binomial random variable with constants n, p and $q = 1 - p$, then $(X - np)/\sqrt{npq}$ has approximately a normal distribution with mean 0 and variance 1 if n is sufficiently large. *Syn.* Gaussian distribution. See BINOMIAL—binomial distribution, BIVARIATE—bivariate normal distribution, F—F

distribution, LOGNORMAL—lognormal distribution, PRECISION—modulus of precision.

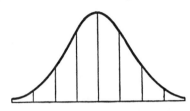

normal equations. A set of equations derived by the method of least squares to obtain estimates of the parameters a and b in the equation $y = a + bx$, where y is a random variable and x is a fixed variate. By minimizing $\sum_{i=1}^{n} [y_i - (a + bx_i)]^2$ with respect to a and to b, where n is the number of paired observations, the following normal equations are derived:

$$\sum y_i - an - b\sum x_i = 0,$$

$$\sum y_i x_i - a\sum x_i - b\sum x_i^2 = 0.$$

The process may be applied to functions with k parameters so that k normal equations are obtained which, when solved, yield the least-squares estimates of the k parameters. See LEAST—method of least squares.

normal extension of a field. See EXTENSION—extension of a field.

normal family of analytic functions. A family of functions of the complex variable z, all analytic in a common domain D, and such that every infinite sequence of functions of the family contains a subsequence which converges uniformly to an analytic function (which might be identically infinite) in every closed region in D.

normal form of an equation. See LINE—equation of a line, PLANE—equation of a plane.

normal frequency function. See PROBABILITY—probability density function.

normal (or normalized) functions. See ORTHOGONAL—orthogonal functions.

normal lines and planes. See PERPENDICULAR—perpendicular lines and planes, and below, normal to a curve or surface.

normal matrix. See MATRIX.

normal number. A number whose decimal expansion is such that all digits occur with equal frequency, and all blocks of the same length occur equally often. *Tech.* Let a real number X be written as an infinite decimal with base r (r not necessarily 10). Let $N(d, n)$ be the number of occurrences of the digit d among the first n digits of X. Then the number X is **simply nor-**

mal with respect to the base r if

$$\lim_{n \to \infty} \frac{N(d, n)}{n} = \frac{1}{r}$$

for each of the values $0, 1, \cdots, r-1$ of d. Now let $N(D_k, n)$ be the number of occurrences of the block D_k of k successive integers in the first n digits of X. Then X is **normal** with respect to the base r if

$$\lim_{n \to \infty} \frac{N(D_k, n)}{n} = \frac{1}{r^k}$$

for each positive integer k and block D_k (actually, it is sufficient to use only an infinite number of values of k). A normal number is irrational (see IRRATIONAL —irrational number), but a simply normal number may be rational (e.g., the *repeating* decimal $0.0123456789012345678 \cdots$). It is not known whether $\sqrt{2}$, π, or e is normal to some base, but the set of all real numbers which *do not* have the property of being normal with respect to every base is of *measure zero*. An example of a normal number with respect to base 10 is given by writing the integers in succession:

$$0.12345678910111213141 5 \cdots .$$

normal order. See ORDER.

normal section of a surface. A plane section made by a plane containing a normal (perpendicular) to the surface. The **principal normal sections** are the normal sections in the principal directions (see CURVATURE —curvature of a surface). *Syn.* perpendicular section of a surface.

normal space. See REGULAR —regular space.

normal stress. See STRESS.

normal subgroup. A subgroup H of a group G for which $gH = Hg$ for each g in G. A subgroup is normal if and only if all its right cosets are also left cosets. A subgroup H is **quasi-normal** if $JH = HJ$ for all subgroups J of G. *Syn.* invariant subgroup, normal divisor. See QUOTIENT —quotient space.

normal to a curve or surface. A **normal line to a curve** at a point is a line perpendicular to the tangent line at the point (see TANGENT —tangent line). For a **plane curve**, its equation is

$$(y - y_1) = \left[-1/f'(x_1) \right](x - x_1),$$

where $f'(x_1)$ is the slope of the curve at the point (x_1, y_1) in which the normal cuts the curve. See DERIVATIVE, and PERPENDICULAR —perpendicular lines and planes. The **normal plane to**

a space curve at a point P is the plane perpendicular to the tangent line at P; the normal lines at P are the lines through P which lie in the normal plane. The *binormal* to a space curve at a point P is the line passing through P and normal to the osculating plane of the curve at P. The positive direction of the binormal is chosen so that its direction cosines are $\rho(y'z'' - z'y'')$, $\rho(z'x'' - x'z'')$, $\rho(x'y'' - y'x'')$, the primes denoting differentiation with respect to the arc length. The **principal normal** to a space curve at a point P is the line perpendicular to the space curve at the point P and lying in the *osculating plane* at P. The positive direction of the principal normal at P is chosen so that the tangent, principal normal, and binormal at P have the same mutual orientation as the positive x, y, and z axes. The **normal line to a surface** at a point is the line perpendicular to the tangent plane at the point (see TANGENT —tangent plane). *Syn.* perpendicular to a curve or surface.

normal transformation. A bounded linear transformation T is normal if it commutes with its *adjoint* ($TT^* = T^*T$). A normal transformation must commute with its adjoint; but if it is not bounded, other conditions are usually imposed (e.g., that it be *closed*). A bounded linear transformation T is normal if and only if $T = A + iB$, where A and B are symmetric transformations for which $AB = BA$. See MATRIX —normal matrix, SPECTRAL —spectral theorem.

polar normal. See POLAR —polar tangent.

principal normal. See above, normal to a curve or surface.

NOR'MAL-IZED, *adj.* **normalized random variable.** See RANDOM —normalized random variable.

NORMED, *adj.* **normed linear (vector) space.** See VECTOR —vector space.

NORTH, *adj.* **north declination.** See DECLINATION —declination of a celestial point.

NO-TA'TION, *n.* Symbols denoting quantities, operations, etc.

continuation notation. See CONTINUATION.

factorial notation. The notation used in writing factorials. See FACTORIAL

functional notation. See FUNCTION.

Plücker's abridged notation. See ABRIDGED.

NOTE, *n.* A signed promise to pay a specified sum of money at a given time, or in partial payments at specified times.

bank note. A note given by a bank and used for currency. Usually has the shape and general appearance of government paper money.

demand note. A note that must be paid when the payee demands payment.

interesting-bearing note. A note containing the words "with interest at—per cent," meaning that interest must be paid when the note is paid or at other times agreed upon by the payee and payer.

noninterest-bearing note. (1) A note on which no interest is paid. (2) A note on which the interest has been paid in advance.

note receivable. *A promissory note*, payment of which is to be made *to* the person under consideration, as contrasted to *note payable* which is to be paid *by* this person. Terms used in bookkeeping and accounting.

promissory note. A note, usually given by an individual, promising to pay a given sum of money at a specified time (or on demand).

NOUGHT, *n.* Same as ZERO; more commonly spelled naught.

NOVIKOV, Serge P. (1938–). Russian geometric and algebraic topologist and Fields Medal recipient (1970). Research on differentiable manifolds, homotopy theory, cobordism, and foliations.

N-TUPLE, *n.* See ORDERED —ordered pair.

NU, *n.* The thirteenth letter of the Greek alphabet: lower case ν; capital, N.

NU'CLE-US, *n.* nucleus of an integral equation. Same as the KERNEL.

NULL, *adj.* (1) Nonexistent; of no value or significance. (2) Quantitatively zero (as *null circle* —a circle of zero area). (3) Not any, empty (as *null set*).

null hypothesis. See HYPOTHESIS —test of a hypothesis.

null matrix. A matrix whose elements are all zero.

null sequence. A sequence whose limit is zero.

null set. The set which is empty—has no members. *Syn.* empty set.

NUM'BER, *n.* (1) The positive integers (see INTEGER). *Number* is still used quite commonly to denote *integer*, in such phrases as *The Theory of Numbers*. (2) The set of all **complex numbers** (see COMPLEX —complex numbers). The **real**

numbers constitute part of the set of complex numbers (the complex numbers which are not real numbers are the numbers $a + bi$ for which a and b are real numbers and $b \neq 0$). The real numbers are of two types, **irrational numbers** and **rational numbers** (see DEDEKIND —Dedekind cut, and RATIONAL). The irrational numbers are of two types, **algebraic irrational numbers** and **transcendental numbers** (see IRRATIONAL — irrational numbers). The rational numbers are of two types, the **integers** and the real numbers that can be represented by **fractions** of the form m/n, where m and n are integers and m is not divisible by n (see INTEGER). (3) See CARDINAL—cardinal number, ORDINAL —ordinal number.

absolute number. See ABSOLUTE.

abstract number. See ABSTRACT.

abundant number. See below, perfect number.

amicable numbers. See AMICABLE.

arithmetic numbers. Real positive numbers; positive integers, fractions, or radicals such as occur in arithmetic.

Bernoulli numbers. See BERNOULLI, James.

cardinal number. See CARDINAL.

Cayley number. See CAYLEY—Cayley algebra.

chromatic number. See FOUR—four-color problem.

concrete number. See CONCRETE.

counting number. See COUNTING.

defective number. See below, perfect number.

denominate number. A number whose unit represents a unit of measure. The **denomination** of a denominate number is the kind of unit to which the number refers, as pounds, feet, gallons, tenths, hundreds, thousands, etc.

extended real-number system. See EXTENDED.

Fermat numbers. See FERMAT.

imaginary number. See COMPLEX—complex number.

Liouville number. See LIOUVILLE.

mixed number. See MIXED.

natural number. Same as POSITIVE INTEGER.

negative number. See POSITIVE—positive number.

normal number. See NORMAL.

number class modulo n. The totality of integers each of which is congruent to a given integer modulo n. Thus if $x \equiv 2 \pmod 3$, then x belongs to the class of integers each of which is congruent to 2 modulo 3. This is the totality of all numbers of type $3k + 2$, where k represents an arbitrary integer.

number field. See FIELD.

number line. A straight line on which points are identified with real numbers, points identified with successive integers usually being spaced at unit distance apart.

number scale See SCALE—number scale.

number sentence. A statement about numbers that is a sentence if its symbols are interpreted in the usual way. Examples are $3 < 7$, $5 - 2 = 3$, $2x + 1 = 17$, $x^2 + 8x = 5$.

number sieve. A mechanical device for factoring large numbers. See ERATOSTHENES.

number system. (1) A method of writing numerals to indicate numbers, as for the *binary number system*, *decimal number system*, and *duodecimal number system*. (2) A mathematical system consisting of a set of objects called numbers, a set of axioms, and some operations that act on the numbers, as for the *real number system*, the *complex number system*, and *Cayley numbers* [see CAYLEY—Cayley algebra].

number theory. See THEORY—number theory.

ordinal number. See ORDINAL.

p-adic number. See *p*-ADIC.

perfect number. An integer which is equal to the sum of all of its divisors except itself; 28 is a perfect number, since $28 = 1 + 2 + 4 + 7 + 14$. If the sum of the divisors of a number (except itself) is less than the number, the number is **defective** (or **deficient**); if its is greater than the number, the number is **abundant**. Every number of type $2^{p-1}(2^p - 1)$ for which p and $2^p - 1$ are primes is perfect, and every even perfect number is of this type with p and $2^p - 1$ prime. If there is an odd perfect number, it is greater than 10^{250}. If n is perfect and $\sigma(n)$ is the sum of all divisors of n, then $\sigma(n) = 2n$. There are numbers n with $\sigma(n) = 3n$; *e.g.*, 120 and 672. A **perfect number of the second kind** is an n with n^2 the product of all divisors of n; *e.g.*, p^3 or pq with p and q different primes. See MERSENNE—Mersenne number.

positive number. See POSITIVE.

Pythagorean numbers. See PYTHAGORAS—Pythagorean triple.

Ramsey number. See RAMSEY—Ramsey theory.

random numbers. See RANDOM—table of random numbers.

square numbers. The numbers $1, 4, 9, 16, 25, \cdots, n^2, \cdots$, which are squares of integers.

transfinite number. See TRANSFINITE.

triangular numbers. The numbers $1, 3, 6, 10, \cdots$. They are called triangular numbers because they are the number of dots employed in making successive triangular arrays of dots. The process is started with one dot, successive rows of dots being placed beneath the first dot, each row having one more dot than the preceding one. The number of dots in the nth array is $(1 + 2 + 3 + \cdots + n) = \frac{1}{2}n(n + 1)$. See ARITHMETIC—arithmetic series.

whole number. A nonnegative integer.

NU′MER-ALS, *n.* Symbols used to denote numbers. See ARABIC, BABYLONIAN, CHINESE, EGYPTIAN, GREEK, ROMAN, VIGESIMAL.

NU′MER-A′TION, *n.* The process of writing or stating numbers in their natural order; the process of numbering.

NU′MER-A′TOR, *n.* The expression above the line in a fraction, the expression which is to be divided by the other expression, called the *denominator*.

NU-MER′I-CAL, *adj.* Consisting of numbers, rather than letters; of the nature of numbers.

numerical analysis. The study of methods of obtaining approximations for solutions of mathematical problems.

numerical determinant. A determinant whose elements are all numerical, rather than literal. See DETERMINANT.

numerical equation. An equation in which the coefficients and constants are numbers rather than letters. The equation $2x + 3 = 5$ is a numerical equation, whereas $ax + b = c$ is a *literal equation*.

numerical phrase. A collection of numbers and associated signs which indicate how the numbers are to be combined to produce a single number. *E.g.*, $3 + 2(7 - 4)$ is a numerical phrase which is equal to 9. An **open numerical phrase** is an expression which becomes a number phrase when values are given to certain variables that occur in the expression. *E.g.*, $2x(3 + y)$ is an open number phrase; when x and y are replaced by 7 and 2, it becomes the numerical phrase $2 \cdot 7(3 + 2)$ which equals 70. See below, numerical sentence.

numerical sentence. A statement about numbers. *E.g.*, $2 + 3 = 5$ is a true numerical sentence and $2 + 5 = 3$ is a false numerical sentence. An **open numerical sentence** is an expression which becomes a numerical sentence when values are given to certain variables. *E.g.*, $2x + 3y = 7$ is an open numerical sentence. An open numerical sentence is a special type of propositional function. See PROPOSITIONAL—propositional function.

numerical value. (1) Same as ABSOLUTE VALUE. (2) Value given as a number, rather than by letters.

O

O, o NOTATION. See MAGNITUDE —order of magnitude.

OB'LATE, *adj.* Resembling a sphere flattened at opposite ends of a diameter.
 oblate ellipsoid of revolution (oblate spheroid). See ELLIPSOID.

OB-LIQUE', *adj.* Neither perpendicular nor horizontal; slanting; changed in direction (as: "having a direction *oblique* to that of a former motion"). An **oblique angle** is an angle not a multiple of 90°; an **oblique triangle** is any triangle (plane or spherical) which does not contain a right angle; an **oblique prism** has bases that are not perpendicular to the lateral edges. **Oblique lines** are lines which are neither parallel nor perpendicular; **oblique coordinates** are coordinates determined from oblique axes (see CARTESIAN—Cartesian coordinates). A line is oblique to a plane if it is neither parallel nor perpendicular to the plane (an **oblique circular cone** is a circular cone whose axis is oblique to its base).

OB-TUSE', *adj.* An **obtuse angle** is an angle greater than a right angle and smaller than a straight angle. An **obtuse triangle** is a triangle one of whose angles is obtuse.

OC'TA-GON, *n.* A polygon having eight sides. A **regular octagon** is an octagon whose angles are all equal and whose sides are all equal.

OC-TA-HE'DRAL, *adj.* **octahedral group.** The group of motions (symmetries) of 3-dimensional space that preserves a regular octahedron. It also is the symmetric group of order 4!; *i.e.*, the set of permutations on four objects. See GROUP, SYMMETRY—group of symmetries.

OC'TA-HE'DRON, *n.* [*pl.* **octahedrons** or **octahedra**]. A polyhedron having eight faces. A **regular octahedron** is an octahedron whose faces are all congruent regular equilateral triangles. There are 257 types of convex octahedrons. See POLYHEDRON.

OC'TAL, *adj.* **octal number system.** A system of numerals representing real numbers for which 8 is the base. *Syn.* octonary number system. See BASE—base of a number system.

OC'TANT, *n.* See CARTESIAN —Cartesian coordinates.

OC-TIL'LION, *n.* (1) In the U.S. and France, the number represented by one followed by 27 zeros. (2) In England, the number represented by one followed by 48 zeros.

OC'TO-NAR-Y, *adj.* **octonary number system.** See OCTAL.

ODD, *adj.* **odd function.** See FUNCTION —odd function.
 odd number. An integer that is not evenly divisible by 2; any number of the form $2n + 1$, where n is an integer; 1, 3, 5, 7 are odd numbers.

ODDS, *n.* (1) *In betting*, the ratio of the wager of one party to that of the other, as to lay or give *odds*, say 2 to 1. (2) The probability or degree of probability in favor of some event on which bets are laid. (3) An equalizing allowance given to a weaker side or player by a stronger, as a piece at chess or points at tennis.

OFF'SET, *n.* A shift from a *given direction* perpendicular to that direction in order to pass an obstruction and yet obtain the total distance through the obstruction in the given direction. The distance across a pond in a given direction can be obtained by adding all the line segments in this direction which appear in a set of offsets (steps) going around the pond.

OHM, Georg Simon (1787–1854). German physicist.
 ohm. A unit of electrical resistance. (1) The **absolute ohm** is defined as the resistance of a conductor which carries a steady current of one absolute ampere when a steady potential difference of one absolute volt is impressed across its terminals. This is equivalent to the statement that the conductor dissipates heat at the rate of one watt when it carries a steady current of one absolute ampere. The absolute ohm has been the legal standard of resistance since 1950. (2) The **international ohm**, the legal standard before 1950, is the resistance offered to a steady electric current by a column of mercury of 14.4521 gm mass, constant cross-sectional area, and a length of 106.300 cm, at 0° C.

$$1 \text{ int. ohm} = 1.000495 \text{ abs. ohm}$$

 Ohm's law. (*Electricity*) Current is proportional to electromotive force divided by resistance. This law applies to a metallic circuit if the electromotive force and the current are constant. See HELMHOLTZ.

O-ME′GA, *n.* The twenty-fourth and last letter of the Greek alphabet: lower case, ω; capital, Ω.

OM′I-CRON, *n.* The fiteenth letter of the Greek alphabet: lower case, o; capital, O.

ONE, *adj.* The cardinal number of a set with a single member; the multiplicative identity for real numbers, *i.e.*, the number 1 for which $1 \cdot x = x \cdot 1 = x$ for all numbers x.

    one-parameter family of curves or surfaces. See FAMILY.

    one-sided surface. See SURFACE —one-sided surface.

    one-to-one. See CORRESPONDENCE—one-to-one correspondence.

    one-valued, *adj.* Same as SINGLE-VALUED.

    one-way classification. See CLASSIFICATION.

ON′TO, *prep.* A map or transformation of a set X which maps points of X into points of Y is said to be a map of X *into* Y; it is a map of X *onto* Y if each point of Y is the image of at least one point of X. *E.g.*, $y = 3x + 2$ is a map of the real numbers *onto* the real numbers; $y = x^2$ is a map of the real numbers *into* the real numbers, or *onto* the nonnegative real numbers. See SURJECTION.

O′PEN, *adj.* **open interval.** See INTERVAL.

    open mapping. A mapping (correspondence, transformation, or function) which associates with each point of a space D a unique point of a space Y, is **open** if the image of each open set of D is open in R; it is **closed** if the image of each closed set is closed. An open mapping might or might not be closed and a closed mapping might or might not be open. Also, a continuous mapping need not be either open or closed, but f^{-1} is continuous if f is one-to-one and either open or closed. See CONTINUOUS —continuous function. *Syn.* interior mapping.

    open-mapping theorem. Let X and Y be Banach spaces (or Fréchet spaces) and T be a linear mapping whose domain is X and whose range is Y (or a second-category subspace of Y). The **closed-mapping theorem** asserts T is continuous if and only if T is *closed* [see CLOSED — closed mapping or transformation (2)]. The **open-mapping theorem** asserts T is *open* if T is continuous. The **inverse-mapping theorem** asserts the inverse T^{-1} is continuous if T is continuous and T is one-to-one (so T^{-1} exists). See FRÉCHET —Fréchet space (2).

    open phrase. See NUMERICAL—numerical phrase.

    open sentence. Same as OPEN STATEMENT. See next column.

open set of points. A set U such that each point of U has a neighborhood all the points of which are points of U; the complement of a closed set. The interior of a circle and the set of all points on one side of a straight line in a plane are open sets. See TOPOLOGICAL —topological space.

    open statement. Same as PROPOSITIONAL FUNCTION, *i.e.*, an **open statement** is a function whose range is a collection of statements. See NUMERICAL—numerical sentence, PROPOSITIONAL—propositional function.

OP′ER-A′TION, *n.* (1) The process of carrying out rules of procedure, such as addition, subtraction, differentiation, taking logarithms, making substitutions or transformations. (2) An operation on a set S is a function whose domain is a set of ordered sequences $(x_1, x_2, \cdots, x_n)$ of members of S and whose range is contained in S. The operation is **unary, binary, ternary,** etc., according as n is $1, 2, 3$, etc. Sometimes such a function is said to be an **internal operation** on S and an **external operation** is a function for which either the values of the function are not in S or not all the independent variables have values in S. *E.g.*, the *vector product* is an internal operation on vectors, but *multiplication by scalars* and the *scalar product* are external operations. See MULTIPLICATION —multiplication of vectors, BINARY —binary operation, TERNARY —ternary operation.

    fundamental operations of arithmetic. See FUNDAMENTAL.

OP′ER-A′TOR, *n.* Same as FUNCTION.

    differential operator. See DIFFERENTIAL—differential operator.

    linear operator. Common designation for a linear transformation on a Hilbert space. As for any normed linear space, the set of bounded linear operators on a Hilbert space H has the **norm** or **uniform topology** (see LINEAR —linear transformation) and the **strong operator topology,** for which a subbase is the collection of all sets of type

$$\{T : \|(T - T_0)x\| < \epsilon\}, \qquad x \text{ in } H,$$

but also has the **weak operator topology** for which a subbase is the collection of all sets of type

$$\{T : |((T - T_0)x, y)| < \epsilon\}, \qquad x \text{ and } y \text{ in } H.$$

OP′PO-SITE, *adj.* An angle and a side of a triangle are **opposite** if the sides of the angle are the other two sides of the triangle. For any polygon with an even number of sides (*e.g.*, a quadrilateral) **opposite vertices** (or **opposite sides**) are two vertices (or sides) separated by the same number of sides whichever way one travels around the polygon.

opposite rays. See PARALLEL —parallel rays.

OP′TI-CAL, *adj.* **optical property of conics.** See *focal property* under ELLIPSE, HYPERBOLA, PARABOLA.

OP′TI-MAL, *adj.* **optimal strategy.** See STRATEGY.

OP′TI-MAL′I-TY, *n.* **principle of optimality.** In dynamic programming, the principle that, regardless of the initial state of the process under consideration and regardless of the initial decision, the remaining decisions must form an optimal policy relative to the state resulting from the first decision. See PROGRAMMING —dynamic programming.

OP′TION, *n.* **option term insurance.** See INSURANCE —life insurance.

OP′TION-AL, *adj.* (*Finance*) Same as CALLABLE. See BOND.

OR′BIT, *n.* Let *G* be a set of functions, each of which maps a given set *S* into *S*. For each *x* in *S*, the **orbit** of *x* is the set of all $g(x)$ for $g \in G$. *E.g.*, if *f* is a function that maps *S* into *S* and $x \in S$, then the sequence $\{x, f(x), f^2(x), \cdots\}$ is an orbit of *x* if *G* is the sequence $\{f, f^2, \cdots\}$, where $f^2 = f[f(x)]$, etc. If *G* is a *group*, then an *equivalence relation* can be defined by letting two points of *S* be equivalent if they belong to the same equivalence class. The resulting *quotient space* is the **orbit space** of *G*. See GROUP, QUOTIENT —quotient space.

OR′DER, *adj.*, *n.* **differences of first, second, third order** See DIFFERENCE.

normal order. An established arrangement of numbers, letters, or objects which is called *normal* relative to all other arrangements. If a, b, c is defined as the normal order for these letters, then b, a, c is an *inversion*. See INVERSION —inversion of a sequence of objects.

of the order of. See INFINITESIMAL —order of an infinitesimal, MAGNITUDE —order of magnitude.

order of an algebra. See ALGEBRA —algebra over a field.

order of an algebraic curve or surface. The degree of its equation; the greatest number of

points (real or imaginary) in which any straight line can cut it.

order of an *a*-point of an analytic function. See ANALYTIC —*a*-point of an analytic function.

order of a branch-point of a Riemann surface. For $k \geq 1$, a **branch-point of order *k*** is a point at which $k + 1$ sheets of a Riemann surface hang together.

order of contact of two curves. A measure of how close the curves lie together in the neighborhood of a point at which they have a common tangent. *Tech.* The order of contact of two curves is one less than the order of the infinitesimal difference of the distances from the two curves to their common tangent, measured along the same perpendicular, relative to the distance from the foot of this perpendicular to their point of contact with the tangent; the order of contact of two curves is *n* when the *n*th-order differential coefficients, from their equations, and all lower orders, are equal at the point of contact, but the $(n + 1)$st-order differential coefficients are unequal. See INFINITESIMAL —order of an infinitesimal.

order of a derivative. See DERIVATIVE —derivative of higher order.

order of a differential equation. See DIFFERENTIAL —differential equation (ordinary).

order of an elliptic function. The sum of the orders of the poles of the function in a primitive period parallelogram. There are no elliptic functions of order 0 or 1. An elliptic function of order *k* takes on every complex value exactly *k* times in a primitive period parallelogram.

order of the fundamental operations of arithmetic. When several of the fundamental operations occur in succession, multiplications and divisions are performed before additions and subtractions, and in the order in which they occur; *e.g.*,

$$3 + 6 \div 2 \times 4 - 7 = 3 + 3 \times 4 - 7$$
$$= 3 + 12 - 7 = 8.$$

order of a group. See GROUP, PERIOD —period of a member of a group.

order of an infinitesimal. See INFINITESIMAL.

order of magnitude. See MAGNITUDE —order of magnitude.

order of a pole of an analytic function. See ISOLATED —isolated singular point of an analytic function.

order properties of real numbers. If $x < y$ is defined to mean that there is a positive number *a* such that $y = x + a$, then this order relation is a *linear ordering*, *i.e.*, it has the two properties: (i) (**Trichotomy**) for any two numbers *x* and *y*, exactly one of $x < y$, $x = y$, and $y < x$ is true;

(ii) (**Transitivity**) if $x < y$ and $y < z$, then $x < z$. Many other order properties can be proved for the real numbers, *e.g.*: (iii) (**Addition**) if $x < y$, then $x + a < y + a$ for all numbers a; (iv) (**Multiplication**) if $x < y$ and $a > 0$, then $ax < ay$; if $x < y$ and $a < 0$, then $ay < ax$; (v) if x and y are positive, then $x < y$ if and only if $x^2 < y^2$; (vi) (**Archimedean**) if x and y are positive, there is a positive integer n such that $x < ny$. The set of positive integers (or any other set of integers that has a smallest member) has the **well-ordering property**, *i.e.*, if S is a subset that is not empty, then S has a least member (a member less than or equal to every other member). See INEQUALITY, ORDERED —ordered set.

order of a radical. Same as the INDEX. See INDEX —index of a radical.

order of units. The place of a digit in a number. The unit in the unit's place is the unit of the *first order*, in ten's place, of the *second order*, etc.

order of a zero point of an analytic function. See ZERO —zero of a function.

OR′DERED, *adj.* **ordered integral domain and field.** See DOMAIN —integral domain, FIELD.

ordered pair. A set with two (possibly equal) terms for which one term is designated as the first and the other as the second. An **ordered triple** and **ordered n-tuple** are defined similarly, the n-tuple $(x_1, x_2, \cdots, x_n)$ having the first term x_1, the second term x_2, etc. An n-tuple can also be described as a function whose domain is the set consisting of the first n positive integers. An ordered pair of numbers has many interpretations; *e.g.*, it can represent a point in a plane, the two numbers being the Cartesian coordinates of the point, or it can represent a vector whose components are the two numbers.

ordered partition. See PARTITION.

ordered set. A **partially ordered** set (or **poset**) is a set which has a relation $x < y$, or "x precedes y", defined for some members x and y and satisfying the conditions: (1) If $x < y$, then $y < x$ is false and x and y are not the same element. (2) If $x < y$ and $y < z$, then $x < z$. Subsets of a set S are partially ordered if one defines $U < V$ for sets U and V to mean that U is a proper subset of V. Positive integers are partially ordered if one defines $a < b$ to mean a is a factor of b and $a \neq b$. A **linearly ordered set** or **totally ordered set** is a partially ordered set which satisfies the following strengthened form of (1) (the *trichotomy property*): "For any two members x and y, exactly one of $x < y$, $x = y$, $y < x$ is true." The set of positive integers (or the set of real numbers) in their natural order is a linearly

ordered set. A linearly ordered set also is called a **chain, ordered set, serially ordered set,** and **simply ordered set.** A **well-ordered set** is a linearly ordered set for which each subset has a first member (one that precedes all other members). The set of positive integers in their natural order is well ordered. The set of all integers in their natural order is not well ordered, since the set itself has no first member; the set of nonnegative real numbers in their natural order is not well ordered, since the set of numbers greater than 3 has no first member. Zermelo proved that every set can be well ordered if it is assumed that in each nonempty subset T one element of T can be chosen (or designated) as a "special" element. This assumption is the **axiom of choice** or **Zermelo's axiom.** See CHOICE, ORDER —order properties of real numbers, ZORN—Zorn's lemma.

OR′DIN-AL, *adj.* **ordinal numbers.** Numbers that denote order of the members of a set as well as the cardinal number property of the set. Two simply ordered sets are said to be **similar** if they can be put into one-to-one correspondence which preserves the ordering. All simply ordered sets which are similar to one another are said to have the same **ordinal type** and to have the same **ordinal number.** The ordinal number of integers $1, 2, \cdots, n$ is denoted by n, of all positive integers by ω, of all negative integers by ω^*, of all integers by π, of the rational numbers with the usual ordering by η, and of the real numbers with the usual ordering by λ. If α and β are the ordinal numbers of simply ordered sets P and Q, then $\alpha + \beta$ is defined as the ordinal number of the set (P, Q) containing all elements of P and Q with the given ordering of P and Q and with the definition that each element of P precedes each element of Q. We have $\omega \neq \omega^*$, $\omega^* + \omega = \pi \neq \omega + \omega^*$. Two ordered sets having the same ordinal number have the same *cardinal number*, but two ordered sets having the same cardinal number do not necessarily have the same ordinal number unless they are finite (e.g., $\omega \neq \omega^*$). Many authors allow only ordinal types which correspond to *well-ordered* sets. With this restriction, any set of ordinal numbers is well-ordered if one defines $\alpha \leq \beta$ to mean that any set of ordinal type α can be put into a one-to-one, order-preserving correspondence with an initial segment of any set of ordinal type β.

OR′DI-NAR′Y, *adj.* **ordinary annuity and life insurance.** See ANNUITY, INSURANCE —life insurance.

ordinary differential equation. See DIFFERENTIAL —differential equation (ordinary).

ordinary point of a curve. See POINT—ordinary point of a curve.

OR′DI-NATE, *n.* The coordinate of a point, in the Cartesian coordinates in the plane, which is the distance from the axis of abscissas (*x*-axis) to the point, measured along a line parallel to the axis of ordinates (*y*-axis).

average (mean) ordinate. See MEAN—mean value of a function.

double ordinate. A line segment between two points on a curve and parallel to the axis of ordinates (in Cartesian coordinates); used with reference to curves that are symmetrical with respect to the axis of abscissas, such as the parabola $y^2 = 2px$ or the ellipse

$$x^2/a^2 + y^2/b^2 = 1.$$

ORDVAC. An automatic digital computing machine built at the University of Illinois and installed at the Aberdeen Proving Ground. ORDVAC is an acronym for *Ordnance Discrete Variable Automatic Computer*.

O′RI-EN-TA′TION, *n.* See ANGLE, COMPLEX—simplicial complex, MANIFOLD, SIMPLEX, SURFACE, TRIHEDRAL.

OR′I-GIN, *n.* **origin of Cartesian coordinates.** The point of intersection of the axes. See CARTESIAN—Cartesian coordinates.

origin of a coordinate trihedral. The point of intersection of the **coordinate planes**. See CARTESIAN—Cartesian coordinates.

origin of a ray. See RAY.

OR′THO-CEN′TER, *n.* **orthocenter of a triangle.** The point of intersection of the three altitudes of the triangle. See PEDAL—pedal triangle.

OR-THOG′O-NAL, *adj.* Right-angled; pertaining to or depending on the use of right angles.

orthogonal basis. See BASIS—basis of a vector space.

orthogonal complement. The orthogonal complement of a vector **v** (or of a subset *S*) of a vector space is the set of all vectors of the space which are orthogonal to **v** (or to each vector of *S*). The orthogonal complement of a vector in three dimensions is the set of all vectors perpendicular to the given vector, *i.e.*, the set of all linear combinations of any two linearly independent vectors perpendicular to the given vector. See VECTOR—vector space.

orthogonal functions. Real functions f_1, $f_2, \cdots$, are said to be *orthogonal* on the interval (a, b) if

$$\int_a^b f_n(x) f_m(x)\, dx = 0$$

for $m \neq n$, and to be **normal**, or **normalized**, or to be an **orthonormal** system if also

$$\int_a^b [f_n(x)]^2\, dx = 1$$

for all *n*. The integral $\int_a^b f_n(x) f_m(x)\, dx$ is the **inner product** (f_n, f_m) of f_n and f_m. A **complete** orthonormal system is an orthonormal system such that

$$(F, F) = \sum_{n=1}^{\infty} (F, f_n)^2$$

for all continuous functions F or, equivalently, $\sum_1^\infty (F, f_n) f_n$ converges in the mean (of order 2) to F. If Lebesgue integration is used and the functions involved are measurable functions whose squares are integrable, then an orthonormal system is complete if and only if $F = 0$ whenever

$$\int_a^b F(x) f_n(x)\, dx = 0$$

for all *n*. The above is valid for integration over more general sets, and also for complex-valued functions, if (F, G) is defined as $\int_a^b \overline{F(x)} G(x)\, dx$. Examples of complete orthonormal systems of continuous functions are: (1) The functions

$$\frac{1}{(2\pi)^{1/2}}, \frac{\cos nx}{\pi^{1/2}}, \frac{\sin nx}{\pi^{1/2}},$$

$n = 1, 2, 3, \cdots$, on the interval $(0, 2\pi)$; (2) the functions

$$e^{nix}/(2\pi)^{1/2} \quad (n = 0, 1, \cdots)$$

on the intervals $(0, 2\pi)$; (3) the functions $[\frac{1}{2}(2n + 1)]^{1/2} P_n(x)$ $(n = 0, 1, 2, \cdots)$ on the interval $(-1, 1)$, where P_n is the *n*th Legendre polynomial. See BESSEL—Bessel's inequality, GRAM—Gram-Schmidt process, PARSEVAL—Parseval's theorem, RIESZ—Riesz-Fischer theorem.

orthogonal matrix. See MATRIX—orthogonal matrix.

orthogonal projection. See PROJECTION— orthogonal projection.

orthogonal substitution or transformation. A substitution which transforms from one set of rectangular coordinates to another.

orthogonal system of curves on a surface. A system of two one-parameter families of curves on a surface S such that through any point of S there passes exactly one curve of each family, and such that at each point P on S the tangents to the two curves of the system through P are mutually orthogonal.

orthogonal trajectory of a family of curves. A curve which cuts all the curves of the family at right angles. Any line through the origin is an *orthogonal trajectory* of the family of circles which has the origin as a common center, and any one of these circles is an orthogonal trajectory of the family of lines passing through the origin. The equation of the orthogonal trajectories of a family of curves may be obtained from the differential equation of the family by replacing dy/dx in that equation by its negative reciprocal, $-dx/dy$, and solving the resulting differential equation.

orthogonal transformation. (1) A linear transformation of the form

$$y_i = \sum_{j=1}^{n} a_{ij} x_j \quad (i = 1, 2, \cdots, n),$$

which leaves the quadratic form $x_1^2 + x_2^2 + \cdots + x_n^2$ invariant; a linear transformation whose matrix is an orthogonal matrix. (2) A transformation of the form $P^{-1}AP$ of a matrix A by an orthogonal matrix P. These two concepts are closely related. For let $A = (a_{ij})$ be the matrix of a linear transformation. Then since $A^T A$ is the identity matrix (I) if A is orthogonal and A^T is the transpose of A,

$$\sum_{i=1}^{n} y_i^2 = (y) I\{y\} = (x) A^T A\{x\} = \sum_{i=1}^{n} x_i^2,$$

where (y) is the one-row matrix $(y_1, \cdots, y_n)$, $\{x\}$ is the similar one-column matrix, and multiplication is matrix multiplication. A real orthogonal transformation is **proper** or **improper** according as the determinant of A is 1 or -1. The rotation $x' = x \cos \phi + y \sin \phi$, $y' = -x \sin \phi + y \cos \phi$ is a proper orthogonal transformation. A proper orthogonal transformation is also called a **rotation**, being the usual rotation in two or three dimensions. If a matrix is symmetric, it can be reduced to diagonal form by an orthogonal transformation. Hence orthogonal transformations are often called *principal-axis transformations* and the eigenvectors of the matrix are called *normal coordinates*. See EQUIVALENT —equivalent matrices, TRANSFORMATIOIN —congruent transformation, UNITARY —unitary transformation.

orthogonal vectors. Two vectors whose scalar product is zero. See MULTIPLICATION —multiplication of vectors, and VECTOR —vector space. For vectors represented by directed line segments in the plane or three-dimensional space, this is equivalent to the vectors (or lines) being perpendicular. See RECIPROCAL —reciprocal systems of vectors.

triply orthogonal system of surfaces. Three families of surfaces which are such that one member of each family passes through each point in space and each surface is orthogonal to every member of the other two families. In the figure, the three triply orthogonal systems of surfaces $x^2 + y^2 = r_0^2$, $y = x \tan \theta_0$ and $z = z_0$ intersect at right angles, for instance at the point P. See CONFOCAL —confocal quadratics, CURVILINEAR —curvilinear coordinates of a point in space.

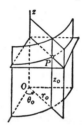

OR'THO-GRAPH'IC, *adj.* **orthographic projection.** Same as ORTHOGONAL PROJECTION. See AZIMUTHAL—azimuthal map, PROJECTION— orthogonal projection.

OR'THO-NOR'MAL, *adj.* An abbreviation of "orthogonal and normalized." See BASIS—basis of a vector space, ORTHOGONAL —orthogonal functions.

OS'CIL-LAT'ING, *adj.* **oscillating series.** See DIVERGENT —divergent series.

OS'CIL-LA'TION, *n.* A single swing from one extreme to another of an object with an oscillating or vibrating motion. *Syn.* vibration. Oscillations with variable amplitude decreasing toward zero as time increases are **damped oscillations**; **forced oscillations** are oscillations imparted to a body by an intermittent or oscillatory force, giving the motion of the body a different amplitude than it would have without such a force; oscillations which are not forced are **free** (a pendulum which has free oscillation very nearly describes simple harmonic motion if the oscillation is small); oscillations that tend toward fixed and well-defined limiting positions are **stable oscillations**. Motion as described by the differential

equation

$$\frac{d^2y}{dx^2} + A\frac{dy}{dx} + By = f(t)$$

has *free oscillations* if $A = 0$, $B > 0$, and $f(t) \equiv 0$; it then is *simple harmonic motion*. The motion has *damped oscillations* if $A > 0$ [if $f(t) \equiv 0$, the motion is oscillatory if $A < \frac{1}{2}\sqrt{B}$]. If $f(t)$ represents an oscillatory force with

$$f(t) = k\sin(\lambda t + \theta),$$

then there are *forced oscillations*. If $\lambda^2 = B$ and there is no damping ($A = 0$), then the general solution is

$$y = -\frac{kt}{2\sqrt{B}}\cos(\sqrt{B}\,t + \theta)$$
$$+ C_1\sin(\sqrt{B}\,t + C_2)$$

and the motion becomes increasingly violent as t increases; this phenomenon is called *resonance* and results when the impressed force is of the same frequency as the free vibrations of the system.

oscillation of a function. The oscillation of a function **on an interval** I is the difference between the least upper bound and the greatest lower bound of values of the function at points of I. The oscillation of a function **at a point** P is the limit of the oscillation of the function on an interval which has P as an interior point, as the length of the interval approaches zero. *Syn.* saltus of a function. See VARIATION —variation of a function.

OS′CU-LAT′ING, *p.* **osculating circle.** The **osculating circle** of a curve C at a point P is the circle in the osculating plane at P which is tangent to C, has radius equal to the reciprocal of the curvature at C, and is on the concave side of the projection of C onto the osculating plane. The osculating circle is the *circle of curvature* of the projection of C onto the osculating plane at P. *Syn.* circle of curvature. See CURVATURE — curvature of a curve.

osculating plane. The **osculating plane** of a curve C at a point P is the plane that contains the unit tangent vector $\mathbf{T}$ at P and contains the principal normal vector $d\mathbf{T}/ds$, where s is distance along the curve (the osculating plane does not exist if $d\mathbf{T}/ds = 0$, *e.g.*, if the curve is a straight line). The osculating plane is the plane in the limiting position, if this exists, of the plane through the tangent to C at the point P, and through a variable point P' on C, as $P' \to P$ along C. See NORMAL—normal to a curve or surface.

osculating sphere of a space curve at a point. The sphere through the osculating circle having contact of highest order (generally third) with the curve C at the point. Its center is on the polar line and its radius r is given by $r^2 = \rho^2 + \left(\tau\frac{d\rho}{ds}\right)^2$, where ρ and τ are the radii of curvature and torsion of C, respectively, and s is the arc length. A **stationary osculating plane** is the osculating plane at a point where the rate of change of each of the direction cosines of the binomial to the curve vanishes.

OS′CU-LA′TION, *n.* **point of osculation.** A point on a curve at which two branches have a common tangent and each branch extends in both directions of the tangent. The curve $y^2 = x^4(1 - x^2)$ has a point of osculation at the origin, the double tangent there being the x-axis. Also called a **tacnode** and a **double cusp**.

POINT OF OSCULATION

OSGOOD, William Fogg (1864–1943). American analyst. Wrote influential testbooks on calculus, mechanics, and complex-variable theory. See RIEMANN —Riemann mapping theorem.

OSTROGRADSKI, Michel Vassilievitch (1801–1862). Russian analyst.
 Ostrogradski's theorem. Same as the DIVERGENCE THEOREM.

OUT′ER, *adj.* **outer automorphism.** See ISOMORPHISM.
 outer measure. Same as EXTERIOR MEASURE. See MEASURE —exterior and interior measure.

OUT′PUT, *adj.* **output component.** In a computing machine, any component that is used in making available the results of the computation; *e.g.*, a typewriter, punched-card machine, or tape might be used for this purpose.

O′VAL, *n.* A curve shaped like a section of a football or of an egg; any curve that is closed and always concave toward the center; a closed curve bounding a convex domain.
 oval of Cassini. See CASSINI.

O′VER-HEAD′, *adj.* **overhead expenses.** All expenses except labor and material.

P

***p*-ADIC,** *adj.* For a fixed prime p, a **p-adic integer** is a sequence of integers $\{x_0, x_1, \cdots\}$ for which $x_n \equiv x_{n-1} \pmod{p^n}$ for all $n \geq 1$ (see CONGRUENCE). Two p-adic integers $\{x_n\}$ and $\{y_n\}$ being *equal* means that $x_n \equiv y_n \pmod{p^{n+1}}$ for all $n \geq 0$. If addition and multiplication are defined by $\{x_n\} + \{y_n\} = \{x_n + y_n\}$, and $\{x_n\}\{y_n\} = \{x_n y_n\}$, then the set of p-adic integers is a *commutative ring*; $\{x_n\}$ is a *unit* if and only if $x_0 \not\equiv 0 \pmod{p}$. A **$p$-adic number** is any fraction of type a/p^k, where a is a p-adic integer, k is a nonnegative integer, and $a/p^m = b/p^n$ means ap^n and bp^m are equal p-adic integers. The **p-adic field** is the set of all p-adic numbers with addition and multiplication defined by

$$\frac{a}{p^m} + \frac{b}{p^n} = \frac{ap^n + bp^m}{p^{m+n}}, \qquad \frac{a}{p^m} \cdot \frac{b}{p^n} = \frac{ab}{p^{m+n}}.$$

Any non-zero p-adic number is representable in the form $p^n e$, where n is an integer and e is a unit. If $V(p^n e) = n$, then V is a **valuation** of the field.

PAIR, *n.* See ORDERED —ordered pair.

PAIRED, *adj.* **paired observations.** See MATCHED —matched samples.

PALEY, Raymond Edward Alan Christopher (1907–1933). Gifted English mathematician killed by an avalanche while skiing near Banff, Canada.

    Paley-Wiener theorem. If $\{x_i\}$ is a basis for a Banach space X, $\{y_i\}$ is a sequence in X, and there is a positive number $\theta < 1$ such that

$$\left\| \sum_{i=1}^{n} a_i (x_i - y_i) \right\| \leq \theta \left\| \sum_{i=1}^{n} a_i x_i \right\|$$

for all numbers $\{a_i\}$, then $\{y_i\}$ is a basis for X and $y_i = T(x_i)$ defines an isomorphism T of X onto X. This theorem was stated by Paley and Wiener for complete orthonormal bases for Hilbert space. One can omit θ and $\{a_i\}$ and replace the inequality by

$$\sum_{i=1}^{\infty} \|x_i - y_i\| \|f_i\| < 1,$$

where $\{f_i\}$ is the sequence of continuous linear functionals such that $f_i(x_j)$ is 1 or 0 as $i = j$ or $i \neq j$. See BASIS (2), ORTHOGONAL —orthogonal functions.

PAN′TO-GRAPH, *n.* A mechanical device for copying figures and at the same time changing the scale by which they are drawn, *i.e.*, for drawing figures similar to given figures. It consists of four graduated bars forming an adjustable parallelogram with the sides extended (see figure). The point P is fixed and, while the point Q traces out the figure, the point S traces out the copy (or vice versa). A, B, and C are free to move.

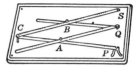

PAPPUS of Alexandria (fl. c. 300 A.D.). Greek geometer.

    theorems of Pappus. I. The area of a surface of revolution, formed by revolving a curve about a line in its plane not cutting the curve, is equal to the product of the length of the generating curve and the circumference of the circle described by the centroid of the curve. II. The volume of a solid of revolution, formed by revolving a plane set about a line in its plane not cutting the set, is equal to the product of the area of the generating set and the circumference of the circle described by the centroid of the set.

PAR, *n.* (1) Value stated in a contract to pay, such as a bond or note. Also called **par value**. *At par, below par,* and *above par* refer to an amount equal to, less than, or greater than the **face value** (the amount stated in the contract). (2) The established value of the monetary unit of one country expressed in the monetary unit of another; called in full *par of exchange, mint par,* or *commercial par.*

PA-RAB′O-LA, *n.* The intersection of a conical surface and a plane parallel to an element of the surface; a plane curve which is the set of all points equidistant from a fixed point and a fixed line in the plane. Its **standard equation** in rectangular Cartesian coordinates is $y^2 = 2px$ (also written $y^2 = 4mx$), where the fixed point is on the positive x-axis at a distance $\frac{1}{2}p$ (or m) from the origin, and the given line is parallel to the y-axis at a distance $\frac{1}{2}p$ to the left of the origin. The given point is the **focus** of the parabola, and the given line is the **directrix**. The axis of symmetry (the x-axis in the standard form given above) is the **axis** of the parabola. The point where the axis cuts the parabola is the **vertex**, and the chord through the focus and perpendicular to the axis is the **latus rectum**. Important **parametric equations** of the parabola are those used, for instance, in determining the trajectory of a projectile. If v_0 is the magnitude of the initial velocity and β the angle the projectile makes with the horizontal plane when it starts, the equations of its path are

$$x = v_0 t \cos \beta, \qquad y = v_0 t \sin \beta - \tfrac{1}{2}gt^2,$$

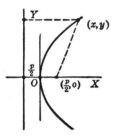

where the parameter t represents the time that has elapsed since the flight of the object started, and g is the acceleration of gravity. These are equations of a parabola. If $\beta = 45°$ (the angle at which a projectile must be thrown to travel farthest, neglecting air resistance) and g is approximated as 32, then the equations reduce to

$$x = \tfrac{1}{2}\sqrt{2}\, v_0 t, \qquad y = \tfrac{1}{2}\sqrt{2}\, v_0 t - 16t^2,$$

from which $y = x - 32x^2/v_0^2$.

cubical parabola. The plane locus of an equation of the form $y = kx^3$. When k is positive the x-axis is an inflectional tangent, the curve passes through the origin and has infinite branches in the 1st and 3rd quadrants, and is concave up in the first and concave down in the 3rd quadrant. When k is negative the curve is the graph of $y = |k|x^3$ reflected in the y-axis.

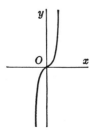

The **semicubical parabola** is the plane locus of the equation $y^2 = kx^3$. It has a cusp of the first kind at the origin, the x-axis being the double tangent. It is the locus of the intersection of a variable chord, perpendicular to the axis of an ordinary parabola, with a line drawn through the vertex of the parabola and perpendicular to the tangent at the end of the chord. The cubical and semicubical parabolas are *not* parabolas.

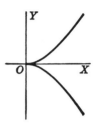

diameter of a parabola. The locus of the midpoints of a set of parallel chords. Any line parallel to the axis of the parabola is a diameter with reference to some set of chords. See DIAMETER—diameter of a conic.

focal property of the parabola. The focal radius to any point P on the parabola, and a line through P parallel to the axis of the parabola, make equal angles with the tangent to the parabola at P. If the parabola is constructed from a polished strip of metal, rays from a source of light at F (see figure) will be reflected from the parabola in rays parallel to the axis of the parabola. Likewise rays parallel to the axis of the parabola will be reflected and brought together at the focus. This property is the **optical or reflection property of the parabola**, and the corresponding property for sound is the **acoustical property of the parabola**.

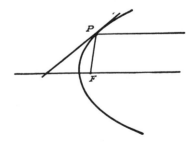

PAR'A-BOL'IC, *adj.* Of, relating to, resembling, or generated by a parabola.

parabolic cable. A cable suspended at both ends and supporting equal weights at equal distances apart horizontally. The curve is a parabola if the weight is uniformly and *continuously* distributed along the horizontal, the cable is flexible, and the weight of the cable is negligible. See CATENARY. A supporting cable of a suspension bridge hangs in a parabolic curve except for the slight modification of the curve due to the weight of the cable and the fact that the load is attached at intervals, not continuously.

parabolic curve. See CURVE—parabolic curve.

parabolic cylinder. See CYLINDER.

parabolic partial differential equation. A real second-order partial differential equation of the form

$$\sum_{t,j=1}^{n} a_{ij} \frac{\partial^2 u}{\partial x_i \partial x_j}$$

$$+ F\left(x_1, \cdots, x_n, u, \frac{\partial u}{\partial x_1}, \cdots, \frac{\partial u}{\partial x_m}\right) = 0$$

for which the determinant $|a_{ij}|$ vanishes. That is, the quadratic form $\sum_{i,j=1}^{n} a_{ij} y_i y_j$ is *singular* (by means of a real linear transformation it can be reduced to the sum of fewer than n squares, not necessarily all of the same sign). A typical example is the *heat equation*. See INDEX—index of a quadratic form.

parabolic point on a surface. A point whose Dupin indicatrix is a pair of parallel straight lines (see DUPIN); a point at which the total curvature vanishes.

parabolic Riemann surface. See RIEMANN—Riemann surface.

parabolic segment. A segment of a parabola which is subtended by a chord perpendicular to the axis of the parabola. Its area is $\frac{2}{3}cd$, where c (the **base**) is the length of the chord, and d (the **altitude**) is the distance from the vertex to the chord.

parabolic space. Same as the (Euclidean) projective plane. See PROJECTIVE.

parabolic spiral. The spiral in which the square of the radius vector is proportional to the vectorial angle. Its equation in polar coordinates is $r^2 = a\theta$. Also called **Fermat's spiral**.

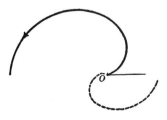

PA-RAB'O-LOID, *n.* A term applied to the *elliptic* and *hyperbolic paraboloids*. The **elliptic paraboloid** is a surface which can be put in a position such that its sections parallel to one of the coordinate planes are ellipses, and parallel to the other coordinate planes are parabolas. When the surface is in the position illustrated, with its axis along the z-axis, its equation is

$$x^2/a^2 + y^2/b^2 = 2cz.$$

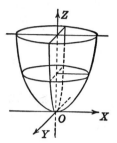

An elliptic paraboloid is a **paraboloid of revolution** if it is formed by revolving a parabola about its axis. This is the special case of the elliptic paraboloid in which the cross sections perpendicular to the axis are circles. The **hyperbolic paraboloid** is a surface which can be put in a position such that its sections parallel to one coordinate plane are hyperbolas, and parallel to the other coordinate planes are parabolas. In the position illustrated, its equation is

$$x^2/a^2 - y^2/b^2 = 2cz.$$

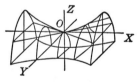

This is a ruled surface, the two families of rulings being

$$x/a - y/b = 1/p, \qquad x/a + y/b = 2pcz,$$

and

$$x/a + y/b = 1/p, \qquad x/a - y/b = 2pcz,$$

where p is an arbitrary parameter. These rulings are called RECTILINEAR GENERATORS, since either set may be used to generate the surface.

similar paraboloids. See SIMILAR—similar hyperboloids and paraboloids.

PAR'A-COM-PACT', *adj.* **paracompact space.** A topological space T which has the property that, for any family F of open sets whose union contains T, there is a *locally finite* family F^* of open sets whose union contains T and which are such that each member of F^* is contained in a member of F; T is **countably paracompact** if F has this property whenever F is countable. A paracompact space is *regular* and *normal*. See COMPACT, METACOMPACT.

PAR'A-DOX, *n.* An argument in which it appears that an obvious untruth has been proved. See BANACH—Banach-Tarski paradox, BURALI-FORTI, GALILEI, HAUSDORFF—Hausdorff paradox, PETERSBURG PARADOX, RUSSELL, ZENO.

PAR'AL-LAC'TIC, *adj.* **parallactic angle of a star.** The angle between the arcs of great circles joining the star and the zenith, and the star and the pole. See HOUR—hour angle and hour circle.

PAR'AL-LAX, *n.* **geodesic parallax of a star.** The plane angle subtended at the star by the radius of the earth.

PAR'AL-LEL, *adj., n.* [Greek *parallēlos*: *para* (side-by-side) and *allēlōn* (of one another)]. See CURVE—parallel curves (in a plane), GRAPH THEORY, and headings below, especially, parallel lines and planes, parallel vectors.
   Euclid's postulate of parallels. If two lines are cut by a transversal and the sum of the interior angles on one side of the transversal is less than a straight angle, the two lines will meet if produced and will meet on that side of the transversal. If suitable other axioms are used, this is logically equivalent to: Only one line can be drawn parallel to a given line through a given point not on this line.
   geodesic parallels on a surface. See GEODESIC.
   parallel-axis theorem. If I is the *moment of inertia* about a line L_0 through the center of mass of an object of mass M, and L_1 is parallel to L_0, then the moment of inertia about L_1 is $I + h^2 M$, where h is the distance between the lines. For a *random variable X* with *variance* σ^2 and *mean* $\overline{X}$, this theorem gives the *expected value* of $(X - a)^2$ to be $\sigma^2 + (\overline{X} - a)^2$.
   parallel circles. See SURFACE—surface of revolution.
   parallel displacement of a vector along a curve. If C is an arbitrarily given curve with parametric

equations

$$x^i = f^i(t) \, (t_0 \le t \le t_1),$$

and if ξ is any given contravariant vector at the point $x^i(t_0)$ on the curve C, then, under suitable restrictions on the metric tensor g_{ij} and on the curve C, the system of differential equations

$$\frac{d\xi^i(t)}{dt} + \Gamma^i_{\alpha\beta}(x^1(t), \cdots, x^n(t)) \xi^\alpha(t) \frac{dx^\beta(t)}{dt}$$

$$= 0,$$

subject to the initial conditions $\xi^i(r_0) = \xi^i_0$, will define a unique contravariant vector $\xi^i(t)$ at each point $x^i(t)$ of the given curve C. The vector $\xi^i(t)$ at the point $x^i(t)$ of the curve C is **parallel** to the given vector ξ^i_0 with respect to the given curve C, and the vector $\xi^i(t)$ is said to be obtained from the given vector ξ^i_0 by a **parallel displacement**. The set of vectors $\xi^i(t)$, as the point $x^i(t)$ on C varies, is a **parallel (contravariant) vector field with respect to the given curve** C. E.g., the tangent vector field $dx^i(s)/ds$ to a geodesic forms a parallel (contravariant) field with respect to the geodesic.
   parallel lines and planes. Two lines are parallel if there is a plane in which they both lie and they do not meet however far they are produced. The analytic condition that two noncoincident lines in a plane be parallel is that the coefficients of the corresponding variables in their rectangular Cartesian equations be proportional, that their slopes be equal, or that the determinant of the coefficients of the variables in their equations be zero. The condition that two noncoincident lines in space be parallel is that they have the same direction cosines (or direction cosines of opposite signs) or that their direction numbers be proportional. **Two planes** are parallel if they do not meet however far produced. The analytic condition that two noncoincident planes be parallel is that the direction numbers of their normals be proportional or that the coefficients of like variables in the Cartesian equations of the planes be proportional. A line and plane which do not meet however far produced are **parallel**. A line is parallel to a plane if, and only if, it is perpendicular to the normal to the plane (see PERPENDICULAR—perpendicular lines and planes). A necessary and sufficient condition that three lines all be parallel to some plane is the vanishing of the third-order determinant whose rows are the direction numbers of the three lines, taken in a fixed order.

parallel rays. Two rays that lie on the same line or on parallel lines. It sometimes also is required that they point in the same direction (then two rays that point in opposite directions are **opposite rays**).

parallel sailing. See SAILING.

parallel surfaces. Surfaces having common normals. The only surfaces parallel to the surface S: $x = x(u,v)$, $y = y(u,v)$, $z = z(u,v)$, are the surfaces whose coordinates are $(x + aX, y + aY, z + aZ)$, where X, Y, Z are the direction cosines of the normal to S, and a is constant.

parallel vectors. Two nonzero vectors **u** and **v** for which there is a nonzero scalar k for which $\mathbf{u} = k\mathbf{v}$. For ordinary three-dimensional vectors, two nonzero vectors **u** and **v** are parallel if and only if their vector product $\mathbf{u} \times \mathbf{v}$ is zero, or if and only if they lie along the same line when they are represented by arrows with the same initial point. Sometimes a more restrictive definition is used, that **parallel vectors** are vectors **u** and **v** for which there is a positive constant k for which $\mathbf{u} = k\mathbf{v}$. Then three-dimensional vectors are parallel if and only if they lie along the same line and point in the same direction when they are represented by arrows with the same initial point (their vector product is zero and their scalar product is positive), and they are **antiparallel** if they point in opposite directions (then $k < 0$, their vector product is zero, and their scalar product is negative).

parallels of latitude. Circles on the earth's surface whose planes are parallel to the plane of the equator.

PAR'AL-LEL'E-PI'PED, *n.* A prism whose bases are parallelograms; a polyhedron, all of whose faces are parallelograms. The faces other than the bases are **lateral faces**; the sum of their areas, the **lateral area** of the prism; and their intersections, the **lateral edges**. A **diagonal** of a parallelepiped is a line segment joining two vertices which are not in the same face. There are four of these, the **principal diagonals**, the other diagonals being the diagonals of the faces. An **altitude** of a parallelepiped is the perpendicular distance from one face (the **base**) to the opposite face; the volume is equal to the product of an altitude and the area of a base. A **right parallelepiped** is a parallelepiped whose bases are perpendicular to its lateral faces. It is a special type of right prism. A **rectangular parallelepiped** is a right parallelepiped whose bases are rectangles. If its edges are of length a, b, c, then its volume is abc and the area of its entire surface is

$$2(ab + bc + ac).$$

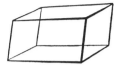

An **oblique parallelepiped** is a parallelepiped whose lateral edges are oblique to its bases.

PAR'AL-LEL'O-GRAM, *n.* A quadrilateral with its opposite sides parallel. The two line segments, AC and BD, which join opposite vertices are the **diagonals**. The **altitude** of a parallelo-

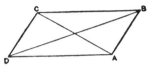

gram is the perpendicular distance between two of its parallel sides. The side to which the altitude is drawn is the **base**. The **area** of a parallelogram is the product of an altitude by the length of the corresponding base, *i.e.*, the product of the length of any side (taken as a base) by the perpendicular distance from that side to the opposite side.

parallelogram law. See SUM—sum of vectors.

parallelogram of forces, velocities, or accelerations. Same as PARALLELOGRAM OF VECTORS, with *forces, velocities,* or *accelerations* substituted for *vectors.* See SUM—sum of vectors.

parallelogram of periods of a doubly periodic function of a complex variable. See PERIOD.

PAR'AL-LEL'O-TOPE, *n.* A parallelepiped the lengths of whose sides are proportional to $1, \frac{1}{2}, \frac{1}{4}$.

Hilbert parallelotope. The set of all points $x = (x_1, x_2, \cdots)$ of Hilbert space for which $|x_n| \leq (\frac{1}{2})^n$ for each n. Any compact metric space is homeomorphic to a subset of the Hilbert parallelotope. The Hilbert parallelotope is homeomorphic to the set of all x for which $|x_n| \leq 1/n$ for each n, this set is sometimes called a Hilbert parallelotope. Each of these sets is sometimes called a **Hilbert cube**.

PA-RAM'E-TER, *n.* An arbitrary constant of a variable in a mathematical expression, which distinguishes various specific cases. Thus in $y = a + bx$, a and b are parameters which specify the particular straight line represented by the equation. Also, the term is used in speaking of any letter, variable, or constant, other than the coordinate variables; *e.g.*, in the parametric equations of a line, $x = at + x_0$, $y = bt + y_0$, $z = ct + z_0$, the variable t is a parameter whose value determines a point on the line; a, b, c are parameters whose values determine a particular line.

The parameters in a *binomial distribution* are the number of trials n and the probability p of success.

conformal parameters. See CONFORMAL—conformal map.

differential parameters. See DIFFERENTIAL.

one-parameter family. See FAMILY.

parameter of distribution of a ruled surface. For a fixed ruling L on a ruled surface S, the value of the parameter of distribution b, to within algebraic sign, is the limit of the ratio, as the variable ruling L' approaches L, of the minimum distance between L and L' to the angle between L and L'. The value of b is positive or negative according as the motion of the tangent plane is left-handed or right-handed as the point of tangency moves along

variation of parameters. See VARIATION—variation of parameters.

PAR-A-MET'RIC, *adj.* **differentiation of parametric equations.** Finding the derivative from parametric equations. If the parametric equations are $y = h(t)$, $x = g(t)$, the derivative is given by

$$\frac{dy}{dx} = \frac{dy}{dt} \div \frac{dx}{dt},$$

provided dx/dt is not zero (in case dx/dt is zero dy/dx might not have a value for this value of x or it may be possible to use some other parametric equation to find it). *E.g.*, if $x = \sin t$ and $y = \cos^2 t$, then

$$\frac{dx}{dt} = \cos t, \qquad \frac{dy}{dt} = -2\sin t \cos t,$$

and

$$\frac{dy}{dx} = \frac{dy}{dt} \div \frac{dx}{dt} = -2\sin t.$$

equidistant system of parametric curves on a surface. A system of parametric curves for the surface such that the first fundamental quadratic form reduces to

$$ds^2 = du^2 + 2F\,du\,dv + dv^2.$$

See SURFACE—fundamental quadratic form of a surface. *Syn.* Chebyshev net of parametric curves on a surface.

parametric curves on a surface. The curves of the families $u = $ const. and $v = $ const. on a surface S: $x = x(u,v)$, $y = y(u,v)$, $z = z(u,v)$; the former are called v-curves, the latter u-curves.

parametric equations. Equations in which coordinates are each expressed in terms of quantities called *parameters*. For a **curve**, parametric equations express each coordinate (two in the plane and three in space) in terms of a single parameter (see CURVE). The curve may be plotted point by point by giving this parameter values, each value of the parameter determining a point on the curve. Any equation can be written in an unlimited number of parametric forms, since the parameter can be replaced by an unlimited number of functions of the parameter. However, the term *parametric equations* sometimes refers to a specific parameter intrinsically related to the curve, as in the **parametric equations of the circle:** $x = r\cos\theta$, $y = r\sin\theta$, where θ is the central angle. See PARABOLA, ELLIPSE, and LINE—equation of a line, for specific parametric equations. The **parametric equations of a surface** are three equations (usually in Cartesian coordinates) giving x, y, z each as a function of two other variables, the parameters (see SURFACE). Elimination of the parameters between the three equations results in a Cartesian equation of the surface. The points determined when one parameter is fixed and the other varies form a curve on the surface, called a **parametric curve**. The parameters are called **curvilinear coordinates**, since a point on the surface is uniquely determined by the intersection of two parametric curves.

PA-REN'THE-SES, *n.* The symbols ( ). They enclose sums or products to show that they are to be taken collectively, *e.g.*, $2(3 + 5 - 4) = 2 \times 4 = 8$. See AGGREGATION, DISTRIBUTIVE.

PAR'I-TY, *n.* If two integers are both odd or both even they are said to have the **same parity**; if one is odd and the other even they are said to have **different parity**.

PARSEVAL DES CHÊNES, Marc Antoine (1755–1836). French mathematician. Forced to flee France after writing poems critical of the Napoleonic government.

Parseval's theorem. (1) If

$$a_k = \frac{1}{\pi} \int_0^{2\pi} f(x)\cos kx\,dx$$

and

$$b_k = \frac{1}{\pi} \int_0^{2\pi} f(x)\sin kx\,dx$$

(for $n = 0, 1, 2, \cdots$) and the numbers A_k, B_k are

defined similarly for F, then

$$\int_0^{2\pi} f(x)F(x)\,dx$$

$$= \pi\left[\tfrac{1}{2}a_0 A_0 + \sum_{n=1}^{\infty}(a_n A_n + b_n B_n)\right].$$

It is necessary to make restrictions on f and F, such that $\int_0^{2\pi} f(x)\,dx$ and $\int_0^{2\pi}|f(x)|^2\,dx$ both exist (and similarly for F), or that f and F are (Lebesgue) measurable and their squares are Lebesgue integrable on $[0, 2\pi]$. For a *complete orthonormal system* of vectors $x_1, x_2, \cdots$ in an infinite-dimensional vector space with an inner product (x, y) defined (such as Hilbert space), the theorem takes the form

$$(u, v) = \sum_{k=1}^{\infty}(u, x_k)\overline{(v, x_k)}.$$

(2) The preceding formulas with $f = F$ and $u = v$. The last one can be written as

$$\|u\|^2 = (u, u) = \sum_{k=1}^{\infty}|(u, x_k)|^2.$$

See BESSEL—Bessel's inequality, VECTOR—vector space.

PAR′TIAL, *adj.* **chain rule for partial differentiation.** See CHAIN—chain rule.

    elliptic, hyperbolic, and parabolic partial differential equations. See ELLIPTIC, HYPERBOLIC, PARABOLIC.

    mixed partial derivative. A partial derivative of the second or higher order for which not all differentiations are with respect to the same variable. Usually it does not matter in what order the differentiations are performed; *e.g.*, the mixed second partial derivative of $x^2 y + x y^4$ is $2x + 4y^3$ whichever order of differentiation is used. If f_{12} denotes the partial derivative with respect to the second variable of the partial derivative f_1 of f with respect to the first variable, and if f_{21} denotes the result of differentiation in the other order, then each of the following is a sufficient condition that $f_{12}(x_0, y_0) = f_{21}(x_0, y_0)$: (1) (x_0, y_0) is in the domain of f_1 and there is a neighborhood of (x_0, y_0) on which f_{21} is continuous; (2) f_{21} is continuous at (x_0, y_0) and there is a neighborhood of (x_0, y_0) that is contained in the intersection of the domain of f_{21} and the domain of f_1. A theorem showing that under suitable conditions $f_{12}(x_0, y_0) = f_{21}(x_0, y_0)$ was proved first by Nikolaus Bernoulli, II.

    partial correlation. See CORRELATION.

    partial derivative. The ordinary derivative of a function of two or more variables with respect to one of the variables, considering the others as constants. If the variables are x and y, the partial derivatives of $f(x, y)$ are written $\partial f(x, y)/\partial x$ and $\partial f(x, y)/\partial y$, $D_x f(x, y)$ and $D_y f(x, y)$, $f_x(x, y)$ and $f_y(x, y)$, $f_1(x, y)$ and $f_2(x, y)$; or (x, y) may be deleted from any of these, leaving $\partial f/\partial x$, f_1, etc. The partial derivative of $x^2 + y$ with respect to x is $2x$; with respect to y it is 1. Geometrically, the partial derivatives of a function f of two variables with respect to x and with respect to y at the point (a, b) are equal to the slopes of the tangents to the curves which are the intersections of the surface $z = f(x, y)$ and the planes whose equations are $y = b$ and $x = a$, respectively. In the figure, the partial derivative with respect to x, evaluated at the point P, is the slope of the line PT which is tangent to the curve AB. A partial derivative of a partial derivative is a partial derivative of **second order**. *E.g.*, if $f(x, y) = x^3 y$, then $f_1(x, y) = 3x^2 y$ and $f_{12}(x, y) = 3x^2$. Partial derivatives of higher order are defined similarly.

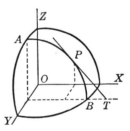

    partial differential. See DIFFERENTIAL.

    partial-differential equations. Equations involving more than one independent variable and partial derivatives with respect to these variables. Such an equation is **linear** if it is of the first degree in the dependent variables and their partial derivatives, *i.e.*, if each term either consists of the product of a known function of the independent variables, and a dependent variable or one of its partial derivatives, or is itself a known function of the independent variables. The **order** of a partial-differential equation is the order of the partial derivative of highest order which occurs in the equation.

    partial fractions. A set of fractions whose algebraic sum is a given fraction. The term *method of partial fractions* is applied to the study of methods of finding these fractions and using them, particularly in integrating certain rational fractions. *E.g.*, it is known that $1/(x^2 - 1) = A/(x - 1) + B/(x + 1)$ for some values of A and B, from which $A = \frac{1}{2}$, $B = -\frac{1}{2}$ are obtained by clearing of fractions and equating coefficients of like powers of x, or by multiplying by $x - 1$ and then letting $x = 1$ to obtain $A = \frac{1}{2}$, and multiplying by $x + 1$ and then letting $x = -1$ to obtain $B = -\frac{1}{2}$. Any quotient of polynomials for which the numerator is of lesser degree than the denominator can be expressed as a sum of fractions of types

$$\frac{A}{x - a}, \quad \frac{B}{(x - a)^n},$$

$$\frac{Cx + D}{x^2 + bx + c}, \quad \frac{Ex + F}{(x^2 + bx + c)^n},$$

where n is a positive integer and all coefficients are real if all coefficients in the original polynomials were real. Indeed, *partial fractions* are usually understood to be fractions of these relatively simple types.

partial product. The product of the multiplicand and one digit of the multiplier, when the latter contains more than one digit.

partial quotient. See FRACTION —continued fraction.

partial remainders. The (detached) coefficients of the quotient in synthetic division. See SYNTHETIC —synthetic division.

partial sum of an infinite series. See SUM.

PAR-TIC′I-PAT′ING, *adj.* **participating insurance policy.** See INSURANCE.

PAR′TI-CLE, *n.* Physically, a minute bit of matter of which a material body is composed. A mathematical idealization of a physical particle is obtained by disregarding its spatial extensions, representing it as a mathematical point, and endowing it with the property of inertia (mass).

PAR-TIC′U-LAR, *adj.* **particular solution (or integral)** of a differential equation. Any solution that does not involve arbitrary constants (constants of integration); a solution obtainable from the general solution by giving special values to the constants of integration. See DIFFERENTIAL —solution of a differential equation.

PAR-TI′TION, *n.* **partition of an integer.** The *number of partitions* $p(n)$ of a positive integer n is the number of ways n can be written as a sum of positive integers, $n = a_1 + a_2 + \cdots + a_k$, where k is any positive integer and $a_1 \geq a_2 \geq \cdots \geq a_k$. If k is restricted so that $k \leq s$, this is called the *number of partitions of n into at most s parts*. Various other types of partitions have been studied. *E.g.*, the number of partitions of n for which the *summands are all different* can be shown to be equal to the number of partitions of n for which the *summands are all odd* (but repetitions are allowed); 5 is equal to 5, 4 + 1, 3 + 2; it is also equal to 5, 3 + 1 + 1, 1 + 1 + 1 + 1 + 1.

partition of an interval. A set of closed intervals $[x_1, x_2], [x_2, x_3], \cdots, [x_n, x_{n+1}]$ for which x_1 and x_{n+1} are the end points of the interval and either $x_i \leq x_{i+1}$ for all i, or $x_i \geq x_{i+1}$ for all i. The **fineness** (or **mesh** or **norm**) of the partition is the greatest of $|x_{i+1} - x_i|$ for $i = 1, 2, \cdots, n$.

partition of a set. A collection of disjoint sets whose union is the given set. If the set has measure (or area, volume, etc.), then it may be required that each member of the partition have measure (or area, volume, etc.). Sometimes the requirement the sets be disjoint is replaced by the requirement that $A \cap B$ have zero measure (or area, volume, etc.) if A and B are members of the partition. The **fineness** of a partition of a metric space is the least upper bound of distances between points p and q with p and q in the same member of the partition. For a partition $\mathscr{P}$ of a set S, any ordered sequence $(A_1, A_2, \cdots)$ whose members are the members of $\mathscr{P}$ is an **ordered partition** of S. A partition $\mathscr{P}$ is **finer** than a partition $\mathscr{Q}$ and $\mathscr{Q}$ is **coarser** than $\mathscr{P}$ if each member of $\mathscr{P}$ is a subset of a member of $\mathscr{Q}$.

PARTS, *n.* **integration by parts.** See INTEGRATION —integration by parts.

PAS-CAL, *n.* The unit of pressure that results from a force of one newton acting over an area of one square meter. See NEWTON.

PASCAL, Blaise (1623–1662). Great French geometer, probabilist, combinatorist, physicist, and philosopher. In correspondence with Fermat, he helped lay the foundations of modern probability theory. Invented and constructed first calculating machine in history. See LIMACON.

Pascal distribution. Same as NEGATIVE BINOMIAL DISTRIBUTION. See under BINOMIAL.

Pascal's theorem. If a hexagon is inscribed in a conic, the three points of intersection of pairs of opposite sides lie on a line. See BRIANCHON —Brianchon's theorem.

Pascal triangle. A triangular array of numbers composed of the coefficients in the expansion of $(x + y)^n$ for $n = 0, 1, 2, 3$, etc. The "triangle" extends down indefinitely, the coefficients in the expansion of $(x + y)^n$ being in the $(n + 1)$st row. As shown, the array is bordered by 1's and the sum of two adjacent numbers in one row is equal to the number in the next row between the two numbers. The array is symmetric about the vertical line through the "vertex." See BINOMIAL —binomial coefficients.

```
              1
            1   1
          1   2   1
        1   3   3   1
      1   4   6   4   1
    1   5  10  10   5   1
       .    .    .
```

   principle of Pascal. The pressure in a fluid is transmitted undiminished in all directions. *E.g.*, if a pipe projects vertically above a closed tank and the tank and pipe are filled with water, then the resulting pressure on the inside surface of the tank is equal to that due to the water in the tank *plus a constant due to the water in the pipe*. That constant is equal to the weight of a column of water the height of the pipe and with unit cross section, regardless of the diameter of the pipe.

PATCH, *n.* **surface patch.**   See SURFACE.

PATH, *n.*   (1) Same as CURVE. However, a path is sometimes defined to be a *piecewise-smooth curve*. See SMOOTH—smooth curve. (2) (*Graph theory*). A sequence of edges that goes from one edge to another through a common node, but each edge in the sequence occurs only once. A path is **closed** if its starting node and ending nodes are the same. The path is (topologically) a **circle** if no node except the starting node occurs more than once. See GRAPH THEORY.
   path connected.   See CONNECTED —arc-wise connected set.
   path curve of a continuous surface deformation.   The locus of a given point of the surface under the deformation.
   path of a projectile.   See PARABOLA —parametric equations of a parabola.

PAY′EE′, *n.*   The person to whom a sum of money is to be paid. Most frequently used in connection with notes and other written promises to pay.

PAY′MENT, *n.*   A sum of money used to discharge a financial obligation, either in part or in its entirety. See DEFAULTED —defaulted payments, INSTALLMENT —installment payments.
   equal-payment method.   See INSTALLMENT —installment payments.

PAY′OFF, *adj., n.*   The amount received by one of the players in a play of a game. For a two-person zero-sum game, the **payoff function** M is the function for which $M(x, y)$ (positive or negative) is the amount paid by the *minimizing player* to the *maximizing player* in case the maximizing player uses pure strategy x and the minimizing player uses pure strategy y. A **payoff matrix** for a finite two-person zero-sum game is a matrix such that the element a_{ij} in the ith row and jth column represents the amount (positive or negative) paid by the *minimizing player* to the *maximizing player* in case the maximizing player uses his ith pure strategy and the minimizing player uses his ith pure strategy. See GAME, PLAYER.

PEANO, Giuseppe (1858–1932).   Italian logician, analyst, and geometer.
   Peano postulates.   See INTEGER.
   Peano space.   A space (or set) S which is a *Hausdorff topological space* and is the image of the *closed unit interval* $[0, 1]$ for a continuous mapping. A Hausdorff topological space is a Peano space if and only if it is *compact, connected, locally connected, metrizable,* and *nonempty* (a Peano space is also *arc-wise connected*). A Peano space is sometimes called a **Peano curve,** although a Peano curve usually is required to be a curve [*i.e.*, to have parametric equations of the form $x = f(t), y = g(t)$, where f and g are continuous and $0 \leq t \leq 1$] which passes through each point of the unit square.

PEARL-REED CURVE.   Same as LOGISTIC CURVE.

PEARSON, Karl (1857–1936).   English statistician. Introduced chi-square test. See CHI—chi-square test, NEYMAN —Neyman–Pearson test.
   Pearson classification of distributions.   The differential equation

$$\frac{dy}{dx} = \frac{x + a}{b + cx + dx^2} y$$

is satisfied by many common *probability-density functions* (*e.g.*, beta, normal, χ^2, t). If a probability density function is known to satisfy an equation of this type, then the values of a, b, c, d, and f itself are determined by the first four moments. Pearson classified the probability-density functions that are solutions of such a differential equation according to the nature of the zeros of $b + cx + dx^2$. If $a = -\mu$, $b = -\sigma^2$, and $c = d = 0$, then the *normal distribution* with mean μ and variance σ^2 results.
   Pearson coefficient.   See CORRELATION —correlation coefficient.

PEAUCELLIER, A. (1832–1913). French engineer and geometer.

Peaucellier cell. See INVERSION —inversion of a point with respect to a circle, INVERSOR.

PED'AL, *adj.* **pedal curve.** The locus of the foot of the perpendicular from a fixed point to a variable tangent to a given curve. *E.g.*, if the given curve is a parabola and the fixed point the vertex, the *pedal curve* is a cissoid.

pedal triangle. The triangle formed within a given triangle by joining the feet of the perpendiculars from any given point to the sides. The triangle *DEF* is the pedal triangle formed within the triangle *ABC* by joining the feet of the altitudes. The figure illustrates the fact that the altitudes of the given triangle bisect the angles of this pedal triangle.

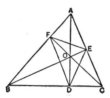

PELL, John (1610–1685). English algebraist, geometer, and astronomer.

Pellian equation. The special Diophantine equation, $x^2 - Dy^2 = 1$, where D is a positive integer not a perfect square.

PEN'CIL, *n.* A collection of geometric objects (*e.g.*, a collection of lines or a collection of spheres) for which pairs of objects have a common property. *E.g.*, the intersection of any two of the objects might be the same as that of any other two. In this case, if two different members of the collection have equations $f(x, y) = 0$ and $g(x, y) = 0$, then equations of members of the pencil are given by

$$hf(x, y) + kg(x, y) = 0,$$

where h and k are parameters that are not both zero; sometimes one parameter is taken to be 1, but this excludes one member of the pencil [if $k = 1$, for instance, there is no value of h which will reduce the equation of the pencil to $f(x, y) = 0$]. See below for special cases.

pencil of circles. All the circles that lie in a given plane and pass through two fixed points. Equations of the members of the pencil of circles through the intersections of $x^2 + y^2 - 4 = 0$ and $x^2 + 2x + y^2 - 4 = 0$ is given by

$$h(x^2 + y^2 - 4) + k(x^2 + 2x + y^2 - 4) = 0,$$

where h and k are arbitrary parameters not both zero. In the figure, $S = 0$ is the equation of one circle and $S' = 0$ is the equation of the other. If the coefficients of x^2 and y^2 are 1 in both equations, then the equation $S - S' = 0$ is the equation of the *radical axis* of any two members of the pencil, *i.e.*, the line through the points of intersection.

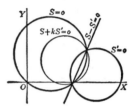

pencil of families of curves on a surface. A one-parameter set of families of curves on a surface such that each two families of the set intersect under constant angle.

pencil of lines through a point. All the lines passing through the given point and lying in a given plane. The point is the **vertex** of the pencil. *E.g.*, equations of members of the pencil through the intersection of the lines $2x + 3y = 0$ and $x + y - 1 = 0$ are given by $h(2x + 3y) + k(x + y - 1) = 0$, where h and k are not both zero.

pencil of parallel lines. All the lines having a given direction; all the lines parallel to a given line. In projective geometry, a pencil of parallel lines (pencil of parallels) is included in the classification of *pencils of lines*; the vertex of the pencil, when the lines are parallel, being an *ideal point*. The notion of ideal point thus unifies the concepts of *pencil of lines* and *pencil of parallels*. The equations of the lines of a parallel pencil can be obtained by holding m (the slope) constant and varying b (the y-intercept) in the slope-intercept form $y = mx + b$ of the equation of a line, except when the pencil is perpendicular to the x-axis, in which case the equation $x = c$ suffices. See above, pencil of lines through a point.

pencil of plane algebraic curves. All curves whose equations are given by assigning particular values to h and k, not both zero, in

$$hf_1(x, y) + kf_2(x, y) = 0,$$

where $f_1 = 0$ and $f_2 = 0$ are of the same order (degree). If n is this order, the family passes through the n^2 points (with complex coordinates) common to $f_1 = 0$ and $f_2 = 0$. A pencil of conics consists of all conics passing through four fixed points, and a pencil of cubics consists of all the cubics passing through nine fixed points. See above, pencil of circles, pencil of lines through a point.

pencil of planes. All the planes passing through a given line. The line is the **axis** of the pencil (the line **AB** in the figure).

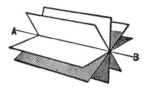

pencil of spheres. All the spheres which pass through a given circle. The plane containing the circle is the **radical plane** of the pencil.

PEN′DU-LUM, *adj., n.* **Foucault's pendulum.** A pendulum with a very long wire and a heavy bob, designed to exhibit the revolution of the earth about its axis. It is supported so as not to be restricted to remain in the same plane relative to the earth.

pendulum property of a cycloid. See CY-CLOID.

simple pendulum. A particle suspended by a weightless rod or cord; a body suspended by a cord whose weight is neglected, the body being treated as if it were concentrated as its center of gravity. The period of a simple pendulum is equal to

$$4\sqrt{\frac{L}{g}}\int_0^{\pi/2}[1-k^2\sin^2 t]^{-1/2}\,dt$$

$$=2\pi\sqrt{\frac{L}{g}}\left[1+\left(\frac{1}{2}\right)^2 k^2+\left(\frac{1\cdot 3}{2\cdot 4}\right)^2 k^4\right.$$

$$\left.+\left(\frac{1\cdot 3\cdot 5}{2\cdot 4\cdot 6}\right)^2 k^6+\cdots\right],$$

where L is the length of the pendulum and $k=\sin\frac{1}{2}\theta$ with θ the largest angle between the pendulum and the vertical. See ACCELERATION —acceleration of gravity.

PEN′TA-DEC′A-GON, *n.* A polygon having fifteen sides. A **regular pentadecagon** is a pentadecagon having all of its sides and interior angles equal. The measure of each interior angle is 156°.

PEN′TA-GON, *n.* A polygon having five sides. A **regular pentagon** is a pentagon whose sides and interior angles are all equal. The measure of each interior angle is 108°.

PEN′TAG′O-NAL, *adj.* **pentagonal-number theorem.** See EULER—Euler pentagonal-number theorem.

pentagonal pyramid. A pyramid whose base is a pentagon.

PEN′TA-GRAM (of Pythagoras). The five-pointed star formed by drawing all the diagonals of a regular pentagon and deleting the sides.

PEN-TA-HE′DRON, *n* A polyhedron with five faces. There are just two types of convex pentahedrons: (1) A pyramid with a quadrilateral base; (2) a "cylindrical type" with three quadrilateral faces and two disjoint triangular faces.

PE-NUM′BRA, *n.* See UMBRA.

PERCENT or PER CENT, *n.* Hundredths; denoted by %; 6% of a quantity is $\frac{6}{100}$ of it.

percent decrease or increase. When the value of something changes from x to y, the **percent increase** is $100(y-x)/x$ if $y>x$, and the **percent decrease** is $100(x-y)/x$ if $y<x$. E.g., if the price of eggs changes from 40¢ to 48¢ per dozen, the percent increase is $100\cdot 8/40$, i.e., 20%; if the price changes from 48¢ to 40¢, the percent decrease is $100\cdot 8/48$, i.e., $16\frac{2}{3}$%.

percent error. See ERROR.

percent profit on cost. The quotient of the selling price minus the cost, by the cost—all multiplied by 100. If an article costs 9 cents and sells for 10 cents, the *percent gain* is $\frac{1}{9}\times 100$, or 11.11%.

percent profit on selling price. The quotient of the selling price minus the cost, by the selling price—all multiplied by 100, or $100(s-c)/s$. The percent gain on the cost price is always greater than the percent gain on the selling price. If an article costs 9 cents and sells for 10 cents, the percent gain on selling price is $\frac{1}{10}\times 100$, or 10%. Compare **percent profit on cost**.

rate percent. Rate in hundredths; same as YIELD.

PER-CENT′AGE, *n.* (1) The result found by taking a certain percent of the base. (2) Parts per hundred. One would say, "A percentage (or percent) of the students are excellent"; but he would say, "Money is worth 6 percent" (never 6 percentage).

PER-CEN′TILE, *n.* One of the set of division points which divides a set of data into one hundred equal parts.

PER′FECT, *adj.* **perfect field.** A field F such that each *irreducible* polynomial with coefficients in F is *separable.* A field is perfect if it has characteristic 0. If it has characteristic $p \neq 0$, it is perfect if and only if, for each a in F, the polynomial $x^p - a$ has a root in F. See FIELD, SEPARABLE —separable polynomial.

perfect number. See NUMBER —perfect number.

perfect power. A number or polynomial which is the exact nth power of some number or polynomial for some positive integer $n > 1$. A **perfect square** is the exact square of another number or polynomial; *e.g.*, 4 is a perfect square, as is also $a^2 + 2ab + b^2$, which equals $(a + b)^2$. A polynomial such as $a^2 + 2ab + b^2$ which is the square of a binomial is a **perfect trinomial square.** A **perfect cube** is a number or polynomial that is an exact third power, such as 8, 27, and $a^3 + 3a^2b + 3ab^2 + b^3$, which is equal to $(a + b)^3$.

perfect set. A set of points (or a set in a metric space) which is identical with its *derived set*; a set which is *closed* and *dense in itself*.

PER′I-GON, *n.* An angle of 360° or 2π radians. *Syn.* round angle.

PER-I-HE′LI-ON, *n.* The point on the orbit of a planet or comet that is nearest the sun. The point farthest from the sun is the **aphelion.**

PER-IM′E-TER, *n.* The length of a closed curve, as the perimeter of a circle, the perimeter of an ellipse, or the sum of the lengths of the sides of a polygon. See ISOPERIMETRIC.

PE′RI-OD, *adj., n.* **conversion period.** The time between two successive conversions of interest. *Syn.* interest period.

parallelogram of periods. For a *doubly periodic function* of the complex variable z, a *parallelogram of periods* is a parallelogram with vertices $z_0, z_0 + \eta, z_0 + \eta + \eta', z_0 + \eta'$, where η and η' are periods (with $\eta \neq k\eta'$ for any real k) but are not necessarily a *primitive-period pair.* See below, primitive-period parallelogram.

period in arithmetic. (1) The number of digits set off by a comma when writing a number. It is customary to set off periods of three digits, as 1,253,689. These periods are called unit period, thousands period, millions period, etc. (2) When using certain methods for extracting roots, periods are set off equal to the index of the root to be extracted. (3) Period of a repeating decimal. The number of digits that repeat. See DECIMAL —repeating decimal.

period of a function. See PERIODIC —periodic function of a real variable, periodic function of a complex variable.

period of a member of a group. The least power of the member which is the identity. Sometimes called the **order.** In the group whose members are the roots of $x^6 = 1$ with multiplication as the group operation, $-\frac{1}{2} + \frac{1}{2}i\sqrt{3}$ is of period 3, since

$$\left(-\tfrac{1}{2} + \tfrac{1}{2}i\sqrt{3} \right)^3 = 1$$

and $(-\frac{1}{2} + \frac{1}{2}i\sqrt{3})^2 \neq 1$. See GROUP.

period region. For a periodic function of a complex variable, a *period region* is a *primitive-period strip* or a *primitive-period parallelogram,* according as the function is simply periodic or doubly periodic.

period of simple harmonic motion. See HARMONIC —simple harmonic motion.

primitive-period pair. Two periods ω and ω' of a *doubly periodic function* such that all periods of the function are of the form $n\omega + n'\omega'$, where n and n' are integers. See PERIODIC —periodic function of a complex variable. *Syn.* fundamental period pair.

primitive-period parallelogram. If ω and ω' form a primitive-period pair for a *doubly periodic function* of the complex variable z, and if z_0 is any point of the finite complex plane, then the parallelogram with vertices $z_0, z_0 + \omega, z_0 + \omega + \omega', z_0 + \omega'$ is a **primitive-period parallelogram** for the function. The vertex z_0 and the adjacent sides of the boundary, exclusive of their other end points, are considered as belonging to the parallelogram, the rest of the boundary being excluded. Thus each point of the finite plane belongs to exactly one parallelogram of a set of congruent primitive-period parallograms paving the entire finite plane. *Syn.* fundamental-period parallelogram.

primitive-period strip. If f is a simply periodic function of the complex variable z in the domain D, and ω is a primitive period, then a region in D bounded by a line C (or a suitable simple curve extending across D) together with an image of C translated by an amount ω is a **primitive-period strip** for f. *Syn.* fundamental-period strip.

select period of a mortality table. See MORTALITY —mortality table.

PE′RI-OD′IC, *adj.* **almost-periodic function.** A continuous function f is (**uniformly**) **almost periodic** if, for any $\epsilon > 0$, the set of all t for which the inequality

$$|f(x+t)-f(x)| < \epsilon$$

is satisfied for all x has the property that there is a number M such that any interval of length M contains at least one such t. *E.g.*, the function

$$f(x) = \sin 2\pi x + \sin 2\pi x\sqrt{2}$$

is uniformly almost periodic: $|f(x+t)-f(x)|$ is small if t is an integer and $\sqrt{2}\,t$ is near an integer. A function f is uniformly almost periodic if and only if there is a sequence of finite trigonometric sums which converges uniformly to f (the terms in the sum are of type $a_r\cos rx$ and $b_r\sin rx$, where r need not be an integer). A number of generalized definitions of almost periodicity have been studied. *E.g.*, the expression $|f(x+t)-f(x)|$ in the above definition might be replaced by the least upper bound (for $-\infty < x < +\infty$) of

$$\left[\frac{1}{k}\int_x^{x+k}|f(x+t)-f(x)|^p\,dx\right]^{1/p},$$

for some specific value of k, or by the limit of this expression as $k \to \infty$. See BOHR.

periodic continued fraction. See FRACTION—continued fraction.

periodic curves. Curves whose ordinates repeat at equal distances on the axis of abscissas; the graph of a periodic function. The loci of $y = \sin x$ and $y = \cos x$ are periodic curves, repeating themselves in every successive interval of length 2π.

periodic decimal. Same as REPEATING DECIMAL. See DECIMAL—decimal number system.

periodic function of a complex variable. The function f, analytic in a domain D, is **periodic** in D if f is not constant and if there is a complex number $\omega \neq 0$ such that if z is in D, then also $z + \omega$ is in D, and $f(z+\omega) \equiv f(z)$. The number ω is a **period** of f. If there is no period of the form $\alpha\omega$, where α is real and $|\alpha| < 1$, then ω is a **primitive period** or **fundamental period** of f. A **simply periodic** (or **singly periodic**) **function** of a complex variable is a function f of the complex variable z having a primitive period ω but having no periods other than $\pm\omega, \pm 2\omega, \cdots$. A **doubly periodic function** of a complex variable is a periodic function of a complex variable which is not *simply periodic*. It can be shown that if a periodic function is not simply periodic then there exist two *primitive periods* ω and ω' such that all periods are of the form $n\omega + n'\omega'$, where n and n' are integers but not both zero. This is

Jacobi's theorem. See ELLIPTIC—elliptic function.

periodic function of a real variable. A function such that the range of the independent variable can be separated into equal subintervals such that the graph of the function is the same in each subinterval. The length of the smallest such equal subintervals is the **period** of the function. *Tech.* If there is a least positive number p for which $f(x+p)=f(x)$ for all x [or $f(x)$ and $f(x+p)$ are both undefined], then p is the **period** of f. The trigonometric function *sine* has period 2π (radians), since

$$\sin(x+2\pi) = \sin x \quad \text{for all } x.$$

The **frequency** of a periodic function in a given interval is the quotient of the length of the interval and the period of the function (*i.e.*, the number of times the function repeats itself in the given interval). If an interval of length 2π is being considered, the frequency of $\sin x$ is 1; of $\sin 2x$, 2; and of $\sin 3x$, 3.

periodic motion. Motion which repeats itself, occurs in cycles. See HARMONIC—simple harmonic motion.

PE′RI-O-DIC′I-TY, *adj.* **periodicity of a function** (**or curve**). The property of having periods or being periodic.

PE-RIPH′ER-Y, *n.* The boundary line or *circumference* of any figure; the surface of a solid.

PER′MA-NENT-LY, *adj.* **permanently convergent series.** See CONVERGENT—permanently convergent series.

PER-MIS′SI-BLE, *adj.* **permissible values of a variable.** The values for which a function under consideration is defined and which lie on the interval or set on which the function is being considered. Zero is not a *permissible* value for x in the function log x, and 4 is not a permissible value for x if log x is being considered on the interval (1, 2). *Permissible* is also used of *any* values in the domain of the function.

PER′MU-TA′TION, *n.* (1) An ordered arrangement or sequence of all or part of a set of things. All possible permutations of the letters a, b, and c are: a, b, c, ab, ac, ba, bc, ca, cb, abc, acb, bac, bca, cab and cba. A **permutation of n things taken all at a time** is an ordered arrangement of all the members of the set. If n is the number of members of the set, the total possible numbers of such permutations is $n!$, for any one of the set can be put in the first place, any one of the remaining $n - 1$ things in the second place, etc., until n places are filled. When **some members of the set are alike** (two permutations obtainable from each other by interchanging like objects are the same permutation) the number of such permutations is the number of permutations of n different things taken all at a time divided by the product of the factorials of the numbers representing the number of repetitions of the various things. The letters a, a, a, b, b, c can be arranged in $6!/(3!2!)$ or 60 different ways (permutations). A **permutation of n things taken r at a time** is a permutation containing only r members of the set. The number of such permutations is denoted by $_nP_r$ and is equal to

$$n(n - 1)(n - 2) \cdots (n - r + 1),$$

or $n!/(n - r)!$, for any one can be put first, any one of the remaining $n - 1$ second, etc., until r places have been filled. A **permutation of n things taken r at a time with repetitions** is an arrangement obtained by putting any member of the set in the first position, any member of the set, including a repetition of the one just used, in the second, and continuing this until there are r positions filled. The total number of such permutations is $n \cdot n \cdot n \cdots$ to r factors, *i.e.*, n^r. The ways in which a, b, c can be arranged two at a time are $aa, ab, ac, ba, bb, bc, ca, cb, cc$. A **circular permutation** is an arrangement of objects around a circle. The total number of circular permutations of n different things taken n at a time is equal to the number of permutations of n things n at a time, divided by n because each arrangement will be exactly like $n - 1$ others except for a shift of the places around the circle. (2) An **operation** which replaces each of a set of objects by itself or another object in the set in a one-to-one manner. The permutation which replaces x_1 by x_2, x_2 by x_1, x_3 by x_4, and x_4 by x_3, is denoted by

$$\begin{pmatrix} 1 & 2 & 3 & 4 \\ 2 & 1 & 4 & 3 \end{pmatrix}$$

or $(12)(34)$. A **cyclic** (or **circular**) **permutation** (or simply a **cycle**) is the advancing of each member of an ordered set of objects one position, the last member taking the position of the first. If the objects are thought of as arranged in order around a circle, a cyclic permutation is effected by rotating the circle; cab is a cyclic permutation of abc, denoted by $\begin{pmatrix} a & b & c \\ c & a & b \end{pmatrix}$ or (acb), for which a maps into c, c into b, and b into a. The **degree** of a cyclic permutation is the number of objects in the set. A cyclic permutation of degree

two is called a **transposition**. Every permutation can be factored into a product of transpositions. *E.g.*, $(abc) = (ab)(ac)$ in the sense that the permutation (abc) has the same effect as the permutation (ab) followed by the permutation (ac). A permutation is **even** or **odd** according as it can be written as the product of an even number or an odd number of *transpositions*. Let $x_1, x_2, \cdots, x_n$ be n indeterminants and let D be the product $(x_1 - x_2)(x_1 - x_3) \cdots (x_{n-1} - x_n)$ of all differences $x_i - x_j$, where $i < j$. A permutation of the subscripts $1, 2, \cdots, n$ is even or odd according as it leaves the sign of D unchanged or changes the sign of D.

   permutation group. A group whose elements are permutations, the product of two permutations being the permutation resulting from applying each in succession. Thus the product of the permutation $p_1 = (abc)$, which takes a into b, b into c, and c into a, and the permutation $p_2 = (bc)$, which takes b into c and c into b, is $p_1 p_2 = (abc)(bc) = (ac)$, which takes a into c and c into a. The group of all permutations on n letters is a group of order $n!$ and **degree** n, called a **symmetric group**. The subgroup of this group (of order $n!/2$ and **degree** n) which contains all *even permutations* is called an **alternating group**. A permutation group of order n on n letters is called **regular**. See GROUP, PERMUTATION (2), SYMMETRY —group of symmetries. *Syn.* substitution group.

   permutation matrix. If a permutation on $x_1, x_2, \cdots, x_n$ carries x_i into $x_{i'}$ for each i, then the *permutation matrix* corresponding to this permutation is the square matrix of order n in which the elements in the ith column (for each i) are all zero except the one in the i'th row, which is unity. Any permutation group is isomorphic with the group of corresponding permutation matrices. In general, a permutation matrix is any square matrix whose elements in any column (or any row) are all zero, except for one element equal to unity.

PER'PEN-DIC'U-LAR, *adj., n. Syn.* normal, orthogonal.

   perpendicular bisector. For a line segment in a plane, the line perpendicular to the segment at its midpoint; for a line segment in space, the plane perpendicular to the segment at its midpoint. In either case, the perpendicular bisector is the set of all points equidistant from the end points of the segment.

   perpendicular lines and planes. Two **straight lines** which intersect so as to form a pair of equal adjacent angles are **perpendicular** (each line is said to be **perpendicular** to the other). The condition (in *analytic geometry*) that two lines be perpendicular is: (1) *In a plane*, that the slope of one of the lines be the negative reciprocal of that of the other, or that one be horizontal and the other vertical; (2) *in space*, that the sum of the products of the corresponding direction numbers (or direction cosines) of the two lines be zero (two **lines in space** are perpendicular if there exist intersecting perpendicular lines, each of which is parallel to one of the given lines). A **common perpendicular** to two or more lines is a line which is perpendicular to each of them. In a plane, the only lines that can have a common perpendicular are parallel lines, and they have any number. In space, any two lines have any number of common perpendiculars (only one of which intersects both lines, unless the lines are parallel). A **line perpendicular to a plane** is a line which is perpendicular to every line through its intersection with the plane. It is sufficient that it be perpendicular to two nonparallel lines in the plane. The condition (in *analytic geometry*) that a line be perpendicular to a plane is that its direction numbers be proportional to those of the normal to the plane; or, what amounts to the same thing, that its direction numbers be proportional to the coefficient of the corresponding variables in the equation of the plane. The **foot** of the perpendicular to a line (or plane) is the point of intersection of the perpendicular with the line (or plane). Two **perpendicular planes** are two planes such that a line in one, which is perpendicular to their line of intersection, is perpendicular to the other; *i.e.*, planes forming a *right dihedral angle*. The condition (in *analytic geometry*) that two planes be perpendicular is that their normals be perpendicular, or that the sum of the products of the coefficients of like variables in their two equations be zero. See NORMAL—normal lines and planes.

PER′PE-TU′I-TY, *n.* An annuity that continues forever. See CAPITALIZED —capitalized cost.

PER-SPEC′TIVE, *adj.* **perspective position.** A pencil of lines and a range of points are in **perspective position** if each line of the pencil goes through the point of the range which corresponds to it. Two pencils of lines are in perspective position if corresponding lines meet in points which lie on a line, the **axis of perspectivity.** Likewise two ranges of points are in **perspective position** provided lines through their corresponding points meet in a point, the **center of perspectivity.** A range of points and an axial pencil (pencil of planes) are in **perspective position** if each plane of the pencil goes through the point which corresponds to it; a pencil of lines and an axial pencil are in **perspective position** if each

line of the pencil lies in the plane to which it corresponds; likewise two axial pencils are in **perspective position** if intersections of corresponding planes lie in a plane. Each of the above relationships is called a **perspectivity.** See PROJECTIVE —projective relation.

PERSPECTIVITY. See PERSPECTIVE.

PETERSBURG PARADOX. Suppose Peter and Paul flip a coin with the understanding that if the first $n - 1$ throws result in heads and the nth in a tail, then Paul pays Peter 2^n dollars and the game ends. Peter has the advantage, no matter how much he pays Paul to play the game. If the number of throws is restricted to n, then Peter should pay Paul $\sum_1^n (\frac{1}{2})^k 2^{k-1} = \frac{1}{2}n$ dollars to play the game. This was proposed by D. Bernoulli in the *Commentarii* of the Petersburg Academy.

PFAFF, Johann Friedrich (1765–1825). German analyst. Friend and teacher of Gauss.
  Pfaffian. An expression of the form

$$u_1\, dx_1 + u_2\, dx_2 + u_3\, dx_3 + \cdots + u_n\, dx_n,$$

where the coefficients $u_1, \cdots, u_n$ are functions of the variables $x_1, \cdots, x_n$.

PHASE, *n.* **phase of simple harmonic motion.** The angle $\phi + kt$ in the equation of simple harmonic motion, $x = a\cos(\phi + kt)$. See HARMONIC —simple harmonic motion.
  initial phase. The phase when $t = 0$, namely ϕ in $x = a\cos(kt + \phi)$.

PHI, *adj., n.* The twenty-first letter of the Greek alphabet: lower case, ϕ or φ; capital, Φ.
  phi coefficient. See COEFFICIENT.
  phi function. See EULER —Euler ϕ-function.

PHRAGMÉN, Lars Edvard (1863–1937). Swedish analyst.
  Phragmén-Lindelöf function. Relative to an entire function f of finite order ρ, the function

$$h(\theta) \equiv \limsup_{r \to \infty} \frac{\log\left| f(re^{i\theta}) \right|}{r^\rho}.$$

The Phragmén-Lindelöf function $h(\theta)$ is a *subsine function* of order ρ. See ENTIRE —entire function.

PI, *n.* The sixteenth letter of the Greek alphabet: lower case, π; capital, Π. The symbol π is used to denote the ratio of the circumference of a circle to its diameter:

$$\pi = 3.14159265358979323846264 3 + \,.$$

The Rhind Papyrus (c. 1650 B.C.) contains calculations that use $\pi = (\frac{16}{9})^2 = 3.1605 - \,.$ In the third century B.C., Archimedes showed that π is between $3\frac{10}{71}$ and $3\frac{1}{7}$, although both the Bible and the Talmud give π the value 3. During the third century, π was calculated in China to six digits by using a polygon with 3072 sides to approximate a circle. In the 17th century, von Ceulen determined the first 35 digits of π. In 1949, 2037 digits were known; in 1961, 100,000 digits; in 1973, one million digits; and in 1989, more than one billion digits. Modern calculations are based on foundations laid by Ramanujan. In 1770, Lambert proved π is irrational. Lindemann proved in 1882 that π is transcendental, which showed the impossibility of solving the ancient Greek problem of ruler-and-compass squaring of the circle. It is known now that π is not a Liouville number and that e^π is transcendental, but it is not known whether any of $\pi + e$, π/e, and $\log \pi$ is irrational, although $e^{\pi i} = -1$. Π denotes a product. See BUFFON, CEULEN, EULER —Euler formula, MACHIN, PRODUCT —infinite product, VIÈTE—Viète formula. WALLIS—Wallis' product for π.

PI′CA, *n. (Printing).* A measure of type body, equal to 12 points in U.S. scale. See POINT (4).

PICARD, Charles Émile (1856–1941). Distinguished French analyst, group theorist, and algebraic geometer.
   Picard's method. An iterative method for solving differential equations. For the differential equation $dy/dx = f(x, y)$, the solution that passes through the point (x_0, y_0) satisfies the equation

$$y(x) = y_0 + \int_{x_0}^{x} f[t, y(t)]\, dt.$$

Starting with an initial function, say the constant y_0, the method consists of making successive substitutions in accordance with the formula

$$y_n(x) = y_0 + \int_{x_0}^{x} f[t, y_{n-1}(t)]\, dt.$$

The method extends to the solution of systems of linear differential equations and to the solution of higher-order linear differential equations and systems of equations.

Picard's theorems. *Picard's first theorem* states that if f is an entire function, and $f(z) \neq$ const., then f takes on every finite complex value with at most one exception. *E.g.,* $f(z) = e^z$ takes on all values except 0. For *Picard's second theorem,* see SINGULAR —isolated singular point of an analytic function.

PICO. A prefix meaning 10^{-12}, as in **picometer.**

PIC′TO-GRAM, *n.* Any figure showing numerical relations, as *bar graphs, broken-line graphs.*

PIECE′WISE, *adj.* **piecewise-continuous function.** A function is piecewise continuous on R if it is defined on R and R can be separated into a finite number of pieces such that the function is continuous on the interior of each piece and such that the function approaches a finite limit as a point moves in the interior of a piece and approaches a boundary point in any way. It is necessary to restrict the nature of the pieces, *e.g.,* that their boundaries be simple closed curves if they are plane regions, or two points if on a straight line. If R is bounded, an equivalent definition is that it be possible to separate R into a finite number of pieces such that the function is uniformly continuous on the interior of each piece.
   piecewise-smooth curve. See SMOOTH— smooth curve.

PIERC′ING, *adj.* **piercing point of a line in space.** See POINT—piercing point of a line in space.

PIG′EON, *adj.* **pigeon-hole principle.** See DIRICHLET —Dirichlet drawer principle.

PITCH, *n.* **pitch of a roof.** The quotient of the **rise** (height from the level of plates to the ridge) by the **span** (length of the plates); one-half of the *slope* of the roof.

PLACE, *n.* **decimal place.** See DECIMAL.
   place value. The value given to a digit by virtue of the place it occupies in the number relative to the units place. In 423.7, 3 denotes merely 3 units, 2 denotes 20 units, 4 denotes 400 units, and 7 denotes $\frac{7}{10}$ of a unit; 3 is in unit's place, 2 in ten's place, 4 in hundred's place, etc. *Syn.* local value.

PLA′NAR, *adj.* **planar graph.** A graph that can be represented in a plane with edges as arcs of simple curves joining corresponding nodes in such a way that two different edges intersect only in nodes. Kuratowski showed that a graph is planar if and only if it has (1) no subgraph that is either the complete graph with five nodes, or (2) a graph with six nodes—all of valence 3 and connected by six edges, as for a hexagon, and with opposite nodes connected by three nonintersecting segments. See GRAPH THEORY.

planar point of a surface. A point of the surface at which $D = D' = D'' = 0$. See SURFACE —fundamental coefficients of a surface. At a planar point, every direction on the surface is an *asymptotic direction*. A surface is a plane if, and only if, all its points are planar points.

PLANE, *adj., n.* A surface such that a straight line joining any two of its points lies entirely in the surface. *Syn.* plane surface.

collinear planes. See COLLINEAR.

complex plane. See COMPLEX.

coordinate plane. See CARTESIAN —Cartesian coordinates.

diametral plane. See DIAMETRAL.

equation of a plane. In three-dimensional Cartesian coordinates, a polynomial equation which is of the first degree. The equation $Ax + By + Cz + D = 0$ with A, B and C not all zero is the **general form** of the equation of a plane. Special cases of the general form are: (1) **Intercept form.** The equation $x/a + y/b + z/c = 1$, where a, b and c are the x, y, z intercepts, respectively. (2) **Three-point form.** The equation of the plane expressed in terms of three points on the plane. The simplest form is obtained by equating to zero the determinant whose rows are $x, y, z, 1$; $x_1, y_1, z_1, 1$; $x_2, y_2, z_2, 1$; and $x_3, y_3, z_3, 1$, where (x_1, y_1, z_1), (x_2, y_2, z_2), and (x_3, y_3, z_3) are the three given points. (3) **Normal form.** The equation

$$lx + my + nz - p = 0,$$

where l, m, and n are the direction cosines of the normal to the plane, directed from the origin to the plane, and p is the length of the normal from the origin to the plane. If l, m, n and the coordinates of a point in the plane, say x_1, y_1, z_1, are given, then $p = lx_1 + my_1 + nz_1$ and the equation of the plane can be written

$$l(x - x_1) + m(y - y_1) + n(z - z_1) = 0.$$

The left member of this equation is the *scalar product* of the vectors (l, m, n) and $(x - x_1, y - y_1, z - z_1)$. Therefore the left member of either of the above equations is the distance from the point (x, y, z) to the plane. An equation, $Ax + By + Cz + D = 0$, of a plane can be reduced to

normal form by dividing by $\pm(A^2 + B^2 + C^2)^{1/2}$, the sign being opposite that of the constant term D.

half-plane. See HALF.

normal (perpendicular) lines or planes to lines, curves, planes, and surfaces. See NORMAL— normal to a curve or surface, PERPENDICULAR — perpendicular lines and planes.

parallel lines and planes. See PARALLEL.

pencil of planes. See PENCIL.

plane angle of a dihedral angle. The angle formed by two intersecting lines, one of which lies in each face and both of which are perpendicular to the edge of the dihedral angle; the plane angle between the intersections of the faces of the dihedral angle with a third plane which is perpendicular to the edge of the dihedral angle. Such a plane angle is said to **measure** its dihedral angle. When the plane angle is *acute, right, obtuse,* etc., its *dihedral* angle is said to be *acute, right, obtuse,* etc.

plane curve (figure, surface, etc.). A curve (figure, surface, etc.) lying entirely in a plane.

plane geometry. See GEOMETRY.

plane sailing. See SAILING.

plane section. The intersection of a plane and a surface or a solid.

principal plane of a quadric surface. See PRINCIPAL —principal plane of a quadric surface.

projection plane. The plane upon which a figure is projected; a plane section of the projection rays of a projection. See PROJECTION.

projective plane. See PROJECTIVE —projective plane.

sheaf of planes. See SHEAF.

shrinking of the plane. See SIMILITUDE — transformation of similitude, STRAIN—one-dimensional strains.

PLA·NIM'E·TER, *n.* A mechanical device for measuring plane areas. Merely requires moving a pointer on the *planimeter* around the bounding curve. A common type is the *polar planimeter*. See INTEGRATOR.

PLAS·TIC'I·TY, *n.* **theory of plasticity.** The theory of behavior of substances beyond their elastic range.

PLATE, *n.* **plate of a building.** A horizontal timber (beam) that supports the lower end of the rafters.

PLATEAU, Joseph Antoine Ferdinand (1801– 1883). Belgium physicist.

Plateau problem. The problem of determining the existence of a *minimal surface* with a given twisted curve as its boundary. It might or might not be required that the minimal surface have minimum area. For several different contours the problem was solved by the physicist Plateau in soap-film experiments.

PLATYKURTIC, *adj.* **platykurtic distribution.** See KURTOSIS.

PLAY, *n.* **play of a game.** Any particular performance, from beginning to end, involved in a game. See GAME, MOVE.

PLAY′ER, *n.* An individual, or group of individuals acting as one, involved in the play of a game. In a two-person zero-sum game, the **maximizing player** is the player to whom all payments are considered as being made by the other player (a payment made to him is considered as a positive payment, while a payment made by him is considered as a negative payment); the **minimizing player** is the player who is considered to be making all payments (a payment made by him is considered as a positive payment, while a payment made to him is considered as a negative payment). See GAME, PAYOFF.

PLOT, *v.* **plot a point.** To locate the point geometrically, either in the plane or in space, when its coordinates are given in some coordinate system. In Cartesian coordinates, a point is plotted by locating it on cross-section paper or by drawing lines on plain paper parallel to indicated axes of coordinates and at a distance from them equal to the corresponding coordinate of the point. See COORDINATE.
point-by-point plotting (graphing) of a curve. Finding an ordered set of points which lie on a curve and drawing through these points a curve which is assumed to resemble the required curve.

PLÜCKER, Julius (1801–1868). German geometer and mathematical physicist.
Plücker's abridged notation. See ABRIDGED.

PLUMB, *adj., n.* A weight attached to a cord.
plumb line. See LINE—plumb line.

PLUS, *adj., n.* Denoted by +. (1) Indicates addition, as $2 + 3$ (3 added to 2). (2) Property of being positive. (3) A little more or in addition to, as $2.35 +$. The symbol + is the **plus sign.**

POINCARÉ, Jules Henri (1854–1912). Great French mathematician, mathematical physicist, astronomer, and philosopher. Called the last mathematical universalist, although Hilbert was about as versatile.

Poincaré-Birkhoff fixed-point theorem. Let a continuous one-to-one transformation map the ring R formed by two concentric circles in such a way that one circle moves in the positive sense and the other in the negative sense and areas are preserved. Then the transformation has at least two fixed points. This theorem was conjectured by Poincaré and proved by G. D. Birkhoff.
Poincaré conjecture. The unproved conjecture that a three-manifold is topologically equivalent to a three-sphere if it is closed, compact, and simply connected. The **general Poincaré conjecture** states that a compact n-dimensional manifold M^n of the same homotopy class as the n-sphere S^n is homeomorphic to S^n; M^n being of the same homotopy class as S^n means that every continuous map of S^k into M^n for $k < n$ can be continuously deformed to a point (see DEFORMATION). The generalized Poincaré conjecture was proved by Smale (1960) for $n > 4$ and by Freedman (1984) for $n = 4$.
Poincaré duality theorem. See DUALITY.
Poincaré recurrence theorem. Let X be a bounded open region of n-dimensional Euclidean space and let T be a homeomorphism of X onto itself that preserves volume, that is, any open set and its transform by T have the same volume (or *measure*). Poincaré proved that there is a set S of *measure zero* in X such that, if x is not in S and U is any open set in X which contains x, then an infinite number of the points $x, T(x), T^2(x), T^3(x), \cdots$ belong to U, where $T^n(x)$ is the result of applying T to x successively n times. The theorem is still true if S is required to be of *first category* as well as of measure zero. Numerous generalizations and modifications of Poincaré's theorem are known. See ERGODIC—ergodic theory.

POINT, *adj., n.* (1) An undefined element of geometry. According to Euclid, it is that which has position but no nonzero dimensions. (2) An element of geometry defined by its coordinates, such as the point $(1, 3)$. (3) An element which satisfies the postulates of a certain space. See POSTULATE—Euclid's postulates, METRIC—metric space. (4) A unit used in measuring bodies of type, leads, etc. It is equal to 0.0138 inches or 0.0351 centimeters, in the U. S. system.
accumulation (or **cluster** or **limit**) **point.** See ACCUMULATION.
antipodal points. Points on a sphere at opposite ends of a diameter.
collinear points. See COLLINEAR.
condensation point. See CONDENSATION.
conjugate points relative to a conic. See CONJUGATE—conjugate points relative to a conic.

decimal point. See DECIMAL.

double point. See below, multiple point.

homologous point. See HOMOLOGOUS.

isolated point. A point in whose neighborhood there is no other point of the set under consideration. The origin is such a point on the graph of a polynomial equation when the lowest-degree homogeneous polynomial in the equation of the curve, referred to a system of Cartesian coordinates having its origin at the given point, vanishes for no values of x and y in a neighborhood of zero, except for both x and y equal to zero. The curve $x^2 + y^2 = x^3$ has an isolated point at the origin, since the equation $x^2 + y^2 = 0$ is satisfied only by the point $(0, 0)$. The lowest-degree homogeneous polynomial which can satisfy the above conditions is a quadratic; hence isolated points are at least double points. *Syn.* Acnode.

material point. See MATERIAL.

multiple (or *n*-**tuple**) **point.** For a curve, a point P which is an interior point of each of n (but not more than n) arcs such that any two of the arcs intersect only at P. The equations of the tangents at an n-tuple point on an algebraic curve can be determined by equating to zero the lowest-degree terms (in this case the terms of the nth degree) in the equation of the curve referred to a Cartesian coordinate system whose origin is at the multiple point. If k arcs through a multiple point P all have the same tangent at P, then P is said to have a multiple (*k*-**tuple**) **tangent** at P. A multiple point for which $n = 2$ is a **double point**. The equations of the tangents at a double point on an algebraic curve can be determined by equating to zero the quadratic terms in the equation of the curve referred to a rectangular Cartesian coordinate system whose origin is at the double point, the linear terms and constant term being zero in this case. This quadratic may be a perfect square, in which case there is a **double tangent** at the point.

ordinary (or **simple**) **point of a curve.** (1) A point that is not a multiple point and is an interior point of an arc on which the curve has a smoothly turning tangent. *Tech.* A point P that is not a multiple point and for which the curve has parametric equations, $x = f(t)$, $y = g(t)$, for which P corresponds to a value of t in a neighborhood of which f' and g' are continuous and not both zero (this definition extends naturally to higher dimensions). If the equation of a plane curve is $f(x, y) = 0$ and the first-order partial derivatives of f are continuous, then a sufficient condition for a point to be an ordinary point is that it not be true that $f_x = f_y = 0$ at the point; if f has continuous second-order partial derivatives, then at a point for which $f_x = f_y = 0$, the

curve has two tangents or no tangent at all, according as $f_{xx}f_{yy} - f_{xy}^2$ is less than, or greater than, zero (if this expression is zero, the curve may have a double tangent, as at a cusp). A point of a curve which is not an ordinary point is a **singular point.** Cusps, crunodes, and multiple points are singular points. (2) The preceding, except that the restriction that f' and g' be continuous is replaced by the restriction that f and g be analytic (see ANALYTIC —analytic function of a real variable).

piercing point of a line in space. Any one of the points where the line passes through one of the coordinate planes.

point-by-point plotting (graphing) of a function. See PLOT.

point charge. See CHARGE.

point circle (null circle) and **point ellipse.** See CIRCLE —null circle, ELLIPSE.

point of contact. See TANGENCY —point of tangency.

point of discontinuity. A point at which a curve (or function) is not continuous. See CONTINUOUS, DISCONTINUITY.

point of division. The point which divides the line segment joining two given points in a given ratio. If the two given points have the Cartesian coordinates (x_1, y_1) and (x_2, y_2) and it is desired to find a point P such that the distance from the first point to the point P, divided by the distance from the point P to the second point, is equal to r_1/r_2, the formulas giving the coordinates x and y of the desired point P are

$$x = \frac{r_2 x_1 + r_1 x_2}{r_1 + r_2}, \qquad y = \frac{r_2 y_1 + r_1 y_2}{r_1 + r_2}.$$

When r_1/r_2 is positive, the point of division lies between the two given points, and the division is **internal**; the point P **divides the line segment internally** in the ratio r_1/r_2. When this ratio is negative, the point of division must lie on the line segment extended, and it **divides the line segment externally** in the ratio $|r_1/r_2|$. When $r_1 = r_2$, the point P bisects the line segment and the above formulas reduce to

$$x = \frac{x_1 + x_2}{2}, \qquad y = \frac{y_1 + y_2}{2}.$$

For **points in space**, the situation is the same as in the plane except that the points now have three coordinates. The formulas for x and y are the same, and the formula for z is

$$z = \frac{r_2 z_1 + r_1 z_2}{r_1 + r_2}.$$

point-finite. See FINITE—locally finite family of sets.

point of inflection. See INFLECTION.

point at infinity. (1) See IDEAL—ideal point. (2) See INFINITY—point at infinity.

point of osculation. See OSCULATION.

point of tangency. See TANGENCY.

point-slope form of the equation of a straight line. See LINE—equation of a line.

power of a point. See POWER—power of a point.

salient point. See SALIENT.

simple point. Same as ORDINARY POINT. See p. 319.

singular point. See above, ordinary point of a curve, and various headings under SINGULAR.

umbilical point on a surface. See UMBILICAL.

POISSON, Siméon Denis (1781–1840). French analyst, probabilist, and applied mathematician.

Poisson differential equation. The partial differential equation

$$\frac{\partial^2 v}{\partial x^2} + \frac{\partial^2 v}{\partial y^2} + \frac{\partial^2 v}{\partial z^2} = -u,$$

or $\nabla^2 v = -u$. See DIRICHLET—Dirichlet characteristic properties of the potential function.

Poisson distribution. A random variable X has a **Poisson distribution** or is a **Poisson random variable** if the range of X is the set of nonnegative integers and there is a number μ for which the *probability function P* satisfies

$$P(n) = \frac{e^{-\mu}\mu^n}{n!} \quad \text{if } n \geq 0;$$

μ is both the *mean* and the *variance* and the *moment generating function* is $M(t) = e^{\mu(e^t - 1)}$. If $n \to \infty$ and $p \to 0$ in the binomial distribution in such a way that $np = \mu$, the *binomial distribution* approaches the Poisson distribution. Thus the Poisson distribution is a useful approximation for the binomial distribution when studying independent events which are very improbable but for which a large number of trials occur, *e.g.*, traffic deaths, accidents, and radioactive emissions. If a *Poisson process* has parameter λ and $\mu = \lambda s$, then X is a Poisson random variable with parameter μ if X is the (random) number of events occurring in an interval of length s.

Poisson integral. The integral

$$\frac{1}{2\pi}\int_0^{2\pi} U(\phi)\frac{a^2 - r^2}{a^2 - 2ar\cos(\theta - \phi) + r^2}\,d\phi$$

or, for $\zeta = ae^{i\phi}$ and $z = re^{i\theta}$,

$$\frac{1}{2\pi}\int_0^{2\pi} \text{Re}\left(\frac{\zeta + z}{\zeta - z}\right)U(\phi)\,d\phi,$$

which gives the value at the point $x = r\cos\theta$, $y = r\sin\theta$ of the function which is harmonic for $x^2 + y^2 < a^2$, continuous for $x^2 + y^2 \leq a^2$, and which coincides with the continuous boundary-value function $U(\phi)$ on $x^2 + y^2 = a^2$. More general boundary-value functions can be considered.

Poisson process. A *stochastic process* $\{X(t): t \in T\}$ is a **Poisson process** if its index set T is an interval of real numbers, $X(t)$ is the number of occurrences of some particular event prior to "time" t, and (i) there is a number λ (the **parameter** or **mean rate** or **intensity**) such that

$$\lim_{h \to 0}\frac{P[X(h) = 1]}{h} = \lambda,$$

i.e., the probability $P[X(h) = 1]$ of the occurrence of exactly one event in an interval of length h is approximately λh if h is small; (ii) $\lim_{h \to 0} P[X(h) \geq 2]/h = 0$; (iii) if $a < b \leq c < d$, the random variables $X(b) - X(a)$ and $X(d) - X(c)$ are independent and have the same distribution whenever $b - a = d - c$. Poisson processes provide useful models for radioactive decay, arrivals at a service counter, flaws in a long tape or wire, etc. See GAMMA—gamma distribution, and above, Poisson distribution.

Poisson ratio. The numerical value of the ratio of the strain in the transverse direction to the longitudinal strain. *E.g.*, a thin elastic rod, subjected to the action of a longitudinal stress T, undergoes a contraction e_1 in the linear dimensions of its cross section, and an extension e_2 in the longitudinal direction. The numerical value of the ratio $\sigma = |e_1/e_2|$ is the **Poisson ratio**. From Hooke's law, $T = Ee_2$, where E is *Young's modulus* of tension, so that $\sigma = -e_1E/T$. For most structural materials, the Poisson ratio has a value between $\frac{1}{4}$ and $\frac{1}{3}$.

PO′LAR, *adj.*, *n.* As a noun, same as POLAR LINE. See p. 321.

polar coordinates in the plane. The system of coordinates in which a point is located by its distance from a fixed point and the angle that the line from this point to the given point makes with a fixed line, called the **polar axis**. The fixed point, O in the figure, is the **pole**; the distance, $OP = r$, from the pole to the given point, is the **radius vector**; the angle θ (taken positive when counter-clockwise) is the **polar angle** or **vectorial angle**. The polar coordinates of the point P are written (r, θ). The polar angle is sometimes called the **amplitude, anomaly** or **azimuth** of the point. From the figure it can be seen that the relations between rectangular Cartesian and polar coordinates are $x = r \cos \theta$, $y = r \sin \theta$. If r is positive (as in the figure), the amplitude, θ, of the point P is any angle (positive or negative) having OX as initial side and OP as terminal side. If r is negative, θ is any angle having OX as initial side and the extension of PO through O as terminal side. The point whose coordinates are $(1, 130°)$ or $(-1, -50°)$ is in the 2nd quadrant; the point whose coordinates are $(-1, 130°)$ or $(1, -50°)$ is in the 4th quadrant. See ANGLE.

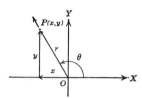

polar coordinates in space. Same as SPHERICAL COORDINATES. See SPHERICAL.

polar distance. See CODECLINATION.

polar equation. An equation in polar coordinates. See CONIC, LINE—equation of a line.

polar form of a complex number. The form a complex number takes when it is expressed in polar coordinates. This form is $r(\cos \theta + i \sin \theta)$, where r and θ are polar coordinates of the point represented by the complex number. The number r is the **modulus** and the angle θ the **amplitude, argument,** or **phase.** *Syn.* trigonometric form (representation) of a complex number. See COMPLEX—complex numbers, DE MOIVRE—De Moivre's theorem, EULER—Euler's formula.

polar line or plane. See POLE—pole and polar of a conic, and pole and polar of a quadric surface.

polar line of a space curve. The line normal to the osculating plane of the curve at the center of curvature. *Syn.* polar.

polar planimeter. See PLANIMETER.

polar of a quadratic form. The bilinear form obtained from a quadratic form

$$Q = \sum_{i,j=1}^{n} a_{ij} x_i x_j \ (a_{ij} = a_{ji}) \text{ by the operator}$$

$$\frac{1}{2} \sum_{i=1}^{n} y_i \frac{\partial}{\partial x_i},$$

i.e., the bilinear form

$$Q' = \sum_{i,j=1}^{n} a_{ij} y_i x_j.$$

If x and y are regarded as points in $n-1$ dimensions with homogeneous coordinates $(x_1, \cdots, x_n)$ and $(y_1, \cdots, y_n)$, then $Q = 0$ is the equation of a quadric and $Q' = 0$ is that of the polar of y with respect to the quadric. See below, pole and polar of a conic.

polar reciprocal curves. See RECIPROCAL—polar reciprocal curves.

polar tangent. The segment of the tangent line to a curve cut off by the point of tangency and a line through the pole perpendicular to the radius vector. The projection of the polar tangent on this perpendicular is the **polar subtangent.** The segment of the normal between the point on the curve and this perpendicular is the **polar normal.** The projection of the polar normal on this perpendicular is the **polar subnormal.**

polar triangle of a spherical triangle. The spherical triangle whose vertices are poles of the sides of the given triangle, the poles being the ones nearest to the vertices opposite the sides of which they are poles. See POLE—pole of an arc of a circle on a sphere.

reciprocal polar figures. See RECIPROCAL.

PO′LAR-I-ZA′TION, *n.* polarization of a complex of charges. See POTENTIAL—concentration method for the potential of a complex.

POLE, *n.* **pole of an analytic function.** See LAURENT—Laurent expansion, SINGULAR—singular point of an analytic function.

pole of the celestial sphere. One of the two points where the earth's axis, produced, pierces the celestial sphere. They are called the *north* and *south celestial poles.* See HOUR—hour angle and hour circle.

pole of a circle on a sphere. A point of intersection of the sphere and the line through the center of the circle and perpendicular to the plane of the circle. The north and south poles are the poles of the equator. The poles of an **arc of a circle** on a sphere are the poles of the circle containing the arc.

pole of geodesic polar coordinates. See GEODESIC—geodesic polar coordinates.

pole and polar of a conic. A point and the line which is the locus of the harmonic conjugates of this point with respect to the two points in which a secant through the given point cuts the conic; a point and the line which is the locus of points conjugate (see CONJUGATE—conjugate points relative to a conic) to the given point. The point is the **pole** of the line and the line is the **polar** of the point. Analytically, the polar of a point is the locus of the equation obtained by replacing the coordinates of the point of contact in the equation of a general tangent to the conic by the coordinates of the given point. See CONIC—tangent to a general conic. *E.g.*, if a **circle** has the equation $x^2 + y^2 = a^2$, the equation of the *polar* of the point (x_1, y_1) is $x_1 x + y_1 y = a^2$. When a point lies so that two tangents can be drawn from it to the conic, the *polar* of the point is the secant through the points of contact of the tangents. The *polar line* of P_1 (in the figure), relative to the ellipse, is the line $P_2 P_3$.

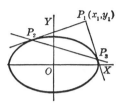

pole and polar of a quadric surface. A point (the **pole** of the plane) and a plane (the **polar** of the point) which is the locus of the harmonic conjugates of the point with respect to the two points in which a variable secant through the pole cuts the quadric. Analytically, the polar plane of a given point is the plane whose equation is that obtained by replacing the coordinates of the point of tangency in the general equation of a tangent plane by the coordinates of the given point. See TANGENT—tangent plane to a quadric surface. If, for instance, the quadric is an ellipsoid whose equation is $x^2/a^2 + y^2/b^2 + z^2/c^2 = 1$, the polar of the point (x_1, y_1, z_1) is the plane $x_1 x/a^2 + y_1 y/b^2 + z_1 z/c^2 = 1$.

pole of stereographic projection. See PROJECTION—stereographic projection of a sphere on a plane.

pole of a system of coordinates. See POLAR—polar coordinates in the plane.

POL′I-CY, *n.* **annuity and insurance policies.** See ANNUITY, INSURANCE.

POLISH, *adj.* **Polish space.** A topological space that is separable, complete, and metrizable.

POL′Y-GON, *n.* A plane figure consisting of n points, $p_1, p_2, p_3, \cdots, p_n$ (the **vertices**), $n \geq 3$, and of the line segments $p_1 p_2, \ p_2 p_3, \cdots,$ $p_{n-1} p_n, p_n p_1$ (the **sides**). In elementary geometry, it is usually required that the sides have no common point except their end points. A polygon of 3 sides is a **triangle**; of 4 sides, a **quadrilateral**; of 5 sides, a **pentagon**; of 6 sides, a **hexagon**; of 7 sides, a **heptagon**; of 8 sides, an **octagon**; of 9 sides, a **nonagon**; of 10 sides, a **decagon**; of 12 sides, a **dodecagon**; of n-sides, an **n-gon**. The plane region enclosed by the sides of the polygon is the **interior** of the polygon. The

(interior) angles of a polygon are the angles made by adjacent sides of the polygon and lying within the polygon. A polygon is **convex** if it lies on one side of any line that contains a side of the polygon, *i.e.*, if each interior angle is less than or equal to 180°. A polygon is **concave** if it is not convex, *i.e.*, if at least one of its interior angles is greater than 180°. A polygon is concave if and only if there is a straight line which passes through the interior of the polygon and cuts the polygon in four or more points. A convex polygon always has an interior. A concave polygon has an interior if no side touches any other side, except at a vertex, and no two vertices coincide (*i.e.*, if it is a *simple closed curve* or a *Jordan curve*). A polygon is **equiangular** if its interior angles are congruent; it is **equilateral** if its sides are congruent. A triangle is equiangular if and only if it is equilateral, but this is not true for polygons of more than three sides. A polygon is **regular** if its sides are congruent and its interior angles are congruent.

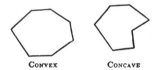

CONVEX   CONCAVE

circumscribed and inscribed polygons. See CIRCUMSCRIBED.

diagonal of a polygon. A line segment joining any two nonadjacent vertices of the polygon.

frequency polygon. See FREQUENCY—frequency curve or diagram.

similar polygons. Polygons having their corresponding angles equal and their corresponding sides proportional. See SIMILAR.

spherical polygon. A portion of a sphere bounded by arcs of great circles.

PO-LYG′ON-AL, *adj.* **polygonal region.** The interior of a polygon together with all, some, or none of the polygon. The region is **closed** or **open** according as it contains all or none of the polygon. See REGION.

POL′Y-HE′DRAL, *adj.* **polyhedral angle.** See ANGLE—polyhedral angle.

polyhedral region. The interior of a polyhedron together with all, some, or none of the polyhedron. The region is **closed** or **open** according as it contains all or none of the polyhedron. See REGION.

POL′Y-HE′DRON, *n.* [*pl.* **polyhedrons** or **polyhedra**]. A solid bounded by plane polygons. The bounding polygons are the **faces**; the intersections of the faces are the **edges**; and the points where three or more edges intersect are the **vertices**. A polyhedron of four faces is a **tetrahedron**; one of five faces, a **pentahedron**; one of six faces, a **hexahedron**; one of seven faces, a **heptahedron**; one of eight faces, an **octahedron**; one of twelve faces, a **dodecahedron**; and one of twenty faces, an **icosahedron**. A **convex polyhedron** is a polyhedron which lies entirely on one side of any plane containing one of its faces, *i.e.*, a polyhedron any plane section of which is a convex polygon. A polyhedron that is not convex is **concave**. For a concave polyhedron, there is at least one plane which contains one of its faces and is such that there is a part of the polyhedron on each side of the plane. A **simple polyhedron** is a polyhedron which is *topologically equivalent* to a sphere, a polyhedron with no "holes" in it. A **regular polyhedron** is a polyhedron whose faces are congruent regular polygons and whose polyhedral angles are congruent (see ARCHIMEDES —Archimedean solid). There are only five regular polyhedrons: the regular **tetrahedron, hexahedron** (or cube), **octahedron, dodecahedron**, and **icosahedron**. These are shown in the figures. Useful in proving that there are only five regular polyhedrons is *Euler's theorem*, which states that, for any simple polyhedron, $V - E + F = 2$, where V, E, and F are the number of vertices, edges, and faces, respectively. More generally, a polyhedron may be an object which is homeomorphic to a set consisting of all the points which belong to simplexes of a *simplicial complex*. See ARCHIMEDES —Archimedean solid, DELTAHEDRON.

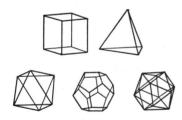

circumscribed and inscribed polyhedrons. See CIRCUMSCRIBED.

diagonal of a polyhedron. See DIAGONAL — diagonal of a polyhedron.

Euler theorem for polyhedrons. See EULER.

similar polyhedrons. Polyhedrons which can be made to correspond in such a way that corresponding faces are similar each to each and similarly placed and such that their corresponding polyhedral angles are congruent.

symmetric polyhedrons. Two polyhedrons each of which is congruent to the mirror image of the other.

POL′Y-HEX, *n.* See POLYOMINO.

POL′Y-NO′MI-AL, *adj., n.* A **polynomial in one variable** (usually called simple a **polynomial**) of **degree** n is a rational integral algebraic expression of the form $a_0 x^n + a_1 x^{n-1} + \cdots + a_{n-1} x + a_n$, where a_i, $i = 0, 1, 2, \cdots, n$, are complex numbers (real or imaginary), and n is a nonnegative integer; constants are then polynomials of degree 0, except that the constant 0 is not assigned a degree. A polynomial is **linear, quadratic, cubic, quartic** (or **biquadratic**), etc., according as its degree is 1, 2, 3, 4, etc. A **polynomial in several variables** is an expression which is the sum of terms, each of which is the product of a constant and various nonnegative powers of the variables. A **polynomial over the integers, rational numbers,** or **real numbers** is a polynomial all of whose coefficients are integers, rational numbers, or real numbers, respectively. The **polynomial form** or **expanded form** of 853 is $8 \cdot 10^2 + 5 \cdot 10 + 3 \cdot 1$. See IRREDUCIBLE— irreducible polynomial.

continuation of sign in a polynomial. See CONTINUATION.

cyclotomic polynomial. See CYCLOTOMIC.

polynomial equation and inequality. See EQUATION —polynomial equation, INEQUALITY.

polynomial function. A function whose values can be computed by substituting the value of the independent variable in a polynomial.

polynomials of Bernoulli, Chebyshev, Hermite, Laguerre, and **Legendre.** See the respective names.

primitive polynomial. See PRIMITIVE.

separable polynomial. See SEPARABLE.

POL-Y-OM′I-NO, *n.* The plane figure formed by joining unit squares along their edges. Polyominos all of which are congruent to a given polyomino that uses 4 or fewer squares can be used as tiles to cover the plane (*i.e.*, **monominos, dominos, trominos,** or **tetrominos**). This is true for the 12 **pentominos** (5 squares) and 35 **hexominos** if some can be turned over, but 4 of the 108 types of **heptominos** and probably 26 of the 369 types of **octominos** are nontilers (even if some are turned over). Any **polyiamond** formed by joining 8 or fewer equilateral triangles along edges is a tiler, except for one case of 7 triangles. Any **polyhex** formed by joining 5 or fewer regular hexagons is a tiler. See TESSELLATION.

POL′Y-TOPE′, *n.* The analogue in n-dimensional space of point, segment, polygon, and polyhedron in spaces of dimensions 0, 1, 2 and 3. A **convex polytope** in n-space is the convex span of a finite set of points that do not all lie in the same hyperplane; thus a convex polytope is a bounded convex subset enclosed by a finite number of hyperplanes. A **face** of a convex polytope K is the empty set or K (the *improper faces*) or any set F for which there is a supporting hyperplane H of K with $F = H \cap K$. A **facet** of a convex polytope is a proper face not contained in any larger face.

PONCELET, Jean Victor (1788–1867). French engineer and projective geometer. Revived study of projective geometry and laid foundations for its modern study; formulated (sharing priority with Gergonne) the principle of duality; introduced points at infinite and theory of transversals.

Poncelet's principle of continuity. The very vague principle that "if one figure can be derived from another by a continuous change and the latter is as general as the former, then any property of the first figure can be asserted for the second figure."

PONTRYAGIN, Lev Semenovich (1908–1988). Soviet mathematician who made important contributions to algebra, topology, differential equations, and control theory. He was permanently blind after an accident at the age of 14.

POOLED, *adj.* **pooled sum of squares.** When several random samples of various sizes are regarded as originating from the same model, S is a **pooled sum of squares** if

$$S = \sum_{j=1}^{k} \sum_{i=1}^{n_j} \left(x_{ij} - \bar{x}_j \right)^2,$$

where $j = 1, \cdots, k$ over k samples and $i = 1, \cdots, n_j$, where n_j is the number of observations in the jth sample and $\bar{x}_j$ is the mean of the jth sample. Then $S \Big/ \left(\sum_{j=1}^{k} n_j \right)$ is a **pooled variance.**

POP'U-LA'TION, *n.* The set of all possible outcomes of an experiment, or the totality of numbers or symbols that describe the outcomes (*i.e.*, all possible values of an associated random variable). Examples of populations are the set {heads, tails} for the toss of a coin; the totality of potential measurements of the length of a rod; the totality of automobile tires produced under prescribed conditions (or the lengths of life of such tires when subjected to some standard test). *Syn.* parent population, target population, universe.

PO'SET, *n.* A partially ordered set. See OR-DERED —ordered set.

PO-SI'TION-AL, *adj.* **positional notation.** See EXPANDED —expanded notation.

POS'I-TIVE, *adj.* **positive angle.** See ANGLE.

positive correlation. See CORRELATION.

positive number. *Positive* and *negative numbers* are used to denote numbers of units taken in opposite directions or opposite senses. If a positive number denotes miles east, a negative number denotes miles west. *Tech.* If a is a number, positive, negative, or zero, then b is the negative of a if $a + b = 0$; a itself is written $+a$, or simply a, while b, the negative of a, is written $-a$. The set of real numbers is an *ordered field.* See FIELD —ordered field.

positive part of a function. Let f be a function whose range is a set of real numbers. Then the **positive part** of f is the function f^+ defined by $f^+(x) = f(x)$ if $f(x) \geq 0$ and $f^+(x) = 0$ if $f(x) < 0$. The negative part of f is the function f^- defined by $f^-(x) = 0$ if $f(x) > 0$ and $f^-(x) = -f(x)$ if $f(x) \leq 0$. For all f, we have $f = f^+ - f^-$ and $|f| = f^+ + f^-$.

positive sign. Same as PLUS SIGN. See PLUS.

POS'TU-LATE, *n.* See AXIOM.

Euclid's postulates. (1) A straight line may be drawn between any two points. (2) Any terminated straight line may be produced indefinitely. (3) About any point as center a circle with any radius may be described. (4) All right angles are equal. (5) (The parallel postulate.) If two straight lines lying in a plane are met by another line, making the sum of the internal angles on one side less than two right angles, then those straight lines will meet, if sufficiently produced, on the side on which the sum of the angles is less than two right angles. (There is not complete agreement on how many of Euclid's assumptions were designated as postulates, but these five are generally so recognized.)

PO'TEN-CY, *n.* **potency of a set.** See CARDINAL —cardinal number.

POTENTIAL, *adj., n.* The work done against a conservative field or its negative (depending on conventions) in bringing a unit of the proper sort from infinity to the point in question, or the value at the given point of a function whose directional derivative is equal in magnitude to the component of the field intensity in that direction at the given point. This concept is so extensive and well developed that it can be described satisfactorily only by enumerating its special instances. See various headings below.

concentration method for the potential of a complex. This method consists of selecting a point O inside the complex and expressing r_i in terms of quantities associated with O, namely, r (the distance from O to the field point), l_i (the distance from O to the charge e_i), and θ_i (the angle between r and l_i). Thus by the law of cosines and the binomial theorem we have

$$r_i^{-1} = \left(r^2 + l_i^2 - 2rl_i \cos \theta_i \right)^{-1/2}$$

$$= \frac{1}{r} + (l_i \cos \theta_i)/r^2$$

$$+ l_i(3 \cos^2 \theta_i - 1)/(2r^3) + \cdots .$$

If we let λ_i be the vector from O to e_i, ρ_1 the unit vector pointing from O toward the field point, and μ_i the vector $e_i \lambda_i$, then $\mu_i \cdot \rho_1 = e_i l_i \cos \theta_i$ and therefore $\Sigma e_i l_i \cos \theta_i = \Sigma \mu_i \cdot \rho_1 = \mu \cdot \rho_1$, where $\mu = \Sigma \mu_i$. The vector μ is the **polarization of the complex**. If we multiply the foregoing equation for r_i^{-1} by e_i, sum on i, and denote the total charge Σe_i by e, we see that the potential assumes the form $e/r + (\mu \cdot \rho_1)/r^2 +$ higher-order terms. If the complex consists of but two charges equal in magnitude but opposite in sign and we call the negative charge e_1, then

$$\mu_1 + \mu_2 = e_2(-\lambda_1 + \lambda_2) = \mu.$$

Now $-\lambda_1 + \lambda_2$ is a vector from the negative charge e_1 to the positive charge e_2. Consequently, the polarization of this special complex (which in one sense is the electrical counterpart of a magnet) is a vector having direction from the negative charge to the positive charge and having magnitude m equal to the magnitude of the positive charge times the distance between charges. A **doublet** or **dipole** is the abstraction obtained by allowing l_1 and l_2 to approach zero while e_2 ($e_2 = -e_1$) approaches infinity in such a way that μ remains constant. This limiting process eliminates the higher-order terms in the expansion of r_i^{-1}. Hence the potential of the doublet is given by the single term $(\mu \cdot \rho_1)r^{-2}$ or $(m \cos \theta)r^{-2}$, where θ is the angle between ρ_1 and μ. Returning to the more general complex of charges, we see that, except for terms involving the third and higher powers of $1/r$, its potential is that due to a single charge of magnitude e and a dipole of moment μ, both located at O.

conductor potential. See CONDUCTOR.

Dirichlet characteristic properties of the potential function. See DIRICHLET.

electrostatic potential. See ELECTROSTATIC.

first, second, and third boundary-value problems of potential theory. See BOUNDARY.

Gauss' mean-value theorem for potential functions. See GAUSS—Gauss' mean-value theorem.

gravitational potential of a complex of particles (Newtonian potential). The function obtained from $\Sigma e_i/r_i$ by replacing e_i with $-Gm_i$, where G denotes the gravitational constant and m_i the mass of the ith particle. Many writers omit the minus sign and compensate for it otherwise. When this is done it is the positive gradient of the potential that gives the field strength or force a unit mass would experience if placed at the point in question. If the minus sign is dropped and G is given the value unity, the *Newtonian potential function* for the set of point-masses is then $\Sigma m_i/r_i$.

kinetic potential. The difference between the kinetic energy and the potential energy. *Syn.* Lagrangian function.

logarithmic potential. See LOGARITHMIC.

potential energy. See ENERGY.

potential function for a double layer of distribution of dipoles on a surface. This potential function U is given by

$$U = \int m \cos \theta \, r^{-2} \, dS.$$

Here m is the moment per unit area of the dipole distribution and θ is the angle between the polarization vector and the vector to the field point (see p. 324, concentration method for the potential of a complex). If m not only is continuous but is of class C and the polarization vector is normal to the surface, then

$$\lim \left. \frac{\partial U}{\partial n} \right|_{\text{at } N} = \lim \left. \frac{\partial U}{\partial n} \right|_{\text{at } M}$$

as M and N approach P. However, in this case U suffers a jump on passing through the surface; for, if m is continuous and M and N are points on the positive and negative sides of the normal through the surface point P, then $\lim U(M) = U(P) + 2\pi m(P)$, while

$$\lim U(N) = U(P) - 2\pi m(P).$$

potential function for a surface distribution of charge or mass. The function U defined by $U = \int \sigma/r \, dS$. Here σ is the surface density of charge or mass if σ is continuous. U is continuous but its normal derivative suffers a jump at the surface. More precisely, if we select a point P on the surface (but not on its edge if it is a surface patch), draw the normal through P, select two points M and N on the normal but on opposite sides of P, and compute the normal derivatives in the sense M to N at both M and

N, then the limit as M and N approach P of

$$\partial U/\partial n|_{\text{at }N} - \partial U/\partial n|_{\text{at }M}$$

is $-4\pi\sigma$.

potential function relative to a given vector-valued function ϕ. A scalar-valued function S such that $\nabla S = \phi$, or $-\nabla S = \phi$, depending on the convention adopted. If ϕ is velocity, then S is the **velocity potential**. See IRROTATIONAL — irrotational vector in a region.

potential function for a volume distribution of charge or mass. If we are given a continuous space distribution of charge or mass (*i.e.*, a density function instead of a discrete collection of points endowed with charge or mass), the potential function is $\int \rho/r\, dV$. Here ρ is a point function, say $\rho(X, Y, Z)$ in Cartesian coordinates, r is the distance from the charge point (X, Y, Z) to the field point (x, y, z), whereas the region of integration is the volume occupied by the charge. Thus in

$$\iiint \rho/r\, dX\, dY\, dZ,$$

the integration variables are the coordinates X, Y, Z, while the letters x, y, z appear as parameters. Consequently,

$$\int \rho/r\, dV$$

is a function of the field point variables x, y, z.

potential in magnetostatics. Work done by the magnetic field in repelling a unit positive pole from the given point to a point at infinity, or to a point which has been selected as a point of zero potential. The potential due to a distribution of magnetic material is essentially that of a similar dipole distribution.

potential theory. The theory of potential functions. From one point of view, it is the theory of Laplace's equation. Every harmonic function can be regarded as a potential function and the Newtonian potential functions are harmonic functions in free space.

spreading method for the potential of a complex. Instead of replacing the complex with a series of fictitious elements located at a single point, the spreading method replaces the set of point charges with a continuous distribution of charge characterized by a density function $\rho(x, y, z)$, or with both a density of charge and a density of polarization. If both charge and polarization are spread, simpler functions will suffice for a given degree of approximation than will if only charge is distributed. The potential in the

two cases is taken to be the volume integral $\int \rho/r\, dV$ if charge alone is spread, and otherwise

$$\int \rho/r\, dV + \int (m\cos\theta)/r^2\, dV.$$

Here m is the absolute value of the polarization per unit volume. If the polarization may be regarded as concentrated on a surface, the last integral should be replaced with the surface integral

$$\int m\cos\theta\, r^{-2}\, dS,$$

in which m is the magnitude of the polarization per unit area.

vector potential relative to a given vector-valued function ϕ. A vector-valued function ψ such that $\nabla \times \psi = \phi$. See SOLENOIDAL— solenoidal vector in a region.

POUND, *n.* A unit of weight; the weight of one mass pound. See MASS and WEIGHT. Since weight varies slightly at different points on the earth, for extremely accurate work one pound of force is taken as the weight of a mass pound at sea level and $45°$ north latitude.

pound of mass. See MASS.

POUND'AL, *n.* A unit of force. See FORCE— unit of force.

POUSSIN. See DE LA VALLÉE POUSSIN.

POW'ER, *adj., n.* See EXPONENT.

Abel theorem on power series. See ABEL.

difference of like powers of two quantities. See DIFFERENTIAL.

differentiation of a power series. See SERIES —differentiation of an infinite series.

integration of a power series. See SERIES— integration of an infinite series.

perfect power. See PERFECT.

power. (*Physics*) The rate at which work is done.

power of a point. (1) *With reference to a circle,* the quantity obtained by substituting the coordinates of the point in the equation of the circle when written with the right-hand side equal to zero and the coefficients of the square terms equal to unity. This is equal to the algebraic product of the distances from the point to the points where any line through the given point cuts the circle (this product is the same for all such lines); it is equal to the square of the length of a tangent from the point to the circle when the point is external to the circle. (2) *With refer-*

ence to a sphere, the power of a point is the power of the point with reference to any circle formed by a plane passing through the point and the center of the sphere. This is equal to the value obtained by substituting the coordinates of the point in the equation of the sphere when written with the right-hand side equal to zero and the coefficients of the square terms equal to unity. It is also the product of the distances from the point to the points where any line through the given point cuts the sphere, or the square of the length of a tangent from the point to the sphere if the point is external to the sphere.

power of a set. See CARDINAL —cardinal number.

power of a test of a hypothesis. See HYPOTH-ESIS —test of a hypothesis.

power residue. See RESIDUE.

power series. See SERIES.

sums of like powers of two quantities. See SUM —sum of like powers.

PRE-CI′SION, *n.* **double precision.** A term applied when, in a computation, two words (or storage positions) of a computing machine are used to denote a single number to more decimal places than would be possible with one storage position.

modulus of precision. When analyzing errors of estimation, the modulus of precision is $1/(\sigma\sqrt{2}\,)$, where σ^2 is the *variance*. The normal distribution with modulus of precision h has probability density function

$$f(t) = \frac{h}{\sqrt{\pi}} e^{-h^2 t^2}.$$

PRE-IMAGE, *n.* See IMAGE.

PRE′MI-UM, *adj., n.* (1) The amount paid for the loan of money, in addition to normal interest. (2) The difference between the selling price and par value of stocks, bonds, notes, and shares, when the selling price is greater (*e.g.*, **premium bonds** are bonds which sell at a premium—for more than par value). Compare DISCOUNT. (3) One kind of currency is at a *premium* when it sells for more than its face value in terms of another; when one dollar in gold sells for $1.10 in paper money, gold is at a *premium* of 10 cents per dollar. (4) The amount paid for insurance. A **net premium** is a premium which does not include any of the company's operating expenses. Explicitly, **net annual premiums** are equal annual payments made at the beginning of each policy year to pay the cost of a policy figured under the following assumptions: all policyholders will die at a rate given by a standard (accepted) mortality table; the insurance company's funds will draw interest at a certain given rate; every benefit will be paid at the close of the policy year in which it becomes due, and there will be no charge for carrying on the company's business; the **net single premium** is the present value of the contract benefits of the insurance policy. The **natural premium** is the net single premium for a one-year term insurance policy at a given age (this is the yearly sum required to meet the cost of insurance each year, not including the company's operating expenses). **Net level premiums** are fixed (equal) premiums (usually annual), which are equivalent over a period of years to the natural premiums over the same period. In the early years, the premiums are greater than the natural premium; in the later years, they are less. **Gross premium** (or **office premium**) is the premium paid to the insurance company; net premium, plus allowances for office expenses, medical examinations, agents' fees, etc., minus deductions due to income. **Installment premiums** are annual premiums payable in installments during the year. The single premium for an insurance policy is the amount which, if paid on the policy date, would meet all premiums on the policy. See RESERVE.

PRES′ENT, *adj.* **present value.** See SURRENDER —surrender value of an insurance policy, VALUE —present value.

PRES′SURE, *n.* A force, per unit area, exerted over the surface of a body. See below, fluid pressure.

center of pressure. See CENTER —center of pressure.

fluid pressure. The force exerted per unit area by a fluid. The fluid pressure on a unit horizontal area (plate) at a depth h is equal to the product of the density of the fluid and h. The total force on a horizontal area at depth k is khA, where A is the area and k the density of the fluid. The total force on a nonhorizontal region is found by dividing the region into infinitesimal regions (*e.g.*, vertical or horizontal strips, if the region is in a vertical plane) and using integration to find the limit of the sum of suitable approximations to the forces on these strips. See ELEMENT —element of integration.

PRE-VAIL′ING, *adj.* **prevailing interest rate for a given investment.** The rate which is generally accepted for that particular type of investment at the time under consideration. *Syn.* income rate, current rate, yield rate.

PRICE, *n.* The quoted sum for which merchandise or contracts (bonds, mortgages, stock, etc.) are offered for sale, or the price for which they are actually sold (the **selling price**). The price recorded in wholesale catalogues and other literature is the **list price** (it is usually subject to a discount to retail merchants). The **net price** is the price after all discounts and other reductions have been made. For bonds, the total payment made for a bond is the **flat price** or **purchase price**; the **quoted price** (or **"and-interest price"**) is the same as the *book value* of the bond (see VALUE). The theoretical value of the purchase price of a bond on a dividend date is the present value of the redemption price (usually face value) plus the present value of an annuity whose payments are equal to the dividends on the bond; between dividend dates, the purchase price is the sum of the price of the bond at the last interest date and the *accrued interest* (the proportionate part of the next coupon which is paid to the seller). The *flat price* is equal to the *quoted price* plus the *accrued interest*. The **redemption price** of a bond is the price that must be paid to redeem the bond. If a bond specifies that it may be redeemed at specified dates prior to maturity, the price at which it may be redeemed on such dates is the **call price**.

PRI′MA-RY, *adj.* **primary infinitesimal and infinite quantity.** See STANDARD —standard infinitesimal and infinite quantity.

PRIME, *adj., n.* An integer *p* which is not 0 or ±1 and is divisible by no integers except ±1 and ±*p*, *e.g.*, ±2, ±3, ±5, ±7, ±11. Sometimes a prime number is required to be positive. There are an infinite number of prime numbers, but no general formula for these primes. More generally, a prime can mean any member of an integral domain that is not a unit and can not be written as the product of two members that are not units. See DIRICHLET —Dirichlet theorem, ERATOSTHENES, FUNDAMENTAL —fundamental theorem of arithmetic, GOLDBACH —Goldbach conjecture, and above, prime-number theorem.

prime direction. An initial directed line; a fixed line with reference to which directions (angles) are defined; usually the positive *x*-axis or the *polar axis*.

prime factor. A prime quantity (number, polynomial) that will exactly divide the given quantity. *E.g.*, (1) the numbers, 2, 3, and 5 are the prime factors of 30; (2) the quantities *x*, $(x + 1)$ and $(x - 1)$ are the prime factors of $x^5 - 2x^3 + x$. See above, PRIME and prime polynomial.

prime meridian. See MERIDIAN.

prime number. See above, PRIME.

prime-number theorem. Let $\pi(n)$ be the number of positive primes not greater than *n*. Then $\pi(n)$ is asymptotic to $n/\log_e n$, *i.e.*,

$$\lim_{n \to \infty} \frac{\pi(n)\log_e n}{n} = 1.$$

This theorem was stated as a conjecture by Gauss in 1792 and proved (independently) for the first time by Hadamard and de la Vallée Poussin in 1896. The first elementary proof (no use of integral calculus) was given by Selberg and Erdös in 1948 and 1949. The prime-number theorem is equivalent to $\lim_{n \to \infty} \pi(n)/Li(n) = 1$, where

$$Li(n) = \lim_{\epsilon \to 0} \left(\int_0^{1-\epsilon} \frac{dx}{\log_e x} + \int_{1+\epsilon}^n \frac{dx}{\log_e x} \right).$$

The difference $\pi(n) - Li(n)$ changes sign infinitely often. S. Skewes proved (1955) that the first value of *n* (**Skewes number**) for which $\pi(n) > Li(n)$ is less than $(10^{10^{10}})^{1000}$ (later it was shown to be less than 10^{370}). If $n = 10^4$, then $\pi(n) = 1,229$ and $Li(n) = 1,246$; If $n = 10^{10}$, then $\pi(n) = 455,052,512$ and $Li(n) = 455,055,614$.

prime polynomial. A polynomial which has no polynomial factors except itself and constants. The polynomials $x - 1$ and $x^2 + x + 1$ are prime. *Tech.* A polynomial which is *irreducible*. See IRREDUCIBLE —irreducible polynomial.

prime (or accent) as a symbol. The symbol (′) placed to the right and above a letter. (1) Used to denote the first derivative of a function: y' and $f'(x)$, *y-prime* and *f-prime*, denote the first derivatives of *y* and $f(x)$. Similarly y'' and $f''(x)$, read *y double prime* and *f double prime*, denote second derivatives. If general, $y^{[n]}$ and $f^{[n]}(x)$ denote the *n*th derivatives. (2) Sometimes used on letters to denote constants, x' denoting a particular value of *x*, (x', y') denoting the particular point whose coordinates are x' and y', in distinction to the variable point (x, y). (3) Used to denote different variables with the same letters, as x, x', x'', etc. (4) Used to denote feet and inches, as $2', 3''$, read two feet and three inches. (5) Used to denote minutes and seconds in circular measurement of angles, as $3° 10' 20''$, read three degrees, ten minutes, and twenty seconds.

relatively prime. Two integers are relatively prime if they have no common factors other than +1 or −1; two polynomials are relatively prime if they have no common factors except constants.

twin primes. See TWIN.

PRIM′I-TIVE, *adj.* (1) A geometrical or analytic form from which another is derived; a quantity whose derivative is a function under consideration. (See DERIVATIVE.) (2) A function which satisfies a differential equation. (3) A curve of which another is the polar or reciprocal, etc.

primitive curve. A curve from which another curve is derived; a curve of which another is the polar, reciprocal, etc.; the graph of the primitive of a differential equation (a member of the family of curves which are graphs of the solutions of the differential equation). See INTEGRAL— integral curves.

primitive of a differential equation. See DIFFERENTIAL —solution of a differential equation.

primitive element of a monogenic analytic function. See MONOGENIC —monogenic analytic function.

primitive nth root of unity. See ROOT—root of unity.

primitive period of a periodic function of a complex variable. See PERIODIC —periodic function of a complex variable, and various headings under PERIOD.

primitive polynomial. A polynomial with integers as coefficients for which the greatest common divisor of the coefficients is 1. If a primitive polynomial ρ is the product of two polynomials r and s with rational coefficients, then there are polynomials f and g with integer coefficients that differ from r and s by constant factors and for which $p = fg$.

PRIN′CI-PAL, *adj., n.* Most important or most significant. In *finance*, money put at interest, or otherwise invested.

principal curvature and radii of curvature. See CURVATURE —curvature of a surface.

principal diagonal. See DETERMINANT, MATRIX, PARALLELEPIPED.

principal ideal See IDEAL, RING—principal ideal ring.

principal meridian. See MERIDIAN.

principal normal. See NORMAL—normal to a curve or surface.

principal part of a function of a complex variable. See LAURENT—Laurent expansion.

principal part of the increment of a function. See INCREMENT —increment of a function.

principal parts of a triangle. The sides and interior angles. The other parts, such as the bisectors of the angles, the altitudes, the circumscribed and inscribed circles, are **secondary parts**.

principal plane of a quadric surface. A plane of symmetry of the quadric surface.

principal root of a number. The positive real root in the case of roots of positive numbers; the negative real root in the case of odd roots of negative numbers. Every nonzero number has two square roots, three cube roots, and in general n nth roots, if complex numbers are used.

principal value of an inverse trigonometric function. See TRIGONOMETRIC— inverse trigonometric functions.

PRIN-CI′PI-A, *n.* One of the greatest scholarly works of all time, written by Sir Isaac Newton, and first printed in London in 1687 under the title *Philosophiae Naturalis Principia Mathematica*. This work lies at the base of all present structure of mechanics of rigid and deformable bodies and mathematical astronomy.

PRIN′CI-PLE, *n.* A general truth or law, either assumed or proved. See AXIOM—axiom of continuity, DUALITY, ENERGY —principle of energy, PROPORTIONAL—proportional parts, SAINT-VENANTS′ PRINCIPLE.

principle of the maximum. The principle that if f is a regular analytic function of the complex variable z in the domain D, and f is not a constant, then $|f(z)|$ does not have a maximum value at any interior point of D.

principle of the minimum. The principle that if f is a regular analytic function of the complex variable z in the domain D, if there is no z in D for which $f(z) = 0$, and if f is not a constant, then $|f(z)|$ does not have a minimum value at any interior point of D. Note that if $f(z) = z$, then $|f(z)|$ does have a minimum at the origin.

PRINGSHEIM, Alfred (1850–1941). German analyst.

Pringsheim's theorem on double series. See SERIES —double series.

PRISM, *n.* A polyhedron with two congruent and parallel faces, the **bases**, whose other faces, the **lateral faces**, are parallelograms formed by joining corresponding vertices of the bases; the intersections of lateral faces are **lateral edges**. A **diagonal** is any line segment joining two vertices that do not lie in the same face or base. The **altitude** is the perpendicular distance between the bases. The **lateral area** is the total area of the lateral faces (equal to an edge times the perimeter of a right section),

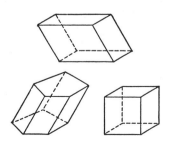

and its volume is equal to the product of its base and its altitude. A prism with a triangle as base is a **triangular prism**; one with a quadrilateral as base is a **quadrangular prism**; etc. An **oblique prism** is a prism whose bases are not perpendicular to the lateral edges. A **right prism** has bases perpendicular to the lateral edges and all lateral faces are rectangles. A **regular prism** is a right prism whose bases are regular polygons. A **truncated prism** is a portion of a prism lying between two nonparallel planes which cut the prism and have their line of intersection outside the prism. A **right truncated prism** is a truncated prism in which one of the cutting planes is perpendicular to a lateral edge.

circumscribed and inscribed prisms. See CIRCUMSCRIBED.

right section of a prism. A plane section perpendicular to the lateral faces of the prism.

PRIS-MAT'IC, *adj.* **prismatic surface.** A surface generated by a moving straight line which always intersects a broken line lying in a given plane and is always parallel to a given line not in the plane. When the broken line is a polygon, the surface is a **closed prismatic surface**.

PRIS'MA-TOID, *adj., n.* A polyhedron whose vertices all lie in one or the other of two parallel planes. The faces which lie in the parallel planes are the **bases** of the prismatoid and the perpendicular distance between the bases is the **altitude**. See PRISMOIDAL —prismoidal formula.

PRIS'MOID, *n.* A prismatoid whose bases are polygons having the same number of sides, the other faces being trapezoids or parallelograms. A prismoid whose bases are congruent is a **prism**.

PRIS-MOI'DAL, *adj.* **prismoidal formula.** The volume of a prismatoid is equal to one-sixth of the altitude times the sum of the areas of the bases and four times the area of a plane section midway between the bases: $V = \frac{1}{6}h(B_1 + 4B_m + B_2)$. This formula also gives the volume of any solid having two parallel plane bases, whose cross-sectional area (by a plane parallel to the bases) is given by a linear, quadratic, or cubic function of the distance of the cross section from one of the bases (an elliptic cylinder, and

quadratic cone, satisfy these conditions). The prismoidal formula is sometimes given as $V = \frac{1}{4}h(B_1 + 3S)$, where S is the area of a section parallel to the base and $\frac{2}{3}$ the distance from B_1 to B_2. This is equivalent to the preceding form. See SIMPSON'S RULE.

PROB'A-BIL'I-TY, *adj., n.* (1) Let n be the number of exhaustive, mutually exclusive, and equally likely cases of an event under a given set of conditions. If m of these cases are known as the event A, then the (**mathematical** or *a priori*) **probability** of event A under the given set of conditions is m/n. E.g., if one ball is to be drawn from a bag containing two white balls and three red balls, and each ball is equally likely to be drawn, then the probability of drawing a white ball is $\frac{2}{5}$ and the probability of drawing a red ball is $\frac{3}{5}$. This definition is circular in that *equally likely* means *equally probable*, but the intuitive meaning is useful. (2) If, in a *random sequence* of n trials of an event with m favorable events, that ratio m/n as n increases indefinitely has the limit P, then P is the **probability** of the event m. This is sometimes amended to state that, if it is practically certain that m/n is approximately equal to P when n is very large, P is the probability of the event m. (3) A third general type of definition is based on certain axiomatic statements and leaves the problem of applying these to empirical situations to the ingenuity of the practicing statistician [see below, probability function]. All the above definitions have either logical or empirical difficulties. They usually give the same numerical values for the usual empirical problems. See below, empirical or *a posteriori* probability.

conditional probability. If A and B are events, then the **conditional probability** of A given B is the probability of A, assuming B holds. If $P(B) \neq 0$, then the conditional probability $P(A|B)$ of A given B is $P(A \text{ and } B)/P(B)$. E.g., the probability at least one of two dice shows a 3 when the sum is 7 is P(at least one 3 and a sum of 7) $\div P$(sum of 7) $= \frac{1}{18}/\frac{1}{6} = \frac{1}{3}$.

convergence in probability. Let $x_1, x_2, x_3, \cdots$ be a sequence of random variables (*e.g.*, the means of samples of size $1, 2, 3, \cdots$). Then x_n **converges in probability** to a constant k if, for any $\epsilon > 0$, the probability of $|x_n - k| > \epsilon$ tends to zero as $n \to \infty$.

empirical or *a posteriori* probability. If in a number of trials an event has occurred n times and failed m times, the *probability* of its occurring in the next trial is $n/(n + m)$. It is assumed, in determining *empirical probability*, that there is no known information relative to the probability of the occurrence of the event other than the past trials. The probability of a man living through any one year, based upon past observations as recorded in a mortality table, is *empirical probability*.

   inverse probability. See BAYES—Bayes' theorem.

   mathematical or *a priori* probability. See above, PROBABILITY (1).

   probability-density function. For a given *probability function P*, a **probability-density function** is a function p such that, for any event represented by a set E, $P(E) = \int_E p(x)\,dx$. When the domain of p is the real line and p is continuous at x, then $p(x)$ is the derivative of the distribution function F defined by $F(x) = P(E_x) = \int_{-\infty}^{x} p(x)\,dx$, where E_x is the set of all ξ for which $\xi \leq x$. Sometimes a probability density function also is called a **relative-frequency function** or merely a **frequency function**. For examples, see *distributions* under CAUCHY, CHI, GAMMA, LOGNORMAL, NORMAL, F, T, UNIFORM.

   probability function. Suppose we are considering various events and that T is a set for which these events are identified with subsets of T, the "certain" event being identified with T itself. Then a **probability function** is a function P whose domain is this collection of events, whose range is contained in the closed interval $[0,1]$, and which satisfies the conditions: (i) $P(T) = 1$; (ii) $P(A \cup B) = P(A) + P(B)$ if A and B are two events for which $A \cap B$ is empty; $P(A_1 \cup A_2 \cup \cdots) = \sum_{1}^{\infty} P(A_n)$ whenever $\{A_1, A_2, \cdots\}$ is a sequence of events and $A_i \cap A_j$ is empty when $i \neq j$. E.g., suppose two dice are thrown and we let T be the set of all ordered pairs (m, n) for which each of m and n is one of the integers $1, 2, 3, 4, 5, 6$. The usual probability function assigns $\frac{1}{36}$ as the probability for any particular ordered pair. Then the event "sum of 8" corresponds to the set $\{(2,6),(3,5),(4,4),(5,3),(6,2)\}$, which has probability $5 \cdot \frac{1}{36}$. More generally, for any set T we can call a function P whose domain is a collection of subsets of T a **probability function** and the members of the domain of P can be called **events** if the range of P is a set of nonnegative real numbers, $P(T) = 1$, $A - B$ and $A \cup B$ are in the domain of P if A and B are in the domain, and $P(A \cup B) = P(A) + P(B)$ if $A \cap B$ is empty [customarily, P is also required to be countably additive, i.e., $A_1 \cup A_2 \cup \cdots$ is in the domain of P and $P(A_1 \cup A_2 \cup \cdots) = \sum_{1}^{\infty} P(A_n)$ whenever each A_i is in the domain of P and $A_i \cap A_j$ is empty when $i \neq j$]. *Syn.*

probability measure. See MEASURE—measure of a set, and above, probability density function.

   probability limit. T is the *probability limit* of the statistic t_n, derived from a random sample of n observations, if the probability of $|t_n - T| < \epsilon$ approaches 1 as a limit as $n \to \infty$, for any $\epsilon > 0$. See above, probability convergence.

   probability measure. Same as PROBABILITY FUNCTION. See above.

   probability in a number of repeated trials. (1) The probability that an event will happen *exactly r times* in n trials, for which p is the probability of its happening and q of its failing in any given trial, is given by the formula $n! p^r q^{n-r} / [r!(n-r)!]$, which is the $(n-r+1)$th term in the expansion of $(p+q)^n$. The probability of throwing exactly two sixes in five throws of a die is

$$5! \left(\tfrac{1}{6}\right)^2 \left(\tfrac{5}{6}\right)^3 / (2!3!) = 0.16 + .$$

(2) The probability that an event will happen *at least r times* in n throws is the probability that it will happen every time plus the probability that it will happen exactly $n-1$ times, $n-2$ times, etc., to exactly r times. This probability is given by the sum of the first $n-r+1$ terms of the expansion of $(p+q)^n$.

   probability paper. Graph paper, one axis of which is scaled so that the graph of the *cumulative frequency* of the *normal-distribution function* forms a straight line.

PROB′A·BLE, *adj.* Likely to be true or to happen.

   probable deviation, or **probable error.** See DEVIATION—probable deviation.

PROB′LEM, *n.* A question proposed for solution; a matter for examination; a proposition requiring an operation to be performed or a construction to be made, as to bisect an angle or find an eighth root of 2. See ACCUMULATION— accumulation problem, APOLLONIUS, DIDO′S PROBLEM, DISCOUNT—discount problem under compound interest, FOUR—four-color problem, MOMENT—moment problem, THREE—three-point problem.

   problem formulation. In numerical analysis, problem formulation is the process of deciding what information the customer really wanted, or should have wanted, and then of writing this in mathematical terms preparatory to programming the problem for machine solution. See PROGRAMMING—programming for a computing machine.

PRO′CEEDS, *n.* (1) The sum of money obtained from a business transaction or enterprise.

The proceeds of a farm for a year is the sum of all the money taken in during the year; the proceeds of a sale of goods is the money received in return for the goods. The amount of money left after deducting all discounts and expenses from the proceeds of a transaction is called the **net proceeds**. (2) The difference between the face of a note, or other contract to pay, and the discount; the balance after interest, in advance, has been deducted from the face of the note.

PRO-DUCE′, *v.* **produce a line.** To continue the line. *Syn.* prolong, extend.

PROD′UCT, *adj., n.* The product of two or more objects is the object which is determined from these objects by a given operation called **multiplication**. See various headings below (particularly *product of real numbers*) and under MULTIPLICATION. Also see COMPLEX—complex numbers, SERIES—multiplication of infinite series.

    Cartesian product. The Cartesian product of two sets A and B is the set (denoted by $A \times B$) of all pairs (x, y) such that x is a member of A and y is a member of B. If multiplication, addition, or multiplication by scalars is defined for each of the sets A and B, then the same operation can be defined for $A \times B$ by

$$(x_1, y_1) \cdot (x_2, y_2) = (x_1 \cdot x_2, y_1 \cdot y_2),$$

$$(x_1, y_1) + (x_2, y_2) = (x_1 + x_2, y_1 + y_2),$$

$$a(x, y) = (ax, ay).$$

If A and B are **groups**, their Cartesian product is a group. A matrix representation of the Cartesian product of two groups is given by the *direct product* of corresponding matrices in representations of the two groups. If a group G has subgroups H_1 and H_2 such that H_1 and H_2 have only the identity in common, each element of G is a product of an element of H_1 and an element of H_2, and each element of H_1 commutes with each element of H_2, then G is isomorphic with the Cartesian product $H_1 \times H_2$. If A and B are **rings**, then $A \times B$ is a ring. If A and B are **vector spaces** (with the same scalar multipliers), then $A \times B$ is a vector space. If A and B are **topological spaces**, then $A \times B$ is a topological space if a set in $A \times B$ is defined to be open if it is a Cartesian product $U \times V$, where U and V are open sets in A and B. If A and B are **topological groups** (or **topological vector spaces**), then $A \times B$ is a topological group (or a topological vector space). If A and B are **metric spaces**, the usual distance relation for $A \times B$ is

$$d[(x_1, y_1), (x_2, y_2)] = \left[d(x_1, x_2)^2 + d(y_1, y_2)^2 \right]^{1/2}.$$

With this definition, the Cartesian product $R \times R$, where R is the space of real numbers, is the two-dimensional space of all points (x, y) with the usual distance of plane geometry. If A and B are **normed vector spaces**, $A \times B$ is a normed vector space if the norm is defined by

$$\|(x, y)\| = \left[\|x\|^2 + \|y\|^2 \right]^{1/2}.$$

Many other definitions are used, such as $\|(x, y)\| = \|x\| + \|y\|$. If A and B are **Hilbert spaces**, then $A \times B$ is a Hilbert space if the norm is defined as above or, equivalently, if the *inner product* of (x_1, y_1) and (x_2, y_2) is defined to be the sum of the inner product of x_1 and x_2 and the inner product of y_1 and y_2. The above definitions can be extended in natural ways to the product of any finite number of spaces. The Cartesian product of sets X_a, where a is a member of an index set A, is the set of all functions x defined on A for which $x(a)$ is a member of X_a for each a of A. This means that a point of the product space is a set consisting of a point chosen from each of the sets X_a, the point $x(a)$ being the *a-th coordinate* of the point x of the product. If each of the sets X_a is a **topological space**, then their Cartesian product is a topological space if an open set is defined to be any set which is a union of sets which are Cartesian products of sets Y_a, where $Y_a = X_a$ for all but a finite number of members of A and Y_a is an open set of X_a for the other members of A. For a Cartesian product of a finite number of topological spaces $X_1, X_2, \cdots, X_n$, a set is open in the product if and only if it is a product of sets $U_1, U_2, \cdots, U_n$, where U_k is open in X_k for each k. With this topology for the Cartesian product, it can be shown that the Cartesian product is compact if and only if each X_a is compact (this is the **Tychonoff theorem**). Sometimes the elements of the Cartesian product of a nonfinite number of spaces are restricted by some convergence requirement. *E.g.*, the Cartesian product of Hilbert spaces $H_1, H_2, \cdots$ is the set of all sequences $h = (h_1, h_2, \cdots)$ for which h_n belongs to H_n for each n and $\|h\|$ is finite, where

$$\|h\| = \left[\|h_1\|^2 + \|h_2\|^2 + \cdots \right]^{1/2}.$$

A Cartesian product is sometimes called a **direct product** or a **direct sum**.

    continued product. See CONTINUED.

    direct product of matrices. The direct product of square matrices A and B (not necessarily of the same order) is the matrix whose elements are the products $a_{ij}b_{mn}$ of elements of A and B, where i, m are row indices and j, n are column indices; the row containing $a_{ij}b_{mn}$ precedes that containing $a_{i'j'}b_{m'n'}$ if $i < i'$, or if $i = i'$ and $m < m'$, and similarly for columns. Other ordering conventions are sometimes used.

infinite product. A product which contains an unlimited number of factors. An infinite product is denoted by a capital pi, Π; *e.g.*, $\Pi[n/(n + 1)] = \frac{1}{2} \cdot \frac{2}{3} \cdot \frac{3}{4} \cdot \frac{4}{5} \cdots$ is an infinite product. An infinite product $u_1 \cdot u_2 \cdots u_n \cdots$ **converges** if it is possible to choose k so that the sequence

$$u_k, u_k \cdot u_{k+1}, u_k \cdot u_{k+1} \cdot u_{k+2}, \cdots$$

converges to some limit which is not zero. If the product becomes infinite, or if the above sequence approaches zero for all k, it is said to diverge. If there is a k such that the sequence above neither approaches a limit nor becomes infinite, it is said to oscillate. Because of certain relations to infinite series, infinite products are frequently written in the form $\Pi(1 + a_n)$. A necessary and sufficient condition for the convergence of $\Pi(1 + a_n)$ and $\Pi(1 - a_n)$, if each $a_n > 0$, is the convergence of Σa_n. If the series Σa_n^2 is convergent, the infinite product converges if, and only if, the series Σa_n converges. An infinite product, $\Pi(1 + a_n)$, is said to **converge absolutely** if $\Sigma |a_n|$ is convergent. An absolutely convergent infinite product is convergent. The factors of a convergent infinite product can be rearranged in any way whatever without changing the limit of the product if, and only if, this product converges absolutely.

inner product. See headings under INNER.

limit of a product. See LIMIT—fundamental theorems on limits.

partial product. See PARTIAL—partial product.

product formulas. See TRIGONOMETRY—identities of plane trigonometry.

product moment. See MOMENT—product moment.

product-moment correlation coefficient. Same as CORRELATION COEFFICIENT.

product of determinants, polynomials, and vectors. See MULTIPLICATION —multiplication of determinants, multiplication of vectors, DISTRIBUTIVE —distributive property of arithmetic and algebra.

product of matrices. The product AB of matrices A and B is the matrix whose elements are determined by the rule that the element c_{rs} in row r and column s is the sum over i of the product of the element a_{ri} in row r and column i of A by the element b_{is} in row i and column s

of B:

$$c_{rs} = \sum_{i=1}^{n} a_{ri}b_{is}.$$

This product is defined only if the number n of columns in A is equal to the number of rows in B. Matrix multiplication is associative, but not commutative. The product of a **scalar c and a matrix A** is the matrix whose elements are the products of c and the corresponding elements of A. The determinant of cA (when A is a square matrix of order n) is equal to the product of c^n and the determinant of A.

product of real numbers. Positive integers (and zero) can be thought of as symbols used to describe the "manyness" of sets of objects (also see PEANO—Peano's postulates). Then the **product of two integers A and B** (denoted by $A \times B$, $A \cdot B$, or AB) is the integer which describes the "manyness" of the set of objects obtained by combining A sets, each of which contains B objects (or by combining B sets, each of which contains A objects; $AB = BA$). *E.g.*,

$$3 \cdot 4 = 4 + 4 + 4 = 3 + 3 + 3 + 3 = 12$$

(see SUM—sum of real numbers). Also, $0 \cdot 3 = 3 \cdot 0 = 0 + 0 + 0 = 0$. The **product of two fractions** $\frac{a}{b}$ and $\frac{c}{d}$ (an integer n may be regarded as the fraction $\frac{n}{1}$) is defined by

$$\frac{a}{b} \cdot \frac{c}{d} = \frac{ac}{bd}.$$

The same rule applies if some of a, b, c, d are fractions. *E.g.*,

$$\frac{3}{5} \cdot \frac{1}{2} = \frac{3}{10}; \quad \frac{\frac{2}{3}}{\frac{1}{5}} \cdot \frac{3}{\frac{1}{2}} = \frac{6/3}{1/10} = 20$$

[the last equality results from multiplying numerator and denominator by 10 and then dividing by 3, which is valid since $\frac{a}{b} = \frac{ak}{bk}$ for any numbers a, b, k (b and k not zero)]. Multiplication by a fraction a/b can be interpreted as dividing the other factor into b equal parts and taking a of these ($\frac{1}{2}$ is equal to the sum of 5 addends, each equal to $\frac{1}{10}$, so $\frac{3}{5} \cdot \frac{1}{2}$ is $\frac{1}{10} + \frac{1}{10} + \frac{1}{10} = \frac{3}{10}$). The **product of mixed numbers** can be obtained by multiplying each term of one number by each term of the other, or by reducing each mixed number to a fraction. *E.g.*,

$$\left(2\tfrac{1}{2}\right)\left(3\tfrac{2}{3}\right) = \left(2+\tfrac{1}{2}\right)\left(3+\tfrac{2}{3}\right) = 6+\tfrac{4}{3}+\tfrac{3}{2}+\tfrac{2}{6} = 9\tfrac{1}{6}$$

$$= \left(\tfrac{5}{2}\right)\left(\tfrac{11}{3}\right) = \tfrac{55}{6}.$$

Two **decimals** can be multiplied by reducing them to fractions, or by ignoring the decimal point and multiplying as if the decimals were whole numbers and then pointing off as many decimal places in this product as there are in both multiplicand and multiplier together. The meaning and application of this rule are shown by the example

$$2.3 \times 0.02 = \tfrac{23}{10} \times \tfrac{2}{100} = \tfrac{46}{1000} = 0.046.$$

When the numbers being multiplied have been given signs, the multiplication is done by multiplying the numerical values of the numbers and making the result positive, if both numbers are positive or both are negative, and making it negative if the numbers have different signs (this is sometimes called **algebraic multiplication**). The rule of signs is: *Like signs give plus, unlike give minus.* *E.g.*, $2 \times (-3) = -6$, $-2 \times 3 = -6$, $-2 \times (-3) = 6$. An explanation of this rule is that $a(-b)$ is the number which added to ab will give zero (*i.e.*, the *negative* or *additive inverse of ab*), since

$$ab + a(-b) = a[b + (-b)] = a \cdot 0 = 0.$$

Likewise, $(-a)(-b)$ is the number which added to $a(-b)$ will give zero, *i.e.*,

$$(-a)(-b) = ab.$$

The **product of irrational numbers** may be left in indicated form after similar terms have been combined, until some specific application indicates the degree of accuracy desired. Such a product as $(\sqrt{2} + \sqrt{3})(2\sqrt{2} - \sqrt{3})$ would be left as $1 + \sqrt{6}$. A product such as $\pi\sqrt{2}$ can be approximated as

$$(3.1416)(1.4142) = 4.443.$$

It is necessary to have a specific definition of irrational numbers before one can specifically define the product of numbers one or more of which is irrational. See DEDEKIND —Dedekind cut.

   product of sets and spaces. See INTERSECTION, and Cartesian product, p. 332.

   product of the sum and difference of two quantities. Such products as $(x + y)(x - y)$ used in factoring, since this product is equal to $x^2 - y^2$.

   products of inertia. See MOMENT—moment of inertia.

   scalar and vector products. See MULTIPLICATION—multiplication of vectors.

PRO'FILE, *adj.* **profile map.** A vertical section of a surface, showing the relative altitudes of the points which lie in the section.

PROF'IT, *n.* The difference between the price received and the sum of the original cost and the selling expenses, when the price received is the larger. The selling expenses include storage, depreciation, labor, and sometimes accumulations in a reserve fund. This is sometimes called the **net profit**; the **gross profit** is the difference between the selling price and the original cost. Also see headings under PERCENT.

PRO'GRAM-MING, *n.* **convex programming.** The particular case of nonlinear programming in which the function to be extremized, and also the constraints, are appropriately convex or concave functions of the x's. See below, linear programming, quadratic programming.

   dynamic programming. The mathematical theory of multistage decision processes.

   linear programming. The mathematical theory of the minimization or maximization of a linear function subject to linear constraints. As often formulated, it is the problem of minimizing the linear form $\sum_{i=1}^{n} a_i x_i$, $x_i \geq 0$, subject to the linear constraints $\sum_{i=1}^{n} b_{ij} x_i = c_j$, $j = 1, 2, \cdots, m$. See TRANSPORTATION —Hitchcock transportation problem. The analogous mathematical theory for which the function to be extremized and the constraints are not all linear is **nonlinear programming.** A **solution** of a linear-programming problem is any set of values x_i that satisfy the m linear constraints; a solution consisting of nonnegative numbers is a **feasible solution**; a solution consisting of m x's for which the matrix of coefficients in the constraints is not singular, and otherwise consisting of zeros, is a **basic solution**; a feasible solution that minimizes the linear form is an **optimal solution.** See below, quadratic programming, and SIMPLEX —simplex method.

   programming for a computing machine. In preparing a problem for machine solution, programming is the process of planning the logical sequence of steps to be taken by the machine. It usually, but not necessarily, involves the preparation of flow charts. This process usually follows problem formulation and precedes coding. See

CODING, CHART—flow chart, PROBLEM—problem formulation.

quadratic programming. The particular case of nonlinear programming in which the function to be extremized, and also the constraints, are quadratic functions of the x's wherein the second-degree terms constitute appropriately semidefinite quadratic forms. See above, convex programming.

PRO-GRES'SION, *n.* **arithmetic, geometric, and harmonic progressions.** See ARITHMETIC— arithmetic sequence, GEOMETRIC —geometric sequence, HARMONIC —harmonic sequence.

PRO-JEC'TILE, *n.* **path of a projectile.** See PARABOLA —parametric equations of a parabola.

PRO-JECT'ING, *adj.* **projecting cylinder.** A cylinder whose elements pass through a given curve and are perpendicular to one of the coordinate planes. There are three such cylinders for any given curve, unless the curve lies in a plane perpendicular to a coordinate plane, and their equations in rectangular Cartesian coordinates can each be obtained by eliminating the proper one of the variables x, y, and z between the two equations which define the curve. The space curve, a circle, which is the intersection of the sphere $x^2 + y^2 + z^2 = 1$ and the plane $x + y + z = 0$ has the three projecting cylinders whose equations are $x^2 + y^2 + xy = \frac{1}{2}$, $x^2 + z^2 + xz = \frac{1}{2}$, and $y^2 + z^2 + yz = \frac{1}{2}$. These are elliptic cylinders.

projecting plane of a line in space. A plane containing the line and perpendicular to one of the coordinate planes. There are three projecting planes for every line in space, unless the line is perpendicular to a coordinate axis. The equation of each projecting plane contains only two variables, the missing variable being the one whose axis is parallel to the plane. These equations can be derived as the three equations given by the double equality of the symmetric space equation of the straight line. See LINE—equation of a line.

PRO-JEC'TION, *n.* **center of projection.** See below, central projection.

central projection. A projection of one configuration (A, B, C, D in the figure) on a given plane (the **plane of projection**) in which the projection in this plane (A', B', C', D') is formed by the intersections, with this plane, of all lines passing through a fixed point (not in the plane) and the various points in the configuration. The image on a photographic film is a projection of the image being photographed, if the lens is considered as a point. The point is the **center of projection** and the lines, or rays, are the **projectors.** When the center of projection is a point at infinity (when the rays are parallel), the projection is a **parallel projection.** See below, orthogonal projection.

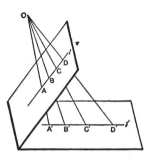

orthogonal projection. A projection of a set S into a given line or plane for which the projection of each point P of S is the foot of the perpendicular from P to the given line or plane, and the projection of S is the set of all projections of points of S. The projection of a **line segment** is the line segment which joins the projections of the end points of the given line segment; the projection of a **vector (force, velocity,** etc.) is the vector whose initial and terminal points are the projections of the initial and terminal points, respectively, of the given vector; the projection of a **directed broken line** into a line is the directed line segment whose initial and terminal points are, respectively, the projections of the initial point of the first segment of the directed broken line and the terminal point of the last segment (sometimes one speaks of the signed length of this projection as the *projection* of the directed broken line).

projection of a vector space. A transformation P of the vector space into itself which is *linear* (*i.e., additive* and *homogeneous*) and *idempotent* ($P \cdot P = P$). If P is a projection of the vector space T, then there are vector spaces M and N contained in T such that each element of T is uniquely representable as the sum of an element of M and an element of N. Explicitly, M is the *range* of P and N is the *null space* of P (the space of all vectors x such that $P(x) = 0$). It is said that P **projects** T onto M along N. If T is a **Banach space**, then P is *continuous* if and only if there is a positive number ϵ such that $\|x - y\| \geq \epsilon$ if x and y are vectors of unit norm (length) belonging to M and N, respectively, or (equivalently) if there is a constant M such that

$\|P(x)\| < M\|x\|$ for each x. If T is a Hilbert space, then P is an **orthogonal projection** (or sometimes simply a **projection**) if $\|P(x)\| \leq \|x\|$ for each x, or (equivalently) if M and N are orthogonal.

stereographic projection of a sphere on a plane. For a given point P, called the **pole**, on the surface of a sphere S, and for a given plane π not passing through P, and perpendicular to a diameter through P, the line joining P with a variable point p on π intersects S in a second point q. This mapping of the points q of the sphere S on the points p of π is a **stereographic projection** of S on π. If an ideal "point at infinity" is adjoined to the plane π, to correspond to P, then the correspondence between the points of S and those of π is one-to-one. The map is a conformal one, much used in the theory of functions of a complex variable. The plane π is often taken as the equatorial plane of S relative to P, or as the tangent plane to S diametrically opposite P.

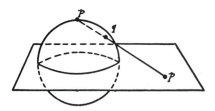

PRO-JEC'TIVE, *adj.* **projective algebraic variety.** See VARIETY.

projective geometry. The study of those properties of geometric configurations which are invariant under projection. See ALBERTI, DESARGUES, PONCELET.

projective plane. (1) The set of all number triples (x_1, x_2, x_3) except $(0,0,0)$ with the convention that $(x_1, x_2, x_3) = (y_1, y_2, y_3)$ if there are two nonzero numbers a and b such that $ax_i = by_i$ for $i = 1, 2, 3$. Points with $x_3 \neq 0$ can be regarded as points of the *Euclidean plane* with abscissa x_1/x_3 and ordinate x_2/x_3; points with $x_3 = 0$ are **points at infinity** or **ideal points** (see IDEAL—ideal point). Each "direction" in the Euclidean plane determines a single point at infinity and the projective plane is topologically equivalent to a disc (a circle and its interior) with the two end points of each diameter identified. It also can be described as the set of all lines through the origin in three-dimensional space (see DIRECTION—direction numbers of a line in space). The projective plane is topologically equivalent to a *sphere with one cross-cap.* See COORDINATE—homogeneous coordinates, and, projective space.

(2) Suppose we have a set of objects called **points,** a set of objects called **lines,** and a concept of a point being **on a line** (or a line **containing a point**). This set of points and lines is called a **projective plane** if (i) any two distinct points are on exactly one line and (ii) for any two distinct lines, there is exactly one point that is on each line. One usually uses additional axioms; *e.g.*, that each line contains at least three points. It is a **finite projective plane of order** n if (iii) each line contains $n + 1$ points and (iv) each point is on $n + 1$ lines. Then there are $n^2 + n + 1$ points and $n^2 + n + 1$ lines. There is a projective plane of order n if and only if there exists a complete set of mutually orthogonal Latin squares of order n. There are no projective planes of orders 6 or 10, but there are four nonequivalent projective planes of order 9. See LATIN—Latin square.

projective plane curve. The set of all points in the projective plane for which $f(x_1, x_2, x_3) = 0$, where f is a homogeneous polynomial. If the gradient $(\partial f/\partial x_1, \partial f/\partial x_2, \partial f/\partial x_3)$ is zero only when $x_1 = x_2 = x_3 = 0$, then the curve is a **smooth projective plane curve.** See CURVE—algebraic plane curve, and above, projective plane (1).

projective space. The n-dimensional projective space over a field F is the set of all $(n + 1)$-tuples $\{x_1, x_2, \cdots, x_{n+1}\}$ of members of F for which not all x_i's are 0 and two $(n + 1)$-tuples are equal if and only if the terms of one are proportional to the terms of the other. Topologically, n-dimensional projective space is equivalent to the solid n-dimensional sphere with each diameter having its end points identified. See ORDERED—ordered pair, and above, projective plane (1).

projective topology. See TENSOR—tensor product of vector spaces.

PRO-JEC'TORS, *n.* See PROJECTION—central projection.

PRO'LATE, *adj.* **prolate cycloid.** A *trochoid* which has loops.

prolate ellipsoid of revolution. See ELLIPSOID.

PRO-LONG', *v.* **prolong a line.** To continue the line. *Syn.* produce.

PROM'IS-SO'RY, *adj.* **promissory note.** A written promise to pay a certain sum on a certain future date, the sum usually drawing interest until paid. The person who signs the note (the one who first guarantees its payment) is the **maker** of the note (also see INDORSE).

PROOF, *n.* (1) The logical argument which establishes the truth of a statement. (2) The process of showing by means of an assumed logical process that what is to be proved follows from certain previously proved or axiomatically accepted propositions. See ANALYTIC—analytic proof, DEDUCTIVE —deductive method or theory, INDUCTION —mathematical induction, INDUCTIVE —inductive methods, SYNTHETIC—synthetic method of proof.

direct and indirect proofs. A **direct proof** uses an argument that makes direct use of the hypotheses and arrives at the conclusion. An **indirect proof** shows that it is impossible for that which is to be proved to be false, because if it is false some accepted facts are contradicted; in other words, it assumes the negation of the proposition to be proved and then shows that this leads to a contradiction. *E.g.*, suppose we accept the axiom that only one line can be drawn through a given point parallel to a given line and we wish to prove that two coplanar lines are parallel if each is parallel to a third line. For a **direct proof**, we could reason that if each of two lines L_1 and L_2 is parallel to a third line M, then L_1 and L_2 do not have a point P in common, since through such a point P there is only one line parallel to M. Therefore, L_1 and L_2 do not intersect, *i.e.*, they are parallel. For an *indirect proof*, we could assume there are coplanar lines L_1 and L_2 that are not parallel, but both are parallel to a third line M. Then L_1 and L_2 have a point of intersection P. This gives a contradiction to our axiom, since L_1 and L_2 are two lines through P parallel to M. For another example of an indirect proof, suppose we wish to prove there are an infinite number of primes. We assume there are a finite number, *i.e.*, the number of primes is a positive integer n. Then we can let $p_1, p_2, \cdots, p_n$ be all the primes. The number $(p_1 p_2 \cdots p_n) + 1$ is a prime, since it is not divisible by any prime, *i.e.*, not divisible by any of the numbers $p_1, \cdots, p_n$. The contradiction is that this new prime is larger than all primes and therefore larger than itself. An indirect proof also is called a *proof by contradiction* or a *reductio ad absurdum proof*.

PROP′ER, *adj.* **proper factor.** See FACTOR —factor of an integer.

proper fraction. See FRACTION.

proper subset. See SUBSET.

PROP′ER-LY, *adv.* **contained properly.** See SUBSET.

properly divergent series. See DIVERGENT —divergent series.

PROP′ER-TY, *n.* **property of finite character.** See CHARACTER.

PRO-POR′TION, *n.* The statement of equality of two ratios; an equation whose members are ratios. Four numbers, a, b, c, d, are in proportion when the ratio of the first pair equals the ratio of the second pair. This is denoted by $a : b = c : d$, or better by $a/b = c/d$; a notation becoming obsolete is $a : b :: c : d$. The numbers a and d are the **extremes**, b and c the **means**, of the proportion. A **continued proportion** is an ordered set of three or more quantities such that the ratio between any two successive ones is the same. This is equivalent to saying that any one of the quantities except the first and last is the *geometric mean* between the previous and succeeding ones, or that the quantities form a *geometric progression*; $1, 2, 4, 8, 16$ form a continued proportion, written

$$1 : 2 : 4 : 8 : 16 \quad \text{or} \quad \tfrac{1}{2} = \tfrac{2}{4} = \tfrac{4}{8} = \tfrac{8}{16}.$$

If four numbers are in proportion, then various other proportions can be derived from this proportion. If $\dfrac{a}{b} = \dfrac{c}{d}$, then

$$\frac{a+b}{b} = \frac{c+d}{d};$$

$$\frac{a+b}{a-b} = \frac{c+d}{c-d}, \quad \text{if } a \neq b;$$

$$\frac{a}{c} = \frac{b}{d}, \quad \text{if } c \neq 0;$$

$$\frac{b}{a} = \frac{d}{c}, \quad \text{if } a \neq 0;$$

$$\frac{a-b}{b} = \frac{c-d}{d}.$$

These five proportions are said to be derived from the given proportion by **addition, addition and subtraction, alternation, inversion,** and **subtraction,** respectively.

PRO-POR′TION-AL, *adj., n.* As a noun, one of the terms in a proportion. A **fourth proportional** for numbers a, b, and c is a number x such that $a/b = c/x$; a **third proportional** for numbers a and b is a number x such that $a/b = b/x$. A **mean proportional** between two numbers a and b is a number x such that $a/x = x/b$. *E.g.*, 10 is a fourth proportional for 1, 2, and 5, since $\tfrac{1}{2} = \tfrac{5}{10}$; 4 is a third proportional for 1 and 2, since $\tfrac{1}{2} = \tfrac{2}{4}$

(also, 2 is a mean proportional between 1 and 4). See PROPORTION.

directly proportional quantities. Same as PROPORTIONAL QUANTITIES. See below.

inversely proportional quantities. Two quantities whose product is constant; two quantities such that one is *proportional* to the reciprocal of the other.

proportional parts. For a positive number n, **proportional parts** are positive numbers whose sum is n and which are in the same proportion as a given set of numbers. The parts of 12 proportional to 1, 2, and 3 are 2, 4, and 6. The use of **proportional parts** is a common method of estimating the value of a function f at a number x between a and b by replacing the graph of f by a straight line between the points $(a, f(a))$ and $(b, f(b))$; *i.e.*, estimating $f(x)$ so that $f(x) - f(a)$ and $f(b) - f(x)$ are in the same proportion as $x - a$ and $b - x$. See INTERPOLATION, LOGARITHM—proportional parts in a table of logarithms.

proportional quantities. Two variable quantities having fixed (constant) ratio.

proportional sample. See RANDOM—stratified random sample.

proportional sets of numbers. Two sets of numbers in one-to-one correspondence for which there exist two numbers m and n, not both zero, such that m times any number of the first set is equal to n times the corresponding number of the second set. The sets $1, 2, 3, 7$, and $4, 8, 12, 28$ are proportional. The numbers $m = 4$ and $n = 1$ suffice for these sets. This definition is more general than if corresponding numbers must have equal quotients, for the sets $\{1, 5, 0, 9, 0\}$ and $\{2, 10, 0, 18, 0\}$ are in proportion, where $m = 2$ and $n = 1$; but some corresponding numbers do not have quotients, because of the impossibility of dividing by zero.

PRO-POR'TION-AL'I-TY, *n.* The state of being in proportion.

factor of proportionality. See VARIATION — direct variation.

PROP'O-SI'TION, *n.* (1) A theorem or problem. (2) A theorem or problem with its proof or solution. (3) Any statement which makes an assertion which is either true or false, or which has been designated as true or false. *Syn.* sentence, statement.

PROP'O-SI'TION-AL, *adj.* **propositional function.** A function whose range is a collection of propositions (statements). The **truth set** of the propositional function p is the set of all members of the domain of p for which the value of p is a true proposition. *E.g.*, the expression "$x < 3$" defines a propositional function which is a *true* proposition if $x = 2$, a *false* proposition if $x = 4$; the truth set is the set of all numbers less than 3. The statement function "$x^2 + 3x = 0$" is true if x is -3 or 0, so its truth set is the set whose members are -3 and 0. For the propositional function "x is isosceles if x is a right triangle," whose domain is the set of all plane triangles, the truth set is the set of all triangles that either are not right triangles or are both isosceles and right. **Equivalent propositional functions** are propositional functions that have the same truth sets. *E.g.*, if p and q are propositional functions with the same domains, then the propositional functions $\sim p \wedge \sim q$ and $\sim (p \vee q)$ are equivalent, where for a given x these propositional functions determine the propositions "$p(x)$ is false and $q(x)$ is false" and "it is not true that at least one of $p(x)$ and $q(x)$ is true." See above, equivalent equations, inequalities, etc., for other examples. *Syn.* open sentence, open statement, sentential function, statement function. See DE MORGAN—De Morgan formulas, QUANTIFIER.

PRO-SPEC'TIVE, *adj.* **prospective method of computing reserves.** See RESERVE.

PRO-TRAC'TOR, *n.* A semicircular plate graduated, usually in degrees, from one extremity of the diameter to the other and used to measure angles.

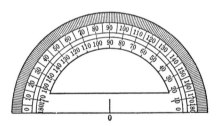

PROVE, *v.* To establish by evidence or demonstration; show the truth of; find a proof of. See PROOF.

PRÜFER, Heinz (1896–1934). German mathematician who contributed to group theory, projective geometry, and the theory of differential equations.

Prüfer substitution. The substitution $py' = r \cos \theta$, $y' = r \sin \theta$, which replaces the differential equation $(py')' + qy = 0$ with dependent variable y by the two equations $\theta' = q \sin^2 \theta + (\cos^2 \theta)/p$, $r' = \frac{1}{2}(-q + 1/p)r \sin 2\theta$, with dependent variables r and θ. This transformation is particularly useful in developing the Sturm-Liouville theory of ordinary differential equations.

PSEU′DO-SPHERE, *n.* The surface of revolution of a tractrix about its asymptote; a pseudospherical surface of revolution of parabolic type.

PSEU-DO-SPHER′I-CAL, *adj.* **pseudospherical surface.** A surface whose total curvature K has the same negative value at all its points. See SPHERICAL —spherical surface. A pseudospherical surface of **elliptic type** is a pseudospherical surface whose linear element is reducible to the form $ds^2 = du^2 + a^2 \sinh^2(u/a)\, dv^2$; the coordinate system is a geodesic polar one. A pseudospherical surface of revolution of elliptic type consists of a succession of hour-glass shaped

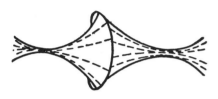

zones with cusps at the maximum parallels. A pseudospherical surface of **hyperbolic type** is a pseudospherical surface whose linear element is reducible to the form $ds^2 = du^2 + \cosh^2(u/a)\, dv^2$; the coordinate system is a geodesic one, with the coordinate geodesics orthogonal to a geodesic, $u = 0$. A *pseudospherical surface of revolution of hyperbolic type* consists of a succession of congruent spool-shaped zones with cusps at the maximum parallels. A pseudospherical surface of

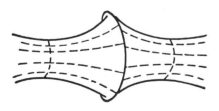

parabolic type is a pseudospherical surface whose linear element is reducible to the form $ds^2 = du^2 + e^{2u/a}\, dv^2$; the coordinate system is a geodesic one, with the coordinate geodesics orthogonal to a curve of constant geodesic curvature. The only pseudospherical surface of revolution of parabolic type is the pseudosphere, or surface of revolution of a tractrix about its asymptote.

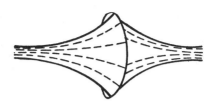

PSI, *n.* The twenty-third letter of the Greek alphabet: lower case, ψ; capital, Ψ.

PTOLEMY (Claudius Ptolemaus) (2nd century A.D.) Alexandrian geometer, astronomer, and geographer.
    Ptolemy's theorem. A necessary and sufficient condition that a convex quadrilateral be inscribable in a circle is that the sum of the products of two pairs of opposite sides be equal to the product of the diagonals.

PUR′CHASE, *adj., n.* **purchase price of a bond.** See PRICE.

PURE, *adj.* **pure geometry.** See SYNTHETIC —synthetic geometry.
    pure-imaginary number. See COMPLEX—complex number.
    pure mathematics. See MATHEMATICS.
    pure projective geometry. Projective geometry employing only geometric methods and bringing up properties other than projective only in a subordinate way. See GEOMETRY —modern analytic geometry.
    pure surd. See SURD.

PYR′A-MID, *n.* A polyhedron with one face a polygon and the other faces triangles with a common vertex. The polygon is the **base** of the pyramid and the triangles are the **lateral faces**. The common vertex of the lateral faces is the **vertex** of the pyramid and the intersections of pairs of lateral faces are **lateral edges**. The **altitude** is the perpendicular distance from the vertex to the base. The **lateral area** is the total area of the lateral faces, and the **volume** is equal to $\frac{1}{3}bh$, where b is the area of the base and h is the altitude. A **regular pyramid** is a pyramid whose base is a regular polygon and whose lateral faces make equal angles with the base (a pyramid with a regular polygon for base and with the foot of its altitude at the center of the base). Its lateral surface area is $\frac{1}{2}SP$, where S is the **slant height** (common altitude of its faces) and P the perimeter of the base.

circumscribed and inscribed pyramids. See CIRCUMSCRIBED.

frustum of a pyramid. The section of a pyramid between the base and a plane parallel to the base. The **bases** of the frustum are the base of the pyramid and the intersection of the pyramid with this plane parallel to the base. The **altitude** is the perpendicular distance between the base and the plane. The volume of a frustum of a pyramid is equal to

$$\tfrac{1}{3}h\left(A + B + \sqrt{AB} \right),$$

where A and B are the areas of the bases, and h is the altitude. If the pyramid is regular, the lateral area of a frustum is $\tfrac{1}{2}S(P_1 + P_2)$, where S is the slant height (altitude of a face) and P_1 and P_2 are the perimeters of the bases.

spherical pyramid. A figure formed by a spherical polygon and planes passing through the sides of the polygon and the center of the sphere. Its volume is

$$\frac{\pi r^3 E}{540},$$

where r is the radius of the sphere and E the *spherical excess* of the base of the pyramid. The polyhedral angle, at the center of the sphere, made by the plane faces of the pyramid is said to **correspond** to the spherical polygon which forms the base of the pyramid.

truncated pyramid. The part of a pyramid between the base and a plane oblique to the base that intersects the base only outside the pyramid. The **bases** are the base of the pyramid and the intersection of the plane and the pyramid.

PY-RAM'I-DAL, *adj.* **pyramidal surface.** A surface generated by a line passing through a fixed point and moving along a broken line in a plane that does not contain the fixed point. It is a **closed pyramidal surface** if the broken line is a polygon.

PYTHAGORAS of Samos (c. 580–c. 500 B.C.). Greek geometer and philosopher. Sought to interpret all things through numbers.

pentagram of Pythagoras. See PENTAGRAM.
Pythagorean identities. See TRIGONOMETRY —identities of plane trigonometry.

Pythagorean relation between direction cosines. The sum of the squares of the direction cosines of a line is equal to unity.

Pythagorean theorem. The sum of the squares of the lengths of the legs of a right triangle is equal to the square of the length of the hypotenuse. The right triangle whose legs are of length 3 and 4, and whose hypotenuse is of length 5, has been used for ages to square corners. Geometrically, this theorem is that the area of *ABGF* (in the figure) is equal to the sum of the areas of *ACDE* and *BCKH*.

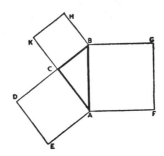

Pythagorean triple. Any set of three positive integers satisfying the equation $x^2 + y^2 = z^2$; *e.g.*, $(3,4,5)$ or $(5,12,13)$. All such triples with y even are given by $x = r - s$, $y = 2\sqrt{rs}$, $z = r + s$, where r and s are positive integers with $r > s$ and rs the square of an integer. All Pythagorean triples with y even and x and y relatively prime are given by $x = m^2 - n^2$, $y = 2mn$, $z = m^2 + n^2$, where m and n are relatively prime positive integers with $m > n$. *Syn.* Pythagorean numbers.

Q

QUAD'RAN'GLE, *n.* A **simple quadrangle** is a plane geometric figure consisting of four points, no three of which are collinear, and the four lines connecting them in a given order. A **complete quadrangle** consists of four coplanar points, no three of which are collinear, and the six lines determined by the points in pairs. See QUADRILATERAL —complete quadrilateral.

QUAD-RAN'GU-LAR, *adj.* Relating to quadrilaterals. For example, a **quadrangular prism** is a prism whose bases are quadrilaterals and a **quadrangular pyramid** is a pyramid whose base is a quadrilaterial.

QUAD-RANT, *adj., n.* **laws of quadrants** for a right spherical triangle. (1) Any angle and the side opposite it are in the same qudarant; (2) when two of the sides are in the same quadrant the third is in the first quadrant, and when two are in different quadrants the third is in the second. (Being in the *first*, *second*, *third*, or *fourth* quadrants means having measure from 0° to 90°, 90° to 180°, 180° to 270°, or 270° to 360°, respectively.)

quadrant angles. Angles are designated as first, second, third, or fourth quadrant angles when the initial side coincides with the positive

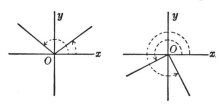

abscissa axis in a system of rectangular coordinates and the terminal side lies in the first, second, third, or fourth quadrants, respectively. The angles in the first figures are in the first and second quadrants; those in the second are in the third and fourth quadrants.

quadrant of a circle. (1) One-half of a semicircle; one-fourth of a circle; the minor arc subtended by two perpendicular radii. (2) The plane area bounded by two perpendicular radii and the minor arc they subtend.

quadrant of a great circle on a sphere. One-fourth of the great circle; the minor arc of the great circle subtended by a right angle at the center of the sphere.

quadrant in a system of plane rectangular coordinates. One of the four compartments into which the plane is separated by the coordinate axes of a Cartesian system of coordinates. They are called first, second, third, and fourth quadrants as counted counterclockwise beginning with the quadrant in which both coordinates are positive. See CARTESIAN —Cartesian coordinates in the plane.

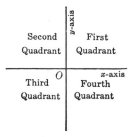

QUAD-RAN′TAL, *adj.* **quadrantal angles.** The angles 0°, 90°, 180°, 270°, or in radians 0, $\pi/2$, π, $3\pi/2$, and all angles having the same terminal

sides as any one of the these, as 2π, $5\pi/2$, 3π, $7\pi/2$, $-\pi/2$, $-\pi$, $\cdots$.

quadrantal spherical triangle. See SPHERICAL.

QUAD-RAT′IC, *adj.* Of the second degree.

discriminant of a quadratic. See DISCRIMINANT—discriminant of a polynomial equation.

quadratic equation. A polynomial equation of the second degree. The **general form** (sometimes called an **affected quadratic**) is $ax^2 + bx + c = 0$. The **reduced form** is

$$x^2 + px + q = 0.$$

A **pure quadratic** equation is an equation of the form $ax^2 + b = 0$.

quadratic form. See FORM.

quadratic formula. A formula giving the roots of a quadratic equation. If the equation is in the form $ax^2 + bx + c = 0$ with $a \neq 0$, the formula is

$$x = \frac{-b \pm \sqrt{b^2 - 4ac}}{2a}.$$

See DISCRIMINANT —discriminant of a polynomial equation.

quadratic function. See below, quadratic polynomial.

quadratic inequality. An inequality of type $ax^2 + bx + c < 0$, or with $<$ replaced by $\leq$, $>$, or $\geq$. The inequality $x^2 + 1 < 0$ has no solutions; but the inequality $-x^2 + 2x - 3 < 0$ is satisfied for all x, since $-x^2 + 2x - 3 = -(x-1)^2 - 2 \leq -2$ for all x. The inequality $x^2 + 2x - 3 < 0$ is equivalent to $(x-1)(x+3) < 0$ and its solution set is the set of all x for which one of $x - 1$ and $x + 3$ is positive and the other is negative, *i.e.*, all x for which $-3 < x < 1$.

quadratic polynomial. A polynomial of the second degree, *i.e.*, a polynomial of the form $ax^2 + bx + c$. A **quadratic function** is a function f whose value $f(x)$ at x is given by a quadratic polynomial. If $f(x) = ax^2 + bx + c$, then the graph of f is the graph of the equation $y = ax^2 + bx + c$ and is a parabola with vertical axis.

quadratic reciprocity law. If p and q are distinct odd primes, then

$$(q|p)(p|q) = (-1)^{(1/4)(q-1)(p-1)}.$$

See LEGENDRE —Legendre symbol.

QUAD′RA-TURE, *n.* The process of finding a square equal in area to the area of a given surface.

quadrature of a circle. See SQUARING — squaring the circle.

QUAD′RE-FOIL, *n.* See MULTIFOIL.

QUAD′RIC, *adj., n.* (1) Of the second degree; quadratic. (2) An expression of the second degree in all its terms; a homogeneous expression of the second degree. A **quadric curve (quadric surface)** is a curve (surface) whose equation in Cartesian coordinates is algebraic and of the second degree (see CONIC). The quadric surfaces are the *ellipsoids, hyperboloids,* and *paraboloids.* See CENTRAL—central quadrics, CONICAL— quadric conical surfaces.

    confocal quadrics. See CONFOCAL.
    quadric quantic. See QUANTIC.

QUAD′RI-LAT′ER-AL, *n.* A polygon having four sides. See PARALLELOGRAM, RECTANGLE, RHOMBUS, TRAPEZOID.

    complete quadrilateral. A figure consisting of four lines in a plane and their six points of intersection.
    quadrilateral inscribable in a circle. See PTOLEMY—Ptolemy's theorem.
    regular quadrilateral. A quadrilateral whose sides and interior angles are all equal; a square.
    simple quadrilateral. A figure consisting of four lines in a plane and their four successive intersections in pairs. Simple as distinguished from a *complete* quadrilateral.

QUAD-RIL′LION, *n.* (1) In the U. S. and France, the number represented by one followed by 15 zeros. (2) In England, the number represented by one followed by 24 zeros.

QUAD′RU-PLE, *adj., n.* Fourfold; consisting of four. An **ordered quadruple** is a set with 4 members, designated as first, second, third, and fourth. An ordered quadruple of numbers can represent a point in four-dimensional space.

QUAN′TIC, *n.* A rational integral homogeneous function of two or more variables; a homogeneous algebraic polynomial in two or more variables. Quantics are classified as *quadric, cubic, quartic,* etc., according as they are of the second, third, fourth, etc., degrees. They are classified as *binary, ternary, quaternary,* etc., according as they contain two, three, four, etc., variables.

QUAN′TI-FI′ER, *n.* Prefixes such as "for every" or "there are " in phrases such as *for every*

$x, y, z, \cdots,$ and *there are* $x, y, z, \cdots$ *for which.* The first type is a **universal quantifier,** the latter an **existential quantifier.** Quantifiers precede a *propositional function* and may be represented symbolically; *e.g.,* "*for any* x, $p(x)$" might be written as $\forall_x[p(x)]$, $A_x[p(x)]$, or $(x)[p(x)]$; "*there is an* x *for which* $p(x)$" as $\exists_x[p(x)]$, $E_x[p(x)]$, or $(\exists x)p(x)$. Such a statement as "there is a man who is disliked by all men" could be written as

$$\exists_x\left[\forall_y(x \text{ is disliked by } y)\right].$$

The negation of $\forall_x[p(x)]$ is the statement $\exists_x[p(x)$ is false]; the negation of $\exists_x[p(x)]$ is $\forall_x[p(x)$ is false].

QUAN′TI-TY, *n.* Any arithmetic, algebraic, or analytic expression which is concerned with value rather than relations between such expressions.

QUAR′TER, *n.* One fourth-part.

QUAR′TIC, *adj., n.* Of the fourth degree; of the fourth order. A **quartic curve** is an algebraic curve of the fourth order (the graph of a fourth-degree equation). A **quartic equation** is a polynomial equation of the fourth degree.

    quartic symmetry. Symmetry like that of a regular octagon, that is, symmetry of a plane figure with respect to four lines through a point, neighboring pairs intersecting at $45°$.
    solution of the quartic. See FERRARI.

QUAR′TILE, *n.* The three points which divide a distribution or a set of data into four equal parts. The middle quartile is the *median* and the other two are the **lower** and **upper quartiles.** For a continuous random variable with probability density function f, the quartiles are the numbers Q_1, Q_2, Q_3 such that

$$\int_{-\infty}^{Q_1} f(x)\,dx = \int_{Q_1}^{Q_2} f(x)\,dx$$

$$= \int_{Q_2}^{Q_3} f(x)\,dx = \int_{Q_3}^{\infty} f(x)\,dx = \tfrac{1}{4}.$$

    quartile deviation. See DEVIATION—quartile deviation.

QUA′SI. quasi-analytic function. See ANALYTIC —quasi-analytic function.

QUA-TER′NA-RY, *adj.* Consisting of four; containing four.

quaternary quantic. See QUANTIC.

QUA-TER′NI-ON, *n.* A symbol of type $x = x_0 + x_1 i + x_2 j + x_3 k$, where x_0 and the coefficients of i, j, k are real numbers. *Scalar multiplication* is defined by

$$cx = cx_0 + cx_1 i + cx_2 j + cx_3 k;$$

the *sum* of x and $y = y_0 + y_1 i + y_2 j + y_3 k$ is

$$x + y = (x_0 + y_0) + (x_1 + y_1)i + (x_2 + y_2)j$$
$$+ (x_3 + y_3)k;$$

the *product* xy is computed by formally multiplying x and y by use of the distributive law and the conventions

$$i^2 = j^2 = k^2 = -1,$$

$$ij = -ji = k, \quad jk = -kj = i, \quad ki = -ik = j.$$

The set of quaternions is a *division ring* and a *skew field*; it satisfies all the axioms for a *field* except the commutative law of multiplication. Quaternions were invented by Wm. R. Hamilton on Oct. 16, 1843. See FROBENIUS —Frobenius theorem, HYPERCOMPLEX.

conjugate quaternions. The **conjugate** of the quaternion $x = x_0 + x_1 i + x_2 j + x_3 k$ is $\bar{x} = x_0 - x_1 i - x_2 j - x_3 k$. In general, $\overline{x + y} = \bar{x} + \bar{y}$, $\overline{x \cdot y} = \bar{x} \cdot \bar{y}$, and $x \cdot \bar{x} = \bar{x} \cdot x = x_0^2 + x_1^2 + x_2^2 + x_3^2 = N(x)$. The number $N(x)$ is the **norm** of x. For any x and y, $N(xy) = N(x)N(y)$.

QUAT′RE-FOIL, *n.* See MULTIFOIL.

QUETELET, Lambert Adolphe Jacque (1796–1874). Belgian statistician and astronomer. See DANDELIN.

QUILLEN, Daniel Grey (1940–   ). American mathematician awarded a Fields Medal (1978) for his contributions to the cohomology of groups and to algebraic K-theory and topological K-theory.

QUIN′TIC, *adj., n.* (1) Of the fifth degree. (2) An algebraic function of the fifth degree. A **quintic curve** is an algebraic curve of the fifth order (the graph of a fifth-degree equation). A **quintic equation** is a polynomial equation of the fifth degree.

quintic quantic. See QUANTIC.

QUIN-TIL′LION, *n.* (1) In the U. S. and France, the number represented by one followed by 18 zeros. (2) In England, the number represented by one followed by 30 zeros.

QUO′TIENT, *adj., n.* The quantity resulting from the division of one quantity by another. The division may have been actually performed or merely indicated; *e.g.*, 2 is the quotient of 6 divided by 3, as is also 6/3. In case the division is not exact, one speaks of the *quotient* and the *remainder*, or simply the *quotient* (meaning the integer obtained plus the indicated division of the remainder), *e.g.*, $7 \div 2$ gives the quotient 3 and the remainder 1, or the *quotient* $3\frac{1}{2}$. See DIVISION —division algorithm.

derivative of a quotient. See DIFFERENTIATION FORMULAS in the appendix.

difference quotient. See DIFFERENCE.

quotient space or **factor space.** Let T be a set for which an *equivalence relation* is defined and let T be divided into *equivalence classes* (see EQUIVALENCE). If certain operations (distance, etc.) are defined for elements of T, then it may be possible to define these operations (distance, etc.) for the equivalence classes in such a way that the set of equivalence classes is a space of the same type as T. In this case, the set of equivalence classes is a **quotient space** or **factor space** of T. *E.g.*, the quotient space of the set C of complex numbers, *modulo* the set R of real numbers, is the set C/R of equivalence classes defined by the equivalence relation $x \equiv y$ if and only if $x - y$ is a real number. The elements of C/R are the set of numbers represented by the horizontal lines in the complex plane, and the sum of two "lines" is the "line" which contains the sum of two numbers, one on each of the given lines (the elements of C/R are also called *residue classes* (modulo R)). The **quotient group** of a group G by an *invariant subgroup H* is the group (denoted by G/H) whose elements are the cosets of H (these cosets are also equivalence classes if one defines x and y to be equivalent if xy^{-1} belongs to H). The unit element of G/H is H and the product of two cosets is the coset containing a product of an element of one coset by an element of the other, the multiplication being in the same order as that of the corresponding cosets. The uniqueness of the product and the group properties of G/H are a consequence of H being an invariant subgroup of G. If G is also a **topological group** and H is a closed set as well as being an invariant subgroup, then G/H is a topological group if one defines a set U^* of elements to be open if an only if U is open in G, where U is the set of all elements of G which belong to a coset of H which is a member of U^*. If G is a metric space, then there is a metric for G which is equivalent to the metric of G and which is right-invariant (*i.e.*, distance as given by the metric satisfies $d(xa, ya) = d(x, y)$ for any elements a, x, y). Then G/H is a metric space if the distance between cosets H_1 and H_2 is defined to be

$$\bar{d}(H_1, H_2) = \text{g.l.b.}\, d(x_1, x_2)$$

for x_1 in H_1 and x_2 in H_2. For the above example of the quotient space C/R of the set of complex numbers modulo the set of real numbers, an open set in C/R is a set of horizontal "lines" in the complex plane whose points form an open set in the plane, while the distance between two elements of C/R is the distance between the corresponding "lines" in the plane. The **quotient ring** of a ring R by an *ideal I* is the ring (denoted by R/I) whose elements are the cosets of I. These cosets are equivalence classes if one defines x and y to be equivalent if $x - y$ belongs to I (they are also called residue classes and R/I a **residue class ring**). The zero element of R/I is I and the sum (product) of two cosets is the coset containing a sum (product) of an element of one coset and an element of the other (the multiplication being in the same order as that of the corresponding cosets). The uniqueness of the sum and product, and the ring properties of R/I, are consequences of I being an ideal. Also, if R is a ring with unit element, or a commutative ring, or an integral domain, then R/I is a set of the same type. Let V be a **vector space** and L be a subset of V which is also a vector space. Let V/L be the set of equivalence classes (or residue classes) defined by the equivalence relation $f \equiv g$ if and only if $f - g$ belongs to L. Then V/L is a vector space if the sum of two equivalence classes F and G is the equivalence class which contains the sum of an element of F and an element of G, and the product of a scalar α and an equivalence class F is the equivalence class which contains the product of α and an element of F. If B is a **Banach space** and L is a subset of B which is also a Banach space, then B/L can be defined in the same way as for vector spaces. If one also defines $\|F\|$ for an equivalence class F to be the greatest lower bound of $\|f\|$ for f belonging to F, then B/L is a Banach space. If H is a **Hilbert space**, then H/L can be defined in the same way as for Banach spaces and is isometric with the orthogonal complement of L in H.

R

RAABE, Josef Ludwig (1801–1859). Swiss analyst.

    Raabe's ratio test. See RATIO—ratio test.

RADEMACHER, Hans Adolph (1892–1969). German mathematician who fled to the U.S. in 1933 because of pacifist views. Made important contributions to analysis and analytic number theory.

    Rademacher functions. The functions $\{r_n\}$ defined on the interval $[0, 1]$ by $r_n(x) = \text{sgn}[\sin(2^n\pi x)]$, where n is a positive integer and $\text{sgn}(x)$ is 1, 0, or -1 when $x > 0$, $x = 0$, or $x < 0$, respectively. The Rademacher functions are orthonormal on the interval $[0, 1]$ and their closed linear span in L^p ($1 \leq p < \infty$) is a subspace isomorphic to Hilbert space and complemented if $p > 1$, where L^p is the Banach space of functions on $[0, 1]$ with $\|f\| = \left[\int_0^1 |f|^p \cdot dx\right]^{1/p}$. See HAAR —Haar functions, LEBESGUE —Lebesgue integral, ORTHOGONAL—orthogonal functions, SIGNUM —signum function, WALSH —Walsh functions.

RA′DI-AL-LY, *adv.* **radially related figures.** Figures which are central projections of each other; figures such that a line drawn from some fixed point to a point of one of them passes through a point of the other, such that the ratio of the distances from the fixed point to the two points is always the same (two similar figures can always be so placed). The fixed point is the **homothetic center**, the **center of similitude,** or **ray center.** The ratio of the two line segments is the **ray ratio, ratio of similitude,** or **homothetic ratio.** Two radially related figures are similar. They are also called **homothetic figures.**

RA′DI-AN, *n.* A central angle subtended in a circle by an arc whose length is equal to the radius of the circle. Thus the **radian measure** of an angle is the ratio of the arc it subtends to the radius of the circle in which it is the central angle (a constant ratio for all such circles); also called circular measure, π measure (rare), natural measure (rare); 2π *radians* $= 360°$, π *radians*

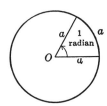

$= 180°$ or 1 *radian* $= (180/\pi)°$; $\frac{1}{4}\pi$ *radians* $=$ $45°$, $\frac{1}{3}\pi$ *radians* $= 60°$, $\frac{1}{2}\pi$ *radians* $= 90°$. See SEXAGESIMAL, MIL.

RA′DI-ATE, *v.* **radiate from a point.** To be a *ray* with the point as origin.

RA′DI-A′TION, *adj., n.* **radiation phenomena.** Wave phenomena in which a disturbance at a single point at time $t = 0$ spreads out with the passage of time. The region into which the disturbance spreads is called the **range of influence.** See DEPENDENCE —domain of dependence.

RAD′I-CAL, *adj., n.* (1) The indicated root of a quantity, as $\sqrt{2}, \sqrt{x}$. (2) The sign indicating a root to be taken, a **radical sign,** the sign $\sqrt{}$ (a modified form of the letter r, the initial of the Latin *radix*, meaning root), placed before a quantity to denote that its root is to be extracted. To distinguish the particular root, a number (the **index**) is written over the sign; thus, $\sqrt[2]{}, \sqrt[3]{}, \sqrt[n]{}$, etc., denote respectively the square root, cube root, nth root, etc. In the case of the square root, the index is omitted, $\sqrt{}$ instead of $\sqrt[2]{}$ being written. The *radical sign* is frequently said to include a bar above the radicand as well as the above sign. This combination is written $\sqrt{}$. See SIMPLIFIED.

radical axis. The radical axis of **two circles** is the locus of the equation resulting from eliminating the square terms between the equations of the circles. When the circles intersect, the radical axis passes through their two points of intersection. The radical axis of two circles is also the line consisting of those points whose *powers* with respect to the two circles are equal (see POWER —power of a point with reference to a circle or a sphere). The radical axis of **three spheres** is the line of intersection of the three radical planes taken with respect to the three possible pairs of spheres. The line is finite if, and only if, the centers of the spheres do not lie on a straight line.

radical center. The radical center of **three circles** is the point in which the three radical axes of the circles, taken in pairs, intersect. This point is finite if, and only if, the centers of the circles do not lie on a line. The radical center of **four spheres** is the point of intersection of the six radical planes formed with respect to the six possible pairs of spheres made up from the given four. The point is finite if, and only if, the centers of the four spheres are not coplanar.

radical of an ideal. For an ideal I of a ring R, the **radical** of I (denoted by $\sqrt{I}$) is the set of all x in R for which there is an n such that $x^n \in I$. See IDEAL.

radical of a ring. The nilpotent ideal that contains all nilpotent ideals. See NILPOTENT — nilpotent ideal. *Syn.* nilradical.

radical plane of two spheres. The locus of the equation resulting from eliminating the square terms between the equations of the two spheres. When the spheres intersect, the radical plane is the plane of their circle of intersection.

RAD′I-CAND′, *n.* The quantity under a radical sign; as 2 in $\sqrt{2}$, or $a + b$ in $\sqrt{a + b}$.

RA′DI-US, *n.* [*pl.* **radii**]. **focal radius.** See FOCAL—focal chord of a conic.

long radius of a regular polygon. The distance from the center to a vertex; the radius of the circumscribed circle.

radius of a circle (or sphere). (1) The distance from the center to the circle (or sphere). (2) A line segment joining the center and a point of the circle (or sphere).

radius of convergence of a power series. The radius of the circle of convergence. See CONVERGENCE—circle of convergence.

radius of curvature. See CURVATURE— curvature of a curve, curvature of a surface, GEODESIC —geodesic curvature, and below, radius of total curvature.

radius of geodesic torsion. The reciprocal of the *geodesic torsion.* See GEODESIC —geodesic torsion of a surface at a point in a given direction.

radius of gyration. The distance from a fixed line (point, or plane) to a point in, or near, a body where all the mass of the body could be concentrated without altering the *moment of inertia* of the body about the line (point, or plane); the square root of the quotient of the moment of inertia by the mass.

radius of torsion of a space curve. See TORSION—torsion of a space curve at a point. *Syn.* radius of second curvature.

radius of total curvature of a surface at a point. The quantity ρ defined by $K = -\dfrac{1}{\rho^2}$, where K is the *total curvature* of the surface at the point. If K is negative, then ρ is real. If the asymptotic lines are taken as parametric curves, so that we have $D = D'' = 0$, then $\dfrac{1}{\rho} = \dfrac{D'}{H}$, where $H = \sqrt{EG - F^2}$.

radius vector [*pl.* **radii vectors; radius vectors**]. See POLAR—polar coordinates, and SPHERICAL —spherical coordinates.

short radius of a regular polygon. The perpendicular distance from the center to a side; the radius of the inscribed circle. *Syn.* apothem.

RA′DIX, *n.* [*pl.* **radices** or **radixes**]. (1) A root. (2) Any number which is made the fundamental number or base of any system of numbers; thus, 10 is the radix of the decimal system of numeration. (3) A name sometimes given to the base of a system of logarithms. In the common system of logarithms the radix is 10; in the natural system it is 2.7182818284 $\cdots$, denoted by e. See BASE—base of a system of numbers.

radix fraction. An indicated sum of fractions, of the form $a/r + b/r^2 + c/r^3 + d/r^4 \cdots$, where the letters $a, b, \cdots$ are all integers less than r (which is also an integer). See BASE—base of a system of numbers.

radix of a mortality table. See MORTALITY — mortality table.

RADÓ, Tibor (1895–1965). Hungarian-American mathematician who worked in complex-variable theory, minimal surfaces, and measure theory. Solved the *Plateau problem* about the same time as J. Douglas.

RADON, Johann Karl August (1887–1956). Austro-German algebraist, analyst, and geometer.

Radon-Nikodým derivative. See below, Radon-Nikodým theorem.

Radon-Nikodým property. See below, Radon-Nikodým theorem.

Radon-Nikodým theorem. Let μ be a σ-*finite measure* defined on a σ-algebra $\mathscr{A}$ of subsets of a set X and let ν be any σ-finite measure defined on $\mathscr{A}$ that is absolutely continuous with respect to μ [*i.e.*, $\nu(A) = 0$ if $\mu(A) = 0$]. Then there exists a nonnegative μ-measurable function ϕ such that

$$\nu(A) = \int_A \phi \, d\mu \quad \text{and} \quad \int_A f \, d\nu = \int_A f\phi \, d\mu$$

if A is in $\mathscr{A}$ and f is μ-measurable. The function ϕ is the **Radon-Nikodým derivative** of ν with respect to μ; two such functions ϕ_1 and ϕ_2 differ only on a set whose μ-measure is zero. The theorem still is true if ν and ϕ are complex-valued. If Bochner integrals are used, the theorem is true if ν and ϕ have values in a finite-dimensional space or in certain Banach spaces (*e.g.*, any reflexive space). Such Banach spaces are said to have the **Radon-Nikodým property.** See INTEGRABLE —integrable function, KREIN—Krein-Milman property, MEASURE —measure of a set.

Radon theorem. If S is a subset of n-dimensional space and S contains at least $n + 2$ points, then S can be represented as the union of two disjoint sets X and Y whose convex spans are disjoint. See related theorems under CARATHÉO-DORY, HELLY, STEINITZ.

RAMANUJAN, Srinivasa (1887–1920). Highly original Indian genius of number theory. Worked with Hardy. See PI.

RAM′PHOID, *n.* **ramphoid cusp.** Same as CUSP OF THE SECOND KIND. See CUSP.

RAMSEY, Frank Plumpton (1902–1930). English mathematician, economist, and philosopher.

Ramsey theory. Intuitively, principles such as "Any large structure contains an orderly substructure," "Complete disorder is not possible," or "Any sufficiently large system of a certain kind contains a large subsystem with a greater degree of organization than the original system." Following are five illustrations of this. (1) **The Ramsey theorem**: "Given a positive integer λ and a set of positive integers $\{\mu_1, \mu_2, \cdots, \mu_k\}$ with each $\mu_i \geq \lambda$, there exists a number N with the following property. Suppose T is a set with N members for which the collection of λ-member subsets has been partitioned into sets $A_1, \cdots, A_k$. Then there is a subset M_i of T with μ_i members such that each λ-member subset of M_i belongs to A_i." The minimal such N is the **Ramsey number** $R_\lambda(\mu_1, \cdots, \mu_k)$. It is known that $R_2(3,3) = 6$, $R_2(3,4) = 9$, $R_2(4,4) = 18$, $R_2(3,5) = 14$, $R_2(3,6) = 18$, $R_2(3,7) = 23$, $R_2(3,9) = 36$. *E.g.*, (i) Suppose a set has $R_2(\mu_1, \cdots, \mu_k)$ members and each pair of members has one of k relations: $R_1, R_2, \cdots, R_k$. Let A_i be the set of all pairs that are R_i-related. Then M_i is a set with μ_i members for which each pair of members of M_i is R_i-related. (ii) Suppose no 3 points of a set of $R_4(p,5)$ points in the plane are collinear. Let a 4-member set L belong to A_1 or A_2 according as the points of L form a convex quadrilateral or do not. If $i = 1$, then M_i is a set of p points that form a convex polygon with p sides; but i cannot be 2, since any set of 5 points (no 3 collinear) contains a set of 4 points that form a convex quadrilateral. (2) **The Van der Waerden theorem**: "Given k and p, there is a number n such that, if each of the first n integers is colored one of k colors, then there exists an arithmetic progression of p terms for which each term has the same color." (3) Any set of positive integers whose *upper density* is positive contains arbitrarily long arithmetic progressions. (4) If I and C are infinite sets and f is a single-valued function with domain I and range in C, then there is an infinite subset S of I such that either all members of S map into the same member of C or f is one-to-one when restricted to S. (5) For any r, there is an n such that, if the first n positive integers are divided into r subsets, then at least one of the subsets has members x, y, and z with $x + y = z$.

RAN′DOM, *adj.* **normalized random variable.** (1) A transformation which converts a random variable X into a normally distributed random variable or approximately so is a **normalizing transformation**, the new random variable being called a **normalized random variable.** (2) See STANDARIZED —standarized random variable.

random sample. Suppose an experiment has outcomes represented by a real number X. If the experiment is repeated in such a way that a particular outcome is not influenced by the other outcomes, then the set of outcomes $\{X_1, X_2, \cdots, X_n\}$ is a random sample of size n. *Tech.* A **random sample of size n** of a random variable X is a set of n independent random variables $\{X_1, X_2, \cdots, X_n\}$ each of which has the same distribution function as X. *E.g.*, suppose an urn contains n balls numbered from 1 to n and that an elementary event is the outcome of drawing a ball, the value of the associated random variable X being the number on the ball. If X_1 and X_2 are the numbers of the balls drawn on the first and second draw, respectively, then the set $\{X_1, X_2\}$ is a random sample of order 2 if the ball drawn on the first draw is replaced before the second draw. See below, stratified random sample.

random sequence. A sequence that is irregular, nonrepetitive, and haphazard. A **sequence of random digits** is a random sequence whose terms are chosen from the ten digits $0, 1, \cdots, 9$ in such a manner that for each position the probability of a particular digit being chosen is $1/10$ and the choices at two different places are independent. A random sample of k objects from n objects can be obtained by numbering the objects from 0 to n and then choosing a sequence of numbers from a sequence of random digits (each with the same number of digits as n), disregarding those greater than n, until k numbers have been chosen. A completely satisfactory definition of random sequence is yet to be discovered. However, tests of randomness can be made; *e.g.*, by subdividing the sequence into blocks and using the chi-square test to analyze the frequencies of occurrence of specified individual integers or of runs consisting of specified digits in a specified order. A table of one million random digits has been published.

random variable. A function X whose range is a set of real numbers, whose domain is the sample space (set of outcomes) S of an experiment, and for which the set of all s in S for which $X(s) \leq x$ is an *event* if x is any real number. It is understood that a probability function is given that specifies the probability X has certain values (or values in certain sets). In fact, one might define a random variable to be simply a probability function P on suitable subsets of a set T, the points of T being "elementary events" and each set in the domain of P an "event" (see PROBABILITY —probability function). A **discrete random variable** is a random variable X that has an associated probability function P and real numbers $\{x_n\}$ such that $\Sigma P(X = x_n) = 1$. Sometimes it is required that each bounded interval contain only finitely many of the numbers $\{x_n\}$, which assures that the distribution function is constant between any two adjacent members of the set $\{x_n\}$ and its only discontinuities are jump discontinuities at points x_n such that $P(X = x_n) > 0$. Any random variable that has only a finite number of values is said to be **discrete.** *E.g.*, if three coins are tossed and h indicates the number of heads, then h is a discrete random variable that can have values 0, 1, 2 or 3. A **continuous random variable** is a random variable X that has an associated probability density function f for which the probability that $a \leq X \leq b$ is given by $\int_a^b f(t)\, dt$, for any numbers $a \leq b$. Often additional restrictions are imposed, such as that f is continuous, that the set of points of discontinuity of P is discrete, or that the distribution function F defined by $F(x) = \int_{-\infty}^x f(t)\, dt$ is differentiable (except possibly at points of a discrete set). Such conditions assure that $F'(x) = f(x)$ for almost all x. For a continuous random variable X and any number x, $P(X = x) = 0$. A random variable need be neither discrete nor continuous. *Syn.* chance variable, stochastic variable, variate.

random walk. A succession of "walks" along line segments in which the direction (and possibly the length) of each walk is determined in a random way. Random walks can be used to obtain probabilistic solutions to mathematical or physical problems. *Tech.* A **random walk** is a sequence $\{S_n\}$, where $S_n = \sum_{i=1}^{n} X_i$ and $\{X_n\}$ is a sequence of independent random variables. *E.g.*, each X_i might have values $+h$ or $-h$ with equal probability and the process could be described as a person making a step of length h every r seconds with each step equally likely to be to the right or to the left; then the probability at time t of being at distance x from where he was at time $t = 0$ can be shown to be given by a function $U(x, t)$ which satisfies the difference equation

$$U(x, t+r) = \tfrac{1}{2}U(x+h, t) + \tfrac{1}{2}U(x-h, t).$$

The function U can be evaluated approximately by letting a computing machine "make" a large number of random walks by reference to a sequence of random numbers. If $h^2 = r$ and $h \to 0$, the limit $u(x, t)$ of $U(x, t)$ satisfies the heat equation

$$\frac{\partial^2 u}{\partial x^2} = 2\frac{\partial u}{\partial t}$$

with the boundary conditions $u(x, 0) = 0$ if $x \neq 0$ and $\int_{-\infty}^{\infty} u(x, t)\, dx = 1$. A random-walk method is a type of Monte Carlo method.

randomized blocks. See BLOCK.

stratified random sample. Let a population be separated into several subpopulations, called **strata**. If from each of these strata random samples are drawn, the resulting pooled sample is a **stratified random sample**. Thus a stratified random sample is basically a group of random samples. Let a population be separated into several strata. Then for that system of classification, the stratified random sample which provides the *minimum-variance unbiased estimate* $\sum p_i \bar{x}_i$ of the mean $\bar{x}$ of the population is the one for which the number of random observations for the ith stratum is proportional to $p_i \sigma_i$, where p_i is the proportion of the entire population in the ith stratum, $\bar{x}_i$ is the average of the sample from the ith stratum, and σ_i is the standard deviation of the ith stratum. If σ_i cannot be estimated, then the sampling procedure which minimizes the variance of the estimates of the mean of the population is one in which the number of observations in the ith stratum is proportional to p_i. This is called a **representative** or **proportional sample**.

table of random numbers. A table listing the members of a random sequence of numbers. See above, random sequence.

vector random variable. A sequence $(X_1, X_2, \cdots, X_n)$ of random variables which are defined on the same sample space of some experiment. For example, if H, W and A denote height, weight and age, and the experiment consists of picking a resident of Chicago, then (H, W, A) is a vector random variable defined on the population of Chicago. If an experiment consists of drawing three numbered balls from a bag and S is the smallest number drawn and L the largest, then (S, L) is a vector random variable. For a vector random variable, the concepts of

discreteness, continuity, and distribution function are defined analogously to what is done for a single random variable. See DISTRIBUTION —distribution function, HYPERGEOMETRIC— hypergeometric distribution, MULTINOMIAL— multinomial distribution, and above, random variable.

RAN'DOM-IZED, *adj.* **randomized blocks.** See BLOCK.

RANGE, *n.* The **range of a function** is the set of values that the function takes on. The range of the function $f(x) = x^2$ is the set of all non-negative real numbers, if the domain of the function is the set of all real numbers. The **range of a variable** is the set of values the variable takes on.

RANK, *n.* **rank of a matrix or quadratic form.** See INDEX —index of a quadratic form, MATRIX.

RARE, *adj.* **rare set.** Same as NOWHERE DENSE SET. See DENSE —dense set.

RATE, *n., v.* (1) Reckoning by comparative values or relations. (2) Relative amount, quantity, or degree; as, the *rate* of interest is 6% (*i.e.*, \$6 for every \$100 for every year); the *rate* per mile of railroad charges; a rapid *rate* of growth. See CORRESPONDING—corresponding rates, DIVIDEND—dividend rate, DEATH—central death rate, INTEREST, MORTALITY —rate of mortality, SPEED, VELOCITY, YIELD.

rate of change of a function at a point. The limit of the ratio of an increment of the function value at the point to that of the independent variable as the increment of the variable approaches zero; the limit of the average rate of change over an interval including the point as the length of the interval approaches zero. This is sometimes called the **instantaneous rate of change** since the rates of change at neighboring points are in general different. The rate of change of a function at a point is the slope of the tangent to the graph of the function, the **derivative at the point**.

RA'TIO, *adj., n.* The quotient of two numbers (or quantities); the relative sizes of two numbers (or quantities). The **inverse ratio** or **reciprocal ratio** of two quantities is their ratio taken in reverse order, *i.e.*, the reciprocal of their ratio. See ANTECEDENT, CONSEQUENT, CORRELATION —correlation ratio, DEFORMATION —deformation ratio, LIKELIHOOD—likelihood ratio, POINT—

point of division, POISSON —Poisson's ratio, PRO-PORTIONAL —proportional sets of numbers.

cross ratio (or **anharmonic ratio**). If A, B, C, D are **four distinct collinear points**, the cross ratio (AB, CD) is defined as the quotient of the ratio in which C divides AB by the ratio in which D divides AB; if the abscissas (or ordinates) of four points are x_1, x_2, x_3, x_4, the cross ratio is

$$\frac{(x_3 - x_1)(x_4 - x_2)}{(x_3 - x_2)(x_4 - x_1)}.$$

If no ordering of the four points gives a harmonic ratio (see below, harmonic ratio) there are, in general, six distinct values of the cross ratio, depending on how the points are ordered. If L_1, L_2, L_3, L_4 are **four distinct concurrent lines** with sloopes equal to m_1, m_2, m_3, m_4, respectively, the cross ratio of the four lines is

$$\frac{(m_3 - m_1)(m_4 - m_2)}{(m_3 - m_2)(m_4 - m_1)}.$$

harmonic ratio. If the cross ratio of four points (or four lines) is equal to -1, it is called a **harmonic ratio** and the last two points are said to divide the first two **harmonically.**

ratio paper. Same as SEMILOGARITHMIC PAPER. See LOGARITHMIC.

ratio of similitude. The ratio of the lengths of corresponding lines of similar figures; the **ray ratio** (see RADIALLY —radially related figures). Also called *homothetic ratio.*

ratio test. Any of several tests for convergence (or divergence) of an infinite series which make use of the ratio of successive terms of the series. The **ordinary ratio test** (or **Cauchy's ratio test**) is that a series converges or diverges according as the absolute value of the limit, as n becomes infinite, of the ratio of the nth to the $(n - 1)$th term is less than or greater than one. If it is equal to one, the test fails. *E.g.*, (1) for the series $1 + 1/2! + 1/3! + \cdots + 1/n! + \cdots$, the ratio of the nth to the $(n - 1)$th term is

$$(1/n!)\big/[1/(n - 1)!] = 1/n,$$

and

$$\lim_{n \to \infty} (1/n) = 0.$$

Hence the series converges. (2) For the harmonic series,

$$1 + \tfrac{1}{2} + \tfrac{1}{3} + \cdots + 1/n + \cdots,$$

the ratio is

$$(1/n)\big/[1/(n - 1)] = (n - 1)/n$$

and

$$\lim_{n \to \infty} (n - 1)/n = 1;$$

hence the test fails. (However, this series diverges, as can be shown by grouping the terms so that each group equals or exceeds $\tfrac{1}{2}$, namely

$$1 + \tfrac{1}{2} + \left(\tfrac{1}{3} + \tfrac{1}{4}\right) + \left(\tfrac{1}{5} + \tfrac{1}{6} + \tfrac{1}{7} + \tfrac{1}{8}\right) + \cdots.)$$

The existence of the limit of the ratio of the nth to the $(n - 1)$th term is not needed. The **generalized ratio test** (also called **d'Alembert's test**) states that a series converges if after some term the absolute value of the ratio of any term to the preceding is always less than a fixed number less than unity; if this ratio is always greater than unity, the series diverges. **Raabe's ratio test**, which is a more refined test, is that if the series is $u_1 + u_2 + u_3 + \cdots + u_n + \cdots$, and $u_{n+1}/u_n = 1/(1 + a_n)$, then the series converges if, after a certain term, the product na_n is always greater than a fixed number which is greater than unity, and it diverges if, after a certain term, the same product is always less than or equal to unity.

RA′TION-AL, *adj.* **rational expression or function.** An algebraic expression which involves no variable in an irreducible radical or under a fractional exponent; a function which can be written as a quotient of polynomials. The expressions $2x^2 + 1$ and $2x + 1/x$ are *rational*, but $\sqrt{x + 1}$ and $x^{3/2} + 1$ are not. See PARTIAL —partial fractions.

rational integral function. A function containing only rational and integral terms in the variable (or variables). A function may be rational and integral in one or more of the variables while it is not in others; *e.g.*, $w + x^2 + 2xy^{1/2} + 1/z$ is rational and integral in x and in w and x together, but not rational in y and not integral in z. See TERM *Syn.* polynomial.

rational number. A number that can be expressed as an integer or as a quotient of integers (such as $\tfrac{1}{2}, \tfrac{4}{3}, 7$). *Tech.* Once integers have been defined (see INTEGER), rational numbers can be defined as the set of all ordered pairs (a, b) for which a and b are integers $(b \neq 0)$ and equality, addition, and multiplication are defined as follows:

$$(a, b) = (c, d) \quad \text{if and only if} \quad ad = bc;$$

$$(a, b) + (c, d) = (ad + bc, bd);$$

$$(a, b) \cdot (c, d) = (ac, bd).$$

The usual practice is to write (a, b) as a/b, in which case the above definitions of equality, addition, and multiplication take the form:

$$\frac{a}{b} = \frac{c}{d} \quad \text{if and only if} \quad ad = bc;$$

$$\frac{a}{b} + \frac{c}{d} = \frac{ad + bc}{bd};$$

$$\frac{a}{b} \cdot \frac{c}{d} = \frac{ac}{bd}.$$

The rational number $(a, 1)$, or $a/1$, is called an integer and usually written simply as a. See IR-RATIONAL —irrational number.

rational operations. Addition, subtraction, multiplication, and division.

rational-root theorem. If a rational number p/q, where p and q have no common factors, is a root of a polynomial equation whose coefficients are integers,

$$a_0 x^n + a_1 x^{n-1} + a_2 x^{n-2} + \cdots$$

$$+ a_{n-1} x + a_n = 0,$$

then a_0 is divisible by q and a_n is divisible by p.

RA′TION·AL·IZE, *v.* To remove radicals without altering the value of an expression or the roots of an equation. To **rationalize an algebraic equation** means to remove the radicals which contain the variable (not always possible). A procedure that sometimes suffices is to isolate the radical in one member of the equation (or if there be more than one radical, to arrange them to the best advantage) and raise both sides to a power equal to the index of the radical (or one of the radicals), repeating this process if necessary. Extraneous roots may be introduced by this procedure. *E.g.*, (1) $\sqrt{x-1} = x - 2$ rationalizes into $x - 1 = x^2 - 4x + 4$, or $x^2 - 5x + 5 = 0$; (2) $\sqrt{x-1} + 2 = \sqrt{x+1}$ is written $\sqrt{x-1} - \sqrt{x+1} = -2$; squaring gives $x - 1 - 2\sqrt{x^2-1} + x + 1 = 4$ or $\sqrt{x^2-1} = x - 2$, whence $x^2 - 1 = x^2 - 4x + 4$ or $4x - 5 = 0$. To **rationalize the denominator of a fraction** means to multiply numerator and denominator by a quantity such that the resulting expression will contain no radicals in the denominator. *E.g.*, if the fraction is $\dfrac{1}{\sqrt{a} + \sqrt{b}}$, a rationalizing factor is $\sqrt{a} - \sqrt{b}$ and we obtain $\dfrac{\sqrt{a} - \sqrt{b}}{a - b}$; if the fraction is $\dfrac{1}{\sqrt[3]{c^2}}$, a rationalizing factor is $\sqrt[3]{c}$ and we obtain $\dfrac{\sqrt[3]{c}}{c}$.

To **rationalize an integral** means to make a substitution (changing variables) so that the radicals in the integrand disappear; the integral $\displaystyle\int \frac{x^{1/2}}{1 + x^{3/4}}\, dx$ is *rationalized* into $\displaystyle\int \frac{4z^5}{1 + z^3}\, dz$ by the substitution $x = z^4$ $(dx = 4z^3\, dz)$.

RAY, *adj., n.* A set consisting of a point P on a straight line and all points of the line on one side of P. The point P is the **initial point** or **origin** of the ray. Often a ray with or without its initial point is called a **half-line**, being a **closed half-line** if it contains the point and an **open half-line** if it does not. The point is the **boundary** of the half-line in either case.

ray center. Same as CENTER OF PROJECTION. See PROJECTION —central projection, RADIALLY —radially related figures.

ray ratio. See RADIALLY —radially related figures.

RAYLEIGH-RITZ METHOD. A method for determining approximate solutions of functional equations through the expedient of replacing them by finite systems of equations. Thus, for example, any function (and its first n derivatives) that is of class $C^{(n)}$ on a closed interval can be approximated arbitrarily closely by a polynomial.

RE·AC′TION, *n.* **law of action and reaction.** See ACTION.

REAL, *adj.* **real-number axis** or **real axis.** A straight line upon which the real numbers are plotted; the horizontal axis is an Argand diagram. See COMPLEX —complex numbers.

real number. Any rational or irrational number (see RATIONAL and IRRATIONAL). The *complex numbers* consist of the real and the imaginary numbers (those numbers $a + bi$ for which a and b are real and $b = 0$ and $b \ne 0$, respectively). The set of all real numbers is the **real number system** or the **real continuum** (see CONTINUUM).

real part of a complex number. The term which does not contain the factor i. If the number is $z = x + iy$ (where x and y are real), the real part is x, denoted by $\mathrm{R}(z)$, $\mathrm{Re}(z)$, or $\Re(z)$.

real plane. A plane in which all points are assigned ordered pairs of real numbers for coordinates, as contrasted to the *complex plane*.

real-valued function. A function whose range is a set of real numbers. See FUNCTION.

real variable. A variable which has only real numbers for its values.

REAM, *n.* A measure of paper; 500 sheets. See DENOMINATE NUMBERS in the appendix.

RE-AR-RANGE′MENT, *n.* **rearrangement of the terms in a series.** See SERIES.

RE-CEIPT′, *n.* (1) The act of receiving payment in money or goods. (2) A statement acknowledging money or goods having been received.

RE-CEIPTS′, *n.* Money or other assets taken in, as contrasted to expenditures.

RE-CEIV′A-BLE, *adj.* See NOTE—note receivable.

RE-CIP′RO-CAL, *adj., n.* The reciprocal of a **number** is the number whose product with the given number is equal to 1; *i.e.,* 1 divided by the number. For a **fraction,** the reciprocal is the fraction formed by interchanging the numerator and denominator in the given fraction. For any set of objects for which multiplication is defined and there is a multiplicative *identity* (whose product with any member x of the set is equal to x), the reciprocal (or **inverse**) of an object x is an object y such that xy and yx are each equal to the identity (provided there is only one y with this property). *E.g.,* the reciprocal of the *polynomial* $x^2 + 1$ is $1/(x^2 + 1)$, since $(x^2 + 1)\dfrac{1}{x^2 + 1} = 1$. See GROUP.

reciprocal curve of a curve. The curve obtained by replacing each ordinate of a given curve by its reciprocal; the graph of the equation derived from the given equation (in Cartesian coordinates) by replacing y by $1/y$. The graphs of $y = 1/x$ and $y = x$ are reciprocals of each other; so are the graphs of $y = \sin x$ and $y = \operatorname{cosec} x$.

reciprocal equation. An equation in one variable whose set of roots remains unchanged if the roots are replaced by their reciprocals; an algebraic equation whose roots are unchanged if the variable is replaced by its reciprocal. *E.g.,* when x is replaced by $1/x$ and the equations are simplified, $x + 1 = 0$ yields $1 + x = 0$, and $x^4 - ax^3 + bx^2 - ax + 1 = 0$ yields $1 - ax + bx^2 - ax^3 + x^4 = 0$.

reciprocal of a matrix. Same as the *inverse* of the matrix. See MATRIX.

reciprocal polar figures in the plane. Two figures made up of lines and their points of intersection are reciprocal polar figures if each point in either one of them is the *pole* of a line in the other with respect to some given conic (see POLE—pole and polar of a conic); **polar-**

reciprocal triangles are two triangles such that the vertices of each of them are the poles of the sides of the other with respect to some conic. **Polar-reciprocal curves** are two curves so related that the *polar*, with respect to a given conic, of every point on one of them is tangent to the other (it then follows that the polars of the points on the latter are tangent to the former).

reciprocal ratio. See RATIO.

reciprocal spiral. See HYPERBOLIC —hyperbolic spiral.

reciprocal substitution. The substitution of a new variable for the reciprocal of the old; a substitution such as $y = 1/x$.

reciprocal system of vectors. Sets of vectors A_1, A_2, A_3 and B_1, B_2, B_3 such that $A_i \cdot B_i = 1$, $i = 1, 2, 3$, and $A_i \cdot B_j = 0$ if $i \neq j$. If the triple scalar product

$$[A_1 A_2 A_3] \neq 0,$$

then the set of vectors reciprocal to A_1, A_2, A_3 is $A_2 \times A_3/[A_1 A_2 A_3]$, $A_3 \times A_1/[A_1 A_2 A_3]$, $A_1 \times A_2/[A_1 A_2 A_3]$. See TRIPLE—triple scalar product.

reciprocal theorems. (1) In *plane geometry,* theorems such that the interchanging of two geometric elements, *e.g.,* angles and sides, points and lines, etc., transfers each of the theorems into the other. Two such theorems are not always simultaneously true or false. (2) In *projective geometry,* same as DUAL THEOREMS.

Volterra reciprocal functions. See VOLTERRA.

REC′TAN′GLE, *n.* A parallelogram with one angle a right angle and therefore all of its angles right angles: a quadrilateral whose angles are all right angles. A **diagonal** of a rectangle is a line joining opposite vertices; if the sides are of length a and b, the length of the diagonal is $\sqrt{a^2 + b^2}$. The **altitude** is the perpendicular distance from one side (designated as the **base**) to the opposite side. The **area** of a rectangle is the product of two adjacent sides. If a rectangle has two sides of length 2 and 3, respectively, its area is 6.

REC-TAN′GU-LAR, *adj.* Like a rectangle; mutually perpendicular.

rectangular axes and **coordinates.** See CARTESIAN.

rectangular form of a complex number. The form $x + yi$, as distinguished from the polar or trigonometric form $r(\cos\theta + i \sin\theta)$.

rectangular graph. Same as BAR GRAPH. See GRAPH.

rectangular hyperbola. See HYPERBOLA—rectangular hyperbola.

rectangular region. See REGION.

rectangular solid. A rectangular parallelepiped. See PARALLELEPIPED.

REC'TI-FI'A-BLE, *adj.* **rectifiable curve.** A curve that has length (*i.e.*, finite length). See LENGTH—length of a curve.

REC'TI-FY-ING, *adj.* **rectifying plane of a space curve at a point.** The plane of the tangent and binormal to the curve at the point. See DEVELOPABLE—rectifying developable of a space curve.

REC'TI-LIN'E-AR, *adj.* (1) Consisting of lines. (2) Bounded by lines.

rectilinear generators. See RULED—ruled surface, HYPERBOLOID—hyperboloid of one sheet, PARABOLOID.

rectilinear motion. Motion along a straight line. See VELOCITY.

RE-CUR'RENCE, *n.* **recurrence theorem.** See POINCARÉ—Poincaré recurrence theorem.

RE-CUR'RING, *adj.* **recurring continued fraction.** See FRACTION—continued fraction.

recurring decimal. Same as REPEATING DECIMAL.

RE-DEEM', *v.* To repurchase, to release by making payments. To redeem a note, bond, or mortgage means to pay the sum it calls for. To redeem property means to get ownership by paying off lapsed liability for which it was security.

RE-DEMP'TION, *n.* The act of redeeming. See PRICE.

RE-DUCED', *p.* **reduced cubic equation.** A cubic equation of the form $y^3 + py + q = 0$; the form that the general cubic, $x^3 + ax^2 + bx + c = 0$, takes when the x^2 term is eliminated by substituting $x - \frac{1}{3}a$ for x. See CARDAN—Cardan solution of the cubic.

reduced differential equation. See DIFFERENTIAL—linear differential equation.

RE-DU'CI-BLE, *adj.* A curve or surface is said to be **reducible** in a given region if it can be shrunk to a point by a continuous deformation without passing outside that region. See DEFORMATION—continuous deformation, and CONNECTED—simply connected region.

reducible polynomial. For some given domain or field, a polynomial that can be written as the product of two polynomials with degrees at least 1 and coefficients in the given domain or field. See IRREDUCIBLE —irreducible polynomial.

reducible set of matrices. A set of matrices which correspond to linear transformations of an n-dimensional vector space V is **reducible** if there is a proper subset V' of V which contains a nonzero element and is such that each point of V' is transformed into a point of V' by any linear transformation corresponding to one of the matrices. See REPRESENTATION —reducible representation of a group.

reducible transformation. A linear transformation T of a **linear space** L into itself is reducible if there are two linear subsets M and N of L such that $T(x)$ belongs to M if x belongs to M, $T(x)$ belongs to N if x belongs to N, and M and N are complementary in the sense that any vector of L can be uniquely represented as the sum of a vector of M and a vector of N. The transformation T can then be completely specified by describing its effect on M and on N. For **Hilbert space**, it is customary to require that M and N be orthogonal complements of each other. Then T is reducible by M and N if and only if T and its *adjoint* T^* map M into M, or if and only if T commutes with the (orthogonal) projection whose range is M.

REDUCTIO AD ABSURDUM PROOF. Same as INDIRECT PROOF. See PROOF—direct and indirect proof.

RE-DUC'TION, *adj., n.* (1) A diminution or decreasing, as a reduction of 10% in the price. (2) The act of changing to a different form, by collecting terms, powering equations, simplifying fractions, making substitutions, etc.

reduction ascending. Changing a denominate number into one of higher order, as feet and inches into yards.

reduction descending. Changing a denominate number into one of lower order, as yards and feet into inches.

reduction formulas in integration. See INTEGRATION —reduction formulas in integration.

reduction formulas of trigonometry. See TRIGONOMETRY —identities of plane trigonometry.

reduction of a common fraction to a decimal. Annexation of a decimal point and zeros to the numerator and dividing (usually approximately) by the denominator. *E.g.*,

$$\frac{1}{4} = \frac{1.00}{4} = 0.25; \qquad \frac{2}{3} = \frac{2.000}{3} = 0.667 - .$$

reduction of a fraction to its lowest terms. The process of dividing all common factors out of numerator and denominator.

RE-DUN′DANT, *adj.* **redundant equation.** See EQUATION —redundant equation.

redundant number. Same as ABUNDANT NUMBER. See NUMBER —perfect number.

RE-EN′TRANT, *adj.* **reentrant angle.** An angle which is an interior angle of a polygon and greater than 180° (angle *HFM* in figure). The other angles of the figure (interior angles of less than 180°) are **salient angles**.

REF′ER-ENCE, *n.* **axis of reference.** One of the axes of a Cartesian coordinate system, or the polar axis in a polar coordinate system; in general, any line used to aid in determining the location of points, either in the plane or in space.

frame of reference. See FRAME.

reference angle. Same as RELATED ANGLE. See RELATED.

RE-FLEC′TION, *adj., n.* The change of direction which, *e.g.*, a ray of light, radiant heat, or sound, experiences when it strikes upon a surface and is thrown back into the same medium from which it came. Reflection follows two laws: (1) the reflected and incident rays are in a plane normal to the surface; (2) the angle of incidence is equal to the angle of reflection (the **angle of incidence** is the angle the incident ray makes with the normal at the point of incidence; the **angle of reflection** is the angle which the reflected ray makes with this normal).

reflection in a line. Replacing each point in the reflected configuration by a point symmetric to the given point with respect to the line. A reflection in a coordinate axis in the plane is defined by one of the transformations $x' = x$, $y' = -y$, or $x' = -x$, $y' = y$. Each given point is replaced by a point symmetric to the given point with respect to the axis in which the reflection is made, the x-axis and y-axis, respectively, in the above transformations.

reflection in the origin. Replacing each point by a point symmetric to the given point with respect to the origin (in the plane, a reflection in the origin is a rotation about the origin through 180°): the result of successive reflections in each axis of a rectangular system of coordinates. See above, reflection in a line.

reflection in a plane. Replacing each point in the reflected configuration by a point symmetric to the given point with respect to the plane; *e.g.*, the reflection of the point (x, y, z) in the (x, y)-plane is the point $(x, y, -z)$.

reflection property of the ellipse, hyperbola, and parabola. See *focal property* under ELLIPSE, HYPERBOLA, PARABOLA.

RE′FLEX, *adj.* **reflex angle.** An angle greater than 180° and less than 360°.

RE-FLEX′IVE, *adj.* **reflexive Banach space.** Let B be a Banach space and B^* and B^{**} be the *first* and *second conjugate spaces* of B (see CONJUGATE —conjugate space). If x_0 is an element of B, then F, defined by $F(f) = f(x_0)$, is a continuous linear functional defined on B^*; B is **reflexive** if every linear functional defined on B^* is of this type, it then following that B and B^{**} are identical if each x_0 is identified with the linear functional F defined by $F(f) = f(x_0)$. However, there exists a nonreflexive Banach space J for which there is an isometric correspondence between J and J^{**}. A Banach space is reflexive if and only if the unit ball is *weakly compact*, or if and only if, for each continuous linear functional f, there is an element $x_0 \neq 0$ such that $f(x_0) = \|f\| \cdot \|x_0\|$, or if and only if each hyperplane of support for the unit ball contains a point of the ball. Hilbert space is a reflexive Banach space, as are all uniformly convex or uniformly non-square Banach spaces. *Syn.* regular Banach space.

reflexive relation. A relation of which it is true that, for any x, x bears the given relation to itself. The relation of equality in arithmetic is *reflexive*, since $x = x$, for all x. A relation such that x does not bear the given relation to itself for any x is **antireflexive** or **irreflexive**. The relation of being greater than is *antireflexive*, since it is not true for any x that $x > x$. A relation such that there is at least one x which does not bear the given relation to itself is **nonreflexive**. The relation of *being the reciprocal of* is *non-reflexive*, since x is the reciprocal of x if x is $+1$ or -1, but otherwise x is not the reciprocal of x.

RE′FLEX-IV′I-TY, *n.* The property of being reflexive. See REFLEXIVE.

RE-FRAC′TION, *n.* A change of direction of rays (as of light, heat, or sound) which are obliquely incident upon and pass through a surface bounding two media in which the ray has different velocities (as light going from air to water). It is found that for isotropic media: (1) When passing into a denser medium, the ray is refracted toward a perpendicular to the surface, and, when passing into a less dense medium, it is bent away from the perpendicular; (2) the incident and refracted rays are in a plane that is perpendicular to the surface; (3) the sines of the angle of incidence and the angle of refraction bear a constant ratio to each other for any two given media (the **angles of incidence** and **refraction** are the angles which the incident and refracted ray make, respectively, with the perpendicular to the surface). If the first medium is air, this ratio is called the **index of refraction** or the **refraction index** of the second medium. The law stated in (3) is known as **Snell's law**.

RE′GION, *n.* A set that is the union of an open connected set and none, some, or all of its boundary points. The set is an **open region** if none of the boundary is included; it is a **closed region** if all of the boundary is included. For example, a **closed triangular region** (or **triangular region**) is a triangle together with its interior; a **closed circular region** (or **circular region**) is a circle together with its interior; a **closed rectangular region** (or **rectangular region**) is a rectangle together with its interior. The interior of a triangle, circle, or rectangle is an **open triangular region**, an **open circular region**, and an **open rectangular region**, respectively. The interior of a sphere is an open region, and the sphere together with its interior is a closed region.

RE-GRES′SION, *adj., n.* In *statistics*, the term *regression line* was first used for the least-squares line in studies estimating the extent by which the height of children of tall parents "regresses" toward the mean height of the population. However, see various headings below.

edge of regression. The tangent surface S of a space curve C generally consists of two sheets which are tangent to one another along C, forming a sharp edge there. C is called the *edge of regression* of S.

linear regression. See below, regression function, regression line.

regression coefficient. Any coefficient of a random variable in a regression equation. See below, regression line.

regression curve. The graph of a regression equation of type $Y = f(X)$. See below, regression function, regression line.

regression function. A function giving the *conditional expected value* of a random variable Y for given values of random variables $X_1, X_2,$

$\cdots, X_n$. If $f(X_1, X_2, \cdots, X_n)$ is the expected value of Y given $X_1, X_2, \cdots, X_n$, then f is the **regression function** and $Y = f(X_1, X_2, \cdots, X_n)$ is the **regression equation**. If $n > 1, f$ is a **multiple regression function**. If f is linear, it is a **linear regression function**. To determine a regression function, it usually is assumed to be of a particular type with unknown parameters, and the method of least squares is used to determine the parameters. *E.g.*, it is not true that the height of an individual in the U.S. is a function of his weight; however, it might be reasonable to assume there is an equation $h(w) = a + bw$ that expresses the *mean* or *expected height* of a person of weight w as a function of w. By use of random samples and the method of least squares, one might then determine a and b. See below, regression line.

regression line. If the conditional expected value of a random variable Y, given the value of a random variable X, is given by an equation $Y = mX + b$, the graph of this equation is the **regression line** of Y on X. If the *regression coefficients* m and b are determined by the method of least squares, the regression line is

$$(Y - \bar{Y})/\sigma_Y = r(X - \bar{X})/\sigma_X,$$

where $\bar{X}$ and $\bar{Y}$ are sample means, σ_X and σ_Y are sample standard deviations, and r is the sample correlation coefficient. See LEAST—method of least squares, and above, regression function.

REGULA FALSI (rule of false position). The method of calculating an unknown (as a root of a number) by making an estimate (or estimates) and working from it and properties of the unknown to secure the value of the latter. If one estimate is used, it is **simple position**; if two, **double position**. Double position is used in approximating irrational roots of an equation and in approximating logarithms of numbers which contain more significant digits than are listed in the tables being used. The method assumes that small arcs are approximately coincident with the chords which join their extremities. This makes the changes in the abscissas proportional to the changes in the corresponding ordinates; *e.g.*, if $y = f(x)$ has the value -4 when x is 2 and the value 8 when x is 3, then the chord joining the points whose coordinates are $(2, -4)$ and $(3, 8)$ crosses the x-axis at a point whose abscissa x is such that $\frac{1}{4}(x - 2) = \frac{1}{12}$ which gives $x = 2\frac{1}{3}$ as an approximate value of a root of $f(x) = 0$. *Newton's method* for approximating roots is an example of *simple position* (see NEWTON).

REG′U-LAR, *adj.* **regular analytic curve.** See ANALYTIC —analytic curve.

regular Banach space. See REFLEXIVE— reflexive Banach space.

regular curve. A curve all of whose points are ordinary points. See POINT—ordinary point on a curve.

regular definition of the sum of a divergent series. A definition which, when applied to convergent series, gives their ordinary sums. **Consistent** is sometimes used to denote the same property. *Regular* is also used to denote not only the above property, but the added property of failing to sum properly divergent series.

regular function of a complex variable. See ANALYTIC —analytic function of a complex variable.

regular permutation group. See PERMUTATION—permutation group.

regular point of a curve. See POINT—ordinary point of a curve.

regular point of a surface. A point of the surface which is not a singular point of the surface. See SINGULAR —singular point of a surface.

regular polygon. See POLYGON.

regular polyhedron. See POLYHEDRON.

regular sequence. (1) A *convergent sequence* See SEQUENCE —convergent sequence. (2) See SEQUENCE —Cauchy sequence.

regular space. A (topological) space such that if U is any neighborhood of a point x of the space, then there is a neighborhood V of x with the closure of V contained in U. A topological space is **normal** if, for any two nonintersecting closed sets P and Q, there are two disjoint open sets, one of which contains P and the other Q; it is **completely normal** if for any sets P and Q, neither of which contains a point of the closure of the other, there are two disjoint open sets one of which contains P and the other Q. A normal space is regular, and a regular space which satisfies the *second axiom of countability* is normal. See METRIC—metric space. A topological space is **completely regular** if, for each x of T and neighborhood U of x, there is a continuous function with values in the interval $[0,1]$ for which $f(x) = 1$ and $f(y) = 0$ if y is not in U. A completely regular T_1 space is sometimes called a **Tychonoff space**. A completely regular space is regular.

RE-LAT′ED, *adj.* **related angle.** The acute angle (angle in the first quadrant) for which the trigonometric functions have the same absolute values as for a given angle in another quadrant, with reference to which the acute angle is called

the *related angle*; 30° is the related angle of 150° and of 210°.

related expressions or functions. Same as DEPENDENT FUNCTIONS, but less commonly used. See DEPENDENT.

RE-LA′TION, *n.* Equality, inequality, or any property that can be said to hold (or not hold) for two objects in a specified order. *Tech.* A relation is a set R of ordered pairs (x, y), it being said that x is *related* to y (sometimes written xRy) if (x, y) is a member of R. E.g., the relation "less than" for real numbers is the set of all ordered pairs (x, y) for which x and y are real numbers with $x < y$; the relation "sister of" is the set of all ordered pairs (x, y) for which x is a person who is the sister of y. The **inverse** of a relation R is the relation R^{-1} for which (x, y) belongs to R if and only if (y, x) belongs to R^{-1}. See COMPOSITION —composition of relations.

antireflexive, irreflexive, nonreflexive, and reflexive relation. See REFLEXIVE —reflexive relation.

antisymmetric, asymmetric, nonsymmetric, and symmetric relations. See SYMMETRIC — symmetric relation.

connected relation. See CONNECTED—connected relation.

equivalence relation. See EQUIVALENCE.

inclusion relation. See INCLUSION.

intransitive, nontransitive, and transitive relations. See TRANSITIVE.

REL′A-TIVE, *adj.* **relative error.** See ERROR.

relative frequency. See FREQUENCY.

relative maximum and minimum. See MAXIMUM.

relative velocity. See VELOCITY.

REL′A-TIVE-LY, *a.* **relatively prime.** See PRIME —relatively prime.

REL′A-TIV′I-TY, *n.* **mathematical theory of relativity.** A special (or restricted) mathematical theory of relativity is based on two postulates: (1) Physical laws and principles can be expressed in the same mathematical form in all reference systems which move relative to one another with constant velocity. (2) The speed of light has the same constant value c (approximately 3×10^{10} cm/sec.) which is independent of the velocity of the source of light. The adoption of these postulates leads to the conclusions that the velocity of an object with nonzero mass must be less than the velocity of light, and that the mass m of the body depends on its velocity and hence on the kinetic energy of the body. It turns out that mass increases with increase in velocity, and this leads to the celebrated mass-energy relation $E = mc^2$. The general theory subsumes that the physical laws and principles are invariant with respect to all possible reference frames. It provides an elegant mathematical formulation of particle dynamics that is essentially geometrical in character. It provides a reasonable explanation of several astronomical phenomena that are not easily explained by Newtonian mechanics, but it fails to provide a satisfactory unified account of electrodynamical phenomena. See EINSTEIN.

RE′LAX-A′TION, *adj., n.* **relaxation method.** In numerical analysis, a method in which the errors, or residuals, resulting from an initial approximation are considered as constraints that are to be relaxed. New approximations are chosen to reduce the worst of the residuals until finally all are within the toleration limit.

RE-LI-A-BIL′I-TY, *n.* Following are some of the many contexts in which the term "reliability" is used. (1) For a method of sampling or measurement, the sample *variance* is a measure of the reliability of the method. (2) The reliability for time t of a device that sometimes works and sometimes does not work is the probability the device will work without any failures for a length of time t. (3) The probability a device will work when it should; *e.g.*, the probability a valve will shut off when a container is full. (4) The reliability for time t of an object that has a limited life is the probability the life of the object is as long as t.

RE-MAIN′DER, *adj., n.* When an **integer** m is divided by a positive integer n, and a quotient q is obtained for which $m = nq + r$ with $0 \leq r < n$, then r is the remainder. When a **polynomial** f is divided by a nonconstant polynomial g, and a quotient q is obtained for which $f(x) \equiv g(x)q(x) + r(x)$ with either $r \equiv 0$ or the degree of r less than the degree of g, then r is the remainder (when the divisor is of the first degree the *remainder* is a constant). See DIVISION — division algorithm, and below, remainder theorem. The minuend minus the subtrahend in subtraction is sometimes called the remainder (*difference* is more common).

    Chinese remainder theorem. See CHINESE.

    remainder of an infinite series after the nth term. (1) The difference R_n between the sum, S, of the series and the sum, S_n, of n terms, *i.e.*, $R_n = S - S_n$, when the series is convergent. (2) The difference between the sum of the first n terms of the series and the quantity (or function)

whose expansion is sought [see TAYLOR —Taylor's theorem, and FOURIER —Fourier series]. The series converges and represents the quantity or function for all values of the independent variable for which the remainder converges to zero.

    remainder in Taylor's theorem. See TAYLOR —Taylor's theorem.

    remainder theorem. When a polynomial in x is divided by $x - h$, the remainder is equal to the number obtained by substituting h for x in the polynomial. More concisely, $f(x) = (x - h)q(x) + f(h)$, where $q(x)$ is the quotient and $f(h)$ the remainder, which is easily verified by substituting h for x. E.g., $(x^2 + 2x + 3) \div (x - 1)$ leaves a remainder of $1^2 + 2 \times 1 + 3$, or 6. If $f(h) = 0$, then it follows from the remainder theorem that $f(x) = (x - h)q(x)$. This is a proof of the *factor theorem*.

RE-MOV′A-BLE, *adj.* **removable discontinuity.** See DISCONTINUITY.

RE-MOV′AL, *n.* **removal of a term of an equation.** Transforming the equation into a form having this term missing. See ROTATION —rotation of axes, TRANSLATION —translation of axes, REDUCED —reduced cubic equation.

RENT, *n.* (1) A sum of money paid at regular intervals in return for the use of property or nonperishable goods. (2) The periodic payments of an annuity. The period between successive payments of rent is the **rent period.**

RENTES, *n.* French perpetuity bonds.

RE-PEAT′ED, *adj.* **repeated root.** See MULTIPLE—multiple root.

RE-PEAT′ING, *adj.* **repeating decimal.** See DECIMAL—decimal number system.

RE-PLACE′MENT, *n.* **replacement cost of equipment.** (1) The cost of new equipment minus the scrap value. (2) The purchase price minus scrap value.

REP′RE-SEN-TA′TION, *n.* **reducible matrix representation of a group.** Let $D_1, D_2, \cdots$ be the matrices of a representation of a group G by square matrices of order n. This representation is *reducible* if there is a matrix M such that, for each i, $M^{-1}D_i M = E_i$ is a matrix whose elements are all zero except in two or more matrices $A_{i1}, A_{i2}, \cdots, A_{ip}$ having main diagonals along the main diagonal of E_i, where A_{im} is of the same order for all i. When the number of such matrices A_{im} is maximal, the set of matrices A_{im} for a fixed value of m is said to be an **irreducible representation** of the group; such a set of matrices is isomorphic to a subset of G which contains the (group) product of any two of its members, and G is the direct product of all such subsets. The number of irreducible representations is equal to the number of distinct *conjugate sets* of elements. For an Abelian group, this number of irreducible representations is the order of the group, and each matrix of the irreducible representations is of order one; *i.e.*, any finite Abelian group can be represented as the direct sum of cyclic subgroups. This definition of irreducible representation is equivalent to that given for a set of matrices (see REDUCIBLE — reducible set of matrices) when the *set* is a group.

representation of a group. (1) A group of a particular type (*e.g.*, a permutation group, or a group of matrices) which is *isomorphic* with the given group. Every finite group can be represented by a permutation group and by a group of matrices. (2) A group H is a **representation** of a group G if there is a homomorphism of G onto H. A set of representations consisting of matrices (or transformations) is a **complete system of representations** of G if for any g of G other than the identity there is a representation for which g does not correspond to the identity matrix (or identity transformation). Any finite group has a complete system of matrix representations and any *locally compact topological group* has a complete system of representations consisting of *unitary transformations* of Hilbert space. The order of the matrices in a matric representation is called the **degree** or **dimension** of the representation. See PERMUTATION —permutation matrix, and above, reducible matrix representation of a group.

spherical representation. See SPHERICAL.

RE-SERVE′, *n.* (*Life Insurance*) The amount an insurance company needs (at a given time) to add to future net premiums and interest to pay all claims expected according to the particular mortality table being used (this is the difference between the present value of future benefits and the present value of future premiums and is a liability, also called the **reinsurance fund** or **self-insurance fund**). The **reserve per policy** (the **net premium reserve**) is called the **initial reserve** when computed at the beginning of a policy year just after the premium has been received, and the **terminal reserve** when computed at the end of a policy year before the premium has been paid. The average of the initial and terminal reserves is called the **mean reserve**. The **prospective method** for computing reserves makes use of

the fact that the reserve (initial or terminal) is the difference between the present value of future benefits and the present value of future premiums. It is also possible to use the past history of the policy to compute the reserve (the **retrospective method**), computing the present value of the past differences of the *net level premiums* and the *natural premiums* (this difference is positive in the early years of the policy and negative in the later years). A **premium deficiency reserve** is an amount equal to the difference between the present value of future net premiums and future gross premiums (required in most states when gross premium is less than net premium).

RE-SID′U-AL, *adj.* **residual set.** See CATEGORY —category of sets.

RES′I-DUE, *adj., n.* If the congruence $x^n \equiv a$ (mod m) has a solution, then a is called a **residue** (in particular, a *power residue* of m of the nth order). If this congruence has no solution, a is called a **nonresidue** of m. Thus 4 is a residue of 5 of the second order, since $3^2 \equiv 4$ (mod 5). The congruence $x^n \equiv a$ (mod m) is solvable if and only if $a^{\phi(m)/d} \equiv 1$ (mod m), where ϕ is the *Euler ϕ-function* and d is the greatest common divisor of n and $\phi(m)$. Thus a is a residue of m of order n if and only if $a^{\phi(m)/d} \equiv 1$ (mod m). This is the **Euler criterion**.

complete residue system modulo n. Any set of integers which has the property that no two of the integers belong to the same number class modulo n is a **complete residue system modulo** n. This set of integers is also said to form a complete system of incongruent numbers modulo n. *E.g.*, $1, 9, 3, -3, 5, -1, 7$ form such a system modulo 7. This set of integers may be also expressed in terms of nonnegative integers each less than 7 by the numbers $0, 1, 2, 3, 4, 5, 6$.

reduced residue system modulo n. A complete residue system modulo n contains some numbers prime to n. This set of numbers is called a reduced residue system modulo n. Thus a reduced residue system modulo 6 is $1, 5$; whereas a complete residue system modulo 6 is $0, 1, 2, 3, 4, 5$. See above, complete residue system modulo n.

residue of an analytic function at an isolated singular point. If f is an analytic function of the complex variable z in the "deleted" neighborhood consisting of all z satisfying $0 < |z - z_0|$ $< \epsilon$, then the residue of f at z_0 is $\dfrac{1}{2\pi i} \int_C f(z)\,dz$, where C is a simple closed rectifiable curve about z_0 in the "deleted" neighborhood. The value of the residue is a_{-1}, where a_{-1} is the coefficient of $(z - z_0)^{-1}$ in the Laurant expansion of $f(z)$ about z_0.

RE-SIST′ANCE, *n*. **electrical resistance.** That property of a conductor which causes the passage of an electric current through it to be accompanied by the transformation of electric energy into heat. See OHM.

RE-SOL′VENT, *adj.*, *n*. **resolvent cubic.** See FERRARI —Ferrari's solution of the quartic.

    resolvent kernel. See VOLTERRA —Volterra reciprocal functions, KERNEL —iterated kernels.

    resolvent of a matrix. The inverse of the matrix $\lambda I - A$, where A is the given matrix, and I is the identity matrix. The resolvent exists for all λ which are not *eigenvalues* of the matrix.

    resolvent set of a transformation. See SPECTRUM —spectrum of a transformation.

RES′O-NANCE, *n*. See OSCILLATION.

RE-SULT′, *n*. The end sought in a computation or proof.

RE-SULT′ANT, *n*. **resultant of a set of polynomial equations.** A relation between coefficients which is obtained by eliminating the variables and which is zero if the equations have a nonempty solution set (the resultant is also called an **eliminant**). In case the equations are linear, this can be accomplished expeditiously by equating to zero the determinant (of order $n + 1$) whose columns are the coefficients of the respective variables and the constant terms. *E.g.*, the result of eliminating x and y from

$$\begin{array}{l} ax + by + c = 0 \\ dx + ey + f = 0 \\ gx + hy + k = 0 \end{array} \quad \text{is} \quad \begin{vmatrix} a & b & c \\ d & e & f \\ g & h & k \end{vmatrix} = 0.$$

The determinant of the coefficients of a system of n homogeneous linear equations in n variables is also called a resultant (or eliminant) of the equations (it is zero if and only if the equations have a *nontrivial* simultaneous solution). For two polynomial equations in one variable,

$$f(x) = a_0 x^m + a_1 x^{m-1} + \cdots + a_m = 0,$$
$$a_0 \neq 0,$$
$$g(x) = b_0 x^n + b_1 x^{n-1} + \cdots + b_n = 0,$$
$$b_0 \neq 0,$$

the resultant is usually taken to be

$$R(f, g) = a_0^n g(r_1) g(r_2) \cdots g(r_m),$$

where $r_1, r_2, \cdots, r_m$ are the roots of $f(x) = 0$. This is also equal to the following determinant, which has n rows containing the coefficients of $f(x)$ and m rows containing the coefficients of $g(x)$:

$$\begin{vmatrix} a_0 & a_1 & a_2 & \cdots & a_m & 0 & \cdots & \cdots & 0 \\ 0 & a_0 & a_1 & a_2 & \cdots & a_m & 0 & \cdots & 0 \\ 0 & 0 & a_0 & a_1 & a_2 & \cdots & a_m & \cdots & 0 \\ \cdots & \cdots & \cdots & \cdots & \cdots & \cdots & \cdots & \cdots & \cdots \\ b_0 & b_1 & b_2 & \cdots & b_n & 0 & \cdots & \cdots & 0 \\ 0 & b_0 & b_1 & b_2 & \cdots & b_n & 0 & \cdots & 0 \\ \cdots & \cdots & \cdots & \cdots & \cdots & \cdots & \cdots & \cdots & \cdots \end{vmatrix}$$

This determinant is the resultant obtained by Sylvester's dialytic method. See DISCRIMINANT —discriminant of a polynomial equation, SYLVESTER.

    resultant of two functions. Same as the CONVOLUTION of the two functions.

    resultant of vectors (forces, velocities, accelerations, etc.). The sum of the vectors. See SUM —sum of vectors.

RE-TRACT′, *n*. A subset X of a topological space T being a **retract** of T means that there is a continuous function f that maps T onto X and satisfies $f(x) = x$ if $x \in X$, *i.e.*, the identity map of X has a continuous extension to T. If X is a retract of T, then each continuous function on X has a continuous extension to T (see TIETZE —Tietze extension theorem). A topological space X being an **absolute retract** means that if T is any normal topological space and X is homeomorphic to a closed subset Y of T, then Y is a retract of T. A *disk* or an *n-ball* is an absolute retract, but a *circle* is not.

RET′RO-SPEC′TIVE, *adj.* **retrospective method of computing reserves.** See RESERVE.

REULEAUX, Franz (1829–1905). German geometer.

    Reuleaux "triangle." A closed curve consisting of three arcs that join vertices of an equilateral triangle; each arc joins two vertices and lies on the circle through these vertices that has its center at the remaining vertex. This "triangle" is a *curve of constant width* in the sense that if r is the radius of the circles, then for any line L the "triangle" lies between two lines parallel to L and separated by distance r.

RE-VERSE′, *adj.* Backward. A series of steps in a computation is taken in *reverse order* when the last is taken first, the next to last second, etc.; a finite sequence of terms is in *reverse order* when the last is made first, etc.

RE-VER′SION, *adj.* **reversion of a series.** The process of expressing x as a series in y, having given y expressed as a series in x.

RE-VER′SION-AR′Y, *adj.* See ANNUITY, INSURANCE—participating insurance policy.

REV′O-LU′TION, *n.* **axis of revolution.** See SURFACE —surface of revolution, and below, solid of revolution.

cone, cylinder, and ellipsoid of revolution. See CONE, CYLINDER, ELLIPSOID.

solid of revolution. A solid generated by revolving a plane area about a line (the **axis of revolution**). The volume of a solid of revolution can be computed without multiple integration in three ways. If a plane perpendicular to the axis of revolution intersects the solid in a region bounded by two circles, and the larger and smaller circles have radii r_2 and r_1, then the element of volume $\pi(r_2^2 - r_1^2)\,dh$ can be used (the **washer method**), where h is measured along the axis of revolution, r_2 and r_1 are functions of h, and

$$\int_{h_1}^{h_2} \pi\left(r_2^2 - r_1^2\right) dh$$

is the volume of the solid (h_1 and h_2 are the least and greatest values of h for which the plane intersects the solid). If the axis of revolution is the x-axis and the area is that bounded by the x-axis, the ordinates corresponding to $x = a$ and $x = b$, and the curve $y = f(x)$, the elements of volume are discs ($r_1 = 0$) and the volume (the **disc method**) is

$$\int_a^b \pi f^2(x)\,dx.$$

If the plane area is all on the positive side of the axis of revolution and a line parallel to the axis of revolution and at distance x from it intersects the plane area in a line segment (or segments) of total length $L(x)$, then the element of volume $2\pi x L(x)\,dx$ is an approximation to the volume generated by revolving a thin strip of width dx and length $L(x)$ about the axis of revolution. The volume by this method (the **shell method**) is

$$\int_{x_1}^{x_2} 2\pi x L(x)\,dx,$$

where x_1 and x_2 are the minimum and maximum distances of the plane area from the axis of revolution. The washer and disc methods are special cases of **volume by slicing**: If a solid is

between the planes $x = a$ and $x = b$ and the area of a plane section in the plane $x = t$ is $A(t)$, then the volume is

$$\int_a^b A(t)\,dt.$$

surface of revolution. See SURFACE —surface of revolution.

RE-VOLVE′, *v.* To rotate about an axis or point. One would speak of revolving a figure in the plane about the origin, through an angle of a given size, or of *revolving* a curve in space about the x-axis with the understanding that the revolution is through an angle of 360° unless otherwise stipulated. See SURFACE —surface of revolution.

RHIND, Alexander Henry (1833–1863). Scottish antiquarian. In 1858 he purchased in an Egyptian resort town the famous papyrus now bearing his name. See AHMES PAPYRUS.

RHO, *n.* The seventeenth letter of the Greek alphabet: lower case, ρ; capital, P.

RHOMB, *n.* Same as rhombus.

RHOMBOHEDRON, *n.* A six-sided prism whose faces are parallelograms.

RHOM′BOID, *n.* A parallelogram with adjacent sides not equal.

RHOM′BUS, *n.* [*pl.* **rhombi** or **rhombuses**]. A parallelogram with adjacent sides equal (all of its sides are then necessarily equal). Some authors require that a rhombus not be a square, but the preference seems to be to call the square a special case of the rhombus.

Rhombus

RHUMB, *adj.* **rhumb line.** The path of a ship sailing so as to cut the meridians at a constant angle; a spiral on the earth's surface, winding around a pole and cutting the meridians at a constant angle. *Syn.* loxodromic spiral.

RICCATI, Count Jacopo Francesco (1676–1754). Italian geometer and analyst.

Riccati equation. A differential equation of type $dy/dx + ay^2 = bx^n$. Daniel Bernoulli determined that n is of form $-4k/(2k \pm 1)$ where k is a positive integer if the equation is integrable in finite form. A differential equation of form $dy/dx + f + yg + y^2 h = 0$ is a **generalized Riccati equation**. The substitution $y = w'/(hw)$ transforms this equation into $w'' + [g - (h'/h)]w' + fhw = 0$. Thus if $h(x) \equiv 1$, then the general solution of $y'' + gy' + fy = 0$ is $y(x) = ce^{\int u(x)\,dx}$, where u is the general solution of the generalized Riccati equation.

RICCI, Curbastro Gregorio (1853–1925). Italian algebraist, analyst, geometer, and mathematical physicist. Originated tensor analysis as a means of studying differential invariants of Riemannian geometry.

Ricci tensor. The *contracted curvature tensor* $R_{ij} = R_{ij\sigma}^{\sigma}$, where R_{ijk}^{p} is the *Riemann-Christoffel curvature tensor*. It is often called the **Einstein tensor** in general relativity theory, since it occurs in the Einstein gravitational equations. The Ricci tensor is a symmetric tensor since

$$\frac{\partial \log \sqrt{g}}{\partial x^i} = \left\{ \begin{matrix} i \\ ij \end{matrix} \right\}.$$

RIEMANN, Georg Friedrich Bernhard (1826–1866). Great innovative German mathematician who made fundamental contributions to geometry and the theory of analytic functions of a complex variable, as well as to number theory, potential theory, topology, and mathematical physics. His Riemannian geometry provides the foundation for modern relativity theory.

covariant Riemann-Christoffel curvature tensor. The covariant tensor field of rank four:

$$R_{i\alpha\beta\gamma}(x^1, \cdots, x^n) = g_{i\sigma} R_{\alpha\beta\gamma}^{\sigma}(x^1, \cdots, x^n).$$

See below, Riemann-Christoffel curvature tensor.

Riemann-Christoffel curvature tensor. The tensor field

$$R_{\alpha\beta\gamma}^{i}(x^1, x^2, \cdots, x^n)$$

$$= \frac{\partial \left\{ \begin{matrix} i \\ \alpha\beta \end{matrix} \right\}}{\partial x^\gamma} - \frac{\partial \left\{ \begin{matrix} i \\ \alpha\gamma \end{matrix} \right\}}{\partial x^\beta}$$

$$+ \left\{ \begin{matrix} \sigma \\ \alpha\beta \end{matrix} \right\} \left\{ \begin{matrix} i \\ \sigma\gamma \end{matrix} \right\} - \left\{ \begin{matrix} \sigma \\ \alpha\gamma \end{matrix} \right\} \left\{ \begin{matrix} i \\ \sigma\beta \end{matrix} \right\},$$

where the summation convention applies and

$\left\{ \begin{matrix} i \\ jk \end{matrix} \right\}$ are the *Christoffel symbols of the second kind* of the n-dimensional Riemannian space with the fundamental differential form $g_{ij}\,dx^i\,dx^j$. $R_{\alpha\beta\gamma}^{i}$ is a tensor field of rank four, contravariant of rank one and covariant of rank three. Many authors define $R_{\alpha\beta\gamma}^{i}$ as the negative of the above. See above, covariant Riemann-Christoffel curvature tensor.

Riemann hypothesis. The zeta function has zeros at $-2, -4, \cdots$. All other complex numbers which are zeros of the zeta function lie in the strip of complex numbers z whose real parts satisfy $0 < \mathrm{Re}(z) < 1$. The Riemann hypothesis is the (unproved) conjecture that these zeros actually lie on the line $\mathrm{Re}(z) = \frac{1}{2}$. It was proved by G. H. Hardy that an infinite number of zeros lie on this line. The first $1.5 \cdot 10^9$ zeros are known, all are simple, and all lie on this line. The proof of the Riemann hypothesis would have extensive consequences in the theory of prime numbers. The Riemann hypothesis is true if and only if

$$\sum_{1}^{\infty} \mu(n)n^{-s}$$

converges for the real part of s greater than $\frac{1}{2}$, where μ is the Möbius function.

Riemann integral. See INTEGRAL —definite integral, generalized Riemann integral.

Riemann-Lebesgue lemma. If f and $|f|$ are integrable on the interval $[a, b]$, then

$$\lim_{t \to +\infty} \int_a^b f(x) \sin tx\,dx = 0$$

and $\lim_{t \to +\infty} \int_a^b f(x)\cos tx\,dx = 0$, or (equivalently)

$$\lim_{t \to +\infty} \int_a^b f(x)\sin(tx + b)\,dx = 0 \text{ for all } b.$$

This lemma is useful when studying convergence of Fourier series. In particular, when t is an integer the lemma expresses the fact that $\lim_{n \to \infty} a_n = \lim_{n \to +\infty} b_n = 0$, where a_n and b_n are the coefficients of $\cos nx$ and $\sin nx$ in the Fourier series for f.

Riemann mapping theorem. Any nonempty simply connected open set in the plane, other than the whole plane, can be mapped one-to-one and conformally onto the interior of a circle. If the circle is $|z| = 1$ and z_0 is in the open set, then there is exactly one such map f for which $f(z_0) = 0$ and $f'(z_0) > 0$. Stated by Riemann in 1851, but with an incorrect "proof." The first successful proof was given by P. Koebe in 1908, although a related theorem from which the Riemann mapping theorem can be derived was proved in 1900 by W. F. Osgood. Other mathematicians have contributed techniques which yielded additional proofs.

Riemann sphere (or **spherical surface**). The surface on a unit sphere corresponding to a (plane) Riemann surface under a stereographic projection.

Riemann-Stieltjes integral. See STIELTJES.

Riemann sum. See INTEGRAL—definite integral.

Riemann surface. The relation between complex numbers z and complex numbers w expressed by the monogenic analytic function $w = f(z)$ might be one-to-one, one-to-many, many-to-one, or many-to-many. Respective examples are $w = (z + 1)/(z - 1)$, $w = z^2$, $w^3 = z$, $w^3 = z^2$. Riemann surfaces furnish a schematic device whereby the relation is considered as one-to-one (between points of the z- and w-Riemann surfaces) in any case. A suitable number, possibly a denumerably infinite number, of sheets is considered over the z-plane and over the w-plane. These might be joined in a variety of ways at branch points. The sheets are distinguished by imaginary branch cuts joining the branch points or extending to infinity. Thus $w^3 = z^2$ gives a one-to-one mapping of a three-sheeted z-surface on a two-sheeted w-surface. Any simply connected Riemann surface can be mapped conformally on exactly one of the following: the interior of the unit circle; the finite plane (the punctured complex plane, excluding the point at infinity); the closed complex plane (including the point at infinity). In these three cases the surface is said to be respectively of **hyperbolic**, **parabolic**, or **elliptic type**.

Riemann zeta function. See ZETA.

Riemannian curvature. The scalar determined by a point and two linearly independent directions (contravariant vectors) ξ_1^i and ξ_2^i at that point:

$$k = \frac{R_{\alpha\beta\gamma\delta}\xi_1^\alpha\xi_2^\beta\xi_1^\gamma\xi_2^\delta}{\left(g_{\alpha\delta}g_{\beta\gamma} - g_{\alpha\gamma}g_{\beta\delta}\right)\xi_1^\alpha\xi_2^\beta\xi_1^\gamma\xi_2^\delta}.$$

The $g_{\alpha\beta}$ is the metric tensor of the *Riemannian space*, and $R_{\alpha\beta\gamma\delta}$ is the covariant Riemann-Christoffel curvature tensor (see above). A geometrical construction leading to the Riemannian curvature k is as follows: Consider the two-parameter family of directions $u\xi_1^\alpha + v\xi_2^\alpha$ at the given point and form the two-dimensional geodesic surface swept out by the geodesics through the point and having directions in the two-parameter family of directions. The *Gaussian curvature* (total curvature) of the geodesic surface at the given point is the Riemannian curvature of the enveloping n-dimensional Rie-

mannian space at the given point and with respect to the given orientation.

Riemannian space. An n-dimensional coordinate manifold, *i.e.*, a space of points ($x_1, x_2 \cdots, x_n$), whose element of arc length ds is given by a symmetric quadratic differential form

$$ds^2 = g_{ij}(x^1, \cdots, x^n)\, dx^i\, dx^j,$$

in which the coefficients g_{ij} have a nonvanishing determinant and the *summation convention* applies. It is also often required that the differential form be positive definite, although this restriction is not imposed in applications to general relativity theory. The g_{ij} are the components of a symmetric covariant tensor, called the **fundamental metric tensor**.

Riemannian space of constant Riemannian curvature. A Riemannian space such that the *Riemannian curvature* is the same throughout space and at the same time is independent of the orientation ξ_1^i, ξ_2^i. The spaces of constant Riemannian curvature k such that $k > 0$, $k < 0$, and $k = 0$ are locally the **Riemann spherical space**, **Lobachevski space**, and **Euclidean space**, respectively.

RIESZ, Frigyes (1880–1956). Innovative Hungarian functional analyst. Introduced subharmonic functions and abstract concept of operator.

Riesz-Fischer theorem. Let a countably additive measure m be defined on a σ-algebra of subsets of a set Ω and let L_2 be the set of all measurable (real or complex) functions f for which $\int |f|^2\, dm$ is finite. The Riesz-Fischer theorem asserts that L_2 is *complete*; *i.e.*, for a sequence $f_1, f_2, \cdots$ of elements of L_2, there is an f in L_2 such that the sequence converges in the mean (of order 2) to f if $\|f_m - f_n\| \to 0$ as m and n become infinite, where $\|f_m - f_n\|^2 = \int |f_m - f_n|^2\, dm$. An immediate consequence of this, also called the Riesz-Fischer theorem, is that if $u_1, u_2, \cdots$ is an orthonormal sequence of functions and $a_1, a_2, \cdots$ is a sequence of complex (or real) numbers for which $\sum |a_n|^2$ is convergent, then there exists a function f belonging to L_2 such that $a_n = \int f(x)\overline{u_n(x)}\, dm$ for each n.

E.g., a trigonometric series $\frac{1}{2}a_0 + \sum_1^\infty (a_n \cos nx + b_n \sin nx)$ is a Fourier series of some function if (and only if) $\sum_1^\infty (a_n^2 + b_n^2)$ is convergent.

RIGHT, *adj., n.* **continuous on the right.** See CONTINUOUS.

limit on the right. See LIMIT—limit on the left or right.

right angle. See ANGLE—right angle.

right circular cone. See CONE.

right coset. See COSET.

right dihedral angle. See PLANE—plane angle of a dihedral angle.

right-handed coordinate system. See COORDINATE—right-handed coordinate system.

right-handed curve. If the torsion of a directed curve *C* at a point *P* is negative, then *C* is said to be right-handed at *P*. See LEFT—left-handed curve. *Syn.* dextrorsum [Latin] or dextrorse curve.

right-handed trihedral. See TRIHEDRAL.

right ideal. See IDEAL.

right identity and inverse. See IDENTITY—identity element, INVERSE—inverse of an element.

right line. A straight line.

right section of a surface. See SECTION—section of a surface.

right triangle. See SPHERICAL—spherical triangle, TRIANGLE.

RIG′ID, *adj.* **rigid body.** An ideal body which is characterized by the property that the distance between every pair of points of the body remains unchanged. Objects which experimentally are not readily deformable approximate rigid bodies.

rigid motion. Moving a configuration into another position, but making no change in its shape or size; a rotational transformation followed by a translation, or the two taken in reverse order or simultaneously. An isometry is a rigid motion if it does not change the orientation of the coordinate axes [see ISOMETRY (2)]. Superposition of figures in plane geometry is a rigid motion.

RI-GID′I-TY, *n.* **modulus of rigidity.** The ratio of the shearing stress to the change in angle produced by the shearing stress. *Syn.* shearing modulus. See LAMÉ—Lamé's constants.

RING, *adj., n.* A set with two binary operations, called addition and multiplication, which have the properties: (1) The set is an *Abelian group* with respect to the operation of addition. (2) Each pair *a, b* of elements determines a unique product *a · b*, multiplication is *associative*, and multiplication is *distributive with respect to addition*; i.e.,

$$a \cdot (b + c) = a \cdot b + a \cdot c$$

and

$$(b + c) \cdot a = b \cdot a + c \cdot a$$

for each *a*, *b*, and *c* of the set. If it is also true that multiplication is *commutative*, the ring is a **commutative ring.** If there is an identity for multiplication (an element 1 for which $1 \cdot x = x \cdot 1 = x$ for all *x*), the ring is a **ring with unit element** (or a **ring with unity**). A commutative ring with unit element is an integral domain if no product of nonzero elements is zero and a field if each nonzero element has a multiplicative inverse. In a commutative ring with unit element, a **unit** is any *x* for which there is a *y* such that $xy = 1$. A **division ring** is a ring whose nonzero members form a group under multiplication. A commutative division ring is a **field.** A division ring that is not commutative is a **skew field.** A **simple ring** is a ring that contains no ideals except the ring itself and the ideal containing only 0. See CHAIN—chain conditions on rings, DOMAIN—integral domain, FIELD, IDEAL.

Euclidean ring. See EUCLIDEAN.

normed vector ring. See ALGEBRA—Banach algebra.

principal ideal ring. A commutative ring in which all ideals are principal.

quotient (or factor) ring. See QUOTIENT—quotient space.

radical of a ring. See RADICAL.

ring of sets. A nonempty class of sets which contains the union and the difference of any two of its members. It is a *σ*-**ring** if it also contains the union of any sequence of its members. A ring of sets is also a *ring* if *symmetric difference* and *intersection* are taken as the *addition* and *multiplication* operations of the ring. For an arbitrary set *S*, the class of all finite subsets of *S* is a ring of sets. Another example of a ring of sets is the class of sets of real numbers which are finite unions of intervals which contain their left endpoints and do not contain their right endpoints. See ALGEBRA—algebra of subsets, MEASURE—measure ring and measure algebra.

ring surface, torus ring. Same as ANCHOR RING. See ANCHOR.

semiring of sets. A class *S* of sets which contains the empty set and the intersection of any two of its members and which is such that if *A* and *B* are members of *S* with $A \subset B$, then there are a finite number of sets $C_1, C_2, \cdots, C_n$ such that $B - A = \cup C_i$, $C_i \cap C_j = 0$ if $i \neq j$, and each C_i is a member of *S*. Every ring of sets is also a semiring of sets.

RISE, *n.* **rise between two points.** The difference in elevation of the two points. See RUN.

rise of a roof. (1) The vertical distance from the plates to the ridge of the roof. (2) The vertical distance from the lowest to the highest point of the roof.

ROBIN, Victor Gustave (1855–1897). French analyst and applied mathematician.

  Robin's function. For a region R with boundary surface S, and for a point Q interior to R, the Robin's function $R_{k,h}(P,Q)$ is a function of the form $R_{k,h}(P,Q) = 1/(4\pi r) + V(P)$, where r is the distance PQ, $V(P)$ is harmonic, and $k\partial R_{k,h}/\partial n + hR_{k,h} = 0$ on S. The solution $U(Q)$ of the third boundary-value problem of potential theory (the Robin problem) can be represented in the form

$$U(Q) = \int_S f(P) R_{k,h}(P,Q)\, d\sigma_P.$$

See GREEN—Green's function, BOUNDARY—third boundary-value problem of potential theory.

ROBINSON, Abraham (1918–1974). Born in Germany, lived in England, Canada, Israel and finally U.S. Logician, metamathematician, algebraist, analyst, functional analyst, and aerodynamicist who created **nonstandard analysis**.

RO-BUST′, *adj.* **robust statistics.** Test procedures usually depend on assumptions such as that certain distributions are normal. If the inferences are little affected by departures from those assumptions, or by small data or procedural errors, *e.g.*, if the *significance level* and the *power* of a test vary little if the population departs quite substantially from normality, the test is said to be **robust**. More generally, a statistical procedure is robust if it is not very sensitive to small departures from the assumptions on which it depends and somewhat larger deviations would not be catastrophic. Statisticians have long considered such questions, but a theory of robustness is relatively new—the term "robustness" was introduced about 1953. See HYPOTHESIS—test of a hypothesis.

RODRIGUES, Benjamin Olinde (1795–1850). French economist and reformer whose early training was in mathematics.

  equations of Rodrigues. The equations

$$dx + \rho\, dX = 0,\ dy + \rho\, dY = 0,\ dz + \rho\, dZ = 0$$

characterizing the lines of curvature of the surface S. The function ρ is the radius of normal curvature in the direction of the line of curvature.

Rodrigues formula. The equation

$$P_n(x) = \frac{1}{2^n n!} \frac{d^n}{dx^n} (x^2 - 1)^n,$$

where P_n is a *Legendre polynomial*.

ROLLE, Michel (1652–1719). French analyst, algebraist, and geometer.

  Rolle's theorem. If a continuous curve crosses the x-axis at two points and has a tangent at all points between these two x-intercepts, it has a tangent parallel to the x-axis at at least one point between the two intercepts. *Tech.* If f is a continuous function for $a \le x \le b$ and vanishes for $x = a$ and $x = b$ and is differentiable at all interior points on (a,b), then f' vanishes at some point between and distinct from a and b. (It may also vanish at a or b or both.) *E.g.*, the sine curve crosses the x-axis at the origin and at $x = \pi$, and has a tangent parallel to the x-axis at $x = \frac{1}{2}\pi$ (radians).

RO′MAN, *adj.* **Roman numerals.** A system of writing integers, used by the Romans, in which I denotes 1; V, 5; X, 10; L, 50; C, 100; D, 500; M, 1000. All integers are then written using the following rules: (1) When a letter is repeated or immediately followed by a letter of lesser value, the values are added. (2) When a letter is immediately followed by a letter of greater value, the smaller is subtracted from the larger. The integers from 1 to 10 are written: I, II, III, IIII or IV, V, VI, VII, VIII, IX, X. The tens are written: X, XX, XXX, XL, L, LX, LXX, LXXX, XC. Hundreds are written C, CC, CCC, CD, D, DC, DCC, DCCC, CM.

ROOT, *n.* **double, equal, simple, triple and multiple roots.** See MULTIPLE—multiple root of an equation.

  infinite root of an equation. An equation of degree $r < n$ which is considered to be an equation of degree n is said to have infinity as a root $n - r$ times. *E.g.*, the equation $ax^2 + bx + c = 0$ has one infinite root if $a = 0$ and $b \ne 0$; two infinite roots if $a = b = 0 \ne c$. If x is replaced by $1/y$ in this quadratic equation one obtains the equation $a + by + cy^2 = 0$, which has the same number of zero roots as the original equation had infinite roots. With this convention, it can be said that a line and a hyperbola always intersect in two points; one or both of these may be a point at infinity. See IDEAL—ideal point.

  rational-root theorem. See RATIONAL—rational-root theorem.

root of a congruence. A number which when substituted in the congruence, expressed in a form $f(x) \equiv 0 \pmod{n}$, makes the left member of the congruence divisible by the modulus n. Thus 8 is a root of the congruence $x + 2 \equiv 0 \pmod 5$, since $8 + 2$ or 10 is divisible by 5. Another root is 3.

root of an equation. A number which, when substituted for the variable in the equation, reduces it to an identity (a root of the equation $x^2 + 3x - 10 = 0$ is 2, since $2^2 + 3 \cdot 2 - 10 = 0$). A root of an equation is said to *satisfy the equation* or to be a *solution* of the equation, but *solution* more often refers to the process of finding the root. There are many ways to approximate a root of an equation (see FALSE —method of false position, GRAEFFE, GRAPHICAL —graphical solution of an equation, HORNER —Horner's method, and NEWTON —Newton's method of approximation). A basic step in approximating a root is to **isolate** the root, *i.e.*, find two numbers between which there is one and only one root of the equation. The following **location principle** is very useful: If a polynomial or other continuous function of one variable has different signs for two values of the variable, it is zero for some value of the variable between these two values; the equation obtained by equating the given function to zero has a root between two values of the variable for which the function has different signs. Geometrically, if the graph of a continuous function of a variable x is for one value of x on one side of the x-axis and for another value on the other side (changes sign), it must cross the axis between the two positions. From a given equation, it is possible to derive new equations whose roots are related to those of the given equation in various ways. One can **change the signs of the roots** by replacing the variable by its negative, giving a new equation whose roots are the negatives of the roots of the original equation. One can **decrease the roots** by the amount a by transforming the equation by the substitution $x = x' + a$, $a > 0$, where x is the variable in the given equation. If the old equation has the root x_1, the new one has a root $x_1' = x_1 - a$. The substitution of $x = x' + 2$ in $x^2 - 3x + 2 = 0$, whose roots are 1 and 2, results in the equation $(x')^2 + x' = 0$ whose roots are -1 and 0. The substitution $x = 1/x'$ transforms an equation so that the given equation has roots which are the **reciprocals of the roots** of the transformed equation (see RECIPROCAL —reciprocal equation). The **roots and coefficients** of a polynomial equation are related in the following ways: For a quadratic equation, the sum of the roots is equal to the negative of the coefficient of the first-degree term and the product is equal to the constant

term, when the coefficient of the square term is 1. In $ax^2 + bx + c = 0$, the sum of the roots is $-b/a$ and the product is c/a. If the equation is of the nth degree and the coefficient of the nth-degree term is unity, the sum of the roots is the negative of the coefficient of x^{n-1}, the sum of the products of the roots taken two at a time in every possible way is the coefficient of x^{n-2}, the sum of the products of the roots taken three at a time is the negative of the coefficient of x^{n-3}, etc.; finally, the product of all the roots is the constant term with a positive or negative sign according as n is even or odd. If $r_1, r_2, \cdots, r_n$ are the roots of

$$x^n + a_1 x^{n-1} + a_2 x^{n-2} + \cdots + a_n = 0,$$

then

$$r_1 + r_2 + \cdots + r_n = -a_1,$$

$$r_1 r_2 + r_1 r_3 + \cdots + r_1 r_n + r_2 r_3 + \cdots$$

$$+ r_{n-1} r_n = a_2,$$

$$\cdots \cdots \cdots \cdots \cdots \cdots \cdots \cdots \cdots$$

$$r_1 r_2 r_3 \cdots r_n = (-1)^n a_n.$$

See CARDAN, EQUATION —polynomial equation, FERRARI, FUNDAMENTAL —fundamental theorem of algebra, MULTIPLE —multiple root of an equation, NEWTON —Newton's method of approximation, QUADRATIC —quadratic formula, RATIONAL —rational-root theorem.

root field. Same as GALOIS FIELD. See GALOIS.

root-mean-square deviation. See DEVIATION —standard deviation.

root of a number. An nth root of a number is a number which, when taken as a factor n times (raised to the nth power), produces the given number. There are n nth roots of any nonzero number (these may be real or imaginary). If n is odd and the number real, there is one real root; *e.g.*, the cube roots of 27 are 3 and $\frac{3}{2}(-1 \pm \sqrt{3}\,i)$. If n is even and the number positive, there are two real roots, numerically equal but opposite in sign; *e.g.*, the 4th roots of 4 are $\pm \sqrt{2}$ and $\pm \sqrt{2}\,i$. A **square root** of a number is a number which, when multiplied by itself, produces the given number. A positive (real) number has two real square roots, a negative number two imaginary square roots. The **positive square root** of a positive number a is denoted by $\sqrt{a}$. A **cube root** of a number is a number whose cube is the given number. Each real number (except zero) has one real cube root and two imaginary cube roots. If a

complex number (which may be a real number) is written in the form

$$r[\cos\theta + i\sin\theta],$$

or the equivalent form

$$r[\cos(2k\pi + \theta) + i\sin(2k\pi + \theta)],$$

its nth roots are the numbers

$$\sqrt[n]{r}\left[\cos\frac{(2k\pi + \theta)}{n} + i\sin\frac{(2k\pi + \theta)}{n}\right],$$

where k takes on the values $0, 1, 2, \cdots, (n-1)$ and $\sqrt[n]{r}$ is an nth root of the nonnegative number r. See DE MOIVRE—De Moivre's theorem, MECHANIC—mechanic's rule, and below, root of unity.

root of unity. Any complex number z such that $z^n = 1$ for some positive integer n (z is an **nth root of unity**). The nth roots of unity are the numbers $\cos\left(\dfrac{k}{n}\cdot 360°\right) + i\sin\left(\dfrac{k}{n}\cdot 360°\right)$ for $k = 0, 1, 2, \cdots, n - 1$ (see DE MOIVRE—De Moivre's theorem). The set of all nth roots of unity is a *group* with multiplication as the group operation; they are n in number and equally spaced around the unit circle in the complex plane. A **primitive nth root of unity** is an nth root of unity which is not a root of unity of a lower order than n. Primitive roots are always imaginary except in the cases $n = 1$ and $n = 2$. The primitive square root of unity is -1; the primitive cube roots are $\frac{1}{2}(-1 \pm \sqrt{3}\,i)$; the primitive fourth roots are $\pm i$.

root test. A series Σa_n with nonnegative terms converges if there is a positive number $r < 1$ and a number N such that $\sqrt[n]{a_n} < r$ if $n > N$; the series diverges if $\sqrt[n]{a_n} \geq 1$ for infinitely many values of n. Consider the series,

$$1 + x + 2x^2 + 3x^3 + \cdots + nx^n + \cdots.$$

Here the nth root of the nth term is $n^{1/n}x$. Since $\lim_{n\to\infty} n^{1/n} = 1$, for any x_0 numerically less than 1, it is possible to choose an N such that $|n^{1/n}x_0| < 1$ for all $n > N$. Hence the series converges if $|x| < 1$. This test serves whenever the ratio test does, but the converse is not true. It follows from the above test that if $\lim_{n\to\infty}(a_n)^{1/n} = r$ for a series Σa_n, then the series converges if $r < 1$ and diverges if $r > 1$. If $r = 1$ no conclusion can be drawn unless $a_n^{1/n} \geq 1$ for infinitely many values of n, in which case the series diverges. *Syn.* Cauchy's root test.

ROSE, *n.* The graph in polar coordinates of $r = a\sin n\theta$, or $r = a\cos n\theta$, where n is a positive integer. It consists of rose petal-shaped loops with the origin a point common to all of them. When n is odd there are n loops; when n is even there are $2n$ loops. The **three-leafed rose** is the graph of the equation $r = a\sin 3\theta$, or $r = a\cos 3\theta$. The curve consists of three loops with their vertices at the pole. The locus of the first equation has the first petal tangent to the positive polar axis, and symmetric about the line $\theta = 30°$, the second petal symmetric about the line $\theta = 150°$, and the third symmetric about the line $\theta = 270°$, each loop thus being tangent to the sides of an angle of $60°$. The locus of the second equation is the same as that of the first rotated $30°$ about the origin. The **four-leafed rose** is the graph of the equation $r = a\sin 2\theta$ or $r = a\cos 2\theta$. The graph of the first equation (shown in the figure) has the four petals symmetric by pairs about

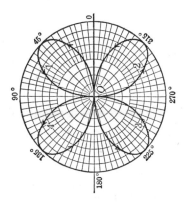

each of the lines $\theta = 45°$ and $\theta = 135°$, and tangent to the coordinate axes in the several quadrants. The graph of the second equation is the same, except that the petals are symmetric about the coordinate axes and tangent to the lines $\theta = 45°$ and $\theta = 135°$.

RO-TA'TION, *n.* See CURL.

rotation about a line. Rigid motion about the line of such a kind that every point in the figure moves in a circular path about the line in a plane perpendicular to the line.

rotation about a point. Rigid motion in a circular path (in a plane) about the point.

rotation angle. See ANGLE.

rotation of axes. A rigid motion which leaves the origin fixed. Such a transformation of axes is convenient in studying curves and surfaces, since it does not alter them intrinsically (preserves size and shape). *E.g.*, by a proper rotation of the coordinate axes in the plane, they can be made parallel to the axes of any given ellipse or hyperbola, or one of them parallel to the axis of any given parabola, thus in each case making the term which contains xy disappear. **In the plane**, the formulas for rotation (rotation formulas) which give the relations between the coordinates (x', y') of a point with reference to a set of axes obtained by rotating a set of rectangular axes through the angle θ, and the coordinates (x, y) relative to the old axes, are

$$x = x' \cos \theta - y' \sin \theta,$$

$$y = x' \sin \theta + y' \cos \theta,$$

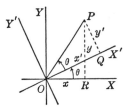

where θ is the angle ROQ. **In space**, a rotation moves the coordinate trihedral in such a way as to leave the origin fixed and the axes in the same relative position. The coordinates of a point are transformed from those referred to one system of rectangular Cartesian axes to coordinates referred to another system of axes having the same origin but different directions and making certain given angles with the original axes. If the direction angles, with respect to the old axes, of the new x-axis (the x'-axis) are A_1, B_1, C_1; of the y'-axis are A_2, B_2, C_2; and of the z'-axis, A_3, B_3, C_3, then the formulas for rotation of axes in space are

$$x = x' \cos A_1 + y' \cos A_2 + z' \cos A_3,$$

$$y = x' \cos B_1 + y' \cos B_2 + z' \cos B_3,$$

$$z = x' \cos C_1 + y' \cos C_2 + z' \cos C_3.$$

See ORTHOGONAL —orthogonal transformation.

ROTH, Klaus Friedrich (1925–   ). English number theorist and Fields Medal recipient (1958). See THUE—Thue-Siegel-Roth theorem.

RO-TUND′, *adj.* **rotund space.** See CONVEX —strictly convex space, uniformly convex space.

ROUCHÉ, Eugène (1832–1910). French algebraist, analyst, geometer, and probabilist.
   Rouché's theorem. If f and g are analytic functions of the complex variable z inside and on a simple rectifiable curve C, and if $f(z) +$ $\lambda g(z) \neq 0$ whenever $0 \leq \lambda \leq 1$ and z is on C, then f and $f + g$ have the same number of zeros in the interior of C. The hypothesis is satisfied if $|f(z)| > |g(z)|$ whenever z is on C. It also is satisfied if $f = \varphi$, $g = -\theta - \varphi$, and $|\varphi(z) + \theta(z)| < |\varphi(z)| + |\theta(z)|$ whenever z is on C. This yields the *symmetric form*: "If $|\varphi(z) + \theta(z)| < |\varphi(z)| + |\theta(z)|$ whenever z is on C, then φ and θ have the same number of zeros inside C."

ROUND, *adj.* **round angle.** An angle of $360°$. *Syn.* perigon.

ROUND′ING, *n.* **rounding off.** Dropping decimals after a certain significant place. When the first digit dropped is less than 5, the preceding digit is not changed; when the first digit dropped is greater than 5, or 5 and some succeeding digit is not zero, the preceding digit is increased by 1; when the first digit dropped is 5, and all succeeding digits are zero, the commonly accepted rule (computer's rule) is to make the preceding digit even, *i.e.*, add 1 to it if it is odd, and leave it alone if it is already even. *E.g.*, 2.324, 2.316, and 2.315 would take the form 2.32, if rounded off to two places.

ROUND-OFF ERROR. An error in computation resulting from the fact that the computation is not exact but instead is carried out to only a specified number of decimal places.

ROW, *n.* An arrangement of terms in a horizontal line. Used with *determinants* and *matrices* to distinguish horizontal arrays of elements from vertical arrays, which are called **columns**. See DETERMINANT.
   row matrix. A matrix with exactly one row. *Syn.* row vector.

RUFFINI, Paolo (1765–1822). Italian algebraist and group theorist. In 1799, published an incomplete proof that the general quintic equation cannot be solved by a finite number of algebraic operations. See ABEL.

RULE, *n.* (1) A prescribed operation or method of procedure; a formula (usually in words, although *rule* is often used synonymously with *formula*). See DESCARTES —Descartes' rule of signs, EMPIRICAL —empirical rule, L'HÔPITAL —L'Hôpital's rule, MECHANIC —mechanic's rule, MERCHANT —merchant's rule, THREE —rule of three. (2) A graduated straight edge. *Syn.* ruler.
   slide rule. See SLIDE.

RULED, *adj.* **conjugate ruled surface of a given ruled surface.** The ruled surface whose rulings are the lines tangent to the given ruled surface S, at the points of the line of striction L of S, and orthogonal to the rulings of S at the corresponding points of L.

ruled paper. Same as CROSS-SECTION PAPER.

ruled surface. A surface that can be generated by a moving straight line. The generating straight line is called the **rectilinear generator**. A **doubly ruled surface** is a ruled surface admitting two different sets of generators. Quadric surfaces are the only doubly ruled surfaces. A **skew ruled surface** is a ruled surface which is not a *developable surface* (see DEVELOPABLE). The various positions of a straight line which generate a ruled surface are the **rulings** of the surface. A **directrix** is any curve that contains at least one point of each ruling and does not have any points not on rulings. The cone, cylinder, hyperbolic paraboloid, and hyperboloid of one sheet are ruled surfaces. See RULING.

RUL′ER, *n.* A straight edge graduated in linear units. If English units are used, the *ruler* is usually a foot long, graduated to fractions of an inch. *Syn.* rule.

RUL′ING, *n.* See RULED—ruled surface.

central plane and point of a ruling. For a fixed ruling L on a ruled surface S, the **central point** is the point in the limiting position of the foot on L of the common perpendicular to L and a variable ruling L' on S, as $L' \to L$. The plane tangent to a ruled surface S at any point of a ruling L on S necessarily contains L. The plane tangent to S at the central point of L is the **central plane** of the ruling L on the ruled surface S.

RUN, *n.* A term sometimes used in speaking of the difference between the abscissas of two points. The *run* from the point whose coordinates are $(2,3)$ to the one whose coordinates are $(5,7)$ is $5-2$, or 3. The distance between the ordinates is sometimes called the **rise**. Thus the *run* squared plus the *rise* squared is equal to the square of the distance between the two points.

RUNGE, Carl David Tolmé (1856–1927). German analyst.

Runge-Kutta method. A method for obtaining an approximate solution of a differential equation of type $dy/dx = f(x, y)$. To determine an approximate solution that passes through the point (x_0, y_0), we let $x_1 = x_0 + h$ and the method

determines a corresponding $y_1 = y_0 + k$ by means of the formulas

$$k_1 = h \cdot f(x_0, y_0), \; k_2 = h \cdot f\left(x_0 + \tfrac{1}{2}h, y_0 + \tfrac{1}{2}k_1\right),$$

$$k_3 = h \cdot f\left(x_0 + \tfrac{1}{2}h, y_0 + \tfrac{1}{2}k_2\right),$$

$$k_4 = h \cdot f(x_0 + h, y_0 + k_3),$$

$$k = \tfrac{1}{6}(k_1 + 2k_2 + 2k_3 + k_4).$$

The process is then repeated, starting with (x_1, y_1). This method, which reduces to *Simpson's rule* if f is a function of x alone, can be extended to the approximate solution of systems of linear differential equations and to the solution of higher-order linear differential equations and systems of equations. The preceding is the classical Runge-Kutta method. It is a natural refinement of the Euler method (see EULER—Euler method). Many variations of this also are called Runge-Kutta methods.

RUSSELL, Bertrand Arthur William (1872–1970). Great English philosopher and logician. With Whitehead, made profound studies in the logical foundation of mathematics.

Russell's paradox. Suppose that all sets are separated into two types: A set M is of the *first type* if it does not contain M itself as a member; and it is of the *second type* if it does contain M itself as a member. **Russell's paradox** is that the set N of all sets of the first type must be of the first type, since otherwise the set N of the second type would be a member of N; but N would then be of the second type, since N itself would be a member of N. It thus appears that the concept of all sets which are not members of themselves is not free from contradiction. See BURALI-FORTI —Burali-Forti paradox.

S

SAD′DLE, *adj.* **saddle point.** A point for which the two first partial derivatives of a function $f(x, y)$ are zero, but which is not a local maximum or a local minimum. If the second-order partial derivatives are continuous in a neighborhood of p, $\dfrac{\partial f}{\partial x} = \dfrac{\partial f}{\partial y} = 0$ and $\left(\dfrac{\partial^2 f}{\partial x \, \partial y}\right)^2 - \dfrac{\partial^2 f}{\partial x^2} \dfrac{\partial^2 f}{\partial y^2} > 0$ at p, then p is a saddle point (see

MAXIMUM). At a saddle point, the tangent plane to the surface $z = f(x, y)$ is horizontal, but near the point the surface is partly above and partly below the tangent plane. Usually there are two "mountains" and two "valleys" as one circles a saddle point, hence the name. A saddle point for which there are three "valleys" and three "mountains" is a monkey saddle, there being a valley for the tail. *Syn.* minimax.

saddle-point method. See STEEPEST —method of steepest descent (2).

saddle point of a game. It is easy to see that, for any finite two-person zero-sum game, the elements a_{ij} of the *payoff matrix* satisfy the relation

$$\max_i \left(\min_j a_{ij} \right) \leq \min_j (\max_i a_{ij}).$$

If the sign of equality holds, then $\max_i(\min_j a_{ij}) = \min_j(\max_i a_{ij}) = v$ and there exist pure strategies i_0 and j_0 for the maximizing and minimizing players, respectively, such that if the maximizing player chooses i_0 then the payoff will be at least v no matter what strategy the minimizing player chooses, and if the minimizing player chooses j_0 then the payoff will be at most v no matter what strategy the maximizing player chooses. Thus

$$v = a_{i_0 j_0} = \max_i a_{ij_0} = \min_j a_{i_0 j}.$$

In this case, the game is said to have a *saddle point* at (i_0, j_0). There might be more than one saddle point at which the value v is taken on. Similar statements hold for an infinite two-person zero-sum game, for which there might or might not be a saddle point. See BOX—three boxes game, MINIMAX —minimax theorem, PAY-OFF.

saddle point of a matrix. Any finite matrix of real numbers, with element a_{ij} in the ith row and jth column, might be considered to be the *payoff matrix* of a finite two-person zero-sum game. If the game has a saddle point at (i_0, j_0), then the matrix is said to have a saddle point at (i_0, j_0). A necessary and sufficient condition that a matrix have a saddle point is that there exist an element that is both the minimum element of its row and the maximum element of its column. See above, saddle point of a game.

SAIL'ING, *n.* **middle-latitude sailing.** Approximating the difference in longitude (DL) of two places from their latitudes (L_1 and L_2) and departure (p) by the formula

$$p \sec \tfrac{1}{2}(L_1 + L_2) = DL$$

measured in minutes.

parallel sailing. Sailing on a parallel of latitude; using the above formula, putting $L_1 = L_2$.

plane sailing. Sailing on a rhumb line. The constant angle which the rhumb line makes with the meridians is called the **ship's course.** Requires solving a plane right triangle.

triangle of plane sailing. See TRIANGLE —triangle of plane sailing.

SAINT-VENANT, Adhémar Jean Claude Barré de (1797–1886). French applied mathematician and engineer who made contributions to the theory of mechanics, elasticity, hydrostatics, and hydrodynamics.

Saint-Venant's compatibility equations. See STRAIN —strain tensor.

Saint-Venant's principle. If some distribution of forces acting on a portion of the surface of a body is replaced by a different distribution of forces acting on the same portion of the body, then the effects of the two different distributions on the parts of the body sufficiently far removed from the region of application of the forces are essentially the same, provided the two distributions of forces have the same resultant force and moment.

SA'LI-ENT, *adj.* **salient angle.** See REENTRANT —reentrant angle.

salient point on a curve. A point at which two branches of a curve meet and stop and have different tangents. The curves

$$y = x/(1 + e^{1/x}) \quad \text{and} \quad y = |x|$$

have *salient points* at the origin.

SAL'I-NON, *n.* A plane figure S bounded by a semicircle C of diameter d, two small semicircles inside C with equal diameters Δ lying along the diameter of C, and another semicircle outside C and between the small semicircles with diameter $d - 2\Delta$ lying along the diameter of C. The area of S is $\tfrac{1}{4}\pi(d - \Delta)^2$. If $\Delta = \tfrac{1}{2}d$, the salinon is an arbilos. See ARBILOS.

SAL'TUS, *n.* **saltus of a function.** See OSCILLA-TION—oscillation of a function.

SAM'PLE, *n.* A finite subset of a population. See RANDOM —random sample, stratified random sample, SYSTEMATIC —systematic sample.

sample mean. See below, sample moment.

sample moment. For a random sample $\{X_1, X_2, \cdots\}$ of outcomes of an experiment, the

kth sample moment is $\sum_{i=1}^{n} (X_i)^k/n$. When $k = 1$, this becomes the **sample mean** $\sum_{i=1}^{n} X_i/n$. See MOMENT—moment of a distribution, RANDOM—random sample.

sample variance. See VARIANCE.

SAM′PLING, *adj.* **sampling error.** See ERROR.

SAT′IS-FY, *v.* (1) To fulfill the conditions of, such as to *satisfy* a theorem, a set of assumptions, or a set of hypotheses. (2) A set of values of the variables which will reduce an equation (or equations) to an identity are said to satisfy the equation (or equations); $x = 1$ *satisfies* $4x + 1 = 5$; $x = 2$, $y = 3$ *satisfy* the simultaneous equations

$$x + 2y - 8 = 0$$

$$x - 2y + 4 = 0.$$

SCA′LAR, *adj.*, *n* **scalar field.** See TENSOR.
  scalar matrix. See MATRIX.
  scalar product. See MULTIPLICATION —multiplication of vectors.
  scalar quantity. (1) The ratio between two quantities of the same kind, a number. (2) A number, as distinguished from a vector, matrix, quaternion, etc. (3) A tensor of order zero. See TENSOR. *Syn.* scalar.

SCALE, *n.* A system of marks in a given order and at known intervals. Used on rulers, thermometers, etc., as an aid in measuring various quantities.
  binary scale. Numerals written with the base two, instead of ten. Numerals in which the second digit to the left indicates the two's; the third, four's, etc.; 1101 with base 2 means $2^3 + 2^2 + 0 \times 2 + 1$ or 13 written with base ten. See BASE—base of a system of numbers.
  diagonal scale. See DIAGONAL—diagonal scale.
  drawing to scale. Making a copy of a drawing with all distances in the same ratio to the corresponding distances in the original; making a copy of a drawing of something with all distances multiplied by a constant factor. *E.g.*, an architect drawing the plan of a house lets feet in the house be denoted by inches, or fractions of an inch, in his drawing, but a bacteriologist might draw at a scale of 4000 to 1.
  logarithmic scale. See LOGARITHMIC —logarithmic coordinate paper.
  natural scale. The section of the number scale which contains positive integers only.

number scale (complete number scale). The scale formed by marking a point 0 on a line, separating the line into equal parts, and labeling the separation points to the right of 0 with the integers $1, 2, 3, \cdots$ and those to the left with the negative integers, $-1, -2, -3, \cdots$.
  scale of imaginaries. The number scale modified by multiplying each of its numbers by i $(= \sqrt{-1})$. In plotting complex numbers, the scale of imaginaries is laid off on a line perpendicular to the line which contains the real number scale. See ARGAND—Argand diagram.
  uniform scale. A scale in which equal numerical values correspond to equal distances.

SCA-LENE′, *adj.* **scalene triangle.** A triangle no two of whose sides are equal (the triangle may be either a plane triangle or a spherical triangle).

SCAT′TER, *adj.* **scatter diagram.** A diagram useful in studying the relation between two random variables with the same domains. An observation consists of values x and y of the two random variables and is plotted as a point (x, y) using ordinary rectangular axes. A set of n observations yields n points and this set of points may suggest a relationship between the random variables. *Syn.* scattergram.

SCHAUDER, Juliusz Pawel (–   ).
  Schauder's fixed-point theorem. See BROUWER—Brouwer's fixed-point theorem.

SCHERK, Heinrich Ferdinand (1798–1885). German algebraist and differential geometer.
  surface of Scherk. See under SURFACE.

SCHLÄFLI, Ludwig (1814–1895). Swiss analyst and geometer.
  Schläfli integral for $P_n(z)$. The integral

$$\frac{1}{2\pi i} \int_C \frac{(t^2 - 1)^n}{2^n (t - z)^{n+1}} \, dt = P_n(z),$$

where P_n is the *Legendre polynomial* of order n and the integration is counterclockwise around a contour C encircling the point z in the complex plane.

SCHLICHT, *adj.* (*German*) **schlicht function.** Same as SIMPLE FUNCTION. See SIMPLE.

SCHLÖMILCH, Oskar Xaver (1823–1901). German analyst.
  Schlömilch form of the remainder for Taylor's theorem. See TAYLOR—Taylor's theorem.

SCHMIDT, Erhard (1876–1959). German analyst. See GRAM—Gram-Schmidt process, HILBERT—Hilbert-Schmidt theory.

SCHNEIDER, Theodor (1911–1988). German mathematician who made important contributions to the theory of Abelian functions and integrals, Diophantine equations, and the geometry of numbers. See GELFOND.

SCHRÖDER, Ernst (1841–1902). German algebraist and logician.

Schröder-Bernstein theorem. If there exists a one-to-one correspondence between a set A and a subset of B, and between B and a subset of A, then there exists a one-to-one correspondence between A and B.

SCHUR, Friedrich Heinrich (1856–1932). German differential geometer.

Schur theorem. If the *Riemannian curvature* k of an n-dimensional ($n \geq 2$) Riemannian space is independent of the orientation ξ_1^i, ξ_2^i, then k does not vary from point to point. With the aid of Schur's theorem it follows that a necessary and sufficient condition that an n-dimensional ($n \geq 2$) Riemannian space be of constant Riemannian curvature k is that the metric tensor g_{ij} satisfy the system of second-order partial differential equations

$$R_{\alpha\beta\gamma\delta} = k(g_{\alpha\beta}g_{\beta\gamma} - g_{\alpha\gamma}g_{\beta\delta}).$$

SCHUR, Issai (1875–1941). German algebraist and number theorist.

Shur lemma. (1) Let S_1 and S_2 be two *irreducible* collections of matrices, corresponding to linear transformations of vector spaces of dimensions n and m, respectively. If there is an $n \times m$ matrix P such that for any A of S_1 there is a B of S_2, and for any B of S_2 an A of S_1, such that $AP = PB$, then either P has all elements zero or P is *square* and *nonsingular*. In the latter case, the two collections S_1 and S_2 are *equivalent* (for any B of S_2 there is an A of S_1 such that $B = P^{-1}AP$). (2) If M is an irreducible module over a ring R and there are members r of R and m of M for which $rm \neq 0$, then the ring of homomorphisms of M into M is a division ring.

SCHWARTZ, Laurent (1915–   ). French functional analyst, topologist, and Fields Medal recipient (1950). Works in mathematical physics and theory of distributions. See GENERALIZED—generalized function.

SCHWARZ, Hermann Amandus (1843–1921). German mathematician who worked in the theory of functions of a complex variable, minimal surfaces, and calculus of variations.

Schwarz inequality. (1) The square of the integral of the product of two real functions over a given interval or region is equal to, or less than, the product of the integrals of their squares over the same interval or region, provided these integrals exist. For complex functions, $f(z)$ and $g(z)$,

$$\left| \int_{z_1}^{z_2} \bar{f}g\, dz \right|^2 \leq \left[\int_{z_1}^{z_2} \bar{f}f\, |dz| \right] \left[\int_{z_1}^{z_2} \bar{g}g\, |dz| \right],$$

where $\bar{f}$ and $\bar{g}$ are the complex conjugates of f and g. This inequality is easily deduced from the *Cauchy inequality* (see CAUCHY). It is also called the *Cauchy-Schwarz inequality* and the *Buniakovski inequality* (Buniakovski called attention to it earlier than Schwarz). (2) For a *vector space* with an *inner product* (x, y) *defined*, the inequality $|(x, y)| \leq \|x\| \cdot \|y\|$ is called Schwarz's inequality. For suitable representations of Hilbert space, this inequality is equivalent to the above inequality and to Cauchy's inequality.

Schwarz's lemma. If the function f of the complex variable z is analytic for $|z| < 1$, with $|f(z)| < 1$ for $|z| < 1$, and $f(0) = 0$, then either $|f(z)| < |z|$ for $0 < |z| < 1$ and $|f'(0)| < 1$, or $f(z) = e^{i\theta}z$, where θ is a real constant.

SCI-EN-TIF′IC, *adj.* **scientific notation.** Decimal numerals written as the product of a power of 10 and a decimal numeral between 1 and 10, all significant digits being written. *E.g.*, 297.2 and 0.00029 would be written as $2.972 \cdot 10^2$ and $2.9 \cdot 10^{-4}$; 697000 would be written as $6.970 \cdot 10^5$ if there are four significant digits.

SCORE, *n.* Twenty.

SCRAP, *n.* **scrap value of equipment.** Its sale value when it is no longer useful. *Syn.* salvage value.

SE′CANT, *adj., n.* (1) A line of unlimited length cutting a given curve. (2) One of the trigonometric functions. See TRIGONOMETRIC —trigonometric functions.

secant curve. The graph of $y = \sec x$. Between $-\frac{1}{2}\pi$ and $\frac{1}{2}\pi$, it is concave up. It is asymptotic to the lines $x = -\frac{1}{2}\pi$ and $x = \frac{1}{2}\pi$, and has its y-intercept equal to unity. Similar arcs appear in other intervals of length π radians, being alternately concave upward and concave downward.

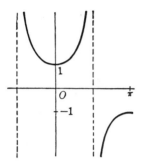

SEC'OND, *adj., n.* **second of angle.** One-sixtieth of a minute and one thirty-six hundredth part of a degree. Denoted by a double accent, as 10″, read ten seconds. See SEXAGESIMAL—sexagesimal measure of an angle.

second derivative. The derivative of the first derivative. See DERIVATIVE—derivatives of higher order.

second mean-value theorem. See MEAN—mean-value theorems for derivatives, mean-value theorems for integrals.

second moment. Same as MOMENT OF INERTIA.

second of time. Approximately $1/86,400$ part of the *mean solar day*; 9,192,631,770 periods of the radiation corresponding to the transition between the two hyperfine levels of the ground state of the cesium-133 atom when unperturbed by exterior fields. See METRIC—metric system, SI, TIME.

SEC'OND-AR'Y, *adj.* **secondary diagonal of a determinant.** See DETERMINANT.

secondary parts of a triangle. Parts other than the sides and interior angles, such as the altitude, exterior angles, and medians. See PRINCIPAL—principal parts of a triangle.

SEC'TION, *n.* **golden and harmonic section of a line.** See GOLDEN, HARMONIC—harmonic division of a line.

method of sections. A method for graphing a surface. Consists of drawing sections of the surface (usually those made by the coordinate planes and planes parallel to them) and inferring the shape of the surface from these sections.

plane section. The plane geometric configuration obtained by cutting any configuration by a plane. A plane section made by a plane containing a normal to the surface is a **normal section**. A **meridian section** of a surface of revolution is a plane section made by a plane containing the axis of revolution. A **right section** of a cylinder is a plane section by a plane perpendicular to the elements of the cylinder, or to the lateral faces of the prism.

section of a polyhedral angle. See ANGLE—polyhedral angle.

SEC'TOR, *n.* **sector of a circle.** A portion of a circle bounded by two radii of the circle, and one of the arcs which they intercept. The shorter arc is the **minor arc**. The longer is the **major arc**. The area of a sector is $\frac{1}{2}r^2\phi$, where r is the radius of the circle and ϕ the angle in radian measure subtended at the center of the circle by the arc of the sector.

spherical sector. A solid generated by rotating a sector of a circle about a diameter. Some writers require that this diameter not lie in the sector, while some require that it contain one of the radii bounding the circular sector. Most writers do not restrict the diameter at all, including both of the above cases as spherical sectors. The figure shows a sector of a circle and the spherical sector resulting from rotating it about a diameter (the dotted line). The volume of a spherical sector is equal to the product of the radius of the sphere and one-third the area of the zone which forms the **base** of the sector; or $\frac{2}{3}\pi r^2 h$, where r is the radius of the sphere and h the altitude of the zone (see ZONE).

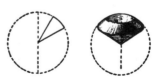

SEC'U-LAR, *adj.* **secular trend.** See TREND.

SE-CU'RI-TY, *n.* Property, or written promises to pay, such as notes and mortgages, used to guarantee payment of a debt. See COLLATERAL.

SEG′MENT, *n.* A part cut off from any figure by a line or plane (or planes). Used most commonly when speaking of a limited piece of a line or of an arc of a curve. See below, segment of a curve, and segment of a line.

addition of line segments. See SUM—sum of directed line segments.

segment of a curve. (1) The part of the curve between two points on it. (2) The region bounded by a chord and the arc of the curve subtended by the chord. A **segment of a circle** is the region between a chord and an arc subtended by the chord. Any chord bounds two segments, which are different in area except when the chord is a diameter. The longer and shorter segments are the **major** and **minor segments**, respectively. The area of a segment of a circle is $\frac{1}{2}r^2(\beta - \sin\beta)$, where r is the radius of the circle and β the angle in radians subtended at the center of the circle by the arc. See figure under SECTOR.

segment of a line or **line segment.** The part of a straight line between two points. The segment may include one or both of the points. A line segment whose end points are identical is a **nil segment.** See DIRECTED—directed line.

spherical segment. The solid bounded by a sphere and two parallel planes intersecting, or tangent to, the sphere, or by a *zone* and the planes (or plane) of its bases (or base). If one plane is tangent to the sphere, the segment is a *spherical segment of one base*; otherwise it is a *spherical segment of two bases.* The **bases** are the intersections of the parallel planes with the solid bounded by the sphere; the **altitude** is the perpendicular distance between these planes. The volume of a spherical segment is equal to

$$\tfrac{1}{6}\pi h\left(3r_1^2 + 3r_2^2 + h^2\right),$$

spherical segment
of one base   of two bases.

where h is the altitude and r_1 and r_2 are the radii of the bases. The formula for the volume of a segment of one base is derived by making one of these r's, say r_2, zero.

SEGRE, Corrado (1863–1924). Italian algebraist and geometer.

Segre characteristic of a matrix. See CANONICAL—canonical form of a matrix.

SELBERG, Atle (1917–   ). Norwegian-American number theorist, analyst, and Fields Medal recipient (1950). Contributed basic results regarding the Riemann zeta function, but proved the prime-number theorem without its use. See PRIME—prime-number theorem.

SE-LECT′, *adj.* **select mortality table.** See MORTALITY—mortality table.

SELF, *pref.* **self-adjoint transformation.** A linear transformation which is its own *adjoint.* For finite-dimensional spaces, a transformation T, which transforms vectors $x = (x_1, x_2, \cdots, x_n)$ into $Tx = (y_1, y_2, \cdots, y_n)$ with $y_i = \sum_j a_{ij}x_j$ for each i, is self-adjoint if and only if the matrix (a_{ij}) of its coefficients is a *Hermitian matrix.* If (x, y) denotes the inner product of elements x and y of a Hilbert space H, then a bounded linear transformation T of H into H is self-adjoint if and only if $(Tx, y) = (x, Ty)$ for any x and y in H. Any bounded linear transformation T of a (complex) Hilbert space (with domain the entire space) can be uniquely written in the form $T = A + iB$, where A and B are self-adjoint transformations. *Syn.* Hermitian transformation. See SPECTRAL—spectral theorem, SYMMETRIC—symmetric transformation.

SELL′ING, *adj.* **selling price.** See PRICE.

percent profit on selling price. See PERCENT.

SEM′I, *pref.* Meaning half; partly; somewhat less than; happening or published twice in an interval or period. A **semicircle** is one-half of a circle (either of the parts of a circle which are cut off by a diameter); a **semicircumference** is one-half of a circumference.

semiannual. Twice a year.

semiaxis. A line segment that has one end point at the center and is half of an axis of an ellipse, ellipsoid, hyperbola, etc. See ELLIPSE, ELLIPSOID, HYPERBOLA.

semicontinuous function. See CONTINUOUS—semicontinuous function.

semicubical parabola. See PARABOLA—cubical parabola.

semigroup. An associative groupoid, *i.e.,* a set G with a binary operation (which we shall call multiplication) whose domain is the set of ordered pairs of members of G, whose range is contained in G, and which is associative, *i.e.,* $a(bc) = (ab)c$ for any members a, b and c. A semigroup is Abelian (or commutative) if $ab = ba$ for any members a and b. Sometimes a cancellation law is assumed (that $x = y$ if there is a member z for which $xz = yz$ or $zx = zy$). A semigroup with a finite number of members satisfies this cancellation law if and only if it is a *group*. A semigroup with an identity element is a *monoid*.

semilogarithmic graphing. Graphing in a plane for which a logarithmic scale is used on one axis and a uniform scale on the other.

semiregular solid. See ARCHIMEDES —Archimedean solid.

semiring. See RING—semiring of sets.

SENSE, *n.* **sense of an inequality.** See INEQUALITY.

SEN-IOR'I-TY, *n.* **law of uniform seniority.** The following law used in evaluating joint life insurance policies: The difference between the age that can be used in computation (instead of the actual ages) and the lesser of the unequal ages is the same for the same difference of the unequal ages, regardless of the actual ages. (The age used in computation instead of the actual ages is the age which two persons of the same age would have if they were given insurance identical with that given those with the different ages.)

SEN'SI-TIV'I-TY, *adj.* **sensitivity analysis.** An analysis of the variation of the solution of a problem with variations in the values of the parameters involved.

SEN'TENCE, *n.* **numerical sentence.** See NUMERICAL.

open sentence. Same as PROPOSITIONAL FUNCTION.

SEP'A-RA-BLE, *adj.* **separable extension of a field.** Let F^* be a field that contains a field F. Then a member c of F^* is **separable** with respect to F if c is a zero of a separable polynomial with coefficients in F. The extension F^* is **separable** if all members of F^* are separable. A **perfect field** is a field all of whose finite extensions are separable, *i.e.*, no irreducible polynomial with coefficients in the field has a multiple zero.

separable polynomial. A polynomial that has no multiple zeros; *i.e.*, it has n distinct roots in the Galois field of the polynomial. A polynomial with coefficients in a field F is separable if and only if the greatest common divisor of f and its *formal derivative* f' is a constant.

separable space. A (topological) space which contains a countable (or finite) set W of points which is *dense* in the space; *i.e.*, every neighborhood of any point of the space contains a point of W. A space which satisfies the *second axiom of countability* is separable. Such a space is

sometimes said to be **completely separable** or **perfectly separable.** A separable metric space is perfectly separable. Hilbert space and Euclidean space of n dimensions are separable.

SEP'A-RA'TION, *n.* **separation of a set.** The separation of the set into two classes. A separation of an ordered set (such as the real numbers or the rational numbers) of the **first kind** is a separation for which each member of one class is less than every member of the other class and the separating number belongs to one or the other of the classes. The number 3 may be thought of as separating all rational numbers into those less than or equal to 3 and those greater than 3. A separation of an ordered set of the **second kind** is a separation for which each member of one class is less than every member of the other and there is no greatest member of the class of lesser objects and no least in the class of greater objects. The separation of the rational numbers into the sets A and B, where x is in A if $x \leq 0$, and each positive x is in A or B according as $x^2 < 2$ or $x^2 > 2$, is of the second kind. See DEDEKIND —Dedekind cut.

separation of variables. See DIFFERENTIAL—differential equation with variables separable.

Sturm separation theorem. See STURM.

SEP'A-RA'TRIX, *n.* Something that separates; a comma that separates a number into periods, as in 234,569; a space that separates a number into periods, as in 234 569. A decimal point is sometimes called a *separatrix*.

SEP-TIL'LION, *n.* (1) In the U.S. and France, the number represented by one followed by 24 zeros. (2) In England, the number represented by one followed by 42 zeros.

SE'QUENCE, *n.* A set of quantities ordered as are the positive integers. The sets

$$\left\{ 1, \frac{1}{2}, \frac{1}{3}, \cdots, \frac{1}{n} \right\}$$

and $\{x, 2x^2, 3x^3, \cdots, nx^n\}$ are sequences. These are **finite sequences**, terminating with the nth term. An **infinite sequence** is nonterminating, there being another term after each term. An infinite sequence often is called a **sequence** and is written

$$\{a_1, a_2, a_3, \cdots, a_n, \cdots\}, \{a_n\}, \quad \text{or} \quad (a_n).$$

Tech. A **finite sequence** with n terms is a function whose domain is the set of integers $\{1, 2, 3, \cdots, n\}$; if a denotes the function, then $a(k)$, or a_k, denotes the kth term. An **infinite sequence** is a function whose domain is the set of positive integers.

accumulation point of a sequence. A point P such that there are an infinite number of terms of the sequence in any neighborhood of P; *e.g.*, the sequence

$$\left\{1, \tfrac{1}{2}, 1, \tfrac{1}{3}, 1, \tfrac{1}{4}, 1, \tfrac{1}{5}, \cdots\right\}$$

has two accumulation points, the numbers 0 and 1. If, for any number M, there are an infinite number of terms of a sequence of real numbers which are greater (less) than M, then $+\infty$ ($-\infty$) is said to be an accumulation point of the sequence. An accumulation point of a sequence is also called a **cluster point**, or **limit point**, of the sequence. For a sequence of real numbers, the largest accumulation point is also called the **limit superior** (or **greatest of the limits**, or **maximum limit**) and is the number L (or $\pm \infty$) which is the largest number such that there are an infinite number of terms of the sequence greater than $L - \epsilon$ for any positive ϵ ($L = +\infty$ or $L = -\infty$ according as for each number c infinitely many of the terms are larger than c, or for each number c only finitely many of the terms are larger than c); the smallest accumulation point is also called the **limit inferior** (or **least of the limits**, or **minimum limit**) and is the number l (or $\pm\infty$) which is the least number such that there are an infinite number of terms of the sequence less than $l + \epsilon$ for any positive ϵ ($l = +\infty$ or $l = -\infty$ according as for each number c only finitely many of the terms are less than c, or for each number c infinitely many of the terms are less than c). The *limit superior* (*limit inferior*) is the limit of the *upper* (*lower*) *bounds* of the numbers in the subsequences

$$\begin{aligned} &a_1, a_2, a_3, \cdots, a_n, \cdots \\ &a_2, a_3, a_4, \cdots, a_{n+1}, \cdots \\ &a_3, a_4, a_5, \cdots, a_{n+2}, \cdots \end{aligned}$$

$$\cdots \cdots \cdots \cdots \cdots \cdots$$

The *limit superior* and *limit inferior* are not always the *least upper* and *greatest lower bounds* of a sequence. The limit superior and limit inferior of the sequence

$$\left\{2, -\tfrac{3}{2}, \tfrac{4}{3}, \cdots, (-1)^{n-1}(1 + 1/n), \cdots\right\}$$

are 1 and -1, while the upper and lower bounds are 2 and $-\tfrac{3}{2}$. The limit superior and limit inferior of any sequence, $\{a_n\}$, are denoted respectively by

$$\overline{\lim_{n \to \infty}} \, a_n \quad \text{and} \quad \underline{\lim_{n \to \infty}} \, a_n,$$

or by $\limsup_{n \to \infty} a_n$ and $\liminf_{n \to \infty} a_n$. Either limit is denoted by

$$\overline{\lim_{n \to \infty}} \, a_n.$$

When these two limits are the same the sequence has a *limit* (see below, limit of a sequence).

arithmetic sequence. See ARITHMETIC.

bound to a sequence. An **upper bound** (**lower bound**) to a sequence of real numbers is a number which is equal to or greater than (equal to or less than) every number in the sequence. A sequence that has both an upper bound and a lower bound is a **bounded sequence**. The smallest upper bound is the **least upper bound** (sometimes simply **the upper bound**) and is the largest term in the sequence if there is a largest, otherwise a number, L, such that there are terms between $L - \epsilon$ and L for every $\epsilon > 0$ but no terms greater than L. The largest lower bound is the **greatest lower bound** (sometimes simply **the lower bound**) and is the least term, or if there is no least, then a number, l, such that there are terms of the sequence between l and $l + \epsilon$ for every $\epsilon > 0$, but no terms less than l. Sometimes **limit** is used in place of bound in the above expressions.

Cauchy sequence. A sequence of points $\{x_1, x_2, \cdots\}$ such that for any $\epsilon > 0$ there is a number N for which $\rho(x_i, x_j) < \epsilon$ if $i > N$ and $j > N$, where $\rho(x_i, x_j)$ is the distance between x_i and x_j. If the points are points of Euclidean space, this is equivalent to the sequence being convergent. If the points are real (or complex) numbers, then $\rho(x_i, x_j)$ is $|x_i - x_j|$ and the sequence is convergent if and only if it is a Cauchy sequence. *Syn.* convergent sequence, fundamental sequence, regular sequence. See CAUCHY — Cauchy's condition for convergence of a sequence, COMPLETE — complete space.

cluster point of a sequence. See above, accumulation point of a sequence.

convergent and divergent sequences. See below, limit of a sequence.

geometric sequence. See GEOMETRIC.

integral of the limit of a sequence. See BOUNDED — bounded convergence theorem, MONOTONE — monotone convergence theorem,

LEBESGUE —Lebesgue convergence theorem, SE-RIES—integration of an infinite series.

limit of a sequence. A sequence of numbers $\{s_1, s_2, s_3, \cdots, s_n, \cdots\}$ has the limit s if, for any prescribed accuracy, there is a position in the sequence such that all terms after this position approximate s within this prescribed accuracy; i.e., for any $\epsilon > 0$ there exists an N such that $|s - s_n| < \epsilon$ for all n greater than N. A sequence of points $\{p_1, p_2, p_3, \cdots\}$ has the limit p if, for each neighborhood U of p, there is a number N such that p_n is in U if $n > N$. A sequence which has a limit is **convergent**; otherwise, it is **divergent**. A sequence of numbers $\{s_1, s_2, \cdots\}$ is convergent if and only if the series

$$s_1 + (s_2 - s_1) + (s_3 - s_2)$$

$$+ \cdots + (s_n - s_{n-1}) + \cdots$$

has a *sum*. See SUM—sum of an infinite series. A **convergent sequence of sets** is a sequence of sets whose inferior and superior limits are equal, i.e., a sequence of sets for which an object belongs to infinitely many of the sets if and only if it belongs to all but a finite number. The **limit of a convergent sequence of sets** is the set consisting of all objects that belong to infinitely many of the sets.

monotonic (or monotone) sequence. See MONOTONIC.

random sequence. See RANDOM.

regular sequence. See above, Cauchy sequence.

SE-QUEN′TIAL, *adj.* **sequential analysis.** The analysis of observations obtained sequentially. Sequential analysis is particularly useful for testing hypotheses, often requiring much less sampling than methods using fixed sample sizes. When testing a hypothesis H_0 versus a hypothesis H_1, the experimenter decides after each observation, and on the basis of predetermined rules, whether to accept H_0, or to accept H_1, or to use another observation. The **sequential probability ratio test** often is used: if the distribution functions are discrete and $f_0(x_i)$ and $f_1(x_i)$ are the probabilities of x_i if H_0 is true or if H_1 is true, respectively, let

$$\lambda_n = \frac{f_1(x_1) f_1(x_2) \cdots f_1(x_n)}{f_0(x_1) f_0(x_2) \cdots f_0(x_n)},$$

where the numerator and denominator are the probabilities of obtaining the observations $\{x_1, x_2, \cdots, x_n\}$ if H_1 is true or if H_0 is true, respectively. For continuous distributions, f_0 and

f_1 should be the *probability-density functions*. If it is decided α is to be the probability of erroneously accepting H_1 and β is to be the probability of erroneously accepting H_0, then there are numbers c and d, usually approximated by $\beta/(1 - \alpha)$ and $(1 - \beta)/\alpha$, such that H_0 is accepted if $\lambda_n \leq c$, H_1 is accepted if $\lambda_n \geq d$, and an additional observation is taken if $c < \lambda_n < d$. See HYPOTHESIS —test of a hypothesis.

sequential compactness. See COMPACT.

SE′RI-AL, *adj.* **serial bond.** See BOND.

serial plan of building and loan association. See BUILDING —building and loan association.

SE′RIES, *n.* The indicated sum of a finite or infinite sequence of terms. It is a **finite** or an **infinite** series according as the number of terms is finite or infinite. An infinite series can be written in the form

$$a_1 + a_2 + a_3 + \cdots + a_n + \cdots,$$

or Σa_n, where a_n is the **general term** or the **nth term.** *Infinite series* is usually shortened to *series*, as in *convergent series, Taylor series*, etc. An infinite series need not have a sum; it is a **convergent series** if it has a sum and a **divergent series** if it does not (see DIVERGENT, SUM—sum of an infinite series, and various headings under CONVERGENCE). A series is **a positive series** (or **a negative series**) if its terms are all positive (or all negative) real numbes. See GEOMETRIC —geometric sequence and series, and various headings below.

Abel theorem on power series. See ABEL.

addition of infinite series. The addition of corresponding terms of the two series. If two convergent series of constant terms,

$$a_1 + a_2 + a_3 + \cdots + a_n + \cdots$$

and

$$b_1 + b_2 + b_3 + \cdots + b_n + \cdots,$$

have the sums S and S', then the series

$$(a_1 + b_1) + (a_2 + b_2) + (a_3 + b_3) + \cdots$$

$$+ (a_n + b_n) + \cdots$$

converges and has the sum $S + S'$. If the series

$$u_1 + u_2 + u_3 + \cdots + u_n + \cdots$$

and the series

$$v_1 + v_2 + v_3 + \cdots + v_n + \cdots,$$

whose terms are functions of x, converge in certain intervals, the term by term sum of these series, namely

$$(u_1 + v_1) + (u_2 + v_2) + (u_3 + v_3) + \cdots$$
$$+ (u_n + v_n) + \cdots,$$

converges in any interval common to the two intervals.

alternating series. See ALTERNATING.

arithmetic series. See ARITHMETIC—arithmetic series.

asymptotic series. See ASYMPTOTIC—asymptotic expansion.

autoregressive series. See AUTOREGRESSIVE.

binomial series. See BINOMIAL—binomial series.

differentiation of an infinite series. The term-by-term differentiation of the series. This is permissible, i.e., the resulting series represents the derivative of the function represented by the given series in the same interval, if the resulting series is uniformly convergent in this interval. This condition is always satisfied by a **power series** in any interval within its interval of convergence; e.g., the series

$$x - \frac{x^2}{2} + \frac{x^3}{3} - \cdots + (-1)^{n+1}\frac{x^n}{n} + \cdots$$

converges for $-1 < x \leq 1$ and represents $\log(1 + x)$ in this interval; the derived series

$$1 - x + x^2 - \cdots \pm x^{n-1} \mp \cdots$$

converges uniformly for $-a < x < a$ if $a < 1$, and represents

$$\frac{1}{1 + x}$$

in any such interval.

division of two power series. The division of the two series as if they were polynomials arranged in ascending powers of the variable. Their quotient converges and represents the quotient of the sums of the series for all values of the variable within a region of convergence common to both their regions and numerically less than the numerically least real or complex number for which the series in the denominator is zero.

double series. Consider an array of the form

$$u_{1,1}\ u_{1,2}\ u_{1,3}\ u_{1,4}\ \cdots$$
$$u_{2,1}\ u_{2,2}\ u_{2,3}\ u_{2,4}\ \cdots$$
$$u_{3,1}\ u_{3,2}\ u_{3,3}\ u_{3,4}\ \cdots\ .$$

Let the sum of the terms in the rectangular array formed by the first n terms of each of the first m rows be denoted by $S_{m,n}$. If $S_{m,n}$ approaches S as m and n increase, then S is the sum of the series. *Tech.* **A convergent double series** is a double series for which there is a number S (the sum) such that for any positive number ϵ there is an integer K for which $|S - S_{m,n}| < \epsilon$ if $m > K$ and $n > K$. If an infinite series is formed from the terms in each row (or column), then the infinite series consisting of the sums of these series is the **sum by rows** (or the **sum by columns**) of the double series. If S exists (as defined above) and the sum by rows and the sum by columns exist, then these three sums are equal. This is known as **Pringsheim's theorem**.

entire series. See ENTIRE.

Euler transformation of series. See EULER.

exponential series. See EXPONENTIAL—exponential series.

factorial series. The series

$$1 + \frac{1}{1!} + \frac{1}{2!} + \frac{1}{3!} + \frac{1}{4!} + \cdots + \frac{1}{n!} + \cdots.$$

The sum of this series is the number e. See e.

Fourier series. See FOURIER.

geometric series. See GEOMETRIC—geometric series.

harmonic series. See HARMONIC—harmonic series.

hypergeometric series. See HYPERGEOMETRIC.

integration of an infinite series. The term-by-term integration (definite integration) of an infinite series. Any series of continuous functions which converges uniformly on an interval may be integrated term by term and the result will converge and equal the integral of the function represented by the original series, provided the limits of integration are finite and lie within the interval of uniform convergence. Any **power series** satisfies this condition in any interval within its interval of convergence and may be integrated term by term provided the limits of integration lie within the interval of convergence. The series

$$1 - x + x^2 - \cdots (-1)^{n+1}x^{n-1} \cdots$$

converges when $|x| < 1$. Hence term by term integration is permissible between the limits 0 and $\frac{1}{2}$, for instance, or between x_1 and x_2, provided $|x_1| < 1$ and $|x_2| < 1$. Actually, one of x_1 or x_2 can be 1. This is a special case of the following more general theorem: Let $S_n(x)$ be the sum of the first n terms of an infinite series for which there is a set of *measure zero* such that, on the complement of this set in the interval $[a, b]$, $|S_n(x)|$ is uniformly bounded and the series is convergent to a sum $S(x)$. If $\int_a^b S(x)\,dx$ and $\int_a^b S_n(x)\,dx$ exist for each n, then $\lim \int_a^b S_n(x)\,dx = \int_a^b S(x)\,dx$. If Lebesgue (instead of Riemann) integration is used, then it is not necessary to assume the existence of $\int_a^b S(x)\,dx$ and the assumption of the existence of $\int_a^b S_n(x)\,dx$ can be replaced by the assumption that each S_n is *measurable*. See BOUNDED — bounded convergence theorem, and LEBESGUE —Lebesgue convergence theorem.

Laurent series. See LAURENT.

logarithmic series. The expansion in a Taylor series of $\log(1 + x)$, namely,

$$x - x^2/2 + x^3/3 - x^4/4 + \cdots$$

$$+ (-1)^{n+1} x^n/n \cdots .$$

From this series is derived the relation

$$\log(n+1)$$

$$= \log n + 2\left[(2n+1)^{-1} + \tfrac{1}{3}(2n+1)^{-3} \right.$$

$$\left. + \tfrac{1}{5}(2n+1)^{-5} + \cdots \right],$$

which is convenient for approximating logarithms of numbers because it converges rapidly.

Maclaurin series. See TAYLOR —Taylor's theorem.

multiplication of infinite series. The multiplication of the series as if they were polynomials, multiplying each term of one series by all the terms of the other. If each series converges absolutely, the terms of the product series have a sum equal to the product of the sums of the given series, whatever the order of the terms in the product series. This need not be the case if one series is conditionally convergent. The **Cauchy product** (usually called the *product*) of two series $a_1 + a_2 + a_3 + \cdots$ and $b_1 + b_2 + b_3 + \cdots$ is the series $c_1 + c_2 + c_3 + \cdots$ for which

$$c_n = a_1 b_n + a_2 b_{n-1} + \cdots + a_n b_1$$

is the sum of all products $a_i b_j$ for which $i + j = n + 1$. For power series, the nth term of a product is the sum of all terms of degree $n - 1$ which are products of a term of one series by a term of the other. If two series are convergent and one (or both) is absolutely convergent, then their Cauchy product is convergent and has a sum which is the product of the sums of the given series. Also, if two series and their Cauchy product are convergent, then the sum of the Cauchy product is the product of the sums of the given series. A **power series** converges absolutely within its interval of convergence; hence two power series can always be multiplied, and the result will be valid within their common interval of convergence.

oscillating series. See DIVERGENT —divergent series.

p series. The series

$$1 + \left(\tfrac{1}{2}\right)^p + \left(\tfrac{1}{3}\right)^p + \cdots + (1/n)^p + \cdots .$$

It is of importance in applying the *comparison test*, since it converges for all values of p greater than one and diverges for p equal to or less than one. When $p = 1$, it is the *harmonic series*.

power series. A series whose terms contain ascending positive integral powers of a variable, a series of the form

$$a_0 + a_1 x + a_2 x^2 + \cdots + a_n x^n + \cdots ,$$

where the a's are constants and x is a variable; or a series of the form

$$a_0 + a_1(x - h) + a_2(x - h)^2 + \cdots$$

$$+ a_n(x - h)^n + \cdots .$$

A **formal power series** is a power series $a_0 + a_1 x + a_2 x^2 + \cdots$ for which one is not concerned about convergence. Such series can be added by adding corresponding terms and multiplied by multiplying each term of one series by each term of the other [see above, multiplication of infinite series]. The set of all formal power series in x is a *commutative ring with unit element*. Any power series F with a nonzero constant term is a *unit*, since one can by formal division obtain a power series F^{-1} for which $FF^{-1} = 1$; thus, two power series F and G are *associates* if there is a power series E such that $F = GE$ and E has a nonzero constant term. These concepts can be extended to several variables, a power series in $x_1, x_2, \cdots, x_n$ being an indicated sum of type $\sum_{p=0}^{\infty} F_p(x_1, \cdots, x_n)$, where F_p is a homogeneous polynomial of degree p in $x_1, \cdots, x_n$. See TAYLOR —Taylor's theorem.

rearrangement of the terms of a series. Defining another series which contains all the terms of the original series, but not necessarily in the same order. *I.e.*, for any n the first n terms of the new series are all terms of the old series, and the first n terms of the old series are all terms of the new series. If a series is absolutely convergent, all the rearrangements have the same sum. If it is conditionally convergent, rearrangements can be made such as to give any arbitrary sum, or to diverge.

reciprocal series. A series whose terms are each reciprocals of the corresponding terms of another series, of which it is the *reciprocal series*.

remainder of an infinite series. See REMAINDER—remainder of an infinite series.

reversion of a series. See REVERSION.

sum of an infinite series. See SUM.

Taylor series. See TAYLOR—Taylor's theorem.

telescopic series. A series such as

$$\frac{1}{k(k+1)} + \frac{1}{(k+1)(k+2)} + \cdots$$

$$+ \frac{1}{(k+n-1)(k+n)} + \cdots,$$

where k is not a negative integer, which can be "telescoped" by writing it in the form

$$\left[\frac{1}{k} - \frac{1}{k+1}\right] + \left[\frac{1}{k+1} - \frac{1}{k+2}\right] + \cdots$$

$$+ \left[\frac{1}{k+n-1} - \frac{1}{k+n}\right] + \cdots$$

which sums to $\dfrac{1}{k}$.

time series. See TIME—time series.

trigonometric series. See TRIGONOMETRIC—trigonometric series.

two-way series. A series of the form

$$\cdots + a_{-2} + a_{-1} + a_0 + a_1 + a_2 + \cdots, \quad \text{or}$$

$$\sum_{n=-\infty}^{\infty} a_n.$$

See LAURENT—Laurent series.

SER'PEN-TINE, *adj.* **serpentine curve.** The curve defined by the equation $x^2 y + b^2 y - a^2 x = 0$. It is symmetric about the origin, passes through the origin, and has the x-axis as an asymptote.

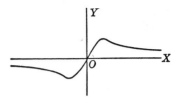

SERRE, Jean-Pierre (1926–   ). French analyst, topologist, and Fields Medal recipient (1954). Works in complex-variable theory in terms of cohomology in a complex-analytic sheaf.

SERRET, Joseph Alfred (1819–1885). French mathematician and astonomer. See FRENET.

SER'VO-MECH'A-NISM, *n.* An amplifying device that effects a certain relation between an input signal and an output signal. Examples are steering devices, automatic stabilizers, and components of computing machines.

hunting of a servomechanism. The output of a servomechanism is designed to follow the instructions of the input. Errors (or deviations) in the output, which ideally should be self-correcting, are **hunting motions**.

SET, *n.* A collection of particular things, as the *set* of numbers between 3 and 5, the *set* of points on a segment of a line, or within a circle, etc. See COMPLEMENT, INTERSECTION, SUBSET, UNION.

bounded set of numbers. A set such that the absolute value of each of its members is less than some constant. All proper fractions consititue a bounded set, for they are all less than 1 in absolute value.

bounded set of points. See under BOUNDED.

F_σ, G_δ, and Borel sets. See BOREL—Borel set.

finite and infinite sets. See FINITE, INFINITE.

fuzzy set. See FUZZY.

ordered set. See ORDERED.

SEX-A-DEC'I-MAL, *adj.* **sexadecimal number system.** A system of numerals for representing real numbers for which sixteen is the base. *Syn.* hexadecimal number system.

SEX-A-GES'I-MAL, *adj.* Pertaining to the number sixty.

sexagesimal measure of an angle. The system in which one complete revolution is divided into 360 parts, written 360° and called **degrees**; one degree into 60 parts, written 60′ and called **minutes**; and one minute into 60 parts, written 60″ and called **seconds**. See RADIAN.

sexagesimal system of numbers. A number system using sixty for a base. See BABYLONIAN, BASE—base of a number system.

SEX'TIC, *adj.* Of the sixth degree; of the sixth order (when speaking of curves or surfaces). A **sextic curve** is an algebraic curve of the sixth order; a **sextic equation** is a polynomial equation of the sixth degree.

SEX-TIL'LION, *n.* (1) In the U.S. and France, the number represented by one followed by 21 zeros. (2) In England, the number represented by one followed by 36 zeros.

SHANNON, Claude Elwood (1916–). American applied mathematician who contributed to theories of Boolean algebra, cryptography, communications, and computing machines. Founded information theory. See IN-FORMATION —information theory.

SHEAF, *n.* **sheaf of planes.** All the planes that pass through a given point. The point is the **center** of the sheaf. The equation of any plane in the sheaf can be found by multiplying the equations of three planes of the sheaf not having a line in common by different parameters (arbitary constants) and adding these equations. See PEN-CIL—pencil of planes. *Syn.* bundle of planes.

SHEAR, *n.* **modulus of shear.** See RIGIDITY—modulus of rigidity.
    simple shear transformation. See TRANSFOR-MATION.

SHEAR'ING, *adj.* **shearing force.** One of two equal forces acting in opposite directions and not in the same line, causing, when acting upon a solid, a distortion known as a *shearing strain*.
    shearing motion. The motion that takes place when a body gives way due to a shearing stress.
    shearing strain and stress. See STRAIN, STRESS.

SHEET, *n.* **sheet of a surface.** A part of the surface such that one can travel from any point on it to any other point on it without leaving the surface. See HYPERBOLOID —hyperboloid of one sheet, hyperboloid of two sheets.
    sheet of a Riemann surface. Any portion of a Riemann surface which cannot be extended without giving a multiple covering of some part of the plane over which the surface lies. Thus for the function $w = z^{1/2}$ a sheet of the Riemann surface of definition consists of the z-plane cut by any simple curve extending from the origin to the point at infinity.

SHEPPARD, William Fleetwood (1863–1936). English probabilist and statistician.
    Sheppard's correction. Suppose values of a random variable are grouped in intervals of length h, frequencies are given for each interval, and all values in a given interval are regarded as being at the midpoint. This produces errors when computing moments, for which corrections were proposed by Sheppard. The corrected moments μ'_i are expressed in terms of moments μ_i computed from the grouped data by $\mu'_1 = \mu_1$, $\mu'_2 = \mu_2 - h^2/12$, $\mu'_3 = \mu_3 - \frac{1}{4}\mu_1 h$, etc.

SHIFT, *n.* **unilateral shift.** Let H be the Hilbert space of all sequences $x = (x_1, x_2, \cdots)$ of complex numbers with $\sum_1^\infty |x_i|^2$ finite. The **unilateral shift** is the bounded linear operator T for which $T(x) = (0, x_1, x_2, \cdots)$. The operator T is an isometry of H onto a proper subspace of H. See ISOMETRY (3).

SHOCK, *adj.* **shock wave.** In fluid dynamics, a discontinuous solution of a nonlinear hyperbolic equation or system of equations, arising from continuous initial and boundary conditions.

SHOE'MAK'ER, *n.* **shoemaker's knife.** See ARBILOS.

SHORT, *adj.* **short arc of a circle.** The shorter of the two arcs subtended by a chord of the circle.
    short division See DIVISION—short division.
    shortest condfidence interval. See CONFI-DENCE —most selective confidence interval.

SHRINK'ING, *n.* **shrinking of the plane.** See SIMILITUDE—transformation of similitude, STRAIN —one-dimensional strain.

SI, *n.* The international system of units for which the basic units are the meter, second, kilogram, ampere, kelvin, candela, and mole. Known in French as *Système International d'Unités*, but abbreviated as SI in all languages.

SIDE, *n.* **side of an angle.** See ANGLE.
    side of a line. See HALF—half-plane.
    side opposite an angle in a triangle or polygon. The side separated from the vertex of the angle by the same number of sides in whichever direction they are counted around the triangle or polygon.
    side of a polygon. Any one of the line segments forming the polygon.

SI-DE′RE-AL, *adj.* Pertaining to the stars.

sidereal clock. A clock that keeps sidereal time.

sideral time. Time as measured by the apparent diurnal motion of the stars. It is equal to the hour angle of the vernal equinox (see HOUR). The sidereal day, the fundamental unit of sidereal time, is assumed to begin and to end with two successive passages over the meridian of the vernal equinox. There is one more sidereal day than mean solar days in a sidereal year.

sidereal year. The time during which the earth makes one complete revolution around the sun with respect to the stars. Its length is 365 days, 6 hours, 9 minutes, 9.5 seconds. See YEAR.

SIEGEL, Carl Ludwig (1896–1981). German-American mathematician known particularly for his work in number theory, functions of one or several complex variables, and differential equations. See THUE—Thue-Siegel-Roth theorem.

SIERPIŃSKI, Waclaw (1882–1969). Polish logician, number and set theorist, and topologist. Leader of modern Polish school of mathematicians in Warsaw.

Sierpiński set. (1) Let G be the class of all uncountable G_δ sets on a line (see BOREL—Borel set). A Sierpiński set is a set S on the line which has the property that both S and its complement contain at least one point from each set belonging to G. Such a set can be shown to exist by using the *well-ordering principle* (or *axiom of choice*) to obtain a well-ordering of G with the property that the set of all predecessors of an element of G has *cardinal number* less than the cardinal number c of the real numbers (G itself has cardinal number c). Zorn's lemma can then be used to choose two points from each set of G, such that for any set neither of the two points chosen from that set were chosen from any previous set. A Sierpiński set can then be formed by choosing one of the members of each of these two-point sets. A Sierpiński set S has the properties that, for every set E, either E is of *measure zero* or one of the intersections of E with S and with the complement of S is *nonmeasurable*, and either E is of *first category* or one of the intersections of E with S and with the complement of S does not have the *property of Baire*. If S is a Sierpiński set and the exterior measure of a set A is $m_e(A)$, the set function $M(A) = m_e(A \cap S)$ defines a measure on a σ-algebra which includes S and all measurable sets. Also, $M(A) = m(A)$ if A is measurable. (2) A set S of points in the plane is a Sierpiński set if S contains at least one point of each closed set of nonzero measure and no three points of S are collinear. Such a set S is not measurable, although no line contains more than two points of S (see FUBINI—Fubini's theorem). The class C of closed sets of nonzero measure has cardinal number c and can be well-ordered so that each member of C has fewer than c predecessors. The set S can then be constructed by use of Zorn's Lemma, choosing a point x_α from each C_α of C in such a way that x_α is not collinear with any two points chosen from previous members of C.

SIEVE, *n.* **number sieve.** See NUMBER—number sieve.

sieve of Eratosthenes. See ERATOSTHENES.

SIG′MA, *adj., n.* The eighteenth letter of the Greek alphabet: lower case, σ; capital, Σ. See SUMMATION—summation sign.

σ-algebra and **σ-ring.** See ALGEBRA—algebra of subsets, RING—ring of sets.

σ-field. Same as σ-ALGEBRA, but becoming archaic.

σ-finite. See MEASURE—measure of a set.

SIGN, *n.* **algebraic sign.** A positive or negative sign.

continuation of a sign in a polynomial. See CONTINUATION.

Descartes' rule of signs. See DESCARTES.

law of signs. In *addition and subtraction,* two adjacent like signs can be replaced by a positive sign, and two adjacent unlike signs can be replaced by a negative sign. We have

$$2 - (-1) = 3,$$

while

$$2 - (+1) = 2 - 1 = 1$$

and

$$2 + (-) = 1.$$

Since, for example, $2 - (-1)$ can be regarded as meaning $2 + (-1)(-1)$, these are special cases of properties of *multiplication and division,* namely, that the product or quotient of two factors with like signs takes a positive sign; with unlike signs, a negative sign. We have

$$(-4)(-2) = 2,$$

and

$$\frac{-4}{2} = \frac{4}{-2} = -2.$$

See PRODUCT—product of real numbers, SUM—sum of real numbers.

sign of aggregation. See AGGREGATION.

summation sign. See SUMMATION.

SIG′NA-TURE, *n.* **signature of a quadratic form, Hermitian form,** or **matrix.** See INDEX—index of a quadratic form.

SIGNED, *adj.* **signed measure.** See MEASURE—measure of a set.

signed numbers. Positive and negative numbers. *Syn.* directed numbers.

SIG-NIF′I-CANCE, *n.* *(Statistics)* A significance test for a hypothesis H is a test of the hypothesis that H is false. Thus observations that are so improbable under the hypothesis as to cause one to believe the difference is not due merely to sampling errors or fluctuations are said to be statistically significant. *E.g.,* when testing a physical theory, an experimental value is significant if it suggests that some other theory is needed. One is primarily concerned with whether the experimental value is *significantly far* from what is expected. See HYPOTHESIS —test of a hypothesis.

significance level of a test. See HYPOTHESIS—test of a hypothesis.

SIG-NIF′I-CANT, *adj.* **significant digit** or **figure.** See DIGIT.

SIG′NUM, *n.* **signum function.** The function whose value is 1 for $x > 0$, -1 for $x < 0$, and 0 for $x = 0$. It is denoted by sgn x or sg x.

SIM′I-LAR, *adj.* Two geometric figures are **similar** if one can be made congruent to the other by using a transformation of similitude [see SIMILITUDE], *i.e.,* if one is a magnification or reduction of the other. If corresponding lengths in two similar figures are in the ratio k, then corresponding areas are in the ratio k^2 and corresponding volumes are in the ratio k^3. See various headings below.

similar decimals. See DECIMAL.

similar ellipses (or hyperbolas). Ellipses (or hyperbolas) are similar if they have the same eccentricity or if their semiaxes are in the same ratio.

similar ellipsoids. Ellipsoids are similar if their principal sections are similar ellipses. Thus the ellipsoids

$$\frac{x^2}{a^2} + \frac{y^2}{b^2} + \frac{z^2}{c^2} = \mu,$$

where μ is a parameter greater than zero, are all similar.

similar fractions. See FRACTION.

similar hyperboloids and paraboloids. Hyperboloids and paraboloids are similar if their principal sections are similar. The hyperboloids whose equations are

$$\frac{x^2}{a^2} + \frac{y^2}{b^2} - \frac{z^2}{c^2} = \mu$$

with μ taking different positive values (different negative values) are similar. The paraboloids whose equations are

$$\frac{x^2}{a^2} + \frac{y^2}{b^2} = \mu z,$$

with μ taking different nonzero values, are similar elliptic paraboloids. The paraboloids whose equations are

$$\frac{x^2}{a^2} - \frac{y^2}{b^2} = \mu z,$$

with μ taking different nonzero values, are similar hyperbolic paraboloids.

similar matrices. Matrices which are transforms of each other by a nonsingular matrix. See TRANSFORMATION —collineatory transformation.

similar polygons. Two polygons are similar if the angles of one are equal to the corresponding angles of the other and the corresponding sides are proportional, or if their vertices are respectively the points of two *similar sets of points.* See below, similar sets of points.

similar sets of points. Points so situated on a pencil of lines (two points on each line) that all the ratios of the distances from the vertex of the pencil to the two points (one in each of the two sets) on a given line are equal. The set of points (one on each line) whose distances from the vertex are the antecedents of the ratios, and the set of points whose distances are the consequents, are called *similar sets* of points, or *similar systems* of points. Two such sets of points are also said to be **homothetic** and any figures formed by joining corresponding pairs of points in each set are **homothetic.** See SIMILITUDE —transformation of similitude.

similar solids. See SOLID.

similar surfaces. Surfaces which can be made to correspond point to point in such a way that the distance between any two points on one surface is always the same multiple of the distance between the two corresponding points on the other. The areas of similar surfaces are to each other as the squares of corresponding distances.

similar terms. Terms which contain the same power (or powers) of the variable. The terms $3x$ and $5x$, ax and bx, axy and bxy, are similar terms. *Syn.* like terms.

similar triangles. Two triangles are similar if and only if corresponding angles are equal. Corresponding sides are then proportional.

SIM′I-LAR′I-TY, *n.* The property of being similar.

general similarity transformation. A transformation (composed possibly of a translation, a rotation, and a homothetic transformation), which transforms figures into similar figures.

SI-MIL′I-TUDE, *n.* **center of similitude.** See RADIALLY —radially related figures.

ratio of similitude. See RATIO.

transformation of similitude. The transformation $x' = kx$, $y' = ky$, in rectangular coordinates. It multiplies the distance between every two points by the same constant k, called the ratio of similitude. If k is less than one, the

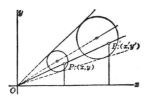

transformation is said to **shrink** the plane. In the figure, the circumference of the larger circle is k times the circumference of the smaller, and the point P' is k times as far from the origin as the point P. This transformation is also called the **homothetic transformation.** See AFFINE —affine transformation, RADIALLY —radially related.

SIM′PLE, *adj.*

simple algebra. See ALGEBRA —algebra over a field.

simple arc. A set of points which is the image of the closed interval $[0, 1]$ under a one-to-one continuous transformation (the transformation necessarily has a continuous inverse). See TOPOLOGICAL —topological transformation. A continuum (of at least two points) such that there are not more than two points whose omission does not destroy the connectedness of the set is a simple arc. See below, simple closed curve.

simple closed curve. A curve such as a circle, an ellipse, or the circumference of a rectangle, that is closed and does not intersect itself. *Tech.*

A set of points which is the image of a circle under a one-to-one continuous transformation (the transformation necessarily has a continuous inverse). See TOPOLOGICAL —topological transformation. A continuum (of at least two points) which is no longer connected if any two arbitrary points are removed is a simple closed curve. *Syn.* Jordan curve. See CURVE, JORDAN —Jordan curve theorem.

simple curve. See CURVE.

simple cusp. See CUSP —cusp of the first kind.

simple elongations and compressions. Same as one-dimensional strains. See STRAIN.

simple event. See EVENT.

simple extension of a field. See EXTENSION —extension of a field.

simple fraction. See FRACTION.

simple function. (1) A simple function of a complex variable in a region D is an analytic function that does not take on any value more than once in D. *Syn.* schlicht function. (2) See INTEGRABLE —integrable function.

simple group. See GROUP —simple group.

simple harmonic motion. See HARMONIC.

simple hexagon. See HEXAGON —simple hexagon.

simple integral. A single integral, as distinguished from multiple and iterated integrals.

simple interest. See INTEREST.

simple pendulum. See PENDULUM.

simple point on a curve. Same as ORDINARY POINT. See POINT.

simple polyhedron. See POLYHEDRON.

simple ring. See RING.

simple root. A root of an equation that is not a repeated root. For an equation $f(x) = 0$, with $f(x)$ either a polynomial or a power series, a simple root is a root r such that $f(x)$ is divisible by the first power of $x - r$ and by no higher power. See MULTIPLE —multiple root of an equation.

SIM′PLEX, *n.* An n-dimensional simplex (or simply an **n-simplex**) is a set which consists of $n + 1$ linearly independent points $p_0, p_1, \cdots, p_n$ of an Euclidean space of dimension greater than n together with all the points of type

$$x = \lambda_0 p_0 + \lambda_1 p_1 + \cdots + \lambda_n p_n,$$

where $\lambda_0 + \lambda_1 + \cdots + \lambda_n = 1$ and $0 \leq \lambda_i$ for each i (see BARYCENTRIC —barycentric coordinates). Such a set is sometimes called a **closed simplex,** while the set of all such points x for which each λ_i is positive in an **open simplex.** A set of points of one of these two types is sometimes called a **degenerate simplex** if the points $p_0, \cdots, p_n$ are not linearly independent (or if two or more of these points coincide). Each of the points $p_0, \cdots, p_n$ is a **vertex** of the simplex, and any simplex whose vertices are $r + 1$ of these points is an **r-dimensional face**, or an **r-face**, of the simplex. An n-simplex is its own n-face, while faces of dimension less than n are **proper faces**. A simplex of dimension 0 is a single point; a simplex of dimension 1 has 2 vertices and consists of the straight line segment joining these vertices (its vertices are its only proper faces); a simplex of dimension 2 has 3 vertices and is a triangle with its interior (its 1-dimensional faces are its sides and its 0-dimensional faces are its vertices); a simplex of dimension 3 has 4 vertices and is a tetrahedron together with its interior (its 2-dimensional faces are triangles). The set of all the vertices of a simplex is its **skeleton**. Any $n + 1$ objects can be called an **abstract n-simplex** (see COMPLEX —simplicial complex). A **topological simplex** is any topological space (such as a solid sphere) which is homeomorphic to a simplex. A simplex is **oriented** if an order has been assigned to its vertices. If $(p_0 p_1 \cdots p_n)$ is an orientation of a simplex with vertices $p_0, \cdots, p_n$, this is regarded as being the same as any orientation obtained from it by an even permutation of the vertices and as the negative of any orientation obtained by an odd permutation of the vertices. E.g., a 2-simplex with vertices p_0 and p_1 has the two orientations $(p_0 p_1)$ and $(p_1 p_0)$. A 3-simplex, or triangle, has the two orientations corresponding to the two directions for enumerating vertices around the triangle. If $(p_0 p_1 \cdots p_n)$ is an orientation of an n-simplex, then this simplex and the $(n - 1)$-simplex $p_0, \cdots, p_{i-1}, p_{i+1}, \cdots, p_n$ obtained by discarding the vertex p_i are **coherently** (or **concordantly**) **oriented** if the orientation of the $(n - 1)$-simplex is $(-1)^i (p_0 \cdots p_{i-1} \cdots p_n)$. E.g., if (ABC) is an orientation of a triangle with vertices A, B, C, then this triangle is coherently oriented with each of its sides if the sides have the orientations $(AB), (BC), -(AC) = (CA)$.

simplex method. A standard finite iterative algorithm for solving a linear programming problem by successively determining basic *feasible solutions*, if any exist, and testing them for optimality. See PROGRAMMING —linear programming.

SIM-PLIC'I-AL, *adj.* **simplicial complex.** See COMPLEX.

simplicial mapping. A mapping of a simplicial complex K_1 into a simplicial complex K_2 for which the images of simplexes of K_1 are simplexes of K_2. If the mapping is one-to-one and the image of K_1 is all of K_2, then K_1 and K_2 are said to be *isomorphic*, or *combinatorially equivalent*.

SIM'PLI-FI-CA'TION, *n.* The process of reducing an expression or a statement to a briefer form, or one easier to work with. See SIMPLIFIED.

SIM'PLI-FIED, *adj.* The simplified form of an expression, quantity, or equation can mean either (1) the briefest, least complex form, or (2) the form best adapted to the next step to be taken in the process of seeking a certain result. One of the most indefinite terms used seriously in mathematics. Its meaning depends upon the operation as well as the expression at hand and its setting. E.g., if one desired to factor $x^4 + 2x^2 + 1 - x^2$, to collect the x^2 terms would be foolish, since it would conceal the factors. Usually a **radical** is said to be in simplified form when there is no fraction under the radical and no factor under the radical which possesses the root indicated by the index; $\sqrt{2}$ and $2\sqrt{3}$ are in simplest form, but $\sqrt{\frac{2}{3}}$ and $\sqrt{12}$ are not. A **fraction** whose numerator and denominator are rational numbers is usually said to be in simplified form when written so that numerator and denominator are integers with no common factors other than ± 1.

SIM'PLY, *adv.* **simply connected set.** See CONNECTED.

  simply ordered set. See ORDERED —ordered set.

SIMPSON, Thomas (1710–1761). English algebraist, analyst, geometer, and probabilist.

  Simpson's rule. A formula for approximating an integral of type $\int_a^b f(x)\, dx$. The formula can be thought of as the result of separating the interval from a to b into an even number of congruent subintervals by points $a = x_0, x_1, x_2, \cdots, x_{2n} = b$, and then approximating the graph of $y = f(x)$ for x between x_{2k} and x_{2k+2} by the parabola through the points on the graph for which x has the values $x_{2k}, x_{2k+1}, x_{2k+2}$. The formula is

$$\frac{(b-a)}{6n} \big[y_a + 4y_1 + 2y_2 + 4y_3 + 2y_4 + \cdots + 4y_{2n-1} + y_b \big]$$

where $y_a, y_1, y_2, \cdots, y_{2n-1}, y_b$ are the respective ordinates at the points $a, x_1, \cdots, x_{2n-1}, b$. The numerical difference between the number given by this formula and the actual integral is not greater than

$$\frac{M(b-a)^5}{180(2n)^4},$$

where M is the least upper bound of the absolute value of the fourth derivative of f on the interval from a to b. If the degree of f is 3 or less, the formula gives the exact value of the integral. If also $f(x) \geq 0$ for each x, then the formula in the form

$$\frac{b-a}{6}[y_a + 4y_1 + y_b],$$

where $n = 1$, gives the area under the graph of f and is the **prismoidal formula** (for area). See NEWTON —Newton's three-eighths rule, TRAPEZOID —trapezoid rule.

SI′MUL-TA′NE-OUS, *adj.* **simultaneous equations.** Two or more equations that are conditions imposed simultaneously on all the variables, but may or may not have common solutions; *e.g.,* $x + y = 2$ and $3x + 2y = 5$, treated as simultaneous equations, are satisfied by $x = 1$, $y = 1$, these values being the coordinates of the point of intersection of the straight lines which are the graphs of the two equations. The number of solutions of two **simultaneous polynomial equations** in two variables is equal to the product of their degrees (provided they have no common factor), infinite values (see homogeneous coordinates) being allowed, and equal solutions being counted to the degree of their multiplicity. *E.g.,* (1) the equations $y = 2x^2$ and $y = x$ have the two common solutions $(0,0)$ and $(\frac{1}{2}, \frac{1}{2})$. (2) the equations $y - 2x^2 = 0$ and $y^2 - x = 0$ have two real and two imaginary common solutions. **Simultaneous linear equations** are simultaneous equations which are linear (of the first degree) in the variables (see CONSISTENCY —consistency of linear equations). *Syn.* system of equations.

simultaneous inequalities. Two or more inequalities that are conditions imposed simultaneously on all the variables, but may or may not have common solutions. The simultaneous inequalities $x^2 + y^2 < 1$, $y > 0$ have as their solution set the set of all points above the x-axis and inside the unit circle about the origin. The interior of a convex polygon or polyhedron is the *graph* (or *solution set*) of suitable simultaneous linear inequalities—in two variables for the polygon, and three variables for the polyhedron. *Syn.* system of inequalities.

SINE, *adj., n.* See TRIGONOMETRIC —trigonometric functions.

exponential values of sin x and cos x. See EXPONENTIAL —exponential values of sin x and cos x.

laws of sines. For a **plane triangle**, the sides of a triangle are proportional to the sines of the opposite angles. If the angles are A, B, C, and the lengths of the sides opposite these angles are a, b, c, this law is

$$\frac{a}{\sin A} = \frac{b}{\sin B} = \frac{c}{\sin C}.$$

For a **spherical triangle**, the sines of the sides are proportional to the sines of the opposite angles.

sine curve. The graph of $y = \sin x$. The curve passes through the origin and all points on the x-axis whose abscissas are multiples of π (radians), is concave toward the x-axis, and the greatest distance from the x-axis to the curve is unity.

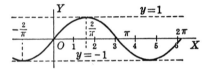

sine of a number or angle. See TRIGONOMETRIC —trigonometric functions.

sine series. See FOURIER —Fourier half-range series.

SIN′GLE, *adj.* **single-address system.** A method of coding problems for machine solution, whereby each separate instruction is restricted to telling what to do with a single item at a specified address or memory position. See MULTIADDRESS.

single premium. See PREMIUM.

single-valued function. See FUNCTION —multiple-valued function.

SIN′GLE-TON, *n.* A set that has exactly one member.

SIN′GU-LAR, *adj.* **singular curve on a surface.** A curve C on a surface S such that every point of C is a singular point of S. See below, singular point of a surface.

singular matrix. A matrix whose determinant is zero. See MATRIX.

singular point of an analytic function. A point at which the function (of a complex variable) is not analytic, but in every neighborhood of which there are points of analyticity. An **isolated singular point** is a point z_0 on the Riemann surface of existence of the function at which it is not analytic, but such that there exists on the surface a neighborhood

$$|z - z_0| < \epsilon$$

of z_0 at each point z ($\neq z_0$) of which $f(z)$ is analytic. An isolated singular point can be of any one of three types. (1) **Removable singular point**. An isolated singular point z_0 such that f can be defined, or redefined, at z_0 in such a way as to be analytic at z_0. E.g., if $f(z) = z$ for $0 < |z| < 1$ and $f(0) = 1$, then f has a removable singular point at 0. (2) **Pole**. An isolated singular point z_0 such that $f(z)$ can be represented by an expression that is of the form $f(z) = \dfrac{\phi(z)}{(z - z_0)^k}$, where k is a positive integer, ϕ is analytic at z_0, and $\phi(z_0) \neq 0$. The integer k is called the **order** of the pole. E.g., $f(z) = (z - 1)/(z - 2)^3$ has a pole of order 3 at 2, with $\phi(z) = z - 1$. (3) **Essential isolated singular point**. An isolated singular point which is neither a removable singularity nor a pole. In any neighborhood of an essential singular point, and for every finite complex number α with the exception of at most one number α, the equation $f(z) - \alpha = 0$ has an infinitude of roots. This is the **second theorem of Picard**. E.g., $f(z) = \sin \dfrac{1}{z}$ has an essential isolated singular point at 0. An **essential singular point** is any singular point which is not a pole and which is not a removable singularity. For example, for $f(z) = \tan \dfrac{1}{z}$, the origin is an essential singular point, but is not an *isolated singular point*, since it is a limit point of poles of f.

singular point of a curve. See POINT—ordinary point of a curve.

singular point of a surface. A point of the surface S: $x = x(u, v)$, $y = y(u, v)$, $z = z(u, v)$, at which $H^2 \equiv EG - F^2 = 0$. See SURFACE—fundamental coefficients of a surface. Since

$$H^2 = \left[\frac{\partial(y, z)}{\partial(u, v)}\right]^2 + \left[\frac{\partial(z, x)}{\partial(u, v)}\right]^2 + \left[\frac{\partial(x, y)}{\partial(u, v)}\right]^2,$$

it follows that, for real surfaces and parameters, the quantity H^2 is nonnegative; and H^2 is positive at a point unless all three Jacobians vanish there. H appears in the denominator of several important formulas in differential geometry. See REGULAR —regular point of a surface.

singular solution of a differential equation. See DIFFERENTIAL —solution of a differential equation.

singular transformation. See LINEAR —linear transformation.

SINISTRORSUM [*Latin*] **or SIN′IS-TRORSE,** *adj.* Same as LEFT-HANDED. See LEFT.

SINK, *n.* A negative source. See SOURCE.

SINKING FUND. See FUND—sinking fund.

SI′NUS-SOID, *n.* The *sine curve*. See SINE.

SIZE, *n.* **size of a test.** See HYPOTHESIS —test of a hypothesis.

SKEL′E-TON, *n.* See COMPLEX—simplicial complex, SIMPLEX.

SKEW, *adj.* **skew field.** See RING.

skew lines. Nonintersecting, nonparallel lines in space. Two lines are skew if and only if they do not lie in a common plane. The **distance** between two skew lines is the length of the line segment joining the lines that is perpendicular to both.

skew quadrilateral. The figure formed by joining four noncoplanar points by line segments, each point being joined to two, and only two, other points.

skew-symmetric determinant. A determinant having its conjugate elements numerically equal but opposite in sign. If the element in the first row and second column is 5, the element in the first column and second row would be -5. A skew-symmetric determinant of odd order is always equal to zero.

skew-symmetric matrix. A matrix which is equal to the negative of its transpose; a square matrix such that $a_{ij} = -a_{ji}$, where a_{ij} is the element in the ith row and jth column. *Syn.* skew matrix.

skew-symmetric tensor. See TENSOR.

SKEWES NUMBER. See PRIME—prime-number theorem.

SKEW′NESS, *n.* Lack of symmetry of a distribution about the mean. There are several measures of skewness. A common measure is μ_3/σ^3, where μ_3 is the third moment about the mean and σ^2 is the variance or second moment about the mean. However, it is possible to have $\mu_3 = 0$ for a distribution that is far from symmetric.

SLANT, *adj.* **slant height.** The slant height of a **right circular cone (cone of revolution)** is the common length of the elements of the cone; the slant height of a **frustum of a right circular cone** is the length of the segment of an element of the cone intercepted by the bases of the frustum. The slant height of a **regular pyramid** is the common altitude of its lateral faces; the slant height of a **frustum of a regular pyramid** is the common altitude of its faces (the perpendicular distance between the parallel edges of the faces).

SLIDE, *adj.* **slide rule.** A mechanical device to aid in calculating by the use of logarithms. It consists essentially of two rules, one sliding in a groove in the other, containing logarithmic scales by means of which products and quotients are calculated by adding and subtracting logarithms. Calculators and computers have replaced the slide rule for doing calculations.

SLOPE, *n.* **angle of slope.** Same as angle of inclination. See ANGLE—angle of inclination.

    point-slope and slope-intercept forms of the equation of a line. See LINE—equation of a line.

    slope of a curve at a point. The slope of the tangent line at that point; the derivative, dy/dx, evaluated at the point. See DERIVATIVE.

    slope of a line. The tangent of the angle that the line makes with the positive x-axis; the rate of change of the ordinate with respect to the abscissa, *i.e.* (in rectangular Cartesian coordinates),

$$\frac{y_2 - y_1}{x_2 - x_1},$$

where (x_1, y_1) and (x_2, y_2) are points on the line. In *calculus* the slope at (x_1, y_1) is the derivative of the ordinate with respect to the abscissa,

$$\lim_{x_2 \to x_1} \frac{y_2 - y_1}{x_2 - x_1} \quad \text{or} \quad \left(\frac{dy}{dx}\right)_{x = x_1}$$

(which is the same for all points on the line). The slope of $y = x$ is 1; of $y = 2x$, 2; of $y = 3x + 1$, 3. See DERIVATIVE. Slope is not defined for lines perpendicular to the x-axis.

SMALE, Stephen (1930–). American differential topologist, global analyst and Fields Medal recipient (1966). Proved that the sphere can be turned inside out and that Poincaré's conjecture is valid for dimensions greater than 4.

SMALL, *adj., n.* **group without small subgroups.** See under GROUP.

    in the small. In the neighborhood of a point. *E.g.*, when studying properties such as curvature of a curve at a point, one is concerned only with the behavior of the curve in the neighborhood of the point. Classical differential geometry is a study **in the small** or **im kleinen.** The study of a geometric object in its entirety, or the study of definite sections of it, or the study of a function in a given fixed interval, is a study **in the large** or **im grossen.** Algebraic geometry is a study in the large.

    small arcs, angles, or line segments. Arcs, lines, or line segments, which are small enough to satisfy certain conditions, such as making the difference between two ordinates of a curve less than a stipulated amount, or the quotient of the sine of an angle by the angle (in radians) differ from 1 by less than a given amount.

    small circle. See CIRCLE—small circle.

SMITH, Henry Lee (1893–1957). See MOORE (E. H.)—Moore-Smith convergence.

SMOOTH, *n.* **smooth curve.** If C is a curve in an Euclidean space, then C is the image of an interval $[a, b]$ under a continuous transformation. Also, x_i is a continuous function on $[a, b]$ if $x_i(t)$ denotes the ith Cartesian coordinate of the point on C that corresponds to t in $[a, b]$. The curve C being **smooth** or **continuously differentiable** means that the first derivative of each such function x_i is continuous on $[a, b]$; **piecewise smooth** means that these first derivatives are continuous except for a finite number of points and at each of these points the function is differentiable on the left and on the right.

    smooth map. See DIFFEOMORPHISM.

    smooth projective plane curve. See PROJECTIVE—projective plane curve.

    smooth surface or **smooth surface element.** (1) A surface which has the property that there is a tangent plane at each point and the direction of the normal is a continuous function of the point of tangency. (2) A set that is the range of a continuous one-to-one transformation T which has the properties that the domain of T is a bounded closed plane set D whose boundary is a rectifiable simple closed curve, and T can be described by parametric equations $x = f(u, v)$, $y = g(u, v)$, $z = h(u, v)$ for which the first-order partial derivatives of f, g, and h are continuous on an open set containing D and there is no point of D at which all the following Jacobians are zero: $\partial(y, z)/\partial(u, v)$, $\partial(z, x)/\partial(u, v)$, $\partial(x, y)/\partial(u, v)$. The *edge* of the surface is the image of the boundary of D. Such a surface has property (1).

SN, *n.* See ELLIPTIC—Jacobian elliptic functions.

SNEDECOR, George Waddel (1881–   ). American statistician and specialist on punch-card equipment. See F—*F* distribution.

SNEL or **SNELL van Roijen** (*also* Snellius), **Willebrord** (1580–1626). Dutch astronomer and mathematician who succeeded his father as professor of mathematics at the University of Leiden in 1613.
  Snell's law. See REFRACTION.

SO′LAR, adj. **solar time.** See TIME.

SO′LE-NOI′DAL, *n.* **solenoidal vector in a region.** A vector-valued function **F** whose domain contains the region and whose integral over every reducible surface S in the region is zero; *i.e.*,
$\int_S \mathbf{F} \cdot \mathbf{n}\, da = 0$, where **n** is the unit vector in the direction of the outer normal to the element of area da. The divergence of a vector is zero at every point in a region if, and only if, the vector is solenoidal in the region, or if, and only if, the vector is the curl of some vector function (vector potential). See EQUATION —equation of continuity.

SOL′ID, *adj., n.* See GEOMETRIC —geometric solid.
  frustum of a solid. See FRUSTUM.
  similar solids. Solids bounded by similar surfaces; solids whose points can be made to correspond in such a way that the distances between all pairs of points of the one are a constant multiple of the distances between corresponding points of the other. Volumes of similar solids are proportional to the cubes of the distances between corresponding points. All spheres are similar solids; so are all cubes.
  solid angle. A surface formed by rays with a common origin, the **vertex** of the solid angle, and passing through a closed curve (a polyhedral angle is a special type of solid angle for which the curve is a polygon). A measure of the solid angle at P (see figure) subtended by a surface S is equal to the area A of the portion of the surface of a sphere of unit radius, with center at P, which is cut by a conical surface with vertex at P and the perimeter of S as a generatrix. The unit solid angle is the **steradian.** The total solid angle about a point is equal to 4π steradians. See SPHERICAL —spherical degree.

solid geometry. See GEOMETRY —solid geometry.
  solid of revolution. See REVOLUTION.

SOL′I-DUS, *n.* A slant line used to indicate division in a fraction, *e.g.*, 3/4 or a/b. Also used in dates, *e.g.*, 7/4/1776.

SO-LU′TION, *n.* (1) The process of finding a required result by the use of certain given data, previously known facts or methods, and newly observed relations. (2) The *result* is also spoken of as the *solution*. *E.g.*, a root of an equation is called a *solution* of the equation it satisfies, although a *solution* of the equation may refer either to the process of finding a root or to the root itself.
  algebraic, analytic, and geometric solutions. See headings under ALGEBRAIC, ANALYTIC, GEOMETRIC.
  solution of a differential equation. See DIFFERENTIAL —solution of a differential equation.
  solution of equations. For a single equation, solution may mean either (1) the process of finding (or approximating) a root of the equation, or (2) a root of the equation. The solution of a set of **simultaneous equations** is the process of finding a set of values of the variables which satisfy all the equations (this set of values of the variables also is called a solution) (see SIMULTA-NEOUS—simultaneous equations). The **geometric** (or **graphical**) solution of an equation $f(x) = 0$ is the process of finding the roots by graphing $y = f(x)$ and estimating where its graph crosses the x-axis (see ROOT—root of an equation). A **solution by inspection** consists of guessing a root and testing it by substitution in the equation. See POLYNOMIAL —polynomial equation, ROOT —root of an equation.
  solution of inequalities. See INEQUALITY — graph of an inequality.
  solution of linear-programming problem. See PROGRAMMING —linear programming.
  solution set. The set of all solutions of a given equation, system of equations, inequality, etc. *E.g.*, the solution set of the equation $x^2 - 2x = 0$ is the set whose members are the numbers 0 and 2; the solution set of $x^2 + y^2 = 4$ is the circle with center at the origin and radius 2; the solution set of the system, $x + y = 1$, $x - y$

= 3, is the set whose only member is the ordered pair $(2, -1)$; the solution set of the inequality $3x + 4y + z < 2$ is the set of all ordered triples (x, y, z) that represent points which are below the plane whose equation is $3x + 4y + z = 2$. *Syn.* truth set. See PROPOSITIONAL —propositional function.

solution of a two-person zero-sum game. See GAME.

solution of a triangle. Finding the remaining angles and sides when sufficient of these have been given. For a **plane right triangle**, it is sufficient to know any two sides, or to know one of the acute angles and one side. The unknown parts are found by use of trigonometric tables and the definitions of the trigonometric functions (see TRIGONOMETRIC): if a, b, c represent the legs and hypotenuse, respectively, and A, B are the angles opposite sides a and b, then $a = b \tan A = c \sin A$, $b = c \cos A$, $A = \tan^{-1} a/b$, $B = 90° - A$. For an **oblique plane triangle**, it is sufficient to know all three sides, two angles and one side, or two sides and one angle (except that when two sides and the angle opposite one of them is given there may be two solutions; see AMBIGUOUS). See SINE—law of sines, COSINE— law of cosines, TANGENT—tangent law, TRIGONOMETRY —half-angle formulas of plane trigonometry, HERON—Hero's formula. For a right spherical triangle, *Napier's rules* supply all the formulas needed. For formulas providing solutions of an **oblique spherical triangle** in cases when solutions exist, see COSINE—law of cosines, GAUSS—Gauss' formulas, NAPIER—Napier's analogies, SINE—law of sines, TRIGONOMETRY — half-angle formulas and half-side formulas of spherical trigonometry. Also see QUAD- RANT—laws of quadrants, SPECIES—law of species.

SOLV′A-BLE, *adj.* **solvable group.** See GROUP.

SOURCE, *n.* In hydrodynamics, potential theory, etc., a point at which additional fluid is considered as being introduced into the region occupied by the fluid. If fluid is being removed at the point, the negative source is called a **sink.**

SOUSLIN (or SUSLIN), Michail Jakovlevich (1894–1919). Russian analyst and topologist.

Souslin's conjecture. The conjecture that a topological space L is topologically equivalent to the real line if L is linearly ordered with no first or last member, the open intervals are a base for the topology of L, L is connected, and there is no uncountable collection of disjoint open intervals in L. It is known that L is topologically equivalent to the real line if L is separable and

satisfies the first three of the preceding four conditions. A **Souslin line** is a set that satisfies the four conditions and is not separable. Souslin's conjecture is false if and only if there exists a Souslin line. However, Souslin's conjecture is undecidable on the basis of the usual axioms of set theory, even with the continuum hypothesis.

Souslin's theorem. See ANALYTIC —analytic set.

SOUTH, *adj.* **south declination.** See DECLINA - TION—declination of a celestial point.

SPACE, *adj., n.* (1) A three-dimensional region. (2) Any *abstract space.* See ABSTRACT.

coordinate in space. See CARTESIAN —Cartesian coordinates, CYLINDRICAL —cylindrical coordinates, SPHERICAL —spherical coordinates.

enveloping space. The space in which a configuration lies. The configuration is then said to be **embedded** in the enveloping space. Thus the circle $x = \cos \theta$, $y = \sin \theta$ is embedded in the two-dimensional Euclidean (x, y)-space.

half-space. See HALF.

orbit space. See ORBIT.

space curves. Curves that may or may not be plane curves; the intersection of two distinct surfaces is usually a space curve. Space curves do not lie in a plane (*i.e.*, they are twisted) except when their *torsion* is zero.

SPAN, *n.* The **span of a set** S is the minimal set that contains S and has some prescribed property. Thus the **convex span** of S is the minimal convex set that contains S; *i.e.*, the intersection of all convex sets which contain S. The **linear span** of S is the minimal linear space that contains S. *Syn.* HULL.

SPE′CIES, *n.* **law of species.** One-half the sum of any two sides of a spherical triangle and one-half the sum of the opposite angles are the same species. Two angles, two sides, or an angle and a side are said to be of the *same species* if they are both acute or both obtuse, and of *different species* if one is acute and one obtuse.

species of a set of points. Let G' be the derived set of a set G, G'' the derived set of G' and in general $G^{(n)}$ the derived set of $G^{(n-1)}$. If one of the sets $G', G'', \cdots$ is the null set (contains no points), then G is of the **first species.** Otherwise G is of the **second species.** The set G of all numbers of the form $m + 1/n$ with m and n integers is of the first species, since $G'' = 0$. The set of all rational numbers is of the second species, since all derived sets consist of all real numbers.

SPE-CIF'IC, *adj.* **specific gravity.** The ratio of the weight of a given volume of any substance to the weight of the same volume of a standard substance. The substance taken as the standard for solids and liquids is water at 4° C, the temperature at which water has the greatest density.

specific heat. (1) The number of calories required to raise the temperature of one gram of a substance 1° C, or the number of B.T.U.'s required to raise one pound of the substance 1° F. Sometimes called *thermal capacity.* (2) The ratio of the quantity of heat necessary to change the temperature of a given mass 1° to the amount necessary to change an equal mass of water 1°.

SPEC'TRAL, *adj.* **spectral measure and integral.** Let H be a Hilbert space and S a set with a specified σ-algebra A of subsets. A **spectral measure** on S is a function which assigns a projection $P(X)$ to each member X of A in such a way that $P(S)$ is the identity transformation on H and $P(\bigcup_1^\infty X_k) = \sum_1^\infty P(X_k)$ for any sequence of pairwise disjoint sets $X_1, X_2, \cdots$ belonging to A. It follows that if $X_1 \subset X_2$, then $P(X_2 - X_1) = P(X_2) - P(X_1)$; also, $P(X_1) \le P(X_2)$ in the sense that the range of $P(X_1)$ is contained in the range of $P(X_2)$, or that $P(X_1) \cdot P(X_2) = P(X_1)$. For any two members X_1 and X_2 of A, $P(X_1 \cup X_2) + P(X_1 \cap X_2) = P(X_1) + P(X_2)$ and $P(X_1 \cap X_2) = P(X_1) \cdot P(X_2)$. If X_1 and X_2 are disjoint, then the ranges of $P(X_1)$ and $P(X_2)$ are orthogonal. If S is the complex plane (or a subset of the complex plane) and A is the σ-algebra of Borel sets, then the spectral measure has the additional property that, for X a member of A, the range of $P(X)$ is the union of the ranges of projections $P(X_a)$ for X_a a compact subset of X. The **spectrum** of a spectral measure is the complement of the union of all the open sets U for which $P(U) = 0$. If the spectrum is bounded and $f(\lambda)$ is a bounded (Borel) measurable function (real or complex valued), then $T = \int f(\lambda) \, dP$ defines a bounded transformation T in the sense that the approximating sums for the integral define operators which converge in norm to T. Also, for any two elements x and y of the Hilbert space, $m(X) = (P(X)x, y)$ defines a complex-valued *measure* on A, and $(Tx, y) = \int f(\lambda) \, dm$. It follows that $\int f \cdot g \, dP = \int f \, dP \cdot \int g \, dP$ and that, if f is continuous, $\|\int f(\lambda) \, dP\|$ is the least upper bound of $|\lambda|$ for λ belonging to the spectrum; the spectrum of the transformation $T = \int \lambda \, dP$ is coincident with the spectrum of the spectral measure. If the spectrum is not bounded, but f is bounded on

bounded sets, then $\int f(\lambda) \, dP$ is the unique transformation which coincides with $\int f_X(\lambda) \, dP$ on the range of the projection $P(X)$ for each bounded X of A, where f_X coincides with f on X and is zero on the complement of X.

spectral theorem. For any *Hermitian, normal,* or *unitary* transformation T defined on a Hilbert space, there is a unique spectral measure defined on the Borel sets of the complex plane for which $T = \int \lambda \, dP$. If T is Hermitian, then $P(X) = 0$ if X does not intersect the real line and $\int \lambda \, dP$ can be regarded as an integral along the real line; if T is unitary, then $P(X) = 0$ if X does not intersect the circle $|z| = 1$ and $\int \lambda \, dP$ can be regarded as an integral around this circle.

SPEC'TRUM, *n.* **spectrum of a transformation.** For a **matrix,** the set of its eigenvalues. More generally, let T be a linear transformation of a vector space L into itself and I be the identity transformation, $I(x) \equiv x$. The **spectrum** of T then consists of three pair-wise disjoint sets: the **point spectrum,** which is the set of numbers λ for which $T - \lambda I$ does not have an inverse (is not one-to-one); the **continuous spectrum,** which is the set of numbers λ for which $T - \lambda I$ has an inverse which is not bounded (*i.e.,* not continuous) and whose domain is dense in L; the **residual spectrum,** which is the set of numbers λ for which $T - \lambda I$ has an inverse whose domain is not dense in L. The set of numbers which do not belong to the spectrum is the **resolvent set** and consists of those numbers λ for which $T - \lambda I$ has a bounded inverse with dense domain. If L is a finite-dimensional vector space and T is the transformation that transforms vectors $\mathbf{x} = (x_1, x_2, \cdots, x_n)$ into vectors $T(\mathbf{x}) = (y_1, y_2, \cdots, y_n)$ with $y_i = \sum_j a_{ij} x_j$, then the point spectrum is the entire spectrum of T and is the set of *eigenvalues* of the **matrix** (a_{ij}). If λ_0 is in the point spectrum of T, then there is a vector $\mathbf{x} \ne 0$ such that $T(\mathbf{x}) = \lambda_0 \mathbf{x}$; λ_0 is an **eigenvalue** of T and $\mathbf{x}$ is an **eigenvector** of T. The linear space of eigenvectors corresponding to λ_0 is the **manifold of eigenvalues** corresponding to λ_0. If L is a Banach space, the spectrum is a nonempty set. If T is a bounded linear transformation and $|\lambda| > \|T\|$, then λ belongs to the resolvent set and the inverse of $T - \lambda I$ is $-\sum_1^\infty \lambda^{-n} T^{n-1}$. If L is a (complex) Hilbert space and λ belongs to the residual spectrum of T, then $\bar{\lambda}$ belongs to the point spectrum of T^*; if λ belongs to the point spectrum of T, then $\bar{\lambda}$ belongs to either the point spectrum or the residual spectrum of T^*. If T is *Hermitian, normal,* or *unitary,* then the residual spectrum of T is empty. If T is Hermitian, all numbers in the spectrum are real; if T is unitary, all numbers in the spectrum are on the circle $|z| = 1$. *E.g.*, let $u_1, u_2, \cdots$ be a complete orthonormal sequence in Hilbert space, $\lambda_1, \lambda_2, \cdots$ a sequence of numbers with limit 1 ($\lambda_n \neq 1$), and T the linear transformation defined by $T\left(\sum_1^\infty a_i u_i\right) = \sum_1^\infty a_i \lambda_i u_i$. Then the numbers $\lambda_1, \lambda_2, \cdots$ constitute the point spectrum. The transformation $T - I$ has an inverse which is not bounded, but has dense domain. Thus 1 belongs to the continuous spectrum. All numbers other than 1 and λ_i ($i = 1, 2, \cdots$) belong to the resolvent set. See ADJOINT —adjoint of a transformation, SPECTRAL —spectral theorem.

SPEED, *n.* Distance passed over per unit of time. Speed is concerned only with the length of the path passed over per unit of time and not with its direction (see VELOCITY). The **average speed** of an object during a given interval of time is the quotient of the distance traveled during this time interval and the length of the time interval. The speed (or **instantaneous speed**) is the limit of the average speed as the time interval approaches zero. If the distance the object has traveled at time t is $h(t)$, then the average speed between times t_0 and t is the absolute value of the ratio

$$\frac{h(t) - h(t_0)}{t - t_0}.$$

The speed at time t_0 is the absolute value of the limit of this ratio as t approaches t_0. For instance, if the distance passed over is equal to the cube of the time, the speed at a time t_0 is the limit of $(t_1^3 - t_0^3)/(t_1 - t_0)$ as t_1 approaches t_0, which is $3t_0^2$. If the distance traveled is represented as a function of the time, speed is the absolute value of the *derivative* of this function with respect to the time.

    angular speed (in a plane). Relative to a point O, the **average angular speed** of a point, during a time interval of length t, is A/t, where A is the measure of the angle through which the line joining O to the point passes during this interval of time. The angular speed (the **instantaneous angular speed**) is the limit of the average speed over an interval of time, as that interval approaches zero. *Tech.* If the angle between some fixed line through O and the line joining O to the point is a function of time, the angular speed is the absolute value of the derivative of this function with respect to time.

    constant speed. See CONSTANT —constant speed and velocity.

SPHERE, *n.* The set of points in space at a given distance from a fixed point. The fixed point is the **center**, the given distance the **radius**. The **diameter** of a sphere is twice the radius (the diameter may be either the segment intercepted by the sphere on a line passing through the center or the length of this segment). The **volume** of a sphere is $\frac{4}{3}\pi r^3$, where r is the radius. The **area** of the surface of a sphere is four times the area of a great circle of the sphere, *i.e.*, $4\pi r^2$. In rectangular coordinates, the equation of a sphere of radius r is

$$x^2 + y^2 + z^2 = r^2$$

when the center is at the origin, and it is

$$(x - a)^2 + (y - b)^2 + (z - c)^2 = r^2$$

when the center is at the point whose coordinates are (a, b, c). (See DISTANCE —distance between two points.) In spherical coordinates, the equation of a sphere is $\rho = r$ when the center is at the pole. Sometimes the set of points at distance not greater than r from a point P is called a sphere, but it is also a **closed ball** of radius r; an **open ball** of radius r is the set of all points at distance less than r from a point P. An **n-sphere** with center $(a_1, a_2, \cdots, a_{n+1})$ and radius r is the set of points $(x_1, x_2, \cdots, x_{n+1})$ of $(n + 1)$-dimensional space for which $\sum_{i=1}^{n+1} (x_i - a_i)^2 = r^2$. A **closed n-ball** with center $(a_1, a_2, \cdots, a_n)$ and radius r is the set of points $(x_1, x_2, \cdots, x_n)$ of n-dimensional space for which $\sum_{i=1}^{n} (x_i - a_i)^2 \leq r^2$; the set of points for which $\sum_{i=1}^{n} (x_i - a_i)^2 < r^2$ is an **open n-ball**. If p is a point in a **metric space** with distance function ρ and r is a positive number, then the **open ball** with center p and radius r is the set of all points x for which $\rho(x, p) < r$ and the **closed ball** is the set of all x for which $\rho(x, p) \leq r$. For **normed linear spaces**, the set of points for which $\|x\| < 1$ is the **open unit ball**; the set of points for which $\|x\| \leq 1$ is the **closed unit ball**; and the set of points for which $\|x\| = 1$ is the **unit sphere**. Balls and spheres with center x_0 and radius r are defined similarly.

    celestial sphere. The spherical surface in which the stars appear to move.

    chord of a sphere. A line segment joining two points on a sphere. Any line cutting a sphere

is a **secant** and the segment cut out of a secant
by the sphere is a chord.

circumscribed and inscribed spheres. See
CIRCUMSCRIBED.

Dandelin sphere. See DANDELIN.

exotic sphere. See EXOTIC.

family of spheres. See FAMILY—one parameter family of surfaces.

secant of a sphere. See above, chord of a
sphere.

SPHER'I-CAL, *adj.* **spherical angle.** See ANGLE
—spherical angle.

spherical cone. See CONE—spherical cone.

spherical coordinates. A system of coordinates in space. The position of any point P (see
the figure) is assigned by its **radius vector** $OP = r$
(*i.e.*, the distance of P from a fixed origin or
pole O), and two angles: the **colatitude** θ, which

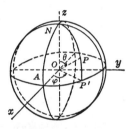

is the angle NOP made by OP with a fixed axis
ON, the **polar axis**; and the **longitude** ϕ, which is
the angle AOP' between the plane of θ and a
fixed plane NOA through the polar axis, the
initial meridian plane. A given radius vector r
confines the point P to the sphere of radius r
about the pole O. The angles θ and ϕ serve to
determine the position of P on this sphere. The
angle θ is always taken between 0 and π radians,
while ϕ can have any value (r being taken as
negative if ϕ is measured to $P'O$ extended). The
relations between the spherical and Cartesian
coordinates are:

$$x = r \sin \theta \cos \phi,$$

$$y = r \sin \theta \sin \phi,$$

$$z = r \cos \theta.$$

Sometimes ρ is used in place of r, and θ and ϕ
are often interchanged. *Syn.* geographical coordinates, polar coordinates in space.

spherical degree. The area of the *birectangular spherical triangle* whose third angle is one
degree. The area of the triangle APB in the
figure is one spherical degree. See SOLID—solid
angle.

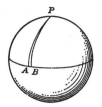

spherical excess. Of a **spherical triangle**: The
difference between the sum of the angles of a
spherical triangle and 180° (the sum of the angles of a spherical triangle is greater than 180°
and less than 540°). Of a **spherical polygon** of n
sides: The difference between the sum of the
angles of the spherical polygon and $(n - 2)180°$
(the sum of the angles of a plane polygon of n
sides).

spherical harmonic. See HARMONIC —spherical harmonic.

**spherical image (or representation) of curves
and surfaces.** For a **curve,** see various headings
under INDICATRIX. The spherical image of a point
on a surface is the extremity of the radius of the
unit sphere parallel to the positive direction of
the normal to the surface at the point. The
spherical representation (or image) of a **surface**
is the locus of the spherical images of points on
the surface. *Syn.* Gaussian representation of a
surface.

spherical polygon. A portion of a spherical
surface bounded by three or more arcs of great
circles. Its area is

$$\frac{\pi r^2 E}{180},$$

where r is the radius of the sphere and E is the
spherical excess of the polygon.

spherical pyramid. See PYRAMID.

spherical sector and segment. See SECTOR
and SEGMENT.

spherical surface. A surface whose total curvature K has the same positive value at all its
points. See PSEUDOSPHERICAL —pseudospherical
surface, and SURFACE —surface of constant curvature. Not all spherical surfaces are spheres,
but all are applicable to spheres. Hence all
spherical surfaces have the same intrinsic properties. A spherical surface is of **elliptic type** if its
linear element is reducible to the form

$$ds^2 = du^2 + c^2 \sin^2(u/a)\, dv^2, \qquad c < a;$$

the coordinate system is a geodesic one.

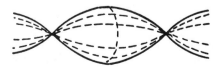

A spherical surface of revolution of elliptic type consists of a succession of congruent spindle-shaped zones. A spherical surface is of **hyperbolic type** if its linear element is reducible to the form

$$ds^2 = du^2 + c^2 \sin^2(u/a)\, dv^2, \qquad c > a;$$

the coordinate system is a geodesic one.

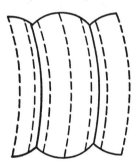

A spherical surface of revolution of hyperbolic type consists of a succession of congruent cheese-shaped zones each of which is bounded by parallels of minimum radius. A spherical surface is of **parabolic type** if its linear element is reducible to the form

$$ds^2 = du^2 + a^2 \sin^2(u/a)\, dv^2;$$

the coordinate system is a geodesic polar one. The only spherical surfaces of revolution of parabolic type are spheres.

   spherical triangle. A spherical polygon with three sides; a portion of a sphere bounded by three arcs of great circles. In the spherical triangle *ABC* (in figure), the **sides** of the triangle are a = angle *BOC*, b = angle *AOC*, and c = angle *AOB*. The **angles** of the triangle are A = angle *B'A'P*, *B*, and C = angle *B'C'P*. A spherical triangle is a **right spherical triangle** if it has at least one right angle (it may have two or three and is **birectangular** if it has two right angles and

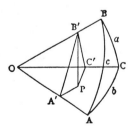

trirectangular if it has three right angles), a **quadrantal spherical triangle** if it has one side equal to 90° (a quadrant), an **oblique spherical triangle** if none of its angles are right angles, an

isosceles spherical triangle if it has two equal sides, a **scalene spherical triangle** if no two sides are equal. The **area** of a spherical triangle is $\pi r^2 E/180$, where r is the radius of the sphere and E is the triangle's spherical excess. See SOLUTION—solution of a triangle.

   spherical trigonometry. The study of spherical triangles—finding unknown sides, angles, and areas by the use of trigonometric functions of the plane angles which measure angles and sides of the triangles. See TRIGONOMETRY.

   spherical wedge. A solid the shape of a slice (from stem to blossom) of a spherical watermelon; the solid bounded by a lune of a sphere and the two planes of its great circles. Its volume is

$$\frac{\pi r^3 A}{270},$$

where r is the radius of the sphere and A is the dihedral angle (in degrees) between the plane faces of the wedge.

SPHE′ROID, *n.* Same as ELLIPSOID OF REVOLUTION. See ELLIPSOID.

SPI′NODE, *n.* Same as CUSP.

SPI′RAL, *adj., n.* See HYPERBOLIC—hyperbolic spiral, LOGARITHMIC—logarithmic spiral, PARABOLIC—parabolic spiral.

   cornu spiral. The plane curve which has parametric equations

$$x = \int_0^s \cos \tfrac{1}{2}\pi\theta^2\, d\theta, \qquad y = \int_0^s \sin \tfrac{1}{2}\pi\theta^2\, d\theta.$$

The curvature of this curve at a point P is πs, where s is the length of the curve from the origin to P. See FRESNEL—Fresnel integrals.

   equiangular spiral. Same as LOGARITHMIC SPIRAL.

   spiral of Archimedes. The plane curve which is the locus of a point that moves with uniform speed, starting at the pole, along the radius vector while the radius vector moves with uniform angular speed. Its polar equation is $r = a\theta$.

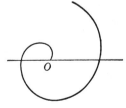

The figure shows the portion of the curve for which r is positive. See POLAR—polar coordinates in the plane.

spiral surface. A surface generated by rotating a curve C about an axis A and simultaneously transforming C homothetically relative to a point of A in such a way that for each point P of C the angle between A and the points on the locus described by P remains constant.

SPLINE, *n.* A function defined on an interval and consisting of pieces which are defined on a set of subintervals, usually as polynomials or some other simple form, and which match up with each other at the end points of the subintervals with a specified degree of accuracy. Given an interval $[a, b]$ and real numbers $a = x_0 < x_1 < \cdots < x_n = b$, a **spline of degree** m with **knots** $\{x_i : 0 \le i \le n\}$ is a function S such that the $(m - 1)$st derivative of S is continuous everywhere and, on each of the intervals $(-\infty, x_0), (x_0, x_1), \cdots, (x_{n-1}, x_n), (x_n, \infty)$, S is identically a polynomial of degree not greater than m. Splines are used for approximation of solutions of differential and integral equations and in many other ways.

SPLIT'TING, *adj.* **splitting field.** Same as GALOIS FIELD. See GALOIS.

SPREAD'ING, *adj.* **spreading method for the potential of a complex.** See POTENTIAL.

SPUR, *n.* (*German*). **spur of a matrix.** The sum of the elements in the principal diagonal. *Syn.* trace.

SQUARE, *adj., n.* In *arithmetic* or *algebra*, the result of multiplying a quantity by itself. In *geometry*, a quadrilateral with equal sides and equal angles; a rectangle with two adjacent sides equal. The area of a square is equal to the square of the length of a side.

difference of squares. See DIFFERENCE —difference of like powers.

magic squares. See MAGIC.

method of least squares. See LEAST.

perfect square. See PERFECT —perfect power.

pooled sum of squares. See POOLED.

square matrix. See MATRIX.

square numbers. Numbers which are the squares of integers, as $1, 4, 9, 16, 25, 36, 49$, etc.

square root. See MECHANIC—mechanic's rule, ROOT—root of a number.

SQUAR'ING, *n.* **squaring the circle.** The problem, using straight edge and compass alone, of constructing a square with the same area as that of a given circle. A solution is impossible, since a segment whose length is a transcendental number cannot be constructed using only straight edge and compass and the number $\sqrt{\pi}$ is tran-

scendental (the edge of the square would be $\sqrt{\pi}$ if the radius of the circle were 1). *Sny.* quadrature of a circle. See HIPPOCRATES OF CHIOS.

STA'BLE, *adj.* **stable oscillations.** See OSCILLATION.

stable point. See CHAOS.

stable system. A physical system described by a system of differential equations

$$\frac{dx_i}{dt} = f_i(x_1, \cdots, x_n); \qquad x_i(t_0) = c_i;$$

$$i = 1, \cdots, n,$$

is said to be **stable** if it returns to a stationary state under perturbations of sufficiently small magnitude. It is said to be **totally stable** if it returns to a stationary state from arbitrary perturbations. See STATIONARY —stationary state.

STAGE, *n.* See LIMIT.

STAND'ARD, *adj.* **standard atmosphere.** See ATMOSPHERE.

standard deviation. See DEVIATION.

standard error. See ERROR—standard error.

standard form of an equation. A form that has been universally accepted by mathematicians as such, in the interest of simplicity and uniformity. *E.g.*, the *standard form* of a rational integral (polynomial) equation of the nth degree in x is

$$a_0 x^n + a_1 x^{n-1} + \cdots + a_n = 0;$$

the *standard form* in rectangular Cartesian coordinates of the equation of an ellipse is

$$\frac{x^2}{a^2} + \frac{y^2}{b^2} = 1.$$

standard (or primary) infinitesimal and infinite quantities. The infinitesimal or infinite quantity relative to which *orders* are defined. If x is the *standard* or *primary infinitesimal*, then x^2 is an infinitesimal of higher (second) order with respect to x. Similarly, if x is becoming large, then x^2 is an infinite quantity of higher (second) order with respect to the *standard* or *primary infinite quantity* x. See INFINITESIMAL—order of an infinitesimal, INFINITY—order of infinities.

standard time. See TIME.

STAND'ARD-IZED', *adj.* **standardized random variable.** Given a random variable X with mean $\overline{X}$ and standard deviation σ, the random variable $(X - \overline{X})/\sigma$ has mean 0 and standard deviation 1 and is said to be a **standardized random variable** (sometimes called a **normalized random variable**).

STAR, *n.* For a member P of a family of sets, the **star** of P consists of all sets which contain P as a subset. The star of a simplex S of a *simplicial complex K* is the set of all simplexes of K for which S is a face (the star of a vertex P is the set of all simplexes which have P as a vertex). *E.g.,* the star of a vertex of a tetrahedron is the set of all edges and faces which contain P.

    star-shaped set. A set B in Euclidean space of any number of dimensions, or in any linear space, is *star-shaped with respect to a point P of B* provided that, for every point Q of B, all points of the linear segment PQ are points of B.

STATE′MENT, *adj., n.* **open statement.** Same as PROPOSITIONAL FUNCTION, *i.e.,* an **open statement** is a function whose range is a collection of statements. See PROPOSITIONAL—propositional function.

    statement function. Same as PROPOSITIONAL FUNCTION.

STAT′IC, *adj.* **static moment.** Same as MOMENT OF MASS.

STAT′ICS, *n.* The branch of mechanics of solids and fluids dealing with those situations wherein the forces acting on a body are so arranged that the body remains at rest relative to a given frame of reference. See FRAME—frame of reference.

STA′TION-AR′Y, *adj.* **stationary point.** A point on a curve at which the tangent is horizontal. For a function of one variable, a point at which the derivative is zero; for a function of several variables, a point at which all partial derivatives are zero.

    stationary state. For a physical system described at time t by a set of state variables $x_1(t), \cdots, x_n(t)$ that vary with time in accordance with a system of differential equations

$$\frac{dx_i}{dt} = f_i(x_1, \cdots, x_n); \qquad x_i(t_0) = c_i;$$

$$i = 1, \cdots, n;$$

a stationary state is a set of values $a_1, \cdots, a_n$ of $x_1, \cdots, x_n$, such that $f_i(a_1, \cdots, a_n) = 0$ for $i = 1, \cdots, n$. See STABLE—stable system.

    stationary value of an integral. See VARIATION.

STA-TIS′TIC, *n.* A function of the terms of a random sample. The *observed value of a statistic* can be computed once a sample has been taken.

For example, the *sample mean* $\sum_{i=1}^{n} X_i/n$ is a statistic. Usually a statistic is an *estimator* of some parameter of the distribution. See ESTIMATOR.

    sufficient statistic. Roughly, a sufficient statistic contains all the information in a sample about a population parameter in that no other information can be obtained from the sample to improve the estimate of the parameter. Let $\{X_1, \cdots, X_n\}$ be a *random sample* for a random variable X whose distribution is known if the value of a parameter θ is determined. A **sufficient statistic** is a statistic $t(X_1, \cdots, X_n)$ which has the property that the conditional distribution of $(X_1, \cdots, X_n)$, given a value of t, is independent of θ; i.e., if $S_1, S_2, \cdots, S_n$ are events, then the probability of $(S_1, \cdots, S_n)$, given a value of t, does not depend on θ. Under suitable regularity conditions, if t is a sufficient statistic for θ, the *maximum-likelihood estimator* for θ is a function of t. If $f(X_1, \cdots, X_n; \theta)$ is the *probability density* for $(X_1, X_2, \cdots, X_n)$ when the parameter has the value θ (or f is the *frequency function* if the distribution is discrete), then t is sufficient if $f(X_1, \cdots, X_n; \theta)$ can be factored as

$$g[t(X_1, \cdots, X_n), \theta] h(X_1, \cdots, X_n).$$

As the notation implies, the value of g is not dependent on $X_1, X_2, \cdots, X_n$, except in so far as they determine t, and the value of h is independent of θ.

    test statistic. See TEST.

STA-TIS′TI-CAL, *adj.* **statistical control.** See CONTROL.

    statistical independence. See EVENT—dependent and independent events, INDEPENDENT—independent random variables.

    statistical significance. See SIGNIFICANCE.

STA-TIS′TICS, *n.* (1) Methods of planning experiments for obtaining data and drawing conclusions or making decisions on the basis of available data. This includes: (a) Inference from samples to population by means of probability (commonly called **statistical inference**); (b) characterizing and summarizing a given set of data without direct reference to inference (called **descriptive statistics**); (c) methods of obtaining samples for statistical inference (called **sampling statistics**). (2) A collection of data (obsolete). (3) Plural of *statistic.*

    robust statistics. See ROBUST.

STEEP′EST, *adj.* **method of steepest descent.** (1) A method for approximating extreme values by use of gradients. *E.g.*, suppose (x_1, y_1) is given and one wishes a better approximation of a point that gives a local minimum for f. Then

$$x_2 = x_1 - tf_x(x_1, y_1), \qquad y_2 = y_1 - tf_y(x_1, y_1),$$

where t is determined by minimizing

$$F(t) = f\left[x_1 - tf_x(x_1, y_1), y_1 - tf_y(x_1, y_1)\right].$$

(2) A method for obtaining asymptotic expansions of functions of type $f(t) = \int_C g(z)e^{th(z)}\, dz$ as $t \to \infty$, where g and h are analytic and C is a contour in the complex plane. A point z_0 where $h'(z_0) = 0$ is called a **saddle point** of the function $e^{th(z)}$. When z moves from z_0 along the ray

$$\arg(z - z_0) = \tfrac{1}{2}\pi - \tfrac{1}{2}\arg[th''(z_0)],$$

$|e^{zh(t)}|$ decreases more rapidly than in any other direction. One then deforms C so as to pass through z_0 in this direction. *E.g.*, this technique can be used to find an asymptotic expansion for the Bessel function

$$J_n(t) = \pi^{-1} i^{-n} \int_0^\pi \cos nz\, e^{it\cos z}\, dz.$$

Syn. saddle-point method.

STEINITZ, Ernest (1871–1928). German algebraist and topologist.

    Steinitz theorem. If x is an interior point of the convex span of a subset S of n-dimensional Euclidean space, then S contains a subset X which has at most $2n$ points and for which x is an interior point of the convex span of X. See related theorems under CARATHEODORY, HELLY, RADON.

STEP, *adj., n.* **step function.** A function which is defined throughout some interval I and is constant on each one of a finite set of nonintersecting intervals whose union is I. See INTEGRABLE —integrable function, and INTERVAL.

STERADIAN, *n.* See SOLID—solid angle.

STERE, *n.* One cubic meter, or 35.3156 cubic feet. Used mostly in measuring wood. See METRIC—metric system.

STER'E-O-GRAPH'IC, *adj.* **stereographic projection.** See PROJECTION.

STEVIN (or STEVINUS), Simon (1548–1620). Flemish algebraist, arithmetist, and engineer. Popularized decimal fractions and discussed limits, anticipating calculus. See SUM—sum of vectors.

STIELTJES, Thomas Jan (1856–1894). French analyst and number theorist. See MOMENT— moment problem.

    Lebesgue-Stieltjes integral. Let a *measurable* function f and a monotonically increasing function ϕ be defined on an interval $[a, b]$. Define $F(\xi)$ for $\phi(a) \leq \xi \leq \phi(b)$ by the relations: (1) $F(\xi) = f(x)$ if there is a point x such that $\xi = \phi(x)$; (2) if $\xi_0 \neq \phi(x)$ for any x, then it follows that there is a unique point of discontinuity x_0 of ϕ such that

$$\phi(x_0 - 0) \leq \xi_0 \leq \phi(x_0 + 0),$$

and $F(\xi_0)$ is defined as $f(x_0)$. If the *Lebesgue integral* $\int_{\phi(a)}^{\phi(b)} F(\xi)\, d\xi$ exists, its value is defined to be the **Lebesgue-Stieltjes integral** of f with respect to ϕ, written

$$\int_a^b f(x)\, d\phi(x).$$

If ϕ is of *bounded variation*, it is the difference $\phi_1 - \phi_2$ of monotonic increasing functions ϕ_1 and ϕ_2, and the Lebesgue-Stieltjes integral $\int_a^b f(x)\, d\phi(x)$ is defined as

$$\int_a^b f(x)\, d\phi_1(x) - \int_a^b f(x)\, d\phi_2(x).$$

If F as defined above is measurable on $[\phi(a), \phi(b)]$, f is measurable on $[a, b]$, and

$$\phi(x) = \int_a^x \theta(x)\, dx$$

for some measurable function θ, then

$$\int_a^b f(x)\theta(x)\, dx = \int_a^b f(x)\, d\phi(x),$$

the former being a Lebesgue integral.

    Riemann-Stieltjes integral. Let $a = x_0, x_1, x_2, \cdots, x_n = b$ be a subdivision of the interval $[a, b]$ and let

$$s_n = \max|x_i - x_{i-1}| \qquad (i = 1, 2, \cdots, n).$$

Let f and ϕ be bounded real-valued functions

defined on $[a, b]$ and

$$S_n = \sum_{i=1}^{n} f(\xi_i)[\phi(x_i) - \phi(x_{i-1})],$$

where ξ_i are arbitrary numbers satisfying $x_{i-1} < \xi_i < x_i$. If $\lim S_n$ exists as n becomes infinite in such a way that s_n approaches zero, and if the limit is independent of the choice of ξ_i and of the manner of the successive subdivisions, then this limit is the **Riemann-Stieltjes integral** of f with respect to ϕ, written

$$\int_a^b f(x)\,d\phi(x).$$

If $\int_a^b f(x)\,d\phi(x)$ exists, then $\int_a^b \phi(x)\,df(x)$ exists and

$$\int_a^b f(x)\,d\phi(x) + \int_a^b \phi(x)\,df(x)$$

$$= f(b)\phi(b) - f(a)\phi(a).$$

If f is bounded on $[a, b]$ and ϕ is of bounded variation on $[a, b]$, then

$$\int_a^b f(x)\,d\phi(x)$$

exists if, and only if, the total variation of ϕ over the set of points of discontinuity of f is zero (*i.e.*, the set of all points $\phi(x)$ for which f is discontinuous at x is of measure zero, $\phi(x)$ being taken as the interval between $\phi(x - 0)$ and $\phi(x + 0)$ if ϕ is discontinuous at x).

STIRLING, James (1692–1770). A Scot who began his career at Oxford, Venice, and London with brilliant academic work in mathematics amid political and religious conflicts and friendship with Newton and Maclaurin, but devoted most of his life after 1735 to industrial management.

Stirling's formula. (1) The formula $(n/e)^n \sqrt{2\pi n}$, which is asymptotic to $n!$; *i.e.*, $\lim_{n\to\infty} n!/[(n/e)^n\sqrt{2\pi n}] = 1$. More precisely, for each n there is θ_n for which

$$n! = (n/e)^n \sqrt{2\pi n}\, e^{\theta_n/(12n)},$$

and $0 < \theta_n < 1$. Stirling's formula can be extended to the asymptotic series

$(n/e)^n \sqrt{2\pi n}\, e^w$, where

$$w = \frac{1}{12n} - \frac{1}{360n^3} + \frac{1}{1260n^5} - \cdots.$$

(2) Maclaurin series, discovered by Stirling, but published first by Maclaurin.

Stirling's series. Either of the two asymptotic expansions:

$$\log \Gamma(x) = \left(x - \tfrac{1}{2}\right) \log x - x + \tfrac{1}{2}\log(2\pi)$$

$$+ \sum_{k=1}^{\infty} \frac{(-1)^{k-1} B_k}{2k(2k-1)x^{2k-1}},$$

$$\Gamma(x) = e^{-x} x^{x-1/2} (2\pi)^{1/2}$$

$$\times \left\{ 1 + \frac{1}{12x} + \frac{1}{288x^2} - \frac{139}{51840x^3} + O\!\left(\frac{1}{x^4}\right) \right\};$$

where $\Gamma(x)$ is the *gamma function*; $B_1, B_2, \cdots$, are the *Bernoulli numbers* $\tfrac{1}{6}, \tfrac{1}{30}, \tfrac{1}{42}, \cdots$; and $O(1/x^4)$ is a function such that $x^4 \cdot O(1/x^4)$ is bounded as $x \to \infty$ (the second series can be extended through successive powers of x^{-1}).

STO-CHAS′TIC, *adj.* **stochastic independence.** Refers to independence of events or random variables. See EVENT—dependent and independent events, INDEPENDENT —independent random variables.

stochastic process. A collection of **random variables** $\{X(t): t \in T\}$, where T is the *index set* and there is a random variable $X(t)$ for each t in T. When T is a discrete set (*e.g.*, a set of integers), the process is a **discrete parameter process.** When T is an interval of real numbers, the process is a **continuous parameter process.** For examples, see MARTINGALE, POISSON —Poisson process, RANDOM —random walk, WIENER —Wiener process.

stochastic variable. Same as RANDOM VARIABLE. See RANDOM.

STOCK, *n.* The capital (or certain assets) of a corporation or company which is in the form of transferable shares. **Preferred stock** is stock upon which a fixed rate of interest (dividend) is paid, after the dividends on bonds have been paid and before any dividends are paid on common stock. If the company fails, the order of redemption is the same as the above order of payments. **Common stock** is stock which receives as dividends its proportionate share of the proceeds after all other demands have been met. A **stock certificate** is a written statement that the owner of the certificate has a certain amount of capital in the corporation which issued the certificate. A **stock company** is a company composed of the purchasers of certain stock. A **stock insurance company** is a stock company whose business is insurance. The policies they sell are nonparticipating. The profits go to the stockholders. However, some stock companies, because of the competition of mutual companies, write participating policies. *Syn.* nonparticipating insurance company.

STOKES, Sir George Gabriel (1819–1903). British analyst and physicist who used mathematics as a tool for his study of hydrodynamics, elasticity, and wave theory.

Stokes' theorem. Let S be an orientable surface and C be the boundary of S. Then the line integral of a vector-valued function **F** in the positive direction around C is equal to the surface integral over S of $(\nabla \times \mathbf{F}) \cdot \mathbf{n}$, where **n** is the unit normal to S (see SURFACE —surface integral). It is necessary to restrict S and **F**. Sufficient conditions are that S be the union of a finite number of smooth surface elements and that the first-order partial derivatives of the components of **F** be continuous on S. If C is piece-wise smooth and each surface element of S can be represented using x and y as parameters, and using y and z as parameters, and using z and x as parameters, then Stokes' theorem means that the integral of $Ldx + Mdy + Ndz$ around C is equal to the integral over S of

$$\left(\frac{\partial N}{\partial y} - \frac{\partial M}{\partial z} \right)(\pm dy\,dz) + \left(\frac{\partial L}{\partial z} - \frac{\partial N}{\partial x} \right)(\pm dz\,dx)$$

$$+ \left(\frac{\partial N}{\partial x} - \frac{\partial L}{\partial y} \right)(\pm dx\,dy),$$

where $\mathbf{F} = L\mathbf{i} + M\mathbf{j} + N\mathbf{k}$ and the signs are determined by whether **n** points toward the positive or the negative side of the corresponding coordinate planes.

STONE, *n.* See DENOMINATE NUMBERS in the appendix.

STONE, Marshall Harvey (1903–1989). American functional analyst, algebraist, logician, and topologist. Son of former chief justice Harlan Stone.

Stone-Čech compactification. See COMPACTIFICATION.

Stone-Weierstrass theorem. Stone's generalization of the *Weierstrass approximation theorem*: Let T be a compact topological space and S be a set of continuous real-valued functions defined on T. Then each continuous real-valued function defined on T can be uniformly approximated by a member of S if S has the following properties: (1) if f and g are members of S and a is a real number, then $af, f + g$, and $f \times g$ are members of S; (2) if x and y are distinct points of T and a and b are real numbers, there is a member f of S for which $f(x) = a$ and $f(y) = b$.

STOR'AGE, *adj.* **storage component.** In a computing machine, any component that is used in storing information for later use. The storage might be permanent or temporary, of quick or slow access, etc. Magnetic drums and tapes, television tubes, mercury delay lines, etc., are used for this purpose. *Syn.* memory component.

STRAIGHT, *adj.* Continuing in the same direction; not swerving or turning; in a *straight line*. See LINE.
    straight angle. See ANGLE—straight angle.
    straight line. See LINE—straight line.

STRAIN, *adj.*, *n.* The change in the relative positions of points in a medium, the change being produced by a deformation of the medium as a result of *stress*.
    coefficient of strain. See below, one-dimensional strains.
    homogeneous strains. The concept in dynamics represented approximately by the *homogeneous affine transformation*; the forces acting internally in an elastic body when it is deformed.
    longitudinal strain. See below, strain tensor.
    one-dimensional strains. The transformations $x' = x$, $y' = Ky$, or $x' = Kx$, $y' = y$. These transformations elongate or compress a configuration in the directions parallel to the axes, according as $K > 1$ or $K < 1$. The constant K is called the *coefficient of the strain*. *Syn.* simple elongations and compressions, one-dimensional elongations and compressions.
    principal directions of strain. At each point of an undeformed medium there exists a set of three mutually orthogonal directions which remain mutually orthogonal after the deformation has taken place. These directions are **principal directions of strain.**
    principal strains. The elongations in the directions of the principal directions of strain (*q.v.*).
    shearing strain. Strain due to the distortion of the angles between the initially orthogonal directions in a medium which has been deformed. See below, strain tensor.
    simple strains. A general name given to *simple elongations* and *compressions*, and *simple shears*.
    strain tensor. In the linear theory of elasticity, a set of six functions $e_{xx}, e_{yy}, e_{zz}, e_{xy}, e_{zy}, e_{xz}$,

related to the displacements u, v, w along the Cartesian axes x, y, z, respectively, by the formulas $e_{xx} = \dfrac{\partial u}{\partial x}$,

$$e_{yy} = \frac{\partial v}{\partial y}, \qquad e_{zz} = \frac{\partial w}{\partial z}, \qquad e_{xy} = \frac{1}{2}\left(\frac{\partial v}{\partial x} + \frac{\partial u}{\partial y}\right),$$

$$e_{zy} = \frac{1}{2}\left(\frac{\partial w}{\partial y} + \frac{\partial v}{\partial z}\right), \qquad e_{xz} = \frac{1}{2}\left(\frac{\partial u}{\partial z} + \frac{\partial w}{\partial x}\right).$$

These six quantities (or alternatively, the set of three principal strains) characterize the state of strain of a body. The quantities e_{xx}, e_{yy}, e_{zz} are the **longitudinal strains** and the remaining ones are the **shearing strains** (*q.v.*). The integrability conditions for the six components of the strain tensor (**Saint-Venant's compatibility equations**) are

$$(e_{ij})_{kl} + (e_{kl})_{ij} - (e_{ik})_{jl} - (e_{jl})_{ik} = 0,$$

where i, j, k, l take on any of the values x, y, z and subscripts on the outside of parentheses indicate partial differentiation.

STRAT′A, *n.* See RANDOM —stratified random sample.

STRAT′E-GY, *n.* **strategy for a game.** A **pure strategy** is any specifically determined plan, covering all possible contingencies but not involving the use of random devices, that a player might make in advance for a complete play of a game. If a player of a game has m possible pure strategies, then any probability vector $X = (x_1, x_2, \cdots, x_m)$, with each $x_i \geq 0$ and with $\sum x_i = 1$, is a **mixed strategy** for the player. If the player chooses this mixed strategy, then with probability x_i as determined by a random device he will employ his ith pure strategy for a given play of the game. Similarly, for continuous games a mixed strategy is a probability distribution over the continuum $[0, 1]$ of pure strategies. Since, for example, the mixed strategy $(1, 0, \cdots, 0)$ is equivalent to the player's first pure strategy, it follows that any pure strategy can be considered as a special mixed strategy. The word **strategy** is often used (when the meaning is clear from the context) to denote a *pure strategy*; it is also sometimes used to denote a *mixed strategy*. A pure strategy is a **dominant strategy** for one player of a game, relative to a second pure strategy for the same player, if the first strategy has, for each pure strategy of the opponent, at least as great *payoff* as the second (it is a **strictly dominant strategy** if its payoff is always greater than that of the second). For a two-person zero-sum game having *value* v, a strategy—either a pure strategy or a mixed strategy given by a probability vector or a probability distribution function—for the *maximizing player* that will make the expected value of the payoff at least v (or for the *minimizing player* that will make the expected value of the payoff at most v), regardless of the strategy chosen by the opponent, is said to be an **optimal strategy**.

STRAT′I-FIED, *adj.* **stratified sample.** See RANDOM —stratified random sample.

STREAM, *adj.* **stream function** and **stream lines.** See FUNCTION —stream function.

STRESS, *n.* A material body is said to be *stressed* when the action of external forces is transmitted to its interior. The average stress $\bar{T}$ is the average force F per unit area a of the planar element passing through a given point in the medium. The actual stress at the point is the limit of the ratio $\bar{T} = F/a$ as the area a containing the point in question is made to approach zero. The magnitude and the direction of the stress vector $\mathbf{T}$ depend not only on the choice of the point in the body but also on the orientation of the planar element at the chosen point. The component $\mathbf{T}_n$ of the stress vector $\mathbf{T}$ in the direction of the normal to the planar element is called the **normal stress**, while the component T in the plane of the element is the **shearing stress**.

    internal stress. The resistance of a physical body to external forces; the unit internal resistance set up by external forces.

STRETCH′ING, *adj.* **stretching and shrinking transformations.** See SIMILITUDE —transformation of similitude.

STRICTLY, *adv.* **strictly convex space.** See CONVEX.

    strictly increasing and **strictly decreasing.** See INCREASING —increasing function, DECREASING —decreasing function.

STRIC′TION, *n.* **line of striction of a ruled surface.** The locus of the central points of the rulings on the surface. See RULING —central plane and point of a ruling.

STRING, *n.* See BRAID.

STRONG, *adj.* **strong law of large numbers.** See LARGE —law of large numbers.

    strong topology. See OPERATOR —linear operator. TOPOLOGY —topology of a space.

STROPH'OID, *n.* The plane locus of a point on a variable line passing through a fixed point when the distance from the describing point to the intersection of the line with the y-axis is equal to the y-intercept. If the coordinates of the fixed point are taken as $(-a, 0)$, the equation of the curve is $y^2 = x^2(x+a)/(a-x)$. In the figure, $P'E = EP = OE$, A is the point through which the line always passes, and the dotted line is the asymptote of the curve.

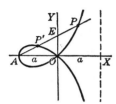

STUDENT (*Pseu.* for **W. S. Gosset**).
    Student's t. See T.

STURM, Jacques Charles François (1803–1855). Swiss-French analyst and mathematical physicist.
    Sturm comparison theorem. Suppose p and p_1 have derivatives that are continuous on an interval I, that q and q_1 are continuous on I, and that $p(x) \geq p_1(x) > 0$ and $q_1(x) \geq q(x)$ for all x in I. Then between any two zeros in I of each function u, which is not identically zero on I and for which $(pu')' + qu = 0$ on I, there is at least one zero of u_1, if $(p_1 u_1')' + q_1 u_1 = 0$ on I and it is not true that $u = cu_1$ for some c, $p = p_1$, and $q = q_1$.
    Strum-Liouville differential equation. A differential equation of the form

$$\frac{d}{dx}\left[p(x)\frac{dy}{dx}\right] + [\lambda\rho(x) - q(x)]y = 0,$$

where $p(x)$ and $\rho(x)$ are positive if x is in a given closed interval $[a, b]$, the functions p', q, and ρ are continuous on $[a, b]$, and λ is a parameter. **A regular Sturm-Liouville system** is such a differential equation together with boundary conditions of type $\alpha y(a) + \beta y'(a) = 0$, $\gamma y(b) + \delta y'(b) = 0$, where not both α and β are zero and not both γ and δ are zero. The operator T, with $T(y)$ defined as $-d[py']/dx + qy$, is *symmetric* on the set of continuously second-differentiable functions satisfying the boundary conditions; there is an increasing sequence of **eigenvalues** $\{\lambda_n\}$ which are real numbers for which $\lim \lambda_n = \infty$; each eigenvalue λ_n has an **eigenfunction** ϕ_n that is unique within scalar multiplication; ϕ_i is orthogonal to ϕ_j if $i \neq j$; ϕ_n has exactly $n - 1$ zeros on (a, b), and the sequence

$\{\phi_n\}$ is a complete orthogonal sequence for the Hilbert space of (Lebesgue) measurable functions f for which $|f|^2$ has a finite integral on $[a, b]$. Similar theorems can be proved for certain **singular systems**, such as the Legendre equation on $[-1, +1]$ for which $p(x)$ is zero at both endpoints and boundary conditions are not used $[p(x) = 1 - x^2]$, and cases for which the interval is not bounded. See PRÜFER.
    Sturm functions. A sequence of functions derived from a given polynomial f; explicitly, the sequence of functions $f_0, f_1, \cdots, f_n$, where $f_0(x) \equiv f(x)$, $f_1(x) \equiv f'(x)$, and $f_2(x)$, $f_3(x)$, etc., are the negatives of the remainders occurring in the process of finding the highest common factor of $f(x)$ and $f'(x)$ by Euclid's algorithm. This sequence is a **sequence of Sturm functions.**
    Sturm separation theorem. If u and v are real linearly independent solutions of the differential equation

$$y'' + p(x)y' + q(x)y = 0$$

on an interval I on which p and q are continuous, then between any two consecutive zeros of u there is exactly one zero of v.
    Sturm's theorem. A theorem determining the number of real roots of an algebraic equation which lie between any two arbitrarily chosen values of the variable. The theorem states that the number of real roots of $f(x) = 0$ between two numbers a and b with $f(a) \neq 0$ and $f(b) \neq 0$ is equal to the difference between the number of variations of sign in the sequence of values of the *Sturm's functions* [derived from f] when $x = a$ and when $x = b$, vanishing terms not being counted and each multiple root being counted exactly once. See VARIATION —variation of sign in an ordered set of numbers.

SUB-AD'DI-TIVE, *adj.* See ADDITIVE— additive function, additive set function.

SUB'BASE, *n.* See BASE—base for a topological space.

SUB'CLASS', *n.* Same as SUBSET. See SUBSET.

SUB-FAC-TO'RI-AL, *n.* **subfactorial of an integer.** If n is the integer, *subfactorial* n is

$$n! \times \left[\frac{1}{2!} - \frac{1}{3!} + \frac{1}{4!} - \cdots \frac{(-1)^n}{n!}\right].$$ This is

equal to $n!E$, where E is the sum of the first $n + 1$ terms in the Maclaurin expansion of e^x

with $x = -1$. E.g., *subfactorial* 4 is equal to

$$4!\left(\frac{1}{2!} - \frac{1}{3!} + \frac{1}{4!}\right) = 24\left(\frac{1}{2} - \frac{1}{6} + \frac{1}{24}\right) = 9.$$

SUB'FIELD, *n.* A subset of a field which is itself a field. *E.g.*, the set of rational numbers is a subfield of the set of real numbers. See FIELD.

SUB'GROUP, *n.* See GROUP.

    invariant or **normal subgroup.** See NORMAL —normal subgroup.

SUB'HAR-MON'IC, *adj.* **subharmonic function.** A real-valued function u whose domain is a two-dimensional domain D is **subharmonic** in D provided u satisfies the following conditions in D: (1) $-\infty \le u(x, y) < +\infty$. (The condition $u(x, y) \not\equiv -\infty$ is sometimes added.) (2) u is upper semicontinuous in D. (3) For any subdomain D' included, together with its boundary B', in D, and for any function h harmonic in D', continuous in $D' + B'$, and satisfying $h(x, y) \ge u(x, y)$ on B', we have $h(x, y) \ge u(x, y)$ in D'. A subharmonic function u which satisfies $u(x, y) \not\equiv -\infty$ necessarily is summable. If a function u satisfies $u(x, y) \not\equiv -\infty$ and is upper semi-continuous in its domain D of definition, then u is subharmonic if and only if it satisfies either of the following mean-value inequalities for each circular disc in D:

$$u(x_0, y_0)$$
$$\le \frac{1}{2\pi}\int_0^{2\pi} u(x_0 + \rho\cos\theta, y_0 + \rho\sin\theta)\, d\theta,$$

$$u(x_0, y_0)$$
$$\le \frac{1}{\pi r^2}\int_0^r\int_0^{2\pi} u(x_0 + \rho\cos\theta,$$

$$y_0 + \rho\sin\theta)\rho\, d\rho\, d\theta.$$

If a function u has continuous second-order partial derivatives in its domain D of definition, then it is subharmonic in D if, and only if, the following differential inequality is satisfied at each point of D:

$$\Delta u = \frac{\partial^2 u}{\partial x^2} + \frac{\partial^2 u}{\partial y^2} \ge 0.$$

The notion of subharmonic function extends immediately to functions of n variables. See CONVEX—convex function.

SUB-NOR'MAL, *n.* The projection on the axis of abscissas (x-axis) of the segment of the normal between the point of the curve and the point of intersection of the normal with the x-axis. The length of the subnormal is $y(dy/dx)$, where y and dy/dx (the derivative of y with respect to x) are evaluated at the given point on the curve. See TANGENT—length of a tangent.

    polar subnormal. See POLAR—polar tangent.

SUB'RE'GION, *n.* A region within a region.

SUB'SCRIPT, *n.* A small number, letter, or symbol written below and to the right or left of a letter or symbol. Often used on a variable to denote a constant value of that variable or to distinguish between variables. The symbols a_1, a_2, etc., denote constants; $D_x f$ denotes the derivative of f with respect to x; (x_0, y_0), (x_1, y_1), etc., denote coordinates of fixed points; $f(x_1, x_2, \cdots, x_n)$ denotes a function of n variables, $x_1, x_2, \cdots, x_n$; $_nC_r$ denotes the number of possible combinations of n things r at a time. **Double subscripts** are used, for example, in writing determinants with general terms (the general term might be denoted by a_{ij}, where the first subscript denotes the row number and the second the column number). See SUPERSCRIPT, and the usage in various headings under TENSOR.

SUB-SE'QUENCE, *n.* A sequence within a sequence; $\frac{1}{2}, \frac{1}{4}, \cdots, 1/(2n), \cdots$, is a subsequence of $1, \frac{1}{2}, \frac{1}{3}, \frac{1}{4}, \cdots, 1/n, \cdots$.

SUB'SET, *n.* If each member of a set A belongs to a set B, then one says that A is contained in B, B contains A, A is a **subset** of B, or B is a **superset** of A. A subset R is a **proper subset** of (or **contained properly** in) a set S if R is a subset of S and $R \ne S$.

SUB'SINE, *adj.* **subsine function of order** ρ. A function f that is dominated by functions of the form $F(x) \equiv A\cos\rho x + B\sin\rho x$ in the way that convex functions are dominated by linear functions. Thus for f to be a subsine function of order ρ on the interval I, if the numbers x_1, x_2 are in I and satisfy $0 < x_2 - x_1 < \pi/\rho$, and the above function F satisfies $F(x_1) = f(x_1)$, $F(x_2) = f(x_2)$, we must have $f(x) \le F(x)$ for $x_1 < x < x_2$. See PHRAGMÉN—Phragmén-Lindelöf function.

SUB'STI-TU'TION, *n.* **elimination by substitution.** See ELIMINATION.

integration by substitution. See INTEGRATION —change of variable in integration.

inverse substitution. The substitution which exactly undoes the effect of a given substitution. For examples, see TRANSFORMATION —inverse transformation.

substitution group. Same as PERMUTATION GROUP.

substitution of one quantity for another. Replacing the one quantity by the other. Substitutions are made in order to simplify equations, simplify integrands (in the calculus), and to change (transform) geometric configurations into other forms or to different positions. See TRANSFORMATION.

trigonometric substitution. See TRIGONOMETRIC.

SUB-TAN'GENT, *n.* The projection on the axis of abscissas (*x*-axis) of the segment of the tangent joining the point of tangency on the curve and the point of intersection of the tangent with the *x*-axis; the segment of the axis of abscissas between the foot of the ordinate at the point on the curve and the *x*-intercept of the tangent. The length of the subtangent is $y(dx/dy)$, where y and dx/dy are evaluated at the point on the curve. See DERIVATIVE, TANGENT —length of a tangent.

SUB-TEND', *v.* To be opposite to, or measure off, as a side of a triangle subtends the opposite angle, and an arc of a circle subtends the central angle of the arc. The angle is also said to be *subtended by* the side of the triangle or arc of the circle.

SUB-TRAC'TION, *adj.*, *n.* The process of finding a quantity which when added to one of two given quantities will give the other. These quantities are, respectively, the **subtrahend** and **minuend**, and the quantity found is the **difference** or **remainder**. *E.g.*, 2 subtracted from 5 is written $5 - 2 = 3$; 5 is the *minuend*, 2 the *subtrahend*, and 3 the *difference* or *remainder*. Subtraction of signed numbers is **algebraic subtraction** (this is equivalent to changing the sign of the subtrahend and adding it to the minuend). *E.g.*, $5 - 7 = 5 + (-7)$, which is -2. See SUM —sum of real numbers.

subtraction formulas. See TRIGONOMETRY —identities of plane trigonometry.

SUB'TRA-HEND', *n.* A quantity to be subtracted from another.

SUC-CES'SIVE, *adj.* Same as CONSECUTIVE. See CONSECUTIVE.

SUC-CES'SOR, *n.* **successor of an integer.** The next integer; the successor of n is $n + 1$. *Syn.* consequent. See INTEGER.

SUF-FI'CIENT, *adj.* **sufficient condition.** See CONDITION.

sufficient statistic. See STATISTIC —sufficient statistic.

SUM, *n.* The **sum** of two or more objects is the object which is determined from these objects by a given operation called **addition**. Usually this addition operation is related (sometimes remotely) to some process of accumulation. *E.g.*, $2 + 3 = 5$ is related to the statement that putting together two sets, one containing 2 things and the other 3 things, will yield a set containing 5; the sum of vectors which represent forces is the vector which represents a force equivalent to all the individual forces operating together. See the various headings below.

algebraic sum. The combination of terms either by addition or subtraction in the sense that adding a negative number is equivalent to subtracting a positive one. The expression $x - y + z$ is an *algebraic sum* in the sense that it is the same as $x + (-y) + z$.

arithmetic sum. The number obtained by adding positive numbers. Five is the sum of two and three, written $2 + 3 = 5$.

limit of a sum. See LIMIT —fundamental theorems on limits.

partial sum of an infinite series. The sum of a finite number of consecutive terms of the series, beginning with the first term. If the series is $a_1 + a_2 + a_3 + \cdots$, then each of the quantities S_n is a partial sum, where

$$S_n = a_1 + a_2 + \cdots + a_n$$

is the sum of the first n terms of the series.

sum of angles. *Geometrically*, the angle determined by a rotation from the initial side through one angle, followed by a rotation, beginning with the terminal side of this angle, through the other angle; *algebraically*, the ordinary algebraic addition of the same kind of measures of the angles (*e.g.*, degrees plus degrees or radians plus radians).

sum of complex numbers. See COMPLEX —complex numbers.

sum of directed line segments. The line segment which extends from the initial point of the first line segment to the terminal point of the last line segment when the line segments are placed so that the terminal point of each is coincident with the initial point of the next. *E.g.*, 5 miles east plus 2 miles west is 3 miles east. This is a special case of the sum of vectors; see below.

sum of an infinite series. The limit of the sum of the first n terms of the series, as n increases. This is not a sum in the ordinary sense of arithmetic, because the terms of an infinite series can never all be added term by term. The sum of the series

$$\tfrac{1}{2} + \tfrac{1}{4} + \tfrac{1}{8} + \cdots + \left(\tfrac{1}{2}\right)^n + \cdots$$

is 1, because that is the limit approached by the sum of the first n of these terms, namely $1 - 1/2^n$, as n becomes infinite. The sum of the series is precisely 1, even though the actual arithmetic sum of a finite number of terms of the series is always less than 1. The series $1 + (-1) + 1 + (-1) + 1 + \cdots$ does not have a sum, since the sum of the first n terms is $+1$ if n is odd and 0 if n is even and therefore the sum of the first n terms does not have a limit as n increases. *Tech.* An infinite series $a_1 + a_2 + a_3 + \cdots$ is **convergent** and its **sum** is S if

$$\lim_{n \to \infty} (a_1 + a_2 + \cdots + a_n)$$

exists and equals S. The series is divergent if this limit does not exist. See SERIES —geometric series, double series, and various headings under CONVERGENCE.

sum of like powers. (1) An algebraic expression of type $x^n + y^n$. Such sums are of interest in factoring, because when n is odd the sum $x^n + y^n$ is divisible by $x + y$. See DIFFERENCE —difference of like powers. (2) See WARING —Waring's problem.

sum of matrices The sum $A + B$ of matrices A and B is the matrix whose elements are formed by the rule that the element in row r and column s is the sum of the elements a_{rs} and b_{rs} in row r and column s of A and B, respectively. This sum is defined only if A and B have the same number of rows and the same number of columns.

sum of order t. For positive numbers a_i, the expression $(\Sigma a_i^t)^{1/t}$. For the definition of the analogous mean of order t, see AVERAGE.

sum of real numbers. Positive integers (and zero) can be thought of as symbols used to describe the "manyness" of the set of objects (also see INTEGER). Then the **sum of two integers** A and B is the integer which describes the "manyness" of the set of objects obtained by combining a set of A objects with a set of B objects. This means that addition of integers is the process of finding the number class which is composed of the number classes denoted by the addends (see CARDINAL —cardinal number). The

sum of fractions is obtained by the same process, after a common denominator (common unit) has been established. *E.g.*, $\tfrac{1}{2}$ and $\tfrac{2}{3}$ are the same as $\tfrac{3}{6}$ and $\tfrac{4}{6}$, and 3 sixths plus 4 sixths is 7 sixths. In general,

$$\frac{a}{b} + \frac{c}{d} = \frac{ad + bc}{bd}.$$

The **sum of mixed numbers** can be determined by adding the integral parts and fractions separately, or by reducing each mixed number to a fraction. *E.g.*, $2\tfrac{1}{2} + 3\tfrac{1}{4} = 2 + 3 + \tfrac{1}{2} + \tfrac{1}{4} = 5\tfrac{3}{4}$, or $2\tfrac{1}{2} + 3\tfrac{1}{4} = \tfrac{10}{4} + \tfrac{13}{4} = \tfrac{23}{4}$. When the numbers being added have been given signs, the addition is done as follows (this is sometimes called **algebraic addition**). Two positive numbers are added as above; two negative numbers are added by adding their numerical values and making the result negative; a positive and a negative number are added by subtracting the lesser numerical value from the greater and giving the difference the sign of the number which has the greater numerical value. *E.g.*, $(-2) + (-3) = -5$; $(-2) + 3 = 1$. The significance of this definition becomes apparent when we let positive numbers denote distances eastward and negative numbers distances westward and think of their sum as the distance from the starting point to the place reached by travelling in succession the paths measured by the addends. *E.g.*, an interpretation of $(-3) + 2 = -1$ is that one would finish 1 mile west of the starting point if one traveled 3 miles west and then 2 miles east. The **sum of irrational numbers** may be left in indicated form, after similar terms have been combined until some specific application indicates the degree of accuracy desired. Such a sum as $(\sqrt{2} + \sqrt{3}) - (2\sqrt{2} - 5\sqrt{3})$ would thus be left in the form $6\sqrt{3} - \sqrt{2}$. A sum such as $\pi + \sqrt{2}$ can be approximated as

$$3.1416 + 1.4142 = 4.5558.$$

It is necessary to have a specific definition of irrational numbers before one can specifically define the sum of numbers, one or more of which is irrational. See DEDEKIND —Dedekind cut.

sum of sets. Same as the UNION.

sum of vectors. *Algebraically*, the vector obtained by the addition of corresponding components. *E.g.*,

$$(2i + 3j) + (i - 2j) = 3i + j,$$

$$(2i + 3j + 5k) + (i - 2j + 3k) = 3i + j + 8k.$$

Geometrically, the sum of vectors can be determined by representing the vectors as arrows joined in sequence with the initial point of each vector coinciding with the terminal point of the preceding one; the sum is then the vector with initial point at the initial point of the first vector and terminal point at the terminal point of the last vector. For two vectors, this gives the **parallelogram law** illustrated in the figure: the sum of the vectors **A** and **B** is the vector **C** along the diagonal of the parallelogram determined by **A** and **B** (the figure illustrates commutativity of vector addition—it also is associative). It was enunciated in 1586 by Stevinus that the resultant of two forces can be represented in this way (the **parallelogram law of forces**). The resultant is that force whose effect is equivalent to the several forces jointly. Many other physical entities can be represented by vectors and combined in this way (*e.g.*, velocity and acceleration). *Syn.* resultant of vectors. See VECTOR.

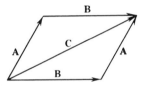

SUM′MA-BLE, *adj.* **absolutely summable series.** Refers to summability by Borel's integral method. A series Σa_n is said to be absolutely summable if the integrals

$$\int_0^\infty e^{-x}|a(x)|\,dx \quad \text{and} \quad \int_0^\infty e^{-x}|a^{(m)}(x)|\,dx$$

all exist, where $m = 1, 2, 3, \cdots$ denotes derivatives of these orders, and $a(x) = a_0 + a_1 x + a_2 x^2/2! + \cdots$.

 summable divergent series. Series to which a sum is assigned by some *regular* definition of the sum of a divergent series. Better usage is to speak of a series as, *e.g.*, Cesàro summable. *I.e.*, indicate the method by which the series is summable. See SUMMATION —summation of divergent series.

 summable function. Same as *integrable function*. See INTEGRABLE.

 uniformly summable series. A series of variable terms is *uniformly summable* on a set S by a given definition of the sum of a divergent series if the sequence which defines the sum converges uniformly on S. The series $\Sigma(-x)^n$ diverges for $x = 1$, but is uniformly summable for $0 \leq x \leq 1$ by any of the standard definitions, such as Hölder's,

Cesàro's, and Borel's. By Hölder's definition, we have for the sum

$$\lim_{n \to \infty}\left[1 + (1 - x) + (1 - x + x^2)\right.$$

$$\left. + \cdots + \sum_{k=0}^{n-1}(-x)^k\right] \div n$$

$$= \lim_{n \to \infty}\left[\frac{1}{n} - x\frac{(n-1)}{n} + x^2\frac{(n-2)}{n}\right.$$

$$\left. - \cdots + (-x)^{n-1}\frac{1}{n}\right],$$

which converges uniformly, with regard to x, on the closed interval $[0, 1]$.

SUM′MAND, *n.* Any one of two or more terms to be added to form a sum.

SUM-MA′TION, *adj.*, *n.* **integration as a summation process.** See INTEGRAL —definite integral.

 summation convention. The convention of letting the repetition of an index (subscript or superscript) denote a summation with respect to that index over its range. *E.g.*, if $\{1, 2, 3, 4, 5, 6\}$ is the range of the index i, then $a_i x^i$ stands for

$$\sum_{i=1}^6 a_i x^i = a_1 x^1 + a_2 x^2 + a_3 x^3 + a_4 x^4$$

$$+ a_5 x^5 + a_6 x^6.$$

The superscript i in x^i is not the ith power of the number x, but merely an index which denotes that x^i is the ith object of the six objects $x^1, x^2, \cdots, x^6$. A repeated index, such as i in $a_i x^i$, is a **dummy index** or an **umbral index**, since the value of the expression does not depend on the symbol used for this index. An index which is not repeated, such as i in $a_{ij}x^j$, is a **free index**.

 summation of divergent series. Attributing sums to divergent series by transforming them into convergent series, or by other devices. *E.g.*, the sum of $1 - 1 + 1 - 1 + \cdots$ can be defined as the limit of the sum of $1 - x + x^2 - x^3 + \cdots$, as

x approaches $+1$ with $x < +1$, or as

$$\lim_{n \to \infty} \frac{S_1 + S_2 + \cdots + S_n}{n}$$

$$= \lim_{n \to \infty} \frac{1 + 0 + 1 + \cdots + \frac{1}{2}\left[1 - (-1)^n\right]}{n},$$

where S_n denotes the sum of the first n terms. In both cases the sum is $\frac{1}{2}$. The former method is an illustration of the use of *convergence factors*, in this instance, $1, x, x^2, \cdots$. The latter method is an illustration of the method of arithmetic means. See ABEL—Abel method of summation, BOREL, CESÀRO—Cesàro summation formula, HÖLDER—Hölder definition of the sum of a divergent series.

summation of an infinite series. The process of finding the sum of the series. See SUM—sum of an infinite series.

summation sign. Sigma, the Greek letter corresponding to the English S, written Σ. When the process of summing includes the first to the nth terms of a set of numbers $a_1, a_2, a_3, \cdots, a_n, \cdots$, the sum is written

$$\sum_{i=1}^{n} a_i, \quad \text{or} \quad \sum_{1}^{n} a_i.$$

When the summation includes infinitely many terms, it is written

$$\sum_{i=1}^{\infty} a_i, \ \sum_{1}^{\infty} a_i, \quad \text{or simply} \ \sum a_i.$$

SUP, *n.* See BOUND.

SU′PER-AD′DI-TIVE, *adj.* **superadditive function.** See ADDITIVE—additive function.

SU′PER-HAR-MON′IC, *adj.* **superharmonic function.** A function that is related to subharmonic functions in the way that concave functions are related to convex functions; *i.e.*, a real function f, of any number of variables, such that $-f$ is subharmonic. See SUBHARMONIC.

SU-PE′RI-OR, *adj., n.* **limit superior.** (1) See SEQUENCE—accumulation point of a sequence. (2) The **limit superior** of a function f at a point x_0 is the largest number L such that for any $\epsilon > 0$ and neighborhood U of x_0 there is a point $x \neq x_0$ of U for which $f(x) > L - \epsilon$; this definition applies to the case $L = +\infty$ if $f(x) > L - \epsilon$ is replaced by $f(x) > \epsilon$, while $L = -\infty$ if for any

$\epsilon > 0$ there is a neighborhood U of x_0 in which $f(x) < -\epsilon$ for each $x \neq x_0$. This limit is denoted by $\limsup_{x \to x_0} f(x)$ or $\overline{\lim}_{x \to x_0} f(x)$. The limit superior of $f(x)$ at x_0 is equal to the limit as $\epsilon \to 0$ of the l.u.b. of $f(x)$ for $|x - x_0| < \epsilon$ and $x \neq x_0$, and may be positively or negatively infinite. (3) The **limit superior** of a sequence of sets $\{U_1, U_2, \cdots\}$ is the set consisting of all objects belonging to infinitely many of the sets U_n. It is equal to the intersection of all unions of the form $U_p \cup U_{p+1} \cup \cdots$; *i.e.*, $\bigcap_{p=1}^{+\infty} \bigcup_{n=p}^{+\infty} U_n$. For sequences of sets, the limit superior is also called the **complete limit.** See INFERIOR—limit inferior. *Syn.* upper limit.

SU′PER-OS′CU-LAT-ING, *adj.* **superosculating curves on a surface.** Normal sections of a surface which are *superosculated* by their circles of curvature. See below, SUPEROSCULATION.

SU′PER-OS′CU-LA-TION, *n.* The property of some pairs of curves or surfaces of having contact of higher order than other pairs, which are said to osculate.

SU′PER-POSE′, *v.* To place one configuration upon another in such a way that corresponding parts coincide. To superpose two triangles which have their corresponding sides equal is to place one upon the other so that corresponding sides coincide.

SU′PER-POS′A-BLE, *adj.* **superposable configurations.** Two configurations which can be superposed. *Syn.* congruent.

SU′PER-PO-SI′TION, *adj., n.* **axiom of superposition.** See AXIOM—axiom of superposition.

superposition principle for electrostatic intensity. See ELECTROSTATIC.

SU′PER-RE-FLEX′IVE, *adj.* A **super-reflexive Banach space** is a Banach space X for which there is no nonreflexive Banach space that is *finitely representable* in X. A Banach space is super-reflexive if and only if it is isomorphic to a uniformly nonsquare space, or if and only if it is isomorphic to a uniformly convex space. See CONVEX—uniformly convex space, FINITELY—finitely representable, NON—nonsquare, REFLEXIVE.

SU′PER-SCRIPT, *n.* A small number, letter, or symbol written above and to the right or left of a letter or symbol. Commonly used to indicate a power, as for x^3 or $7^{\frac{1}{2}}$. Often used on a variable to denote a constant value of that variable or to distinguish between variables. See EXPONENT, PRIME—prime (or accent) as a symbol, SUBSCRIPT, and the usage in various headings under TENSOR.

SU′PER-SET, *n.* See SUBSET.

SUP′PLE-MEN′TAL, *adj.* **supplemental chords of a circle.** The chords joining a point on the circumference to the two extremities of a diameter. Supplemental chords are perpendicular.

SUP′PLE-MEN′TA-RY, *adj.* **supplementary angles.** Two angles whose sum is 180°; two angles whose sum is a straight angle. The angles are **supplements** of each other.

SUP-PORT′, *adj., n.* **line of support.** Relative to a convex region B in the plane, a line of support is a line containing at least one point of B but such that one of the two open half planes determined by the line contains no point of B. The equations of such a line can be written in the form

$$x \cos \theta + y \sin \theta = S(Q),$$

where Q is the point with coordinates $(\cos \theta, \sin \theta)$ and $S(Q)$ is the *normalized support function*. The function $S(Q)$ is a *subsine function* of the angle θ. For a convex or concave function, a line of support can be defined similarly in terms of the graph of the function.

   plane (and hyperplane) of support. Relative to a *convex set B* in three-dimensional space, a plane of support is a plane containing at least one point of B, but such that one of the two open half-spaces determined by the plane contains no point of B. For a *normed vector space T* and a convex set B contained in T, a hyperplane of support is a hyperplane H whose distance from B is zero and which is the separating hyperplane between two open half spaces, one of which contains no points of B. This means that H is a hyperplane of support of B if and only if there is a *continuous linear functional f* and a constant c for which $f(P) \le c$ if P belongs to B and H is the set of all P with $f(P) = c$. An *arbitrary Banach space* is *reflexive* if and only if the distance between H and B being zero implies H contains a point of B, for any bounded closed convex set B and hyperplane of support H. Given a space with an inner product, for a bounded closed convex subset B a hyperplane of support must contain a point of B; also, there is a point P for which the hyperplane of support

consists of all those points Q for which $(P, Q) = S(P)$, where S is the support function. See below, support function.

   support function. Relative to a bounded closed convex set B in any space with a real *inner product* (*e.g.*, a *Euclidean space* of any dimension, or a real *Hilbert space*), the support function S is defined (for all points P of the space other than $P = 0$) by letting

$$S(P) = \max(P, Q)$$

for Q in B, where (P, Q) is the inner product of P and Q. Thus for each point Q of B we have $(P, Q) \le S(P)$, with the sign of equality holding for some point Q_0 of B. All of B lies in one of the two closed half spaces bounded by the hyperplane consisting of all points R for which $(P, R) = S(P)$. The function $S(P)$ is a convex function of P. The support function satisfies the relation $S(kP) = kS(P)$ for $k \ge 0$. Accordingly, $S(P)$ is completely determined by its values $S(Q)$ on the unit sphere consisting of all points Q with $(Q, Q) = 1$. With its independent variable thus restricted, the function $S(Q)$ is a **normalized support function** of B. See MINKOWSKI —Minkowski distance function, and above, plane of support.

   support of a function. The closure of the set of points at which the function is not zero. A function has **compact support** if its support is a compact set.

SU-PREM′UM, *n.* See BOUND.

SURD, *n.* A sum with one or more irrational indicated roots as addends. Sometimes used for *irrational number.* A surd of one term is **quadratic, cubic, quartic, quintic,** etc., according as the index of the radical is two, three, four, five, etc. It is an **entire surd** if it does not contain a rational factor or term (*e.g.*, $\sqrt{3}$ or $\sqrt{2} + \sqrt{3}$); a **mixed surd** if it contains a rational factor or term (*e.g.*, $2\sqrt{3}$ or $5 + \sqrt{2}$); a **pure surd** if each term is a surd (*e.g.*, $3\sqrt{2} + \sqrt{5}$). A **binomial surd** is a binomial, at least one of whose terms is a surd, such as $2 + \sqrt{3}$ or $\sqrt[3]{2} - \sqrt{3}$. **Conjugate binomial** surds are two binomial surds of the form $a\sqrt{b} + c\sqrt{d}$ and $a\sqrt{b} - c\sqrt{d}$, where a, b, c and d are rational and $\sqrt{b}$ and $\sqrt{d}$ are not both rational. The product of two conjugate binomial surds is rational, *e.g.*, $(a + \sqrt{b})(a - \sqrt{b}) = a^2 - b$. A **trinomial surd** is a trinomial at least two of whose terms are surds which cannot be expressed as a surd of one term; $2 + \sqrt{2} + \sqrt{3}$ and $3 + \sqrt{5} + \sqrt[3]{2}$ are trinomial surds.

SUR'FACE, *adj., n.* A surface is the geometric figure consisting of those points whose coordinates satisfy any equation such as $z = f(x, y)$, or $F(x, y, z) = 0$, or parametric equations $x = x(u, v)$, $y = y(u, v)$, $z = z(u, v)$, with conditions such as continuity or nonvanishing of a Jacobian imposed to ensure nondegeneracy [see SMOOTH —smooth surface]. *E.g.*, the surface of the sphere of radius 2 with center at $(0, 0, 0)$ has the equation $x^2 + y^2 + z^2 = 4$; it also has the parametric equations $x = 2 \sin \phi \cos \theta$, $y = 2 \sin \phi \sin \theta$, $z = 2 \cos \phi$. *Tech.* In a rather restrictive sense, a *closed surface* might be defined as a connected, compact metric space which is homogeneous in the sense that each point has a neighborhood which is homeomorphic with the interior of a circle in the plane; a *surface with boundary curves* would then be defined by changing the neighborhood condition so that each point on a boundary curve has a neighborhood which is homeomorphic with half of a 2-cell with the diameter included and lying along the boundary curve. An equivalent definition is that a surface is a geometric figure which can be subdivided into a finite number of "triangles" (each of which has vertices and edges and is homeomorphic with a plane triangle) such that (i) if two triangles intersect, their intersection is a side of each of them; (ii) no side belongs to more than two triangles; (iii) for any two triangles R and S, there is a sequence of triangles $T_1, T_2, \cdots, T_n$ such that $T_1 = R$, $T_n = S$, and any two adjacent triangles, T_i and T_{i+1}, have a side in common. Such a surface is **closed** if each side of a triangle also belongs to another triangle; otherwise the surface has a finite number of closed boundary curves. A surface is **orientable** if the above triangles can be oriented so that a direction around the perimeter is prescribed in such a way that two intersecting triangles assign opposite orientations (directions) to their common side (this is equivalent to requiring that the surface not contain a Möbius strip, or that a small oriented "circle" can not be moved in the surface so as to return to its initial position with its orientation reversed). See GENUS—genus of a surface.

algebraic surface. A surface which admits a parametric representation such that the coordinate functions are algebraic functions of the parameters u, v.

applicable surfaces. Surfaces such that there exists a length-preserving map of one on the other. See ISOMETRY.

canal surface. The envelope of a one-parameter family of spheres of equal radii having their centers on a given space curve. For any point on the curve, the *characteristic* is the great circle in the plane normal to the curve at the point.

curvature of a surface. See various headings under CURVATURE.

curved surface. A surface no part of which is a plane surface.

cylindrical surface. See CYLINDRICAL —cylindrical surface.

equation of a surface. See PARAMETRIC — parametric equations, SURFACE.

fundamental coefficients of a surface. The **fundamental coefficients of the first order** of a surface are the coefficients E, F, G of the first fundamental quadratic form of the surface. *Syn.* fundamental quantities of the first order of a surface. The **fundamental coefficients of the second order** of a surface are the coefficients D, D', D'' of the second fundamental quadratic form of the surface. See below, fundamental quadratic forms of a surface.

fundamental quadratic forms of a surface. The expression $E \, du^2 + 2F \, du \, dv + G \, dv^2$ is the **first fundamental quadratic form of a surface.** Also written $g_{\alpha\beta} \, du^\alpha \, du^\beta$ in tensor notation. See LINEAR —linear element. The **second fundamental quadratic form of a surface** is the expression $\Phi = D \, du^2 + 2D' \, du \, dv + D'' \, dv^2$. Also written

$$\Phi = L \, du^2 + 2M \, du \, dv + N \, dv^2,$$

and $\Phi = e \, du^2 + 2f \, du \, dv + g \, dv^2$, and, in tensor notation, $d_{\alpha\beta} \, du^\alpha \, du^\beta$. See DISTANCE —distance from a surface to a tangent plane. The **third fundamental quadratic form of a surface** is the first fundamental quadratic form of the spherical representation of the surface.

fundamental quantities of the first order of a surface. See above, fundamental coefficients of a surface.

Gaussian representation of a surface. Same as spherical representation of a surface. See SPHERICAL.

imaginary surface. See IMAGINARY —imaginary curve (surface).

material surface. See MATERIAL.

minimal surface. See MINIMAL.

molding surface. A surface generated by a plane curve whose plane rolls without slipping over a cylinder. If the cylinder is a line, the molding surface is a surface of revolution. See below, surface of Monge.

normal to a surface. See NORMAL—normal to a curve or surface.

one-sided surface. A surface which has only one side, in the sense that any two bugs on the surface can get to each other without going around an edge, regardless of where put. See MÖBIUS —Möbius strip, KLEIN—Klein bottle, and MINIMAL —double minimal surface. *Tech.* A surface is one-sided if it is *nonorientable*; it is nonorientable if and only if it contains a Möbius strip (see also SURFACE, above).

parallel surfaces. See PARALLEL.

plane surface. A plane.

principal direction on a surface. See DIRECTION.

pseudospherical surface. See PSEUDOSPHERICAL.

quadric surface. See QUADRIC.

ruled surface. See RULED.

similar surfaces. See SIMILAR—similar surfaces.

smooth surface. See SMOOTH.

spherical representation of a surface. See SPHERICAL.

spherical surface. See SPHERICAL.

surface area. Care must be taken in approximating the area of a surface as the sum of areas of plane sets. For example, by selecting appropriate triplets of points evenly distributed over a given surface as vertices of plane triangles, a crinkled surface can be made that approximates the given surface. The area of the crinkled surface is the sum of the areas of its triangular parts. It might seem reasonable that as more and more points are chosen closer and closer together, the areas of the crinkled surfaces would approach the area of the given surface. However, if the surface is a right circular cylinder, the points can be chosen so that not only do the areas of the crinkled surfaces not approach the area of the cylinder, but there is no upper bound to the areas of the crinkled surfaces. The following methods of determining area avoid this difficulty by using tangential plane sets to approximate the surface. (For surfaces that can be rolled out on a plane, such as cylinders and cones, the area can be determined directly as the area of the resulting plane region.): (1) Area of a surface is the limit of the sum of the areas of the polygons formed by the intersections of tangent planes at neighboring points distributed over the entire surface, as the area of the largest of these polygons approaches zero. Each of these plane tangential areas is obtainable by projecting some area lying in one of the coordinate planes onto the tangent plane (2) Let a plane P be such that no line perpendicular to P cuts a given surface S in more than one point, and let A be the projection of S into P. Let a partition of A be chosen and let each member A_k of the partition be projected into a tangent plane along lines perpendicular to P, where the tangent plane is tangent to S at a point on a line perpendicular to P at a point of A_k. The area of S is the limit of the sum of the areas of these projections as the fineness of the partition approaches zero. If P is the (x, y)-plane, and if β is the angle between the tangent plane and the xy-plane, the area of

S is equal to the integral of $(\sec \beta)\, dx\, dy$ (the **element of area** or **differential of area**) over the projection of the surface into the (x, y)-plane. If the equation of the surface is in the form $z = f(x, y)$, then

$$\sec \beta = \left[1 + (D_x z)^2 + (D_y z)^2\right]^{1/2},$$

where $D_x z$ and $D_y z$ denote the partial derivatives of z with respect to x and y. If the equation of the surface is $f(x, y, z) = 0$ and subscripts denote the partial derivatives, then $\sec \beta = (f_x^2 + f_y^2 + f_z^2)^{1/2}/f_z$. It is assumed that $\sec \beta$ is finite, *i.e.*, none of the tangent planes is perpendicular to the (x, y)-plane. (3) If S is a smooth surface with the position vector $\mathbf{P}$ whose domain is a set D in the (u, v)-plane, then the area of S can be defined as

$$\int_D \left| \frac{\partial \mathbf{P}}{\partial u} \times \frac{\partial \mathbf{P}}{\partial v} \right| dA.$$

For a surface generated by rotating the graph of $y = f(x)$ ($y \geq 0$) around the x-axis, this formula becomes $\int 2\pi f(x)\{1 + [f'(x)]^2\}^{1/2}\, dx$ or $\int 2\pi f(x)\, ds$. In more advanced theory, many definitions of area have been given, a frequently used one being that given by Lebesgue: The area of a surface is the least value which is the limit of the sum of the areas of polyhedrons converging (in the sense of Fréchet) to the surface. See below, surface integral.

surface of constant curvature. A surface whose total curvature is the same at all its points. These are *developable surfaces*, for $K = 0$; *spherical surfaces* (not just the sphere), for $K > 0$; and *pseudospherical surfaces*, for $K < 0$. See PSEUDOSPHERICAL, SPHERICAL.

surface of Enneper. The real minimal surface for which $\phi(u) = $ const. See WEIERSTRASS — equations of Weierstrass. If we take $\phi(u) = 3$, and let $u = s + it$, the parametric curves are the lines of curvature and the coordinate functions are

$$x = 3s + 3st^2 - s^3,$$

$$y = 3t + 3s^2t - t^3,$$

$$z = 3s^2 - 3t^2;$$

the *map* is a conformal one, and the coordinate functions are harmonic.

surface harmonic. See HARMONIC —surface harmonic.

surface of Henneberg. The real minimal surface for which $\phi(u) = 1 - 1/u^4$. See WEIERSTRASS—equations of Weierstrass. The surface of Henneberg is a *double minimal surface*. See MINIMAL.

surface integral. The integral of some function f over a surface S: $\iint_S f(x, y, z) \, d\sigma$. Suppose a partition of the surface is chosen and the sum of the products of the area of each member A of the partition by a value of f at a point in A is formed. Then the integral over the surface is the limit of this sum as the number of members of the partition increases in such a way that the fineness of the partition approaches zero. If S is a smooth surface with the position vector $\mathbf{P}$ whose domain is a set D in the (u, v)-plane and n is a continuous vector-valued function whose domain is D and whose values are all of unit length and perpendicular to S, then the surface integral of a vector-valued function $\mathbf{F}$ over S is the integral

$$\int_S (\mathbf{F} \cdot \mathbf{n}) \, d\sigma = \int_D (\mathbf{F} \cdot \mathbf{n}) \left| \frac{\partial \mathbf{P}}{\partial u} \times \frac{\partial \mathbf{P}}{\partial v} \right| dA,$$

which is equal to $\int_D \mathbf{F} \cdot \left(\dfrac{\partial \mathbf{P}}{\partial u} \times \dfrac{\partial \mathbf{P}}{\partial v} \right) dA$ if $\mathbf{n}$ is chosen as the vector $(\partial \mathbf{P}/\partial u) \times (\partial \mathbf{P}/\partial v)$ divided by its length. If D is in the (x, y)-plane, the equation of the surface becomes $z = g(x, y)$ and the integral becomes

$$\int_D f[x, y, g(x, y)] \sec \beta \, dx \, dy,$$

where f denotes $\mathbf{F} \cdot \mathbf{n}$ and β is the angle between $\mathbf{n}$ and the unit vector $\mathbf{k}$ perpendicular to the (x, y)-plane. See STOKES—Stokes' theorem, and above, surface area.

surface of Joachimsthal. A surface such that all the members of one of its two families of lines of curvature are plane curves, and such that all these planes are coaxial.

surface of Liouville. A surface which admits a parametric representation such that the first fundamental quadratic form reduces to

$$ds^2 = [f(u) + g(v)][du^2 + dv^2].$$

surface of Monge. A surface generated by a plane curve whose plane rolls without slipping over a developable surface. See above, molding surface.

surface patch. A surface, or part of a surface, bounded by a closed curve, in contradistinction to a surface of infinite extent or a closed surface such as a sphere.

surface of revolution. A surface which can be generated by revolving a plane curve about an axis in its plane. Sections of a surface of revolution perpendicular to this axis are circles, called **parallel circles** or simply **parallels**; sections containing the axis are **meridian sections**, or simply **meridians**. The earth is a surface of revolution which can be generated by revolving a meridian about the line through the north and south poles. A surface of revolution can also be generated by a circle moving always perpendicular to a fixed line with its center on the fixed line and expanding or contracting so as to continually pass through a curve which lies in a plane with the straight line. The **element of area** of a surface of revolution can be taken as $2\pi r \, ds$, where r is the distance from the axis of revolution of any point in the element of arc, ds, of the curve which is rotated to form the surface. If the curve $y = f(x)$ is revolved about the x-axis (as in the figure), then $2\pi r \, ds = 2\pi f(x) \, ds$ and the area of the surface of revolution, between the values a and b of x, is

$$\int_a^b 2\pi f(x) \sqrt{1 + (dy/dx)^2} \, dx.$$

From the figure, it can be seen that $2\pi r \, \Delta s$ is the area derived by rotating the arc $BC = \Delta s$ about the x-axis, and hence $2\pi r \, ds$ is an approximation to this area (see ELEMENT—element of integration, and above, surface area).

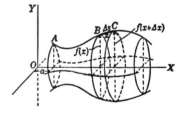

surface of Scherk. The real minimal surface for which $\phi(u) = \dfrac{2}{1 - u^4}$. See WEIERSTRASS—equations of Weierstrass. The surface of Scherk is *doubly periodic*.

surface of translation. A surface admitting a representation of the form $x = x_1(u) + x_2(v)$, $y = y_1(u) + y_2(v)$, $z = z_1(u) + z_2(v)$. It might be considered as being generated by translating the curve C_1: $x = x_1(u)$, $y = y_1(u)$, $z = z_1(u)$ parallel to itself in such a way that each point of C_1 describes a curve congruent to C_2: $x = x_2(v)$, $y = y_2(v)$, $z = z_2(v)$; or equally well by a translation of C_2 parallel to itself in such a way that each point of C_2 describes a curve congruent to C_1. The loci described by the points of C_1 (or of

C_2) are called the **generators** of the surface. *Syn.* translation surface.

surface of Voss. A surface with a conjugate system of geodesics.

surfaces of center relative to a given surface. The loci of the centers of principal curvature of the given surface. See CURVATURE —curvature of a surface. The surfaces of center of S are also surfaces of center of any surface parallel to S. See above, parallel surfaces, and COMPLEMENTARY—surface complementary to a given surface.

traces of a surface. See TRACE —traces of a surface

Weingarten surface. See WEINGARTEN.

SUR-JEC′TION, *n.* A **surjection** from a set A to a set B is a function whose domain is A and whose range is B, *i.e.*, a function from A *onto* B. *Syn.* surjective function. See BIJECTION, INJECTION.

SUR-JEC′TIVE, *adj.* See SURJECTION.

SUR-REN′DER, *adj.* **surrender charge.** (*Insurance*) The deduction that is made from the terminal reserve to determine the cash surrender value (not over $2\frac{1}{2}\%$ of the terminal reserve is allowed by law in most states).

surrender value of an insurance policy. The amount the insurance company is willing to pay the insured for the return (cancellation) of the policy; the difference between the terminal reserve and the surrender charge. Same as *present value* or *cash equivalent* of the insurance policy. See VALUE—present value.

SUR′TAX′, *n.* Tax, additional to the normal tax, levied on incomes above a certain level.

SUR-VEY′OR, *n.* **surveyor's measure.** See DENOMINATE NUMBERS in the appendix.

SYL′LO-GISM, *n.* A logical statement that involves three propositions, usually called the *major premise*, *minor premise*, and *conclusion*, the conclusion necessarily being true if the premises are true. *E.g.*, the three propositions might be: "John likes fishing or likes singing"; "John does not like fishing"; "John likes singing." A **hypothetical syllogism** is a particular type of syllogism which relates three implications (p, q, r) and states: "If p implies q, and q implies r, then p implies r." This is frequently written as: $[(p \rightarrow q)$ and $(q \rightarrow r)] \rightarrow (p \rightarrow r)$. A **categorical syllogism** relates implications with universal quantifiers, an example of which is: If the propositions "*For any quadrilateral T, if T is a square, then T is a rectangle*" and "*For any quadrilateral T, if T is a rectangle, then T is a parallelogram*" are true, then the proposition "*For any quadrilateral T, if T is a square, then T is a parallelogram*" is true. See IMPLICATION.

SYLOW, Peter Ludvig (1832–1918). Norwegian group theorist.

Sylow's theorem. As proved by Sylow, the theorem is that if p is a prime and G is a group whose order is divisible by p^n but not by p^{n+1}, then there is an integer k such that G contains $1 + kp$ subgroups of order p^n. Later, Frobenius proved that the number of subgroups of order p^n is $1 + kp$ even if the order of G is divisible by a higher power of p than p^n.

SYLVESTER, James Joseph (1814–1897). English algebraist, combinatorist, geometer, number theorist, and poet. Cofounder with Cayley of the theory of algebraic invariants (anticipated to some extent by Boole and Lagrange). Spent two periods in the U.S., where he was a stimulant to mathematical research.

Sylvester's dialytic method. A method of eliminating a variable from two algebraic equations. It consists essentially of multiplying each of the equations by the variable, thus getting two more equations and only one higher power of the variable, doing the same with the two new equations, etc., until the number of equations is one greater than the number of powers of the variable, then eliminating the various powers of the variable between these equations as if the powers were different variables (see ELIMINATION —elimination of an unknown...). *Sylvester's method* is equivalent to the procedure illustrated by the following example, which does not require determinants: It is desired to eliminate x from

$$(1) \qquad x^2 + ax + b = 0$$

and

$$(2) \qquad x^3 + cx^2 + dx + e = 0.$$

Multiply equation (1) by x and subtract the result from equation (2). This results in an equation of the second degree. Eliminate x^2 between this equation and equation (1), and so on. Finally one reaches two linear equations, subtraction of which eliminates the variable entirely. See RESULTANT—resultant of a set of polynomial equations.

Sylvester's law of inertia. See INDEX—index of a quadratic form.

SYM′BOL, *n.* A letter or mark of any sort representing quantities, relations, or operations. See the appendix for a list of mathematical symbols.

algebraic symbols. Symbols representing numbers, and algebraic combinations and operators with these numbers.

SYM-MET′RIC. SYM-MET′RI-CAL, *adj.* Possessing symmetry. See various headings below and under SYMMETRY.

cyclosymmetric function. A function which remains unchanged under a *cyclic* change of the variables. See below, symmetric function.

elementary symmetric functions. For *n* variables $x_1, x_2, \cdots, x_n$, the elementary symmetric functions are

$$\sigma_1 = x_1 + x_2 + \cdots + x_n,$$

$$\sigma_2 = x_1 x_2 + x_1 x_3 + \cdots + x_{n-1} x_n,$$

. . . .

$$\sigma_n = x_1 x_2 x_3 \cdots x_n,$$

where σ_k is the sum of all products of k of the variables $x_1, x_2, \cdots, x_n$. Any symmetric polynomial in *n* variables can be written in exactly one way as a polynomial in the elementary symmetric functions. For a polynomial $p(x)$ of degree *n* with roots $a_1, a_2, \cdots, a_n$,

$$p(x) = (x - a_1)(x - a_2)(x - a_3) \cdots (x - a_n)$$

$$= x^n - \sigma_1 x^{n-1} + \sigma_2 x^{n-2} - \sigma_3 x^{n-3}$$

$$+ \cdots + (-1)^n \sigma_n,$$

where $\sigma_1, \cdots, \sigma_n$ are the elementary symmetric functions of $a_1, \cdots, a_n$.

skew-symmetric determinants and matrices. See SKEW.

symmetric determinant. A determinant having all its *conjugate elements* equal, a determinant which is symmetric about the principal diagonal.

symmetric difference. See DIFFERENCE—difference of two sets.

symmetric distribution. See DISTRIBUTION.

symmetric dyadic. See DYAD.

symmetric form of the equations of a line in space. See LINE—equation of a straight line.

symmetric function. A function of two or more variables which remains unchanged under every interchange of two of the variables; $xy + xz + yz$ is a symmetric function of x, y, and z. Such a function is sometimes called **absolutely**

symmetric; a function which remains unchanged under cyclic changes of the variables is **cyclosymmetric**. The word absolute is usually omitted, symmetry and cyclosymmetry being sufficient. The function

$$abc + a^2 + b^2 + c^2$$

has absolute symmetry, whereas

$$(a - b)(b - c)(c - a)$$

has only cyclosymmetry.

symmetric geometric configurations. (1) A geometric configuration (curve, surface, etc.) is said to be symmetric (have symmetry) with respect to a point, a line, or a plane, when for every point on the configuration there is another point of the configuration such that the pair is symmetric with respect to the point, line, or plane. The point is the **center of symmetry**; the line is the **axis of symmetry**, and the plane is the **plane of symmetry**. See p. 411, symmetric points. For a **plane curve**, the following are tests for symmetry: (1) In Cartesian coordinates, if its equation is unaltered when the variables are replaced by their negatives, it is symmetric with respect to the origin; if its equation is unaltered when y is replaced by $-y$, it is symmetric with respect to the x-axis (in this case the equation contains only even powers of y, if it is rational in y); if its equation is unaltered when x is replaced by $-x$ it is symmetric with respect to the y-axis (in this case the equation contains only even powers of x, if it is rational in x). If it is symmetric with respect to both axes, it is symmetric with respect to the origin, but the converse is not true. (2) In polar coordinates, (r, θ), a curve is symmetric with respect to the origin if its equation is unchanged when r is replaced by $-r$ ($r^2 = \theta$ is symmetric with respect to the origin); it is symmetric about the polar axis if its equation is unchanged when θ is replaced by $-\theta$ ($r = \cos \theta$ is symmetrical with respect to the polar axis); and it is symmetrical about the line $\theta = \frac{1}{2}\pi$ if its equation is unchanged when θ is replaced by $180° - \theta$ ($r = \sin \theta$ is symmetrical about $\theta = \frac{1}{2}\pi$). The conditions for polar coordinates are sufficient, but not necessary. Similar tests for symmetry of other geometric configurations can be made. A plane figure has **two-fold symmetry** with respect to a point if after being revolved, in its plane, about the point through 180° it forms the same figure as before. If the angle through which it is revolved is 120°, it has **three-fold symmetry**; if the angle is $360°/n$, it has **n-fold symmetry** with respect to the point. A regular polygon of *n* sides has *n*-fold symmetry about its center. (2) Two geometric configurations are symmetric with respect to a point, line, or plane, if for each point of either configuration there is a point of the other configuration such that the pair of points is symmetric with respect to the point, line, or plane. One of the geometric configurations is then said to be the *reflection* of the other through the point, line, or plane.

symmetric group. See PERMUTATION —permutation group.

symmetric matrix. A matrix which is equal to its *transpose*; a square matrix which is symmetric about the principal diagonal. See ORTHOGONAL —orthogonal transformation.

symmetric pair of equations. A pair of equations which remains unchanged as a pair, although the equations may be interchanged, when the variables are interchanged. The equations $x^2 + 2x + 3y - 4 = 0$ and $y^2 + 2y + 3x - 4 = 0$ are a symmetric pair.

symmetric points. (1) Two points are said to be symmetric (have *symmetry*) with respect to a third point (the **center of symmetry**) if the third point bisects the line joining the points. (2) Two points are said to possess symmetry with respect

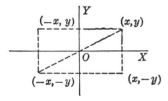

to a line or plane (the **axis** or **plane of symmetry**) if the line, or plane, is the perpendicular bisector of the line segment joining the two points. The pairs of points whose coordinates are (x, y) and $(-x, -y)$ are symmetric with respect to the origin; the points (x, y) and $(x, -y)$ are symmetric with respect to the x-axis; the points (x, y) and $(-x, y)$ are symmetric with respect to the y-axis.

symmetric relation. A relation which has the property that if a is related to b, then b is related in like manner to a. The equals relation of algebra is symmetric, since if $a = b$, then $b = a$. A relation is **asymmetric** if there are no pairs (a, b) such that a is related to b and b is related to a. The property of *being older than* is *asymmetric*; if a is older than b, then b is not older than a. A relation is **nonsymmetric** if there is at least one pair (a, b) such that a is related to b, but b is not related to a. The relation of *love* is *nonsymmetric*, since if a loves b, b may, or may not, love a. A relation is **antisymmetric** if a related to b and b related to a imply $a = b$. The relation $a \leq b$ for real numbers is *antisymmetric*.

symmetric spherical triangles. Spherical triangles whose corresponding sides and corresponding angles are equal, but appear in opposite order when viewed from the center of the sphere. The triangles are not superposable.

symmetric tensor. See TENSOR.

symmetric transformation. A transformation T defined on a **Hilbert space** H is symmetric if the inner products (Tx, y) and (x, Ty) are equal for every x and y in the domain of T. If, also, the domain of T is dense in H, then the second *adjoint* T^{**} of T is a symmetric transformation which is also *closed*. Any bounded symmetric transformation has an extension which is self-adjoint. A symmetric transformation whose domain (or range) is all of H is bounded and self-adjoint. For finite-dimensional spaces, a transformation T, which transforms vectors $x = (x_1, x_2, \cdots, x_n)$ into $Tx = (y_1, y_2, \cdots, y_n)$ with $y_i = \sum_j a_{ij} x_j$ for each i, is symmetric if and only if the matrix (a_{ij}) of its coefficients is a Hermitian matrix. See SELF—self-adjoint transformation.

symmetric trihedral angles. See TRIHEDRAL —trihedral angle.

SYM′ME-TRY, *adj.*, *n.* See below and various headings under SYMMETRIC.

axial symmetry. Symmetry with respect to a line. See SYMMETRIC —symmetric geometric configurations. *Syn.* line symmetry.

axis, center, and plane of symmetry. See SYMMETRIC—symmetric geometric configurations.

central symmetry. Symmetry with respect to a point. See SYMMETRIC —symmetric geometric configurations.

cyclosymmetry. See SYMMETRIC —symmetric function.

group of symmetries. The set of all rigid motions that transform a given geometric figure onto itself. For example, the symmetries of a circle consist of all rotations around the center and of all rotations of 180° around a diameter. There are eight symmetries of a square: rotations in the plane of the square around its center of 0°, 90°, 180°, and 270°; rotations of 180° around the two diagonals; and rotations of 180° around the two perpendicular bisectors of pairs of opposite sides. The symmetries of a geometric figure form a group if the product of two symmetries S_1 and S_2 is defined as the symmetry obtained by first applying S_1 and then applying S_2. The symmetries of a polygon or a polyhedron can be described as permutations of the vertices, so that such groups of symmetries are subgroups of permutation groups [see PERMUTATION]. See DIHE-

DRAL—dihedral group, FOUR—four-group, ICOSAHEDRAL —icosahedral group, OCTAHEDRAL —octahedral group, TETRAHEDRAL —tetrahedral group.

SYN-THET′IC, *adj.* **synthetic division.** Division of a polynomial in one variable, say *x*, by *x* minus a constant (positive or negative), making use of detached coefficients and a simplified arrangement of the work. Consider the division of $2x^2 - 5x + 2$ by $x - 2$. Using ordinary long division, the process would be written

$$
\begin{array}{r|l}
2x^2 - 5x + 2 & \,x - 2 \\ \cline{2-2}
2x^2 - 4x & 2x - 1 \\ \cline{1-1}
-x + 2 & \\
-x + 2 & \\ \cline{1-1}
\end{array}
$$

Noting that the coefficient in the quotient is always the coefficient of the first term in the dividend, that it is useless to write down the $-x$, and that by changing the sign of -2 in the quotient one could add instead of subtract, we can put the process in the *synthetic division* form

$$
\begin{array}{r|l}
2 - 5 + 2 & \underline{\,2} \\
4 - 2 & \\ \hline
2 - 1 + 0 &
\end{array}
$$

The detached coefficients of the quotient, 2 and -1, are **partial remainders**, while the last term, here 0, is the **remainder**.

synthetic geometry. The study of geometry by synthetic and geometric methods. See below, synthetic method of proof. Synthetic geometry usually refers to projective geometry. *Syn.* pure geometry.

synthetic method of proof. A method of proof involving a combining of propositions into a whole or system; involving reasoning by advancing to a conclusion from principles established or assumed and propositions already proved; the opposite of *analysis. Syn.* deductive method of proof.

synthetic substitution. Same as SYNTHETIC DIVISION. The latter is more commonly used.

SYS′TEM, *n.* (1) A set of quantities having some common property, such as the *system* of even integers, the *system* of lines passing through the origin, etc. (2) A set of principles concerned with a central objective, as, a *coordinate system*, a *system* of notation, etc.

coordinate system. See COORDINATE.

decimal system. See DECIMAL—decimal system.

dense system of numbers. See DENSE.

duodecimal system. See DUODECIMAL.

logarithmic system. Logarithms using a certain base, as the *Briggs system* (which uses 10 for a base), or the *natural system* (which uses $e = 2.71828 \cdots$).

metric system. See METRIC.

number system. See NUMBER—number system.

system of circles. Sometimes used for *family of circles.* See CIRCLE—family of circles.

system of equations or inequalities. Same as SIMULTANEOUS EQUATIONS OR INEQUALITIES. See SIMULTANEOUS.

SYS-TE-MAT′IC, *adj.* **systematic sample.** A sample selected from a statistical population by choosing a first member randomly and then choosing successive members located at even intervals, *e.g.*, every tenth member in an alphabetical listing, or an object from an assembly line every five minutes. It is important that the subpopulation from which the sample is drawn is representative of the total population—that the sampling intervals not be related to some cyclical pattern in the population. See RANDOM—random sample.

T

T, *n.* **t distribution.** A random variable X has a **t distribution** or is a **t random variable** with n degrees of freedom if it has a *probability density function f* for which

$$
f(x) = \frac{\Gamma\left[\frac{1}{2}(n + 1)\right]}{\sqrt{n\pi}\,\Gamma\left(\frac{1}{2}n\right)} \left(1 + \frac{x^2}{n}\right)^{-\frac{1}{2}(n+1)},
$$

where Γ is the *gamma function.* The *mean* is 0 if $n > 1$ and the *variance* is $n/(n - 2)$ if $n > 2$. A random variable is a *t* random variable if it is symmetrically distributed about 0 and its square is an *F random variable* with $(1, n)$ degrees of freedom. Equivalently, $(X - \mu)\sqrt{n}\,/s$ is a *t* random variable if X is a normal random variable with mean μ and $s = \left[\sum_{i=1}^{n} (X_i - \mu)^2\right]^{1/2}$, where $(X_1, \cdots, X_n)$ is a random sample of X. For large values of n, the *t* distribution is approximately the normal distribution with mean 0 and variance 1. The *t test* and *t distribution* were developed by "Student" (W. S. Gosset), although the distribution had been noted earlier by Helmert. *Syn.* Student's *t.* See CAUCHY—Cauchy distribution, CHI—chi-square distribution, *F—F* distribution.

t **test.** Suppose X is a normal random variable with unknown mean and variance. To test (with level of significance α) the hypothesis that the mean is μ_0, the *t test* uses the random variable

$$T = \frac{(n-1)^{1/2}(\mu - \mu_0)}{s},$$

where μ is the mean of a random sample of size n and s is $\left[\sum_{i=1}^{n} (X_i - \mu)^2 / n \right]^{1/2}$; when the hypothesis is true, T has the *t distribution* with $n-1$ degrees of freedom and the hypothesis is rejected if $|T| > t_\alpha$, where $F(t_\alpha) = 1 - \frac{1}{2}\alpha$ with F the distribution function for this *t* distribution. The *t* distribution can be used for many other tests of hypotheses.

TA′BLE, *n.* A systematic listing of results already worked out, which reduces the labor of computors and investigators or forms a basis for future predictions. See headings under ACCURACY, COMMUTATION, CONTINGENCY, CONVERSION, MORTALITY.

TAB′U-LAR, *adj.* **tabular differences.** The differences between successive values of a function, as recorded in a table. The tabular differences of a table of *logarithms* are the differences between successive mantissas, usually recorded in a column of their own. The tabular differences of a *trigonometric table* are the differences between successive recorded values of a trigonometric function.

TAC′NODE, *n.* Same as POINT OF OSCULATION. See OSCULATION.

TAC-POINT, *n.* A **tac-point** of a family of curves is a point where two different members of the family intersect and have a common tangent; a **tac-locus** is a set of tac-points. *E.g.,* for the family of circles of radius one which are tangent to the *x*-axis, the lines $y = 1$ and $y = -1$ are tac-loci. See DISCRIMINANT —discriminant of a differential equation.

TAN′GEN-CY, *n.* **point of tangency.** The point in which a line tangent to a curve meets the curve, or the point in which a line or a plane tangent to a surface meets the surface. *Syn.* point of contact.

TAN′GENT, *adj., n.* **length of a tangent.** The distance from the point of contact to the intersection of the tangent line with the *x*-axis. In the figure, the length of the tangent at P_1 is TP_1; the length of the **normal** at P_1 is NP_1; the **subtangent** at P_1 is TM_1; and the **subnormal** at P_1 is NM_1.

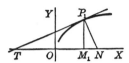

polar tangent. See POLAR—polar tangent.

tangent circles. Two circles are tangent at the point Q if Q is the only point on both circles. The circles are **internally tangent** if one is inside the other; otherwise they are **externally tangent**. The line through the centers of the circles passes through Q and the perpendicular to this line at Q is tangent to both circles. See below, tangent lines and curves.

tangent cone. See CONE—tangent cone of a quadric surface.

tangent curve. The graph of $y = \tan x$. The graph has a point of inflection at the origin, the branch that goes through the origin is asymptotic to the lines $x = -\frac{1}{2}\pi$ and $x = \frac{1}{2}\pi$, and this branch is separated by the origin into two curves each of which is convex toward the *x*-axis. The graph is periodic with period π; *i.e.,* it is duplicated in each successive interval of length π. See TRIGONOMETRIC —trigonometric functions.

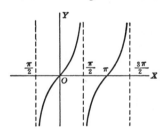

tangent formulas of spherical trigonometry. Same as HALF-ANGLE and HALF-SIDE FORMULAS. See TRIGONOMETRY —half-angle and half-side formulas of spherical trigonometry.

tangent function. See TRIGONOMETRIC —trigonometric functions.

tangent law, or law of tangents. A relation between two sides and the opposite angles of a plane triangle, which is adapted to calculations by logarithms. If A and B are two angles of a triangle, and a and b the sides opposite A and B, respectively, then the law is

$$\frac{a-b}{a+b} = \frac{\tan\frac{1}{2}(A-B)}{\tan\frac{1}{2}(A+B)}.$$

tangent lines and curves. A tangent line to a circle or a sphere is a line that contains exactly one point of the circle or sphere. For more complicated curves or surfaces, more careful definitions must be made. In general, a tangent to a curve C at a point P (the **point of tangency** or **point of contact**) is a line that approximates C very closely near P (it may cross C at P). More precisely, the tangent line at P is the limiting position, if this exists, of the secant line through a fixed point P on C and a variable point P' on C as $P' \rightarrow P$ along C. This means that if there is a tangent line at a point P on a curve, this is the line L which passes through P and has the property that for any positive number ϵ there is a positive number δ such that, if Q is any point on the curve for which the distance from P to Q is less than δ, then the angle between the line L and the line through P and Q is less than ϵ. For the plane curve in the figure, the tangent line at P_1 is the limiting position of a secant ($P_1 P_2$ in the figure) when P_2 approaches P_1. The tangent is the line $P_1 T$. The equation of the tangent at a point on a plane curve is obtained by substituting the coordinates of the point and the slope of the curve at the point in the point-slope form of the equation of a line. If the curve has the equation $y = f(x)$, then the derivative $f'(x_0)$ is the slope of the tangent at x_0. If a curve has the parametric equations $x = f(t)$, $y = g(t)$, $z = h(t)$, and if f, g, and h are differentiable at t_0 and not all these derivatives are zero at t_0, then the curve has a tangent at the corresponding point on the curve and the vector $f'(t_0)\mathbf{i} + g'(t_0)\mathbf{j} + h'(t_0)\mathbf{k}$ is parallel to the tangent. **Two curves** being tangent at a point P means that the two curves pass through P and have the same tangent line at P. A curve or a line is **tangent to a surface** at a point P if the curve (or line) is tangent to a curve on the surface at the point P. See CONIC —tangent to a general conic.

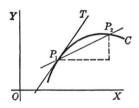

tangent plane. The tangent plane to a surface at a point P is the plane which is such that each line in the plane which passes through P is tangent to the surface at P. If the first-order partial derivatives of f are continuous in a neighborhood of (x_0, y_0, z_0) and not all zero, then direction numbers of the normal line of the plane tangent to the surface whose equation is $f(x, y, z) = 0$, at the point (x_0, y_0, z_0), are the partial derivatives of f with respect to x, y, and z, respectively, evaluated at the point. Hence an equation of the tangent plane is

$$f_1(x_0, y_0, z_0)(x - x_0)$$
$$+ f_2(x_0, y_0, z_0)(y - y_0)$$
$$+ f_3(x_0, y_0, z_0)(z - z_0) = 0,$$

awhere f_1, f_2, and f_3 are the partial derivatives of f with respect to x, y, and z, respectively. If a surface has the equation $z = f(x, y)$ and (x_0, y_0) is an interior point of the domain of f, then f is *differentiable* at (x_0, y_0) if and only if the surface has a tangent plane at $[x_0, y_0, f(x_0, y_0)]$ and the tangent plane is not parallel to the z-axis; the equation of this plane is $z - z_0 = f_1(x_0, y_0)(x - x_0) + f_2(x_0, y_0)(y - y_0)$. See PLANE—equation of a plane, and PARTIAL—partial derivative. The tangent plane to a **cone** or **cylinder** at a point is the plane determined by the element through the point and the tangent to the directrix at its intersection with this element. The tangent plane to a **sphere** at a point P is the plane through P which touches the sphere only at P (the plane which is perpendicular to the radius which terminates at P). If the equation of a **general quadric surface** is $ax^2 + by^2 + cz^2 + 2dxy + 2exz + 2fyz + 2gx + 2hy + 2kz + l = 0$, then the equation of the tangent plane at the point (x_1, y_1, z_1) can be derived by replacing x^2 by xx_1, y^2 by yy_1, etc., $2xy$ by $(xy_1 + x_1 y)$, etc., $2x$ by $(x + x_1)$, etc.

tangent surface of a space curve. The envelope of the family of *osculating planes* of the space curve; the totality of points on lines tangent to the space curve. See DEVELOPABLE —developable surface, OSCULATING—osculating plane.

TAN-GEN′TIAL, *adj.* **tangential acceleration.** See ACCELERATION.

TAR′IFF, *n.* Duties, considered collectively. Sometimes used in the same sense as *duty*.

TARSKI, Alfred (1902–1983). Polish-American algebraist, analyst, logician, and metamathematician.
Banach-Tarski paradox See BANACH.

TARTAGLIA, Niccolò (*real name:* **Niccolò Fontana**) (c. 1500–1557). Italian linguist, mathematician, and physicist. About 1541, he learned how to solve the reduced cubic equation in one variable, possibly after receiving a hint from an earlier secret solution by Ferro of $x^3 + mx = n$

for m and n positive. See CARDAN—Cardan's solution of the cubic, FERRO.

TAU, n. The nineteenth letter of the Greek alphabet: lower case, τ; capital T.

TAUBER, Alfred (1866–1942). Austrian analyst. Prof. Univ. Wien, 1919–1933.

Tauberian theorem. A theorem which establishes some type of limit for a specified class of functions, one of the assumptions being that the limit can be obtained by some stronger limit process. *E.g.*, this includes any theorem which establishes a sufficient condition for convergence of a series which is known to be summable by some (regular) method of summation. The *Tauberian theorem* of this type states that if $f(x) = \sum_0^\infty a_n x^n$, where $\lim_{n \to \infty} n a_n = 0$ and $f(x) \to S$ as $x \to 1$ (with $x < 1$), then $\sum_0^\infty a_n$ converges and has the sum S. See ABEL—Abel method of summation.

TAU′TO-CHRONE, n. (1) An isochronous curve, *i.e.*, a cycloid. (2) For a family of curves through a point P and a positive number c, a curve that intersects each of the given curves at the point which a particle sliding along the given curve from P would reach in time c.

TAX, n. A charge levied for the support of the government.

direct tax. Tax levied upon the person who actually pays it, such as tax levied on real estate, personal property, etc.

indirect tax. A tax ultimately paid by a person other than the one upon whom it is levied, such as taxes on industry which are paid by the consumer in the form of increased prices.

poll tax. Tax levied on individuals, usually on voters only.

TAYLOR, Brook (1685–1731). English analyst, geometer, and philosopher. Published books on perspective and calculus of finite differences.

Taylor formula. The formula in TAYLOR'S THEOREM.

Taylor's theorem. A theorem which describes approximating polynomials for rather general functions and provides estimates for errors. For a function f of one variable, *Taylor's theorem* can be written as

$$f(x) = f(a) + f'(a)(x - a) + f''(a)(x - a)^2/2!$$
$$+ f'''(a)(x - a)^3/3! + \cdots$$
$$+ f^{[n-1]}(a)(x - a)^{n-1}/(n - 1)! + R_n,$$

where R_n is the **remainder**, or the remainder after n terms. The remainder has been put in several different forms, the usefulness of the particular form depending on the particular type of function being expanded. Four of these forms are:

(1) $R_n = \dfrac{1}{(n-1)!} \displaystyle\int_a^x (x - t)^{n-1} f^{[n]}(t)\, dt$;

(2) **Lagrange form,**

$$R_n = \frac{h^n}{n!} f^{[n]}(a + \theta h);$$

(3) **Cauchy form,**

$$R_n = \frac{h^n (1 - \theta)^{n-1}}{(n - 1)!} f^{[n]}(a + \theta h);$$

(4) **Schloemilch form,**

$$R_n = \frac{h^n}{p(n - 1)!} (1 - \theta)^{n-p} f^{[n]}(a + \theta h).$$

In the last three of these, θ is some number between 0 and 1, and $h = x - a$. When $p = 1$, or $p = n$, the Schloemilch form becomes the Cauchy form and Lagrange form, respectively. If n is allowed to increase without limit in the polynomial obtained by *Taylor's theorem*, the result is a **Taylor series.** The sum of such a series represents the expanded function if, and only if, the limit of R_n as n becomes infinite is zero. If $a = 0$ in a *Taylor series* in one variable, the series is called a **Maclaurin series,** though Maclaurin himself stated the series is just a special case of Taylor series. The binomial expansion of $(x + a)^n$ is a Maclaurin series, R_{n+1} being zero when n is an integer. In fact, if any function can be expressed as a power series, such as $c_0 + c_1(x - a) + c_2(x - a)^2 + \cdots + c_n(x - a)^n + \cdots$, that series is a *Taylor series.* Obviously a function cannot be expanded in a *Taylor series* which represents the function in the above sense unless it processes derivatives of all orders on the interval under consideration. Expansions in Taylor series were known to Gregory and, in essence, to John Bernoulli, long before known by Taylor! **For functions of two variables,** *Taylor's theorem* states that

$$f(x,y) = f(a,b) + \left[(x-a)\frac{\partial}{\partial x} + (y-b)\frac{\partial}{\partial y} \right]$$

$$\times f(a,b) + \cdots$$

$$+ \left[(x-a)\frac{\partial}{\partial x} + (y-b)\frac{\partial}{\partial y} \right]^{n-1}$$

$$\times \frac{f(a,b)}{(n-1)!} + R_n,$$

where $f(a,b)$ following the brackets means that the partials within the brackets are to operate upon f at the point (a,b), and the brackets indicate that the expansion of the quantity within is to be a binomial expansion except that

$$\left(\frac{\partial}{\partial x} \right)^h \left(\frac{\partial}{\partial y} \right)^k$$

is to be replaced by

$$\frac{\partial^{h+k}}{\partial x^h \partial y^k}, \quad \text{and} \quad \left(\frac{\partial}{\partial x} \right)^0 \quad \text{and} \quad \left(\frac{\partial}{\partial y} \right)^0$$

are to be replaced by 1; $|R_n|$ is equal to or less than the numerically greatest value of all the nth partial derivatives, multiplied by $(|x-a| + |y-b|)^n$. Explicitly,

$$R_n = \left[(x-a)\frac{\partial}{\partial x} + (y-b)\frac{\partial}{\partial y} \right]^n \frac{f(x_n, y_n)}{n!},$$

for certain x_n and y_n such that

$$x_n = a + \theta(x-a), \qquad y_n = b + \theta(y-b),$$

where $0 < \theta < 1$. If an unlimited (infinite) number of terms of the form used above is taken [it being necessary that all the partial derivatives of f exist] the result is a **Taylor series in two variables**. The series represents the function from which it was derived if, and only if, the remainder approaches zero as the number of terms becomes infinite. Similarly, *Taylor's theorem* and series can be extended to any number of variables. Taylor's theorem also is called the TAYLOR FORMULA and sometimes the EXTENDED or GENERALIZED mean-value theorem, although the latter two are sometimes used for the second mean-value theorem. See LAURENT—Laurent series.

TCHEBYCHEFF. See CHEBYSHEV.

TEM′PO-RAR′Y, *adj.* **temporary annuity.** See ANNUITY.

temporary life insurance. Same as TERM LIFE INSURANCE. See INSURANCE —life insurance.

TEN, *n.* **ten's place.** See PLACE—place value.

TEN′SION, *n.* Any force which tends to extend a body lengthwise, in contrast to a **compression** which is a force that tends to shorten or compress it. A weight hanging on a cord causes a *tension* in the cord, while a weight resting on a stool causes a *compression* in the legs of the stool.

modulus in tension. See HOOKE—Hooke's law, MODULUS —Young's modulus.

TEN′SOR, *adj., n.* An abstract object having a definitely specified system of components in every coordinate system under consideration and such that, under transformations of coordinates, the components of the object undergo a transformation of a certain nature. Explicitly, let

$$A^{pq \cdots t}_{jk \cdots m}$$

be one of a set of functions of the variables x^i $(i = 1, 2, \cdots, n)$, where each index can take on the values $1, 2, \cdots, n$ and the number of superscripts is r, the number of subscripts s. Then these n^{r+s} quantities are the x-components of a **tensor of order** $r + s$, provided its components in any other system

$$x'^i \qquad (i = 1, 2, \cdots, n)$$

are given by

$$A'^{pq \cdots t}_{jk \cdots m} = A^{ab \cdots d}_{ef \cdots h} \frac{\partial x'^p}{\partial x^a} \cdots \frac{\partial x'^t}{\partial x^d} \frac{\partial x^e}{\partial x'^j} \cdots \frac{\partial x^h}{\partial x'^m}$$

where the summation convention is to be applied to the indices $a, b, \cdots, d$ and $e, f, \cdots, h$. (See SUMMATION —summation convention.) Such a tensor is said to be *contravariant of order r* and *covariant of order s*. The superscripts are **contravariant indices**, the subscripts **covariant indices**. See *contravariant tensor* and *covariant tensor* (below) for examples of tensors. When it is desired to distinguish between an abstract object of the above type whose domain of definition is a single point (in each coordinate system) and one whose domain of definition is a region, it is customary to call the former a **tensor** and the latter a **tensor field.** The above is also called an **absolute tensor field** to distinguish it from a *relative tensor field.* A **scalar field** or **invariant** is a tensor field which is contravariant and covariant of order zero (*i.e.*, it has only one component, and this has the same value in all coordinate systems). See below, relative tensor field of weight *w*.

addition and subtraction of tensors. The sum of two tensors $A^{i_1 \cdots i_p}_{j_1 \cdots j_q}$ and $B^{i_1 \cdots i_p}_{j_1 \cdots j_q}$, which have the same number of contravariant indices and the same number of covariant indices, is the tensor

$$T^{i_1 \cdots i_p}_{j_1 \cdots j_q} = A^{i_1 \cdots i_p}_{j_1 \cdots j_q} + B^{i_1 \cdots i_p}_{j_1 \cdots j_q},$$

and their difference is the tensor

$$S^{i_1 \cdots i_p}_{j_1 \cdots j_q} = A^{i_1 \cdots i_p}_{j_1 \cdots j_q} - B^{i_1 \cdots i_p}_{j_1 \cdots j_q},$$

associated tensors. A tensor is said to be **associated with** the tensor $T^{i_1 \cdots i_p}_{j_1 \cdots j_q}$ if it can be obtained from $T^{i_1 \cdots i_p}_{j_1 \cdots j_q}$ by raising or lowering any number of the indices by means of a series of *inner multiplications* of the form $g^{i\sigma} T^{i_1 \cdots i_p}_{j_1 \cdots \sigma \cdots j_q}$ or $g_{j\sigma} T^{i_1 \cdots \sigma \cdots i_p}_{j_1 \cdots j_q}$, where g_{ij} is the *fundamental metric tensor* and g^{ij} is $1/g$ times the cofactor of g_{ji} in the determinant g which has g_{ij} in the *i*th row and *j*th column.

components of the stress tensor. See COMPONENT.

composition (or inner multiplication of) tensors. See INNER—inner product of tensors.

contraction of a tensor. See CONTRACTION.

contravariant tensor. A tensor which has only contravariant indices. If there are *r* indices, it is a **contravariant tensor of order *r*.** If the variables are x^1, x^2, x^3, the differentials dx^1, dx^2, and dx^3 are the components of a contravariant tensor of order one (*i.e.*, a *contravariant vector*), since

$$dx'^i = \frac{\partial x'^i}{\partial x^j} dx^j$$

$$= \frac{\partial x'^i}{\partial x^1} dx^1 + \frac{\partial x'^i}{\partial x^2} dx^2 + \frac{\partial x'^i}{\partial x^3} dx^3,$$

for *i* = 1, 2, or 3.

covariant tensor. A tensor which has only covariant indices. If there are *s* indices, it is a **covariant tensor of order *s*.** The *gradient* of a function is a covariant tensor of order one (*i.e.*, a *covariant vector*). If the function is

$$f(x^1, x^2, x^3),$$

the components of the tensor are

$$\frac{\partial f}{\partial x^i} \quad (i = 1, 2, 3),$$

and we have

$$\frac{\partial f}{\partial x'^i} = \frac{\partial f}{\partial x^j} \frac{\partial x^j}{\partial x'^i}$$

$$= \frac{\partial f}{\partial x^1} \frac{\partial x^1}{\partial x'^i} + \frac{\partial f}{\partial x^2} \frac{\partial x^2}{\partial x'^i} + \frac{\partial f}{\partial x^3} \frac{\partial x^3}{\partial x'^i}.$$

covariant and contravariant derivatives of a tensor. See CONTRAVARIANT, COVARIANT.

divergence of a tensor. See DIVERGENCE.

Einstein tensor. See RICCI—Ricci tensor.

fundamental metric tensor. See RIEMANN—Riemannian space.

mixed tensor. A tensor which has both contravariant and covariant indices.

multiple-point tensor field. A generalized tensor field whose components depend on the coordinates of two or more points. *E.g.*, the distance (say in the Euclidean plane) between two variable points in the plane is a two-point scalar field.

numerical tensor. A tensor which has the same components in all coordinate systems. The *Kronecker delta* δ^i_j and the *generalized Kronecker delta* are numerical tensors.

product of tensors. The product of two tensors $A^{a_1 \cdots a_n}_{i_1 \cdots i_m}$ and $B^{b_1 \cdots b_p}_{j_1 \cdots j_q}$ is the tensor $C^{a_1 \cdots a_n b_1 \cdots b_p}_{i_1 \cdots i_m j_1 \cdots j_q} = A^{a_1 \cdots a_n}_{i_1 \cdots i_m} B^{b_1 \cdots b_p}_{j_1 \cdots j_q}$. This is also called the **outer product.** See INNER—inner product of tensors.

relative tensor field of weight *w*. Its definition differs from that of a tensor field by the presence of the *Jacobian* $\left| \dfrac{\partial x^i}{\partial x'^j} \right|$ to the *w*th power as a factor in the right-hand side of the transformation law of the components of the tensor field. A relative tensor field of weight 1 is a **tensor density.** The *epsilon symbol* $\epsilon^{i_1 i_2 \cdots i_n}$ is a tensor density. The components of a *scalar field of weight one* (a *scalar density*) are related by

$$s'(x'^1, \cdots, x'^n) = \left| \frac{\partial x^i}{\partial x'^j} \right| s(x^1, \cdots, x^n).$$

If t_{ij} are the components of a covariant tensor field and if $t = |t_{ij}|$ is the *n*-rowed determinant with element t_{ij} in the *i*th row and *j*th column, then $\sqrt{t}$ is a *scalar density*.

Ricci tensor. See RICCI.

Riemann-Christoffel curvature tensor. See RIEMANN.

skew-symmetric tensor. When the interchange of two contravariant (or covariant) indices changes only the sign of each component, the tensor is **skew-symmetric with respect to these indices**. A tensor is **skew-symmetric** if it is skew-symmetric with respect to every two contravariant and every two covariant indices.

strain tensor. See STRAIN.

symmetric tensor. When the interchange of two or more contravariant (or covariant) indices in the components of a tensor does not change any of the components of the tensor, the tensor is **symmetric with respect to these indices**. A tensor is **symmetric** if it is symmetric with respect to every two contravariant and every two covariant indices.

tensor analysis. The study of objects with components possessing characteristic laws of transformation under transformation of coordinates. The subject is intimately connected with the various Riemannian and non-Riemannian geometries, including the theory of surfaces in Euclidean and non-Euclidean spaces.

tensor density. See above, relative tensor field of weight w.

tensor field. See above, TENSOR.

tensor product of vector spaces. If X and Y are vector spaces over a field F, the **tensor product** $X \otimes Y$ is the *conjugate* (or *dual*) of the space $L(X, Y)$ of bilinear functions from X and Y to F. If X and Y have dimensions m and n, then $X \otimes Y$ has dimension mn. For members x and y of X and Y, the member z of $X \otimes Y$ defined by $z(\varphi) = \varphi(x, y)$ for each bilinear function φ is denoted by $z = x \otimes y$. If X and Y are locally convex topological vector spaces, then the **projective topology** on $X \otimes Y$ is the finest locally convex topology for which the map F defined by $F(x, y) = x \otimes y$ is continuous.

TER'A. A prefix meaning 10^{12}, as in **terameter**.

TERM, *adj., n.* (1) A term of a **fraction** is the numerator or denominator of the fraction (see LOWEST—fraction in lowest terms). A term of a **proportion** is any one of the *extremes* or *means*. See PROPORTION. (2) A term of an **equation** or an **inequality** is the entire quantity on one (or the other) side of the sign of equality or inequality. *Member* is better usage. (3) For an expression which is written as the sum of several quantities, each of these quantities is called a term of the expression, *e.g.*, in $xy^2 + y \sin x - \dfrac{x+1}{y-1} - (x+y)$, the terms are xy^2, $y \sin x$, $-\dfrac{x+1}{y-1}$ and $-(x+y)$. In the **polynomial** $x^2 - 5x - 2$, the

terms are x^2, $-5x$, and -2; in $x^2 + (x+2) - 5$, the terms are x^2, $(x+2)$, and -5. A **constant** (or **absolute**) term is a term that does not contain any of the variables. An **algebraic term** is a term containing only algebraic symbols and numbers; *e.g.*, $7x$, $x^2 + 3ay$, and $\sqrt{3x^2 + y}$. An algebraic term is **rational and integral** if the variable or variables do not appear under any radical sign, in any denominator, or with fractional or negative exponents. The terms $2xy$ and $2x^2y^2$ are rational and integral in x and y. A term which is not algebraic is **transcendental**. Examples of transcendental terms are **trigonometric terms**, which contain only trigonometric functions and constants (sometimes a term is called trigonometric when it also contains algebraic factors); **exponential terms**, which contain the variable only in exponents (sometimes terms are called exponential when they contain both exponential and algebraic factors); **logarithmic terms**, which contain the variable or variables affected by logarithms, such as $\log x$, $\log(x+1)$ (sometimes terms are called logarithmic when they also contain algebraic factors, as $x^2 \log x$). **Like** (or **similar**) terms are terms that contain the same variables, each variable of the same kind being raised to the same power; $2x^2yz$ and $5x^2yz$ are **like terms**.

general term. A term containing parameters in such a way that the parameters can be given specific values so that the term itself reduces to any special term of some set under consideration. See BINOMIAL—binomial theorem. The general term in the general algebraic equation of the nth degree $a_0 x^n + a_1 x^{n-1} + \cdots + a_n = 0$ would be written $a_i x^{n-i}$ $(i = 0, \cdots, n)$.

term life insurance. See INSURANCE —life insurance.

TER'MI-NAL, *adj.* **terminal form of commutation columns.** See COMMUTATION —commutation tables.

terminal point. See CURVE, DIRECTED —directed line.

terminal side of an angle. See ANGLE.

TER'MI-NAT'ING, *adj.* Coming to an end; limited; expressed in a finite number of figures or terms. *E.g.*, the decimal 3.147 is a **terminating decimal**, while the repeating decimal $7.414141\cdots$ is **nonterminating**. Infinite sequences and series are nonterminating.

terminating continued fraction. See FRACTION—continued fraction.

terminating plan of building and loan association. See BUILDING —building and loan association.

TER′NA-RY, *adj.* **ternary number system.** A system of notation for real numbers that uses the base 3 instead of 10. (See BASE—base of a system of numbers). *E.g.,* the number $38\frac{5}{27}$ in the decimal system would be 1102.012 when written with base 3, since with base 3 the number 1102.012 is equal to $1 \cdot 27 + 1 \cdot 9 + 0 \cdot 3 + 2 + 0 \cdot \frac{1}{3} + 1 \cdot \frac{1}{9} + 2 \cdot \frac{1}{27}$.

    ternary operation. An operation which is applied to three objects. Examples are the operation of determining the *median* of three numbers, and the operation on the numbers x, y, and z that produces the number $x(y + z)$. *Tech.* A **ternary operation** is a function f whose domain is a set of ordered triples of members of a set S. See BINARY—binary operation.

TER-RES′TRI-AL, *adj.* **terrestrial triangle.** A spherical triangle on the earth's surface (considered as a sphere) having for its vertices the north pole and two points whose distance apart is being found.

TES-SEL-LA′TION, *n.* A covering of the plane by congruent polygons or of space by congruent polyhedrons. See POLYOMINO.

TES′SE-RACT, *n.* See CUBE.

TES′SER-AL, *adj.* **tesseral harmonic.** See HARMONIC—surface harmonic.

TEST, *adj.* **test statistic.** A statistic upon which a *test of a hypothesis* is based. *Syn.* test (random) variable. See HYPOTHESIS —test of a hypothesis.

    test function. See GENERALIZED—generalized function.

TET′RA-HE′DRAL, *adj.* **tetrahedral angle.** A polyhedral angle having four faces. See ANGLE—polyhedral angle.

    tetrahedral group. The group of motions (symmetries) of 3-dimensional space that preserve the regular tetrahedron. It also is the alternating group of order 12 that is the set of even permutations on 4 objects. See GROUP, SYMMETRY—group of symmetries.

    tetrahedral surface. A surface admitting parametric representation of the form

$$x = A(u - a)^{\alpha}(v - a)^{\beta},$$

$$y = B(u - b)^{\alpha}(v - b)^{\beta},$$

$$z = C(u - c)^{\alpha}(v - c)^{\beta},$$

where a, b, c, A, B, C, α, and β are constants.

TET′RA-HE′DRON, *n.* [*pl.* **tetrahedrons** or **tetrahedra**]. A four-faced polyhedron. *Syn.* triangular pyramid. A **regular tetrahedron** is a tetrahedron having all of its faces equilateral triangles. *Syn.* triangular pyramid. See POLYHEDRON.

THE′O-REM, *n.* A statement that has been proved to be true if certain hypotheses (axioms) are true. However, such a statement might not be labeled as a theorem unless it is believed to be worthy of special attention. A proof of a theorem may use previously proved theorems and not make explicit use of the hypotheses. A statement that follows "easily" from a theorem is often said to be a **corollary** of that theorem. A theorem that is proved primarily because of its use in proving another theorem is said to be a **lemma**.

THE′O-RY, *n.* The principles concerned with a certain concept, and the facts postulated and proved about it.

    function theory. The theory of functions of a real variable, the theory of functions of a complex variable, etc.

    group theory, or theory of groups. See GROUP.

    linear theory. See ELASTICITY.

    number theory, or theory of numbers (elementary). The study of integers and relations between them.

    theory of equations. The study of methods of solving and the possibility of solving polynomial equations, and of the relations between roots and between roots and coefficients of equations.

THE′TA, *n.* The eighth letter of the Greek alphabet: lower case, θ or ϑ; capital, Θ.

    theta functions. Let $q = e^{\pi i \tau}$, where τ is a constant complex number whose imaginary part is positive. The four theta functions (usually written without explicitly indicating the dependence on τ) are

$$\vartheta_1(z) = 2\sum_{0}^{\infty}(-1)^n q^{(n+1/2)^2}\sin(2n+1)z,$$

$$\vartheta_2(z) = 2\sum_{0}^{\infty} q^{(n+1/2)^2}\cos(2n+1)z,$$

$$\vartheta_3(z) = 1 + 2\sum_{1}^{\infty} q^{n^2}\cos 2nz,$$

$$\vartheta_4(z) = 1 + 2\sum_{1}^{\infty}(-1)^n q^{n^2}\cos 2nz.$$

Other notations are also used for these functions

$(e.g., \vartheta$ for ϑ_4 and $\vartheta_i(\pi z)$ for $\vartheta_i(z)$). It can be shown that

$$\vartheta_1(z) = -\vartheta_2\left(z + \tfrac{1}{2}\pi\right)$$
$$= \left(-iq^{1/4}e^{iz}\right)\vartheta_3\left(z + \tfrac{1}{2}\pi + \tfrac{1}{2}\pi\tau\right)$$
$$= \left(-iq^{1/4}e^{iz}\right)\vartheta_4\left(z + \tfrac{1}{2}\pi\tau\right).$$

Each of the theta functions satisfies a relation analogous to

$$\vartheta_4(z + \pi) = \vartheta_4(z) = \left(-qe^{2iz}\right)\vartheta_4(z + \pi\tau),$$

and are said to be *quasi doubly periodic*. The theta functions are entire functions.

THOM, René (1923–). French differential topologist and Fields Medal recipient (1958). Created theory of cobordism.

THOMPSON, John Griggs (1932–). English mathematician and Fields Medal recipient (1970). With Walter Feit, proved in 1963 that all noncyclic finite simple groups have even order (the *Feit-Thompson theorem*, which was conjectured by William Burnside in 1911). Determined the minimal simple finite groups, that is, the simple finite groups whose proper subgroups are solvable.

THOU'SAND, *n.* Ten hundred (1000).

THREE, *adj., n.* **divisibility by three.** See DIVISIBLE.

rule of three. The rule that the product of the *means* of a proportion equals the product of the *extremes* (see PROPORTION). This can be applied to such problems as: "If 6 apples cost 90¢, what will 10 apples cost?" Three pieces of information are given. To find the cost c of 10 apples, we can equate the costs per apple, calculated in two ways: $90/6 = c/10$, and obtain $6 \cdot c = 900$ or $c = 150$¢. The following quote is from "The Mad Gardener's Song" in Lewis Carroll's book SYLVIE AND BRUNO (1889).

> "He thought he saw a Garden-Door
> That opened with a key:
> He looked again, and found it was
> A Double Rule of Three:
> 'And all its mystery,' he said,
> 'Is clear as day to me!' "

The text Carroll used in school indicates that the **double rule of three** refers to problems such as the preceding, but involving more information, as: "If $100 earns $7 simple interest in 14 months, what principal will earn $18 in 9 months?" To find the principal P, compute the rate of interest in two ways [rate = interest/(time × principal)]:

$$\frac{7}{14 \cdot 100} = \frac{18}{9 \cdot P} \quad \text{or} \quad P = \$400.$$

three-circles theorem. See HADAMARD.

three-dimensional geometry. The study of figures in three (as well as two) dimensions. See GEOMETRY —solid geometry, and DIMENSION.

three-point form of the equation of a plane. See PLANE—equation of a plane.

three-point problem. Given three collinear points, A, B, C, with the distances AB and BC known, and a fourth point S, with the angles ASB and BSC known, to find the distance SB. This is the problem of finding the distance from a ship S to a point B on the shore.

three-squares theorem. A positive integer n is the sum of the squares of three integers if and only if there are no nonnegative integers r and s for which $n = 4^r(8s + 7)$. See WARING —Waring's problem.

THUE, Axel (1863–1922). Norwegian number theorist.

Thue-Siegel-Roth theorem. For an irrational number α, let $\overline{\mu}(\alpha)$ be the least upper bound of all μ for which there are infinitely many rational numbers p/q such that

$$\left|\frac{p}{q} - \alpha\right| < q^{-\mu}.$$

Then for all α, $\overline{\mu}(\alpha) \geq 2$. It has been proved that if α is an algebraic number of degree n, then $\overline{\mu}(\alpha) \leq n$ (Liouville, 1844), $\overline{\mu}(\alpha) \leq \tfrac{1}{2}n + 1$ (Thue, 1908), $\overline{\mu}(\alpha) \leq 2\sqrt{n}$ (Siegel, 1921), $\overline{\mu}(\alpha) \leq \sqrt{2n}$ (Dyson, 1947), $\overline{\mu}(\alpha) = 2$ (Roth, 1955).

THURSTON, William P. (1946–). American mathematician awarded a Fields Medal (1983) for his contributions to the theory of geometric structures of three-dimensional manifolds.

TIETZE, Heinrich Franz Friedrich (1880–1964). Austro-German analyst and topologist.

Tietze extension theorem. If T is a Hausdorff topological space, each of the following is a necessary and sufficient condition for T to be *normal*: (a) For each closed subset X and each continuous function f that maps X into the closed interval $[0,1]$, there is a continuous function F that maps T into $[0,1]$ and satisfies $F(x) = f(x)$ if $x \in X$. (b) For each closed subset X and each continuous function f that maps X

into the set of real numbers, there is a continuous function F that maps T into the set of real numbers and satisfies $F(x) = f(x)$ if $x \in X$. See RETRACT. *Syn.* Tietze-Urysohn extension theorem.

TILE, *n.* See POLYOMINO.

TIME, *adj., n.* Continuous existence as indicated by some sequence of events, such as the hours indicated by a clock or the rotation of the earth about its axis; the experience of duration or succession. **Mean solar time** (or **astronomical time**) is the average time between successive passages of the sun over the meridian of a place, the time that would be shown by a sun dial if the sun were always on the celestial equator (in the plane of the earth's equator) and moving at a uniform rate. **Apparent solar time** is the time indicated by the sun dial, which divides each day into 24 hours; the hour-angle of the apparent or true sun (see HOUR) plus 12 hours. The hours are not exactly the same length, because of the inclination of the earth's axis to the plane of the ecliptic (the plane of the earth's orbit) and to the eccentricity (elliptic shape) of the earth's orbit. Also see SIDEREAL —sidereal time. **Standard time** is a uniform system of measuring time, originated for railroad use in the United States and Canada and now in common use. The United States is divided into four belts, each extending through approximately 15° of longitude, designated as *Eastern, Central, Mountain,* and *Pacific.* The time in each belt is the mean solar time of its central meridian. For instance, 7 a.m. *Central Time* is 8 a.m. *Eastern Time,* 6 a.m. *Mountain Time,* and 5 a.m. *Pacific Time. Tech.* Standard time is the mean solar time of a standard meridian, a meridian whose longitude differs by a certain multiple of 15° from the longitude at Greenwich, 15° being equivalent to one hour. See SECOND.

    time discount. See DISCOUNT —time discount.

    time rate. See SPEED, VELOCITY.

    time series. Data taken at time intervals, such as the temperature or the rainfall taken at a certain time each day for a succession of days.

TON, *n.* See DENOMINATE NUMBERS in the appendix.

TON-TINE', *adj.* **tontine annuity.** An annuity purchased by a group with the *benefit of survivorship* (i.e., the share of each member who dies is divided among the others, the last survivor getting the entire annuity during the balance of his life).

    tontine fund. A fund accumulated by investments of withheld annuity payments.

tontine insurance. Insurance in which all benefits except those due to death, including such as dividends and cash surrender values, are allowed to accumulate until the end of a certain period and then are divided among those who have carried this insurance throughout the period.

TOP'O-LOG'I-CAL, *adj.* **linear topological space.** See VECTOR —vector space.

    topological dimension. See dimension.

    topological group. An abstract group which is also a topological space and in which the group operations are continuous. Continuity of the group operations means that, for any elements x and y, (1) if W is a neighborhood of xy, then there exist neighborhoods U and V of x and y such that uv belongs to W if u and v belong to U and V, respectively; (2) if V is any neighborhood of x^{-1} (the inverse of x), then there exists a neighborhood U of x such that u^{-1} belongs to V if u belongs to U. The set of all real numbers is a topological group, as is also the group of all nonsingular square matrices of a certain order with multiplication as the group operation and a neighborhood of a matrix A as the set of all matrices B such that the *norm* of $A - B$ is less than some fixed number ϵ.

    topological manifold. See MANIFOLD.

    topological property. Any property of a geometrical figure A that holds as well for every figure into which A may be transformed by a topological transformation. Examples are the properties of *connectedness* and *compactness,* of subsets being *open* or *closed,* and of points being *accumulation points.*

    topological space. A set X which has associated with it a collection $\mathcal{T}$ of subsets for which the empty set and X belong to $\mathcal{T}$, $U \cap V$ belongs to $\mathcal{T}$ if U and V belong to $\mathcal{T}$, and the union of any collection of sets belonging to $\mathcal{T}$ belongs to $\mathcal{T}$. The members of $\mathcal{T}$ are **open sets.** The plane is a topological space if open sets are those sets U with the property that for any x in U there is a number $\epsilon > 0$ such that U contains the disk with radius ϵ and center at x; similarly, any metric space is a topological space if open sets are defined similarly. There are many special types of topological spaces: A T_0-**space** (or **Kolmogorov space**) has the property that if $x \neq y$ then either there is an open set that contains x and does not contain y or there is an open set that contains y and does not contain x; a T_1-**space** (or **Fréchet space**) has the property that if $x \neq y$ then there is an open set that contains x and does not contain y; a T_2-**space** (or **Hausdorf space**) has the property that if $x \neq y$ then there are disjoint open sets U and V that contain x

and y, respectively; a **T_3-space** is a T_1-space that is regular; a **T_4-space** is a T_1-space that is *normal*; a **T_5-space** is a T_1-space that is completely normal. A **Tychonoff space** (or $T_{3\frac{1}{2}}$-space) is a T_1-space that is completely regular. See REGULAR—regular space.

topological transformation. A one-to-one correspondence between the points of two geometric figures A and B which is continuous in both directions; a one-to-one correspondence between the points of A and B such that open sets in A correspond to open sets in B, and conversely (or analogously for closed sets). If one figure can be transformed into another by a topological transformation, the two figures are said to be **topologically equivalent**. Continuous deformations are examples of topological transformations (see DEFORMATION). Any two "knots" formed by looping and interlacing a piece of string and then joining the ends together are topologically equivalent, but cannot necessarily be continuously deformed into each other. *Syn.* homeomorphism.

TOP′O-LOG′I-CAL-LY, *adv.* **topologically complete.** See COMPLETE —complete space.

TO-POL′O-GY, *n.* That branch of geometry which deals with the *topological properties* of figures. **Combinatorial topology** is the branch of topology which is the study of geometric forms by decomposing them into the simplest geometric figures (simplexes) which adjoin each other in a regular fashion (see COMPLEX —simplicial complex, SURFACE). **Algebraic topology** includes the fields of topology which use algebraic methods (especially group theory) to a large extent (see HOMOLOGY —homology group). **Point-set topology** is the study of sets as accumulations of points (as contrasted to combinatorial methods of representing an object as a union of simpler objects) and describing sets in terms of topological properties such as being open, closed, compact, normal, regular, connected, etc. See ANALYSIS—analysis situs.

discrete topology. See DISCRETE.

topology of a space. The set of all *open subsets* of the space (the space must be a *topological space*). A set can be assigned a topology by specifying a family of subsets with the properties that any union and any finite intersection of members of the family is a member of the family (see BASE—base for a topology). A given set of objects together with a topology is a topological space (see TOPOLOGICAL —topological space). For a normed linear space, the topology defined by the norm is sometimes called the **strong topology**

to distinguish it from the weak topology (see WEAK—weak topology).

trivial topology. See TRIVIAL.

uniform topology. See UNIFORM.

TORQUE, *n.* See MOMENT—moment of force.

TOR′SION, *adj., n.* geodesic torsion. See GEODESIC.

torsion coefficients of a group. If G is a commutative group with a finite set of generators, then G is a *Cartesian product* of infinite cyclic groups $F_1, F_2, \cdots, F_m$ and cyclic groups $H_1, H_2, \cdots, H_n$ of finite order. The number m and the orders $r_1, r_2, \cdots, r_n$ of $H_1, H_2, \cdots, H_n$ form a complete system of invariants. The numbers $r_1, r_2, \cdots, r_n$ are **torsion coefficients** of G (G is **torsion free** if $n = 0$).

torsion of a space curve at a point. If P is a fixed point, and P' a variable point, on a directed space curve C, Δs the length of arc on C from P to P', and $\Delta\psi$ the angle between the positive directions of the binormals of C at P and P', then the torsion $1/\tau$ of C at P is defined, to within sign, by $1/\tau = \lim\limits_{\Delta s \to 0} \pm \dfrac{\Delta\psi}{\Delta s}$. The sign of $1/\tau$ is chosen so that we have $d\gamma/ds = \beta/\tau$. See FRENET—Frenet-Serret formulas. The torsion may be taken as a measure of the rate at which C is turning out of its osculating plane relative to the arc length s. We have

$$1/\tau = -\rho^2 \begin{vmatrix} x' & y' & z' \\ x'' & y'' & z'' \\ x''' & y''' & z''' \end{vmatrix},$$

the primes denoting differentiation with respect to the arc length. The reciprocal of the torsion is the **radius of torsion**. Some authors use the symbol τ rather than $1/\tau$ to denote the torsion.

TO′RUS, *n.* Same as ANCHOR RING.

TO′TAL, *adj.* **total curvature.** See CURVATURE —curvature of a surface.

total differential. See DIFFERENTIAL.

TO′TAL-LY, *adv.* **totally bounded** and **totally disconnected.** See BOUNDED—bounded set of points, DISCONNECTED —disconnected set.

totally ordered set. See ORDERED—ordered set.

TO′TIENT, *n.* **totient of an integer.** Same as *Euler ϕ-function* of the integer (see EULER —Euler ϕ-function); the number of *totitives* of the integer.

TOT′I-TIVE, *n.* **totitive of an integer.** A positive integer not greater than the given positive integer and relatively prime to it ($+1$ being the only common factor that is a positive integer); 1, 3, 5, and 7 are the totitives of 8. Each integer less than a given prime is a totitive of the prime.

TRACE, *n.* **trace of a line in space.** (1) A point at which it pierces a coordinate plane. (2) Its projection in a coordinate plane; the intersection of a *projecting plane* of the line with the corresponding coordinate plane. When trace is used in the latter sense, the point of definition (1) is called the *piercing point.*

    trace of a matrix. The sum of the elements of the principal diagonal (German "Spur").

    traces of a surface. The curves in which it cuts the coordinate planes.

TRAC′ING, *n.* **curve tracing.** See CURVE—curve tracing.

TRAC′TRIX, *n.* The involute of a catenary; a curve the lengths of whose *tangents* are all equal; the path of one end (*P* in the figure) of a rod *PQ* of fixed length *a* attached to a point *Q* which moves along the *x*-axis from 0 to $\pm\infty$, the initial position of the rod being *OA* and the rod moving in such a way as to always be tangent to the path described by *P*. Its equation is

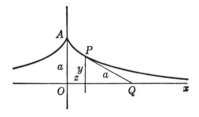

$$x = a \log \frac{\left(a \pm \sqrt{a^2 - y^2}\right)}{y} \mp \sqrt{a^2 - y^2}.$$

TRA-JEC′TO-RY, *n.* (1) The path of a moving particle or a celestial body. (2) A curve which cuts all curves (or surfaces) of a given family at the same angle. An **orthogonal trajectory** is a curve which cuts all the members of a given family of curves (or surfaces) at right angles. (3) A curve or surface which fits some given law such as passing through a given set of points.

TRAN′SCEN-DEN′TAL, *adj.* **transcendental curves.** Graphs of *transcendental functions.*

    transcendental functions. Functions which cannot be expressed algebraically in terms of the variable (or variables) and constants. Contains terms involving trigonometric functions, logarithms, exponentials, etc. *Tech.* A transcendental function is any function which is not an *algebraic function*. An entire function is transcendental if it is not a polynomial. See FUNCTION—algebraic function.

    transcendental number. See ALGEBRAIC—algebraic number, IRRATIONAL—irrational number.

TRANS-FI′NITE, *adj.* **transfinite induction.** The principle that, if some statement is true for the first member of a *well-ordered set S*, and the statement is true for a member *a* of *S* if it is true for each member preceding *a*, then the statement is true for every member of *S*. This is easy to show. Let *F* be the set of all members of *S* for which the statement is false. If *F* is not empty, then it follows from *S* being well-ordered that *F* has a first member *α*. Since the statement is true for the first member of *S*, *α* is not the first member of *S*. But then the statement is true for all members of *S* that precede *α*, which implies the statement is true for *α*. This contradiction implies *F* is empty and the statement is true for all members of *S*. See INDUCTION—mathematical induction, ORDERED—ordered set.

    transfinite number. A cardinal or ordinal number which is not an integer. See CARDINAL.

TRANS-FORM′, *n.* **transform of an element of a group.** The transform of an element *A* by an element *X* is the element $B = X^{-1}AX$. The set of all transforms of *A* by elements of the group is the set of **conjugates** of *A* and is a **conjugate set** (or **class**) of elements of the group. The set of different subgroups obtained by transforming a given subgroup by all the elements of the group is a **conjugate set of subgroups**; any two of these subgroups are **conjugate** to each other. See GROUP, NORMAL—normal subgroup.

    transform of a matrix. A matrix *B* related to a given matrix *A* by $B = P^{-1}AP$, where *P* is a nonsingular matrix.

TRANS-FOR-MA′TION, *adj., n.* A passage from one figure or expression to another, as: (1) the changing of one algebraic expression to another one having different form (*e.g.,* see below, congruent transformation); (2) the changing of an equation or algebraic expression by substituting for the variables their values in terms of another set of variables; (3) a *function.* See FUNCTION, LINEAR—linear transformation (2).

    affine transformation. See AFFINE.

    adjoint transformation. See ADJOINT.

collineatory transformation. (1) A nonsingular linear transformation of $(n - 1)$-dimensional Euclidean space of the form $y_i = \sum_{j=1}^{n} a_{ij}x_j$ $(i = 1, 2, \cdots, n)$ in homogeneous coordinates; a transformation which takes collinear points into collinear points. (2) A transformation B of the form $P^{-1}AP$ of a matrix A by a nonsingular matrix P; A and B are then said to be **similar** and are **transforms** of each other. These two concepts are closely related. For let the coordinates of two points x and y be related by $y_i = \sum_{j=1}^{n} a_{ij}x_j$ $(i = 1, 2, \cdots, n)$ or symbolically by $y = Ax$, where x is thought of as a one-column matrix (vector) with elements (components) $x_1, x_2, \cdots, x_n$ (and likewise for y). If P is the matrix of a nonsingular linear transformation, with $y = Py'$ and $x = Px'$, then $Py' = APx'$, or $y' = P^{-1}APx' = Bx'$, in the new frame of reference introduced by the linear transformation defined by P. See EQUIVALENT —equivalent matrices. *Syn.* collineation.

congruent transformation. A transformation of the form $B = P^{T}AP$ of a matrix A by a nonsingular matrix P, where P^{T} is the transpose of P. B is said to be **congruent** to A. Let the quadratic form $Q = \sum_{i,j=1}^{n} a_{ij}x_ix_j$ be written symbolically as $Q = (x)A\{x\}$, where (x) is the one-row matrix $(x_1, \cdots, x_n)$, $\{x\}$ is the similar one-column matrix, and multiplication is matrix multiplication. If

$$x_i = \sum_{j=1}^{n} p_{ij}y_j \quad (i = 1, 2, \cdots, n),$$

or symbolically $\{x\} = P\{y\}$, then

$$(x) = (y)P^{T}$$

and $Q = (y) \cdot P^{T}AP \cdot \{y\}$. Thus the matrix A is transformed into a congruent matrix under a linear transformation of the variables. Every symmetric matrix is congruent to a diagonal matrix, and hence every quadratic form can be changed to a form of type $\Sigma k_i x_i^2$ by a linear transformation. See DISCRIMINANT—discriminant of a quadratic form, ORTHOGONAL —orthogonal transformation.

conjunctive transformation. See EQUIVALENT —equivalent matrices. A conjunctive transformation is related to *Hermitian forms* in the same way that a *congruent transformation* is related to *quadratic forms* except that P^{T} is replaced by the Hermitian conjugate of P. See above, con-

gruent transformation. Every Hermitian matrix can be made diagonal by a conjunctive transformation, and hence every Hermitian form can be reduced to the form $\sum_{i=1}^{n} a_i z_i \bar{z}_i$ by a linear transformation, where a_i is real for each i.

division transformation. See DIVISION.

equiangular transformation. Same as ISOGONAL TRANSFORMATION.

Euler transformation. See EULER—Euler transformation of series.

factoring of a transformation. See FACTORIZATION —factorization of a transformation.

Hermitian transformation. See HERMITIAN —Hermitian transformation.

homogeneous transformation. A transformation whose equations are algebraic and whose terms are all of the same degree. Rotation of axes, reflection in the axes, stretching, and shrinking are homogeneous transformations.

homothetic transformation. See SIMILITUDE —transformation of similitude.

identity transformation. See IDENTITY—identity function.

inverse transformation. The transformation which exactly undoes the effect of a given transformation. That is, if T is a transformation, then T^{-1} is its inverse if $T^{-1}T = I$, where I is the identity transformation. For the nonzero real (or complex) numbers, the transformation $T(x) = 1/x$ is its own inverse, because two reciprocals of a number return it to its original value. If T is a one-to-one transformation of a set X onto a set Y, then the **inverse** of T is the transformation T^{-1} which maps a point y of Y to a point x of X if T maps x to y. A transformation has an inverse if and only if it is one-to-one. See IMAGE, and, since a transformation is a function, see INVERSE —inverse of a function.

isogonal transformation. See ISOGONAL—isogonal transformation.

linear transformation. See LINEAR —linear transformation.

matrix of a linear transformation. See MATRIX—matrix of a linear transformation.

normal transformation. See NORMAL.

orthogonal transformation. See ORTHOGONAL.

product of two transformations. The transformation resulting from the successive application of the two given transformations. Such a product may not be commutative, *i.e.*, the product may depend on the order in which the transformations are applied. *E.g.*, the transformation $x = x' + a$ and $x = (x')^2$ are not commutative, since replacing x by $x' + a$ and the new x (x' with the prime dropped) by $(x')^2$ gives $x =$

$(x')^2 + a$ as the product transformation, while reversing the order gives $x = (x' + a)^2$.

rational transformation. The replacement of the variables of an equation, or function, by other variables which are each rational functions of the first. The transformations $x' = x + 2$, $y' = y + 3$ and $x' = x^2$, $y' = y^2$ are rational transformations.

reducible transformation. See REDUCIBLE.

simple shear transformation. A transformation which represents a shearing motion for which each point of a coordinate axis in the plane, or of a coordinate plane in space, is mapped onto itself; in a plane, it is a transformation of the form $x' = x$, $y' = kx + y$, or $x' = ky + x$, $y' = y$.

symmetric transformation. See SYMMETRIC.

topological transformation. See TOPOLOGICAL.

transformation of coordinates. Changing the coordinates of a point to another set which refers to a new system of coordinates, either of the same type or of another type. Examples are affine transformations, linear transformations, translation of axes, rotation of axes, and transformations between Cartesian and polar or spherical coordinates.

transformation group. A set of transformations which form a group. *Syn.* group of transformations. See GROUP, and above, product of two transformations, and inverse transformation.

transformation of similitude. See SIMILITUDE.

unitary transformation. See UNITARY—unitary transformation.

TRANS'IT, *n.* An instrument for measuring angles. Consists essentially of a small telescope which rotates horizontally and vertically, the angles through which it rotates being indicated on graduated scales.

TRAN'SI-TIVE, *adj.* **transitive relation.** A relation which has the property that if A bears the relation to B and B bears the same relation to C, then A bears the relation to C. Equality in arithmetic is transitive, since if $A = B$ and $B = C$, then $A = C$ [also see ORDER—order properties of real numbers]. A relation which is not transitive is **intransitive** or **nontransitive** according as there does not, or does, exist a set of objects A, B, C such that A bears the relation to B, B bears the relation to C, and A bears the relation to C. The relation of *being the father of* is *intransitive*, since if A is the father of B, and B is the father of C, then A is not the father of C. The relation of *being a friend of* is *nontransitive*, since if A is a friend of B, and B is a friend of C, then A may or may not be a friend of C. See EQUIVALENCE —equivalence relation.

TRANS-LA'TION, *adj.*, *n.* **translation of axes.** Changing the coordinates of points to coordinates referred to new axes parallel to the old. Used to change the form of equations so as to aid in the study of their loci; *e.g.*, one may desire to translate the axes so that the new origin is on the curve, which yields an equation with zero constant term, or to translate the axes until they are coincident with the axes of symmetry when these are parallel to the axes, as in the case of conics, thus obtaining an equation lacking first-degree terms. **Translation formulas** are formulas expressing a translation of axes analytically. In the *plane*, these formulas are

$$x = x' + h, \qquad y = y' + k,$$

where h and k are the coordinates of the origin of the x', y' system with reference to the x, y system; *i.e.*, when $x' = y' = 0$, $x = h$ and $y = k$. In *space*, if the new origin has the coordinates (h, k, l) with respect to the old axes, and a point has coordinates (x', y', z') with respect to the new axes, and coordinates (x, y, z) with respect to the old axes, then $x = x' + h$, $y = y' + k$, $z = z' + l$.

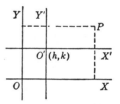

translation and rotation. A transformation which both translates and rotates the axes. Used, for instance, in the study of the general quadratic in x and y to obtain an equation with zero coefficients of the xy and x and y terms. The transformation formulas are

$$x = x' \cos \theta - y' \sin \theta + h,$$

$$y = x' \sin \theta + y' \cos \theta + k,$$

where h and k are the coordinates of the new origin relative to the old and θ is the angle through which the positive x-axis is rotated to be parallel to the positive x'-axis.

translation surface. See SURFACE —surface of translation.

TRANS'POR-TA'TION, *adj.*, *n.* **Hitchcock transportation problem.** The linear programming problem of minimizing the total cost of moving ships between ports. Thus if there are a_i ships in port A_i $(i = 1, 2, \cdots, n)$ and it is desired to deliver a total of b_j of the ships to port B_j ($j = 1, 2, \cdots, m$) with $\sum_{i=1}^{n} a_i = \sum_{j=1}^{m} b_j$, and the cost of moving one ship from A_i to B_j is c_{ij}, then we want to choose nonnegative integers x_{ij} that will minimize $\sum_{i,j=1}^{n,m} c_{ij} x_{ij}$, subject to the constraints $\sum_{j=1}^{m} x_{ij} = a_i$, $\sum_{i=1}^{n} x_{ij} = b_j$. See PROGRAMMING—linear programming.

TRANS-POSE, *n.*, *v.* To move a term from one member of an equation to the other and change its sign. This is equivalent to subtracting the term from both members. The equation $x + 2 = 0$ yields $x = -2$ after *transposing* the 2.

transpose of a matrix. The matrix resulting from interchanging the rows and columns in the given matrix.

TRANS'PO-SI'TION, *n.* (1) The act of transposing terms from one side of an equation to the other. See TRANSPOSE. (2) The interchange of two objects; a cyclic permutation of two objects. See PERMUTATION —cyclic permutation.

TRANS-VER'SAL, *adj.* (1) A line intersecting a system of lines. See ANGLE—angles made by a transversal. (2) See TRANSVERSALITY —transversality condition.

TRANS'VER-SAL'I-TY, *adj.* **transversality condition.** A condition generalizing the fact that the shortest line segment joining a point (x_1, y_1) to a curve C must be orthogonal to C at the point (x_2, y_2) where the segment meets C. For a curve C with parametric equations $x = X(t)$, $y = Y(t)$, the transversality condition is

$$\left(f - y' f_{y'} \right) X_t + f_{y'} Y_t = 0,$$

which must be satisfied at the point (x_2, y_2) if the function y is to minimize the integral $I = \int_{x_1}^{x_2} f(x, y, y') \, dx$, where (x_1, y_1) is fixed and (x_2, y_2) is constrained only to lie on C. A curve which satisfies the transversality condition relative to a curve C and an integral $I = \int_{x_1}^{x_2} f(x, y, y') \, dx$ and which minimizes this integral for (x_2, y_2) on C is a **transversal**. See FOCAL—focal point.

TRANS-VERSE', *adj.* **transverse axis of a hyperbola.** See HYPERBOLA.

TRA-PE'ZI-UM, *n.* A quadrilateral, none of whose sides are parallel.

TRAP'E-ZOID, *adj.*, *n.* A quadrilateral which has two parallel sides. It is sometimes required that the other sides be nonparallel. The parallel sides are the **bases** of the trapezoid and the **altitude** is the perpendicular distance between the bases. An **isosceles trapezoid** is a trapezoid in which the nonparallel sides are of equal length. The **area** of a trapezoid is the product of its altitude and one-half the sum of the lengths of its bases, written

$$A = h \frac{(b_1 + b_2)}{2}.$$

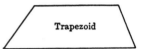

Trapezoid

trapezoid rule. A formula for approximating an integral of type $\int_a^b f(x) \, dx$. The formula can be thought of as the result of separating the interval from a to b into congruent subintervals by points $a = x_0, x_1, x_2, \cdots, x_n = b$, and then approximating the graph of $y = f(x)$ for x between x_k and x_{k+1} by the line segment joining the points on the graph for which x has the values x_k and x_{k+1}. The formula is

$$\frac{b-a}{n} \left[\tfrac{1}{2}(y_0 + y_n) + \sum_{i=1}^{n-1} y_i \right],$$

where $y_k = f(x_k)$ for each k. The numerical difference between the number given by this formula and the actual integral is not greater than

$$\frac{M(b-a)^3}{12n^2},$$

where M is the least upper bound of the absolute value of the second derivative of f on the interval from a to b. *Syn.* **trapezoid formula.** See SIMPSON —Simpson's rule.

TREE, *n.* A connected nonempty graph that contains no closed paths. See GRAPH THEORY, PATH.

TRE'FOIL, *n.* See MULTIFOIL.

TREND, *adj.*, *n.* The general drift, tendency, or bent of a set of data; such, for instance, as the price of steel over a long period of time. Particular data will generally fluctuate from the trend. Trend extending over long periods of time that is the result of slowly changing, long-lasting forces is called **secular trend**. Trend is usually represented by a smooth mathematical function, such as the trend line, but see AVERAGE —moving average.

TRI′AL, *n.*   Same as EXPERIMENT.

TRI′AN′GLE, *n.*   (a) The figure formed by connecting three noncollinear points (the **vertices**) by line segments. (2) The figure described in (1) together with the points in the same plane and interior to the figure. Six kinds of triangles are illustrated. As indicated in the figures below, an **acute triangle** is a triangle whose interior angles are all acute; an **obtuse triangle** is a triangle that contains an obtuse interior angle; a **scalene triangle** is a triangle with no two sides equal; an **isosceles triangle** is a triangle with two equal sides (the third side is called the **base** and the angle opposite it the **vertex**); a **right triangle** is a triangle one of whose angles is a right angle (the side opposite the right angle is called the **hypotenuse** and the other two sides the **legs** of the right triangle); an **equilateral triangle** is a triangle with all three sides equal (it must then also be equiangular, *i.e.*, have its three interior angles equal). An **oblique triangle** is a triangle which contains no right angles. The **altitude** of a triangle is the perpendicular distance from a vertex to the opposite side, which has been designated as the **base**. The **area** of a triangle is one-half the product of the base and the corresponding altitude. The area is equal to one-half the determinant whose first column consists of the abscissas of the vertices, the second of the ordinates (in the same order), and the third entirely of ones (this is positive if the points are taken around the triangle in counterclockwise order).

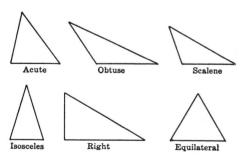

Acute   Obtuse   Scalene

Isosceles   Right   Equilateral

astronomical triangle.   The spherical triangle on the celestial sphere which has for its vertices the nearer celestial pole, the zenith, and the

celestial body under consideration. See HOUR—hour angle and hour circle.

congruent triangles.   See CONGRUENT —congruent figures.

excenter, incenter, and orthocenter of a triangle.   See EXCENTER, INCENTER, ORTHOCENTER.

Pascal's triangle.   See PASCAL.

pedal triangle.   See PEDAL.

polar triangle.   See POLAR —polar triangle.

solution of a triangle.   See SOLUTION.

spherical triangle.   See SPHERICAL.

terrestrial triangle.   See TERRESTRIAL.

triangle inequality.   An inequality of type $|x + y| \leq |x| + |y|$. For x and y real or complex numbers or vectors with three or fewer components, this inequality can be proved by using the fact that the length of one side of a triangle is less than or equal to the sum of the lengths of the other two. The triangle inequality for normed vector spaces is $\|x + y\| \leq \|x\| + \|y\|$, where $\|x\|$ is the *norm* of the member x of the space. See INNER —inner-product space, and VECTOR —vector space.

triangle of plane sailing.   The right spherical triangle (treated as a plane triangle) which has for legs the difference in latitude and the departure of two places, and for its hypotenuse the rhumb line between the two places.

TRI-AN′GU-LAR, *adj.*   Like a triangle; having three angles or three sides.

triangular number.   See NUMBER —triangular numbers.

triangular prism.   A prism with triangular bases.

triangular pyramid.   A pyramid whose base is a triangle. *Syn.* tetrahedron.

triangular region.   See REGION.

TRI-AN′GU-LA′TION, *n.*   A triangulation of a topological space T is a *homeomorphism* of T onto a polyhedron consisting of the points belonging to simplexes of a simplicial complex (see COMPLEX —simplicial complex). A **triangulable space** is a space that is homeomorphic to a simplicial complex. *E.g.*, the surface of an ordinary sphere is triangulable since it is homeomorphic to the surface of an inscribed regular tetrahedron, the homeomorphism consisting of projections of points of the sphere onto the tetrahedron along radii (and of points of the tetrahedron onto the sphere along radii). The surface of a regular tetrahedron is a simplicial complex whose simplexes are triangles. This mapping of the tetrahedron onto the sphere separates the sphere into four spherical triangles corresponding to the four faces of the tetrahedron.

TRI-CHOT′O-MY, *n.* **trichotomy property.** The property sometimes assumed for an ordering that exactly one of the statements $x < y$, $x = y$, and $y < x$ is true for any x and y. See ORDER —order properties of real numbers, and ORDERED — ordered set.

TRI′DENT, *n.* **trident of Newton.** The cubic curve defined by the equation $xy = ax^3 + bx^2 + cx + d$ $(a \neq 0)$. It cuts the x-axis in 1 or 3 points and is asymptotic to the y-axis if $d \neq 0$. If $d = 0$, the equation factors into $x = 0$ (the y-axis) and $y = ax^2 + bx + c$ (a parabola).

TRIG′O-NO-MET′RIC, *adj.* **inverse trigonometric function.** A trigonometric function f is a *relation* and has an inverse f^{-1} which also is a relation (sometimes called a *multiple-valued function*). E.g., the inverse sine of x is the relation denoted by arc sine, or $\sin^{-1}$, for which the values of $\arcsin x$ [*i.e.*, the numbers (or angles) y for which x is related to y] are the numbers (or angles) whose sine is x. The numbers to which a number A is related by the inverse of a trigonometric function are denoted either by $\sin^{-1} A$, $\cos^{-1} A$, $\tan^{-1} A$, etc., or by $\arcsin A$, $\arccos A$, $\arctan A$, etc. The graphs of the inverse trigonometric relations are simply graphs of the trigonometric functions with the positive x-axis and y-axis interchanged, or, what is the same, the graphs of the trigonometric functions reflected in the line $y = x$. For a trigonometric function f to have an inverse which is a function, the domain of f (or the range of f^{-1}) must be restricted. It is customary to choose the domain of the sine to be the closed interval $[-\frac{1}{2}\pi, \frac{1}{2}\pi]$; the numbers in this interval are **principal values of arc sine** and this interval is the range of the function arc sine. The **principal values of arc cosine** are the numbers in the closed interval $[0, \pi]$, and the **principal values of arc tangent** are the numbers in the open interval $(-\frac{1}{2}\pi, \frac{1}{2}\pi)$. The **principal values of arc cotangent** usually are taken in the interval $(0, \pi)$, but sometimes in $(-\frac{1}{2}\pi, \frac{1}{2}\pi]$ with zero deleted. There is no uniformity in definitions of the principal values of arc secant and arc cosecant, the most usual choices being to use $[0, \pi]$ with $\frac{1}{2}\pi$ deleted, or $[-\pi, -\frac{1}{2}\pi) \cup [0, \frac{1}{2}\pi)$, for the arc secant, and $[-\frac{1}{2}\pi, \frac{1}{2}\pi]$ with 0 deleted, or $(-\pi, -\frac{1}{2}\pi] \cup (0, \frac{1}{2}\pi]$, for the arc cosecant. For the inverse of each trigonometric function f, the principal value of $f^{-1}(x)$ is in the interval $[0, \frac{1}{2}\pi]$ if $x \geq 0$ and there is any real y for which $x = f(y)$. *Syn.* antitrigonometric functions. See ARC SINE, ARC COSINE, etc., INVERSE —inverse of a function, RELATION.

trigonometric cofunctions. Trigonometric functions f and g for which f has the same value at x as g has at y whenever x and y are complementary. The *sine* and *cosine* are cofunctions, as are also the *tangent* and *cotangent*, and the *secant* and *cosecant*. E.g., $\sin 30° = \cos 60°$, $\tan 15° = \text{ctn } 75°$, and $\sec -10° = \csc 100°$.

trigonometric curves. Graphs of the trigonometric functions in rectangular coordinates. See under SINE, COSINE, TANGENT, COTANGENT, SECANT, COSECANT. The term *trigonometric curve* is also applied to the graphs of any function involving only trigonometric functions, such as $\sin 2x + \sin x$ or $\sin x + \tan x$.

trigonometric equation. An equation that contains the variable in the argument of trigonometric functions, as for the equations $\cos x - \sin x = 0$ and $\sin^2 x + 3x = \tan(x + 2)$.

trigonometric form of a complex number. Same as POLAR FORM. See POLAR.

trigonometric functions. For **acute angles**, the trigonometric functions of the angles are certain ratios of the sides of a right triangle containing the angle. If A is an angle in a right triangle with hypotenuse of length c, side opposite A of length a, and side adjacent of length b, then the values of the trigonometric functions of A are

$$\frac{a}{c}, \ \frac{b}{c}, \ \frac{a}{b}, \ \frac{b}{a}, \ \frac{c}{b}, \ \text{ and } \ \frac{c}{a}.$$

They are named, respectively, sine A, cosine A, tangent A, cotangent A, secant A, and cosecant A, and written $\sin A$, $\cos A$, $\tan A$, $\text{ctn } A$ (or $\cot A$), $\sec A$, and $\csc A$ (or $\operatorname{cosec} A$). Other trigonometric functions not so commonly

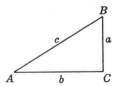

used are versed sine of $A = 1 - \cos A$ = versine of A, written vers A; coversed sine of $A = 1 - \sin A$ = versed cosine of A, written covers A; exsecant of $A = \sec A - 1$, written exsec A; haversine of $A = \frac{1}{2}$ vers A, written hav A. Now let A be **any positive or negative angle** and let OP be a line segment with O the origin of a system of rectangular coordinates and such that A is the angle from the positive x-axis to OP (see ANGLE). Let P have coordinates (x, y). If

OP has length r, then

$$\sin A = \frac{y}{r}, \quad \cos A = \frac{x}{r}, \quad \tan A = \frac{y}{x},$$

$$\operatorname{ctn} A = \frac{x}{y}, \quad \sec A = \frac{r}{x}, \quad \csc A = \frac{r}{y},$$

or

$$\sin A = (\text{ordinate})/r,$$

$$\cos A = (\text{abscissa})/r$$

$$\tan A = (\text{ordinate})/(\text{abscissa}),$$

$$\operatorname{ctn} A = (\text{abscissa})/(\text{ordinate}),$$

$$\sec A = r/(\text{abscissa}),$$

$$\csc A = r/(\text{ordinate}),$$

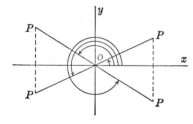

and other trigonometric functions are defined in terms of these basic six functions as for the case A is an acute angle. Although each of these six functions has the same sign in two quadrants, if the signs of two functions which are not reciprocals are given, the quadrant is uniquely determined; *e.g.*, if $\sin A$ is positive and $\cos A$ is negative, then the terminal side OP of A is in the second quadrant. The variation of the trigonometric functions as A varies from $0°$ to $360°$ often is described by stating the values for the angles $0°$, $90°$, $180°$, and $270°$, since each function has its greatest or least value (or is undefined) at certain of these points. For the sine they are $0, 1, 0, -1$; for the cosine $1, 0, -1, 0$; for the tangent $0, \infty, 0, \infty$; for the cotangent, $-\infty, 0, -\infty, 0$; for the secant, $1, \infty, -1, -\infty$; for the cosecant, $-\infty, 1, \infty, -1$. The signs ∞ and $-\infty$ here mean that the function increases or decreases, respectively, without limit as the angle approaches the given angle in a counterclockwise direction. The opposite signs (on ∞) would result if the direction were clockwise. The following are

fundamental identities relating the trigonometric functions:

$$\sin x = \frac{1}{\csc x}, \quad \cos x = \frac{1}{\sec x},$$

$$\tan x = \frac{1}{\operatorname{ctn} x}, \quad \tan x = \frac{\sin x}{\cos x},$$

$$\sin^2 x + \cos^2 x = 1, \quad \tan^2 x + 1 = \sec^2 x,$$

$$\operatorname{ctn}^2 x + 1 = \csc^2 x.$$

The first four of these identities can be derived directly from the definitions of the functions; the last three make use of the Pythagorean theorem and are called **Pythagorean identities**. In most fields of mathematics, it is customary to speak of trigonometric functions of **numbers** rather than angles. A trigonometric function of a number x has a value equal to that of the given trigonometric function of an angle whose radian measure is equal to x. The sine and cosine of a number x can also be defined by means of the series:

$$\sin x = x - \frac{x^3}{3!} + \frac{x^5}{5!} - \cdots,$$

$$\cos x = 1 - \frac{x^2}{2!} + \frac{x^4}{4!} - \cdots.$$

Then the other trigonometric functions can be defined by the above fundamental relations. For a **complex number** z, the sine and cosine of z may be defined in terms of the exponential function,

$$\sin z = \frac{e^{iz} - e^{-iz}}{2i}, \quad \cos z = \frac{e^{iz} + e^{-iz}}{2},$$

or by series,

$$\sin z = z - \frac{z^3}{3!} + \frac{z^5}{5!} - \cdots,$$

$$\cos z = 1 - \frac{z^2}{2!} + \frac{z^4}{4!} - \cdots.$$

The other trigonometric functions of z are defined in terms of these in the usual way.

    trigonometric identities. See TRIGONOMETRY —identities of plane trigonometry.

trigonometric series. A series of the form

$$a_0 + (a_1 \cos x + b_1 \sin x)$$

$$+ (a_2 \cos 2x + b_2 \sin 2x) + \cdots$$

$$= a_0 + \sum (a_n \cos nx + b_n \sin nx),$$

where the a's and b's are constants. See FOURIER —Fourier series.

trigonometric substitutions. Substitutions used to rationalize quadratic surds of the forms

$$\sqrt{a^2 - x^2}, \qquad \sqrt{x^2 + a^2}, \qquad \sqrt{x^2 - a^2}.$$

The substitutions $x = a \sin u$, $x = a \tan u$, and $x = a \sec u$ reduce these surds to $|a \cos u|$, $|a \sec u|$, $|a \tan u|$, respectively. The quadratic surd $\sqrt{x^2 + px + q}$ can always be put into one of the above forms by completing the square. See INTEGRATION —change of variables in integration.

TRIG'O-NOM'E-TRY, *n.* The name "trigonometry" is derived from two Greek words, combining to mean measurement of triangles. While the solution of triangles forms an important part of modern trigonometry, it is by no means the only part or even the most important part. In the development of methods for the solution of triangles by computation, certain **trigonometric functions** occur (see TRIGONOMETRIC). The study of the properties of these functions and their applications to various mathematical problems, including the solution of triangles, constitutes the subject matter of trigonometry. Trigonometry has applications in surveying, navigation, construction work and many branches of science. It is particularly essential for most branches of mathematics and physics. In **plane trigonometry,** the solution of plane triangles is considered; **spherical trigonometry** treats of the solution of spherical triangles. See various headings under SPHERICAL, TRIGONOMETRIC, TRIGONOMETRY.

half-angle formulas of plane trigonometry. (1) Formulas for the solution of plane triangles which give relations between the sides and one of the angles of a triangle; used in place of the *cosine law* because they are better adapted to calculation by logarithms. The half-angle formulas are

$$\tan \tfrac{1}{2}A = r/(s-a),$$

$$\tan \tfrac{1}{2}B = r/(s-b),$$

$$\tan \tfrac{1}{2}C = r/(s-c),$$

where A, B, C are the angles of the triangle, a, b, c the lengths of the sides opposite A, B, C, respectively, $s = \tfrac{1}{2}(a + b + c)$, and

$$r = \sqrt{(s-a)(s-b)(s-c)/s}.$$

(2) See below, identities of plane trigonometry.

half-angle and half-side formulas of spherical trigonometry. The **half-angle formulas** are formulas giving the tangents of half of an angle of a spherical triangle in terms of functions of the sides. If α, β, γ are the angles, a, b, c, respectively, the opposite sides, and $s = \tfrac{1}{2}(a + b + c)$, then

$$\tan \tfrac{1}{2}\alpha = \frac{r}{\sin(s-a)},$$

where

$$r = \sqrt{\frac{\sin(s-a)\sin(s-b)\sin(s-c)}{\sin s}}.$$

Formulas for $\tan \tfrac{1}{2}\beta$ and $\tan \tfrac{1}{2}\gamma$ are obtained from this formula by a cyclic change between a, b, c. The **half-side formulas** give the tangents of one-half of each of the sides in terms of the angles. If $S = \tfrac{1}{2}(\alpha + \beta + \gamma)$, then

$$\tan \tfrac{1}{2}a = R \cos(S - \alpha),$$

where

$$R = \sqrt{\frac{-\cos S}{\cos(S-\alpha)\cos(S-\beta)\cos(S-\gamma)}}.$$

Formulas for $\tan \tfrac{1}{2}b$ and $\tfrac{1}{2}c$ are obtained by a cyclic change between α, β and γ.

identities of plane trigonometry. Equations which express relations among trigonometric functions that are valid for all values of the variables for which the functions involved are defined. The *fundamental relations* among the trigonometric functions are the simplest trigonometric identities (see TRIGONOMETRIC —trigonometric functions). The following are some more trigonometric identities: The **reduction formulas** are identities that can be used to express the value of a trigonometric function of any angle in terms of a trigonometric function of an angle A such that $0 \le A < 90°$, or $0 \le A < 45°$. Some formulas for the sine, cosine, and tangent are:

$$\sin(90° \pm A) = \cos A,$$

$$\sin(180° \pm A) = \mp \sin A,$$

$$\sin(270° \pm A) = -\cos A,$$

$$\cos(90° \pm A) = \mp \sin A,$$

$$\cos(180° \pm A) = -\cos A,$$

$$\cos(270° \pm A) = \pm \sin A,$$

$$\tan(90° \pm A) = \mp \cot A,$$

$$\tan(180° \pm A) = \pm \tan A,$$

$$\tan(270° \pm A) = \mp \cot A,$$

where, in each formula, either the upper signs, or the lower, are to be used throughout. The **Pythagorean identities** are:

$$\sin^2 + \cos^2 x = 1,$$

$$\tan^2 x + 1 = \sec^2 x,$$

$$1 + \cot^2 x = \csc^2 x.$$

The **addition and subtraction formulas (identities)** are formulas expressing the sine, cosine, tangent, etc., of the sum or difference of two angles in terms of functions of the angles. They answer the need created by the fact that functions of sums and differences are not distributive; i.e., $\sin(x \pm y) \neq \sin x \pm \sin y$. The most important of these formulas are

$$\sin(x \pm y) = \sin x \cos y \pm \cos x \sin y,$$

$$\cos(x \pm y) = \cos x \cos y \mp \sin x \sin y,$$

and

$$\tan(x \pm y) = \frac{\tan x \pm \tan y}{1 \mp \tan x \tan y}.$$

(The upper signs, and the lower signs, are to be taken together throughout.) The **double-angle formulas (identities)** are formulas expressing the sine, cosine, tangent, etc., of twice an angle in terms of functions of the angle. These are easily obtained by replacing y by x in the above *addition formulas*. The most important are

$$\sin 2x = 2 \sin x \cos x,$$

$$\cos 2x = \cos^2 x - \sin^2 x,$$

and

$$\tan 2x = 2 \tan x / (1 - \tan^2 x).$$

The **half-angle formulas (identities)** are formulas expressing trigonometric functions of half an angle in terms of functions of the angle. They are easily derivable from the *double-angle formulas* (with the double angle, $2x$, replaced by A and x replaced by $\frac{1}{2}A$). The most important half-angle formulas are

$$\sin \tfrac{1}{2}A = \pm \sqrt{(1 - \cos A)/2},$$

$$\cos \tfrac{1}{2}A = \pm \sqrt{(1 + \cos A)/2},$$

and

$$\tan \tfrac{1}{2}A = \sin A / (1 + \cos A),$$

$$= (1 - \cos A)/\sin A.$$

The **product formulas** (identities) are the formulas

$$\sin x \cos y = \tfrac{1}{2}[\sin(x + y) + \sin(x - y)],$$

$$\cos x \sin y = \tfrac{1}{2}[\sin(x + y) - \sin(x - y)],$$

$$\cos x \cos y = \tfrac{1}{2}[\cos(x + y) + \cos(x - y)],$$

$$\sin x \sin y = \tfrac{1}{2}[\cos(x - y) - \cos(x + y)].$$

These can be easily derived from the *addition* and *subtraction formulas*.

spherical trigonometry. See SPHERICAL.

TRI-HE′DRAL, *adj., n.* (1) A figure formed by three noncoplanar lines which intersect in a point. A trihedral formed by three directed lines is a **directed trihedral**. The trihedrel formed by the three axes of a system of Cartesian coordinates in space is a directed trihedral, called the **coordinate trihedral**. (2) The union of three noncoplanar rays with the same initial point. An **oriented trihedral** is a trihedral for which one of the rays is designated as the first, another as the second, and the remaining one as the third. An oriented trihedral is **left-handed** if, when the thumb of the left hand extends from the initial point along the first ray, the fingers fold in the direction in which the second ray could be rotated through an angle of less than 180° to coincide with the third, it is **right-handed** if, when the thumb of the right hand extends from the initial point along the first ray, the fingers fold in the direction in which the second ray could be rotated through an angle of less than 180° to coincide with the third. If vectors **u**, **v**, and **w** are represented by arrows with initial points at the initial points of the rays of an oriented trihedral, **u** along the first ray, **v** along the second ray, and **w** along the third ray, then the trihedral is **positively oriented** or **negatively oriented** according as the *triple scalar product* $\mathbf{u} \cdot (\mathbf{v} \times \mathbf{w})$ is positive or negative. An oriented trihedral is positively oriented if and only if it and the coordinate trihedral are both right-handed or both left-handed. A **trirectangular trihedral** is a trihedral formed by three mutually perpendicular rays. If the rays forming an oriented trihedal are L_1, L_2, L_3, a necessary and sufficient condition that the trihedral be trirectangular is that the absolute value of the determinant whose rows are $l_1, m_1, n_1; l_2, m_2, n_2$; and l_3, m_3, n_3 (in these orders) be 1, where $l_1, m_1, n_1; l_2, m_2, n_2$; and l_3, m_3, n_3 are, respectively, the direction cosines of the rays L_1, L_2, and L_3, *i.e.*, components of unit vectors along the rays. The determinant is positive if and only if the trihedral is positively oriented.

moving trihedral of space curves and surfaces. The *moving trihedral* (or *trihedral*) of a directed space curve C is the configuration consisting of the tangent, principal normal, and binormal of C at a variable point of C. The moving trihedral of a **surface** relative to a directed curve on the surface is defined as follows: For a point P of a directed curve C on a surface S, let α be the unit vector from P in the positive direction of the tangent to C at P, let γ be the unit vector from P in the positive direction of the normal to S at P, and let β be the unit vector from P in the tangent plane to S at P, and such that α, β, γ have the same mutual orientation as the x, y, z axes. Then axes directed along the vectors α, β, γ form the moving trihedral of S relative to C. The **rotations of the moving trihedral** of a surface are a certain set of six particular functions determining the orientation of the *moving trihedral* of the surface, but not its position in space.

trihedral angle. A polyhedral angle having three faces. Two trihedral angles are **symmetric** if they have their face angles equal in pairs but arranged in opposite order. Such trihedral angles are not superposable.

TRIL′LION, *n.* (1) In the U. S. and France, the number represented by one followed by 12 zeros. (2) In England, the number represented by one followed by 18 zeros.

TRI-NO′MI-AL, *n.* A polynomial of three terms, such as $x^2 - 3x + 2$.

TRI′PLE, *adj.*, *n.* Threefold; consisting of three. An **ordered triple** is a set with 3 members for which one member is designated as the first, another as second, and the remaining one as third. A number triple (a, b, c) can be used to designate a vector with components a, b, and c, or to represent any object which can be determined in some specified way when three real numbers are given; *e.g.*, the point with spherical coordinates (a, b, c) or the circle with radius a and center (b, c).

triple integral. See INTEGRAL—iterated integral, multiple integral.

triple of conjugate harmonic functions. Three functions, $x(u, v), y(u, v), z(u, v)$, harmonic in a common domain D and there satisfying the relations $E = G$, $F = 0$. Such functions give conformal maps of D on minimal surfaces.

triple root of an equation. A root which occurs three times. See MULTIPLE—multiple root.

triple (scalar) product of three vectors. The scalar $\mathbf{A} \cdot (\mathbf{B} \times \mathbf{C})$, written $(\mathbf{ABC})$ or $[\mathbf{ABC}]$. The dot indicates *scalar multiplication* and the cross *vector multiplication*. If **A**, **B**, and **C** are written in the forms $\mathbf{A} = a_1\mathbf{i} + a_2\mathbf{j} + a_3\mathbf{k}$, $\mathbf{B} = b_1\mathbf{i} + b_2\mathbf{j} + b_3\mathbf{k}$, $\mathbf{C} = c_1\mathbf{i} + c_2\mathbf{j} + c_3\mathbf{k}$, the triple scalar product is the determinant of the coefficients of $\mathbf{i}, \mathbf{j}, \mathbf{k}$. It is clear from this determinant that any cyclic change between the vectors in a triple product does not alter its value. The numerical value of the product is equal to the volume of the parallelepiped of which the vectors of the triple product are coterminal sides.

TRI′PLY, *adv.* Containing a property three times; repeating an operation three times.

triply orthogonal system of surfaces. See ORTHOGONAL.

TRI′REC-TAN′GU-LAR, *adj.* Having three right angles. *E.g.*, a spherical triangle can have three right angles; such a triangle is trirectangular. A trirectangular trihedral angle is a trihedral angle whose three dihedral angles are right angles.

TRI-SEC′TION, *n.* The process of separating into three equal parts.

trisection of an angle. The problem of trisecting any angle with straightedge and compasses, alone. It was proved impossible by P. L. Wantzel in 1847. Any angle can be trisected, however, in several ways, for instance, by the use of a *protractor*, the *limaçon* of Pascal, the *conchoid* of Nicodemes, or the *trisectrix* of Maclaurin. See TRISECTRIX.

TRI-SEC′TRIX, *n.* The locus of the equation $x^3 + xy^2 + ay^2 - 3ax^2 = 0$. The curve is symmetric with respect to the *x*-axis, passes through the origin, and has the line $x = -a$ as an asymptote. It is of interest in connection with the problem of trisecting a given angle, or if a line having an angle of inclination $3A$ is drawn through the point $(2a, 0)$, then the line passing through the origin and the point of intersection of this line with the trisectrix has an angle of inclination A. Also called the trisectrix of Maclaurin.

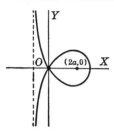

TRIV′I-AL, *adj.* **trivial solutions of a set of homogeneous linear equations.** Zero values for all the variables, *trivial* because they are a solution of any system of homogeneous equations. Solutions in which at least one of the variables has a value different from zero are **nontrivial solutions.** See CONSISTENCY —consistency of linear equations.

trivial topology. For a set *S*, the **trivial topology** (or **indiscrete topology**) is the topology for which the only open sets are *S* and the empty set ∅. then the only closed sets are *S* and ∅; each point in *X* has only one neighborhood (*S* itself); and if *A* is a nonempty subset of *S*, then *S* is the closure of *A*. See TOPOLOGY —topology of a space.

TRO′CHOID, *n.* (*Greek*: TROCHOS = wheel + EIDOS = form). The plane locus of a point on the radius of a circle, or on the radius produced, as the circle rolls (in a plane) on a fixed straight line. If *a* is the radius of the rolling circle, *b* the distance from the center of the circle to the point describing the curve, and θ the angle (in radians) subtended by the arc which has contacted the fixed line in getting to the point under consideration, then the parametric equations of the *trochoid* are

$$x = a\theta - b\sin\theta \quad \text{and} \quad y = a - b\cos\theta.$$

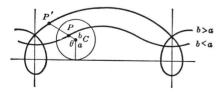

When *b* is greater than *a*, the curve has a loop between every two arches, nodes at $\theta = \theta_1 + n\pi$, where $0 < \theta_1 < \pi$ and $a\theta_1 - b\sin\theta_1 = 0$; it is then a **prolate cycloid.** If *b* is less than *a*, the curve never touches the base line; it is then a **curtate cycloid** (the names prolate and curtate are sometimes interchanged). As *b* approaches zero, the curve tends to smooth out nearer to the straight line described by the center of the circle. When $b = a$ the curve is a **cycloid.**

TROY, *adj.* **troy weight.** A system of weights having a pound of 12 ounces as its basic unit. It is used mostly for weighing fine metals. See DENOMINATE NUMBERS in the appendix.

TRUN′CAT-ED, *adj.* See CONE, PRISM, PYRAMID.

TRUTH, *adj.* **truth set.** For a propositional function *p*, the **truth set** is the set of all objects in the domain of *p* for which the value of *p* is a true proposition. Also called a **solution set,** particularly if the propositional function is described by means of equations or inequalities. See PROPOSITIONAL —propositional function, SOLUTION—solution set.

TUKEY, John Wilder (1915–   ). American operations analyst, statistician, and topologist.
Tukey's lemma. See ZORN—Zorn's lemma.

TU′PLE, *n.* See ORDERED —ordered pair.

TURN′ING, *adj.* **turning point.** A point on a curve at which the ordinates of the curve cease increasing and begin decreasing, or vice versa; a maximum or minimum point.

TWELVE, *adj.* **twelve-color theorem.** See FOUR —four-color problem.

TWIN, *adj.* **twin primes.** Pairs of prime numbers, such as $(3, 5)$, $(5, 7)$, $(17, 19)$, that differ by 2. It is not known whether there is an infinite number of twin primes.

TWIST′ED, *adj.* **twisted curve.** See CURVE.

TWO, *adj.* **two-dimensional geometry.** The study of figures in a plane. See GEOMETRY —plane geometry.

two-point form of the equation of a line. See LINE—equation of a line.

TYCHONOFF (or TIHONOV), Andrei Nikolaevich (1906–   ). Russian geophysicist, mathematical physicist, and topologist.

Tychonoff space See REGULAR—regular space.

Tychonoff theorems. See BROUWER—Brouwer's fixed-point theorem, PRODUCT—Cartesian product.

TYDAC. An imaginary computing machine having representative characteristics of existing machines. TYDAC is an acronym for *Typical Digital Automatic Computer*.

TYPE, *n.* **problem of type.** The problem of determining the type of a given simply connected Riemann surface. See RIEMANN—Riemann surface.

TYPE I or **TYPE II ERROR.** See ERROR, HYPOTHESIS—test of a hypothesis.

U

UL′TRA-FIL′TER, *n.* See FILTER.

UM-BIL′IC, *n.* Same as UMBILICAL POINT.

UM-BIL′I-CAL, *adj.* **umbilical geodesic of a quadratic surface.** A geodesic lying on the surface S and passing through an umbilical point of S.

umbilical point on a surface. A point of a surface S which is either a *circular point* or a *planar point* of S. A point of S is an umbilical point of S if, and only if, its first and second fundamental quadratic forms are proportional. The normal curvature of S is the same in all directions on S at an umbilical point of S. All points on a sphere or plane are umbilical points. The points where an ellipsoid of revolution cuts the axis of revolution are umbilical points. *Syn.* umbilic point.

UM′BRA, *n.* [*pl.* **umbrae**]. The part of the shadow of an object from which all direct light is excluded. For the sun and the earth, the part of the shadow in the cone tangent to the earth and sun is in complete shadow, and is known as the **umbra**, while the outer region (the **penumbra**) shades from full illumination at A and D, through partial illumination at B and C, to complete shadow at the midpoint.

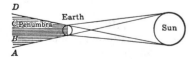

U′NA-RY, *adj.* **unary operation.** A unary operation on a set S is a function whose domain is S and whose range is contained in S. See BINARY—binary operation, OPERATION.

UN-BI′ASED, or **UN-BI′ASSED,** *adj.* **unbiased estimator** (*Statistics*). An estimator Φ for a parameter ϕ such that the expected value $E(\Phi)$ of Φ is ϕ. The difference $E(\Phi) - \phi$ is the **bias** of Φ; Φ is **biased** if the bias of Φ is not zero. *E.g.*, $\sum_{i=1}^{n} X_i/n$ is an unbiased estimator for the *mean* of any distribution (also see VARIANCE). An estimator Φ for a parameter ϕ is **asymptotically unbiased** if $\lim_{n \to \infty} E[\Phi(X_1, X_2, \cdots, X_n)] = \phi$. An unbiased estimator whose variance is less than or equal to that of any other unbiased estimator is a minimum-variance unbiased estimator. Minimum-variance unbiased estimators are unique for many of the parameters that are important in applied statistics. See CONSISTENT—consistent estimator.

unbiased test. See HYPOTHESIS—test of a hypothesis.

UN-BOUND′ED, *adj.* **unbounded function.** A function which is not bounded; a function such that for any number M there is a value of the function whose numerical value is larger than M. *Tech.* f is unbounded on the set S if for any number M there is a point x_m of S such that $|f(x_m)| > M$. The function $1/x$ is unbounded on the interval $0 < x \leq 1$; $\tan x$ is unbounded on the interval $0 \leq x < \frac{1}{2}\pi$.

UN′CON-DI′TION-AL, *adj.* **unconditional inequality.** See INEQUALITY.

UN′DE-FINED′, *adj.* **undefined term.** A term used without specific mathematical definition, whatever meaning it has being purely psychological; a term which satisfies certain axioms, but is not otherwise defined.

UN′DE-TER′MINED, *adj.* **undetermined coefficients.** Unknowns to be determined so as to satisfy certain conditions. *E.g.*, if it is desired to factor $x^2 - 3x + 2$, the factors may be taken to be $x + a$ and $x + b$, where a and b are to be determined so as to make the product of these two factors equal to the original expression; *i.e.*,

$$x^2 + (a + b)x + ab \equiv x^2 - 3x + 2,$$

whence $a + b = -3$ and $ab = 2$, from which $a = -1$, $b = -2$, or $a = -2$, $b = -1$. The **method of undetermined coefficients** used for differential equations is illustrated by the following examples: To find a particular solution of the equation $y'' + 2y' - 5y = 5 \sin x$, substitute $y = A \sin x + B \cos x$, to obtain $A = -\frac{3}{4}$, $B = -\frac{1}{4}$; to find a particular solution of $y'' - 3y' + 2y = 27x^2 e^x$, substitute

$$y = x^2(Ax^2 e^x + Bxe^x + Ce^x),$$

to obtain $A = \frac{3}{4}$, $B = -1$, $C = 1$ [the factor x^2 was introduced because $y'' - 3y' + 2y = (D + 2)(D - 1)^2 y$ and e^x is a solution of $(D - 1)y = 0$; the terms in $Ax^2 e^x + Bxe^x + Ce^x$ represent all types obtainable by successive differentiation of $27x^2 e^x$]. See PARTIAL—partial fractions.

U-NI-CUR′SAL, *adj.* **unicursal curve.** A curve that has parametric equations $x = \theta(t)$, $y = \phi(t)$ for which θ and ϕ are rational functions of t.

U′NI-FORM, *adj.* **uniform acceleration.** Acceleration in which there are equal changes in the velocity in equal intervals of time. *Syn.* constant acceleration.

    uniform-boundedness principle. See BANACH —Banach-Steinhaus theorem.

    uniform circular motion. Motion around a circle with constant speed.

    uniform continuity. A function whose domain and range are sets of real numbers is **uniformly continuous** on a set S contained in its domain if for any positive number ϵ there exists a positive number δ such that $|f(r) - f(s)| < \epsilon$ whenever $|r - s| < \delta$ and r and s are in S. If I is a bounded closed interval, then f is uniformly continuous on I if f is continuous on I. If f is differentiable on an interval I and there is a positive number M such that $|f'(x)| < M$ if x is in I, then f is uniformly continuous on I. The function x^2 is uniformly continuous on the interval $[0, 1]$, but is not uniformly continuous on the set of real numbers. The function $1/x$ is continuous and not uniformly continuous on the open interval $(0, 1)$, but is uniformly continuous on all closed intervals $[r, 1]$ for which $0 < r < 1$. More generally, let F be a function whose domain is a set S, where S may be a set of real numbers, of points in the plane or space, or of points in a metric space; let the range of F be a set of real numbers, of points in the plane or space, or of points in a metric space. Then F is *uniformly continuous* on S if for any positive number ϵ there exists a number δ such that the numerical distance between $F(x_1)$ and $F(x_2)$ is less than ϵ whenever the numerical distance between x_1 and x_2 is less than δ. If S is compact, then F is

uniformly continuous if F is continuous at each point of S.

    uniform convergence of a series. See CONVERGENCE.

    uniform convergence of a set of functions. Convergence such that the difference between each function and its limit can be made less than the same arbitrary positive number for a common interval of values of the argument. *Tech.* If the set of functions is such that each function f_i has a limit L_i, as x approaches x_0, then they *converge uniformly*, as x approaches x_0, if for any positive number ϵ there exists a positive δ such that

$$|f_i(x) - L_i| < \epsilon$$

for all i, when $|x - x_0| < \delta$. See ASCOLI —Ascoli's theorem.

    uniform distribution. A random variable X has a uniform distribution or is a **uniform random variable** on the interval $[a, b]$ if its probability density function f is such that $f(x) = 0$ if $x \notin [a, b]$ and $f(x) = 1/(b - a)$ if $a < x < b$. The *mean* is $\frac{1}{2}(a + b)$, the *variance* is $\frac{1}{12}(b - a)^2$, and the *moment generating function* is

$$M(t) = \frac{e^{tb} - e^{ta}}{t(b - a)}.$$

If F is the *distribution function*, then $F(t) = 0$ if $t \leq a$, $F(t) = (t - a)/(b - a)$ if $a \leq t \leq b$, and $F(t) = 1$ if $b \leq t$.

    uniform scale. See LOGARITHMIC —logarithmic coordinate paper.

    uniform speed and velocity. See CONSTANT —constant speed and velocity.

    uniform topology. The topology of a topological space T is a **uniform topology** if there is a family F of subsets of the Cartesian product $T \times T$ such that a subset A of T is *open* if and only if, for any x of A, there is a member V of F for which the set of all y with (x, y) in V is a subset of A—in addition, the family F must have the properties: (i) each member of F contains each (x, x) for x in T; (ii) for each V of F, V^{-1} belongs to F (V^{-1} is the set of all (x, y) with (y, x) in V); (iii) for each V of F there is a V^* of F which has the property that V contains all (x, z) for which there is a y such that (x, y) and (y, z) belong to V^*; (iv) the intersection of two members of F is a member of F; (v) a subset of $T \times T$ is a member of F if it contains a member of F. A family of sets which satisfies conditions (i)–(v) is said to be a **uniformity**, or a **uniform structure** for T. A family F of subsets of $T \times T$ which satisfies (i), (ii), and (iii) is sometimes said to be a uniformity (it can be shown that the set of all finite intersections of members of such a family is a *base* for a uniformity which satisfies (i)–(v), a base B of a uniformity U being a subset of U which has the property that each member of U contains a member of B). A topological space with a uniform topology is *metrizable* if and only if it is a Hausdorff topological space and its uniformity has a countable base. If T is a metric space, then T has a uniformity which is the family of all subsets V of $T \times T$ for which there is a positive number ϵ such that V contains all (x, y) for which $d(x, y) < \epsilon$.

U′NI-FORM′LY, *adv.* **uniformly continuous function.** See UNIFORM —uniform continuity.

   uniformly most powerful test. See HYPOTHESIS—test of a hypothesis.

U′NI-LAT′ER-AL, *adj* **unilateral shift.** See SHIFT. **unilateral surface.** Same as ONE-SIDED SURFACE. See SURFACE.

U′NI-MOD′U-LAR, *adj.* **unimodular matrix.** A square matrix whose determinant is equal to 1.

UN′ION, *n.* The union of a collection of sets is the set whose members are those objects that belong to at least one of the given sets. For example, the union of the set with members A, B, C and the set with members B and D is the set with members A, B, C, and D. The union of two sets U and V usually is denoted by $U \cup V$. *Syn.* join, sum.

U-NIQUE′, *adj.* Leading to one and only one result; consisting of one, and only one. The product of two integers is unique; the square root of an integer is not unless the integer is 0.

   unique factorization. See DOMAIN—integral domain, FUNDAMENTAL —fundamental theorem of arithmetic, IRREDUCIBLE —irreducible polynomial.

U-NIQUE′LY, *adv.* **uniquely defined.** A concept so defined that it is the only concept that fits the definition.

U-NIQUE′NESS, *adj.* **uniqueness theorem.** A theorem that asserts there is at most one object of some specified type. Examples are: (1) "Given a point not in a plane, there is at most one plane containing the point and parallel to the given plane"; (2) "If f, g, and h are continuous on the closed interval $[a, b]$ and y_0 and y_1 are two real numbers, then there is at most one solution y of the differential equations $y'' + f(x)y' + g(x)y = h(x)$ for which y'' is continuous on $[a, b]$, $y(a) =$ y_0, and $y'(a) = y_1$." An argument that establishes an uniqueness theorem is an **uniqueness proof.** See EXISTENCE —existence theorem.

U′NIT, *n.* A standard of measurement such as an inch, a foot, a centimeter, a pound, or a dollar; a single one of a number, used as the basis of counting or calculating. The **unit (real) number** is the number 1, a **unit complex number** is a complex number of unit absolute value (*i.e.,* a complex number of type $\cos \theta + i \sin \theta$), and the **unit imaginary number** is the number i. A **unit vector** is a vector of unit length.

   unit circle and **unit sphere.** A circle (or sphere) whose radius is one unit of distance. The circle (sphere) of unit radius having its center at the origin of coordinates is spoken of **as the unit circle (the unit sphere).**

   unit in a domain, groupoid, ring, or field. See GROUPOID, RING.

   unit fraction. See FRACTION.

   unit matrix. See MATRIX.

   unit square and **unit cube.** A square (cube) with its edges of length equal to one unit of distance.

U′NI-TAR′Y, *adj.* Undivided; relating to a unit or units.

   unitary analysis. See ANALYSIS.

   unitary matrix. See MATRIX.

   unitary space. See INNER—inner product space.

   unitary transformation. (1) A linear transformation whose adjoint is also its inverse. For **finite-dimensional spaces,** a linear transformation, which transforms $x = (x_1, x_2, \cdots, x_n)$ into $Tx = (y_1, y_2, \cdots, y_n)$ with

$$y_i = \sum_{j=1}^{n} a_{ij} x_j \qquad (i = 1, 2, \cdots, n),$$

is unitary if and only if the matrix (a_{ij}) is unitary, or if and only if it leaves the Hermitian form

$$x_1 \bar{x}_1 + x_2 \bar{x}_2 + \cdots + x_n \bar{x}_n$$

invariant. If (x, y) denotes the inner product of elements of a **Hilbert space** H, then a transformation T of H into H is unitary if $(Tx, Ty) = (x, y)$ for each x and y, or if T is an *isometric mapping* of H into H $[(Tx, Tx) = (x, x)$ for each $x]$. A unitary transformation is also a normal transformation. (2) A unitary transformation of a **matrix** A is a transformation of the form $P^{-1}AP$, where P is a unitary matrix. The concepts of unitary transformation of a finite-dimensional space and of a matrix are related in the same way as for *orthogonal transformations* except that the transpose A^T is replaced by the Hermitian conjugate of A. A Hermitian matrix can be reduced to diagonal form by a unitary transformation; hence every Hermitian form can be reduced to the type $\sum_{i=1}^{\infty} p_i x_i \bar{x}_i$ by a unitary transformation such as described above. See ORTHOGONAL —orthogonal transformation, SPECTRAL—spectral theorem.

UNITED STATES RULE. A rule for evaluating a debt upon which partial payments have been made: Apply each payment first to the interest, any surplus being deducted from the principal. If a payment is less than all interest due, the balance of the interest must be added to the principal, and cannot draw interest. This is the legal rule.

U′NI-TY, *n.* Same as ONE or 1.
root of unity. See ROOT—root of unity.

UNIVAC. An automatic digital computing machine manufactured by the Sperry Rand Corporation. UNIVAC is an acronym for *Universal Automatic Computer*.

U′NI-VER′SAL, *adj.* **universal quantifier.** See QUANTIFIER.
universal set. The set of all objects admissible in a particular problem or discussion. *Syn.* universe.

U′NI-VERSE′, *n.* In statistics, same as POPULATION.

UN-KNOWN′, *ad., n.* **unknown quantity.** (1) A letter or literal expression whose numerical value is implicit in certain given conditions by means of which this value is to be found. It is used chiefly in connection with equations. In the equation $x + 2 = 4x + 5$, x is the *unknown*. (2) More properly, a symbol as described in (1) is a **variable**, and "unknown" refers to a member of the solution set. Thus, for the equation $x^2 - 5x + 6 = 0$, x is a variable and the unknowns are 2 and 3, the values of x that make the equation a true statement.

UP′PER, *adj.* **upper bound of a set of numbers.** See BOUND.
upper density. See DENSITY—density of a sequence of integers.
upper limit of integration. See INTEGRAL—definite integral.

UP′SI-LON, *n.* The twentieth letter of the Greek alphabet: lower case, v; capital, Υ.

URYSOHN, Paul Samuilovich (1898–1924). Soviet analyst and topologist. Drowned at age 26.
Urysohn's lemma. If P and Q are two nonintersecting closed sets (in a normal topological space T), there exists a real function f defined and continuous in T and such that $0 \leq f(p) \leq 1$ for all p, with $f(p) = 0$ for p in P, and $f(p) = 1$ for p in Q. See METRIC—metric space, TIETZE —Tietze extension theorem.

V

VA′LENCE, *n.* **valence of a node.** See GRAPH THEORY.

VAL′U-A′TION, *n.* Act of finding or determining the value of.
valuation of a field. A map V of the field F to an *ordered ring* such that, for all x and y in F, (i) $V(x) \geq 0$ and $V(x) = 0$ if and only if $x = 0$; (ii) $V(xy) = V(x)V(y)$; (iii) $V(x + y) \leq V(x) + V(y)$. See p-ADIC.
valuation of bonds. See BOND.

VAL′UE, *n.* **absolute value.** See ABSOLUTE —absolute value of a real number, MODULUS —modulus of a complex number, VECTOR —absolute value of a vector.
accumulated value of an annuity. See ACCUMULATED.
assessed value. See ASSESSED.
book value. The book value of a **bond** is the purchase price minus the amount accumulated for amortization of the premium, or plus the amount of the accumulation of the discount, according as the bond has been bought at a premium or discount. When a bond is purchased at a premium, each dividend provides both for interest on the investor's principal and for a return of part of the premium, the *book value* at a given time being the purchase price minus the accumulation of the parts of the dividends used for amortization of the premium (the book value at maturity is equal to the face value). When the yield rate is greater than the dividend rate, the bond sells at a discount. The income is then greater than the dividend for any period, so the difference is added to the purchase price of the bond to bring it back to par value at maturity. These increasing values are the *book values* of

the bond. The book value of a **debt** is equal to the difference between the face value of the debt and the sinking fund set up to pay the debt; if the debt is being amortized, the book value is the amount which, with interest, would equal the amount of the debt at the time it is due. The book value of **depreciating assets** (such as equipment) is the difference between the cost price and the accumulated depreciation charges at the date under consideration.

face value. See PAR.

line value of a trigonometric function. A line segment whose length is equal to the absolute value of the function, usually taken as the segment whose length is the numerator in the definition of the function with the segment whose length is the denominator chosen to be of unit length. The vertex of the angle is frequently placed at the center of a unit circle.

market value. See MARKET.

mean-value theorems. See headings under MEAN.

numerical value. Same as ABSOLUTE VALUE.

par value. See PAR.

permissible value. See PERMISSIBLE.

place value. See PLACE.

present value. A sum of money which, with accrued interest, will equal a specified sum at some specified future time, or equal several sums at different times, as in the case of the present value (cost) of an annuity. The present value of a sum of money A_n due in n years at interest rate i is, at simple interest,

$$P = A_n/(1 + ni).$$

At compound interest, it is

$$P = A_n/(1 + i)^n = A_n(1 + i)^{-n}.$$

The present value of an **ordinary annuity** is

$$A = R\frac{(1 + i)^n - 1}{i(1 + i)^n} = \frac{R}{i}\left[1 - (1 + i)^{-n}\right],$$

where A is the present value, R the periodic payment, n the number of periods, and i the rate of interest per period. The present value of an **annuity due** is

$$A = R\frac{(1 + i)^n - 1}{i(1 + i)^{n-1}} = \frac{R}{i}\left[1 + i - (1 + i)^{-n+1}\right].$$

principal value of the inverse of a trigonometric function. See TRIGONOMETRIC—inverse trigonometric function.

scrap or salvage value. See SCRAP.

surrender value of an insurance policy. See SURRENDER —surrender value of an insurance policy.

value of an expression. The result that would be obtained if the indicated operations were carried out. The value of $\sqrt{9}$ is 3; the value of

$$\int_a^b 2x\,dx \quad \text{is} \quad b^2 - a^2.$$

The value of the polynomial $x^2 - 5x - 7$ is -1 when $x = 6$. See EVALUATE.

value of a function. Any member of the range of the function. For a particular value (or values) of the independent variable (or variables), the value of the function is the corresponding member of the range. See FUNCTION.

value of an insurance policy. The difference between the expectation of the future benefit and the expectation of the future net premiums; the difference between the accumulated premiums and the accumulated losses. *Syn.* terminal reserve.

wearing value. Same as REPLACEMENT COST.

VANDERMONDE, Alexandre Théophile (1735–1796). French algebraist. Gave first logical exposition of the theory of determinants.

Vandermonde determinant or matrix. See under DETERMINANT, MATRIX.

VAN DER WAERDEN, Bartel Leendert (1903–). **Van der Waerden theorem**. See RAMSEY —Ramsey theory.

VANDIVER, Harry Schultz (1882–1973). Self-educated American algebraist and number theorist. Best known for his contributions toward the solution of Fermat's last theorem.

VAN'ISH, *v.* To become zero; to take on the value zero.

VAN'ISH-ING, *adj.* Approaching zero as a limit; taking on the value zero.

VAR'I-A-BIL'I-TY, *n.* (*Statistics*) Same as DISPERSION.

measures of variability. The *range, quartile deviation, average deviation*, and *standard deviation* are common usages.

VAR'I-A-BLE, *n.* A symbol used to represent an unspecified member of some set. A variable is a "place holder" or a "blank" for the name of some member of the set. Any member of the set is a **value** of the variable and the set itself is the **range** of the variable. If the set has only one member, the variable is a **constant**. The symbols x and y in the expression $x^2 - y^2 = (x - y)(x + y)$ are variables that represent unspecified numbers in the sense that the equality is true whatever numbers may be put in the places held by x and y. See FUNCTION.

chance, random, or stochastic variable. See RANDOM—random variable.

change of variable in differentiation and integration. See CHAIN—chain rule, INTEGRATION—change of variables in integration.

dependent and independent variables. See FUNCTION.

separation of variables. See DIFFERENTIAL—differential equation with variables separable.

VAR′I-ANCE, *adj., n.* The **variance** of a random variable X (or of the associated distribution function) is the *second moment* of X about the mean, *i.e.*, the *expected value* of $(X - \mu)^2$, where μ is the mean. It is customary to denote the variation by σ^2. For a discrete random variable with values $\{x_n\}$ and probability function p,

$$\sigma^2 = \sum (x_i - \mu)^2 p(x_i),$$

if this sum converges; for a continuous random variable with probability density function f,

$$\sigma^2 = \int_{-\infty}^{\infty} (x - \mu)^2 f(x)\, dx,$$

if this integral converges. The variance minimizes the second moment, since

$$E\left[(X - c)^2\right] = \sigma^2 + (\mu - c)^2$$

for any c and this is least when $c = \mu$ (see PARALLEL—parallel-axis theorem). For a sample $\{x_1, \cdots, x_n\}$ whose mean is $\bar{x}$, the **sample variance** either is defined to be $\sum_{i=1}^{n} (x_i - \bar{x})^2/n$, which is a maximum-likelihood estimator of the variance of the distribution if the distribution is normal, or is defined to be $\sum_{i=1}^{n}(x_i - \bar{x})^2/(n - 1)$, which is an *unbiased estimator* in general (also a minimum-variance estimator if the distribution is normal). If the mean μ of the distribution is known, then $\sum_{i=1}^{n} (x_i - \mu)^2/n$ is an unbiased estimator (also a maximum-likelihood, minimum-variance estimator if the distribution is normal). See DEVIATION—standard deviation, EXPECTED

—expected value, MOMENT—moment of a distribution.

analysis of variance. Any technique for analyzing the *variation* (the sum of squares of differences from the mean) of a set of observations composed of samples from several populations by representing the variation as the sum of components; usually one component is associated with the actual population variability and another is associated with variations in sample means. More generally, analysis of variance refers to any process of partitioning a sum of squares of (usually normal) random variables into a sum of squares of other independent (usually normal) random variables. Analysis-of-variance techniques provide methods of analyzing differences between means of sets of samples, where the differences between means are due to factors whose influence on the mean of a group is to be investigated.

VA′RI-ATE, *n.* Same as RANDOM VARIABLE. See RANDOM—random variable.

VAR′I-A′TION, *n.* (*Calculus of Variations*). A *variation* δy of a function y is a function δy which is added to y to give a new function $y + \delta y$. The name *calculus of variations* was adopted as a result of this notation, introduced by Lagrange in about 1760 when comparing the value of an integral along an arc with its value along a neighboring arc. The **first variation of an integral**

$$I = \int_a^b f(x, y, y')\, dx$$

is

$$\delta I = \left[\frac{d}{d\epsilon} \int_a^b f(x, y + \epsilon\phi, y' + \epsilon\phi')\, dx \right]_{\epsilon = 0},$$

if it exists for suitably restricted functions ϕ. If $\phi(a) = \phi(b) = 0$, then

$$\delta I = \int_a^b \phi \left[\frac{\partial f}{\partial y} - \frac{d}{dx}\left(\frac{\partial f}{\partial y'} \right) \right] dx.$$

A function y is said to make I **stationary**, or I is said to have a **stationary value** at y, if the first variation of I is zero at y for all *admitted* functions ϕ such that $\phi(a) = \phi(b) = 0$, a function ϕ being *admitted* if it satisfies certain conditions (*e.g.*, that it be rectifiable, or that it have a continuous derivative). A necessary condition that y make I have a (relative) maximum or minimum is that it give a stationary value of I. The **nth variation** $\delta^n I$ is given by

$$\delta^n I = \left[\frac{d^n}{d\epsilon^n} \int_a^b f(x, y + \epsilon\phi, y' + \epsilon\phi')\, dx\right]_{\epsilon = 0}.$$

See CALCULUS —calculus of variations.

calculus of variations. See CALCULUS.

coefficient of variation. The quotient of the *standard deviation* and the *mean* of a distribution, sometimes multiplied by 100.

combined variation. One quantity varying as some combination of other quantities, such as z varying directly as y and inversely as x.

direct variation. When two variables are so related that their ratio remains constant, one of them is said to *vary directly* as the other, or they are said to *vary proportionately*; *i.e.*, when $y/x = c$, or $y = cx$, where c is a constant, y is said to *vary directly* as x. This is sometimes written: $y \propto x$. The number c is the **constant of proportionality** (or **factor of proportionality** or **constant of variation**). *E.g.*, when velocity is constant, the distance s traveled per unit of time is proportional to the time; *i.e.*, $s = kt$, where k is the constant of proportionality.

fundamental lemma of the calculus of variations. See FUNDAMENTAL.

inverse variation. When the ratio of one variable to the reciprocal of the other is constant (*i.e.*, when the product of the two variables is constant) one of them is said to *vary inversely* as the other, *i.e.*, if $y = c/x$, or $xy = c$, y is said to *vary inversely* as x, or x to *vary inversely* as y.

joint variation. When one variable *varies directly* as the product of two variables, the one variable is said to *vary jointly* as the other two, *i.e.*, when $x = kyz$, x varies jointly as y and z. When $x = kyz/w$, x is said to vary jointly as y and z and inversely as w.

variation of a function in an interval (a, b). The least upper bound of the sum of the oscillations in the closed subintervals (a, x_1), $(x_1, x_2), \cdots, (x_n, b)$ $(a < x_1 < x_2 < \cdots < b)$ of (a, b), for all possible such subdivisions. A function with finite variation in (a, b) is of **bounded variation** or **limited variation** in (a, b) and can be expressed as the sum of two monotone functions.

variation of a function on a surface. On a surface S: $x = x(u, v)$, $y = y(u, v)$, $z = z(u, v)$, the rate of change of a given function $f(u, v)$ at a point P of S varies with the direction from P on S. It vanishes in the direction of the tangent to

the curve $f = $ const., and has its greatest absolute value in the direction on S perpendicular to $f = $ const.; this latter value is given by

$$\left|\frac{df}{ds}\right| = \frac{\left[E\left(\frac{\partial f}{\partial v}\right)^2 - 2F\frac{\partial f}{\partial u}\frac{\partial f}{\partial v} + G\left(\frac{\partial f}{\partial u}\right)^2\right]^{1/2}}{[EG - F^2]^{1/2}}.$$

See GRADIENT —gradient of a function, SURFACE —fundamental coefficients of a surface.

variation of parameters. A method for finding a particular solution of a linear differential equation when the general solution of the reduced equation is known. *E.g.*, suppose it has been determined that the general solution of $(x - 1)y'' - xy' + y = 0$ is $y = Ax + Be^x$. To solve

$$(x - 1)y'' - xy' + y = 1 - x,$$

the parameters A and B are replaced by functions A and B to be determined so that $y = xA(x) + e^x B(x)$ is a particular solution. By imposing the condition $xA'(x) + e^x B'(x) = 0$, one obtains $y' = A(x) + e^x B(x)$. Then substitution in the differential equation gives $A'(x) + e^x B'(x) = -1$, and the desired solution can be obtained by solving the equations

$$xA'(x) + e^x B'(x) = 0$$

and

$$A'(x) + e^x B'(x) = -1$$

for $A'(x)$ and $B'(x)$, and then integrating to find $A(x)$ and $B(x)$. A similar procedure can be used for an equation of order n, obtaining $n - 1$ equations in n unknown functions by equating to zero the terms involving derivatives of the unknown functions in the first $n - 1$ derivatives of y and then obtaining the nth equation by substitution in the differential equation. A variation of this procedure can be used when some solutions of the reduced equation are known, but the general solution is not known.

variation of sign in a polynomial. A change of sign between two successive terms when arranged in descending order. The polynomial $x - 2$ has one *variation of sign*, while $x^3 - x^2 + 2x - 1$ has three. See DESCARTES —Descartes' rule of signs.

variation of sign in an ordered set of numbers. A change of sign between two successive numbers; *e.g.*, the sequence $\{1, 2, -3, 4, -5\}$ has three variations of sign.

VA-RI'E-TY, *n.* Let V be an n-dimensional vector space with scalars in some field F. If A is a subset of V that is the set of all points $(x_1, \cdots, x_n)$ which satisfy each of a set of polynomial equations $\{p_k(x_1, \cdots, x_n) = 0\}$ with coefficients in F, then A is an (affine) **algebraic variety**. Given a field F, an **affine algebraic variety** (or **affine variety**) is a subset of an n-dimensional affine space F_n over F if it is the set of common zeros of a set of polynomials $\{p_k(x_1, \cdots, x_n)\}$ with coefficients in F. A **projective algebraic variety** (or **projective variety**) is defined similarly, but with F_n the n-dimensional projective space over F and each p_k a homogeneous polynomial. See AFFINE —affine space, PROJECTIVE —projective space.

VAR'Y, *v.* **vary directly** or **inversely.** See VARIATION —direct variation, inverse variation.

VEBLEN, Oswald (1880–1960). American analyst, projective geometer, and topologist. Active in founding Institute for Advanced Study in Princeton, N.J. See JORDAN —Jordan curve theorem.

VEC'TOR, *adj.*, *n.* In three-dimensional Euclidean space, an entity that can be described by a directed line segment and is subject to certain operations of addition and multiplication [see MULTIPLICATION —multiplication of vectors, and SUM —sum of vectors]; an ordered triple of numbers subject to corresponding operations. Any set of vectors whose sum is a given vector are **components** of this vector, although the component of a vector in a given direction is understood to be the projection of the vector onto a line in that direction. If the unit vectors in the directions of the x, y and z axes are denoted by $\boldsymbol{i}$, $\boldsymbol{j}$, and $\boldsymbol{k}$, then the components parallel to the axes are of the form $x\boldsymbol{i}$, $y\boldsymbol{j}$, and $z\boldsymbol{k}$, and the vector can be written as $x\boldsymbol{i} + y\boldsymbol{j} + z\boldsymbol{k}$, or (x, y, z). The vector $\boldsymbol{R} = x\boldsymbol{i} + y\boldsymbol{j} + z\boldsymbol{k}$ is shown in the figure. See ACCELERATION, FORCE —force vector, VELOCITY, and below, vector space.

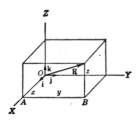

absolute value of a vector. The numerical length of the vector (without regard to direction);

the square root of the sum of the squares of its components along the axes. The absolute value of $2\boldsymbol{i} + 3\boldsymbol{j} + 4\boldsymbol{k}$ is $\sqrt{29}$; in general, the absolute value of $a\boldsymbol{i} + b\boldsymbol{j} + c\boldsymbol{k}$ is $\sqrt{a^2 + b^2 + c^2}$. *Syn.* numerical value.

addition and multiplication of vectors. See MULTIPLICATION —multiplication of vectors, SUM —sum of vectors.

contravariant vector field. A contravariant tensor field of order one. See TENSOR —contravariant tensor.

covariant vector field. A covariant tensor field of order one. The gradient of a scalar field is a covariant vector field, while $t_i(x^1, x^2, \cdots, x^n)$ is locally the gradient of some scalar field if the conditions

$$\frac{\partial t_i}{\partial x^j} = \frac{\partial t_j}{\partial x^i}$$

are satisfied for all i and j in a region where the partial derivatives of t_i exist and are continuous. See TENSOR —covariant tensor.

derivative of a vector. See DERIVATIVE —derivative of a vector.

dominant vector. See DOMINANT.

irrotational vector. See IRROTATIONAL.

orthogonal vectors. See ORTHOGONAL.

parallel (contravariant) vector field. See PARALLEL —parallel displacement of a vector along a curve.

position vector. A vector from the origin to a point under consideration. If the point has Cartesian coordinates x, y, and z, the *position vector* is $\boldsymbol{R} = x\boldsymbol{i} + y\boldsymbol{j} + z\boldsymbol{k}$ (see the illustration under VECTOR).

radius vector. See POLAR —polar coordinates, SPHERICAL —spherical coordinates.

reciprocal system of vectors. See RECIPROCAL.

solenoidal vector. See SOLENOIDAL.

vector analysis. The study of vectors, relations between vectors, and their applications.

vector potential. See POTENTIAL.

vector product. See MULTIPLICATION —multiplication of vectors.

vector space. (1) A space of vectors, as ordinary vectors in three dimensions, or (in general) a space of elements called *vectors* described by n components $(x_1, x_2, \cdots, x_n)$. The space is a **real vector space** if the component are real numbers. The sum of vectors $x = (x_1, x_2, \cdots, x_n)$ and $y = (y_1, y_2, \cdots, y_n)$ is defined as $x + y = (x_1 + y_1, x_2 + y_2, \cdots, x_n + y_n)$ and ax as $(ax_1, ax_2, \cdots)$. The **scalar product** (or **inner product** or **dot**

product) of x and y is $\sum_{i=1}^{n} \bar{x}_i y_i$, and the length or **norm** of the vector x is $\left[\sum_{i=1}^{n} |x_i|^2 \right]^{1/2}$. The vectors may have an infinite number of components, as for Hilbert space. (2) A set V of elements, called *vectors*, such that any vectors x and y determine a unique vector $x + y$ and any vector x and scalar a have a product ax in V, with the properties: (i) V is an Abelian group with addition as the group operation; (ii) $a(x + y) = ax + ay$ and $(a + b)x = ax + bx$; (iii) $(ab)x = a(bx)$ and $1 \cdot x = x$; properties (ii) and (iii) hold for any vectors x and y and any scalars a and b. The scalars may be real numbers, complex numbers, or elements of some other field. *Syn.* linear space, linear vector space. A vector space is a **linear topological space** (or **topological vector space**) if it is a topological group and scalar multiplication is continuous (*i.e.*, for any neighborhood W of $a \cdot x$ there is a neighborhood U of a and a neighborhood V of x such that $b \cdot y$ is in W if b and y belong to U and V, respectively. A vector space is a **normed vector space** (or **normed linear space**) if there is a real number $\|x\|$ (called the **norm** of x) associated with each "vector" x and $\|x\| > 0$ if $x \neq 0$, $\|ax\| = |a| \|x\|$, $\|x + y\| \leq \|x\| + \|y\|$. A normed vector space is also a linear topological space. See BANACH, BASIS, FRÉCHET —Fréchet space, HILBERT, INNER —inner-product space, ORTHOGONAL —orthogonal vectors.

vector-valued function. A function whose range is a subset of a vector space. See FUNCTION.

VEC-TO′RI-AL, *adj.* **vectorial angle.** See POLAR—polar coordinates in the plane.

VE-LOC′I-TY, *n.* Directed speed. The velocity of an object at time t is the limit of the **average velocity** as the time interval (Δt) approaches zero (see below, average velocity). Velocity is sometimes called **instantaneous velocity** to distinguish it from average velocity. If an object is moving along a straight line, its velocity is spoken of as **linear** (or **rectilinear**) **velocity** and is equal in magnitude to the speed. If an object is moving along a curve, its velocity is sometimes called **curvilinear velocity** (its direction is that of the tangent to the curve, its magnitude the speed of the object). Velocity is said to be **absolute**, or **relative**, according as it is computed relative to a coordinate system considered stationary, or relative to a moving coordinate system. If the coordinates of a moving point are x, y, and z, its *position vector* is $R = xi + yj + zk$, and its *vector*

velocity is $dR/dt = (dx/dt)i + (dy/dt)j + (dz/dt)k$. See SPEED.

angular velocity. (1) If a particle is moving in a plane, its *angular velocity* about a point in the plane is the rate of change of the angle between a fixed line and the line joining the moving particle to the fixed point. (As far as angular velocity is concerned, the particle may as well be considered as moving on a circle.) (2) If a rigid body is rotating about an axis, its *angular velocity* is (is represented by) a vector directed along the axis in the direction a right-hand screw would advance if subject to the given rotation and having magnitude equal to the angular speed of rotation about the axis (*i.e.*, the number of degrees or radians through which the body rotates per unit of time).

average velocity. The difference between the position vectors of a body at the ends of a given time interval, divided by the length of the interval. If $R = x(t)i + y(t)j + z(t)k$ is the *position vector* of a particle at time t, the average velocity over the time interval Δt is

$$\frac{x(t + \Delta t) - x(t)}{\Delta t}i + \frac{y(t + \Delta t) - y(t)}{\Delta t}j$$

$$+ \frac{z(t + \Delta t) - z(t)}{\Delta t}k,$$

or $\dfrac{R(t + \Delta t) - R(t)}{\Delta t}$, which is the resultant of the average velocities along the x, y, and z axes. See VECTOR.

constant (or uniform) velocity. See CONSTANT—constant speed and velocity.

VERSED, *adj.* **versed sine and versed cosine.** See TRIGONOMETRIC —trigonometric functions.

VER-SI-E′RA, *n* See WITCH.

VER′TEX, *n.* [*pl.* **vertices**]. See ANGLE—polyhedral angle, APEX, CONE, CONICAL —conical surface, ELLIPSE, HYPERBOLA, PARABOLA, PARABOLOID, PENCIL—pencil of lines through a point, POLYGON, POLYHEDRON, PYRAMID, TRIANGLE.

opposite vertex. See OPPOSITE.

VER′TI-CAL, *adj.* **vertical angles.** Two angles such that each side of one is a prolongation through the vertex of a side of the other.

vertical line. (1) A line perpendicular to a horizontal line. The horizontal line is usually thought of as being directed from left to right and the vertical line as being directed upward, when they are coordinate axes. (2) A line perpendicular to the plane of the horizon. (3) A line from the observer to his zenith, *i.e.*, the plumb line.

VI'BRAT-ING, *p.* **equation of vibrating string.** The equation

$$\frac{\partial^2 y}{\partial t^2} = \frac{T}{\rho} \frac{\partial^2 y}{\partial x^2},$$

where x is the direction in which the string is stretched, y denotes displacement, and t is the time variable; T is the tension in the string and ρ is the density (mass per unit length) of the string. The boundary conditions are usually taken to be of the form $y = f(x)$ at $t = 0$ and $\partial y / \partial t = g(x)$ at $t = 0$, where $g = 0$ if the string is at rest at time 0. It is assumed that the string is perfectly flexible, that T is constant, and that T is large enough that gravity can be neglected in comparison with it.

VI-BRA'TION, *n.* A periodic motion; a motion which is approximately periodic. *Syn.* oscillation. See OSCILLATION.

VIÈTE, François (Franciscus Vieta) (1540–1603). Most influential French mathematician of his time. Algebraist, arithmetist, cryptanalyst, and geometer. Used letters to represent both constants and variables, but rejected negative numbers. Gave trigonometric solution of general cubic equation in one variable.

Viète formula. The formula

$$\frac{2}{\pi} = \cos \frac{\pi}{4} \cos \frac{\pi}{8} \cos \frac{\pi}{16} \cdots,$$

which is equivalent to

$$\frac{2}{\pi} = \frac{2^{1/2}}{2} \frac{\left(2 + 2^{1/2}\right)^{1/2}}{2}$$

$$\times \frac{\left[2 + \left(2 + 2^{1/2}\right)^{1/2}\right]^{1/2}}{2} \cdots.$$

VI-GES'I-MAL, *adj.* Having to do with twenty. The **vigesimal number system** is the system of notation for real numbers for which 20 is the base. It is suggested by the English *score* and the French *vingt* (20). The Aztecs and Mayas used such a system, except that the Mayan system used $18 \cdot 20 = 360$ instead of 20^2 in the third

place—perhaps because the Mayan year had 360 days. See BASE—base of a number system.

VIN'CU-LUM, *n.* See AGGREGATION.

VI'TAL, *adj.* **vital statistics.** Statistics relating to the length of life and the number of persons dying during certain years, the kind of statistics from which mortality tables are constructed.

VITALI, Giuseppe (1875–1932). Italian analyst and set theorist.

Vitali covering. Let S be a set in n-dimensional Euclidean space and J be a class of sets such that, for each point x of S, there is a positive number $\alpha(x)$ and a sequence of sets $U_1, U_2, \cdots$, which belong to J, each of which contains x, and which have the properties that their *diameters* approach zero and for each integer n there is a *cube* C_n such that $m(U_n) \geq \alpha(x) m(C_n)$ and C_n contains U_n [$m(U_n)$ and $m(C_n)$ are the *measures* of U_n and C_n]. Such a class of sets is said to **cover S in the sense of Vitali**.

Vitali covering theorem. Let S be a set in n-dimensional Euclidean space. The Vitali covering theorem states that, if a class J of closed sets is a *Vitali covering* of S, then there is a finite or denumerably infinite sequence of pairwise disjoint sets belonging to J whose union contains all of S except a set of *measure zero*.

Vitali set. A set of real numbers such that no two numbers of the set have a difference which is a rational number and each real number is equal to a rational number plus a member of the set. Such a set can be formed by choosing exactly one element from each *coset* of the rational numbers, considered as a subgroup of the additive group of real numbers. A Vitali set is nonmeasurable and an intersection of a Vitali set with an interval either is of measure zero or is nonmeasurable. See SIERPINSKI —Sierpinski set.

VOLTA, Allesandro Giuseppe Antonio Anastasio (1745–1827). Italian physicist and inventor of electric battery. *Volt* named in his honor.

VOLT, *n.* A unit of measure of electromotive force. (1) The **absolute volt** is the steady potential difference which must exist across a conductor which carries a steady current of one absolute ampere and which dissipates thermal energy at the rate of one watt. The absolute volt has been the legal standard of potential difference since 1950. (2) The **international volt**, the legal standard prior to 1950, is the steady potential difference which must be maintained across a conductor which has a resistance of one international ohm and which carries a steady current of one international ampere.

$$1 \text{ int. volt} = 1.000330 \text{ abs. volts.}$$

VOLTERRA, Vito (1860–1940). Italian analyst and physicist. Pioneer in the development of integral-differential equations and functional analysis.

Volterra integral equations. The *Volterra integral equation of the first kind* is the equation

$$f(x) = \int_a^x K(x, t) y(t) \, dt,$$

and the *Volterra integral equation of the second kind* is $y(x) = f(x) + \lambda \int_a^x K(x, t) y(t) \, dt$, in which f and K are two given functions, and y is the unknown function. The function K is the **kernel** or **nucleus** of the equation. Volterra's equation of the second kind is **homogeneous** if $f(x) \equiv 0$. For an example, see ABEL—Abel's problem.

Volterra reciprocal functions. Two functions $K(x, y)$ and $k(x, y; \lambda)$ for which

$$K(x, y) + k(x, y; \lambda) = \lambda \int_a^b k(x, t; \lambda) K(t, y) \, dt.$$

If the *Fredholm determinant* $D(\lambda) \neq 0$ and $K(x, y)$ is continuous in x and y, then

$$k(x, y; \lambda) = -D(x, y; \lambda) / [\lambda D(\lambda)],$$

where $D(x, y; \lambda)$ is the *first Fredholm minor*. If g is a solution of

$$g(x) = f(x) + \lambda \int_a^b K(x, t) g(t) \, dt,$$

then f is a solution of

$$f(x) = g(x) + \lambda \int_a^b k(x, t; \lambda) f(t) \, dt,$$

and conversely. The function $k(x, y; \lambda)$ is the **resolvent kernel**. The above is sometimes modified by letting $\lambda = 1$. See KERNEL—iterated kernels.

Volterra solutions of the Volterra integral equations. If f and K are continuous functions of x in $a \leq x \leq b$ and of x and t in $a \leq t \leq x \leq b$,

respectively, then the Volterra integral equation of the second kind,

$$y(x) = f(x) + \lambda \int_a^x K(x, t) y(t) \, dt,$$

has a unique continuous solution given by the formula

$$y(x) = f(x) + \int_a^x k(x, t; \lambda) f(t) \, dt,$$

where $k(x, t; \lambda)$ is the *resolvent kernel* of the given *kernel* $K(x, y)$ and is a continuous function of x and t for $a \leq t \leq x \leq b$. The *Volterra integral equation of the first kind*,

$$f(x) = \lambda \int_a^x K(x, t) y(t) \, dt,$$

can be reduced to an equation of the second kind by differentiation, giving

$$f'(x) = \lambda K(x, x) y(x) + \lambda \int_a^x \frac{\partial K(x, t)}{\partial x} y(t) \, dt,$$

it being assumed that $\dfrac{\partial K(x, t)}{\partial x}$ exists and is continuous. The above is sometimes modified by letting $\lambda = 1$.

VOLUME, *n.* A number describing the three-dimensional extent of a set. A cube with side 1 has *unit* volume. A rectangular parallelepiped with adjacent edges of lengths a, b, and c has volume abc. The volume of any bounded set is the least upper bound α of the sum of the volumes of a finite collection of nonoverlapping rectangular parallelepipeds contained in the set, or the greatest lower bound β of the sum of the volumes of a finite collection of rectangular parallelepipeds which together completely cover the set, provided $\alpha = \beta$ (if $\alpha = \beta = 0$, the set has volume and the volume is zero; if $\alpha \neq \beta$, the set does not have volume). An unbounded set with volume is an unbounded set S for which there is a number m such that $R \cap S$ has volume not greater than m whenever R is a cube; the volume of S is then the least upper bound of the volumes of $R \cap S$ for cubes R. This definition can be used to prove the usual formulas for volume (see CONE, CYLINDER, SPHERE, etc.). For example, the volume of a tetrahedron can be shown to be $\frac{1}{3}$ the product of the altitude and the area of the corresponding base. Any polyhedron can be represented as the union of nonoverlapping tetrahederons and its volume is the sum of the volumes of such tetrahedrons. For many solids (*e.g.*, spheres), the volume is the least upper bound of the volumes of polyhedrons contained in the solid. Calculus is very useful for computing volumes (see INTEGRAL —multiple integral, REVOLUTION —solid of revolution). *Syn.* three-dimensional content. See CAVALIERI, CONTENT, EXHAUSTION, MEASURE —measure of a set, PAPPUS.

coefficient of volume expansion. See COEFFICIENT.

differential (or **element**) **of volume.** See ELEMENT—element of integration.

volumes of similar solids. See SOLID—similar solids.

VON NEUMANN, John (1903–1957) Born in Budapest and taught at Berlin and Hamburg before coming to U.S. Pure and applied mathematician and economist. One of the most versatile and innovative mathematicians of the twentieth century. Created theory of games and made important contributions to mathematical economics, quantum theory, ergodic theory, theory of operators on Hilbert space, computer theory and design, linear programming, theory of continuous groups, logic, probability.

von Neumann algebra. A subset A of the algebra of bounded linear operators on a complex Hilbert space is a $*$-algebra if A is an algebra and also contains the adjoint of each of its members. A **von Neumann algebra** is a $*$-algebra that contains the identity operator and is closed in the strong operator topology. *Syn.* operator ring, W^*-algebra. See ALGEBRA—algebra over a field, OPERATOR —linear operator.

VOSS, Aurel Edmund (1845–1931). German differential geometer. See SURFACE —surface of Voss.

VUL′GAR, *adj.* **vulgar fraction.** See FRACTION.

W

W. W-surface, *n.* Same as WEINGARTEN SURFACE. See WEINGARTEN.

WALK, *n.* **random walk.** See RANDOM.

WALLIS, John (1616–1703). English algebraist, analyst, cryptographer, logician, and theologian. He was the most accomplished English mathematician before Newton. His and Barrow's work on infinitesimal analysis strongly influenced

Newton. He was perhaps first (1685) to introduce graphical representation of complex numbers, though he did not explicitly employ an axis of imaginaries. See ARGAND—Argand diagram, COMPLEX —complex plane, WESSEL.

Wallis' formulas. Formulas giving the values of the definite integrals from 0 to $\frac{1}{2}\pi$ of each of the functions $\sin^m x, \cos^m x$, and $\sin^m x \cos^n x$, for m and n any positive integers. See formula 359 in the INTEGRAL TABLES in the appendix.

Wallis' product for π. The infinite product

$$\frac{\pi}{2} = \frac{2}{1}\frac{2}{3}\frac{4}{3}\frac{4}{5}\frac{6}{5}\frac{6}{7} \cdots \frac{2k}{2k-1}\frac{2k}{2k+1} \cdots.$$

WALSH, Joseph Leonard (1895–1973). American analyst. Worked with polynomial, rational, harmonic, and orthogonal functions, particularly in relation to approximation theory.

Walsh functions. The functions $\{w_n\}$ defined on the interval $[0,1]$ by $w_1 \equiv 1$ and $w_{n+1} = r_{n_1+1}r_{n_2+1} \cdots r_{n_k+1}$, where n is $2^{n_1} + 2^{n_2} + \cdots + 2^{n_k}$ and $n_1 > n_2 > \cdots > n_k \geq 0$. The sequence of Walsh functions is orthonormal on the interval $[0,1]$ and contains the Rademacher functions. Their closed linear span in L^p ($1 \leq p < \infty$) is L^p. See RADEMACHER —Rademacher functions.

WANTZEL, Pierre Laurent (1814–1848). French algebraist and geometer. See TRISECTION —trisection of an angle.

WARING, Edward (1734–1798). English algebraist and number theorist.

Waring's problem. To prove the assertion made by Waring in 1770 that, for any integer n, there is a least integer $g(n)$ such that any integer can be represented as the sum of not more than $g(n)$ numbers, each of which is an nth power of an integer. Waring's problem was solved by Hilbert (1909). Lagrange showed in 1770 that any integer can be represented as the sum of not more than 4 squares (any prime of the form $4n + 1$ can be represented as the sum of 2 squares in exactly one way). Euler conjectured that $g(n) = 2^n + A - 2$ if $n \geq 2$, where A is the greatest integer less than $(\frac{3}{2})^n$. The Euler conjecture fails for at most finitely many values of n, all greater than 471,600,000. *E.g.*, $g(3) = 9$, $g(4) = 19$, $g(5) = 37$, $g(6) = 73$, and $g(7) = 143$. If $G(n)$ is the least integer such that all but at most finitely many integers can be represented as the sum of not more than $G(n)$ numbers, each of which is an nth power of an integer, then $G(3) \leq 7$ and $G(4) = 16$. See THREE—three-squares theorem.

WAST′ING, *adj.* **wasting assets.** Same as DE-PRECIATING ASSETS. See ASSETS.

WATT, James (1736–1819). British engineer and inventor.

watt. A metric unit of measure of power; the power required to keep a current of one ampere flowing under a potential drop of one volt; about $\frac{1}{736}$ of one horsepower (English and American). The **international watt** is defined in terms of the *international ampere* and the *international volt* and differs slightly from the **absolute watt,** which is equivalent to 10^7 ergs (one joule) of work per second.

watt-hour. A unit of measure of electric energy; the work done by one watt acting for one hour. It equals $36 \cdot 10^9$ ergs.

WAVE, *adj.* **wave equation.** The partial differential equation

$$\frac{\partial^2 \psi}{\partial x^2} + \frac{\partial^2 \psi}{\partial y^2} + \frac{\partial^2 \psi}{\partial z^2} = \frac{1}{c^2} \frac{\partial^2 \psi}{\partial t^2}.$$

In the theory of sound, this equation is satisfied by the velocity potential (in a perfect gas); in the theory of elastic vibrations, it is satisfied by each component of the displacement; and, in the theory of electric or electromagnetic waves, it is satisfied by each component of the electric or magnetic (force) vector. The constant c is the velocity of propagation of the periodic disturbance.

wave length of motion as represented by a trigonometric function. The period of the trigonometric function. See PERIODIC—periodic function of a real variable.

WEAK, *adj.* **weak compactness.** Compactness relative to the weak topology [see COMPACT]. A set S contained in a normed linear space N is *weakly compact* if and only if each sequence of elements of S has a subsequence which *converges weakly* to a point of S. A Banach space has the property that each bounded closed subset is weakly compact if and only if the space is reflexive.

weak completeness. Completeness relative to the weak topology [see COMPLETE—complete space]. A weakly complete normed linear space is complete (and is a Banach space). A reflexive Banach space is weakly complete, but the space l^1 [of sequences $x = (x_1, x_2, \cdots)$ for which

$\|x\| = \Sigma |x_i|$ is finite] is weakly complete and not reflexive.

weak convergence. A sequence of elements $\{x_1, x_2, \cdots\}$ of a linear topological space N is *weakly convergent* (or is a *weakly fundamental sequence*) if $\lim f(x_n)$ exists for each continuous linear functional defined on N. If $\lim f(x_n) = f(x)$ for each f, then the sequence **converges weakly to x** (and x is a **weak limit** of the sequence). A continuous linear functional f is a **weak*** **limit** (or **w*-limit**) of a sequence $f_1, f_2, \cdots$ of continuous linear functionals if $\lim f_i(x) = f(x)$ for each x of N. Weak* is reads as **weak-star.** See below, weak topology.

weak law of large numbers. See LARGE—law of large numbers.

weak operator topology. See OPERATOR —linear operator.

weak topology. The **weak topology** of a linear topological space N is generated by the set of neighborhoods defined as follows: For each positive number ϵ, each element x_0 of N, and each finite set $\{f_1, f_2, \cdots, f_n\}$ of continuous linear functionals defined on N, the set U of all x for which $|f_k(x) - f_k(x_0)| < \epsilon$ for each k is a neighborhood of x_0. Open sets of the topology are then unions of such neighborhoods. The linear space with its weak topology is a *Hausdorff space* if and only if for each x and y with $x \neq y$ there is a continuous linear functional f such that $f(x) \neq f(y)$; this is true for all normed linear spaces. The **weak*** **topology** (or **w*-topology**) of the *first conjugate space* N^* of a linear topological space N is generated by the set of neighborhoods defined as follows: For each positive number ϵ, element f_0 of N^*, and finite set $\{x_1, x_2, \cdots, x_n\}$ of elements of N, the set V of all f for which $|f(x_k) - f_0(x_k)| < \epsilon$ for each k is a neighborhood of f_0. If N is a normed linear space, then the unit ball of N^* (the set of all f with $\|f\| \leq 1$) is compact in the weak* topology. For a reflexive Banach space, the weak* topology of B^* and the weak topology of B^* are identical.

WEAR′ING, *adj.* **wearing value of equipment.** The difference between the purchase price and the scrap value. Also called **original wearing value.** The difference between the book value and scrap value is the **remaining wearing value.**

WEDDERBURN, Joseph Henry Maclagan (1882–1948). Scottish-American algebraist. Worked on matrices, hypercomplex numbers, and algebraic structures.

Wedderburn theorem on finite division rings. All finite division rings are fields.

Wedderburn's structure theorems. Theorems such as the following: (1) If F is a field and A is a simple algebra over F, then there is a unique positive integer n for which there is a division algebra D over F such that A is isomorphic to the algebra of n by n matrices with elements in D. (2) A ring R satisfies the descending chain condition on right ideals and contains no ideals except the zero ideal that consists entirely of nilpotent members if and only if R is the direct sum of a finite number of ideals each of which is isomorphic to the ring of matrices with elements in a division ring.

WEDDLE, Thomas (1817–1853). English analyst and geometer.
   Weddle's rule. An alternative to Simpson's rule for approximating an integral of type $\int_a^b f(x)\,dx$. The interval (a, b) is divided into $6n$ equal parts, and the formula is

$$\frac{b-a}{20n}\left[y_a + 5y_1 + y_2 + 6y_3 + y_4 + 5y_5\right.$$

$$\left. + y_6 + \cdots + 5y_{6n-1} + y_{6n}\right].$$

See NEWTON —Newton's three-eighths rule, SIMPSON—Simpson's rule.

WEDGE, n. elliptic wedge. Suppose a straight line L, a plane P, and an ellipse have the properties that L is not parallel to P and the plane of the ellipse is parallel to L but does not contain L. An **elliptic wedge** is the union of all line segments that are parallel to P and have one end point on L and the other on the ellipse.
   spherical wedge. See SPHERICAL.

WEIERSTRASS, Karl Theodor Wilhelm (1815–1897). Powerful German analyst. First recognized in 1854 when as a secondary-school teacher he published a paper for which he was given an honorary doctorate at the University of Königsberg and a position at the University of Berlin in 1856. He defined analytic functions of a complex variable by means of power series and contributed to notions of continuity and the real number system, real- and complex-variable theory, Abelian and elliptic functions, and the calculus of variations.
   equations of Weierstrass. Integral equations for the coordinate functions of all real minimal surfaces in isothermic representation:

$$x = R \int (1 - u^2)\phi(u)\,du,$$

$$y = R \int i(1 + u^2)\phi(u)\,du,$$

$$z = R \int 2u\phi(u)\,du,$$

where R denotes the real part of any function. E.g., the right helicoid is obtained by setting $\phi(u) = ik/2u^2$, where k is a real constant. The equations of Weierstrass can be obtained from those of Enneper by letting u and v, and ϕ and ψ, be conjugate imaginaries. See ENNEPER — equations of Enneper. The functions x, y, z are harmonic in accordance with a theorem of Weierstrass that a necessary and sufficient condition for a surface given in isothermic representation to be a minimal surface is that the coordinate functions be harmonic. See SURFACE — surface of Enneper, surface of Henneberg, surface of Scherk.
   theorem of Weierstrass. See above, equations of Weierstrass.
   Weierstrass approximation theorem. A continuous function may be approximated over a closed interval by a polynomial, to any assigned degree of accuracy. *Tech.* For every function f continuous on a closed interval $[a, b]$ and any positive number ϵ there exists a polynomial P such that

$$|f(x) - P(x)| < \epsilon$$

for every x in $[a, b]$. See BERNSTEIN —Bernstein polynomials, STONE—Stone-Weierstrass theorem.
   Weierstrass elliptic functions. See ELLIPTIC.
   Weierstrass M-test for uniform convergence: If $|f_1(x)|, |f_2(x)|, |f_3(x)|, \cdots$ are bounded for x on the interval (a, b) by the corresponding terms of the sequence $M_1, M_2, M_3, \cdots$, and ΣM_n converges, then Σf_n converges uniformly on (a, b). E.g., the terms of the sequence $x, x^2, x^3, \cdots$ are bounded for x on the interval $(0, \frac{1}{2})$ by the corresponding terms of $\frac{1}{2}, (\frac{1}{2})^2, (\frac{1}{2})^3, \cdots$, and $\Sigma(\frac{1}{2})^n$ converges. Hence Σx^n converges uniformly on $(0, \frac{1}{2})$.
   Weierstrass necessary condition. A condition that must be satisfied if the function y is to minimize

$$\int_{x_1}^{x_2} f(x, y, y')\,dx,$$

namely, the condition $E(x, y, y', Y') \geq 0$ for all admissible $(x, y, Y') \neq (x, y, y')$, where

$$E = f(x, y, Y') - f(x, y, y')$$

$$-(Y' - y')f_{y'}(x, y, y').$$

Legendre's necessary condition, $f_{y'y'}(x, y, y') \geq 0$, follows from this. See CALCULUS—calculus of variations, EULER—Euler equation, LEGENDRE—Legendre necessary condition.

Weierstrass preparation theorem. Suppose $F(x_1, \cdots, x_n)$ is a *formal power series* in $x_1, \cdots, x_n$ that does not have a constant term and contains a term in x_1 alone, the least degree of such a term being k. Then there is a formal power series E with a constant term and a unique expression

$$G = x_1^k + x_1^{k-1}G_1 + x_1^{k-2}G_2 + \cdots + G_k$$

for which each G_i is a formal power series in $x_2, x_3, \cdots, x_n$ without a constant term and $F = GE$.

WEIGHT, *n.* (1) The gravitational pull on a body. See POUND. (2) See AVERAGE.

apothecaries' weight. The system of weights used by druggists. The pound and the ounce are the same as in troy weight, but the subdivisions of the ounce are different. See DENOMINATE NUMBERS in the appendix.

avoirdupois weight. The system of weights which uses a pound of 16 ounces as its basic unit. See the appendix.

pound of weight. See POUND.

troy weight. The system of weights using a pound consisting of 12 ounces. Used mostly for weighing metals. See DENOMINATE NUMBERS in the appendix.

WEIGHT′ED, *adj.* **weighted mean.** See MEAN.

WEINGARTEN, Johannes Leonard Gottfried Julius (1836–1910). German applied mathematician and differential geometer.

Weingarten surface. A surface such that each of the principal radii is a function of the other. *E.g.*, surfaces of constant total curvature and surfaces of constant mean curvature are Weingarten surfaces. *Syn.* *W*-surface.

WELL, *a.* **well-order property.** See ORDER—order property of real numbers, ORDERED—ordered set.

WESSEL, Caspar (1745–1818). Danish mathematician. One of the first to publish (1798) an account of the graphical representation of complex numbers. See ARGAND—Argand diagram, GAUSS—Gauss plane, WALLIS.

WEYL, Hermann (1885–1955). German mathematician and philosopher. Did basic work on group representations and Riemann surfaces, as well as working in algebra, number theory, quantum mechanics, relativity theory, topology, and the foundations of mathematics. He supported the intuitionism of Brouwer. Spent the last 22 years of his life at the Institute for Advanced Study in Princeton, N.J.

WHIRLWIND. An automatic digital computing machine at the Mass. Inst. of Tech.

WHITEHEAD, Alfred North (1861–1947). English algebraist, analyst, applied mathematician, logician, and philosopher. Made important contributions to the philosophy of mathematics. See RUSSELL.

WHOLE, *adj.* **whole-life insurance.** See INSURANCE—life insurance.

whole number. (1) One of the integers $0, 1, 2, 3, \cdots$. (2) A positive integer; *i.e.*, a natural number. (3) An integer, positive, negative, or zero.

WHYBURN, Gordon Thomas (1904–1969). American analytical topologist. Worked especially on compact, monotone, open, and quotient maps.

WIDTH, *n.* For a convex set in a plane, the greatest lower bound of numbers w such that the set is between two parallel lines separated by distance w. The same definition often is given for convex sets in n-dimensional space, except that "parallel lines" is replaced by "parallel hyperplanes." However, there are other usages: *e.g.*, a box with edges $a < b < c$ is said to have width b and length c, as "a sheet of steel 24 cm long, 4 cm wide, and $\frac{1}{2}$ cm thick."

WIENER, Norbert (1894–1964). American analyst and applied mathematician. Made important contributions to probability and potential theories, Fourier integrals and transforms, and automatic computers and feedback analysis. Founded science of cybernetics. See CYBERNETICS.

Wiener process. A *stochastic process* $\{X(t): t \geq 0\}$ is a **Wiener process** (or **Brownian-motion**

process) if (i) $X(0) = 0$; (ii) for each t, $X(t)$ is a normal random variable with zero mean; (iii) if $a < b \leq c < d$, the random variables $X(b) - X(a)$ and $X(d) - X(c)$ are independent and have the same distribution whenever $b - a = d - c$. Each Wiener process is a *Martingale* and has a constant B such that if $0 \leq t_1 \leq t_2$, then $X(t_2) - X(t_1)$ is *normal* with mean 0 and variance $B(t_2 - t_1)$. Wiener processes have applications to the study of Brownian motion, price movements in stock markets, quantum mechanics, etc.

WILSON, John (1741–1793). English number theorist.

   Wilson's theorem. The number $[(n - 1)! + 1]$ is divisible by n if, and only if, n is a prime. *E.g.*, $4! + 1 = 25$ is divisible by 5, but $5! + 1 = 121$ is not divisible by 6.

WIND'ING, *adj.* **winding number.** The number of times a closed curve in the plane passes around a designated point of the plane in the counterclockwise (positive) direction. *Tech.* Let C be a closed curve in the plane which is the image of a circle for a continuous transformation. This means that C has parametric equations of type $x = u(t)$, $y = v(t)$ for $0 \leq t \leq 1$, where u and v are continuous functions with $u(0) = u(1)$ and $v(0) = v(1)$; or (equivalently) an equation of type $w = f(z)$ for $|z| = 1$, where z and w are complex numbers and f is a continuous function. If P is chosen as a point which is not on C, the numbers $\{t_i\}$ satisfy $t_0 = 0 < t_1 < t_2 < \cdots < t_n = 1$, and Q_i is the point for which the parameter t is equal to t_i ($i = 1, 2, \cdots, n$), then there is a positive number E such that the quantity $\dfrac{1}{2\pi} \sum_1^n \theta_i$ has a value $n(C, P)$ which is independent of the choice of the numbers $\{t_i\}$ provided $(t_i - t_{i-1}) < E$ for each i, where θ_i is the angle (in radian measure) from the line PQ_{i-1} to the line PQ_i for each i. This number $n(C, P)$ is an integer and is the **winding number** of C relative to P, or the **index** of P relative to C. The winding number of a curve is not changed by a continuous deformation of the curve for which the curve does not pass through the point P. *E.g.*, if $p(z)$ is a polynomial of degree n with $p(0) \neq 0$, and C_K is the image of the circle $|z| = K$ for the mapping $w = p(z)$, then for sufficiently large K the winding number of C_K relative to the origin is n, and for sufficiently small K the winding number of C_K relative to the origin is 0. Since these curves can be continuously deformed into each other (by letting K vary continuously), there must be a value of K for which the curve C_K passes through the origin and therefore a value of z for which

$p(z) = 0$. This gives a proof of the *fundamental theorem of algebra*. If P is the complex number a and the curve C is defined by $w = f(z)$, where f is *piecewise differentiable*, then

$$n(C, a) = \frac{1}{2\pi i} \int_C \frac{dz}{z - a}.$$

WITCH, *n.* A plane cubic curve defined by drawing a circle of radius a, tangent to the x-axis at the origin, then drawing a line through the origin and forming a right triangle with its hypotenuse on this line and one leg parallel to the x-axis, the other parallel to the y-axis, and passing, respectively, through

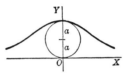

the points of intersection of this line with the circle and the line $y = 2a$. The witch is then the locus of the intersection of the legs of all such triangles. Its equation in rectangular coordinates is $x^2 y = 4a^2(2a - y)$. The *witch* is usually called the **witch of Agnesi**, after Donna Marie Gaetana Agnesi, who discussed the curve. *Syn.* versiera.

WITTEN, Edward (1951–   ). American mathematician awarded a Fields Medal (1990) for his contributions to applying advanced mathematical tools in theoretical physics and to applying physical ideas to discover new and beautiful mathematics.

WORD, *n.* In a digital computing machine, the content of one storage position; an ordered set of digits considered as a unit, regardless of whether they designate a number, an address code, an instruction to the machine, or a combination of these.

WORK, *n.* For a constant force acting on an object moving in the direction of the force, the **work** done on the object by the force is the product of the magnitude of the force and the distance the object moves. *E.g.*, the work necessary to lift a 25 pound bag of flour from the floor to the top of a 3 foot counter is 75 foot-pounds. When the force is constant but at an angle θ with the line along which the object moves, then the work W is $(F \cos \theta)s$, where F is the magnitude of the force, $F \cos \theta$ is the magnitude of the component of the force in the direction of motion, and s is the distance the object moves. More generally, if the object moves along a curve C, then the work is either of the line integrals, $W = \int_C F_t \, ds = \int_C \mathbf{F} \cdot d\mathbf{P}$, where $\mathbf{F}$ is the force (a vector), F_t is the magnitude of the component of the force along the tangent to C in the direction of motion, and $\mathbf{P}$ is the position vector of points on C [see INTEGRAL—line integral]. If C is described by parametric equations using time t as a parameter, then $W = \int_C \mathbf{F} \cdot \mathbf{V} \, dt$, where $\mathbf{F} \cdot \mathbf{V}$ is the scalar product of $\mathbf{F}$ and the velocity vector $\mathbf{V} = d\mathbf{P}/dt$. The total work done on a particle by all forces acting on it is equal to the change in kinetic energy.

WRONSKI (*also* **Höené-Wronski**), **Josef Maria** (1778–1853). Analyst, combinatorialist, philosopher, and physicist. Born in Poland, lived in France.
  Wronskian. The *Wronskian* of n functions $u_1, u_2, \cdots, u_n$ is the determinant of order n which has these functions as the elements of the first row, and their kth derivatives as the elements of the $(k+1)$st row $(k = 1, 2, \cdots, n-1)$. If the Wronskian of n functions is not identically zero, the functions are *linearly independent*, while n functions are *linearly dependent* on an interval (a, b) if their Wronskian is identically zero on (a, b), it being assumed that their first $n-1$ derivatives are continuous and they are solutions of a differential equation of the form

$$p_0 \frac{d^n y}{dx^n} + p_1 \frac{d^{n-1}y}{dx^{n-1}} + \cdots + p_{n-1}\frac{dy}{dx} + p_n y = 0,$$

where the functions p_i are continuous on (a, b) and p_0 is not zero at any point of (a, b). Indeed, the n solutions of this differential equation are linearly dependent and their Wronskian is identically zero on (a, b), if their Wronskian vanishes at a single point of (a, b).

X

X. (1) The letter most commonly used to denote an **unknown number** or **variable**. (2) Used to denote one of the axes in a system of Cartesian coordinates. See CARTESIAN.
  *x***-axis.** See CARTESIAN—Cartesian coordinates.

XI, *n.* The fourteenth letter of the Greek alphabet: lower case, ξ; capital, Ξ.

Y

Y. *y***-axis.** See CARTESIAN —Cartesian coordinates.

YARD, *n.* An English unit of linear measure equal to 3 feet; the distance between two lines on a specially prepared and carefully preserved bar, at a temperature of 62° F. See DENOMINATE NUMBERS in the appendix.

YATES, Frank (1902–). English statistican noted for contributions to experimental design and analysis, theory of sampling surveys, and applications of computers.
  Yates' correction for continuity. The estimate of χ^2 for a 2×2 table needs a correction for small frequencies. The following formula for χ^2 contains a correction which results in a reasonably close approximation to a χ^2 distribution when the expected number of cases in each cell of the 2×2 table is small:

$$\chi^2 = \sum_{i=1}^{4} \frac{\left(|x_i - m_i| - \frac{1}{2}\right)^2}{m_i},$$

where x_i is the observed frequency and m_i is the expected frequency in the ith cell. See CHI—chi-square test.

YAU, Shing-Tung (1949–). American mathematician, born in China. Awarded a Fields Medal (1983) for his work in differential geometry and partial differential equations.

YAW, *adj.* **yaw angle.** In exterior ballistics, the angle between the direction of the axis of a shell and the direction of its velocity vector.

YEAR, *n.* All definitions of year depend on the revolution of the earth about the sun. The **sidereal year** is the time during which the earth makes one complete revolution around the sun with respect to the stars. (Its length in mean solar days is 365 days, 6 hours, 9 minutes, 9.5 seconds). The **tropical year** (also called the **astronomical**, **equinoctial**, **natural**, or **solar year**) is the time required for the earth (or apparently the sun) to pass from the vernal equinox back to the vernal equinox (its length is 365 days, 5 hours, 48 minutes, 46 seconds). Because of the precession of the equinoxes, this is 20 minutes, 23.5 seconds shorter than the sidereal year. The tropical year is the basis of practically all ancient and modern calendars. The **anomalistic year** is the time required for the earth to pass from some point in its elliptic orbit back to the same point again and is 365 days, 6 hours, 13 minutes, 53 seconds in length, differing from the other two because of the fact that the major axis of the earth's orbit is slowly moving at the rate of $11''$ per year. The **civil year** (also called the **calendar** or **legal year**) is 365 days (except on **leap years**, when it is 366 days). A **commercial year** is 360 days, as used in computing simple interest.

YIELD, *n.* The rate per cent which gives a certain profit. *Syn.* rate per cent yield, yield rate, investor's (investment) rate. Thus the **yield of a bond** is the effective rate of interest which a person realizes on an investment in this type of bonds (the **approximate yield** is the average interest per period divided by the average capital invested).

YOUNG, Thomas (1773–1829). English physician, physicist, and Egyptologist. Instrumental in deciphering Egyptian hieroglyphics of the Rosetta Stone.
  Young's modulus. See MODULUS.

YOUNG, William Henry (1863–1942). British analyst. Worked on theory of integration and of orthogonal series.
  Young's inequality. Let the function f be continuous and strictly increasing for $x \geq 0$, with $f(0) = 0$, let g be the inverse function, and let $a \geq 0$ and $b \geq 0$ be numbers in the domains of f and g, respectively. Then Young's inequality is

$$ab \leq \int_0^a f(x)\,dx + \int_0^b g(y)\,dy,$$

the sign of equality holding if and only if $b = f(a)$. The result, which becomes intuitively clear when a figure is drawn showing the three cases $b < f(a)$, $b = f(a)$, and $b > f(a)$, has many applications in the theory of inequalities.

Z

Z. Fisher's *z*. See FISHER.
  ***z*-axis.** See CARTESIAN—Cartesian coordinates.

ZE′NITH, *n.* **zenith distance of a star.** The angular distance from the zenith to the star, measured along the great circle through the zenith, the nadir, and the star. It is the complement of the altitude. See HOUR—hour angle and hour circle.
  zenith of an observer. The point on the celestial sphere directly above the observer; the point where a plumb line, extended upward, would pierce the celestial sphere.

ZENO of Elea (c. 490–c. 435 B.C.). Greek philosopher and mathematician. Propounded paradoxes to point out the problem of the relation of the discrete to the continuous. It is not clear whether or not Zeno took a position regarding the resolution of his paradoxes, but because of them he is usually thought of as the ancient version of Brouwer and Kronecker. See BROUWER, KRONECKER.
  Zeno's paradox of Achilles and the tortoise. A tortoise has a head start on Achilles equal to the distance from a to b and both start running, Achilles after the tortoise. Although Achilles runs faster than the tortoise, he would never catch up with the tortoise, since, while Achilles goes from a to b, the tortoise goes from b to c, and while Achilles goes from b to c, the tortoise goes from c to d, etc., this process never ending. The explanation of the fallacy is that motion is measured by space intervals per unit of time, not by numbers of points. If Achilles takes time $t_1, t_2, t_3, \cdots$ to go from a to b, b to c, c to $d, \cdots$, then Achilles will catch the tortoise in time equal to $\sum_1^\infty t_i$, if this sum is finite. If the tortoise travels 10 feet per second and Achilles 20 feet and the tortoise starts 10 feet in advance, Achilles will catch him at the end of the first second, since then $t_1 = \frac{1}{2}$, $t_2 = \frac{1}{4}, \cdots, t_n = 1/2^n, \cdots$ and $\frac{1}{2} + \frac{1}{4} + \frac{1}{8} + \cdots = 1$. However, if Achilles always runs faster than the tortoise, but the tortoise gradually increases his speed so that $t_1 = 1$, $t_2 = \frac{1}{2}$, $t_3 = \frac{1}{3}, \cdots, t_n = 1/n, \cdots$, then $\sum_1^n t_i$ becomes large beyond all bounds as n increases and Achilles will never catch the tortoise.

ZERMELO, Ernst Friedrich Ferdinand (1871–1953). German analyst and set theorist.
  Zermelo's axiom. See CHOICE—axiom of choice, ZORN—Zorn's lemma.

ZE′RO, *adj., n.* In arithmetic, the *identity element* of addition, *i.e.*, the number 0 for which $x + 0 = x$ and $0 + x = x$ for all numbers x. Zero is also the cardinal number of the empty set. See CARDINAL —cardinal number.

division by zero. See DIVISION.

division of zero. The quotient of zero and any nonzero number is zero; $0/k = 0$ for all k not zero, since $0 = k \times 0$. See DIVISION.

divisor of zero. See DOMAIN—integral domain.

factorial zero. Defined as equal to 1. See FACTORIAL.

multiplication by zero. The product of zero and any other number is zero, *i.e.*, $0 \times k = k \times 0 = 0$ for all k. See MULTIPLICATION —multiplication property of one and zero.

zero in a category. See CATEGORY (2).

zero of a function. A value of the argument for which the function is zero. A **real zero** is a real number for which the function is zero. If the function f has only real number values for real number values of x (*e.g.*, if f is a **polynomial** with real numbers as coefficients), then the real zeros of f are the values of x for which the curve $y = f(x)$ meets the x-axis. See ROOT—root of an equation. If z_0 is a zero (also called **zero point**) of an **analytic function** f of a complex variable, then there is a positive integer k such that $f(z) \equiv (z - z_0)^k \phi(z)$, where ϕ is analytic at z_0 and $\phi(z_0) \neq 0$. The integer k is the **order** of the zero.

zero-sum game. See GAME.

zero vector. A vector of zero length; a vector all of whose components are zero. For vectors of type $\mathbf{V} = a\mathbf{i} + b\mathbf{j} + c\mathbf{k}$, the zero vector is the vector $\mathbf{O} = 0\mathbf{i} + 0\mathbf{j} + 0\mathbf{k}$. The zero vector is the zero element for vector addition, meaning that $\mathbf{V} + \mathbf{O} = \mathbf{O} + \mathbf{V} = \mathbf{V}$ for all $\mathbf{V}$.

ZE′TA, *adj., n.* The sixth letter of the Greek alphabet: lower case, ζ; capital, Z.

Riemann zeta function. The zeta function ζ of complex numbers $z = x + iy$ is defined for $x > 1$ by the series

$$\zeta(z) = \sum_{n=1}^{\infty} n^{-z} = \sum_{n=1}^{\infty} e^{-z \log n},$$

where $\log n$ is real. The function can be defined by analytic continuation for all finite z. It is a meromorphic function having a simple pole at $z = 1$. See RIEMANN —Riemann hypothesis.

ZON′AL, *adj.* **zonal harmonic.** See HARMONIC —zonal harmonic.

ZONE, *n.* A portion of a sphere bounded by the two intersections of two parallel planes with the sphere. One of the planes may be a tangent plane in which case one of the circular intersections is a point and the zone is said to be a **zone of one base.** A **base** of a zone is an intersection with the sphere of one of the planes forming the zone. The **altitude** of a zone is the perpendicular distance between the planes cutting the zone out of the sphere. The **area** of a zone is equal to the product of its altitude and the perimeter of a great circle of the sphere, *i.e.*, $2\pi rh$, where r is the radius of the sphere and h the altitude of the zone.

zone of a surface of revolution. The portion of the surface contained between two planes normal to the axis of revolution.

ZORN, Max August (1906–   ). German-American algebraist, group theorist and analyst.

Zorn's lemma. The maximal principal: If T is *partially ordered* and each *linearly ordered* subset has an upper bound in T, then T contains at least one *maximal element* (an element x such that there is no y of T with $x < y$). Other alternative forms of this principle are: (1) **(Kuratowski's lemma)** Each simply ordered subset of a partially ordered set is contained in a maximal linearly ordered subset. (2) If a collection A of sets has the property that for each *nest* in A there is a member of A which contains each member of the nest, then there is a maximal member of A. (3) **(Hausdorff maximal principle)** If A is a collection of sets and N is a nest in A, then there is a nest N^* that contains N and is not contained in any larger nest. (4) **(Tukey's lemma)** A collection of sets which is of *finite character* has a *maximal member*. (5) Any set can be well ordered (see ORDERED —ordered set). (6) The axiom of choice (see CHOICE). If the finite axiom of choice is assumed, all of the above principles are logically equivalent. See FILTER.

DENOMINATE NUMBERS

LENGTH

1   point (printer's) = 0.013837 inches
1   inch (in.) = 2.54 centimeters
1   hand = 4 inches
1   palm = 3 inches (sometimes 4 in.)
1   span = 9 inches
12   in. = 1 foot (ft)
1   (military) pace = $2\frac{1}{2}$ ft
3   ft = 1 yard (yd)
$16\frac{1}{2}$ ft = $5\frac{1}{2}$ yd = 1 rod (rd)
40   rods = 1 furlong = 666 ft
1   mile (*statute* or *land*) = 5280 ft
     = 320 rd = 1.609344 kilometers
3   miles = 1 league
terameter (Tm) = 10^{12} m
gigameter (Gm) = 10^{9} m
megameter (Mm) = 10^{6} m
myriameter (mym) = 10^{4} m
kilometer (km) = 10^{3} m = 0.62137+ miles
hectometer (hm) = 100 m
decameter (or dekameter) (dam) = 10 m
meter (m) = 39.3701− in. = 1.09361+ yd
decimeter (dm) = $\frac{1}{10}$ m
centimeter (cm) = 10^{-2} m
millimeter (mm) = 10^{-3} m
micrometer (or micron) (μm) = 10^{-6} m
nanometer (nm) = 10^{-9} m
picometer (pm) = 10^{-12} m
femtometer (fm) = 10^{-15} m
attometer (am) = 10^{-18} m
astronomical unit = approx. 92,955,807
    miles = mean distance between the sun
    and the earth.
light year = approx. $5.8785 \cdot 10^{12}$ miles = the
    distance traveled by light in 1 year.
1 parsec = the distance of a star at which the
    angle subtended by the radius of the
    earth's orbit is 1″ (about 3.3 light years).

SURVEYORS' MEASURE

7.92 inches = 1 link (li)
25   links = 1 rod (rd)
4   rods = 1 (Gunter's) chain = 66 feet
100   links = 1 (Gunter's) chain
80   chains = 1 mile
    (An engineers' chain, or measuring tape,
     is usually 100 feet long.)
625   square links = 1 square rod
16   square rods = 1 square chain
10   square chains = 1 acre
36   square miles = 1 township

MARINERS' MEASURE

6   feet = 1 fathom
120   fathoms = 1 cable lgth. (or cable)
$7\frac{1}{2}$ cable lgths. = 1 nautical mile
     = 6076.11549 feet
     = 1.852 kilometers
3   nautical miles = 1 marine league

CIRCULAR MEASURE

60   seconds (″) = 1 minute (′)
60   minutes = 1 degree (°)
360   degrees = 1 circumference
1   radian = 57° 17′ 44.806″
1   degree = 0.01745329 radians
1   mil = 1/6400 of a circumference
69   miles = 1 degree of latitude, approximately.

AREA

144   square inches = 1 square foot
9   square feet = 1 square yard
$30\frac{1}{4}$ square yards = 1 square rod
40   sq. rods = 1 rood
43,560   square ft = 160 square rds. = 1 acre
640   acres = 1 square mile
100   square meters = 1 are
     = 119.599 sq. yds.
100   ares = 1 hectare = 2.471 acres
100   hectares = 1 square kilometer

VOLUME

1728   cubic inches = 1 cubic foot
27   cubic feet = 1 cubic yard
$24\frac{3}{4}$ cubic feet = 1 perch
1   teaspoonful = $\frac{1}{4}$ tablespoonful
           (medical)
        = $\frac{1}{3}$ tablespoonful
           (cooking)
1   tablespoonful = $\frac{1}{2}$ fluid ounce
4   gills = 1 pint (liquid measure)
2   pints = 1 quart (liquid measure)
4   quarts = 1 gallon
1   gallon = 231 cubic inches
1   cubic foot = 7.48 gallons
$31\frac{1}{2}$ gallons = 1 barrel
63   gallons = 1 hogshead
2   pints = 1 quart (dry measure)
8   quarts (dry measure) = 1 peck
     = 537.605 cubic inches
4   pecks = 1 bushel
In the U.S. one *bushel* contains 2150.42 cubic inches;
    in Great Britain, 2218.2. In the U.S. one *heaped*
    bushel is 2,747.715 cubic inches.
kiloliter (stere) = 1,000 liters
hectoliter = 100 liters
decaliter = 10 liters
liter = 1 cu. dm = 1.0567− quarts
deciliter = 1/10 liter
centiliter = 1/100 liter
millileter = 1/1000 liter
(One *liter* equals 0.908 qt *dry measure* and 1.0567−
    qt *liquid measure*.)

APOTHECARIES' FLUID MEASURE

60   minims = 1 fluid dram = 0.2256 cu. in.
8   fluid drams = 1 fluid ounce
16   fluid ounces = 1 pint
8   pints = 1 gallon = 231 cu. in.

AVOIRDUPOIS WEIGHT

(used in weighing all articles except drugs, gold, silver, and precious stones).

1	grain = 64.800 milligrams
$27\frac{11}{32}$	grains (gr) = 1 dram (dr)
16	drams = 1 ounce (oz) = 28.3495 grams
16	ounces = 1 pound (lb) = 453.52+ grams
14	pounds = 1 stone (English)
25	pounds = 1 quarter
4	quarters = 1 hundredweight (cwt)
20	hundredweight = 1 ton (T)
2000	pounds = 1 ton (*net ton* or *short ton*)
2240	pounds = 1 long ton, or gross ton
2352	pounds = 1 Cornish mining ton

TROY WEIGHT

1	grain = 64.800− milligrams
24	grains = 1 pennyweight
20	pennyweight = 1 ounce
12	ounces = 1 pound = 373.24+ grams
1	pound = 5760 grains
1	carat = 3.168 grains

APOTHECARIES' WEIGHT

1	grain = 64.800− milligrams
20	grains = 1 scruple
3	scruples = 1 dram
8	drams = 1 ounce
12	ounces = 1 pound = 373.24+ grams

The pound, ounce and grain have the same weight as those of Troy Weight.

METRIC WEIGHT

mil′lier′ = 1,000,000 gr = 2,204.623 lb
    = 1,000 kg = 1 metric ton
quin′tal = 100,000 gr = 220.46 lb
myr′i-a-gram = 10,000 gr = 22.046 lb
kil′o-gram = 1,000 gr = 2.205 lb
hec′to-gram = 100 gr = 3.527 oz
dec′a-gram = 10 gr = 0.353 oz
gram = 1 gr = 15.432 grains = 0.035 oz
dec′i-gram = 1/10 gr = 1.543 grains
cen′ti-gram = 1/100 gr = 0.154 grains
mil′li-gram = 1/1000 gr = 0.015 grains

BOARD AND WOOD MEASURE

16 cubic feet = 1 cord foot
8 cord feet or 128 cubic feet = 1 cord
1 ft B. M. = 1 piece 1 ft square and 1 inch, or less, thick. For pieces more than 1 inch thick, the B. M. is the product of the area in feet by the nominal thickness in inches.

PAPER MEASURE

24	sheets = 1 quire
20	sheets = 1 quire of outsides
25	sheets = 1 printer's quire
20	quires = 1 ream
$21\frac{1}{2}$	quires = 1 printer's ream
2	reams = 1 bundle
4	reams = 1 printers' bundle
10	reams = 1 bale
60	skins = 1 roll of parchment
480	sheets = 1 short ream
500	sheets = 1 long ream

TIME

60	seconds = 1 minute
60	minutes = 1 hour
24	hours = 1 day
7	days = 1 week
365	days = 1 common year
366	days = 1 leap year
12	months = 1 year
360	days = 1 commercial year
1	sidereal year = 365.256+ days (mean solar)
1	tropical (equinoctial) year = 365.242+ days
100	years = 1 century
10	years = 1 decade

UNITED STATES MONEY

10	mills (m.) = 1 cent (ct. or ¢)
10	cents = 1 dime
10	dimes = 1 dollar ($)
10	dollars = 1 eagle
5	cents = 1 nickel
5	nickels = 1 quarter
2	quarters = 1 half dollar

ENGLISH MONEY

(coins in the left column are in use)

1	halfpenny = $\frac{1}{2}$ penny
1	penny = $\frac{1}{12}$ shilling
1	shilling = 12 pence
1	florin = 2 shillings
$\frac{1}{2}$	crown = $2\frac{1}{2}$ shillings
1	crown = 5 shillings
1	pound (sterling) = 20 shillings

MISCELLANEOUS

12 units = 1 dozen
12 dozen = 1 gross
12 gross = 1 great gross
20 units = 1 score
1 cu ft water = 62.425 lb (max. density)
1 gallon water = 8.337 lb
1 cu ft air = 0.0807 lb (32° F)
g (acceleration of gravity) = 32.174 feet per sec per sec
g (legal) = 980.665 cm/sec^2
1 horse power (mech) = 550 ft-lb/sec
    = 745.70 watts
1 horse power (elec) = 746.00 watts
1 BTU/min = 17.580 watts (abs)
    = 778.0 ft-lb/min
1 kw-hr = 3413.0 BTU
1 watt-sec = 1 (abs) joule = 1 newton-meter
    = $1 \cdot 10^7$ ergs
1 joule (international) = 1.000165 joules (abs)
vel. of sound in air = 1,088 ft/sec
vel. of sound in water (20° C) = 4,823 ft/sec
speed of light = 186,282.397 miles/sec
e = Naperian Base = 2.71828 18284 59045+
π = 3.14159 26535 89793 23846 2643+
M = $\log_{10}$ e = 0.43429 44819 03252−
1/M = 2.30258 50929 94045 68402−

MATHEMATICAL SYMBOLS

Arithmetic, Algebra, Number Theory

$+$ Plus; positive.

$-$ Minus; negative.

$\pm$ Plus or minus; positive or negative.

$\mp$ Minus or plus; negative or positive.

ab, $a \cdot b$, $a \times b$ a times b; a multiplied by b.

a/b, $a \div b$, $a:b$ a divided by b; the ratio of a to b.

$a/b = c/d$ or $a:b::c:d$ A proportion: a is to b as c is to d (the second form is seldom used).

$=$, $::$ Equals (the symbol $::$ is practically obsolete).

$\equiv$ Is identically equal to; is identical with.

$\neq$ Does not equal.

$\cong$ or $\equiv$ Congruent; approximately equal (not common).

$\sim$ or $\backsim$ Equivalent; similar.

$>$ Is greater than.

$<$ Is less than.

$\geqq$ or $\geq$ Is greater than or equal to.

$\leqq$ or $\leq$ Is less than or equal to.

a^n $aaa \cdots$ to n factors.

$\sqrt{a}$, $a^{1/2}$ The positive square root of a, for positive a.

$\sqrt[n]{a}$, $a^{1/n}$ The nth root of a, usually the *principal* nth root.

a^0 The number 1 (if $a \neq 0$).

a^{-n} The reciprocal of a^n; $1/a^n$.

$a^{m/n}$ The nth root of a^m.

$(\ )$ Parentheses.

$[\ ]$ Brackets.

$\{\ \}$ Braces.

$\overline{}$ Vinculum (used as a symbol of aggregation).

$a \propto b$ a varies directly as b; a is directly proportional to b (seldom used).

i (or j) Square root of -1; $\sqrt{-1}$; j is used in physics, where i denotes current, but i is almost universally used in mathematics.

$\omega_1, \omega_2, \omega_3$ or $1, \omega, \omega^2$ The three cube roots of unity.

a' a prime.

a'' a double prime; a second.

$a^{[n]}$ a with n primes.

a_n a sub n, a subscript n.

xRy x is in the relation R to y.

$f(x)$, $F(x)$, $\phi(x)$, etc. The value at x of the function f, F, ϕ, etc.

$f^{-1}(a)$ If f has an inverse function, the value of the inverse at a; otherwise, the set of x such that $f(x) = a$.

$|z|$ Absolute value of z; numerical value of z; modulus of z.

$\bar{z}$ or $\mathbf{conj}\, z$ Conjugate of z.

$\mathbf{arg}\, z$ Argument, amplitude, or phase of z.

$\mathbf{R}(z)$, $\Re(z)$, $\mathrm{Re}(z)$ Real part of z; $\mathbf{R}(z) = x$, if $z = x + iy$ and x and y are real.

$\mathrm{I}(z)$, $\Im(z)$, $\mathrm{Im}(z)$ Imaginary part of z; $\mathrm{I}(z) = y$, if $z = x + iy$ and x and y are real.

$n!$ (or $\underline{|n}$) Factorial n; n factorial: $0! = 1, n! = 1 \cdot 2 \cdot 3 \cdots n$ if $n \geq 1$.

$P(n, r)$, $_nP_r$ The number of permutations of n things taken r at a time; $n!/(n-r)! = n(n-1)(n-2) \cdots (n-r+1)$.

$\binom{n}{r}$, $_nC_r$, C_r^n, or $C(n, r)$. The number of combinations of n things taken r at a time; $n!/[r!(n-r)!]$; the $(r+1)$st binomial coefficient.

$|a_{ij}|$ The determinant whose element in the ith row and jth column is a_{ij}.

$\|a_{ij}\|$ or (a_{ij}) The matrix whose element in the ith row and the jth column is a_{ij}.

$|abc \cdots|$ The determinant $\begin{vmatrix} a_1 & a_2 & \cdots \\ b_1 & b_2 & \cdots \\ \cdots & \cdots & \cdots \end{vmatrix}$

$\|abc \cdots\|$ or $(abc \cdots)$ The matrix

$$\begin{Vmatrix} a_1 & a_2 & \cdots \\ b_1 & b_2 & \cdots \\ \cdots & \cdots & \cdots \end{Vmatrix} \quad \text{or} \quad \begin{pmatrix} a_1 & a_2 & \cdots \\ b_1 & b_2 & \cdots \\ \cdots & \cdots & \cdots \end{pmatrix}$$

$\begin{pmatrix} a & b & c & \cdots \\ b & c & d & \cdots \end{pmatrix}$ or $(abcd \cdots)$ The permutation which replaces a by b, b by c, c by d, etc.

$\mathbf{adj}\, A$, $[A_{ji}]$, (A) Adjoint of the matrix $A = [a_{ij}]$.

$\bar{A}$ Complex conjugate of the matrix A.

I An identity matrix.

A^{-1} Inverse of the matrix A.

A^* Hermitian conjugate of the matrix A.

A', A^T Transpose of the matrix A.

A_{ij} Cofactor of the element a_{ij} in the matrix $[a_{ij}]$.

$\|A\|$ The norm of the matrix A.

$\oplus$ An operation in a postulational algebraic system, with $x \oplus y$ called the sum of x and y.

$\otimes$ An operation in a postulational algebraic system, with $x \otimes y$ called the product of x and y.

$\circ$, $*$ An operation in a postulational algebraic system, with $x \circ y$, or $x * y$, an element of the system.

G.C.D. or g.c.d. Greatest common divisor.

L.C.D. or l.c.d. Least common denominator.

L.C.M. or l.c.m. Least common multiple.

(a, b) The G.C.D. of a and b; the open interval from a to b.

$[a, b]$ The L.C.M. of a and b; the closed interval from a to b.

$a|b$ a divides b.

$a \nmid b$ a does not divide b.

$x \equiv a \pmod{p}$ $x - a$ is divisible by p, read: x is congruent to a modulus p, or modulo p.

$[x]$ The greatest integer not greater than x.

$\phi(n)$ Euler's ϕ-function of n (the number of positive integers prime to n and not greater than n).

$p(n)$ The number of partitions of n.

$\tau(n)$, $d(n)$ The number of positive divisors of n.

$\sigma(n)$ Sum of positive divisors of n; if n is a perfect number, $\sigma(n) = 2n$.

$\sigma_k(n)$ Sum of the kth powers of the positive divisors of n.

$\omega(n)$, $\nu(n)$ The number of different primes which divide n.

$\Omega(n)$ The number of prime factors of n. E.g., $\Omega(12) = 3$, $\Omega(32) = 5$.

$\pi(n)$ The number of primes which are not greater than n.

$\lambda(n)$ Liouville function.

$\mu(n)$ Möbius function.

F_n Fermat number.

M_p Mersenne number.

Trigonometry and Hyperbolic Functions

$a°$ a degrees (angle).

a' a minutes (angle).

a'' a seconds (angle).

$a^{(r)}$ a radians, unusual.

s One-half the sum of the lengths of the sides of a triangle (plane or spherical).

S, σ One-half the sum of the angles of a spherical triangle.

E Spherical excess.

$s.a.s.$ Side, angle, side.

$s.s.s.$ Side, side, side.

sin   Sine.

cos   Cosine.

tan   Tangent.

ctn   (or **cot**) Cotangent.

sec   Secant.

csc   Cosecant (or **cosec**).

covers   Coversed sine or coversine.

exsec   Exsecant.

gd (or **amh**)   Gudermannian (or hyperbolic amplitude).

hav   Haversine.

vers   Versed sine or versine.

$\sin^{-1} x$ (or **arc sin** x)   The principal value of the angle whose sine is x (when x is real); antisine x; inverse sine x.

$\sin^2 x, \cos^2 x$, etc. $(\sin x)^2, (\cos x)^2$, etc.

sinh   Hyperbolic sine.

cosh   Hyperbolic cosine.

tanh   Hyperbolic tangent.

ctnh (or **coth**)   Hyperbolic cotangent.

sech   Hyperbolic secant.

csch   Hyperbolic cosecant.

$\sinh^{-1} x$ (or **arc sinh** x)   The number whose hyperbolic sine is x; antihyperbolic sine of x; inverse hyperbolic sine of x.

Geometry

$\angle$ Angle.

$\angle\!s$ Angles.

$\perp$ Perpendicular; is perpendicular to.

$\perp\!s$ Perpendiculars.

$\parallel$ Parallel; is parallel to.

$\parallel\!s$ Parallels.

$\nparallel$ Not parallel.

$\cong, \equiv$ Congruent; is congruent to.

$\sim$ Is similar to.

$\therefore$ Therefore; hence.

$\triangle$ Triangle.

$\triangle\!\!\!\!s$ Triangles.

$\diagup\!\!\!\!\square$ Parallelogram.

$\square$ Square.

$\bigcirc$ Circle.

ⓢ   Circles.

π The ratio of the circumference of a circle to the diameter, the Greek letter pi, equal to 3.1415926536-.

O Origin of a coordinate system.

(x, y) Rectangular coordinates of a point in a plane.

(x, y, z) Rectangular coordinates of a point in space.

(r, θ) Polar coordinates.

χ The angle from the radius vector to the tangent to a curve.

(ρ, θ, ϕ) or (r, θ, ϕ) Spherical coordinates of a point in space.

(r, θ, z) Cylindrical coordinates.

$\cos \alpha, \cos \beta, \cos \gamma$ Direction cosines.

l, m, n Direction numbers.

e Eccentricity of a conic.

p Half of the latus rectum of a parabola (usage general in U.S.).

m Slope.

$\overline{AB}$ or AB The line segment between A and B.

$\overrightarrow{AB}$ The directed line segment from A to B; the ray from A through B.

$\overarc{AB}$ The arc between A and B.

$P(x, y)$ or $P{:}(x, y)$ Point P with coordinates x and y in the plane.

$P(x, y, z)$ or $P{:}(x, y, z)$ Point P with coordinates x, y, z in space.

(AB, CD) or $(AB|CD)$ The cross ratio of the elements (points, lines, etc.) A, B, C, and D, the quotient of the ratio in which C divides AB by the ratio in which D divides AB.

$[A]\overline{\wedge}[B]$ Indicates that there is a perspective correspondence between the ranges $[A]$ and $[B]$.

$[A]\wedge[B]$ Indicates a projective correspondence between ranges $[A]$ and $[B]$.

i, j, k Unit vectors along the coordinate axes.

$a \cdot b, (a, b), \mathbf{S}ab, (ab)$ Scalar product, or dot product, of the vectors a and b.

$a \times b, \mathbf{V}ab, [ab]$ Vector product, or cross product, of the vectors a and b.

$[abc]$ The *scalar triple product* of the vectors a, b and c: $(a \times b) \cdot c$, $a \cdot (b \times c)$, or $b \cdot (c \times a)$.

Calculus and Analysis

(a, b) The open interval $a < x < b$.

$[a, b]$ The closed interval $a \le x \le b$.

$(a, b]$ The interval $a < x \le b$.

$[a, b)$ The interval $a \le x < b$.

$\{a_n\}, [a_n], (a_n)$ The sequence whose terms are $a_1, a_2, \cdots, a_n, \cdots$.

$\displaystyle\sum_1^n$ or $\displaystyle\sum_{i=1}^n$ Sum to n terms, one for each positive integer from 1 to n.

$\displaystyle\sum$ Sum of certain terms, the terms being indicated by the context or as in $\displaystyle\sum_{i=1}^n X_i$ or $\displaystyle\sum_{a \in A} X_a$.

$\sum_{i=1}^{\infty} X_i$ The infinite series $x_1 + x_2 + \cdots$; the sum of this series.

$\prod_1^n$ or $\prod_{i=1}^n$ Product of n terms, one for each positive integer from 1 to n.

$\prod$ Product of certain terms, the terms being indicated by the context or by added notation, as in $\prod_{i=1}^n X_i$ or $\prod_{a \in A} X_a$.

$\prod_{i=1}^{\infty} x_i$ The infinite product $x_1 x_2 x_3 \cdots$; $\lim_{n \to \infty} \prod_{i=1}^n x_i$.

I Moment of inertia.

k Radius of gyration.

$\bar{x}, \bar{y}, \bar{z}$ Coordinates of the center of mass.

s (or σ)   Length of arc.

ρ Radius of curvature.

κ Curvature of a curve.

τ Torsion of a curve.

l.u.b. or **sup**   Least upper bound; supremum.

g.l.b. or **inf**   Greatest lower bound; infimum.

$\lim_{x \to a} y = b$, or $\lim_{x = a} y = b$ The limit of y as x approaches a is b.

$\overline{\lim}_{n \to \infty} t_n$ The greatest of the accumulation points of the sequence (t_n); limit superior of (t_n).

$\underline{\lim}_{n \to \infty} t_n$ The least of the accumulation points of the sequence (t_n); limit inferior of (t_n).

$\to$ Approaches, or implies.

lim sup or $\overline{\lim}$ Limit superior.

lim inf or $\underline{\lim}$ Limit inferior.

e The base of the system of natural logarithms;

$$\lim_{n \to \infty} (1 + 1/n)^n = 2.7182818285\text{-}.$$

$\log_a x$ Logarithm (base a) of x.

$\log a$, $\log_{10} a$ Common (Briggsian) logarithm of a: $\log a$ is used for $\log_{10} a$ when the context shows that the base is 10.

ln a, $\log a$, $\log_e a$ Natural logarithm of a.

antilog   Antilogarithm.

colog   Cologarithm.

exp x e^x, where e is the base of the natural system of logarithms $(2.718 \cdots)$.

xRy x is in the relation R to y.

$f(x)$, $F(x)$, $\phi(x)$, etc.   The value at x of the function f, F, ϕ, etc.

$f^n(x)$ The result of n applications of f on x; $f^2(x) = f[f(x)]$.

$f^{-1}(a)$ If f has an inverse function, the value of the inverse at a; otherwise, the set of x such that $f(x) = a$.

$|z|$ Absolute value of z; numerical value of z; modulus of z.

$\bar{z}$ or **conj** z Conjugate of z.

arg z Argument, amplitude, or phase of z.

$R(z)$, $\Re(z)$, $Re(z)$ Real part of z; $R(z) = x$, if $z = x + iy$ and x and y are real.

$I(z)$, $\Im(z)$, $Im(z)$ Imaginary part of z; $I(z) = y$, if $z = x + iy$ and x and y are real.

$f(a + 0)$, $f(a +)$, $\lim_{x \downarrow a} f(x)$, or $\lim_{x \to a+} f(x)$ The limit on the right of f at a.

$f(a - 0)$, $f(a -)$, $\lim_{x \uparrow a} f(x)$, or $\lim_{x \to a-} f(x)$ The limit on the left of f at a.

$f'(a +)$ The derivative on the right of f at the number a.

$f'(a -)$ The derivative on the left of f at the number a.

Δy An increment of y.

∂y A variation in y; an increment of y.

dy Differential of y.

$\dot{s}$, ds/dt, v The derivative of s with respect to t; speed.

$\ddot{s}$, dv/dt, d^2s/dt^2, a The second derivative of s with respect to the time t; acceleration.

ω, α Angular speed and angular acceleration, respectively.

$\dfrac{dy}{dx}$, $\dfrac{df(x)}{dx}$, y', $f'(x)$, $D_x y$ The derivative of y with respect to x, where $y = f(x)$.

D The operator $\dfrac{d}{dx}$.

$\dfrac{d^n y}{dx^n}$, $y^{(n)}$, $f^{(n)}(x)$, $D_x^n y$ The nth derivative of y, with respect to x, where $y = f(x)$.

$\dfrac{\partial u}{\partial x}$, u_x, $f_x(x, y)$, $f_1(x, y)$, $D_x u$ The partial derivative of $u = f(x, y)$ with respect to x.

$\dfrac{\partial^2 u}{\partial y \, \partial x}$, u_{xy}, $f_{xy}(x, y)$, $f_{12}(x, y)$, $D_y(D_x u)$ The second partial derivative of $u = f(x, y)$, taken first with respect to x, and then with respect to y.

D_i, D_{ij}, etc.   Partial differentiation operators

$$\left(e.g., \; D_{ij} = \frac{\partial^2}{\partial x_i \, \partial x_j} \right).$$

$D_s f$ Directional derivative of f in the direction s.

E The operator defined by $Ef(x) = f(x + h)$, for a specified constant h.

Δ The operator defined by $\Delta f(x) = f(x + h) - f(x)$, for a specified constant h (also see below, ∇^2 or Δ).

∇ Del (or nabla).

$$\left(\mathbf{i} \frac{\partial}{\partial x} + \mathbf{j} \frac{\partial}{\partial y} + k \frac{\partial}{\partial z} \right).$$

∇u or **grad** u Gradient of u:

$$\left(\mathbf{i} \frac{\partial u}{\partial x} + \mathbf{j} \frac{\partial u}{\partial y} + k \frac{\partial u}{\partial z} \right).$$

$\nabla \cdot \mathbf{v}$ or **div** $\mathbf{v}$ Divergence of $\mathbf{v}$.

$\nabla \times \mathbf{F}$ Curl (or rotation) of $\mathbf{F}$.

∇^2 or Δ The Laplacian operator:

$$\frac{\partial^2}{\partial x^2} + \frac{\partial^2}{\partial y^2} + \frac{\partial^2}{\partial z^2}.$$

δ_j^i Kronecker delta.

$\delta_{j_1, j_2, \cdots, j_k}^{i_1, i_2, \cdots, i_k}$ The generalized Kronecker delta.

$\epsilon^{i_1,i_2,\cdots,i_k}$, $\epsilon_{i_1,i_2,\cdots,i_k}$ The epsilon symbols.

g_{ij}, g^{ji} The components of the fundamental metric tensor of a *Riemannian space*,

$$ds^2 = g_{ij}\, dx^i\, dx^j = g^{ij}\, dx_i\, dx_j.$$

E, F, G The coefficients in the first fundamental quadratic form of a surface.

$F(x)]_a^b$ $F(b) - F(a)$.

$\int f(x)\, dx$ The indefinite integral or antiderivative of f with respect to x.

$\int_a^b f(x)\, dx$ The definite integral of f between the limits a and b.

$\overline{\int}_a^b$ The upper Darboux integral.

$\underline{\int}_a^b$ The lower Darboux integral.

$m_e(S)$, $m^*(S)$, $\mu^*(S)$ Exterior measure of S.
$m_i(S)$, $m_*(S)$, $\mu_*(S)$ Interior measure of S.
$m(S)$, $\mu(S)$ Measure of S.
a.e.   Almost everywhere; except for a set of measure zero.

G_δ set; F_σ set   See BOREL—Borel set.

BV Of bounded variation.

$T_f(I)$, $V_f(I)$, or $V(f, I)$ Total variation of f on the interval I.

$\Omega_f(I)$, $\omega_f(I)$, or $o_f(I)$ Oscillation of f on I.

$\omega_f(x)$, $o_f(x)$ Oscillation of f at the point x.

(f, g) Inner product of the functions f and g.

$\|f\|$ Norm of the function f; i.e., $(f, f)^{1/2}$.

$f * g$ Convolution of f and g.

$W(u_1, u_2, \cdots, u_n)$ Wronskian of $u_1, u_2, \cdots, u_n$.

$\dfrac{\partial(f_1, f_2, \cdots, f_n)}{\partial(x_1, x_2, \cdots, x_n)}$, $\dfrac{D(f_1, f_2, \cdots, f_n)}{D(x_1, x_2, \cdots, x_n)}$, or

$J\left(\dfrac{f_1, f_2, \cdots, f_n}{x_1, x_2, \cdots, x_n}\right)$ Jacobian of the functions

$f_i(x_1, x_2, \cdots, x_n)$.

C_n, $C^{(n)}$ See FUNCTION—function of class C_n.
L_p, $L^{(p)}$ See FUNCTION—function of class L_p.

$f(x) \sim \sum\limits_0^\infty A_n$ The series is an asymptotic expansion of the function $f(x)$.

$x_n \sim y_n$ Limit $x_n/y_n = 1$; x_n and y_n are *asymptotically equal*.

$u_n = O(v_n)$ u_n is *of the order of* v_n (u_n/v_n is bounded).

$u_n = o(v_n)$ $\lim u_n/v_n = 0$.

summable C_k, or (Ck) Summable by Cesàro's method of summation of order k.

γ Euler's constant.

$B_1, B_2, B_3, \cdots$ The Bernoulli numbers. The Bernoulli numbers are also sometimes $B_1, B_3, B_5, \cdots$.

$\operatorname*{Res}\limits_{z=a} f(z)$ Residue of f at a.

$\Gamma(z)$ The Gamma function.

$(\gamma a, x)$; $\Gamma(a, x)$ Incomplete gamma functions.

$B_n(x)$ The Bernoulli polynomial of degree n.

$F(a, b; c; z)$ A hypergeometric function.

$H_n(x)$ The Hermite polynomial of degree n.

$J_n(p, q; x)$ A Jacobi polynomial.

$J_n(x)$ The nth Bessel function.

$I_n(z)$; $K_n(z)$ Modified Bessel functions.

$H_n^{(1)}(z)$; $H_n^{(2)}(z)$ Hankel functions.

$N_p(z)$; $Y_n(z)$ Neumann functions.

$\beta(m, n)$, $B(m, n)$ The beta function.

$B_x(m, n)$ Incomplete beta function.

$ber(z)$, $bei(z)$, $ker(z)$, $kei(z)$ See BER.

$J(\tau)$; $\lambda(\tau)$, $f(\tau)$, $g(\tau)$, $h(\tau)$ Modular functions.

$\vartheta_1(z)$, etc.   Theta functions.

$\vartheta_1, \vartheta_2, \cdots$; $\vartheta_1', \cdots$ Theta functions and their derivatives with zero argument.

$\zeta(z)$ Riemann zeta function.

$Erf(x)$ $\int_0^x e^{-t^2}\, dt = \frac{1}{2}\gamma(\frac{1}{2}, x^2)$; see ERROR—error function.

$Erfc(x)$ $\int_x^\infty e^{-t^2}\, dt = \frac{1}{2}\pi^{1/2} - Erf(x) = \frac{1}{2}\Gamma(\frac{1}{2}, x^2)$.

$Erfi(x)$ $\int_0^x e^{t^2}\, dt = -i\, Erf(ix)$.

$L_n(x)$ The Laguerre polynomial of degree n.

$L_n^k(x)$ An associated Laguerre polynomial.

$P_n(x)$ The Legendre polynomial of degree n.

$P_n^m(x)$ An associated Legendre function.

$T_n(x)$ The Tchebycheff polynomial of degree n.

$ce_n(x)$, $se_n(x)$ Mathieu functions.

$sn\, z, cn\, z, dn\, z$ The Jacobian elliptic functions.

$p(z)$, $p'(z)$ The Weierstrass elliptic functions.

Logic and Set Theory

$\therefore$ Therefore.

$\ni$ Such that.

$\sim p$, $-p$, $\bar{p}$, p' Not p.

$p \wedge q$, $p \cdot q$, $p \& q$ Both p and q; p and q.

$p \vee q$ At least one of p and q; p or q.

$p|q$, p/q Not both p and q; not p or not q.

$p \downarrow q$, $p \vartriangle q$ Neither p nor q.

$p \to q$, $p \Rightarrow q$, $p \subset q$ If p, then q; p only if q.

$\leftrightarrow$, $\Leftrightarrow$, $\equiv$, $\sim$, **iff**   If and only if.

$\vee, I, 1$ The universal class (containing all the members of some specific class, such as the set of all real numbers).

$\varnothing$, $\wedge$, Λ, 0 The null class; the class containing no members.

., :, :., etc.   Dots used in place of parentheses, n dots being stronger than $n - 1$ dots, and n dots with a sign such as $\vee$, $\to$, or $\leftrightarrow$ being equivalent to $n + 1$ dots.

xRy x has the relation R to y; (x, y) belongs to the equivalence class R.

G/H Quotient space. See QUOTIENT—quotient space or factor space.

(x), Π_x, A_x, $\forall_x$ For all x.

$A_{x,y,\cdots}$; $\forall_{x,y,\cdots}$ For all $x, y, \cdots$

$\exists$ There exists.

$(\exists x)$, (Ex), Σ_x There is an x such that.

$E_{x,y,\cdots}$ There exist $x, y, \cdots$ such that.

E_x, $\hat{x}$, $\mathbf{C}_x$, $[x|]$, $[x:]$ The class of all objects x which satisfy the condition stated after the symbol (or after the vertical bar or colon of the last symbols).

$x \in M$ The point x belongs to the set M.

$x \notin M$ The point x does not belong to the set M.

$M = N$ The sets M and N coincide.

$M \subset N$ Each point of M belongs to N; M is a subset of N. Sometimes (but rarely) understood to mean that M is a proper subset of N.

$M \not\subset N$ Some point of M does not belong to N.

$M \subseteq N$ Each point of M belongs to N; M is a subset of N.

$M \supset N$ Each point of N belongs to M; M contains N as a subset. Sometimes (but rarely) understood to mean that N is a proper subset of M.

$M \not\supset N$ Some point of N does not belong to M.

$M \supseteq N$ Each point of N belongs to M; M contains N as a subset.

$M \cap N$, $M \cdot N$ The intersection of M and N.

$M \cup N$, $M + N$ The join (or sum) of M and N.

$\bigcap_{\alpha \in A} M_\alpha$, $\prod_{\alpha \in A} M_\alpha$ The set of all points which belong to M_α for all α of A.

$\bigcup_{\alpha \in A} M_\alpha$, $\sum_{\alpha \in A} M_\alpha$ The set of all points which belong to M_α for some α of A.

$\sim M$, $C(M)$, $\overline{M}$, $\tilde{M}$, M' The complement of M.

$M - N$, $M \sim N$ The complement of N in M; all points of M not in N.

$M \sim N$ The sets M and N can be put into one-to-one correspondence.

$\aleph$ Aleph, the first letter of the Hebrew alphabet.

$\aleph_0$ Aleph-null, or aleph-zero. The cardinal number of the set of positive integers.

c The cardinal number of the set of all real numbers.

$\aleph_\alpha$ An infinite cardinal number, the least being $\aleph_0$, the next $\aleph_1$, the next $\aleph_2$, etc. The first cardinal number greater than $\aleph_\alpha$ is denoted by $\aleph_{\alpha+1}$.

$M \simeq N$ M and N are of the same ordinal type.

ω The ordinal number of the positive integers in their natural order.

ω^*, $^*\omega$ The ordinal number of the negative integers in their natural order.

π The ordinal number of all integers in their natural order.

η The ordinal number of the rational numbers in the open interval $(0, 1)$.

θ The ordinal number of the real numbers of the closed interval $[0, 1]$.

α^*, $^*\alpha$ The ordinal number of a simply ordered set whose ordering is exactly reversed from that of a set of ordinal type α.

Q.E.D. Quod erat demonstrandum (L., which was to be proved).

Topology and Abstract Spaces

$\overline{M}$ The closure of M.

M' The derived set of M.

$d(x, y)$, $\delta(x, y)$, $\rho(x, y)$, (x, y) Distance from x to y.

$M \times N$ The Cartesian product of spaces M and N.

M/N The quotient space of M by N.

$\prec$, $<$; $\succ$, $>$ Symbols denoting an order relation.

T_0-**space** A topological space such that for distinct x and y there is either a neighborhood of x not containing y or a neighborhood of y not containing x.

T_1-**space** A topological space such that for distinct x and y there is a neighborhood of x not containing y.

T_2-**space** A Hausdorff topological space.

T_3-**space** A T_2-space which is regular.

T_4-**space** A T_3-space which is normal.

T_5-**space** A T_4-space which is completely normal.

F_σ, G_δ See BOREL—Borel set.

E_n, E^n, R_n, R^n Real n-dimensional Euclidean space.

Z_n, C_n Complex n-dimensional space.

H, $\mathfrak{H}$ Hilbert space.

(x, y) Inner product of the elements x and y of a vector space.

$\|x\|$ Norm of x (see VECTOR—vector space).

(B)-**space** A Banach space

(C), C The space of all continuous real-valued functions on some specified compact set, as on the closed interval $[0, 1]$ (then sometimes denoted by $C[0, 1]$), with $\|f\|$ defined as $\sup|f(x)|$.

(M), M The space of bounded functions on some set (particularly the interval $[0, 1]$), with $\|f\|$ defined as $\sup|f(x)|$.

ℓ_∞, ℓ^∞, (m), m The space of all bounded sequences $x = (x_1, x_2, \cdots)$, with $\|x\|$ defined as $\sup|x_i|$.

(c), c The space of all convergent sequences $x = (x_1, x_2, \cdots)$, with $\|x\|$ defined as $\sup|x_i|$.

(c_0), c_0 The space of all sequences $x = (x_1, x_2, \cdots)$ with $\lim_{n \to \infty} x_n = 0$ and $\|x\|$ defined as $\sup|x_i|$.

l_p, $l^{(p)}$ The space of all sequences $x = (x_1, x_2, \cdots)$ with $\sum|x_i|^p$ convergent $(p \geq 1)$ and $\|x\|$ defined as $[\sum|x_i|^p]^{1/p}$.

L_p, $L^{(p)}$ The space of all measurable functions f on a specified set S with $|f(x)|^p$ integrable $(p \geq 1)$ and

$$\|f\| = \left[\int_S |f(x)|^p \, dx \right]^{1/p};$$

S is frequently taken as the interval $[0, 1]$.

p The genus of an orientable surface (sometimes the number of "handles," whether the surface is orientable or not—see SURFACE).

q The number of cross-caps on a non-orientable surface (see SURFACE).

r The number of boundary curves on a surface.

χ Euler characteristic.

∂S, ΔS, $d(S)$ Boundary of the set S.

B_m^s An s-dimensional Betti group modulo m (m a prime).

B_0^s An s-dimensional Betti group relative to the group of integers.

R_m^s An s-dimensional Betti number modulo m (m a prime).

R_0^s An s-dimensional Betti number relative to the group of integers.

Mathematics of Finance

% Percent.

$ Dollar, dollars.

¢ Cent; cents.

@ At.

P Principal; present value.

$j_{(p)}$ Nominal rate (p conversion periods per year).

i, j, r Rate of interest.

s Compound amount of $1 for n periods;

$$s = (1 + i)^n.$$

n Number of periods or years, usually years.

v^n Present value of $1 ($n$ periods);

$$v^n = 1/(1 + i)^n.$$

l_x Number of persons living at age x (mortality table).

d_x Number of deaths per year of persons of age x (mortality table).

p_x Probability of a person of age x living one year.

q_x Probability of a person of age x dying within one year.

D_x $v^x l_x$.

C_x $v^{x+1} d_x$.

N_x $D_x + D_{x+1} + D_{x+2} + \cdots$ to table limit.

M_x $C_x + C_{x+1} + C_{x+2} + \cdots$ to table limit.

a_x Present value of a life annuity of $1 at age x; N_{x+1}/D_x.

$a_x, \bar{a}_x$ Present value of a (life) annuity due of $1 at age x; $1 + a_x$.

u_x D_x/D_{x+1}.

k_x C_x/D_{x+1}.

A_x Net single premium for $1 of whole-life insurance taken out at age x.

P_x Net annual premium for an insurance of $1 at age x on the ordinary life plan.

$_nA_x$ Net single premium for $1 of term insurance for n years for a person aged x.

$_nP_x$ Premium for a limited payment life policy of $1 with a term of n years at age x.

$_nV_x$ Terminal reserve, at the end of n years after the policy was issued, on an ordinary life policy of $1 for a life of age x.

$_nE_x$ The present value of a pure endowment to be paid in n years to a person of age x.

R Annual rent.

$s_{\overline{n}|}$ or $s_{\overline{n}|i}$ Compound amount of $1 per annum for n years at interest rate i; $[(1+i)^n - 1]/i$.

$s_{\overline{n}|}^{(p)}$ or $s_{\overline{n}|i}^{(p)}$ Amount of $1 per annum for n years at interest rate i when payable in p equal installments at intervals of $(1/p)$th part of a year.

$a_{\overline{n}|}$ or $a_{\overline{n}|i}$ Present value of $1 per annum for n years at interest rate i; $[1 - (1+i)^{-n}]/i$.

$a_{\overline{n}|}^{(p)}$ or $a_{\overline{n}|i}^{(p)}$ Present value of $1 per annum for n years at interest rate i if payable in p installments at intervals of $(1/p)$th part of a year.

$A_{x:\overline{n}|}$ Net single premium for an endowment insurance for $1 for n years at age x.

$P_{x:\overline{n}|}$ Net annual premium for an n-year endowment policy for $1 taken out at age x.

$A_{x:\overline{n}|}^1$ or $_nA_x$ Net single premium for an endowment policy of $1 for n years taken out at age x.

$P_{x:\overline{n}|}^1$ or $_nP_x$ Net annual premium for a term insurance of $1 for n years at age x.

Statistics

χ^2 Chi-square.

d.f. Degrees of freedom.

F F ratio.

i Width of a class interval.

P.E. Probable error (same as probable deviation).

r Correlation coefficient (Pearson product moment correlation coefficient between two variables).

$r_{12 \cdot 34 \cdots n}$ Partial correlation coefficient between variables 1 and 2 in a set of n variables.

$r_{1 \cdot 234 \cdots n}$ Multiple correlation coefficient between variable 1 and remainder of a set of n variables.

s Standard deviation (from a sample).

σ_x Standard deviation of x.

$\sigma_{x \cdot y}$ Standard error of estimate; also standard deviation of x for given value of y.

σ_x^2 Variance of x.

$\sigma_{x, y}^2$ Covariance of x and y.

t Students' "t" statistic.

V Coefficient of variation.

$\bar{x}$ Arithmetic average of the variable x (from a sample).

μ Arithmetic mean of a population.

$\mu_2 = \sigma^2$ Second moment about the mean.

μ_r The rth moment about the mean.

$\beta_1 = \dfrac{\mu_3}{\sigma^3}$ Coefficient of skewness.

$\beta_2 = \dfrac{\mu_4}{\sigma^4}$ Coefficient of kurtosis.

$\beta_{12 \cdot 34}$ Multiple-regression coefficient in terms of standard-deviation units.

η Correlation ratio.

z Fisher's z statistic.

Q_1 First quartile.

Q_3 Third quartile.

$E(x)$ Expected value of x.

$E(x|y)$ Conditional expectation of x given y.

$P(x_i)$ Probability that x assumes the value x_i.

DIFFERENTIATION FORMULAS

(See Derivative)

In the following formulas, u, v and y are differentiable functions, a, c, and n are constants, and $\ln u = \log_e u$.

$$\frac{dc}{dx} = 0.$$

$$\frac{d}{dx} x = 1.$$

$$\frac{d}{dx}(cv) = c\frac{dv}{dx}.$$

$$\frac{d}{dx} x^n = nx^{n-1}, \ n > 1 \text{ if } x = 0.$$

$$\frac{dy}{dx} = \frac{dy}{du} \cdot \frac{du}{dx}.$$

$$\frac{d}{dx}(u+v) = \frac{du}{dx} + \frac{dv}{dx}.$$

$$\frac{d}{dx}(uv) = u\frac{dv}{dx} + v\frac{du}{dx}.$$

$$\frac{d}{dx}\left(\frac{u}{v}\right) = \frac{v\frac{du}{dx} - u\frac{dv}{dx}}{v^2}.$$

$$\frac{du^n}{dx} = nu^{n-1}\frac{du}{dx}, \ n > 1 \text{ if } u = 0.$$

$$\frac{d}{dx}(\sin u) = \cos u \cdot \frac{du}{dx}.$$

$$\frac{d}{dx}(\cos u) = -\sin u \frac{du}{dx}.$$

$$\frac{d}{dx}(\tan u) = \sec^2 u \frac{du}{dx}.$$

$$\frac{d}{dx}(\cot u) = -\csc^2 u \frac{du}{dx}.$$

$$\frac{d}{dx}(\sec u) = \sec u \tan u \frac{du}{dx}.$$

$$\frac{d}{dx}(\csc u) = -\csc u \cot u \frac{du}{dx}.$$

$$\frac{d}{dx}(\sinh x) = \cosh x.$$

$$\frac{d}{dx}(\cosh x) = \sinh x.$$

$$\frac{d}{dx}(\tanh x) = \operatorname{sech}^2 x.$$

$$\frac{d}{dx}(\operatorname{ctnh} x) = -\operatorname{csch}^2 x.$$

$$\frac{d}{dx}(\operatorname{sech} x) = -\operatorname{sech} x \tanh x.$$

$$\frac{d}{dx}(\operatorname{csch} x) = -\operatorname{csch} x \operatorname{ctnh} x.$$

$$\frac{d}{dx}(\ln u) = \frac{\frac{du}{dx}}{u}.$$

$$\frac{d}{dx}(\log_a u) = \log_a e \cdot \frac{\frac{du}{dx}}{u}, \ a > 0 \text{ and } a \neq 1.$$

$$\frac{d}{dx}(e^u) = e^u \cdot \frac{du}{dx}.$$

$$\frac{d}{dx}(a^u) = \ln a \cdot a^u \cdot \frac{du}{dx}, \ a > 0.$$

$$\frac{d}{dx}(\arcsin u) = \frac{\frac{du}{dx}}{\sqrt{1-u^2}}, \ |u| < 1.$$

$$\frac{d}{dx}(\arccos u) = -\frac{\frac{du}{dx}}{\sqrt{1-u^2}}, \ |u| < 1.$$

$$\frac{d}{dx}(\arctan u) = \frac{\frac{du}{dx}}{1+u^2}.$$

$$\frac{d}{dx}(\operatorname{arccot} u) = -\frac{\frac{du}{dx}}{1+u^2}.$$

$$\frac{d}{dx}(\operatorname{arcsec} u) = \frac{\frac{du}{dx}}{u\sqrt{u^2-1}}.$$

$$-\pi < \sec^{-1} u < -\frac{\pi}{2}, \ 0 < \sec^{-1} u < \frac{\pi}{2},$$

$$\frac{d}{dx}(\operatorname{arccsc} u) = -\frac{\frac{du}{dx}}{u\sqrt{u^2-1}},$$

$$-\pi < \csc^{-1} u < -\frac{\pi}{2}, \ 0 < \csc^{-1} u < \frac{\pi}{2}.$$

$$\frac{d}{dx}(\sinh^{-1} x) = \frac{1}{\sqrt{x^2+1}}.$$

$$\frac{d}{dx}(\cosh^{-1} x) = \frac{1}{\sqrt{x^2-1}}, \ x > 1.$$

$$\frac{d}{dx}(\tanh^{-1} x) = \frac{1}{1-x^2}, \ |x| < 1.$$

$$\frac{d}{dx}(\operatorname{ctnh}^{-1} x) = \frac{1}{1-x^2}, \ |x| > 1.$$

$$\frac{d}{dx}(\operatorname{sech}^{-1} x) = \frac{-1}{x\sqrt{1-x^2}}, \ 0 < x < 1.$$

$$\frac{d}{dx}(\operatorname{csch}^{-1} x) = \frac{-1}{|x|\sqrt{1+x^2}}, \ x \neq 0.$$

INTEGRAL TABLES*

In the following tables, the constant of integration, **C**, is omitted but should be added to the result of every integration. The letter **x** represents any variable; **u** represents any function of **x**; the remaining letters represent arbitrary constants, unless otherwise indicated; all angles are in radians. **Unless otherwise mentioned $\log_e u \equiv \log u$.**

Short Table of Integrals.

1. $\int df(x) = f(x).$

2. $d \int f(x)\,dx = f(x)\,dx.$

3. $\int 0 \cdot dx = C.$

4. $\int af(x)\,dx = a \int f(x)\,dx.$

5. $\int (u \pm v)\,dx = \int u\,dx \pm \int v\,dx.$

6. $\int u\,dv = uv - \int v\,du.$

7. $\int \frac{u\,dv}{dx}\,dx = uv - \int v\,\frac{du}{dx}\,dx.$

8. $\int f(y)\,dx = \int \frac{f(y)\,dy}{\frac{dy}{dx}}.$

9. $\int u^n\,du = \frac{u^{n+1}}{n+1}, \quad n \neq -1.$

10. $\int \frac{du}{u} = \log u.$

11. $\int e^u\,du = e^u.$

12. $\int b^u\,du = \frac{b^u}{\log b}.$

13. $\int \sin u\,du = -\cos u.$

14. $\int \cos u\,du = \sin u.$

15. $\int \tan u\,du = \log \sec u = -\log \cos u.$

16. $\int \operatorname{ctn} u\,du = \log \sin u = -\log \csc u.$

17. $\int \sec u\,du = \log(\sec u + \tan u) = \log \tan\left(\frac{u}{2} + \frac{\pi}{4}\right).$

18. $\int \csc u\,du = \log(\csc u - \operatorname{ctn} u) = \log \tan \frac{u}{2}.$

19. $\int \sin^2 u\,du = \frac{1}{2}u - \frac{1}{2}\sin u \cos u.$

20. $\int \cos^2 u\,du = \frac{1}{2}u + \frac{1}{2}\sin u \cos u.$

21. $\int \sec^2 u\,du = \tan u.$

22. $\int \csc^2 u\,du = -\operatorname{ctn} u.$

23. $\int \tan^2 u\,du = \tan u - u.$

24. $\int \operatorname{ctn}^2 u\,du = -\operatorname{ctn} u - u.$

25. $\int \frac{du}{u^2 + a^2} = \frac{1}{a}\tan^{-1}\frac{u}{a}.$

26. $\int \frac{du}{u^2 - a^2} = \frac{1}{2a}\log\left(\frac{u-a}{u+a}\right) = -\frac{1}{a}\operatorname{ctnh}^{-1}\left(\frac{u}{a}\right), \quad \text{if } u^2 > a^2,$

$\qquad = \frac{1}{2a}\log\left(\frac{a-u}{a+u}\right) = -\frac{1}{a}\tanh^{-1}\left(\frac{u}{a}\right), \quad \text{if } u^2 < a^2.$

27. $\int \frac{du}{\sqrt{a^2 - u^2}} = \sin^{-1}\left(\frac{u}{a}\right).$

28. $\int \frac{du}{\sqrt{u^2 \pm a^2}} = \log\left(u + \sqrt{u^2 \pm a^2}\right).$

*These tables were taken, with permission, from "HANDBOOK OF MATHEMATICAL TABLES AND FORMULAS," *Handbook Publishers, Inc.,* Sandusky, Ohio, edited by RICHARD S. BURINGTON.

29. $\displaystyle\int \frac{du}{\sqrt{2au - u^2}} = \cos^{-1}\left(\frac{a-u}{a}\right).$

30. $\displaystyle\int \frac{du}{u\sqrt{u^2 - a^2}} = \frac{1}{a}\sec^{-1}\left(\frac{u}{a}\right).$

31. $\displaystyle\int \frac{du}{u\sqrt{a^2 \pm u^2}} = -\frac{1}{a}\log\left(\frac{a + \sqrt{a^2 \pm u^2}}{u}\right)^{*}.$

32. $\displaystyle\int \sqrt{a^2 - u^2}\cdot du = \frac{1}{2}\left(u\sqrt{a^2 - u^2} + a^2\sin^{-1}\frac{u}{a}\right).$

33. $\displaystyle\int \sqrt{u^2 \pm a^2}\, du = \frac{1}{2}\left[u\sqrt{u^2 \pm a^2} \pm a^2\log\left(u + \sqrt{u^2 \pm a^2}\right)\right]^{*}.$

34. $\displaystyle\int \sinh u\, du = \cosh u.$

35. $\displaystyle\int \cosh u\, du = \sinh u.$

36. $\displaystyle\int \tanh u\, du = \log(\cosh u).$

37. $\displaystyle\int \operatorname{ctnh} u\, du = \log(\sinh u).$

38. $\displaystyle\int \operatorname{sech} u\, du = \sin^{-1}(\tanh u).$

39. $\displaystyle\int \operatorname{csch} u\, du = \log\left(\tanh\frac{u}{2}\right).$

40. $\displaystyle\int \operatorname{sech} u\cdot\tanh u\cdot du = -\operatorname{sech} u.$

41. $\displaystyle\int \operatorname{csch} u\cdot\operatorname{ctnh} u\cdot du = -\operatorname{csch} u.$

Expressions Containing $(ax + b)$.

42. $\displaystyle\int (ax + b)^n\, dx = \frac{1}{a(n+1)}(ax + b)^{n+1}, \quad n \neq -1.$

43. $\displaystyle\int \frac{dx}{ax + b} = \frac{1}{a}\log, (ax + b).$

$^{*}\log\left(\dfrac{u + \sqrt{u^2 + a^2}}{a}\right)$

$= \sinh^{-1}\left(\dfrac{u}{a}\right); \quad \log\left(\dfrac{a + \sqrt{a^2 - u^2}}{u}\right) = \operatorname{sech}^{-1}\left(\dfrac{u}{a}\right);$

$\log\left(\dfrac{u + \sqrt{u^2 - a^2}}{a}\right)$

$= \cosh^{-1}\left(\dfrac{u}{a}\right); \quad \log\left(\dfrac{a + \sqrt{a^2 + u^2}}{u}\right) = \operatorname{csch}^{-1}\left(\dfrac{u}{a}\right).$

44. $\displaystyle\int \frac{dx}{(ax + b)^2} = -\frac{1}{a(ax + b)}.$

45. $\displaystyle\int \frac{dx}{(ax + b)^3} = -\frac{1}{2a(ax + b)^2}.$

46. $\displaystyle\int x(ax + b)^n\, dx = \frac{1}{a^2(n+2)}(ax + b)^{n+2}$
$$- \frac{b}{a^2(n+1)}(ax + b)^{n+1}, \quad n \neq -1, -2$$

47. $\displaystyle\int \frac{x\, dx}{ax + b} = \frac{x}{a} - \frac{b}{a^2}\log(ax + b).$

48. $\displaystyle\int \frac{x\, dx}{(ax + b)^2} = \frac{b}{a^2(ax + b)} + \frac{1}{a^2}\log(ax + b).$

49. $\displaystyle\int \frac{x\, dx}{(ax + b)^3} = \frac{b}{2a^2(ax + b)^2} - \frac{1}{a^2(ax + b)}$

50. $\displaystyle\int x^2(ax + b)^n\, dx = \frac{1}{a^3}\left[\frac{(ax + b)^{n+3}}{n+3}\right.$
$$\left. - 2b\frac{(ax + b)^{n+2}}{n+2} + b^2\frac{(ax + b)^{n+1}}{n+1}\right], \quad n \neq -1, -2, -3.$$

51. $\displaystyle\int \frac{x^2\, dx}{ax + b} = \frac{1}{a^3}\left[\frac{1}{2}(ax + b)^2 - 2b(ax + b) + b^2\log(ax + b)\right].$

52. $\displaystyle\int \frac{x^2\, dx}{(ax + b)^2} = \frac{1}{a^3}\left[(ax + b) - 2b\log(ax + b) - \frac{b^2}{ax + b}\right].$

53. $\displaystyle\int \frac{x^2\, dx}{(ax + b)^3} = \frac{1}{a^3}\left[\log(ax + b) + \frac{2b}{ax + b} - \frac{b^2}{2(ax + b)^2}\right].$

54. $\displaystyle\int x^m(ax + b)^n\, dx$
$$= \frac{1}{a(m + n + 1)}\left[x^m(ax + b)^{n+1} - mb\int x^{m-1}(ax + b)^n\, dx\right]$$
$$= \frac{1}{m + n + 1}\left[x^{m+1}(ax + b)^n + nb\int x^m(ax + b)^{n-1}\, dx\right],$$
$$m > 0, \quad m + n + 1 \neq 0.$$

55. $\displaystyle\int \frac{dx}{x(ax + b)} = \frac{1}{b}\log\frac{x}{ax + b}.$

56. $\displaystyle\int \frac{dx}{x^2(ax + b)} = -\frac{1}{bx} + \frac{a}{b^2}\log\frac{ax + b}{x}.$

57. $\displaystyle\int \frac{dx}{x^3(ax + b)} = \frac{2ax - b}{2b^2x^2} + \frac{a^2}{b^3}\log\frac{x}{ax + b}.$

58. $\displaystyle\int \frac{dx}{x(ax + b)^2} = \frac{1}{b(ax + b)} - \frac{1}{b^2}\log\frac{ax + b}{x}.$

59. $\displaystyle\int \frac{dx}{x(ax + b)^3} = \frac{1}{b^3}\left[\frac{1}{2}\left(\frac{ax + 2b}{ax + b}\right)^2 + \log\frac{x}{ax + b}\right].$

60. $\int \dfrac{dx}{x^2(ax+b)^2} = -\dfrac{b+2ax}{b^2 x(ax+b)} + \dfrac{2a}{b^3}\log\dfrac{ax+b}{x}$.

61. $\int \sqrt{ax+b}\,dx = \dfrac{2}{3a}\sqrt{(ax+b)^3}$.

62. $\int x\sqrt{ax+b}\,dx = \dfrac{2(3ax-2b)}{15a^2}\sqrt{(ax+b)^3}$.

63. $\int x^2\sqrt{ax+b}\,dx = \dfrac{2(15a^2x^2-12abx+8b^2)\sqrt{(ax+b)^3}}{105a^3}$.

64. $\int x^3\sqrt{ax+b}\,dx$

$$= \dfrac{2(35a^3x^3-30a^2bx^2+24ab^2x-16b^3)\sqrt{(ax+b)^3}}{315a^4}.$$

65. $\int x^n\sqrt{ax+b}\,dx = \dfrac{2}{a^{n+1}}\int u^2(u^2-b)^n\,du, \quad u=\sqrt{ax+b}$.

66. $\int \dfrac{\sqrt{ax+b}}{x}\,dx = 2\sqrt{ax+b} + b\int \dfrac{dx}{x\sqrt{ax+b}}$.

67. $\int \dfrac{dx}{\sqrt{ax+b}} = \dfrac{2\sqrt{ax+b}}{a}$.

68. $\int \dfrac{x\,dx}{\sqrt{ax+b}} = \dfrac{2(ax-2b)}{3a^2}\sqrt{ax+b}$.

69. $\int \dfrac{x^2\,dx}{\sqrt{ax+b}} = \dfrac{2(3a^2x^2-4abx+8b^2)}{15a^3}\sqrt{ax+b}$.

70. $\int \dfrac{x^3\,dx}{\sqrt{ax+b}} = \dfrac{2(5a^3x^3-6a^2bx^2+8ab^2x-16b^3)}{35a^4}\sqrt{ax+b}$.

71. $\int \dfrac{x^n\,dx}{\sqrt{ax+b}} = \dfrac{2}{a^{n+1}}\int (u^2-b)^n\,du, \quad u=\sqrt{ax+b}$.

72. $\int \dfrac{dx}{x\sqrt{ax+b}} = \dfrac{1}{\sqrt{b}}\log\dfrac{\sqrt{ax+b}-\sqrt{b}}{\sqrt{ax+b}+\sqrt{b}}, \quad$ for $b>0$.

73. $\int \dfrac{dx}{x\sqrt{ax+b}} = \dfrac{2}{\sqrt{-b}}\tan^{-1}\sqrt{\dfrac{ax+b}{-b}}, \quad b<0,$

$\qquad$ or $\dfrac{-2}{\sqrt{b}}\tanh^{-1}\sqrt{\dfrac{ax+b}{b}}, \quad b>0$.

74. $\int \dfrac{dx}{x^2\sqrt{ax+b}} = -\dfrac{\sqrt{ax+b}}{bx} - \dfrac{a}{2b}\int \dfrac{dx}{x\sqrt{ax+b}}$.

75. $\int \dfrac{dx}{x^3\sqrt{ax+b}} = -\dfrac{\sqrt{ax+b}}{2bx^2}$

$\qquad + \dfrac{3a\sqrt{ax+b}}{4b^2x} + \dfrac{3a^2}{8b^2}\int \dfrac{dx}{x\sqrt{ax+b}}$.

76. $\int \dfrac{dx}{x^n(ax+b)^m} = -\dfrac{1}{b^{m+n-1}}\int \dfrac{(u-a)^{m+n-2}\,du}{u^m},$

$\qquad u = \dfrac{ax+b}{x}$.

77. $\int (ax+b)^{\pm\frac{n}{2}}\,dx = \dfrac{2(ax+b)^{\frac{2\pm n}{2}}}{a(2\pm n)}$.

78. $\int x(ax+b)^{\pm\frac{n}{2}}\,dx = \dfrac{2}{a^2}\left[\dfrac{(ax+b)^{\frac{4\pm n}{2}}}{4\pm n} - \dfrac{b(ax+b)^{\frac{2\pm n}{2}}}{2\pm n}\right]$.

79. $\int \dfrac{dx}{x(ax+b)^{\frac{n}{2}}} = \dfrac{1}{b}\int \dfrac{dx}{x(ax+b)^{\frac{n-2}{2}}} - \dfrac{a}{b}\int \dfrac{dx}{(ax+b)^{\frac{n}{2}}}$.

80. $\int \dfrac{x^m\,dx}{\sqrt{ax+b}} = \dfrac{2x^m\sqrt{ax+b}}{(2m+1)a} - \dfrac{2mb}{(2m+1)a}\int \dfrac{x^{m-1}\,dx}{\sqrt{ax+b}}$.

81. $\int \dfrac{dx}{x^n\sqrt{ax+b}} = \dfrac{-\sqrt{ax+b}}{(n-1)bx^{n-1}} - \dfrac{(2n-3)a}{(2n-2)b}\int \dfrac{dx}{x^{n-1}\sqrt{ax+b}}$.

82. $\int \dfrac{(ax+b)^{\frac{n}{2}}}{x}\,dx = a\int (ax+b)^{\frac{n-2}{2}}\,dx$

$\qquad + b\int \dfrac{(ax+b)^{\frac{n-2}{2}}}{x}\,dx$.

83. $\int \dfrac{dx}{(ax+b)(cx+d)} = \dfrac{1}{bc-ad}\log\dfrac{cx+d}{ax+b}, \quad bc-ad\neq 0$.

84. $\int \dfrac{dx}{(ax+b)^2(cx+d)}$

$\qquad = \dfrac{1}{bc-ad}\left[\dfrac{1}{ax+b} + \dfrac{c}{bc-ad}\log\left(\dfrac{cx+d}{ax+b}\right)\right], \quad bc-ad\neq 0$.

85. $\int (ax+b)^n(cx+d)^m\,dx = \dfrac{1}{(m+n+1)a}$

$\qquad \cdot\left[(ax+b)^{n+1}(cx+d)^m\right.$

$\qquad \left. - m(bc-ad)\int (ax+b)^n(cx+d)^{m-1}\,dx\right]$.

86. $\int \dfrac{dx}{(ax+b)^n(cx+d)^m} = \dfrac{-1}{(m-1)(bc-ad)}$

$\qquad \times\left[\dfrac{1}{(ax+b)^{n-1}(cx+d)^{m-1}} + a(m+n-2)\right.$

$\qquad \left.\times\int \dfrac{dx}{(ax+b)^n(cx+d)^{m-1}}\right], \quad m>1, \quad n>0, \quad bc-ad\neq 0$.

464

87. $\displaystyle \int \frac{(ax+b)^n}{(cx+d)^m}\,dx = -\frac{1}{(m-1)(bc-ad)}$

$$\times \left[\frac{(ax+b)^{n+1}}{(cx+d)^{m-1}} + (m-n-2)a \int \frac{(ax+b)^n\,dx}{(cx+d)^{m-1}} \right],$$

$$= \frac{-1}{(m-n-1)c}\left[\frac{(ax+b)^n}{(cx+d)^{m-1}} + n(bc-ad)\int \frac{(ax+b)^{n-1}}{(cx+d)^m}\,dx \right].$$

88. $\displaystyle \int \frac{x\,dx}{(ax+b)(cx+d)}$

$$= \frac{1}{bc-ad}\left[\frac{b}{a}\log(ax+b) - \frac{d}{c}\log(cx+d) \right], \quad bc-ad \neq 0.$$

89. $\displaystyle \int \frac{x\,dx}{(ax+b)^2(cx+d)} = \frac{1}{bc-ad}\left[-\frac{b}{a(ax+b)} \right.$

$$\left. -\frac{d}{bc-ad}\log\frac{cx+d}{ax+b} \right], \quad bc-ad \neq 0.$$

90. $\displaystyle \int \frac{cx+d}{\sqrt{ax+b}}\,dx = \frac{2}{3a^2}(3ad-2bc+acx)\sqrt{ax+b}\,.$

91. $\displaystyle \int \frac{\sqrt{ax+b}}{cx+d}\,dx = \frac{2\sqrt{ax+b}}{c}$

$$-\frac{2}{c}\sqrt{\frac{ad-bc}{c}}\,\tan^{-1}\sqrt{\frac{c(ax+b)}{ad-bc}}\,, \quad c>0, \quad ad>bc.$$

92. $\displaystyle \int \frac{\sqrt{ax+b}}{cx+d}\,dx = \frac{2\sqrt{ax+b}}{c}$

$$+\frac{1}{c}\sqrt{\frac{bc-ad}{c}}\,\log\frac{\sqrt{c(ax+b)}-\sqrt{bc-ad}}{\sqrt{c(ax+b)}+\sqrt{bc-ad}}\,, \quad c>0, \quad bc>ad.$$

93. $\displaystyle \int \frac{dx}{(cx+d)\sqrt{ax+b}}$

$$= \frac{2}{\sqrt{c}\,\sqrt{ad-bc}}\,\tan^{-1}\sqrt{\frac{c(ax+b)}{ad-bc}}\,, \quad c>0, \quad ad>bc.$$

94. $\displaystyle \int \frac{dx}{(cx+d)\sqrt{ax+b}}$

$$= \frac{1}{\sqrt{c}\,\sqrt{bc-ad}}\,\log\frac{\sqrt{c(ax+b)}-\sqrt{bc-ad}}{\sqrt{c(ax+b)}+\sqrt{bc-ad}}\,, \quad c>0, \quad bc>ad.$$

Expressions Containing ax^2+c, ax^n+c, $x^2 \pm p^2$, and $p^2 - x^2$.

95. $\displaystyle \int \frac{dx}{p^2+x^2} = \frac{1}{p}\tan^{-1}\frac{x}{p}, \quad \text{or} \quad -\frac{1}{p}\operatorname{ctn}^{-1}\left(\frac{x}{p}\right).$

96. $\displaystyle \int \frac{dx}{p^2-x^2} = \frac{1}{2p}\log\frac{p+x}{p-x}, \quad \text{or} \quad \frac{1}{p}\tanh^{-1}\left(\frac{x}{p}\right).$

97. $\displaystyle \int \frac{dx}{ax^2+c} = \frac{1}{\sqrt{ac}}\tan^{-1}\left(x\sqrt{\frac{a}{c}}\right), \quad a \text{ and } c>0.$

98. $\displaystyle \int \frac{dx}{ax^2+c} = \frac{1}{2\sqrt{-ac}}\log\frac{x\sqrt{a}-\sqrt{-c}}{x\sqrt{a}+\sqrt{-c}}, \quad a>0, \quad c<0.$

$$= \frac{1}{2\sqrt{-ac}}\log\frac{\sqrt{c}+x\sqrt{-a}}{\sqrt{c}-x\sqrt{-a}}, \quad a<0, \quad c>0.$$

99. $\displaystyle \int \frac{dx}{(ax^2+c)^n} = \frac{1}{2(n-1)c}\cdot\frac{x}{(ax^2+c)^{n-1}}$

$$+\frac{2n-3}{2(n-1)c}\int \frac{dx}{(ax^2+c)^{n-1}}, \quad n \text{ a positive integer, } n>1.$$

100. $\displaystyle \int x(ax^2+c)^n\,dx = \frac{1}{2a}\frac{(ax^2+c)^{n+1}}{n+1}, \quad n \neq -1.$

101. $\displaystyle \int \frac{x}{ax^2+c}\,dx = \frac{1}{2a}\log(ax^2+c).$

102. $\displaystyle \int \frac{dx}{x(ax^2+c)} = \frac{1}{2c}\log\frac{x^2}{ax^2+c}.$

103. $\displaystyle \int \frac{dx}{x^2(ax^2+c)} = -\frac{1}{cx} - \frac{a}{c}\int \frac{dx}{ax^2+c}.$

104. $\displaystyle \int \frac{x^2\,dx}{ax^2+c} = \frac{x}{a} - \frac{c}{a}\int \frac{dx}{ax^2+c}.$

105. $\displaystyle \int \frac{x^n\,dx}{ax^2+c} = \frac{x^{n-1}}{a(n-1)} - \frac{c}{a}\int \frac{x^{n-2}\,dx}{ax^2+c}, \quad n \neq 1.$

106. $\displaystyle \int \frac{x^2\,dx}{(ax^2+c)^n} = -\frac{1}{2(n-1)a}\cdot\frac{x}{(ax^2+c)^{n-1}}$

$$+\frac{1}{2(n-1)a}\int \frac{dx}{(ax^2+c)^{n-1}}.$$

107. $\displaystyle \int \frac{dx}{x^2(ax^2+c)^n} = \frac{1}{c}\int \frac{dx}{x^2(ax^2+c)^{n-1}} - \frac{a}{c}\int \frac{dx}{(ax^2+c)^n}.$

108. $\displaystyle \int \sqrt{x^2 \pm p^2}\,dx = \frac{1}{2}\left[x\sqrt{x^2 \pm p^2} \pm p^2\log\left(x+\sqrt{x^2 \pm p^2}\right) \right].$

109. $\displaystyle \int \sqrt{p^2-x^2}\,dx = \frac{1}{2}\left[x\sqrt{p^2-x^2} + p^2\sin^{-1}\left(\frac{x}{p}\right) \right]$

110. $\displaystyle \int \frac{dx}{\sqrt{x^2 \pm p^2}} = \log\left(x+\sqrt{x^2 \pm p^2}\right).$

111. $\displaystyle \int \frac{dx}{\sqrt{p^2-x^2}} = \sin^{-1}\left(\frac{x}{p}\right) \quad \text{or} \quad -\cos^{-1}\left(\frac{x}{p}\right).$

112. $\displaystyle \int \sqrt{ax^2+c}\,dx = \frac{x}{2}\sqrt{ax^2+c}$

$$+\frac{c}{2\sqrt{a}}\log\left(x\sqrt{a}+\sqrt{ax^2+c}\right), \quad a>0.$$

465

113. $\int \sqrt{ax^2 + c}\, dx = \frac{x}{2}\sqrt{ax^2 + c} + \frac{c}{2\sqrt{-a}}\sin^{-1}\left(x\sqrt{\frac{-a}{c}}\right),$

$$a < 0.$$

114. $\int \frac{dx}{\sqrt{ax^2 + c}} = \frac{1}{\sqrt{a}}\log\left(x\sqrt{a} + \sqrt{ax^2 + c}\right),\quad a > 0.$

115. $\int \frac{dx}{\sqrt{ax^2 + c}} = \frac{1}{\sqrt{-a}}\sin^{-1}\left(x\sqrt{\frac{-a}{c}}\right),\quad a < 0.$

116. $\int x\sqrt{ax^2 + c}\cdot dx = \frac{1}{3a}(ax^2 + c)^{\frac{3}{2}}.$

117. $\int x^2\sqrt{ax^2 + c}\, dx = \frac{x}{4a}\sqrt{(ax^2 + c)^3} - \frac{cx}{8a}\sqrt{ax^2 + c}$

$$- \frac{c^2}{8\sqrt{a^3}}\log\left(x\sqrt{a} + \sqrt{ax^2 + c}\right),\quad a > 0.$$

118. $\int x^2\sqrt{ax^2 + c}\, dx = \frac{x}{4a}\sqrt{(ax^2 + c)^3} - \frac{cx}{8a}\sqrt{ax^2 + c}$

$$- \frac{c^2}{8a\sqrt{-a}}\sin^{-1}\left(x\sqrt{\frac{-a}{c}}\right),\quad a < 0.$$

119. $\int \frac{x\, dx}{\sqrt{ax^2 + c}} = \frac{1}{a}\sqrt{ax^2 + c}.$

120. $\int \frac{x^2\, dx}{\sqrt{ax^2 + c}} = \frac{x}{a}\sqrt{ax^2 + c} - \frac{1}{a}\int \sqrt{ax^2 + c}\, dx.$

121. $\int \frac{\sqrt{ax^2 + c}}{x}\, dx = \sqrt{ax^2 + c} + \sqrt{c}\log\frac{\sqrt{ax^2 + c} - \sqrt{c}}{x},\quad c > 0.$

122. $\int \frac{\sqrt{ax^2 + c}}{x}\, dx = \sqrt{ax^2 + c} - \sqrt{-c}\tan^{-1}\frac{\sqrt{ax^2 + c}}{\sqrt{-c}},\quad c < 0.$

123. $\int \frac{dx}{x\sqrt{p^2 \pm x^2}} = -\frac{1}{p}\log\left(\frac{p + \sqrt{p^2 \pm x^2}}{x}\right).$

124. $\int \frac{dx}{x\sqrt{x^2 - p^2}} = \frac{1}{p}\cos^{-1}\left(\frac{p}{x}\right),\quad \text{or } -\frac{1}{p}\sin^{-1}\left(\frac{p}{x}\right).$

125. $\int \frac{dx}{x\sqrt{ax^2 + c}} = \frac{1}{\sqrt{c}}\log\frac{\sqrt{ax^2 + c} - \sqrt{c}}{x},\quad c > 0.$

126. $\int \frac{dx}{x\sqrt{ax^2 + c}} = \frac{1}{\sqrt{-c}}\sec^{-1}\left(x\sqrt{\frac{-a}{c}}\right),\quad c < 0.$

127. $\int \frac{dx}{x^2\sqrt{ax^2 + c}} = -\frac{\sqrt{ax^2 + c}}{cx}.$

128. $\int \frac{x^n\, dx}{\sqrt{ax^2 + c}} = \frac{x^{n-1}\sqrt{ax^2 + c}}{na}$

$$- \frac{(n-1)c}{na}\int \frac{x^{n-2}\, dx}{\sqrt{ax^2 + c}},\quad n > 0.$$

129. $\int x^n\sqrt{ax^2 + c}\, dx = \frac{x^{n-1}(ax^2 + c)^{\frac{3}{2}}}{(n+2)a}$

$$- \frac{(n-1)c}{(n+2)a}\int x^{n-2}\sqrt{ax^2 + c}\cdot dx,\quad n > 0.$$

130. $\int \frac{\sqrt{ax^2 + c}}{x^n}\, dx = -\frac{(ax^2 + c)^{\frac{3}{2}}}{c(n-1)x^{n-1}}$

$$- \frac{(n-4)a}{(n-1)c}\int \frac{\sqrt{ax^2 + c}}{x^{n-2}}\, dx,\quad n > 1.$$

131. $\int \frac{dx}{x^n\sqrt{ax^2 + c}} = -\frac{\sqrt{ax^2 + c}}{c(n-1)x^{n-1}}$

$$- \frac{(n-2)a}{(n-1)c}\int \frac{dx}{x^{n-2}\sqrt{ax^2 + c}},\quad n > 1.$$

132. $\int (ax^2 + c)^{\frac{3}{2}}\, dx = \frac{x}{8}(2ax^2 + 5c)\sqrt{ax^2 + c}$

$$+ \frac{3c^2}{8\sqrt{a}}\log\left(x\sqrt{a} + \sqrt{ax^2 + c}\right),\quad a > 0.$$

133. $\int (ax^2 + c)^{\frac{3}{2}}\, dx = \frac{x}{8}(2ax^2 + 5c)\sqrt{ax^2 + c}$

$$+ \frac{3c^2}{8\sqrt{-a}}\sin^{-1}\left(x\sqrt{\frac{-a}{c}}\right),\quad a < 0.$$

134. $\int \frac{dx}{(ax^2 + c)^{\frac{3}{2}}} = \frac{x}{c\sqrt{ax^2 + c}}.$

135. $\int x(ax^2 + c)^{\frac{3}{2}}\, dx = \frac{1}{5a}(ax^2 + c)^{\frac{5}{2}}.$

136. $\int x^2(ax^2 + c)^{\frac{3}{2}}\, dx = \frac{x^3}{6}(ax^2 + c)^{\frac{3}{2}} + \frac{c}{2}\int x^2\sqrt{ax^2 + c}\, dx.$

137. $\int x^n(ax^2 + c)^{\frac{3}{2}}\, dx = \frac{x^{n+1}(ax^2 + c)^{\frac{3}{2}}}{n+4}$

$$+ \frac{3c}{n+4}\int x^n\sqrt{ax^2 + c}\, dx.$$

138. $\int \frac{x\, dx}{(ax^2 + c)^{\frac{3}{2}}} = -\frac{1}{a\sqrt{ax^2 + c}}.$

139. $\int \frac{x^2\, dx}{(ax^2 + c)^{\frac{3}{2}}} = -\frac{x}{a\sqrt{ax^2 + c}}$

$$+ \frac{1}{a\sqrt{a}}\log\left(x\sqrt{a} + \sqrt{ax^2 + c}\right),\quad a > 0.$$

140. $\displaystyle\int \frac{x^2\, dx}{(ax^2+c)^{\frac{3}{2}}} = -\frac{x}{a\sqrt{ax^2+c}}$

$\displaystyle\qquad\qquad + \frac{1}{a\sqrt{-a}}\sin^{-1}\!\left(x\sqrt{\frac{-a}{c}}\right),\quad a<0.$

141. $\displaystyle\int \frac{x^2\, dx}{(ax^2+c)^{\frac{3}{2}}} = -\frac{x^2}{a\sqrt{ax^2+c}} + \frac{2}{a^2}\sqrt{ax^2+c}\,.$

142. $\displaystyle\int \frac{dx}{x(ax^n+c)} = \frac{1}{cn}\log\frac{x^n}{ax^n+c}\,.$

143. $\displaystyle\int \frac{dx}{(ax^n+c)^m} = \frac{1}{c}\int \frac{dx}{(ax^n+c)^{m-1}} - \frac{a}{c}\int \frac{x^n\, dx}{(ax^n+c)^m}\,.$

144. $\displaystyle\int \frac{dx}{x\sqrt{ax^n+c}} = \frac{1}{n\sqrt{c}}\log\frac{\sqrt{ax^n+c}-\sqrt{c}}{\sqrt{ax^n+c}+\sqrt{c}}\,,\quad c>0.$

145. $\displaystyle\int \frac{dx}{x\sqrt{ax^n+c}} = \frac{2}{n\sqrt{-c}}\sec^{-1}\sqrt{\frac{-ax^n}{c}}\,,\quad c<0.$

146. $\displaystyle\int x^{m-1}(ax^n+c)^p\, dx$

$\displaystyle\quad = \frac{1}{m+np}\left[x^m(ax^n+c)^p + npc\int x^{m-1}(ax^n+c)^{p-1}\, dx\right]$

$\displaystyle\quad = \frac{1}{cn(p+1)}\left[-x^m(ax^n+c)^{p+1} + (m+np+n)\right.$

$\displaystyle\qquad\qquad\left. \times \int x^{m-1}(ax^n+c)^{p+1}\, dx\right]$

$\displaystyle\quad = \frac{1}{a(m+np)}\left[x^{m-n}(ax^n+c)^{p+1} - (m-n)c\right.$

$\displaystyle\qquad\qquad\left. \times \int x^{m-n-1}(ax^n+c)^p\, dx\right]$

$\displaystyle\quad = \frac{1}{mc}\left[x^m(ax^n+c)^{p+1} - (m+np+n)a\int x^{m+n-1}(ax^n+c)^p\, dx\right].$

147. $\displaystyle\int \frac{x^m\, dx}{(ax^n+c)^p} = \frac{1}{a}\int \frac{x^{m-n}\, dx}{(ax^n+c)^{p-1}} - \frac{c}{a}\int \frac{x^{m-n}\, dx}{(ax^n+c)^p}\,.$

148. $\displaystyle\int \frac{dx}{x^m(ax^n+c)^p} = \frac{1}{c}\int \frac{dx}{x^m(ax^n+c)^{p-1}}$

$\displaystyle\qquad\qquad - \frac{a}{c}\int \frac{dx}{x^{m-n}(ax^n+c)^p}\,.$

Expressions Containing $(ax^2 + bx + c)$.

149. $\displaystyle\int \frac{dx}{ax^2+bx+c}$

$\displaystyle\quad = \frac{1}{\sqrt{b^2-4ac}}\log\frac{2ax+b-\sqrt{b^2-4ac}}{2ax+b+\sqrt{b^2-4ac}}\,,\quad b^2>4ac.$

150. $\displaystyle\int \frac{dx}{ax^2+bx+c} = \frac{2}{\sqrt{4ac-b^2}}\tan^{-1}\frac{2ax+b}{\sqrt{4ac-b^2}}\,,\quad b^2<4ac.$

151. $\displaystyle\int \frac{dx}{ax^2+bx+c} = -\frac{2}{2ax+b}\,,\quad b^2=4ac.$

152. $\displaystyle\int \frac{dx}{(ax^2+bx+c)^{n+1}} = \frac{2ax+b}{n(4ac-b^2)(ax^2+bx+c)^n}$

$\displaystyle\qquad\qquad + \frac{2(2n-1)a}{n(4ac-b^2)}\int \frac{dx}{(ax^2+bx+c)^n}\,.$

153. $\displaystyle\int \frac{x\, dx}{ax^2+bx+c} = \frac{1}{2a}\log(ax^2+bx+c)$

$\displaystyle\qquad\qquad - \frac{b}{2a}\int \frac{dx}{ax^2+bx+c}\,.$

154. $\displaystyle\int \frac{x^2\, dx}{ax^2+bx+c} = \frac{x}{a} - \frac{b}{2a^2}\log(ax^2+bx+c)$

$\displaystyle\qquad\qquad + \frac{b^2-2ac}{2a^2}\int \frac{dx}{ax^2+bx+c}\,.$

155. $\displaystyle\int \frac{x^n\, dx}{ax^2+bx+c} = \frac{x^{n-1}}{(n-1)a} - \frac{c}{a}\int \frac{x^{n-2}\, dx}{ax^2+bx+c}$

$\displaystyle\qquad\qquad - \frac{b}{a}\int \frac{x^{n-1}\, dx}{ax^2+bx+c}\,.$

156. $\displaystyle\int \frac{x\, dx}{(ax^2+bx+c)^{n+1}} = \frac{-(2c+bx)}{n(4ac-b^2)(ax^2+bx+c)^n}$

$\displaystyle\qquad\qquad - \frac{b(2n-1)}{n(4ac-b^2)}\int \frac{dx}{(ax^2+bx+c)^n}\,.$

157. $\displaystyle\int \frac{x^m\, dx}{(ax^2+bx+c)^{n+1}} = -\frac{x^{m-1}}{a(2n-m+1)(ax^2+bx+c)^n}$

$\displaystyle\qquad - \frac{n-m+1}{2n-m+1}\cdot\frac{b}{a}\int \frac{x^{m-1}\, dx}{(ax^2+bx+c)^{n+1}}$

$\displaystyle\qquad + \frac{m-1}{2n-m+1}\cdot\frac{c}{a}\int \frac{x^{m-2}\, dx}{(ax^2+bx+c)^{n+1}}\,.$

158. $\displaystyle\int \frac{dx}{x(ax^2+bx+c)} = \frac{1}{2c}\log\frac{x^2}{ax^2+bx+c}$

$\displaystyle\qquad\qquad - \frac{b}{2c}\int \frac{dx}{(ax^2+bx+c)}\,.$

159. $\displaystyle\int \frac{dx}{x^2(ax^2+bx+c)} = \frac{b}{2c^2}\log\!\left(\frac{ax^2+bx+c}{x^2}\right)$

$\displaystyle\qquad\qquad - \frac{1}{cx} + \left(\frac{b^2}{2c^2}-\frac{a}{c}\right)\int \frac{dx}{(ax^2+bx+c)}\,.$

160. $\displaystyle\int \frac{dx}{x^m(ax^2+bx+c)^{n+1}} = -\frac{1}{(m-1)cx^{m-1}(ax^2+bx+c)^n}$

$\displaystyle\qquad -\frac{(n+m-1)}{m-1}\cdot\frac{b}{c}\int \frac{dx}{x^{m-1}(ax^2+bx+c)^{n+1}}$

$\displaystyle\qquad -\frac{(2n+m-1)}{m-1}\cdot\frac{a}{c}\int \frac{dx}{x^{m-2}(ax^2+bx+c)^{n+1}}.$

161. $\displaystyle\int \frac{dx}{x(ax^2+bx+c)^n} = \frac{1}{2c(n-1)(ax^2+bx+c)^{n-1}}$

$\displaystyle\qquad -\frac{b}{2c}\int \frac{dx}{(ax^2+bx+c)^n} + \frac{1}{c}\int \frac{dx}{x(ax^2+bx+c)^{n-1}}.$

162. $\displaystyle\int \frac{dx}{\sqrt{ax^2+bx+c}}$

$\displaystyle\qquad = \frac{1}{\sqrt{a}}\log\left(2ax+b+2\sqrt{a}\sqrt{ax^2+bx+c}\right),\quad a>0.$

163. $\displaystyle\int \frac{dx}{\sqrt{ax^2+bx+c}} = \frac{1}{\sqrt{-a}}\sin^{-1}\frac{-2ax-b}{\sqrt{b^2-4ac}},\quad a<0.$

164. $\displaystyle\int \frac{x\,dx}{\sqrt{ax^2+bx+c}} = \frac{\sqrt{ax^2+bx+c}}{a}$

$\displaystyle\qquad -\frac{b}{2a}\int \frac{dx}{\sqrt{ax^2+bx+c}}.$

165. $\displaystyle\int \frac{x^n\,dx}{\sqrt{ax^2+bx+c}} = \frac{x^{n-1}}{an}\sqrt{ax^2+bx+c} - \frac{b(2n-1)}{2an}$

$\displaystyle\qquad \times \int \frac{x^{n-1}\,dx}{\sqrt{ax^2+bx+c}} - \frac{c(n-1)}{an}\int \frac{x^{n-2}\,dx}{\sqrt{ax^2+bx+c}}.$

166. $\displaystyle\int \sqrt{ax^2+bx+c}\;dx = \frac{2ax+b}{4a}\sqrt{ax^2+bx+c}$

$\displaystyle\qquad +\frac{4ac-b^2}{8a}\int \frac{dx}{\sqrt{ax^2+bx+c}}.$

167. $\displaystyle\int x\sqrt{ax^2+bx+c}\;dx = \frac{(ax^2+bx+c)^{\frac{3}{2}}}{3a}$

$\displaystyle\qquad -\frac{b}{2a}\int \sqrt{ax^2+bx+c}\;dx.$

168. $\displaystyle\int x^2\sqrt{ax^2+bx+c}\;dx = \left(x-\frac{5b}{6a}\right)\frac{(ax^2+bx+c)^{\frac{3}{2}}}{4a}$

$\displaystyle\qquad +\frac{(5b^2-4ac)}{16a^2}\int \sqrt{ax^2+bx+c}\;dx.$

169. $\displaystyle\int \frac{dx}{x\sqrt{ax^2+bx+c}}$

$\displaystyle\qquad = -\frac{1}{\sqrt{c}}\log\left(\frac{\sqrt{ax^2+bx+c}+\sqrt{c}}{x}+\frac{b}{2\sqrt{c}}\right),\quad c>0.$

170. $\displaystyle\int \frac{dx}{x\sqrt{ax^2+bx+c}} = \frac{1}{\sqrt{-c}}\sin^{-1}\frac{bx+2c}{x\sqrt{b^2-4ac}},\quad c<0.$

171. $\displaystyle\int \frac{dx}{x\sqrt{ax^2+bx}} = -\frac{2}{bx}\sqrt{ax^2+bx},\quad c=0.$

172. $\displaystyle\int \frac{dx}{x^n\sqrt{ax^2+bx+c}} = -\frac{\sqrt{ax^2+bx+c}}{c(n-1)x^{n-1}} + \frac{b(3-2n)}{2c(n-1)}$

$\displaystyle\qquad \times \int \frac{dx}{x^{n-1}\sqrt{ax^2+bx+c}} + \frac{a(2-n)}{c(n-1)}\int \frac{dx}{x^{n-2}\sqrt{ax^2+bx+c}}$

173. $\displaystyle\int \frac{dx}{(ax^2+bx+c)^{\frac{3}{2}}} = -\frac{2(2ax+b)}{(b^2-4ac)\sqrt{ax^2+bx+c}},$

$\displaystyle\qquad\qquad\qquad\qquad\qquad\qquad b^2\neq 4ac.$

174. $\displaystyle\int \frac{dx}{(ax^2+bx+c)^{\frac{3}{2}}} = -\frac{1}{2\sqrt{a^3}(x+b/2a)^2},\quad b^2=4ac.$

Miscellaneous Algebraic Expressions.

175. $\displaystyle\int \sqrt{2px-x^2}\;dx$

$\displaystyle\qquad = \frac{1}{2}\left[(x-p)\sqrt{2px-x^2}+p^2\sin^{-1}[(x-p)/p]\right].$

176. $\displaystyle\int \frac{dx}{\sqrt{2px-x^2}} = \cos^{-1}\left(\frac{p-x}{p}\right).$

177. $\displaystyle\int \frac{dx}{\sqrt{ax+b}\cdot\sqrt{cx+d}} = \frac{2}{\sqrt{-ac}}\tan^{-1}\sqrt{\frac{-c(ax+b)}{a(cx+d)}}$

$\displaystyle\qquad\qquad \text{or}\; \frac{2}{\sqrt{ac}}\tanh^{-1}\sqrt{\frac{c(ax+b)}{a(cx+d)}}.$

178. $\displaystyle\int \sqrt{ax+b}\cdot\sqrt{cx+d}\;dx$

$\displaystyle\qquad = \frac{(2acx+bc+ad)\sqrt{ax+b}\cdot\sqrt{cx+d}}{4ac}$

$\displaystyle\qquad -\frac{(ad-bc)^2}{8ac}\int \frac{dx}{\sqrt{ax+b}\cdot\sqrt{cx+d}}.$

468

179. $\displaystyle\int \sqrt{\frac{cx+d}{ax+b}}\,dx = \frac{\sqrt{ax+b}\cdot\sqrt{cx+d}}{a}$

$\displaystyle\qquad + \frac{(ad-bc)}{2a}\int \frac{dx}{\sqrt{ax+b}\cdot\sqrt{cx+d}}.$

180. $\displaystyle\int \sqrt{\frac{x+b}{x+d}}\,dx = \sqrt{x+d}\cdot\sqrt{x+b}$

$\displaystyle\qquad + (b-d)\log\left[\sqrt{x+d} + \sqrt{x+b}\,\right].$

181. $\displaystyle\int \sqrt{\frac{1+x}{1-x}}\,dx = \sin^{-1}x - \sqrt{1-x^2}.$

182. $\displaystyle\int \sqrt{\frac{p-x}{q+x}}\,dx = \sqrt{p-x}\cdot\sqrt{q+x} + (p+q)\sin^{-1}\sqrt{\frac{x+q}{p+q}}.$

183. $\displaystyle\int \sqrt{\frac{p+x}{q-x}}\,dx = -\sqrt{p+x}\cdot\sqrt{q-x} - (p+q)\sin^{-1}\sqrt{\frac{q-x}{p+q}}.$

184. $\displaystyle\int \frac{dx}{\sqrt{x-p}\cdot\sqrt{q-x}} = 2\sin^{-1}\sqrt{\frac{x-p}{q-p}}.$

Expressions Containing $\sin ax$.

185. $\displaystyle\int \sin ax\,dx = -\frac{1}{a}\cos ax.$

186. $\displaystyle\int \sin^2 ax\,dx = \frac{x}{2} - \frac{\sin 2ax}{4a}.$

187. $\displaystyle\int \sin^3 ax\,dx = -\frac{1}{a}\cos ax + \frac{1}{3a}\cos^3 ax.$

188. $\displaystyle\int \sin^4 ax\,dx = \frac{3x}{8} - \frac{3\sin 2ax}{16a} - \frac{\sin^3 ax\cos ax}{4a}.$

189. $\displaystyle\int \sin^n ax\,dx = -\frac{\sin^{n-1} ax\cos ax}{na}$

$\displaystyle\qquad + \frac{n-1}{n}\int \sin^{n-2} ax\,dx, \quad (n \text{ pos. integer}).$

190. $\displaystyle\int \frac{dx}{\sin ax} = \frac{1}{a}\log\tan\frac{ax}{2} = \frac{1}{a}\log(\csc ax - \operatorname{ctn} ax).$

191. $\displaystyle\int \frac{dx}{\sin^2 ax} = \int \csc^2 ax\,dx = -\frac{1}{a}\operatorname{ctn} ax.$

192. $\displaystyle\int \frac{dx}{\sin^n ax} = -\frac{1}{a(n-1)}\frac{\cos ax}{\sin^{n-1} ax}$

$\displaystyle\qquad + \frac{n-2}{n-1}\int \frac{dx}{\sin^{n-2} ax}, \quad n \text{ integer} > 1.$

193. $\displaystyle\int \frac{dx}{1\pm\sin ax} = \mp\frac{1}{a}\tan\left(\frac{\pi}{4}\mp\frac{ax}{2}\right).$

194. $\displaystyle\int \frac{dx}{b+c\sin ax}$

$\displaystyle\qquad = \frac{-2}{a\sqrt{b^2-c^2}}\tan^{-1}\left[\sqrt{\frac{b-c}{b+c}}\,\tan\left(\frac{\pi}{4}-\frac{ax}{2}\right)\right], \quad b^2 > c^2.$

195. $\displaystyle\int \frac{dx}{b+c\sin ax}$

$\displaystyle\qquad = \frac{-1}{a\sqrt{c^2-b^2}}\log\frac{c+b\sin ax + \sqrt{c^2-b^2}\,\cos ax}{b+c\sin ax}, \quad c^2 > b^2.$

196. $\displaystyle\int \sin ax\sin bx\,dx = \frac{\sin(a-b)x}{2(a-b)} - \frac{\sin(a+b)x}{2(a+b)}, \quad a^2 \neq b^2.$

197. $\displaystyle\int \sqrt{1+\sin x}\,dx = \pm 2\left(\sin\frac{x}{2} - \cos\frac{x}{2}\right); \quad \text{use } + \text{ sign}$

$\qquad$ when $(8k-1)\dfrac{\pi}{2} < x \leq (8k+3)\dfrac{\pi}{2}$, otherwise $-$, k an integer.

198. $\displaystyle\int \sqrt{1-\sin x}\,dx = \pm 2\left(\sin\frac{x}{2} + \cos\frac{x}{2}\right); \quad \text{use } + \text{ sign}$

$\qquad$ when $(8k-3)\dfrac{\pi}{2} < x \leq (8k+1)\dfrac{\pi}{2}$, otherwise $-$, k an integer.

Expressions Involving $\cos ax$.

199. $\displaystyle\int \cos ax\,dx = \frac{1}{a}\sin ax.$

200. $\displaystyle\int \cos^2 ax\,dx = \frac{x}{2} + \frac{\sin 2ax}{4a}.$

201. $\displaystyle\int \cos^3 ax\,dx = \frac{1}{a}\sin ax - \frac{1}{3a}\sin^3 ax.$

202. $\displaystyle\int \cos^4 ax\,dx = \frac{3x}{8} + \frac{3\sin 2ax}{16a} + \frac{\cos^3 ax\sin ax}{4a}.$

203. $\displaystyle\int \cos^n ax\,dx = \frac{\cos^{n-1} ax\sin ax}{na}$

$\displaystyle\qquad + \frac{n-1}{n}\int \cos^{n-2} ax\,dx, \quad (n \text{ pos. integer}).$

204. $\displaystyle\int \frac{dx}{\cos ax} = \frac{1}{a}\log\tan\left(\frac{ax}{2}+\frac{\pi}{4}\right) = \frac{1}{a}\log(\tan ax + \sec ax).$

205. $\displaystyle\int \frac{dx}{\cos^2 ax} = \frac{1}{a}\tan ax.$

206. $\displaystyle\int \frac{dx}{\cos^n ax} = \frac{1}{a(n-1)}\frac{\sin ax}{\cos^{n-1} ax} + \frac{n-2}{n-1}\int \frac{dx}{\cos^{n-2} ax},$

$\qquad n \text{ integer} > 1.$

207. $\int \dfrac{dx}{1+\cos ax} = \dfrac{1}{a}\tan\dfrac{ax}{2}$, $\int \dfrac{dx}{1-\cos ax} = -\dfrac{1}{a}\operatorname{ctn}\dfrac{ax}{2}$.

221. $\int \dfrac{\sin ax}{b+c\cos ax}\,dx = -\dfrac{1}{ac}\log(b+c\cos ax)$.

208. $\int \sqrt{1+\cos x}\cdot dx = \pm\sqrt{2}\int \cos\dfrac{x}{2}\,dx = \pm 2\sqrt{2}\,\sin\dfrac{x}{2}$.

Use $+$ when $(4k-1)\pi < x \leqq (4k+1)\pi$, otherwise $-$, k an integer.

222. $\int \dfrac{dx}{b\sin ax + c\cos ax} = \dfrac{1}{a\sqrt{b^2+c^2}}\left[\log\tan\tfrac{1}{2}\left(ax+\tan^{-1}\dfrac{c}{b}\right)\right]$.

209. $\int \sqrt{1-\cos x}\cdot dx = \pm\sqrt{2}\int \sin\dfrac{x}{2}\,dx = \mp 2\sqrt{2}\,\cos\dfrac{x}{2}$.

Use top signs when $4k\pi < x \leqq (4k+2)\pi$, otherwise bottom signs.

223. $\int \dfrac{dx}{b+c\cos ax + d\sin ax} = \dfrac{-1}{a\sqrt{b^2-c^2-d^2}}\sin^{-1}U$.

$U = \left[\dfrac{c^2+d^2+b(c\cos ax + a\sin ax)}{\sqrt{c^2+d^2}\,(b+c\cos ax + d\sin ax)}\right]$;

210. $\int \dfrac{dx}{b+c\cos ax} = \dfrac{1}{a\sqrt{b^2-c^2}}\tan^{-1}\left(\dfrac{\sqrt{b^2-c^2}\cdot\sin ax}{c+b\cos ax}\right)$,

or $= \dfrac{1}{a\sqrt{c^2+d^2-b^2}}\log V$,

$b^2 > c^2$.

211. $\int \dfrac{dx}{b+c\cos ax} = \dfrac{1}{a\sqrt{c^2-b^2}}\tanh^{-1}\left[\dfrac{\sqrt{c^2-b^2}\cdot\sin ax}{c+b\cos ax}\right]$,

$V = \left[\dfrac{c^2+d^2+b(c\cos ax + d\sin ax)}{\,+\sqrt{c^2+d^2-b^2}\,(c\sin ax - d\cos ax)}\middle/ \sqrt{c^2+d^2}\,(b+c\cos ax + d\sin ax)\right]$

$c^2 > b^2$.

$b^2 \neq c^2 + d^2$, $-\pi < ax < \pi$.

212. $\int \cos ax \cdot \cos bx\,dx = \dfrac{\sin(a-b)x}{2(a-b)} + \dfrac{\sin(a+b)x}{2(a+b)}$, $a^2 \neq b^2$.

224. $\int \dfrac{dx}{b+c\cos ax + d\sin ax}$

$= \dfrac{1}{ab}\left[\dfrac{b-(c+d)\cos ax + (c-d)\sin ax}{b+(c-d)\cos ax + (c+d)\sin ax}\right]$, $b^2 = c^2 + d^2$.

Expressions Containing $\sin ax$ and $\cos ax$.

213. $\int \sin ax \cos bx\,dx = -\dfrac{1}{2}\left[\dfrac{\cos(a-b)x}{a-b} + \dfrac{\cos(a+b)x}{a+b}\right]$,

225. $\int \dfrac{\sin^2 ax\,dx}{b+c\cos^2 ax} = \dfrac{1}{ac}\sqrt{\dfrac{b+c}{b}}\tan^{-1}\left(\sqrt{\dfrac{b}{b+c}}\tan ax\right) - \dfrac{x}{c}$

$a^2 \neq b^2$.

226. $\int \dfrac{\sin ax \cos ax\,dx}{b\cos^2 ax + c\sin^2 ax} = \dfrac{1}{2a(c-b)}\log(b\cos^2 ax + c\sin^2 ax)$.

214. $\int \sin^n ax \cos ax\,dx = \dfrac{1}{a(n+1)}\sin^{n+1}ax$, $n \neq -1$.

227. $\int \dfrac{dx}{b^2\cos^2 ax - c^2\sin^2 ax} = \dfrac{1}{2abc}\log\dfrac{b\cos ax + c\sin ax}{b\cos ax - c\sin ax}$.

215. $\int \cos^n ax \sin ax\,dx = -\dfrac{1}{a(n+1)}\cos^{n+1}ax$, $n \neq -1$.

228. $\int \dfrac{dx}{b^2\cos^2 ax + c^2\sin^2 ax} = \dfrac{1}{abc}\tan^{-1}\left(\dfrac{c\tan ax}{b}\right)$.

216. $\int \dfrac{\sin ax}{\cos ax}\,dx = -\dfrac{1}{a}\log\cos ax$.

229. $\int \sin^2 ax \cos^2 ax\,dx = \dfrac{x}{8} - \dfrac{\sin 4ax}{32a}$.

217. $\int \dfrac{\cos ax}{\sin ax}\,dx = \dfrac{1}{a}\log\sin ax$.

230. $\int \dfrac{dx}{\sin ax \cos ax} = \dfrac{1}{a}\log\tan ax$.

218. $\int (b+c\sin ax)^n \cos ax\,dx = \dfrac{1}{ac(n+1)}(b+c\sin ax)^{n+1}$,

$n \neq -1$.

231. $\int \dfrac{dx}{\sin^2 ax \cos^2 ax} = \dfrac{1}{a}(\tan ax - \operatorname{ctn} ax)$.

219. $\int (b+c\cos ax)^n \sin ax\,dx = -\dfrac{1}{ac(n+1)}(b+c\cos ax)^{n+1}$,

$n \neq -1$.

232. $\int \dfrac{\sin^2 ax}{\cos ax}\,dx = \dfrac{1}{a}\left[-\sin ax + \log\tan\left(\dfrac{ax}{2} + \dfrac{\pi}{4}\right)\right]$.

220. $\int \dfrac{\cos ax\,dx}{b+c\sin ax} = \dfrac{1}{ac}\log(b+c\sin ax)$.

233. $\int \dfrac{\cos^2 ax}{\sin ax}\,dx = \dfrac{1}{a}\left[\cos ax + \log\tan\dfrac{ax}{2}\right]$.

470

234. $\displaystyle\int \sin^m ax \cos^n ax\,dx = -\frac{\sin^{n-1} ax \cos^{n+1} ax}{a(m+n)}$

$\displaystyle\qquad +\frac{m-1}{m+n}\int \sin^{m-2} ax \cos^n ax\,dx, \quad m,n > 0.$

235. $\displaystyle\int \sin^m ax \cos^n ax\,dx = \frac{\sin^{m+1} ax \cos^{n-1} ax}{a(m+n)}$

$\displaystyle\qquad +\frac{n-1}{m+n}\int \sin^m ax \cos^{n-2} ax\,dx, \quad m,n > 0.$

236. $\displaystyle\int \frac{\sin^m ax}{\cos^n ax}\,dx = \frac{\sin^{m+1} ax}{a(n-1)\cos^{n-1} ax}$

$\displaystyle\qquad -\frac{m-n+2}{n-1}\int \frac{\sin^m ax}{\cos^{n-2} ax}\,dx, \quad m,n > 0,\; n \neq 1.$

237. $\displaystyle\int \frac{\cos^n ax}{\sin^m ax}\,dx = \frac{-\cos^{n+1} ax}{a(m-1)\sin^{m-1} ax}$

$\displaystyle\qquad +\frac{m-n-2}{(m-1)}\int \frac{\cos^n ax}{\sin^{m-2} ax}\,dx, \quad m,n > 0,\; m \neq 1.$

238. $\displaystyle\int \frac{dx}{\sin^m ax \cos^n ax} = \frac{1}{a(n-1)}\frac{1}{\sin^{m-1} ax \cos^{n-1} ax}$

$\displaystyle\qquad +\frac{m+n-2}{(n-1)}\int \frac{dx}{\sin^m ax \cos^{n-2} ax}.$

239. $\displaystyle\int \frac{dx}{\sin^m ax \cos^n ax} = -\frac{1}{a(m-1)}\frac{1}{\sin^{m-1} ax \cos^{n-1} ax}$

$\displaystyle\qquad +\frac{m+n-2}{(m-1)}\int \frac{dx}{\sin^{m-2} ax \cos^n ax}.$

240. $\displaystyle\int \frac{\sin^{2n} ax}{\cos ax}\,dx = \int \frac{\left(1-\cos^2 ax\right)^n}{\cos ax}\,dx.$

(Expand, divide, and use 203).

241. $\displaystyle\int \frac{\cos^{2n} ax}{\sin ax}\,dx = \int \frac{\left(1-\sin^2 ax\right)^n}{\sin ax}\,dx.$

(Expand, divide, and use 189).

242. $\displaystyle\int \frac{\sin^{2n+1} ax}{\cos ax}\,dx = \int \frac{\left(1-\cos^2 ax\right)^n}{\cos ax}\sin ax\,dx.$

(Expand, divide, and use 215).

243. $\displaystyle\int \frac{\cos^{2n+1} ax}{\sin ax}\,dx = \int \frac{\left(1-\sin^2 ax\right)^n}{\sin ax}\cos ax\,dx.$

(Expand, divide, and use 214).

Expressions Containing $\tan ax$ or $\operatorname{ctn} ax$
($\tan ax = 1/\operatorname{ctn} ax$).

244. $\displaystyle\int \tan ax\,dx = -\frac{1}{a}\log\cos ax.$

245. $\displaystyle\int \tan^2 ax\,dx = \frac{1}{a}\tan ax - x.$

246. $\displaystyle\int \tan^3 ax\,dx = \frac{1}{2a}\tan^2 ax + \frac{1}{a}\log\cos ax.$

247. $\displaystyle\int \tan^n ax\,dx = \frac{1}{a(n-1)}\tan^{n-1} ax - \int \tan^{n-2} ax\,dx,$

$\qquad\qquad\qquad\qquad n$ integer $> 1.$

248. $\displaystyle\int \operatorname{ctn} u\,du = \log\sin u, \text{ or } -\log\csc u,$

where u is any function of x.

249. $\displaystyle\int \operatorname{ctn}^2 ax\,dx = \int \frac{dx}{\tan^2 ax} = -\frac{1}{a}\operatorname{ctn} ax - x.$

250. $\displaystyle\int \operatorname{ctn}^3 ax\,dx = -\frac{1}{2a}\operatorname{ctn}^2 ax - \frac{1}{a}\log\sin ax.$

251. $\displaystyle\int \operatorname{ctn}^n ax\,dx = \int \frac{dx}{\tan^n ax} = -\frac{1}{a(n-1)}\operatorname{ctn}^{n-1} ax$

$\qquad\qquad -\int \operatorname{ctn}^{n-2} ax\,dx, \quad n$ integer $> 1.$

252. $\displaystyle\int \frac{dx}{b+c\tan ax} = \int \frac{\operatorname{ctn} ax\,dx}{b\operatorname{ctn} ax + c}$

$\displaystyle\qquad = \frac{1}{b^2+c^2}\left[bx + \frac{c}{a}\log(b\cos ax + c\sin ax)\right].$

253. $\displaystyle\int \frac{dx}{b+c\operatorname{ctn} ax} = \int \frac{\tan ax\,dx}{b\tan ax + c}$

$\displaystyle\qquad = \frac{1}{b^2+c^2}\left[bx - \frac{c}{a}\log(c\cos ax + b\sin ax)\right].$

254. $\displaystyle\int \frac{dx}{\sqrt{b+c\tan^2 ax}}$

$\displaystyle\qquad = \frac{1}{a\sqrt{b-c}}\sin^{-1}\left(\sqrt{\frac{b-c}{b}}\sin ax\right), \quad b \text{ pos.}, b^2 > c^2.$

Expressions Containing $\sec ax = 1/\cos ax$ or
$\csc ax = 1/\sin ax$.

255. $\displaystyle\int \sec ax\,dx = \frac{1}{a}\log\tan\left(\frac{ax}{2} + \frac{\pi}{4}\right).$

256. $\displaystyle\int \sec^2 ax\,dx = \frac{1}{a}\tan ax.$

257. $\displaystyle\int \sec^3 ax\,dx = \frac{1}{2a}\left[\tan ax \sec ax + \log\tan\left(\frac{ax}{2} + \frac{\pi}{4}\right)\right].$

258. $\displaystyle\int \sec^n ax\,dx = \frac{1}{a(n-1)}\frac{\sin ax}{\cos^{n-1}ax}$

$\displaystyle\qquad + \frac{n-2}{n-1}\int \sec^{n-2}ax\,dx, \quad n \text{ integer} > 1.$

259. $\displaystyle\int \csc ax\,dx = \frac{1}{a}\log\tan\frac{ax}{2}.$

260. $\displaystyle\int \csc^2 ax\,dx = -\frac{1}{a}\operatorname{ctn} ax.$

261. $\displaystyle\int \csc^3 ax\,dx = \frac{1}{2a}\left[-\operatorname{ctn} ax \csc ax + \log\tan\frac{ax}{2}\right].$

262. $\displaystyle\int \csc^n ax\,dx = -\frac{1}{a(n-1)}\frac{\cos ax}{\sin^{n-1}ax}$

$\displaystyle\qquad + \frac{n-2}{n-1}\int \csc^{n-2}ax\,dx, \quad n \text{ integer} > 1.$

Expressions Containing tan ax and sec ax or ctn ax and csc ax.

263. $\displaystyle\int \tan ax \sec ax\,dx = \frac{1}{a}\sec ax.$

264. $\displaystyle\int \tan^n ax \sec^2 ax\,dx = \frac{1}{a(n+1)}\tan^{n+1}ax, \quad n \neq -1.$

265. $\displaystyle\int \tan ax \sec^n ax\,dx = \frac{1}{an}\sec^n ax, \quad n \neq 0.$

266. $\displaystyle\int \operatorname{ctn} ax \csc ax\,dx = -\frac{1}{a}\csc ax.$

267. $\displaystyle\int \operatorname{ctn}^n ax \csc^2 ax\,dx = -\frac{1}{a(n+1)}\operatorname{ctn}^{n+1}ax, \quad n \neq -1.$

268. $\displaystyle\int \operatorname{ctn} ax \csc^n ax\,dx = -\frac{1}{an}\csc^n ax, \quad n \neq 0.$

269. $\displaystyle\int \frac{\csc^2 ax\,dx}{\operatorname{ctn} ax} = -\frac{1}{a}\log\operatorname{ctn} ax.$

Expressions Containing Algebraic and Trigonometric Functions.

270. $\displaystyle\int x \sin ax\,dx = \frac{1}{a^2}\sin ax - \frac{1}{a}x\cos ax.$

271. $\displaystyle\int x^2 \sin ax\,dx = \frac{2x}{a^2}\sin ax + \frac{2}{a^3}\cos ax - \frac{x^2}{a}\cos ax.$

272. $\displaystyle\int x^3 \sin ax\,dx = \frac{3x^2}{a^2}\sin ax - \frac{6}{a^4}\sin ax$

$\displaystyle\qquad - \frac{x^3}{a}\cos ax + \frac{6x}{a^3}\cos ax.$

273. $\displaystyle\int x \sin^2 ax\,dx = \frac{x^2}{4} - \frac{x\sin 2ax}{4a} - \frac{\cos 2ax}{8a^2}.$

274. $\displaystyle\int x^2 \sin^2 ax\,dx = \frac{x^3}{6} - \left(\frac{x^2}{4a} - \frac{1}{8a^3}\right)\sin 2ax - \frac{x\cos 2ax}{4a^2}.$

275. $\displaystyle\int x^3 \sin^2 ax\,dx = \frac{x^4}{8} - \left(\frac{x^3}{4a} - \frac{3x}{8a^3}\right)\sin 2ax$

$\displaystyle\qquad - \left(\frac{3x^2}{8a^2} - \frac{3}{16a^4}\right)\cos 2ax.$

276. $\displaystyle\int x \sin^3 ax\,dx = \frac{x\cos 3ax}{12a} - \frac{\sin 3ax}{36a^2} - \frac{3x\cos ax}{4a} + \frac{3\sin ax}{4a^2}.$

277. $\displaystyle\int x^n \sin ax\,dx = -\frac{1}{a}x^n\cos ax + \frac{n}{a}\int x^{n-1}\cos ax\,dx,$

$\qquad n > 0.$

278. $\displaystyle\int \frac{\sin ax\,dx}{x} = ax - \frac{(ax)^3}{3\cdot 3!} + \frac{(ax)^5}{5\cdot 5!} - \cdots.$

279. $\displaystyle\int \frac{\sin ax\,dx}{x^m} = \frac{-1}{(m-1)}\frac{\sin ax}{x^{m-1}} + \frac{a}{(m-1)}\int \frac{\cos ax\,dx}{x^{m-1}}.$

280. $\displaystyle\int x \cos ax\,dx = \frac{1}{a^2}\cos ax + \frac{1}{a}x\sin ax.$

281. $\displaystyle\int x^2 \cos ax\,dx = \frac{2x}{a^2}\cos ax - \frac{2}{a^3}\sin ax + \frac{x^2}{a}\sin ax.$

282. $\displaystyle\int x^3 \cos ax\,dx = \frac{(3a^2x^2-6)\cos ax}{a^4} + \frac{(a^2x^2-6x)\sin ax}{a^3}.$

283. $\displaystyle\int x \cos^2 ax\,dx = \frac{x^2}{4} + \frac{x\sin 2ax}{4a} + \frac{\cos 2ax}{8a^2}.$

284. $\displaystyle\int x^2 \cos^2 ax\,dx = \frac{x^3}{6} + \left(\frac{x^2}{4a} - \frac{1}{8a^3}\right)\sin 2ax + \frac{x\cos 2ax}{4a^2}.$

285. $\int x^3 \cos^2 ax\, dx = \dfrac{x^4}{8} + \left(\dfrac{x^3}{4a} - \dfrac{3x}{8a^3} \right) \sin 2ax$

$$+ \left(\dfrac{3x^2}{8a^2} - \dfrac{3}{16a^4} \right) \cos 2ax.$$

286. $\int x \cos^3 ax\, dx = \dfrac{x \sin 3ax}{12a} + \dfrac{\cos 3ax}{36a^2} + \dfrac{3x \sin ax}{4a} + \dfrac{3 \cos ax}{4a^2}.$

287. $\int x^n \cos ax\, dx = \dfrac{1}{a} x^n \sin ax - \dfrac{n}{a} \int x^{n-1} \sin ax\, dx, \quad n \text{ pos.}$

288. $\int \dfrac{\cos ax\, dx}{x} = \log ax - \dfrac{(ax)^2}{2 \cdot 2!} + \dfrac{(ax)^4}{4 \cdot 4!} - \cdots.$

289. $\int \dfrac{\cos ax}{x^m} dx = - \dfrac{1}{(m-1)} \cdot \dfrac{\cos ax}{x^{m-1}} - \dfrac{a}{(m-1)} \int \dfrac{\sin ax\, dx}{x^{m-1}}.$

Expressions Containing Exponential and Logarithmic Functions.

290. $\int e^{ax}\, dx = \dfrac{1}{a} e^{ax}, \quad \int b^{ax}\, dx = \dfrac{b^{ax}}{a \log b}.$

291. $\int x e^{ax}\, dx = \dfrac{e^{ax}}{a^2} (ax - 1), \quad \int x b^{ax}\, dx = \dfrac{x b^{ax}}{a \log b} - \dfrac{b^{ax}}{a^2 (\log b)^2}.$

292. $\int x^2 e^{ax}\, dx = \dfrac{e^{ax}}{a^3} (a^2 x^2 - 2ax + 2).$

293. $\int x^n e^{ax}\, dx = \dfrac{1}{a} x^n e^{ax} - \dfrac{n}{a} \int x^{n-1} e^{ax}\, dx, \quad n \text{ pos.}$

294. $\int x^n e^{ax}\, dx = \dfrac{e^{ax}}{a^{n+1}} \big[(ax)^n - n(ax)^{n-1}$

$$+ n(n-1)(ax)^{n-2} - \cdots + (-1)^n n! \big], \quad n \text{ pos. integ.}$$

295. $\int x^n e^{-ax}\, dx = - \dfrac{e^{-ax}}{a^{n+1}} \big[(ax)^n + n(ax)^{n-1}$

$$+ n(n-1)(ax)^{n-2} + \cdots + n! \big], \quad n \text{ pos. integ.}$$

296. $\int x^n b^{ax}\, dx = \dfrac{x^n b^{ax}}{a \log b} - \dfrac{n}{a \log b} \int x^{n-1} b^{ax}\, dx, \quad n \text{ pos.}$

297. $\int \dfrac{e^{ax}}{x} dx = \log x + ax + \dfrac{(ax)^2}{2 \cdot 2!} + \dfrac{(ax)^3}{3 \cdot 3!} + \cdots.$

298. $\int \dfrac{e^{ax}}{x^n} dx = \dfrac{1}{n-1} \left[- \dfrac{e^{ax}}{x^{n-1}} + a \int \dfrac{e^{ax}}{x^{n-1}} dx \right], \quad n \text{ integ.} > 1.$

299. $\int \dfrac{dx}{b + ce^{ax}} = \dfrac{1}{ab} \big[ax - \log(b + ce^{ax}) \big].$

300. $\int \dfrac{e^{ax}\, dx}{b + ce^{ax}} = \dfrac{1}{ac} \log(b + ce^{ax}).$

301. $\int \dfrac{dx}{be^{ax} + ce^{-ax}} = \dfrac{1}{a\sqrt{bc}} \tan^{-1} \left(e^{ax} \sqrt{\dfrac{b}{c}} \right), \quad b \text{ and } c \text{ pos.}$

302. $\int e^{ax} \sin bx\, dx = \dfrac{e^{ax}}{a^2 + b^2} (a \sin bx - b \cos bx).$

303. $\int e^{ax} \sin bx \sin cx\, dx = \dfrac{e^{ax} [(b-c) \sin(b-c)x + a \cos(b-c)x]}{2 [a^2 + (b-c)^2]}$

$$- \dfrac{e^{ax} [(b+c) \sin(b+c)x + a \cos(b+c)x]}{2 [a^2 + (b+c)^2]}.$$

304. $\int e^{ax} \cos bx\, dx = \dfrac{e^{ax}}{a^2 + b^2} (a \cos bx + b \sin bx).$

305. $\int e^{ax} \cos bx \cos cx\, dx$

$$= \dfrac{e^{ax} [(b-c) \sin(b-c)x + a \cos(b-c)x]}{2 [a^2 + (b-c)^2]}$$

$$+ \dfrac{e^{ax} [(b+c) \sin(b+c)x + a \cos(b+c)x]}{2 [a^2 + (b+c)^2]}.$$

306. $\int e^{ax} \sin bx \cos cx\, dx$

$$= \dfrac{e^{ax} [a \sin(b-c)x - (b-c) \cos(b-c)x]}{2 [a^2 + (b-c)^2]}$$

$$+ \dfrac{e^{ax} [a \sin(b+c)x - (b+c) \cos(b+c)x]}{2 [a^2 + (b+c)^2]}.$$

307. $\int e^{ax} \sin bx \sin(bx + c)\, dx$

$$= \dfrac{e^{ax} \cos c}{2a} - \dfrac{e^{ax} [a \cos(2bx + c) + 2b \sin(2bx + c)]}{2 (a^2 + 4b^2)}.$$

308. $\int e^{ax} \cos bx \cos(bx + c)\, dx$

$$= \dfrac{e^{ax} \cos c}{2a} + \dfrac{e^{ax} [a \cos(2bx + c) + 2b \sin(2bx + c)]}{2 (a^2 + 4b^2)}.$$

309. $\int e^{ax} \cos bx \sin(bx + c)\, dx$

$$= \dfrac{e^{ax} \sin c}{2a} + \dfrac{e^{ax} [a \sin(2bx + c) - 2b \cos(2bx + c)]}{2 (a^2 + 4b^2)}.$$

310. $\int e^{ax} \sin bx \cos(bx + c)\, dx$

$$= - \dfrac{e^{ax} \sin c}{2a} + \dfrac{e^{ax} [a \sin(2bx + c) - 2b \cos(2bx + c)]}{2 (a^2 + 4b^2)}.$$

311. $\int x e^{ax} \sin bx\, dx = \dfrac{x e^{ax}}{a^2 + b^2} (a \sin bx - b \cos bx)$

$$- \dfrac{e^{ax}}{(a^2 + b^2)^2} \big[(a^2 - b^2) \sin bx - 2ab \cos bx \big].$$

473

312. $\displaystyle\int xe^{ax}\cos bx\,dx = \frac{xe^{ax}}{a^2+b^2}(a\cos bx + b\sin bx)$

$$-\frac{e^{ax}}{(a^2+b^2)^2}\left[(a^2-b^2)\cos bx + 2ab\sin bx\right].$$

313. $\displaystyle\int e^{ax}\cos^n bx\,dx = \frac{e^{ax}\cos^{n-1}bx(a\cos bx + nb\sin bx)}{a^2+n^2b^2}$

$$+\frac{n(n-1)b^2}{a^2+n^2b^2}\int e^{ax}\cos^{n-2}bx\,dx.$$

314. $\displaystyle\int e^{ax}\sin^n bx\,dx = \frac{e^{ax}\sin^{n-1}bx(a\sin bx - nb\cos bx)}{a^2+n^2b^2}$

$$+\frac{n(n-1)b^2}{a^2+n^2b^2}\int e^{ax}\sin^{n-2}bx\,dx.$$

315. $\displaystyle\int \log ax\,dx = x\log ax - x.$

316. $\displaystyle\int x\log ax\,dx = \frac{x^2}{2}\log ax - \frac{x^2}{4}.$

317. $\displaystyle\int x^2\log ax\,dx = \frac{x^3}{3}\log ax - \frac{x^3}{9}.$

318. $\displaystyle\int (\log ax)^2\,dx = x(\log ax)^2 - 2x\log ax + 2x.$

319. $\displaystyle\int (\log ax)^n\,dx = x(\log ax)^n - n\int (\log ax)^{n-1}\,dx,\quad n\text{ pos.}$

320. $\displaystyle\int x^n\log ax\,dx = x^{n+1}\left[\frac{\log ax}{n+1} - \frac{1}{(n+1)^2}\right],\quad n\neq -1.$

321. $\displaystyle\int x^n(\log ax)^m\,dx = \frac{x^{n+1}}{n+1}(\log ax)^m$

$$-\frac{m}{n+1}\int x^n(\log ax)^{m-1}\,dx.$$

322. $\displaystyle\int \frac{(\log ax)^n}{x}\,dx = \frac{(\log ax)^{n+1}}{n+1},\quad n\neq -1.$

323. $\displaystyle\int \frac{dx}{x\log ax} = \log(\log ax).$

324. $\displaystyle\int \frac{dx}{x(\log ax)^n} = -\frac{1}{(n-1)(\log ax)^{n-1}}.$

325. $\displaystyle\int \frac{x^n\,dx}{(\log ax)^m} = \frac{-x^{n+1}}{(m-1)(\log ax)^{m-1}}$

$$+\frac{n+1}{m-1}\int \frac{x^n\,dx}{(\log ax)^{m-1}},\quad m\neq 1.$$

326. $\displaystyle\int \frac{x^n\,dx}{\log ax} = \frac{1}{a^{n+1}}\int \frac{e^y\,dy}{y},\quad y = (n+1)\log ax.$

327. $\displaystyle\int \frac{x^n\,dx}{\log ax} = \frac{1}{a^{n+1}}\left[\log|\log ax| + (n+1)\log ax\right.$

$$\left.+\frac{(n+1)^2(\log ax)^2}{2\cdot 2!} + \frac{(n+1)^3(\log ax)^3}{3\cdot 3!} + \cdots\right].$$

328. $\displaystyle\int \frac{dx}{\log ax} = \frac{1}{a}\left[\log|\log ax| + \log ax\right.$

$$\left.+\frac{(\log ax)^2}{2\cdot 2!} + \frac{(\log ax)^3}{3\cdot 3!} + \cdots\right].$$

329. $\displaystyle\int \sin(\log ax)\,dx = \frac{x}{2}\left[\sin(\log ax) - \cos(\log ax)\right].$

330. $\displaystyle\int \cos(\log ax)\,dx = \frac{x}{2}\left[\sin(\log ax) + \cos(\log ax)\right].$

331. $\displaystyle\int e^{ax}\log bx\,dx = \frac{1}{a}e^{ax}\log bx - \frac{1}{a}\int \frac{e^{ax}}{x}\,dx.$

Expressions Containing Inverse Trigonometric Functions.

332. $\displaystyle\int \sin^{-1}ax\,dx = x\sin^{-1}ax + \frac{1}{a}\sqrt{1-a^2x^2}.$

333. $\displaystyle\int (\sin^{-1}ax)^2\,dx = x(\sin^{-1}ax)^2 - 2x + \frac{2}{a}\sqrt{1-a^2x^2}\sin^{-1}ax.$

334. $\displaystyle\int x\sin^{-1}ax\,dx = \frac{x^2}{2}\sin^{-1}ax - \frac{1}{4a^2}\sin^{-1}ax + \frac{x}{4a}\sqrt{1-a^2x^2}.$

335. $\displaystyle\int x^n\sin^{-1}ax\,dx = \frac{x^{n+1}}{n+1}\sin^{-1}ax$

$$-\frac{a}{n+1}\int \frac{x^{n+1}\,dx}{\sqrt{1-a^2x^2}},\quad n\neq -1.$$

336. $\displaystyle\int \frac{\sin^{-1}ax\,dx}{x} = ax + \frac{1}{2\cdot 3\cdot 3}(ax)^3 + \frac{1\cdot 3}{2\cdot 4\cdot 5\cdot 5}(ax)^5$

$$+\frac{1\cdot 3\cdot 5}{2\cdot 4\cdot 6\cdot 7\cdot 7}(ax)^7 + \cdots,\quad a^2x^2 < 1.$$

337. $\displaystyle\int \frac{\sin^{-1}ax\,dx}{x^2} = -\frac{1}{x}\sin^{-1}ax - a\log\left|\frac{1+\sqrt{1-a^2x^2}}{ax}\right|.$

338. $\displaystyle\int \cos^{-1}ax\,dx = x\cos^{-1}ax - \frac{1}{a}\sqrt{1-a^2x^2}.$

339. $\int \left(\cos^{-1} ax\right)^2 dx = x\left(\cos^{-1} ax\right)^2 - 2x$

$$-\frac{2}{a}\sqrt{1 - a^2 x^2}\,\cos^{-1} ax.$$

340. $\int x \cos^{-1} ax\, dx = \frac{x^2}{2}\cos^{-1} ax$

$$-\frac{1}{4a^2}\cos^{-1} ax - \frac{x}{4a}\sqrt{1 - a^2 x^2}.$$

341. $\int x^n \cos^{-1} ax\, dx = \frac{x^{n+1}}{n+1}\cos^{-1} ax$

$$+\frac{a}{n+1}\int \frac{x^{n+1}\,dx}{\sqrt{1 - a^2 x^2}},\quad n \neq -1.$$

342. $\int \frac{\cos^{-1} ax\, dx}{x} = \frac{\pi}{2}\log|ax| - ax - \frac{1}{2 \cdot 3 \cdot 3}(ax)^3$

$$-\frac{1 \cdot 3}{2 \cdot 4 \cdot 5 \cdot 5}(ax)^5 - \frac{1 \cdot 3 \cdot 5}{2 \cdot 4 \cdot 6 \cdot 7 \cdot 7}(ax)^7 - \cdots,\quad a^2 x^2 < 1.$$

343. $\int \frac{\cos^{-1} ax\, dx}{x^2} = -\frac{1}{x}\cos^{-1} ax + a\log\left|\frac{1 + \sqrt{1 - a^2 x^2}}{ax}\right|.$

344. $\int \tan^{-1} ax\, dx = x\tan^{-1} ax - \frac{1}{2a}\log(1 + a^2 x^2).$

345. $\int x^n \tan^{-1} ax\, dx = \frac{x^{n+1}}{n+1}\tan^{-1} ax - \frac{a}{n+1}\int \frac{x^{n+1}\,dx}{1 + a^2 x^2},$

$$n \neq -1.$$

346. $\int \frac{\tan^{-1} ax\, dx}{x^2} = -\frac{1}{x}\tan^{-1} ax - \frac{a}{2}\log\left(\frac{1 + a^2 x^2}{a^2 x^2}\right).$

347. $\int \text{ctn}^{-1} ax\, dx = x\,\text{ctn}^{-1} ax + \frac{1}{2a}\log(1 + a^2 x^2).$

348. $\int x^n \text{ctn}^{-1} ax\, dx = \frac{x^{n+1}}{n+1}\text{ctn}^{-1} ax + \frac{a}{n+1}\int \frac{x^{n+1}\,dx}{1 + a^2 x^2},$

$$n \neq -1.$$

349. $\int \frac{\text{ctn}^{-1} ax\, dx}{x^2} = -\frac{1}{x}\text{ctn}^{-1} ax + \frac{a}{2}\log\left(\frac{1 + a^2 x^2}{a^2 x^2}\right).$

350. $\int \sec^{-1} ax\, dx = x\sec^{-1} ax - \frac{1}{a}\log\left(ax + \sqrt{a^2 x^2 - 1}\right).$

351. $\int x^n \sec^{-1} ax\, dx = \frac{x^{n+1}}{n+1}\sec^{-1} ax \pm \frac{1}{n+1}\int \frac{x^n\,dx}{\sqrt{a^2 x^2 - 1}},$

$$n \neq -1.$$

Use + sign when $\frac{\pi}{2} < \sec^{-1} ax < \pi$; − sign when $0 < \sec^{-1} ax < \frac{\pi}{2}$.

352. $\int \csc^{-1} ax\, dx = x\csc^{-1} ax + \frac{1}{a}\log\left(ax + \sqrt{a^2 x^2 - 1}\right).$

353. $\int x^n \csc^{-1} ax\, dx = \frac{x^{n+1}}{n+1}\csc^{-1} ax \pm \frac{1}{n+1}\int \frac{x^n\,dx}{\sqrt{a^2 x^2 - 1}},$

$$n \neq -1.$$

Use + sign when $0 < \csc^{-1} ax < \frac{\pi}{2}$; − sign when $-\frac{\pi}{2} < \csc^{-1} ax < 0$.

Definite Integrals

354. $\int_0^\infty \frac{a\,dx}{a^2 + x^2} = \frac{\pi}{2}$, if $a > 0$; $\ 0$, if $a = 0$; $\ \frac{-\pi}{2}$, if $a < 0$.

355. $\int_0^\infty x^{n-1} e^{-s}\,dx = \int_0^1 \left[\log_e \frac{1}{x}\right]^{n-1} dx = \Gamma(n),$

$\Gamma(n+1) = n \cdot \Gamma(n)$, if $n > 0$. $\quad \Gamma(2) = \Gamma(1) = 1.$

$\Gamma(n+1) = n!$, if n is an integer. $\Gamma(\tfrac{1}{2}) = \sqrt{\pi}$.

356. $\int_0^\infty e^{-zx} \cdot z^n \cdot x^{n-1}\,dx = \Gamma(n),\quad z > 0.$

357. $\int_0^1 x^{m-1}(1 - x)^{n-1}\,dx = \int_0^\infty \frac{x^{m-1}\,dx}{(1 + x)^{m+n}} = \frac{\Gamma(m)\Gamma(n)}{\Gamma(m+n)}.$

358. $\int_0^\infty \frac{x^{n-1}}{1 + x}\,dx = \frac{\pi}{\sin n\pi},\quad 0 < n < 1.$

359. $\int_0^{\frac{\pi}{2}} \sin^n x\, dx = \int_0^{\frac{\pi}{2}} \cos^n x\, dx$

$$= \frac{1}{2}\sqrt{\pi} \cdot \frac{\Gamma\left(\frac{n}{2} + \frac{1}{2}\right)}{\Gamma\left(\frac{n}{2} + 1\right)},\quad \text{if } n > -1;$$

$$= \frac{1 \cdot 3 \cdot 5 \cdots (n-1)}{2 \cdot 4 \cdot 6 \cdots (n)} \cdot \frac{\pi}{2},\quad \text{if } n \text{ is an even integer;}$$

$$= \frac{2 \cdot 4 \cdot 6 \cdots (n-1)}{1 \cdot 3 \cdot 5 \cdot 7 \cdots n},\quad \text{if } n \text{ is an odd integer.}$$

360. $\int_0^\infty \frac{\sin^2 x}{x^2}\,dx = \frac{\pi}{2}.$ **361.** $\int_0^\infty \frac{\sin ax}{x}\,dx = \frac{\pi}{2}$, if $a > 0$.

362. $\int_0^\infty \frac{\sin x \cos ax}{x}\,dx = 0$, if $a < -1$, or $a > 1$;

$$= \frac{\pi}{4},\quad \text{if } a = -1, \text{ or } a = 1;$$

$$= \frac{\pi}{2},\quad \text{if } -1 < a < 1.$$

363. $\int_0^\pi \sin^2 ax\, dx = \int_0^\pi \cos^2 ax\, dx = \frac{\pi}{2}.$

364. $\int_0^{\pi/a} \sin ax \cdot \cos ax\, dx = \int_0^\pi \sin x \cdot \cos x\, dx = 0.$

475

365. $\int_0^\pi \sin ax \sin bx\, dx = \int_0^\pi \cos ax \cos bx\, dx = 0, \quad a \neq b,\ a \text{ and}$
$$b \text{ integers.}$$

366. $\int_0^\pi \sin ax \cos bx\, dx = \dfrac{2a}{a^2 - b^2}, \quad \text{if } a - b \text{ is odd};$
$$= 0, \quad \text{if } a - b \text{ is even; } a \text{ and } b \text{ unequal}$$
$$\text{integers.}$$

367. $\int_0^\infty \dfrac{\sin ax \sin bx}{x^2}\, dx = \dfrac{1}{2}\pi a, \quad \text{if } 0 \leq a < b.$

368. $\int_0^\infty \cos(x^2)\, dx = \int_0^\infty \sin(x^2)\, dx = \dfrac{1}{2}\sqrt{\dfrac{\pi}{2}}\,.$

369. $\int_0^\infty e^{-a^2 x^2}\, dx = \dfrac{\sqrt{\pi}}{2a} = \dfrac{1}{2a}\Gamma\left(\dfrac{1}{2}\right), \quad \text{if } a > 0.$

370. $\int_0^\infty x^n \cdot e^{-ax}\, dx = \dfrac{\Gamma(n+1)}{a^{n+1}}, \text{if } a > 0.$

371. $\int_0^\infty x^{2n} e^{-ax^2}\, dx = \dfrac{1 \cdot 3 \cdot 5 \cdots (2n-1)}{2^{n+1} a^n}\sqrt{\dfrac{\pi}{a}}\,.$

372. $\int_0^\infty \sqrt{x}\, e^{-ax}\, dx = \dfrac{1}{2a}\sqrt{\dfrac{\pi}{a}}\,.$ 373. $\int_0^\infty \dfrac{e^{-ax}}{\sqrt{x}}\, dx = \sqrt{\dfrac{\pi}{a}}\,.$

374. $\int_0^\infty e^{(-x^2 - a^2/x^2)}\, dx = \dfrac{1}{2}e^{-2a}\sqrt{\pi}, \quad \text{if } a > 0.$

375. $\int_0^\infty e^{-as}\cos bx\, dx = \dfrac{a}{a^2 + b^2}, \quad \text{if } a > 0.$

376. $\int_0^\infty e^{-ax}\sin bx\, dx = \dfrac{b}{a^2 + b^2}, \quad \text{if } a > 0.$

377. $\int_0^\infty \dfrac{e^{-ax}\sin x}{x}\, dx = \operatorname{ctn}^{-1} a, \quad a > 0.$

378. $\int_0^\infty e^{-a^2 x^2}\cos bx\, dx = \dfrac{\sqrt{\pi}\cdot e^{-b^2/(4a^2)}}{2a}, \quad \text{if } a > 0.$

379. $\int_0^1 (\log x)^n\, dx = (-1)^n \cdot n!, \quad n \text{ pos. integ.}$

380. $\int_0^1 \dfrac{\log x}{1-x}\, dx = -\dfrac{\pi^2}{6}.$ 381. $\int_0^1 \dfrac{\log x}{1+x}\, dx = -\dfrac{\pi^2}{12}.$

382. $\int_0^1 \dfrac{\log x}{1-x^2}\, dx = -\dfrac{\pi^2}{8}$ 383. $\int_0^1 \dfrac{\log x}{\sqrt{1-x^2}}\, dx = -\dfrac{\pi}{2}\log 2.$

384. $\int_0^1 \log\left(\dfrac{1+x}{1-x}\right)\cdot\dfrac{dx}{x} = \dfrac{\pi^2}{4}.$ 385. $\int_0^\infty \log\left(\dfrac{e^x + 1}{e^x - 1}\right) dx = \dfrac{\pi^2}{4}.$

386. $\int_0^1 \dfrac{dx}{\sqrt{\log(1/x)}} = \sqrt{\pi}\,.$

387. $\int_0^{\frac{\pi}{2}} \log\sin x\, dx = \int_0^{\frac{\pi}{2}} \log\cos x\, dx = -\dfrac{\pi}{2}\cdot\log_e 2.$

388. $\int_0^\pi x \log\sin x\, dx = -\dfrac{\pi^2}{2}\cdot\log_e 2.$

389. $\int_0^1 \log|\log x|\, dx = \int_0^\infty e^{-x}\log x\, dx = -\gamma = -0.5772157\cdots.$

390. $\int_0^1 \left(\log\dfrac{1}{x}\right)^{\frac{1}{2}} dx = \dfrac{\sqrt{\pi}}{2}.$

391. $\int_0^1 \left(\log\dfrac{1}{x}\right)^{-\frac{1}{2}} dx = \sqrt{\pi}\,.$

392. $\int_0^1 x^m \left(\log\dfrac{1}{x}\right)^n dx = \dfrac{\Gamma(n+1)}{(m+1)^{n+1}}, \quad \text{if } m+1 > 0,\, n+1 > 0.$

393. $\int_0^\pi \log(a \pm b\cos x)\, dx = \pi\log\left(\dfrac{a + \sqrt{a^2 - b^2}}{2}\right), \quad a \geq b.$

394. $\int_0^\pi \dfrac{\log(1 + \sin a\cos x)}{\cos x}\, dx = \pi a.$

395. $\int_0^1 \dfrac{x^b - x^a}{\log x}\, dx = \log\dfrac{1+b}{1+a}.$

396. $\int_0^\pi \dfrac{dx}{a + b\cos x} = \dfrac{\pi}{\sqrt{a^2 - b^2}}, \quad \text{if } a > b > 0.$

397. $\int_0^{\frac{\pi}{2}} \dfrac{dx}{a + b\cos x} = \dfrac{\cos^{-1}\left(\dfrac{b}{a}\right)}{\sqrt{a^2 - b^2}}, \quad a > b.$

398. $\int_0^\infty \dfrac{\cos ax\, dx}{1 + x^2} = \dfrac{\pi}{2}\cdot e^{-a}, \text{ if } a > 0; \quad = \dfrac{\pi}{2}e^a, \text{ if } a < 0.$

399. $\int_0^\infty \dfrac{\cos x\, dx}{\sqrt{x}} = \int_0^\infty \dfrac{\sin x\, dx}{\sqrt{x}} = \sqrt{\dfrac{\pi}{2}}\,.$

400. $\int_0^\infty \dfrac{e^{-ax} - e^{-bx}}{x}\, dx = \log\dfrac{b}{a}.$

401. $\int_0^\infty \dfrac{\tan^{-1} ax - \tan^{-1} bx}{x}\, dx = \dfrac{\pi}{2}\log\dfrac{a}{b}.$

402. $\int_0^\infty \dfrac{\cos ax - \cos bx}{x}\, dx = \log\dfrac{b}{a}.$

403. $\int_0^{\frac{\pi}{2}} \dfrac{dx}{a^2\cos^2 x + b^2\sin^2 x} = \dfrac{\pi}{2ab}.$

404. $\int_0^{\frac{\pi}{2}} \dfrac{dx}{(a^2\cos^2 x + b^2\sin^2 x)^2} = \dfrac{\pi(a^2 + b^2)}{4a^3 b^3}.$

405. $\int_0^\pi \dfrac{(a - b \cos x)\, dx}{a^2 - 2ab \cos x + b^2} = 0, \quad \text{if } a^2 < b^2;$

$$= \frac{\pi}{a}, \quad \text{if } a^2 > b^2;$$

$$= \frac{\pi}{2a}, \quad \text{if } a = b.$$

406. $\int_0^1 \dfrac{1 + x^2}{1 + x^4}\, dx = \dfrac{\pi}{4}\sqrt{2}\,.$

407. $\int_0^1 \dfrac{\log(1 + x)}{x}\, dx = \dfrac{1}{1^2} - \dfrac{1}{2^2} + \dfrac{1}{3^2} - \dfrac{1}{4^2} + \cdots = \dfrac{\pi^2}{12}.$

408. $\int_{+\infty}^1 \dfrac{e^{-xu}}{u}\, du = \gamma + \log x - x + \dfrac{x^2}{2 \cdot 2!} - \dfrac{x^3}{3 \cdot 3!} + \dfrac{x^4}{4 \cdot 4!} - \cdots,$

where $\gamma = \lim\limits_{t \to \infty} \left(1 + \dfrac{1}{2} + \dfrac{1}{3} + \cdots + \dfrac{1}{t} - \log t\right) = 0.5772157 \cdots.$

409. $\int_{+\infty}^1 \dfrac{\cos xu}{u}\, du = \gamma + \log x - \dfrac{x^2}{2 \cdot 2!} + \dfrac{x^4}{4 \cdot 4!} - \dfrac{x^6}{6 \cdot 6!} + \cdots,$

where $\gamma = 0.5772157 \cdots.$

410. $\int_0^1 \dfrac{e^{xu} - e^{-xu}}{u}\, du = 2\left(x + \dfrac{x^3}{3 \cdot 3!} + \dfrac{x^4}{5 \cdot 5!} + \cdots\right).$

411. $\int_0^1 \dfrac{1 - e^{-xu}}{u}\, du = x - \dfrac{x^2}{2 \cdot 2!} + \dfrac{x^3}{3 \cdot 3!} - \dfrac{x^4}{4 \cdot 4!} + \cdots.$

412. $\int_0^{\frac{\pi}{2}} \dfrac{dx}{\sqrt{1 - K^2 \sin^2 x}} = \dfrac{\pi}{2}\left[1 + \left(\dfrac{1}{2}\right)^2 K^2 + \left(\dfrac{1 \cdot 3}{2 \cdot 4}\right)^2 K^4 \right.$

$$\left. + \left(\dfrac{1 \cdot 3 \cdot 5}{2 \cdot 4 \cdot 6}\right)^2 K^6 + \cdots\right], \quad \text{if } K^2 < 1.$$

413. $\int_0^{\frac{\pi}{2}} \sqrt{1 - K^2 \sin^2 x}\, dx = \dfrac{\pi}{2}\left[1 - \left(\dfrac{1}{2}\right)^2 K^2 - \left(\dfrac{1 \cdot 3}{2 \cdot 4}\right)^2 \dfrac{K^4}{3} \right.$

$$\left. - \left(\dfrac{1 \cdot 3 \cdot 5}{2 \cdot 4 \cdot 6}\right)^2 \dfrac{K^6}{5} - \cdots\right], \quad \text{if } K^2 < 1.$$

414. $\int_0^\infty e^{-ax} \cosh bx\, dx = \dfrac{a}{a^2 - b^2}, \quad 0 \le |b| < a.$

415. $\int_0^\infty e^{-ax} \sinh bx\, dx = \dfrac{b}{a^2 - b^2}, \quad a > 0,\, a^2 \ne b^2.$

416. $\int_0^\infty x e^{-ax} \sin bx\, dx = \dfrac{2ab}{(a^2 + b^2)^2}, \quad a > 0.$

417. $\int_0^\infty x e^{-ax} \cos bx\, dx = \dfrac{a^2 - b^2}{(a^2 + b^2)^2}, \quad a > 0.$

418. $\int_0^\infty x^2 e^{-ax} \sin bx\, dx = \dfrac{2b(3a^2 - b^2)}{(a^2 + b^2)^3}, \quad a > 0.$

419. $\int_0^\infty x^2 e^{-ax} \cos bx\, dx = \dfrac{2a(a^2 - 3b^2)}{(a^2 + b^2)^3}, \quad a > 0.$

420. $\int_0^\infty x^3 e^{-ax} \sin bx\, dx = \dfrac{24ab(a^2 - b^2)}{(a^2 + b^2)^4}, \quad a > 0.$

421. $\int_0^\infty x^3 e^{-ax} \cos bx\, dx = \dfrac{6(a^4 - 6a^2 b^2 + b^4)}{(a^2 + b^2)^4}, \quad a > 0.$

422. $\int_0^\infty x^n e^{-ax} \sin bx\, dx = \dfrac{i \cdot n!\left[(a - ib)^{n+1} - (a + ib)^{n+1}\right]}{2(a^2 + b^2)^{n+1}},$

$$a > 0.$$

423. $\int_0^\infty x^n e^{-ax} \cos bx\, dx = \dfrac{n!\left[(a - ib)^{n+1} + (a + ib)^{n+1}\right]}{2(a^2 + b^2)^{n+1}},$

$$a > 0.$$

GREEK ALPHABET

Letters	Names		Letters	Names		Letters	Names
A α	Alpha		I ι	Iota		P ρ	Rho
B β	Beta		K κ	Kappa		Σ σ ς	Sigma
Γ γ	Gamma		Λ λ	Lambda		T τ	Tau
Δ δ	Delta		M μ	Mu		Υ υ	Upsilon
E ϵ	Epsilon		N ν	Nu		Φ ϕ φ	Phi
Z ζ	Zeta		Ξ ξ	Xi		X χ	Chi
H η	Eta		O o	Omicron		Ψ ψ	Psi
Θ θ ϑ	Theta		Π π	Pi		Ω ω	Omega

French—English Index

Abaque. Abacus
Abscisse. Abscissa
Accélération. Acceleration
Accélération angulaire. Angular acceleration
Accélération centripète. Centripetal acceleration
Accélération tangentielle. Tangential acceleration
Accolade. Brace
Accumulateur. Accumulator
Acnode. Acnode
Acre. Acre
Action centrifuge. Centrifugal force
Action réciproque. Interaction
Actives. Assets
Addende. Addend
Addition. Addition
Adiabatique. Adiabatic
Adjoint d'une matrice. Adjoint of a matrix
Agent de... Broker
Agnésienne. Witch of Agnesi
Aire. Area
Aire-conservateur. Equiareal (or area-preserving)
Aire de superficie. Surface area
Aire de surface. Surface area
Aire latérale. Lateral area
Ajouteur. Adder
Ajustement des courbes. Curve fitting
Aleph-nul. Aleph-null (or aleph zero)
Aleph zéro. Aleph-null (or aleph zero)
Algébrique. Algebraic
Algèbre. Algebra
Algèbre homologique. Homological algebra
Algorisme. Algorithm
Algorithme. Algorithm
Allongement. Dilatation
Altitude. Altitude
Amortissement. Amortization
Amortisseur. Buffer (in a computing machine)
Amplitude d'un nombre complexe. Amplitude of a complex number
An. Year
Analogie. Analogy
Analyse. Analysis
Analyse de sensitivité. Sensitivity analysis
Analyse des facteurs. Factor analysis
Analyse des vecteurs. Vector analysis
Analyse infinitesimale. Infinitesimal analysis
Analyse tensorielle. Tensor analysis
Analyse vectorielle. Vector analysis
Analysis situs combinatoire. Combinatorial topology
Analyticité. Analiticity
Analytique. Analytical
Angle. Angle
Angle aigu. Acute angle
Angle central. Central angle
Angle dièdre. Dihedral angle
Angle directeur. Direction angle
Angle excentrique d'un ellipse. Eccentric angle of an ellipse
Angle extérieur. Exterior angle
Angle horaire. Hour angle
Angle intérieur. Interior angle
Angle obtus. Obtuse angle
Angle parallactique. Parallactic angle

Angle polyèdre. Polyhedral angle
Angle polyédrique. Polyhedral angle
Angle quadrantal. Quadrantal angles
Angle rapporteur. Protractor
Angle réflex. Reflex angle
Angle relatif. Related angle
Angle rentrant. Reentrant angle
Angle solide. Solid angle
Angle tétraédral. Tetrahedral angle
Angle trièdre. Trihedral angle
Angle vectoriel. Vectorial angle
Angles alternes. Alternate angles
Angles complémentaires. Complementary angles
Angles conjugués. Conjugate angles
Angles correspondants. Corresponding angles
Angles coterminals. Coterminal angles
Angles supplémentaires. Supplementary angles
Angles verticaux. Vertical angles
Anneau circulaire. Annulus
Anneau de cercles. Annulus
Anneau de mesure. Measure ring
Anneau des nombres. Ring of numbers
Année. Year
Annihilateur. Annihilator
Annuité. Annuity
Annuité abregée. Curtate annuity
Annuité contingente. Contingent annuity
Annuité différée. Deferred annuity
Annuité diminuée. Curtate annuity
Annuité fortuite. Continent annuity
Annuité suspendue. Deferred annuity
Annuité tontine. Tontine annuity
Anomalie d'un point. Anomaly of a point
Anse sur une surface. Handle on a surface
Antilogarithme. Antilogarithm
Antiautomorphisme. Antiautomorphism
Anticommutatif. Anticommutative
Antiisomorphisme. Antiisomorphism
Antisymétrique. Antisymmetric
Aphélie. Aphelion
Apothème. Apothem
Appareil chiffreur. Digital device
Application contractante. Contraction mapping
Application d'un espace. Mapping of a space
Application inessentielle. Inessential mapping
Application lisse. Smooth map
Application nonexpansive. Nonexpansive mapping
Approximation. Approximation
Arbélos. Arbilos
Arbre. Tree
Arc-cosécante. Arc-cosecant
Arc-cosinus. Arc-cosine
Arc-cotangente. Arc-cotangent
Arc gothique. Ogive
Arc gradué. Protractor
Arc-sécante. Arc-secant
Arc-sinus. Arc-sine
Arc-tangente. Arc-tangent
Arête d'un solide. Edge of a solid
Arête multiple d'un graphe. Multiple edge in a graph
Argument d'un nombre complexe. Amplitude of a complex number
Argument d'une fonction. Argument of a function

Arithmétique. Arithmetic
Arithmomètre. Arithmometer
Arpenteur. Surveyor
Arrondissage des nombres. Rounding off numbers
Ascension. Grade of a path
Assurance. Insurance
Assurance à vie entière (toute). Whole life insurance
Assurance de vie. Life insurance
Astroïde. Astroid
Asymétrie. Skewness
Asymétrique. Asymmetric
Asymptote. Asymptote
Atmosphère. Atmosphere
Atôme. Atom
Automorphisme. Automorphism
Automorphisme intérieur. Inner automorphism
Autre hypothèse. Alternative hypothesis
Avoir-dupoids. Avoirdupois
Axe. Axis
Axe mineur. Minor axis
Axe principale. Major axis
Axe radicale. Radical axis
Axe transverse. Transverse axis
Axes rectangulaires. Rectangular axes
Axiome. Axiom
Azimut. Azimuth

Barre, bar. Bar
Barre oblique. Solidus
Barycentre. Barycenter
Base. Base
Base. Basis
Base de filtre. Filter base
Base rétrécissante (= base "shrinking"). Shrinking basis
Bei-fonction. Bei function
Bénéficiaire. Beneficiary
Ber-fonction. Ber function
Bicompactum. Bicompactum
Biennal. Biennial
Bijection. Bijection
Bilinéaire. Bilinear
Billion. Billion
Bimodale. Bimodal
Binarie. Binary
Binôme. Binomial (n)
Binormale. Binormal
Biquadratique. Biquadratic
Biréctange. Birectangular
Bissecteur. Bisector
Bon. Bond
Bon de série. Serial bond
Borne. Bound
Borne d'un ensemble. Boundary of a set
Borne d'une suite. Boundary of a set
Borne inferieure. Lower bound
Borne superieure. Upper bound
Borne superieure la moindre. Least upper bound
Borné essentiellement. Essentially bounded
Boule ouverte. Open ball
Bourbaki. Bourbaki
Bout d'une courbe. End point of a curve
Brachistochrone. Brachistochrone
Brachystochrone. Brachistochrone
Branche de la courbe. Branch of curve

Bras de levier. Lever arm
Brasse. Cord

Calcul. Calculation; calculus
Calcul automatique. Automatic computation
Calcul des variations. Calculus of variations
Calcul intégral. Integral calculus
Calculateur analogique. Analogue computer
Calculateur arithmétique. Arithmometer
Calculatoir. Calculating machine
Calorie. Calory
Cancellation. Cancellation
Candela. Candela
Cap-croix. Cross-cap
Caractère. Digit
Caractéristique de logarithme. Characteristic of a logarithm
Cardioïde. Cardioid
Carré. Square
Carré magique. Magic square
Carré parfait. Perfect square
Carte de flux du procedé technologique. Flow chart
Carte profile. Profile map
Cas mutuellement exclusifs. Mutually exclusive events
Catégorie. Category
Catégorique. Categorical
Caténaire. Catenary
Caténoïde. Catenoid
Cathète. Leg of a right triangle
Céleste. Celestial
Cent. Hundred
Centaine. Hundred
Centième part d'un nombre. Hundredth part of a number
Centième partie d'un nombre. Hundredth part of a number
Centigramme. Centigram
Centimètre. Centimeter
Centre de cercle circonscrit à triangle. Circumcenter of a triangle
Centre de cercle inscrit dans un triangle. Incenter of a triangle
Centre de conversion. Fulcrum
Centre de groupe. Central of agroup
Centre de gravité. Barycenter
Centre de gravité. Centroid
Centre de masse. Center of mass
Centre de rayon. Ray center
Centre d'un cercle. Center of a circle
Centre d'une droite. Midpont of a line segment
Cercle. Circle
Cercle auxiliaire. Auxiliary circle
Cercle circonscrit. Circumcircle
Cercle circonscrit. Circumscribed circle
Cercle de convergence. Circle of convergence
Cercle des sommets d'une hyperbole. Auxiliary circle of an hyperbola
Cercle d'unité. Unit circle
Cercle exinscrit. Excircle
Cercle inscrit dans un triangle. Incircle
Cercle vertical. Auxiliary circle
Cercle vicieux. Circular argument
Cercles coaxials. Coaxial circles
Cercles concentriques. Concentric circles
Cercles écrits. Escribed circle

Chaîne des simplexes. Chain of simplexes
Chaînette. Catenary
Chaleur spécifique. Specific heat
Chances. Odds
Changement de base. Change of base
Chaos. Chaos
Charge de dépréciation. Depreciation charge
Cheval-vapeur. (C.V. ou H.P.) Horsepower
Chi-carré. Chi-square
Chiffre. Cipher
Chiffre. Digit
Chiffre signifiant. Significant digit
Chiffre significatif. Significant digit
Cinématique. Kinematics
Cinétique. Kinetics
Cinq. Five
Circonférence. Circumference
Circonférence. Girth
Circuit flip-flop. Flip-flop circuit
Circulant. Circulant
Ciseau contrainte. Shearing strain
Ciseau transformation. Shear transformation
Classe d'équivalence. Equivalence class
Cloture d'ensemble. Closure of a set
Coder à calculateur. Coding for a computing machine
Coefficient. Coefficient
Coefficient binomial. Binomial coefficient
Coefficient de corrélation. Correlation coefficient
Coefficient de corrélation bisériale. Biserial correlation coefficient
Coefficient de régression. Regression coefficient
Coefficient principal. Leading coefficient
Coefficients détachés. Detached coefficients
Coefficients indéterminés. Undetermined coefficients
Cofacteur. Cofactor
Cofonction. Cofunction
Coin. Wedge
Coincident. Coincident
Collinéation. Collineation
Cologarithme. Cologarithm
Coloration de graphes. Graph coloring
Combinaison d'ensemble d'objets. Combination of a set of objects
Combinaison d'une suite d'objets. Combination of a sequence of objects
Combinaison linéaire. Linear combination
Commensurable. Commensurable
Commissionnaire. Broker
Commutateur. Commutator
Commutatif. Commutative
Compactification. Compactification
Compas. Compass
Compas. Dividers
Complément d'ensemble. Complement of a set
Complément de facteur. Cofactor
Complément de latitude. Colatitude
Compléter un carré parfait. Completing the square
Complex simplicieux. Simplicial complex
Composant d'inclusion. Input component
Composant d'une force. Component of a force
Composant de productivité. Output component
Compte. Score
Compter par deux. Count by twos
Compteur du calculateur. Counter of a computing machine

Computation. Computation
Comultiple. Common multiple
Concavité. Concavity
Conchoïde. Conchoid
Conclusion statistique. Statistical inference
Concorde. Union
Condition de chaîne ascendante. Ascending chain condition
Condition de chaîne descendante. Descending chain condition
Condition nécessaire. Necessary condition
Condition suffisante. Sufficient condition
Cône. Cone
Cône circulaire. Circular cone
Cône d'ombre. Umbra
Cône directeur. Director cone
Cône tronqué. Truncated cone
Confiance. Reliability
Configuration. Configuration
Configuration en deux variables. Form in two variables
Configurations superposables. Superposable configurations
Confondu. Coincident
Congru. Coincident
Congruence. Congruence
Conicoïde. Conicoid
Conique. Conic
Conique dégénérée. Degenerate conic
Coniques confocales. Confocal conics
Conjecture de Bieberbach. Bieberbach conjecture
Conjecture de Mordell. Mordell conjecture
Conjecture de Poincaré. Poincaré conjecture
Conjecture de Souslin. Souslin's conjecture
Conjonction. Conjunction
Connexion. Bond; connectivity
Conoïde. Conoid
Consistance des equation. Consistency of equations
Constante d'intégration. Constant of integration
Constante essentielle. Essential constant
Constante littérale. Literal constant
Contenu d'ensemble. Content of a set
Continu. Continuum
Continuation de signe. Continuation of sign
Continuité. Continuity
Continuité uniforme. Uniform continuity
Contour. Contour lines
Contraction. Contraction mapping
Contraction de tenseur. Contraction of a tensor
Convergence absolue. Absolute convergence
Convergence conditionnelle. Conditional convergence
Convergence de série. Convergence of a series
Convergence de suite. Convergence of a sequence
Convergence faible. Weak convergence
Convergence uniforme. Uniform convergence
Convergent de fraction continue. Convergent of a continued fraction
Converger à limite. Converge to a limit
Conversion d'un théorème. Converse of a theorem
Convolution de deux fonctions. Convolution of two functions
Coopératif; coopérative. Cooperative
Coordonnées barycentriques. Barycentric coordinates
Coordonnées cartésiennes. Cartesian coordinates

Coordonnée d'un point. Coordinate of a point
Coordonnées géographiques. Geographic coordinates
Coordonnée polaires. Polar coordinates
Coordonnées sphériques. Spherical coordinates
Corde. Chord
Corde; cordage. Cord
Corde. String
Corde focale. Focal chord
Cordes supplémentaires. Supplemental chords
Corollaire. Corollary
Corps algébriquement complet. Algebraically complete field
Corps convex d'ensemble. Convex hull of a set
Corps de Galois. Galois field
Corps de Galois. Splitting field
Corps parfait. Perfect field
Corrélation illusoire. Illusory correlation
Correspondence bi-univoque. One-to-one correspondence
Cosécante d'angle. Cosecant of angle
Cosinus d'angle. Cosine of angle
Cotangente d'angle. Cotangent of angle
Côté d'un polygone. Side of a polygon
Côté d'un solide. Edge of a solid
Côté initiale d'un angle. Initial side of an angle
Côté terminale d'un angle. Terminal side of an angle
Côtés opposés. Opposite sides
Coup en jeu. Move of a game
Coup personnel. Personal move
Courbe caractéristique. Characteristic curve
Courbe close. Close curve
Courbe convexe. Convex curve
Courbe croisée. Cruciform curve
Courbe dans le plan projectif. Projective plane curve
Courbe de fréquence. Frequency curve
Courbe de la probabilité. Probability curve
Courbe de sécante. Secant curve
Courbe de sinus. Sine curve
Courbe des valeurs cumulaires. Ogive
Courbe du quatrième ordre. Quartic
Courbe empirique. Empirical curve
Courbe épitrochoïde. Epitrochoidal curve
Courbe d'espace. Space curve
Courbe exponentielle. Exponential curve
Courbe fermée. Closed curve
Courbe filetée à gauche. Left-handed curve
Courbe isochrone. Isochronous curve
Courbe lisse sur le plan projectif. Smooth projective plane curve
Courbe logarithmique. Logarithmic curve
Courbe logarithmique à base quelconque. Logistic curve
Courbe logistique. Logistic curve
Courbe méridienne. Meridian curve
Courbe ogive. Ogive
Courbe pédale. Pedal curve
Courbe quartique. Quartic curve
Courbe rectifiable. Rectifiable curve
Courbe réductible. Reducible curve
Courbe serpentine. Serpentine curve
Courbe simple. Simple curve
Courbes supérieure plan. Higher plane curve
Courbe tordue. Twisted curve
Courbe torse. Twisted curve
Courbe unicursale. Unicursal curve

Courbes superosculantes sur une surface. Superosculating curves on a surface
Courbure. Kurtosis
Courbure d'une courbe. Curvature of a curve
Course (distance) entre deux points. Run between two points
Courtier. Broker
Couteau du cordonnier. Shoemaker's knife
Covariance. Covariance
Coversinus. Coversed sine (coversine)
Crible. Sieve
Crochet. Bracket
Croisé de référence. Frame of reference
Crunode. Crunode
Cube. Cube
Cubique bipartite. Bipartite cubic
Cuboctaèdre. Cuboctahedron
Cuboïde. Cuboid
Cumulants. Cumulants
Cuspe. Cusp
Cybernétique. Cybernetics
Cycle. Cycle
Cyclides. Cyclides
Cycloïde. Cycloid
Cylindre. Cylinder
Cylindre hyperbolique. Hyperbolic cylinder
Cylindre parabolique. Parabolic cylinder
Cylindroïde. Cylindroid

Dans le sens contraire des aiguilles d'une montre. Counter-clockwise
De six mois. Biannual
Décagone. Decagon
Décalage unilatéral. Unilateral shift
Décamètre. Decameter
Décimale répétante. Repeating decimal
Décimale terminée. Terminating decimal
Décimètre. Decimeter
Déclinaison. Declination
Déclinaison norde. North declination
Déclinaison sud. South declination
Décomposable aux facteurs. Factorable
Décomposer aux facteurs. Factorization
Décomposition en facteurs uniques. Unique factorization
Décomposition spectrale. Spectral decomposition
Dédoubler. Bisect
Déduction statistique. Statistical inference
Défini uniquement. Uniquely defined
Déformation d'un objet. Deformation of an object
Degré d'un polynôme. Degree of a polynomial
Degré d'un sommet. Valence of a node
Degré d'une trajectoire. Grade of a path
Del. Del
Deltaèdre. Deltahedron
Deltoïde. Deltoid
Demi-angle formules. Half-angle formulas
Démonstration indirecte. Indirect proof
Démontrer une théorème. Prove a theorem
Dénombrabilité. Countability
Dénombrablement compact. Countably compact
Dénombrer par deux. Count by two
Dénominateur. Denominator
Densité. Density
Densité asymptotique. Asymptotic density

Densité supérieure. Upper density
Dépôt composant. Storage component
Dérivée covariant. Covariant derivative
Dérivée directrice. Directional derivative
Dérivée d'ordre supérieur. Derivative of higher order
Dérivée d'une distribution. Derivative of a distribution
Dérivée d'une fonction. Derivative of a function
Dérivée formelle. Formal derivative
Dérivée normale. Normal derivative
Dérivée partielle. Partial derivative
Dérivée suivant un vecteur. Directional derivative
Descent. Grade of a path
Dessiner par composition. Graphing by composition
Désunion. Disjunction
Déterminant. Determinant
Déterminant antisymétrique. Skew-symmetric determinant
Deux. Two
Deuxième dérivée. Second derivative
Développante d'une courbe. Involute of a curve
Développée d'une courbe. Evolute of a curve
Développement. Evolution
Développement asymptotique. Asymptotic expansion
Développement d'un déterminant. Expansion of a determinant
Devenir égaux. Equate
Déviation. Deviation
Déviation probable. Probable deviation
Déviation quartile. Quartile deviation
Diagonale d'un déterminant. Diagonal of a determinant
Diagonale principale. Principal diagonal
Diagonale secondaire. Secondary diagonal
Diagonaliser. Diagonalize
Diagramme. Diagram
Diagramme de barres. Bar graph
Diagramme de dispersement. Scattergram
Diagramme de dispersion. Scattergram
Diagramme d'une équation. Graph of an equation
Diagramme des rectangles. Bar graph
Diamètre d'un cercle. Diameter of a circle
Dichotomie. Dichotomy
Difféomorphisme. Diffeomorphism
Différence de deux carrés. Difference of two squares
Différence tabulaire. Tabular differences
Différencier une fonction. Differencing a function
Différentiation d'une fonction. Differentiation of a function
Différentiation implicite. Implicit differentiation
Différentielle complète. Total differential
Différentielle d'une fonction. Differential of a function
Différentielle entière. Total differential
Différentielle totale. Total differential
Dilatation. Dilatation
Dimension. Dimension
Dimension fractale. Fractal dimension
Dimension de Hausdorff. Hausdorff dimension
Dimension topologique. Topological dimension
Dipôle. Dipole; doublet
Direction asymptotique. Asymptotic direction
Direction d'aiguille. Clockwise
Direction de montre. Clockwise
Directrice d'une conique. Directrix of a conic

Discontinuité. Discontinuity
Discontinuité amovible. Removable discontinuity
Discontinuité insurmontable. Nonremovable discontinuity
Discontinuité pas écartante. Nonremovable discontinuity
Discriminant d'un polynôme. Discriminant of a polynomial
Disjonction. Disjunction
Dispersion. Dispersion
Dispersiongramme. Scattergram
Disproportionné. Disproportionate
Disque. disc (or disk)
Distance de deux points. Distance between two points
Distance de zenith. Coaltitude
Distance polaire. Codeclination
Distribution bêta. Beta distribution
Distribution leptocurtique. Leptokurtic distribution
Distribution lognormale. Lognormal distribution
Distribution mésocurtique. Mesokurtic distribution
Distribution normale bivariée. Bivariate normal distribution
Distribution par courbure haute. Leptokurtic distribution
Distribution par une courbe aplatie. Platikurtic distribution
Distribution par une moyenne courbure. Mesokurtic distribution
Distribution platicurtique. Platykurtic distribution
Divergence d'une série. Divergence of a series
Diverger à partir d'un point. Radiate from a point
Dividende aux un bon. Dividend on a bond
Divine proportion. Golden section
Diviser. Divide
Diviser en deux parties égales. Bisect
Diviseur. Divisor
Diviseur exact. Exact divisor
Divisibilité. Divisibility
Divisibilité par onze. Divisibility by eleven
Division. Division
Division brève. Short division
Division synthétique. Synthetic division
Dix. Ten
Dodécaèdre. Dodecahedron
Dodécagone. Dodecagon
Domaine. Domain
Domaine connecté multiplement. Multiply connected region
Domaine conservatif de pouvoir (force). Conservative field of force
Domaine des nombres. Field of numbers
Domaine de recherche. Field of study
Domaine d'examen. Field of study
Domaine d'investigation. Field of study
Domaine du nombre. Number field
Domaine préservatif de pouvoir (force). Conservative field of force
Domaine simplement connexe. Simply connected region
Domino. Domino
Double règle de trois. Double rule of three
Douze. Twelve
Dualité. Duality
Dualité. Dyad
Duel muet. Silent duel

Duel silencieux. Silent duel
Duel tumultueux. Noisy duel
Duplication du cube. Duplication of the cube
Dyade. Dyad
Dyadique. Dyadic
Dynamique. Dynamics
Dyne. Dyne

Écart-type. Standard deviation
Échangeur. Alternant
Échantillon. Sample
Échelle des imaginaires. Scale of imaginaries
Echelle de température Celsius. Celsius temperature
  scale
Écliptique. Ecliptic
Écrancher. Cancel
Effacer. Cancel
Égal asymptotiquement. Asymptotically equal
Égaler. Equate
Égaliser. Equate
Égalité. Equality
Égalité. Parity
Élargissement. Dilatation
Élasticité. Elasticity
Élément d'intégration. Element of integration
Élément linéaire. Lineal element
Élévation. Altitude
Élévation entre deux points. Rise between two points
Éliminant. Eliminant
Élimination par substitution. Elimination by substitu-
  tion
Ellipse. Ellipse
Ellipsoïde. Ellipsoid
Ellipsoïde aplati. Oblate ellipsoid
Ellipsoïde étendu. Prolate ellipsoid
Élongation. Elongation
Émaner à partir d'un point. Radiate from a point
Emprunt. Loan
Endomorphisme. Endomorphism
Énergie cinétique. Kinetic energy
Ensemble. Manifold, set
Ensemble absorbant. Absorbing set
Ensemble analytique. Analytic set
Ensemble borélien. Borel set
Ensemble borné. Bounded set
Ensemble compact. Compact set
Ensemble connexe. Connected set
Ensemble connexe par arcs. Arc-wise connected set
Ensemble de Julia. Julia set
Ensemble de Mandelbrot. Mandelbrot set
Ensemble dénombrable. Countable set
Ensemble dense. Dense set
Ensemble disconnexe. Disconnected set
Ensemble de vérité. truth set
Ensemble discret. Discrete set
Ensemble énumérable. Countable set
Ensemble fermé. Closed set
Ensemble fini. Finite set
Ensemble flou. Fuzzy set
Ensemble mesurable. Measurable set
Ensemble net. Crisp set
Ensemble ordonné. Ordered set
Ensemble ordonné par série. Serially ordered set
Ensemble ouvert. Open set
Ensemble rare. Rare set

Ensemble secondaire de sous-groupe. Coset of a
  group
Ensemble totalement ordonné. Totally ordered set
Ensemble vide. Empty set
Ensembles disjoints. Disjoint sets
Entier cyclotomique. Cyclotomique integer
Entier naturel. Counting number
Entropie. Entropy
Énumérabilité. Countability
Énumérer par deux. Count by twos
Enveloppe d'une famille des courbes. Envelope of a
  family of curves
Épicycloïde. Epicycloid
Épitrochoïde. Epitrochoid
Épreuve de rapport. Ratio test
Épreuve rapport généralisé. Generalized ratio test
Épuisement de la correlation. Attenuation of correla-
  tion
Équateur. Equator
Équateur célestiel. Celestial equator
Équation aux différences. Difference equation
Équation caractéristique de matrice. Characteristic
  equation of a matrix
Équation cubique réduite. Reuced cubic equation
Équation cyclotomique. Cyclotomic equation
Équation d'ondulation. Wave equation
Équation d'une courbe. Equation of a curve
Équation dépressée. Depressed equation
Équation dérivée. Derived equation
Équation différentielle. Differential equation
Équation différentielle exacte. Exact differential
  equation
Équation homogène. Homogeneous equation
Équation intégrale. Integral equation
Équation monique. Monic equation
Équation polynomiale. Polynomial equation
Équation quadratique. Quadratic equation
Équation quarrée. Quadratic equation
Équation sextique. Sextic equation
Équations consistantes. Consistent equations
Équations dépendantes. Dependent equations
Équations différentielles complètes. Exact differential
  equations
Équations paramétriques. Parametric equations
Équations réciproques. Reciprocal equations
Équations simultanées. Simultaneous equations
Équi-aire. Equiareal (or area-preserving)
Équicontinu pour la topologie de la convergence sim-
  ple. Point-wise equicontinuous
Équicontinu uniformément. Uniformly equicontinu-
  ous
Équidistant. Equidistant
Équilibre. Equilibrium
Équinoxe. Equinox
Érg. Erg
Erreur absolue. Absolute error
Erreur de rond. Round-off error
Erreur d'échantillonnage. Sampling error
Erreur par cent. Percent error
Escompte. Discount
Espace. Space
Espace abstrait. Abstract space
Espace affine. Affine space
Espace bicompact. Bicompact space
Espace compact. Compact space

Espace complet. Complete space

Espace complet topologiquement. Topologicaly complete space

Espace conjugué. Adjoint (or conjugate) space

Espace de Baire. Baire space

Espace de Fréchet. Fréchet space

Espace de Hardy. Hardy space

Espace des orbites. Orbit space

Espace uniformément convex. Uniformly convex space

Espace lacunaire. Lacunary space

Espace métacompact. Metacompact space

Espace métrique. Metric space

Espace métrisable. Metrizable space

Espace métrisable et compact. Compactum

Espace non carré. Nonsquare space

Espace normé. Normed space

Espace paracompact. Paracompact space

Espace projectif. Projective space

Espace qui on peut mettre métrique. Metrizable space

Espace quotient. Quotient space

Espace séparable. Separable space

Espaces séparé. Hausdorff space

Espèce d'un ensemble des points. Species of a set of points

Espèce d'une suite des points. Species of a set of points

Espérance. Expected value

Essais successifs. Successive trials

Estimation impartiale. Unibased estimate

Estimation d'une quantité. Estimate of a quantity

Étendu. Width

Étendu d'un variable. Range of a variable

Éternité. Perpetuity

Étoile circumpolaire. Circumpolar star

Étoile d'un complex. Star of a complex

Évaluation. Evaluation

Évaluer. Evaluate

Évasement. Dilatation

Événements indépendants. Independent events

Évolute d'une courbe. Evolute of a curve

Évolution. Evolution

Excentre. Excenter

Excentricité d'une hyperbole. Eccentricity of a hyperbola

Excès des neuves. Excess of nines

Exercise. Exercise

Expectation de la vie. Expectation of life

Exposant. Exponent

Exposant fractionel. Fractional exponent

Exsécante. Exsecant

Extension. Dilatation

Extension d'un corps. Extension of a field

Extirper. Cancel

Extrapolation. Extrapolation

Extrêmement discontinu. Extremally disconnected

Extrêmes. Extreme terms (or extremes)

Extrémité d'un ensemble. Bound of a set

Extrémité d'une courbe. End point of a curve

Extrémité d'une suite. Bound of a sequence

Face d'un polyèdre. Face of a polyhedron

Facette. Facet

Facteur d'un polynôme. Factor of a polynomial

Facteur intégrant. Integrating factor

Factorielle d'un nombre entier. Factorial of an integer

Faiblement compact. Weakly compact

Faire la preuve de théorème. Prove a theorem

Faire le programme dynamique. Dynamic programming

Faire un programme. Programming

Faire un programme linéaire. Linear programming

Faire une programme non-linéaire. Nonlinear programming

Faisceau des cercles. Pencil of circles

Faisceau des plans. Sheaf of planes

Famille des courbes. Family of curves

Fibré en plans. Bundle of planes

Figure plane. Plane figure

Figure symétrique. Symmetric figure

Figures affines radialement. Radially related figures

Figures congruentes. Congruent figures

Figures homothétiques. Homothetic figures

Figures homotopes. Homotopic figures

Fil à plomb. Plumb line

Filtre. Filter

Finesse d'une partition. Fineness of a partition

Finiment représentable. Finitely representable

Focale d'une parabole. Focus of a parabola

Folium de Descartes. Folium of Descartes

Foncteur. Functor

Fonction absolument continue. Absolutely continuous function

Fonction additive. Additive function

Fonction analytique. Analytic function

Fonction analytique monogène. Monogenic analytic function

Fonction arc-hyperbolique. Arc-hyperbolic function

Fonction automorphe. Automorphic function

Fonction bei. Bei function

Fonction ber. Ber function

Fonction bessélienne. Bessel functions

Fonction caractéristique. Characteristic function

Fonction complémentaire. Cofunction

Fonction composée. Composite function

Fonction continuée. Continuous function

Fonction continue par morceaux. Piecewise continuous function

Fonction croissante. Increasing function

Fonction de classe C^n. Function of class C^n

Fonction kei. Kei function

Fonction ker. Ker function

Fonction de Cantor. Cantor function

Fonction décroissante. Decreasing function

Fonction de Koebe. Koebe function

Fonction delta de Dirac. Dirac delta function

Fonction de payement. Payoff function

Fonction digamma. Digamma function

Fonction d'incidence. Incidence function

Fonction discontinue. Discontinuous function

Fonction disparaissante. Vanishing function

Fonction distributive. Distribution function

Fonction en escalier. Step function

Fonction entière. Entire function

Fonction explicite. Explicit function

Fonction Gamma. Gamma function
Fonction généralisée. Generalized function
Fonction holomorphe. Holomorphic function
Fonction illimite. Unbounded function
Fonction implicite. Implicit function
Fonction injective. Injective function
Fonction intégrable. Integrable function
Fonction localement intégrable. Locally integrable function
Fonction méromorphe. Meromorphic function
Fonction modulaire. Modular function
Fonction monotone. Monotone function
Fonction multiforme. Many valued function
Fonction orthogonale. Orthogonal function
Fonction positive. Positive function
Fonction potentielle. Potential function
Fonction presque périodique. Almost periodic function
Fonction propositionnelle. Propositional function
Fonction propre. Eigenfunction
Fonction sans bornes. Unbounded function
Fonction semi-continue. Semicontinuous function
Fonction sommable. Summable function
Fonction sous-additive. Subadditive function
Fonction sous-harmonique. Subharmonic function
Fonction strictement croissante. Strictly increasing function
Fonction Thêta. Theta function
Fonction trigonométrique inverse. Inverse trigonometric function
Fonction univalente. Schlict function
Fonction univoque. Single valued function
Fonction Zêta. Zeta function
Fonctions de Rademacher. Rademacher functions
Fonctions équicontinues. Equicontinuous functions
Fonctions trigonométriques. Trigonometric functions
Fonds. Capital stock
Fonds d'amortissement. Sinking fund
Force centrifuge. Centrifugal force
Force de mortalité. Force of mortality
Force électromotrice. Electromotive force
Forme canonique. Canonical form
Forme en deux variables. Form in two variables
Forme indéterminée. Indeterminate form
Formule. Formula
Formule de doublement. Duplication formula
Formule de prismoïde. Prismoidal formula
Formule de Viète. Viète formula
Formule par réduction. Reduction formula
Formules par soustraction. Subtraction formulas
Fractal. Fractal
Fraction. Fraction
Fraction continue. Continued fraction
Fraction ordinaire. Common fraction
Fraction partielle. Partial fraction
Fraction propre. Proper fraction
Fraction pure. Proper fraction
Fraction simplifiée. Simplified fraction
Fraction vulgaire. Common fraction
Fraction vulgaire. Vulgar fraction
Fréquence cumulative. Cumulative frequency
Fréquence de classe. Class frequency
Friction. Friction
Frontière d'un ensemble. Frontier of a set
Frontière d'une suite. Frontier of a set

Frustrum d'un solide. Frustrum of a solid

Gamma fonction. Gamma function
Garantie complémentaire. Collateral security
Garantie supplémentaire. Collateral security
Générateur (génératrice) d'une surface. Generator of a surface
Générateurs rectilignes. Rectilinear generators
Génératrice. Generatrix
Gentre d'un ensemble des points. Species of a set of points
Genre d'une suite des points. Species of a set of points
Genre d'une surface. Genus of a surface
Géoïde. Geoid
Géométrie. Geometry
Géométrie à deux dimensions. Two-dimensional geometry
Géométrie à trois dimensions. Three-dimensional geometry
Géométrie projective. Projective geometry
Googol. Googol
Gradient. Gradient
Gradient. Grade of a path
Gramme. Gram
Grandeur d'une étoile. Magnitude of a star
Grandeur inconnue. Unknown quantity
Grandeur scalaire. Scalar quantity
Grandeurs égales. Equal quantities
Grandeurs identiques. Identical quantities
Grandeurs proportionnelles. Proportional quantities
Graphe biparti. Bipartite graph
Graphe complet. Complete graph
Graphe eulérien. Eulerian graph
Graphe hamiltonien. Hamiltonian graph
Graphe planaire. Planar graph
Gravitation. Gravitation
Gravité. Gravity
Grillage. Lattice
Groupe alternant. Alternating group
Groupe alterné. Alternating group
Groupe commutatif. Commutative group
Groupe contrôle, -lant. Control group
Groupe de homologie. Homology group
Groupe de Klein. Four-group
Groupe de l'icosaèdre. Icosahedral group
Groupe de l'octaèdre. Octahedral group
Groupe des nombres. Group of numbers
Groupe des transformations. Transformation group
Groupe diédral. Dihedral group
Groupe diédrique. Dihedral group
Groupe du tétraèdre. Tetrahedral group
Groupe homologue. Homology group
Groupe icosaédral. Icosahedral group
Groupe icosaédrique. Icosahedral group
Groupe octaédral. Octahedral group
Groupe octaédrique. Octahedral group
Groupe résoluble. Solvable group
Groupe tétraédral. Tetrahedral group
Groupe tétraédrique. Tetrahedral group
Groupe topologique. Topological group
Groupement des termes. Grouping terms
Groupoïde. Groupoid
Gudermanienne. Gudermannian
Gyration. Gyration

Harmonique tesséral. Tesseral harmonic
Harmonique zonal. Zonal harmonic
Haut oblique. Slant height
Hauteur. Altitude
Hélice. Helix
Hélicoïde. Helicoid
Hémisphère. Hemisphere
Heptaèdre. Heptahedron
Heptagone. Heptagon
Hexaèdre. Hexahedron
Hexagone. Hexagon
Histogramme. Histogram
Hodographe. Hodograph
Homeomorphisme de deux ensembles. Homeomorphism of two sets
Homogénéité. Homogeneity
Homologique. Homologous
Homologue. Homologous
Homomorphisme de deux ensembles. Homomorphism of two sets
Homos édastique. Homoscedastic; *i.e.*, having equal variance
Horizon. Horizon
Horizontal, -e. Horizontal
Huit. Eight
Hyperplan. Hyperplane
Hyperbole. Hyperbola
Hyperboloïde à une nappe. Hyperboloid of one sheet
Hypersurface. Hypersurface
Hypervolume. Hypervolume
Hypocycloïde. Hypocycloid
Hypoténuse. Hypotenuse
Hypothèse. Hypothesis
Hypothèse admissible. Admissible hypothesis
Hypotrochoïde. Hypotrochoid

Icosaèdre. Icosahedron
Idéal contenu dans un anneau. Ideal contained in a ring
Idéal nilpotent. Nilpotent ideal
Idemfacteur. Idemfactor
Identité. Identity
Image d'un point. Image of a point
Implication. Implication
Impôt. Tax
Impôt supplémentaire. Surtax
Impôt sur le revenu. Income tax
Inch. Inch
Inclinaison. Grade of a path
Inclinaison d'une droite. Inclination of a line
Inclinaison d'un toit. Pitch of a roof
Incrément d'une fonction. Increment of a function
Indicateur, -trice d'un nombre. Indicator of an integer
Indicateur d'un nombre entier. Totient of an integer
Indicatrice d'une courbe. Indicatrix of a curve
Indice d'un radical. Index of a radical
Induction. Induction
Induction incomplète. Incomplete induction
Induction mathématique. Mathematical induction
Induction transfinie. Transfinite induction
Inégalité. Inequality
Inégalité de Bienaymé-Tchebitchev. Chebyshev inequality

Inégalité sans condition. Unconditional inequality
Inégalité sans réserve. Unconditional inequality
Inertie. Inertia
Inférence. Inference
Infinité. Infinity
Insérer dans un espace. Imbed in a space
Insertion d'un ensemble. Imbedding of a set
Insertion d'une suite. Imbedding of a set
Instrument chiffreur. Digital device
Intégrale de Bochner. Bochner integral
Intégrale définie. Definite integral
Intégrale d'énergie. Energy integral
Intégrale de Riemann généralisée. Generalized Riemann integral
Intégrale de surface. Surface integral
Intégrale double. Double integral
Intégrale d'une fonction. Integral of a function
Intégrale impropre. Improper integral
Intégrale indéfinie. Antiderivative
Intégrale indéfinie. Indefinite integral
Intégrale itérée. Iterated integral
Intégrale multiple. Multiple integral
Intégrale particulière. Particular integral
Intégrale simple. Simple integral
Intégrande. Integrand
Intégraphe. Integraph
Intégrateur. Integraph
Intégrateur. Integrator
Intégration mécanique. Mechanical integration
Intégration par parties. Integration by parts
Intensité lumineuse. Candlepower
Intercalation d'un ensemble. Imbedding of a set
Intercalation d'une suite. Imbedding of a set
Intercaler dans un espace. Imbed in a space
Intercepte par une axe. Intercept on an axis
Intérêt composé. Compound interest
Intérêt effectif. Effective interest rate
Intérêt réel. Effective interest rate
Intermédiaire. Average
Interpolation. Interpolation
Intersection. Cap
Intersection de courbes. Intersection of curves
Intersection de deux ensembles. Intersection of two sets
Intervalle de certitude. Confidence interval
Intervalle de confiance. Confidence interval
Intervalle de convergence. Interval of convergence
Intervalle fermé. Closed interval
Intervalle ouvert. Open interval
Intervalles nid en un à l'autre. Nested intervals
Intuitionisme. Intuitionism
Invariant d'une équation. Invariant of an equation
Inverse d'une opération. Inverse of an operation
Inversible. Invertible
Inversion d'un point. Inversion of a point
Inverseur. Inversor
Inversion d'un théorème. Converse of a theorem
Investissement. Investment
Involution sur une droite (ligne). Involution on a line
Isohypses. Level lines
Isolé d'une racine. Isolate a root
Isolement. Disiunction
Isomorphisme de deux ensembles. Isomorphism of two sets

Isothère (ligne d'égale température d'un moyen été). Isothermal line
Isotherme. Isotherm

Jeu à deux personnes. Two-person game
Jeu absolument mélangé. Completely mixed game
Jeu absolument mêlé. Completely mixed game
Jeu absolument mixte. Completely mixed game
Jeu concavo-convexe. Concave-convex game
Jeu coopératif. Cooperative game
Jeu de Banach-Mazur. Mazur-Banach game
Jeu de hex. Game of hex
Jeu de Morra. Morra (a game)
Jeu de Nim. Game of nim
Jeu de position. Positional game
Jeu de somme null. Zero-sum game
Jeu des paires des pieces. Coin-matching game
Jeu entièrement mélangé. Completely mixed game
Jeu entièrement mêlé. Completely mixed game
Jeu fini. Finite game
Jeu entièrement mixte. Completely mixed game
Jeu parfaitement mélangé. Completely mixed game
Jeu parfaitement mêlé. Completely mixed game
Jeu parfaitement mixte. Completely mixed game
Jeu séparable. Separable game
Jeu totalement mélangé. Completely mixed game
Jeu totalement mêlé. Completely mixed game
Jeu totalement mixte. Completely mixed game
Jeu tout à fait mélangé. Completely mixed game
Jeu tout à fait mêlé. Completely mixed game
Jeu tout à fait mixte. Completely mixed game
Joueur d'un jeu. Play of a game
Joueur qui augmente jusqu'à maximum. Maximizing player
Joueur qui augmente jusqu'à minimum. Minimizing player
Joule. Joule

Kappa courbe. Kappa curve
Kei fonction. Kei function
Ker fonction. Ker function
Kilogramme. Kilogram
Kilomètre. Kilometer
Kilowatt. Kilowatt

Lacet. Loop of a curve
Lame. Lamina
Largeur. Breadth
Largeur. Width
Latitude d'un point. Latitude of a point
Lemme. Lemma
Le plus grand commun diviseur. Greatest common divisor
Le problème des ponts de Königsberg. Königsberg bridge problem
Lemniscate. Lemniscate
Lexicographiquement. Lexicographically
Lien. Bond
Lieu. Locus
Lieu-tac. Tac-locus
Ligne brisée. Broken line
Ligne centrale. Bisector
Ligne de tendre. Trend line
Ligne diamétrale. Diametral line
Ligne directée. Directed line

Ligne droite. Straight line
Ligne isotherme. Isothermal line
Ligne isothermique. Isothermal line
Ligne nodale. Nodal line
Ligne orientée. Directed line
Ligne verticale. Vertical line
Lignes antiparallèles. Antiparalleled lines
Lignes concourantes. Concurrent lines
Lignes des contoures. Contour lines
Lignes coplanaires. Coplanar lines
Lignes courantes. Stream lines
Lignes de niveau. Level lines
Lignes obliques. Skew lines
Lignes parallèles. Parallel lines
Lignes perpendiculaires. Perpendicular lines
Limaçon. Limacon
Limite d'un ensemble. Bound of a set
Limite d'une fonction. Limit of a function
Limite inférieure. Inferior limit
Limite inférieure. Lower bound
Limite le moindre supérieure. Least upper bound
Limite supérieure. Superior limit
Limite supérieure. Upper bound
Limité essentialement. Essentially bounded
Limites probables. Fiducial limits
Litre. Liter
Lituus. Lituus
Livre. Pound
Localement compact. Locally compact
Localement connexe par arcs. Locally arc-wise connected
Logarithme d'un nombre. Logarithm of a number
Logarithme naturel. Natural logarithm
Logarithmes ordinaires. Common logarithms
Logique floue. Fuzzy logic
Logistique. Logistic curve
Loi associatif. Associative law
Loi des éxposants. Law of exponents
Loi distributif. Distributive law
Loi du khi carré. Chi-square distribution
Longueur d'un arc. Arc length
Longueur d'une courbe. Length of a curve
Longitude. Longitude
Loxodromie. Loxodromic spiral
Lune. Lune
Lunules d'Hippocrate. Lunes of Hippocrates

Machine à calculer. Computing machine
Mantisse. Mantissa
Marche en jeu. Move in a game
Masse. Mass
Mathématique, -s. Mathematics
Mathématiques abstraites. Abstract mathematics
Mathématiques appliquées. Applied mathematics
Mathématiques constructives. Constructive mathematics
Mathématiques discrètes. Discrete mathematics
Mathématiques du fini. Finite mathematics
Mathématiques pures. Pure mathematics
Matière isotrope. Isotropic matter
Matière isotropique. Isotropic matter
Matrice augmentée. Augmented matrix
Matrice de coéfficients. Matrix of coefficients
Matrice de Vandermonde. Vandermonde matrix

Matrice échelon. Echelon matrix
Matrice hermitienne. Hermitian matrix
Matrice unimodale. Unimodular matrix
Matrice unitaire. Unitary matrix
Matrices conformables. Conformable matrices
Matrices correspondantes. Conformable matrices
Matrices équivalentes. Equivalent matrices
Maximum d'une fonction. Maximum of a function
Mécanique de fluides. Mechanics of fluids
Mécanique de liquides. Mechanics of liquids
Mécanisme chiffreur. Digital device
Médiane. Bisector
Membre d'une equation. Member of an equation
Mémoire component. Memory component
Mensuration. Mensuration
Méridien sur la terre. Meridian on the earth
Mesure d'un ensemble. Measure of a set
Mesure zéro. Measure zero
Méthode de la plus grande pente. Methode of steepest descent
Méthode de simplex. Simplex method
Méthode des moindres carrés. Method of least squares
Méthode d'exhaustion. Method of exhaustion
Méthode dialytique de Sylvester. Dialytic method
Méthode du point-selle. Saddle-point method
Méthode heuristique. Heuristic method
Méthode inductive. Inductive method
Mètre. Meter
Mètre cubique. Stere
Mettre au même niveau que… Equate
Mil. Mil
Mille. Mile
Mille. Thousand
Mille nautique. Nautical mile
Mille naval. Nautical mile
Millimètre. Millimeter
Million. Million
Mineur d'un déterminant. Minor of a determinant
Minimum d'une fonction. Minimum of a function
Minuende. Minuend
Minus. Minus
Minute. Minute
Mode. Mode
Modèle. Sample
Module. Module
Module de la compression. Bulk modulus
Module d'une congruence. Modulus of a congruence
Moitié de cône double. Nappe of a cone
Moitié de rhombe solide. Nappe of a cone
Mole. Mole
Moment d'inertie. Moment of inertia
Moment d'une force. Moment of a force
Moment statique. Static moment
Momentume. Momentum
Monôme. Monomial
Monômial, -e. Monomial
Morphisme. Morphism
Mouvement curviligne. Curvilinear motion
Mouvement harmonique. Harmonic motion
Mouvement périodique. Periodic motion
Mouvement raide. Rigid motion
Mouvement rigide. Rigid motion
Moyenne. Average

Moyenne de deux nombres. Mean (or average) of two numbers
Moyenne géométrique. Geometric average
Moyenne pondérée. Weighted mean
Multiple commun. Common multiple
Multiple d'un nombre. Multiple of a number
Multiplicande. Multiplicand
Multiplicateur. Multiplier
Multiplication de vecteurs. Multiplication of vectors
Multiplicité. Manifold
Multiplicité d'une racine. Multiplicity of a root
Multiplier deux nombres. Multiply two numbers
Myriade. Myriad

Nadir. Nadir
Nappe d'une surface. Sheet of a surface
Négation. Negation
Nerf d'un système des ensembles. Nerve of a system of sets
Neuf. Nine
Newton. Newton
n-ième racine primitive. Primitive nth root
Nilpotente. Nilpotent
Niveler. Equate
Nœud. Loop of a curve
Nœud (dans topologie). Knot in topology
Nœud de distance. Knot of distance
Nœud d'une courbe. Node of a curve
Noeud en astronomie. Node in astronomy
Nombre. Cipher
Nombre. Number
Nombre à ajouter. Addend
Nombre à soustraire. Subtrahend
Nombre abondant. Abundant number
Nombre abondant. Redundant number
Nombre arithmétique. Arithmetic number
Nombre caractéristique d'une matrice. Eigenvalue of a matrice
Nombre cardinal. Cardinal number
Nombre chromatique. Chromatic number
Nombre complexe. Complex number
Nombre complexe conjugué. Conjugate complex numbers
Nombre composé. Composite number
Nombre concret. Denominate number
Nombre défectif. Defective (or deficient) number
Nombre défectueux. Defective (or deficient) number
Nombre déficient. Deficient number
Nombre dénommé. Denominate number
Nombre d'or. Golden section
Nombre de Ramsey. Ramsey number
Nombre entier. Integer
Nombre impair. Odd number
Nombre imparfait. Defective (or deficient) number
Nombre incomplet. Defective (or deficient) number
Nombre irrationnel. Irrational number
Nombre mixte. Mixed number
Nombre négatif. Negative number
Nombre ordinal. Ordinal number
Nombre p-adique. p-adic number
Nombre pair. Even number
Nombre positif. Positive number
Nombre premier. Prime number
Nombre rationnel. Rational number

Nombre rationnel dyadique. Dyadic rational
Nombre réel. Real number
Nombre tordu. Winding number
Nombre tortueux. Winding number
Nombre transcendant. Transcendental number
Nombres algébriques. Signed numbers
Nombres avec signes. Signed numbers
Nombres amiables. Amicable numbers
Nombres amicals. Amicable numbers
Nombres babyloniens. Babylonian numerals
Nombres de Catalan. Catalan numbers
Nombres égyptiens. Egyptian numerals
Nombres grecs. Greek numerals
Nombres hypercomplexes. Hypercomplex numbers
Nombres hyperréels. Hyperreal numbers
Nombres incommensurables. Incommensurable numbers
Nombres non standards. Nonstandard numbers
Nombres premiers jumeaux. Twin primes
Nombres sino-japonais. Chinese-Japanese numerals
Nomogramme. Nomogram
Non biaisé asymptotiquement. Asymptotically unbiased
Non coopératif. Noncooperative
Non résidu. Nonresidue
Nonagone. Nonagon
Normale d'une courbe. Normal to a curve
Norme d'une matrice. Norm of a matrix
Notation. Notation
Notation factorielle. Factorial notation
Notation fonctionnelle. Functional notation
Notation scientifique. Scientific notation
Noyau de Dirichlet. Dirichlet kernel
Noyau de Féjer. Féjer kernel
Noyau d'une équation intégrale. Nucleus (or kernel) of an integral equation
Noyau d'un homomorphisme. Kernel of a homomorphism
Numérateur. Numerator
Numération. Numeration
Numéraux. Numerals

Obligation. Bond
Obligation. Liability
Octaèdre. Octahedron
Octagone. Octagon
Octant. Octant
Ogive. Ogive
Ohme. Ohm
Onze. Eleven
Opérateur. Operator
Opérateur linéaire. Linear operator
Opérateur nabla. Del
Opération. Operation
Opérations élémentaires. Elementary operations
Opération unaire. Unary operation
Orbite. Orbit
Ordonnée d'un point. Ordinate of a point
Ordre de contact. Order of contact
Ordre d'un groupe. Order of a group
Orientation. Orientation
Orienté cohérentement. Coherently oriented
Orienté d'une manière cohérente. Coherently oriented
Orienté en conformité. Concordantly oriented

Orienté en connexion. Coherently oriented
Origine des coordonnées. Origin of coordinates
Orthocentre. Orthocenter
Oscillation d'une fonction. Oscillation of a function

Pantographe. Pantograph
Papiers de valeurs négociables. Negotiable papers
Parabole. Parabola
Parabole cubique. Cubical parabola
Paraboloïde de révolution. Paraboloid of revolution
Paraboloïde hyperbolique. Hyperbolic paraboloid
Paradoxe. Paradox
Paradoxe de Banach-Tarski. Banach-Tarski paradox
Paradoxe de Hausdorff. Hausdorff paradox
Paradoxe de Petersburg. Petersburg paradox
Parallax d'une étoile. Parallax of a star
Parallélépipède. Parallelepiped
Parallèles de latitude. Parallels of latitude
Parallèles géodésiques. Geodesic parallels
Parallélogramme. Parallelogram
Parallélotope. Parallelotope
Paramètre. Parameter
Parenthèse. Parenthesis
Parité. Parity
Partage en deux. Bisect
Partie imaginaire d'un nombre. Imaginary part of a number
Partition d'un nombre entier. Partition of an integer
Partition plus grossière. Coarser partition
Pascal. Pascal
Pavage. Tesselation
Payement en acompte (s). Installment paying
Payement par annuité. Installment paying
Payement par termes. Installment paying
Pendule. Pendulum
Pénombre. Penumbra
Pentadécagone. Pentadecagon
Pentagone. Pentagon
Pentagramme. Pentagram
Pentaèdre. Pentahedron
Pente. Grade of a path
Pente d'un toit. Pitch of a roof
Pente d'une courbe. Slope of a curve
Percentage. Percentage
Percentile. Percentile
Périgone. Perigon
Périhélie. Perihelion
Périmètre. Perimeter
Période d'une fonction. Period of a function
Périodicité. Periodicity
Périphérie. Periphery
Permutation cyclique. Cyclic permutation
Permutation de n objets. Permutation of n things
Permutation droite. Even permutation
Permutation groupe. Permutation group
Permutation paire. Even permutation
Permuteur. Alternant
Perpendiculaire à une surface. Perpendicular to a surface
Perspectivité. Perspectivity
Pharmaceutique. Apothecary
Phase de movement harmonique simple. Phase of simple harmonic motion
Pictogramme. Pictogram
Pied d'une perpendiculaire. Foot of a perpendicular

Pinceau de cercles. Pencil of circles
Plan projectant. Projecting plane
Plan projectif fini. Finite projective plane
Plan rectificant. Rectifying plane
Plan tangent. Tangent plane
Plan tangent à une surface. Plane tangent to a surface
Planimètre. Planimeter
Plans concourants. Copunctal planes
Plans des coordonnées. Coordinate planes
Plasticité. Plasticity
Plus. Plus sign
Poids. Weight
Poids de troy. Troy weight
Point adhérent. Adherent point
Point bissecteur. Bisecting point
Point d'accumulation. Accumulation point
Point d'amas. Cluster point
Point d'appui. Fulcrum
Point d'inflexion. Inflection point
Point de bifurcation. Bifurcation point
Point de condensation. Condesnation point
Point de discontinuité. Point of discontinuity
Point de la courbure. Bend point
Point de la flexion. Bend point
Point de selle. Saddle point
Point de ramification. Branch point
Point de rebroussement. Cusp
Point de tour. Turning point
Point décimal flottant. Floating decimal point
Point décimal mutable. Floating decimal point
Point double. Crunode
Point ellipse. Point ellipse
Point fixe. Fixed point
Point isolé. Acnode
Point limite. Limit point
Point médian. Median point
Point nodal d'une courbe. Node of a curve
Point ombilic. Umbilical point
Point ordinaire. Ordinary point
Point perçant. Piercing point
Point planaire. Planar point
Point saillant. Salient point
Point singulaire. Singular point
Point stable. Stable point
Point stationnaire. Stationary point
Point transperçant. Piercing point
Pointe. Cusp
Points antipodaux. Antipodal points
Points collinéaires. Collinear points
Points concycliques. Concyclic points
Polaire d'une forme quadratique. Polar of a quadratic form
Polarisation. Polarization
Pôle d'un cercle. Pole of a circle
Polyèdre. Polyhedron
Polygone. Polygon
Polygone concave. Concave polygon
Polygone inscrit (dans un cercle, ellipse ...) Inscribed polygon
Polygone régulier. Regular polygon
Polygone régulier avec côtés courbes. Multifoil
Polyhex. Polyhex
Polynôme de Legendre. Polynomial of Legendre
Polyomino. Polyomino

Polytope. Polytope
Population. Population
Possession par temps illimité. Perpetuity
Poste. Addend
Postulate. Postulate
Potentiel électrostatique. Electrostatic potential
Poundale. Poundal
Poutre console. Cantilever beam
Pouvoir centrifuge. Centrifugal force
Pression. Pressure
Preuve. Proof
Preuve déductive. Deductive proof
Preuve indirecte. Indirect proof
Preuve par la descente. Proof by descent
Preuve par neuf. Casting out nines
Prime. Bonus
Prime. Premium
Primitif d'une équation différentielle. Primitive of a differential equation
Principe. Principle
Principe de la borne uniforme. Uniform boundedness principle
Principe de la meilleuse. Principle of optimality
Principe de la plus advantage. Principle of optimality
Principe de localisation. Localization principle
Principe d'optimalité. Principle of optimality
Principe des boîtes. Pidgeon-hole principle
Principe des tiroirs. Pidgeon-hole principle
Principe des tiroirs de Dirichlet. Dirichlet drawer principle
Principe de superposition. Superposition principle
Prismatoïde. Prismatoid
Prisme. Prism
Prisme hexagonale. Hexagonal prism
Prisme hexagone. Hexagonal prism
Prisme quadrangulaire. Quadrangular prism
Prismoïde. Prismoid
Prix. Bonus
Prix. Premium
Prix de rachat. Redeemption price
Prix fixe. Flat price
Prix vente. Selling price
Probabilité d'événement. Probability of occurrence
Probe à comparison. Comparison test
Problème. Exercise
Problème. Problem
Problème à quatre couleurs. Four-color problem
Problème de fermeture-complémentation de Kuratowski. Kuratowski closure-complementation problem
Problème de Kakeya. Kakeya problem
Problème de la valeur au bord. Boundary-value problem
Problème isopérimétrique. Isoperimetric problem
Produit. Yield
Produit cartésien. Cartesian product
Produit de Blaschke. Blaschke product
Produit des nombres. Product of numbers
Produit direct. Direct product
Produit-espace. Product space
Produit infini. Infinite product
Produit interne. Inner product
Produit scalaire. Dot product
Produit tensoriel d'espaces vectoriels. Tensor product of vector spaces

Profit. Profit
Profit brut. Gross profit
Profit net. Net profit
Programme d'Erlangen. Erlangen program
Progression. Progression
Projection d'un vecteur. Projection of a vector
Projection stéréographique. Stereographic projection
Projectivité. Projectivity
Prolongation. Dilatation
Prolongement de signe. Continuation of sign
Proportion. Proportion
Proportion composée. Composition in a proportion
Proportion de déformation. Deformation ratio
Proportionalité. Proportionality
Proposition. Proposition
Propriété d'absorption. Absorption property
Propriété d'approximation. Approximation property
Propriété de bon ordre. Well-ordering property
Propriété de caractère finite. Property of finite character
Propriété de Krein-Milman. Krein-Milman property
Propriété de réflexion. Reflection property
Propriété de trichotomie. Trichotomy property
Propriété globale. Global property
Propriété idempotente. Idempotent property
Propriété intrinsèque. Intrinsic property
Propriété invariante. Invariant property
Propriété locale. Local property
Prouver un théorème. Prove a theorem
Pseudosphère. Pseudosphere
Puissance d'un ensemble. Potency of a set
Puissance d'un nombre. Power of a number
Pyramide. Pyramid
Pyramide pentagonale. Pentagonal pyramid
Pyramide triangulaire. Triangular pyramid

Quadrangle. Quadrangle
Quadrant d'un cercle. Quadrant of a circle
Quadrature d'un cercle. Quadrature of a circle
Quadrifolium. Quadrefoil
Quadrilatéral. Quadrilateral
Quadrilatère. Quadrilateral
Quadrillion. Quadrillion
Quadrique. Quadric
Quantificateur. Quantifier
Quantificateur effectif. Existential quantifier
Quantificateur universal. Universal quantifier
Quantique. Quantic
Quantique quaternaire. Quaternary quantic
Quantité. Quantity
Quantité inconnue. Unknown quantity
Quantité scalaire. Scalar quantity
Quantités égales. Equal quantities
Quantités identiques. Identical quantities
Quantités inversement proportionelles. Inversely proportional quantities
Quantités linéairement dépendantes. Linearly dependent quantities
Quantités proportionnelles. Proportional quantities
Quart. Quarter
Quartier. Quarter
Quaternion. Quaternion
Quatre. Four
Quintillion. Quintillion
Quintique. Quintic

Quotient de deux nombres. Quotient of two numbers

Rabais. Discount
Raccourcissement de la plan. Shrinking of the plane
Racine. Radix
Racine caractéristique d'une matrice. Characteristic root of a matrix
Racine carrée. Square root
Racine cubique. Cube root
Racine d'une équation. Root of an equation
Racine étrangère. Extraneous root
Racine extraire. Extraneous root
Racine irréductible. Irreducible radical
Racine simple. Simple root
Radian. Radian
Radical. Radical
Radical d'un idéal. Radical of an ideal
Radicande. Radicand
Radier à partir d'un point. Radiate from a point
Raison extérieure. External ratio
Rame. Ream
Rangée d'un déterminant. Row of a determinant
Rapidité. Speed
Rapidité constante. Constant speed
Rapport. Ratio
Rapport anharmonique. Anharmonic ratio
Rapport de similitude. Ratio of similitude
Rapport extérieur. External ratio
Rapport interne. Internal ratio
Rarrangement de termes. Rearrangement of terms
Rationnel. Commensurable
Rayon d'un cercle. Radius of a circle
Rebroussement. Cusp
Récepteur de payement. Payee
Réciproque d'un nombre. Reciprocal of a number
Recouvrement d'ensemble. Covering of a set
Rectangle. Rectangle
Rectification d'un cercle. Squaring a circle
Réduction de tenseur. Contraction of a tensor
Réduction d'une fraction. Reduction of a fraction
Réflexibilité. Reflection property
Réflexion dans une ligne. Reflection in a line
Réfraction. Refraction
Région. Domain
Région de confiance. Confidence region
Réglage à une surface. Ruling on a surface
Règle. Ruler
Règle de calcul. Slide rule
Règle de conjointe. Chain rule
Règle de mécanicien. Mechanic's rule
Règle du trapèze. Trapezoid rule
Règle des signes. Rule of signs
Relation. Relation
Relation antisymétrique. Antisymmetric relation
Relation connexe. Connected relation
Relation d'inclusion. Inclusion relation
Relation intransitive. Intransitive relation
Relation réflexive. Reflexive relation
Relation transitive. Transitive relation
Rendement. Yield
Rendre rationnel un dénominateur. Rationalize a denominator
Rente. Annuity
Rente abrégée. Curtate annuity
Rente contingente. Contingent annuity

Rente différée. Deferred annuity
Rente fortuite. Contingent annuity
Rente diminuée. Curtate annuity
Rente suspendue. Deferred annuity
Rente tontine. Tontine annuity
Répandu également. Homoscedastic
Représentation d'un groupe. Representation of a group
Représentation ternaire de nombres. Ternary representation of numbers
Résidu d'une fonction. Residue of a function
Résidu d'une série infinie. Remainder of an infinite series
Résolution graphique. Graphical solution
Résolvante d'une matrice. Resolvent of a matrix
Responsabilité. Liability
Résultante des fonctions. Resultant of functions
Retardation. Deceleration
Rétracte. Retract
Rétrécissement de la plan. Shrinking of the plane
Rétrécissement de tenseur. Contraction of a tensor
Réunion d'ensembles. Union of sets
Revenu net. Net profit
Réversion des séries. Reversion of a series
Révolution d'une courbe à la ronde d'un axe. Revolution of a curve about an axis
Rhombe. Rhombus
Rhomboèdre. Rhombohedron
Rhomboïde. Rhomboid
Rhumb. Rhumb line; bearing of a line
Rosace à trois feuilles. Rose of three leafs
Rotation des axes. Rotation of axes
Rumb. Rhumb line

Saltus d'une fonction. Saltus of a function
Satisfaire une équation. Satisfy an equation
Saut d'une fonction. Jump discontinuity
Schème au hazard. Random device
Schème mnémonique. Mnemonic device
Sécante d'un angle. Secant of an angle
Secteur d'un cercle. Sector of a circle
Section cylindrique. Section of a cylinder
Section d'or. Golden section
Section dorée. Golden section
Section du cylindre. Section of a cylinder
Segment d'une courbe. Segment of a curve
Segment d'une ligne. Line segment
Salinon. Salinon
Salinon d'Archimède. Salinon
Semestriel, -le. Biannual
Semi-cercle. Semicircle
Semi-sinus-versus. Haversine
Sens d'une inégalité. Sense of an inequality
Séparation d'un ensemble. Separation of a set
Sept. Seven
Septillion. Septillion
Série, Séries (*pl.*). Series
Série arithmétique. Arithmetic series
Série autorégressive. Autoregressive series
Série convergente. Convergent series
Série de nombre. Series of numbers
Série de puissances. Power series
Série de puissances formelle. Formal power series
Séries divergentes décidées. Properly divergent series
Séries géométriques. Geometric series

Séries hypergéométriques. Hypergeometric series
Séries infinis. Infinite series
Séries oscillatoires. Oscillating series
Séries sommables. Summable series
Servomécanisme. Servomechanism
Sextillion. Sextillion
Shift unilatéral. Unilateral shift
Signe de sommation. Summation sign
Signe d'un nombre. Sign of a number
Signification d'une déviation. Significance of a deviation
Signum fonction. Signum function
Similitude. Similitude
Simplement équicontinu. Point-wise equicontinuous
Simplex. Simplex
Simplification. Simplification
Singularité-pli. Fold singularity
Sinus d'un angle. Sine of an angle
Sinus verse. Versed sine
Sinusoïde. Sinusoid
Six. Six
Solide d'Archimède. Archimedean solid
Solide de révolution. Solid of revolution
Solides élastiques. Elastic bodies
Solide semi-régulier. Semi-regular solid
Solution d'une équation. Solution of an equation
Solution graphique. Graphical solution
Solution insignificante. Trivial solution
Solution simple. Simple solution
Solution triviale. Trivial solution
Solution vulgaire. Trivial solution
Sommation des séries. Summation of series
Somme des nombres. Sum of numbers
Sommet. Apex
Sourd. Surd
Sous-corps. Subfield
Souscrit. Subscript
Sous-ensemble. Subset
Sous-ensemble definitif complément. Cofinal subset
Sous-ensemble limité complément. Cofinal subset
Sous-groupe. Subgroup
Sous-groupe quasi-distingué. Quasi-normal subgroup
Sous-groupe quasi-invariant. Quasi-normal subgroup
Sous-groupe quasi-normal. Quasi-normal subgroup
Sous-groupes conjuguées. Conjugate subgroups
Sous-normal. Subnormal
Sous-suite. Subsequence
Sous-suite definitive complémente. Cofinal subsequence
Sous-suite limitée complémente. Cofinal subsequence
Sous-tangente. Subtangent
Soustendre un angle. Subtend an angle
Soustraction des nombres. Subtraction of numbers
Spécimen. Sample
Spécimen stratifié. Stratified sample
Spectre d'une matice. Spectrum of a matrix
Spectre résiduel. Residual spectrum
Sphère. Sphere
Sphère exotique. Exotic sphere
Sphères de Dandelin. Dandelin spheres
Sphéroïde. Spheroid
Spinode. Spinode
Spirale équiangle. Equiangular spiral
Spirale sphérique. Loxodromic spiral
Spline. Spline

Squelette d'un complex. Skeleton of a complex
Statique. Statics
Statistique. Statistic
Statistiques. Statistics
Statistiques avec erreurs systématiques. Biased statistics
Statistiques de la vie. Vital statistics
Statistiques robustes. Robust statistics
Stéradiane. Steradian
Stère. Stere
Stock. Stock
Stock. Capital stock
Stratégie dominante. Dominant strategy
Stratégie d'un jeu. Strategy of a game
Stratégie la meilleuse. Optimal strategy
Stratégie la plus avantageuse. Optimal strategy
Stratégie pure. Pure strategy
Stratégie strictement dominant. Strictly dominant strategy
Strophoïde. Strophoid
Substitution dans une équation. Substitution in an equation
Suite arithmétique. Arithmetic sequence
Suite au hazard. Random sequence
Suite autorégressive. Auto-regressive sequence
Suite convergente. Convergent sequence
Suite dense. Dense sequence
Suite des nombres. Sequence of numbers
Suite divergente. Divergent sequence
Suites généralisée de points partiellement ordonnés. Net of partially ordered points
Suite géométrique. Geometric sequence
Suite orthonormale. Orthonormal sequence
Suites disjointes. Disjoint sequences
Suivant de rapport. Consequent in a ratio
Superficie prismatique. Prismatic surface
Superosculation. Superosculation
Superposer deux configurations. Superpose two configurations
Super-réflexif. Super-reflexive
Support d'une fonction. Support of a function
Surensemble. Superset
Surface conique. Conical surface
Surface convexe d'un cylindre. Cylindrical surface
Surface cylindrique. Cylindrical surface
Surface de révolution. Surface of revolution
Surface développable. Developable surface
Surface du quatrième ordre. Quartic
Surface élliptique. Elliptic surface
Surface équipotentielle. Equipotential surface
Surface minimale. Minimal surface
Surface prismatique. Prismatic surface
Surface pseudosphérique. Pseudospherical surface
Surface pyramidale. Pyramidal surface
Surface réglée. Ruled surface
Surface spirale. Spiral surface
Surface translatoire. Translation surface
Surface unilatérale. Unilateral surface
Surfaces isométriques. Isometric surfaces
Surjection. Surjection
Suscrite. Superscript
Syllogisme. Syllogism
Symbole. Symbol
Symboles cunéiformes. Cuneiform symbols
Symétrie axiale. Axial symmetry

Symétrie cyclique. Cyclosymmetry
Symétrie de l'axe. Axial symmetry
Symétrie d'une fonction. Symmetry of a function
Système centésimal de mésure des angles. Centesimal system of measuring angles
Système d'addresse seule. Single address system
Système d'adresse simple. Single address system
Système de courbes isothermes. Isothermic system of curves
Système de courbes isothermiques. Isothermic system of curves
Système décimal. Decimal system
Système de numération hexadécimale. Hexadesimal number system
Système de numération octale. Octal number system
Système de numération sexagésimale. Sexagésimal number system
Système des équations. System of equations
Système duodecimal des nombres. Duodecimal system of numbers
Système international d'unités. International system of units
Système multiadresse. Multiaddress system
Système polyadresse. Multiaddress system
Système sexagésimal des nombres. Sexagesimal system of numbers
Système triplement orthogonal. Triply orthogonal system

Table d'éventualité. Continency table
Table de hazard. Contingency table
Table de mortalité. Mortality table
Table de mortalité choisi. Select mortality table
Table des logarithmes. Table of logarithms
Table du change. Conversion table
Tamis. Sieve
Tangence. Tangency
Tangente d'un angle. Tangent of an angle
Tangente à un cercle. Tangent to a circle
Tangente commune à deux cercles. Common tangent of two circles
Tangente de rebroussement. Inflexional tangent
Tangente d'inflexion. Inflectional tangent
Tangente extérieur à deux cercles. External tangent of two circles
Tangente interne à deux cercles. Internal tangent of two circles
Tantième. Bonus
Tarif. Tariff
Taux (d'intérêts) pour cent. Interest rate
Taux (d'intérêts) pour cent nominale. Nominal rate of interest
Taxe. Tax
Taxe supplémentaire. Surtax
Temps. Time
Temps astral. Sidereal time
Temps nivelé. Equated time
Temps régulateurs. Standard time
Temps sidéral. Sidereal time
Temps solaire. Solar time
Tenseur. Tensor
Tenseur contraindre. Strain tensor
Tenseur contrevariant. Contravariant tensor
Tenseur tendre. Strain tensor
Tension d'une substance. Stress of a body

Terme. Summand
Terme d'une fraction. Term of a fraction
Terme non defini. Undefined term
Termes dissemblables. Dissimilar terms
Termes divers. Dissimilar terms
Termes extrêmes. Extreme terms (or extremes)
Termes hétérogènes. Dissimilar terms
Termes pas ressemblants. Dissimilar terms
Tessélation. Tesselation
Tesseract. Tesseract
Tétraèdre. Tetrahedron
Thème. Exercise
Théorème. Theorem
Théorème de Bezout. Bezout's theorem
Théorème de la récurrence. Recurrence theorem
Théorème de la sous-base d'Alexander. Alexander's subbase theorem
Théorème de la valeur moyenne. Mean-value theorem
Théorème de minimax. Minimax theorem
Théorème de monodrome. Monodromy theorem
Théorème de Pythagore. Pythagorean theorem
Théorème de Radon-Nikodým. Radon-Nikodým theorem
Théorème des douze couleurs. Twelve-color theorem
Théorème des trois carrés. Three-squares theorem
Théorème de Tauber. Tauberian theorem
Théorème de valeur intérmediaire. Intermediate value theorem
Théorème d'existence. Existence theorem
Théorème d'extension de Tietze. Tietze extension theorem
Théorème du minimax. Minimax theorem
Théorème d'unicité. Uniqueness theorem
Théorème du nombre pentagonal d'Euler. Euler pentagonal-number theorem
Théorème du point fixe. Fixed-point theorem
Théorème du point fixe de Banach. Banach fixed-pont theorem
Théorème du résidu. Remainder theorem
Théorème du sandwich au jambon. Ham-sandwich theorem
Théorème étendue de la moyenne. Extended mean value theorem
Théorème fondamental d'algèbere. Fundamental theorem of algebra
Théorème pythagoréen. Theorem of Pythagoras
Théorème pythagoricien. Theorem of Pythagoras
Théorème pythagorique. Theorem of Pythagoras
Théorème réciproque. Dual theorems
Théorie de la rélativité. Relativity theory
Théorie des catastrophes. Catastrophe theory
Théorie des equations. Theory of equations
Théorie des fonctions. Function theory
Théorie des graphes. Graph theory
Théorie ergodique. Ergodic theory
Thermomètre centigrade. Centigrade thermometer
Titres valeurs négociables. Negotiable paper
Toise. Cord
Tonne. Ton
Topographe. Surveyor
Topologie. Topology
Topologie combinatore. Combinatorial topology
Topologie discrète. Discrete topology

Topologie grossière. Indiscrete topology
Topologie projective. Projective topology
Topologie triviale. Trivial topology
Tore. Torus
Torque. Torque
Torsion d'une courbe. Torsion of a curve
Totient d'un nombre entier. Totient of an integer
Totitif d'un nombre entier. Totitive of an integer
Tourbillon de vecteur. Curl of a vector
Trace d'une matrice. Spur of a matrix; trace of a matrix
Tractrice. Tractrix
Trajectoire. Trajectory
Trajectoire d'un projectile. Path of a projectile
Transformation affine. Affine transformation
Transformation auto-adjoint. Self-adjoint transformation
Transformation collinéaire. Collineatory transformation
Transformation conformale. Conformal transformation
Transformation de Fourier rapide. Fast Fourier transform
Transformation des coordonnées. Transformation of coordinates
Transformation étendante. Stretching transformation
Transformation isogonale. Isogonal transformation
Transformation linéaire. Linear transformation
Transformation non singulaire. Nonsingular transformation
Transformation orthogonale. Orthogonal transformation
Transformation par similarité. Similarity transformation
Transformation subjonctive. Conjunctive transformation
Transforme d'une matrice. Transform of a matrix
Transormée de Fourier discrète. Discrete Fourier transform
Transit. Transit
Translation des axes. Translation of axes
Translation unilatérale. Unilateral shift
Transporter un terme. Transpose a term
Transposée d'une matrice. Transpose of a matrix
Transposer un terme. Transpose a term
Transposition. Transposition
Transversale. Transversal
Transverse. Transversal
Trapèze. Trapezium
Trapézoïde. Trapezoid
Travail. Work
Trèfle. Trefoil
Treize. Thirteen
Tresse. Braid
Triangle. Triangle
Triangle équilatéral. Equilateral triangle
Triangle équilatère. Equilateral triangle
Triangle isocèle. Isosceles triangle
Triangle oblique. Oblique triangle
Triangle rectangulaire. Right triangle
Triangle scalène. Scalene triangle
Triangle sphérique trirectangle. Trirectangular spherical triangle
Triangle terrestre. Terrestrial triangle
Triangle similaires. Similar triangles

Triangulation. Triangulation
Trident de Newton. Trident of Newton
Trièdre formé par trois lignes. Trihedral formed by three lines
Trigonométrie. Trigonometry
Trillion. Trillion
Trinôme. Trinomial
Triple intégrale. Triple integral
Triple racine. Triple root
Triplet pythagoréen. Pythagorean triple
Trisection d'un angle. Trisection of an angle
Trisectrice. Trisectrix
Trochoïde. Trochoid
Trois. Three
Tronc d'un solide. Frustum of a solid
Tuile. Tile

Ultrafiltre. Ultrafilter
Ultrafiltre non trivial. Free ultrafilter
Un, une. One
Union. Cup, union
Unité. Unity
Unité astronomique. Astronomical unit

Valeur absolue. Absolute value
Valeur accumulée. Accumulated value
Valeur à livre. Book value
Valeur capitalisée. Capitalized cost
Valeur courante. Market value
Valeur critique. Critical value
Valeur de laitier. Scrap value
Valeur de place. Place value
Valeur de rendre. Surrender value
Valeur d'une police d'assurance. Value of an insur-acne policy
Valeur future. Future value
Valeur locale. Local value
Valeur nominale. Par value
Valeur numéraire. Numerical value
Valeur présente. Present value
Valeur propre. Eigenvalue
Valuation d'un corps. Valuation of a field
Variabilité. Variability
Variable. Variable
Variable dépendant. Dependant variable
Variable indépendant. Independent variable

Variable stochastique. Stochastic variable
Variate. Variate
Variate normalé. Normalized variate
Variation. Variance
Variation des paramètres. Variation of parameters
Variation d'une fonction. Variation of a function
Variété. Manifold
Variéte algébrique affine. Affine algebraic variety
Variété exotique de dimension quatre. Exotic four space
Vecteur. Vector
Vecteur de la force. Force vector
Vecteur non-rotatif. Irrotational vector
Vecteur propre. Eigenvector
Vecteur solenoïdal. Solenoidal vector
Verification de solution. Check on a solution
Versement à compet. Installment payments
Vertex. Apex
Vertex d'un angle. Vertex of an angle
Vibration. Vibration
Vie annuité commune. Joint life annuity
Vie rente commune. Joint life annuity
Vinculé. Vinculum
Vingt. Score, twenty
Vitesse. Speed
Vitesse. Velocity
Vitesse-constante. Constant speed
Vitesse instantanée. Instantaneous velocity
Vitesse relative. Relative velocity
Voisinage d'un point. Neighborhood of a point
Volte. Volt
Volume d'un solide. Volume of a solid

Watt. Watt
Wronskienne. Wronskian

X-Axe. X-axis

Yard de distance. Yard of distance
Y-Axe. Y-axis

Zenith distance. Zenith distance
Zenith d'un observateur. Zenith of an observer
Zéro. Zero
Zêta-fonction. Zeta function
Zone. Zone
Zone interquartile. Interquartile range

German—English Index

Abbildung eines Raumes. Mapping of a space
Abgekürzte Division. Short division
Abgeleitete Gleichung. Derived equation
Abgeplattetes Rotationsellipsoid. Oblate ellipsoid
Abgeschlossene Kurve. Closed curve
Abgeschlossene Menge. Closed set
Abhängige Gleichungen. Dependent equations
Abhängige Veränderliche, abhängige Variable. Dependent variable
Ableitung (Derivierte) einer Funktion. Derivative of a function
Ableitung einer Distribution. Derivative of a distribution
Ableitung höherer Ordnung. Derivative of higher order
Ableitung in Richtung der Normalen. Normal derivative
Ablösungsfond, Tilgungsfond. Sinking fund
Abrundungsfehler, Rundungsfehler. Round-off error
Abschreibungsaufschlag, Abschreibungsposten. Depreciation charge
Abschwächung einer Korrelation. Attenuation of correlation
Absolut stetige Funktion. Absolutely continuous function
Absolute Konvergenz. Absolute convergence
Absoluter Fehler. Absolute error
Absorbierende Menge. Absorbing set
Absorptionseigenschaft. Absorption property
Absteigende Kettenbedingung. Descending chain condition
Abstrakte Mathematik. Abstract mathematics
Abstrakter Raum. Abstract space
Abszisse. Abscissa
Abszissenzuwachs zwischen zwei Punkten. Run between two points
Abundante Zahl. Abundant number, Redundant number
Abweichung, Fehler. Deviation
Abzählbar kompakt. Countably compact
Abzählbare Menge. Denumerable set; Countable set
Abzählbarkeit. Countability
Achse. Axis
Achsenabschnitt. Intercept on an axis
Achsendrehung. Rotation of axes
Achsentranslation. Translation of axes
Acht. Eight
Achteck. Octagon
Acker (= 40.47 a). Acre
Adder, Addierer. Adder
Addition. Addition
Additive Funktion. Additive function
Adiabatisch. Adiabatic
Adjungierte einer Matrix. Adjoint of a matrix
Adjungierter Raum, Dualer Raum, Raum der Linearformen. Adjoint (or conjugate) space
Aegyptisches Zahlensystem (mit ägyptischen Symbolen). Egyptian numerals
Aequinoktium (Tag-und Nachtgleiche). Equinox
Affine algebraische Varietät. Affine algebraic variety
Affine Transformation. Affine transformation
Affiner Raum. Affine space
Ähnliche Dreiecke. Similar triangles

Ähnliche Figuren. Homothetic figures; Radially related figures
Ähnlichkeit. Similitude
Ähnlichkeitstransformation. Similarity transformation
Ähnlichkeitsverhältnis. Ratio of similitude
Aktien. Stock
Aktienkapital. Capital stock
Aktiva, Vermögen. Assets
Alef-Null. Aleph-null (or aleph zero)
Alexanderscher Subbasissatz. Alexander's subbase theorem
Algebra, hyperkomplexes System. Algebra
Algebraisch. Algebraic
Algebraisch abgeschlossener Körper. Algebraically complete field
Algebraische Gleichung sechsten Grades. Sextic equation
Algebraische Kurve höherer als zweiten Ordnung. Higher plane curve
Algebraisches Komplement, Adjunkte. Cofactor
Algorithmus. Algorithm
Allgemeine Unkosten. Overhead expenses
Alternierend. Alternant
Alternierende Gruppe. Alternating group
Amerik. Tonne (= 907,18 kg). Ton
Amortisation. Amortization
Amplitude (Arcus) einer komplexen Zahl. Amplitude of a complex number
Analogie. Analogy
Analog-Rechner. Analogue computer
Analysis. Analysis
Analytische Funktion. Analytic function
Analytische Menge. Analytic set
Analytisches Gebilde. Analytic function
Analytizität. Analyticity
Anbeschriebener Kreis. Inscribed circle
Anfangsstrahl eines Winkels. Initial side of an angle
Angewandte Mathematik. Applied mathematics
Ankreis. Excircle
Anlage. Investment
Annihilator, Annullisator. Annihilator
Anstieg zwischen zwei Punkten. Rise between two points
Antiautomorphismus. Antiautomorphism
Antiisomorphismus. Antiisomorphism
Antikommutativ. Anticommutative
Antipodenpaar. Antipodal points
Antisymmetrisch. Antisymmetric
Antisymmetrische Relation. Antisymmetric relation
Anwartschaftsrente, aufgeschobene Rente. Deferred annuity
Anzahl der primen Restklassen. Indicator of an integer
Aphel. Aphelion
Apotheker. Apothecary
Approximation (Annäherung). Approximation
Approximationseigenschaft. Approximation property
Äquator. Equator
Äquipotentialfläche. Equipotential surface
Äquivalente Matrizen. Equivalent matrices
Äquivalenzklasse. Equivalence class
Arbeit. Work

Arbeitsgebiet. Field of study
Arcus cosekans. Arc-cosecant
Arcus cosinus. Arc-cosine
Arcus cotangens. Arc-cotangent
Arcusfunktion, Zyklometrische Funktion. Inverse trigonometric function
Arcus sekans. Arc-secant
Arcus sinus. Arc-sine
Arcus tangens. Arc tangent
Argument einer Funktion. Argument of a function
(Argument)bereich. Domain
Arithmetik. Arithmetic
Arithmetische Reihe. Arithmetic series
Associatives Gesetz. Associative law
Astroide. Astroid
Astronomische Einheit. Astronomical unit
Asymmetrisch. Asymmetric
Asymptote. Asymptote
Asymptotisch gleich. Asymptotically equal
Asymptotisch unverfälscht. Asymptotically unbiased
Asymptotische Dichte. Asymptotic density
Asymptotische Entwicklung. Asymptotic expansion
Asymptotische Richtung. Asymptotic direction
Atmosphäre. Atmosphere
Atom. Atom
(Aufeinander) senkrechte Geraden. Perpendicular lines
Aufeinanderfolgende Ereignisse. Successive trials
Auflösbare Gruppe. Solvable group
Aufsteigende Kettenbedingung. Ascending chain condition
Aufzählbare Menge. Enumerable set
Ausgangskomponente, Entnahme. Output component
Ausrechnen, den Wert bestimmen. Evaluate
Ausrechnung. Evaluation
Aussage. Proposition (in logic)
Aussagefunktion, Relation, Prädikat (Hilbert-Ackermann). Propositional function
Ausschöpfungsmethode. Method of exhaustion
Aussenglieder. Extreme terms (or extremes)
Aussenwinkel. Exterior angle
Äussere Algebra. Exterior algebra
Äussere Tangente zweier Kreise. External tangent of two circles
Äusseres Teilverhältnis. External ratio
Auswahlaxiom. Axiom of choice
Auszahlungsfunktion. Payoff function
Automatische Berechnung. Automatic computation
Automorphe Funktion. Automorphic function
Automorphismus. Automorphism
Autoregressive Folge. Autoregressive series
Axiale Symmetrie. Axial symmetry
Axiom. Axiom
Azimut. Azimuth

Babylonisches Zahlensystem (mit babylonischen Symbolen). Babylonian numerals
Bahn. Orbit
Bahnenraum. Orbit space
Bairescher Raum. Baire space
Balkendiagramm. Bar graph
Banach-Tarski-Paradoxon. Banach-Tarski paradox
Banachscher Fixpunktsatz. Banach fixed-point theorem

Barwert. Present value
Baryzentrische Koordinaten. Barycentric coordinates
Basis. Base; Basis
Basis wechsel. Change of base
Baum. Tree
Bedeutsame Ziffer, geltende Stelle. Significant digit
Bedingte Konvergenz. Conditional convergence
Befreundete Zahlen. Amicable numbers
Begebbares Papier. Negotiable paper
Benannte Zahl. Denominate number
Berechnung, Rechnung. Computation
Bereich einer Variable. Range of a variable
Berührender Doppelpunkt. Osculation
Berührpunkt. Tangency
Berührung dritter Ordnung. Superosculation
Berührungspunkt. Adherent point
Beschleunigung. Acceleration
Beschränkte Menge. Bounded set
Bestimmt divergente Reihe. Properly divergent series
Bestimmtes Integral. Definite integral
Beta-Verteilung. Beta distribution
Betrag (Absolutwert). Absolute value
Bevölkerung, statistische Gesamtheit, Gesamtmasse, Personengesamtheit. Population
Bewegung. Rigid motion
Beweis. Proof
Beweis durch Abstieg. Proof by descent
Bewertungsring. Valuation ring
Bewichtetes Mittel. Weighted mean
Bezeichnung, Notation. Notation
Bezeichnung der Fakultät. Factorial notation
Bezeichnung mit Funktionssymbolen. Functional notation
Bézoutscher Satz. Bézout's theorem
Biasfreie Schätzung, erwartungstreue Schätzung. Unbiased estimate
Bieberbachsche Vermutung. Bieberbach conjecture
Bijektion. Bijection
Bikompakter Raum. Bicompact space
Bikompaktum. Bicompactum
Bild eines Punktes. Image of a point
Bilinear. Bilinear
Billion. Trillion
Bimodal. Bimodal
Binomial. Binomial
Binomialkoeffizienten. Binomial coefficients
Binormale. Binormal
Biquadratisch. Biquadratic
Biquadratische Kurve. Quartic curve
Blaschke-Produkt. Blaschke product
Blatt einer Riemannschen Fläche. Sheet of a Riemann surface
Bochner-Integral. Bochner integral
Bogenlänge. Arc length
Borelmenge. Borel set
Bourbaki. Bourbaki
Brachistochrone. Brachistochrone
Brechung, Refraktion. Refraction
Breite. Breadth
Breite eines Punktes (geogr.). Latitude of a point
Breitenkreise. Parallels of latitude
Brennpunkt einer Parabel. Focus of a parabola
Brennpunktssehne. Focal chord
Bruch. Fraction
Bruttogewinn. Gross profit

Buchstabenkonstante, d.i. Mitteilungsvariable für Objekte. Literal constant

Buchwert. Book value

Candela (photometrische Einheit für Lichtstärke). Candela

Cantorsche Funktion. Cantor function

Cap (Symbol für das Schneiden von Mengen: ∩). Cap

Cartesisches Produkt. Cartesian product

Catalansche Zahl. Catalan numbers

Celsius-Temperaturskala. Celsius temperature scale

Chancen. Odds

Charakteristische Gleichung einer Matrix. Characteristic equation of a matrix

Charakteristische Kurven (Charakteristken). Characteristic curves

Chaos. Chaos

Chinesisch-Japanisches Zahlensystem (mit entsprechenden Symbolen). Chinese-Japanese numerals

Chi-Quadrat-Verteilung. Chi-square distribution

Chiquadrat, χ^2. Chi-square

Chromatische Zahl. Chromatic number

Cosekans eines Winkels. Cosecant of an angle

Cosinus eines Winkels. Cosine of an angle

Cotangens eines Winkels. Cotangent of an angle

Counting number. Counting number

Crisp set. Crisp set

Cup (Symbol für die Vereinigung von Mengen: ∪). Cup

Dandelinsche Kugeln. Dandelin spheres

Darlehen, Anleihe (in der Versicherung: Policendarlehen). Loan

Darstellung einer Gruppe. Representation of a group

Deduktiver Beweis. Deductive proof

Defiziente Zahl. Defective number

Defiziente Zahl. Deficient number

Deformation (Verformung) eines Objekts. Deformation of an object

Deformationsverhältnis. Deformation ratio

Dehnungstransformation. Stretching transformation

Deklination. Declination

Deltaeder. Deltahedron

Deltoid. Deltoid

Descartes'sches Blatt. Folium of Descartes

Determinante. Determinant

Dezimalsystem. Decimal system

Dezimeter. Decimeter

Diagonale einer Determinante. Diagonal of a determinant

Diagonalisieren. Diagonalize

Diagramm. Diagram

Dialytische Methode. Dialytic method

Dichotomie. Dichotomy

Dichte. Density

Dichte Menge. Dense set

Diedergruppe. Dihedral group

Differential einer Funktion. Differential of a function

Differentialgleichung. Differential equation

Differentiation einer Funktion. Differentiation of a function

Differenzen einer Funktion nehmen. Differencing a function

Differenz zweier Quadrate. Difference of two squares

Differenzengleichung. Difference equation

Diffeomorphismus. Diffeomorphism

Dilatation, Streckung. Dilatation

Dimension. Dimension

Dipol. Dipole; Doublet

Diracsche Distribution. Dirac δ-function

Direktes Produkt. Direct product

Direktrix eines Kegelschnittes, Leitlinie eines Kegelschnittes. Directrix of a conic

Dirichletscher Kern. Dirichlet kernel

Dirichletsches Schubfachprinzip. Dirichlet drawer principle

Disjunkte Mengen. Disjoint sets

Disjunktion. Disjunction

Diskrete Fouriertransformation. Discrete Fourier transform

Diskrete Mathematik. Discrete mathematics

Diskrete Menge. Discrete set

Diskrete Topologie. Discrete topology

Diskriminante eines Polynoms. Discriminant of a polynomial

Dispersion. Dispersion

Distributives Gesetz. Distributive law

Divergente Folge. Divergent sequence

Divergenz einer Vektorfunktion. Divergence of a vectorfunction

Divergenz von Reihen. Divergence of a series

Division. Division

Divisor. Consequent in a ratio

Dodekaeder. Dodecahedron

Dominierende Strategie. Dominant strategy

Dominostein. Domino

Doppelintegral. Double integral

Doppelpunkt. Crunode

Doppelte Sicherstellung. Collateral security

Doppelverhältnis. Anharmonic ratio (modern: cross ratio)

Drehimpuls. Angular momentum

Drehmoment. Torque

Drehpunkt, Stützpunkt. Fulcrum

Drei. Three

Dreibein. Trihedral formed by three lines

Dreiblatt. Trefoil

Dreiblättrige Rose. Rose of three leafs

Dreidimensionale Geometrie. Three-dimensional geometry

Dreieck. Triangle

Dreifache Wurzel. Triple root

Dreifaches Integral. Triple integral

Dreiquadratesatz. Three-squares theorem

Dreisatz. Double rule of three

Dreiteilung eines Winkels. Trisection of an angle

Dreizehn. Thirteen

Druck. Pressure

Druckeinheit. Bar

Druck(spannung). Compression

Duale Theoreme, duale Sätze. Dual theorems

Dualität. Duality

Duodezimalsystem der Zahlen. Duodecimal system of numbers

Durchmesser. Diametral line

Durchmesser eines Kreises. Diameter of a circle

Durchschnitt. Average

Durchschnitt von Mengen. Intersection of sets

Durchschnitt zweier Mengen. Meet of two sets
Durchschnittlicher Fehler. Mean deviation
Durchstosspunkt (einer Geraden), Spurpunkt. Piercing point
Dyade. Dyad
Dyadisch. Dyadic
Dyadische rationale Zahl. Dyadic rational
Dyn. Dyne
Dynamik. Dynamics
Dynamisches Programmieren. Dynamic programming

Ebene. Plane
Ebene Figur. Plane figure
Ebene projektive Kurve. Projective plane curve
Ebenenbündel. Bundle of planes
Ebenenbündel. Copunctal planes; Sheaf of planes
Ebenenschrumpfung. Shrinking of the plane
Echt steigende Funktion. Strictly increasing function
Echter Bruch. Proper fraction
Ecke (einer Kurve). Salient point
Effectiver Zinsfuss. Effective interest rate
Eigenfunktion. Eigenfunction
Eigenschaft finiten Charakters. Property of finite character
Eigenvektor. Eigenvector
Eigenwert einer Matrix. Characteristic root of a matrix, Eigenvalue of a matrix
Eilinie, Oval. Oval
Einbeschriebenes Polygon. Inscribed polygon
Einbettung einer Menge. Imbedding of a set
Eindeutig definiert. Uniquely defined
Eindeutige Funktion. Single valued function
Eindeutige Zerlegung (in Primelemente). Unique factorization
Eindeutigkeitssatz. Uniqueness theorem
Eineindeutige Entsprechung, umkehrbar eindeutige Entsprechung. One-to-one correspondence
Einfach geschlossene Kurve. Simple closed curve
Einfach zusammenhängendes Gebiet. Simply connected region
Einfache Lösung. Simple solution
Einfache Wurzel. Simple root
Einfaches Integral. Simple integral
Eingabe Komponente, Eingang. Input component
Einheitsdyade (bei skalarer Multiplikation). Idemfactor
Einheitselement. Unity
Einheitskreis. Unit circle
Einhüllende (Enveloppe) einer Familie von Kurven. Envelope of a family of curves
Einkommensteuer. Income tax
Ein, Eins. One
Einschaliges Hyperboloid. Hyperboloid of one sheet
Einschliessungssatz. Ham-sandwich theorem
Einseitige Fläche. Unilateral surface
Einseitige Verschiebung. Unilateral shift
Einstellige Operation. Unary operation
Ekliptik. Ecliptic
Elastische Körper. Elastic bodies
Elastizität. Elasticity
Elektromotorische Kraft. Electromotive force
Elecktrostatisches Potential. Electrostatic potential
Elementare Operationen. Elementary operations
Elf. Eleven

Elimination durch Substitution. Elimination by substitution
Ellipse. Ellipse
Ellipsoid. Ellipsoid
Elliptische Fläche. Elliptic surface
Elongation. Elongation
Empfindlichkeitsanalyse. Sensitivity analysis
Empirisch eine Kurve bestimmen. Curve fitting
Empirische Kurve. Empirical Curve
Endlich darstellbar. Finitely representable
Endliche projektive Ebene. Finite projective plane
Endliches Spiel. Finite game
Endomorphismus. Endomorphism
Endpunkt einer Kurve. End point of a curve
Endstrahl eines Winkels. Terminal side of an angle
Endwert. Accumulated value; Future value
Energieintegral. Energy integral
Entarteter Kegelschnitt. Degenerate conic
Entfernung zwischen zwei Punkten. Distance between two points
Entropie. Entropy
Entsprechende Winkel. Corresponding angles
Entwicklung einer Determinante. Expansion of a determinant
Epitrochoidale Kurve. Epitrochoidal curve
Epitrochoide. Epitrochoid
Epizykloide. Epicycloid
Erdmeridian. Meridian on the earth
Ereigniswahrscheinlichkeit. Probability of occurrence
Erg. Erg
Ergänzungswinkel. Conjugate angle
Ergodentheorie. Ergodic theory
Erlanger Programm. Erlangen program
Erwartungswert. Expected value
Erweiterte Matrix (eines linearen Gleichungssystems). Augmented matrix
Erzeugende. Generatrix
Erzeugende einer Fläche. Generator of a surface
Erzeugende Gerade einer Fläche. Ruling on a surface
Erzeugende Geraden (einer Regelfläche). Rectilinear generators
Eulerscher Graph. Eulerian graph
Evolute einer Kurve. Evolute of a curve
Evolvente einer Kurve. Involute of a curve
Exakte Differentialgleichung. Exact differential equation
Existenzsatz. Existence theorem
Exotische Sphäre. Exotic sphere
Exotischer vierdimensionaler Raum. Exotic four-space
Explizite Funktion. Explicit function
Exponent. Exponent
Exponentenregel. Law of exponents
Exponentialkurve. Exponential curve
Extrapolation. Extrapolation
Extremal unzusammenhängend. Extremally disconnected
Extremalpunkt. Bend point
Extrempunkt. Turning point
Exzentrizität einer Hyperbel. Eccentricity of a hyperbola

Facette. Facet
Fächergestell. Scattergram

Faktor. Multiplier
Faktor eines Polynoms. Factor of a polynomial
Faktoranalyse. Factor analysis
Faktorisierbar, zerlegbar. Factorable
Faktorisierung, Zerlegung. Factorization
Fakultät einer ganzen Zahl. Factorial of an integer
Fallende Funktion. Decreasing function
Faltsingularität. Fold singularity
Faltung zweier Funktionen. Convolution of two functions; Resultant of two functions
Färbung von Graphen. Graph coloring
Faserbund. Fiber bundle
Faserraum. Fiber space
Fast periodisch. Almost periodic
Feinere Partition. Finer partition
Feinheit einer Partition. Fineness of a partition
Fejérscher Kern. Fejér kernel
Feldmesser, Gutachter. Surveyor
Filter. Filter
Filterbasis. Filter base
Finite Mathematik. Finite mathematics
Fixpunkt. Fixed point
Fixpunktsatz. Fixed-point theorem
Flächeninhalt. Area
Flächentren. Equiareal (or area-preserving)
Flugbahn. Path of a projectile
Fluss. Flux
Flussdiagramm. Flow chart
Folgenkompakt. Weakly compact
Folgenkorrelations Koeffizient. Biserial correlation coefficient
Form in zwei Variablen. Form in two variables
Formale Ableitung. Formal derivative
Formale Potenzreihe. Formal power series
Formel. Formula
Fraktal. Fractal
Fraktale Dimension. Fractal dimension
Fréchet-Raum. Fréchet space
Freier Ultrafilter. Free ultrafilter
Freitragender Balken. Cantilever beam
Fundamentalsatz der Algebra. Fundamental theorem of algebra
Fünf. Five
Fünfeck. Pentagon
Fünfeckzahlsatz von Euler. Euler pentagonal-number theorem
Fünfflächner. Pentahedron
Fünfseitige Pyramide. Pentagonal pyramid
Fünfzehneck. Pentadecagon
Funktion der Differentiations Klasse C^n. Function of class C^n
Funktionentheorie. Function theory
Funktor. Functor
Für einen Rechenautomaten verschlüsseln. Coding for a computing machine
Fusspunkt einer Senkrechten. Foot of a perpendicular
Fusspunktkurve. Pedal curve
Fuzzy logic. Fuzzy logic
Fuzzy set. Fuzzy set

Galoiskörper. Galois field
Gammafunktion. Gamma function
Ganze Funktion. Entire function
Ganze Vielfache rechter Winkel. Quadrantal angles

Ganze Zahl. Integer
Ganzen komplexen Zahlen. Gaussian integers
(Ganzes) Vielfaches einer Zahl. Multiple of a number
Garbe. Sheaf
Gebrochene Linie. Broken line
Gebrochener Exponent. Fractional exponent
Gebundene Variable. Bound variable
Gedächtnisstütze. Mnemonic device
Gegen den Uhrzeigersinn. Counterclockwise
Gegen einen Grenzwert konvergieren. Converge to a limit
Gegenhypothese. Alternative hypothesis
Gegenüberliegende Seiten. Opposite sides
Gemeinsame Tangente zweier Kreise. Common tangent of two circles
Gemeinsames Vielfaches. Common multiple
Gemischte Versicherung. Endowment insurance
Gemischter Bruch. Mixed number
Geodätische Parallelen. Geodesic parallels
Geoid (leicht abgeplattete Kugel). Geoid
Geometrie. Geometry
Geometrisches Mittel. Geometric average
Geometrische Reihe. Geometric series
Geometrischer Orf. Locus
Geordnete Menge, Verein (auch: teilweise geordnete Menge). (Partially) ordered set
Gerade. Straight line
Gerade Permutation. Even permutation
Gerade Zahl. Even number
Gerade Zahlen zählen. Count by twos
Geraden derselben Ebene. Coplanar lines
Geradenabschnitt. Line segment
Geradenbüschel. Concurrent lines
Gerichtete Gerade. Directed line
Gerüst eines Komplexes. Skeleton of a complex
Geschlecht einer Fläche. Genus of a surface
Geschweifte Klammer. Brace
Geschwindigkeit. Speed; Velocity
Gesicherheit einer Abweichung. Significance of a deviation
Gewicht. Weight
Gewichte zum Wägen von Edelmetallen. Troy weight
Gewinn. Profit
Gewöhnliche Logarithmen. Common logarithms
Gewöhnlicher Bruch. Vulgar fraction; Common fraction
Gitter. Lattice (in physics)
Glatte Abbildung. Smooth map
Glatte ebene projektive Kurve. Smooth projective plane curve
Gleichartige Grössen. Equal quantities
Gleichgewicht. Equilibrium
Gleichgradig stetige Funktionen. Equicontinuous functions
Gleichheit. Equality
Gleichmässig gleichgradig stetig. Uniformly equicontinuous
Gleichmässig konvexer Raum. Uniformly convex space
Gleichmässige Konvergenz. Uniform convergence
Gleichmässige Stetigkeit. Uniform continuity
Gleichschenkliges Dreieck. Isosceles triangle
Gleichseitiges Dreieck. Equilateral triangle
Gleichsetzen. Equate

Gleichung einer Kurve. Equation of a curve
Gleichungssystem. System of equations
Gleitendes Komma. Floating decimal point
Glied eines Bruches. Term of a fraction
Globale Eigenschaft. Global property
Goldener Schnitt. Golden section
Googol (10 hoch 100 oder sehr grosse Zahl). Googol
Grad Celsius Thermometer. Centigrade thermometer
Grad eines Polynoms. Degree of a polynomial
Gradient. Gradient
Gramm. Gram
Graph, graphische Darstellung. Pictogram
Graph einer Gleichung. Graph of an equation
Graphentheorie. Graph theory
Graphische Lösung. Graphical solution
Gravitation, Schwerkraft. Gravitation
Grenzwert, Limes. Limit point
Grenzwert einer Funktion, Limes einer Funktion.
    Limit of a function
Griechisches Zahlensystem (mit griechischen Sym-
    bolen). Greek numerals
Gröbere Partition. Coarser partition
Grösse. Quantity
Grosse Disjunktion, Existenzquantor, Partikularisator.
    Existential quantifier
Grösse (Helligkeit) eines Sternes. Magnitude of a star
Grosse Konjunktion, Allquantor, Generalisator. Uni-
    versal quantifier
Grösster gemeinsamer Teiler. Greatest common divi-
    sor
Gruppoid. Groupoid
Gütefunktion. Power function

Halbieren. Bisect
Halbierungspunkt. Bisecting point
Halbjährlich. Biannual
Halbkreis. Semicircle
Halbregulärer Körper. Archimedean solid
Halbregulärer Körper. Semi-regular solid
Halbschaffen. Penumbra
Halbstetige Funktion. Semicontinuous function
Halbwinkelformeln. Half-angle formulas
Halm einer Garbe. Stalk of a sheaf
Hamiltonscher Graph. Hamiltonian graph
Handelsgewicht. Avoirdupois weight
Hardyscher Raum. Hardy space
Harmonische Bewegung. Harmonic motion
Harmonische Funktion. Harmonic function
Häufigkeitskurve. Frequency curve
Häufungspunkt. Cluster point; Accumulation point
Hauptachse. Major axis
Hauptdiagonale. Principal diagonal
Hauptidealring. Principal ideal ring
Haussdorff-Dimension. Hausdorff dimension
Haussdorffsches Paradoxon. Hausdorff paradox
Hebbare Unstetigkeit. Removable discontinuity
Hebelarm. Lever arm
Hellebardenspitze. Cusp of first kind
Hemisphäre, Halbkugel. Hemisphere
Henkel an einer Fläche. Handle on a surface
Hermitesche Matrix. Hermitian matrix
Heuristische Methode. Heuristic method
Hex-Spiel. Game of hex
Hexäder. Hexahedron
Hexadezimalsystem. Hexadecimal number system

Hilfskreis. Auxiliary circle
Himmels-. Celestial
Himmelsäquator. Celestial equator
Hinreichende Bedingung. Sufficient condition
Höchster Koeffizient. Leading coefficient
Hodograph. Hodograph
Höhe. Altitude
Holomorphe Funktion. Holomorphic function
Holzmass. Cord (of wood)
Homogene Gleichung. Homogeneous equation
Homogenes Polynom. Quantic
Homogenes Polynom in vier Variablen. Quarternary
    quantic
Homogenität. Homogeneity
Homolog. Homologous
Homologiegruppe. Homology group
Homologische Algebra. Homological algebra
Homomorphismus zweier algebraischer Strukturen.
    Homomorphism of two algebraic structures
Homöomorphismus zweier Räume. Homeomorphism
    of two spaces
Homotope Figuren. Homotopic figures
Horizont. Horizon
Horizontal, waagerecht. Horizontal
Hülle einer Menge, (abgeschlossene Hülle einer
    Menge). Closure of a set
Hundert. Hundred
Hundertster Teil einer Zahl. Hundredth part of a
    number
Hydromechanik. Mechanics of fluids
Hyperbel. Hyperbola
Hyperbolischer Zylinder. Hyperbolic cylinder
Hyperbolisches Paraboloid. Hyperbolic paraboloid
Hypereben . Hyperplane
Hyperfläche. Hypersurface
Hypergeometrische Reihe. Hypergeometric series
Hyperkomplexe Zahlen. Hypercomplex numbers
Hyperreelle Zahlen. Hyperreal numbers
Hypervolumen. Hypervolume
Hypotenuse. Hypotenuse
Hypothese. Hypothesis
Hypotrochoide. Hypotrochoid
Hypozykloide. Hypocycloid

Idempotent. Idempotent
Idempotenzeigenschaft. Idempotent property
Identische Grössen. Identical quantities
Identität. Identity
Ikosäder. Icosahedron
Ikosaedergruppe. Icosahedral group
Imaginäre Zahlengerade. Scale of imaginaries
Imaginärteil der modifizierten Besselfunktion. Kei
    function
Imaginärteil einer Zahl. Imaginary part of a number
Implikation. Implication
Implizite Differentiation. Implicit differentiation
Implizite Funktion. Implicit function
Impuls. Momentum
Im Gegenuhrzeigersinn. Counter-clockwise
Im Uhrzeigersinn. Clockwise
Im wesentlichen beschränkt. Essentially bounded
In einen Raum einbetten. Imbed in a space
In einem Ring enthaltenes Ideal. Ideal contained in a
    ring
In Raten rückkäufliches Anlagepapier. Serial bond

Indikatrix einer quadratischen Form. Indicatrix of a quadratic form
Indirekter Beweis. Indirect proof
Indiskrete Topologie. Indiscrete topology
Induktion. Induction
Induktive Methode. Inductive method
Ineinandergeschachtelte Intervalle. Nested intervals
Infimum, grösste untere Schranke. Greatest lower bound
Infinitesimalrechnung, Analysis. Calculus
Infinitesimalrechnung. Infinitesimal analysis
Inhalt einer Menge. Content of a set
Injektive Funktion. Injective function
Inklusionsrelation. Inclusion relation
Inkommensurabel Zahlen. Incommensurable numbers
Inkreis. Incircle
Inkreismittelpunkt eines Dreiecks. Incenter of a triangle
Inkreisradius (eines Polygons). Apothem
Innenwinkel. Interior angle
Innenwinkel eines Polygons, grösser als π. Reentrant angle
Innere Eigenschaft. Intrinsic property
Innere Tangente zweier Kreise. Internal tangent of two circles
Innerer Automorphismus. Inner automorphism
Inneres Produkt, Skalarprodukt. Inner product
Inneres Teilverhältnis. Internal ratio
Insichdicht. Dense-in-itself
Integral einer Funktion. Integral of a function
Integralgleichung. Integral equation
Integralrechnung. Integral calculus
Integrand. Integrand
Integraph. Integraph
Integrationselement. Element of integration
Integrationskonstante. Constant of integration
Integrator. Integrator
Integrierbare Funktion. Integrable function
Integrierender Faktor. Integrating factor
Integritätsbereich. Integral domain
Interpolation. Interpolation
Internationales Einheitensystem. International system of units
Intervallschachtelung. Nest of intervals
Intransitive Relation. Intransitive relation
Intuitionismus. Intuitionism
Invariante Eigenschaft. Invariant property
Invariante einer Gleichung. Invariant of an equation
Inverse der charakteristischen Matrix. Resolvent of a matrix
Inversion (eines Punktes an einem Kreis). Inversion of a point
Inversor. Inversor
Invertierbar. Invertible
Involution auf einer Geraden. Involution on a line
Inzidenzfunktion. Incidence function
Irrationale algebraische Zahl. Surd
Irrationalzahl. Irrational number
Irreduzible Wurzel. Irreducible radical
Isochrone. Isochronous curve
Isolierter Punkt. Acnode
Isomorphismus. Isomorphism
Isoperimetrisches Problem. Isoperimetric problem
Isotherme. Isotherm

Isotherme. Isothermal line
Isotherme Kurvenschar. Isothermic system of curves
Isotrope Materie. Isotropic matter

Jahr. Year
Joule. Joule
Julia-Menge. Julia set

Kalorie. Calorie
Kanonische Form. Canonical form
Kanonischer Representant einer primen Restklasse einer ganzen Zahl. Totitive of an integer
Kante eines Körpers. Edge of a solid
Kapitalisierte Kosten. Capitalized cost
Kappakurve. Kappa curve
Kardinalzahl. Cardinal number
Kardioide, Herzkurve. Cardioid
Katastrophentheorie. Catastrophe theory
Kategorie. Category
Kategorisch. Categorical
Katenoid, Drehfläche der Kettenlinie. Catenoid
Kathete eines rechtwinkligen Dreiecks. Leg of a right triangle
Kaufpreis. Flat price
Kegel. Cone
Kegelfläche. Conical surface
Kegelschnitt, konisch. Conic
Kegelstumpf. Truncated cone
Keil. Wedge
Keilschriftsymbole. Cuneiform symbols
Keim von Funktionen. Germ of functions
Kennziffer eines Logarithmus. Characteristic of a logarithm
Kern eines Homomorphismus. Kernel of a homomorphism
Kern einer Integralgleichung. Nucleus (or kernel) of an integral equation
Kettenbruch. Continued fraction
Kettenkomplex. Chain complex
Kettenlinie. Catenary
Kettenregel. Chain rule
Kilogramm (Masse). Kilogram (mass unit)
Kilometer. Kilometer
Kilopond (Kraft). Kilogram (force unit)
Kilowatt. Kilowatt
Kinematik. Kinematics
Kinetik. Kinetics
Kinetische Energie. Kinetic energy
Kippschalter. Flip-flop circuit
Klammer, eckige Klammer. Bracket
Klassenhäufigkeit. Class frequency
Knoten. Knot in topology
Knoten. Knot of velocity
Knotenlinie. Nodal line
Knotenpunkt einer Kurve. Node of a curve
Knotenpunkt in der Astronomie. Node in astronomy
Koaxiale Kreise. Coaxial circles
Koebe-Funktion. Koebe function
Koeffizient. Coefficient
Koeffizientenmatrix. Matrix of coefficients
Kofinale Untermenge. Cofinal subset
Kofunktion, komplementäre Funktion. Cofunction
Kohärent, zusammenhängend orientiert. Coherently oriented
Koinzidierend, Koinzident. Coincident

Kollineare Transformation. Collineatory transformation

Kollinease Punkte. Collinear points

Kollineation. Collineation

Kombination einer Menge von Objekten. Combination of a set of objects

Kombinatorische Topologie. Combinatorial topology

Kommensurabel. Commensurable

Kommutativ. Commutative

Kommutative Gruppe, Abelsche Gruppe. Commutative group

Kommutator. Commutator

Kompakte Menge. Compact set

Kompakter Träger. Compact support

Kompaktifizierung. Compactification

Kompaktum. Compactum

Komplement einer Menge. Complement of a set

Komplementwinkel. Complementary angles

Kompletter Körper. Complete field

Komplexe Zahl. Complex number

Konchoide. Conchoid

Kondensationspunkt. Condensation point

Konfiguration, Stellung. Configuration

Konfokale Kegelschnitte. Confocal conics

Konforme Transformation. Conformal transformation

Kongruente Figuren. Congruent figures

Kongruente Konfigurationen. Superposable configurations

Kongruenz. Congruence

Königsberger Brückenproblem. Königsberg bridge problem

Konjugierte komplexe Zahlen. Conjugate complex numbers

Konjugierte Untergruppen. Conjugate subgroups

Konjunktion. Conjunction

Konjunktive Transformation. Conjunctive transformation

Konkaves Polygon. Concave polygon

Konkav-konvexes Spiel. Concave-convex game

Konkavsein. Concavity

Konnexe Relation. Connected relation

Konoid. Conoid

Konservatives Kraftfeld. Conservative field of force

Konsistente Gleichungen. Consistent equations

Konsistenz (Widerspruchsfreiheit) von Gleichungen. Consistency of equations

Konstante Geschwindigkeit. Constant speed

Konstruktion. Construction

Konstruktive Mathematik. Constructive mathematics

Kontakttransformation, Berührungstransformation. Contact transformation

Kontingenztafel. Contingency table

Kontinuum. Continuum

Kontrahierende Abbildung. Contraction mapping

Kontraktion (Verdünnung) eines Tensors. Contraction of a tensor

Kontravarianter Tensor. Contravariant tensor

Kontrollgruppe. Control group

Kontrollierte Stichprobe, Gruppenauswahl. Stratified sample

Konvergente Folge. Convergent sequence

Konvergenz eines Kettenbruchs. Convergence of a continued fraction

Konvergenz einer Reihe. Convergence of a series

Konvergenzintervall. Interval of convergence

Konvergenzkreis. Circle of convergence

Konvexe Hülle einer Menge. Convex hull of a set

Konvexe Kurve. Convex curve

Konzentrische Kreise. Concentric circles

Konzyklische Punkte (Punkte auf einem Kreis). Concyclic points

Kooperativ, Konsumverein. Cooperative

Koordinate eines Punktes. Coordinate of a point

Koordinaten transformation. Transformation of coordinates

Koordinatenebenen. Coordinate planes

Koordinatennetz, Bezugssystem. Frame of reference

Kopf und Adler. Coin-matching game

Korollar. Corollary

Körperbewertung. Valuation of a field

Körpererweiterung. Extension of a field

Korrelationskoeffizient. Correlation coefficient

Kovariante Ableitung. Covariant derivative

Kovarianz. Covariance

Kraftkomponente. Component of a force

Kraftvektor. Force vector

Krein-Milmansche Eigenschaft. Krein-Milman property

Kreis. Circle

Kreisausschnitt. Sector of a circle

Kreisbüschel. Pencil of circles

Kreiskegel. Circular cone

Kreispunkt. Umbilical point

Kreisring. Annulus

Kreisscheibe. Disc (or disk)

Kreisteilungsgleichung. Cyclotomic equation

Kreuzförmige Kurve. Cruciform curve

Kreuzhaube. Cross-cap

Kritischer Wert. Critical value

Krummlinige Bewegung. Curvilinear motion

Krümmung einer Kurve. Curvature of a curve

Kubikmeter. Stere

Kubikwurzel. Cube root

Kubische Kurve. Cubic curve

Kubische Parabel. Cubical parabola

Kubische Resolvente. Resolvent cubic

Kubooktaeder. Cuboctahedron

Kugelzone. Zone

Kummulanten. Cumulants

Kummulative Häufigkeit. Cumulative frequency

Kuratowskisches Abschluss- und Komplementierungsproblem. Kuratowski closure-complementation problem

Kurtosis. Kurtosis

Kurvenbogen. Segment of a curve

Kurvenlänge. Length of a curve

Kurvenschar. Family of curves

Kürzen. Cancel

Kürzung. Cancellation

Kybernetik. Cybernetics

Ladung. Charge

Länge (geogr.). Longitude

Längentreu aufeinander abbildbare Flächen. Isometric surfaces

Lebenserwartung. Expectation of life

Lebenslängliche Rente. Perpetuity

Lebenslängliche Verbindungsrente. Joint life annuity

Lebensstatistik. Vital statistics

Lebensversicherung. Life insurance
Legendresches Polynom. Polynomial of Legendre
Lehre von den Gleichungen. Theory of equations
Leitfähigkeit. Conductivity
Lemma, Hilfssatz. Lemma
Lemniskate. Lemniscate
Lexikographisch. Lexicographically
Lichtintensität in Candelas. Candlepower
Lineal. Ruler
Linear abhängige Grössen. Linearly dependent quantities
Lineare Programmierung. Linear programming
Lineare Transformation. Linear transformation
Linearer Operator. Linear operator
Linearkombination. Linear combination
Linienelement. Lineal element
Linksgewundene Kurve. Left-handed curve
Liter. Liter
Lituus, Krummstab. Lituus
Logarithmentafel. Table of logarithms
Logarithmische Kurve. Logarithmic curve
Logarithmische Spirale. Equiangular spiral; Logistic spiral
Logarithmus des Reziproken einer Zahl. Cologarithm
Logarithmus einer Zahl. Logarithm of a number
Lognormalverteilung. Lognormal distribution
Lokal integrierbare Funktion. Locally integrable function
Lokal wegzusammenhängend. Locally arc-wise connected
Lokale Eigenschaft. Local property
Lokalisationsprinzip. Localization principle
Lokalkompakt. Locally compact
Loopraum, Raum der geschlossenen Wege. Loop space
Losgelöste Koeffizienten. Detached coefficients
Lösung einer Differentialgleichung. Primitive of a differential equation
Lösung einer Gleichung. Solution of an equation
Lösungsmenge. Truth set
Lot. Plumb line
Loxodrome. Loxodromic spiral
Loxodrome. Rhumb line

Mächtigkeit einer Menge. Potency of a set
Magisches Quadrat. Magic square
Makler. Broker
Mandelbrot-Menge. Mandelbrot set
Mannigfaltigkeit. Manifold
Mantelfläche. Lateral area
Mantisse. Mantissa
Marktwert. Market value
Mass einer Menge. Measure of a set
Mass Null. Measure zero
Masse. Mass
Massenmittelpunkt. Center of mass; Centroid
Mathematik. Mathematics
Mathematische Induktion. Mathematical induction
Matrix in Staffelform. Echelon matrix
Maximisierender Spieler. Maximizing player
Maximum einer Funktion. Maximum of a function
Mazur-Banach-Spiel. Mazur-Banach game
Mechanik der Deformierbaten. Mechanics of deformable bodies

Mechanische Integration. Mechanical integration
Mehradressensystem. Multiaddress system
Mehrfach zusammenhängendes Gebiet. Multiply connected region
Mehrfaches Integral. Iterated integral
Mehrfaches Integral. Multiple integral
Mehrfachkante eines Graphen. Multiple edge in a graph
Mehrwertige Funktion. Many valued function
Meile. Mile
Meridianlinie. Meridian curve
Meromorphe Funktion. Meromorphic function
Messbare Menge. Measurable set
Messung. Mensuration
Metakompakter Raum. Metacompact space
Meter. Meter
Methode der kleinsten Fehlerquadrate. Method of least squares
Methode des steilsten Abstiegs. Method of steepest descent
Metrischer Raum. Metric space
Metrisierbarer Raum. Metrizable space
Milliarde. Billion
Millimeter. Millimeter
Million. Million
Minimalfläche. Minimal surface
Minimax-Satz. Minimax theorem
Minimaxtheorem. Minimax theorem
Minimisierender Spieler. Minimizing player
Minimum einer Funktion. Minimum of a function
Minor einer Determinante. Minor of a determinant
Minuend. Minuend
Minus. Minus
Minute. Minute
Mittel zweier Zahlen. Mean (or average) of two numbers
Mittelpunkt eines Ankreises. Excenter
Mittelpunkt (Zentrum) eines Kreises. Center of a circle
Mittelpunkt einer Strecke. Midpoint of a line segment
Mittelpunktswinkel, Zentriwinkel. Central angle
Mittelwertsatz. Mean-value theorem
Mittlerer Fehler, Standardabweichung, mittlere quadratische Abweichung. Standard deviation
Mit zwei rechten Winkeln. Birectangular
Modifizierte Besselfunktionen. Modified Bessel functions
Modul. Module
Modul einer Kongruenz. Modulus of a congruence
Modulfunktion. Modular function
Modulo 2π gleiche Winkel. Coterminal angles
Modus (einer Wahrscheinlichkeitsdichte). Mode
Mol. Mole
Moment einer Kraft. Moment of a force; static moment
Momentangeschwindigkeit. Instantaneous velocity
Möndchen des Hyppokrates. Lunes of Hippocrates
Monodromiesatz. Monodromy theorem
Monom. Monomial
Monotone Funktion. Monotone function
Mordellsche Vermutung. Mordell conjecture
Morphismus. Morphism
Multifolium. Multifoil
Multinom. Multinomial

Multiplikand, Faktor. Multiplicand
Multiplizierbare Matrizen. Conformable matrices
Myriade. Myriad

Nabla Operator. Del
Nachbarschaft eines Punktes. Neighborhood of a point
Nadir. Nadir
Näherungsregel zur Bestimmung von Quadratwurzeln. Mechanic's rule
Natürliche Logarithmen. Natural logarithms
Nebenachse. Minor axis
Nebendiagonale. Secondary diagonal
Nebenklassen einer Untergruppe. Coset of a subgroup
Negation, Verneinung. Negation
Negative Imaginärteil der Besselfunktion. Bei function
Negative Zahl. Negative number
Neigung einer Geraden. Inclination of a line
Nenner. Denominator
Nennwert, Nominalwert. Redemption price
Nerv eines Mengensystems. Nerve of a system of sets
Neugradsystem zur Winkelmessung. Centesimal system of measuring angles
Neun. Nine
Neuneck. Nonagon
Neunerprobe. Casting out nines
Neunerrest. Excess of nines
Newton. Newton
Newtons Tridens, Cartesische Parabel. Trident of Newton
Nicht beschränkte Funktion. Unbounded function
Nicht erwartungstreue Stichprobenfunktion, nicht reguläre. Biased statistic
Nicht proportionell. Disproportionate
Nicht-kooperativ. Noncooperative
Nichtausgeartete Fläche zweiter Ordnung. Conicoid
Nichtexpansive Abbildung. Nonexpansive mapping
Nichthebbare Unstetigkeit, unbestimmte Unstetigkeit. Nonremovable discontinuity
Nichtlineare Programmierung. Nonlinear programming
Nichtquadratischer Raum. Nonsquare space
Nichtrest. Nonresidue
Nichtsingulärer Punkt, regulärer Punkt. Ordinary point
Nichtsinguläre Transformation. Nonsingular transformation
Nichtstandardzahlen. Nonstandard numbers
Nilpotent. Nilpotent
Nilpotentes ideal. Nilpotent ideal
Nirgends dicht. Nowhere dense
Nirgends dichte Menge. Rare set
Niveaulinien, Höhenlinien. Level lines
Nomineller Zinsfuss. Nominal rate of interest
Nomogramm. Nomogram
Nördliche Deklination. North declination
Norm einer Matrix. Norm of a matrix
Normale einer Kurve. Normal to a curve
Normalzeit. Standard time
Normierter Raum. Normed space
Notwendige Bedingung. Necessary condition
Null. Cipher
Null, Nullelement. Zero

Nullellipse. Point ellipse
Nullmenge, leere Menge. Null set
Numerierung. Numeration
Numerischer Wert. Numerical value
Numerus. Antilogarithm
Nutznießer. Beneficiary

Obere Dichte. Upper density
Obere Grenze, kleinste obere Schranke, Supremum. Least upper bound
Obere Schranke. Upper bound
Oberfläche, Flächeninhalt. Surface area
Oberflächenintegral, Flächenintegral. Surface integral
Obermenge. Superset
Obligation, Anlagepapier. Bond
Offene Kugel. Open ball
Offenes Intervall. Open interval
Ohm. Ohm
Oktaeder, Achtflach. Octahedron
Oktaedergruppe. Octahedral group
Oktales Zahlensystem. Octal number system
Oktant. Octant
Operation. Operation
Operator. Operator
Optimale Strategie. Optimal strategy
Ordinalzahlen. Ordinal numbers
Ordinate eines Punktes. Ordinate of a point
Ordnung der Berührung. Order of contact
Ordnung einer Gruppe. Order of a group
Orientierung. Orientation
Orthogonale Funktionen. Orthogonal functions
Orthonormale Folge. Orthonormal sequence
Oszillierende Reihe. Oscillating series

p-Adische Zahl. p-adic number
Paarer Graph. Bipartite graph
Pantograph. Pantograph
Papiermass. Ream
Parabel. Parabola
Parabolischer Punkt. Parabolic point
Parabolischer Zylinder. Parabolic cylinder
Paradoxie, Paradoxon. Paradox
Parakompakter Raum. Paracompact space
Parallaktischer Winkel. Parallactic angle
Parallaxe eines Sternes. Parallax of a star
Parallele Geraden. Parallel lines
Parallelepipedon. Parallelepiped
Parallelogramm. Parallelogram
Parallelotop. Parallelotope
Parameter. Parameter
Parametergleichungen. Parametric equations
Parität. Parity
Parkettierung oder Pflasterung. Tessellation
Parkettierungselement. Tile
Partialbrüche. Partial fractions
Partie eines Spiels. Play of a game
Partielle Ableitung. Partial derivative
Partielle Integration. Integration by parts
Partikuläres Integral. Particular integral
Pascal. Pascal
Pendel. Pendulum
Pentagramm, Fünfstern. Pentagram
Perfekte Menge. Perfect set
Perfekter Körper. Complete field

Perihel. Perihelion
Periode einer Funktion. Period of a function
Periodische Bewegung. Periodic motion
Periodischer Dezimalbruch. Repeating decimal
Periodizität. Periodicity
Peripherie, Rand. Periphery
Permutation von n Dingen. Permutation of n things
Permutationsgruppe. Permutation group
Persönlicher Zug. Personal move
Perspektivität. Perspectivity
Perzentile. Percentile
Petersburger Paradoxon. Petersburg paradox
Pferdestärke. Horsepower
Pfund. Pound
Phase einer einfach harmonischen Bewegung. Phase of simple harmonic motion
Planarer oder plättbarer Graph. Planar graph
Planimeter. Planimeter
Plastizität. Plasticity
Pluszeichen. Plus sign
Pointcarésche Vermutung. Poincaré conjecture
Pol eines Kreises (auf einer Kugelfläche). Pole of a circle
Polare einer quadratischen Form. Polar of a quadratic form
Polarisation. Polarization
Polarkoordinaten. Polar coordinates
Polarwinkel. Anomaly of a point
Polarwinkel. Vectorial angle
Poldistanz. Codeclination
Poldistanz (auf der Erde). Colatitude
Polyeder. Polyhedron
Polygon, Vieleck. Polygon
Polyhex. Polyhex
Polynomische Gleichung, Polynomgleichung. Polynomial equation
Polyomino (ebene Figuren bestehehnd aus aneinandergefügten Einheitsquadraten). Polyomino
Polytop. Polytope
Positionsspiel. Positional game
Positive reelle Zahl. Arithmetic number
Positive Zahl. Positive number
Postulat, Forderung. Postulate
Potentialfunktion. Potential function
Potenz einer Zahl. Power of a number
Potenzlinie. Radical axis
Potenzreihe. Power series
Praedikatensymbol. Predicate
Prämie. Premium
Prämie, Dividende. Bonus
Prämienreserve. Value of an insurance policy
Primitive n-te Einheitswurzel. Primitive nth root of unity
Primzahl. Prime number
Primzahlpaar. Twin primes
Primzahlzwilling. Twin primes
Prinzip, Grundsatz. Principle
Prinzip der gleichmässigen Beschränktheit. Uniform boundedness principle
Prinzip der Optimalität. Principle of optimality
Prisma. Prism
Prismatische Fläche. Prismatic surface
Prismoidformel. Prismoidal formula
Prismoid, Prismatoid. Prismoid, Prismatoid
Probe auf das Ergebnis machen. Check on a solution

Problem. Problem
Problem von Kakeya. Kakeya problem
Produkt von Zahlen. Product of numbers
Produktraum. Product space
Programmierung, Programmgestaltung. Programming
Progression, Reihe. Progression
Projektion eines Vektors. Projection of a vector
Projektionszentrum. Ray center
Projektive Geometrie. Projective geometry
Projektive Topologie. Projective topology
Projektiver Raum. Projective space
Projektivität. Projectivity
Projizierende Ebene. Projecting plane
Proportion, Verhältnis. Proportion
Proportionale Grössen. Proportional quantities
Proportionalität. Proportionality
Prozentischer Fehler. Percent error
Prozentsatz. Percentage
Pseudosphäre. Pseudosphere
Pseudosphärische Fläche. Pseudospherical surface
Psi-Funktion. Digamma function
Punktweise gleichgradig stetig. Point-wise equicontinuous
Pyramide. Pyramid
Pyramidenfläche. Pyramidal surface
Pythagoräischer Lehrsatz, Satz von Pythagoras. Pythagorean theorem
Pythagoreisches Tripel. Pythagorean triple

Quader. Cuboid
Quadrant eines Kreises. Quadrant of a circle
Quadrant. Square
Quadratische Ergänzung. Completing the square
Quadratische Gleichung. Quadratic equation
Quadratur eines Kreises. Quadrature of a circle
Quadratwurzel. Square root
Quadrik. Quadric
Quadrillion. Septillion
Quandtoren. Quantifier
Quartile. Quartile
Quasinormalteiler. Quasi-normal subgroup
Quaternion. Quaternion
Quellenfreies Wirbelfeld. Solenoidal vector field
(Quer)schnitt eines Zylinders. Section of a cylinder
Querstrich. Bar
Quotient zweier Zahlen. Quotient of two numbers
Quotientenkriterium. Ratio test
Quotientenkriterium. Generalized ratio test
Quotientenraum, Faktorraum. Quotient space

Rabatt. Discount
Rademacher-Funktion. Rademacher functions
Radiant. Radian
Radikal. Radical
Radikal eines Ideals. Radical of an ideal
Radikal eines Rings. Radical of a ring
Radikand. Radicand
Radius eines Kreises, Halbmesser eines Kreises. Radius of a circle
Radizierung. Evolution
Ramsey-Zahl. Ramsey number
Rand einer Menge. Boundary of a set; Frontier of a set
Randwertproblem. Boundary value problem
Ratenzahlungen. Installment payments

Rationale Zahl. Rational number
Raum. Space
Raumkurve. Space curve
Raumwinkel. Solid angle
Raumwinkel eines Polyeders. Polyhedral angle
Realteil der Besselfunktion. Ber function
Realteil der modifizierten Besselfunktion. Ker function
Rechenbrett, Abakus. Abacus
Rechenmaschine. Arithmometer; calculating machine
Rechenmaschine, Rechenanlage. Computing machine
Rechenschieber. Slide rule
Rechenwerk, Zählwerk einer Rechenmaschine. Counter of a computing machine
Rechnen, Berechnen. Calculate
Rechteck. Rectangle
Rechtwinklige Achsen. Rectangular axes
Rechtwinkliges Dreieck. Right triangle
Reduktion eines Bruches, Kürzen eines Bruches. Reduction of a fraction
Reduktionsformeln. Reduction formulas
Reduzible Kurve. Reducible curve
Reduzierte Gleichung nach Abspaltung eines Linearfaktors. Depressed equation
Reduzierte kubische Gleichung. Reduced cubic equation
Reelle Zahl. Real number
Reflexionseigenschaft. Reflection property
Reflexive Relation. Reflexive relation
Regelfläche. Ruled surface
Regressionskoeffizient. Regression coefficient
Reguläres Polygon, regelmässiges Vieleck. Regular polygon
Reibung. Friction
Reihensummation. Summation of series
Reihe von Zahlen. Series of numbers
Reine Mathematik. Pure mathematics
Reine Strategie. Pure strategy
Reinverdienst, Nettoverdienst. Net profit
Rektaszension. Right ascension
Rektifizierbare Kurve. Rectifiable curve
Rektifizierende Ebene, Streckebene. Rectifying plane
Relation, Beziehung. Relation
Relativgeschwindigkeit. Relative velocity
Relativitätstheorie. Relativity theory
Reliefkarte. Profile map
Residualspektrum. Residual spectrum
Residuum einer Funktion. Residue of a function
Restglied einer unendlichen Reihe. Remainder of an infinite series
Restklasse. Residue class
Resultante. Eliminant
Resultante (eines Gleichungssystems). Resultant of a set of equations
Retrakt. Retract
Reziproke einer Zahl. Reciprocal of a number
Rhomboeder. Rhombohedron
Rhomboid. Rhomboid
Rhombus, Raute. Rhombus
Richtung einer Ungleichung. Sense of an inequality
Richtungsableitung. Directional derivative
Richtungskegel. Director cone
Richtungswinkel. Direction angles
Robuste Statistik. Robust statistics

Rotation einer Kurve um eine Achse. Revolution of a curve about an axis
Rotation eines Vektors, Rotor eines Vektors. Curl of a vector
Rotationsellipsoid. Spheroid
Rotationsfläche. Surface of revolution
Rotationskörper. Solid of revolution
Rückkauf. Redemption
Rückkaufswert. Surrender value
Rückkehrpunkt. Cusp
Runde Klammern, Parenthesen. Parentheses

Säkulartrend. Secular trend
Sammelwerk. Accumulator
Sattelpunkt. Saddle point
Sattelpunktmethode. Saddle-point method
Satz. Proposition (theorem)
Satz von Radon-Nykodým. Radon-Nikodým theorem
Schätzung einer Grösse. Estimate of a quantity
Scheinkorrelation (eigentlich Scheinkausalität). Illusory correlation
Scheitel eines Winkels. Vertex of an angle
Scherungsdeformation. Shearing strain
Scherungstransformation. Shear transformation
Schichtlinien, Isohypsen. Contour lines
Schiebfläche. Translation surface
Schiefe. Skewness
Schiefer Winkel. Oblique triangle
Schiefkörper. Skew field
Schiefkörper. Division ring
Schiefsymmetrische Determinante. Skew-symmetric determinant
Schlagschatten. Umbra
Schleife einer Kurve. Loop of a curve
Schlichte Funktion. Schlicht function
Schluss, Folgerung. Inference
Schmiegebene. Osculating plane
Schnabelspitze. Cusp of second kind
Schnelle Fourier-Transformation. Fast Fourier transform
Schnittfläche. Cross section
Schnittpunkt der Höhen eines Dreiecks. Orthocenter
Schnittpunkt der Seitenhalbierenolen. Median point
Schnittpunkt von Kurven. Intersection of curves
Schrägstrich (für Brüche). Solidus
Schranke einer Menge. Bound of a set
Schraubenfläche. Helicoid
Schraubenlinie. Helix
Schrottwert. Scrap value
Schrumpfende Basis. Shrinking basis
Schubfachprinzip (Dirichletsches). Pigeon-hole principle
Schustermesser (begrenzt durch 3 Halbkreise). Arbilos
Schustermesser (begrenzt durch 3 Halbkreise). Salinon
Schustermesser. Shoemaker's knife
Schwache Konvergenz. Weak convergence
Schwankung einer Funktion (auf einem abgeschlossenen Intervall). Oscillation of a function
Schwere. Gravity
Schwerpunkt. Barycenter
Schwingung. Vibration
Seemeile. Nautical mile
Sechs. Six

Sechseck. Hexagon
Sechseckiges Prisma. Hexagonal prism
Sehne. Chord
Seite, Seitenfläche eines Polyeders. Face of a polyhedron
Seite einer Gleichung. Member of an equation
Seite eines Polygons. Side of a polygon
Seitenhöhe. Slant height
Sekans eines Winkels. Secant of an angle
Sekanskurve. Secant curve
Selbstadjungierte Transformation. Self-adjoint transformation
Selektionstafel, (Sterblichkeitstafel unter Berücksichtigung der Selektionswirkung). Select mortality table
Senkrecht auf einer Fläche. Perpendicular to a surface
Separable Raum. Separable space
Serpentine. Serpentine curve
Sexagesimalsystem (Basis 60). Sexagesimal number system
Sexagesimalsystem der Zahlen. Sexagesimal system of numbers
Sich gegenseitig ausschliessende Ereignisse. Mutually exclusive events
Sieb. Sieve
Sieben. Seven
Siebeneck. Heptagon
Siebenflächner. Heptahedron
Simplex. Simplex
Simplexkette. Chain of simplexes
Simplexmethode. Simplex method
Simplizialer Komplex. Simplicial complex
Simultane Gleichungen. Simultaneous equations
Singulärer Punkt. Singular point
Sinus einer Zahl. Sine of a number
Sinuskurve. Sine curve
Sinuskurve. Sinusoid
Skalare Grösse. Scalar quantity
Skalarprodukt, Inneres Produkt. Dot product
Sonnenzeit. Solar time
Spalte einer Matrix. Column of a matrix
Spannung. Voltage
Spannungszustand eines Körpers. Stress of a body
Speicherkomponente. Memory component
Speicherkomponente. Storage component
Spektrale Zerlegung. Spectral decomposition
Spektrum einer Matrix. Spectrum of a matrix
Spezifische Wärme. Specific heat
Spezifischer Widerstand. Resistivity
Sphäre, Kugelfläche. Sphere
Sphärische Koordinaten, Kugelkoordinaten. Spherical coordinates
Sphärische Polarkoordinaten, Kugelkoordinaten. Geographic coordinates
Sphärisches Dreieck mit drei rechten Winkeln. Trirectangular spherical triangle
Spiegelung an einer Geraden. Reflection in a line
Spiel mit Summe null. Zero-sum game
Spiralfläche. Spiral surface
Spitze. Cusp; Spinode; Apex
Spitzer Winkel. Acute angle
Spline. Spline
Sprung(grösse) einer Funktion. Saltus of a function
Sprungstelle. Jump discontinuity

Spur einer Matrix. Spur of a matrix; Trace of a matrix
Stabiler Punkt. Stable point
Standardisierte Zufallsvariable. Normalized variate
Statik. Statics
Stationärer Punkt. Stationary point
Statistik. Statistics
(Statistische) Grösse, stochastische Variable. Statistic
Statistischer Schluss. Statistical inference
Stechzirkel. Dividers
Steigende Funktion. Increasing function
Steigung einer Kurve. Slope of a curve
Steigung eines Weges. Grade of a path
Stelle. Place
Stellenwert. Local value
Stellenwert. Place value
Steradiant. Steradian
Sterblichkeitsintensität. Force of mortality
Sterblichkeitstafel, Sterbetafel, Absterbeordnung. Mortality table
Stereografische Projektion. Stereographic projection
Stern eines Komplexes. Star of a complex
Sternzeit. Sidereal time
Stetige Funktion. Continuous function
Stetige Teilung. Golden section
Stetigkeit. Continuity
Steuer. Tax
Steuerzuschlag. Surtax
Stichprobe. Sample
Stichprobenfehler. Sampling error
Stichprobenstreuung, Verlässlichkeit. Reliability
Strategie eines Spiels. Strategy of a game
Streng dominierende Strategie. Strictly dominant strategy
Strich als verbindende Überstreichung. Vinculum
String. String
Strom. Current
Stromlinien. Stream lines
Strophoide. Strophoid
Stuckweis stetige Funktion. Piecewise continuous function
Stumpfer Winkel. Obtuse angle
Stundenwinkel. Hour angle
Subadditive Funktion. Subadditive function
Subharmonische Funktion. Subharmonic function
Subnormale. Subnormal
Substitution in eine Gleichung. Substitution in an equation
Subtangente. Subtangent
Subtrahend. Subtrahend
Subtraktionsformeln. Subtraction formulas
Subtraktion von Zahlen. Subtraction of numbers
Südliche Deklination. South declination
Summand. Addend
Summand. Summand
Summationszeichen. Summation sign
Summe von Zahlen. Sum of numbers
Summierbare Funktion. Summable function
Summierbare Reihe. Summable series
Super-reflexiv. Super-reflexive
Superpositionsprinzip. Superposition principle
Supplementsehnen. Supplemental chords
Supplementwinkel. Supplementary angles
Surjektive Abbildung. Surjection
Suslinsche Vermutung. Souslin's conjecture

Syllogismus, Schluss. Syllogism
Symbol, Zeichen. Symbol
Symmetrie einer Funktion. Symmetry of a function
Symmetrische Figur. Symmetric figure

Tafeldifferenzen. Tabular differences
Tangens eines Winkels. Tangent of an angle
Tangente an einen Kreis. Tangent to a circle
Tangentenbild einer Kurve. Indicatrix of a curve
Tangentialbeschleunigung. Tangential acceleration
Tangentialebene. Tangent plane
Tangentialebene an eine Fläche. Plane tangent to a surface
Tauberscher Satz. Tauberian theorem
Tausend. Thousand
Tausend Billionen. Quadrillion
Tausend Trillionen. Sextillion
Teil eines Körpers zwischen zwei parallelen Ebenen, Stumpf. Frustum of a solid
Teilbarkeit. Divisibility
Teilbarkeit durch elf. Divisibility by eleven
Teilen, dividieren. Divide
Teiler. Divisor
Temporäre Leibrente. Curtate annuity
Temporäres Speichersystem. Buffer (in a computing machine)
Tenäre Darstellung von Zahlen. Ternary representation of numbers
Tensor. Tensor
Tensoranalysis. Tensor analysis
Tensorprodukt von Vektorräumen. Tensor product of vector spaces
Terme gruppieren. Grouping terms
Terrestrisches Dreieck. Terrestrial triangle
Tesserale harmonische Funktion. Tesseral harmonic
Teträder. Triangular pyramid
Teträderwinkel. Tetrahedral angle
Tetraedergruppe. Tetrahedral group
Theodolit. Transit
Theorem, Hauptsatz. Theorem
Thetafunktion. Theta function
Tietzscher Erweiterungssatz. Tietze extension theorem
Todesfallversicherung. Whole life insurance
Topologie. Topology
Topologisch vollständiger Raum. Topologically complete space
Topologische Dimension. Topological dimension
Topologische Gruppe. Topological group
Torse, Abwickelbare Fläche. Developable surface
Torsion einer Kurve, Windung einer Kurve. Torsion of a curve
Torsionskurve. Twisted curve
Torus, Ringfläche. Torus
Totales Differential. Total differential
Totalgeordnete Menge. Serially ordered set
Totalgeordnete Menge, Kette. (Totally) ordered set
Träger einer Funktion. Support of a function
Trägheit. Inertia
Trägheitsmoment. Moment of inertia
Trägheitsradius. Radius of gyration
Trajektorie. Trajectory
Traktrix, Hundekurve. Tractrix
Transfinite Induktion. Transfinite induction
Transformationsgruppe. Transformation group

Transformierte Matrix. Transform of a matrix
Transitive Relation. Transitive relation
Transponierte Matrix. Transpose of a matrix
Transposition. Transposition
Transversal. Transversal
Transzendentale Zahl. Transcendental number
Trapez. Trapezoid
Trapezregel. Trapezoid rule
Trendkurve. Trend line
Trennungsaxiome. Separation axioms
Treppenfunktion. Step function
Triangulation. Triangulation
Trichotomie-Eigenschaft. Trichotomy property
Triederwinkel. Trihedral angle
Trigonometrie. Trigonometry
Trigonometrische Funktionen, Winkelfunktionen. Trigonometric functions
Trillion. Quintillion
Trinom. Trinomial
Trisektrix. Trisectrix
Triviale Lösung. Trivial solution
Triviale Topologie. Trivial topology
Trochoide. Trochoid
Tschebyscheffsche Ungleichung. Chebyshev inequality

Überdeckung einer Menge. Cover of a set
Übereinstimmend orientiert. Concordantly oriented
Überflüssige Wurzel. Extraneous root
Überlagerungsfläche. Covering space
Überlebensrente. Contingent annuity
Übung, Aufgabe. Exercise
Ultrafilter. Ultrafilter
Umbeschriebener Kreis. Circumscribed circle
Umdrehungsparaboloid. Paraboloid of revolution
Umfang. Perimeter; Girth
Umfang, Peripherie. Circumference
Umgebung eines Punktes. Neighborhood of a point
Umgekehrt proportionale Grössen. Inversely proportional quantities
Umkehrung einer hyperbolischen Funktion, Areafunktion. Arc-hyperbolic function
Umkehrung einer Operation. Inverse of an operation
Umkehrung einer Reihe. Reversion of a series
Umkehrung eines Theorems. Converse of a theorem
Umkreis. Circumcircle
Umkreismittelpunkt eines Dreiecks. Circumcenter of a triangle
Umordnung von Gliedern. Rearrangement of terms
Umwandlungstabelle. Conversion table
Unabhängige Ereignisse. Independent events
Unabhängige Variable. Independent variable
Unbedingte Ungleichheit. Unconditional inequality
Unbekannte Grösse. Unknown quantity
Unbestimmte Ausdrücke. Indeterminate forms
Unbestimmte Koeffizienten. Undetermined coefficients
Unbestimmtes Integral. Antiderivative; indefinite integral
Undefinierter Term, undefinierter Ausdruck. Undefined term
Uneigentliches Integral. Improper integral
Unendlich, Unendlichkeit. Infinity
Unendliche Reihen. Infinite series
Unendliches Produkt. Infinite product

Ungerade Zahl. Odd number
Ungleichartige Terme. Dissimilar terms
Ungleichheit. Inequality
Ungleichseitiges Dreieck. Scalene triangle
Unimodulare Matrix. Unimodular matrix
Unitäre Matrix. Unitary matrix
Unstetige Funktion. Discontinuous function
Unstetigkeit. Discontinuity
Unstetigkeitsstelle. Point of discontinuity
Untere Grenze. Greatest lower bound
Untere Schranke. Lower bound
Unterer Index. Subscript
Untergruppe. Subgroup
Unterkörper. Subfield
Untermenge. Subset
Unvollständige Induktion. Incomplete induction
Unwesentliche Abbildung. Inessential mapping
Unzusammenhängende Menge. Disconnected set
Ursprung eines Koordinatensystems. Origin of a co-
ordinate system

Vandermondsche Matrix. Vandermonde matrix
Variabilität. Variability
Variabel, Veränderliche. Variable
Varianz, Streuung. Variance
Variation einer Funktion. Variation of a function
Variation von Parametern. Variation of parameters
Variationsrechnung. Calculus of variations
Vektor. Vector
Vektoranalysis. Vector analysis
Vektormultiplikation. Multiplication of vectors
Verallgemeinerte Funktion. Generalized function
Verallgemeinerter Mittelwertsatz; Satz von Taylor.
Extended mean-value theorem
Verallgemeinertes Riemann-Integral. Generalized
Riemann integral
Verband. Lattice (in mathematics)
Verdoppelungsformel. Duplication formula
Vereinfachter Bruch. Simplified fraction
Vereinfachung. Simplification
Vereinigung von Mengen. Join of sets, Union of sets
Vergleichskriterium. Comparison test
Verkaufspreis. Selling price
Verschwindende Funktion. Vanishing function
Versicherung. Insurance
Vertängerts Rotationsellipsoid. Prolate ellipsoid of
revolution
(Verteilungen) mit gleicher Varianz. Homoscedastic
Verteilungsfunktion. Distribution function
Vertikale, Senkrechte. Vertical line
Vertrauensbereich. Confidence region
Vertrauensgrenzen. Fiducial limits
Vertrauensintervall, Konfidenzintervall. Confidence
interval
Verzerrungstensor. Strain tensor
Verzögerung. Deceleration
Verzweigungspunkt. Bifurcation point
Verzweigungspunkt, Windungspunkt. Branch point
Vielfachheit einer Wurzel. Multiplicity of a root
Vier. Four
Vierblattkurve. Quadrefoil
Vierdimensionaler Würfel. Tesseract
Viereck. Quadrangle
Vierergruppe. Four-group
Vierfarbenproblem. Four-color problem

Vierseit. Trapezium; Quadrilateral
Vierseitiges Prisma. Quadrangular prism
Viertel. Quarter
Viètascher Lehrsatz. Viète formula
Vollkommener Körper. Perfect field
Vollständig gemischtes Spiel. Completely mixed game
Vollständig normal. Perfectly normal
Vollständiger Graph. Complete graph
Vollständiger Körper. Complete field
Vollständiger Raum. Complete space
Vollständiges Quadrat. Perfect square
Vollwinkel. Perigon
Volt. Volt
Volumelastizitätsmodul. Bulk modulus
Volumen eines Körpers. Volume of a solid
Von einem Punkt ausgehen. Radiate from a point
Von gleicher Entfernung. Equidistant
Von zwei Grosskreishälften begrenztes Stück einer
Kugelfläche. Lune
Vorzeichen einer Zahl. Sign of a number

Wahrheitsmenge. Truth set
Wahrscheinlicher Fehler. Probable deviation
Wahrscheinlichkeitskurve. Probability curve
Watt. Watt
(Wechsel)inhaber. Payee
Wechselwinkel. Alternate angles
Wechselwirkung. Interaction
Wegzusammenhängende Menge. Arc-wise connected
set
Weite. Width
Wellengleichung. Wave equation
Wendepunkt. Inflection point
Wendetangente. Inflectional tangent
Wertigkeit eines Knotens. Valence of a node
Wesentliche Konstante. Essential constant
Widerstand. Resistance
Wiederkehrsatz. Recurrence theorem
Windschiefe Geraden. Skew lines
Windungszahl. Winding number
Winkel in den Parametergleichungen einer Ellipse in
Normalform. Eccentric angle of an ellipse
Winkel. Angle
Winkelbeschleunigung. Angular acceleration
Winkelhalbierende. Bisector of an angle
Winkel zweier Ebenen. Dihedral angle
Winkelmesser. Protractor
Winkeltreue Transformation. Isogonal transforma-
tion
Wirbelfreies Vektorfeld. Irrotational vector field
Wissenschaftliche Schreibweise. Scientific notation
Wohlordnungseigenschaft. Well-ordering property
Wronskische Determinante. Wronskian
Würfel. Cube
Würfelgruppe. Octahedral group
Würfelverdopplung. Duplication of the cube
Wurzel. Radix; root
Wurzel einer Gleichung. Root of an equation
Wurzelexponent. Index of a radical

X-Achse. X-axis

Y-Achse. Y-axis
Yard (= 91,44 cm). Yard of distance

Zahl, Nummer. Number
Zahl der primen Restklassen einer ganzen Zahl. Totient of an integer
Zahlen abrunden. Rounding off numbers
Zahlen mit Vorzeichen. Signed numbers
Zähler. Numerator
Zahlfolge. Sequence of numbers
Zahlgruppe. Group of numbers
Zahlkörper. Field of numbers, number field
Zahlmenge. Set of numbers
Zahlring. Ring of numbers
Zahlzeichen, Zahlwörter. Numerals
Zehn. Ten
Zehn Meter. Decameter
Zehneck. Decagon
Zeichenregel. Rule of signs
Zeile einer Determinante. Row of a determinant
Zeit. Time
Zeitrente, Rente. Annuity
Zelle. Cell
Zenit eines Beobachters. Zenith of an observer
Zenitdistanz. Zenith distance; Colatitude
Zentigramm. Centigram
Zentimeter. Centimeter
Zentrifugalkraft. Centrifugal force
Zentripedalbeschleunigung. Centripetal acceleration
Zentrum einer Gruppe. Center of a group
Zerfällungskörper. Splitting field
Zerlegung einer ganzen Zahl (in Primfaktoren). Partition of an integer
Zetafunktion. Zeta function
Ziffein-Rechner. Digital device (computer)
Ziffer. Digit
Zinsen eines Anlagepapiers. Dividend of a bond
Zinseszins. Compound interest
Zinsfuss. Interest rate
Zirkel. Circle, (pair of) compasses
Zirkelschluss. Circular argument
Zirkumpolarstern. Circumpolar star
Zoll. Inch
Zoll, Tarif. Tariff
Zonale harmonische Funktion. Zonal harmonic
Zopf. Braid
Zufallsfolge. Random sequence
Zufallsvariable. Variate

Zufallsvariable, zufällige Variable, aleatorische Variable. Stochastic variable
Zufallsvorrichtung. Random device
Zufallszug. Chance move
Zug in einem Spiel. Move of a game
Zug(spannung). Tension
Zugehöriger Winkel (beider Reduktion von Winkelfunktionen in den ersten Quadranten). Related angle
Zulässige Hypothese. Admissible hypothesis
Zusammengesetzte Funktion. Composite function
Zusammengesetzte Zahl. Composite number
Zusammenhang. Connectivity
Zusammenhängende Menge. Connected set
Zuwachs einer Funktion. Increment of a function
Zwanzig. Twenty, Score
Zwei. Two
Zwei Konfigurationen superponieren. Superpose two configurations
Zwei paarweis senkrechte Geraden (relativ zu zwei gegebenen Geraden). Antiparallel lines
Zwei-Personen-Spiel. Two-person game
Zweidimensionale Geometrie, eben Geometrie. Two-dimensional geometry
Zweidimensionale Normalverteilung. Bivariate normal distribution
Zweig einer Kurve. Branch of a curve
Zweijährlich, alle zwei Jahre. Biennial
Zwei Zahlen multiplizieren. Multiply two numbers
Zweistellig. Binary
Zweite Ableitung. Second derivative
Zwischenwertsatz. Intermediate value theorem
Zwölf. Twelve
Zwölfeck. Dodecagon
Zwölffarbensatz. Twelve-color theorem
Zykel, Zyklus. Cycle
Zykliden. Cyclides
Zyklische Permutation. Cyclic permutation
Zykloide. Cycloid
Zyklotomische ganze Zahl. Cyclotomic integer
Zylinder. Cylinder
Zylindrische Fläche. Cylindrical surface
Zylindroid. Cylindroid

Russian—English Index

Абак. Abacus

Абстрактное пространство. Abstract space

Абсолютная величина. Absolute value

Абсолютное значение. Absolute value

Абсолютная погрешность, абсолютная ошибка. Absolute error

Абсолютная сходимость. Absolute convergence

Абсолютно - непрерывная функция. Absolutely continuous function

Абстрактная математика. Abstract mathematics

Абсцисса. Abscissa

Автоматическое вычисление. Automatic computation

Автоморфизм. Automorphism

Автоморфная функция. Automorphic function

Авторегрессивные ряды (серии). Autoregressive series

Аддитивная функция. Additive function

Адиабатный. Adiabatic

Азимут. Azimuth

Акр. Acre

Аксиома. Axiom, postulate

Активы. Assets

Акции. Capital stock

Акционерный капитал. Stock

Алгебра. Algebra

Алеф - нуль, алеф - нулевое. Aleph-null (or aleph zero)

Алгебра, основанная на теории гомологии. Homological algebra

Алгебраич - ный, -еский. Algebraic

Алгебраически - заполненное поле. Algebraically complete field

Алгорифм. Algorithm

Алтернант. Alternant

Альтернативное предположение, гипотеза. Alternative hypothesis

Амортизационный капитал. Sinking fund

Амортизация. Amortization

Анализ. Analysis

Анализ точности. Sensitivity analysis

Анализ чувствительности. Sensitivity analysis

Аналитическая последовательность, ряд, множество. Analytic set

Аналитическая функция. Analytic function

Аналитичность. Analyticity

Аналогия. Analogy

Английская система мер веса. Avoirdupois

Аннигилятор. Annihilator

Аномалия точки. Anomaly of a point

Антиавтоморфизм. Antiautomorphism

Антиизоморфизм. Antiisomorphism

Антикоммутативный. Anticommutative

Антилогарифм. Antilogarithm

Антипараллельные линии. Antiparallel lines

Антипроизводная. Antiderivative

Антисимметричное отношение. Antisymmetric relation

Антисимметричный. Antisymmetric

Апофема. Apothem

Аптекарский. Apothecary

"Арбилос", особая геометрическая фигура, описанная Апхимедом. Arbilos

Аргумент комплексного числа. Amplitude of a complex number

Аргумент функции. Argument of a function

Арифметика. Arithmetic

Арифметический ряд. Arithmetic series

Арифметическое число. Arithmetic number

Арифмометр. Calculating machine

Аркгиперболическая функция. Arc-hyperbolic function

Арккосеканс. Arc-cosecant

Арккосинус. Arc-cosine

Арккотангенс. Arc-cotangent

Арксеканс. Arc-secant

Арксинус. Arc-sine

Арктангенс. Arc-tangent

Апхимедово твёрдое тело. Archimedean solid

Асимметричный. Asymmetric

Асимметрия распределения. Skewness

Асимптота. Asymptote

Асимптотическое направление. Asymptotic direction

Асимптотическое растяжение. Asymptotic expansion

Асимптотическое расширение. Asymptotic expansion

Ассимптотная плотность. Asymptotic density

Ассоциативный закон. Associative law

Астроида. Asteroid

Астрономическая величина. Astronomical unit

Атмосфера. Atmosphere

Атом. Atom

Аффинная трансформация. Affine transformation

Аффинное преобразование. Affine transformation

Базарная цена. Market value

Базис. Basis

Бар, столбик гистограммы. Bar

Барицентр. Barycenter

Барицентрические координаты. Barycentric coordinates

Без систематической ошибки в ассимптотах. Asymptotically unbiased

Безусловное неравенство. Unconditional inequality

Безусловный. Categorical

Бесконечная последовательность. Infinite series

Бесконечное произведение. Infinite product

Бесконечность. Infinity

Бесконечность. Perpetuity

Бесконечный ряд. Infinite series

Бета - распределение. Beta distribution

Биквадратный. Biquadratic

Бикомпактное. Bicompact

Билинейный. Bilinear

Бимодальный. Bimodal

Бинарный. Binary

Биномиальные коэффициенты. Binomial coefficients

Бижекция (вид функции), взаимно - однозначное соответствие. Bijection

Биссектрисса. Bisector

Боковая площадь. Lateral area

Боковая поверхность. Lateral surface

Большее множество полученное в результате растленения. Coarser partition
Боны. Bond
Брахистохрона. Brachistochrone
Будущая ценность. Future value
Буквенная постоянная. Literal constant
Бурбаки, Николя. Bourbaki
Буфер. Buffer (in a computing machine)
Быстрое преобразование Фурье. Fast Fourier transform

Вавилонские цифры. Babylonian numerals
Валентность узла, вершины (графа). Valence of a node
Валовая прибыль. Gross profit
Валовой доход. Gross profit
Вариация параметров. Variation of parameters
Вариация функции. Variation of a function
Вариационное исчисление. Calculus of variations
Ватт. Watt
Вводный элемент. Input component
Ведущий коэффициент. Leading coefficient
Ведьма агнези. Witch of Agnesi
Вековое направление. Secular trend
Вектор. Vector
Вектор силы. Force vector
Векторное исчисление. Vector analysis
Векторный угол. Vectorial angle
Величина звезды. Magnitude of a star
Вероятность события. Probability of occurrence
Версинус. Versed sine
Вертикальная линия. Vertical line
Вертикальные углы. Vertical angles
Вершина. Apex
Вершина угла. Vertex of an angle
Верхний предел. Superior limit
Верхняя грань. Upper bound
Верхняя плотность (распределения). Upper density
Верхушка. Apex
Вес. Weight
Ветвь кривой. Branch of a curve
Вечность. Perpetuity
Взаимно исключающиеся события (случаи). Mutually exclusive events
Взаимодействие. Interation
Взимнооднозначное соответствие. One-to-one correspondence
Взвешенное среднее. Weighted mean
Вибрация. Vibration
Видоизмененные бесселевы функции. Modified Bessel functions
Винкуль. Vinculum
Вклад. Investment
Вихрь. Curl
Вкладное страхование. Endowment insurance
Включение во множестве. Imbedding of a set
Вложение. Investment
В направление часовой стрелки. Clockwise
Внешние члены. Extreme terms (or extremes)
Внешний угол. Exterior angle
Внешняя касательная к двум окружностям. External tangent of two circles
Внешняя пропорция. External ratio

Внутреннее отношение. Internal ratio
Внутреннее произведение. Inner product
Внутреннее свойство. Intrinsic property
Внутренние накрест лежащие углы. Alternate angles
Внутренний автоморфизм. Inner automorphism
Внутренний угол. Interior angle
Внутренняя касательная к двум окружностям. Internal tangent of two circles
Внутренняя пропорция. Internal ratio
Внутреродность. Endomorphism
Вовлечение. Implication
Вогнутая поверхность. Concave surface
Вогнуто-выпуклая игра. Concave-convex game
Вогнутость. Concavity
Вогнутый многоугольник. Concave polygon
Возведение в степени на линии. Involution on a line
Возвратная точка совмещенная с точкой перехода. Flecnode
Возвышенность. Altitude
Возможное отклонение. Probable deviation
Возрастающая функция. Increasing function
Вольт. Volt
Волшебный квадрат. Magic square
Восемь. Eight
Восьмая часть круга. Octant
Восьмигранная (октаэдральная) группа. Octahedral group
Восьмигранник. Octahedron
Восьмиугольник. Octagon
Вписанный (в окружность) многоугольник. Inscribed polygon
Вполне смешанная игра. Completely mixed game
Вращательное движение. Rotation
Вращение. Rotation
Вращение вокруг оси. Revolution about an axis
Вращение кривой вокруг оси. Revolution of a curve about an axis
Вращение осей. Rotation of axes
Время. Time
Всеобщее (не локальное) свойство. Global property
Всеобщий квантор. Universal quantifier
Вспомогательный круг. Auxiliary circle
Вставленные (внутри) промежутки. Nested intervals
Вторая диагональ определителя. Secondary diagonal
Вторая производная. Second derivative
Второй член пропорции. Consequent in a ratio
Входной угол. Reentrant angle
Выбор-ка (в статистике). Sample
Быборочная погрешность. Sampling error
Бывод. Inference
Быделять корень (числа, уравнения). Isolate a root
Быпрямляющаяся плоскость. Rectifying plane
Выпуклая кривая. Convex curve
Выпуклая оболочка множества. Convex hull of a set
Выражать в числах. Evaluate
Вырождающаяся конусная поверхность. Degenerate conic
Выражение во второй степени. Quadric
Высота. Altitude

Бысота уклона. Slant height
Бытянутый эллипсоуд. Prolate ellipsoid
Выход. Yield
Вычеркивание. Cancellation
Вычеркнуть. Cancel
Вычерчиване пространства. Mapping of a space
Вычет функции. Residue of a function
Вычисление. Computation
Вычисление промежуточных значений функции. Interpolation
Вычисление разностей функции. Differencing a function
Вычислительная машина. Computing machine
Вычислительный прибор. Calculating machine
Вычислять. Calculate
Вычислять. Cipher (*v.*)
Вычитаемое. Subtrahend
Бычитание чисел. Subtraction of numbers
Бычитательные формулы. Subtraction formulas

Гаверсинус. Haversine
Гамма функция. Gamma function
Гармоническая функция. Harmonic function
Гармоническое движение. Harmonic motion
Гексаэдрон. Hexahedron
Геликоида. Helicoid
Генеральная совокупность. Population
Генератриса. Generatrix
Географические координаты. Geographic coordinates
Геодезические параллели. Geodesic parallels
Геоид (вид элипсоида). Geoid
Геометрическая средняя. Geometric average
Геометрические последовательности. Geometric series
Геометрический ряд (-ы). Geometric series
Геометрическое место (траектория) двойных точек кривой. Tac-locus
Геометрическое место точек. Locus
Геометрия. Geometry
Геометрия двух измерений. Two-dimensional geometry
Геометрия трех измерений. Three-dimensional geometry
Гибкость. Flexibility
Гипер-вещественные числа. Hyperreal numbers
Гипербола. Hyperbola
Гиперболический параболоид. Hyperbolic paraboloid
Гиперболический цилиндр. Hyperbolic cylinder
Гиперболоид одного листа. Hyperboloid of one sheet
Гипергеометрические последовательности. Hypergeometric sequences
Гипергеометричекие ряды. Hypergeometric series
Гипер-комплексные числа. Hypercomplex numbers
Гипер-объём (свойство множества в Эвклидовом пространством). Hypervolume
Гиперплоскость. Hyperplane
Гипер-поверхность (вид подмножества). Hypersurface
Гипотеза. Hypothesis
Гипотенуза. Hypotenuse
Гипоциклоида. Hypocycloid

Гистограмма. Bar graph
Гистограмма. Histogram
Главная диагональ. Principal diagonal
Главная ось. Major axis
Гладкая кривая в проецируемой плоскости. Smooth projective plane curve
Гладкое отображение (карта). Smooth map
Год. Year
Голоморфная функция. Holomorphic function
Гомеоморфизм двух множеств. Homeomorphism of two sets
Гомоморфизм двух множеств. Homomorphism of two sets
Гомотетичные фигуры. Homothetic figures
Горизонт. Horizon
Горизонтальный. Horizontal
Грам. Gram
Граница множества. Boundary of a set
Граница множества. Bound of a set
Границы изменения переменного. Range of a variable
Грань многогранника. Face of a polyhedron
Грань (полипота). Facet
Грань пространственной фигуры. Edge of a solid
Граф Гамилтона. Hamiltonian graph
Граф Эйлера. Eulerian graph
Графика по составлению. Graphing by composition
График уравнения. Graph of an equation
Графическое решение. Graphical solution
Греческие цифры. Greek numerals
Группа гомологии. Homology group
Группа контролирующая. Control group
Группа отображения. Homology group
Группа перемещений. Commutative group
Группа перестановок. Permutation group
Группа преображения (трансформации). Transformation group
Группа четвёртого порядка. Four-group
Группа чисел. Group of numbers
Группоид (вид множества). Groupoid

Давление. Pressure
Два (две). Two
Два десятка, двадцать. Score
Двадцатигранная (икозаэдральная) группа. Icosahedral group
Двенадцатигранник. Dodecahedron
Двенадцатиугольник. Dodecagon
Двадцать. Twenty
Дважды в год. Biannual
Двенадцатиричная система счисления—(нумерации). Duodecimal system of numbers
Двенадцать. Twelve
Движение неменяющее фигуру. Rigid motion
Движущая сула. Momentum
Движущийся на окружности (вокруг). Circulant
Двойная нормаль. Binormal
Двойник. Doublet
Двойно прямоугольный. Birectangular
"Двойное правило, основанное на трёх данных" (из книги Л. Кэролла). Double rule of three
Двойной интеграл. Double integral
Двойной счет. Count by two

Двудольный граф. Bipartite graph

Двумерное нормальное распределение. Bivariate normal distribution

Двусериальный коэффициент корреляции. Biserial correlation coefficient

Двухгранная (диэдральная) группа. Dihedral group

Двухгранный угол. Dihedral angle

Двучлен. Binomial (n.)

Девятиугольник. Nonagon

Девять. Nine

Дедуктивное доказательство. Deductive proof

Действие. Operation

Действие в нгре. Play of a game

Действительная норма процента, Действительная процентная ставка. Effective interest rate

Действующий. Operator

Декагон. Decagon

Декаметр. Decameter

Декартовы координаты. Cartesian coordinates

Декартобо произведение. Cartesian product

Деление. Division

Деленное пространство. Quotient space

Делимость на одиннадцать. Divisibility by eleven

Делители. Dividers

Делитель. Divisor

Делительность. Divisibility

Дельта-функция Дирака. Dirac δ-function

Дельтаэдр. Deltahedron

Дельтоид. Deltoid

"Дерево" (в теории графов). Tree

Десятичная система нумерации. Decimal system

Десять. Ten

Детерминант Вронского. Wronksian

Деформапия. Deformation

Дециметр. Decimeter

Джоуль (единица измерения энергии или работы). Joule

Дзета. Zeta

Диагональ определителя. Diagonal of a determinant

Диаграмма. Diagram

Диада. Dyad

Диадный рационал, вещественное число, получаемое в результате определённой комбинации целых чисел. Dyadic rational

Диалитическая метода Сильвестра. Dialytic method

Диаметр в конической кривой. Diametral line

Диаметр круга (окружности). Diameter of a circle

Дивергенция. Divergence

Дивиденд облигации. Dividend on a bond

Дина. Dyne

Динамика. Dynamics

Динамическое программирование. Dynamic programming

Диполь. Dipole

Директрисса конической кривой. Directrix of a conic

Диск. Disc (or disk)

Дисконт. Discount

Дискретная математика. Discrete mathematics

Дискретная топология. Discrete topology

Дискретное множество. Discrete set

Дискретное преобразованние Фурье. Discrete Fourier transform

Дискриминант многочлена. Discriminant of a polynomial

Дисперсия. Variance

Диффеоморфизм, прямое отобпажение для дифференциируем, гладких функций. Diffeomorphism

Дифференциальное уравнение. Differential equation

Дифференциал функции. Differential of a function

Дифференцирование функции. Differentiation of a function

Дихотомия. Dichotomy

Длина дуги. Arc length

Длина кривой. Length of a curve

Добавля-ющийся, -емое. Addend

Добавочные углы. Supplementary angles

Добавочный налог. Surtax

Добытое уравнение. Derived equation

Доверие. Reliability

Доверительные пределы. Fiducial limits

Доверительный интервал. Confidence interval

Додекагон. Dodecagon

Додекаэдр. Dodecahedron

Доказательство. Proof

Доказательство от противного. Indirect proof

Доказательство го выводу. Deductive proof

Доказательство с помощью "спуска", построения от общего к частному. Proof by descent

Доказать теорему. Prove a theorem

Долг. Liability

Долг. Loan

Долгота. Longitude

Доминирующая стратегия. Dominant strategy

Домино (геометрическая фигура). Domino

Дополнение для наклонения (до 90°). Codeclination

Дополнение для широты (до 90°). Colatitude

Дополнение множества. Complement of a set

Дополнительная часть конечной части множества. Cofinal subset

Дополнительная функция. Cofunction

Дополнительная широта (до 90°). Colatitude

Дополнительное обезпечение. Collateral security

Дополнительные хорды. Supplemental chords

Дополнительный угол. Complementary angle

Дополнить квадрат. Completing the square

Достаточное условие (положение). Sufficient condition

Доходный налог. Income tax

Дробный показатель степени. Fractional exponent

Дробь. Fraction

Дружные числа. Amicable numbers

Дуальность, двойственность. Duality

"Дуга", символ для обозначения пересечения или нижней грани. Cap

Дюйм. Inch

Египетские цифровые иероглифы. Egyptian numerals

Единица. Unity

Единица. One

Единообразная цена. Flat price

Единственное разложение на множители. Unique factorization

Ежегодная рента. Annuity
Естественное следствие. Corollary

Жизненое страхование. Life insurance

Завертывание на линии. Involution on a line
Зависимая переменная. Dependent variable
Зависимые уравнения, система уравнений. Dependent equations
Задача. Problem
Задача Какеи. Kakeya problem
Задача "Кенигсбергского моста". Königsberg bridge problem
Задача Куратовского по замыканию и дополнению. Kuratowski closure-complementation problem
Задача о граничных значениях. Boundary-value problem
Задача о четырех красках (закрашиваний). Four-color problem
Задолженность. Liability
Заключение. Conclusion
Заключительная сторона угла. Terminal side of an angle
Закон. Principle
Закон показателей степеней. Law of exponents
Закон распределения. Distributive law
Замечание. Note
Замкнутая кривая. Closed curve
Замкнутое множество. Closed set
Замыкание множества. Closure of a set
Зашифровать. Cipher (v.)
Звезда комплекса. Star of a complex
Звезда, приближённая к полюсу. Circumpolar star
Звездное время. Sidereal time
Землемер. Surveyor
Земной меридиан. Meridian on the earth
Зенит наблюдателя. Zenith of an observer
Зенитное расстояние. Zenith distance
Зета функция. Zeta function
Змеиная кривая. Serpentine curve
Знак. Symbol
Знак дробного деления. Solidus
Знак корня. Radical
Знак плюс. Plus-sign
Знак сложения. Summation sign
Знак числа. Sign of a number
Знаковая (сигнус) функция. Signum function
Знаменатель. Denominator
Значащий цифр. Significant digit
Значение места в числе. Place value
Значимость отклонения. Significance of a deviation
"Золотой" отрезок, участок, секция. Golden section
Зона. Zone
Зональная гармоника. Zonal harmonic
Зэта функция. Zeta function

Игра в "две монетки". Coin-matching game
Игра в "шестёрки" ("шестиугольники"). Game of hex
Игра двух лиц. Two-person game
Игра Мазура-Банаха. Mazur-Banach game
Игра морра. Morra (a game)
Игра Ним. Game of Nim

Игра с нулевой суммой. Zero-sum game
Идеал в кольце. Ideal contained in a ring
Идеальное поле. Perfect field
Идемфактор. Idemfactor
Избыточное число. Abundant number
Извлечение корня. Evolution
Изменение. Change
Изменение основания (логарифмов). Change of base
Измениние параметров. Variation of parameters
Изменение порядка членов. Rearrangement of terms
Изменчивость. Variability
Изменяющийся. Variate
Измерение. Mensuration
Измерение. Dimension
Измерение Хаусдорфа. Hausdorff dimension
Измеримое множество. Measurable set
Изображение знаками, буквами, цифрами. Notation
Изогональная трансформация. Isogonal transformation
Изогональное преобразование. Isogonal transformation
Изолированная точка. Acnode
Изолировать корень (числа, уравнения). Isolate a root
Изометрические поверхности. Isometric surfaces
Изоморфизм двух множеств. Isomorphism of two sets
Изотерма. Isothermal line
Изохронная кривая. Isochronous curve
Икосаэдр. Icosahedron
Импульс. Momentum
Имущество. Assets
Инвариант уравнения. Invariant of an equation
Инверсия точки. Inversion of a point
Инверсор. Inversor
Индикатриса пространственной кривой. Indicatrix of a curve
Индуктивный метод. Inductdive method
Индукция. Induction
Инерция. Inertia
Интеграл Бохнера. Bochner integral
Интеграл по поверхности. Surface integral
Интеграл функции. Integral of a function
Интеграл энергии. Energy integral
Интегральная однородная функция от четырех переменных. Quaternary quantic
Интегральное исчисление. Integral calculus
Интегральное уравнение. Integral equation
Интегратор. Integrator
Интеграф. Integraph
Интегрирование по частям. Integration by parts
Интегрирующий множитель. Integrating factor
Интегрируемая функция. Integrable function, summable function
Интервал. Interval
Интервал доверия. Confidence interval
Интервал сходимости. Interval of convergence
Интерквартильная зона. Interquartile range
Интерполяция. Interpolation
"Интуиционизм" (философско - математическая доктрина. Intuitionism
Иррациональное число. Irrational number

Иррациональное число. Surd
Исключение. Elimination
Испытание делимости на девять. Casting out nines
Испытание по сравнению. Comparison test
Истинное множество, истинная совокупность объектов. Truth set
Исходить из точки. Radiate from a point
Исчезающаяся функция. Vanishing function
Исчисление. Calculus
Исчисление. Evaluation
Исчисление бесконечно малых. Infinitesimal analysis
Исчизление факторов. Factor analysis
Исчизлять. Evaluate
Итеративный интеграл. Iterated integral

Каждые два года. Biennial
Калория. Calory
Каноническая форма. Canonial form
Капитализированная стоимость (цена). Capitalized cost
Капиталовложение. Investment
Каппа кривая. Kappa curve
Кардинальное число. Cardinal number
Кардиоида. Cardiod
Карта технологического процесса. Flow chart
Касание. Tangency
Касательная плоскость. Tangent plane
Касающаяся к кругу. Tangent of a circle
Каталонские цифры. Catalan numbers
Категорический. Categorical
Категория. Category
Катеноида. Catenoid
Катет (прямоугольного трехугольника). Leg of a right triangle
Качество ограничения. Property of finite character
Качество ("тонкость") расчленения множества. Fineness of a partition
Квадратное уравнение. Quadratic equation
Квадрантные углы. Quadrantal angles
Квадрат. Square
Квадратнчное уравнение. Quadratic equation
Квадратный корень. Square root
Квадратура круга. Quadrature of a circle
Квадратура круга. Squaring a circle
Квадратурная кривая. Rectifiable curve
Квадриллион. Quadrillion
Квази-нормальная подгруппа. Quasi-normal subgroup
Квантика. Quantic
Квантор. Quantifier
Кватернион. Quaternion
Квинтиллион. Quintillion
Кибернетика. Cybernetics
Киловатт. Kilowatt
Килограмм. Kilogram
Километр. Kilometer
Кинематика. Kinematics
Кинетика. Kinetics
Кинетическая знергия. Kinetic energy
Китайско-Ьпонские цифры. Chinese-Japanese numerals
Класс. Class or set
Класс эквивалентности. Equivalence class

Клин. Wedge
Клинопись, клинописные символы. Cuneiform symbols
Ковариантное производная. Covariant derivative
Коверсинус. Covered sine (coversine)
Колебание. Vibration
Колебание функции. Oscillation of a function.
Колебающиеся ряд(ы). Oscillating series
Колинейные преобразования. Collineatory transformations
Количество. Quantity
Количество движения. Momentum
Кологарифм. Cologarithm
Кольцо. Ring, annulus
Кольцо мер. Measure ring
Комбинаторная топология. Combinatorial topology
Комиссионер. Broker
Коммутативная группа. Commutative group
Коммутатор. Commutator
Компактизация. Compactification
Компактное множество. Compact set
Компактум. Compactum
Компас. Compass
Комплексное число. Complex number
Компонента памяти. Memory component
Компонента силы. Component of a force
Компонента хранения. Storage component
Конгруентные фигуры (тела). Congruent figures
Конгруенция. Congruence
Конечная десятичная дробь. Terminating decimal
Конечная игра. Finite game
Конечная проецируемая плоскость. Finite projective plane
Конечная точка кривой. End point of a curve
Конечное множество. Finite set
Коникоида. Conicoid
Коническая поверхность. Conical surface
Конические кривые с общими фокусами. Confocal conics
Конический, Коническая кривая. Conic
Коноида. Conoid
Консопьная балка. Cantilever beam
Констркция. Construction
Континуум. Continuum
Контравариантный тензор. Contravariant tensor
Контрольная группа. Control group
Контурные линии. Contour lines
Конус. Cone
Конусная поверхноць разделенная верхушкой конуса. Nappe of a cone
Конусхая поверхноць. Conical surface
Конусообразная поверхноць. Conical surface
Конфигурация. Configuration
Конформная транцформация. Conformal transformation
Конформное преобразование. Conformal transformation
Конхоида. Conchoid
Концентричные круги. Concentric circles
Концентричные окружности. Concentric circles
Концентрические круги. Concentric circles
Кооперативнбая игра. Cooperative game
Координата точки. Coordinate of a point
Координатная плоскость. Coordinate plane
Корд. Cord

Корень. Radix
Корень третьей степени. Cube root
Корень уравнения. Root of an equation
Косая высота. Slant height
Косвенное доказательство. Indirect proof
Косеканс угла. Cosecant of an angle
Косинус угла. Cosine of an angle
Косо-симметричный определитель. Skew-symmetric determinant
Косой треугольник. Oblique triangle
Косые линии. Skew lines
Котангенс угла. Cotangent of an angle
Кофункция. Cofunction
Коциклические точки, точки принадлежащие одной общей окружности. Concyclic points
Коэффициент. Coefficient
Коэффициент. Multiplier
Коэффициент деформации. Deformation ratio
Коэффициент корреляции. Correlation coefficient
Коэффициент обьема, массы. Bulk modulus
Коэффициент отношения подобности. Ratio of similitude
Коэффициент регрессии. Regression coefficient
Кратное двух чисел. Quotient of two numbers
Кратное пространство. Quotient space
Кратное числа. Multiple of a number
Кратные (параллельные) рёбра графа. Multiple edge in a graph
Кратный интеграл. Iterated integral
Крестообразная кривая. Cruciform curve
Кривая Агнези. Witch of Agnesi
Кривая в проецируемой плоскости. Projective plane curve
Кривая вероятности. Probability curve
Кривая возрастания. Logistic curve
Кривая движения векторов скорости. Hodograph
Кривая кратчайшего спуска. Brachistochrone
Кривая полета снаряда. Trajectory
Кривая Пэрл-Рида. Logistic curve
Кривая разделения угла на три части. Trisectrix
Кривая распределения частост. Frequency curve
Кривая (линия) с одним направлением. Unicursal curve
Кривая секанса. Secant curve
Кривая синуса. Sine curve
Кривая третьей степени. Cubic curve
Кривая третьей степени с двумя отдельными частями. Bipartite cubic
Кривая четвертой степени. Quartic curve
Кривизна. Curvature
Кривой квадрат. Quatrefoil
Криволинейное движение. Curvilinear motion
Криволинейно-четырехугольная гармоническая кривая. Tesseral harmonic
Критическое значение. Critical value
Кросс-кап. Cross-cap
Круг. Circle
Круг. Cycle
Круг с радиусом равным единице. Unit circle
Круг (кружок) сходимости. Circle of convergence
Круги с общей осью. Coaxial circles
Круглый конус. Circular cone
Круговая перестановка. Cyclic permutation
Круговая симметрия переменных. Cyclosymmetry
Крутость крыши. Pitch of a roof

Крунода. Crunode
Крученная кривая. Twisted curve
Куб. Cube
Кубическая кривая. Cubic curve
Кубическая парабола. Cubical parabola
Кубический корень. Cube root
Кубоид, прямоугольный параллелипипед. Cuboid
Кусочно-непрерывная функция. Piecewise continuous function

Лакунарное пространство. Lacunary space
Леворучная кривая. Left-handed curve
Лексиграфически—(упорядоченная последовательность). Lexicographically
Лемма. Lemma
Лемниската. Lemniscate
Линейка. Rule
Линейная комбинация. Linear combination
Линейно зависимые количества. Linearly dependent quantities
Линейное преобразование. Linear transformation
Линейное прогпаммирование. Linear programming
Линейные формы на поверхности. Ruling on a surface
Линейный оператор. Linear operator
Линейный элемент в дифференциальном уравнении. Linear element
Линейно-связанное множество. Arc-wise connected set
Линейчатая поверхноуть. Ruled surface
Линии лежащие в одной и той же плоскости. Coplanar lines
Линии потока. Stream lines
Линии течения. Stream lines
Линия общего направления. Trend line
Лист. Lamina
Лист Декарта. Folium of Descartes
Лист поверхности. Sheet of a surface
Литр. Liter
Литуус. Lituus
Лицо получающее плату по страховой полиси. Beneficiary
Личный ход. Personal move
Логарифм числа. Logarithm of a number
Логарифмическая кривая. Logarithmic curve
Логарифмическая линейка. Slide rule
Логарифмически-нормальное распределение. Lognormal distribution
Логарифмические таблицы. Table of logarithms
Локально (местно)-интегрируемая функция. Locally integrable function
Lokalьno kompaktnuй. Locally compact
Локально (место) связанные (соединённые) линейно. Locally arc-wise connected
Локальное (местное) свойство. Local property
Локсодромная спираль. Loxodromic spiral
Локус. Locus
Лошадиная сила. Horsepower
Луночка. Lune
"Луны" Гиппократа. Lunes of Hippocrates
Лучевой центр. Ray center
Любое число очень большой величины. Googol

Маклер. Broker
Максимизирующий игрок. Maximizing player
Максимум функции. Maximum of a function
Мантисса. Mantissa
Масса. Mass
Математика. Mathematics
Математика, базирующаяся на методах конструктивизма. Constructive mathematics
Математическая индукция. Mathematical induction
Матрица Вандермонде. Vandermonde matrix
Матрица коэффициентов. Matrix of coefficients
Матрица Эрмита. Hermitian matrix
Маятник. Pendulum
Мгновенная скорость. Instantaneous velocity
Медианная точка. Median point
Межа множества. Boundary of a set
Междуквартильный размах. Interquartile range
Международная система единиц. International system of units
Меньшая граница. Lower bound
Меньшая ось. Minor axis
Меньшее множество полученное в результате расчленения. Finer partition
Меньший предел. Inferior limit
Мера концентрации распределения—куртосис. Kurtosis
Мера множества. Measure of a set
Меридианная кривая. Meridian curve
Мероморфная функция. Meromorphic function
Местная ценность. Local value
Место точек. Locus
Метасжатое (метауплотнённое, метакомпактное) пространство. Metacompact space
Метод наименьших квадратов. Method of least squares
Метод полного перебора (вариантов). Method of exhaustion
Метод резкого "спуска". Method of steepest descent
Метод седловой точки. Saddle-point method
Метр. Meter
Метризуемое пространство. Metrizable space
Метрическое пространство. Metric space
Механика жидкостей. Mechanics of fluids
Механическое интегрирование. Mechanical integration
Мил. Mil
Миллиард. Billion (10^9)
Миллиметр. Millimeter
Миллион. Million
Миля. Mile
Минимальная поверхность. Minimal surface
Минимум функции. Minimum of a function
Минор определителя. Minor of a determinant
Минута. Minute
Минус. Minus
Мириада. Myriad
Мнемоническая схема. Mnemonic device
Мнимая корреляция. Illusory correlation
Мнимая часть числа. Imaginary part of a number
Многоадресная система. Multiaddress system
Многогранник. Polyhedron
Многогранный угол. Polyhedral angle
Многократный интеграл. Multiple integral
Многозначная функция. Many-valued function

Многолистник. Multifoil
Многообразие. Manifold
Многоугольник. Polygon
Многочисленность корня. Multiplicity of a root
Многочлен. Multinomial
Множественно связанные области. Multiply connected regions
Множество. Set
Множество Бореля. Borel set
Множество Джулии. Julia set
Множество Мандельброта. Mandelbrot set
Множество чисел. Set of numbers
Множимое. Multiplicand
Множитель. Multiplier
Множитель многочлена. Factor of a polynomial
Множить два числа. Multiply two numbers
Модулирующая машина. Analog computer
Модуль. Module
Модуль конгруентности. Modulus of a congruence
Модуль объема, массы. Bulk modulus
Модульная функция. Modular function
Моль. Mole
Момент вращения. Torque
Момент инерции. Moment of inertia
Момент силы. Moment of a force
Момент скручивания. Torque
Монетный вес. Troy weight
Моническое уравнение. Monic equation
Моногеническая аналитическая функция. Monogenic analytic function
Монотонная функция. Monotone function
Морская миля. Nautical mile
Морской узел. Knot of distance
Морфизм. Morphism
Мощность множества. Potency of a set

Набла. Del, nabla
Награда. Premium
Надир. Nadir
Надпись (сверху). Superscript
Наиболее благоприятный маневр. Optimal strategy
Наиболее благоприятная стратеяия. Optimal strategy
Наибольший общий делитель. Greatest common divisor
Наименьшая верхная грань. Least upper bound
Накладные расходы. Overhead expenses
Накладываемые конфигурации (формы). Superposable configuration
Накладывать две конфигурации (формы). Superpose two configurations
Наклон дороги. Grade of a path
Наклон кривой. Slope of a curve
Наклон линии. Inclination of a line
Наклонение. Declination
Наклонный треугольник. Oblique triangle
Наклонный угол. Gradient
Накопленная ценность. Accumulated value
Накопленная частота. Cumulative frequence
Накопленное значение. Accumulated value
Накопители. Cumulants
Налог. Tax
Направленная линия. Directed line

Направляющий конус. Director cone
Напряжение. Tension
Напряжение резки. Shearing strain
Напряжение тела. Stress of a body
Нарицательное число. Denominate number
Нарушение симметрии. Assymetry
Настоящая ценность. Present value
Натуральное число. Natural number
Натуральный логарифм. Natural logarithms
Натяжение. Tension
Находиться в пространстве. Imbed in a space
Начало координатных осей. Origin of coordinates
Начинающая сторона угла. Initial side of an angle
Небесный. Celestial
Небесный экватор. Celestial equator
Невращающийся вектор. Irrotational vector
Невыырожденное преобразование. Nondegenerate transformation
Негармоническая частость (пропорция). Anharmonic ratio
Недвижущаяся точка. Stationary point
Неединственное преобразование. Nonsingular transformation
Независимое переменное. Independent variable
Независнмые события. Independent events
Незначительное решение. Trivial solution
Неизвестное количество. Unknown quantity
Некоопероտивная угра. Noncooperative game
Нелинейное программирование. Nonlinear programming
Неменяющий множитель. Idemfactor
Неограниченная функция. Unbounded function
Неопределённые коэффициенты. Undetermined coefficients
Неопределённые формы. Indeterminant forms
Неопределённый интеграл. Antiderivative; indefinite integral
Неопределённый член. Undefined term
Необходимая постоянная. Essential constant
Необходимое условие. Necessary condition
Неограниченная функция. Unbounded function
Неособое преобразование. Nonsingular transformation
Неособое отображение. Inessential mapping
Неостаток. Nonresidue
Непереходная зависимость. Intransitive relation
Непереходная связь. Intransitive relation
Непереходное отношение. Intransitive relation
Неперовский логарифм. Natural logarithm
Неповоротимый вектор. Irrotational vector
Неподобные члены. Dissimilar terms
Неполная индукция. Incomplete induction
Неполное число (противоположность избыточному числу). Deficient number
Непостоянство. Variability
Неправильный четыреугольник. Trapezium
Непрерывная дробь. Continued fraction
Непрерывная (недискретная) топология сети. Indiscrete topology
Непрерывная (недискретная, "тривиальная") топология. Trivial topology
Непрерывная функция. Continuous function
Непрерывно деформируемые из одной в другую фигуры. Homotopic figures

Непрерывное многообразие (множество). Continuum
Непрерывное пространство. Normal space
Непрерывность. Continuity
Неприводимый корень (числа). Irreducible radical
Непропорциональный. Disproportionate
Непрямоугольное пространство. Nonsquare space
Неравенство. Inequality
Неравенство. Odds
Неравенство без ограничений. Unconditional inequality
Неравенство Чебышева. Chebyshev inequality
Неправомерное доказательство. Circular argument
Нерв системы множеств. Nerve of a system of sets
Неротативный вектор. Irrotational vector
Несвободное поле силы. Conservative field of force
Несвоеобразное преобразование. Nonsingular transformation
Несвязное множество. Disconnected set
Несмещенная оценка. Unbiased estimate
Несобственный интеграл. Improper integral
Несовершенное число. Defective (or deficient) number
Несоизмеримые числа. Incommensurable numbers
Нестандартные (гипер-вещественные) числа. Nonstandard numbers
Несущественный разрыв. Removable discontinuity
Нечетное число. Odd number
Нечёткая (размытая) логика. Fuzzy logic
Нечёткое (размытое) множество. Fuzzy set
Неявная дифференциация. Implicit differentiation
Неявная функция. Implicit function
Неявное дифференцирование. Implicit differentiation
Нивелировочные линии. Level lines
Нижний предел. Inferior limit
Нижняя грань. Lower bound
Нильпотентный. Nilpotent
Номинальная норма процентов—номинальная процентная ставка. Nominal rate of interest
Номограмма. Nomogram
Норма матрицы. Norm of a matrix
Норма процента. Interest rate
Нормализованное переменное. Normalized variate
Нормаль кривой. Normal to a curve
Нормальная производная. Normal derivative
Номинальная стоимость. Par value
Нормальное время. Standard time
Нормальное пространство. Normal space
Нулевая мера. Measure zero
Нуль. Zero
Нуль. Cipher (*n.*)
Нуль-потентный идеал. Nilpotent ideal
Нумерация. Numeration
Ньютон. Newton
Ньютоново уравнение третьей степени (тридента). Trident of Newton

Обеспечение функции. Support of a function
Обеспечение (функции) сжатием, компактное обеспечение (функции). Compact support
Обесценивать. Discount
Область. Domain
Область изучения. Field of study

Область исследования. Field of study

Область стопроцентной вероятности (определённости). Confidence region

Область учения. Field of study

Облигация. Bond

Обобщённая функция. Generalized function

Обобщённый интеграл Риманна. Generalized Riemann integral

Обобщенное коши признак сходимости (рядов). Generalized ratio test

Обозначение. Notation

Оболочка множества. Covering of a set

Образ точки. Image of a point

Образующая. Generatrix

Образующая поверхности. Generator of a surface

Обратная теорема. Converse of a theorem

Обпатная тригонометрическая функрция. Inverse trigonometric function

Обратно - пропорциональные величины. Inversely proportional quantities

Обратно - пропорциональные количества. Inversely proportional quantities

Обратное уравнение. Reciprocal equation

Обратное число. Reciprocal of a number

Обратный оператор. Inverse of an operator

Обсоблять корень (числа, уравнения). Isolate a root

Общая касательная к двум окружностям. Common tangent of two circles

Общая пожизненная годовая рента. Joint-life annuity

Общий множитель. Common multiple

Объединение множеств. Join of sets

Объем. Volume of a solid

Обыкновенные логарифмы. Common logarithms

Обязательство. Liability

Овал. Oval

Огибающая семейства кривых. Envelope of a family of curves

Огива. Ogive

Ограниченное множество. Bounded set

Ограниченный по существу. Essentially bounded

Одна сотая част числа. Hundredth part of a number

Один, одна. One

Одинаковое (во всех направлениях) вещество. Isotropic matter

Одиннадцать. Eleven

Одно-адресная система. Single-address system

Одновременные уравнения. Simultaneous equations

Однозначная функция. Single-valued function

Однозначно определенный. Uniquely defined

Одно-однозначное соответствие. One-to-one correspondence

Однородно - выпуклое пространство. Uniformly convex space

Однородно - эквинепрерывная (совокупность функций). Uniformly equicontinuous

Однородность. Homogeneity

Однородность двух множеств. Isomorphism of two sets

Однородные уравнения. Homogeneous equation

Одно-связная область. Simply-connected region

Односторонная поверхность. Unilateral surface

Одностороннее смещение (граничный линейный оператор). Unilateral shift

Одночлен. Monomial

Ожидаемая вероятностная величина. Expected value

Окончивающая десятичная дробь. Terminating decimal

Окрестность точки. Neighborhood of a point

Округление чисел. Rounding off numbers

Окружность. Circle

Окружность (круга). Circumference

Окружность. Periphery

Окружность бписцанная в треугольник. Incircle

Окружность описанная около треугольника касающаяся к одной стороне и к продолжениям двух других сторон. Excircle, escribed circle

Октагон. Octagon

Октант. Octant

Октаэдр. Octahedron

Ом. Ohm

Оператор. Operator

Описанная окружность (круг) вокруг многоугольника. Circumcircle

Описанный круг вокруг многоугольника. Circumscribed circle

Определенная точка. Fixed point

Определенный интеграл. Definite integral

Определитель. Determinant

Определитель Вронского. Wronksian

Определитель Гудермана. Gudermannian

Опрокидывающая схема. Flip-flop circuit

Оптимальная стратегия. Optimal strategy

Опытная кривая. Empirical curve

Орбита. Orbit

Ордината точки. Ordinate of a point

Ориентировка, Ориентация, Ориентирование. Orientation

Ортогональные функции. Orthogonal functions

Ортонормальная последовательность. Orthonormal sequence

Ортоцентр. Orthocenter

Освобождение знаменателя дроби от иррациональности. Rationalize a denominator

Осевая симметрия. Axial symmetry

Основание. Base

Основание. Basis

Основание перпендикуляра. Foot of a perpendicular

Основание системы исчисления. Radix

Основание системы логарифмов. Radix

Основная теорема алгебры. Fundamental theorem of algebra

Основание (база) фильтра. Filter base

Особая точка. Singular point

Особость (сингулярность) сгиба. Fold singularity

Особый интеграл. Particular integral

Остаток бесконечной последовательности. Remainder of an infinite series

Остаток от бесконечного ряда. Remainder of an infinite series

Остаток при кратном после деления на девять. Excess of nines

Остаточный спектр. Residual spectrum

Острый треугольник. Scalene triangle

Острый угол. Acute angle

Осуществлять диагональную трансформацию матрицы. Diagonalize
Ось. Axis
Ось абсцис. X-axis
Ось ординат. Y-axis
Ось пересечения. Transverse axis
Отборная статистическая таблица смертности. Select mortality table
Отвесная линия. Plumb-line
Отвесная линия. Vertical line
Ответственность. Liability
Отвлеченная математика. Abstract mathematics
Отделение множества. Separation of a set
Отделенная точка. Acnode
Отделнные коэффициенты. Detached coefficients
Отделимое пространство. Separable space
Отделять корень (числа, уравнения). Isolate a root
Отклонение. Deviation
Открытый промежуток. Open interval
"Открытый шар" (для нормализованных линейных пространств). Open ball
Отмена. Cancellation
Отменить. Cancel
Относительная скорость. Relative velocity
Относительность. Relativity
Отношение включения. Inclusion relation
Отношение связи. Connected relation
Отнять. Subtract
Отображение в линии. Reflection in a line
Отображение сжатости (отображение Липшица). Contraction mapping
Отображенный. Homologous
Отображенный угол. Reflex angle
Отрезок. Segment
Отрезок на оси. Intercept on an axis
Отрезочная трансформация. Shear transformation
Отрезочное преобразование. Shear transformation
Отрицание. Negation
Отрицательное число. Negative number
Отсроченный платеж по ежегодной ренте. Deferred annuity
Охват. Girth
Оценивание. Evaluation
Оценивать. Evaluate
Оценка. Evaluation
Оценка величины. Estimate of a quantity
Оценка по (бухгалтерским) книгам. Book value
Оценка (определение знатения) поля. Valuation of a field
Очертание. Configuration
Ошибка процентновая, ошибка данная в процентах. Percent error
Оширенная теорема о среднем значении функции. Extended mean-value theorem

Пантограф. Pantograph
Пара простых чисел с разницей в 2. Twin primes
Парабола. Parabola
Парабола третьей степени. Cubic parabola
Параболический цилиндр. Parabolic cylinder
Параболоидвращения. Paraboloid of revolution
Парадокс. Paradox
Парадокс Банаха-Тарского. Banach-Tarski paradox
Парадокс Хаусдорфа. Hausdorff dimension

Параллакс звезды. Parallax of a star
Параллаксный угол. Parallactic angle
Параллелепипед. Parallelepiped
Параллели широты. Parallels of latitude
Параллелограм. Parallelogram
Параллелотоп. Parallelotope
Параллельные линии. Parallel lines
Параметр. Parameter
Параметрические уравнения. Parametric equations
Паскаль. Pascal
Педальная кривая. Pedal curve
Пентагон. Pentagon
Пентаграмма. Pentagram
Пентадекагон, 15-ти-сторонный многоугольник. Pentadecagon
Первообразная. Antiderivative
Переводная таблица. Conversion table
Перегибная касательная. Inflectional tangent
Перекрестная точка возврата кривой. Crunode
Перемежающаяся группа. Alternating group
Перемена параметров. Variation of parameters
Переменная группа. Alternating group
Переменная. Variable
Переменный. Alternant
Переместительный. Commutative
Переместить член. Transpose a term
Перемещение. Displacement
Перемещение осей. Translation of axes
Пересекающая (линия, поверхность). Transversal
Пересекающая ось. Transverse axis
Пересечение двух множеств. Meet of two sets
Перестановка n вещей. Permutation of n things
Пересчитываемость. Countability
Переход(ка) осей. Translation of axes
Переходное родство. Transitive relation
Перигелион (точка ближайшая к солнцу). Perihelion
Перигон. Perigon
Периметр, длина всех сторон многоугольника. Perimeter
Период функции. Period of a function
Периодическое движение. Periodic motion
Периодичность. Periodicity
Периферия. Periphery
Перпендикуляр. Vertical line
Перпендикуляр к поверхности. Perpendicular to a surface
Перпендикуляр к точке касания касательной к кривой. Normal to a curve
Перпендикулярные линии. Perpendicular lines
Перспективность. Perspectivity
Петербургский парадокс. Petersburg paradox
Петля кривой. Loop of a curve
Пиктограма. Pictogram
Пирамида. Pyramid
Пирамидная поверхность. Pyramidal surface
Плавающая запятая. Floating decimal point
Планарный (плоский) граф. Planar graph
Планиметр. Planimeter
Планиметрия. Two-dimensional geometry
Планиметрические кривые высшего порядка. Higher plane curves
Пластичность. Plasticity
Плечо рычага. Lever arm
Плоская фигура. Plane figure

Плоскости с общей точкой. Copunctal planes
Плоскостная точка поверхности. Planar point
Плотное множество. Dense set
Плотность. Density
Площадь. Area
Площадь поверхности. Surface area
Поверхностный интеграл. Surface integral
Поверхность вращения. Surface of revolution
Поверхность перемещения. Translation surface
Поворотная точка. Turning point
Повторение одного и того же алгебраического знака. Continuation of sign
Повторный интеграл. Iterated integral
Повторяющаяся десятичная дробь. Repeating decimal
Погашение долга. Amortization
Поглощающая способность, свойство поглощения. Absorption property
Поглущающий массив, множество. Absorbing set
Погрешность округления числа. Round off error
Подбазиеная теорема Александера. Alexander's subbase theorem
Подгруппа. Subgroup
Поддающиеся матрицы. Conformable matrices
Подинтегральная функция. Integrand
Подкоренное число. Radicand
Подмножество. Subset
Поднормаль. Subnormal
Подобие. Similitude
Подобная трансформация. Similarity transformation
Подобное преобразование. Similarity transformation
Подобные треугольники. Similar triangles
Подпись (снизу). Subscript
Подполе. Subfield
Подразумеваемое. Inference
Подразумевание. Implication
Подстановка в уравнении. Substitution in an equation
Подсчитывать. Calculate
Подтангенс. Subtangent
Пожизненная рента. Annuity (life)
Пожизненная пента. Perpetuity
Пожизненная пента прерываемая со смертыо получающаго её. Curtate annuity
Позиционная игра. Positional game
Показатель корня. Index of a radical
Показатель степени. Exponent
Показательная кривая. Exponential curve
Поле Галуа. Galois field
Поле разделения, расщепления. Splitting field
Полигон. Polygon
Полином Лежандра. Legendre polynomial
Полиомино. Polyomino
Политоп. Polytope
Полигекс (геометрическая фигура). Polyhex
Полиэдр. Polyhedron
Полное пространство. Complete space
Полностью (совершенно) упорядоченное множество. Totally ordered set
Полный граф. Complete graph
Полный дифференциал. Total differential
Положительное число. Positive number
Положительный знак. Plus sign

Полуаддитивная функция. Subadditive function
Полукруг. Semicircle
Полунепрерывная функция. Semicontinuous function
Полуокружность. Semicircle
Полурегулярное твёрдое тело (Архимедово твёрдое тело). Semi-regular solid
Полусфера. Hemisphere
Полутень. Penumbra
Получатель денег. Payee
Полюс круга. Pole of a circle
Поляр квадратной формы. Polar of a quadratic form
Поляризация. Polarization
Полярные координаты. Polar coordinates
Популяция. Population
Порода множества точек,Ророда точечного множества. Species of a set of points
Порядковое число. Ordinal number
Порядок группы. Order of a group
Порядок касания. Order of contact
Последовательно ориентированный. Coherently oriented
Последовательно-упорядоченное множество. Serially ordered set
Последовательность чисел. Sequence of numbers
Последовательные испытания (пробы, опыты). Successive trials
Последовательные трансформации пропорции. Composition in a proportion
Постоянная интегрирования. Constant of integration
Постоянная поворотхая точка кривой. Spinode
Постоянная скорость. Constant speed
Постоянно-выпуклое пространство. Uniformly convex space
Постоянный член интеграции. Constant of integration
Построение. Construction
Постулат. Postulate (n.)
Потенциальная функция. Potential function
Поундаль. Poundal
Почти периодический. Almost periodic
Почтиплотное пространство. Paracompact space
Почтисжатое пространство. Paracompact space
Правило. Principle
Правило знаков. Rule of signs
Правило механика. Mechanic's rule
Правило трапеции. Trapezoid rule
Правильно расходящиеця ряды. Properly divergent series
Правильный многоугольник (полигон). Regular polygon
Предел функции. Limit of a function
Предельная точка. Limit point
Предложение (для доказательства). Proposition
Предложительная функция. Propositional function
Предположительные цифры остающейся жизни статистически выведенные для любого возраста. Expectation of life
Представимая в конечном виде. Finitely representable
Представление группы. Representation of a group
Предъявитель чека, векселя. Payee
Преимущество. Odds

Премия. Bonus
Премия. Premium
Преобразование координат. Transformation of coordinates
Преобразование точек в точки, прямых в прямые и т.д. Collineation
Преобразованная матрица. Transform of a matrix
Преобразователя. Commutator
Прибавление. Addition
Прибавля-ющийся, -емое. Addend
Приближение. Approximation
Приблизительность. Approximation
Прибор осуществляющий преобразование инверсии. Inversor
Прибыток. Profit
Приведение дроби. Reduction of a fraction
Приемлемая гипотеза, допустимое предположение. Admissible hypothesis
Призма. Prism
Призматическая поверхность. Prismatic surface
Призматоид. Prismatoid
Призмоида. Prismoid
Призмоидная формула, Призмоидное правило. Prismoidal formula
Прикладная математика. Applied mathematics
Примитивный корень n-ой степени. Primitive nth root
Принцип. Principle
Принцип локализации. Localization principle
Принцип наложения. Superposition principle
Принцип однородной граничности (теорема Банаха-Штанхауза). Uniform boundedness principle
Принцип "ящика стола". Pigeon hole principle
Принцип "ящика стола" Дирихле. Dirichlet drawer principle
Приравнять. Equate
Приращение функции. Increment of a function
Принцип оптимальности. Principle of optimality
Присоединенная матрица. Adjoint matrix
Притяжение. Gravitation
Приходный налог. Income tax
Проба отношением. Ratio test
Проверка решения. Check on a solution
Программирование. Programming
Прогрессия. Progression
Продажная цена. Selling price
Продление (расширение) поля. Extension of a field
Проективная геометрия. Projective geometry
Проективность. Projectivity
Проектируемая плоскость. Projecting plane
Проекция вектора. Projection of a vector
Проекция шара на плоскость или плоскости на шар. Stereographic projection
Проецируемая топология. Projective topology
Проецируемое пространство. Projective space
Произведение Блашке. Blaschke product
Произведение чисел. Product of numbers
Произведенное уравнение. Derived equation
Производная вероятности. Derivative of a distribution
Производная высшего порядка. Derivative of higher order
Производная от натурального логарифма гамма-функции. Digamma function

Производная по направлению. Directional derivative
Производная функции. Derivative of a function
Промежуток сходимости. Interval of convergence
Пропорциональность. Proportionality
Пропорциональные величины. Propostional quantities
Пропорция. Proportion
Прорез цилиндра. Section of a cylinder
Простая дробь. Common fraction
Простая дробь. Vulgar fraction
Простая кривая, Простая закрытая кривая. Simple curve
Простая точка. Ordinary point
Простая функция. Schlicht function
Простое решение. Simple solution
Простое число. Prime number
Простой интеграл. Simple integral
Простой неповторяющийся корень уравнения. Simple root
Просто-связная область. Simply connected region
Пространственная кривая. Space curve
Пространственная спираль. Helix
Пространственная фигура вращения. Solid of revolution
Пространственная фигура с шестью гранями. Hexahedron
Пространственный угол. Solid angle
Пространство. Space
Пространство Бэйра. Baire space
Пространство орбиты. Orbit space
Пространство произведения. Product space
Пространство с открытыми областями. Lacunary space
Пространство Фреше. Fréchet space
Против движения часовой стрелки. Counterclockwise
Против часовой стрелки. Counter-clockwise
Противолежащий угол. Alternate angle
Противопараллельные линии. Antiparallel lines
Противоположные стороны. Opposite sides
Противоположные, противолежащие точки. Antipodal points
Профильная карта. Profile map
Процент. Interest rate
Процент. Percentage
Процентная квантиля. Percentile
Процентная ставка. Interest rate
Процентное отношение. Percentage
Прямая (линия). Straight line
Прямое произведение (груп, матриц). Direct product
Прямой треугольник. Right triangle
Прямолинейные образователи. Rectilinear generators
Прямоугольник. Rectangle
Прямоугольные (координатные) оси. Rectangular axes
Псевдосфера. Pseudosphere
Псевдосферическая поверхность. Pseudospherical surface
Псевдошар. Pseudosphere
Пупочная точка. Umbilical point
Пустое множество. Null set
Пучок кругов. Pencil of circles

Пучковая точка. Cluster point
Пучок плоскостей. Sheaf of planes
Пятиугольная пирамида. Pentagonal pyramid
Пятиугольник. Pentagon
Пятиугольник Пифагора. Pentagram
Пятигпанник, пентаэдр. Pentahedron

Работа. Work
Равенство. Equality
Равенство. Parity
Равновеликие. Equiareal
Равнобедренный треугольник. Isosceles triangle
Равновесие. Equilibrium
Равнодействующая. Resultant
Равноизмененный. Homoscedastic
Равнонепрерывные функции. Equicontinuous functions
Равноотстоящие. Equidistant
Равномерная непрерывность. Uniform continuity
Равномерная сходимость. Uniform convergence
Равномерно - выпуклое пространство. Uniformly convex space
Равносторонний треугольник. Equiliateral triangle
Равнотемпературная линия. Isothermal line
Равноугольная спираль. Equiangular spiral
Равноугольная трансформация. Isogonal transformation
Равноугольное преобразование. Isogonal transformation
Равноценность. Parity
Равные в ассимптотах, ассимптотически - равные. Asymptotically equal
Равные количества. Equal quantities
Радиан. Radian
Радикал. Radical
Радикал идеала. Radical of an ideal
Радикал кольца (кольцевой сети). Radical of a ring
Радикальная ось. Radical axis
Радиус круга (окружности). Radius of a circle
p-адическое число (в теории целых чисел). p-adic number
Развертка кривой. Involute of a curve
Развертка на линии. Involution on a line
Развертывающаяся поверхность. Developable surface
Разделение. Disjunction
Разделение множества. Separation of a set
Разделение на множители. Factorization
Разделение угла на три (равные) части. Trisection of an angle
Разделение целого числа. Partition of an integer
Разделимый на множители. Factorable
Разделители. Dividers
Разделить. Divide
Разделить пополам. Bisect
Разделы математики, не включающие вычисли теляные аспекты высшей математики и изучение иределов. Finite mathematics
Разделяющ-ий (-ая) пополам. Bisector
Размеры выработки. Yield
Разница. Odds
Разница между абсциссами двух точек. Run between two points
Разнообразие корня. Multiplicity of a root

Разностное уравнение. Difference equation
Разность двух квадратов. Difference of two squares
Разобщение. Disjunction
Разобщение множества. Separation of a set
Разрезать пополам. Bisect
Разрешающая группа. Solvable group
Разрыв со скачком. Jump discontinuity
Разрывность. Discontinuity
Разряд. Category
Разъединение. Disjuntion
Раскрытие определителя. Expansion of a determinant
Распределение с большой концентрацией около средней. Leptokurtic distribution
Распределение с малой концентрацией около средней. Platykurtic distribution
Распределение слабо сгущенное около средней. Mesokurtic distribution
Распределение χ^2 (хи - квадратное). Chi-square distribution
Распространение. Dilatation
Рассеяние. Dispersion
Рассроченная плата, уплата, Рассрочунный платеж. Installment payment
Расстояние. Distance
Рассчитывать. Calculate
Растягиваемое преобразование. Stretching transformation
Растяжение. Elongation
Расхождение рядов (последовательностей). Divergence of series
Расходящаяся последовательность. Divergent sequence
Расцветивание графов (в теории графов). Graph coloring
Расчётно - компактное (пространство, интервал). Countably compact
Расчлененные множества. Disjoint sets
Расширение. Dilatation
Рациональное число. Rational number
Реверсия последовательностей. Reversion of series
Регулярное пространство. Normal space
Редкое множество. Rare set
Режим. Mode
Резольвента матрицы. Resolvent of a matrix
Резольвентное уравнение третьей степени. Resolvent cubic
Результат. Result
Ректификация. Rectification
Рефракция. Refraction
Решетка. Lattice
Решето. Sieve
Решительный. Categorical
Род множества точек, Род точечного мхожества. Species of a set of points
Родственность. Relation
Родственный угол. Related nagle
Розетка из трех листов. Rose of three leafs
Ромб. Rhombus
Ромбовая призма. Rhombohedron
Ромбоид. Rhomboid
Ротор. Curl
Румбовая линия. Rhumb line
Ручка на поверхности. Handle on a surface
Ручка рычага. Lever arm

Руночная цена. Market value
Ряд чисел. Series of numbers

Салинон (геометрическая фигура). Salinon
Салтус функции. Saltus of a function
Самосопряженное преобразование. Self-adjoint transformation
Сантиграмм. Centigram
Сантиметр. Centimeter
"Сапожничий нож", арбелос (геометрическая фигура). Shoemaker's knife
Сверх-рефлективный. Super-reflexive
Сверхсоприкосновающиеся кривые на поверхности. Superosculating curves on a surface
Сверхсоприкосновение. Superosculation
Сверхтрохоида. Hypotrochoid
Сверхфильтр. Ultrafilter
Световая интенсивность измеряемая в свечах. Candlepower
Свеча (единица световой интенсивности). Candela
Свивание кривой. Torsion of a curve
Свободный ультрафильтр (вид фильтра). Free ultrafilter
Свойство идемпотентности. Idempotent property
Свойство инвариантности. Invariant property
Свойство Крейна-Мильмана. Krein-Milman property
Свойство отображения. Reflection property
Свойство регулярного (формального) упорядочения (множества). Well-ordering property
Связная трансформация. Conjunctive transformation
Связно ориентированный. Coherently oriented
Связное множество. Connected set
Связное преобразование. Conjunctive transformation
Связность. Connectivity
Связь. Bond
Связь. Brace
Связь. Conjunction
Сглаживание кривых. Curve fitting
Северное наклонение. North declination
Сегмент кривой (линии). Segment of a curve (line)
Седловая точка. Saddle point
Секанс угла. Secant of an angle
Секстиллион. Sextillion
Сектор круга. Sector of a circle
Семигранник, гептаэдр. Heptahedron
Семиугольник. Heptagon
Семь. Seven
Семья кривых. Family of curves
Сепарабельная игра. Separable game
Септиллион. Septillion
Сервомеханизм, Серво. Servomechanism
Сериальная облигация. Serial bond
Сериально-упорядоченное множество. Serially ordered set
Сеть неполно упорядоченных множеств. Net of partially ordered points
Сжатая трансформация. Contact transformation
Сжатие, отображение Липшица. Nonexpansive mapping
Сжатие тензора. Contraction of a tensor
Сжатое множесво. Compact set

Сжатое преобразование. Contact transformation
Сжимание. Compactification
Сжимающийся (сокращающийся) базис. Shrinking basis
Сила смертности. Force of mortality
Силлогизм. Symbol
Символ. Symbol
Символ, обозначающий связь, принадлежность или наименьший верхний предел. Cup
Симметричная фигура. Symmetric figure
Симметрия функции. Symmetry of a function
Симплекс. Simplex
Симплекс-метод. Simplex method
Симплициальное множество. Simplicial complex
Синус угла. Sine of an angle
Синтетическое деление. Synthetic division
Синусоида. Sinusoid
Система восьмеричных чисел. Octal number system
Система равнотемпературных кривых. System of isothermal curves
Система уравнений. System of equations
Скала мнимых чисел. Scale of imaginaries
Скалярное количество. Scalar quantity
Скелет комплекса. Skeleton of a complex
Скидка. Discount
Складыва-ющийся, -емое. Addend
Скобка. Bracket
Скобки (круглые). Parentheses
Скорость. Speed
Скорость. Velocity
Скорость движения. Momentum
Скрученность кривой. Torsion of a curve
Скручивание кривой. Torsion of a curve
Скручивающее усилие. Torque
Слабая сходимость. Weak convergence
Слагаемое. Summand
Слагающая силы. Component of a force
Слабо компактный. Weakly compact
Слабо сжатый. Weakly compact
Сложение. Addition
След матрицы. Spur of a matrix
Следствие. Corollary
Сложная функция. Composite function
Сложность корня. Multiplicity of a root
Сложные проценты. Compound interest
Сломанная линия. Broken line
Случайная последовательность. Random sequence
Случайное отклонение. Probable deviation
Случайный ход. Chance move
Смешанное число. Mixed number
Смещающаяся поверхность. Translation surface
Смещенная статистика. Biased statistic
Сморщивание плоскости. Shrinking of the plane
Смысл неравенства. Sense of an inequality
Собственная функция. Eigenfunction
Собственное значение. Eigenvalue
Собственный вектор. Eigenvector
Совершенно (экстремально) разъединённые (разобщённые) множества. Externally disconnected
Совершенно смешанная игра. Completely mixed game
Совместимость уравнений. Consistency of equations

Совместимые уравнения.   Consistent equations
Совместная пожизненная годовая рента.   Joint life annuity
Совокупность плоскостей.   Bundle of planes
Совпадающие линии.   Concurrent lines
Совпадающие фигуры (тела).   Congruent figures
Совпадающиеся углы, но различающиеся на 360°.   Coterminal angles
Совпадающий.   Coincident
Совпадение.   Congruence
Совпадение.   Conjunction
Согласно ориентированные.   Concordantly oriented
Согласно расположенные.   Concordantly oriented
Согласование.   Congruence
Согласованность уравнений.   Consistency of equations
Согласованные уравнения.   Consistent equations
Содержание множества.   Content of a set
Соединение.   Conjunction
Соединение (с свиванием) двух функций.   Convolution of two functions
Соединение множеств.   Join of sets
Соединение множеств.   Union of sets
Соединение множества предметов.   Combination of a set of objects
Соединение членов.   Grouping terms
Соизменимое производное.   Covariant derivative
Соизменимость.   Covariance
Соизмеримый.   Commensurable
Сокращать.   Cancel
Сокращение.   Cancellation
Сокращение (в топологии).   Retract
Сокращение тензора.   Contraction of a tensor
Соленоидный вектор.   Solenoidal vector
Солнечное время.   Solar time
Сомножество подгруппы.   Coset of a subgroup
Сомножитель.   Cofactor
Соответственные углы.   Corresponding angles
Соответствие.   Congruence
Соответствие.   Parity
Соответствующие матрицы.   Conformable matrices
Соответствующий.   Coincident
Соприкасающаяся плоскость.   Osculating plane
Соприкосновение.   Osculation
Сопряженное пространство.   Adjoint (or conjugate) space
Сопряженные комплексные числа.   Conjugate complex numbers
Сопряженные подгруппы.   Conjugate subgroups
Сопряженные углы.   Conjugate angles
Сорт поверхности.   Genus of a surface
Соседство точки.   Neighborhood of a point
Составляющая силы.   Component of a force
Составная функция.   Composite function
Составная часть.   Component
Составное число.   Composite number
Составной элемент.   Component
Сотная система меры углов.   Centesimal system of measuring angles
Сочетание.   Conjunction
Сочетание множества предметов.   Combination of a set of objects
Сплайн, полиномная кривая, кусочно-полиномиальное приближение, сплайн приближение.   Spline

Спектр матрицы.   Spectrum of a matrix
Спектральный анализ.   Spectral analysis
Специальная точка.   Singular point
Спиральная поверхность.   Spiral surface
Спиральное число.   Winding number
Сплющенный эллипсоид.   Oblate ellipsoid
Способность (свойство) приближения.   Approximation property
Способный к инверсии (обратному преобразованию).   Invertible
Среднее.   Average
Среднее двух чисел.   Mean (or average) of two numbers
Средняя точка.   Median point
Средняя точка отрезка линии.   Midpoint of a line segment
Ставить условием.   Postulate ($v.$)
Стандартное время.   Standard time
Стандартное отклонение.   Standard deviation
Статистика.   Statistics
Статистика рождаемости, смертности, и т.д.   Vital statistics
(Статистическая) таблица смертности.   Mortality table
Статистический вывод.   Statistical inference
Статистическое данное.   Statistic
Статистическое заключение.   Statistical inference
Статика.   Statics
Статический момент.   Static moment
Степенная кривая.   Power curve
Степенные ряды, серии.   Power series
Степень полинома (многочлена).   Degree of a polynomial
Степень числа.   Power of a number
Стерадиан.   Steradian
Стере—Кубический метр.   Stere
Стереографическая проекция.   Stereographic projection
Стереометрия.   Three-dimensional geometry
Сто.   Hundred
Стоградусный термометр.   Centigrade thermometer
Стоимость амортизации (изнашивания).   Depreciation charge
Стопа (бумаги).   Ream
Сторона многоугольника.   Side of a polygon
Стохастическая переменная.   Stochastic variable
Стратегия в игре.   Strategy of a game
Страхование.   Insurance
Страхование всей жизни.   Whole life insurance
Строго возрастающая функция.   Strictly increasing function
Строго-доминирующая стратегия.   Strictly dominant strategy
Строка определителя.   Row of a determinant
Строка, цепочка, последовательность.   String
Строка, цепочка элементов.   Braid
Строфоида.   Strophoid
Ступенчатая функция.   Step function
Стягивание тензора.   Contraction of a tensor
Стягивать угол.   Subtend an angle
Стяжатель.   Accumulator
Субгармоническая функция.   Subharmonic function
Субгруппа.   Subgroup
Субтангенс.   Subtangent

Суженная трансформация. Contact transformation

Суженное преобразование. Contact transformation

Сумма причитающаяся отказавшемуся от страхово го полиса. Surrender value

Сумма чисел. Sum of numbers

Суммирование ряда. Summation of a series

Суммируемая функция. Summable function

Супермножество, множество множесть. Superset

Суржекция (функция). Surjection

Существенно ограниченный. Essentially bounded

Существенное свойство. Intrinsic property

Существенный квантор. Existential quantifier

Существенный разрыв. Non-removable discontinuity

Сфера. Sphere

Сферические оси (координаты). Spherical coordinates

Сферический прямоугольник с тремя прямыми углам. Trirectangular spherical triangle

Сфероид. Spheroid

Сферы Дандлена. Dandelin spheres

Схема комплекса. Skeleton of a complex

Схема случайности (беспорядочности). Random device

Сходимость бесконечного ряда. Convergence of a series

Сходство. Analogy

Сходиться к пределу. Converge to a limit

Сходящееся последование. Convergent sequence

Сходящийся знаменатель цепной дроби. Convergent of a continued fraction

Сходящийся ряд. Convergent series

Сцепленно ориентированный. Coherently oriented

Счет. Numeration

Счет. Score

Счет с основаним два. Binary

Счетная линейка. Slide rule

Счетная машина. Calculating machine, computing machine

Счетное множество. Countable set, enumerable set, denumerable set

Счетность. Countability

Счетчик на вычитательной машине (счетной машине). Counter of a computing machine

Счеты. Abacus

Счётное число. Counting number

Счислять. Calculate, count

Считать. Compute, add, count

Считать по два. Count by twos

Считать по двойкам. Count by twos

Таблица логарифмов. Table of logarithms

Таблицы возможности, случайности, условности. Contingency table

Табличная разность. Tabular differences

Тангенс угла. Tangent of a angle

Тариф. Tariff

Температурная шкала Цельсия. Celsius temperature scale

Тензорное произведение векторных пространств. Tensor product of vector spaces

Тензор. Tensor

Тензор напряжения. Stress tensor

Тензорное исчисление. Tensor analysis

Тензорный анализ. Tensor analysis

Тень. Umbra

Теорема. Theorem

Теорема Безо́. Bézeut's theorem

Теорема Дарбу об аналитическом продолжении. Monodromy theorem

Теорема двойственности. Duality theorem

Теорема единственности. Uniqueness theorem

"Теорема о двенадцати цветах". Twelve-color theorem

Теорема о минимаксе. Minimax theorem

Теорема о неподвижной точке. Fixed-point theorem

Теорема о постоянной точке Банаха. Banach fixed point theorem

Теорема о пределах функций ("теорема бутерброда с ветчиной"). Ham sandwich theorem

Теорема пятиугольных (пентагональных) чисел Эйлера. Euler pentagonal-number theorem

Теорема о промежуточной величине. Intermediate value theorem

Теорема о промежуточном значении. Intermediate value theorem

Теорема Радона-Никодима. Radon-Nikodým theorem

Теорема расширения Титце. Tietze extension theorem

Теорема рекурентности (рекурсии). Recurrence theorem

Теорема о средней величине. Mean-value theorem

Теорема об остатке. Remainder theorem

Теорема Пифагора. Pythagorean theorem

Теорема Ролля. Mean-value theorem

Теорема существования. Existence theorem

Теорема Таубера. Tauberian theorem

Теорема трёх квадратов. Three-squares theorem

Теория графов. Graph theory

Теория катастроф. Catastrophe theory

Теория относительности. Relativity theory

Теория функций. Function theory

Теория уравнений. Theory of equations

Теперешняяценность. Present value

Термометр Шельсия. Centigrade thermometer

Тесселяция, покрытие плоскости многоугольяниками или заполнение пространства многогранниками. Tessellation

Трение, фрикция. Friction

Тетраедр. Tetrahedron

Тихая дуэль. Silent duel

Тождественное алгебраигеское множество. Affine algebraic variety

Тождественное преобразование. Affine transformation

Тождественное пространство. Affine space

Тождественные количества. Identical quantities

Тождество. Identity

Тонкая пластинка. Lamina

Тонко-полосный выбор. Stratified sample

Тонна. Ton

Тонтинная ежегодная рента. Tontine annuity

Топологическая группа. Topological group

Топологически-заполненное пространство. Topologically complete space

Топологическое измерение, габаритное поле (в графопо строителях). Topological dimension

Топологическое преобразование двух множеств. Homeomorphism of two sets

Топология. Toplogy

Тор. Torus

Точечная диаграмма. Scattergram

Точечно-эквинепрерывный. Point-wise equicontinuous

Точечное произведение. Dot product

Точечный эллипс. Point ellipse

Точка ветвления. Branch point

Точка вращения рычага. Fulcrum

Точка делящая пополам. Bisecting point

Точка изгибания. Bend point

Точка конденсации. Condensation point

Точка, наиболее удалённая от Солнца (в астрономии). Aphelion

Точка накопления. Accumulation point

Точка опоры рычага. Fulcrum

Точка перегиба. Inflection point

Точка перерыва (прерывания). Point of discontinuity

Точка пересечения высот треугольника. Orthocenter.

Точка пересечения двух кривых с разными касательными. Salient point

Точка пересечния орбиты объекта с эклиптикой (в астрономии). Node in astronomy

Точка поворота. Turning point

Точка приложения силы. Fulcrum

Точка, присущая (напр. прямой, плоскости и т.д.). Adherent point

Точка пронизывания. Piercing point

Точка разветвления. Branch point

Точка разветвления, разъединения. Bifurcation point

Точка сгущения. Condensation point

Точка уплотнения. Condensation point

Точка устойчивости, устойчивая точка. Stable point

Точки лежащие на одной и той же линии. Collinear points

Точки равноденствия (в астрономии). Equinox

Точное дифференциальное уравнение. Exact differential equation

Точный делитель. Exact divisor

Точный квадрат (числа). Perfect square

Траектория. Trajectory

Траектория снаряда. Path of a projectile

Трактриса. Tractrix

Транзит-(ный телескоп). Transit

Транзитивное родство. Transitive relation

Транспозиция. Transposition

Транспозиця матрицы. Transpose of a matrix

Транспортир. Protractor

Трансфинитная индукция. Transfinite induction

Трансформация координат. Transformation of coordinates

Трансцендентное число. Transcendental number

Трапеция. Trapezoid

Треугольная пирамида. Tetrahedron

Треугольник. Triangle

Треугольник на земном шаре. Terrestrial triangle

Трехгранный (поверхностый) угол. Trihedral angle

Трехгранный угол образованный тремя линиями. Trihedral formed by three lines

Трехсторонняя пирамида. Triangular pyramid

Трехсторонний угол. Tetrahedral angle

Трехсторонний угол образованный тремя линиями. Trihedral formed by three lines

Трехчлен, Трехчленное выражение. Trinomial

Три. Three

Триангуляция. Triangulation

Тривиальное решение. Trivial solution

Тригонометрические функции. Trigonometric functions

Тригонометрия. Trigonometry

Трижды ортогональная система. Triply orthogonal system

Трилистик. Trefoil

Триллион. Trillion

Тринадцать. Thirteen

Трисектрисса. Trisectrix

Трисекция угла. Trisection of an angle

Трихотомическое свойство. Trichotomy property

Троичное представление чисел. Ternary representation of numbers

Тройка. Triplet

Тройка (целых гисел) Пифагора. Pythagorean triple

Тройной корень (уравнения). Triple root

Тройной интеграл. Triple integral

Трохоида. Trochoid

Тупой угол. Obtuse angle

Тысяча. Thousand

Тэта функция. Theta function

Убавить. Discount

Убавлять. Discount

Убывающаяся функция. Decreasing function

Углы направления. Direction angles

Угол. Angle

Угол вектора. Vectorial angle

Уголм ежду линией и североюжной линией. Bearing of a line

Угловое ускорение. Angular acceleration

Угломер. Protractor

Удвоение куба. Doubling of the cube

Удельная теплота. Specific heat

Удлинение. Elongation

Удлиняемое преобразование. Stretching transformation

Удовлетворить уравнение. Satisfy an equation

Узел. Cusp

Узел (в топологии). Knot in topology

Узел на кривой. Node of a curve

Узловая кривая. Nodal curve

Узловая линия. Nodal line

Укорочение тензора. Contraction of a tensor

Укороченное деление. Short division

Улитка. Limaçon

Ультра-фильтр. Ultra-filtre

Уменьшаемая (до точки) кривая. Reducible curve

Уменьшающееся. Minuend

Уменьшение корреляции. Attenuation of correlation

Уменьшение скорости. Deceleration

Уменьшение тензора. Contraction of a tensor

Умножение векторов. Multiplication of vectors

Умножить два числа. Multiply two numbers

Унарная операция, операция с одним операндом. Unary operation

Уникурсальная кривая. Unicursal curve
Унимодальная матрица. Unimodal matrix
Унитарная матрица. Unitary matrix
Уничтожитель. Annihilator
Уплотнение. Compactification
Упорядоченное множество. Ordered set
Упражнение. Exercise
Упрощение. Simplification
Упрощенная дробь. Simplified fraction
Упрощенное деление. Short division
Упрощенное кубическое уравнение (третьей сте
пени). Reduced cubic equation
Уравнение пятой степени. Quintic
Уравнение волны. Wave equation
Уравнение второй степени. Quadratic equation
Уравнение кривой. Equation of a cruve
Уравнение с уменьшенным числом корней. De-
pressed equation
Уравнение шестой степени. Sextic equation
Уравненное время. Equated time
Уравнить. Equate
Усеченная пространственная фигура. Frustum of a
solid
Усеченный конус. Truncated cone
Ускорение. Acceleration
Ускорение по тангенсу. Tangential acceleration
Условие возрастающей цепочки. Ascending chain
condition
Условие нисходящей цепочки. Descending chain
condition
Условия. Conditions
Условная ежегодная рента. Contingent annuity
Условная сходимость. Conditional convergence
"Устойчивая" статистика. Robust statistics
Учет векселей. Discount
Утверждение Бибербаха. Bieberbach conjecture
Утверждение Морделла. Mordell conjecture
Утверждение Пуанкаре́. Poincaré conjecture
Утверждение Суслина. Souslin's conjecture

Фаза простого гармонического движения. Phase
of simple harmonic motion
Факториал целого числа. Factorial of an integer
Факториальная нотация. Factorial notation
Факториальное исчисление. Factor analysis
Фигуры родственные по центральным проекциям.
Radially related figures
Фильтр. Filter
Фокус параболы. Focus of a parabola
Форма. Configuration
Форма с двумя переменными. Form in two vari-
ables
Формальная производная. Formal derivative
Формальный степенной ряд. Formal power series
Формула. Formula
Формула Виета. Viéte formula
Формула удвоеняя Лежандра. Duplication formula
Формулы длф вычитания. Subtraction formulas
Формулы для половины угла. Half-angle formulas
Формулы приведения (в тригонометрии). Redu-
uction formulas
Фрактал. Fractal
Фрактальное измерение (измерение Мандель
брота). Fractal dimension

Функтор, функциональный элемент. Functor
Функции Радемахера. Rademacher functions
Функция инжекции. Injective function
Функция инцидентности (вершин в графе). Inci-
dence function
Функция Кантора. Cantor function
Функция Кёбе. Koebe function
Функция класса C^n. Function of class C^n
Функция распределения. Distribution function
Функция платежа. Payoff function
Функция шлихта. Schlicht function
Функциональное обозначение. Functional notation
Фунт. Pound

Хаос. Chaos
Характеристика логарифма. Characteristic of a
logarithm
Характеристический корень матрицы. Character-
istic root of a matrix
Характеристическое уравнение матрицы. Charac-
teristic equation of a matrix
Характерные кривые на поверхности. Characteris-
tic curves on a surface
Хардиево пространство. Hardy space
Хи. Chi
Хи квадрат. Chi square
Ход в игре. Move of a game
Хорда. Chord
Хорда проходящая через фокус. Focal chord
Хроматическое число. Chromatic number

Целая функция. Entire function
Целое число. Integer
Цена выкупления. Redemption price
Цена лома. Scrap value
Ценная бумага. Negotiable paper
Ценность страховой полиси. Value of an insurance
policy
Центр круга (окружности). Center of a circle
Центр круга описанного около трехугольника.
Circumcenter of a triangle
Центр луча. Ray center
Центр массы. Center of mass
Центр окрыжности вписанной в треугольнике. In-
center of a triangle
Центр окружности описанной около треугольника.
Excenter of a triangle
Центр проектирования. Ray center
Централь группы. Central of a group
Центральный (в круге) угол. Central angle
Центробежная сила. Centifugal force
Центроида. Centroid
Центростремительное ускорение. Centripetal ac-
celeration
Цепная дробь. Continued fraction
Цепная кривая (линия). Catenoid
Цепная линия. Catenary
Цепное правило. Chain rule
Цепь симплексов. Chain of simplexes
Цикл. Cycle
Циклиды. Cyclides
Циклическая перестановка. Cyclic permutation
Циклоида. Cycloid
Циклотомное уравнение. Cyclotomic equation

Циклотомное целое число.   Cyclotomic integer
Цилиндр.   Cylinder
Цилиндрическая поверхность.   Cylindrical surface
Цилиндроид—цилиндрическая поверхность с сечениями перпендикулярными к эллипсам. Cylindroid
Циркуль.   Dividers
Цифра.   Digit
Цифровая машина.   Digital device

Часовой угол.   Hour angle
Частичные дроби.   Partial fractions
Частная производная.   Partial derivative
Частный интеграл.   Particular integral
Частота.   Periodicity
Частота класса.   Class frequency
Часть кривой (линии).   Segment of a curve
Часть премии возвращаемая отказавшемуся от страхового полиса.   Surrender value
Чередующийся.   Alternant
"Черепец", тайл, совокупность плоских фигур (полиомино).   Tile
Четверть.   Quarter
Четверть круга.   Quadrant of a circle
Четверть окружности.   Quadrant of a circle
Четверичное отклонение.   Quartile deviation
Четное размещение.   Even permutation
Четное число.   Even number
Четность.   Parity
Четыре.   Four
Четырехсторонная призма.   Quadrangular prism
Четырехчлен.   Quaternion
Четыреугольник.   Quadrangle
Четыреугольник.   Quadrilateral
Четырёхмерный параллелипипед, куб, тессеракт. Tesseract
Четырёхгранная (тетраэдральная) группа.   Tetrahedral group
Четырёхгранник, кубоктаэдр.   Cuboctahedron
Чёткое (неразмытое) множество.   Crisp set
Числа с их знаками.   Signed numbers
Числитель (дроби).   Numerator
Число.   Number
Число из которого корень извлекается.   Radicand
Число относительно простое данного числа и меньшее данного числа.   Totive of an integer
Число Рамсея.   Ramsey number
Число, целое.   Integer
Число чисел относительно простых к данного числа.   Totient of an integer
Числовая величина.   Numerical value
Числовое значение.   Numerical value
Числовое кольцо.   Ring of numbers
Числовое поле.   Field of numbers
Чистая дробь.   Proper fraction
Чистая математика.   Pure mathematics
Чистая прибыль.   Net profit
Чистая стратегия.   Pure strategy
Член дроби.   Term of a fraction
Член уравнения.   Member of an equation
Чрезмерное число.   Redundant number

Шансы.   Odds
Шар.   Sphere

Шаровые оси (координаты).   Spherical coordinates
Шестидесятая система нумерации (числения). Sexagesimal system of numbers
Шестиугольная призма.   Hexagonal prism
Шестнадцатиричная система исчисления.   Sexagesimal number system
Шестнадцатиричная система исчисления.   Hexadesimal number system
Шесть.   Six
Ширина.   Breadth
Ширина.   Width
Ширина (положения точки на сфере).   Latitude of a point
Шифр.   Cipher (n.)
Шифрование для вычислительной машины.   Coding for a computation machine
Шпур матрицы.   Spur of a matrix
Шпур матрицы.   Trace of a matrix
Шумный поединок.   Noisy duel

Эвольвента кривой.   Involute of a curve
Эволюта кривой.   Evolute of a curve
Эвристический метод.   Heuristic method
Эйлера Ф-функция.   Indicator of an integer
Эквивалентные матрицы.   Equivalent matrices
Экватор.   Equator
Эквипотенциальная поверхность.   Equipotential surface
"Экзотическая" сфера (вид множества).   Exotic sphere
"Экзотическое" четырёхмерное пространство (вид четырёхмерного множества).   Exotic four-space
Эклиптика.   Ecliptic
Эксекант угла.   Exsecant
Экспоненциальное представление чисел в виде мантиссы и порядка.   Scientific notation
Экстраполяция.   Extrapolation
Энтропия.   Entropy
Эксцентриситет гиперболы.   Eccentricity of a hyperbola
Эксцентрический угол эллипса.   Eccentric angle of an ellipse
Эластичность.   Elasticity
Эластичные фигуры.   Elastic bodies
Электродвижущая сила.   Electromotive force
Электростатичный потенциал.   Electrostatic potential
Элемент интеграции.   Element of integration
Элемент памяти.   Memory component
Элемент происходящий но прямой линии.   Lineal element
Элемент хранения.   Storage component
Элементарные операции.   Elementary operations
Элиминант.   Eliminant
Эллипс.   Ellipse
Эллипсоид.   Ellipsoid
Эллиптическая поверхность.   Elliptic surface
Эмпирическая кривая.   Empirical curve
Эндоморфизм.   Endomorphism
Эпитрохоида.   Epitrochoid
Эпитрохоидная кривая.   Epitrochoidal curve
Эпициклоида.   Epicycloid
Эрг.   Erg
Эргодическая теорема.   Ergodic theorem

Эрлангенская программа Кляйна. Erlangen program

Эффективная норма процента. Effective interest rate

Эффективная процентная ставка. Effective interest rate

"Эшелонная" матрица. Echelon matrix

Южное наклонение. South declination

Явная функция. Explicit function

Ядро гомоморфизма. Kernel of a homomorphism

Ядро (уравнения) Дирихле. Dirichlet kernel

Ядро интегрального уравнения. Nucleus (or kernel) of an integral equation

Ядро (уравнемия) Фежёра. Fejér kernel

Ярд расстояния. Yard of distance

Ясно ориентированный. Coherently oriented

Spanish—English Index

Abaco. Abacus
Abscisa. Abscissa
Aceleración. Acceleration
Aceleración angular. Angular acceleration
Aceleración centripeta. Centripetal acceleration
Aceleración tangencial. Tangential acceleration
Acotado esencialmente. Essentially bounded
Acre. Acre
Acumulador. Accumulator
Adiabático. Adiabatic
Adjunta de una matriz. Adjoint of a matrix
Adoquinado. Tessellation
Afelio. Aphelion
Agrupando términos. Grouping terms
Aislar una raiz. Isolate a root
Alargamiento. Elongation
Alef cero. Aleph-null (or aleph zero)
Álgebra. Algebra
Álgebra homológica. Homological algebra
Algebráico. Algebraic
Algoritmo. Algorithm
Alternante. Alternant
Altitud (altura). Altitude
Altura sesgada. Slant height
Amortiguador (en una máquina calculadora). Buffer (in a computing machine)
Amortización. Amortization
Amplitud de un número complejo. Amplitude of a complex number
Analicidad. Analyticity
Análisis. Analysis
Análisis de sensibilidad. Sensitivity analysis
Análisis factorial. Factor analysis
Análisis infinitesimal. Infinitesimal analysis
Análisis tensorial. Tensor analysis
Análisis vectorial. Vector analysis
Analogía. Analogy
Ancho. Width; breadth
Angulo. Angle
Angulo agudo. Actue angle
Angulo central. Central angle
Angulo cosecante. Arc-cosecant
Angulo coseno. Arc-cosine
Angulo diedro. Dihedral angle
Angulo excéntrico de una elipse. Eccentric angle of an ellipse
Angulo exterior. Exterior angle
Angulo horario. Hour angle
Angulo interno. Interior angle
Angulo obtuso. Obtuse angle
Angulo paraláctico. Parallactic angle
Angulo polar de un punto. Anomaly of a point
Angulo poliedro. Polyhedral angle
Angulo reentrante. Reentrant angle
Angulo reflejo. Reflex angle
Angulo relacionado. Related angle
Angulo secante. Arc-secant
Angulo seno. Arc-sine
Angulo sólido. Solid angle

Angulo tangente. Arc-tangent
Angulo tetraedro. Tetrahedral angle
Angulo triedro. Trihedral angle
Angulo vectorial. Vectorial angle
Angulo alternos. Alternate angles
Angulos complementarios. Complementary angles
Angulos conjugados. Conjugate angles
Angulos correspondientes. Corresponding angles
Angulos coterminales. Coterminal angles
Angulos cuadrantes. Quadrantal angles
Angulos directores. Direction angles
Angulos suplementarios. Supplementary angles
Angulos verticales. Vertical angles
Anillo de medidas. Measure ring
Anillo de números. Ring of numbers
Aniquilador. Annihilator
Año. Year
Anotación. Score
Antiautomorfismo. Antiautomorphism
Anticonmutativo. Anticommutative
Antiderivada. Antiderivative
Antiisomorfismo. Antiisomorphism
Antilogaritmo. Antilogarithm
Antípodas. Antipodal points
Antisimétrico. Antisymmetric
Anualidad acortada. Curtate annuity
Anualidad contingente. Contingent annuity
Anualidad ó renta vitalicia. Annuity
Anualidad postergada. Deferred annuity
Anualidad tontina. Tontine annuity
Apice. Apex
Apotema. Apothem
Aproximación. Approximation
Arbelos. Arbilos
Árbol. Tree
Area. Area
Area de una superficie. Surface area
Area lateral. Lateral area
Argumento circular. Circular argument
Argumento (ó dominio) de una función. Argument of a function
Arista de un sólido. Edge of a solid
Arista múltiple de una gráfica. Multiple edge in a graph
Aritmética. Arithmetic
Aritmómetro. Arithmometer
Armónica teseral. Tesseral harmonic
Armónica zonal. Zonal harmonic
Artificio mnemotécnico. Mnemonic device
Asa en una superficie. Handle on a surface
Asimétrico. Asymmetric
Asintota. Asymptote
Asintóticamente igual. Asymptotically equal
Asintóticamente insesgado. Asymptotically unbiased
Atenuación de una correlación. Attenuation of correlation
Atmósfera. Atmosphere
Átomo. Atom
Automorfismo. Automorphism

Automorfismo interno. Inner automorphism
Avoir dupois (Sistema de pesos). Avoirdupois
Axioma. Axiom
Azimut. Azimuth

Baria (unidades de presión). Bar
Baricentro. Barycenter
Barra (símbolo); bar. Bar
Base. Base
Base. Basis
Base. Radix
Base de filtro. Filter base
Base retractante. Shrinking basis
Beneficiario. Beneficiary
Bicompacto. Bicompactum
Bicuadrática. Biquadratic
Bienal. Biennial
Bienes. Assets
Bilineal. Bilinear
Billón. Billion
Bimodal. Bimodal
Binario. Binary
Binomio. Binomial
Binormal. Binormal
Birectángular. Birectangular
Bisectar. Bisect
Bisectriz. Bisector
Biyección. Bijection
Bola abierta. Open ball
Bonos seriados. Serial bonds
Bourbaki. Bourbaki
Braquistócrona. Brachistochrone
Brazo de palanca. Lever arm
Bruja de Agnesi. Witch of Agnesi

Caballo de fuerza. Horsepower
Cadena de simplejos. Chain of simplexes
Calculadora análoga. Analogue computer
Calcular. Calculate
Cálculo. Calculus
Cálculo (cómputo). Computation
Cálculo automático. Automatic computation
Cálculo de variaciones. Calculus of variations
Cálculo integral. Integral calculus
Calor específico. Specific heat
Caloría. Calory
Cambio de base. Change of base
Campo algebraicamente completo. Algebraically
 complete field
Campo de estudios. Field of study
Campo de extensión. Splitting field
Campo de fuerza conservador. Conservative field of
 force
Campo de Galois. Galois field
Campo de números. Field of numbers
Campo numérico. Number field
Campo perfecto. Perfect field
Cancelación. Cancellation
Cancelar. Cancel
Candela. Candela
Cantidad. Quantity
Cantidad de movimiento. Momentum
Cantidad escalar. Scalar quantity
Cantidad incógnita. Unknown quantity

Cantidades idénticas. Identical quantities
Cantidades iguales. Equal quantities
Cantidades inversamente proporcionales. Inversely
 proportional quantities
Cantidades linealmente dependientes. Linearly de-
 pendent quantities
Cantidades proporcionales. Proportional quantities
Caos. Chaos
Capital comercial. Capital stock
Capital comercial ó acciones. Stock
Cápsula convexa de un conjunto. Convex hull of a set
Cara de un poliedro. Face of a polyhedron
Característica de un logaritmo. Characteristic of a
 logarithm
Cardioide. Cardioid
Cargo por depreciación. Depreciation charge
Casi periódica. Almost periodic
Categoría. Category
Categórico. Categorical
Catenaria. Catenary
Catenoide. Catenoid
Cedazo. Sieve
Celestial. Celestial
Centésima parte de un número. Hundredth part of a
 number
Centígramo. Centigram
Centímetro. Centimeter
Centro de masa. Center of mass
Centro de proyección. Ray center
Centro de un círculo. Center of a circle
Centro de un grupo. Central of a group
Centroide. Centroid
Centroide de un triángulo. Median point
Cero. Zero
Cerradura de un conjunto. Closure of a set
Cibernética. Cybernetics
Cíclides. Cyclides
Ciclo. Cycle
Cicloide. Cycloid
Cien, ciento. Hundred
Cifra. Cipher
Cifra decimal. Place value
Cilindro. Cylinder
Cilindro hiperbólico. Hyperbolic cylinder
Cilindro parabólico. Parabolic cylinder
Cilindroide. Cylindroid
Cinco. Five
Cinemática. Kinematics
Cinética. Kinetics
Circulante. Circulant
Círculo. Circle
Círculo auxiliar. Auxiliary circle
Círculo circunscrito. Circumscribed circle
Círculo de convergencia. Circle of convergence
Círculo excrito. Escribed circle
Círculo unitario. Unit circle
Círculos coaxiales. Coaxial circles
Círculos concéntricos. Concentric circles
Circuncentro de un triángulo. Circumcenter of a tri-
 angle
Circuncírculo. Circumcircle
Circunferencia. Circumference
Clase de equivalencia. Equivalence class
Clase residual de un subgrupo. Coset of a subgroup

Clave para una máquina calculadora. Coding for a computation machine

Coaltitud. Coaltitude

Cociente de dos números. Quotient of two numbers

Codeclinación. Codeclination

Coeficiente. Coefficient

Coeficiente de correlación. Correlation coefficient

Coeficiente de correlación biserial. Biserial correlation coefficient

Coeficiente de regresión. Regression coefficient

Coeficiente principal. Leading coefficient

Coeficientes binomiales. Binomial coefficients

Coeficientes indeterminados. Undetermined coefficients

Coeficientes separados. Detached coefficients

Cofactor. Cofactor

Cofunción. Cofunction

Coherentemente orientado. Coherently oriented

Coincidente. Coincident

Colatitud. Colatitude

Colineación. Collineation

Cologaritmo. Cologarithm

Coloración de una gráfica. Graph coloring

Combinación de un conjunto de objetos. Combination of a set of objects

Combinación lineal. Linear combination

Compactificación. Compactification

Compactum. Compactum

Compás. Compass

Compás divisor, compás de puntas. Dividers

Complejo simplicial. Simplicial complex

Complemento de un conjunto. Complement of a set

Completar cuadrados. Completing the square

Componente de almacenamiento. Storage component

Componente de consumo. Input component

Componente de la memoria. Memory component

Componente de rendimiento de trabajo. Output component

Componente de una fuerza. Component of a force

Composición en una proporción. Composition in a proportion

Comprobación de una solución. Check on a solution

Comprobación por regla de los nueves. Casting out nines

Concavidad. Concavity

Concoide. Conchoid

Concordantemente orientado. Concordantly oriented

Condición de la cadena ascendente. Ascending chain condition

Condición de la cadena descendente. Descending chain condition

Condición necesaria. Necessary condition

Condición suficiente. Sufficient condition

Conectividad. Connectivity

Configuración. Configuration

Configuraciones superponibles. Superposable configurations

Congruencia. Congruence

Cónica. Conic

Cónica degenerada. Degenerate conic

Cónicas confocales. Confocal conics

Conicoide. Conicoid

Conjetura de Bieberbach. Bieberbach's conjecture

Conjetura de Mordell. Mordell conjecture

Conjetura de Poincaré. Poincaré conjecture

Conjetura de Souslin. Souslin's conjecture

Conjunción. Conjunction

Conjunto absorbente. Absorbing set

Conjunto acotado. Bounded set

Conjunto analítico. Analytic set

Conjunto bien definido. Crisp set

Conjunto borroso. Fuzzy set

Conjunto cerrado. Closed set

Conjunto compacto. Compact set

Conjunto conexo. Connected set

Conjunto conexo por trayectorias. Arc-wise connected set

Conjunto de Borel. Borel set

Conjunto de Julia. Julia set

Conjunto de Mandelbrot. Mandelbrot set

Conjunto de números. Set of numbers

Conjunto denso. Dense set

Conjunto desconectado. Disconnected set

Conjunto difuso. Fuzzy set

Conjunto discreto. Discrete set

Conjunto finito. Finite set

Conjunto mensurable (ó medible). Measurable set

Conjunto nítido. Crisp set

Conjunto numerable. Countable set; enumerable set; denumerable set

Conjunto ordenado. Ordered set

Conjunto ordenado en serie. Serially ordered set

Conjunto raro. Rare set

Conjunto totalmente ordenado. Totally ordered set

Conjunto vacío. Null set

Conjuntos ajenos. Disjoint sets

Conmensurable. Commensurable

Conmutador. Commutator

Conmutativo. Commutative

Cono. Cone

Cono circular. Circular cone

Cono director. Director cone

Cono truncado. Truncated cone

Conoide. Conoid

Consecuente en una relación. Consequent in a ratio

Consistencia de ecuaciones. Consistency of equations

Constante de integración. Constante of integration

Constante esencial. Essential constant

Construcción. Construction

Construcción de una gráfica por composición. Graphing by composition

Contador de una máquina calculadora. Counter of a computing machine

Contenido de un conjunto. Content of a set

Continuación de signo. Continuation of sign

Continuidad. Continuity

Continuidad uniforme. Uniform continuity

Contínuo. Continuum

Contra las manecillas del reloj. Counterclockwise

Contracción. Contraction mapping

Contracción de un tensor. Contraction of a tensor

Contracción del plano. Shrinking of the plane

Convergencia absoluta. Absolute convergence

Convergencia condicional. Conditional convergence

Convergencia de una serie. Convergence of a series

Convergencia debil. Weak convergence

Convergencia uniforme. Uniform convergence

Convergente de una fracción continua. Convergent of a continued fraction

Converger a un limite. Converge to a limit

Convolución de dos funciones. Convolution of two functions
Coordenada de un punto. Coordinate of a point
Coordenadas baricéntricas. Barycentric coordinates
Coordenadas cartesianas. Cartesian coordinates
Coordenadas esféricas. Spherical coordinates
Coordenadas geográficas. Geographic coordinates
Coordenadas polares. Polar coordinates
Corolario. Corollary
Corona circular (anillo). Annulus
Corredor. Broker
Correlación espúrea. Illusory correlation
Correspondencia biunívoca. One-to-one correspondence
Corrimiento unilateral. Unilateral shift
Cosecante de un ángulo. Cosecant of an angle
Coseno de un ángulo. Cosine of an angle
Costo capitalizado. Capitalized cost
Cota de un conjunto. Bound of a set
Cota inferior. Lower bound
Cota superior. Upper bound
Cotangente de un ángulo. Cotangent of an angle
Covariancia. Covariance
Coverseno. Coversed sine (coversine)
Criterio de la razón (convergencia de series). Ratio test
Criterio de razón generalizado (Criterio de D'Alembert). Generalized ratio test
Cuadrado. Square
Cuadrado mágico. Magic square
Cuadrado perfecto. Perfect square
Cuadrángulo. Quadrangle
Cuadrante de un círculo. Quadrant of a circle
Cuadratura de un círculo. Quadrature of a circle
Cuadratura del círculo. Squaring a circle
Cuádrica. Quadric
Cuadrilátero. Quadrilateral
Cuántica. Quantic
Cuántica cuaternaria. Quaternary quantic
Cuantificador. Quantifier
Cuantificador existencial. Existential quantifier
Cuantificador universal. Universal quantifier
Cuarto. Quarter
Cuaternio. Quaternion
Cuatrillón. Quadrillion
Cuatro. Four
Cúbica bipartita. Bipartite cubic
Cubierta de un conjunto. Covering of a set
Cubo. Cube
Cuboctaedro. Cuboctahedron
Cubo tetradimensional. Tesseract
Cuchillo del zapatero. Shoemaker's knife
Cuenta de dos en dos. Count by twos
Cuerda. Chord
Cuerda. Cord
Cuerda. String
Cuerda focal. Focal chord
Cuerdas suplementarias. Supplemental chords
Cuerpos elásticos. Elastic bodies
Cumulantes. Cumulants
Cuña. Wedge
Curso entre dos puntos. Run between two points
Curtosis. Kurtosis
Curva cerrada. Closed curve
Curva convexa. Convex curve

Curva cruciforme. Cruciform curve
Curva cuártica. Quartic curve
Curva cúbica. Cubic curve
Curva de frecuencia. Frequency curve
Curva de probabilidad. Probability curve
Curva empírica. Empirical curve
Curva en el espacio. Space curve
Curva epitrocoide. Epitrochoidal curve
Curva exponencial. Exponential curve
Curva isócrona. Isochronous curve
Curva izquierda (ó alabeada). Left-handed curve
Curva kapa. Kappa curve
Curva logarítmica. Logarithmic curve
Curva meridiana. Meridian curve
Curva pedal. Pedal curve
Curva plana de grado superior. Higher plane curve
Curva proyectiva plana. Projective plane curve
Curva proyectiva plana suave. Smooth projective plane curve
Curva rectificable. Rectifiable curve
Curva reducible. Reducible curve
Curva secante. Secant curve
Curva senoidal. Sine curve
Curva serpentina. Serpentine curve
Curva simple. Simple curve
Curva torcida. Twisted curve
Curva unicursal. Unicursal curve
Curvas características. Characteristic curves
Curvas superosculantes en una superficie. Superosculating curves on a surface
Curvatura de una curva. Curvature of a curve
Cúspide. Spinode; cusp

Débilmente compacto. Weakly compact
Decágono. Decagon
Decámetro. Decameter
Deceleración. Deceleration
Decimal finito. Terminating decimal
Decímetro. Decimeter
Declinación. Declination
Declinación norte. North declination
Declinación sur. South declination
Definido de una manera única. Uniquely defined
Deformación de un objeto. Deformation of an object
Deltaedro. Deltahedron
Deltoide. Deltoid
Demostración. Proof
Demostración indirecta. Indirect proof
Demostración por deducción. Deductive proof
Demostrar un teorema. Prove a theorem
Denominador. Denominator
Densidad. Density
Densidad asintótica. Asymptotic density
Densidad superior. Upper density
Derivada covariante. Covariant derivative
Derivada de orden superior. Derivative of higher order
Derivada de una distribución. Derivative of a distribution
Derivada de una función. Derivative of a function
Derivada direccional. Directional derivative
Derivada formal. Formal derivative
Derivada logarítmica de la función gama. Digamma function

Derivada normal. Normal derivative
Derivada parcial. Partial derivative
Desarollo de un determinante. Expansion of a deter-
minant
Descomposición espectral. Spectral decomposition
Descuento. Discount
Desigualdad. Inequality
Desigualdad de Chebyshev. Chebyshev inequality
Desigualdad incondicional. Unconditional inequality
Desproporcionado. Disproportionate
Desviación. Deviation
Desviación cuartil. Quartile deviation
Desviación probable. Probable deviation
Desviación standard. Standard deviation
Determinación de curvas empíricas. Curve fitting
Determinante. Determinant
Determinante antisimétrico. Skew-symmetric deter-
minant
Diada. Dyad
Diádico. Dyadic
Diagonal de un determinante. Diagonal of a determi-
nant
Diagonalizar. Diagonalize
Diagonal principal. Principal diagonal
Diagonal secundaria. Secondary diagonal
Diagonal (símbolo). Solidus
Diagrama. Diagram
Diagrama de dispersión. Scattergram
Diagrama de flujo. Flow chart
Diámetro de un círculo. Diameter of a circle
Diámetro de una partición. Fineness of a partition
Dicotomía. Dichotomy
Diez. Ten
Difeomorfismo. Diffeomorphism
Diferencia de dos cuadrados. Difference of two
squares
Diferenciación de una función. Differentiation of a
function
Diferenciación implícita. Implicit differentiation
Diferencial de una función. Differential of a function
Diferencial total. Total differential
Diferencias sucesivas de una función. Differencing of
a function
Diferencias tabulares. Tabular differences
Dígito. Digit
Dígito significativo. Significant digit
Digno de confianza (fidedigno). Reliability
Dilatación. Dilatation
Dimensión. Dimension
Dimensión de Hausdorff. Hausdorff dimension
Dimensión fractal. Fractal dimension
Dimensión topológica. Topological dimension
Dina. Dyne
Dinámica. Dynamics
Dipolo. Dipole
Dirección asintótica. Asymptotic direction
Directriz de una cónica. Directrix of a conic
Disco. Disc (or disk)
Discontinuidad. Discontinuity
Discontinuidad irremovible. Nonremovable disconti-
nuity
Discontinuidad removible. Removable discontinuity
Discriminante de un polinomio. Discriminant of a
polynomial
Dispersión. Dispersion

Distancia entre dos puntos. Distance between two
points
Distancia zenital. Zenith distance
Distribución beta. Beta distribution
Distribución ji cuadrada. Chi-square distribution
Distribución leptocúrtica. Leptokurtic distribution
Distribución lognormal. Lognormal distribution
Distribución mesocúrtica. Mesokurtic distribution
Distribución normal bivariada. Bivariate normal dis-
tribution
Distribución platocúrtica. Platykurtic distribution
Disyunción. Disjunction
Divergencia de una serie. Divergence of a series
Dividendo de un bono. Dividend on a bond
Dividir. Divide
Divisibilidad. Divisibility
Divisibilidad por once. Divisibility by eleven
División. Division
División corta. Short division
División sintética. Synthetic division
Divisor. Divisor
Divisor exacto. Exact divisor
Doble regla de tres. Double rule of three
Doblete. Doublet
Doce. Twelve
Dodecaedro. Dodecahedron
Dodecágono. Dodecagon
Dominio. Domain
Dominio de una variable. Range of a variable
Dominio de verdad. Truth set
Dominó. Domino
Dos. Two
Dualidad. Duality
Duelo ruidoso. Noisy duel
Duelo silencioso. Silent duel
Duplicación del cubo. Duplication of the cube

Eclíptica. Ecliptic
Ecuación característica de una matriz. Characteristic
equation of a matrix
Ecuación ciclotómica. Cyclotomic equation
Ecuación cuadrática. Quadratic equation
Ecuación cúbica reducida. Reduced cubic equation
Ecuación de diferencias. Difference equation
Ecuación de grado reducido (al dividir por una raíz).
Depressed equation
Ecuación de onda. Wave equation
Ecuación de sexto grado. Sextic equation
Ecuación de una curva. Equation of a curve
Ecuación derivada. Derived equation
Ecuación diferencial. Differential equation
Ecuación diferencial exacta. Exact differential equa-
tion
Ecuación homogénea. Homogeneous equation
Ecuación integral. Integral equation
Ecuación mónica. Monic equation
Ecuación polinomial. Polynomial equation
Ecuación recíproca. Reciprocal equation
Ecuaciones consistentes. Consistent equations
Ecuaciones dependientes. Dependent equations
Ecuaciones paramétricas. Parametric equations
Ecuaciones simultáneas. Simultaneous equations
Ecuador. Equator
Ecuador celeste. Celestial equator

Eje. Axis
Eje x. X-axis
Eje y. Y-axis
Eje mayor. Major axis
Eje menor. Minor axis
Eje radical. Radical axis
Eje transversal. Transverse axis
Ejercicio. Exercise
Ejes rectangulares. Rectangular axes
Elasticidad. Elasticity
Elemento de integración. Element of integration
Elemento lineal. Lineal element
Elevación entre dos puntos. Rise between two points
Eliminación por substitución. Elimination by substitution
Eliminante. Eliminant
Elipse. Ellipse
Elipse degenerada. Point ellipse
Elipsoide. Ellipsoid
Elipsoide achatado por los polos. Oblate ellipsoid
Elipsoide alargado hacia los polos. Prolate ellipsoid
En el sentido de las manecillas del reloj. Clockwise
Endomorfismo. Endomorphism
Energía cinética. Kinetic energy
Ensayos sucesivos. Successive trials
En sentido opuesto a las manecillas del reloj. Counter-clockwise
Entero. Integer
Entero ciclotómico. Cyclotomic integer
Entropía. Entropy
Envolvente de una familia de curvas. Envelope of a family of curves
Epicicloide. Epicycloid
Epitrocoide. Epitrochoid
Equidistante. Equidistant
Equilibrio. Equilibrium
Equinoccio. Equinox
Ergio. Erg
Error absoluto. Absolute error
Error al redondear un número. Round-off error
Error de muestreo. Sampling error
Escala de imaginarios. Scale of imaginaries
Escala de temperatura Celsius. Celsius temperature scale
Esfera. Sphere
Esfera exótica. Exotic sphere
Esferas de Dandelin. Dandelin spheres
Esferoide. Spheroid
Esfuerzo de un cuerpo. Stress of a body
Espacio. Space
Espacio abstracto. Abstract space
Espacio adjunto (ó conjugado). Adjoint (or conjugate) space
Espacio afín. Affine space
Espacio bicompacto (compacto). Bicompact space
Espacio cociente. Quotient space
Espacio completo. Complete space
Espacio de Baire. Baire space
Espacio de Banach no cuadrado. Nonsquare space
Espacio de cuatro dimensiones exótico. Exotic four-space
Espacio de Frechet. Frechet space
Espacio de Hardy. Hardy space
Espacio de órbitas. Orbit space
Espacio lagunar. Lacunary space

Espacio metacompacto. Metacompact space
Espacio métrico. Metric space
Espacio metrizable. Metrizable space
Espacio normado. Normed space
Espacio paracompacto. Paracompact space
Espacio producto. Product space
Espacio proyectivo. Projective space
Espacio separable. Separable space
Espacio topológicamente completo. Topologically complete space
Espacio uniformemente convexo. Uniformly convex space
Especie de un conjunto de puntos. Species of a set of points
Espectro de una matriz. Spectrum of a matrix
Espectro residual. Residual spectrum
Espiral equiángular. Equiangular spiral
Espiral lituiforme. Lituus
Espiral logarítmica. Logistic curve
Espiral loxodrómica. Loxodromic spiral
Esplain. Spline
Esqueleto de un complejo. Skeleton of a complex
Estadística. Statistic
Estadística. Statistics
Estadística bias. Biased statistic
Estadística vital. Vital statistics
Estadísticas robustas. Robust statistics
Estática. Statics
Esteradian. Steradian
Estéreo. Stere
Estimación de una cantidad. Estimate of a quantity
Estimación sin bias. Unbiased estimate
Estrategia de un juego. Strategy of a game
Estrategia dominante. Dominant strategy
Estrategia estrictamente dominante. Strictly dominant strategy
Estrategia óptima. Optimal strategy
Estrategia pura. Pure strategy
Estrella circumpolar. Circumpolar star
Estrella de un complejo. Star of a complex
Estrofoide. Strophoid
Evaluación. Evaluation
Evaluar. Evaluate
Eventos independientes. Independent events
Eventos que se excluyen mutuamente. Mutually exclusive events
Evolución. Evolution
Evoluta de una curva. Evolute of a curve
Exaedro. Hexahedron
Exágono. Hexagon
Excedente de nueves. Excess of nines
Excentricidad de una hipérbola. Eccentricity of a hyperbola
Excentro. Excenter
Excírculo. Excircle
Expansión asintótica. Asymptotic expansion
Expectativa de vida. Expectation of life
Exponente. Exponent
Exponente fraccionario. Fractional exponent
Exsecante. Exsecant
Extensión de un campo. Extension of a field
Extrapolación. Extrapolation
Extremalmente disconexo. Extremally disconnected
Extremos (o términos extremos). Extreme terms (or extremes)

Faceta. Facet
Factor de un polinomio. Factor of a polynomial
Factor integrante. Integrating factor
Factorial de un entero. Factorial of an integer
Factorizable. Factorable
Factorización. Factorization
Factorización única. Unique factorization
Familia de curvas. Family of curves
Farmacéutico. Apothecary
Fase de movimiento armónico simple. Phase of simple harmonic motion
Fibrado de planos. Bundle of planes
Figura plana. Plane figure
Figura simétrica. Symmetric figure
Figuras congruentes. Congruent figures
Figuras homotéticas. Homothetic figures
Figuras homotópicas. Homotopic figures
Figuras radialmente relacionadas. Radially related figures
Filtro. Filter
Finitamente representable. Finitely representable
Flécnodo. Flecnode
Foco de una parábola. Focus of a parabola
Folio de Descartes. Folium of Descartes
Fondo de amortización. Sinking fund
Forma canónica. Canonical form
Forma en dos variables. Form in two variables
Formas indeterminadas. Indeterminate forms
Fórmula. Formula
Fórmula de duplicación. Duplication formula
Fórmula de Vieta. Viète formula
Fórmula prismoidal. Prismoidal formula
Fórmulas de reducción. Reduction formulas
Fórmulas de resta. Subtraction formulas
Fórmulas del ángulo medio. Half-angle formulas
Fracción. Fraction
Fracción común. Common fraction
Fracción contínua. Continued fraction
Fracción decimal periódica. Repeating decimal
Fracción propia. Proper fraction
Fracción simplificada. Simplified fraction
Fracción vulgar. Vulgar fraction
Fracciones parciales. Partial fractions
Fractal. Fractal
Frecuencia acumulativa. Cumulative frequency
Frecuencia de la clase. Class frequency
Fricción. Friction
Frontera de un conjunto. Frontier of a set; boundary of a set
Fuerza centrífuga. Centrifugal force
Fuerza cortante. Shearing strain
Fuerza de mortandad. Force of mortality
Fuerza electromotriz. Electromotive force
Fulcro. Fulcrum
Función absolutamente continua. Absolutely continuous function
Función aditiva. Additive function
Función analítica. Analytic function
Función analítica biunívoca. Schlicht function
Función analítica monogénica. Monogenic analytic function
Función ángulo-hiperbólica. Arc-hyperbolic function
Función armónica. Harmonic function
Función automorfa. Automorphic function
Función bei. Bei function

Función ber. Ber function
Función compuesta. Composite function
Función continua. Continuous function
Función continua por arcos. Piecewise continuous function
Función creciente. Increasing function
Función de Cantor. Cantor function
Función de clase C^n. Function of class C^n
Función δ de Dirac. Dirac δ-function
Función de distribución. Distribution function
Función de Koebe. Koebe function
Función de núcleo. Ker function
Función de pago. Payoff function
Función de signo. Signum function
Función decreciente. Decreasing function
Función desvaneciente. Vanishing function
Función diferenciable. Smooth map
Función discontinua. Discontinuous function
Función entera. Entire function
Función escalonada. Step function
Función estrictamente creciente. Strictly increasing function
Función explícita. Explicit function
Función gama. Gamma function
Función holomorfa. Holomorphic function
Función implícita. Implicit function
Función integrable. Integrable function
Función inyectiva. Injective function
Función kei. Kei function
Función lisa. Smooth map
Función localmente integrable. Locally integrable function
Función meromorfa. Meromophic function
Función modular. Modular function
Función monótona. Monotone function
Función no acotada. Unbounded function
Función 1-Lipshitz. Nonexpansive mapping
Función polivalente. Many valued function
Función potencial. Potential function
Función propia (ó eigen-función). Eigenfunction
Función proposicional. Propositional function
Función semicontinua. Semicontinuous function
Función (mapeo) suave. Smooth map
Función subaditiva. Subadditive function
Función subarmónica. Subharmonic function
Función sumable. Summable function
Función teta. Theta function
Función trigonométrica inversa. Inverse trigonometric function
Función univalente. Single-valued function
Función zeta. Zeta function
Funciones de Bessel modificadas. Modified Bessel functions
Funciones de Rademacher. Rademacher functions
Funciones equicontínuas. Equicontinuous functions
Funciones generalizadas. Generalized functions
Funciones ortogonales. Orthogonal functions
Funciones trigonométricas. Trigonometric functions
Funtor. Functor

Ganancia bruta. Gross profit
Gastos generales fijos. Overhead expenses
Generador de una superficie. Generator of a surface
Generadores rectilíneos. Rectilinear generators

Generatriz. Generatrix
Geoide. Geoid
Geometría. Geometry
Geometría bidimensional. Two-dimensional geometry
Geometría proyectiva. Projective geometry
Geometría tridimensional. Three-dimensional geometry
Girar alrededor de un eje. Revolve about an axis
Giro. Gyration
Googol. Googol
Gradiente. Gradient
Grado de un polinomio. Degree of a polynomial
Gráfica bipartita. Bipartite graph
Gráfica completa. Complete graph
Gráfica de barras. Bar graph
Gráfica de una ecuación. Graph of an equation
Gráfica euleriana. Eulerian graph
Gráfica hamiltoniana. Hamiltonian graph
Gráfica plana. Planar graph
Gramo. Gram
Gravedad. Gravity
Gravitación. Gravitation
Grupo alternante. Alternating group
Grupo conmutativo (abeliano). Commutative group
Grupo de control. Control group
Grupo de homologia. Homology group
Grupo de Klein. Four-group
Grupo de números. Group of numbers
Grupo de permutaciones. Permutation group
Grupo de transformaciones. Transformation group
Grupo del icosaedro. Icosahedral group
Grupo del octaedro. Octahedral group
Grupo del tetraedro. Tetrahedral group
Grupo diédrico. Dihedral group
Grupo soluble Solvable group
Grupo topológico. Topological group
Grupoide. Groupoid
Gudermaniano. Gudermannian

Haverseno (ó medio verseno). Haversine
Haz de planos. Bundle of planes
Haz de planos. Sheaf of planes
Hélice. Helix
Helicoidal (helicoide). Helicoid
Hemisferio. Hemisphere
Heptaedro. Heptahedron
Heptágono. Heptagon
Hipérbola. Hyperbola
Hiperboloide de una hoja. Hyperboloid of one sheet
Hiperplano. Hyperplane
Hipersuperficie. Hypersurface
Hipervolumen. Hypervolume
Hipocicloide. Hypocycloid
Hipocicloide de cuatro cúspides. Astroid
Hipotenusa. Hypotenuse
Hipótesis. Hypothesis
Hipótesis admisible. Admissible hypothesis
Hipótesis alternativa. Alternative hypothesis
Hipotrocoide. Hypotrochoid
Histograma. Histogram
Hodógrafo. Hodograph
Hoja de una superficie. Sheet of a surface
Hoja de una superficie cónica. Nappe of a cone

Homeomorfismo entre dos conjuntos. Homeomorphism of two sets
Homogeneidad. Homogeneity
Homólogo. Homologous
Homomorfismo entre dos conjuntos. Homomorphism of two sets
Homosedástico. Homoscedastic
Horizontal. Horizontal
Horizonte. Horizon

Icosaedro. Icosahedron
Ideal contenido en un anillo. Ideal contained in a ring
Ideal nilpotente. Nilpotent ideal
Idemfactor. Idemfactor
Idenpotente. Idempotent
Identidad. Identity
Igualdad. Equality
Imagen de un punto. Image of a point
Implicación. Implication
Impuesto. Tax
Impuesto adicional. Surtax
Impuesto sobre la renta. Income tax
Incentro de un triángulo. Incenter of a triangle
Incidencia. Incidence function
Incírculo. Incircle
Inclinación de una recta. Inclination of a line
Incremento de una función. Increment of a function
Indicador de un entero. Totient of an integer
Indicador de un número. Indicator of an integer
Indicativo de un entero. Totitive of an integer
Indicatriz de una curva. Indicatrix of a curve
Indice de un radical. Index of a radical
Indice superior. Superscript
Inducción. Induction
Inducción incompleta. Incomplete induction
Inducción matemática. Mathematical induction
Inducción transfinita. Transfinite induction
Inercia. Inertia
Inferencia. Inference
Inferencia estadística. Statistical inference
Infinito. Infinity
Inmergir (ó sumergir) en un espacio. Imbed in a space
Inmersión de un conjunto. Imbedding of a set
Integración mecánica. Mechanical integration
Integración por partes. Integration by parts
Integral de Bochner. Bochner integral
Integral de la energía. Energy integral
Integral de superficie. Surface integral
Integral de una función. Integral of a function
Integral definida. Definite integral
Integral de Riemann generalizada. Generalized Riemann integral
Integral doble. Double integral
Integral impropia. Improper integral
Integral indefinida. Indefinite integral
Integral iterada. Iterated integral
Integral múltiple. Multiple integral
Integral particular. Particular integral
Integral simple. Simple integral
Integral triple. Triple integral
Integrador. Integrator
Integrador gráfico. Integraph

Integrando. Integrand
Intensidad luminosa. Candlepower
Interacción. Interaction
Interés compuesto. Compound interest
Interpolación. Interpolation
Intersección de curvas. Intersection of curves
Intersección de dos conjuntos. Meet of two sets
Intervalo abierto. Open interval
Intervalo de confidencia. Confidence interval
Intervalo intercuartil. Interquartile range
Intervalos encajados. Nested intervals
Intuicionismo. Intuitionism
Invariante de una ecuación. Invariant of an equation
Inversión. Investment
Inversión de un punto. Inversion of a point
Inverso de una operación. Inverse of an operation
Inversor. Inversor
Invertible. Invertible
Involución de una recta. Involution on a line
Involuta de una curva. Involute of a curve
Irracional. Surd
Isomorfismo entre dos conjuntos. Isomorphism of two sets
Isotermo. Isotherm

Ji-cuadrado. Chi-square
Joule. Joule
Juego completamente mixto. Completely mixed game
Juego cóncavo convexo. Concave-convex game
Juego de dos personas. Two-person game
Juego de hex. Game of hex
Juego de Mazur-Banach. Mazur-Banach game
Juego de nim. Game of nim
Juego de posición. Positional game
Juego de suma nula. Zero-sum game
Juego del polígono. Game of hex
Juego en cooperativa. Cooperative game
Juego finito. Finite game
Juego no cooperativo. Noncooperative game
Juego separable. Separable game
Jugada personal. Personal move
Jugador maximalizante. Maximizing player
Jugador minimalizante. Minimizing player

Kérnel de Dirichlet. Dirichlet kernel
Kérnel de Fejér. Fejér kernel
Kilogramo. Kilogram
Kilómetro. Kilometer
Kilovatio. Kilowatt

Lado de un polígono. Side of a polygon
Lado inicial de un ángulo. Initial side of an angle
Lado terminal de un ángulo. Terminal side of an angle
Lados opuestos. Opposite sides
Lámina. Lamina
Látice. Lattice
Latitud de un punto. Latitude of a point
Lazo de una curva. Loop of a curve
Lema. Lemma
Lemniscato. Lemniscate
Lexicográfico. Lexicographical
Ley asociativa. Associative law
Ley de exponentes. Law of exponents

Ley distributiva. Distributive law
Libra. Pound
Ligadura. Bond
Limazón de Pascal. Limaçon
Límite de una función. Limit of a function
Límite inferior. Inferior limit
Límite superior. Superior limit
Límites fiduciarios (ó fiduciales). Fiducial limits
Línea de curso. Trend line
Línea de rumbo. Rhumb line
Línea diametrical. Diametral line
Línea dirigida. Directed line
Línea isoterma. Isothermal line
Línea nodal. Nodal line
Línea quebrada. Broken line
Línea recta. Straight line
Línea vertical. Vertical line
Líneas antiparalelas. Antiparallel lines
Líneas concurrentes. Concurrent lines
Líneas coplanares. Coplanar lines
Líneas de contorno. Contour lines
Líneas de flujo. Stream lines
Líneas de nivel. Level lines
Líneas (ó rectas) oblícuas. Skew lines
Líneas paralelas. Parallel lines
Literal. Literal constant
Litro. Liter
Llave (ó corchete). Brace
Localmente compacto. Locally compact
Localmente conexo por trayectorias. Locally arc-wise connected
Logaritmo de un número. Logarithm of a number
Logaritmos comunes. Common logarithms
Logaritmos naturales. Natural logarithms
Lógica borrosa. Fuzzy logic
Lógica difusa. Fuzzy logic
Longitud. Longitude
Longitud de arco. Arc length
Longitud de una curva. Length of a curve
Lugar. Locus
Lunas de Hipócrates. Lunes of Hippocrates
Lúnula. Lune

Magnitud de una estrella. Magnitude of a star
Mantisa. Mantissa
Maps (transformaciones) no esenciales. Inessential mapping
Máquina calculadora. Calculating machine; computing machine
Marco de referencia. Frame of reference
Masa. Mass
Matemática abstracta. Abstract mathematics
Matemática aplicada. Applied mathematics
Matemáticas. Mathematics
Matemáticas constructivas. Constructive mathematics
Matemáticas discretas. Discrete mathematics
Matemáticas finitas. Finite mathematics
Matemáticas puras. Pure mathematics
Materia isotrópica. Isotropic matter
Matrices conformes. Conformable matrices
Matrices equivalentes. Equivalent matrices
Matriz aumentada. Augmented matrix
Matriz de coeficientes. Matrix of coefficients
Matriz de Vandermonde. Vandermonde matrix

Matriz escalón. Echelon matrix
Matriz hermitiana. Hermitian matrix
Matriz unimodular. Unimodular matrix
Matriz unitaria. Unitary matrix
Máximo común divisor. Greatest common divisor
Máximo de una función. Maximum of a function
Mecánica de fluidos. Mechanics of fluids
Mecanismo aleatorio. Random device
Mecanismo digital. Digital device
Media (ó promedio) de dos números. Mean (or average) of two numbers
Medición. Mensuration
Medida cero. Measure zero
Medida de un conjunto. Measure of a set
Menor de un determinante. Minor of a determinant
Menos. Minus
Meridiano terrestre. Meridian on the earth
Método de agotamiento de Eudoxo. Method of exhaustion
Método de comparación (de una serie). Comparison test
Método de descenso infinito. Proof by descent
Método de mínimos cuadrados. Method of least squares
Método del gradiente. Method of the steepest descent
Método del punto silla. Saddle-point method
Método del "simplex" (ó simplejo). Simplex method
Método dialítico. Dialytic method
Método heurístico. Heuristic method
Método inductivo. Inductive method
Metro. Meter
Miembro de una ecuación. Member of an equation
Mil. Thousand
Milímetro. Millimeter
Milla. Mile
Milla náutica. Nautical mile
Millón. Million
Mínima cota superior. Least upper bound
Mínimo de una función. Minimum of a function
Minuendo. Minuend
Minuto. Minute
Miríada. Myriad
Moda. Mode
Módulo. Module
Módulo de una congruencia. Modulus of a congruence
Módulo de volumen. Bulk modulus
Mol. Mole
Monomio. Monomial
Momento de inercia. Moment of inertia
Momento de una fuerza. Moment of a force
Momento estático. Static moment
Morfismo. Morphism
Morra (juego). Morra (a game)
Mosaico. Tile
Movimiento al azar. Chance move
Movimiento armónico. Harmonic motion
Movimiento curvilíneo. Curvilinear motion
Movimiento periódico. Periodic motion
Moviemiento rigido. Rigid motion
Muestra. Sample
Muestra estratificada. Stratified sample
Multilobulado. Multifoil
Multinomio. Multinomial

Multiplicación de vectores. Multiplication of vectors
Multiplicador. Multiplier
Multiplicando. Multiplicand
Multiplicidad de una raíz. Multiplicity of a root
Multiplicar dos números. Multiply two numbers
Múltiplo común. Common multiple
Múltiplo de un número. Multiple of a number
Multivibrador biestable. Flip-flop circuit

Nabla. Del
Nadir. Nadir
Negación. Negation
Nervio de un sistema de conjuntos. Nerve of a system of sets
Newton. Newton
Nilpotente. Nilpotent
Nodo (término astronómico). Node in astronomy
Nodo con tangentes distintas. Crunode
Nodo de una curva. Node of a curve
Nomograma. Nomogram
Nonágono. Nonagon
Normal a una curva. Normal to a curve
Norma de una matriz. Norm of a matrix
Notación. Notation
Notación científica. Scientific notation
Notación factorial. Factorial notation
Notación funcional. Functional notation
Núcleo de Dirichlet. Dirichlet kernel
Núcleo de Fejér. Fejér kernel
Núcleo de un homomorfismo. Kernel of a homomorphism
Núcleo de una ecuaciór integral. Nucleus (or kernel) of an integral equation
Nudo de distancia. Knot of distance
Nudo en topología. Knot in topology
Nueve. Nine
Numerablemente compacto. Countably compact
Numerador. Numerator
Numerabilidad. Countability
Numeración. Numeration
Número. Number
Número abundante. Abundant number
Número aritmético. Arithmetic number
Número cardinal. Cardinal number
Número complejo. Complex number
Número compuesto. Composite number
Número cromático. Chromatic number
Número deficiente. Defective (or deficient) number
Número de Ramsey. Ramsey number
Número de vueltas. Winding number
Número denominado. Denominate number
Número idempotente. Idempotent number
Número impar. Odd number
Número irracional. Irrational number
Número mixto. Mixed number
Número negativo. Negative number
Número ordinal. Ordinal numbers
Número par. Even number
Número peádico. p-adic number
Número positivo. Positive number
Número primo. Prime number
Número que se usa para contar. Counting number
Número racional. Rational number
Número real. Real number

Número redundante. Redundant number
Número trascendente. Transcendental number
Números. Numerals
Números amigables. Amicable numbers
Números babilonios. Babylonian numerals
Números chino-japoneses. Chinese-Japanese numerals
Números complejos conjugados. Conjugate complex numbers
Números de Catalan. Catalan numbers
Números dirigidos. Signed numbers
Números egipcios. Egyptian numerals
Números griegos. Greek numerals
Números hipercomplejos. Hypercomplex numbers
Números hiperreales. Hyperreal numbers
Números inconmensurables. Incommensurable numbers
Números no estándar. Hyperreal numbers
Números no estándar. Nonstandard numbers

Ocho. Eight
Octaedro. Octahedron
Octágono. Octagon
Octante. Octant
Ohmio. Ohm
Ojiva. Ogive
Once. Eleven
Operación. Operation
Operación unaria. Unary operation
Operaciones elementales. Elementary operations
Operador. Operator
Operador lineal. Linear operator
Órbita. Orbit
Orden de contacto (de dos curvas). Order of contact
Ordenada al origen. Intercept on an axis
Ordenada de un punto. Ordinate of a point
Orden de un grupo. Order of a group
Orientación. Orientation
Origen de coordenadas. Origin of coordinates
Ortocentro. Orthocenter
Oscilación de una función. Oscillation of a function
Osculación. Osculation
Ovalo. Oval

Pago a plazos. Installment payments
Papel negociable. Negotiable paper
Par (torque). Torque
Parábola. Parabola
Parábola cúbica. Cubical parabola
Paraboloide de revolución. Paraboloid of revolution
Paraboloide hiperbólico. Hyperbolic paraboloid
Paradoja. Paradox
Paradoja de Banach-Tarsky. Banach-Tarsky paradox
Paradoja de Hausdorff. Hausdorff paradox
Paradoja de Petersburgo. Petersburg paradox
Paralaje de una estrella. Parallax of a star
Paralelas geodésicas. Geodesic parallels
Paralelepípedo. Parallelipiped
Paralelepípedo rectangular. Cuboid
Paralelogramo. Parallelogram
Paralelos de latitud. Parallels of latitude
Paralelotopo. Parallelotope
Parámetro. Parameter
Paréntesis. Parentheses

Paréntesis rectangular. Bracket
Paridad. Parity
Parte imaginaria de un número. Imaginary part of a number
Partición de un entero. Partition of an integer
Partición más fina. Finer partition
Partición más gruesa. Coarser partition
Partida de un juego. Play of a game
Partida de un juego. Move of a game
Pascal. Pascal
Pasivo. Liability
Pantógrafo. Pantograph
Pendiente de un tejado. Pitch of a roof
Pendiente de una curva. Slope of a curve
Pendiente de una trayectoria. Grade of a path
Péndulo. Pendulum
Pentadecágono. Pentadecagon
Pentaedro. Pentahedron
Pentágono. Pentagon
Pentagrama. Pentagram
Penumbra. Penumbra
Percentil. Percentile
Periferia. Periphery
Perígono. Perigon
Perihelio. Perihelion
Perímetro de una sección transversal. Girth
Perpendicular de una superficie. Perpendicular of a surface
Perpetuidad. Perpetuity
Perímetro. Perimeter
Periodicidad. Periodicity
Período de una función. Period of a function
Permutación cíclica. Cyclic permutation
Permutación de n objetos. Permutation of n things
Permutación par. Even permutation
Perspectiva. Perspectivity
Peso. Weight
Peso troy. Troy weight
Pictograma. Pictogram
Pie de un triángulo rectángulo. Leg of a right triangle
Pié de una perpendicular. Foot of a perpendicular
Pirámide. Pyramid
Pirámide pentagonal. Pentagonal pyramid
Pirámide triangular. Triangular pyramid
Planímetro. Planimeter
Plano osculador. Osculating plane
Plano proyectante. Projecting plane
Plano proyectivo finito. Finite projective plane
Plano rectificador. Rectifying plane
Plano tangente. Tangent plane
Planos coordenados. Coordinates planes
Planos incidentes en un punto. Copunctal planes
Plasticidad. Plasticity
Plomada. Plumb line
Población. Population
Polar de una forma cuadrática. Polar of a quadratic form
Polarización. Polarization
Poliedro. Polyhedron
Polígono. Polygon
Polígono concavo. Concave polygon
Polígono inscrito. Inscribed polygon
Polígono regular. Regular polygon
Polihex. Polyhex

Polinomio de Legendre. Polynomial of Legendre
Poliominó. Polyomino
Politopo. Polytope
Polo de un círculo. Pole of a circle
Poner en ecuación. Equate
Porcentaje. Pencentage
Porciento de error. Percent error
Porciento de interés. Interest rate
Porciento de interés efectivo. Effective interest rate
Porciento de interés nominal. Nominal rate of interest
Postulado. Postulate
Potencia de un conjunto. Potency of a set
Potencia de un número. Power of a number
Potencial electrostático. Electrostatic potential
Poundal. Poundal
Precio corriente. Market value
Precio de compra. Flat price
Precio de lista. Book value
Precio de rescate (ó desempeño). Redemption price
Precio de venta. Selling price
Preservante de área. Equiareal (or area-preserving)
Presión. Pressure
Préstamo. Loan
Prima. Bonus
Prima. Premium
Primitiva de una ecuación, diferencial. Primitive of a differential equation
Primos gemelos. Twin-primes
Principio. Principle
Principio de las casillas. Dirichlet drawer principle
Principio de las casillas. Pigeon-hole principle
Principio de localización. Localization principle
Principio de superposición. Superposition principle
Principio del acotamiento uniforme. Uniform boundedness principle
Principio del óptimo. Principle of optimality
Prisma. Prism
Prisma cuadrangular. Quadrangular prism
Prisma exagonal. Hexagonal prism
Prismatoide. Prismatoid
Prismoide. Prismoid
Probabilidad. Odds
Probabilidad de ocurrencia. Probability of occurrence
Problema. Problem
Problema con valor en la frontera. Boundary value problem
Problema de Kakeya. Kakeya problem
Problema de la cerradura y complementación de Kuratowsky. Kuratowsky closure-complementation problem
Problema de los cuatro colores. Four-color problem
Problema de los puentes de Koenigsberg. Königsberg bridge problem
Problema isoperimétrico. Isoperimetric problem
Producto cartesiano. Cartesian product
Producto de Blaschke. Blaschke product
Producto de números. Product of numbers
Producto directo. Direct product
Producto infinito. Infinite product
Producto interior. Inner product
Producto punto (producto escalar). Dot product
Producto tensorial de espacios vectoriales. Tensor product of vector spaces
Programación. Programming

Programación dinámica. Dynamic programming
Programación lineal. Linear programming
Programación no lineal. Nonlinear programming
Programa de Erlangen. Erlangen program
Progresión. Progression
Promedio. Average
Promedio geométrico. Geometric average
Promedio pesado. Weighted mean
Propiedad de absorción. Absorption property
Propiedad de aproximación. Approximation property
Propiedad de caracter finito. Property of finite character
Propiedad de Krein-Milman. Krein-Milman property
Propiedad de reflexión. Reflection property
Propiedad de tricotomía. Trichotomy property
Propiedad del buen orden. Well-ordering property
Propiedad global. Global property
Propiedad intrínseca. Intrinsic property
Propiedad invariante. Invariant property
Propiedad local. Local property
Proporción. Proportion
Proporcionalidad. Proportionality
Proposición. Proposition
Proyección de un vector. Projection of a vector
Proyección estereográfica. Stereographic projection
Proyectividad. Projectivity
Pseudoesfera. Pseudosphere
Pulgada. Inch
Punto aislado. Acnode
Punto de acumulación. Accumulation point; cluster point
Punto de adherencia. Adherent point
Punto de bifurcación. Bifurcation point
Punto de cambio. Turning point
Punto de condensación. Condensation point
Punto de curvatura. Bend point
Punto de discontinuidad. Point of discontinuity
Punto de inflexión. Inflection point
Punto de la cerradura. Adherent point
Punto de penetración. Piercing point
Punto de ramificación. Branch point
Punto decimal flotante. Floating decimal point
Punto en puerto. Saddle point
Punto en un plano. Planar point
Punto estable. Stable point
Punto estacionario. Stationary point
Punto extremo de una curva. End point of a curve
Punto fijo. Fixed point
Punto límite. Limit point
Punto medio de un segmento rectilíneo. Midpoint of a line segment
Punto medio ó punto bisector. Bisecting point
Punto ordinario. Ordinary point
Punto saliente. Salient point
Punto singular. Singular point
Punto umbilical. Umbilical point
Puntos colineales (alineados). Collinear points
Puntos concíclicos. Concyclic points
Puntualmente equicontinuo. Point-wise Equicontinuous
Quíntica. Quintic
Quintillón. Quintillion

Racional diádico. Dyadic rational
Racionalizar un denominador. Rationalize a denominator
Radiante. Radian
Radiar desde un punto. Radiate from a point
Radical. Radical
Radical de un anillo. Radical of a ring
Radical de un ideal. Radical of an ideal
Radical irreducible. Irreducible radical
Radicando. Radicand
Radio de un círculo. Radius of a circle
Raíz ajena (ó extraña). Extraneous root
Raíz característica de una matriz. Characteristic root of a matrix
Raíz cuadrada. Square root
Raíz cúbica. Cube root
Raíz de una ecuación. Root of an equation
Raíz enésima primitiva. Primitive nth root
Raíz simple. Simple root
Raíz triple. Triple root
Rama de una curva. Branch of a curve
Rearreglo de términos. Rearrangement of terms
Recíproco de un número. Reciprocal of a number
Recíproco de un teorema. Converse of a theorem
Rectángulo. Rectangle
Rectas perpendiculares. Perpendicular lines
Redondear un número. Rounding off numbers
Reducción de una fracción. Reduction of a fraction
Reflexión en una línea. Reflection in a line
Refracción. Refraction
Región de confianza. Confidence region
Región multiconexa. Multiply connected region
Región simplemente conexa. Simply connected region
Regla. Ruler
Regla de cálculo. Slide rule
Regla de cadena. Chain rule
Regla de signos. Rule of signs
Regla del trapezoide. Trapezoid rule
Regla mecánica para extraer raices cuadradas. Mechanic's rule
Reglado en una superficie. Ruling on a surface
Relación. Relation
Relación anarmónica. Anharmonic ratio
Relación antisimétrica. Antisymmetric relation
Relación de deformación. Deformation ratio
Relación de inclusión. Inclusion relation
Relación de semejanza. Ratio of similitude
Relación externa. External ratio
Relación interna. Internal ratio
Relación intransitiva. Intransitive relation
Relación reflexiva. Reflexive relation
Relación total. Connected relation
Relación transitiva. Transitive relation
Rombo. Rhombus
Romboedro. Rhombohedron
Romboide. Rhomboid
Rendir (producir, ceder). Yield
Renglón de un determinante. Row of a determinant
Renta vitalicia conjunta. Joint life annuity
Representación de un grupo. Representation of a group
Representación ternaria de un número. Ternary representation of numbers
Residuo de una función. Residue of a function

Residuo de una serie infinita. Remainder of an infinite series
Residuo nulo. Nonresidue
Resma. Ream
Resolvente cúbica. Resolvent cubic
Resolvente de una matriz. Resolvent of a matrix
Resta de números. Subtraction of numbers
Resultante de una función. Resultant of functions
Retícula de puntos parcialmente ordenados. Net of partially ordered points
Retracto. Retract
Reversión de una serie. Reversion of a series
Revolución de una curva alrededor de un eje. Revolution of a curve about an axis
Rosa de tres hojas. Rose of three leafs
Rotación de ejes. Rotation of axes
Rotacional de un vector. Curl of a vector
Rumbo de una línea. Bearing of a line

Salinón. Salinon
Salto de discontinuidad. Jump discontinuity
Salto de una función. Saltus of a function
Satisfacer una ecuación. Satisfy an equation
Secante de un ángulo. Secant of an angle
Sección de perfil. Profile map
Sección de un cilindro. Section of a cylinder
Sección (ó capa) cruzada. Cross-cap
Sección dorada. Golden section
Sector de un círculo. Sector of a circle
Segmento de una curva. Segment of a curve
Segmento rectilíneo. Line segment
Segunda derivada. Second derivative
Segundo teorema del valor medio. Extended mean value theorem
Seguridad colateral. Collateral security
Seguro. Insurance
Seguro de vida. Life insurance
Seguro de vida contínuo. Whole life insurance
Seguro dotal. Endowment insurance
Seis. Six
Semejanza. Similitude
Semestral. Biannual
Semicírculo. Semicircle
Seno de un número. Sine of a number
Sentido de una desigualdad. Sense of an inequality
Separación de un conjunto. Separation of a set
Septillón. Septillion
Serie aritmética. Arithmetic series
Serie autoregresiva. Autoregressive series
Serie de números. Series of numbers
Serie de oscilación. Oscillation series
Serie de potencias. Power series
Serie de potencias formal. Formal power series
Serie geométrica. Geometric series
Serie hipergeométrica. Hypergeometric series
Serie infinita. Infinite series
Serie propiamente divergente. Properly divergent series
Serie sumable. Summable series
Servo mecanismo. Servomechanism
Sesgo. Skewness
Sextillón. Sextillion
Siete. Seven

Significado de una desviación. Significance of a deviation

Signo de suma. Summation sign

Signo de un número. Sign of a number

Signo más. Plus sign

Silogismo. Syllogism

Símbolo. Symbol

Símbolo de intersección. Cap

Símbolo de unión. Cup

Símbolos cuneiformes. Cuneiform symbols

Simetría axial. Axial symmetry

Simetría cíclica. Cyclosymmetry

Simetría de una función. Symmetry of a function

Simplejo. Simplex

Simplificación. Simplification

Singularidad de doblez. Fold singularity

Singularidad de pliegue. Fold singularity

Sinusoide. Sinusoid

Sistema centesimal para medida de ángulos. Centesimal system of measuring angles

Sistema de curvas isotérmicas. Isothermic system of curves

Sistema de ecuaciones. System of equations

Sistema decimal. Decimal system

Sistema hexadecimal. Hexadecimal number system

Sistema internacional de unidades. International system of units

Sistema monodireccional. Single address system

Sistema multidireccional. Multiaddress system

Sistema numérico duodecimal. Duodecimal system of numbers

Sistema octal. Octal number system

Sistema sexagesimal. Sexagesimal number system

Sistema sexagesimal de números. Sexagesimal system of numbers

Sistema triortogonal. Triply orthogonal system

Sólido de Arquímedes. Archimedean solid

Sólido de Arquímedes. Semi-regular solid

Sólido de revolución. Solid of revolution

Sólido truncado. Frustum of a solid

Solución de una ecuación. Solution of an equation

Solución gráfica. Graphical solution

Solución simple. Simple solution

Solución trivial. Trivial solution

Sombra (cono de sombra). Umbra

Soporte compacto. Compact support

Soporte de una función. Support of a function

Spline. Spline

Subcampo. Subfield

Subconjunto. Subset

Subconjunto cofinal. Cofinal subset

Subgrupo. Subgroup

Subgrupo cuasinormal. Quasi-normal subgroup

Subgrupos conjugados. Conjugate subgroups

Subíndice. Subscript

Subnormal. Subnormal

Substitución en una ecuación. Substitution in an equation

Subtangente. Subtangent

Subtender un ángulo. Subtend an angle

Sucesión aleatoria. Random sequence

Sucesión convergente. Convergent sequence

Sucesión de números. Sequence of numbers

Sucesión divergente. Divergent sequence

Sucesión ortonormal. Orthonormal sequence

Suma (adición). Addition

Suma de números. Sum of numbers

Suma de una serie. Summation of series

Sumador. Adder

Sumando. Addend

Sumando. Summand

Superconjunto. Superset

Superficie cilíndrica. Cylindrical surface

Superficie cónica. Conical surface

Superficie de revolución. Surface of revolution

Superficie desarrollable. Developable surface

Superficie de traslación. Translation surface

Superficie elíptica. Elliptic surface

Superficie equipotencial. Equipotential surface

Superficie espiral. Spiral surface

Superficie mínima. Minimal surface

Superficie piramidal. Pyramidal surface

Superficie prismática. Prismatic surface

Superficie pseudoesférica. Pseudospherical surface

Superficie reglada. Ruled surface

Superficie unilateral. Unilateral surface

Superficies isométricas. Isometric surfaces

Superosculación. Superosculation

Superponer dos configuraciones. Superpose two configurations

Superreflexivo. Super-reflexive

Suprayección. Surjection

Sustraendo. Subtrahend

Tabla de contingencia. Contingency table

Tabla de conversión. Conversion table

Tabla de logaritmos. Table of logarithms

Tabla de mortandad. Mortality table

Tabla selectiva de mortandad. Select mortality table

Tangencia. Tangency

Tangente a un círculo. Tangent to a circle

Tangente común a dos círculos. Common tangent of two circles

Tangente de inflexión. Inflection tangent

Tangente de un ángulo. Tangent of an angle

Tangente exterior a dos círculos. External tangent of two circles

Tangente interior a dos círculos. Internal tangent of two circles

Tarifa. Tariff

Tendencia secular (Estadistica). Secular trend

Tenedor. Payee

Tensión. Tension

Tensor. Tensor

Tensor contravariante. Contravariant tensor

Tensor de esfuerzo. Strain tensor

Teorema. Theorem

Teorema de Bézout. Bézout's theorem

Teorema de existencia. Existence theorem

Teorema de extensión de Tietze. Tietze extension theorem

Teorema de la subbase de Alexander. Alexander's subbase theorem

Teorema de los doce colores. Twelve-color theorem

Teorema de los números pentagonales de Euler. Euler pentagonal-number theorem

Teorema de los tres cuadrados. Three-squares theorem

Teorema de monodromía. Monodromy theorem

Teorema de Pitágoras. Pythagorean theorem
Teorema de punto fijo. Fixed-point theorem
Teorema de punto fijo de Banach. Banach fixed-point theorem
Teorema de Radon-Nykodim. Radon-Nikodým theorem
Teorema de recurrencia de Poincaré. Recurrence theorem
Teorema de unicidad. Uniqueness theorem
Teorema del mini-max. Minimax theorem
Teorema del residuo. Remainder theorem
Teorema del sandwich. Ham-sandwich theorem
Teorema del valor intermedio. Intermediate-value theorem
Teorema del valor medio. Mean-value theorem
Teorema fundamental del álgebra. Fundamental theorem of algebra
Teorema minimax. Minimax theorem
Teorema tauberiano. Tauberian theorem
Teoremas duales. Dual theorems
Teoría de catástrofes. Catastrophe theory
Teoría de ecuaciones. Theory of equations
Teoría de funciones. Function theory
Teoría de gráficas. Graph theory
Teoría de la relatividad. Relativity theory
Teoría ergódica. Ergodic theory
Término de una fracción. Term of a fraction
Término indefinido. Undefined term
Términos no semejantes. Dissimilar terms
Termómetro centígrado. Centigrade thermometer
Terna pitagórica. Pythagorean triple
Tetraedro. Tetrahedron
Tetrafolio. Quadrefoil
Tiempo. Time
Tiempo ecuacionado. Equated time
Tiempo sideral. Sidereal time
Tiempo solar. Solar time
Tiempo standard. Standard time
Tonelada. Ton
Topógrafo. Surveyor
Topología. Topology
Topología combinatoria. Combinatorial topology
Topología discreta. Discrete topology
Topología indiscreta. Indiscrete topology
Topología indiscreta. Trivial topology
Topología proyectiva. Projective topology
Topología trivial. Trivial topology
Toro. Torus
Torsión de una curva. Torsion of a curve
Trabajo. Work
Tractriz. Tractrix
Transformación afín. Affine transformation
Transformación auto adjunta. Self-adjoint transformation
Transformación colineal. Collineatory transformation
Transformación conforme. Conformal transformation
Transformación conjuntiva. Conjunctive transformation
Transformación de alargamiento. Stretching transformation
Transformación de coordenadas. Transformation of coordinates
Transformación de deslizamiento. Shear transformation

Transformación de semejanza. Similarity transformation
Transformación isógona. Isogonal transformation
Transformación lineal. Linear transformation
Transformación (mapeo) de un espacio. Mapping of a space
Transformación no singular. Nonsingular transformation
Transformación ortogonale. Orthogonal transformation
Transformada de una matriz. Transform of a matrix
Transformada discreta de Fourier. Discrete Fourier transform
Transformada rápida de Fourier. Fast Fourier transform
Transito. Transit
Translación de ejes. Translation of axes
Transponer un término. Transpose a term
Transportador. Protractor
Transposición. Transposition
Transpuesta de una matriz. Transpose of a matrix
Transversal. Transversal
Trapecio. Trapezium
Trapezoide. Trapezoid
Trayectoria. Trajectory
Trayectoria de un proyectil. Path of a projectile
Traza de una matriz. Trace of a matrix
Traza de una matriz. Spur of a matrix
Trébol. Trefoil
Trece. Thirteen
Trenza. Braid
Tres. Three
Triangulación. Triangulation
Triángulo. Triangle
Triángulo equilátero. Equilateral triangle
Triángulo escaleno. Scalene triangle
Triángulo esférico trirectángular. Trirectangular spherical triangle
Triángulo isósceles. Isosceles triangle
Triángulo oblicuo. Oblique triangle
Triángulo rectángulo. Right triangle
Triángulo terrestre. Terrestrial triangle
Triángulos semejantes. Similar triangles
Tridente de Newton. Trident of Newton
Triedro formado por tres líneas. Trihedral formed by three lines
Trigonometría. Trigonometry
Trillón. Trillion
Trinomio. Trinomial
Trisección de un ángulo. Trisection of an angle
Trisectriz. Trisectrix
Trocoide. Trochoid

Ultrafiltro. Ultrafilter
Ultrafiltro libre. Free ultrafilter
Unidad. Unity
Unidad astronómica. Astronomical unit
Uniformemente equicontinuo. Uniformly equicontinuous
Union de conjuntos. Union of sets; join of sets
Uno. One (the number)
Utilidad. Profit
Utilidad neta. Net profit

Valencia de un nodo. Valence of a node
Valor absoluto. Absolute value
Valor actual. Present value
Valor acumulado. Accumulated value
Valor a la par. Par value
Valor crítico. Critical value
Valor de entrega. Surrender value
Valor de una póliza de seguro. Value of an insurance
  policy
Valor depreciado. Scrap value
Valor esperado. Expected value
Valor futuro. Future value
Valor local. Local value
Valor númerico. Numerical value
Valor propio (ó eigen valor) de una matriz. Eigen-
  value of a matrix
Valuación de un campo. Valuation of a field
Variabilidad. Variability
Variable. Variable
Variable dependiente. Dependent variable
Variable estadística. Variate
Variable estocástica. Stochastic variable
Variable estadística normalizada. Normalized variate
Variable independiente. Independent variable
Variación de una función. Variation of a function
Variación de parámetros. Variation of parameters
Variancia. Variance
Variedad. Manifold
Variedad algebraica afín. Affine algebraic variety

Vatio. Watt
Vecindad de un punto. Neighborhood of a point
Vector. Vector
Vector fuerza. Force vector
Vector irrotacional. Irrotational vector
Vector propio (eigen vector). Eigenvector
Vector solenoidal. Solenoidal vector
Veinte. Twenty
Veintena. Score
Velocidad. Speed, velocity
Velocidad constante. Constant speed
Velocidad instantánea. Instantaneous velocity
Velocidad relativa. Relative velocity
Verseno (seno verso). Versed sine
Vértice de un ángulo. Vertex of an angle
Vibración. Vibration
Viga cantilever. Cantilever beam
Vínculo. Vinculum
Voltio. Volt
Volumen de un sólido. Volume of a solid

Wronskiano. Wronskian

Yarda de distancia. Yard of distance

Zenit de un observador. Zenith of an observer
Zona. Zone